BIOLOGY

concepts and applications

FIFTH EDITION

CECIE STARR

LISA STARR
Biology Illustrator

BROOKS/COLE

THOMSON LEARNING™

Australia • Canada • Mexico • Singapore • Spain
United Kingdom • United States

BROOKS/COLE
THOMSON LEARNING

BIOLOGY PUBLISHER: Jack C. Carey

DEVELOPMENT EDITOR: Mary Arbogast

ASSISTANT EDITOR: Mark Andrews, Suzannah Alexander

EDITORIAL ASSISTANT: Karen Hansten

MEDIA PROJECT MANAGER: Pat Waldo

TECHNOLOGY PROJECT MANAGER: Keli Amann

MARKETING MANAGER: Tom Ziolkowski

MARKETING ASSISTANT: Maureen Griffin

ADVERTISING PROJECT MANAGER: Linda Yip

SENIOR PROJECT MANAGER, EDITORIAL/PRODUCTION: Teri Hyde

PRINT/MEDIA BUYER: Karen Hunt

PERMISSIONS EDITOR: Bob Kauser

PRODUCTION SERVICE: Lachina Publishing Services, Inc.

TEXT AND COVER DESIGN: Gary Head, Gary Head Design

ART EDITOR AND PHOTO RESEARCHER: Myrna Engler

ILLUSTRATORS: Lisa Starr, Gary Head

COVER IMAGE: From the African savannah, lioness and cub at sunrise. ©Minden Pictures

COVER PRINTER: Phoenix Color Corp (MD)

COMPOSITOR: Angela Harris, John Becker, Preface, Inc.

FILM HOUSE: Tom Anderson, H&S Graphics

PRINTER: Quebecor/World, Versailles

BOOKS IN THE BROOKS/COLE BIOLOGY SERIES

Printed in the United States of America

1 2 3 4 5 6 7 06 05 04 03 02

For more information about our products, contact us at:
Thomson Learning Academic Resource Center
1-800-423-0563
For permission to use material from this text, contact us by:
Phone: 1-800-730-2214 **Fax:** 1-800-730-2215
Web: http://www.thomsonrights.com

Library of Congress Cataloging-in-Publication Data
Starr, Cecie.
    Biology : concepts and applications / Cecie Starr; Lisa Starr, biology
       illustrator.—5th ed. p. cm.- -(Brooks/Cole biology series)
    Includes bibliographical references (p.).
    ISBN 0-534-38549-4 (hardback)
    1. Biology I. Title. II. Series.
    QH307.2 .S73 2002
    570—dc21

                          2001 0 49 945

Student Paperback Edition: 0-534-38558-3
Annotated Instructor's Edition: 0-534-38555-9

Brooks/Cole—Thomson Learning
10 Davis Drive
Belmont, CA 94002
USA

Asia
Thomson Learning
60 Albert Street, #15-01
Albert Complex
Singapore 189969

Australia
Nelson Thomson Learning
102 Dodds Street
South Melbourne, Victoria 3205
Australia

Canada
Nelson Thomson Learning
1120 Birchmount Road
Toronto, Ontario M1K 5G4
Canada

Europe/Middle East/Africa
Thomson Learning
Berkshire House
168-173 High Holborn
London WC1 V7AA
United Kingdom

CONTENTS IN BRIEF

DETAILED CONTENTS

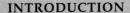

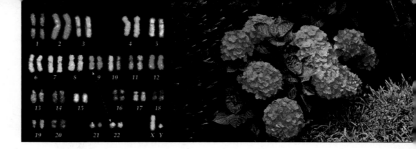

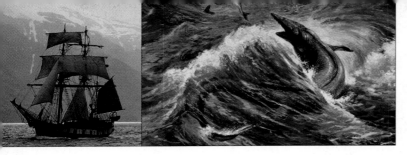

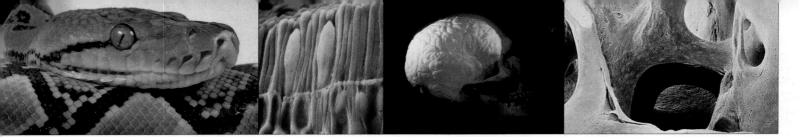

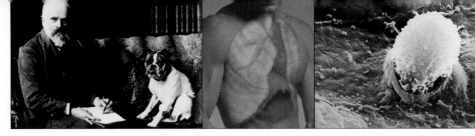

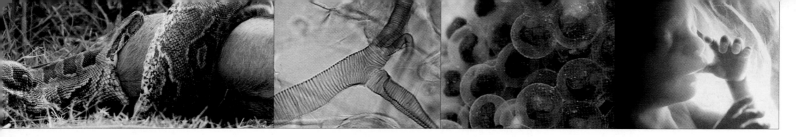

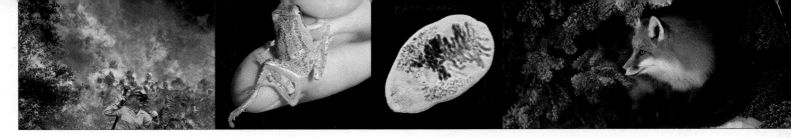

Preface

Teachers of introductory biology know all about the Red Queen effect, whereby one runs as fast as one can to stay in the same place. New and refined information from hundreds of fields of inquiry piles up daily, and somehow teachers are expected to distill it into Biology Lite, a zip through the high points, and help students deepen their understanding of a world of unbelievable richness.

Restricting textbook content runs the risk of splintering understanding of that world, as when an emphasis on human biology inadvertently reinforces archaic notions that everything on Earth is here in the service of Us; or when a molecular focus excludes knowledge of whole organisms. Here I am reminded of the well-intentioned but lethal blanketing of habitats with DDT.

We offer this book as a coherent account of the sweep of life's diversity and its underlying unity. Through its examples of problem solving and experiments, it shows the power of thinking critically about the natural world. It highlights key concepts, current understandings, and research trends for major fields of biological inquiry. It explains the structure and function of a broad sampling of organisms in enough detail so that students can develop a working vocabulary about life's parts and processes.

We start with an overview of the basic concepts and scientific methods. Next are three units on the principles of biochemistry, inheritance, and evolution—a conceptual framework for exploring life's unity and diversity. Units on comparative anatomy and physiology of plants, then animals, follow. The last unit focuses on the patterns and consequences of organisms interacting with one another and with the environment. This conceptual organization parallels the levels of biological organization.

CONCEPT SPREADS Feedback from teachers of more than 3 million students helped us refine our approach. We keep the story line in focus for students by subscribing to the question "How do you eat an elephant?" and its answer, "One bite at a time." We organize descriptions, art, and supporting evidence for each concept on two facing pages at most. Each *concept spread* starts with a numbered tab and concludes with boldfaced summaries of key points (*see below*). On-page summaries allow students to check their understanding of one concept before turning to another.

The clear conceptual organization within chapters also offers teachers flexibility in assigning text material to fit course requirements. For example, those who spend little time on photosynthesis may choose to bypass the spreads on properties of light and the chemiosmotic theory of ATP formation. They may or may not assign Focus essays (one on the global impact of photosynthesis, another on how photosynthesis may have started in the first place). All of the concept spreads flow as parts of the same story, but some clearly offer depth that can be treated as optional.

Concept spreads are not gimmicks. Ongoing feedback guides decisions about when to add depth and when to loosen core material with applications. Within spreads,

---

Numbered tabs indicate the start of a new concept as the chapter's story unfolds. The gold tabs identify basic chapter concepts. Blue tabs identify Focus essays that enrich the basics with examples of experiments (to demonstrate the power of critical thinking), of the nature of life, and of applying the basics to issues of human interest.

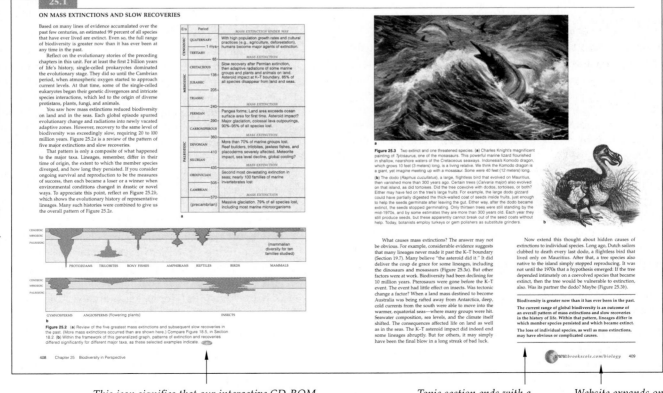

This icon signifies that our interactive CD-ROM further explores the concept being illustrated.

Topic section ends with a summary of key concepts.

Website expands on the section's topic.

headings and subheadings help students keep track of the hierarchy of information. Transitions between spreads help them keep the greater story in focus and discourage memorization for its own sake. To avoid disrupting the basic story line while still attending to interested students, we include some details in optional, enriching illustrations.

Our organization helps students find assigned topics easily and translates into a tangible outcome—improved test scores. Its underlying logic also helps them develop enough confidence to dig deeper into biological science.

BALANCING CONCEPTS WITH APPLICATIONS Chapters start with a lively or sobering application and an advance organizer—a list of key concepts. Essays parallel the core material to help maintain student interest in the basics. The essays afford more depth on medical, environmental, and social issues without interrupting conceptual flows. Briefer applications are integrated in the text. The final pages index all applications separately for fast reference.

FOUNDATIONS FOR CRITICAL THINKING To help students sharpen their capacity for critical thinking, we walk them through experiments that yielded evidence in favor of or against hypotheses. The main index lists all of the experiments we selected for the book (see the entries *Experiment, examples*, and *Test, observational*).

We use certain chapter introductions as well as whole chapters to show students some productive results of critical thinking. The introductions to Mendelian genetics (Chapter 10), DNA structure and function (12), speciation (17), immunology (35), and behavior (41) are examples. Also, each chapter has a set of *Critical Thinking* questions, most from Katherine Denniston. Daniel Fairbanks created many of the *Genetics Problems*, which help students grasp the principles of inheritance (Chapters 10 and 11).

VISUAL OVERVIEWS OF CONCEPTS We simultaneously develop text and art as inseparable parts of the same story. We give visual learners a means to work their way through a visual overview of major processes before reading the corresponding text. Students repeatedly let us know how much they appreciate this art. The overview illustrations include step-by-step descriptions of biological parts and processes. Instead of "wordless" diagrams, we break down information into a series of illustrated callouts. In Figure 13.10, for example, callouts integrated with the art walk students through the stages by which a mature mRNA transcript becomes translated.

Many anatomical drawings are integrated overviews of structure/function. Students need not jump back and forth from text, to tables, to illustrations, and back again to see how an organ system is put together and what its parts do. We hierarchically arrange descriptions of parts to reflect a system's structural and functional organization.

ZOOM SEQUENCES Many illustrations progress from macroscopic to microscopic views of a system or process. Figure 6.3, for example starts with a plant leaf and ends with reaction sites in the chloroplast. Figures 33.15 and 33.16 start with a ballerina's biceps and move down through levels of skeletal muscle contraction.

COLOR CODES Consistent use of colors for molecules, cell structures, and processes helps students track what is going on. We use these colors throughout the book:

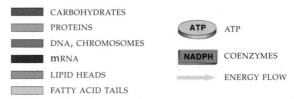

CARBOHYDRATES	
PROTEINS	ATP ATP
DNA, CHROMOSOMES	
mRNA	NADPH COENZYMES
LIPID HEADS	ENERGY FLOW
FATTY ACID TAILS	

ICONS Small icons next to an illustration help students relate a topic to the big picture. For instance, a cell icon reminds them of the location of the plasma membrane relative to the cytoplasm. Some icons relate reactions and processes to cell locations and to one another. Others are reminders of evolutionary relationships among groups, as in Chapters 23 and 24. A multimedia icon directs students to art in the CD-ROM packaged in its own envelope at the back of their book. Others direct them to supplemental material on the Web and to InfoTrac® College Edition, an online database of full-length articles from hundreds of top academic journals and popular sources.

MAJOR CONTENT REVISIONS New to this edition is a chapter on biodiversity and mass extinctions (25). The chapters on principles of metabolism (5), photosynthesis (6), gene control (14), macroevolution (18), immunity (35), principles of animal reproduction/development (39), and ecosystems (43) are significantly reorganized. Like the other chapters, they are now more succinct and accessible owing to line-by-line snipping, as indirectly evidenced by the more open page layouts. Users will find major new art, crisper definitions, new end-of-chapter material, and new or updated material throughout, as highlighted here:

Unit I. Principles of Cellular Life Chapter 2 has new atomic and molecular models. Chapter 3 shows the non-ionized and ionized forms of carboxyl and amino groups (a point of confusion for students, who see both in the literature); its hemoglobin model is new. Chapter 4 has new text and art on the cytoskeleton and, as elsewhere, distinguishes between prokaryotic cell and bacterial cell. Chapter 5 has a new introduction on enzymes and aging, and weaves in fascinating experiments with *Caenorhabditis elegans*. The chapter has a new Critical Thinking question on cyanide poisoning. Revised and reorganized Chapter 6 has livelier writing, as in Section 6.1, and a condensed Summary. It has new text and art on photosystems; the function of electron transport systems is stated more clearly. Art comparing carbon-fixing strategies has accurate cross-sections of C3 and C4 leaves. There are new Critical Thinking questions on the evolution of photosynthesis and on bioluminescence. Chapter 7, a tough topic, has more user-friendly art.

Unit II. Principles of Inheritance In Chapter 8, we moved the overview of chromosome structure forward, added a bit on the cell cycle, refined the mitosis art, and added photographs of Henrietta Lacks and HeLa cells. Micrographs accompany the meiosis diagram in Chapter 9; we finally found some uncomplicated ones that won't confuse students. Chapter 10 has refined definitions of monohybrid and dihybrid crosses, and new text and art on experiments that revealed environmental effects on *Achillea millefolium* phenotype. Chapter 11 has a CML update, the crossing over sketch is moved to the overview section, its karyotyping essay is simplified, and it has a new human example of an outcome of crossing over and genetic recombination. More human genetic disorders are covered (see Table 11.1). We expanded the Summary and added a Critical Thinking question on perceptions of genetic disorders. Chapter 12 includes a new model for DNA and mentions new evidence of the function of DNA's structure (page 195). The updated DNA cloning essay includes concerns about random mistakes in cloning processes. We added a Critical Thinking question on cloning extinct mammals. Updated Chapter 13 includes a recently deduced molecular model for the eukaryotic ribosome; text and art on translation are modified accordingly and show the rRNA catalytic site. The transposon material is updated here and elsewhere; the distinction between ionizing and non-ionizing radiation is clarified. The summary is tighter.

Chapter 14 has a shorter introductory essay on cancer and the nature of gene control; details are in an updated essay on the cell cycle and cancer. A new section gives an overview of control mechanisms. The lactose operon section also describes lactose intolerance. The point of X chromosome inactivation (dosage compensation) is clearly stated. A new invertebrate example (ecdysone control) is included. Chapter 15 includes a recent genetic engineering experiment on aspens plus updates on the mammalian cloning issue, the Human Genome Initiative, and gene therapy for bubble boys. We have an extended Critical Thinking question on genetically engineered food, and another on minimal organisms. We moved description and illustration of bacterial conjugation to Section 20.3.

Unit III. Principles of Evolution *Archaeopteryx* is now in Chapter 18. We refined the text on stabilizing and disruptive selection and balanced polymorphism and added Self-Quiz questions plus a Critical Thinking question on the rise of TB cases. Chapter 17 has new photographs for mechanical isolation and sympatric speciation. Chapter 18 has a new introduction based on insights into the biblical flood story as a way to put catastrophism and evolutionary theories in a cultural context. Next is a new section on the fossil record, a new focus essay on radiometric dating and the geologic time scale, and revised sections on biogeography and comparative morphology with new art. A revised comparative embryology section connects transposons to primate evolution; the comparative biochemistry section is updated and has a better explanation of how molecular clocks are calibrated. A new section on taxonomy and classification follows; it has a simple description of how to construct a cladogram, and considers the pros and cons of the three-domain scheme. A new essay uses *Archaeopteryx* to show how evidence of evolution can be interpreted and misinterpreted. The Summary is revised.

Unit IV. Evolution and Biodiversity Chapter 19 has updates on the RNA world, the selective advantage of DNA structure, the Cambrian "explosion," Paleozoic and Mesozoic mass extinctions, and new reconstructions of early life (Figures

19.14 and 19.16). We moved the visual summary of Earth and life history to Chapter 18, but a concise version also is at the chapter's end. Chapter 20 has stunning new micrographs and art, more on archaebacteria and Lyme disease, a bit on bacterial conjugation, an update on prions and BSE, a bit on toxoplasmosis, a new table on the eight deadliest infectious diseases, and a look at food poisoning. The section on single-celled algae is revised, with a nod to The Cell From Hell. Two new Critical Thinking questions consider unsterilized needles and foot-and-mouth disease. Chapter 21 has new ascomycetes micrographs, a bit on household molds (page 336), an update on some imperfect fungi, and a better section on symbionts. Chapter 22 has new opening photos and more concise sections on evolutionary trends and the rise of seed-bearing plants. Chapter 23 has revised bits on cnidarians, roundworms, spiders and scorpions, and insects, and new Critical Thinking questions. Lobe-finned fishes are back with the fishes in Chapter 24, which also has an example of frog deformities, a cutaway turtle shell and skeleton, and more on birds and mammals, as in new Critical Thinking questions with art. Its sections on primates and human evolution have updated text and art, including new skull reconstructions. The summary is extensively revised.

New Chapter 25 puts biodiversity and extinction crises in evolutionary perspective. It starts with the Easter Island story and the nature of mass extinctions, slow recoveries, and individual species losses. It looks at threatened and endangered species, including their global distribution. It has a case study of coral reefs, an essay on Rachel Carson, and a section on conservation biology, including the role of systematics and bioeconomic analysis. It has two positive examples of recent attempts to reconcile biodiversity with human population growth and economic pressures (strip logging in the tropics and preservation of riparian zones).

Unit V. Plant Structure and Function Line-by-line text changes for clarity and a number of art refinements have improved this unit, already recognized as exceptionally accurate and clear thanks to Nancy Dengler, John Jackson, Cleon Ross, Thomas Rost, and Frank Salisbury.

Unit VI. Animal Structure and Function Chapter 29 has new micrographs and art, and an update on lab-grown tissues. Chapter 30 has a tighter introduction, some new art, sections rewritten for clarity (e.g., 30.1, 30.2, and 30.9), more on multiple sclerosis, and updates on drug addiction. The Summary is revised; we added new Critical Thinking questions. Chapter 31 has revised sections on hearing and vision; Section 31.6 has brief updates (umami, and human pheromones). Chapter 32 has a tighter introduction, an update on signaling molecules, and refined passages on cortisol, thyroid hormone, and prostaglandin secretions. A new section correlates frog deformities, hot-spot habitats, and thyroid malfunctioning. Chapter 33 has updates on vitamin D, a comparison of vertebrate skeletons, a rewrite on calcium metabolism, and new art on muscle contraction.

Chapter 34 has important new art on blood cell lineages, cardiac muscle cells, veins, hemostasis, and lymph nodes, and refined text on blood pressure. Chapter 35, a significant revision, has updates on blood cell lineages and immune functions, autoimmunity, immune deficiencies, and immunoglobulins; new molecular models; easier-to-follow text and art; and an AIDS update. Chapter 36 has new art and text on the respiratory cycle and the respiratory membrane. Chapter 37 has new text on food pyramids and protein-energy malnutrition and dietary excess, plus new art on villus fine structure. Chapter 38 has crisper definitions. Chapter 39 has major updates as well as text rewrites, especially on pattern formation and aging.

Unit VII. Ecology and Behavior　Chapter 40 has updates on human population demographics, TFRs, age structure art, and California's rolling blackouts. The behavior chapter (41) is earlier in the unit to better correspond with levels of biological organization. The text is tighter; some overlaps between sections are eliminated. The example on altruism among termites is now a Critical Thinking question.

Chapter 42 has tie-ins to Chapter 25 on biodiversity. It has updated material on the Canadian lynx/snowshoe hare puzzle, a new table on some species introductions, and a Criticial Thinking question on frog deformities.

Chapter 43 (ecosystems) has major new sections with stunning art on the nature of food chains and food webs, using tallgrass prairie as the key example, followed by an essay on biological magnification based on the original DDT study. It has new art on seasonal shifts in an omnivore's diet, rewritten text for the carbon cycle, and updated text and art for the essay on global warming. It describes the effect of fertilizers on soil ion exchange.

Chapter 44 has new art on environmental gradients and on climate zones, new icon maps for biomes, new text and photographs of tundra, and a revised wetlands section. The coral reef section is now in Chapter 25. Chapter 45 has updates on ozone thinning and nuclear energy, and some changes to the essay on tropical rain forests.

SUPPLEMENTS　The Instructors' Examination copy lists the comprehensive package of print and multimedia supplements to this book, including online resources.

A COMMUNITY EFFORT　This book is the current version of an educational effort that started twenty-six years ago. About 2,000 teachers and researchers have contributed to its ongoing refinement, and along the way we have helped more than 3 million students gain insights into the nature of biology. Much of our art and text, which is accurate and pedagogically sound as an outcome of years of research, brainstorming, and classroom testing, is now standard fare in many textbooks. The individuals listed at right deserve recognition for their commitment to quality in education.

We entrusted this edition to Gary Head, Lisa Starr, Teri Hyde, Grace Davidson, Myrna Engler, and Angela Harris —an exceptionally dedicated publishing team. Each is a superior talent in publishing. Heidi Marschner, Diane Kimmel, and Karen Hunt helped out. Pat Waldo, Chris Evers, and Steve Bolinger created a terrific multimedia package. Susan Badger and Jack Carey especially, and Kathie Head and Sean Wakely, thank you so much for your creative, enlightened management.

Cecie Starr, June 2001

General Advisors/Contributors

JOHN ALCOCK　*Arizona State University*
GEORGE COX　*San Diego State University*
KATHERINE DENNISTON　*Towson State University*
MELANIE DEVORE　*Georgia College and State University*
DANIEL FAIRBANKS　*Brigham Young University*
TOM GARRISON　*Orange Coast College*
PAUL HERTZ　*Barnard College*
JOHN JACKSON　*North Hennipin Community College*
EUGENE KOZLOFF　*University of Washington*
KAREN MESSLEY　*Rock Valley College*
CLEON ROSS　*Colorado State University*
LAURALEE SHERWOOD　*West Virginia University*
STEPHEN WOLFE　*University of California, Davis*

Fifth Edition Reviewers

BENIVENGA, STEPHEN　*University of Wisconsin, Oshkosh*
BORGESON, CHARLOTTE　*University of Nevada*
BUTTON, JERRY　*Portland Community College*
CLARK, DEBORAH C.　*Middle Tennessee University*
DABNEY, MICHAEL W.　*Hawaii Pacific University*
DEGROOTE, DAVID K.　*St. Cloud University*
FENSTER, EUGENE J.　*Longview Community College*
HEIM, WERNER G.　*The Colorado College*
JOHNSTON, TIMOTHY　*Murray State University*
JUILLERAT, FLORENCE　*Indiana University – Purdue University*
KILBURN, KERRY S.　*Old Dominion University*
LYNG, R. DOUGLAS　*Indiana University – Purdue University*
MASON, ROY B.　*Mt. San Jacinto College*
MENDELSON, JOSEPH R.　*Utah State University*
MORENO, JORGE A.　*University of Colorado, Boulder*
NADLER, KENNETH D.　*Michigan State University*
NAPLES, VIRGINIA　*Northern Illinois University*
POLCYN, DAVID M.　*California State University, San Bernardino*
RUETER, JOHN　*Portland State University*

Contributors of Influential Reviews

ALDRIDGE, DAVID　*North Carolina Agricultural/Technical State University*
ARMSTRONG, PETER　*University of California at Davis*
BAJER, ANDREW　*University of Oregon*
BAKKEN, AIMEE　*University of Washington*
BARBOUR, MICHAEL　*University of California, Davis*
BARHAM, LINDA　*Meridian Community College*
BARKWORTH, MARY　*Utah State University*
BELL, ROBERT A.　*University of Wisconsin, Stevens Point*
BENDER, KRISTEN　*California State University, Long Beach*
BINKLEY, DAN　*Colorado State University*
BLEEKMAN, GEORGE　*American River College*
BRENGELMANN, GEORGE　*University of Washington*
BRINSON, MARK　*East Carolina University*
BROWN, ARTHUR　*University of Arkansas*
BUCKNER, VIRGINIA　*Johnson County Community College*
CALVIN, CLYDE　*Portland State University*
CASE, CHRISTINE　*Skyline College*
CASE, TED　*University of California, San Diego*
CHRISTIANSEN, A. KENT　*University of Michigan*
COLAVITO, MARY　*Santa Monica College*
CONKEY, JIM　*Truckee Meadows Community College*
CROWCROFT, PETER　*University of Texas at Austin*
DANIELS, JUDY　*Washtenau Community College*
DAVIS, JERRY　*University of Wisconsin, La Crosse*
DELCOMYN, FRED　*University of Illinois, Urbana*
DEMMANS, DANA　*Finger Lakes Community College*
DEMPSEY, JEROME　*University of Wisconsin*

DENETTE, PHIL *Delgado Community College*
DENGLER, NANCY *University of California, Davis*
DESAIX, JEAN *University of North Carolina*
DETHIER, MEGAN *University of Washington*
DEWALT, R. EDWARD *Louisiana State University*
DIBARTOLOMEIS, SUSAN *Millersville University of Pennsylvania*
DIEHL, FRED *University of Virginia*
DLUZEN, DEAN *Northeastern Ohio Universities College of Medicine*
DONALD-WHITNEY, CATHY *Collin County Community College*
DOYLE, PATRICK *Middle Tennessee State University*
DUKE, STANLEY H. *University of Wisconsin, Madison*
DYER, BETSEY *Wheaton College*
EDLIN, GORDON *University of Hawaii, Manoa*
EDWARDS, JOAN *Williams College*
ELMORE, HAROLD W. *Marshall University*
ENDLER, JOHN *University of California, Santa Barbara*
ENGLISH, DARREL *Northern Arizona University*
ERWIN, CINDY *City College of San Francisco*
EWALD, PAUL *Amherst College*
FALK, RICHARD *University of California, Davis*
FISHER, DAVID *University of Hawaii, Manoa*
FISHER, DONALD *Washington State University*
FLESSA, KARL *University of Arizona*
FONDACARO, JOSEPH *Hoechst Marion Roussel, Inc.*
FRAILEY, CARL *Johnson County Community College*
FRISBIE, MALCOLM *Eastern Kentucky University*
FROEHLICH, JEFFREY *University of New Mexico*
FULCHER, THERESA *Pellissippi State Technical Community College*
GAGLIARDI, GRACE S. *Bucks County Community College*
GENUTH, SAUL M. *Mt. Sinai Medical Center*
GHOLZ, HENRY *University of Florida*
GIBSON, THOMAS *San Diego State University*
GOODMAN, H. MAURICE *University of Massachusetts Medical School*
GOSZ, JAMES *University of New Mexico*
GREGG, KATHERINE *West Virginia Wesleyan College*
GUTSCHICK, VINCENT *New Mexico State University*
HARRIS, JAMES *Utah Valley Community College*
HARTNEY, KRISTINE BEHRENTS *California State University, Fullerton*
HASSAN, ASLAM *Univeristy of Illinois College of Veterinary Medicine*
HELGESON, JEAN *Collin County Community College*
HESS, WILFORD M. *Brigham Young University*
HUFFMAN, DAVID *Southwest Texas State University*
INEICHER, GEORGIA *Hinds Community College*
INGRAHAM, JOHN L. *University of California, Davis*
JENSEN, STEVEN *Southwest Missouri State University*
JOHNSON, LEONARD R. *University of Tennessee College of Medicine*
JUILLERAT, FLORENCE *Indiana University, Purdue University*
KAREIVA, PETER *University of Washington*
KAUFMAN, JUDY *Monroe Community College*
KAYE, GORDON I. *Albany Medical College*
KAYNE, MARLENE *Trenton State College*
KELLNER, CHRIS *Arkansas Tech University*
KENDRICK, BRYCE *University of Waterloo*
KEYES, JACK L. *Linfield College, Portland Campus*
KILLIAN, JOELLA C. *Mary Washington College*
KIRKPATRICK, LEE A. *Glendale Community College*
KREBS, CHARLES *University of British Columbia*
KREBS, JULIA E. *Francis Marion University*
KUTCHAI, HOWARD *University of Virginia Medical School*
LANZA, JANET *University of Arkansas, Little Rock*
LASSITER, WILLIAM *University of North Carolina*
LEVY, MATTHEW *Mt. Sinai Medical Center*
LEWIS, LARRY *Salem State College*
LITTLE, ROBERT *Medical College of Georgia*
LOHMEIER, LYNNE *Mississippi Gulf Coast Community College*
LOPO, ALINA C. *University of California-Los Angeles Medical Center*
LUMSDEN, ANN *Florida State University*
MACKLIN, MONICA *Northeastern State University*
MANN, ALAN *University of Pennsylvania*
MARTIN, JAMES *Reynolds Community College*
MARTIN, TERRY *Kishwaukee College*
MATSON, RONALD *Kennesaw State College*
MATTHEWS, ROBERT *University of Georgia*
MAXWELL, JOYCE *California State University, Northridge*
MCCLURE, JERRY *Miami University*

MCEDWARD, LARRY *University of Florida*
MCKEAN, HEATHER *Eastern Washington State University*
MCKEE, DOROTHY *Auburn University, Montgomery*
MCNABB, ANN *Virginia Polytechnic Institute and State University*
MICKLE, JAMES *North Carolina State University*
MILLER, G. TYLER *Wilmington, North Carolina*
MINORKSY, PETER V. *Western Connecticut State University*
MOISES, HYLAN C. *University of Michigan Medical School*
MOORE–LANDECKER, ELIZABETH *Glassboro State University*
MORRISON–SHETLER, ALLISON *Georgia State University*
MORTON, DAVID *Frostburg State University*
MURPHY, RICHARD *University of Virginia Medical School*
MYRES, BRIAN *Cypress College*
NAGARKATTI, PRAKASH *Virginia Polytechnic Institute & State University*
NELSON, RILEY *University of Texas at Austin*
NORRIS, DAVID *University of Colorado*
PEARCE, FRANK *West Valley College*
PECHENIK, JAN *Tufts University*
PECK, JAMES H. *University of Arkansas, Little Rock*
PERRY, JAMES *University of Wisconsin, Center-Fox Valley*
PETERSON, GARY *South Dakota State University*
PIPERBERG, JOEL *Millersville University*
PLETT, HAROLD *Fullerton College*
REESE, R. NEIL *South Dakota State University*
REEVE, MARIAN *Emeritus, Merritt Community College*
REID, BRUCE *Kean College of New Jersey*
RENFROE, MICHAEL *James Madison University*
REZNICK, DAVID *University of California, Riverside*
RICKETT, JOHN *University of Arkansas, Little Rock*
ROBBINS, ROBERT *National Science Foundation*
ROIG, MATTIE *Broward Community College*
ROSE, GREIG *West Valley College*
ROST, THOMAS *University of California, Davis*
RUIBAL, RODOLFO *University of California, Riverside*
SALISBURY, FRANK *Utah State University*
SCHAPIRO, HARRIET *San Diego State University*
SCHLESINGER, WILLIAM *Duke University*
SCHNEIDEWENT, JUDY *Milwaukee Area Technical College*
SCHNERMANN, JURGEN *University of Michigan School of Medicine*
SCHREIBER, FRED *California State University, Fresno*
SELLERS, LARRY *Louisiana Tech University*
SHEGAL, PREM *East Carolina University*
SHONTZ, NANCY *Grand Valley State University*
SHOPPER, MARILYN *Johnson County Community College*
SLOBODA, ROGER *Dartmouth College*
SMITH, JERRY *St. Petersburg Junior College, Clearwater Campus*
SMITH, MICHAEL E. *Valdosta State College*
SMITH, ROBERT L. *West Virginia University*
SOLOMON, NANCY *Miami University*
STEARNS, DONALD *Rutgers University*
STEELE, KELLY P. *Appalachian State University*
STEINERT, KATHLEEN *Bellevue Community College*
SUMMERS, GERALD *University of Missouri*
SUNDBERG, MARSHALL D. *Louisiana State University*
SWANSON, ROBERT *North Hennepin Community College*
SWEET, SAMUEL *University of California, Santa Barbara*
TAYLOR, JANE *Northern Virginia Community College*
TERHUNE, JERRY *Jefferson Community College, University of Kentucky*
TIZARD, IAN *Texas A&M University*
TROUT, RICHARD E. *Oklahoma City Community College*
TYSER, ROBIN *University of Wisconsin, LaCrosse*
WAALAND, ROBERT *University of Washington*
WAHLERT, JOHN *City University of New York, Baruch College*
WALSH, BRUCE *University of Arizona*
WARING, RICHARD *Oregon State University*
WARNER, MARGARET R. *Purdue University*
WEBB, JACQUELINE F. *Villanova University*
WEIGL, ANN *Winston-Salem State University*
WEISS, MARK *Wayne State University*
WELKIE, GEORGE W. *Utah State University*
WENDEROTH, MARY PAT *University of Washington*
WHITE, EVELYN *Alabama State University*
WHITENBERG, DAVID *Southwest Texas State University*
WINICUR, SANDRA *Indiana University, South Bend*
YONENAKA, SHANNA *San Francisco State University*

Introduction

Current configurations of the Earth's oceans and land masses—the geologic stage upon which life's drama continues to unfold. This composite satellite image reveals global energy use at night by the human population. Just as biological science does, it invites you to think more deeply about the world of life—and about our impact upon it.

CONCEPTS AND METHODS IN BIOLOGY

Why Biology?

Leaf through a newspaper on any given Sunday and you might get an uneasy feeling that the natural world is spinning out of control. Northern forests are burning fiercely (Figure 1.1). The Arctic ice cap is melting and Greenland's glaciers are becoming lakes. Heat waves and droughts are breaking records everywhere, and maybe the whole atmosphere is warming up. The last individual of an ancient primate lineage bit the dust just as the human population neared the 6.1 billion mark. Geneticists decoded yet another garbled message in human DNA that causes another horrible disease. People are still arguing about when, exactly, the mass of human embryonic cells busily dividing inside a woman's womb is "alive." In a bizarre application of genetic engineering, somebody decided to make an "artistic statement" by producing a glow-in-the-dark bunny. Every bit of the live bunny, eyes included, turns fluorescent green in black light.

It's enough to make you throw down the paper and go sit in a park. It's enough to make you wish you lived in the good old days when things were so much simpler.

Of course, read up on the good old days and you'll discover they weren't so good. For example, in 1918, maybe when your grandparents were alive, the Spanish flu swept around the world. Infected people often died within hours of the first symptoms. Before the pandemic subsided, *30 to 40 million* people died. Unlike most flu viruses, which mainly endanger infants, the elderly,

Figure 1.1 Biology starts with the premise that any aspect of nature—including this forest fire in Montana—has one or more underlying causes. By giving us a way of thinking critically about life, biology helps us understand how to deal with nature, and with our place in it.

and the very ill, the Spanish flu virus tended to take down healthy, young adults, the ones who actually support most of the social infrastructure. Then, as today, many people thought the world was coming apart at the seams. They felt helpless and hopeless before a force of nature on the rampage.

What it boils down to is this: For a couple of million years at least, humans and their immediate ancestors have been trying to make sense of the natural world. We observe, we come up with ideas about it, we test the ideas. However, the more pieces of the puzzle we fit together, the bigger the puzzle gets. We are now smart enough to know that it's almost overwhelmingly big.

You might choose to walk away from the challenge and simply let others tell you what to think. Or you might choose to develop your own understanding of the puzzle. Maybe you're interested in the pieces that affect your health, your home, the food you eat, and your offspring, should you choose to reproduce. Maybe you just find the connections among organisms and their environment fascinating. Regardless of the focus, **biology**—the scientific study of life—can help you deepen your perspective on the nature of life.

Start with a question that seems simple enough: *What is life?* Offhandedly, you might say you know it when you see it. Yet the question opens up a story that began at least 3.8 billion years ago and has been unfolding in countless directions ever since.

From a biological perspective, "life" is an outcome of ancient events by which lifeless matter—atoms and molecules—became organized into the first living cells. "Life" is a way of capturing and using energy and raw materials. "Life" is a way of sensing and responding to the environment. "Life" is a capacity to reproduce. And "life" evolves, which means the traits that characterize individuals of a population change from one generation to the next.

Throughout this book, you will come across many examples of how organisms are constructed, how they function, where they live, what they do. The examples support concepts which, when taken together, convey what "life" is. This chapter introduces you to the basic concepts. It sets the stage for forthcoming descriptions of scientific observations, experiments, and tests that help show how you can develop, modify, and refine your views of life. As you continue your reading, you may find it useful to return to this simple overview as a way to reinforce your grasp of the details.

Key Concepts

1. Unity underlies the world of life, for all organisms are alike in key respects. They consist of one or more cells made of the same kinds of substances, put together in the same basic ways. Their activities require inputs of energy, which they must get from their surroundings. All organisms sense and respond to changing conditions in their environment. They all have a capacity to grow and reproduce, based on instructions contained in DNA.

2. The world of life shows immense diversity. Many millions of different kinds of organisms, or species, now inhabit the Earth. Many millions more lived in the past. Each species is unique in some of its traits—that is, in some aspects of its body plan, body functioning, and behavior.

3. Theories of evolution, especially a theory of evolution by natural selection as formulated by Charles Darwin, help explain life's diversity. The theories unite all fields of biological inquiry into a single, coherent whole.

4. Biology, like other branches of science, is based on systematic observations, hypotheses, predictions, and observational and experimental tests. The external world, not internal conviction, is the testing ground for scientific theories.

DNA, ENERGY, AND LIFE

Nothing Lives Without DNA

DNA AND THE MOLECULES OF LIFE   Picture a frog on a rock, busily croaking. Even without thinking about it, you know the frog is alive and the rock is not. Could you explain why? After all, both consist only of protons, electrons, and neutrons. But these units of matter are building blocks for diverse atoms—the building blocks for the larger and more diverse bits of matter called molecules. It is at the molecular level that differences between living and nonliving things start to emerge.

You will never, ever find a rock made of nucleic acids, proteins, carbohydrates, and lipids. In nature, only *cells* build these molecules, which you will read about later. All living things are composed of one or more cells, the smallest units of matter having a capacity for life. And the signature molecule of cells is a nucleic acid known as **DNA**. No chunk of granite or quartz has it.

Encoded in DNA's structure are the instructions for assembling a dazzling array of proteins from a limited number of smaller building blocks, the amino acids. By analogy, if you follow suitable instructions and invest energy in the task, you might organize a heap of a few kinds of ceramic tiles (representing amino acids) into diverse patterns (representing proteins), as in Figure 1.2.

Enzymes are among the proteins. When these worker molecules get an energy boost, they swiftly build, split, and rearrange the molecules of life. As described later, some work with nucleic acids called RNAs to carry out DNA's protein-building instructions. Think of this as a flow of information, from *DNA to RNA to protein*. This molecular trinity is central to our understanding of life.

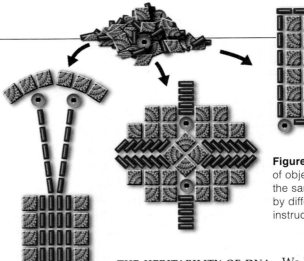

Figure 1.2 Examples of objects built from the same materials by different assembly instructions.

THE HERITABILITY OF DNA   We humans tend to think we all enter the world abruptly and leave it the same way. But we are much more than this. *We and all other organisms are part of a journey that began at least 3.8 billion years ago, with the origin of the first living cells.*

Under present-day conditions on Earth, new cells arise only from cells that already exist. They do so by **inheritance**, which is an acquisition of traits by way of transmission of DNA from parents to offspring. Why do baby storks look like storks and not pelicans? Because they inherited stork DNA, which isn't exactly the same as pelican DNA in its molecular details.

Reproduction refers to actual mechanisms by which parents produce offspring. Often it starts when a sperm fertilizes an egg. However, a fertilized egg would never form if the sperm and egg had not formed earlier. And both formed from DNA instructions passed on countless times—from cell to cell—according to the principles of inheritance and reproduction.

For frogs and humans and other large organisms, DNA also guides **development**, a series of stages that transform a single cell into a multicelled adult, typically with tissues and organs specialized for certain tasks. For instance, a moth is the adult stage of a winged insect (Figure 1.3). It starts as a fertilized egg that develops

Figure 1.3   "The insect"—a series of stages of development. Different adaptive properties emerge at each stage. Shown here, a silkworm moth, from the egg (**a**), to a larval stage (**b**), to a pupal stage (**c**), and on to the winged form of the adult (**d,e**).

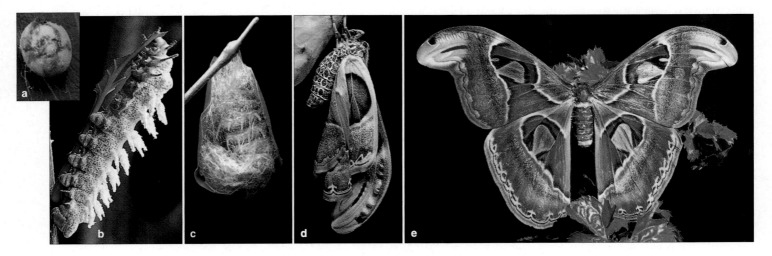

Figure 1.4 Response to signals from pain receptors, activated by a lion cub flirting with disaster.

into an immature larval stage. That stage, a caterpillar, eats leaves and grows rapidly until an internal alarm clock goes off. Then, its tissues become remodeled into a different stage—a pupa. In time an adult emerges. An adult form is adapted to reproduce. Adult moths make sperm or eggs, and the color, pattern, and fluttering frequency of its wings are adapted to attract a mate.

And so "the insect" is a series of organized stages, each of which must be completed before the next begins. Instructions for completing each stage were written into moth DNA long before each moment of reproduction—and so the ancient moth story continues.

Nothing Lives Without Energy

ENERGY DEFINED Moths and everything else in the universe have some amount of **energy**, which we define as the capacity to do work. And nothing—absolutely nothing—happens in the universe without a *transfer* of energy. For instance, suppose some undisturbed atom in a cell absorbs enough energy from the sun to get its electrons jumping. A molecule nearby grabs an excited electron and transfers it to machinery that converts the extra energy into a form the cell can use. In such ways, atoms give up, share, and grab electrons; and molecules form, are rearranged, and are split apart. Such molecular work allows cells to stay alive, grow, and reproduce.

METABOLISM DEFINED Each living cell has the capacity to (1) obtain and convert energy from its surroundings and (2) use energy to maintain itself, grow, and make more cells. We call this capacity **metabolism**. Think of one cell in a leaf that makes food by photosynthesis. It intercepts sunlight energy and converts it to chemical energy, in the form of ATP molecules. ATP helps drive hundreds of events by transferring energy to metabolic workers—in this case, enzymes that put together sugar molecules. ATP also forms by aerobic respiration. This process can release energy that cells have tucked away in sugars and other kinds of molecules.

SENSING AND RESPONDING TO ENERGY It's often said that only organisms respond to the environment. Yet even a rock shows responsiveness, as when it yields to the force of gravity and tumbles down a hill or changes its shape slowly under the repeated battering of wind, rain, or tides. The difference is this: *Organisms can sense changes in their surroundings and then make compensatory, controlled responses to the changes.* How? Each organism has **receptors**, which are molecules and structures that detect stimuli. A **stimulus** is a specific form of energy that a receptor can detect. Examples are sunlight energy,

heat energy, a hormone molecule's chemical energy, and the mechanical energy of a bite (Figure 1.4).

Cells adjust metabolic activities in response to signals from receptors. Each cell (and organism) can withstand only so much heat or cold. It must rid itself of harmful substances. It requires certain foods, in certain amounts. Yet temperatures do shift, harmful substances might be encountered, and food is sometimes plentiful or scarce.

Finish a snack, and simple sugars leave your gut and enter your blood. Blood, along with the tissue fluid that bathes your cells, makes up your *internal* environment. Unchecked, too much or too little sugar in the blood can cause diabetes and other problems. When the level rises, the pancreas (a glandular organ) secretes more insulin. Most of your body cells have receptors for this hormone, which stimulates cells to take up sugar. When enough cells do so, the blood sugar level returns to normal.

Organisms respond so exquisitely to energy changes that their internal operating conditions remain within tolerable limits. We call this a state of **homeostasis**. It is one of the key defining features of life.

All organisms consist of one or more cells, the smallest units of life. Under present-day conditions, new cells form only when cells that are already alive reproduce.

DNA, the molecule of inheritance, encodes protein-building instructions, which RNAs help carry out. Many proteins are enzymes that speed up cellular work, which includes building all the complex molecules characteristic of life.

Cells live only for as long as they engage in metabolism. They acquire and transfer energy that is used to assemble, break down, stockpile, and dispose of materials in ways that promote survival and reproduction.

Single cells and multicelled organisms sense and respond to environmental conditions in ways that help maintain their internal operating conditions.

ENERGY AND LIFE'S ORGANIZATION

Levels of Biological Organization

Taken as a whole, the metabolic activities of single cells and multicelled organisms maintain a great pattern of organization in nature, as in Figure 1.5. Consider the hierarchy. Life's properties emerge when DNA and other molecules are organized into cells. A **cell** is the smallest unit of organization having a capacity to survive and reproduce on its own, given DNA instructions, suitable conditions, building blocks, and energy inputs. This is obvious for amoebas and other free-living cells. Yet the definition also fits **multicelled organisms** composed of many specialized, interdependent cells that are typically organized in tissues and organs. Do you question this? After all, your own cells could never live all alone in nature, because body fluids must continuously bathe them. Yet even isolated human cells remain alive under controlled conditions in laboratories around the world. Researchers routinely maintain isolated human cells for important experiments, as in cancer studies.

Cells and multicelled organisms usually are part of a **population**: a group of organisms of the same kind. A zebra herd is an example. The next level of organization is the **community**: all populations of all species living in the same area, such as the African savanna's bacteria, grasses, trees, zebras, lions, and so on. The next level is the **ecosystem**: a community *together with* its physical and chemical environment. The **biosphere**, the highest level, includes all parts of the Earth's crust, waters, and atmosphere in which organisms live. Astoundingly, *this globe-spanning organization begins with the convergence of energy, certain materials, and DNA in tiny, individual cells.*

BIOSPHERE
All regions of the Earth's crust, waters, and atmosphere that sustain life

ECOSYSTEM
Community and its physical environment

COMMUNITY
Populations of all species occupying the same area

POPULATION
Group of individuals of the same kind (that is, the same species) occupying the same area

MULTICELLED ORGANISM
Individual consisting of interdependent cells typically organized in tissues, organs, and organ systems

ORGAN SYSTEM
Two or more organs interacting chemically, physically, or both in ways that contribute to organism's survival

ORGAN
Structural unit in which tissues, combined in specific amounts and patterns, perform a common task

TISSUE
Organized aggregation of cells and substances functioning together in a specialized activity

CELL
Smallest unit with the capacity to live and reproduce, independently or as part of multicelled organism

ORGANELLE
Membrane-bound internal compartment for specialized reactions (most prokaryotic cells have none)

MOLECULE
Unit of two or more bonded-together atoms of the same element or different elements

ATOM
Smallest unit of an element (a fundamental substance) that still retains the properties of that element

SUBATOMIC PARTICLE
Electron, proton, neutron, or some other fundamental unit of matter

community (all of the populations living in the same area)

populations of shrubs and trees

populations of grasses

population of zebras

ecosystem (community together with its physical environment)

Figure 1.5 Levels of organization in nature.

multicelled individual

once-live multicelled individual, about to revert to molecules and atoms

beetle larva

Figure 1.6 Example of the one-way flow of energy and the cycling of materials through the biosphere.

(a) Plants of a warm, dry grassland called the African savanna capture energy from the sun and use it to build plant parts. Some of the energy ends up inside plant-eating organisms, such as this adult male elephant. He eats huge quantities of plants to maintain his eight-ton self. He produces piles of solid wastes—dung—that still hold some unused nutrients and energy. Although most organisms might not recognize it as such, elephant dung is an exploitable food source.

(b) And so we next have little dung beetles scrambling to the scene almost simultaneously with the uplifting of an elephant tail. Working rapidly, they carve fragments of moist dung into round balls, which they roll off and bury in burrows. In those balls the beetles lay eggs— a reproductive behavior that will help assure their forthcoming offspring (**c**) of a compact food supply.

Thanks to beetles, dung does not pile up and dry out into rock-hard mounds in the intense heat of the day. Instead, the surface of the land is tidied up, the beetle offspring get fed, and the leftover dung accumulates in beetle burrows—there to enrich the soil that nourishes the plants that feed (among others) the elephants.

Within the diagram:

Producers capture, convert, and use or store some energy from the sun.

PRODUCERS

NUTRIENT CYCLING

CONSUMERS, DECOMPOSERS

ONE-WAY FLOW OF ENERGY

Energy gets transferred from one organism to another; in time, all flows back to the environment.

Interdependencies Among Organisms

A great flow of energy into the world of life starts with **producers**, which are plants and other organisms that make their own food. Animals are **consumers**. Directly or indirectly, they depend on food energy that's stored in tissues of producers. For example, some energy gets transferred to zebras after they've browsed on plants. It gets transferred again when a lion devours a zebra, as in Figure 1.5. And it is transferred again when fungal and bacterial decomposers extract energy from tissues and remains of lions, elephants, or any other organism. **Decomposers** break down sugars and other biological molecules to simpler materials, some of which is cycled back to the producers. In time, all of the energy that the plants originally captured from the sun's rays returns to the environment, but that's another story.

For now, keep in mind that organisms connect with one another by a one-way flow of energy *through* them and a cycling of materials *among* them, as in Figure 1.6. Their interconnectedness affects the structure, size, and composition of populations and communities. It affects ecosystems, even the biosphere. Understand the extent of their interactions and you will gain insight into food shortages, cholera epidemics, acid rain, global warming, biodiversity losses, and other modern-day problems.

Nature shows levels of organization. The characteristics of life emerge at the level of single cells and extend through populations, communities, ecosystems, and the biosphere.

A one-way flow of energy through organisms and a cycling of materials among them organizes life in the biosphere. In nearly all cases, energy flow starts with energy from the sun.

IF SO MUCH UNITY, WHY SO MANY SPECIES?

So far, we have focused on life's unity, on characteristics that all living things have in common. Think of it! They consist of the same "lifeless" materials. They can stay alive by metabolism— by ongoing energy transfers at the cellular level. They interact in their requirements for energy and raw materials. They have a capacity to sense and respond to their environment in highly specific ways. They all have the capacity to reproduce based on instructions encoded in DNA. They all inherited DNA from individuals of a preceding generation.

Superimposed on the common heritage is immense diversity. You share the planet with many millions of different kinds of organisms, or **species**. Many millions more preceded us during the past 3.8 billion years, but their lineages vanished; they are extinct. For centuries, scholars have tried to make sense of the confounding diversity. One of them, Carolus Linneaus, came up with a classification scheme that assigns a two-part name to each newly identified species. The first part designates the **genus** (plural, genera). Each genus encompasses all species that seem closely related by way of their form, functioning, and ancestry. The second part of the name designates a particular species within a genus.

For instance, *Quercus alba* is the white oak's scientific name, and *Q. rubra* the red oak's name. (Once you spell out the genus name in a document, you can abbreviate it thereafter.) Use such names as passports when you visit the web, and enter worlds of information.

Biologists also classify life's diversity by assigning species to groups at more encompassing levels. Among other things, they group genera that apparently share a common ancestor into the same *family*, related families into the same *order*, related orders into the same *class*, and related classes into the same *phylum* (plural, phyla). At a higher level, they assign related phyla to the same *kingdom*. They are still refining the groups. For instance, scholars once recognized only two kingdoms—animals and plants. Later, biologists recognized five kingdoms. Today, most agree that there should be at least six: the **Archaebacteria, Eubacteria, Protista, Fungi, Plantae,** and **Animalia** (Figure 1.7). We adhere to a six-kingdom scheme in this book. At this point in your reading, it is enough simply to become acquainted with a few of the defining features of their members.

Archaebacteria and eubacteria are all single cells, of a type called *prokaryotic*. The word means they don't have a nucleus. (In other cell types, this membrane-bound sac keeps DNA separate from the rest of the cell's interior.)

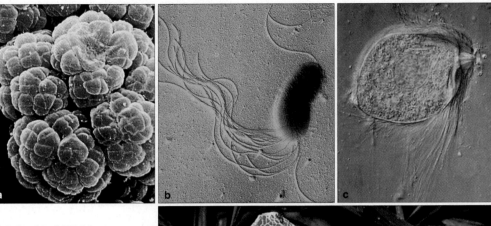

Figure 1.7 A few representatives of life's diversity.

KINGDOM ARCHAEBACTERIA. (**a**) From the muck of an anaerobic (oxygen-free) habitat, a colony of cells (*Methanosarcina*).

KINGDOM EUBACTERIA. (**b**) A "true" bacterium sporting a number of bacterial flagella. Flagella are used for motility.

KINGDOM PROTISTA. (**c**) A trichomonad that lives in a termite's gut. Compared to bacteria, most protistans are much larger and have greater internal complexity. They range from microscopically small single cells, such as this one, to kelps as tall as trees.

KINGDOM FUNGI. (**d**) A stinkhorn fungus. Some species of fungi are parasites and some cause diseases, but the vast majority are decomposers. Without decomposers, communities would gradually become buried in their own wastes.

KINGDOM PLANTAE. (**e**) Trunk of a redwood growing near the coast of California. Like nearly all plants, redwoods produce their own food by photosynthesis. (**f**) Flower of a plant from the family of composites. Its colors and patterning guide bees to nectar. The bees get food. The plant gets help reproducing. Like many other organisms, they interact in a mutually beneficial way.

KINGDOM ANIMALIA. (**g,h**) Male bighorn sheep competing for females and so displaying a characteristic of the kingdom— they actively move about in their environment.

Different kinds are producers, consumers, and decomposers. Of all kingdoms, theirs show the most metabolic diversity.

Diverse archaebacteria live in hot springs, salt lakes, and other habitats as harsh as the ones that prevailed when cells first evolved. The eubacteria are more common and widespread than archaebacteria. Their name means "true bacteria," so in this book we'll just refer to them as bacteria (singular, bacterium).

Characterizing protistans is not easy. Like the plants, fungi, and animals, they are all *eukaryotic*, meaning their DNA is inside a nucleus. Most are larger and show far more internal complexity than prokaryotic cells. Beyond this generalization are confoundingly diverse single-celled producers, consumers, and decomposers, as well as giant multicelled "seaweeds." The food webs of most aquatic provinces start with photoautotrophic protistans.

Most fungi, including the common field mushrooms sold in grocery stores, are multicelled. These eukaryotic decomposers and consumers feed in a distinctive way. They secrete enzymes that digest food outside of the fungal body, then their cells absorb the digested bits.

Both plants and animals are multicelled, eukaryotic organisms. Nearly all plants are photosynthetic, with cellular pipelines for water and solutes that extend through roots, stems, and leaves. Most live on land and are the major producers for communities. Animals are consumers that ingest other organisms or their tissues. Among their ranks are herbivores (which graze on the photosynthesizers), carnivores (meat eaters), parasites, and scavengers. Unlike plants, all animals actively move about during at least some stage of their life.

Pulling this together, you start to get a sense of what it means when someone says life has unity *and* diversity.

Unity threads through the world of life, for all organisms are alike in important ways. They are composed of the same substances, which are assembled in the same basic ways. They engage in metabolism, and they sense and respond to their environment. They all have a capacity to reproduce, based on heritable instructions encoded in their DNA.

Immense diversity threads through the world of life, for organisms differ enormously in body form, in the functions of their body parts, and in their behavior.

To make the study of life's diversity more manageable, we group organisms into six kingdoms—the archaebacteria, eubacteria, protistans, fungi, plants, and animals.

AN EVOLUTIONARY VIEW OF DIVERSITY

Given that organisms are so much alike, what could account for their great diversity? One key explanation is called evolution by means of natural selection. A few simple examples will be enough to introduce you to its premises, which build on the simple observation that individuals of a population vary in their traits.

Mutation—Original Source of Variation

DNA has two striking qualities. Its instructions work to ensure that offspring will resemble their parents, yet they also permit variations in the details of most traits. As an example, having five fingers on each hand is a human trait. Yet some humans are born with six fingers on each hand instead of five. This is an outcome of a **mutation**—a heritable change in DNA. Mutations are the original source of variations in heritable traits.

Many mutations are harmful. A change in even a bit of DNA may be enough to sabotage the body's growth, development, or functioning. Yet some variations are harmless or beneficial. A classic case is a mutation that results in dark-colored moths in a population of light-colored moths. Moths fly at night and rest in the day, when birds that eat them are active. Birds tend to miss light-colored moths resting on light-colored tree trunks. These moths are camouflaged; they "hide in the open."

Suppose people build coal-burning factories nearby. In time, soot-laden smoke darkens all the tree trunks. Dark moths on darkened trunks aren't as conspicuous to bird predators (Figure 1.8), so they are more likely to survive and reproduce. Where soot accumulates, the variant (dark) form of this trait is more adaptive than the light form. An **adaptive trait** is any form of a trait that helps an individual survive and reproduce under prevailing environmental conditions.

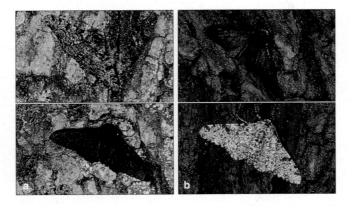

Figure 1.8 How different forms of a trait (body surface coloration) are adaptive to two different conditions. (**a**) Light-colored moths (*Biston betularia*) on a nonsooty tree trunk are well hidden from predators, but dark ones stand out. (**b**) The dark color is more adaptive in places where soot darkens the tree trunks.

Figure 1.9 *Facing page:* One outcome of artificial selection practices: just a few of the 300+ varieties of domesticated pigeons. Breeders started with variant forms of traits in captive populations of wild rock doves.

WILD ROCK DOVE

Evolution Defined

Think about that population of light-colored moths in a sooty forest. At some point, a DNA mutation arose and resulted in a moth of a darker color. When the mutated individual reproduced, some of its offspring inherited the trait. Birds saw and ate many light-colored moths, but most of the dark ones escaped detection and lived long enough to reproduce. So did their dark offspring. So did *their* offspring, and so on. Said another way, the frequency of one form of the trait increased relative to the frequency of the other. In time, the dark form may even become the most common form. People may end up referring to "the population of dark-colored moths."

Evolution is under way. In biology, the word means genetically based change in a line of descent over time. Like moths, individuals of most populations typically show different forms of many (or most) of their traits. The relative frequencies of the different forms typically change over successive generations.

Natural Selection Defined

Long ago, the naturalist Charles Darwin used pigeons to explain a conceptual connection between evolution and variation in traits. Domesticated pigeons differ in their size, feather color, and other traits (Figure 1.9). As Darwin knew, pigeon breeders select certain forms of traits. For example, if they are after black, curly-edged tail feathers, they will let only individual pigeons with the most black and most curl in their tail feathers mate and reproduce. So "black" and "curly" will be the most common forms of tail feathers in the captive pigeon population. Different forms of these traits will become less common or will be eliminated.

Pigeon breeding is a case of **artificial selection**, for the selection among different forms of a trait is taking place in an "artificial" environment—under contrived, manipulated conditions. Yet Darwin saw the practice as a simple model for *natural* selection, a favoring of some forms of traits over others in nature.

Whereas breeders are "selective agents" promoting the reproduction of some individuals but not others in captive populations, different selective agents operate across the range of variation among pigeons in the wild. Pigeon-eating peregrine falcons are such agents. Swifter or more effectively camouflaged individuals of a pigeon

population have a far better chance of escaping from falcons and living long enough to reproduce, compared to the not-so-swift or too-conspicuous individuals among them. What Darwin identified as **natural selection** is simply a difference in *which* individuals of a population survive and reproduce in a given generation. The difference is an outcome of some individuals having more adaptive forms of traits.

Unless you happen to be a pigeon breeder, variation among pigeons may not be a burning issue in your life. So think about something closer to home. **Antibiotics** are toxic secretions of certain soil-dwelling bacteria and fungi; they work against bacterial and fungal competitors for nutrients. In the 1940s, we learned to use antibiotics to kill disease-causing bacteria. Doctors prescribed these "wonder drugs" even for mild infections. In some cases antibiotics were added to toothpaste and chewing gum.

As it turned out, antibiotics are powerful agents of natural selection. Consider streptomycin, an antibiotic that shuts down some essential proteins in bacteria that are its targets. However, certain strains of bacteria have mutated in ways that help them resist streptomycin. The mutations slightly altered the molecular form of the proteins, so streptomycin can't bind to them. Unlike nonmutated cells, the mutants survive and reproduce. Said another way, an antibiotic acts against the bacteria susceptible to it but *favors* variant strains that resist it!

Antibiotic-resistant strains are making it difficult to treat bacterial diseases, including tuberculosis, typhoid, gonorrhea, and staph infections. A few superbugs can no longer be eliminated from the most susceptible people.

When resistance to antibiotics evolves by selection processes, our strategies for using antibiotics must also evolve to overcome the new defenses. That's why drug companies try to modify parts of antibiotic molecules. Such molecular changes in the laboratory might yield more effective antibiotics. But these may work only until new generations of ever more resistant superbugs enter the deadly evolutionary competition for nutrients.

Later in the book, we will consider mechanisms by which populations of moths, pigeons, bacteria, and all other organisms evolve. Meanwhile, keep in mind the following points about natural selection. They are central to biological inquiry, for they have consistently proved useful in explaining a great deal about nature.

1. Individuals of a population vary in form, function, and behavior. Much of this variation is heritable; it can be transmitted from parents to offspring.

2. Some forms of heritable traits are more adaptive to prevailing conditions. They improve an individual's chance of surviving and reproducing, as by helping it secure food, a mate, hiding places, and so on.

3. Natural selection is the outcome of differences in survival and reproduction among variant individuals in a given generation.

4. Natural selection leads to a better fit with prevailing environments. The adaptive forms of traits tend to become more common than other forms. Thus the population's characteristics change; it evolves.

In the evolutionary view, then, *life's diversity is the sum total of variations in traits that have accumulated in different lines of descent generation after generation, as by natural selection and other processes of change.*

Mutations in DNA introduce variations in heritable traits.

Although many mutations are harmful, some give rise to variations in form, function, or behavior that are adaptive under prevailing environmental conditions.

Natural selection is a result of differences in survival and reproduction among individuals of a population that vary in one or more heritable traits. The process helps explain evolution—changes in lines of descent over the generations.

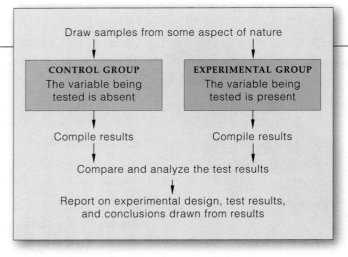

Figure 1.10 Generalized sequence of steps involved in an experimental test of a prediction based on a hypothesis.

THE NATURE OF BIOLOGICAL INQUIRY

The preceding sections sketched out major concepts in biology. Now consider approaching this or any other collection of "facts" with a critical attitude. *"Why should I accept that they have merit?"* The answer requires insight into how biologists make inferences about observations and test the predictive power of the inferences against actual experience in nature or the laboratory.

Observations, Hypotheses, and Tests

To get a sense of "how to do science," start by following some practices that pervade scientific research:

1. Observe some aspect of nature, carefully check what others have found out about it, then frame a question or identify a problem related to your observation.

2. Develop **hypotheses**, or educated guesses, about possible answers to questions or solutions to problems.

3. Using hypotheses as a guide, make a **prediction**—that is, a statement of what you should observe in the natural world if you were to go looking for it. This is often called the "if–then" process. (*If* gravity does not pull objects toward Earth, *then* it should be possible to observe apples falling up, not down, from a tree.)

4. Devise ways to **test** the accuracy of your predictions, as by making systematic observations, building models, and conducting experiments. **Models** are theoretical, detailed descriptions or analogies that help us visualize something that has not yet been directly observed.

5. If tests do not confirm the prediction, check to see what might have gone wrong. For example, maybe you overlooked some factor that influenced the test results. Or maybe the hypothesis is not a good one.

6. Repeat the tests or devise new ones—the more the better, for hypotheses that withstand many tests are likely to have a higher probability of being useful.

7. Objectively analyze and report the test results and the conclusions drawn from them.

You might hear someone refer to these practices as "the scientific method," as if all scientists march to the drumbeat of an absolute, fixed procedure. They do not. Many observe, describe, and report on some subject, then leave it to others to hypothesize about it. Some are lucky; they stumble onto information they aren't even looking for. (Of course, chance does favor a mind that's prepared to know what the information means.) It is not one single method scientists have in common. Rather, it is a critical attitude about being shown rather than told and taking a logical approach to problem solving.

Logic encompasses thought patterns by which an individual draws a conclusion that does not contradict evidence used to support it. Lick a cut lemon and you notice it's mouth-puckeringly sour. Lick ten more. Each time you notice the same thing, so you conclude that all lemons are mouth-puckeringly sour. You correlated one specific (lemon) with another (sour). By this pattern of thinking, called **inductive logic**, an individual derives a general statement from specific observations.

Express the generalization in "if–then" terms, and you have a hypothesis: "If you lick any lemon, then you will get an extremely sour taste in your mouth." By this pattern of thinking, called **deductive logic**, an individual makes inferences about specific consequences or specific predictions that must follow from a hypothesis.

You decide to test the hypothesis by sampling lemons in your area. One variety, the Meyer lemon, is mellow for a lemon. You also find that some people can't taste anything. So you modify the hypothesis: "If most people lick any lemon *except* the Meyer lemon, they will get a very sour taste in their mouth." Suppose you sample all known lemon varieties in the world and conclude the modified hypothesis is a good one. You'll never prove it beyond all shadow of a doubt, for there may be lemon trees in places we don't even know about. You *can* say the hypothesis has a high probability of not being wrong.

Comprehensive observations are a logical means to test the predictions that flow from hypotheses. So are **experiments**. These tests simplify observation in nature or the laboratory by manipulating and controlling the conditions under which observations are made. When suitably designed, observational and experimental tests allow you to predict that something will happen if a hypothesis isn't wrong (or won't happen if it *is* wrong). Figure 1.10 gives a general idea of the steps involved.

AN ASSUMPTION OF CAUSE AND EFFECT Experiments start from the premise that any aspect of nature has one or more underlying causes. With this premise, science is distinct from faith in the supernatural (meaning "beyond nature"). Experiments deal with potentially falsifiable hypotheses. Hypotheses of this sort can be tested in the natural world in ways that might disprove them.

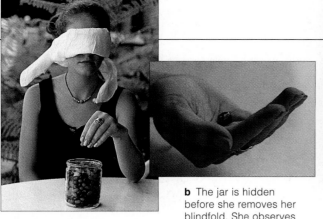

a Natalie, blindfolded, randomly plucks a jellybean from a jar of 120 green and 280 black jellybeans. That's a ratio of 30 to 70 percent.

b The jar is hidden before she removes her blindfold. She observes a single green jellybean in her hand and assumes the jar holds only green jellybeans.

c Still blindfolded, Natalie randomly picks 50 jellybeans from the jar and ends up with 10 green and 40 black ones.

d The larger sample leads her to assume one-fifth of the jar's jellybeans are green and four-fifths are black (a ratio of 20 to 80). Her larger sample more closely approximates the jar's green-to-black ratio. The more times Natalie repeats the sampling, the greater the chance she will come close to knowing the actual ratio.

Figure 1.11 A simple demonstration of sampling error.

EXPERIMENTAL DESIGN To get conclusive test results, experimenters rely on certain practices. They refine test designs by searching the literature for information that may relate to their inquiry. They design experiments to test one prediction of a hypothesis at a time. Each time, they set up a **control group**: a standard for comparison with one or more experimental groups. Ideally, a control group is identical with the experimental group *except* for the one variable being investigated. **Variables** are specific aspects of objects or events that may differ or change over time and among individuals.

Section 1.6 shows the design of a recent experiment. As you will see, the experimenters directly manipulated a single variable in an attempt to support or disprove a prediction. They also tried to hold constant any other variables that could influence the results.

SAMPLING ERROR Rarely can experimenters observe *all* individuals of a group. Rather, they must use subsets or samples of populations, events, and other aspects of nature. They must avoid **sampling error**, or risking tests with groups that by chance are not representative of the whole. Generally, distortion in test results is less likely when the samplings are large and when they are repeated (Figure 1.11).

About the Word "Theory"

Suppose no one has disproved a hypothesis after years of rigorous tests. Suppose scientists use it to interpret more data or observations, which could involve more hypotheses. When a hypothesis meets these criteria, it may become accepted as a **scientific theory**.

You may hear someone apply the word "theory" to a speculative idea, as in the expression "It's only a theory." But a scientific theory differs from speculation for this reason: *Researchers have tested its predictive power many times and in many ways in the natural world and have yet to find evidence that disproves it*. This is why the theory of natural selection is respected. We use it successfully to explain diverse issues, such as how life originated, how plant toxins relate to plant-eating animals, what the sexual advantage of feather color might be, why certain cancers run in families, or why antibiotics are losing effectiveness. Together with the record of Earth history, the theory has even helped explain how life evolved.

Maybe an exhaustively tested theory is as close to the truth as scientists can get with the evidence available. For example, Darwin's theory stands, with only minor modification, after more than a century's worth of many thousands of different scientific tests. We can't prove it holds under all possible conditions; an infinite number of tests would be required to do so. As for any theory, we can only say it has a *high or low probability* of being a good one. So far, biologists haven't found evidence that refutes Darwin's theory. Yet they still keep their eyes open for any new information and new ways of testing that might disprove its premises.

This last point gets us back to the value of thinking critically. Scientists must keep asking themselves: *Will observations or experiments show that a hypothesis is false?* They expect one another to put aside pride or bias by testing ideas even in ways that may prove them wrong. Even if an individual doesn't or won't do this, others will—for science proceeds as a community that is both cooperative and competitive. Ideally, its practitioners share ideas, knowing it's just as useful to expose errors as to applaud insights. Individuals can and do change their mind when presented with contradictory evidence. This is a strength of science, not a weakness.

A scientific approach to studying nature is based on asking questions, formulating hypotheses, making predictions, devising tests, and objectively reporting the results.

A scientific theory is a testable explanation about the cause or causes of a broad range of related phenomena. It remains open to tests, revision, and tentative acceptance or rejection.

The Power of Experimental Tests

BIOLOGICAL THERAPY EXPERIMENTS Worry about the new strains of pathogenic (disease-causing) bacteria that resist antibiotics, and you won't be alone. Bruce Levin of Emory University and Jim Bull of the University of Texas have been looking for possible alternatives to antibiotic therapy. They studied literature on a *biological therapy* that enlists bacteriophages—"bacteria eaters"—to fight infections. Bacteriophages, a class of viruses, attack a narrow range of bacterial strains. When they contact a target cell, they inject enzymes and genetic material into it. What happens next is a hostile takeover of the cell's metabolic machinery. The cell makes many new viral particles, then dies after viral enzymes rupture its outer membrane. Virus particles escape and infect new cells.

The biologists wondered, as others had decades ago, whether injections of bacteriophages could help people fight bacterial infections. The discovery of antibiotics in the 1940s had diverted attention away from that idea, at least in the West. Now the bacteria eaters were starting to look good again. Levin and Bull focused on earlier experiments conducted by two British researchers,

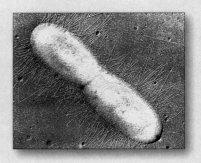

Figure 1.12 *Escherichia coli* cell dividing in two. This is a harmless strain.

H. Williams Smith and Michael Huggins. Would they be able to duplicate the results of those researchers?

Levin and Bull selected 018:K1:H7, which is a strain of *Escherichia coli*. Like a harmless *E. coli* strain that lives in the intestines of humans and other mammals (Figure 1.12), it leaves the body in feces. The bacteriophages used in laboratory work usually ignore 018:K1:H7. So the biologists looked for some *E. coli* killers in samples from a sewage treatment plant in Atlanta. They isolated two kinds of bacteriophages and named them *H* (Hero, an effective killer) and *W* (the less effective Wimp).

HYPOTHESIS: If bacteriophages target and kill cells of *E. coli* 018:K1:H7 in petri dishes, then they will also kill them in laboratory mice infected by that strain.

PREDICTION: Laboratory mice injected with a preparation containing more than 10^7 particles of *H* bacteriophage will not die after being injected with *E. coli* strain 018:K1:H7.

EXPERIMENTAL TEST OF PREDICTION:

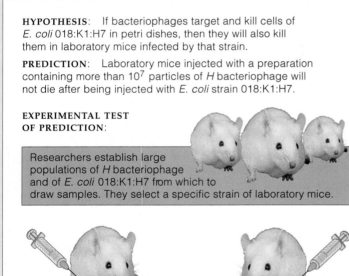

Researchers establish large populations of *H* bacteriophage and of *E. coli* 018:K1:H7 from which to draw samples. They select a specific strain of laboratory mice.

CONTROL GROUP	EXPERIMENTAL GROUP
E. coli injected into right thigh of 15 laboratory mice. No bacteriophage are injected.	*E. coli* is injected into 15 other mice. Left thigh gets bacteriophage injection.

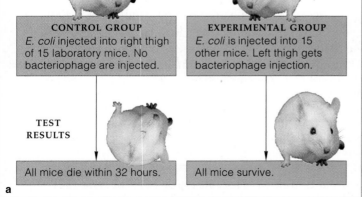

TEST RESULTS

All mice die within 32 hours.	All mice survive.

a

ANOTHER PREDICTION, DERIVED FROM SAME HYPOTHESIS: When treating mice infected with 018:K1:H7, *H* bacteriophages that target and kill 018:K1:H7 will be more effective at killing them compared to single doses of the antibiotic streptomycin.

EXPERIMENTAL TEST OF PREDICTION:

Researchers inject 48 laboratory mice with 018:K1:H17. They divide the 48 into four groups of 12 mice. Eight hours later,

TEST RESULTS

CONTROL GROUP 12 mice receive an injection of saline solution only.	All control mice die.
EXPERIMENTAL GROUP 1 12 mice receive a single injection of *H* bacteriophage.	Of 12 mice, 11 survive.
EXPERIMENTAL GROUP 2 12 mice receive a single dose of 60 micrograms/gram streptomycin.	Of 12 mice, 11 survive.
EXPERIMENTAL GROUP 3 12 mice receive a single dose of 100 micrograms/gram streptomycin.	Of 12 mice, 3 survive.

b

Figure 1.13 Two experiments to compare the effectiveness of bacteriophage injections against antibiotics for treating a bacterial infection.

With graduate student Terry DeRouin and laboratory technician Nina Moore Walker, the biologists grew a large population of 018:K1:H7 in a culture flask. They selected a specific strain of laboratory mice, all females of the same age. Fifteen mice of one experimental group were each injected with 018:K1:H7 in one thigh and more than 10^7 particles of *H* bacteriophage in the other. All mice survived. Fifteen mice of a control group were injected only with 018:K1:H7. None survived. Figure 1.13*a* outlines this experimental test.

Figure 1.13*b* outlines a second experimental test of another prediction derived from the same hypothesis. The results reinforced Smith and Huggins's conclusion that certain bacteriophages are as good as or better than antibiotics at stopping specific bacterial infections.

Each type of bacteriophage targets specific strains of one or at most a few bacterial species. It now takes too long to discover which bacterium is infecting a patient, so the right bacteriophage might not be enlisted until it's too late. New procedures may speed up identification.

IDENTIFYING IMPORTANT VARIABLES In nature, many factors can influence the outcome of an infection. For example, genetic differences among infected individuals lead to differences in immune responses to an infection. Age, nutrition, and health at the time of infection also may affect the outcome. Some pathogens are deadlier than others, and so on.

That's why researchers work to simplify and control variables in experiments. Variables, recall, are specific aspects of objects or events that may differ over time and among individuals. All *E. coli* cells in Levin and Bull's experiments were descended from the same parent cell, and they were raised on the same nutrients at the same temperature in a flask. They could be expected to respond the same way to the bacteriophage attack. The mice were the same age and sex, and they were raised under identical laboratory conditions. Each mouse in each experimental group received the same amount of bacteriophages or streptomycin. Each mouse in a control group received the same injection of saline solution. The focus was on *one variable*—a specific bacteriophage versus a specific antibiotic—in a simple, controlled, artificial situation.

BIAS IN REPORTING RESULTS Whether intentionally or not, experimenters run a risk of interpreting their data in terms of what they want to prove or dismiss. A few have even been known to fake measurements or nudge findings to reinforce their bias. That is why science emphasizes presenting test results in *quantitative* terms—that is, with actual counts or some other precise form. Doing so will allow other experimenters to check or test the results readily and systematically, as Levin and Bull did. At this writing, they are assembling a detailed report of their own experiments for publication in a science journal.

THE LIMITS OF SCIENCE

The call for objective testing strengthens the theories that emerge from scientific studies. It also puts limits on the kinds of studies that can be carried out. Beyond the realm of science, some events remain unexplained. Why do we exist, for what purpose? Why does any one of us have to die at a particular moment? Such questions lead to *subjective* answers. These come from within, as an outcome of all the experiences and mental connections that shape human consciousness. Because people differ vastly in this regard, subjective answers do not readily lend themselves to scientific analysis and experiments.

This is not to say subjective answers are without value. No human society can function for long unless its members share a commitment to certain standards for making judgments, even subjective ones. Moral, aesthetic, philosophical, and economic standards vary from one society to the next. But they all guide people in deciding what is important and good, and what is not. All attempt to give meaning to what we do.

Every so often, scientists stir up controversy when they happen to explain some part of the world that was considered to be beyond natural explanation—that is, as belonging to the supernatural. This is commonly the case when a society's moral codes are interwoven with religious narratives. Exploring a longstanding view of the natural world from a scientific perspective might be misinterpreted as questioning morality, even though the two are not the same thing.

As one example, centuries ago in Europe, Nicolaus Copernicus studied the planets and concluded the Earth circles the sun. Today this seems obvious enough. Back then, it was heresy. The prevailing belief was that the Creator made the Earth—and, by extension, humans—the immovable center of the universe. Later a respected scholar, Galileo Galilei, studied the Copernican model of the solar system, thought it was a good one, and said so. He was forced to retract his statement publicly, on his knees, and put the Earth back as the fixed center of things. (Word has it that when he stood up he muttered, "Even so, it *does* move.") Later still, Darwin's theory of evolution ran up against the same prevailing belief.

Today, as then, society has sets of standards. Those standards might be questioned when some new, natural explanation runs counter to supernatural beliefs. This doesn't mean that scientists who raise the questions are less moral, less lawful, less sensitive, or less caring than anyone else. It simply means one more standard guides their work: *The external world, not internal conviction, must be the testing ground for scientific beliefs.*

Systematic observations, hypotheses, predictions, tests— in all these ways, science differs from systems of belief that are based on faith, force, or simple consensus.

SUMMARY   *Gold* indicates text section

1. There is unity in the living world, for all organisms share these characteristics: They consist of one or more cells. They are assembled from the same kinds of atoms and molecules according to the same laws of energy. They engage in metabolism, and they also sense and respond to specific aspects of the environment. Their DNA gives them a capacity to survive and reproduce. Table 1.1 lists these characteristics. *1.1*

2. The hierarchy of biological organization includes cells, multicelled organisms, populations, communities, ecosystems, and the biosphere. *1.2*

3. Many millions of species (kinds of organisms) exist. Many millions more existed in the past but are extinct. Classification schemes are ever more inclusive groupings of species, starting with genus and proceeding through family, order, class, and phylum, to kingdom. *1.3*

4. Life's diversity arises through mutations: heritable changes in the structure of DNA molecules. The changes are the basis for variation in heritable traits. These are traits that parents bestow on their offspring, including most details of body form and function. *1.4*

5. Darwin's theory of evolution by natural selection is central to biological inquiry. Its key premises are: *1.4*

  a. Individuals of a population differ in the details of their shared heritable traits. Variant forms of traits may affect the ability to survive and reproduce.

  b. Natural selection is the outcome of differences in survival and reproduction among individuals that differ in one or more traits. Adaptive forms of a trait tend to become more common; less adaptive ones become less common or vanish. Thus the defining traits may change over successive generations; the population may evolve.

6. The methods of scientific inquiry are diverse. These terms are important aspects of the methods: *1.5*

  a. Theory: Explanation of a broad range of related phenomena supported by numerous tests. An example is Darwin's theory of evolution by natural selection.

  b. Hypothesis: A proposed explanation of a specific phenomenon. Sometimes called an educated guess.

  c. Prediction: A claim about what can be expected in nature, based on premises of a theory or hypothesis.

  d. Test: An attempt to produce actual observations that match predicted or expected observations.

  e. Conclusion: A statement about whether a theory or hypothesis should be accepted, modified, or rejected, based on tests of predictions derived from it.

7. Logic is a pattern of thought by which an individual draws a conclusion that does not contradict evidence used to support the conclusion. Inductive logic means individuals derive a general statement from specific observations. Deductive logic means individuals make inferences about particular consequences or predictions

Table 1.1 *Summary of Life's Key Characteristics*

SHARED CHARACTERISTICS THAT REFLECT LIFE'S UNITY

1. Organisms consist of one or more cells.

2. Organisms are constructed of the same kinds of atoms and molecules according to the same laws of energy.

3. Organisms engage in metabolism; they acquire and use energy and materials to survive and reproduce.

4. Organisms sense and make controlled responses to their internal and external environments.

5. Heritable instructions encoded in DNA give organisms their capacity to grow and reproduce. DNA instructions also guide the development of complex multicelled organisms.

6. Characteristics that define a population of organisms can change through the generations; the population can evolve.

FOUNDATIONS FOR LIFE'S DIVERSITY

1. Mutations (heritable changes in the structure of DNA) give rise to variation in heritable traits, including most details of body form, functioning, and behavior.

2. Diversity is the sum total of variations that accumulated in different lines of descent over the past 3.8 billion years, as by natural selection and other processes of evolution.

that must follow from a hypothesis. Such a pattern of thinking is often expressed in "if–then" terms. *1.5*

8. Predictions that flow from hypotheses can be tested by comprehensive observations or by experiments in nature or the laboratory. *1.5*

9. Experimental tests simplify observations in nature or the laboratory because conditions under which the observations are made can be precisely manipulated and controlled. They are based on a premise that any aspect of nature has one or more underlying causes, whether obvious or not. With such an assumption of cause and effect, only those hypotheses that can be tested in ways that might disprove them are scientific. *1.5, 1.6*

10. A control group is a standard against which one or more experimental groups (test groups) are compared. Ideally, it is the same as each experimental group in all variables except the one being investigated. *1.5, 1.6*

11. A variable is a specific aspect of an object or event that might differ over time and between individuals. Experimenters directly manipulate the variable they are studying to support or disprove a prediction. *1.5, 1.6*

12. Test results may be distorted by sampling error— chance differences between a population, event, or some other aspect of nature and the samples that were chosen to represent it. Distortion is less likely when samplings are large and when they are repeated. *1.5, 1.6*

13. Scientific theories arise from systematic observations, hypotheses, predictions, and experimental testing. The external world is the testing ground for theories. *1.7*

Review Questions

For this chapter and subsequent chapters, *italics* after a review question identify the section where you can find answers. They include section numbers and *CI* (for Chapter Introduction).

1. Why is it difficult to formulate a simple definition of life? *CI*

2. Name the molecule of inheritance in cells. *1.1*

3. Write out simple definitions of the following terms: *1.1*
   a. cell   b. metabolism   c. energy   d. ATP

4. How do organisms sense changes in their surroundings? *1.1*

5. Study Figure 1.5. Then, on your own, arrange and define the levels of biological organization. *1.2*

6. Study Figure 1.6. Then, on your own, make a sketch of the one-way flow of energy and the cycling of materials through the biosphere. To the side of the sketch, write out definitions of producers, consumers, and decomposers. *1.2*

7. List the shared characteristics of life. *CI, 1.3*

8. What are the two parts of the scientific name for each kind of organism? *1.3*

9. List the six kingdoms of species as outlined in this chapter. Also list some of their general characteristics. *1.3*

10. Define mutation and adaptive trait. Explain the connection between mutation and the diversity of life. *1.4*

11. Write brief definitions of evolution, artificial selection, and natural selection. *1.4*

12. Define and distinguish between: *1.5*
   a. hypothesis and prediction
   b. observational test and experimental test
   c. inductive and deductive logic
   d. speculation and scientific theory

13. With respect to experimental tests, define variable, control group, and experimental group. *1.5*

14. What does sampling error mean? *1.5*

Self-Quiz   ANSWERS IN APPENDIX III

1. _____ is the capacity of cells to extract energy from sources in their environment, and to transform and use energy to grow, maintain themselves, and reproduce.

2. _____ is a state in which the internal environment is being maintained within tolerable limits.

3. The _____ is the smallest unit of life.

4. If a form of a trait improves chances for surviving and reproducing in a given environment, it is a(n) _____ trait.

5. The capacity to evolve is based on variations in heritable traits, which originally arise through _____ .

6. You have some number of traits that also were present in your great-great-great-great-grandmothers and great-great-great-great-grandfathers. This is an example of _____ .
   a. metabolism   c. a control group
   b. homeostasis   d. inheritance

7. DNA molecules _____ .
   a. contain instructions for traits
   b. undergo mutation
   c. are transmitted from parents to offspring
   d. all of the above

8. For many years in a row, a dairy farmer allowed his best milk-producing cows but not the poor producers to mate. Over many generations, milk production increased. This outcome is an example of _____ .
   a. natural selection   c. evolution
   b. artificial selection   d. both b and c

9. A control group is _____ .
   a. a standard against which experimental groups are compared
   b. identical to experimental groups except for one variable
   c. a standard with several variables against which an experimental group is compared
   d. both a and b

10. A specific aspect of an object or event that may change over time or change among individuals is a _____ .
   a. control group   c. variable
   b. experimental group   d. sampling error

11. The fewer the number of individuals from a population that are chosen at random for an experimental group, _____ .
   a. the greater the chance of sampling error
   b. the smaller the chance of sampling error
   c. the less likely that differences among them will distort the test results

12. Match the terms with the most suitable descriptions.
   _____ adaptive   a. statement of what you should find in
         trait   nature if you were to go looking for it
   _____ natural   b. proposed explanation; educated guess
         selection   c. improves chances to survive and
   _____ scientific   reproduce in prevailing environment
         theory   d. related set of hypotheses that form a
   _____ hypothesis   broadly useful, testable explanation
   _____ prediction   e. outcome of differences in survival and
                           reproduction among individuals that
                           differ in details of one or more traits

Critical Thinking

1. Some spiders (*Dolomedes*) that feed on insects around ponds occasionally capture tadpoles and small fishes, as in Figure 1.14, and isn't that fun to think about? While they are immature, the female spiders confine themselves to a small patch of vegetation next to the pond. When sexually mature, they mate and store sperm that fertilize the eggs. Only then do they move out and occupy larger areas around the pond. Develop hypotheses to explain what might cause the spiders to live in different places at different times. Design an experiment to test each hypothesis.

Figure 1.14 A spider (*Dolomedes*) and its prey: a tiny minnow. The spider delivered paralyzing venom as well as digestive enzymes into its captive and is now sucking predigested juices from it. Such spiders can move about below the water's surface. Hairs on their body trap oxygen (for aerobic respiration) during hunting expeditions.

2. A scientific theory about some aspect of nature rests upon inductive logic. The assumption is that, because an outcome of some event has been observed to happen with great regularity, it will happen again. However, we can't know this for certain, because there is no way to account for all possible variables that may affect the outcome. To illustrate this point, Garvin McCain and Erwin Segal offer a parable:

> Once there was a highly intelligent turkey. It lived in a pen, attended by a kind, thoughtful master, and had nothing to do but reflect on the world's wonders and regularities. It observed some major regularities. Morning always began with the sky getting light, followed by the clop, clop, clop of its master's friendly footsteps, then by the appearance of delicious food. Other things varied—sometimes the morning was warm and sometimes cold—but food always followed footsteps. The sequence of events was so predictable that it became the basis of the turkey's theory about the goodness of the world. One morning, after more than one hundred confirmations of the theory, the turkey listened for the clop, clop, clop, heard it, and had its head chopped off.

The turkey learned the hard way that explanations about the world only have a high or low probability of not being wrong. Today, some people take this uncertainty to mean that "facts are irrelevant—facts change." If that is so, should we just stop doing scientific research? Why or why not?

3. Witnesses in a court of law are asked to "swear to tell the truth, the whole truth, and nothing but the truth." What are some of the problems inherent in the question? Can you think of a better alternative?

4. Many popular magazines publish an astounding number of articles on diet, exercise, and other health-related topics. Some authors recommend a specific diet or dietary supplement. What kinds of evidence do you think the articles should include so that you can decide whether to accept the recommendations?

5. Although scientific information is often used when making a decision, it can't tell an individual what's "right" or "wrong." Give an example of this from your own experience.

6. As the old saying goes, Everybody complains about the weather, but nobody does anything about it. Maybe you can start thinking about how we humans might change at least one aspect of local weather conditions. Two researchers at Arizona State University found that, at least for their investigation of the northeast coast of North America, it really does rain more on weekends, when we'd rather be out having fun!

As R. Cerveny and R. Balling Jr. reasoned, if the amount of rain falling each day of the week is a random event, then rules of probability should apply. (*Probability* means the chance that each possible outcome of an event will occur is proportional to the number of ways it can be reached.) Thus, each day of the week should get one-seventh (14.3 percent) of the total rainfall for the week. But it turns out Monday is driest (13.1 percent). Days of the week are wetter, and Saturday is the wettest of all (with 16 percent of the total). Sunday is a bit above average.

What causes this effect? As the researchers hypothesized, if *air pollution* is greater on weekdays than weekends, then human activities might influence regional patterns of rainfall. Monday through Friday, coal-burning factories and gas-powered vehicles release quantities of tiny particles into the air. The particles can promote updrafts and act as "platforms" for the formation of water droplets. Air pollution must build up during the week and extend into Saturday. Over the weekend, fewer factories operate and fewer commuters are on the road. Therefore, by Monday, the air must be cleaner—and drier.

What kind of evidence do you suppose Cerveny and Balling gathered to test the hypothesis? Jot down a few ideas and check them against the researchers' article in the 6 August 1998 issue of *Nature*. Whether or not you continue in biology, this exercise will be good practice in searching through scientific literature for actual data that you can evaluate on topics of interest to you.

7. Scientists devised experiments to shed light on whether different fish species of the same genus compete in their natural habitat. They constructed twelve ponds, identical in chemical composition and physical characteristics. Then they released the following individuals in each pond:

Ponds 1, 2, 3:	Species A	300 individuals each pond
Ponds 4, 5, 6:	Species B	300 individuals each pond
Ponds 7, 8, 9:	Species C	300 individuals each pond
Ponds 10, 11, 12:	Species A, B, C	300 individuals of each species in each pond

Does this experimental design take into consideration all factors that can affect the outcome? If not, how would you modify it?

Selected Key Terms

For this chapter and subsequent chapters, these are the **boldface** terms that appear in the text, in the sections indicated here by *italic* numbers (or *CI*, short for *Chapter Introduction*). As a study aid, make a list of the terms, write a definition for each, then check it against the one in the text. You will use these terms later on. Becoming familiar with each one will help give you a foundation for understanding the material in later chapters.

adaptive trait *1.4*	ecosystem *1.2*	mutation *1.4*
Animalia *1.3*	energy *1.1*	natural selection *1.4*
antibiotic *1.4*	Eubacteria *1.3*	Plantae *1.3*
Archaebacteria *1.3*	evolution *1.4*	population *1.2*
artificial selection *1.4*	experiment *1.5*	prediction *1.5*
biology *CI*	Fungi *1.3*	producer *1.2*
biosphere *1.2*	genus *1.3*	Protista *1.3*
cell *1.2*	homeostasis *1.1*	receptor *1.1*
community *1.2*	hypothesis *1.5*	reproduction *1.1*
consumer *1.2*	inductive logic *1.5*	sampling error *1.5*
control group *1.5*	inheritance *1.1*	species *1.3*
decomposer *1.2*	metabolism *1.1*	stimulus *1.1*
deductive logic *1.5*	model *1.5*	test, scientific *1.5*
development *1.1*	multicelled	theory, scientific *1.5*
DNA *1.1*	organism *1.2*	variable *1.5*

Readings

Carey, S. 1994. *A Beginner's Guide to the Scientific Method.* Belmont, California: Wadsworth. Paperback.

Committee on the Conduct of Science. 1989. *On Being a Scientist.* Washington, D.C.: National Academy of Sciences. Paperback.

McCain, G., and E. Segal. 1988. *The Game of Science.* Fifth edition. Pacific Grove, California: Brooks/Cole. Paperback.

Moore, J. 1993. *Science as a Way of Knowing.* Cambridge, Massachusetts: Harvard University Press.

On-Line readings at Student Guide for InfoTrac: www.brookscole.com/biology

I Principles of Cellular Life

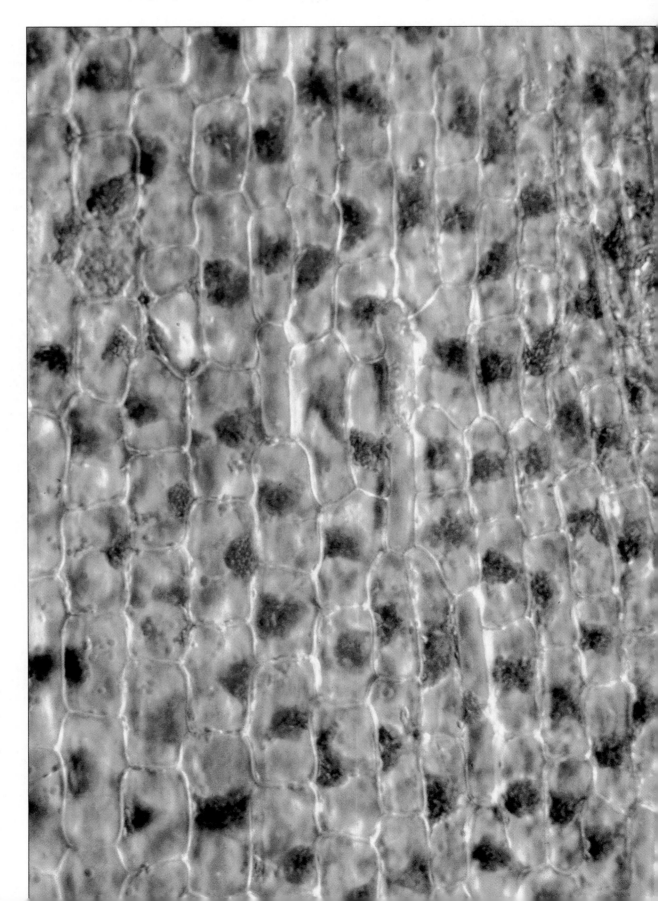

Living cells of one kind of plant (Elodea), observed with the aid of a microscope. Each of these rectangular cells contains efficient chemical factories called chloroplasts (the round, bright green parts inside).

2

CHEMICAL FOUNDATIONS FOR CELLS

Checking Out Leafy Clean-Up Crews

Right now you are breathing in oxygen. You would die without it. Two centuries ago, no one had a clue to what oxygen is, where it comes from, and how it helps keep people alive. Then researchers started to unlock the secrets of this chemical substance and others. As their knowledge of chemistry deepened, they conjured up such amazing things as synthetic fertilizers, nuclear power, nylon stockings, lipsticks, aspirin, antibiotics, and plastic parts of refrigerators, computers, television sets, jet planes, and cars.

Today, chemical "magic" brings us benefits *and* problems. For instance, feeding the human population would be impossible without widespread applications of synthetic fertilizers that boost crop yields. But crop plants don't take up every last bit of fertilizer. Nitrogen, phosphorus, and other nutrients in runoff from fields enter lakes, rivers, and seas—where they "feed" the protistans that cause huge fish kills (Section 20.11).

What about diverting the runoff to holding ponds, where water can evaporate? Farmers commonly do this. However, evaporation ponds are like magnets that pull selenium from the soil. High levels of selenium are toxic to grazing animals and waterfowl.

In 1996, Norman Terry and others at the University of California, Berkeley, grew cattails and other grasses in ten experimental plots contaminated by agricultural runoff (Figure 2.1). As they knew, plants can take up

selenium and convert some of it to a far less toxic gas. They measured how much selenium settled in mud, entered plant tissues, and escaped into the air. As they predicted, before runoff trickled away from the plots, the plants significantly reduced the selenium levels.

Terry's research is an example of **bioremediation**: the use of living organisms to withdraw harmful substances from the environment. Another example is the use of sunflowers to clean a pond at Chernobyl, in the Ukraine. Following a total meltdown at a nuclear power plant, strontium 90, cesium 137, and other extremely nasty radioactive elements contaminated the surroundings, including the pond (Section 45.8).

Fertilizers, wetlands, people, pumpkins, the air you breathe—*every solid, liquid, or gaseous substance in and around you is a collection of one or more elements.* Think of **elements** as fundamental forms of matter that have mass and take up space. Here on Earth, we can't break an element apart into something else except in physics laboratories. Suppose you break a chunk of copper into smaller and smaller bits. The smallest bit you end up with, even a single atom, is still copper.

Ninety-two elements occur naturally on Earth. Four of these—oxygen, hydrogen, carbon, and nitrogen— are the most abundant kinds in your body, as they are in all other living things (Figure 2.2). Organisms also incorporate phosphorus, potassium, sulfur, calcium,

Figure 2.1 At large experimental wetlands in Corcoran, California, cattails, bulrushes, *Spartina*, and other grasses help clean up the water from agricultural runoff by taking up selenium.

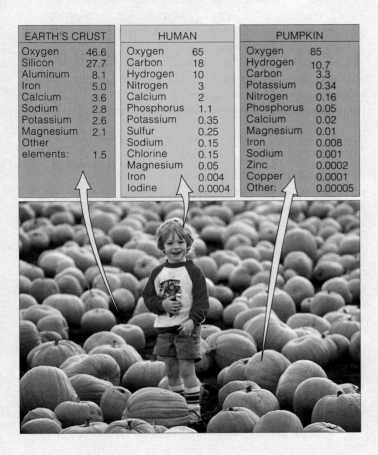

EARTH'S CRUST		HUMAN		PUMPKIN	
Oxygen	46.6	Oxygen	65	Oxygen	85
Silicon	27.7	Carbon	18	Hydrogen	10.7
Aluminum	8.1	Hydrogen	10	Carbon	3.3
Iron	5.0	Nitrogen	3	Potassium	0.34
Calcium	3.6	Calcium	2	Nitrogen	0.16
Sodium	2.8	Phosphorus	1.1	Phosphorus	0.05
Potassium	2.6	Potassium	0.35	Calcium	0.02
Magnesium	2.1	Sulfur	0.25	Magnesium	0.01
Other		Sodium	0.15	Iron	0.008
elements:	1.5	Chlorine	0.15	Sodium	0.001
		Magnesium	0.05	Zinc	0.0002
		Iron	0.004	Copper	0.0001
		Iodine	0.0004	Other:	0.00005

Figure 2.2 Proportions of elements in a human body and the fruit of pumpkin plants, compared to proportions of elements in the materials of the Earth's crust. How are the proportions similar? How do they differ?

sodium, and chlorine, as well as trace elements such as iodine. A *trace* element simply is one that represents less than 0.01 percent of body weight.

Each organism's structure and functions depend on availability of certain kinds and amounts of elements. For example, your muscles will become sore and weak, and your brain simply will not work properly, without daily intakes of magnesium. Similarly, older leaves of plants droop, turn yellow, and die without magnesium.

As you might deduce from this story, safeguarding the environment, our food supplies, and our health starts with knowledge of chemistry. So do efforts to minimize side effects of its applications. Learn about the structure and behavior of chemical substances, and you'll be better equipped to assess risks as well as benefits associated with them.

Key Concepts

1. All substances consist of one or more elements, such as hydrogen, oxygen, and carbon. Each element is composed of atoms, which are the smallest units of matter that still display the element's properties. The atoms of elements are composed of protons, electrons and, except for the hydrogen atom, neutrons.

2. The atoms that make up each kind of element have the same number of protons and electrons. They may differ slightly from one another in their number of neutrons. Variant forms of an element's atoms are called isotopes.

3. Atoms have no overall electric charge unless they become ionized—that is, unless they lose electrons or acquire more of them. An ion is an atom or molecule that has lost or gained one or more electrons. With an extra electron, the charge becomes positive; with an electron loss, the charge becomes negative.

4. Whether a given atom will interact with other atoms depends on how many electrons it has and how they are arranged. When energetic interactions unite two or more atoms, this is a chemical bond.

5. The molecular organization and activities of living things arise largely from ionic, covalent, and hydrogen bonds between atoms.

6. Life probably originated in water and is exquisitely adapted to its properties. Foremost among its properties are temperature-stabilizing effects, cohesiveness, and a capacity to dissolve or repel a variety of substances.

7. All organisms depend on the controlled formation, use, and disposal of hydrogen ions (H^+). The pH scale is a measure of the concentration of H^+ ions in solutions.

REGARDING THE ATOMS

Structure of Atoms

What are the smallest particles that retain the properties of an element? **Atoms**. A line of about a million of them would fit in the period ending this sentence. Small as atoms are, physicists have split them into more than a hundred kinds of smaller particles. The only subatomic particles you will need to consider in this book are the ones called **protons**, **electrons**, and **neutrons**.

All atoms have one or more protons, which carry a positive electric charge (p^+). Except for hydrogen, atoms also have one or more neutrons, which are uncharged. Protons and neutrons make up the atom's core region, or atomic nucleus (Figure 2.3). Zipping about the nucleus and occupying most of the atom's volume are one or more electrons, which carry a negative charge (e^-). Each atom has just as many electrons as protons. This means an atom carries no net charge, overall.

Each element has a unique *atomic* number, which is the number of protons in its atoms. The atomic number

for the hydrogen atom, with its one proton, is 1. For the the carbon atom, with six protons, it is 6 (Table 2.1).

Each element also is assigned a *mass* number, which is the combined number of protons and neutrons in the atomic nucleus. For example, the carbon atom, with six protons and six neutrons, has a mass number of 12.

Why bother with atomic and mass numbers? They can give you a sense of whether and how substances will interact. *This information can help you predict how substances might behave in individual cells, in multicelled organisms, and in the environment, under many conditions.*

Isotopes—Variant Forms of Atoms

Athough all atoms of an element have the same number of protons, they do not all have the same number of neutrons. Atoms that vary in neutron number are called **isotopes**. Most naturally occurring elements have two or more isotopes. Carbon has three, nitrogen has two, and so on. A superscript number to the left of the symbol for an element signifies the isotope. Thus carbon's three isotopes are ^{12}C (or carbon 12, the most common form, with six protons, six neutrons), ^{13}C (six protons, seven neutrons), and ^{14}C (six protons, eight neutrons).

All isotopes of an element interact with other atoms in the same way. This means cells can use any isotope of an element for metabolic activities.

What about radioactive isotopes (radioisotopes)? A physicist, Henri Becquerel, discovered them in 1896 after he put a heavily wrapped rock on top of an unexposed photographic plate inside a desk drawer. The rock held energy-emitting isotopes of uranium. A few days after the plate was exposed to the emissions, a faint image of the rock appeared on it. Marie Curie, a coworker, gave the name "radioactivity" to this chemical behavior.

As we now know, a **radioisotope** is an isotope that has an unstable nucleus and becomes more stabilized by spontaneously emitting energy and particles. (These particles are much smaller than protons, electrons, and neutrons.) The process, radioactive decay, transforms a radioisotope into an atom of a different element at a known rate. For instance, over a predictable time span, carbon 14 becomes nitrogen 14, as Section 18.2 describes.

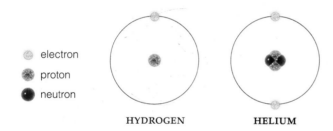

electron
proton
neutron

HYDROGEN   HELIUM

Figure 2.3 The hydrogen atom and helium atom according to one model for atomic structure. The model is highly simplified; the nucleus of these two representative atoms really would be an invisible speck at the scale employed here.

Elements are forms of matter that occupy space, have mass, and cannot be degraded to something else by ordinary means.

Atoms, the smallest particles unique to each element, have one or more positively charged protons, negatively charged electrons, and (except for hydrogen) neutrons.

Most elements have two or more isotopes, which are atoms that differ in the number of neutrons. A radioisotope has an unstable nucleus and stabilizes itself by spontaneously emitting particles and energy at a known rate.

Table 2.1	*Atomic Number and Most Common Mass Number of Selected Elements*		
Element	Symbol	Atomic Number	Mass Number
Hydrogen	H	1	1
Carbon	C	6	12
Nitrogen	N	7	14
Oxygen	O	8	16
Sodium	Na	11	23
Magnesium	Mg	12	24
Phosphorus	P	15	31
Sulfur	S	16	32
Chlorine	Cl	17	35
Potassium	K	19	39
Calcium	Ca	20	40
Iron	Fe	26	56
Iodine	I	53	127

Using Radioisotopes to Track Chemicals and Save Lives

Radioisotopes make splendid tracers. Like a shipping label, a **tracer** is any substance that has a radioisotope attached to it. Researchers track tracers after delivering them into a cell, a multicelled body, an ecosystem, or some other system. Devices detect emissions from a tracer and follow it through a pathway or pinpoint its destination.

For example, Melvin Calvin and other botanists used tracers to track specific steps of photosynthesis. As they knew, all isotopes of an element have the same number of electrons, so all must interact with other atoms the same way. Plants, they hypothesized, should be able to use any carbon isotope when they build carbohydrates. By putting plant cells in a medium enriched with a tracer (^{14}C rather than ^{12}C), they tracked the uptake of carbon through each reaction step in the formation of sugars and starches.

Botanists also use a radioisotope of phosphorus (^{32}P) as a tracer to identify how plants take up and use soil nutrients and synthetic fertilizers. They use such findings in work to improve crop yields.

Besides being research tools, radioisotopes also are medical diagnostic tools. Clinicians use kinds that swiftly decay into harmless elements. Consider how they might analyze a human thyroid. This is the only gland of ours that uses iodine to build thyroid hormones, which greatly influence our body's growth and functions. If a patient's symptoms point to abnormal output of thyroid hormones, clinicians may inject a trace amount of a radioisotope of iodine (^{123}I) into the patient's blood. Then they may use a photographic imaging device to scan the gland. Figure 2.4 shows examples of the type of images they get.

PET (for *Positron-Emission Tomography*) also uses radioisotopes to form images of body tissues. Suppose clinicians attach a tracer to glucose or another molecule. They inject the labeled glucose into the patient, who is moved into a PET scanner (Figure 2.5*a*). Because all cells must use glucose, cells throughout the body take up the labeled glucose. Uptake is greater in some tissues than others, depending on their metabolic activity at the time the patient is being examined. Laboratory devices detect the radioactive emissions and use them to form an image. Such images may reveal variations or abnormalities in metabolic activity, as in Figure 2.5.

Radioisotopes energetic enough to kill cells have uses, also. In *radiation therapy*, radioisotopes kill or impair the activity of their targets: abnormal body cells. For example, emissions from a source of radium 226 or cobalt 60 may destroy small, localized cancers. As another example, the emissions from plutonium 238 drive artificial pacemakers that smooth out irregular heartbeats. The radioisotope is sealed in a case so its emissions won't damage body cells.

normal thyroid

enlarged

cancerous

Figure 2.4 Location of the human thyroid relative to the trachea (windpipe), and scans of the thyroid gland from three patients.

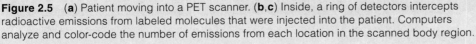
detector ring inside PET scanner   only body section inside ring

The ring intercepts emissions from the labeled molecules

Figure 2.5 (**a**) Patient moving into a PET scanner. (**b,c**) Inside, a ring of detectors intercepts radioactive emissions from labeled molecules that were injected into the patient. Computers analyze and color-code the number of emissions from each location in the scanned body region.

(**d**) Different colors in a brain scan signify differences in metabolic activity. Cells of this brain's left half absorbed and used the labeled molecules at expected rates. However, cells in the right half showed little activity. The patient was diagnosed as having a neurological disorder.

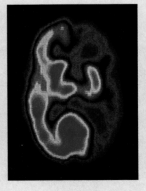
d PET scan (*red* indicates areas of greatest activity; *blue*, of least activity)

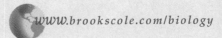

WHAT HAPPENS WHEN ATOM BONDS WITH ATOM?

Electrons and Energy Levels

Cells stay alive because energy inherent in all electrons makes things happen (Figure 2.6). Countless atoms in cells acquire, share, and donate electrons. Atoms of some elements do this easily; others don't. What determines whether an atom will interact with others? *The outcome depends on the number and arrangement of its electrons.*

Tinker with magnets and you can get a sense of the attractive force between unlike charges (+ −) as well as the repulsive force between like charges (++ or − −). Electrons carry a negative charge. In atoms, the positive charge of protons attracts them but they repel each other. They respond to the pulls and pushes by moving about in different orbitals (Figure 2.7). Orbitals are volumes of space around the atomic nucleus in which electrons are likely to be at any instant. Visualize three preschoolers circling a cookie jar, not yet expert in the art of sharing. Each is drawn to the cookies but dreads being shoved away by the others. Two might bob about on opposite sides of the jar to avoid a direct hit. But all three never, ever occupy the same space at the same time.

Each orbital can house one or at most two electrons. Atoms differ in how many occupied orbitals they have because they differ in their number of electrons.

Hydrogen is the simplest atom. It has a lone electron in a spherical orbital, closest to the nucleus. The orbital corresponds to the *lowest available energy level.* In other atoms, two electrons fill the first orbital, and two more electrons occupy a second spherical orbital around the first. Larger atoms have even more electrons occupying orbitals farther from the nucleus, at *higher energy levels.*

An electron at any energy level has a specific amount of energy. On its own, it cannot move to a different level. But suppose an atom absorbs extra energy when sunlight hits it.

Its first orbital has one electron.

Its atomic nucleus has one proton.

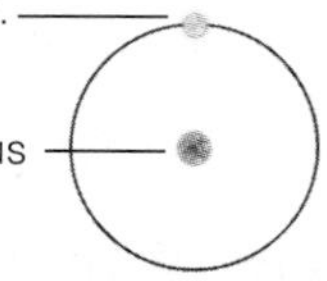

a Shell model of a hydrogen atom

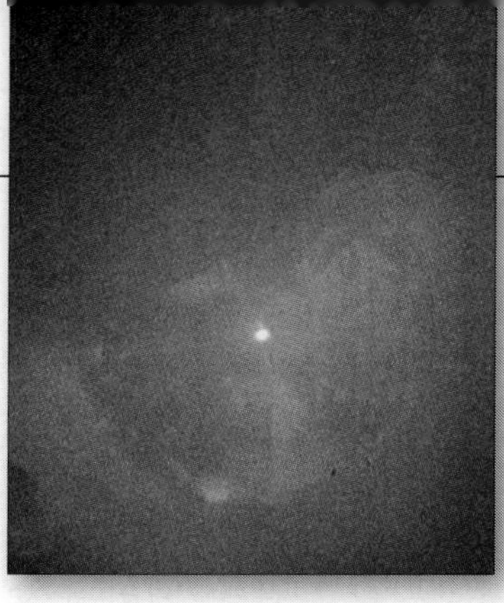

Figure 2.6 How even a lone electron can make things happen. Physicists confined electrons inside a bubble of liquid helium, then used fluctuating sound waves to pop the bubble. When an electron escaped, it made the white-centered red flash shown here.

If the energy input is just the right amount, the electron will become excited enough to reach a higher energy level. Left to itself, it will quickly return to the lower energy level by emitting the extra energy. Most often, its emission is in the form of light. You will read about this event in the sections on photosynthesis.

Think "Shells"

The **shell model** is a simple way to show how electrons are distributed in an atom. By this model, a number of "shells" enclose all orbitals available to electrons, as in Figure 2.8. The first, spherical orbital is inside the first shell. Enclosing the first shell is a second shell at a higher energy level. The second shell has four more available orbitals. More orbitals fit in a third shell, a fourth shell, and so on through large, complex atoms of the heavier elements. Appendix VI lists all of these elements.

From Atoms to Molecules

How can you predict whether two atoms will interact? Check for electron vacancies in their outermost shell. Vacancies mean an atom might give up, gain, or share electrons under suitable conditions, which of course will change the distribution or number of its electrons. You can use the shell model to picture what goes on here.

Figure 2.8 shows the electron distribution for some atoms. Electrons are assigned to an energy level (a shell). Count the vacancies—one or more unfilled orbitals in

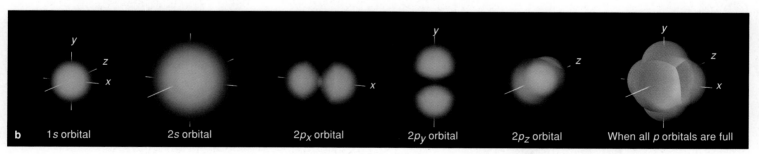

b | 1s orbital | 2s orbital | 2px orbital | 2py orbital | 2pz orbital | When all p orbitals are full

Figure 2.7 Electron arrangements in atoms, which are viewed in terms of three axes. Each axis (*x*, *y*, or *z*) is perpendicular to the other two. (**a**) A hydrogen atom's electron occupies a spherical 1s orbital, at the lowest energy level. (**b**) Every atom has a 1s orbital occupied by one or two electrons. Other orbitals include the 2s and p orbitals at the second energy level. Each can hold two electrons.

Figure 2.8 Examples of the shell model for the distribution of electrons in atoms. Successive shells correspond to higher energy levels from the nucleus. Hydrogen, carbon, and other atoms with electron vacancies (unfilled orbitals) inside their outermost shell tend to give up, accept, or share electrons. Helium and other atoms that have no electron vacancies in their outermost shell are inert; they show little if any tendency to interact with other atoms.

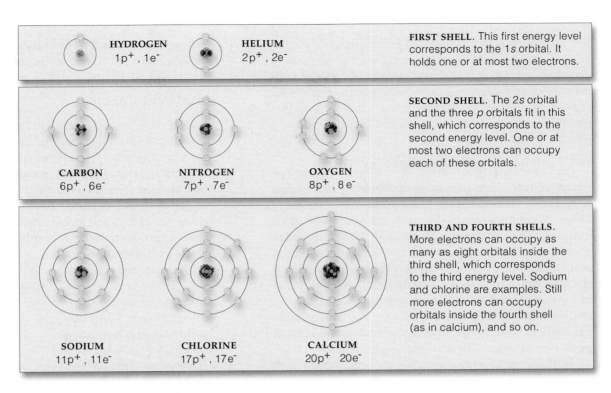

HYDROGEN
$1p^+$, $1e^-$

HELIUM
$2p^+$, $2e^-$

FIRST SHELL. This first energy level corresponds to the $1s$ orbital. It holds one or at most two electrons.

CARBON
$6p^+$, $6e^-$

NITROGEN
$7p^+$, $7e^-$

OXYGEN
$8p^+$, $8e^-$

SECOND SHELL. The $2s$ orbital and the three p orbitals fit in this shell, which corresponds to the second energy level. One or at most two electrons can occupy each of these orbitals.

SODIUM
$11p^+$, $11e^-$

CHLORINE
$17p^+$, $17e^-$

CALCIUM
$20p^+$ $20e^-$

THIRD AND FOURTH SHELLS. More electrons can occupy as many as eight orbitals inside the third shell, which corresponds to the third energy level. Sodium and chlorine are examples. Still more electrons can occupy orbitals inside the fourth shell (as in calcium), and so on.

Figure 2.9 Chemical bookkeeping. We use symbols for elements when writing *formulas*, which identify the composition of compounds. For instance, water has the formula H_2O. The subscript number tells you two hydrogen (H) atoms are present for each oxygen (O) atom. Use such symbols and formulas when writing *chemical equations:* representations of the reactions among atoms and molecules. Substances entering a reaction (reactants) are to the left of the reaction arrow, and products are to the right, as shown by this chemical equation for photosynthesis.

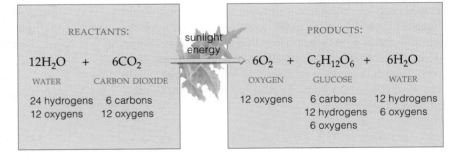

REACTANTS:

$$12H_2O \quad + \quad 6CO_2$$

WATER   CARBON DIOXIDE

24 hydrogens   6 carbons
12 oxygens   12 oxygens

sunlight energy

PRODUCTS:

$$6O_2 \quad + \quad C_6H_{12}O_6 \quad + \quad 6H_2O$$

OXYGEN   GLUCOSE   WATER

12 oxygens   6 carbons   12 hydrogens
           12 hydrogens   6 oxygens
           6 oxygens

each atom's outermost shell. As you can see, helium is one of the atoms with no vacancies. It is an *inert* atom; it shows little tendency to enter chemical reactions.

Now look again at Figure 2.2, which lists the most abundant of the elements making up a typical organism (a human). These elements include hydrogen, oxygen, carbon, and nitrogen. Atoms of all four elements have electron vacancies. Because of this, *they tend to fill the vacancies by forming bonds with other atoms.*

Each **chemical bond** is a union between the electron structures of atoms. Take a moment to study Figure 2.9. It summarizes a few conventions we use when talking about metabolic reactions. A **molecule** forms when two or more atoms join up. Some molecules consist of only one element. Molecular nitrogen (N_2), with two nitrogen atoms, is like this. So is molecular oxygen (O_2).

The molecules of **compounds** consist of two or more different elements in proportions that never vary. Water is an example. In each molecule of water you find one oxygen atom bonded to two hydrogen atoms. Water

molecules in rainclouds, the ocean, a Siberian lake, your bathtub, the petals of a leaf or flower, or anywhere else always have twice as many hydrogen as oxygen atoms.

In a **mixture**, two or more elements intermingle in proportions that can vary (and usually do). For instance, you make a mixture by swirling water with the sugar sucrose, a compound of carbon, hydrogen, and oxygen.

Electrons occupy orbitals, or volumes of space around an atom's nucleus. By a simplified model, orbitals are arranged as a series of shells enclosing the nucleus. Successive shells correspond to levels of energy, which become higher with distance from the nucleus.

One or at most two electrons occupy any orbital. Atoms with unfilled orbitals in their outermost shell tend to interact with other atoms; those with no vacancies do not.

In molecules of an element, all atoms are the same kind. In molecules of a compound, atoms of two or more elements are bonded together, in unvarying proportions.

IMPORTANT BONDS IN BIOLOGICAL MOLECULES

Eat your peas! Drink your milk! For as long as you've been alive, you've been eating complex carbohydrates, proteins, and other "biological molecules." Only living cells make and use these molecules, which consist of a few kinds of atoms joined by a few kinds of bonds—mainly ionic, covalent, and hydrogen bonds.

Ion Formation and Ionic Bonding

An atom, recall, has just as many electrons as protons, so it carries no net charge. That balance can change for atoms having a vacancy—an unfilled orbital—in their outermost shell. For example, a chlorine atom has such a vacancy and can acquire another electron. A sodium atom has a lone electron in its outermost shell, and that electron can be pulled or knocked out. An atom that has either lost or gained one or more electrons is an **ion**. The balance between its protons and its electrons has shifted, so now the atom is ionized; it has become positively or negatively charged (Figure 2.10a).

In living cells, neighboring atoms commonly accept or donate electrons among themselves. When one loses an electron and another grabs it, both become ionized. Depending on cellular conditions, the two ions might not separate; they might stay together as a result of the mutual attraction of opposite charges. An association of two ions that have opposing charges is known as an **ionic bond**. You see one outcome of ionic bonding in Figure 2.10b, which shows a portion of a crystal of table salt, or NaCl. In such crystals, sodium ions (Na^+) and chloride ions (Cl^-) interact through ionic bonds.

Covalent Bonding

Suppose two atoms, each with an unpaired electron in its outermost shell, meet up. Each exerts an attractive force on the other's unpaired electron but not enough to yank it away. Each becomes more stable by *sharing* its unpaired electron with the other. A sharing of a pair of electrons is a **covalent bond**. Example: A hydrogen atom partly fills the electron vacancy in its outermost shell when it's covalently bonded to another hydrogen atom, as shown at right.

In structural formulas, a single line between two atoms represents a *single* covalent bond. Molecular hydrogen (H_2) has such a bond (H—H). Also, two

Two hydrogen atoms, each with a single proton, share two electrons

MOLECULAR HYDROGEN

atoms share two electron pairs of electrons in a *double* covalent bond, as in molecular oxygen (O=O). Two atoms share three pairs of electrons in a *triple* covalent bond, as in molecular nitrogen (N≡N). These examples happen to be gaseous molecules. Each time you breathe in, a stupendous number of H_2, O_2, and N_2 molecules flow through your nose and down toward your lungs.

Covalent bonds are nonpolar or polar. In a *nonpolar* covalent bond, the atoms are exerting the same pull on

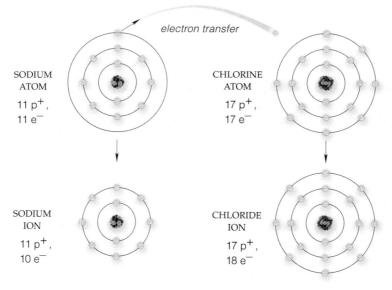

SODIUM ATOM
11 p$^+$,
11 e$^-$

electron transfer

CHLORINE ATOM
17 p$^+$,
17 e$^-$

SODIUM ION
11 p$^+$,
10 e$^-$

CHLORIDE ION
17 p$^+$,
18 e$^-$

a Formation of a sodium ion and a chloride ion

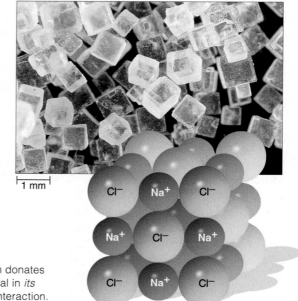

Cl$^-$ Na$^+$ Cl$^-$

Na$^+$ Cl$^-$ Na$^+$

Cl$^-$ Na$^+$ Cl$^-$

b Crystals of sodium chloride (NaCl)

Figure 2.10 (**a**) Ionization by way of an electron transfer. In this case, a sodium atom donates the single electron in its outermost shell to a chlorine atom, which has an unfilled orbital in *its* outermost shell. A sodium ion (Na^+) and a chloride ion (Cl^-) are the outcome of this interaction. (**b**) In each crystal of table salt, or NaCl, many sodium and chloride ions remain together owing to the mutual attraction of opposite charges. Their interaction is a case of ionic bonding.

the electrons and are sharing them equally. "Nonpolar" means there's no difference in charge between the two ends of the bond—that is, at its two poles. Molecular hydrogen has a nonpolar bond. Its two H atoms, each with one proton, attract the shared electrons equally.

In a *polar* covalent bond, atoms of different elements (which have different numbers of protons) don't exert the same pull on shared electrons. The more attractive atom ends up with a slight negative charge. We say it's "electronegative." However, its effect is balanced out by the other atom, which ends up with a slight positive charge. Thus, taken together, the two atoms of a polar covalent bond have no *net* charge.

Consider a water molecule. It has two polar covalent bonds: H—O—H. The electrons are less attracted to the two hydrogens than to the oxygen, which has more protons. The water molecule has no *net* charge overall. However, as you will see shortly, its polarity weakly attracts neighboring polar molecules and ions.

Hydrogen Bonding

The patterns of electron sharing in covalent bonds hold atoms together in specific arrangements in molecules. Some of the patterns also give rise to weak attractions and repulsions between charged functional groups of molecules, as well as between molecules and ions. Like interacting skydivers, such interactions form and break easily (Figure 2.11). Yet they have important roles in the structure and functioning of biological molecules.

For example, a **hydrogen bond** is a weak attraction between an electronegative atom (such as an oxygen or nitrogen atom taking part in a polar covalent bond) and

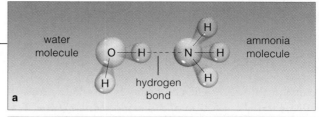
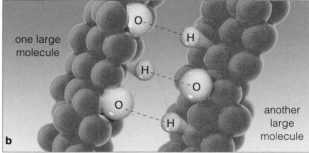
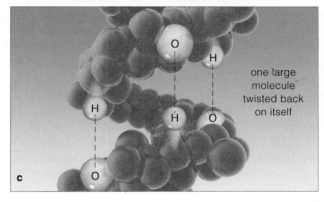

Figure 2.12 Three examples of hydrogen bonds. Compared to covalent bonds, a hydrogen bond is easier to break. Collectively, however, extensive hydrogen bonding has a major role in water, DNA, proteins, and many other substances.

a hydrogen atom taking part in a second polar covalent bond. Hydrogen's slight positive charge weakly attracts the atom with the slight negative charge (Figure 2.12).

Hydrogen bonds often form between different parts of a molecule that has twisted and folded back on itself. They often form between two or more molecules. For example, many hydrogen bonds form between the two strands of a DNA molecule. Although individually easy to break, they collectively stabilize the DNA. Similarly, hydrogen bonds between individual water molecules contribute to water's life-sustaining properties, which is the topic of the next section.

In an ionic bond, two ions of opposite charge attract each other and stay together. Ions form when atoms gain or lose electrons and so acquire a net positive or negative charge.

In a covalent bond, atoms share a pair of electrons. When atoms share the electrons equally, the bond is nonpolar. When the sharing is not equal, the bond itself is polar— slightly positive at one end, slightly negative at the other.

In a hydrogen bond, a covalently bound atom showing a slight negative charge weakly interacts with a covalently bound hydrogen atom showing a slight positive charge.

Figure 2.11 Like skydivers who briefly clasp hands to form an orderly pattern, weak attractions within and between molecules (and between ions and molecules) can form and break easily.

PROPERTIES OF WATER

No sprint through basic chemistry is complete unless it leads us to the collection of molecules called water. Life originated in water. Many organisms still live in it. The ones that don't cart water around with them, in cells and tissue spaces. Many metabolic reactions require water as a reactant. Cell shape and internal structure depend on it. These topics will repeatedly occupy our attention in the book, so you may find it useful to become familiar with the following points about water's properties.

Polarity of the Water Molecule

A water molecule, remember, has no net charge, but the charge that it does carry is unevenly distributed. As a result of its electron arrangements and bond angles, the water molecule's oxygen "end" is a bit negative and its hydrogen end is a bit positive. Figure 2.13*a* is a simple way to think about this charge distribution. Because of the resulting polarity, one water molecule attracts and hydrogen-bonds with others (Figures 2.13*b* and 2.14).

Sugars and other polar molecules easily hydrogen-bond with a water molecule; its polarity attracts them. They are **hydrophilic** (water-loving) **substances**. That same polarity repels oils and other nonpolar molecules, which are **hydrophobic** (water-dreading) **substances**. Observe this for yourself. Shake a bottle full of water and salad oil, and then set it on a table. New hydrogen bonds soon replace the ones that broke when you shook the bottle. As water molecules reunite, they push the oil molecules aside, forcing them to cluster as oil droplets or as an oily film at the water's surface.

A thin, oily membrane separates the water inside a cell from the water outside it. Membrane organization —hence life itself—starts with hydrophilic as well as hydrophobic interactions, as Section 4.1 describes.

Figure 2.14 Arctic habitat of polar bears, which starts with the hydrogen bonding pattern of ice. Below 0°C, each water molecule hydrogen-bonds to four others in a three-dimensional lattice. The molecules are spaced farther apart than they would be in liquid water at room temperature, when molecular motion is greater and not as many bonds form. Ice floats because it has fewer molecules than the same volume of liquid water; its lattice is less dense. Today the Arctic Ocean's ice cap is melting. At current rates of melting, it will be gone in fifty years, and so will the polar bears. Already their seal-hunting season is shorter; the bears are thinner and producing fewer cubs.

Figure 2.13 Water—a substance vital for life. (**a**) Polarity of the water molecule. (**b**) Hydrogen bonding between water molecules in liquid water. Dashed lines signify hydrogen bonds. The photograph shows a human at play in the liquid domain of aquatic organisms.

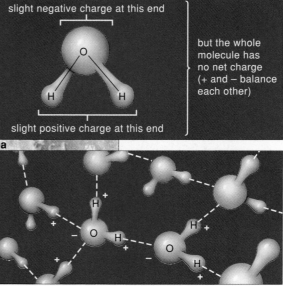

slight negative charge at this end

but the whole molecule has no net charge (+ and − balance each other)

O

H   H

slight positive charge at this end

a

b

Figure 2.15 Water's cohesion. (a) Water strider, which uses its fine, water-resistant leg hairs and water's high surface tension to scoot across water. (b) Cohesion, and evaporation from leaves, pulls water from roots to leaves.

Figure 2.16 Two spheres of hydration.

Water's Temperature-Stabilizing Effects

Cells consist mainly of water. They release a lot of heat energy by metabolism. If it weren't for water's hydrogen bonds, they might cook in their own juices. To see why, start with these facts: Each molecule vibrates nonstop, and its motion increases as it absorbs heat. **Temperature** is a measure of a substance's molecular motion. Compared with most fluids, water absorbs much more heat energy before its temperature increases measurably. Why? Much of the added energy disrupts hydrogen bonding *between* water molecules instead of increasing motion inside them. Hydrogen bonds in water can buffer drastic temperature shifts in aquatic habitats and in the body of a multicelled organism.

Even when liquid water's temperature is not shifting much, hydrogen bonds constantly break, but they form again just as fast. By contrast, large energy inputs can increase molecular motion so much that the hydrogen bonds remain broken, so individual molecules at the water's surface escape into air. By this process, called **evaporation**, heat energy converts liquid water to the gaseous state. As large numbers of molecules break free, they carry away some energy and lower the water's surface temperature. That's how evaporative water loss helps cool you and some other mammals when you work up a sweat on hot and dry days. Sweat—which is about 99 percent water—evaporates from your skin.

Below 0°C, hydrogen bonds resist breaking and lock water molecules in the latticelike bonding pattern of ice (Figure 2.14). Ice is less dense than water. During winter freezes, ice sheets may form near the surface of ponds, lakes, and streams. Like a blanket, ice "insulates" the liquid water beneath it and helps protect many fishes, frogs, and other aquatic organisms against freezing.

Water's Cohesion

Life also depends on water's cohesion. **Cohesion** means something has a capacity to resist rupturing when placed under tension—that is, stretched—as by the weight of a bug's legs (Figure 2.15a). Think of a lake or some other body of liquid water. Uncountable hydrogen bondings

exert a continual inward pull on molecules of water at or near the surface. The bonding causes a high surface tension. You can see evidence of this when, say, flying insects splat against water and float on it.

Cohesion works inside organisms, also. Plants, for example, require nutrient-laden water for growth and metabolism. Largely because of cohesion, narrow columns of liquid water move up through the pipelines of vascular tissues from roots to all leaves and other growing parts of plants. On sunny days, water evaporates from leaves as molecules break free (Figure 2.15b). Hydrogen bonding pulls up more molecules into leaf cells as replacements, in ways you'll read about in Section 27.3.

Water's Solvent Properties

Finally, water is a fine solvent; ions and polar molecules easily dissolve in it. Any dissolved substance is called a **solute**. In general, we say a substance is *dissolved* after water molecules cluster around ions or molecules of it and keep them dispersed in fluid. Such clusters are "spheres of hydration." This is what happens to solutes in cellular fluid, the sap of maple trees, blood, fluid in your gut, and every other fluid associated with life.

Watch this happen when you pour some table salt (NaCl) into a cup of water. After a while, the crystals of salt separate into Na^+ and Cl^-. Each Na^+ attracts the negative end of some water molecules at the same time that Cl^- attracts the positive end of others (Figure 2.16). The spheres of hydration formed in this way keep the ions dispersed in the fluid.

A water molecule has no net charge, but it is slightly polar owing to an uneven distribution of charge. Its oxygen "end" is a bit negative and its hydrogen "end" is a bit positive.

Polarity allows water molecules to hydrogen-bond to each other and to other polar (hydrophilic) substances. Water molecules tend to repel nonpolar (hydrophobic) substances.

Water has temperature-stabilizing effects, internal cohesion, and a capacity to dissolve many substances. These properties influence the structure and functioning of organisms.

ACIDS, BASES, AND BUFFERS

A great variety of ions dissolved in the fluids inside and outside cells influence cell structure and functioning. Among the most influential are **hydrogen ions**, or H^+. Hydrogen ions are the same thing as free (unbound) protons. They have far-reaching effects largely because they are chemically active and there are so many of them.

The pH Scale

At any instant in liquid water, some water molecules are breaking apart into hydrogen ions and **hydroxide ions** (OH^-). This ionization of water is the basis of the **pH scale**, as in Figure 2.17. Biologists use this scale when measuring the H^+ concentration of seawater, tree sap, blood, and other fluids. Pure water (not rainwater or tapwater) always has just as many H^+ as OH^- ions. This condition, which also may occur in other fluids, signifies neutrality. We assign neutrality a value of 7 at the midpoint of the pH scale, which ranges from 0 (the highest H^+ concentration) to 14 (the lowest). *The greater the H^+ concentration, the lower the pH.*

Starting at neutrality, each change by one unit of the pH scale corresponds to a tenfold increase or decrease in H^+ concentration. An easy way to sense that range is to dip your tongue in baking soda (pH 9), water (7), then lemon juice (2.3).

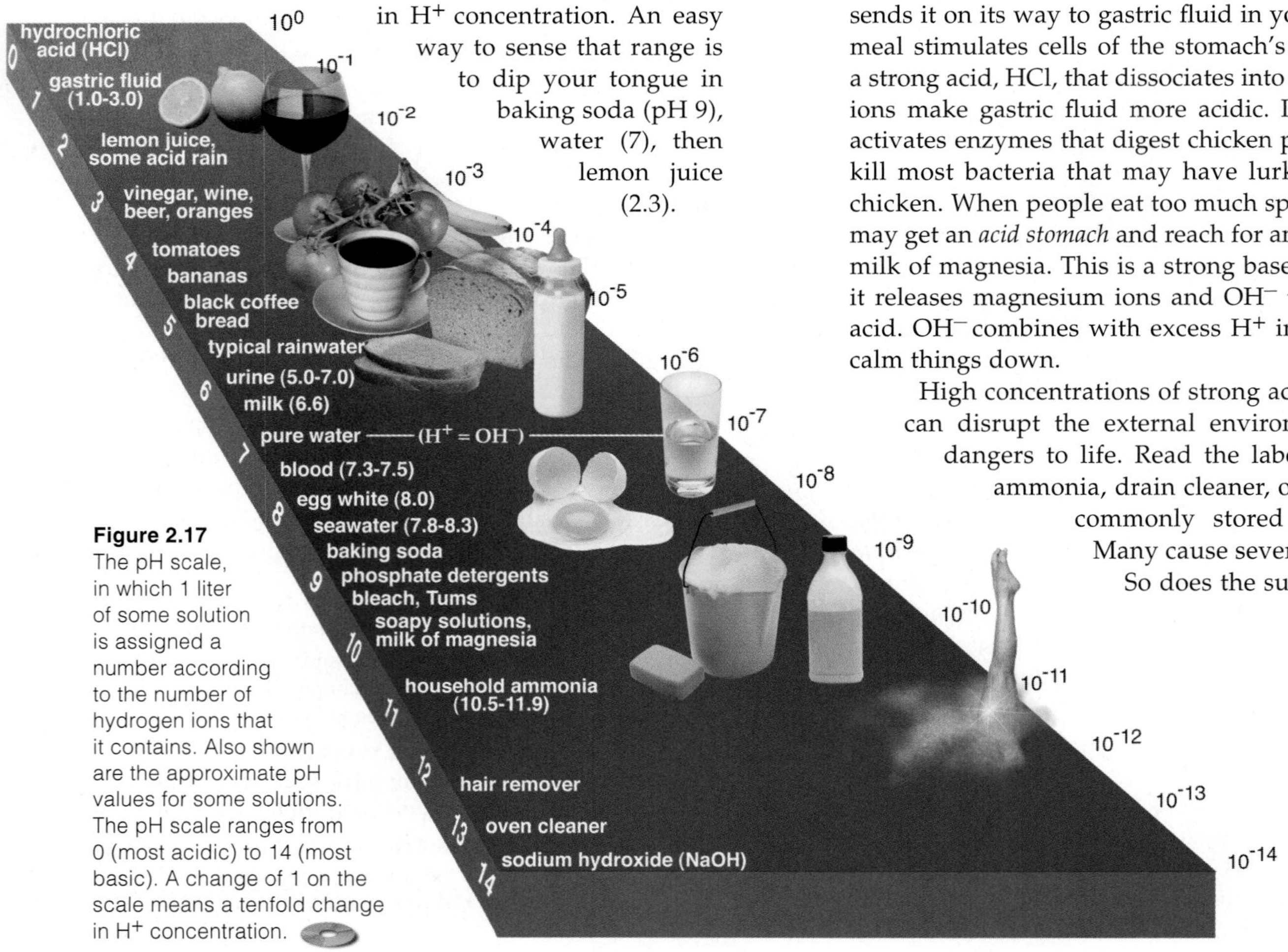

Figure 2.17
The pH scale, in which 1 liter of some solution is assigned a number according to the number of hydrogen ions that it contains. Also shown are the approximate pH values for some solutions. The pH scale ranges from 0 (most acidic) to 14 (most basic). A change of 1 on the scale means a tenfold change in H^+ concentration.

hydrochloric acid (HCl)
gastric fluid (1.0-3.0)
lemon juice, some acid rain
vinegar, wine, beer, oranges
tomatoes
bananas
black coffee
bread
typical rainwater
urine (5.0-7.0)
milk (6.6)
pure water —— ($H^+ = OH^-$) ——
blood (7.3-7.5)
egg white (8.0)
seawater (7.8-8.3)
baking soda
phosphate detergents
bleach, Tums
soapy solutions, milk of magnesia
household ammonia (10.5-11.9)
hair remover
oven cleaner
sodium hydroxide (NaOH)

10^0, 10^{-1}, 10^{-2}, 10^{-3}, 10^{-4}, 10^{-5}, 10^{-6}, 10^{-7}, 10^{-8}, 10^{-9}, 10^{-10}, 10^{-11}, 10^{-12}, 10^{-13}, 10^{-14}

How Do Acids Differ From Bases?

When they dissolve in water, the substances called **acids** *donate* protons (H^+) to other solutes or water molecules. By contrast, the substances called **bases** *accept* H^+ when dissolved in water; OH^- directly or indirectly forms after they do this. *Acidic* solutions, such as lemon juice, gastric fluid, and coffee, release more H^+ than OH^-; their pH is below 7. *Basic* solutions, such as seawater, baking soda, and egg white, release more OH^- than H^+. They are also called "alkaline" fluids, with a pH above 7.

The fluid inside most human cells is about 7 on the pH scale. Most fluids bathing these cells have values between 7.3 and 7.5. This also is the case for the fluid portion of human blood. Seawater is more alkaline than the body fluids of organisms that live in it.

Most acids are weak or strong. Weak ones such as carbonic acid (H_2CO_3) are reluctant H^+ donors. They easily accept H^+ after giving it up, depending on pH, so they alternate between acting as an acid and a base. By contrast, strong acids completely give up H^+ when they dissociate in water. Examples are hydrochloric acid (HCl), nitric acid (HNO_3), and sulfuric acid (H_2SO_4).

Imagine sniffing and eating spicy chicken. Swallowing sends it on its way to gastric fluid in your stomach. The meal stimulates cells of the stomach's lining to secrete a strong acid, HCl, that dissociates into H^+ and Cl^-. The ions make gastric fluid more acidic. Increased acidity activates enzymes that digest chicken proteins and help kill most bacteria that may have lurked in or on the chicken. When people eat too much spicy chicken, they may get an *acid stomach* and reach for an antacid, such as milk of magnesia. This is a strong base. As it dissolves, it releases magnesium ions and OH^- to neutralize the acid. OH^- combines with excess H^+ in gastric fluid to calm things down.

High concentrations of strong acids or bases also can disrupt the external environment and pose dangers to life. Read the labels on bottles of ammonia, drain cleaner, or other products commonly stored in households. Many cause severe *chemical burns*. So does the sulfuric acid in car

Figure 2.18 Sulfur dioxide emissions from a coal-burning power plant. Airborne pollutants such as sulfur dioxide dissolve in water vapor to form acidic solutions. They are a major component of acid rain.

batteries. Fossil fuel burning—as by power plants and cars—and nitrogen fertilizers release strong acid that alter the pH of rain (Figure 2.18). Certain regions are sensitive to the pH of this *acid rain* because of their soil type and vegetation cover. The modified chemistry of habitats in such regions can harm organisms. We will return to this topic in Section 45.1.

Buffers Against Shifts in pH

Metabolic reactions are sensitive to even slight shifts in pH, for H^+ and OH^- can combine with many different molecules and alter their functions. Normally, control mechanisms minimize unsuitable shifts in pH, as they do when HCl enters the stomach in response to a meal. Many of the controls involve buffer systems.

A **buffer system** is a partnership between a weak acid and the base that forms when the acid dissolves in water. The two work like a pair to counter slight shifts in pH. Remember, the OH^- level rises when any strong base enters a fluid. But a weak acid can neutralize part of the incoming OH^- by combining with it. The weak acid's partner (a base) forms as a result of its action. Later, if a strong acid floods in, the base will accept H^+ from it, thus becoming the buffer system's acid.

Bear in mind, the action of a buffer system cannot make new hydrogen ions or eliminate ones that already are present. It can only bind or release them.

In all complex, multicelled organisms, diverse buffer systems operate in the internal environment—in blood and tissue fluids. For example, metabolic reactions in the vertebrate lungs and kidneys help control the acid–base balance of this environment, at levels suitable for life (Chapters 37 and 38). For now, simply think of what happens when the blood level of H^+ decreases and the blood is not as acidic as it should be. At such times, carbonic acid that is dissolved in blood releases H^+ and so becomes the partner base, bicarbonate:

$$H_2CO_3 \longrightarrow HCO_3^- + H^+$$
CARBONIC ACID   BICARBONATE

When blood becomes more acidic, more H^+ becomes bound to the base, thus forming the partner acid:

$$HCO_3^- + H^+ \longrightarrow H_2CO_3$$
BICARBONATE   CARBONIC ACID

Uncontrolled shifts in pH have drastic outcomes. If blood's pH (7.3–7.5) declines even to 7, this invites a *coma*, a sometimes irreversible state of unconsciousness. An increase to 7.8 can lead to *tetany*, a potentially lethal condition in which skeletal muscles start contracting uncontrollably. In *acidosis*, carbon dioxide builds up in blood, too much carbonic acid forms, and blood pH plummets. *Alkalosis* is an uncorrected increase in blood pH. Both conditions can be lethal.

Salts

Salts are compounds that release ions *other than* H^+ and OH^- in solutions. Salts and water often form when a strong acid and strong base interact. Depending on a solution's pH value, salts can form and dissolve easily. Consider how sodium chloride forms, then dissolves:

$$HCl \text{ (acid)} + NaOH \text{ (base)} \longrightarrow NaCl \text{ (salt)} + H_2O$$
HYDROCHLORIC   SODIUM   SODIUM CHLORIDE
ACID   HYDROXIDE

$$Na^+ \quad Cl^- \text{ (ionization)}$$

Many salts dissolve into ions that play key roles in cells. For example, nerve cell activity depends on ions of sodium, potassium, and calcium. Your muscles contract with the help of calcium ions. And water absorption by plant cells depends largely on potassium ions.

Hydrogen ions (H^+) and other ions dissolved in the fluids inside and outside cells affect cell structure and function.

When dissolved in water, acidic substances release H^+, and basic (alkaline) substances accept them. Certain acid–base interactions help maintain the pH value of a fluid—that is, its H^+ concentration—by a buffering action.

A buffer system counters slight shifts in pH by releasing hydrogen ions when their concentration is too low or by combining with them when the concentration is too high.

Salts are compounds that release ions other than H^+ and OH^-, and many of those ions have key roles in cell functions.

SUMMARY   *Gold* indicates text section

1. Chemistry helps us understand the nature of all the substances that make up cells, organisms, and the Earth, its waters, and the atmosphere. Each substance consists of one or more elements. Of the ninety-two naturally occurring elements, the most common in organisms are oxygen, carbon, hydrogen, and nitrogen. Organisms also have lesser amounts of many other elements, such as calcium, phosphorus, potassium, and sulfur. *CI*

2. Table 2.2 summarizes some key chemical terms that you will encounter throughout this book. *CI, 2.1–2.6*

3. Elements consist of atoms. An atom has one or more positively charged protons plus an equal number of negatively charged electrons and (except for hydrogen atoms) one or more uncharged neutrons. The protons and neutrons occupy the core region, or the atomic nucleus. Most elements have isotopes, which are two or more forms of atoms that have the same number of protons but different numbers of neutrons. *2.1*

4. Whether an atom interacts with others depends on the number and arrangement of its electrons, which occupy orbitals (volumes of space) in a series of shells around the atomic nucleus. When an atom has unfilled orbitals in its outermost shell, it tends to bond with atoms of other elements. *2.3*

5. An atom has no net charge. But it may lose or gain one or more electrons and so become an ion, which has an overall positive or negative charge. *2.4*

6. Generally, a chemical bond is a union between the electron structures of atoms. *2.3*

   a. In an ionic bond, a positive ion and negative ion stay together because of the mutual attraction of their opposite charges. *2.4*

   b. Atoms often share one or more pairs of electrons in single, double, and triple covalent bonds. Electron sharing is equal in nonpolar covalent bonds, and it is unequal in polar covalent bonds. The interacting atoms have no net charge, but the bond is slightly negative at one end and slightly positive at the other, overall. *2.4*

   c. In a hydrogen bond, one covalently bonded atom (such as oxygen and hydrogen) with a slight negative charge is weakly attracted to the slight positive charge of a hydrogen atom that is taking part in a different polar covalent bond. *2.4*

7. Polar covalent bonds join together three atoms in a water molecule (two hydrogens and one oxygen). The polarity of a water molecule invites hydrogen bonding between water molecules. Such hydrogen bonding is the basis of liquid water's ability to resist temperature changes more than other fluids do, display internal cohesion, and easily dissolve polar or ionic substances. The properties greatly influence the metabolic activity, structure, shape, and internal organization of cells. *2.5*

8. By the pH scale, a solution is assigned a number that reflects its H^+ concentration. This ranges from 0 (highest concentration) to 14 (lowest). At pH 7, the H^+ and OH^- concentrations are equal. Acids release H^+ in water, and bases combine with them. Buffer systems help maintain pH values of blood, tissue fluids, and the fluid inside cells. *2.6*

Table 2.2 *Summary of Important Players in the Chemical Basis of Life*

ELEMENT	Fundamental form of matter that occupies space, has mass, and cannot be broken apart into a different form of matter by ordinary physical or chemical means.
ATOM	Smallest unit of an element that still retains the characteristic properties of that element.
Proton (p^+)	Positively charged particle of the atomic nucleus. All atoms of an element have the same number of protons, which is the atomic number. A proton without an electron zipping around it is a hydrogen ion (H^+).
Electron (e^-)	Negatively charged particle that can occupy a volume of space (orbital) around an atomic nucleus. All atoms of an element have the same number of electrons. Electrons can be shared or transferred among atoms.
Neutron	Uncharged particle of the nucleus of all atoms except hydrogen. For a given element, the mass number is the number of protons and neutrons in the nucleus.
MOLECULE	Unit of matter in which two or more atoms of the same element, or different ones, are bonded together.
Compound	Molecule composed of two or more different elements in unvarying proportions. Water is an example.
Mixture	Intermingling of two or more elements in proportions that can and usually do vary.
ISOTOPE	One of two or more forms of atoms of an element that differ in their number of neutrons.
Radioisotope	Unstable isotope, having an unbalanced number of protons and neutrons, that emits particles and energy.
Tracer	Molecule of a substance to which a radioisotope is attached. In conjunction with tracking devices, it can be used to follow the movement or destination of that substance in a metabolic pathway, the body, etc.
ION	Atom that has gained or lost one or more electrons, thus becoming positively or negatively charged.
SOLUTE	Any molecule or ion dissolved in some solvent.
Hydrophilic substance	Polar molecule or molecular region that can readily dissolve in water.
Hydrophobic substance	Nonpolar molecule or molecular region that strongly resists dissolving in water.
ACID	Substance that donates H^+ when dissolved in water.
BASE	Substance that accepts H^+ when dissolved in water; OH^- forms directly or indirectly afterward.
SALT	Compound that releases ions other than H^+ or OH^- when dissolved in water.

Review Questions

1. What is an element? Name four elements (and their symbols) that make up more than 95 percent of the body weight of all living organisms. *CI*

2. Define atom, isotope, and radioisotope. *2.1*

3. How many electrons can occupy each orbital around an atomic nucleus? Using the shell model, explain how the orbitals available to electrons are distributed in an atom. *2.3*

4. Define molecule, compound, and mixture. *2.3*

5. Distinguish between:
    a. ionic and hydrogen bonds *2.4*
    b. polar and nonpolar covalent bonds *2.4*
    c. hydrophilic and hydrophobic interactions *2.5*

6. If a water molecule has no net charge, then why does it attract polar molecules and repel nonpolar ones? *2.5*

7. Label the atoms in each water molecule in the sketch below. Indicate which parts of each molecule carry a slight positive charge (+) and which carry a slight negative charge (−). *2.5*

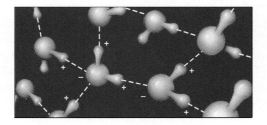

8. Define acid and base. Then describe the behavior of a weak acid in solutions having a high or low pH value. *2.6*

Self-Quiz   ANSWERS IN APPENDIX III

1. Electrons carry a (a) _____ charge.
    a. positive   b. negative   c. zero

2. Atoms share electrons unequally in a(n) _____ bond.
    a. ionic   c. polar covalent
    b. nonpolar covalent   d. hydrogen

3. An individual water molecule shows _____ .
    a. polarity   d. spherelike hydration
    b. hydrogen-bonding capacity   e. a and b
    c. notable heat resistance   f. all of the above

4. In liquid water, spheres of hydration form around _____ .
    a. nonpolar molecules   d. solvents
    b. polar molecules   e. b and c
    c. ions   f. all of the above

5. Hydrogen ions (H^+) are _____ .
    a. the basis of pH values   d. dissolved in blood
    b. unbound protons   e. both a and b
    c. targets of certain buffers f. all of the above

6. When dissolved in water, a(n) _____ donates H^+; however, a(n) _____ accepts H^+.

7. Match the terms with the most suitable descriptions.
    _____ trace element   a. weak acid and its partner base
    _____ buffer system   work as a pair to counter pH shifts
    _____ chemical bond   b. union between electron structures
    _____ temperature   of two atoms
                             c. less than 0.01% of body weight
                             d. measure of molecular motion
                                in some defined region

Critical Thinking

1. An ionic compound forms when calcium combines with chlorine. Referring to Figure 2.8, give the compound's formula. (*Hint:* Be sure the outermost shell of each atom is filled.)

2. David, an inquisitive three-year-old, poked his fingers into warm water inside a metal pan on the stove, then touched the pan and got a nasty burn. Explain why water in a metal pan heats up far more slowly than the pan itself.

3. When molecules absorb microwaves, which are a form of electromagnetic radiation, they move more rapidly. Explain why a microwave oven can heat foods.

4. From what you know about cohesion, devise a hypothesis to explain why water forms droplets.

5. Edward is trying to study a chemical reaction that an enzyme catalyzes (speeds up). H^+ forms in the reaction, but the enzyme is destroyed at low pH. What can he include in his reaction mix to protect the enzyme while he studies the reaction?

6. Through interactions with water and ions, a soluble protein becomes dispersed in cellular fluid. An electrically charged cushion around it makes this happen. Using the sketch at right as a guide, explain the chemical interactions through which this cushion forms. Start with the major bonds in the protein molecule.

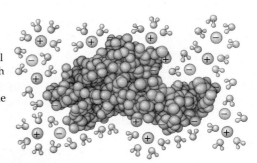

7. Reactants and products often are expressed in moles. A *mole* is a certain number of atoms or molecules of any substance, just as "a dozen" can refer to any twelve cats, roses, and so forth. Molar weight, in grams, equals the total atomic weight of all atoms making up that substance. For example, the atomic weight of carbon is 12, so one mole of carbon weighs 12 grams. A mole of oxygen (atomic weight 16) weighs 16 grams.

   State why a mole of water (H_2O) weighs 18 grams, and why a mole of glucose ($C_6H_{12}O_6$) weighs 180 grams.

Selected Key Terms

acid *2.6*	evaporation *2.5*	mixture *2.3*
atom *2.1*	hydrogen bond *2.4*	molecule *2.3*
base *2.6*	hydrogen ion (H^+) *2.6*	neutron *2.1*
bioremediation *CI*	hydrophilic	pH scale *2.6*
buffer system *2.6*	substance *2.5*	proton *2.1*
chemical bond *2.3*	hydrophobic	radioisotope *2.1*
cohesion *2.5*	substance *2.5*	salt *2.6*
compound *2.3*	hydroxide ion (OH^-) *2.6*	shell model *2.3*
covalent bond *2.4*	ion *2.4*	solute *2.5*
electron *2.1*	ionic bond *2.4*	temperature *2.5*
element *CI*	isotope *2.1*	tracer *2.2*

Readings

Ritter, P. 1996. *Biochemistry: A Foundation*. Pacific Grove, California: Brooks/Cole. Good, easy-to-read introduction to biochemistry, with plenty of human-interest applications.

On-Line readings at Student Guide for InfoTrac:
www.brookscole.com/biology

3

CARBON COMPOUNDS IN CELLS

Carbon, Carbon, In the Sky—Are You Swinging Low and High?

High in the mountains of the Pacific Northwest, vast forests of conifers have endured another murderously cold winter (Figure 3.1). Like other organisms, these evergreen, cone-bearing trees can't grow or reproduce without liquid water. Yet through the winter months, water in their habitat is locked away from them in the form of snow and ice.

The trees get by anyway. Their needle-shaped leaves have an epidermis, a layer of interconnected cells with a thick, waxy wall that restricts water loss. About the only way precious water escapes is through small gaps across the epidermis that close or open in response to changing conditions. During autumn's cool, dry days, the trees go dormant; their metabolic activities idle and growth ceases. Even so, water conserved inside them is enough to keep their leaf cells alive.

When spring arrives, rising temperatures and the water from melting snow invite renewed growth. Tree roots soak up the mineral-laden water. Carbon dioxide, a gaseous molecule consisting of one carbon and two oxygen atoms, moves from the air into each leaf. And now the trees engage in photosynthetic magic, turning water and carbon dioxide into sugars, starches, and other carbon-based compounds. They are the premier producers of the northern forests.

Producers, recall, are organisms that use self-made organic compounds as structural materials and packets of energy. One way or another, compounds made by producers of forests and other ecosystems all over the world nourish every consumer and decomposer.

Today, plants of the northern forests and plains of Canada, the United States, and elsewhere are breaking dormancy earlier than they did just two decades ago. Why? No one knows for sure, but the change in their life cycles may be one outcome of a long-term change in the global climate.

Figure 3.1 A forest of conifers beneath the first snows of winter on Silver Star Mountain. The form, function, and very survival of these trees—and of people, bears, and all other organisms—start with the carbon atom and its diverse molecular partners in organic compounds.

For example, researchers studying carbon dioxide in the air have long known that its concentration shifts seasonally. It declines in the spring and summer, when photosynthesizers use up huge quantities of carbon dioxide. It rises at other times of year, when population sizes of decomposers burgeon. (Decomposers release this gas as a metabolic by-product.) The researchers discovered that the spring decline starts a full week earlier compared to the mid-1970s. Also, the seasonal swings are more dramatic—up to 20 percent greater in Hawaii and a whopping 40 percent in Alaska.

Swings in the atmospheric concentration of carbon dioxide were greatest in 1981 and 1990. Intriguingly, atmospheric *temperatures* were uncommonly high in 1981 and 1990. Is this possible evidence of a long-term rise in the world's atmospheric temperature, a *global warming*? If so, could global warming be promoting the longer growing season and wider seasonal swings?

The picture gets more intricate. We humans burn gasoline, coal, and other fossil fuels for energy (Figure 3.1). Fossil fuels are rich in carbon. As you will read in later chapters, burning them may be releasing enough carbon—as carbon dioxide—to add to global warming.

For now, the point to keep in mind is this: *Carbon permeates the world of life—from the energy-requiring activities and structural organization of cells, to physical and chemical conditions that span the globe and influence ecosystems everywhere.*

With this chapter, we turn to life-giving properties that emerge from the molecular structure of carbon-rich compounds. Study the chapter well, including the summary in Section 3.8. It will serve as a foundation for understanding how different organisms put such compounds together, how they use them, and how the effects of those uses ripple through the biosphere.

Key Concepts

1. Organic compounds have a backbone of one or more carbon atoms to which hydrogen, oxygen, nitrogen, and other atoms are attached. We define cells partly by their capacity to assemble the organic compounds known as carbohydrates, lipids, proteins, and nucleic acids.

2. Cells assemble large biological molecules from their pools of smaller organic compounds, including simple sugars, fatty acids, amino acids, and nucleotides.

3. Glucose and other simple sugars are carbohydrates. So are organic compounds of two or more covalently bonded sugar units. The most complex carbohydrates that cells assemble are polysaccharides, many of which consist of hundreds or thousands of sugar units.

4. Lipids are greasy or oily compounds that show little tendency to dissolve in water but that easily dissolve in nonpolar compounds, such as other lipids. Neutral fats (triglycerides), phospholipids, waxes, and sterols are important lipids.

5. Cells use carbohydrates and lipids as building blocks and as major sources of energy.

6. Proteins have diverse roles. Many serve as structural materials. Many are enzymes, a type of molecule that greatly increases the rate of specific metabolic reactions. Other kinds transport cell substances, contribute to cell movements, trigger changes in cell activities, and defend the body against injury and disease.

7. ATP and other kinds of nucleotides have essential roles in metabolism. DNA and RNA, strandlike nucleic acids assembled from nucleotide units, are the basis of inheritance and reproduction.

PROPERTIES OF ORGANIC COMPOUNDS

The Molecules of Life

Under present-day conditions in nature, *only living cells synthesize carbohydrates, lipids, proteins, and nucleic acids.* These are the molecules characteristic of life. Different classes of biological molecules serve as instant energy sources, structural materials, metabolic workers, cell-to-cell signals, and libraries of hereditary information.

The molecules of life are **organic compounds**, with hydrogen and often other elements covalently bonded to carbon atoms. This term is a holdover from a time when chemists thought "organic" substances were the ones they got from animals and vegetables, as opposed to "inorganic" substances they got from minerals. The term persists even though scientists now make organic compounds in laboratories. And it persists even though we now have reason to believe that organic compounds were present on Earth before organisms were.

Carbon's Bonding Behavior

By far, organisms consist mainly of oxygen, hydrogen, and carbon (Figure 2.2). The oxygen and hydrogen are mainly in the form of water. Put aside the water, and carbon makes up more than half of what's left.

Carbon's importance in life arises from its versatile bonding behavior. As shown in the sketch below, *each carbon atom can share pairs of electrons with as many as four other atoms.* Each covalent bond formed this way is relatively stable. Such bonds join carbon atoms together in chains. The chains are a backbone to which many other elements, such as hydrogen, oxygen, and nitrogen, become attached.

single covalent bond

carbon atom

The bonding arrangement is the start of wonderful three-dimensional shapes of organic compounds. A chain of carbon atoms, bonded covalently one after another, forms a backbone from which other atoms can project:

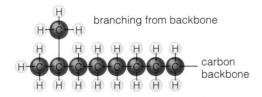

branching from backbone

carbon backbone

Many carbon backbones coil back on themselves in a ring structure, which we can diagram in such ways as:

or

carbon rings

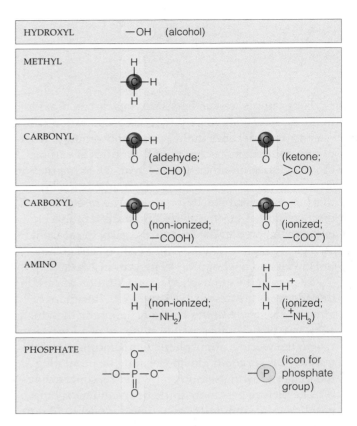

HYDROXYL	—OH (alcohol)	
METHYL	(methyl structure with C bonded to three H's and backbone)	
CARBONYL	(aldehyde; —CHO)	(ketone; >CO)
CARBOXYL	—OH (non-ionized; —COOH)	—O⁻ (ionized; —COO⁻)
AMINO	—N—H with H (non-ionized; —NH₂)	—N—H⁺ with H (ionized; —⁺NH₃)
PHOSPHATE	—O—P—O⁻	(icon for phosphate group)

Figure 3.2 Examples of functional groups.

Functional Groups

A carbon backbone with only hydrogen atoms bonded covalently to it is a **hydrocarbon**, which is a very stable structure. Besides hydrogen, biological molecules have **functional groups**: various atoms or clusters of them covalently bonded to the backbone (Figure 3.2).

To get a sense of their importance, think of **alcohols**: sugars and other organic compounds with one or more hydroxyl groups (—OH). Alcohols dissolve fast because water molecules hydrogen-bond with —OH groups. Or think of a protein. Its shape starts at bonding patterns between amino and carboxyl groups of its backbone. Amino groups also serve as buffers against declines in pH. Other functional groups contribute to differences between males and females of many species (Figure 3.3).

How Do Cells Build Organic Compounds?

Using carbon (from carbon dioxide), water, and sunlight (for energy), the photosynthetic cells described earlier make sugars. The simple sugars are one of four major families of small organic compounds you'll find in every cell; the other three are fatty acids, amino acids, and nucleotides. Cells use these to build all of the biological molecules they require for their structure and functions.

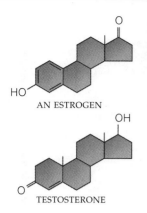

AN ESTROGEN

TESTOSTERONE

FEMALE WOOD DUCK   MALE WOOD DUCK

Figure 3.3 Observable differences in traits between male and female wood ducks (*Aix sponsa*). Two sex hormones have key roles in the development of feather color and other traits that help males and females recognize each other (and therefore influence reproductive success). Both hormones—testosterone and one of the estrogens—have the same carbon ring structure. As you can see at *left*, however, they differ in a single functional group (—OH) attached to the ring.

Figure 3.4
Examples of metabolic reactions by which most biological molecules are put together, rearranged, and broken apart.

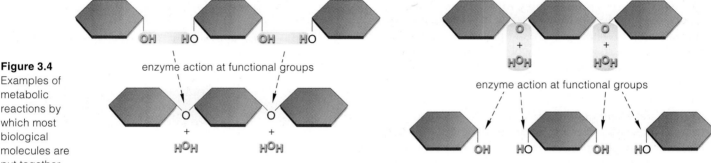

enzyme action at functional groups

enzyme action at functional groups

a Two condensation reactions. Enzymes remove an —OH group and H atom from two molecules, which covalently bond as a larger molecule. Two water molecules form.

b Hydrolysis, a water-requiring cleavage reaction. Enzyme action splits a molecule into three parts, then attaches an —OH group and an H atom derived from a water molecule to each exposed site.

How do they do it? It will take more than one chapter to sketch out answers (and best guesses) to the question. At this point, simply be aware that reactions by which cells assemble, rearrange, and split apart organic compounds require more than energy inputs. They require **enzymes**, a class of proteins that make specific metabolic reactions proceed faster than they would on their own. Different kinds mediate different reactions. In chapters to come, you will read about five categories of reactions:

1. *Functional-group transfer.* One molecule gives up a functional group, which another molecule accepts.

2. *Electron transfer.* One or more electrons stripped from one molecule are donated to another molecule.

3. *Rearrangement.* A juggling of its internal bonds converts one type of organic compound into another.

4. *Condensation.* Through covalent bonding, two molecules combine to form a larger molecule.

5. *Cleavage.* A molecule splits into two smaller ones.

To sense what goes on, think about this **condensation reaction**: Enzymes split off an —OH group from one molecule and an H atom from another, a covalent bond forms between the two at their exposed sites, and the discarded atoms may form water, or H_2O (Figure 3.4*a*).

Starch and other polymers form by many condensation reactions. **Polymers** are molecules of three to millions of subunits that may or may not be identical. Individual subunits of larger molecules are known as **monomers**.

Another example: A type of cleavage reaction called **hydrolysis** is like condensation in reverse (Figure 3.4*b*). Enzymes split molecules at specific groups, then attach one —OH group and a hydrogen atom derived from a water molecule to the exposed sites. With hydrolysis, cells can cleave polymers into smaller molecules when these are required for building blocks or energy.

Organic compounds have diverse, three-dimensional shapes and functions that start with their carbon backbone and the bonding arrangements that arise from it. Functional groups covalently bonded to such backbones add enormously to the structural and functional diversity.

Carbohydrates, lipids, proteins, and nucleic acids are the main biological molecules, the organic compounds that only living cells can assemble under conditions that now occur in nature.

Cells assemble, rearrange, and degrade organic compounds mainly through enzyme-mediated reactions involving the transfer of functional groups or electrons, rearrangement of internal bonds, and a combining or splitting of molecules.

CARBOHYDRATES

Which biological molecules are most abundant? The **carbohydrates**. Most consist of carbon, hydrogen, and oxygen in a 1:2:1 ratio $(CH_2O)_n$. Their functional groups easily dissolve in water, and cells use them as structural materials, transportable forms of energy, or storage forms of energy. The three main classes are **monosaccharides**, **oligosaccharides**, and **polysaccharides**.

The Simple Sugars

"Saccharide" comes from a Greek word meaning sugar. A *mono*saccharide (meaning one sugar monomer) is the simplest carbohydrate. It has at least two —OH groups bonded to the carbon backbone plus an aldehyde or a ketone group. Most readily dissolve in water. Common types have a backbone of five or six carbon atoms that tends to form a ring structure when dissolved in cells or body fluids. Ribose and deoxyribose, the sugar unit of RNA and DNA, respectively, have five carbon atoms. Glucose has six (Figure 3.5a). Most organisms use glucose as the main energy source. Glucose also is a precursor (parent molecule) of many compounds and a building block for larger carbohydrates. Vitamin C (a well-known sugar acid) and glycerol (an alcohol that has three —OH groups) are other compounds with sugar monomers.

Short-Chain Carbohydrates

Unlike the simple sugars, an *oligo*saccharide is a short chain of covalently bonded sugar monomers. (*Oligo—* means "a few.") The ones called *di*saccharides consist of only two sugar units. Lactose, sucrose, and maltose are examples. Lactose, a sugar in milk, has one glucose and one galactose unit. Sucrose, the most plentiful sugar in nature, has a glucose and a fructose unit (Figure 3.5c). Plants convert larger carbohydrates to sucrose, which is easily transported through leaves, stems, and roots. Table sugar is sucrose crystallized from sugarcane and sugar beets. The backbone of some proteins and other large molecules often bears oligosaccharide side chains.

Complex Carbohydrates

The "complex" carbohydrates, or *poly*saccharides, are straight or branched chains of many sugar monomers (often hundreds or thousands) of the same or different types. Cellulose, starch, and glycogen, the most common polysaccharides, consist only of glucose. Cellulose is a structural component of plant cell walls (Figure 3.6a). Like steel rods in reinforced concrete, fibers of cellulose are tough, insoluble, and resistant to weight loads and mechanical stress. Plants store their photosynthetically produced sugars as large starch molecules (Figure 3.6b). Enzymes can readily hydrolyze starch to glucose units.

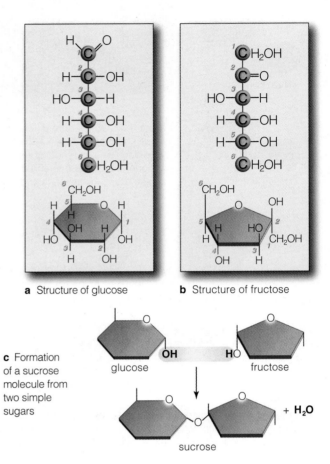

a Structure of glucose   **b** Structure of fructose

c Formation of a sucrose molecule from two simple sugars

glucose   fructose

sucrose $+ \, H_2O$

Figure 3.5 Straight-chain and ring forms of (**a**) glucose and (**b**) fructose. For reference purposes, carbon atoms of simple sugars are commonly numbered in sequence, starting at the end closest to the molecule's aldehyde or ketone group. (**c**) Condensation of two monosaccharides into a disaccharide.

If both cellulose and starch consist of glucose, why are their properties so different? The answer starts with differences in covalent bonding patterns between their glucose units, which are joined together in chains.

In cellulose, many glucose chains stretch out side by side and hydrogen-bond to one another at —OH groups (Figure 3.7a). This bonding arrangement stabilizes the chains into a tightly bundled pattern that resists being digested, at least by most enzymes.

In starch, the pattern of covalent bonding puts each glucose unit at an angle relative to the next unit in line. The chain ends up coiling like a spiral staircase (Figure 3.7b). The coils are not especially stable. In starches that have branched chains, they are even less stable. Many —OH groups project outward from the coiled chains, and this makes them readily accessible to enzymes.

In animals, glycogen is the sugar-storage equivalent of starch in plants. Muscle and liver cells store a lot of it. When the level of sugar in blood declines, liver cells degrade glycogen, so glucose is released and enters the

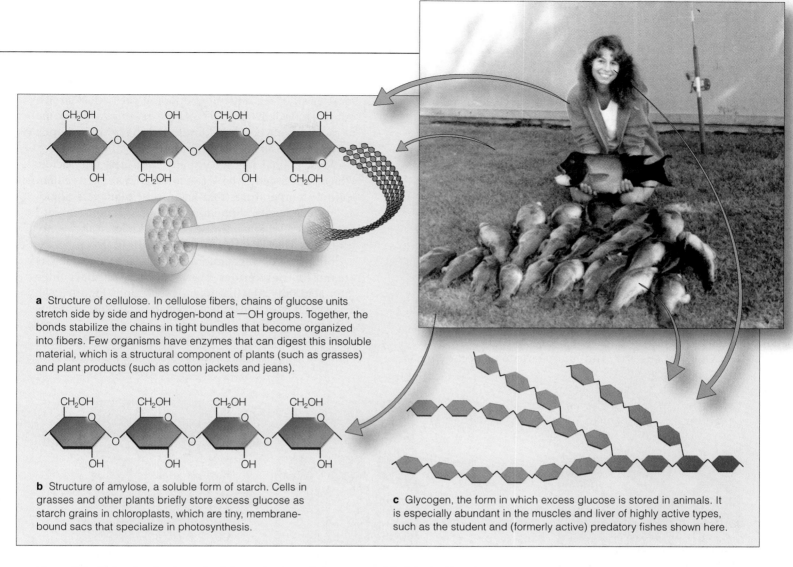

a Structure of cellulose. In cellulose fibers, chains of glucose units stretch side by side and hydrogen-bond at —OH groups. Together, the bonds stabilize the chains in tight bundles that become organized into fibers. Few organisms have enzymes that can digest this insoluble material, which is a structural component of plants (such as grasses) and plant products (such as cotton jackets and jeans).

b Structure of amylose, a soluble form of starch. Cells in grasses and other plants briefly store excess glucose as starch grains in chloroplasts, which are tiny, membrane-bound sacs that specialize in photosynthesis.

c Glycogen, the form in which excess glucose is stored in animals. It is especially abundant in the muscles and liver of highly active types, such as the student and (formerly active) predatory fishes shown here.

Figure 3.6 Molecular structure of cellulose, starch, and glycogen, and their typical locations in a few organisms. All three carbohydrates consist only of glucose units.

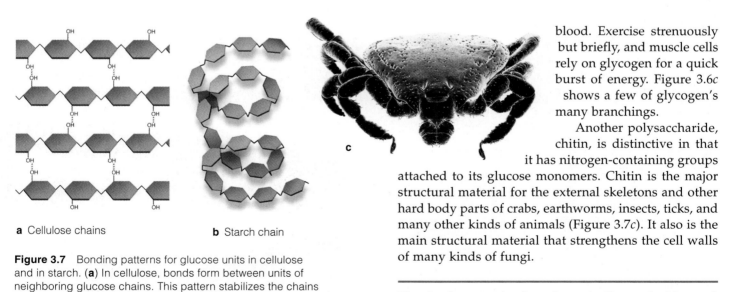

a Cellulose chains

b Starch chain

Figure 3.7 Bonding patterns for glucose units in cellulose and in starch. (**a**) In cellulose, bonds form between units of neighboring glucose chains. This pattern stabilizes the chains and lets them become tightly bundled. (**b**) In amylose, a form of starch, bonding between the units makes the chains coil, which orients the bonds in a way that is accessible to enzymes. (**c**) Scanning electron micrograph of a tick. Its body covering is a protective cuticle reinforced with chitin.

blood. Exercise strenuously but briefly, and muscle cells rely on glycogen for a quick burst of energy. Figure 3.6c shows a few of glycogen's many branchings.

Another polysaccharide, chitin, is distinctive in that it has nitrogen-containing groups attached to its glucose monomers. Chitin is the major structural material for the external skeletons and other hard body parts of crabs, earthworms, insects, ticks, and many other kinds of animals (Figure 3.7c). It also is the main structural material that strengthens the cell walls of many kinds of fungi.

The simple sugars (such as glucose), oligosaccharides, and polysaccharides (such as starch) are carbohydrates. Every cell requires carbohydrates as structural materials, stored forms of energy, and transportable packets of energy.

LIPIDS

If something is greasy or oily to the touch, you can bet it is a lipid or has lipid components. **Lipids** are nonpolar hydrocarbons. They resist dissolving in water but easily dissolve in nonpolar substances (think butter into warm cream sauce). Cells use different lipids as energy stores, as structural materials (for example, in cell membranes and surface coatings), and as signaling molecules. Let's look first at the kinds having fatty acid components—fats, phospholipids, and waxes. Then we'll consider the sterols, each with a backbone of four carbon rings.

Fats and Fatty Acids

Lipids known as **fats** have one, two, or three fatty acids attached to glycerol. Each **fatty acid** has a backbone of as many as thirty-six carbon atoms, a carboxyl group (—COOH) at one end, and hydrogen atoms occupying most or all of the remaining bonding sites. Most stretch out like a flexible tail. Tails that are *unsaturated* have one or more double bonds. *Saturated* tails have single bonds only. Figure 3.8a shows examples.

Most animal fats have many saturated fatty acids, which pack together by weak interactions. They remain solid at temperatures that keep most plant fats liquid, as "vegetable oils." Similar packing interactions in plant fats aren't as stable because of rigid kinks in their fatty acid tails. That is why vegetable oils flow freely.

Butter, lard, vegetable oils, and other natural fats are mostly **triglycerides**. These "neutral" fats have three fatty acid tails attached to a glycerol unit (Figure 3.8b). Triglycerides are the body's most abundant lipids and its richest energy source. Gram for gram, they yield more than twice as much energy when degraded, compared to complex carbohydrates such as starches. Quantities of

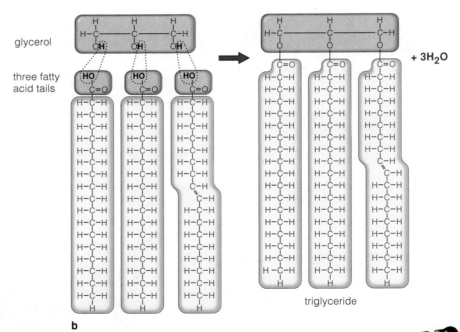

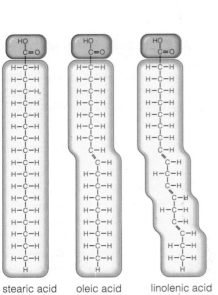

stearic acid   oleic acid   linolenic acid

a

Figure 3.8 *Above:* (**a**) Structural formulas for three fatty acids. In stearic acid, the carbon backbone is fully saturated with hydrogen atoms. Oleic acid, with a double bond in its backbone, is an unsaturated fatty acid. Linolenic acid, with its three double bonds, is a polyunsaturated fatty acid. (**b**) Condensation of fatty acids and a glycerol molecule into a triglyceride.

Figure 3.9 Triglyceride-protected penguins taking the plunge.

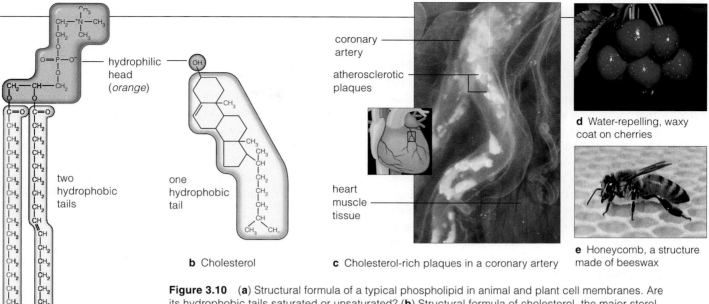

a One of the phospholipids

two hydrophobic tails

hydrophilic head (*orange*)

b Cholesterol

one hydrophobic tail

coronary artery

atherosclerotic plaques

heart muscle tissue

c Cholesterol-rich plaques in a coronary artery

d Water-repelling, waxy coat on cherries

e Honeycomb, a structure made of beeswax

Figure 3.10 (**a**) Structural formula of a typical phospholipid in animal and plant cell membranes. Are its hydrophobic tails saturated or unsaturated? (**b**) Structural formula of cholesterol, the major sterol of animal tissues. (**c**) Your liver synthesizes enough cholesterol for the body. A fat-rich diet may result in excessively high cholesterol levels in blood and the formation of abnormal masses of material in certain blood vessels (arteries). Such *atherosclerotic plaques* may clog the arteries that deliver blood to the heart. (**d**) Demonstration of the water-repelling attribute of a cherry cuticle. (**e**) Honeycomb, a structural material constructed of a firm, water-repellent, waxy secretion called beeswax.

triglycerides are stored as droplets in the cells of body fat (adipose tissue) in every vertebrate. A thick layer of triglycerides under the skin helps penguins and some other animals survive (Figure 3.9). It insulates the body against the extreme temperatures of icy habitats.

One fatty acid is a precursor of eicosanoids. These local signaling molecules include prostaglandins. As you will see, they have roles in contraction, defense, message flow in the nervous system, and other key processes.

Phospholipids

A **phospholipid** has a glycerol backbone, two fatty acid tails, and a hydrophilic "head" with a phosphate group and another polar group (Figure 3.10*a*). Phospholipids are the main materials of cell membranes, which have two layers of lipids. Heads of one layer are dissolved in the cell's fluid interior, and heads of the other layer are dissolved in the surroundings. Sandwiched between the two are all the fatty acid tails, which are hydrophobic.

Sterols and Their Derivatives

Sterols are among the many lipids with no fatty acids. Sterols differ in the number, position, and type of their functional groups, but all have a rigid backbone of four fused-together carbon rings, as below. Eukaryotic cells have sterols in their membranes. Cholesterol is the most common type in tissues of animals. Figure 3.10*b* shows its structure. Also, cholesterol gets remodeled into

sterol backbone

compounds such as vitamin D (necessary for good bones and teeth), steroids, and bile salts. The steroids include sex hormones of the sort shown in Figure 3.3. These govern gamete formation and secondary sexual traits, such as feather color and hair distribution. Bile salts have roles in the digestion of fats in the small intestine.

Waxes

Waxes have long-chain fatty acids tightly packed and linked to long-chain alcohols or carbon rings. All have a firm consistency; all repel water. The cuticle covering aboveground plant parts consists mostly of waxes and another lipid, cutin (Figure 3.10*d*). It restricts water loss and fends off certain parasites. Waxy secretions protect, lubricate, and impart pliability to skin and to hair. Birds secrete waxes, fatty acids, and fats from preen glands to waterproof feathers. Bees use beeswax to construct honeycomb, which houses not only honey but new bee generations (Figure 3.10*e*). Tiny organisms drifting in the seas even use waxes as their main energy source.

Being largely hydrocarbon, lipids can dissolve in nonpolar substances, but they resist dissolving in water.

Triglycerides, or neutral fats, have a glycerol head and three fatty acid tails. They are the body's major energy reservoirs. Phospholipids are the main components of cell membranes.

Sterols such as cholesterol serve as membrane components and precursors of steroid hormones and other compounds. Waxes are firm yet pliable components of water-repelling and lubricating substances.

AMINO ACIDS AND THE PRIMARY STRUCTURE OF PROTEINS

Have you ever wondered how a permanent wave works (Figure 3.11)? By the end of this section and the next, you'll have an idea of how the process alters individual hairs. You will see how the structure of hair and many other body parts begins with the structure of proteins.

Of all large biological molecules, **proteins** are the most diverse. Proteins called enzymes make reactions proceed faster than they would on their own. Structural proteins are the stuff of spider webs, butterfly wings, feathers, bone, and many other body parts and products. Some proteins transport substances across membranes and through fluids. Nutritious proteins abound in eggs and seeds. Protein hormones are signals for change in cell activities. Many proteins act as weapons against pathogens. Amazingly, cells build thousands of diverse proteins from only twenty kinds of amino acids!

Amino Acid Structure

Every **amino acid** is a small organic compound that consists of an amino group, a carboxyl group (which is an acid), a hydrogen atom, and one or more atoms called its R group. Figure 3.12 shows the structure of some amino acids you'll be considering later on in the book. As you can see, their parts generally are bonded covalently to the same carbon atom, as in Figure 3.13.

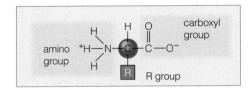

Figure 3.13 Generalized structural formula for amino acids.

What Is a Protein's Primary Structure?

When a cell synthesizes a protein, enzymes link amino acids, one after the other, by peptide bonds. This type of covalent bond forms between the amino group ($-NH_3^+$) of one amino acid and the carboxyl group ($-COO^-$) of the next amino acid in line, as Figure 3.14 shows.

Figure 3.11 Appearance of the hair of actress Nicole Kidman (**a**) before and (**b**) after a permanent wave. The difference, as you will see from this section and the next, starts with strings of amino acids in polypeptide chains.

Figure 3.12
Structural formulas for eight of twenty common amino acids. The *green* boxes highlight their R groups, which are side chains that have functional groups. Each type of side chain contributes in a major way to the distinctive properties of each amino acid.

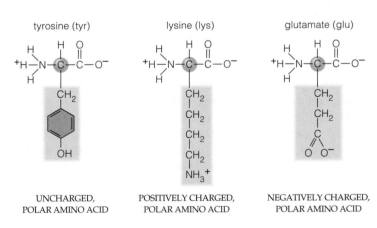

FIVE OF THE NONPOLAR AMINO ACIDS

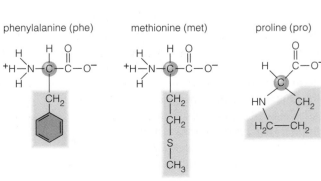

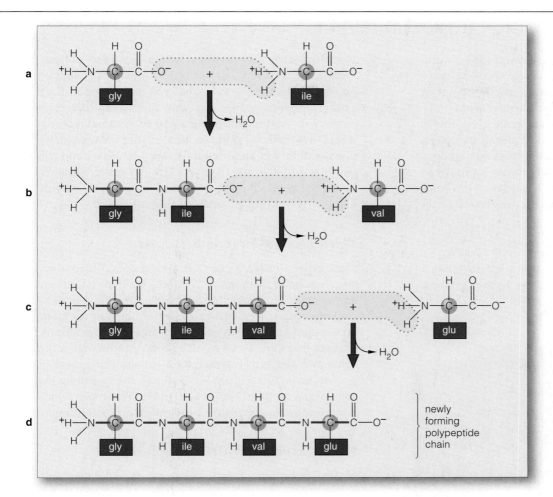

Figure 3.14 Peptide bond formation during protein synthesis.

(**a**) The first two amino acids shown are glycine (gly) and isoleucine (ile). They are at the start of the sequence for a polypeptide chain.

(**b**) A peptide bond forms between glycine and isoleucine during a condensation reaction. A water molecule forms as a by-product.

(**c**) A peptide bond forms between isoleucine and another amino acid, valine (val). Water again forms.

(**d**) Remember, DNA specifies the order in which different kinds of amino acids follow one another in a growing polypeptide chain. In this case, glutamate (glu) is the fourth amino acid specified.

In this case, the outcome will be one of the two polypeptide chains of an insulin molecule, as shown in Figure 3.15. Chapter 14 details steps by which DNA instructions are translated into proteins.

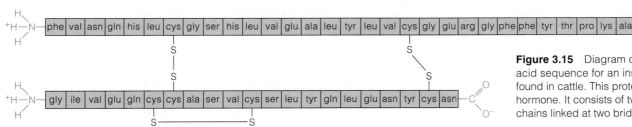

Figure 3.15 Diagram of the amino acid sequence for an insulin molecule found in cattle. This protein serves as a hormone. It consists of two polypeptide chains linked at two bridges (—S—S—).

When peptide bonds join three or more amino acids, we have a **polypeptide chain**. In such chains, the carbon backbone has nitrogen atoms positioned in this regular pattern: —N—C—C—N—C—C—.

For each kind of protein, different amino acids are selected one at a time from the twenty kinds available. Instructions encoded in the cell's DNA determine the selection. Overall, the protein's amino acid sequence is unique and is its *primary* structure (Figures 3.14 and 3.15).

Different cells make thousands of different proteins. Many are fibrous, with polypeptide chains organized as strands or sheets. Collectively, these fibers contribute to a cell's shape and organization. Globular proteins have one or more chains folded in compact, rounded shapes. Most enzymes are globular proteins. So are actin and a few other proteins that help cells, and cell parts, move.

Regardless of the type of protein, its shape and its function arise from the primary structure—that is, from information encoded in its amino acid sequence. That sequence influences which parts of a polypeptide chain will coil, bend, or interact with other chains. The type and arrangement of atoms in the coiled, stretched-out, and folded parts dictate whether the protein will function as, say, an enzyme, a transporter, a receptor, or even an unlucky target for a bacterium or virus.

Each protein consists of one or more polypeptide chains of amino acids. The amino acid sequence (which kind of amino acid follows another in the chain) is unique for each kind of protein and gives rise to its unique structure, chemical behavior, and function.

HOW DOES A THREE-DIMENSIONAL PROTEIN EMERGE?

The preceding section gave you a sense of how amino acids form a polypeptide chain, which is the primary structure of proteins. Now consider a few examples of how that structure dictates the final protein shape.

For the most part, the primary structure gives rise to a protein's shape in two ways. First, it allows hydrogen bonds to form between different amino acids along a polypeptide chain's length. Second, it puts R groups in positions that make them interact. The interactions force the chain to bend and twist in specific ways.

Second Level of Protein Structure

Hydrogen bonds form at regular, short intervals along a new polypeptide chain. They give rise to the protein's

secondary structure—a coiled or an extended pattern. Picture the chain as a set of rigid playing cards joined by sticks (covalent bonds) that can swivel a bit. Each card is a peptide group, and some of its atoms can form bonds with nearby atoms. For instance, hydrogen bonds form between every third peptide group in many chains. The bonding pattern forces the groups to swivel and coil like a spiral staircase (Figure 3.16a). In other proteins, hydrogen bonds hold two or more chains side by side in an extended, sheetlike array (Figure 3.16b).

Third Level of Protein Structure

Most polypeptide chains that have a coiled secondary structure bend even more. Why? Along their length are amino acids, including the bulky proline, that make the chain bend and loop outward. Some R groups far apart on the chain interact to hold loops in specific positions. A chain that folded as a result of bend-producing amino acids and R-group interactions is at the third level of protein structure. Figure 3.17a shows one example: the compact structure of the protein globin. Hydrogen bonds between functional groups brought about its folding.

Fourth Level of Protein Structure

Now picture many bonds forming between *four* globin molecules. Then put an iron-containing functional group called heme close to the center of each one. You end up with **hemoglobin**, an oxygen-transporting, respiratory protein (Figure 3.17b). As you read this, each mature red blood cell inside your body is busily transporting a billion molecules of oxygen. All that oxygen is bound to 250 million hemoglobin molecules.

one peptide group

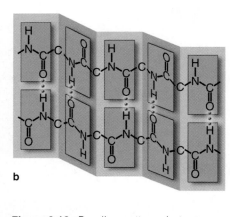

Figure 3.16 Bonding patterns between peptide groups of polypeptide chains. Extensive hydrogen bonding (*dotted* lines) can bring about (**a**) coiling of one chain or (**b**) sheetlike arrays of two or more chains.

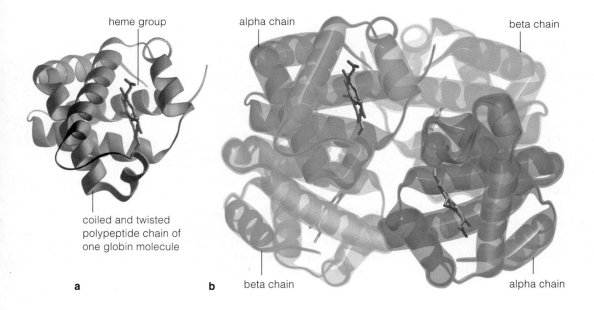

heme group   alpha chain   beta chain

coiled and twisted polypeptide chain of one globin molecule

a   b   beta chain   alpha chain

Figure 3.17 (**a**) Globin molecule. This coiled polypeptide chain associates with a heme group, an iron-containing functional group that strongly attracts oxygen. There is a whole family of globins, identical in most parts of their amino acid sequences but unique in other parts.

(**b**) Hemoglobin, which transports oxygen in blood. This protein consists of four polypeptide chains (globin molecules) and four heme groups. Two chains, designated *alpha*, differ slightly from the other two (*beta* chains) in their amino acid sequence. We sketched the coiled parts of each chain as if they were enclosed in a transparent tube to help you visualize all four globin molecules.

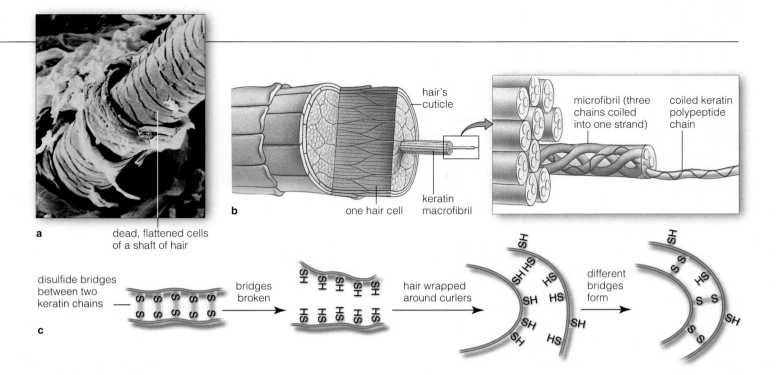

a dead, flattened cells of a shaft of hair

b one hair cell   hair's cuticle   keratin macrofibril

microfibril (three chains coiled into one strand)   coiled keratin polypeptide chain

c disulfide bridges between two keratin chains → bridges broken → hair wrapped around curlers → different bridges form

Hemoglobin is a classic example of the fourth level of protein structure. In all proteins at this level, hydrogen bonds and sometimes covalent bonds between R groups join two or more polypeptide chains into one molecule.

Hemoglobin and most enzymes are examples of the **globular proteins**; their chains are folded in a rounded shape. **Fibrous proteins** are long strands or sheets of polypeptide chains. Examples are keratin (Figure 3.18) and collagen, the most common animal protein. Skin, bone, corneas, arteries, and many other animal body parts depend on collagen's strength.

Glycoproteins and Lipoproteins

Some proteins have other organic compounds attached to their polypeptide chains. For example, **lipoproteins** form when certain proteins circulating in blood join up with cholesterol, triglycerides, and phospholipids that were absorbed from the gut after a meal. Similarly, most **glycoproteins** are proteins that have linear or branched oligosaccharides attached. Almost every protein at the surface of animal cells are glycoproteins. So are most proteins secreted from cells and many proteins in blood.

Structural Changes by Denaturation

Breaking weak bonds of a protein or any other large molecule disrupts its three-dimensional shape, an event called **denaturation**. For example, weak hydrogen bonds are sensitive to increases and decreases in temperature and pH. If the temperature or pH exceeds a protein's range of tolerance, its polypeptide chains will unwind or change shape, and the protein will lose its function. Consider the protein albumin, concentrated in the "egg white" of uncooked chicken eggs. When you cook eggs,

Figure 3.18 Structure of hair. (**a,b**) Hair cells develop from modified skin cells. They synthesize polypeptide chains of the protein keratin. Disulfide bridges link three chains into fine fibers, which become bundled into larger, cable-like fibers. The fibers nearly fill the cells, which eventually die off. (**a**) Dead, flattened cells form a tubelike cuticle around the developing hair shaft.

(**c**) For a permanent wave, hair is exposed to chemicals that break disulfide bridges. When hairs wrap around curlers, their polypeptide chains are held in new positions. Exposure to a different chemical causes disulfide bridges to form again. But, compared to before, the new ones form between different sulfur-bearing amino acids. The displaced bonding locks the hair in curled positions (compare Figure 3.11).

the heat does not disrupt the strong covalent bonds of albumin's primary structure. But it destroys weaker bonds contributing to the three-dimensional shape. For some proteins, denaturation might be reversed when normal conditions are restored—but albumin isn't one of them. There is no way to uncook a cooked egg.

Proteins have a primary structure, which is the sequence of different kinds of amino acids along a polypeptide chain. An individual's DNA specifies that sequence.

Proteins have a secondary structure: a coiled pattern or an extended, sheetlike pattern that arises by hydrogen bonding at short, regular intervals along a polypeptide chain.

At the third level of protein structure, bonding at certain amino acids makes a coiled chain bend and loop at certain angles, in certain directions. Interactions among R groups in the chain hold the loops in characteristic positions.

At the fourth level of protein structure, many hydrogen bonds and other interactions join two or more polypeptide chains. Many proteins are this structurally complex.

NUCLEOTIDES AND NUCLEIC ACIDS

Nucleotides are small organic compounds with a sugar, one or more phosphate groups, and a base. Their sugar, unit, either ribose or deoxyribose, has a five-carbon ring structure. All the bases have a single or double carbon ring structure that incorporates nitrogen.

One nucleotide, **ATP** (adenosine triphosphate), has a string of three phosphate groups attached to its sugar:

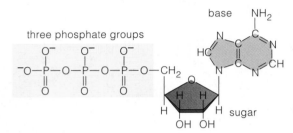

ATP easily transfers a phosphate group to many other molecules, thereby energizing them to enter a reaction. That's why ATP is central to metabolism (Section 5.2).

Some of the nucleotides are part of **coenzymes**. These enzyme helpers transfer hydrogen atoms plus electrons stripped from molecules to other reaction sites. NAD^+ is an example (Sections 5.2 and 7.1). Other coenzymes are chemical messengers between cells and inside cells.

Nucleic acids are composed of nucleotides. **DNA** (deoxyribonucleic acid) is a double-stranded molecule of four kinds of nucleotides that differ in their bases. The bases are adenine, guanine, thymine, or cytosine (Figure 3.19). Their sequential order in a DNA strand encodes information on making all proteins that cells need to survive and reproduce. Bases in some parts of

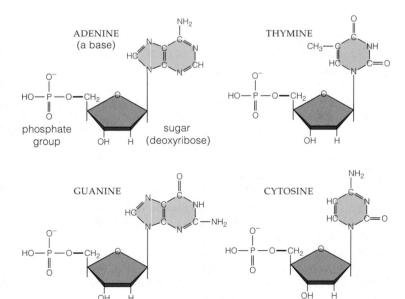

Figure 3.19 Four kinds of nucleotides that function as components of DNA. Two nucleotide bases, adenine and guanine, have a double ring structure. Ttwo other bases, thymine and cytosine, have a single ring structure.

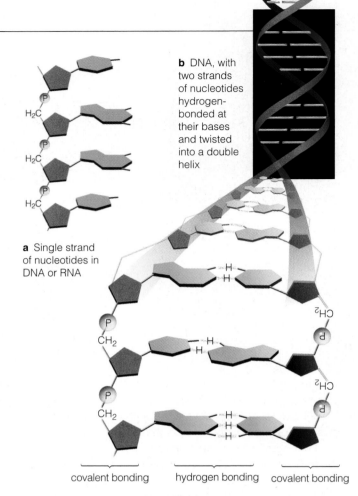

b DNA, with two strands of nucleotides hydrogen-bonded at their bases and twisted into a double helix

a Single strand of nucleotides in DNA or RNA

covalent bonding   hydrogen bonding   covalent bonding

Figure 3.20 (**a**) A series of covalently bonded nucleotides in a single strand of DNA or RNA. (**b**) DNA double helix.

the sequence are unique to each species. As Figure 3.20 shows, hydrogen bonds between bases join the two DNA strands, which twist spirally into a double helix.

Like DNA, **RNAs** (ribonucleic acids) are assembled from four kinds of nucleotides. Unlike DNA, the bases are adenine, guanine, cytosine, and *uracil*. Also unlike DNA, most RNA molecules are single nucleotide strands. RNAs do not encode protein-building instructions, but different classes carry out these instructions in cells. One kind even catalyzes peptide bond formation. Ever since the first cells emerged, RNAs have served in processes by which genetic information is used to build proteins. We return to DNA and RNA in Chapters 12, 13, and 19.

Some nucleotides function as energy carriers or chemical messengers. Others are subunits of coenzymes and nucleic acids. The nucleic acids are single or double strands of covalently bonded nucleotides.

DNA has two helically coiled, hydrogen-bonded strands of nucleotides. Its sequence of bases encodes information about all proteins a cell requires to survive and reproduce. Different classes of single-stranded nucleic acids called RNAs carry out DNA's protein-building instructions.

Food Production and a Chemical Arms Race

Next time you shop for groceries, reflect on what it takes to give you a daily supply of organic compounds. For example, lettuce typically grows in fertilized cropland. Weeds typically compete with lettuce for the fertilizer's nutrients. Each year, these food pirates and others ruin or gobble up nearly half of what people try to grow.

Most plants aren't entirely defenseless. They evolved under intense selection pressure of attacks by insects and other organisms. In nature, they often repel the attackers with toxins. A **toxin** is an organic compound, a normal metabolic product of one species, but its chemical effects harm or kill individuals of a different species that contact it. Humans, too, encounter traces of natural toxins, even in such familiar edibles as hot peppers, potatoes, figs, celery, rhubarb, and alfalfa sprouts. Still, we do not die in droves, so we must have chemical defenses against these toxins.

A few thousand years ago, farmers used sulfur, lead, arsenic, and mercury to kill insects. They freely dispensed these exceedingly toxic metals until the late 1920s, when someone figured out they were poisoning people. Traces of the toxic metals still turn up in contaminated lands.

Also, early farmers took a cue from plants and used *organic* compounds extracted from leaves, flowers, and roots as natural pesticides. In 1945, scientists started to develop synthetic toxins and to understand mechanisms by which toxins attack their targets. *Herbicides* kill weeds by disrupting metabolism and growth. Most *insecticides* clog a target insect's airways, disrupt nerves and muscles, or block reproduction. *Fungicides* work against harmful fungi, including a mold that makes aflatoxin, one of the deadliest poisons. By 1995, people in the United States were spraying or spreading more than 1.25 billion pounds of toxins through fields, gardens, homes, and industrial and commercial sites (Figure 3.21).

Some pesticides also kill birds and other predators that help control pest population sizes. And targeted pests are developing resistance to the chemical arsenal, for reasons given earlier (Chapter 1). Pesticides can't be haphazardly released, for they are easily inhaled, ingested with food, or absorbed by skin. Certain types are active for weeks or years. Some trigger rashes, headaches, hives, asthma, and joint pain in millions of people or trigger life-threatening allergic reactions in abnormally sensitive individuals.

DDT and other long-lived pesticides are now banned in the United States (Section 43.3). Even rapidly degradable types, such as malathion and other organophosphates (Figure 3.21d), must meet rigorous application standards and safety tests. Still, many farmers argue that limiting pesticide applications will invite devastating crop losses, and then how will we feed the hungry multitudes?

b Atrazine. Farmers spray this herbicide directly on leaves of corn and other crop plants. It kills weeds within a few days but doesn't hurt crop plants. Other herbicides that don't persist long are glyphosate (Roundup), daminozide (Alar), and alachlor (Lasso).

c Dichlorodiphenyltrichloroethane, or DDT. This nerve cell poison breaks down two to fifteen years after its application. Chlordane is another type of insecticide that persists for a long time in the environment.

d Malathion. Like other organophosphates, this nerve cell poison lasts one to twelve weeks. Organophosphates break down faster than chlorinated hydrocarbons but are far more toxic. About four dozen kinds are cheap and effective. They account for half of all insecticide uses in the United States. The Environmental Protection Agency wants to minimize pesticide residues on food. At this writing methyl parathion is banned for food crops; applications of azenphos methyl must end at least three weeks before harvest time. Farmers are contesting these actions and want the EPA to consider economic and trade issues as well as human health.

Figure 3.21 (**a**) Lettuce, a common crop plant, and a low-flying crop duster with its rain of pesticides. (**b–d**) A few of the major pesticides, some more toxic than others.

SUMMARY   *Gold* indicates text section

1. Organic compounds consist of carbon, hydrogen, and often other elements. The carbon atoms are often bonded covalently to form a linear or ring-shaped backbone, and functional groups are attached to the backbone. *3.1*

2. In nature, only living cells assemble large organic compounds called biological molecules (Table 3.1). Cells build these complex carbohydrates, lipids, proteins, and nucleic acids from simple sugars, fatty acids, amino acids, nucleotides, and other small organic molecules. *3.1*

3. Cells assemble, rearrange, and break down organic compounds by different enzyme-mediated reactions, as when they transfer functional groups or electrons. *3.1*

4. Cells use the simple sugars (such as glucose) and the oligosaccharides for energy and as building blocks. They use complex carbohydrates (such as starch, cellulose, and glycogen) as structural materials and as energy storage forms. They use lipids (especially triglycerides) as energy storage forms, structural components of cell membranes, and precursors of other compounds. *3.2–3.3*

5. Proteins consist of polypeptide chains of amino acids. Proteins, the most diverse biological molecules, serve in enzyme activity, structural support, transport, changes in cell shape and activities, movements, and defense against disease. Nucleotides function as energy carriers (especially ATP), coenzymes, and subunits for strands of the nucleic acids DNA and RNA, which interact and form the basis of inheritance and reproduction. *3.4–3.6*

Table 3.1 *Summary of the Main Organic Compounds in Living Things*

Category	Main Subcategories	Some Examples and Their Functions	
CARBOHYDRATES . . . contain an aldehyde or a ketone group, and one or more hydroxyl groups	**Monosaccharides** (simple sugars)	Glucose	Energy source
	Oligosaccharides	Sucrose (a disaccharide)	Most common form of sugar transported through plants
	Polysaccharides (complex carbohydrates)	Starch, glycogen Cellulose	Energy storage Structural roles
LIPIDS . . . are largely hydrocarbon; generally do not dissolve in water but dissolve in nonpolar substances, such as other lipids	**Lipids with fatty acids** *Glycerides:* One, two, or three fatty acid tails attached to glycerol backbone	Fats (e.g., butter), oils (e.g., corn oil)	Energy storage
	Phospholipids: Phosphate group, one other polar group, and (often) two fatty acids attached to glycerol backbone	Phosphatidylcholine	Key component of cell membranes
	Waxes: Long-chain fatty acid tails attached to alcohol	Waxes in cutin	Conservation of water in plants
	Lipids with no fatty acids *Sterols:* four carbon rings; the number, position, and type of functional groups differ among various sterols	Cholesterol	Component of animal cell membranes; precursor of many steroids and of vitamin D
PROTEINS . . . are one or more polypeptide chains, each with as many as several thousand covalently linked amino acids	**Fibrous proteins** Long strands or sheets of polypeptide chains; often tough, water-insoluble	Keratin Collagen	Structural component of hair, nails Structural component of bone
	Globular proteins One or more polypeptide chains folded and linked into globular shapes; many roles in cell activities	Enzymes Hemoglobin Insulin Antibodies	Great increase in rates of reactions Oxygen transport Control of glucose metabolism Tissue defense
NUCLEIC ACIDS (AND NUCLEOTIDES) . . . are chains of units (or individual units) that each consist of a five-carbon sugar, phosphate, and a nitrogen-containing base	**Adenosine phosphates**	ATP cAMP (Section 32.2)	Energy carrier Messenger in hormone regulation
	Nucleotide coenzymes	NAD^+, $NADP^+$, FAD	Transport of electrons, protons (H^+), from one reaction site to another
	Nucleic acids Chains of thousands to millions of nucleotides	DNA, RNAs	Storage, transmission, translation of genetic information

Review Questions

1. Define organic compound. Name the type of chemical bond that predominates in the backbone of such a compound. *3.1*

2. Define hydrocarbon. Also define functional group and give an example. *3.1*

3. Name the molecules of life and the families of small organic compounds from which they are built. Do they break apart most easily at their hydrocarbon portion or at functional groups? *3.1*

4. Describe the difference between a condensation reaction and hydrolysis. *3.1*

5. Select one of the carbohydrates, lipids, proteins, or nucleic acids described in this chapter. Speculate on how its functional groups and bonds between the carbon atoms in its backbone contribute to its final shape and function. *3.2 through 3.8*

6. Which item listed includes all of the other items listed? *3.3*
   a. triglyceride   c. wax   e. lipid
   b. fatty acid   d. sterol   f. phospholipid

7. Explain how hemoglobin's three-dimensional shape arises, starting with the primary structure of its four chains. *3.4, 3.5*

Self-Quiz   ANSWERS IN APPENDIX III

1. Each carbon atom can share pairs of electrons with as many as _____ other atoms.
   a. one   b. two   c. three   d. four

2. Hydrolysis is a (an) _____ reaction.
   a. functional group transfer   d. condensation
   b. electron transfer   e. cleavage
   c. rearrangement   f. both b and d

3. _____ is a simple sugar (monosaccharide).
   a. Glucose   c. Ribose   e. both a and b
   b. Sucrose   d. Chitin   f. both a and c

4. In unsaturated fats, fatty acid tails have one or more _____ .
   a. single covalent bonds   b. double covalent bonds

5. _____ are to proteins as _____ are to nucleic acids.
   a. Sugars; lipids   c. Amino acids; hydrogen bonds
   b. Sugars; proteins   d. Amino acids; nucleotides

6. A denatured protein or DNA molecule has lost its _____ .
   a. hydrogen bonds   c. function
   b. shape   d. all of the above

7. _____ are nucleotides.
   a. ATP   c. RNA subunits
   b. DNA subunits d. all of the above

8. Match each molecule with the most suitable description.
   _____ long sequence of amino acids   a. carbohydrate
   _____ the main energy carrier   b. phospholipid
   _____ glycerol, fatty acids, phosphate   c. protein
   _____ two strands of nucleotides   d. DNA
   _____ one or more sugar monomers   e. ATP

Critical Thinking

1. Jack decided to celebrate summer by making a crab salad and a peach pie for some friends. He had such a good time he forgot to put the cap on the bottle of olive oil after making the salad. He also didn't tighten the lid on the shortening can after making the pie crust.
   A few weeks later, the oil was rancid but the shortening still smelled okay. Both substances are fats. Both were stored in a dark cupboard at room temperature. Yet one substance spoiled and the other stayed fresh. Speculate on which molecular bonds were the starting point for the difference.

2. In the following list, identify which is the carbohydrate, the fatty acid, the amino acid, and the polypeptide:
   a. $^+NH_3$—CHR—COO$^-$ c. (glycine)$_{20}$
   b. $C_6H_{12}O_6$ d. $CH_3(CH_2)_{16}COOH$

3. A clerk in a health-food store tells you that certain "natural" vitamin C tablets extracted from rose hips are better for you than synthetic vitamin C tablets. Given your understanding of the structure of organic compounds, what would be your response? Design an experiment to test whether the vitamins differ.

4. Xanthophyll is a yellow pigment. Rabbits that eat green, leafy plants with xanthophyll accumulate this yellow molecule in body fat but not in muscles. What chemical properties of the molecule might cause this selective accumulation?

5. Cows can digest cellulose in grasses, but people can't. A cow's four-chambered stomach holds populations of microbes that are not normal residents of the human stomach. What kind of metabolic reactions do you think the microbes carry out? What do you think might happen to a cow being treated with an antibiotic that kills the microbes?

6. A Gary Larson cartoon states that sheep with steel wool have no natural enemies. You might laugh at the thought without knowing what wool is. Wool is the keratin-rich, soft undercoat of sheep, Angora goats, llamas, and other hairy mammals. Keratin is a strandlike, fibrous protein. Many hydrogen bonds hold it in a helical shape. Visualize three such chains coiled together, with many disulfide bridges in their end regions that stabilize them into tight, ropelike fibers. Each hair has linear groups of the fibers, which resist stretching (Figure 3.18). Given keratin's structure, speculate why, when you run a wool sweater through the hot cycle of a dryer, it shrinks pathetically and permanently.

7. Reflect on the component atoms of the different classes of organic compounds. Now look at the structural formulas for the pesticides shown in Figure 3.21. Which one is a glycoprotein? An organophosphate? A chlorinated hydrocarbon?

8. Many farmers believe crop losses will be disastrous if pesticides are banned without sufficient research and effective alternatives. Many environmental activists are pushing for rapid bans to protect people and the environment. What do you think?

Selected Key Terms

alcohol *3.1*	fibrous protein *3.5*	oligosaccharide *3.2*
amino acid *3.4*	functional group *3.1*	organic compound *3.1*
ATP *3.6*	globular protein *3.5*	phospholipid *3.3*
carbohydrate *3.2*	glycoprotein *3.5*	polymer *3.1*
coenzyme *3.6*	hemoglobin *3.5*	polypeptide chain *3.4*
condensation	hydrocarbon *3.1*	polysaccharide *3.2*
reaction *3.1*	hydrolysis *3.1*	protein *3.4*
denaturation *3.5*	lipid *3.3*	RNA *3.6*
DNA *3.6*	lipoprotein *3.5*	sterol *3.3*
enzyme *3.1*	monomer *3.1*	toxin *3.7*
fat *3.3*	monosaccharide *3.2*	triglyceride *3.3*
fatty acid *3.3*	nucleic acid *3.6*	wax *3.3*
	nucleotide *3.6*	

Readings

Ritter, P. 1996. *Biochemistry: An Introduction*. Pacific Grove, California: Brooks/Cole. Great human applications.

On-Line readings at Student Guide for InfoTrac:
www.brookscole.com/biology

4

CELL STRUCTURE AND FUNCTION

Animalcules and Cells Fill'd With Juices

Early in the seventeenth century, a scholar by the name of Galileo Galilei put two glass lenses inside a cylinder. With this instrument he happened to look at an insect, and later he described the stunning geometric patterns of its tiny eyes. Thus Galileo, who wasn't a biologist, was one of the first to record a biological observation made through a microscope. The study of the cellular basis of life was about to begin. First in Italy, then in France and England, scholars set out to explore a world whose existence had not even been suspected.

At midcentury Robert Hooke, Curator of Instruments for the Royal Society of England, was at the forefront of those studies. When Hooke first turned a microscope to thinly sliced cork from a mature tree, he observed tiny compartments (Figure 4.1c). He gave them the Latin name *cellulae*, meaning small rooms—hence the origin of the biological term "cell." They actually were the interconnecting walls of dead plant cells, which is what cork is made of, but Hooke didn't think of them as being dead because neither he nor anyone else at the time knew cells could be alive. In other plant tissues, he observed cells "fill'd with juices" but didn't have a clue to what they represented.

Given the simplicity of their instruments, it's just amazing that the pioneers in microscopy observed as much as they did. Antony van Leeuwenhoek, a Dutch shopkeeper, had exceptional skill in constructing lenses and possibly the keenest vision of all (Figure 4.1a). By the late 1600s, he was discovering natural wonders everywhere, including "many very small animalcules, the motions of which were very pleasing to behold," in scrapings of tartar from his teeth. Elsewhere he observed protistans, sperm, and even a bacterium—an organism so small it would not be seen again for another two centuries!

a

b

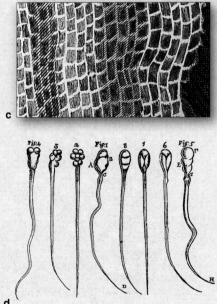

c

d

Figure 4.1 Early glimpses into the world of cells. (**a**) Antony van Leeuwenhoek, with microscope in hand. (**b**) Robert Hooke's compound microscope and (**c**) his drawing of cell walls from cork tissue. (**d**) One of van Leeuwenhoek's early sketches of sperm cells. (**e**) Cartoon evidence of the startling impact of microscopic observations on nineteenth-century London.

In the 1820s, improvements in lenses brought cells into sharper focus. Robert Brown, a botanist, noticed an opaque spot in a variety of cells and called it a nucleus. In 1838 another botanist, Matthias Schleiden, wondered if the nucleus had something to do with development. As he hypothesized, each plant cell might develop as an independent unit even though it's part of the plant.

By 1839, after years of studying animal tissues, the zoologist Theodor Schwann had this to say: Animals as well as plants consist of cells and cell products—and even though the cells are part of a whole organism, to some extent they have an individual life of their own.

A decade later a question remained: Where do cells come from? Rudolf Virchow, a physiologist, completed his own studies of a cell's growth and reproduction—that is, its division into two daughter cells. Every cell, he reasoned, comes from a cell that already exists.

And so, by the middle of the nineteenth century, microscopic analysis had yielded three generalizations, which together constitute the **cell theory**. *First, every organism is composed of one or more cells. Second, the cell is the smallest unit having the properties of life. Third, the continuity of life arises directly from the growth and division of single cells.* All three insights still hold true.

This chapter is not meant for memorization. Read it simply to gain an overview of current understandings of cell structure and function. In later chapters, you can refer back to it as a road map through the details. With its images from microscopy, this chapter and others in the book can transport you into spectacular worlds of juice-fill'd cells and animalcules.

Key Concepts

1. All organisms consist of one or more cells. The cell is the smallest unit that still retains the characteristics of life. Since the origin of life, each new cell has arisen from a preexisting cell. These are the three generalizations of the cell theory.

2. All cells have a plasma membrane. This outermost, double-layered membrane separates a cell's interior from its surroundings. Also, all cells contain cytoplasm, an organized internal region where energy conversions, protein synthesis, movements of cell parts, and other required activities proceed. In eukaryotic cells only, the DNA is enclosed within a membrane-bound nucleus. In prokaryotic cells (archaebacteria and eubacteria), the DNA simply is concentrated in part of the cell interior.

3. The plasma membrane and internal cell membranes consist of lipids and proteins. The lipids are organized as two adjacent layers—that is, as a bilayer. They give a membrane its basic structure and prevent water-soluble substances from freely crossing it. Diverse proteins are embedded in the bilayer or positioned at its surfaces. They carry out many functions, such as transporting water-soluble substances across the membrane.

4. Organelles such as the nucleus are membrane-bound compartments within eukaryotic cells. They physically separate metabolic reactions and allow them to proceed in orderly ways. Prokaryotic cells do not have comparable organelles.

5. When cells are growing, they increase faster in volume than in surface area. This physical constraint on growth influences their size and shape.

6. Different microscopes modify light rays or accelerated beams of electrons in ways that allow us to form images of incredibly small specimens. They are the foundation for our current understanding of cell structure and function.

e

BASIC ASPECTS OF CELL STRUCTURE AND FUNCTION

Inside your body and at its moist surfaces, trillions of cells live in interdependency. In northern forests, pollen grains—four-celled structures—escape from pine trees. In scummy pondwater, a single-celled amoeba moves on its own. For humans, pines, amoebas, and all other organisms, the **cell** is the smallest unit that lives on its own or has the potential to do so (Figure 4.2). Each cell is structurally organized for metabolism. It senses and responds to its environment. And heritable instructions in its DNA give it the potential to reproduce.

Structural Organization of Cells

Cells differ enormously in size, shape, and activities, as you might gather by comparing a tiny bacterium with one of your relatively giant liver cells. Yet they are alike in three respects: They all start out life with a plasma membrane, a region of DNA, and a region of cytoplasm:

1. **Plasma membrane.** This thin, outermost membrane maintains the cell as a distinct entity. By doing so, it allows metabolic events to proceed apart from random events in the environment. A plasma membrane does not isolate the cell interior. Substances and signals continually move across it in highly controlled ways.

2. **Nucleus** or **nucleoid.** Depending on the species, the DNA occupies a membrane-bound sac (nucleus) in the cell or simply a region of the cell interior (nucleoid).

3. **Cytoplasm.** Cytoplasm is everything between the plasma membrane and region of DNA. It consists of a semifluid matrix and other components, such as **ribosomes** (structures on which proteins are built).

This chapter introduces two fundamentally different kinds of cells. **Eukaryotic cells** have organelles, tiny sacs having one or more outer membranes, in the cytoplasm. One, the nucleus, is their defining feature. **Prokaryotic cells** don't have a nucleus; nothing intervenes between their DNA and cytoplasm. The only prokaryotic cells are the archaebacteria and eubacteria. All other organisms—from amoebas to peach trees to puffball mushrooms to whales and zebras—consist of eukaryotic cells.

Fluid Mosaic Model of Cell Membranes

Fluid bathes the two surfaces of a cell membrane and is vital for its functions. The membrane itself has a fluid quality; it isn't a rigid barrier between cytoplasm and extracellular fluid. Why is this so? The answer starts with phospholipids. A **phospholipid** has a hydrophilic head (a phosphate group, another polar group, and a glycerol backbone) and two hydrophobic tails (fatty acids). The head dissolves in water; the tails don't. Immerse many phospholipids in water. They may spontaneously cluster

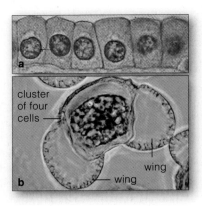

Figure 4.2 (a) A row of cells at the surface of a tiny tube inside a human kidney. Each cell has a nucleus (the sphere stained darker red). (b) A winged pollen grain. One of its cells will give rise to a sperm that may meet up with an egg and help start a new pine tree.

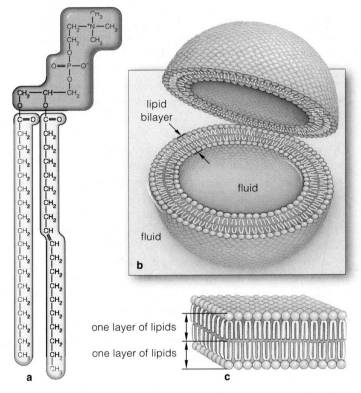

Figure 4.3 Lipid bilayer organization of cell membranes. (a) One of the phospholipids, the most abundant molecules of membranes. (b,c) These and other lipids are arranged into two layers. Their hydrophobic tails are sandwiched between the hydrophilic heads. In cells, the heads are dissolved in cytoplasm on one side of the bilayer and in extracellular fluid on the other side.

as a sheet at the surface. Or they may form two-layered spheres beneath the surface, with all tails sandwiched between all the heads. This **lipid bilayer** arrangement is the structural basis of cell membranes (Figure 4.3).

Figure 4.4 shows a bit of membrane corresponding to the **fluid mosaic model.** By this model, cell membranes have a mixed composition—a *mosaic*—of many diverse phospholipids, sterols, and proteins. Phospholipids are the most abundant components. Cholesterol is the most

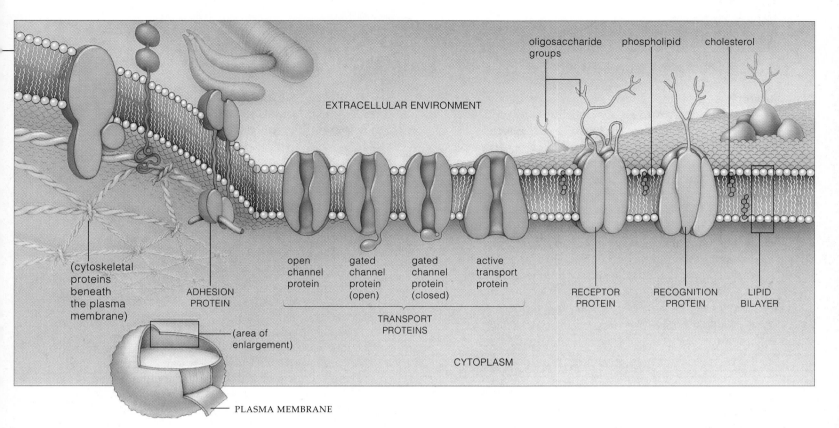

EXTRACELLULAR ENVIRONMENT

oligosaccharide groups   phospholipid   cholesterol

(cytoskeletal proteins beneath the plasma membrane)

ADHESION PROTEIN

open channel protein

gated channel protein (open)

gated channel protein (closed)

active transport protein

RECEPTOR PROTEIN

RECOGNITION PROTEIN

LIPID BILAYER

TRANSPORT PROTEINS

(area of enlargement)

CYTOPLASM

PLASMA MEMBRANE

Figure 4.4 Cutaway view of part of a plasma membrane, based on the fluid mosaic model. In addition to the specialized proteins shown, enzymes also are associated with cell membranes.

common sterol in animal cell membranes; phytosterols are their equivalent in plants. Also by this model, a cell membrane has a *fluid* quality owing to molecular motions and interactions. Hydrophobic interactions give rise to most of the membrane's structure, but they are weaker than covalent bonds. Thus, most phospholipids are free to drift sideways in a membrane, spin about their long axis, and flex their tails. The motions prevent molecules from packing together as a solid layer. Short or kinked (unsaturated) tails add to the membrane fluidity.

The fluid mosaic model is a good starting point for exploring membranes. But bear in mind, cell membranes differ in composition and molecular arrangements. They aren't even the same on both surfaces. Oligosaccharides, for instance, are covalently bonded to certain proteins and lipids of a plasma membrane, but only on its outer surface (Figure 4.3). Also, these carbohydrates and others differ in number and kind among species, even among different cells of the same individual.

Overview of Membrane Proteins

The proteins embedded in a lipid bilayer or attached to it have diverse functions. Many are *enzymes*. Others are *transport* proteins. They span the bilayer and let water-soluble substances move through their interior. They bind molecules or ions on one side of the membrane, then release them on the other side.

Receptor proteins bind extracellular substances, such as hormones, that trigger changes in cell activities. For example, enzymes that crank up the machinery for cell division are switched on when molecules of a hormone, somatotropin, dock at receptors for them. Different cells have different combinations of receptors.

Different *recognition* proteins at the cell surface are like molecular fingerprints; their oligosaccharide chains identify a cell as being of a specific type. For example, the plasma membrane of your cells bristles with "self" proteins. Certain white blood cells chemically recognize self proteins and leave your cells alone, but they attack invading bacterial cells with "nonself" proteins at their surface. *Adhesion* proteins of multicelled organisms help cells of the same type locate and stick to one another and stay in the proper tissues. These are glycoproteins, with oligosaccharides attached. The adhesion sites may become a type of cell junction, as Section 4.11 describes.

All cells have an outermost plasma membrane, an internal region of cytoplasm, and an internal region of DNA. Only eukaryotic cells have a nucleus and other organelles.

A cell membrane has a lipid bilayer, mainly of phospholipid molecules. Hydrophobic parts of the lipids are sandwiched between the hydrophilic parts, which are dissolved in the fluid surroundings. A lipid bilayer imparts structure to the membrane and is a barrier to water-soluble substances.

Proteins embedded in the bilayer or positioned at one of its surfaces carry out diverse membrane functions. Many are enzymes, transporters of substances across the bilayer, or receptors for extracellular substances and signals. Other types function in cell-to-cell recognition or adhesion.

CELL SIZE AND CELL SHAPE

You may be wondering how small cells really are. Can any be observed with the unaided human eye? There are a few, including the "yolks" of bird eggs, cells in the red part of watermelons, and the fish eggs we call caviar. However, most cells cannot be observed without microscopes. To give you a sense of cell sizes, red blood cells are about 8 millionths of a meter across. You could fit a string of 2,000 of them across your thumbnail!

Why are most cells so small? The **surface-to-volume ratio** constrains increases in cell size. By this physical relationship, the volume of an object increases with the cube of the diameter, but the surface area increases only with the square. As Figure 4.5 shows, *if a cell expands in diameter during growth, then its volume will increase more rapidly than its surface area will.*

diameter (cm):	0.5	1.0	1.5
surface area (cm^2):	0.79	3.14	7.07
volume (cm^3):	0.06	0.52	1.77
surface-to-volume ratio:	13.17:1	6.04:1	3.99:1

Figure 4.5 Example of the surface-to-volume ratio.

Imagine that you can make a round cell grow four times wider than normal. Its volume increases 64 times (4^3). But the surface area increases only 16 times (4^2). As a result, each unit of plasma membrane must now serve four times as much cytoplasm as it did before. Moreover, beyond a certain point, the inward flow of nutrients and the outward flow of wastes won't be fast enough, so you'll end up with a dead cell.

A large, round cell also would have trouble moving materials *through* its cytoplasm. In small cells, random, tiny motions of molecules easily distribute materials. If a cell isn't small, you usually can expect it to be long and thin or have outfoldings and infoldings that increase its surface relative to its volume. *The smaller or narrower or more frilly-surfaced the cell, the more efficiently materials cross its surface and become distributed through the interior.*

We also see evidence of surface-to-volume constraints on body plans of multicelled species. For example, cells attach end to end in strandlike algae; each one interacts directly with its surroundings. Muscle cells are thin, but each is as long as the muscle of which it is part.

During their growth, cells increase faster in volume than in surface area. This surface-to-volume ratio is a physical constraint on increases in cell size. It also influences cell shape and the body plans of multicelled organisms.

Microscopes: Gateways to Cells

Modern microscopes are the gateways to amazing worlds. Some even provide glimpses into the very structure of molecules. Different kinds use wavelengths of light or accelerated electrons. The micrographs in Figures 4.6 and 4.7 only hint at the types of details that are now being observed. A **micrograph** simply is a photograph of something that came into view with the aid of a microscope.

LIGHT MICROSCOPES Picture a series of waves moving across an ocean. Each **wavelength** is the distance from one wave's peak to the peak of the wave behind it. Light also travels as waves from sources such as the sun and illuminated specimens. In a *compound light microscope*, such as the one above, two or more sets of glass lenses bend light emanating from a cell or some other specimen in ways that form an enlarged image of it.

A living cell must be small or thin enough for light to pass through. It helps if cell parts differ in color and density from the surroundings, but most have little if any color and appear uniformly dense. To get around this, microscopists *stain* cells—expose them to dyes that react with some parts of a specimen but not others. Staining may alter and kill cells. Dead cells break down fast, so cells typically are pickled or preserved before staining.

Suppose you use the best glass lens system. When you magnify a specimen's diameter by 2,000 times or more, you find that cell parts appear larger but not clearer. To understand why, think about the distance between the two crests of a wavelength of light. To give two examples, the distance is about 750 nanometers for red light and 400 nanometers for violet. Wavelengths of all other colors fall in between. If a cell structure is less than one-half of a wavelength long, that structure won't be able to disturb the rays of light streaming past, and it won't be visible.

ELECTRON MICROSCOPES Better resolution of extremely fine details is possible with the assistance of electrons. Remember, electrons are particles of matter, but they also behave like waves. In electron microscopy, streams of electrons are accelerated to wavelengths of about 0.005 nanometer—about 100,000 times shorter than wavelengths of visible light. The electrons cannot pass through glass lenses, but a magnetic field can bend them from their path and focus them.

In a *transmission electron microscope*, a magnetic field is the "lens." Accelerated electrons are directed through a specimen, focused into an image, and magnified. With *scanning electron microscopes*, a narrow beam of electrons moves back and forth across a specimen to which a thin coat of metal was applied. The metal responds by emitting some of its own electrons. A detector tied into electronic circuitry then transforms the energy of electrons into an image of the specimen's surface on a television screen. Most scanning images have fantastic depth (Figure 4.7*d*).

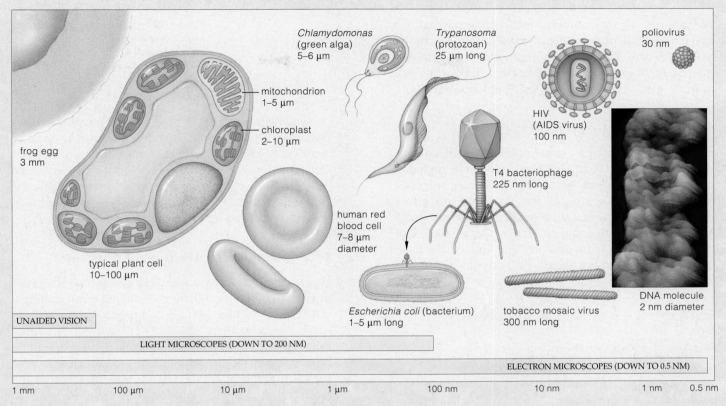

Chlamydomonas (green alga) 5–6 μm

Trypanosoma (protozoan) 25 μm long

poliovirus 30 nm

frog egg 3 mm

mitochondrion 1–5 μm

chloroplast 2–10 μm

HIV (AIDS virus) 100 nm

T4 bacteriophage 225 nm long

human red blood cell 7–8 μm diameter

typical plant cell 10–100 μm

Escherichia coli (bacterium) 1–5 μm long

tobacco mosaic virus 300 nm long

DNA molecule 2 nm diameter

UNAIDED VISION

LIGHT MICROSCOPES (DOWN TO 200 NM)

ELECTRON MICROSCOPES (DOWN TO 0.5 NM)

| 1 mm | 100 μm | 10 μm | 1 μm | 100 nm | 10 nm | 1 nm | 0.5 nm |

Figure 4.6 Units of measure used in microscopy. A *scanning tunneling microscope* (magnifications up to 100 million) gave us the photomicrograph of part of a DNA molecule. The tip of the scope's needlelike probe has a single atom. When voltage is applied between the tip and an atom at a specimen's surface, it tunnels into the electron orbitals. As the tip moves over a specimen's contours, a computer analyzes the tunneling motion and creates a three-dimensional view of the surface atoms.

1 centimeter	(cm)	= 1/100 meter, or 0.4 inch
1 millimeter	(mm)	= 1/1,000 meter
1 micrometer	(μm)	= 1/1,000,000 meter
1 nanometer	(nm)	= 1/1,000,000,000 meter

$$1 \text{ meter} = 10^2 \text{ cm} = 10^3 \text{ mm} = 10^6 \text{ μm} = 10^9 \text{ nm}$$

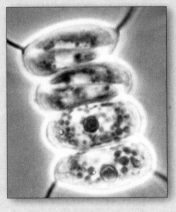

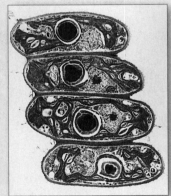

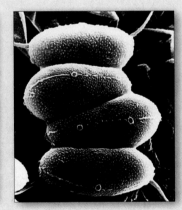

a Light micrograph (phase-contrast process)

b Light micrograph (Nomarski process)

c Transmission electron micrograph, thin section

d Scanning electron micrograph

10 μm

Figure 4.7 How different microscopes can reveal different aspects of the same organism—in this case, a green alga (*Scenedesmus*). Images of all four specimens are at the same magnification. The phase-contrast and Nomarski processes mentioned in (**a**) and (**b**) create optical contrasts without staining cells. Both processes enhance the usefulness of light micrographs. (**c**) Transmission electron microscopes offer finer resolution of details, but specimen preparation is more difficult. As for other micrographs in the book, the short horizontal bar below the micrograph in (**d**) offers a visual reference for size. A micrometer (μm) is 1/1,000,000 of a meter. Using the scale bar, can you estimate the length and width of *Scenedesmus*?

DEFINING FEATURES OF EUKARYOTIC CELLS

We turn now to organelles and other structural features that are typical of the cells of plants, animals, fungi, and protistans. We define an **organelle** as an internal, membrane-bound sac or compartment that serves one or more specialized functions inside eukaryotic cells.

Major Cellular Components

Study micrographs of a typical eukaryotic cell, such as the ones in the preceding section, and you'll probably notice that the nucleus is one of the most conspicuous features. Remember, any cell that starts out life with a nucleus is a eukaryotic cell; it houses a "true nucleus." Many other organelles and structures also are typical of these cells, although the numbers and kinds differ from one cell type to the next. Table 4.1 lists common features.

Figure 4.8 *Facing page:* Generalized sketches showing some of the features of (**a**) a typical photosynthetic plant cell and (**b**) a typical animal cell.

Organelle membranes have another function besides physically separating incompatible reactions. They allow compatible and interconnected reactions to proceed at different times. For instance, a plant's photosynthetic cells produce starch molecules in an organelle called a chloroplast, then store and later release starch for use in different reactions inside the same organelle.

Which Organelles Are Typical of Plants?

Figure 4.8*a* can start you thinking about the location of organelles in a typical plant cell. Bear in mind, calling a cell "typical" is like calling a cactus or a water lily or an apple tree a "typical" plant. As is the case for animal cells, variations on the basic plan are mind-boggling. With this qualification in mind, also take a close look at the micrograph in Figure 4.9, on page 58. It shows the locations of organelles and structures you are likely to observe in many specialized plant cells.

Which Organelles Are Typical of Animals?

Now start thinking about organelles of a typical animal cell, such as the one in Figures 4.8*b* and 4.10, on page 59. Like plant cells, it has a nucleus, mitochondria, and the other features listed in Table 4.1. *These structural similarities correlate with basic functions that are necessary for survival, regardless of the cell type.* We'll be returning to to this concept throughout the book.

Comparing Figures 4.8 through 4.10 also will give you an initial idea of how plant and animal cells differ in their structure. For example, you will never observe an animal cell surrounded by a cell wall. (You might see many different kinds of fungal and protistan cells with one, however.) What other differences can you identify?

Table 4.1 *Common Features of Eukaryotic Cells*

ORGANELLES AND THEIR MAIN FUNCTIONS	
Nucleus	Localizing the cell's DNA
Endoplasmic reticulum	Routing and modifying newly formed polypeptide chains; also, synthesizing lipids
Golgi body	Modifying polypeptide chains into mature proteins; sorting and shipping proteins and lipids for secretion or for use inside the cell
Various vesicles	Transporting or storing a variety of substances; digesting substances and structures in the cell; other functions
Mitochondria	Producing many ATP molecules in highly efficient fashion
NON-MEMBRANOUS STRUCTURES AND THEIR FUNCTIONS	
Ribosomes	Assembling polypeptide chains
Cytoskeleton	Imparting overall shape and internal organization to the cell; moving the cell and its internal structures

Think about Table 4.1 and you might find yourself asking: What is the advantage of partitioning the cell interior with such organelles? *The compartmentalization allows a large number of activities to occur simultaneously in very limited space.* Consider a photosynthetic cell in a leaf. It can put together starch molecules by one set of reactions and break them apart by another set. Yet the cell would gain nothing if the synthesis and breakdown reactions proceeded at the exact same time on the same starch molecule. Without membranes of organelles, the balance of different chemical activities that helps keep eukaryotic cells alive would spiral out of control.

Eukaryotic cells have a number of organelles. These are internal, membrane-bound sacs and compartments with specific metabolic functions.

Organelles physically separate chemical reactions, many of which are incompatible. They also separate different reactions in time, as when certain molecules are assembled, stored, then used later in other reaction sequences.

All eukaryotic cells contain certain organelles (such as the nucleus) and structures (such as ribosomes) that perform functions essential for survival. Specialized cells also may incorporate additional kinds of organelles and structures.

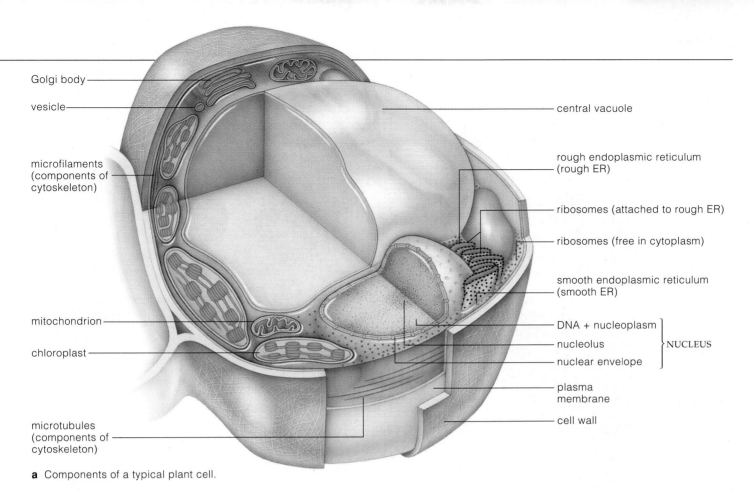

Golgi body

vesicle

microfilaments
(components of
cytoskeleton)

mitochondrion

chloroplast

microtubules
(components of
cytoskeleton)

central vacuole

rough endoplasmic reticulum
(rough ER)

ribosomes (attached to rough ER)

ribosomes (free in cytoplasm)

smooth endoplasmic reticulum
(smooth ER)

DNA + nucleoplasm

nucleolus   NUCLEUS

nuclear envelope

plasma
membrane

cell wall

a Components of a typical plant cell.

nuclear envelope

nucleolus   NUCLEUS

DNA + nucleoplasm

vesicle

lysosome

rough ER

ribosomes
(attached to
rough ER
and free in
cytoplasm)

smooth ER

vesicle

Golgi body

pair of
centrioles

components of
cytoskeleton
  microfilaments

  microtubules

plasma
membrane

mitochondrion

b Components of a typical animal cell.

MICROGRAPHS OF PLANT AND ANIMAL CELLS

The sketches in the preceding section cannot do justice to eukaryotic cell structures. Here are specific examples to give you a better sense of their exquisite details.

Figure 4.9 Transmission electron micrograph of a photosynthetic cell from a blade of Timothy grass (*Phleum pratense*), in cross-section. The sketches highlight key organelles.

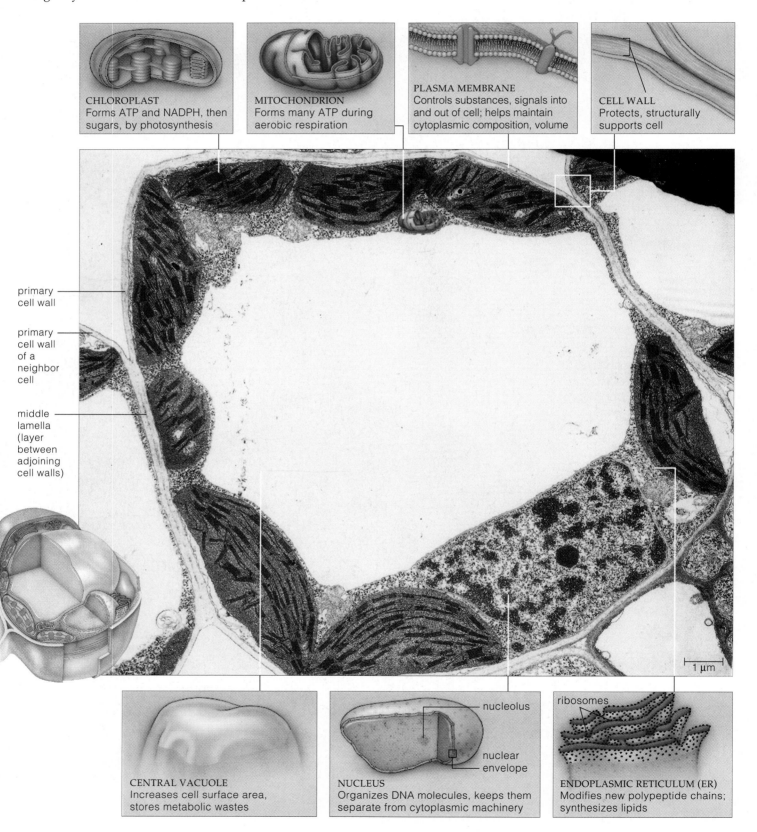

CHLOROPLAST
Forms ATP and NADPH, then sugars, by photosynthesis

MITOCHONDRION
Forms many ATP during aerobic respiration

PLASMA MEMBRANE
Controls substances, signals into and out of cell; helps maintain cytoplasmic composition, volume

CELL WALL
Protects, structurally supports cell

primary cell wall

primary cell wall of a neighbor cell

middle lamella (layer between adjoining cell walls)

1 μm

CENTRAL VACUOLE
Increases cell surface area, stores metabolic wastes

NUCLEUS
Organizes DNA molecules, keeps them separate from cytoplasmic machinery

nucleolus

nuclear envelope

ribosomes

ENDOPLASMIC RETICULUM (ER)
Modifies new polypeptide chains; synthesizes lipids

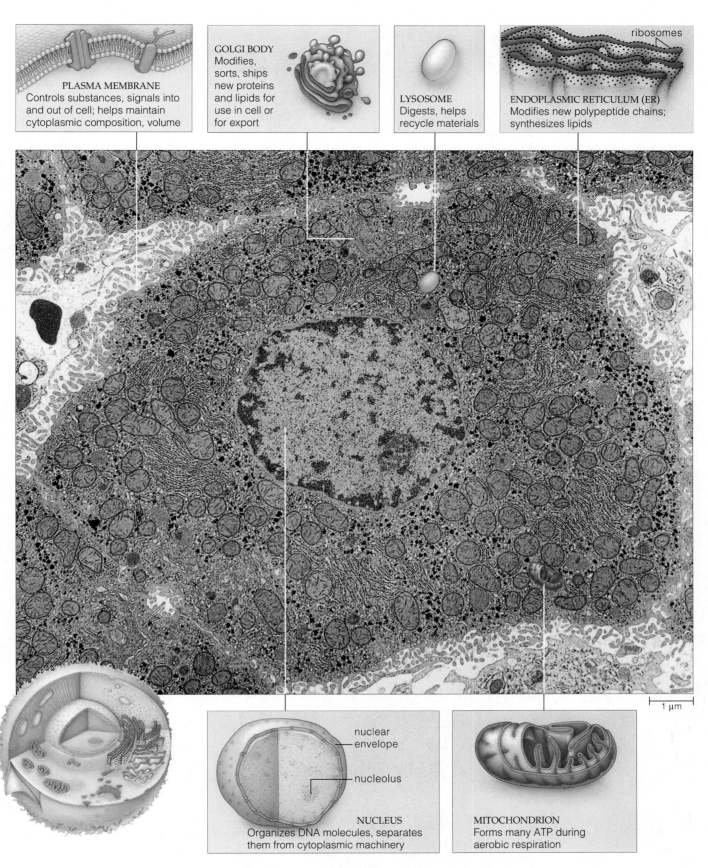

PLASMA MEMBRANE Controls substances, signals into and out of cell; helps maintain cytoplasmic composition, volume

GOLGI BODY Modifies, sorts, ships new proteins and lipids for use in cell or for export

LYSOSOME Digests, helps recycle materials

ENDOPLASMIC RETICULUM (ER) Modifies new polypeptide chains; synthesizes lipids

ribosomes

1 μm

nuclear envelope

nucleolus

NUCLEUS Organizes DNA molecules, separates them from cytoplasmic machinery

MITOCHONDRION Forms many ATP during aerobic respiration

Figure 4.10 Transmission electron micrograph of an animal cell, cross-section. The sketches highlight key organelles. The specimen is a cell from the liver of a rat.

THE NUCLEUS

Constructing, operating, and reproducing cells simply cannot be done without carbohydrates, lipids, proteins, and nucleic acids. It takes a class of proteins—enzymes—to build and use these molecules. Said another way, a cell's structure and function begin with proteins. *And instructions for building proteins are located in DNA.*

Unlike bacteria, eukaryotic cells have their genetic instructions distributed among several to many DNA molecules of different lengths. Each human body cell, for instance, has forty-six DNA molecules. End to end,

the molecules would extend outward for a meter or so. Similarly, stretch a frog cell's twenty-six DNA molecules end to end and they would extend ten meters. We have very little idea what frogs are doing with so much of it, but that's a lot of DNA!

Eukaryotic cells, recall, protect their DNA inside the nucleus. Figure 4.11 shows this organelle's structure. A nucleus has two functions. *First*, it physically separates all of the DNA molecules from the cytoplasm's complex metabolic machinery. Localizing DNA this way makes it easier for a parent cell to copy its hereditary instructions before the time comes for it to divide. Why? DNA molecules can be sorted into parcels —one for each of the two daughter cells that form later. *Second*, the outer membranes of the nucleus serve as a boundary where cells can control the passage of substances and signals to and from the cytoplasm. Table 4.2 lists the components of the nucleus.

Table 4.2 *Components of the Nucleus*	
Nuclear envelope	Pore-riddled double-membrane system that selectively controls the passage of various substances into and out of the nucleus
Nucleoplasm	Fluid interior portion of the nucleus
Nucleolus	Dense cluster of RNA and proteins that will be assembled into subunits of ribosomes
Chromosome	One DNA molecule and many proteins that are intimately associated with it
Chromatin	Total collection of all DNA molecules and their associated proteins in the nucleus

Nuclear Envelope

Unlike cells, a nucleus has two outer membranes, one wrapped around the other. Its double-membrane system is the **nuclear envelope**. It has not one but two lipid bilayers in which many molecules of proteins are embedded (Figure 4.12). The nuclear envelope surrounds the nucleoplasm, the fluid portion of a nucleus. The innermost surface contains attachment sites for strandlike proteins that anchor DNA to the envelope and also help keep it organized. On the outer surface is a profusion of ribosomes. Proteins are built on membrane-bound ribosomes or on other ribosomes positioned in the cytoplasm.

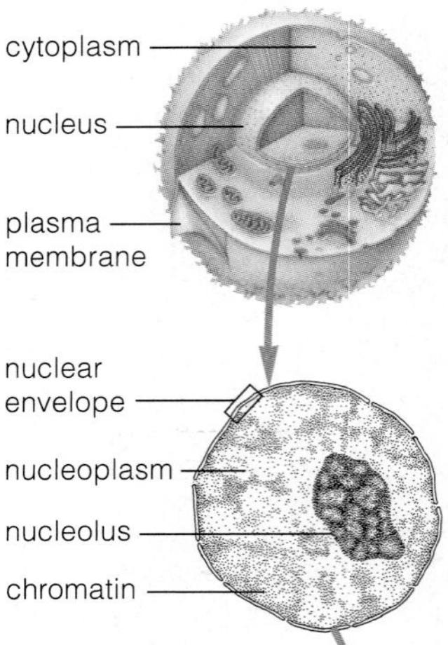

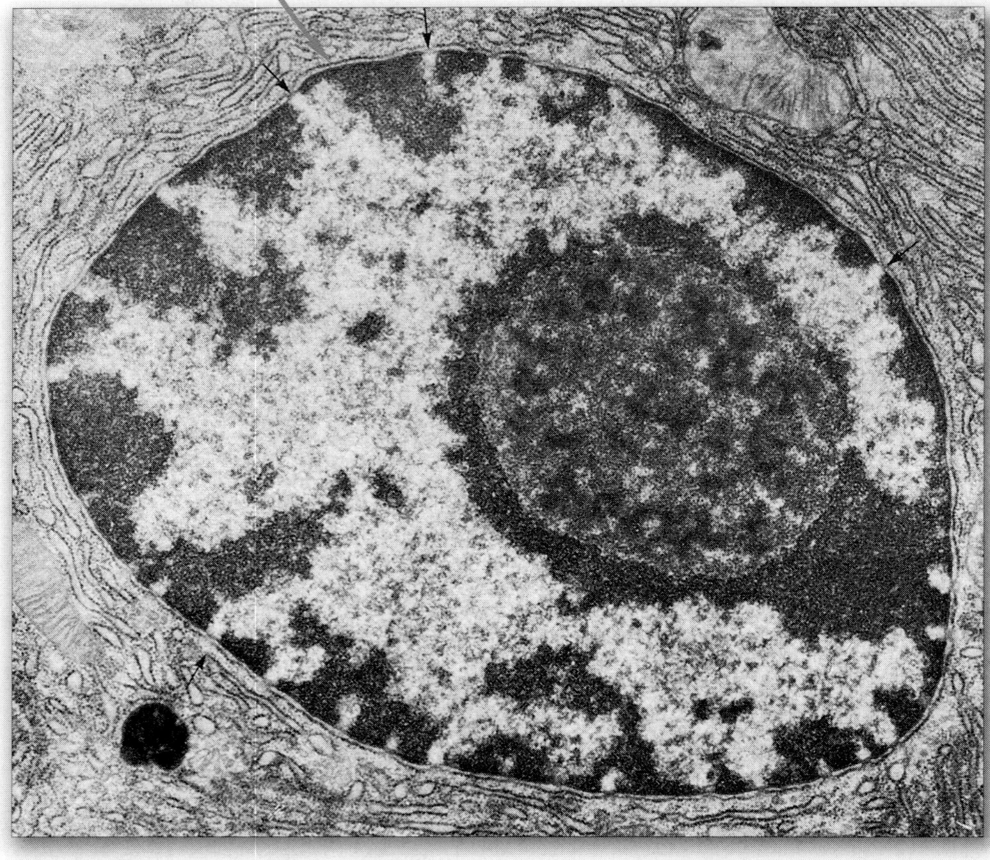

Figure 4.11 Pancreatic cell nucleus. The small arrows on this transmission electron micrograph point to pores where control systems operate to restrict or permit the passage of specific substances across the nuclear envelope.

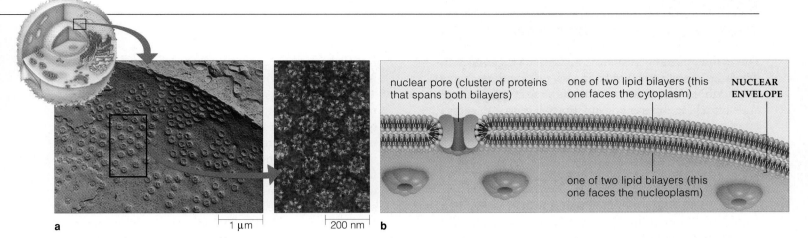

nuclear pore (cluster of proteins that spans both bilayers)

one of two lipid bilayers (this one faces the cytoplasm)

NUCLEAR ENVELOPE

one of two lipid bilayers (this one faces the nucleoplasm)

a   1 μm   200 nm   b

Figure 4.12 (**a**) Part of the outer surface of a nuclear envelope. *Left:* This specimen was fractured to show the intimate layering of its two lipid bilayers. *Right:* Nuclear pores. Each pore across the envelope is an organized array of membrane proteins. The pore permits the selective transport of substances into and out of the nucleus. (**b**) Sketch of the nuclear envelope's structure.

Like all cell membranes, a nuclear envelope's lipid bilayers prevent water-soluble substances from moving freely into and out of a nucleus. However, many pores, each a clustering of proteins, span both bilayers. Small, water-soluble molecules and ions can cross the nuclear envelope at these pores. As you will see later in the book, large molecules (including ribosomal subunits) move across the bilayers at pores in highly controlled ways.

Nucleolus

Take another look at the nucleus in Figure 4.11. Inside is a somewhat spherical, dense mass of material. What is it? When eukaryotic cells grow, one or more of these masses form inside the nucleus. Each is a **nucleolus** (plural, *nucleoli*), a construction site for many molecules of structural proteins and of RNAs in charge of peptide bond formation during protein synthesis. Some RNAs and proteins combine to form ribosomal subunits, then pass through nuclear pores into the cytoplasm. There, two subunits converge as an intact ribosome.

Chromosomes

When eukaryotic cells are not busy dividing, their DNA looks like thin threads. Many proteins are attached to the threads, like beads on a string. The beaded threads have a grainy appearance except at high magnification (Figure 4.11). When a cell prepares to divide, however, it duplicates its DNA molecules so each daughter cell will get all required hereditary instructions. Then each molecule folds into a compact structure, proteins and all.

Early microscopists bestowed the name *chromatin* on the seemingly grainy substance and *chromosomes* on

the condensed structures. We now define **chromatin** as the cell's collection of DNA, together with all proteins associated with it (Table 4.2). Each **chromosome** is one DNA molecule and its associated proteins, regardless of whether it is in threadlike or condensed form:

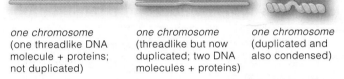

one chromosome (one threadlike DNA molecule + proteins; not duplicated)

one chromosome (threadlike but now duplicated; two DNA molecules + proteins)

one chromosome (duplicated and also condensed)

In other words, the appearance of "the chromosome" changes over the life of a eukaryotic cell! In chapters to come, you will look at different aspects of chromosomes, so you may find it useful to remember this point.

What Happens to the Proteins Specified by DNA?

Outside of the cell nucleus, new polypeptide chains for proteins are assembled on ribosomes. What happens to them? Many are stockpiled in the cytoplasm or get used at once. Many others enter a cytomembrane system. As you will read shortly, this system includes endoplasmic reticulum, Golgi bodies, and a variety of vesicles.

Thanks to DNA's instructions, many proteins take on particular, final forms in the cytomembrane system. Lipids also are packaged and assembled in the system by enzymes and other proteins (which also were built according to DNA's instructions). As you will see next, vesicles deliver the proteins and lipids to specific sites within the cell or to the plasma membrane, for export.

The nucleus, an organelle with two outer membranes, keeps the DNA molecules of eukaryotic cells separated from the metabolic machinery of the cytoplasm.

The localization makes it easier to organize the DNA and to copy it before a parent cell divides into daughter cells.

Pores across the nuclear envelope help control the passage of many substances between the nucleus and cytoplasm.

THE CYTOMEMBRANE SYSTEM

The **cytomembrane system** is a series of organelles in which lipids are assembled and new polypeptide chains are modified into final proteins. Its products are sorted and shipped to different destinations. Figure 4.13 shows how its organelles—the ER, Golgi bodies, and various vesicles—functionally interconnect with one another.

Endoplasmic Reticulum

The functions of the cytomembrane system begin with **endoplasmic reticulum**, or **ER**. In animal cells, the ER is continuous with the nuclear envelope, and it extends through cytoplasm. Its membrane regions appear rough or smooth, depending mainly on whether ribosomes are attached to the membrane facing the cytoplasm.

We typically observe *rough* ER arranged into stacks of flattened sacs with many ribosomes attached (Figure 4.14*a*). Every new polypeptide chain is synthesized on ribosomes. But only the newly forming chains having a built-in signal can enter the space within rough ER or become incorporated into ER membranes. (That signal is a sequence of fifteen to twenty specific amino acids.) Once the chains are in rough ER, enzymes may attach oligosaccharides and other side chains to them. Many specialized cells secrete the final proteins. Rough ER is abundant in such cells. For example, in your pancreas, ER-rich gland cells make and secrete enzymes that end up in the small intestine and help digest your meals.

Smooth ER is free of ribosomes and curves through cytoplasm like connecting pipes (Figure 4.14*b*). Many cells assemble most lipids inside the pipes. Smooth ER is well developed in seeds. In liver cells, certain drugs as well as toxic metabolic wastes are inactivated in it. Sarcoplasmic reticulum, a type of smooth ER in skeletal muscle cells, functions in contraction.

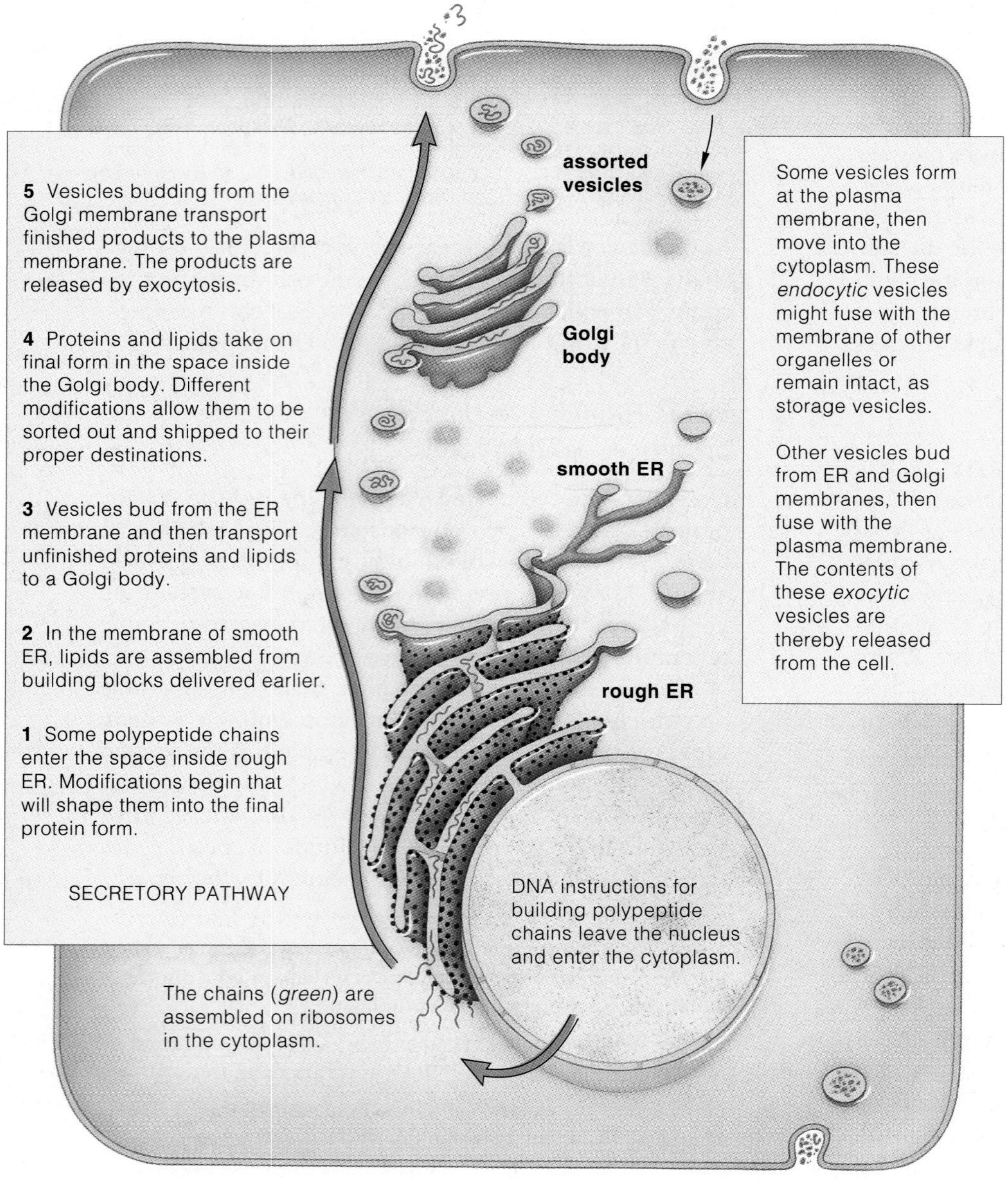

Golgi Bodies

In **Golgi bodies**, enzymes put the finishing touches on proteins and lipids, sort them out, then package them inside vesicles for shipment to specific locations. For example, an enzyme in one Golgi region might attach a phosphate group to a new protein, thereby giving it a mailing tag to its proper destination.

Commonly, a Golgi body looks rather like a stack of pancakes; it is composed of a series of flattened, membrane-bound sacs (Figure 4.15). In functional terms, the last portion of a Golgi body corresponds to the top pancake. Here, vesicles form as patches of the membrane bulge out, then break away into the cytoplasm.

5 Vesicles budding from the Golgi membrane transport finished products to the plasma membrane. The products are released by exocytosis.

4 Proteins and lipids take on final form in the space inside the Golgi body. Different modifications allow them to be sorted out and shipped to their proper destinations.

3 Vesicles bud from the ER membrane and then transport unfinished proteins and lipids to a Golgi body.

2 In the membrane of smooth ER, lipids are assembled from building blocks delivered earlier.

1 Some polypeptide chains enter the space inside rough ER. Modifications begin that will shape them into the final protein form.

SECRETORY PATHWAY

assorted vesicles

Golgi body

smooth ER

rough ER

Some vesicles form at the plasma membrane, then move into the cytoplasm. These *endocytic* vesicles might fuse with the membrane of other organelles or remain intact, as storage vesicles.

Other vesicles bud from ER and Golgi membranes, then fuse with the plasma membrane. The contents of these *exocytic* vesicles are thereby released from the cell.

DNA instructions for building polypeptide chains leave the nucleus and enter the cytoplasm.

The chains (*green*) are assembled on ribosomes in the cytoplasm.

Figure 4.13 Cytomembrane system, a membrane system in the cytoplasm that synthesizes, modifies, packages, and ships proteins and lipids. *Green* arrows highlight a secretory pathway by which some proteins and lipids are packaged and released from many types of cells, including gland cells that secrete mucus, sweat, and digestive enzymes.

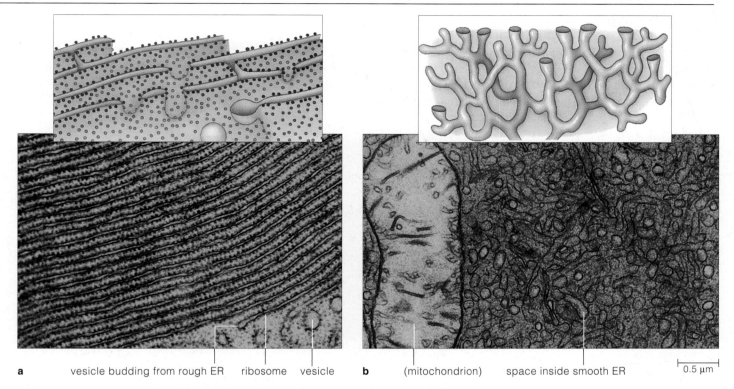

a   vesicle budding from rough ER   ribosome   vesicle   b   (mitochondrion)   space inside smooth ER   0.5 µm

Figure 4.14 Transmission electron micrographs and sketches of endoplasmic reticulum. (**a**) Many ribosomes dot the flattened surfaces of rough ER that face the cytoplasm. (**b**) This section reveals the diameters of many interconnected, pipelike regions of smooth ER.

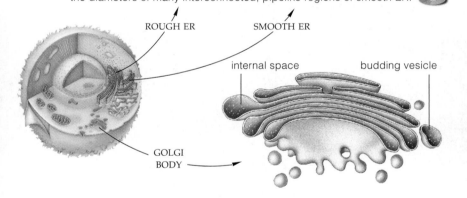

ROUGH ER   SMOOTH ER

internal space   budding vesicle

GOLGI BODY

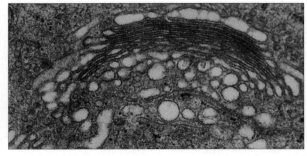

0.25 µm

Figure 4.15 Sketch and micrograph of a Golgi body from an animal cell.

A Variety of Vesicles

Vesicles are tiny, membranous sacs that move through the cytoplasm or take up positions in it. A common type, the lysosome, buds from Golgi membranes of animal cells and some fungal cells. Lysosomes are organelles of intracellular digestion. They hold enzymes that digest complex carbohydrates, proteins, nucleic acids, and some lipids. Often they fuse with vesicles that formed from patches of plasma membrane that surrounded bacteria, molecules, and other items that docked at membrane receptors. Lysosomes even digest whole cells or cell parts. For example, when a tadpole is developing into an adult frog, its tail gradually disappears. Lysosomal enzymes are responding to developmental signals and are helping to destroy cells that make up the tail.

Peroxisomes, another type, are sacs of enzymes that break down fatty acids and amino acids. A product of the reactions, hydrogen peroxide, is toxic, as the Chapter 5 introduction describes. But enzyme action converts it to water and oxygen or delivers it to reaction sites that break down alcohol. Drink alcohol, and peroxisomes of liver and kidney cells will degrade nearly half of it.

Many proteins take on final form and lipids are synthesized in the ER and Golgi bodies of the cytomembrane system.

Lipids, proteins (such as enzymes), and other items become packaged in vesicles destined for export, storage, membrane building, intracellular digestion, and other cell activities.

MITOCHONDRIA

Recall, from Section 3.6, that ATP is an energy carrier. At many different reaction sites, it delivers energy—in the form of phosphate groups—that drives nearly all cell activities. Many ATP molecules form when organic compounds are fully broken down to carbon dioxide and water in a **mitochondrion** (plural, mitochondria).

Only eukaryotic cells contain these organelles. Figure 4.16 gives an idea of their structure. ATP-forming reactions that proceed in mitochondria extract more energy from organic compounds than can be done by any other means. They cannot run to completion without plenty of oxygen. Like all other land-dwelling vertebrates, every time you breathe in, you take in oxygen mainly for the mitochondria in cells—in your case, many trillions of individual cells.

Each mitochondrion has a double-membrane system. The outer membrane faces the cytoplasm. Most commonly, the inner one is repeatedly folded.

What's the point of such an intricate system? It creates two compartments. Enzymes and other proteins of electron transport systems stockpile hydrogen ions in the outer compartment. Then the ions flow out from the compartment in controlled ways. Energy inherent in the flow drives ATP formation, as Chapter 7 describes. Oxygen binds to the spent electrons to form water, an end product of the reactions. Its removal keeps the transport systems clear for operation.

All eukaryotic cells contain one or more mitochondria. You may find only one in a single-celled yeast. You might find one thousand or more in cells that demand a lot of energy, such as those of muscles. Reflect on the profusion of mitochondria in Figure 4.10, which is a micrograph of merely one thin section from a liver cell. It alone tells you that the liver is a notably active, energy-demanding organ.

In size and biochemistry, mitochondria are a lot like bacteria. They even contain their own DNA and some ribosomes. They divide on their own. Did they evolve from ancient prokaryotic cells? Maybe. One idea is that a predatory, amoebalike cell engulfed other cells, which managed to avoid being digested. Perhaps the engulfed cells and their descendants were still able to reproduce. Suppose they became permanent, protected residents. If so, then some structures and functions they formerly needed to maintain independent life would no longer

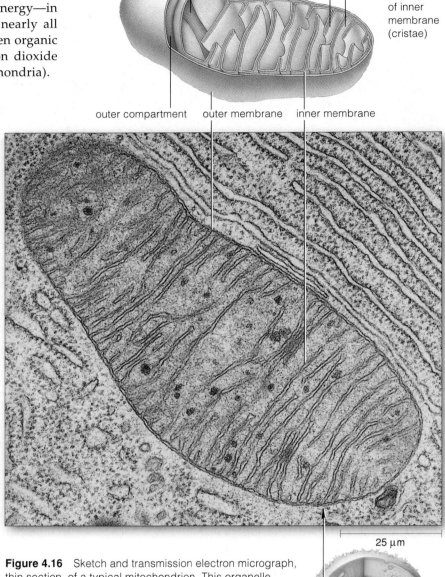

be necessary and might have been lost. In time the descendants may have evolved into mitochondria. We return to this topic in Section 19.4.

Figure 4.16 Sketch and transmission electron micrograph, thin section, of a typical mitochondrion. This organelle specializes in the production of large quantities of ATP, the main energy-carrying molecule between different reaction sites in cells. The process that produces ATP in mitochondria can't proceed without free oxygen.

inner compartment

repeated foldings of inner membrane (cristae)

outer compartment outer membrane inner membrane

25 µm

The organelles called mitochondria are the ATP-producing powerhouses of all eukaryotic cells.

Energy-releasing reactions proceed at the compartmented, internal membrane system of mitochondria. The reactions, which require oxygen, produce far more ATP than can be made by any other cellular reactions.

SPECIALIZED PLANT ORGÁNELLES

Chloroplasts and Other Plastids

Many plant cells contain plastids, a general category of organelles that specialize in photosynthesis or storage. Three types are common in different tissues of plants: chloroplasts, chromoplasts, and amyloplasts.

Of all eukaryotic cells, only the photosynthetic ones have **chloroplasts**. These organelles use sunlight energy to drive the formation of ATP and NADPH. These compounds are then used to assemble sugars and some other organic compounds.

Most chloroplasts have oval or disklike shapes. The semifluid interior, the stroma, is enclosed in two outer membrane layers. In the stroma is a third membrane, called the thylakoid membrane. It's a system of folded, interconnecting, disk-shaped compartments. Commonly, these compartments stack one on top of the other as in Figure 4.17. Botanists call each stack a granum (plural, grana).

The first stage of photosynthesis starts where many light-trapping pigments, enzymes, and other proteins are embedded in the thylakoid membrane. It ends with formation of ATP and NADPH. The ATP and NADPH are used at sites in the stroma where sugars, then starch and other organic compounds are built from carbon dioxide and water. Clusters of new molecules of starch (starch grains) may briefly accumulate in the stroma.

The most abundant photosynthetic pigments are the chlorophylls, which reflect or transmit green light. Others include the carotenoids, which reflect or transmit yellow, orange, and red light. The relative abundances of such pigments contribute to the colors of plant parts.

In many ways, chloroplasts resemble photosynthetic bacteria. Like mitochondria, they might have evolved from bacteria that were engulfed by predatory cells but escaped digestion and became permanent residents in them. We return to this idea in Section 19.4.

Unlike chloroplasts, chromoplasts lack chlorophylls. They have an abundance of carotenoids. They are the source of red-to-yellow colors of many flowers, autumn leaves, ripening fruits, and carrots and other roots. The pigment colors commonly attract animals that pollinate plants or disperse seeds. Amyloplasts lack pigments. Often they store starch grains and are abundant in cells of stems, potato tubers (underground stems), and seeds.

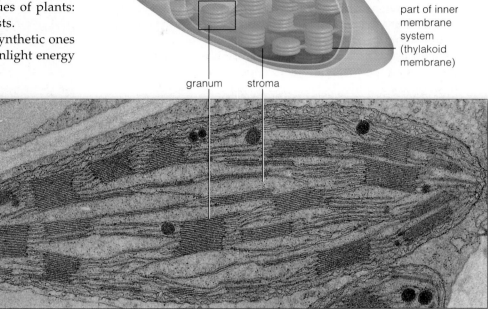

outermost membrane layers (two)

part of inner membrane system (thylakoid membrane)

granum   stroma

0.5 µm

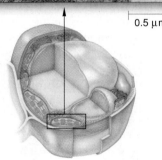

Figure 4.17 Generalized sketch of a chloroplast, the key defining feature of every photosynthetic eukaryotic cell. The transmission electron micrograph shows a thin section of a chloroplast from corn (*Zea mays*).

Central Vacuole

Many mature, living plant cells have a **central vacuole** (Figure 4.8). This fluid-filled organelle stores amino acids, sugars, ions, and toxic wastes. As it enlarges, it causes fluid pressure to build up inside the cell and so forces the cell's still-pliable cell wall to enlarge. Hence the cell itself enlarges. As the cell surface area increases, so does the rate at which water and other substances can be absorbed across the plasma membrane.

In most cases, the central vacuole increases so much in volume that it takes up 50 to 90 percent of the cell's interior. The cytoplasm ends up as a very narrow zone between the central vacuole and plasma membrane.

Photosynthetic eukaryotic cells have chloroplasts and other plastids that function in food production and storage.

Many plant cells have a central vacuole. When this storage vacuole enlarges during growth, cells are forced to enlarge, and this increases the surface area available for absorption.

THE CYTOSKELETON

An interconnected system of fibers, threads, and lattices extends between the nucleus and plasma membrane of eukaryotic cells. This system—the **cytoskeleton**—gives cells their internal organization, shape, and capacity to move. Some elements reinforce the plasma membrane and nuclear envelope. Others are scaffolds for proteins in membranes and cytoplasmic regions. Others are like railroad tracks on which organelles are shipped! Many elements are permanent; others appear only at certain times in a cell's life. Before a cell divides, for instance, many new microtubules form a "spindle" structure that moves chromosomes, then disassemble when the task is done. Energy inputs, as from ATP, trigger the motions.

Figure 4.18 shows **microtubules** and **microfilaments** of an animal cell's cytoskeleton. These two classes of cytoskeletal elements, acting singly or collectively, are responsible for most eukaryotic cell movements. Certain animal cells also have ropelike **intermediate filaments**, which lend mechanical strength to cells and tissues.

Microtubules are long, hollow cylinders of tubulin monomers (Figure 4.19a). They are vital to cell division and some aspects of cell shape and movement. As you might imagine, they are targets in chemical warfare. For instance, the autumn crocus (*Colchicum*) synthesizes colchicine. This poison blocks assembly and promotes disassembly of microtubules in the cells of animals that browse on plants. Taxol, a poison of the western yew (*Taxus brevifolia*), also blocks cell division; doctors use it it to stop uncontrolled cell divisions of some tumors.

Microfilaments are the thinnest elements. Each one consists of two polypeptide chains of actin subunits, as in Figure 4.19b. Actin has roles in many movements.

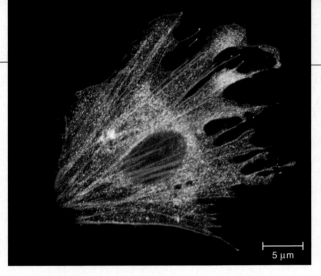

Figure 4.18 Part of the cytoskeleton of a human epithelial cell. These actin filaments (tinted *red*) function in cell motility. The micrographic image also reveals proteins (tinted *green* and *blue*) that work together with cytoskeletal elements to help new cells migrate to their proper locations in developing tissues.

Both tubulin and actin have been highly conserved through time. How do we know this? The microtubules and microfilaments of all eukaryotic cells are built from them. In spite of the basic uniformity, they have other proteins attached to them. The attachments let cells use the same microtubules and microfilaments for a variety of tasks. *Motor* proteins, such as myosin, are examples. Myosin is vital for muscle contraction. Another example: *crosslinking* proteins splice microfilaments together, as in a cell's cortex. The **cell cortex** is a three-dimensional mesh beneath the plasma membrane that reinforces the cell surface and helps the cell move.

Intermediate filaments, the most stable elements of some cytoskeletons, mechanically strengthen and help maintain the shape of cells and cell parts (Figure 4.19c). For example, desmins help organize muscle cells, and cytokeratins reinforce cells that give rise to hairs.

Mechanisms of Cell Movements

If a cell lives, it moves. This is obvious when you study free-living single cells, including the sperm and ciliated cells in Figure 4.20. But even cells with fixed positions in tissues of multicelled organisms move. Microfilaments, microtubules, or both serve in most aspects of cellular movements. They do so by three mechanisms.

First, *the length of a microtubule or microfilament can grow or diminish by the controlled assembly or disassembly of its subunits.* When either lengthens or shortens at one end, a chromosome or some other structure attached to the other end is pushed or dragged through cytoplasm.

Some cells crawl around on protrusions that form as microfilaments assemble beneath their surface. *Amoeba proteus,* a soft-bodied protistan, has **pseudopods** ("false feet"): temporary, lobelike protrusions from the body. Inside each lobe, microfilaments grew rapidly in length and dragged along the plasma membrane attached to it.

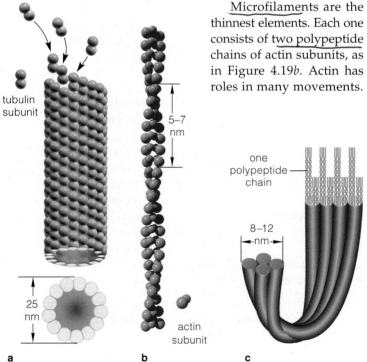

tubulin subunit

5–7 nm

one polypeptide chain

8–12 nm

25 nm

actin subunit

a   b   c

Figure 4.19 Structural arrangement of subunits in (**a**) microtubules, (**b**) microfilaments, and (**c**) one of the intermediate filaments.

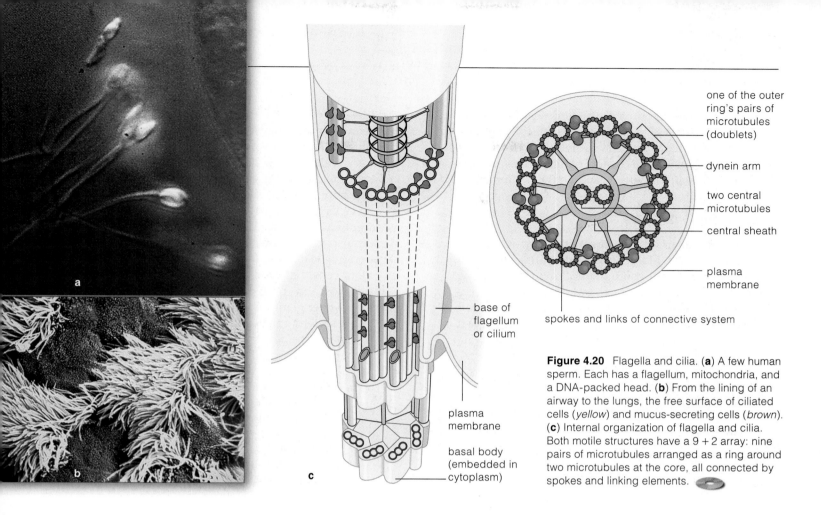

one of the outer ring's pairs of microtubules (doublets)

dynein arm

two central microtubules

central sheath

plasma membrane

spokes and links of connective system

base of flagellum or cilium

plasma membrane

basal body (embedded in cytoplasm)

Figure 4.20 Flagella and cilia. (**a**) A few human sperm. Each has a flagellum, mitochondria, and a DNA-packed head. (**b**) From the lining of an airway to the lungs, the free surface of ciliated cells (*yellow*) and mucus-secreting cells (*brown*). (**c**) Internal organization of flagella and cilia. Both motile structures have a 9 + 2 array: nine pairs of microtubules arranged as a ring around two microtubules at the core, all connected by spokes and linking elements.

Similarly, the animal cell in Figure 4.18 was migrating on sheetlike extensions of microfilaments.

Second, *parallel rows of microfilaments or microtubules actively slide in specific directions.* For instance, a muscle cell is divided into contractile units. In each unit, many actin and myosin filaments are arrayed in parallel. ATP energizes the myosin, which binds, moves, and lets go of the actin. Short, repeated strokes make the actin slide over the myosin, so the unit shortens (Section 33.6).

Third, *microtubules or microfilaments shunt organelles or parts of the cytoplasm from one location to another.* For example, chloroplasts move to new light-intercepting positions as the angle of the sun changes overhead. They are attached to myosin filaments that "walk" on (bind, slide, and let go of) microfilaments. This shunt mechanism also is the basis of cytoplasmic streaming: a dynamic, often rapid flow of some components of the cytoplasmic gel. Observe a living plant cell with a light microscope, and you may see pronounced streaming.

Flagella and Cilia

The **flagellum** (plural, flagella) and the **cilium** (plural, cilia) are examples of structures for cell motility. Both have a ring of nine pairs of microtubules and a central pair. A system of spokes and links stabilizes this "9 + 2 array." The array arises from a **centriole**. This barrel-shaped structure is a microtubule-producing center. It

remains at the base of the completed array, as in Figure 4.20c. In that location it is known as a **basal body**.

How do flagella and cilia differ? Flagella typically are longer and less profuse. Sperm and other free-living, motile cells employ them as whiplike tails for swimming (Figure 4.20a). In numerous multicelled species, ciliated epithelial cells stir air or fluid. Many thousands line the airways inside your chest (Figure 4.20b). Their beating directs mucus-trapped particles away from the lungs.

Both flagella and cilia beat by a sliding mechanism. Extending from each pair of microtubules in the outer ring are short arms of motor proteins (dynein). When energized by ATP, the arms attach to the pair in front of them, tilt in a downward-directed stroke, then release their hold. Repeated strokes make the pairs slide down in sequence. Because spokes and links connect all the pairs, the sliding is converted to a bending motion.

A cytoskeleton of protein filaments and accessory proteins is present in each eukaryotic cell. It is the basis of its shape, internal structure, and capacity for movement. By different mechanisms, the proteins assemble or disassemble, slide past one another, and shunt components to new locations.

Microtubules help organize the cytoskeleton and move cell structures. Microfilaments also help move the cell. They help form and maintain cell shape. Intermediate filaments structurally reinforce some animal cells.

CELL SURFACE SPECIALIZATIONS

Our survey of eukaryotic cells concludes with a look at cell walls and some other specialized surface structures. Many of these architectural marvels are constructed of cell secretions. Others are clusters of membrane proteins that connect neighboring cells and let them interact.

Eukaryotic Cell Walls

Single-celled eukaryotic species are directly exposed to their surroundings. Many have a **cell wall**, a structural component that wraps continuously around the plasma membrane. A cell wall protects and physically supports its owner. The wall is porous, so water and solutes can easily move to and from the plasma membrane. A great variety of protistans, as in Figure 4.21, have a cell wall. So do plant cells and many types of fungal cells.

For instance, young cells in actively growing parts of plants secrete gluelike polysaccharides (such as pectin), glycoproteins, and cellulose. The

Figure 4.21 From a freshwater habitat, one of the single-celled, walled protistans (*Ceratium*, a dinoflagellate).

cellulose molecules form ropelike strands that become embedded in the gluey matrix. All secretions combined form the cell's **primary wall** (Figure 4.22). Being sticky, primary walls cement abutting cells together. They are thin and pliable, so the cell surface area can continue to enlarge under the pressure of incoming water.

At cell surfaces exposed to air, waxes and other cell secretions accumulate. The deposits form a cuticle. This semitransparent, protective surface cover helps restrict evaporative water loss from plants (Figure 4.23*a*).

Plant cells that produce only a thin wall retain the capacity to divide or to change shape as they grow and develop. Other plant cells stop enlarging as they mature and start secreting material on the primary wall's inner surface. The deposits form a rigid, **secondary wall** that reinforces cell shape (Figure 4.22*e*). Whereas cellulose makes up less than 25 percent of the cell's primary wall, the secondary wall deposits are more extensive and contribute more to its structural support.

In woody plants, up to 25 percent of the secondary wall is made of lignin. Lignin consists of a six-carbon ring to which a three-carbon chain and an oxygen atom are attached. It makes plant parts more waterproof, less inviting to plant-attacking organisms, and stronger.

Figure 4.22 Examples of cell walls in (**a**) flax plants. (**b**) Primary cell wall of young cells in flower petals. Cell secretions form the middle lamella, a layer between the walls of adjoining cells. The layer is thickest in adjoining corners. Plasmodesmata, membrane-lined channels across the adjacent walls, connect the cytoplasm of neighboring cells. (**c**,**d**) Part of the lustrous fibers in a cell from a flax stem. We make linen from such fibers, which are three times stronger than cotton fibers. (**e**) In flax fibers, as in many other types of plant cells, more layers become deposited inside the primary wall. The layers stiffen the wall and help maintain its shape. Later, the cell dies, leaving the stiffened walls behind. This also happens in water-conducting pipelines that thread through most plant tissues. Interconnected, stiffened walls of dead cells form the tubes.

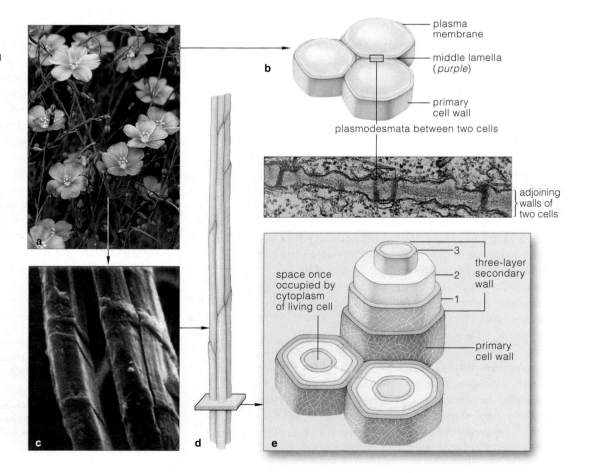

plasma membrane

middle lamella (*purple*)

primary cell wall

plasmodesmata between two cells

adjoining walls of two cells

space once occupied by cytoplasm of living cell

three-layer secondary wall

primary cell wall

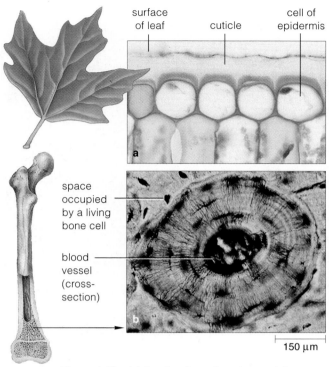

surface
of leaf   cuticle   cell of
epidermis

a

space
occupied
by a living
bone cell

blood
vessel
(cross-
section)

b

150 μm

Figure 4.23 (**a**) Section through a plant cuticle, an outer surface layer of cell secretions. (**b**) Cross-section through compact bone tissue, stained for microscopy.

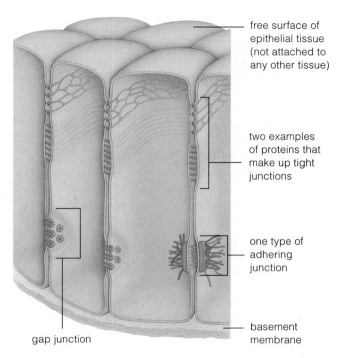

free surface of
epithelial tissue
(not attached to
any other tissue)

two examples
of proteins that
make up tight
junctions

one type of
adhering
junction

basement
membrane

gap junction

Figure 4.24 Composite drawing of the most common types of cell junctions in animals. These are cells from epithelial tissue. See also Section 29.1.

Matrixes Between Animal Cells

Animal cells have no walls. But intervening between many of these cells are matrixes made of cell secretions and materials taken up from the surroundings. Think of cartilage at the knobby ends of your leg bones. Cartilage consists of scattered cells plus collagen or elastin fibers formed from their secretions, all embedded in a ground substance of polysaccharides (Section 29.2). Or think of bone tissue, in which an extensive matrix separates the living bone cells (Figure 4.23*b* and Section 33.4).

Cell-to-Cell Junctions

Even when a wall or some other structure imprisons a cell in its own secretions, the only contact that cell has with the outside world is through its plasma membrane. In multicelled species, membrane components project into adjoining cells and into the surrounding medium. Among these components are junctions where the cell sends and receives signals and materials, and where it recognizes and cements itself to cells of the same type.

In plants, for instance, many tiny channels cross the adjacent primary walls of living cells and interconnect their cytoplasm. Figure 4.23*b* shows a few. Each channel is a plasmodesma (plural, plasmodesmata). The plasma membranes of adjoining cells merged during growth. Now they line the channels, so there is an uninterrupted

flow of substances between the cells. That is how living cells in the plant body engage in quick exchanges.

In most animal tissues, three categories of cell-to-cell junctions are common (Figure 4.24). *Tight* junctions link the cells of most tissues, including all epithelia that line the body's outer surface and internal cavities and organs. They seal abutting cells so water-soluble substances can't leak between them. That is why gastric fluid can't leak across the stomach's lining and damage internal tissues. *Adhering* junctions join cells in tissues of the skin, heart, and other organs subjected to stretching. *Gap* junctions link the cytoplasm of neighboring cells. They are open channels for a rapid flow of signals and substances.

We will be returning to the cell walls, intercellular substances, and cell-to-cell interactions in later chapters. For now, these are the points to remember:

A variety of protistan, plant, and fungal cells have a porous but protective wall that surrounds the plasma membrane. The cells themselves secrete wall-forming materials.

The combined secretions of many cells form the cuticle that covers the surface of plants and certain animals, as well as the extracellular matrixes of animal tissues.

In multicelled organisms, coordinated cell activities rely on cell-to-cell junctions. The junctions are protein complexes or cytoplasmic bridges that serve as physical links and sites of communication between cells.

4.12

PROKARYOTIC CELLS

We turn now to prokaryotic cells. Unlike the cells you've considered so far, a nucleus does not enclose their DNA. *Prokaryotic* means "before the nucleus." Biologists chose the word to remind us that this type of cell appeared on Earth before the nucleus evolved in the forerunners of eukaryotic cells.

These are the smallest known cells. Most are not much more than a micrometer wide; even rod-shaped species are only a few micrometers long. In structural terms, they are the simplest cells to think about. Most have a semirigid or rigid cell wall that surrounds the plasma membrane, supports the cell, and imparts shape to it (Figure 4.25a). As you will see later, their wall differs from walls of eukaryotic cells. Dissolved substances can move freely to and from the plasma membrane because the wall is permeable. Commonly, sticky polysaccharides enclose a wall and help cells attach to interesting surfaces, such as river rocks, teeth, and the vagina. Many pathogenic (disease-causing) eubacteria have a capsule made of thick, jellylike polysaccharides. It surrounds and protects the wall.

As in eukaryotic cells, the plasma membrane of prokaryotic cells mediates the flow of substances to and from the cytoplasm. It, too, has proteins that serve as channels, transporters, and receptors for diverse signals and substances. It, too, incorporates built-in machinery for metabolic reactions, such as the breakdown of energy-rich compounds. In most of the photosynthetic prokaryotic cells, organized arrays of proteins in the plasma membrane harness light energy, then convert it to chemical energy of ATP.

Prokaryotic cells are so small they have only a tiny volume of cytoplasm. But they have many ribosomes on which polypeptide chains are built. The cytoplasm is distinct from that of eukaryotic cells; it is continuous with an irregularly shaped region of DNA. Membranes don't enclose this region, which is the nucleoid (Figure 4.25c). A circular DNA molecule, also called a bacterial chromosome, occupies the region. Given their tiny size and internal simplicity, the cells don't need a complex cytoskeleton, but actin-like protein filaments do form a scaffold that helps shape the sturdier cell wall.

Projecting out from the surface of many prokaryotic cells are one or more bacterial flagella. These threadlike motile structures differ from eukaryotic flagella; they do not have a 9 + 2 array of microtubules. Bacterial flagella help cells move rapidly through fluid environments. Other surface projections include pili (singular, pilus), the main protein filaments that help many types of cells attach to various surfaces, even to one another.

There are two kingdoms of prokaryotic cells: the Archaebacteria and Eubacteria (Sections 1.3 and 18.7).

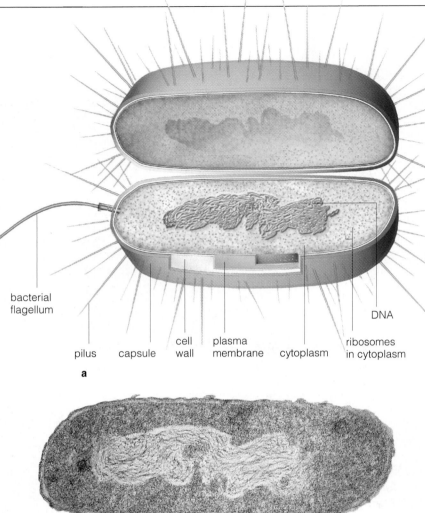

bacterial flagellum

pilus   capsule   cell wall   plasma membrane   cytoplasm   ribosomes in cytoplasm

DNA

a

b   0.5 µm

Figure 4.25 (**a**) Generalized sketch of a typical prokaryotic body plan. (**b**) Micrograph of the bacterium *Escherichia coli*.

The gut of humans and other mammals holds a huge population of a normally harmless strain of *E. coli*. Some harmful strains can contaminates meat sold commercially. One strain also can taint cider. Cooking meat thoroughly or pasteurizing cider kills these pathogens. Where this was not done, people who ate the meat or drank the cider became sick. Some died.

Facing page: (**c**) Researchers manipulated this *E. coli* cell to release its single, circular molecule of DNA. (**d**) Cells of various bacterial species are shaped like balls, rods, or corkscrews. Ball-shaped cells of *Nostoc*, a photosynthetic eubacterium, stick together in a thick, jellylike sheath of their own secretions. Chapter 20 gives other examples. (**e**) Like this *Pseudomonas marginalis* cell, many species have one or more bacterial flagella that propel the cell body in fluid environments.

Together, the species in these kingdoms are the most metabolically diverse organisms of all. Different kinds have managed to exploit energy and raw materials in just about every kind of environment, ranging from hot springs and snow fields to hot, dry deserts. In addition, ancient prokaryotic cells gave rise to all the protistans,

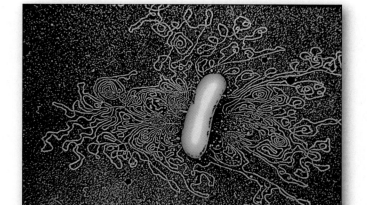

c

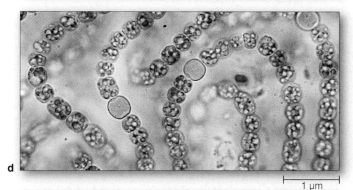

d

1 µm

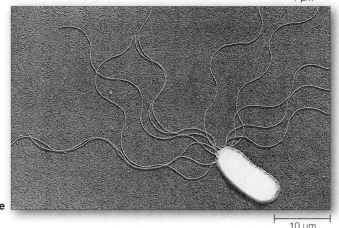

e

10 µm

plants, fungi, and animals ever to appear on Earth. The evolution, structure, and functions of these remarkable cells are topics of later chapters.

Archaebacteria and eubacteria are prokaryotic cells; their DNA is not housed inside a nucleus. Most species have a cell wall around the plasma membrane. Generally, they do not have organelles comparable to those of eukaryotic cells.

These are the simplest cells, but as a group they show the most metabolic diversity. Their metabolic activities proceed at the plasma membrane and within the cytoplasm.

SUMMARY   *Gold* indicates text section

1. Three generalizations constitute the cell theory: *CI*
   a. All living things are composed of one or more cells.
   b. The cell is the smallest entity that still retains the properties of life. That is, it lives independently or has a heritable, built-in capacity to do so.
   c. New cells arise only from cells that already exist.

2. All cells start out life with a plasma membrane, a region of DNA, and cytoplasm. The plasma membrane (a thin, outer membrane) maintains a cell as a distinct entity and lets metabolic activities proceed apart from random events in the environment. Many substances and signals cross it in controlled ways. The cytoplasm consists of ribosomes, structural elements, fluids, and (only in eukaryotic cells) organelles between the region of DNA and the plasma membrane. *4.1*

3. Membranes of lipids and proteins are essential for cell structure and function. The lipids form two layers, with their hydrophobic tails sandwiched between their hydrophilic heads (which are dissolved in fluid). This bilayer gives the membrane its basic structure. Water-soluble substances cannot cross it. Proteins embedded in the bilayer or attached to it have diverse roles. Many are channels or pumps where water-soluble substances cross the bilayer, others are receptors, and so on. *4.1*

4. The surface-to-volume ratio constrains cell size, cell shape, and the body plans of multicelled organisms. *4.2*

5. Modern microscopes harness wavelengths of light or focused beams of electrons to reveal cell structures. *4.3*

6. Membranes divide the cytoplasm of eukaryotic cells into functional compartments (organelles). Prokaryotic cells don't have comparable organelles. *4.4, 4.5*

7. Organelle membranes separate metabolic reactions inside the cytoplasm and let different kinds proceed in orderly fashion. (Many similar reactions proceed at the plasma membrane of prokaryotic cells.) *4.4*
   a. The nuclear envelope functionally separates DNA from the metabolic machinery of the cytoplasm. *4.6*
   b. The cytomembrane system includes the ER, Golgi bodies, and vesicles. Many new proteins are modified into final form and lipids are assembled in this system. Finished products are packaged and then shipped off to destinations inside or outside the cell. *4.7*
   c. Mitochondria specialize in producing numerous ATP molecules by oxygen-requiring reactions. *4.8*
   d. Chloroplasts specialize in capturing energy from the sun and using it to build organic compounds. *4.9*

8. The cytoskeleton of eukaryotic cells functions in cell shape, internal organization, and movement. Many cells have walls and other surface specializations. *4.10, 4.11*

9. Table 4.3 on the next page summarizes the defining features of prokaryotic and eukaryotic cells. *4.1, 4.12*

Table 4.3 *Summary of Typical Components of Prokaryotic and Eukaryotic Cells*

Cell Component	Function	PROKARYOTIC Archaebacteria, Eubacteria	EUKARYOTIC Protistans	Fungi	Plants	Animals
Cell wall	Protection, structural support	✓*	✓*	✓	✓	None
Plasma membrane	Control of substances moving into and out of cell	✓	✓	✓	✓	✓
Nucleus	Physical separation and organization of DNA	None	✓	✓	✓	✓
DNA	Encoding of hereditary information	✓	✓	✓	✓	✓
RNA	Transcription, translation of DNA messages into polypeptide chains of specific proteins	✓	✓	✓	✓	✓
Nucleolus	Assembly of subunits of ribosomes	None	✓	✓	✓	✓
Ribosome	Protein synthesis	✓	✓	✓	✓	✓
Endoplasmic reticulum (ER)	Initial modification of many of the newly forming polypeptide chains of proteins; lipid synthesis	None	✓	✓	✓	✓
Golgi body	Final modification of proteins, lipids; sorting and packaging them for use inside cell or for export	None	✓	✓	✓	✓
Lysosome	Intracellular digestion	None	✓	✓*	✓*	✓
Mitochondrion	ATP formation	**	✓	✓	✓	✓
Photosynthetic pigments	Light–energy conversion	✓*	✓*	None	✓	None
Chloroplast	Photosynthesis; some starch storage	None	✓*	None	✓	None
Central vacuole	Increasing cell surface area; storage	None	None	✓*	✓	None
Bacterial flagellum	Locomotion through fluid surroundings	✓*	None	None	None	None
Flagellum or cilium with 9 + 2 microtubular array	Locomotion through or motion within fluid surroundings	None	✓*	✓*	✓*	✓
Complex cytoskeleton	Cell shape; internal organization; basis of cell movement and, in many cells, locomotion	Rudimentary***	✓*	✓*	✓*	✓

* Known to be present in cells of at least some groups.
** Many groups use oxygen-requiring (aerobic) pathways of ATP formation, but mitochondria are not involved.
*** Protein filaments form a simple scaffold that helps support cell wall in at least some species.

Review Questions

1. State the three key points of the cell theory. *CI*

2. Suppose you wish to observe the three-dimensional surface of an insect's eye. Would you see more details with the aid of a compound light microscope, transmission electron microscope, or scanning electron microscope? *4.3*

3. Label the organelles in the two diagrams below of a plant cell and an animal cell. What are the main differences between the two kinds of cells? In what ways are they similar? *4.4*

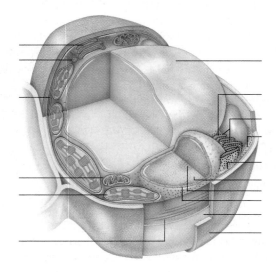
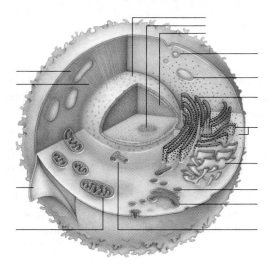

4. Label the parts of this bacterial cell. *4.12*

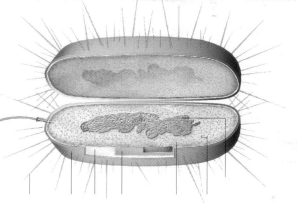

Figure 4.26 Daily chest thumping for a child who is affected by cystic fibrosis.

5. Describe three features that all cells have in common. After reviewing Table 4.3, write a paragraph on the key differences between eukaryotic and prokaryotic cells. *4.1, 4.4, 4.11, 4.12*

6. Briefly characterize the structure and function of the cell nucleus, nuclear envelope, and nucleolus. *4.6*

7. Define chromosome and chromatin. Do chromosomes always have the same appearance during a cell's life? *4.6*

8. Which organelles are part of the cytomembrane system? *4.7*

9. Is this statement true or false: Plant cells have chloroplasts, but not mitochondria. Explain your answer. *4.8, 4.9*

10. What are the functions of the central vacuole? *4.9*

11. Define cytoskeleton. How does it aid in cell functions? *4.10*

12. What gives rise to the 9 + 2 array of cilia and flagella? *4.10*

13. Cell walls are typical of which organisms: bacteria, protistans, fungi, plants, animals? Are the walls impermeable? *4.11*

14. In certain plant cells, is a secondary wall deposited inside or outside the surface of the primary wall? *4.11*

15. In multicelled organisms, coordinated interactions depend on linkages and communications between cells. What types of junctions occur between adjacent animal cells? Plant cells? *4.11*

Self-Quiz ANSWERS IN APPENDIX III

1. Cell membranes consist mainly of a _____ .
   a. carbohydrate bilayer and proteins
   b. protein bilayer and phospholipids
   c. lipid bilayer and proteins
   d. none of the above

2. Organelles _____ .
   a. are membrane-bound compartments
   b. are typical of eukaryotic cells, not prokaryotic cells
   c. separate chemical reactions in time and space
   d. all of the above are features of the organelles

3. Cells of many protistans, plants, and fungi, but not animals, commonly have _____ .
   a. mitochondria   c. ribosomes
   b. a plasma membrane   d. a cell wall

4. Is this statement true or false: The plasma membrane is the outermost component of all cells. Explain your answer.

5. Unlike eukaryotic cells, prokaryotic cells _____ .
   a. lack a plasma membrane   c. do not have a nucleus
   b. have RNA, not DNA   d. all of the above

6. Match each cell component with its function.
   ____ mitochondrion   a. synthesis of polypeptide chains
   ____ chloroplast   b. initial modification of new
   ____ ribosome   polypeptide chains
   ____ rough ER   c. final modification of proteins;
   ____ Golgi body   sorting, shipping tasks
                         d. photosynthesis
                         e. formation of many ATP

Critical Thinking

1. Why is it likely that you will never encounter a predatory two-ton living cell on the sidewalk?

2. In compound light microscopes with blue filters, lenses transmit only blue light. Look at the spectrum of visible light in Figure 6.5. Why is blue light so efficient for viewing objects at high magnification?

3. Your professor shows you an electron micrograph of a cell with large numbers of mitochondria and Golgi bodies. You notice that this particular cell also contains a great deal of rough endoplasmic reticulum. What kinds of cellular activities would require such an abundance of the three kinds of organelles?

4. Each year in the United States alone, there are about 40,000 reported cases of *cystic fibrosis* (CF). This is a fatal genetic disorder that, at this writing, is incurable. On average, CF individuals can expect to live no longer than thirty years.

   Affected glands secrete more than they should, which has far-reaching effects. In time, after digestive enzymes clog a duct between the pancreas and small intestine, food can't be digested properly. Even when food intake increases, malnutrition results. Cysts form in the pancreas, which degenerates and becomes fibrous (hence the disorder's name). Thick mucus builds up in the respiratory tract. CF individuals have trouble expelling airborne bacteria and particles that enter the lungs (Figure 4.26).

   The disorder may arise from a defective protein in the plasma membrane of gland cells that secrete mucus, digestive enzymes, and sweat. Review Section 4.7, then name the organelles involved in the secretory pathway in those cells.

Selected Key Terms

basal body *4.10*	ER (endoplasmic	nucleoid *4.1*
cell *4.1*	reticulum) *4.7*	nucleolus *4.6*
cell theory *CI*	eukaryotic cell *4.1*	nucleus *4.1, 4.6*
cell wall *4.11*	flagellum *4.10*	organelle *4.4*
central vacuole *4.9*	fluid mosaic	phospholipid *4.1*
centriole *4.10*	model *4.1*	plasma membrane *4.1*
chloroplast *4.9*	Golgi body *4.7*	primary wall *4.11*
chromatin *4.6*	intermediate	prokaryotic cell *4.1*
chromosome *4.6*	filament *4.10*	pseudopod *4.10*
cilium *4.10*	lipid bilayer *4.1*	ribosome *4.1*
cytomembrane	microfilament *4.10*	secondary wall *4.11*
system *4.7*	micrograph *4.3*	surface-to-volume
cytoplasm *4.1*	microtubule *4.10*	ratio *4.2*
cytoskeleton *4.10*	mitochondrion *4.8*	vesicle *4.7*
	nuclear envelope *4.6*	wavelength *4.3*

Readings

Alberts, B., et al. 1994. *Molecular Biology of the Cell*. Third edition. New York: Garland. See Chapter 16: The Cytoskeleton.

On-Line readings at Student Guide for InfoTrac:
www.brookscole.com/biology

GROUND RULES OF METABOLISM

Growing Old With Molecular Mayhem

Somewhere in those slender strands of DNA inside your cells are snippets of instructions for making two kinds of enzymes: superoxide dismutase and catalase (Figure 5.1). Both are *antioxidants*. Like vitamins E and C, they help keep you from growing old before your time by cleaning house, so to speak. Working as a team, they neutralize many potentially toxic wastes by using oxygen (O_2) to strip hydrogen atoms from molecules. Many other kinds of organisms, ranging from bacteria to roundworms to plants, also depend on them.

Stripping hydrogen atoms from molecules happens to release electrons, and O_2 normally picks them up. Sometimes it picks up only one. That one electron isn't enough to complete the reaction. But it's enough to give the oxygen a negative charge (O_2^-).

Like other unbound molecular fragments with the wrong number of electrons, O_2^- is a **free radical**. Free radicals slip away from a variety of enzyme-catalyzed reactions, including the digestion of fats and amino acids. They also form when x rays and other forms of ionizing radiation bombard water and other molecules. They even escape from electron transport chains.

Free radicals are highly reactive. When they dock with a molecule, they alter its structure and function.

They even attack molecules that aren't open to just any reaction. Such molecules include DNA, cell membrane lipids, and membrane receptors which, when activated, tell cells it's time to die (Section 14.4).

Enter superoxide dismutase. Under its biochemical prodding, two rogue oxygen molecules combine with hydrogen ions. Hydrogen peroxide (H_2O_2) and O_2 are the outcome. Hydrogen peroxide, a normal by-product of some reactions, is lethal to cells when it accumulates. Enter catalase. Under its prodding, two molecules of hydrogen peroxide react and split into water and oxygen: $2H_2O_2 \longrightarrow 2H_2O + O_2$. This crucial reaction normally occurs before hydrogen peroxide does major damage.

As we age, our cells make copies of enzymes in ever diminishing numbers, in crippled form, or both. When this happens to superoxide dismutase and catalase, free radicals and hydrogen peroxide accumulate. Like loose cannons, they careen through cells and blast away at the structural integrity of proteins, DNA, lipids, and other molecules. Cells suffer or die.

Those brown "age spots" on an older person's skin are evidence of free radical assaults (Figure 5.2a). Each spot is a mass of brownish-black pigments that build up in cells when free radicals take over—all for the want

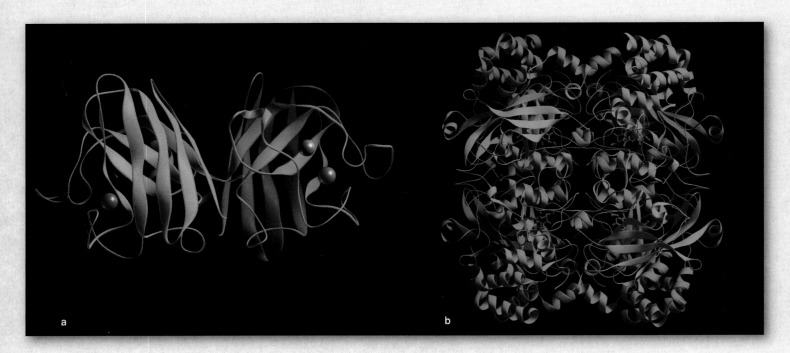

a
b

Figure 5.1 Computer models of the molecular structure of (**a**) superoxide dismutase and (**b**) catalase. Both enzymes help keep free radicals inside the body in check.

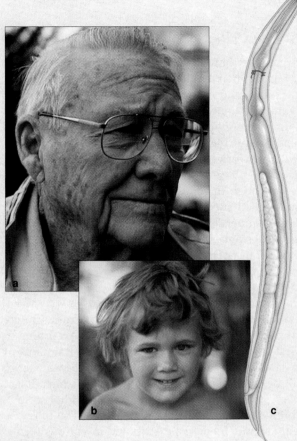

Figure 5.2 (**a**) Owner of skin with a spattering of age spots, evidence of free radicals on the loose. At one time he, like the boy in (**b**), had many smoothly functioning molecules of superoxide dismutase and catalase. (**c**) The worm of choice for aging studies, *Caenorhabditis elegans*.

of specific enzymes. Worse yet, free radicals also may usher in heart disease and other disorders.

Knowing this, Simon Melov, Bernard Malfroy, and their colleagues did experiments with the roundworm *Caenorhabditis elegans* (Figure 5.2*c*). These tiny worms don't live more than a month. However, when synthetic superoxide dismutase and catalase were added to their diet, their life span increased greatly. In some cases it actually doubled! The researchers also fed enzymes to worms that were genetically engineered to succumb prematurely to free radicals. Those worms lived just as long as the unaltered ones.

With this example, we turn to **metabolism**, the cell's capacity to acquire energy and use it to build, degrade, store, and release substances in controlled ways. Each extra day in a worm's life reminds us that cells must continually take in energy-rich solutes, store things, build membranes, replenish enzymes, and check out their DNA. A constant supply of energy drives every one of their activities. At times, the activities may seem remote from your interests. But they help define who you are and who you will become, age spots and all.

Key Concepts

1. Cells engage in metabolism, or chemical work. They use energy to stockpile, build, rearrange, and break apart substances. Cells use energy for mechanical work, as when they move flagella. They also channel energy into electrochemical work, as when they move ions into or out of an organelle compartment.

2. All organisms secure energy from outside sources. Sunlight energy is the primary source for the web of life. Another source is chemical bond energy of molecules in the physical environment or in other, edible organisms.

3. ATP is the main carrier of energy in cells. It couples reactions that release energy with reactions that require it. ATP activates molecules—primes them for chemical change—by transferring a phosphate group to them.

4. Many metabolic reactions involve oxidation–reduction reactions, or electron transfers among substances. Such transfers often are made in electron transport systems.

5. Chemical reactions proceed far too slowly on their own to sustain life. In cells, specific enzymes greatly increase the rate of specific reactions. Coenzymes, such as NAD^+, assist enzymes in the reactions or transfer electrons and hydrogen released during the reactions to other sites.

6. Many enzymes mediate reactions in orderly sequences called metabolic pathways. The coordinated operation of these pathways maintains, increases, and decreases the relative amounts of substances in cells.

7. Much of metabolism relies on concentration gradients. These are differences in the number of molecules or ions of a substance between two regions. Molecules tend to move to the region of lower concentration, a behavior called diffusion.

8. Cells work with and against concentration gradients across the plasma membrane and internal membranes. Specific solutes cross membranes by passive and active transport. Water crosses by osmosis. Materials also cross the plasma membrane by endocytosis and exocytosis.

ENERGY AND THE UNDERLYING ORGANIZATION OF LIFE

Defining Energy

If you've ever watched a house cat stalk a mouse, you know it "freezes" its position to avoid detection before springing at its unsuspecting prey. Like everything else in the universe that's stationary, the cat has a store of **potential energy**—a capacity to do work, simply owing to its position in space and the arrangement of its parts. When that cat springs, some of its potential energy is transformed into **kinetic energy**, the energy of motion.

Energy on the move does work when it imparts motion to other things. In skeletal muscle cells within the cat, ATP gave up some of its potential energy to molecules of contractile units and set them in motion. The combined motions in many muscle cells resulted in the movement of whole muscles. The transfer of energy from ATP also resulted in the release of another form of energy called **heat**, or *thermal* energy.

The potential energy of molecules has its own name: **chemical energy**. It is measurable, as in kilocalories. A kilocalorie is the same thing as 1,000 calories, which is the amount of energy it takes to heat 1,000 grams of water from 14.5°C to 15.5°C at standardized pressure.

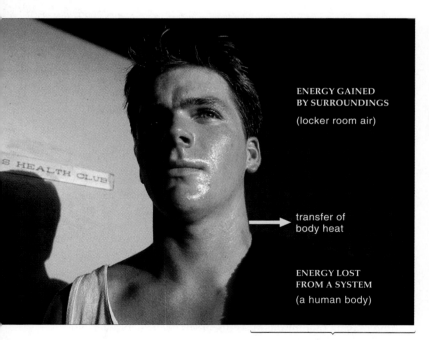

ENERGY GAINED
BY SURROUNDINGS

(locker room air)

transfer of
body heat

ENERGY LOST
FROM A SYSTEM

(a human body)

NET ENERGY CHANGE = 0

Figure 5.3 Example of how the total energy content of any system, *together with its surroundings*, remains constant.

"System" means all matter in a specific region, such as a human body, a plant, a DNA molecule, or a galaxy. "Surroundings" can be a small region in contact with the system or as vast as the whole universe. The system shown (a human male) is giving off heat to the surroundings (a locker room) by evaporative water loss from sweat. What one region loses, the other region gains, so the total energy content of both does not change.

What Can Cells Do With Energy?

All organisms have specific adaptations for securing energy from their environment. Some capture energy from the sun, and others extract energy from inorganic or organic substances in the environment. Regardless of the source, energy inputs become coupled to thousands of energy-requiring processes in cells. Cells use energy for *chemical* work—to stockpile, build, rearrange, and break apart substances. They channel it into *mechanical* work—to move flagella and other cell structures and (in multicelled species) the whole body or portions of it. They channel it into *electrochemical* work—to move charged substances into or out of the cytoplasm or an organelle compartment.

How Much Energy Is Available?

Like single cells, we cannot create energy from scratch; we must get it from someplace else. Why? According to the **first law of thermodynamics**, the total amount of energy in the universe remains constant. More energy cannot be created; existing energy cannot vanish. It can only be converted from one form to some other form.

Think about what the first law means. The universe has only so much energy, distributed in various forms. One form may be converted to another, as when corn plants absorb energy from the sun and convert it to the chemical energy of starch. After you eat and digest corn, your cells extract energy from starch and convert it to other forms, such as mechanical energy for movement.

With each metabolic conversion, a bit of the energy escapes to the surroundings, as heat. Even when you "do nothing," your body gives off about as much heat as a 100-watt lightbulb because of conversions in your cells. The energy being released is transferred to atoms and molecules making up the air, and the conversion of thermal to kinetic energy "heats up" the surroundings (Figure 5.3). The kinetic energy increases the number of ongoing, random collisions among molecules in the air. And with each collision, a bit more energy is released as heat. However, none of the energy ever vanishes.

The One-Way Flow of Energy

Energy available for conversions in cells resides mainly in covalent bonds. Glucose, glycogen, starches, fatty acids, and other organic compounds have organized arrangements of many of these bonds and are said to have a high energy content. When the compounds enter metabolic reactions, bonds break or become rearranged. During the molecular commotion, some amount of heat energy is lost to the surroundings. In general, cells can't recapture energy lost as heat.

Figure 5.4 Example of the one-way flow of energy into the world of life that compensates for the one-way flow of energy out of it. The sun continually loses energy, much of it in the form of wavelengths of light (Section 6.2). Living cells intercept some of the energy and convert it to useful forms of energy, stored in bonds of organic compounds. Each time a metabolic reaction proceeds in cells, stored energy is released—and some inevitably is lost to the surroundings, mostly as heat.

The lower photograph shows green, water-dwelling, photosynthetic cells (*Volvox*). They live in tiny, spherical colonies. The orange cells serve in reproduction. They form new colonies inside the parent sphere.

ENERGY LOST
one-way flow of energy from sun to Earth's environment

ENERGY GAINED
one-way flow of energy from environment to organisms

For example, your cells release usable energy from glucose when they break all of its covalent bonds. After many steps, six molecules of carbon dioxide and six of water remain. Compared with glucose, these leftovers have more stable arrangements of atoms, but chemical energy in their bonds is much less than the overall chemical energy of glucose. Why? *Some energy was lost at each step leading to their formation.* Said another way, glucose is a better source of usable energy.

What about the heat that was transferred from cells to their surroundings when carbon dioxide formed? That heat just isn't useful. Cells can't convert it to other forms, so they can't use it to do work.

Bad news for cells of the remote future: The amount of "low-quality" energy in the universe is increasing. No energy conversion can ever be 100 percent efficient; even highly efficient ones lose heat. So the total amount of energy in the universe is spontaneously flowing from forms rich in energy to forms having less of it. Billions of years from now, energy may not be available for conversions; all of it may be dissipated in space.

Without energy inputs to maintain it, any organized system tends to become more and more disorganized over time. **Entropy** is a measure of the degree of a system's disorder. Think of the Egyptian pyramids—once highly organized, presently crumbling, and many thousands of years from now, dust. It seems the ultimate destination of pyramids and everything else in the universe is a state of maximum entropy. And that, basically, is the point to remember about the **second law of thermodynamics**.

Can life be one glorious pocket of resistance to the depressing flow toward maximum entropy? After all, in every new organism, new bonds form and hold atoms together in precise arrays. So molecules become more organized and have a richer store of energy, not poorer!

Yet a simple example will show that the second law does indeed apply to life on Earth. The primary energy source for life is the sun—which has been releasing energy since it first formed. Plants can capture sunlight

Producer organisms harness sun's energy, use it to build organic compounds from simple raw materials available in their environment.

All organisms tap potential energy stored in organic compounds to drive energy conversions that keep them alive. Some energy is lost with each conversion.

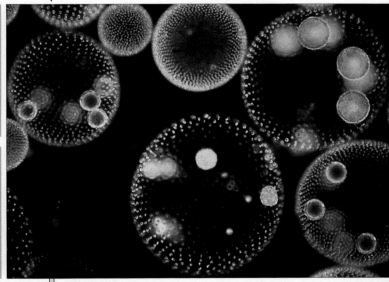

ENERGY LOST
one-way flow of energy from organisms back to the environment

energy, convert it to other forms, then lose energy to other organisms that feed, directly or indirectly, on the plants. At each energy transfer from the sun onward, some energy is lost as heat and joins the universal pool. *Overall, energy still flows in one direction.* The world of life maintains its amazing degree of organization only because it is being resupplied with energy that's being lost from someplace else (Figure 5.4).

The amount of energy in the universe remains constant. Energy can undergo conversion from one form to another, but it cannot be created out of nothing or destroyed.

The total amount of energy in the universe is spontaneously flowing from usable to nonusable forms.

A steady flow of sunlight energy into the interconnected web of life compensates for the steady flow of energy leaving it.

ENERGY CHANGES AND CELLULAR WORK

Energy Inputs, Energy Outputs

When cells convert one form of energy to another, the amount of potential energy available to them changes. The greater the amount a cell taps into, the larger the energy change, and the more work can be done.

Imagine a Martian watching the NASA Rover scoot around her planet. She decides to push it to the top of a rocky hill, which requires tapping into potential energy stored in her muscle cells (Figure 5.5a). Once the Rover is precariously perched on top, it has enough potential energy (owing to its position) to roll to the base on its own. The higher the Rover is relative to its final position at the hill's base, the greater the energy change (Figure 5.5b). This same rule operates in the cellular world, in which temperature and pressure remain fairly constant: *Energy changes in living cells tend to proceed spontaneously in the direction that results in a decrease in usable energy.*

Recall that glucose ($C_6H_{12}O_6$) is built from carbon dioxide ($6CO_2$) and water ($12H_2O$). Each substance has potential energy in chemical bonds, but the bond energy of glucose surpasses those of the other two substances combined. Thus, the assembly of glucose from carbon dioxide and water is not something that will proceed spontaneously. Picture carbon dioxide and water at the base of an energy hill. On their own, they simply don't have enough energy for an uphill run to make glucose.

In photosynthetic cells, *energy inputs* from the sun drive reactions by which glucose is built. The result is a net increase in usable energy for the cell (Figure 5.5c). Said another way, the reaction sequence that produces glucose is **endergonic** (meaning "energy in").

Now picture the reactions running in reverse, from glucose (at the top of the energy hill) to carbon dioxide and water (at the base). Energetically, a downhill run is favorable. The reaction sequence proceeds spontaneously and ends with a net loss in energy; it is **exergonic**, which means "energy out" (Figure 5.5d). All cells release usable energy from glucose and other large organic molecules.

The Role of ATP

Cells survive by *coupling* energy inputs to energy outputs. **ATP** (adenosine triphosphate) is the main coupling agent. This nucleotide consists of a five-carbon sugar (ribose), a nucleotide base (adenine), and a tail of three phosphate groups. Hundreds of different enzymes can split off the tail's outer phosphate group and attach it to another molecule, which thereby is primed to enter a reaction. **Phosphorylation** is the name for this type of phosphate-

Energy input required to push Rover uphill

Potential energy released by the downhill run (but not captured to do useful work)

Figure 5.5 Simple examples of energy changes involved in (**a**,**b**) mechanical work and (**c**,**d**) chemical work.

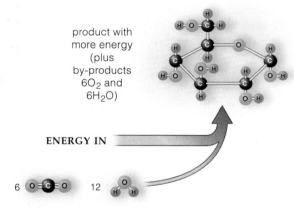

product with more energy (plus by-products $6O_2$ and $6H_2O$)

ENERGY IN

6 $O=C=O$ 12

c energy-poor starting substances

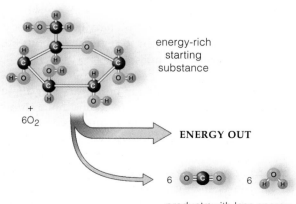

energy-rich starting substance

+ $6O_2$

ENERGY OUT

6 $O=C=O$ 6

d

products with less energy

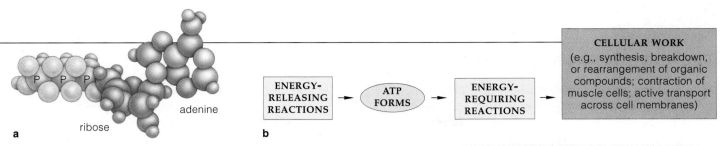

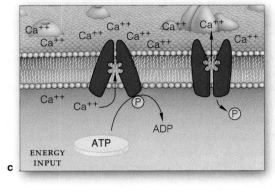

Figure 5.6 (**a**) Three-dimensional model for ATP. (**b**) A key energy relationship in cells. ATP couples exergonic (energy-releasing) reactions with hundreds of endergonic (energy-requiring) ones. (**c**) Example of cellular work initiated by ATP. At a plasma membrane, ATP transfers a phosphate group to a transport protein, which changes shape and pumps calcium ions across the membrane.

group transfers. In this respect ATP's role in cells is like currency in an economy. Cells earn ATP by exergonic reactions and then spend it in endergonic reactions that drive hundreds of activities (Figure 5.6). That's why we often use a cartoon coin to symbolize ATP.

The Role of Electron Transfers

Electron transfers, or **oxidation–reduction reactions**, are part of many energy changes. *Oxidation* means removing electrons from substances; *reduction* means adding them to substances. Cells conserve energy with such transfers.

For example, when a cell runs low on energy, it uses electrons from glucose to make ATP. Glucose is energy-rich yet unstable, compared to what's left at the end of aerobic respiration. It yields the most energy when its carbon and hydrogen combine with oxygen. There is so much oxygen in air that CO_2 is the most stable form of carbon, and H_2O the most stable form of hydrogen.

The more stable forms can't be reached all at once in cells. To see why, toss some glucose into a wood-burning fire. Its carbon and hydrogen atoms swiftly let go of one another and combine with oxygen in the air, but all the

released energy is lost as heat (Figure 5.7). Compared to fire, cells release glucose's energy more efficiently by stripping electrons from it and sending them through **electron transport systems**. Such systems are enzymes and other molecules organized to accept and give up electrons in series at a cell membrane. The electrons are at a higher energy level as they enter a transport system than when they leave. Think of the electrons as moving down a staircase and releasing energy at each step.

Those coenzymes mentioned in Section 3.6 assist in electron transfers. In aerobic respiration, the coenzymes abbreviated NAD^+ and FAD transfer hydrogen as well as electrons stripped from glucose *to* a transport system. (We refer to the reduced forms as NADH and $FADH_2$.) Controlled transfers of hydrogen *at* the system as well as electrons *through* it provide energy to make ATP.

Photosynthesis also uses a coenzyme. Electrons and hydrogen stripped from water molecules enter transport systems, which release energy used to make ATP. At the end of the systems, the coenzyme $NADP^+$ accepts electrons—which attract hydrogen. Hence the reduced coenzyme's name, NADPH. Then the reduced coenzyme transfers electron and hydrogen to reaction sites where they combine with carbon and oxygen to form glucose.

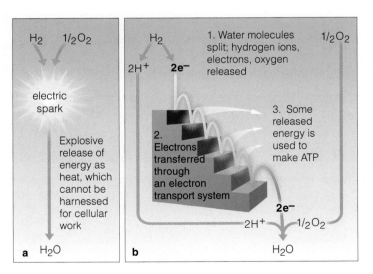

Figure 5.7 Uncontrolled versus controlled energy release. (**a**) Hydrogen and oxygen exposed to an electric spark react and release energy all at once. (**b**) In cells, the same reaction occurs in tiny steps that harness some of the released energy.

On their own, energy changes in cells spontaneously run in the direction that results in a decrease in usable energy.

ATP, an energy carrier in all living cells, couples energy-releasing reactions with energy-requiring ones. It does so by phosphate-group transfers, which release enough usable energy to activate molecules—to prime them to react.

In aerobic respiration, transfers of electrons and hydrogen from glucose drive ATP formation. In photosynthesis, similar transfers drive ATP formation, then the electrons and hydrogen combine with carbon and oxygen to form glucose.

THE DIRECTIONAL NATURE OF METABOLISM

How cells use energy is only one aspect of metabolism. Another aspect is the accumulation, use, and disposal of materials in highly directional ways. The directional nature of metabolism starts with forces that influence the kinds and amounts of substances moving into and out of the cell, and across internal cell membranes.

Concentration Gradients and Diffusion

Molecules or ions of a substance move constantly, and they collide and bounce off one another at random. They collide more frequently in regions where they are more concentrated. When the number in one region is not the same as in an adjoining region, we call this condition a *gradient*. Putting this together, a **concentration gradient** is a difference in the number of molecules or ions of a substance between adjoining regions.

In the absence of other forces, any substance moves from a region where it is more concentrated to a region where it is less concentrated. *The energy inherent in its individual molecules, which keeps them in constant motion, drives the directional movement.* Although the molecules collide randomly and career back and forth millions of times per second, the *net* movement is away from the place of greater concentration (and the most collisions).

Diffusion is the name for the net movement of like molecules or ions down a concentration gradient. It is a factor in how substances move across membranes and through cytoplasmic fluid. In multicelled species, it also moves substances between regions and between the body and its environment. For instance, if oxygen builds up in leaf cells, it may diffuse into air inside the leaf, then into the outside air, where its concentration is lower.

Like other substances, oxygen tends to diffuse in a direction established by its *own* concentration gradient, not those of other substances dissolved in the same fluid. You can see the outcome of this tendency by squeezing a drop of dye into water. Dye molecules diffuse to the region where they are less concentrated.

At the same time, water molecules move to the region where *they* are not as concentrated (Figure 5.8).

The rate and direction of diffusion depend partly on the concentration gradient's steepness. More molecules are moving out from the region of greater concentration than are moving in. Molecules are still moving when the gradient vanishes, but the total number moving one way or the other in a given interval is about the same.

Temperature, molecular size, pressure gradients, and electric gradients also affect diffusion. Molecules move faster and collide more often at higher temperatures, so they diffuse faster than they do in a cooler, adjoining region. With respect to size, smaller molecules tend to move down a gradient faster than larger molecules do.

A *pressure* gradient is a difference in pressure being exerted in adjoining regions. An *electric* gradient is a difference in electric charge between adjoining regions. For example, fluids bathing cell membranes hold many kinds of dissolved ions, each contributing to the overall charge. Opposite charges attract, so the fluid with the more negative charge overall tends to exert the greatest pull on positively charged substances, such as sodium ions. Many events, such as ATP formation, involve the combined force of electric and concentration gradients.

Which Way Will a Reaction Run?

Concentration gradients influence reversible reactions—from starting substances to products, or from products back to starting substances (Figure 5.9). If the energy

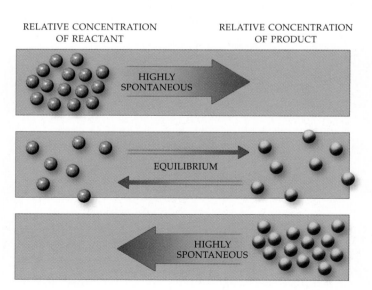

RELATIVE CONCENTRATION OF REACTANT   RELATIVE CONCENTRATION OF PRODUCT

HIGHLY SPONTANEOUS

EQUILIBRIUM

HIGHLY SPONTANEOUS

Figure 5.9 Chemical equilibrium. When the concentration of reactant molecules is high, a reaction runs most strongly in the forward direction (to products). When the concentration of product molecules is high, it runs most strongly in reverse. At equilibrium, rates of the forward and reverse reactions are the same.

Figure 5.8 Example of diffusion. A drop of dye enters a bowl of water. The dye molecules *and* the water molecules slowly become evenly dispersed, because each substance shows a net movement down its own concentration gradient.

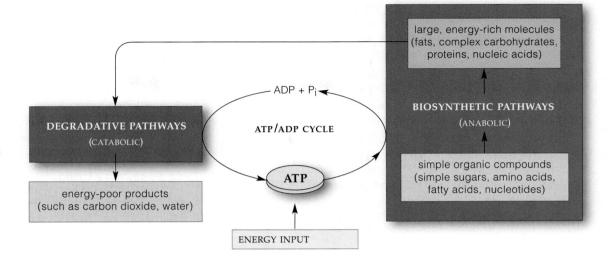

Figure 5.10 Energy relationship between the main biosynthetic and degradative pathways. ATP forms by energy inputs into cells and by catabolic pathways, such as glucose breakdown. In anabolic pathways, ATP drives the synthesis of large, energy-rich biological molecules from simple precursors. Cells renew ATP with an ATP/ADP cycle.

content and concentration of the starting substance are high, this is an energetically favorable state. The reaction tends to run forward (to product). When enough product molecules or ions are available, the reaction tends to run in reverse. Left to themselves, reversible reactions might reach chemical equilibrium, the time when a reaction is running at about the same pace in both directions.

Metabolic Pathways

The concentrations of thousands of substances change continually within the confines of a living cell. Most of the substances enter or leave orderly, enzyme-mediated sequences we call **metabolic pathways**. In *biosynthetic* pathways, small molecules are used to build larger ones of higher bond energies, such as complex carbohydrates, complex lipids, and proteins. These pathways will not run without energy inputs (Figure 5.10). In *degradative* (energy-releasing) pathways, large molecules are broken down to smaller products that have lower bond energies. Photosynthesis is the main biosynthetic pathway, and aerobic respiration is the main degradative pathway.

The participants of metabolic pathways go by these names: *Substrates* are substances that enter a reaction.

They are also called reactants or precursors. Between the start and conclusion of a pathway, any substance that forms is an *intermediate.* Substances left at the end of a reaction or a pathway are *end products.* ATP and a few other compounds are *energy carriers*, which can activate enzymes and other molecules through phosphate-group transfers. ATP is the coupling agent between the main biosynthetic and degradative pathways. Most *enzymes* are proteins that speed specific reactions. (A few RNAs also display enzyme activity.) *Cofactors* are coenzymes (NAD^+ and some other organic compounds) and metal ions. They assist enzymes or pick up electrons, atoms, or functional groups from one reaction site and taxi them to a different site. *Transport proteins*, recall, help substances across membranes. Controls over transport proteins help adjust concentrations on both sides of a membrane and thereby influence metabolic reactions.

Many pathways advance step by step in a straight line from substrates to end products. Others are cyclic; the steps are completed in a circle, with an end product serving as a reactant (Figure 5.11). Intermediates or end products of one pathway may enter different metabolic pathways. Let's turn now to some of the molecules that make all of these events happen.

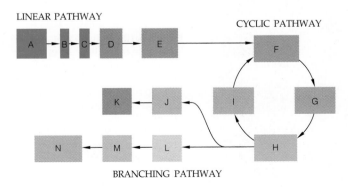

Figure 5.11 Types of reaction sequences—linear, cyclic, or branched—that occur in different metabolic pathways.

Molecules or ions of a substance constantly collide because of their inherent energy of motion. The collisions result in diffusion, a net outward movement of a substance from one region into an adjoining region where it is less concentrated.

A concentration gradient is a form of energy. It can drive the movement of substances into, through, and out of cells.

Metabolic reactions depend on the chemical energy inherent in concentrated amounts of molecules and ions. Cells are able to increase, maintain, and decrease those concentrations by coordinating thousands of metabolic reactions.

Metabolic pathways are enzyme-mediated reaction sequences from substrates to end products. ATP couples the major energy-releasing pathways and energy-requiring pathways.

ENZYME STRUCTURE AND FUNCTION

Without enzymes, the dynamically steady state called "you" would quickly cease to exist. Reactions simply wouldn't proceed fast enough for your body to process food, build and tear down hemoglobin and other vital molecules, send signals to and from brain cells, make muscles contract, and do everything else to stay alive.

To see how enzymes work, return to these concepts: *Cells adjust the concentrations of substances just inside the plasma membrane with respect to the concentrations outside. And eukaryotic cells also control the concentrations across* *organelle membranes.* Remember, molecules or ions of any substance are in constant, random motion that puts them on collision courses. The more concentrated they are, the more often they collide. Energy associated with such a collision might be enough to cause a metabolic reaction—to make a molecule combine with something else, split into smaller parts, or change its shape.

Four Features of Enzymes

By definition, **enzymes** are catalytic molecules; *they speed the rate at which reactions approach equilibrium.* Again, nearly all enzymes are proteins. All share four features. First, enzymes do not make anything happen that could not happen on its own, but they usually make it happen hundreds to millions of times faster. Second, reactions do not permanently alter or use up enzyme molecules; the same enzymes may act repeatedly. Third, the same type of enzyme usually works for the forward and the reverse directions of a reaction. Fourth, each type of enzyme is very picky about substrates. Its substrates are specific substances that it can chemically recognize, bind, and modify in precise ways. For instance, thrombin is one of the enzymes necessary to clot blood. It only recognizes a side-by-side arrangement of arginine and glycine (two amino acids) in a protein molecule. When it does so, it cleaves the peptide bond between them.

Enzyme–Substrate Interactions

Take a look at Figure 5.12. A metabolic reaction occurs when participating molecules collide—provided they collide with some minimum amount of energy called the *activation* energy. The collision may be spontaneous

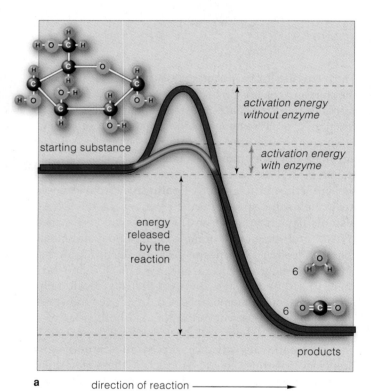

activation energy without enzyme

activation energy with enzyme

starting substance

energy released by the reaction

6

6

products

a   direction of reaction

ENZYME!

Figure 5.12 Activation energy. (**a**) Before reactants can enter a metabolic reaction, an energy input must activate them. Only then do reactants spontaneously proceed to end products. (**b**) An enzyme enhances the rate of a reaction by lowering the amount of activation energy required to boost specific reactants to the transition state. Going back to the Figure 5.5 analogy, enzymes make the hill smaller (they lower the energy barrier).

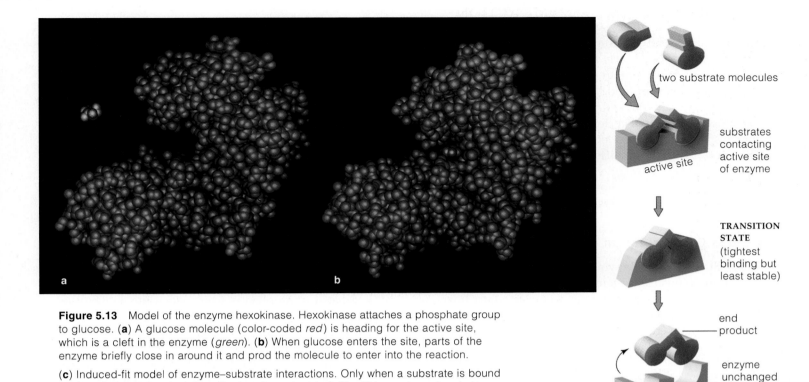

Figure 5.13 Model of the enzyme hexokinase. Hexokinase attaches a phosphate group to glucose. (**a**) A glucose molecule (color-coded *red*) is heading for the active site, which is a cleft in the enzyme (*green*). (**b**) When glucose enters the site, parts of the enzyme briefly close in around it and prod the molecule to enter into the reaction.

(**c**) Induced-fit model of enzyme–substrate interactions. Only when a substrate is bound in place is an enzyme's active site complementary to it. The fit is most precise during the transition state of a reaction. However, the enzyme–substrate complex is short-lived. Why? Usually, the attractive forces holding it together are weak.

(labels in figure c:) two substrate molecules; active site; substrates contacting active site of enzyme; TRANSITION STATE (tightest binding but least stable); end product; enzyme unchanged by the reaction

or enzymes may promote it; this makes no difference. The activation energy is an energy barrier—something like a hill. And that barrier must be surmounted *one way or another* before a reaction will proceed.

An enzyme makes the energy barrier smaller, so to speak. How? Every enzyme contains one or more **active sites**. These are crevices in the molecule's surface where the enzyme interacts with its substrates and catalyzes a reaction. Figure 5.13 shows the active site of a specific enzyme that primes a glucose molecule for reaction by attaching a phosphate group to it.

According to Daniel Koshland's **induced-fit model**, part of each substrate's surface has chemical groups that are almost but not quite complementary to the chemical groups in an active site. When substrates first settle into the site, the contact strains some of their bonds. Strained bonds are easier to break, so they promote formation of new bonds (in products). Also, charged or polar groups in the active site often cause the electric charge to shift in substrates, which primes them for conversion.

When substrates fit most precisely in an active site, they are in an activated, *transition* state and will react (Figure 5.13c). Now the reaction must proceed, just as the NASA Rover must roll down the Martian hill when something pushes it over the crest (Figure 5.12).

What induces the transition state or gets substrates over the energy barrier once that state is reached? The following are examples of the mechanisms involved.

1. *Helping substrates get together.* Substrate molecules rarely collide when there aren't many of them around. Snag them at an active site, and it's like boosting their local concentration. The reaction rate increases by 10,000 to 10,000,000 times, depending on the reaction.

2. *Orienting substrates in positions that favor reaction.* Without enzymes, substrates collide from random directions. Weak but extensive bonding at an active site puts reactive chemical groups on precise collision courses much more often.

3. *Promoting acid–base reactions.* In many cases, acidic or basic side groups of amino acids that are part of the enzyme's active site are poised to donate or accept hydrogen atoms from substrates. The loss or addition destabilizes a substrate's covalent bonds and makes them easier to break. Hydrolysis works this way.

Depending on the type of enzyme, mechanisms of the sort just described work alone or in combination with one another to bring about the transition state.

Enzymes catalyze (speed) the rate at which specific reactions reach equilibrium. They do so by lowering the amount of activation energy necessary to make substrates react.

Enzymes change the rate, not the outcome, of a reaction. They act only on specific substrates. They may catalyze the same reaction repeatedly, as long as substrates are available.

FACTORS AFFECTING ENZYME ACTION

You probably don't get much done when you feel too hot or cold or out of sorts because you ate too many sour plums or salty potato chips. When the cupboard is bare, you focus on food. Maybe you call a friend to go shopping with you, and when you drive too fast to the grocery store, police tend to slow you down. In such respects, you have a lot in common with enzymes. They, too, respond to shifts in temperature, pH, and salinity, and to the relative abundances of particular substances. Many depend on helpers for specific tasks. And all normal enzymes respond to metabolic police.

How Is Enzyme Action Controlled?

Each cell controls its enzyme activity. By coordinating control mechanisms, it maintains, lowers, or increases the concentrations of substances. Controls adjust how fast enzyme molecules are built and become available, or they work on enzymes that are already synthesized.

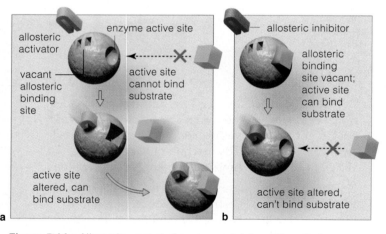

Figure 5.14 Allosteric control of enzymes. (**a**) An active site is unblocked when an activator protein binds to a vacant allosteric site. (**b**) An active site is blocked when an inhibitor protein binds to a vacant allosteric site.

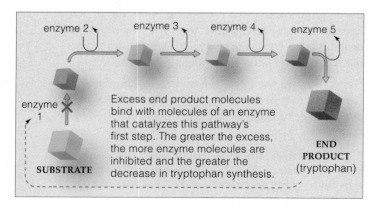

Figure 5.15 One example of feedback inhibition of a metabolic pathway. Five kinds of enzymes act in sequence. They convert a substrate to the end product, tryptophan.

For example, with *allosteric* control, some molecule binds to an enzyme at a site other than the active site. (*Allo*– means different, *steric* means structure, or state.) The active site changes shape in a way that allows or prevents enzyme action. Figure 5.14 shows models for the binding, which is reversible. Study these models, then picture a bacterium synthesizing tryptophan (and other amino acids) used to build proteins. Soon, enough proteins are built. But tryptophan synthesis won't stop until the concentration of tryptophan molecules triggers **feedback inhibition**. By this feedback mechanism, a change caused by an activity *shuts down that activity*. In this pathway, a feedback loop starts and ends with an allosteric enzyme. When tryptophan accumulates, the unused molecules bind with the allosteric site. Binding shuts down the enzyme, hence the rest of the pathway.

What if tryptophan is scarce when the demand for it steps up? The enzyme molecules are not shut down, so tryptophan will be built. Such feedback loops quickly adjust concentrations of many substances (Figure 5.15).

In humans and other multicelled organisms, control of enzyme activity is just amazing. Cells not only work to stay alive, they work with other cells in ways that benefit the whole body! For example, this vast cellular work relies on hormones, a type of signaling molecule. Specialized cells release hormones into tissues. Any cell having receptors for a given hormone responds to it, then its program for building a protein or some other activity changes. The hormone trips cell controls into action—and the activities of certain enzymes change.

Helpers in Enzyme Action

Enzyme action often requires coenzymes or metal ions. These help the reaction or transfer electrons, atoms, or functional groups to a different reaction site. You'll be reading about NAD^+, $NADP^+$, and FAD, all coenzymes derived from vitamin precursors. One of the metal ions, ferrous iron (Fe^{++}), is a helper in cytochrome action. Cytochromes are proteins in cell membranes, including the inner membrane of chloroplasts and mitochondria. They are part of major electron transport systems.

Does the Environment Influence Enzymes?

Temperature, recall, is a measure of molecular motion. A rise in temperature boosts reaction rates by making substrates collide more often with active sites. Yet past some temperature (which differs among enzymes), the increased motion disrupts the weak bonds holding the enzyme in its three-dimensional shape. Substrates no longer bind to the active site, so reaction rates decline sharply, as Figure 5.16*a* shows. When temperatures rise above or fall below the range of tolerance, metabolism

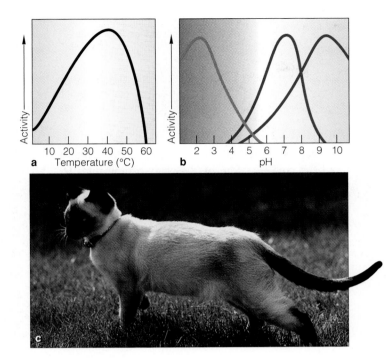

Figure 5.16 How the action of one type of enzyme shifts when (**a**) temperatures and (**b**) pH values fall outside the range of tolerance. (**c**) Siamese cats show observable effects of change in temperature. Fur on the ears and paws has more of a dark brown pigment, melanin, than the rest of the body does. A heat-sensitive enzyme controlling melanin production is less active in warmer parts of the body, which end up with lighter fur.

is disrupted. That's what happens with extremely high fevers. People usually do not survive when the body's internal temperature reaches 44°C (112°F).

pH values that extend beyond an enzyme's range of tolerance also take a toll (Figure 5.16b). Most enzymes work best when the pH is between 6 and 8. Trypsin, for example, is active in the small intestine, where pH is 8 or so. Pepsin, a protein-digesting enzyme, is one of the exceptions. It functions in gastric fluid, a highly acidic liquid (pH of about 1–2) that denatures most enzymes.

In addition, most enzymes do not work well when fluids are saltier than usual. High salinity (extremely high ion concentrations) disrupts the interactions that help hold enzymes in their three-dimensional shapes.

Control mechanisms govern the synthesis of new enzymes and stimulate or inhibit the activity of existing enzymes. By controlling enzymes, cells control the concentrations and kinds of substances available to them.

Coenzymes and metal ions help enzymes catalyze reactions or transfer electrons and atoms to other reaction sites.

Enzymes function best when the cellular environment stays within limited ranges of temperature, pH, and salinity. The actual ranges differ from one type of enzyme to the next.

METABOLISM AND CELL MEMBRANES

All molecules contain energy in the form of chemical bonds. Cell membranes help make that energy available for cell functioning. How? *They selectively concentrate reactants and end products in amounts most favorable for specific reactions.* Think about the water bathing a cell membrane. Plenty of substances are dissolved in it. But the kinds and amounts are not the same next to the two membrane surfaces. The membrane itself established and is maintaining these differences, which are vital for metabolism. How? Each cell membrane shows **selective permeability**: Because of its molecular structure, a cell membrane permits some substances but not others to cross it in certain ways, at certain times (Figure 5.17).

For instance, being largely nonpolar, the hydrocarbon chains of the lipid bilayer passively let carbon dioxide, molecular oxygen, and other small, nonpolar molecules cross the membrane. Some water molecules cross, also. They are polar, but they slip through gaps that open up when the chains flex and bend.

By contrast, large, polar molecules such as glucose seldom move freely across the bilayer. Neither do ions. Such water-soluble substances can cross it passively or actively through the interior of transport proteins. Water molecules in which they are dissolved cross with them.

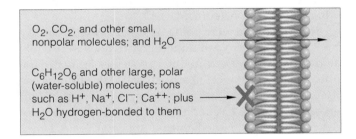

O_2, CO_2, and other small, nonpolar molecules; and H_2O ⟶

$C_6H_{12}O_6$ and other large, polar (water-soluble) molecules; ions such as H^+, Na^+, Cl^-; Ca^{++}; plus H_2O hydrogen-bonded to them ⟶ ✗

Figure 5.17 Selective permeability of cell membranes. Small, nonpolar molecules and some water molecules cross the lipid bilayer. Ions and large, polar, water-soluble molecules and the water dissolving them cannot cross the bilayer; transport proteins spanning the bilayer must help them across.

Membrane structure also allows cells to import and export substances in bulk across the plasma membrane. The processes are called endocytosis and exocytosis.

At any time in a cell's life, all of these mechanisms are operating simultaneously. Let's take a look at how they work, because they will occupy our attention in many chapters to follow.

Metabolic reactions depend on the chemical energy inherent in concentrated amounts of molecules and ions. Cells have mechanisms for increasing or decreasing those concentrations across the plasma membrane and internal cell membranes.

PROTEIN-MEDIATED TRANSPORT

Different kinds of proteins span the lipid bilayer of cell membranes. They let water-soluble molecules and ions diffuse across the membrane, through their tunnel-like interior (Figure 5.18). When a solute enters the tunnel's opening on one side of the membrane, it weakly binds to the protein. The protein's folded shape changes, so that the opening closes behind the bound solute and opens in front of it. The solute is exposed to the fluid bathing the other side of the membrane. The binding site reverts to its former state and the solute is released.

Passive Transport

Many of the proteins simply let solutes diffuse freely across the membrane. **Passive transport** is the name for an unassisted flow of solutes through the interior of a transport protein, down their concentration gradients. Energetically, the flow costs only what the cell already spent to produce and maintain the gradients. Passive transport itself adds nothing more to the energy cost. The *net* direction of movement during a given interval depends on how many molecules or ions of the solute are randomly contacting the transport proteins (Figure 5.18). Encounters are more frequent on the side of the membrane where the solute concentration is greatest. Therefore, the solute's *net* movement is in the direction where it is less concentrated.

If nothing else were going on, the passive, two-way transport would proceed until concentrations were equal across the membrane. But other processes influence the outcome. For instance, nearly all cells take up glucose from the blood, then use it for biosynthesis and instant energy. When the level of glucose in blood is high, they can maintain the gradient even if uptake is rapid. How? As fast as glucose molecules diffuse into the cells, others are entering metabolic reactions. In other words, when the cells use glucose, they help maintain a concentration gradient that favors the uptake of *more* glucose.

Active Transport

Only in a dead cell have solute concentrations become equal on both sides of membranes. Living cells never stop expending energy to pump potassium and other solutes to and from their interior. In **active transport**, energy-driven mechanisms called "membrane pumps" make solutes cross cell membranes *against* concentration gradients, so that their net movement is in the direction where they are more concentrated.

In most cases, ATP molecules deliver the energy for active transport. When it gives up one of its phosphate groups to a transport protein, the chemical fit between the solute and the protein binding site improves on one side of the membrane (Figure 5.19*a,b*).

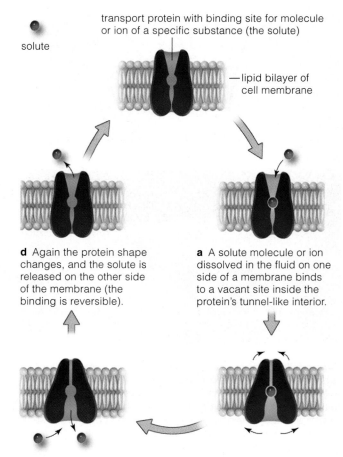

transport protein with binding site for molecule or ion of a specific substance (the solute)

solute

—lipid bilayer of cell membrane

d Again the protein shape changes, and the solute is released on the other side of the membrane (the binding is reversible).

a A solute molecule or ion dissolved in the fluid on one side of a membrane binds to a vacant site inside the protein's tunnel-like interior.

c Once it is exposed to fluid on the other side of the membrane, the binding site releases the solute. The newly vacated binding site attracts another molecule or ion of the solute.

b Binding alters the shape of the protein. The once-open part of the protein molecule closes behind the solute and another part opens in front of it.

Figure 5.18 Passive transport across a cell membrane. Solutes are able to move in both directions through transport proteins. In passive transport, *net* movement will be down the concentration gradient (from higher to lower) until concentrations are the same on both sides of the membrane.

One type of active transport system, the calcium pump, helps keep the calcium concentration inside a cell at least a thousand times lower than outside. Another, the sodium–potassium pump, moves potassium ions (K^+) across the plasma membrane. Activation of this protein helps bind a sodium ion (Na^+) on one side of a cell membrane. After the Na^+ makes the crossing, it is released. This release facilitates the binding of K^+ at a different binding site on the protein, which reverts to its original shape after K^+ is delivered to the other side.

Through operation of such active transport systems, concentration and electric gradients can be maintained across membranes. The gradients are vital to many cell activities and physiological processes, including muscle contraction and nerve cell (neuron) function. We return

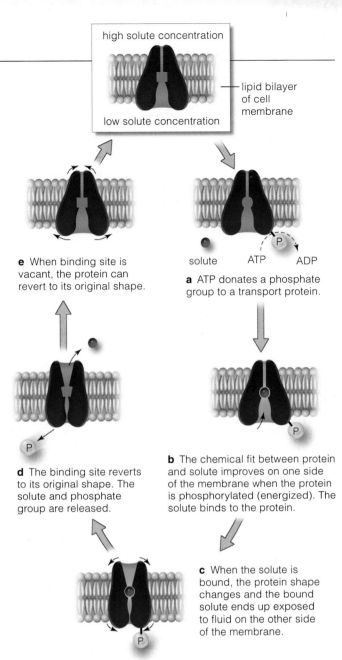

e When binding site is vacant, the protein can revert to its original shape.

a ATP donates a phosphate group to a transport protein.

solute   ATP   ADP

d The binding site reverts to its original shape. The solute and phosphate group are released.

b The chemical fit between protein and solute improves on one side of the membrane when the protein is phosphorylated (energized). The solute binds to the protein.

c When the solute is bound, the protein shape changes and the bound solute ends up exposed to fluid on the other side of the membrane.

high solute concentration

lipid bilayer of cell membrane

low solute concentration

Figure 5.19 Active transport across a cell membrane. ATP transfers a phosphate group to a transport protein. The transfer sets in motion reversible changes in the protein's shape that result in a greater net movement of solute particles *against* the concentration gradient.

to the mechanisms of active transport in later chapters. For now, keep these points in mind:

Transport proteins span the lipid bilayer of cell membranes. They bind solutes on one side of the bilayer and reversibly change shape. This shunts the solute through their interior.

With passive transport, a solute simply diffuses through the protein; its net movement is down its concentration gradient.

With active transport, the net diffusion of a solute is uphill, against its concentration gradient. The transporting protein must be activated, as by ATP energy, to counter the energy inherent in the gradient.

BULK TRANSPORT ACROSS MEMBRANES

Transport proteins are fine for moving a few small ions and molecules at a time. But cells use vesicles to expel or take in big items or lots of them. By **exocytosis**, a vesicle fuses with the plasma membrane, and its contents are released outside (Figure 5.20a). By **endocytosis**, the cell secures items at its outer surface. An indentation forms at the plasma membrane, balloons inward, and pinches off. The resulting endocytic vesicle transports or stores its contents in the cytoplasm (Figure 5.20b).

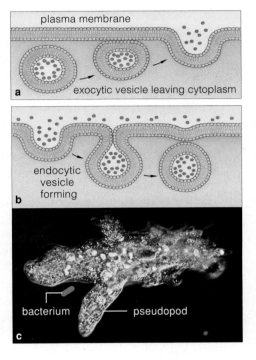

Figure 5.20 (**a**) Exocytosis. Cells release substances when an exocytic vesicle's membrane and the plasma membrane fuse. (**b**) Endocytosis. A patch of plasma membrane balloons inward beneath water and solutes outside. It pinches off as an endocytic vesicle that moves into the cytoplasm. (**c**) A phagocyte, *Amoeba proteus*. Its pseudopods (lobes of cytoplasm) are about to form an endocytic vesicle around a target bacterium.

plasma membrane

a   exocytic vesicle leaving cytoplasm

endocytic vesicle forming   b

bacterium   pseudopod   c

In all cells, endocytic vesicles form around extracellular fluid regardless of which substances are dissolved in it. Sometimes cell receptors chemically recognize and bind specific substances, such as hormones and lipoproteins. **Phagocytosis** (meaning "cell eating") is an active form of endocytosis. Some white blood cells engulf viruses, bacteria, and other pathogens. Amoebas and some other protistans engulf food this way. These phagocytes form pseudopods, which flow over a target and converge to make a vesicle (Figure 5.20c). The vesicle then fuses with lysosomes, the enzymes of which digest the contents.

Whereas transport proteins in a cell membrane deal only with ions and small molecules, exocytosis and endocytosis move larger packets of material across the plasma membrane.

By exocytosis, a cytoplasmic vesicle fuses with the plasma membrane, so that its contents are released outside the cell. By endocytosis, a small patch of the plasma membrane sinks inward and seals back on itself, forming a vesicle inside the cytoplasm. Membrane receptors often mediate this process.

MOVEMENT OF WATER ACROSS MEMBRANES

By far, more water diffuses across cell membranes than any other substance, so the key factors that influence its directional movement deserve special attention.

Osmosis

Turn on a faucet or watch a waterfall. The moving water demonstrates **bulk flow**—the mass movement of one or more substances in response to pressure, gravity, or some other external force. Bulk flow accounts for some movement of water through plants and animals. With each beat, your heart creates fluid pressure that drives a large volume of blood, which is mainly water, through interconnected blood vessels. Sap flows in conducting tissues that thread through maple trees, and this, too, is an example of bulk flow.

What about the movement of water into and out of cells or organelles? A membrane intervening between two regions permits the small, polar water molecules to cross but restricts the passage of ions and large polar molecules. **Osmosis** is the name for the diffusion of water molecules in response to a water concentration gradient between two regions separated by a selectively permeable membrane.

Concentrations of solutes in water on the two sides of a membrane influence osmosis, as you can see from Figure

2% sucrose solution

1 liter of distilled water

1 liter of 10% sucrose solution

1 liter of 2% sucrose solution

a

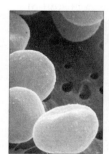

HYPOTONIC CONDITIONS

Water diffuses into red blood cells, which swell up

b

HYPERTONIC CONDITIONS

Water diffuses out of the cells, which shrink

ISOTONIC CONDITIONS

No net movement of water, no change in cell size or shape

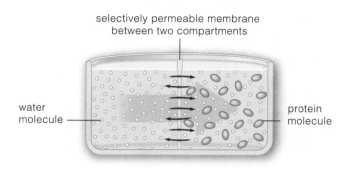
selectively permeable membrane between two compartments

water molecule

protein molecule

Figure 5.21 Demonstration of how a solute concentration gradient influences osmotic movement. Start with a container divided by a membrane that water but not proteins can cross. Pour water into the left compartment. Pour the same volume of a protein-rich solution into the right compartment. There, proteins occupy some of the space. The net diffusion of water in this example is from left to right (large *gray* arrow).

5.21. *The side with more solute particles has a lower water concentration.* Why is this so? Imagine dissolving some glucose in water. Compared to an equivalent volume of water, the glucose solution has fewer water molecules, because each molecule of glucose occupies some of the space formerly occupied by molecules of water.

It's mainly the *total number* of molecules or ions, not the type of solute, that dictates water's concentration. Dissolve some amount of an amino acid or urea in one liter of water, and the concentration of water changes about as much as it did in the glucose solution. Add some sodium chloride (NaCl) to one liter of water, and it dissociates into equal numbers of sodium ions and chloride ions. There are now twice as many particles of solute as there were in the glucose solution—and so the water concentration has decreased proportionately.

Effects of Tonicity

Given that water molecules tend to move osmotically to a region where water is *less* concentrated, the direction of their movement tends to be toward a region where solutes are *more* concentrated. Figure 5.22 illustrates this tendency. Suppose you decide to make a simple observational test of this statement. You construct three sacs from a membrane that water but not sucrose can cross. You fill each sac with a solution that's 2 percent

Figure 5.22 Effect of tonicity on water movement. (**a**) Arrow widths show the direction and relative amounts of water movement. (**b**) The micrographs correspond to the sketches. They show shapes that human red blood cells assume when you put them in fluids of higher, lower, and equal solute concentrations. Normally, solutions inside and outside red blood cells are in balance. This type of cell does not have any built-in mechanisms that would help it adjust to drastic changes in solute levels in its fluid surroundings.

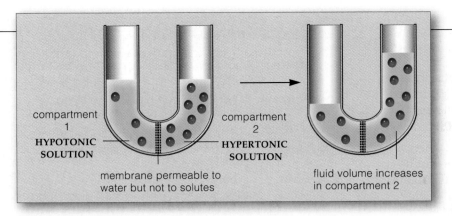

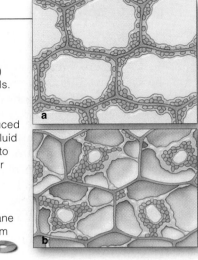

Figure 5.24 (**a**) Young plant cells. (**b**) This is what plasmolysis (an osmotically induced loss of internal fluid pressure) does to such cells. Their cytoplasm and central vacuole shrink and the plasma membrane moves away from the cell wall.

Figure 5.23 Increase in fluid volume owing to osmosis. In time, the net diffusion across a membrane separating two compartments is equal, but the fluid volume in compartment 2 is greater because the membrane is impermeable to solutes.

sucrose. Next you immerse one sac in a liter of distilled water (which has no solutes), one in a solution that's 10 percent sucrose. You immerse the third sac in a solution that's 2 percent sucrose. Each time, tonicity dictates the extent and direction of water movement (Figure 5.22).

Tonicity refers to the relative solute concentrations of two fluids. When two fluids on opposing sides of a membrane differ in solute concentration, the one having fewer solutes is the **hypotonic solution**. The one with more is the **hypertonic solution**. Water tends to diffuse from hypotonic to hypertonic fluids. **Isotonic solutions** have the same solute concentrations, so water shows no net osmotic movement from one to the other.

Normally, fluids on the inside and outside of your cells are isotonic. If a tissue fluid becomes drastically hypotonic, too much water would diffuse into the cells and they'd burst. If the fluid became too hypertonic, an outward diffusion of water would shrivel the cells. Most cells have built-in mechanisms that adjust to shifts in tonicity. Red blood cells don't. (Figure 5.22 shows what happened to them in an observational test of the effects of tonicity differences.) That is why severely dehydrated patients are given infusions of a solution isotonic with blood. Such solutions move by bulk flow from a bottle mounted on a rack above the patient, through a tube, and directly into an incised vein.

Effects of Fluid Pressure

Animal cells generally avoid bursting by engaging in ongoing selective transport of solutes across the plasma membrane. Cells of plants and many protistans, fungi, and bacteria also avoid that unpleasant prospect with the help of pressure exerted on their cell walls.

Pressure differences as well as solute concentrations influence the osmotic movement of water. Take a look at Figure 5.23. It shows how water continues to diffuse across a membrane between a hypotonic solution and a hypertonic solution until the water concentration is the same on both sides. As you can see, the *volume* of the

formerly hypertonic solution has increased (because its solutes cannot diffuse out). Any volume of fluid exerts **hydrostatic pressure**, or a force directed against a wall, membrane, or some other structure enclosing the fluid. The greater the fluid's solute concentration, the greater will be the hydrostatic pressure it exerts.

Living cells cannot increase in volume indefinitely (Section 4.2). At some point, hydrostatic pressure that builds up in the cell counters water's inward diffusion. That point is the osmotic pressure, the amount of force preventing further increase in a solution's volume.

Think of a young plant cell, with its pliable primary wall. As it grows, water diffuses into it and hydrostatic pressure increases against the wall. The wall expands, so the cell volume can increase. The thin walls are strong enough for the cell's internal fluid pressure to develop to the point where it counterbalances water uptake.

Plant cells are vulnerable to water losses, which can happen when the soil dries or becomes too salty. Water stops diffusing in and starts diffusing out, so internal fluid pressure falls. Such osmotically induced shrinkage of cytoplasm is called plasmolysis (Figure 5.24). Plants adjust somewhat to the loss of pressure, as when they actively take up potassium ions against a concentration gradient by mechanisms outlined in Section 5.7.

As you will read in Chapters 34 and 36, hydrostatic and osmotic pressure also influence the distribution of water in the blood, tissue fluid, and cells of animals.

Osmosis is the net diffusion of water between two solutions that differ in water concentration and that are separated by a selectively permeable membrane. The greater the number of molecules and ions dissolved in a solution, the lower its water concentration will be.

Water tends to move osmotically to regions of greater solute concentration (from hypotonic to hypertonic solutions). There is no net diffusion between isotonic solutions.

The fluid pressure that a solution exerts against a membrane or wall also influences the osmotic movement of water.

SUMMARY

1. Cells store, break down, and dispose of substances by acquiring and using energy and raw materials from outside sources. Metabolism, the sum of these energy-driven activities, underlies their survival. *CI*

2. Two laws of thermodynamics affect life. *First*, energy undergoes conversion from one form to another, but its total amount neither increases nor decreases; the total amount in the universe holds constant. *Second*, energy spontaneously flows in one direction, from usable forms to forms that are less and less usable. *5.1*

3. All matter has potential energy (as measured by the capacity to do work) owing to its position in space and the arrangement of its parts. Cells use potential energy to do mechanical, chemical, and electrical work. *5.1*

4. Without energy, a cell (like all organized systems) gets disorganized. It loses chemical potential energy in each metabolic reaction, mainly as heat. It stays alive (organized) by balancing energy outputs and inputs. The sun is life's primary energy source. Plants and other photosynthesizers convert sunlight energy to chemical bond energy of organic compounds. Plants, as well as organisms that feed on plants and each other, use energy in organic compounds to do cellular work. *5.1*

5. Cells conserve energy by coupling energy-releasing reactions with energy-requiring ones. ATP is the main coupling agent; it primes molecules to react by giving up a phosphate group to them. *5.2*

6. In aerobic respiration, coenzymes (NAD^+ and FAD) transfer electrons and hydrogen from glucose to transport systems, where electron transfers yield energy to drive ATP formation. In photosynthesis, electron transfers also drive ATP formation. The coenzyme $NADP^+$ transfers electrons and hydrogen to sites where they will combine with carbon and oxygen to form glucose. *5.2*

7. Much of metabolism is directional; it works with and against concentration gradients. Movement of the ions or molecules of a substance from a region of higher to lower concentration is called diffusion. Diffusion rates are influenced by concentration, electric, and pressure gradients, and by temperature and molecular size. *5.3*

8. Metabolic pathways are orderly sequences of enzyme-mediated reactions (Table 5.1). Photosynthesis and other biosynthetic pathways build energy-rich molecules from small molecules having less energy. Aerobic respiration and other degradative pathways break down energy-rich molecules to small ones of lower energy content. *5.3*

9. Enzymes are catalysts; they greatly enhance reaction rates for specific substrates by lowering the activation energy needed to start the reaction. They make bonds easier to break by binding substrates at their active site. Nearly all are proteins; some RNAs are catalytic. *5.4*

Table 5.1	*Participants in Metabolic Pathways*
SUBSTRATE	Substance that enters a metabolic reaction or metabolic pathway; also called a reactant
INTERMEDIATE	Substance formed between reactants and end products of a pathway
END PRODUCT	Substance left at end of reaction or pathway
ENZYME	Usually a protein that enhances reaction rates; some RNAs also do this
COFACTOR	Coenzyme (such as NAD^+) or metal ion; assist enzymes or taxis electrons, hydrogen, or functional groups between reaction sites
ATP	Main energy carrier in cells; couples energy-releasing reactions with energy-requiring ones
TRANSPORT PROTEIN	Protein that passively assists substances across a cell membrane or actively pumps them across

10. Different controls over enzyme action influence the kinds and amounts of substances available at any time in the cell. Enzyme action is best within a limited range of temperature, pH, and salinity. *5.5*

11. Oxygen, carbon dioxide, and other small nonpolar molecules diffuse across a membrane's lipid bilayer. Ions and large, polar molecules such as glucose cross it with the passive or active help of transport proteins. Water moves through proteins and across the bilayer. *5.6*

12. Transport proteins move water-soluble substances across membranes by reversibly changing shape. Passive transport doesn't require energy inputs; solutes simply diffuse through a protein's interior. Active transport uses energy boosts, as from ATP. The protein pumps a solute against its concentration gradient (Figure 5.25). *5.7*

13. By exocytosis, the membrane of a vesicle fuses with the plasma membrane; the vesicle contents are released outside (Figure 5.25). By endocytosis, a patch of plasma membrane forms a vesicle by way of receptor recognition of specific solutes, by indiscriminate uptake of solutes dissolved in extracellular fluid, or by phagocytosis ("cell eating," an engulfment of cells and substances). *5.8*

14. Osmosis is a special case of diffusion by which water moves across a selectively permeable membrane. *5.9*

Review Questions

1. State the first and second laws of thermodynamics. Does life violate the second law? *5.1*

2. Define and give a few examples of potential energy. *5.1*

3. How does ATP prime a molecule to enter a reaction? *5.2*

4. Define and describe four key features of enzymes. *5.4*

5. If all transport proteins work by changing shape, then how do passive transporters differ from active transporters? *5.7*

6. Define diffusion in terms of concentration gradients, electric gradients, and pressure gradients. What is osmosis? *5.3, 5.9*

7. Distinguish between exocytosis and endocytosis. *5.8*

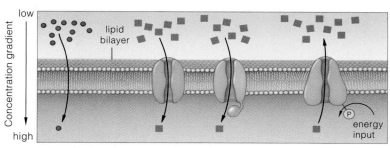

DIFFUSION ACROSS
LIPID BILAYER

Lipid-soluble
molecules and
water molecules
diffuse across

PASSIVE TRANSPORT
(facilitated diffusion)

Water-soluble molecules
and ions diffuse through
interior of transport proteins.
No energy boost required.

ACTIVE
TRANSPORT

Specific solutes pumped
across, through transport
proteins against gradient;
requires energy boost

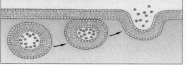

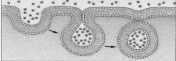

EXOCYTOSIS

Vesicle moves to plasma membrane,
fuses with it; contents released outside

ENDOCYTOSIS

Vesicle forms from patch of plasma
membrane sinking into cytoplasm

Figure 5.25 Summary of membrane crossing mechanisms.

Self-Quiz   ANSWERS IN APPENDIX III

1. Electron transport systems involve _____ .
   a. enzymes, coenzymes   c. cell membranes
   b. electron transfers   d. all of the above

2. Immerse a living red blood cell in a hypotonic solution,
and water will _____ .
   a. move into the cell   c. show no net movement
   b. move out of the cell   d. move in by endocytosis

3. _____ can readily diffuse across a lipid bilayer.
   a. Glucose   b. Oxygen   c. Carbon dioxide   d. b and c

4. Sodium ions cross a membrane through transport proteins
that receive an energy boost. This is an example of _____ .
   a. passive transport   c. facilitated diffusion
   b. active transport   d. a and c

5. Match each substance with the most suitable description.
   ____ coenzyme or metal ion   a. reactant or substrate
   ____ selective membrane crossing b. enzyme
   ____ substance entering a reaction c. enzyme helper
   ____ formed during a reaction   d. intermediate
   ____ substance at end of reaction e. end product
   ____ enhances reaction rate   f. energy carrier
   ____ mainly ATP   g. transport protein

Critical Thinking

1. *Cyanide poisoning* results after cyanide, a toxic compound,
binds to an enzyme of electron transport systems. Binding stops
the enzyme from donating electrons in the system. What effect
will this have on ATP production? From what you know of ATP
function, what effect will this have on a person's health?

2. *Pyrococcus furiosus*, a bacterium that grows in volcanic vents,
thrives at 100°C, the boiling point of water. Enzymes isolated
from *P. furiosus* cells don't function well below 100°C. Which
aspect of enzyme structure keeps them stable and active at such
extreme temperatures? (*Hint:* Review Section 3.5's summary of
interactions that maintain protein structure.)

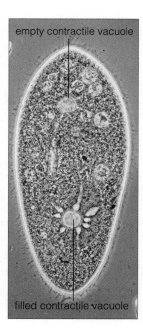

Figure 5.26 Contractile
vacuoles of *Paramecium*.

3. Water moves osmotically
into *Paramecium*, a protistan of
aquatic habitats. If unchecked,
the influx will bloat the cell
and then rupture its plasma
membrane. A mechanism
that requires energy involves
contractile vacuoles to expel
the excess (Figure 5.26). Water
enters tubelike extensions of
this organelle and collects in
a central space in the vacuole.
A filled vacuole contracts and
squirts excess water out of a
pore that opens to the outside.
Are the fluid surroundings
hypotonic, hypertonic, or
isotonic relative to this cell's
cytoplasm?

4. *Vibrio cholerae*, a bacterium,
causes *cholera*, which causes severe diarrhea and losses of up
to twenty liters of fluid a day. Drink contaminated water, and
V. cholerae cells adhere to the intestinal lining and secrete a toxic
metabolic product. In turn, cells of the lining secrete chloride ions
(Cl^-). Sodium ions (Na^+) follow Cl^- into the intestinal fluid.
Correlate the massive fluid loss with these ion movements.

5. Many cultivated fields in California are heavily irrigated.
Over the years, most of the water has evaporated from the soil,
leaving behind all of the solutes in the irrigation water. What
kinds of problems will the altered soil conditions cause plants?

6. Imagine you're a juvenile shrimp living in an *estuary*, where
fresh water draining from the land mixes with saltwater from
the sea. Many people own homes around a lake and want boat
access to the sea. They ask their city for permission to build a
canal to your estuary. If they succeed, what may happen to you?

Selected Key Terms

active site *5.4*
active transport *5.7*
ATP *5.2*
bulk flow *5.9*
chemical energy *5.1*
concentration gradient *5.3*
diffusion *5.3*
electron transport system *5.2*
endergonic reaction *5.2*
endocytosis *5.8*
entropy *5.1*
enzyme *5.4*
exergonic reaction *5.2*
exocytosis *5.8*
feedback inhibition *5.5*
first law of thermodynamics *5.1*
free radical *CI*
heat *5.1*

hydrostatic pressure *5.9*
hypertonic solution *5.9*
hypotonic solution *5.9*
induced-fit model *5.4*
isotonic solution *5.9*
kinetic energy *5.1*
metabolic pathway *5.3*
metabolism *CI*
osmosis *5.9*
oxidation–reduction
   (redox) reaction *5.2*
passive transport *5.7*
phagocytosis *5.8*
phosphorylation *5.2*
potential energy *5.1*
second law of
   thermodynamics *5.1*
selective permeability *5.6*

Readings

Adams, S. 22 October 1994. "No Way Back." *New Scientist*.
A refreshing look at the second law of thermodynamics.

Ritter, P. 1996. *Biochemistry: A Foundation*. Brooks/Cole:
Pacific Grove, California.

On-Line readings at Student Guide for InfoTrac:
www.brookscole.com/biology

6 HOW CELLS ACQUIRE ENERGY

Sunlight and Survival

Think about the last time you were hungry and craved a bit of apple, lettuce, chicken, pizza, or any other food. Where did it come from? For the answers, look past the refrigerator, the market or restaurant, or even the farm. Look to individual plants, the starting point for nearly all edibles you put into your mouth. Plants use environmental sources of energy and raw materials to make glucose and other organic compounds. Organic compounds, recall, are built on a framework of carbon atoms. So the questions become these:

1. *Where does the carbon come from in the first place?*

2. *Where does the energy come from to drive the synthesis of carbon-based compounds?*

Answers vary, because they depend on an organism's mode of nutrition.

Plants generally are "self-nourishing" organisms, or **autotrophs**. Their carbon source is carbon dioxide (CO_2), a gaseous compound in the air and dissolved in water. Plants, some bacteria, and many protistans are *photo*autotrophs, which means they use sunlight for **photosynthesis**. By this process, energy from the sun drives the formation of ATP and a coenzyme, NADPH. ATP delivers energy to sites where glucose and other carbohydrates are built. NADPH delivers electrons and hydrogen building blocks to the same sites.

Many other organisms are **heterotrophs**, meaning they must feed on autotrophs, one another, and organic wastes. (*Hetero*– means other, as in "being nourished by other organisms.") That is how most bacteria, many protistans, and all fungi and animals stay alive. Unlike plants, heterotrophs can't feed themselves with energy and raw materials from the physical environment.

How do we know such things? We didn't have a clue until the mid-seventeenth century. Before then, most people assumed that plants got raw materials to make food from soil. By 1882 a few chemists had an inkling that plants use sunlight, water, and something in the air to make food. A botanist, T. Englemann, wondered: What parts of sunlight do plants favor? As he knew, plants and algae release oxygen during photosynthesis. He also knew that, like many other organisms, some free-living bacteria use oxygen for aerobic respiration.

Englemann hypothesized: If bacterial cells need oxygen, then they will gather in places where the most photosynthesis is going on. He put bacterial cells in a water droplet, then put them on a microscope slide with a green alga (*Spirogyra*). He used a crystal prism to break up a beam of sunlight and cast a spectrum of colors across the slide. Bacteria congregated mostly where violet and red light fell on the green alga. The algal cells released more oxygen in the part illuminated by light of those colors, which happens to be the best light for photosynthesis (Figure 6.1).

Such observations showed us how photosynthesis works. They also helped reveal one of nature's great patterns, *for photosynthesis is the main pathway by which carbon and energy enter the web of life.* Once a cell makes or takes up organic compounds, it uses or stores them. *All* autotrophs and heterotrophs store energy in organic compounds and release it by other processes. Aerobic respiration, the main energy-releasing process, requires oxygen. Figure 6.2 previews the chemical links between photosynthesis and aerobic respiration—the focus of this chapter and the next.

Such processes might seem far removed from your daily interests. But remember, the food that nourishes you and most other organisms can't be produced or used without them. We will return to this point in later chapters. It provides perspective on many issues, such as nutrition, dieting, agriculture, human population growth, genetic engineering, and the impact of pollution on our sources of food—hence on our survival.

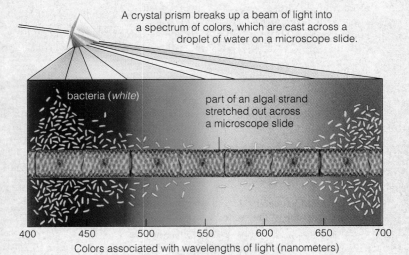

A crystal prism breaks up a beam of light into a spectrum of colors, which are cast across a droplet of water on a microscope slide.

bacteria (*white*)

part of an algal strand stretched out across a microscope slide

400   450   500   550   600   650   700
Colors associated with wavelengths of light (nanometers)

Figure 6.1 Results from T. Englemann's observational test that correlated portions of visible light with photosynthesis in *Spirogyra*, a strandlike green alga. Many oxygen-requiring bacterial cells moved to the colors where algal cells released the most oxygen, which is a by-product of their photosynthetic activity.

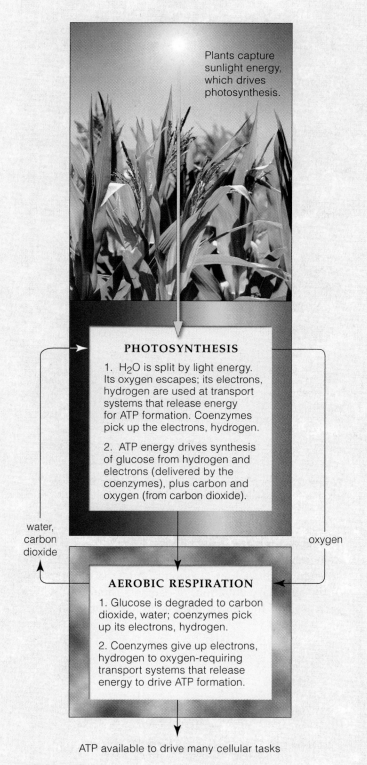

PHOTOSYNTHESIS

Plants capture sunlight energy, which drives photosynthesis.

1. H_2O is split by light energy. Its oxygen escapes; its electrons, hydrogen are used at transport systems that release energy for ATP formation. Coenzymes pick up the electrons, hydrogen.

2. ATP energy drives synthesis of glucose from hydrogen and electrons (delivered by the coenzymes), plus carbon and oxygen (from carbon dioxide).

water, carbon dioxide

oxygen

AEROBIC RESPIRATION

1. Glucose is degraded to carbon dioxide, water; coenzymes pick up its electrons, hydrogen.

2. Coenzymes give up electrons, hydrogen to oxygen-requiring transport systems that release energy to drive ATP formation.

ATP available to drive many cellular tasks

Figure 6.2 Links between photosynthesis—the main energy-requiring process in the world of life—and aerobic respiration, the main energy-releasing process.

Key Concepts

1. Plants, some bacteria, and many protistans use sunlight energy, carbon dioxide, and water to make glucose and other carbohydrates, which have a backbone of carbon atoms. The metabolic process by which they do this is called photosynthesis.

2. Photosynthesis is the main process by which carbon and energy become available to the web of life.

3. In the first stage of photosynthesis, sunlight energy is trapped and converted to chemical bond energy in ATP molecules. Typically, water molecules are split, their electrons and hydrogen atoms are picked up by the coenzyme $NADP^+$ to form NADPH, and their oxygen atoms are released as a by-product.

4. In the second stage of photosynthesis, ATP delivers energy for biosynthesis. NADPH delivers electrons and hydrogen, which combine with carbon and oxygen (both from carbon dioxide) to form glucose.

5. In the photosynthetic cells of plants and in many protistans, both stages of the reactions proceed inside organelles called chloroplasts.

6. Often we summarize photosynthesis this way:

$$12H_2O + 6CO_2 \xrightarrow{\text{LIGHT ENERGY}} 6O_2 + C_6H_{12}O_6 + 6H_2O$$

WATER   CARBON DIOXIDE   OXYGEN   GLUCOSE   WATER

PHOTOSYNTHESIS—AN OVERVIEW

Sit outdoors on a warm, cloudless day and you'll get hot but you'll never build your own food. How do plants do it? Think of it! They have to plug into the sun's energy, but not so much that it might cook them. They have to do magic and turn the sun's energy into ATP's chemical bond energy. They have to get carbon, oxygen, and hydrogen from somewhere and combine them into carbohydrates. As you might imagine, these multiple tasks can't all be done in the same place. Different functions, different structures.

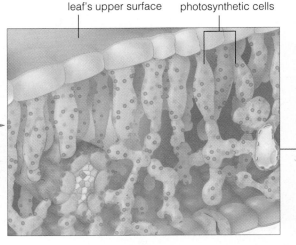
leaf's upper surface   photosynthetic cells

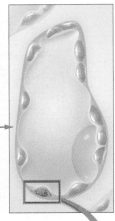

a Sow thistle (*Sonchus*) leaf

b Cutaway of a small section from the leaf. Its upper and lower surfaces enclose many photosynthetic cells.

c One of the photosynthetic cells, with green chloroplasts.

Where the Reactions Take Place

Plants and protistans called algae carry out the tasks in **chloroplasts**: organelles specialized for photosynthesis (Section 4.7). Figure 6.3 shows the structure of one type. All chloroplasts have two outermost membranes that surround a mostly fluid interior, the **stroma**. Another membrane, in the stroma, is highly folded into flattened sacs. The sacs, **thylakoids**, interconnect by channels. It makes no difference how many sacs a chloroplast has or whether they are stacked as in Figure 6.3*d,e*. The inner membrane always forms one functional compartment.

The capture and conversion of sunlight energy to chemical energy takes place at thylakoids. With these *light-dependent* reactions, plants get the electrons and hydrogen for making food. Amazingly, sunlight splits water molecules. Water's oxygen diffuses away. But its electrons flow through transport systems embedded in the thylakoid membrane. (Remember Section 5.2?) As electrons flow through the systems, hydrogen moves inside the thylakoid compartment. ATP forms when it flows out. Waste not, want not: the helpful coenzyme $NADP^+$ picks up the spent electrons and hydrogen.

Making glucose doesn't require light. These *light-independent* reactions proceed elsewhere, in the stroma. This is where ATP gives up energy, the coenzyme gives up electrons and hydrogen, and carbon dioxide (CO_2) is dismantled for its carbon and oxygen atoms.

Overall, you can think about all the reactants and products just described in terms of a simple equation:

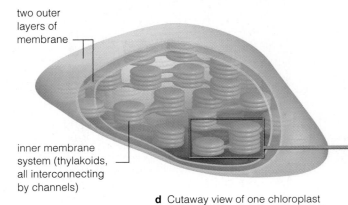
two outer layers of membrane

inner membrane system (thylakoids, all interconnecting by channels)

d Cutaway view of one chloroplast

Figure 6.3 Zooming in on sites of photosynthesis inside a leaf from a typical plant.

In case you're wondering how we know what we know: Remember the description of tracers in Section 2.2? By attaching radioisotopes to atoms of the reactants shown in the summary equation, researchers found out where each atom of the reactants ends up in the products:

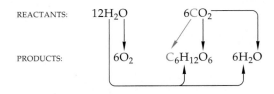

REACTANTS: $12H_2O$ $6CO_2$

PRODUCTS: $6O_2$ $C_6H_{12}O_6$ $6H_2O$

But Things Don't <u>Really</u> End With Glucose

Notice how we show *glucose* as the end product of the reactions. We do so to keep the chemical bookkeeping simple. But you'll find little glucose in a chloroplast.

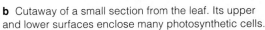

$$12H_2O + 6CO_2 \xrightarrow{\text{LIGHT ENERGY}} 6O_2 + C_6H_{12}O_6 + 6H_2O$$

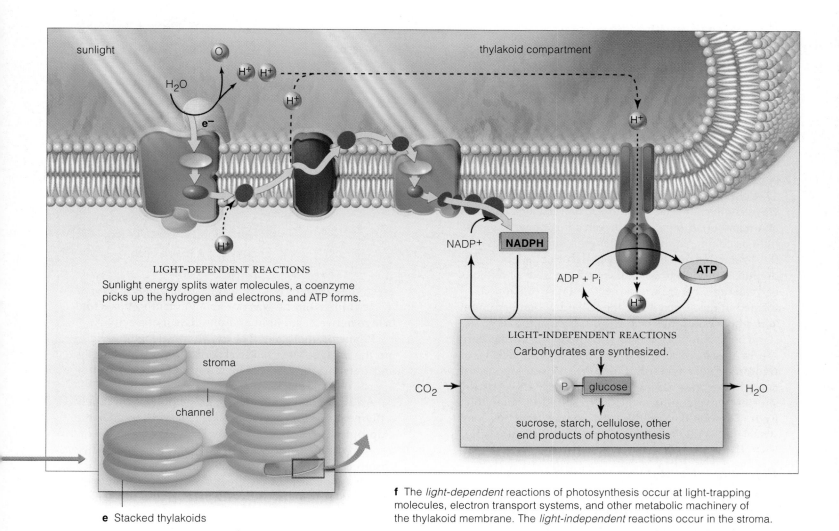

sunlight

O

H_2O

H^+ H^+

H^+

thylakoid compartment

e^-

H^+

H^+

NADPH

NADP$^+$

ADP + P$_i$

ATP

H^+

LIGHT-DEPENDENT REACTIONS
Sunlight energy splits water molecules, a coenzyme
picks up the hydrogen and electrons, and ATP forms.

LIGHT-INDEPENDENT REACTIONS
Carbohydrates are synthesized.

CO_2 →

P glucose

→ H_2O

sucrose, starch, cellulose, other
end products of photosynthesis

stroma

channel

e Stacked thylakoids

f The *light-dependent* reactions of photosynthesis occur at light-trapping
molecules, electron transport systems, and other metabolic machinery of
the thylakoid membrane. The *light-independent* reactions occur in the stroma.

Each newly formed glucose molecule has an attached
phosphate group; it is primed to react with something
else. Almost always, it is used at once to form sucrose,
starch, and cellulose. These are the main end products
of photosynthesis even though we "stop" with glucose.

The sketch below is a simple version of Figure 6.3.
We repeat it in sections that follow to help reinforce
your grasp of how the details fit into the big picture:

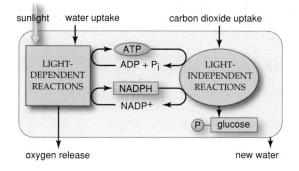

sunlight   water uptake   carbon dioxide uptake

ATP
LIGHT-   ADP + P$_i$ LIGHT-
DEPENDENT   INDEPENDENT
REACTIONS   NADPH   REACTIONS
                NADP$^+$
                                 P glucose

oxygen release   new water

We turn next to the details. Before we do, think about
this: Two thousand chloroplasts, lined single file, would
be no wider than a dime. Imagine all of the chloroplasts
in just one corn or rice plant—each a tiny factory for
producing sugars and starch—and you may get an idea of
the magnitude of the metabolic events required to feed
you and every other organism on this planet.

**Chloroplasts, the organelles of photosynthesis in plants
and many protistans, specialize in food production.**

**In the first stage of photosynthesis, sunlight energy drives
the formation of ATP and NADPH, and oxygen is released.
In chloroplasts, this stage occurs at an inner membrane
system of interconnected, flattened sacs called thylakoids.**

**The second stage occurs in the stroma. Energy from ATP
drives glucose formation. For these synthesis reactions,
carbon dioxide provides the carbon and oxygen atoms, and
NADPH provides the electrons and hydrogen atoms.**

SUNLIGHT AS AN ENERGY SOURCE

Properties of Light

Photoautotrophs are the start of a one-way flow of energy through nearly all webs of life (Figure 6.4). That flow is only a tiny fraction of the energy that is continually released by the sun's thermonuclear reactions. All of the radiant energy undulates across space, something like waves crossing a sea. The horizontal distance between crests of every two successive waves is a **wavelength** (Figure 6.5a). The **electromagnetic spectrum** is the range of all wavelengths of radiant energy (Figure 6.5).

Collectively, different photoautotrophs absorb the wavelengths between 380 and 750 nanometers. Humans and many other organisms perceive this range of visible light as colors. Shorter wavelengths, such as ultraviolet (UV) radiation, are energetic enough to break bonds of organic compounds and kill cells. Hundreds of millions of years ago, before the protective layer of ozone (O_3) formed in the atmosphere, the lethal UV bombardment kept photoautotrophs below the surface of the seas.

When something absorbs it, energy of visible light can be measured as if it were organized in packets, or **photons**. Every type of photon has a fixed amount of energy. The ones having the most energy travel as the shortest wavelengths, which correspond to blue-violet light. Photons having the least energy travel as long wavelengths that correspond to red light (Figure 6.5b).

Pigments—The Rainbow Catchers

Collectively, wavelengths of visible light look white to us. A crystal prism intercepting white light can sort it out into its component colors by bending the different wavelengths by different degrees. In the 1800s, recall, T. Englemann used a prism as part of an experiment to identify which wavelengths drive photosynthesis in a

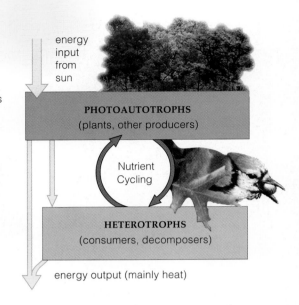

Figure 6.4 Model for how photoautotrophs are the basis of a one-way flow of energy through nearly all ecosystems, or webs of life, starting with energy inputs from the sun.

energy input from sun

PHOTOAUTOTROPHS (plants, other producers)

Nutrient Cycling

HETEROTROPHS (consumers, decomposers)

energy output (mainly heat)

green alga (Figure 6.1). But molecular biology was far in the future, and he was not able to identify the actual bridge between sunlight and photosynthetic activity.

Pigments, molecules that absorb wavelengths of light, are the bridge. Most kinds absorb certain wavelengths and transmit the rest. A few absorb so many that they appear dark or black.

An absorption spectrum is a graph of how effectively a pigment absorbs different wavelengths. Consider the **chlorophylls**. *These are the main pigments that absorb the wavelengths most efficient at driving photosynthesis.* They absorb all wavelengths of visible light. Of course, they absorb only a little of the green and yellow-green ones, and they transmit or reflect the rest (Figure 6.6). That's why plant parts rich with chlorophylls look green to us.

Other, *accessory* pigments extend the effective range of wavelengths that drive photosynthesis. Chlorophyll *a*, a grass-green pigment, absorbs the blue-violet and red wavelengths. It is a key player in the light-dependent

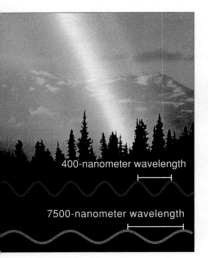

400-nanometer wavelength

7500-nanometer wavelength

a

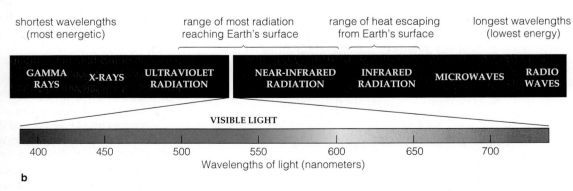

shortest wavelengths (most energetic)

range of most radiation reaching Earth's surface

range of heat escaping from Earth's surface

longest wavelengths (lowest energy)

| GAMMA RAYS | X-RAYS | ULTRAVIOLET RADIATION | NEAR-INFRARED RADIATION | INFRARED RADIATION | MICROWAVES | RADIO WAVES |

VISIBLE LIGHT

400   450   500   550   600   650   700

Wavelengths of light (nanometers)

b

Figure 6.5 (**a**) Two examples of wavelengths, the horizontal distance between crests of successive waves. (**b**) Electromagnetic spectrum. Visible light is one of many forms of electromagnetic radiation in the spectrum.

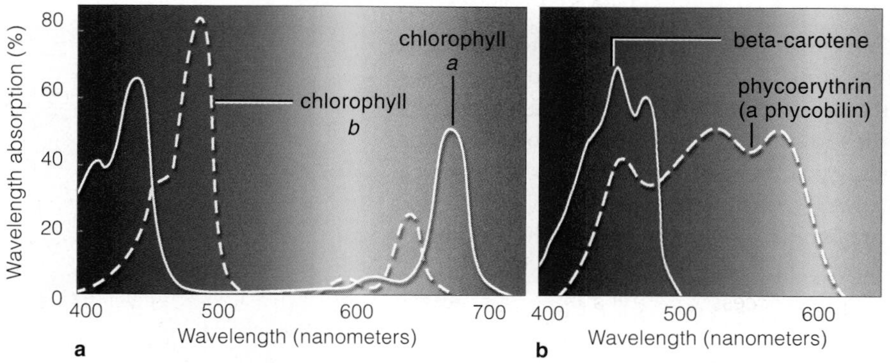

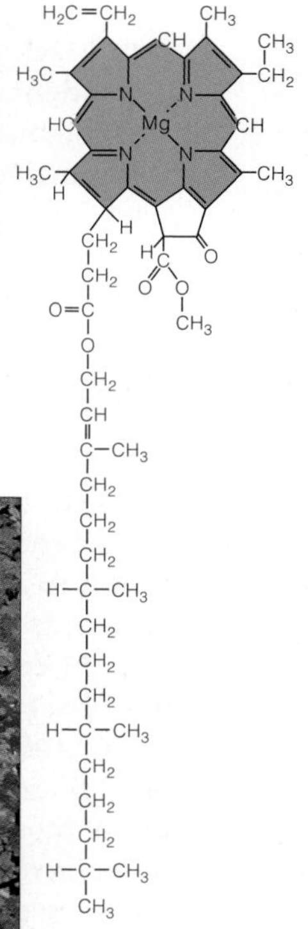

Figure 6.6 **(a)** Absorption spectra showing the efficiency with which chlorophylls a and b respond to wavelengths. **(b)** More absorption spectra, for beta-carotene (one of the carotenoids) and a phycobilin. *Far right:* Structural formula for chlorophyll a. Its light-catching array is color-coded *green;* its hydrocarbon backbone readily dissolves in the lipid bilayers of cell membranes.

Figure 6.7 Leaf color. In intense-green leaves, photosynthetic cells continually make chlorophyll, which masks accessory pigments. In fall, chlorophyll synthesis lags behind its breakdown in many species. Other pigments are unmasked, and more colors show through. Water-soluble anthocyanins accumulate in leaf cells. They appear red when plant fluids are slightly acidic, blue when they are basic (alkaline), or colors in between in fluids of intermediate pH. Soil conditions contribute to pH values.

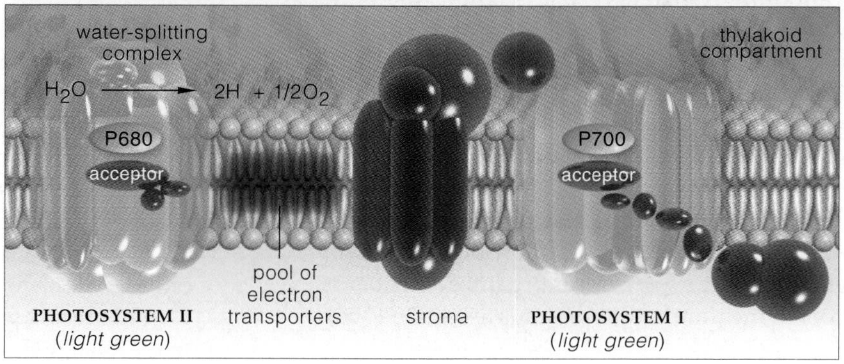

Figure 6.8 Arrangement of the two types of photosystems, II and I, in a chloroplast's thylakoid membranes. Components of neighboring electron transport systems are coded *dark green.* One transport system extends from photosystem II to photosystem I. A second transport system gives up the spent electrons to the coenzyme NADP+.

reactions (Figure 6.6). Chlorophyll b absorbs blue and red-orange wavelengths. In chloroplasts, this pigment harvests wavelengths that chlorophyll a misses. Plants, green algae, and a few photoautotrophic bacteria have this bluish-green pigment.

Carotenoids, other accessory pigments, absorb blue-violet and blue-green wavelengths that the chlorophylls miss. They reflect red, orange, and yellow wavelengths. Thus you can expect the carotenoids to be abundant in flowers, fruits, and vegetables in this color range. They are less abundant than chlorophylls in green leaves but often show up in autumn (Figure 6.7). Each year for three weeks in New England, tourists spend a billion dollars to see the demise of chlorophyll in deciduous trees.

Other accessory pigments include the red and blue anthocyanins and phycobilins. Anthocyanins also help color many flowers as well as leaves. Phycobilins are the signature pigments of red algae and cyanobacteria.

Where Are Photosynthetic Pigments Located?

We find photosynthetic pigments in prokaryotic cells of ancient lineages. Some archaebacteria use a purple one. Cyanobacteria use chlorophyll a and other pigments in some internal foldings of their plasma membrane.

In thylakoid membranes, pigments are organized in many clusters called **photosystems.** Each photosystem includes 200 to 300 pigment molecules, and each is near an electron transport system (Figure 6.8). How the two systems interact is the topic of the next two sections.

Chlorophylls are the main light-absorbing pigments in photoautotrophs. Carotenoids and other accessory pigments help absorb wavelengths of light used in photosynthesis.

Pigment clusters (photosystems) are the functional partners of electron transport systems in thylakoid membranes.

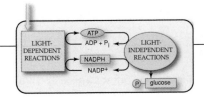

6.3

THE LIGHT-DEPENDENT REACTIONS

Let's track three events that are central to the **light-dependent reactions**, the first stage of photosynthesis. *First*, pigments capture sunlight energy, which also splits water molecules into oxygen, hydrogen, and electrons. *Second*, the energy input is converted to the chemical bond energy of ATP. *Third*, the coenzyme $NADP^+$ picks up the electrons, and hydrogen tags along with them.

What Happens to the Absorbed Energy?

When a photosynthetic pigment absorbs photons, some of its electrons move into an unstable arrangement at a higher energy level (Section 2.2). The electrons quickly return to a lower energy level, and they emit the extra energy as fluorescent light and heat. (*Fluorescent* refers to light emitted when a destabilized molecule reverts to stable form.) If nothing else were around to intercept it, all of the emitted energy would escape as light and heat. But remember, pigments are clustered together in the photosystems of the thylakoid membrane. Most of them merely harvest energy from photons. Rather than releasing excitation energy as a fluorescent afterglow, a harvester randomly transfers it to a neighbor pigment, which randomly passes it to a neighbor pigment, and so on. Figure 6.9 shows this "random walk" of energy.

With each transfer, some energy is lost as heat. Very soon, the energy that's left corresponds to a wavelength that only the photosystem's **reaction center**—a special chlorophyll *a* molecule—can trap. The reaction center passes the electrons not to another pigment but to an acceptor molecule near an electron transport system.

Electron transport systems are organized arrays of enzymes, coenzymes, and other proteins associated with

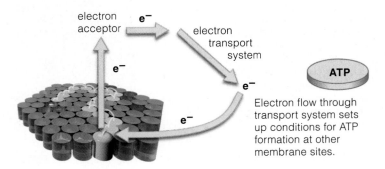

Figure 6.10 Cyclic pathway of ATP formation.

a cell membrane (Section 5.2). Electrons are transferred step-by-step through the array, and some energy escapes at each step. In chloroplasts, much of the energy drives the formation of ATP and NADPH.

Cyclic and Noncyclic Electron Flow

Without NADPH, we wouldn't have big plants. After all, in ancient bacteria, photosynthesis evolved as a way to form ATP, not build organic compounds. Hydrogen and electrons for the synthesis reactions were extracted from inorganic compounds. Energy from the sun drove the electrons in a cycle: from a photosystem, to a transport system, then back to the photosystem (Figure 6.10). This *cyclic* pathway still runs in all photoautotrophs. It uses only the ancient photosystem, which we call type I.

That flow of electrons doesn't pack enough punch to make NADPH. But more than 2 billion years ago, the energy for electron transfers increased. It was as if two batteries became hooked together. Another photosystem, type II, entered the picture. When photosystems II and I operate together, they acquire enough energy to strip electrons from water molecules.

As you can see from Figure 6.11*a*, the more recent path of electron flow is *noncyclic*. Electrons flow from water to photosystem II, through a transport system to photosystem I, then on through yet another transport system. From there, $NADP^+$ accepts two electrons and a hydrogen ion (H^+) to become NADPH.

The energy input to photosystem II is also the exact amount necessary to trigger **photolysis**. By this reaction, water molecules split into oxygen, hydrogen which gets ionized (H^+), and electrons. When photosystem II gives up electrons, new electrons from water replace them.

ATP still forms as it does in the cyclic pathway, as an outcome of what happens at the first electron transport system. The mechanism is a bit complicated, so we'll reserve that part of the story for Section 6.4. For now,

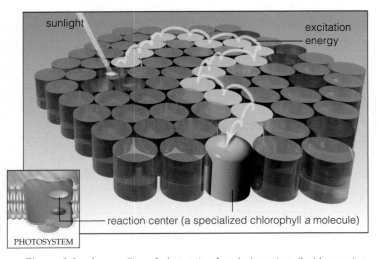

Figure 6.9 A sampling of pigments of a photosystem that harvest photon energy. Notice the random flow of energy among them. The energy of excitation quickly reaches a reaction center which, when suitably activated, releases electrons for photosynthesis.

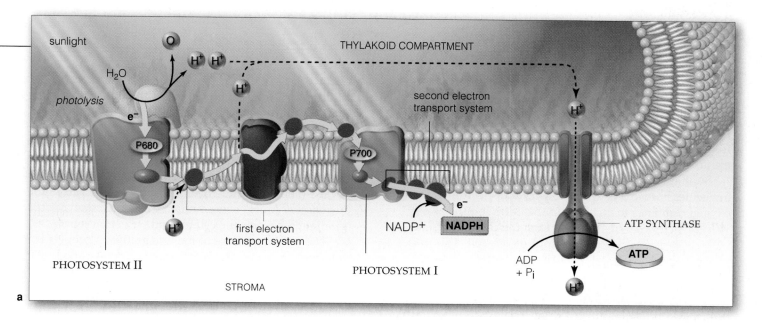

sunlight

THYLAKOID COMPARTMENT

H_2O

photolysis

P680

first electron transport system

second electron transport system

P700

H^+

$NADP^+$ **NADPH**

e^-

ATP SYNTHASE

ADP + P_i

ATP

PHOTOSYSTEM II

PHOTOSYSTEM I

STROMA

a

Figure 6.11 (**a**) Filling in details for Figure 6.3—ATP and NADPH formation in the noncyclic pathway of photosynthesis. The *yellow* arrows show electron flow. Electrons released from water molecules split by photolysis move through two photosystems (*light green*) and two transport systems (*dark green*). The joint operation of the two photosystems boosts electrons to an energy level high enough to drive NADPH formation. Also, as you will read shortly, hydrogen ions move across the thylakoid membrane in ways that lead to ATP formation at proteins called ATP synthases.

(**b**) Diagram the energy changes associated with the noncyclic pathway, and you will end up with this "Z scheme."

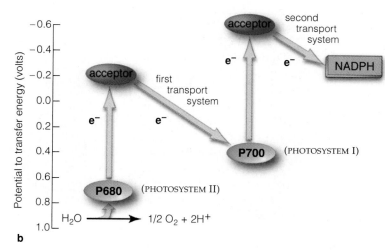

Potential to transfer energy (volts)

acceptor

first transport system

e^-

e^-

acceptor

second transport system

e^- e^-

NADPH

P700   (PHOTOSYSTEM I)

e^- e^-

P680   (PHOTOSYSTEM II)

$H_2O \longrightarrow 1/2\ O_2 + 2H^+$

b

simply note that when electrons leave the first transport system, they still have a bit of extra energy. They get even more upon their arrival at photosystem I, which sunlight also happens to be bombarding. They are able to enter the second transport system only because the energy boost puts them at a higher energy level.

Today, both the cyclic and noncyclic pathways of electron flow operate in all photosynthetic organisms. Which one dominates at a given time depends on the organism's metabolic demands for ATP and NADPH.

The Legacy—A New Atmosphere

On sunny days, you can see bubbles of oxygen escaping from aquatic plants (Figure 6.12). When the noncyclic pathway first evolved, the oxygen it released dissolved in seawater, wet mud, and similar prokaryotic habitats. By about 1.5 billion years ago, enormous quantities of

Figure 6.12 Visible evidence of photosynthesis—bubbles of oxygen escaping from the leaves of *Elodea*, a plant that lives in freshwater habitats.

dissolved oxygen were escaping to what had been an oxygen-free atmosphere. The atmosphere now changed forever. And so did the world of life. The vast, global increase in free oxygen favored the evolution of aerobic respiration. This became the most efficient pathway for releasing usable energy from organic compounds. And so the emergence of the noncyclic pathway allowed you and every other animal to be around today, breathing the oxygen that helps keep your cells alive.

In the light-dependent reactions, sunlight energy drives the release of electrons from photosystems in thylakoid membranes. The electron flow through nearby transport systems results in the formation of ATP, NADPH, or both.

ATP alone forms in a cyclic pathway that uses one type of photosystem. Both ATP and NADPH form by a noncyclic pathway in which electrons flow from water, through two types of photosystems, and finally to $NADP^+$.

All photosynthetic species employ one or both pathways in response to the cell's changing needs for ATP and NADPH.

Oxygen, a by-product of the noncyclic pathway, changed the early atmosphere and made aerobic respiration possible.

CASE STUDY: A CONTROLLED RELEASE OF ENERGY

Spill some sugar into a campfire and you'll release its energy all at once, but not in a form that does any good. Events in the chloroplast control the release of energy from incoming sunlight—and they direct it into doing useful cellular work.

Sunlight prods enzymes to repeatedly split water molecules into oxygen, hydrogen, and electrons. The oxygen combines to form O_2. In this form it diffuses out from the chloroplast and out of the cell. The free hydrogen (H^+) does not leave. It accumulates in the thylakoid, as shown in Figure 6.13a.

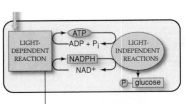

electrons.) Certain components of the transport system shunt the ions across the membrane, as in Figure 6.13b.

Therefore, *by a combination of photolysis and electron transport, H^+ has now become more concentrated in the thylakoid compartment than in the stroma.* In addition, the unequal distribution of these positively charged ions means there is a difference in electric charge across the membrane. An H^+ concentration gradient and an electric gradient have become established.

The combined force of the two gradients propels H^+ into the stroma, through transport proteins called ATP synthases. This type of protein spans the thylakoid membrane (Figure 6.13c). ATP synthases have built-in machinery that responds to the ion flow. They catalyze

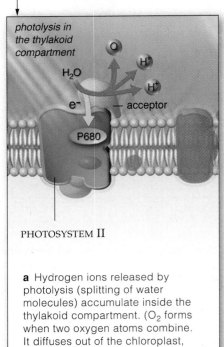

a Hydrogen ions released by photolysis (splitting of water molecules) accumulate inside the thylakoid compartment. (O_2 forms when two oxygen atoms combine. It diffuses out of the chloroplast, then out of the photosynthetic cell.)

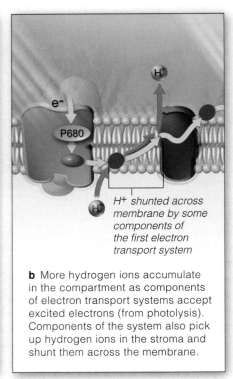

b More hydrogen ions accumulate in the compartment as components of electron transport systems accept excited electrons (from photolysis). Components of the system also pick up hydrogen ions in the stroma and shunt them across the membrane.

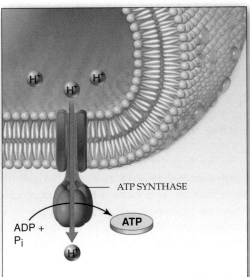

c Ions follow the concentration and electric gradients across the thylakoid membrane. They flow to the stroma, through the interior of ATP synthases that span the membrane. The flow drives the formation of ATP from the cell's pool of ADP and P_i.

Figure 6.13 From photosynthesis, an example of how energy is released and directed to cellular tasks. Sunlight energy sets in motion events that build concentration and electric gradients, which are tapped to make ATP. It also leads to the formation of NADPH.

At the same time, an acceptor molecule picks up the electrons and transfers them to an electron transport system that is positioned next to it in the thylakoid membrane. The membrane has many, many transport systems. While all the systems are accepting electrons, they're also picking up hydrogen ions from the stroma. (These ions are attracted to the negative charge of the

the attachment of unbound phosphate to a molecule of ADP in the stroma, thus forming an ATP molecule.

The sequence of events just described is called the chemiosmotic model for ATP formation in chloroplasts. As you will read in the next chapter, the same model applies to ATP formation in mitochondria.

In chloroplasts, ongoing photolysis and transfers of electrons in transport systems set up H^+ concentration and electric gradients across the thylakoid membrane. Ion flow from the thylakoid compartment to the stroma drives ATP formation.

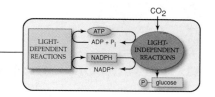

THE LIGHT-INDEPENDENT REACTIONS

The **light-independent reactions** are the "synthesis" part of photosynthesis. They occur as a cyclic pathway called the **Calvin–Benson cycle**. The reactions require energy from ATP, hydrogen and electrons from NADPH, and carbon and oxygen from carbon dioxide (CO_2), which is present in the air (or water) that surrounds photosynthetic cells. We say these reactions are light-independent because they do not depend directly on sunlight. They proceed just as well in the dark, as long as ATP and NADPH are available.

How Do Plants Capture Carbon?

Let's track a CO_2 molecule that diffuses into air spaces inside a leaf and ends up next to a photosynthetic cell. It diffuses into the cell, then into a chloroplast's stroma. An enzyme attaches the carbon atom of CO_2 to **RuBP** (ribulose bisphosphate), a compound with a backbone of five carbon atoms. **Rubisco** (for RuBP carboxylase) is the enzyme's name. Its action produces an unstable six-carbon intermediate that splits into two molecules of **PGA** (phosphoglycerate). PGA is a stable molecule with a three-carbon backbone. Incorporating a carbon atom from CO_2 into a stable organic compound is called **carbon fixation**. Quite simply, food can't be produced without this first step of the Calvin–Benson cycle.

The cycle yields phosphorylated glucose, and it also regenerates the RuBP. For our purposes, we can focus just on the carbon atoms of the substrates, intermediates, and end products, as in Figure 6.14.

How Do Plants Build Glucose?

Each PGA accepts a phosphate group from ATP, plus hydrogen and electrons from NADPH. The resulting intermediate is called **PGAL** (phosphoglyceraldehyde). To build *one* six-carbon sugar phosphate, carbon atoms from six CO_2 must be fixed, and twelve PGAL must form. Most of the PGAL becomes rearranged to form new RuBP, which can be used to fix more carbon. But two of the PGAL combine, thus forming glucose with a phosphate group attached to its six-carbon backbone.

When phosphorylated this way, glucose is primed to enter other reactions. Plants use it as a building block for their main carbohydrates—sucrose, cellulose, and starch. *The synthesis of these organic compounds by other pathways marks the end of the light-independent reactions.*

With six turns of the cycle, enough RuBP molecules form to replace the ones used in carbon fixation. The ADP, NADP+, and phosphate remaining diffuse through the stroma, back to the light-dependent reaction sites. There they may be converted again to NADPH and ATP.

Photosynthetic cells convert phosphorylated glucose to sucrose or starch during daylight hours. Sucrose is

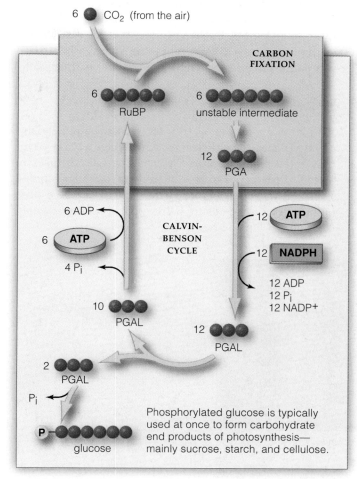

Figure 6.14 Light-independent reactions of photosynthesis. *Red* circles signify carbon atoms of key molecules. All intermediates have one or two phosphate groups attached; for clarity, we show only the phosphate group on glucose. Many water molecules that formed in the light-dependent reactions enter this pathway; six remain when it's over. Appendix V (Figure C) shows details.

the most readily transportable carbohydrate in plants. Starch is the most common storage form. The cells also convert excess PGAL to starch. They briefly store starch (as starch grains) inside the stroma. After the sun goes down, they convert starch to sucrose for export to other living cells in leaves, stems, and roots. Ultimately, the products and intermediates of photosynthesis end up as energy sources and as building blocks for all of the lipids, amino acids, and other organic compounds that plants require for growth, survival, and reproduction.

In the light-independent reactions (the Calvin–Benson cycle), carbon is "captured" from carbon dioxide, glucose forms during reactions that require ATP and NADPH, and RuBP (necessary to capture the carbon) is regenerated.

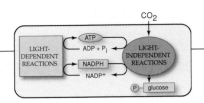

FIXING CARBON—SO NEAR, YET SO FAR

If sunlight intensity, air temperature, rainfall, and soil composition never varied, photosynthesis might be the same in all plants. But environments differ—so details of photosynthesis differ. Comparing a few carbon-fixing adaptations to stressful environments makes this point.

C4 Plants

All plants must take up CO_2 for growth. But CO_2 isn't always plentiful inside leaves, which have a waxy cover that helps plants conserve water. The CO_2 diffuses in and O_2 diffuses out mainly at **stomata** (singular, stoma). These microscopic openings across the leaf surface close on hot, dry days. Closure helps a plant conserve water,

but then CO_2 can't get into leaves. If photosynthetic cells are busy, oxygen accumulates. A high O_2 concentration in a leaf triggers *photorespiration*. This process wastes fixed CO_2 and lowers the plant's sugar-making capacity. Remember rubisco, the enzyme that attaches carbon to RuBP in the Calvin–Benson cycle? Well, rubisco also can attach *oxygen* to it. When CO_2 gets used up and O_2 builds up in the leaf because it can't diffuse out, only one (not two) PGA forms. So does one glycolate molecule, which is later degraded.

The many **C3 plants** include sunflowers and beans. "C3" refers to *three*-carbon PGA, the first intermediate of its carbon-fixing pathway (Figure 6.15a). The first one to form when corn and many other plants fix carbon is *four*-carbon oxaloacetate. Hence *their* name, **C4 plants**.

When C4 plants close stomata on hot, dry days, they still maintain enough CO_2 in leaves by *fixing carbon twice*, in two types of photosynthetic cells. Mesophyll cells, the first type, use CO_2 to form oxaloacetate—which is transferred to bundle-sheath cells at leaf veins (Figure

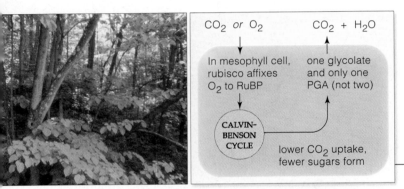

a C3 plants. With low CO_2 / high O_2, photorespiration predominates.

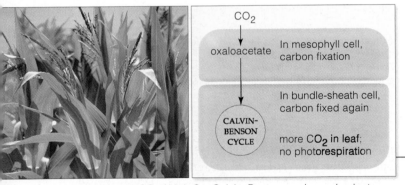

b C4 plants. With low CO_2 / high O_2, Calvin–Benson cycle predominates.

c CAM plants. With low CO_2 / high O_2, Calvin–Benson cycle predominates.

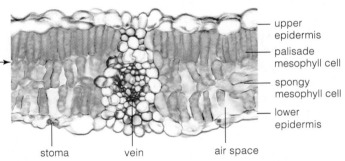

Leaf of basswood (*Tilia*), a typical C3 plant, cross-section

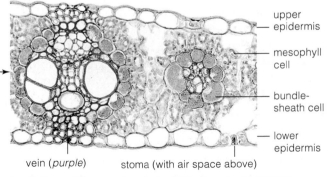

Leaf of corn (*Zea mays*), a typical C4 plant, cross-section

Figure 6.15 Three ways of fixing carbon in hot, dry weather, when there is too little CO_2 and too much O_2 in leaves. (**a**) The C3 pathway is common in evergreens and many nonwoody plants of temperate zones, such as basswood and Kentucky bluegrass. (**b**) The C4 pathway is common in grasses and other plants that evolved in the tropics and that fix CO_2 twice. Corn, sugarcane, and crabgrass are examples. (**c**) CAM plants open stomata and fix carbon at night. They include orchids, pineapples, and cacti and many other succulents.

6.15b). There, CO_2 is released and its local concentration rises, so rubisco doesn't use oxygen. In mesophyll cells, CO_2 gets fixed again in the Calvin–Benson cycle. With this pathway, C4 plants can get by with small stomata, lose less water, and make more glucose than C3 plants can when conditions are hot, bright, and dry.

Experiments show that photorespiration lowers the photosynthetic efficiency of many C3 plants. Example: Grow hothouse tomatoes at CO_2 levels high enough to block photorespiration and you see a dramatic rise in growth rates. So why hasn't natural selection eliminated this wasteful process? Look to rubisco, the enzyme that starts the process. This enzyme evolved long ago, when the atmosphere had little O_2 and abundant CO_2. Maybe some gene coding for its structure can't mutate without having bad effects on carbon-fixing activity. Or maybe the pathway that degrades glycolate to carbon dioxide has proved to be so adaptive that it can't be eliminated. After all, glycolate is toxic at high concentrations.

The C4 pathway evolved separately in many lineages of flowering plants over the past 50 to 60 million years. Before then, when atmospheric CO_2 levels were higher, C3 plants had the selective advantage. Which pathway will be most adaptive in the future? Atmospheric CO_2 levels have been increasing for decades and may double in the next fifty years. If so, photorespiration will again decline—and many vital crop plants may benefit.

CAM Plants

We see a carbon-fixing adaptation to desert conditions in cacti. A cactus is one of the succulents; it has juicy, water-storing tissues and thick surface layers that restrict water loss. It cannot open stomata on hot days without losing precious water. It opens them and fixes CO_2 *at night*. Its cells store the resulting intermediate in their central vacuoles, then use it for photosynthesis the next day when stomata are closed. Many plants are adapted this way. They are **CAM plants** (short for *Crassulacean Acid Metabolism*). Unlike C4 species, CAM plants do not fix carbon twice in different types of cells. They fix it in the same cells but at different times (Figure 6.15c).

Many plants die during prolonged droughts, but some CAM plants survive by closing stomata even at night. They keep fixing the CO_2 from aerobic respiration. Not much forms, but it's enough to maintain low metabolic rates. CAM plants do grow slowly. Try growing a cactus plant in Seattle or other places with a mild climate, and it will compete poorly with C3 and C4 plants.

C4 plants and CAM plants both have modified ways of fixing carbon for photosynthesis. The modifications counter stresses imposed by hot, dry conditions in their environments.

Light in the Deep Dark Sea

During the late 1980s a graduate student, Cindy Lee Van Dover, was puzzling over an eyeless shrimp. Divers had discovered it at a hydrothermal vent deep in the Atlantic Ocean. **Hydrothermal vents** are fissures in the seafloor where molten rock rises and mixes with cold seawater. Mineral-rich water, superheated by 650 degrees, spews into the perpetually dark surroundings. The sides of the vents teem with life, including shrimps and bacteria.

When Van Dover saw videotapes of shrimps clinging to the vent, she noticed a pair of bright strips on their back. By studying specimens in the lab, she found that the strips are connected to a nerve. Could eyeless shrimps have a novel sensory organ?

EYELESS SHRIMP

Steven Chamberlain, a neuroscientist, confirmed the strips are a sensory organ with light-absorbing pigmented cells (photoreceptors). Later, Ete Szuts, a pigment expert, isolated the shrimp pigment. Its absorption spectrum is the same as that for rhodopsin, a visual pigment in eyes as complex as yours.

Van Dover had a hunch: If the strips serve as an eye, then we should find *light* on the seafloor. On later dives, special cameras confirmed her hypothesis. Like coils in a toaster oven, vents release heat (infrared radiation). They also release faint radiation at the low end of the visible spectrum—light up to nineteen times brighter than expected for infrared. A billion billion photons per square inch per second strike typical sunlit trees. Studies of photoautotrophic bacteria living 72 meters (240 feet) below the Black Sea's surface show that they encounter a trillion photons per square inch per second. *Just as many photons are available to hydrothermal vent organisms.*

About this time Van Dover casually asked a colleague, "Hey, what if there's enough light for photosynthesis?" The response was, "What a stupid idea."

Euan Nisbet, who studies ancient environments, thought about this. The first cells arose around 3.8 billion years ago. As Nisbet and Van Dover hypothesized, What if they arose at hydrothermal vents? They could have used inorganic compounds such as hydrogen sulfide (for their hydrogen and electrons) and carbon dioxide (for carbon). Survival probably depended on being able to move away from light at vents and thereby avoid being boiled alive. Millions of years later, some bacterial descendants were evolving in hot springs near the ocean's surface. In time they used their light-detecting machinery to absorb light from the sun. They were adapted to a new energy source; they had become photosynthetic.

Some observations support the intriguing hypothesis that light-sensing machinery of deep-sea bacteria became modified for shallow-water photosynthesis. A few clues: Absorption spectra for the chlorophyll in evolutionarily ancient photosynthetic bacteria correspond to the light measured at hydrothermal vents. Also, photosynthetic machinery contains iron, sulfur, manganese, and other minerals—which are abundant at hydrothermal vents.

Autotrophs, Humans, and the Biosphere

We conclude this chapter with a story that reinforces how photosynthesizers and other autotrophs fit in the world of living things. It is a story of mind-boggling numbers of single-celled and multicelled species on land and in the sunlit waters of the Earth.

Each spring, you sense renewed growth of autotrophs on land as leaves unfurl and lawns and fields turn green. You might not be aware that uncountable numbers of single-celled autotrophs also drift through the surface waters of the world ocean. You can't see them without a microscope; a row of 7 million cells of one aquatic species would be less than a quarter-inch long. In some regions, a cupful of seawater might hold 24 million cells of one species. And that wouldn't include cells of all the other aquatic species.

a Photosynthetic activity in winter.

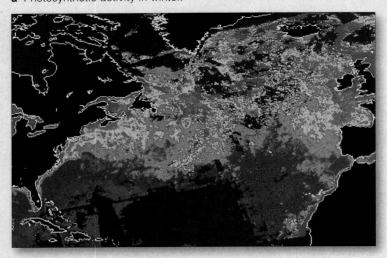

b Photosynthetic activity in spring.

Figure 6.16 Two satellite images that help convey the magnitude of photosynthetic activity during springtime in the North Atlantic portion of the world ocean. In these color-enhanced images, *red-orange* shows where chlorophyll is most concentrated.

Most of the drifters are photoautotrophic prokaryotic cells and protistans. Together they are the "pastures of the seas," producers that ultimately feed most other marine organisms. The pastures "bloom" in the spring, when nutrient inputs sustain rapid reproduction. At that time seawater becomes warmer and enriched with nutrients that winter currents churn up from the deep.

Until NASA gathered data from satellites in space, we had no idea of their numbers and distribution. Figure 6.16a shows visual evidence of their activities one winter in the surface waters of the North Atlantic Ocean. Figure 6.16b shows a springtime bloom stretching from North Carolina all the way past Spain!

Collectively, the cells help shape the global climate, for they deal with staggering numbers of reactant and product molecules. For instance, they sponge up nearly half the carbon dioxide that we humans release each year, as when we burn fossil fuels or burn vast tracts of forests to clear land for farming. Without the aquatic photoautotrophs, atmospheric carbon dioxide would accumulate more rapidly and possibly contribute to global warming, as described in Section 43.10. If the atmosphere warms by only a few degrees, then sea levels will rise and all lowlands near the coasts of continents and islands will become submerged.

Although such global change is a real possibility, tons of industrial wastes, raw sewage, and fertilizers in runoff from croplands still drain into the ocean each day. The pollutants seriously alter the chemical composition of seawater. How long will the marine photoautotrophs be able to function in this chemical brew? The answer will have impact on your own life in more ways than one.

Other autotrophs also influence your life in ways you might not expect. At hydrothermal vents, in hot springs, in waste heaps of coal mines are diverse prokaryotic cells classified as *chemo*autotrophs. They, too, use carbon dioxide as a carbon source. But they use inorganic compounds in their environment as energy sources.

For example, some archaebacteria living on the sides of hydrothermal vents strip hydrogen and electrons from hydrogen sulfide. Other kinds of chemoautotrophs living in soil obtain energy from nitrogen-containing wastes and remains of animals and other organisms.

Although they, too, are microscopically small, the chemoautotrophs also exist in monumental numbers. They influence the cycling of nitrogen, phosphorus, and other elements through the biosphere. We will consider their impact on the environment in later chapters. In this unit, we turn next to pathways by which cells can release the chemical bond energy of glucose and other biological molecules—the chemical legacy of autotrophs everywhere.

SUMMARY

1. Cell structure and functions are based on organic compounds, the synthesis of which depends on sources of carbon and energy. Plants and other autotrophs get carbon from carbon dioxide and energy from the sun or inorganic compounds. Animals and other heterotrophs aren't self-nourishing; they get carbon and energy from organic compounds synthesized by autotrophs. *CI*

2. Photosynthesis is the main process by which carbon and energy enter the web of life. Its two stages are the light-dependent and light-independent reactions. This equation and Figure 6.17 summarize the process: *6.1*

$$12H_2O + 6CO_2 \xrightarrow{\text{LIGHT ENERGY}} 6O_2 + C_6H_{12}O_6 + 6H_2O$$

WATER CARBON   OXYGEN GLUCOSE WATER
       DIOXIDE

3. Chloroplasts are organelles of photosynthesis in algae and plants. Two outer membranes enclose its semifluid interior (stroma). A third membrane in the stroma forms a system of connected sacs (thylakoids) and channels. The light-dependent reactions occur at thylakoids. The light-independent reactions occur in the stroma. *6.1*

4. The light-dependent reactions start at photosystems that each include 200 to 300 pigments. *6.2–6.4*

   a. Chlorophyll *a*, the main photosynthetic pigment, absorbs all wavelengths of visible light but only a little of the yellow-green and green ones. Accessory pigments (e.g., carotenoids) absorb other wavelengths.

   b. A cyclic pathway forms ATP only. Photosystem I gives up excited electrons to a transport system, which sends the "spent" ones back to the photosystem.

   c. The noncyclic pathway forms ATP and NADPH. Water molecules are split into hydrogen, electrons, and oxygen. Excited electrons enter photosystem II, then a transport system, photosystem I, and another transport system. NADP$^+$ picks up the electrons and hydrogen to form NADPH. By 1.5 billion years ago, oxygen released by this pathway had accumulated in the atmosphere. It ultimately made possible aerobic respiration.

5. ATP energy drives the light-independent reactions (by phosphate-group transfers). Hydrogen and electrons from NADPH, plus carbon and oxygen (from carbon dioxide), are used as building blocks for glucose. Most glucose is used at once to synthesize starch, cellulose, and other end products of photosynthesis. *6.5*

   a. Cyclic reactions (Calvin–Benson cycle) start when rubisco, an enzyme, affixes carbon from CO_2 to the five-carbon RuBP. The resulting intermediate splits into two PGA that ATP phosphorylates. Hydrogen and electrons from NADPH help form two PGAL.

   b. For every six carbon atoms that enter the cycle, twelve PGAL form. Two are used to form a six-carbon sugar phosphate. The rest are used to regenerate RuBP.

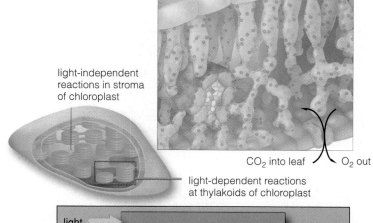

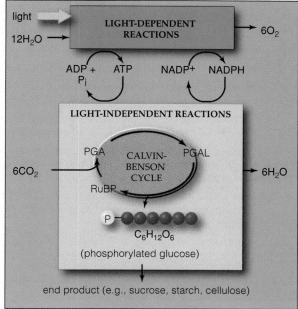

Figure 6.17 Summary of photosynthesis. In the light-dependent reactions, sunlight energy is converted to chemical bond energy of ATP. In a noncyclic pathway, water molecules are split by photolysis. NADP$^+$ conserves the released electrons and hydrogen (as NADPH). Oxygen is released as a by-product.

ATP energy drives the light-independent reactions. Hydrogen and electrons from NADPH, plus carbon and oxygen from carbon dioxide, are used to form glucose. In the Calvin–Benson cycle, the RuBP on which the cycle turns is regenerated, and new water molecules form. *Each turn* of the cycle requires one CO_2, three ATP, and two NADPH. Because each glucose molecule has a backbone of six carbon atoms, its formation requires six turns of the cycle.

6. Rubisco, a carbon-fixing enzyme, evolved when air had far more CO_2 and less O_2. Today, when CO_2 is used up and O_2 builds up in leaves, rubisco attaches oxygen (not carbon) to RuBP. This process, photorespiration, is wasteful: one PGA and glycolate form. Glycolate can't be used to form sugars and is later degraded. *6.6*

7. Photorespiration predominates in C3 plants such as sunflowers under hot, dry conditions. Stomata close, so O_2 from photosynthesis builds up in leaves to levels higher than CO_2 levels. Corn and other C4 plants raise the CO_2 level by fixing carbon twice, in two cell types. CAM plants such as cacti raise the CO_2 level by fixing carbon at night, when stomata are open. *6.6*

Review Questions

1. A cat eats a bird, which earlier ate a caterpillar that chewed on a weed. Which organisms are autotrophs? Which are the heterotrophs? *CI*

2. Summarize the photosynthesis reactions as an equation. Name where each stage takes place inside a chloroplast. *6.1*

3. What is the function of ATP in photosynthesis? What is the function of NADPH? *6.1*

4. Which of the following pigments are most visible in a maple leaf in summer? Which become the most visible in autumn? *6.3*
   a. chlorophylls   c. anthocyanins
   b. phycobilins   d. carotenoids

5. How does chlorophyll *a* differ in function from accessory pigments during the light-dependent reactions? *6.2*

6. Fill in Figure 6.18 for the light-dependent reactions. *6.3*

7. With respect to the light-dependent reactions, how do the cyclic and noncyclic pathways of electron flow differ? *6.3*

8. What substance does *not* take part in the Calvin–Benson cycle: ATP, NADPH, RuBP, carotenoids, O_2, CO_2, or enzymes? *6.5*

9. Fill in the blanks for Figure 6.19. Which substances are the original sources of carbon atoms and hydrogen atoms used in the synthesis of glucose in the Calvin–Benson cycle? *6.1, 6.5*

10. On hot, dry days, oxygen from photosynthesis accumulates in leaves. Explain why, and explain what happens in C3 plants versus C4 plants. *6.6*

11. Are photoautotrophs the only "self feeders"? If not, give examples and where you might find them. *6.7, 6.8*

Self-Quiz   ANSWERS IN APPENDIX III

1. Photosynthetic autotrophs use _____ from the air as a carbon source and _____ as their energy source.

2. In plants, light-dependent reactions proceed at the _____ .
   a. cytoplasm   c. stroma
   b. plasma membrane   d. thylakoid membrane

3. In the light-*dependent* reactions, _____ .
   a. carbon dioxide is fixed
   b. ATP and NADPH form
   c. CO_2 accepts electrons
   d. sugar phosphates form

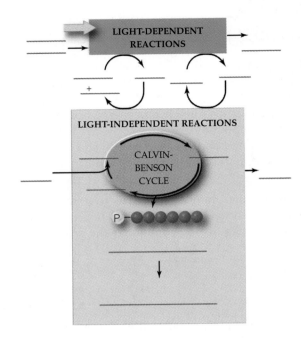

Figure 6.19   Review Figure 6.17. Then, on your own, fill in the blanks (*red* lines) with the names of the key reactants, intermediates, and products of photosynthesis.

4. The light-*independent* reactions proceed in the _____ .
   a. cytoplasm
   b. plasma membrane
   c. stroma

5. When a photosystem absorbs light, _____ .
   a. sugar phosphates are produced
   b. electrons are transferred to ATP
   c. RuBP accepts electrons
   d. light-dependent reactions begin

6. Identify which of the following substances accumulates inside the thylakoid compartment of chloroplasts during the light-dependent reactions:
   a. glucose   c. chlorophyll   e. hydrogen ions
   b. carotenoids   d. fatty acids

Figure 6.18
Review Figure 6.11, which shows the light-dependent reactions of photosynthesis. Then, on your own, fill in the blanks (*red* lines) with the names of key components and activities.

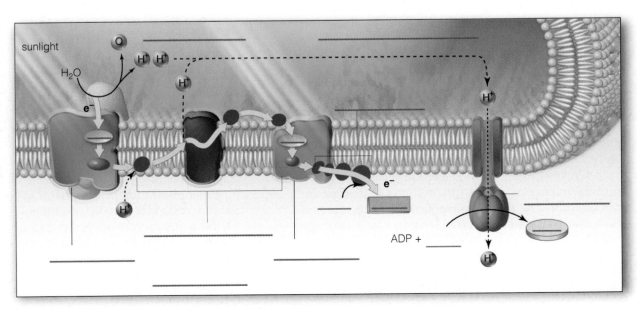

7. The Calvin–Benson cycle starts when _____ .
   a. light is available
   b. light is not available
   c. carbon dioxide is attached to RuBP
   d. electrons leave a photosystem

8. ATP phosphorylates _____ in the light-independent reactions.
   a. RuBP   b. $NADP^+$ c. PGA   d. PGAL

9. Match each event with its most suitable description.
   _____ photon absorption
   _____ NADPH formation
   _____ CO_2 fixation
   _____ PGAL formation
   _____ ATP formation only

   a. rubisco required
   b. ATP, NADPH required
   c. electrons cycled back to photosystem I
   d. energy matches amount required to excite electrons
   e. photolysis required

Critical Thinking

1. Suppose a garden in your neighborhood is filled with red, white, and blue petunias. Explain each of the three floral colors in terms of which wavelengths of light it is absorbing and which wavelengths it is transmitting or reflecting.

2. About 200 years ago, Jan Baptista van Helmont performed an experiment on the nature of photosynthesis. He wanted to know where growing plants acquire the raw materials necessary to increase in size. For his experiment, he planted a tree seedling weighing 5 pounds in a barrel filled with 200 pounds of soil. He watered the tree regularly.

    Five years passed. Van Helmont again weighed the tree and the soil. The tree weighed 169 pounds, 3 ounces. The soil weighed 199 pounds, 14 ounces. Because the tree weight had increased so much and soil weight had decreased so little, he concluded the tree had gained weight after absorbing the water he had added to the barrel.

    Given what you know about the composition of biological molecules, why was van Helmont's conclusion misguided? Reflect on the current model of photosynthesis and then give a more plausible explanation of his results.

3. Melanie is using radioactively labeled carbon atoms ($^{14}CO_2$) in the laboratory. The photosynthesizing plants she is studying take them up. Identify the compound in which the labeled carbon will appear first: NADPH, PGAL, pyruvate, or PGA.

4. The green alga (*Codium*) in Figure 6.20*a* grows in sunlit seawater. Its main pigments absorb red light, and its accessory pigments help harvest photon energy in the sunlit waters. Some of its pigments even function as shields against ultraviolet radiation. Other species of green algae live in freshwater ponds and lakes, just beneath the soil surface, and on rocks. Some survive in snow.

    The red alga in Figure 6.20*b* was found on a tropical reef in clear, warm water. Its phycobilins, which mask the color of chlorophyll, are good at absorbing wavelengths of green and blue-green light. Those wavelengths can penetrate deep water. Some of its relatives live in very deep, dimly lit waters, and their pigments are nearly black.

    The variation in pigments invites a question: If trapping energy is so vital to life, then why doesn't each photosynthetic pigment go after the whole rainbow? In other words, why aren't all pigments black?

    The earliest photoautotrophs evolved in the seas, so their pigments evolved in the seas, also. Speculate on how natural selection may have favored the evolution of different pigments at different depths, starting with the first kinds that may have evolved at hydrothermal vents (Section 6.7).

Figure 6.20 (**a**) A green alga (*Codium*) that grows in coastal waters. (**b**) A red alga from a tropical reef.

5. Fireflies emit light when enzymes called luciferases convert chemical energy to light energy. The reactions start when ATP transfers a phosphate group to a substance called luciferin. Luciferases, with a little help from oxygen, convert activated luciferin to a different molecule. Their action boosts electrons of the molecule to a higher energy level. The excited electrons quickly return to a lower energy level, and they release energy as fluorescent light (Section 6.2). When organisms flash with it, this is called *bioluminescence*. Electrons also get an energy boost in the light-dependent reactions of photosynthesis. So why aren't plants naturally bioluminescent?

Selected Key Terms

autotroph *CI*	PGA *6.5*
C3 plant *6.6*	PGAL *6.5*
C4 plant *6.6*	photolysis *6.3*
Calvin–Benson cycle *6.5*	photon *6.2*
CAM plant *6.6*	photosynthesis *CI*
carbon fixation *6.5*	photosystem *6.2*
carotenoid *6.2*	pigment *6.2*
chlorophyll *6.2*	reaction center *6.3*
chloroplast *6.1*	rubisco *6.5*
electromagnetic spectrum *6.2*	RuBP *6.5*
electron transport system *6.3*	stoma (stomata) *6.6*
heterotroph *CI*	stroma *6.1*
hydrothermal vent *6.7*	thylakoid *6.1*
light-dependent reactions *6.3*	wavelength *6.2*
light-independent reactions *6.5*	

Readings

Bazzaz, F., and E. Jajer. January 1992. "Plant Life in a CO_2-Rich World." *Scientific American*, 266:68–74.

Zimmer, C. November 1996. "The Light at the Bottom of the Sea." *Discover*, 63–73.

HOW CELLS RELEASE STORED ENERGY

The Killers Are Coming! The Killers Are Coming!

Thanks to selective breeding experiments gone wrong, descendants of "killer" bees that flew out of South America a few decades ago buzzed across the border between Mexico and Texas. By 1995, they had invaded 13,287 square kilometers of southern California and were busily setting up colonies. By 1998, when nectar-rich flowers bloomed profusely in the deserts after heavy El Niño storms, they moved even farther west and north than biologists had predicted.

When provoked, the bees are terrifying. To give an example, when a construction worker started a tractor a few hundred yards from a hive, thousands of agitated bees flew into action. They entered a nearby subway station, where they stung passengers on the platform and in the trains. They killed one person and injured a hundred more.

Where did the bees come from? In the 1950s, some queen bees were shipped over from Africa to Brazil for breeding experiments. Why? Honeybees are big business. They are the source of nutritious honey. Also, they are rented to commercial orchards, where their collective pollinating activities may significantly enhance fruit production. Imagine yourself enclosing an orchard tree in a cage that keeps out all pollinators. Without pollinators to spread pollen about, less than 1 percent of that tree's flowers will set fruit. Now imagine putting a hive of honeybees inside the cage with the tree. As many as 40 percent of the flowers will set fruit.

Compared to their relatives in Africa, the bees of Brazil are sluggish pollinators and honey producers. By cross-breeding the two strains, researchers thought they might come up with a strain of mild-mannered but zippier bees. So they put some local bees and imported ones together inside netted enclosures, complete with artificial hives. Then they let nature take its course.

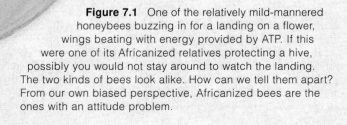

Figure 7.1 One of the relatively mild-mannered honeybees buzzing in for a landing on a flower, wings beating with energy provided by ATP. If this were one of its Africanized relatives protecting a hive, possibly you would not stay around to watch the landing. The two kinds of bees look alike. How can we tell them apart? From our own biased perspective, Africanized bees are the ones with an attitude problem.

Twenty-six African queen bees escaped. That was bad enough. Then beekeepers got wind of preliminary experimental results. After learning that the first few generations of offspring were more energetic but not overly aggressive, they imported hundreds of African queens and encouraged them to mate with the locals. And they set off a genetic time bomb.

Before long, African bees became established in commercial hives—and in wild bee populations. Their traits became dominant. The "Africanized" bees do everything other bees do, but they do more of it faster. Their eggs develop into adults more quickly. Adults fly more rapidly, outcompete other bees for nectar, and even die sooner.

When something disturbs their hives or swarms, Africanized bees become extremely agitated. They can remain that way for as long as eight hours. Whereas a mild-mannered honeybee might chase an intruding animal fifty yards or so, a squadron of Africanized bees will chase it a quarter of a mile. If they catch up with it, they collectively can sting it to death.

Doing things faster means having a continuous supply of energy and efficient ways of using it. An Africanized bee's stomach can hold thirty milligrams of sugar-rich nectar—which is enough fuel to fly sixty kilometers. That's more than thirty-five miles! Besides this, compared to other kinds of bees, an Africanized bee's flight muscle cells have larger mitochondria. These organelles specialize in releasing a great deal of energy from sugars and other organic compounds, then converting it to the energy of ATP.

Whenever they tap into the stored energy of organic compounds, Africanized bees reveal their biochemical connection with other organisms. Study a primrose or puppy, a mold growing on stale bread, an amoeba in pondwater, or a bacterium living on your skin, and you will discover that their energy-releasing pathways differ in some details. But all of the pathways require characteristic starting materials. They yield predictable products and by-products. And they yield the universal energy currency of life—ATP.

In fact, throughout the biosphere, organisms put energy and raw materials to use in amazingly similar ways. *At the biochemical level, we find undeniable unity among all forms of life.* We will return to this idea in the concluding section of the chapter.

Key Concepts

1. The cells of all organisms can release energy stored in glucose and other organic compounds, then use it in ATP production. The energy-releasing pathways differ from one another. But the main types all start with the breakdown of glucose to pyruvate.

2. The initial breakdown reactions, known as glycolysis, can proceed in the presence of oxygen or in its absence. Said another way, glycolysis can be the first stage of either aerobic or anaerobic pathways.

3. Two kinds of energy-releasing pathways are entirely anaerobic, from start to finish. We call them fermentation and anaerobic electron transport. They proceed only in the cytoplasm, and none yields more than a small amount of ATP for each glucose molecule metabolized.

4. Another pathway, aerobic respiration, also starts in the cytoplasm. But this one runs to completion in organelles called mitochondria. Compared to the other pathways, aerobic respiration releases far more energy from glucose.

5. Aerobic respiration has three stages. First, pyruvate forms from glucose (during glycolysis). Second, different reactions break down the pyruvate to carbon dioxide. These reactions liberate electrons and hydrogen, which coenzymes deliver to electron transport systems. Third, stepwise electron transfers through the systems help set up the conditions that favor ATP formation. Free oxygen accepts the electrons at the end of the line and combines with hydrogen, thereby forming water.

6. Over evolutionary time, photosynthesis and aerobic respiration became linked on a global scale. The oxygen-rich atmosphere, a long-term outcome of photosynthetic activity, sustains aerobic respiration, which has become the dominant energy-releasing pathway. And most kinds of photosynthesizers use carbon dioxide and water from aerobic respiration as raw materials when they synthesize organic compounds:

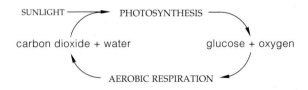

7.1

HOW DO CELLS MAKE ATP?

Organisms stay alive by taking in energy. Plants and all other organisms that engage in photosynthesis get energy from the sun. Animals get energy secondhand, thirdhand, and so on, by eating plants and one another. Regardless of its source, energy must be in a form that can drive thousands of life-sustaining reactions. Energy that becomes converted into the chemical bond energy of adenosine triphosphate—ATP—serves that function.

Plants make ATP during photosynthesis and use it to make glucose and other carbohydrates. *But they and every other organism also produce ATP by breaking down glucose and other carbohydrates, lipids, and proteins.* The breakdown reactions strip electrons and hydrogen from intermediates. Transfers of electrons through transport systems, of the sort described in Section 5.2, are central to the energy-releasing pathways.

Comparison of the Main Types of Energy-Releasing Pathways

The first energy-releasing pathways evolved 3.8 billion years ago, when conditions were different on Earth. The atmosphere had little free oxygen, so the pathways must have been *anaerobic*; they could proceed without using oxygen. Many prokaryotic cells and protistans still live in places where oxygen is absent or not always available. To make ATP, they use anaerobic pathways (fermentation or anaerobic electron transport). Some cells in your body use an anaerobic route for short periods, but only when they aren't getting enough oxygen. Your cells, like most others, rely more on **aerobic respiration**, an oxygen-dependent pathway of ATP formation. Each breath you take provides your actively respiring cells with a fresh supply of oxygen.

Make note of this point: *The main energy-releasing pathways all start with the same reactions in the cytoplasm.* During this initial stage of reactions, called **glycolysis**, enzymes cleave and rearrange a glucose molecule into two molecules of **pyruvate**, an organic compound with a backbone of three carbon atoms. Once glycolysis is over, the energy-releasing pathways differ. Most importantly, only the aerobic pathway continues in a mitochondrion (Figure 7.2). There, oxygen serves as the final acceptor

of electrons used during the reactions. The anaerobic pathways start and end in the cytoplasm. A substance other than oxygen is the final electron acceptor.

As you examine the energy-releasing pathways in sections to follow, keep in mind that the reaction steps do not proceed by themselves. Enzymes catalyze each step, and the intermediate molecules formed at one step serve as substrates for the next enzyme in the pathway.

Overview of Aerobic Respiration

Of all energy-releasing pathways, aerobic respiration gets the most ATP for each glucose molecule. Whereas anaerobic routes typically have a net yield of two ATP molecules, the aerobic route commonly yields thirty-six or more. If you were a bacterium, you wouldn't require much ATP. Being far larger, more complex, and highly active, you rely absolutely on the aerobic route's high yield. When a glucose molecule is the starting material, aerobic respiration can be summarized this way:

$$C_6H_{12}O_6 + 6O_2 \longrightarrow 6CO_2 + 6H_2O$$

GLUCOSE   OXYGEN   CARBON DIOXIDE   WATER

However, as you can see, the summary equation only tells us what the substances are at the start and finish of the pathway. In between are three reaction stages.

Figure 7.2 Where the aerobic and anaerobic pathways of ATP formation start and finish up.

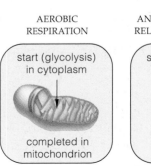

AEROBIC RESPIRATION

start (glycolysis) in cytoplasm

completed in mitochondrion

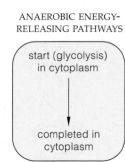

ANAEROBIC ENERGY-RELEASING PATHWAYS

start (glycolysis) in cytoplasm

completed in cytoplasm

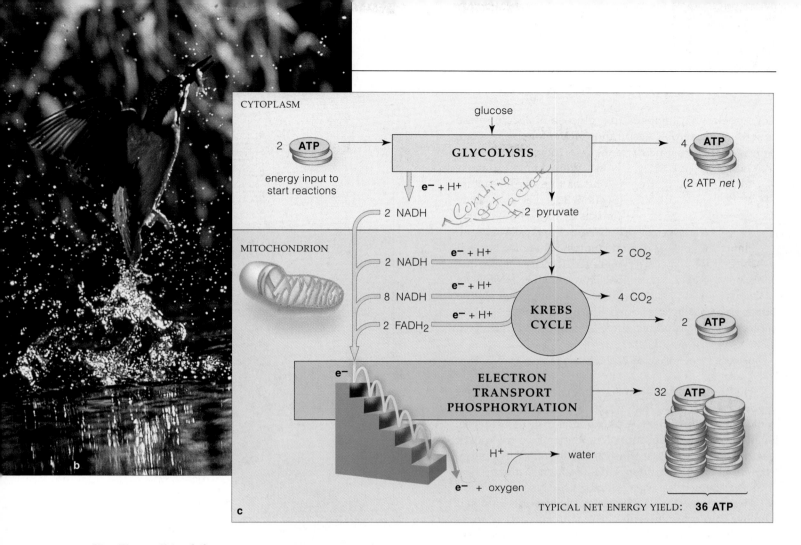

CYTOPLASM

glucose

2 ATP
energy input to start reactions

GLYCOLYSIS

4 ATP
(2 ATP *net*)

e^- + H+

2 NADH

~~Combine get lactose~~

2 pyruvate

MITOCHONDRION

2 NADH e^- + H+ → 2 CO_2

8 NADH e^- + H+ → 4 CO_2

2 $FADH_2$ e^- + H+

KREBS CYCLE

2 ATP

e^-

ELECTRON TRANSPORT PHOSPHORYLATION

32 ATP

H+ → water

e^- + oxygen

TYPICAL NET ENERGY YIELD: **36 ATP**

Figure 7.3 Overview of aerobic respiration, which has three stages. From start to finish, the typical net energy yield from each glucose molecule is thirty-six ATP. **(a,b)** Only this pathway delivers enough ATP to construct and maintain redwoods and all other large, multicelled organisms. It alone delivers enough ATP for highly active animals, including bees, humans, and kingfishers and other birds.

(c) In the first stage (glycolysis), enzymes partially break down glucose to pyruvate. In the second stage, which is mainly the Krebs cycle, enzymes completely break down pyruvate to carbon dioxide. NAD+ and FAD pick up electrons and hydrogen stripped from intermediates at both stages. In the final stage (electron transport phosphorylation), the reduced coenzymes (NADH and $FADH_2$) give up electrons and hydrogen to transport systems. Energy released during the flow of electrons through the systems drives ATP formation. Oxygen accepts the electrons at the end of the third stage.

Use Figure 7.3 while we outline the reactions. Again, glycolysis is the first stage. The second stage is mainly a cyclic pathway, the **Krebs cycle**. Enzymes break down pyruvate to carbon dioxide and water, thereby releasing electrons and hydrogen.

NAD+ (for nicotinamide adenine dinucleotide) and **FAD** (flavin adenine dinucleotide) function in glycolysis and the Krebs cycle. Remember, these coenzymes assist enzymes by accepting electrons and hydrogen stripped from intermediates. The unbound hydrogen atoms are naked protons, or hydrogen ions (H+). They tag along with the oppositely charged electrons. In reduced form, the coenzymes are abbreviated NADH and $FADH_2$.

Little ATP forms in glycolysis or the Krebs cycle. The large energy harvest comes in the third stage, after the coenzymes give up the electrons and hydrogen to many electron transport systems. These systems are the machinery for **electron transport phosphorylation**. It sets up H+ concentration and electric gradients, which drive ATP formation at nearby membrane proteins. It is during this final stage that so many ATP molecules are produced. As it ends, oxygen inside the mitochondrion accepts the "spent" electrons from the last component of each transport system. Oxygen picks up hydrogen at the same time and thereby forms water.

Cells drive nearly all metabolic activities by releasing energy from glucose and other organic compounds and converting it to chemical bond energy of ATP. All of the main energy-releasing pathways start inside the cytoplasm with glycolysis, a stage of reactions that break down glucose to pyruvate.

The most common anaerobic pathways, which include the fermentation routes, end in the cytoplasm. Each has a net energy yield of two ATP.

Aerobic respiration, an oxygen-dependent pathway, runs to completion in the mitochondrion. From start (glycolysis) to finish, it commonly has a net energy yield of thirty-six ATP.

GLYCOLYSIS: FIRST STAGE OF ENERGY-RELEASING PATHWAYS

Let's track what happens to a glucose molecule in the first stage of aerobic respiration. Remember, the same things happen to glucose in the anaerobic pathways.

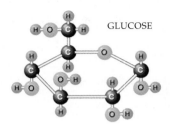

GLUCOSE

As you know, glucose is one of the simple sugars (Section 3.2). Each molecule consists of six carbon, twelve hydrogen, and six oxygen atoms, all covalently bonded together. Its carbons are the backbone. In the cytoplasm, glucose or some other simple sugar is partly broken down during glycolysis to two molecules of pyruvate, a three-carbon compound:

glucose ⟶ glucose–6–phosphate ⟶ 2 pyruvate

The first steps of glycolysis are *energy-requiring*. As Figure 7.4 indicates, they proceed only when two ATP molecules each transfer a phosphate group to glucose. Such phosphate-group transfers, remember, are known as phosphorylations. In this case, they raise the energy content of glucose to a level high enough to allow entry into the *energy-releasing* steps of glycolysis.

The first energy-releasing step cleaves the activated glucose into two molecules. We can call each of them PGAL (phosphoglyceraldehyde). Each PGAL becomes converted to an unstable intermediate that gives up a phosphate group to ADP, and so ATP forms. The next intermediate in the sequence does the same thing.

Thus, a total of four ATP form by **substrate-level phosphorylation**. We define this metabolic event as the direct transfer of a phosphate group from a substrate of a reaction to some other molecule—in this case, ADP. Remember, though, two ATP were invested to jump-start the reactions. So the *net* energy yield is two ATP.

Meanwhile, the coenzyme NAD$^+$ picks up electrons and hydrogen atoms liberated from each PGAL, thus becoming NADH. When NADH delivers its cargo to a different reaction site, it reverts to NAD$^+$. Said another way, NAD$^+$—like other coenzymes—is reusable.

In sum, glycolysis converts energy stored in glucose to a transportable form of energy, in ATP. NAD$^+$ picks up electrons and hydrogen stripped from glucose. These have roles in the next stage of reactions. So do the end products of glycolysis—the two molecules of pyruvate.

Glycolysis is a stage of reactions that partially break down glucose or some other carbohydrate to two molecules of pyruvate, thereby releasing energy.

Two NADH and four ATP form. However, when we subtract the two ATP required to start the reactions, the *net* energy yield of glycolysis is two ATP per glucose molecule.

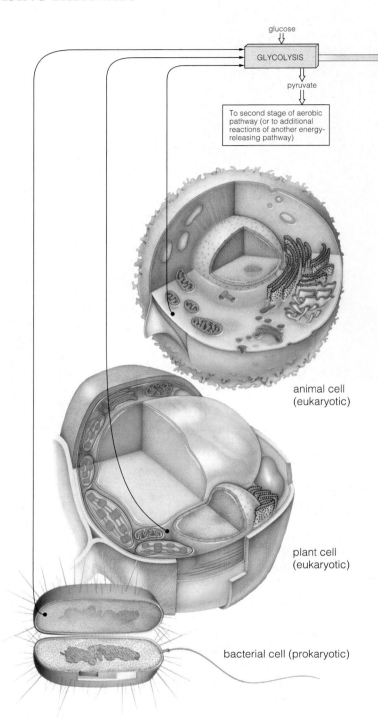

glucose

GLYCOLYSIS

pyruvate

To second stage of aerobic pathway (or to additional reactions of another energy-releasing pathway)

animal cell (eukaryotic)

plant cell (eukaryotic)

bacterial cell (prokaryotic)

Figure 7.4 Glycolysis, the first stage of the main energy-releasing pathways. All prokaryotic and eukaryotic cells use glycolysis, which occurs in the cytoplasm. In this example, glucose is the starting material. Two pyruvate, two NADH, and four ATP form. Because cells invest two ATP to start glycolysis, the *net* energy yield is two ATP. Appendix V (Figure A) gives more details.

Depending on the type of cell and environmental conditions, the pyruvate may enter the second set of reactions of the aerobic pathway, including the Krebs cycle. Or it may be used in other reactions, such as those of fermentation.

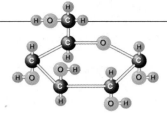

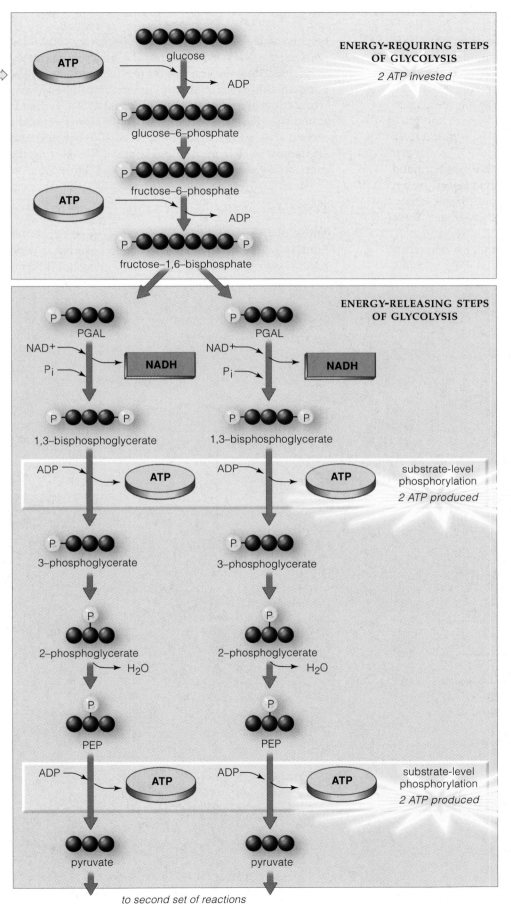

ENERGY-REQUIRING STEPS OF GLYCOLYSIS

2 ATP invested

ATP → ADP

glucose

glucose–6–phosphate

fructose–6–phosphate

ATP → ADP

fructose–1,6–bisphosphate

ENERGY-RELEASING STEPS OF GLYCOLYSIS

PGAL

NAD+
P_i → NADH

1,3–bisphosphoglycerate

ADP → ATP

substrate-level phosphorylation
2 ATP produced

3–phosphoglycerate

2–phosphoglycerate → H_2O

PEP

ADP → ATP

substrate-level phosphorylation
2 ATP produced

pyruvate

PGAL

NAD+
P_i → NADH

1,3–bisphosphoglycerate

ADP → ATP

3–phosphoglycerate

2–phosphoglycerate → H_2O

PEP

ADP → ATP

pyruvate

to second set of reactions

a This diagram tracks the six carbon atoms (*red* spheres) of the glucose molecule shown above. Glycolysis starts with an energy investment of two ATP.

b One ATP makes a phosphate-group transfer to glucose, and some atoms in glucose are rearranged in response.

c A phosphate-group transfer from the second ATP causes rearrangements that form fructose–1,6–bisphosphate. This intermediate can be split easily.

d It splits at once into two molecules, each with a three-carbon backbone. We can call these two PGAL.

e Two NADH form when each PGAL gives up two electrons and a hydrogen atom to NAD+.

f One ATP forms as each PGAL also combines with inorganic phosphate (P_i) and transfers a phosphate group to ADP.

g *Thus, two ATP formed by the direct transfer of a phosphate group from two intermediates of the reactions.* The original energy investment of two ATP has been paid off.

h In the next two reactions, the two intermediates each release a hydrogen atom and an —OH group, which then combine to form water.

i Two 3–phosphoenolpyruvate (PEP) molecules result. Each PEP transfers a phosphate group to ADP.

j *Once again, two ATP have formed by substrate-level phosphorylation.*

k In sum, the net energy yield from glycolysis is two ATP for each glucose molecule entering the reactions. Two molecules of pyruvate, the end product, may enter the next set of reactions in an energy-releasing pathway.

SECOND STAGE OF THE AEROBIC PATHWAY

Suppose two pyruvate molecules formed by glycolysis leave the cytoplasm and enter a **mitochondrion** (plural, mitochondria). In this organelle, the second and third stages of the aerobic pathway are completed. Figure 7.5 shows its structure and functional zones.

Preparatory Steps and the Krebs Cycle

It is during the second stage that the glucose is finally broken down completely to carbon dioxide and water. Two ATP form, but the most striking part of the second stage is the transfer of electrons and hydrogen from intermediates to many coenzymes.

In a few preparatory reactions, an enzyme removes a carbon atom from each pyruvate molecule. An enzyme helper, coenzyme A, becomes **acetyl–CoA** by combining with the two-carbon fragment. The fragment is passed to **oxaloacetate**, the entry point for the Krebs cycle. The name of this cyclic pathway honors Hans Krebs, who worked out many of its details in the 1930s. It also is called the citric acid cycle. *Six* carbon atoms enter the second stage of reactions (three in each pyruvate). And *six* depart, in six carbon dioxide molecules, during the preparatory steps and the cycle proper (Figure 7.6).

Functions of the Second Stage

Think of the second stage as having three functions. First, it loads electrons and hydrogen onto NAD^+ and FAD, resulting in NADH and $FADH_2$. Second, it forms

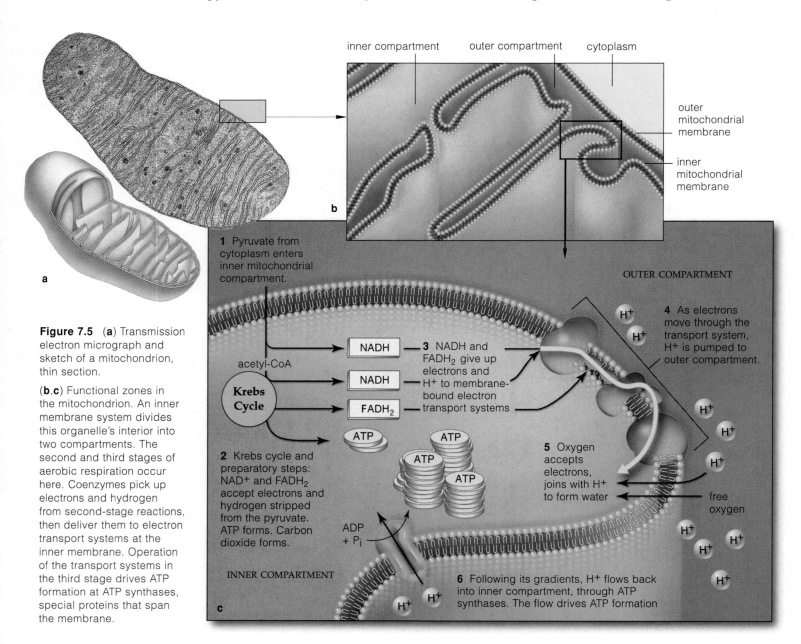

Figure 7.5 (**a**) Transmission electron micrograph and sketch of a mitochondrion, thin section.

(**b,c**) Functional zones in the mitochondrion. An inner membrane system divides this organelle's interior into two compartments. The second and third stages of aerobic respiration occur here. Coenzymes pick up electrons and hydrogen from second-stage reactions, then deliver them to electron transport systems at the inner membrane. Operation of the transport systems in the third stage drives ATP formation at ATP synthases, special proteins that span the membrane.

inner compartment   outer compartment   cytoplasm

outer mitochondrial membrane

inner mitochondrial membrane

b

a

1 Pyruvate from cytoplasm enters inner mitochondrial compartment.

OUTER COMPARTMENT

acetyl-CoA

Krebs Cycle

NADH

NADH

$FADH_2$

3 NADH and $FADH_2$ give up electrons and H^+ to membrane-bound electron transport systems

4 As electrons move through the transport system, H^+ is pumped to outer compartment.

H^+

H^+

ATP

ATP

ATP

ATP

ATP

2 Krebs cycle and preparatory steps: NAD^+ and $FADH_2$ accept electrons and hydrogen stripped from the pyruvate. ATP forms. Carbon dioxide forms.

5 Oxygen accepts electrons, joins with H^+ to form water

H^+

H^+

free oxygen

H^+

ADP + P_i

H^+ H^+

INNER COMPARTMENT

c

H^+ H^+

6 Following its gradients, H^+ flows back into inner compartment, through ATP synthases. The flow drives ATP formation

H^+

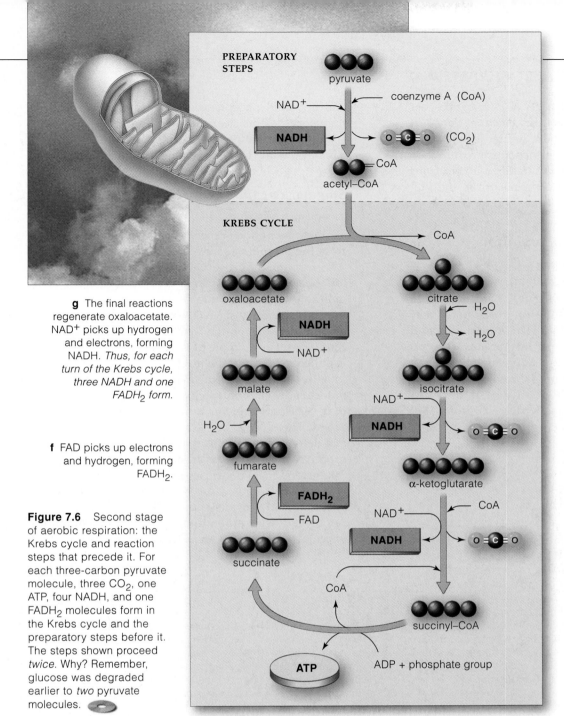

PREPARATORY STEPS

pyruvate

NAD$^+$ — coenzyme A (CoA)

NADH

O=C=O (CO$_2$)

=CoA
acetyl–CoA

KREBS CYCLE

CoA

oxaloacetate

citrate

H$_2$O
H$_2$O

NADH

NAD$^+$

malate

isocitrate

NAD$^+$

H$_2$O

NADH

O C O

fumarate

α-ketoglutarate

FADH$_2$

FAD

NAD$^+$ — CoA

NADH

O C O

succinate

NADH

CoA

succinyl–CoA

ATP

ADP + phosphate group

a Pyruvate from glucose enters a mitochondrion. A carbon atom is released, as CO$_2$. A coenzyme binds with the two-carbon fragment, becoming acetyl–CoA. NADH forms as NAD$^+$ picks up hydrogen and electrons. *Thus, one NADH forms during a few steps preceding the Krebs cycle.*

b The four-carbon oxaloacetate is the entry point into the Krebs cycle. Acetyl–CoA transfers two carbons to it, forming citrate, a six-carbon compound. Adding and removing water changes citrate to another intermediate.

c Another carbon atom is released, as CO$_2$. NADH forms when NAD$^+$ picks up hydrogen and electrons.

d Another carbon atom is released as CO$_2$, another NADH forms, and a coenzyme A molecule binds to the intermediate. *At this point, for each turn of the cycle, three carbon atoms have been released.* This balances out the three carbons that entered the mitochondrion (in pyruvate).

e A phosphate group replaces the coenzyme A and is attached to ADP. *Thus, for each turn of the Krebs cycle, one ATP forms by substrate-level phosphorylation.*

g The final reactions regenerate oxaloacetate. NAD$^+$ picks up hydrogen and electrons, forming NADH. *Thus, for each turn of the Krebs cycle, three NADH and one FADH$_2$ form.*

f FAD picks up electrons and hydrogen, forming FADH$_2$.

Figure 7.6 Second stage of aerobic respiration: the Krebs cycle and reaction steps that precede it. For each three-carbon pyruvate molecule, three CO$_2$, one ATP, four NADH, and one FADH$_2$ molecules form in the Krebs cycle and the preparatory steps before it. The steps shown proceed *twice*. Why? Remember, glucose was degraded earlier to *two* pyruvate molecules.

two ATP by substrate-level phosphorylations. Third, it rearranges Krebs cycle intermediates into oxaloacetate. This is important; cells have only so much oxaloacetate, which must be regenerated to keep the reactions going.

The two ATP do not add much to the small yield from glycolysis. But the reactions load ten coenzymes with electrons and hydrogen. So far, then, the final stage of the aerobic pathway will get all of these coenzymes:

Glycolysis:	2 NADH
Pyruvate conversion before Krebs cycle:	2 NADH
Krebs cycle:	2 FADH$_2$ + 6 NADH
Coenzymes sent to third stage:	2 FADH$_2$ + 10 NADH

Overall, these are the key points to remember about the second stage of aerobic respiration:

During the second stage of aerobic respiration, two pyruvate molecules from glycolysis enter a mitochondrion.

All of pyruvate's carbon atoms (originally from glucose) are released, in the form of carbon dioxide. Two ATP form. The oxaloacetate that is the entry point for the second-stage cyclic reactions is regenerated.

The breakdown of pyruvate releases enough electrons and hydrogen to reduce ten coenzymes. With the two coenzymes that formed in glycolysis, they will deliver electrons and hydrogen to sites of the final stage of the aerobic pathway.

THIRD STAGE OF THE AEROBIC PATHWAY

In the aerobic pathway's third stage, ATP formation goes into high gear. This stage requires electron transport systems and ATP synthases, both located in the inner membrane that divides the mitochondrion into two compartments (Figure 7.7). Both interact with electrons and hydrogen, delivered by coenzymes from the first two stages of the aerobic pathway.

Electron Transport Phosphorylation

The electrons enter transport systems, and hydrogen is released and so becomes ionized (H^+). As the electrons are transferred through the systems, hydrogen ions in the inner compartment are shuttled to the outer one. These repeated shuttlings set up H^+ concentration and electric gradients across the inner membrane.

The only way ions can follow the gradient and flow back to the inner compartment is through the interior of ATP synthases (Figure 7.7). Their flow through the transport proteins drives formation of ATP from ADP and unbound phosphate. Free oxygen helps keep the transport systems clear for operation. It withdraws the spent electrons at the end of the transport systems and combines with H^+ to form water, a by-product.

Summary of the Energy Harvest

Thirty-two ATP typically form during the third stage of aerobic respiration. When you add these to the yield from the preceding stages, the net harvest is thirty-six ATP from one glucose molecule (Figure 7.8). That's a lot! Anaerobic pathways may use up eighteen glucose molecules to get the same net yield.

Think of thirty-six ATP as a typical yield only. The amount depends on the type of cell and the prevailing conditions. For example, it is less when an intermediate is pulled away from the reactions and used elsewhere.

Also, NADH from the cytoplasm can't even enter a mitochondrion. It only delivers electrons and hydrogen to it; transport proteins in the membrane shuttle these across. Then NAD^+ or FAD already inside accept them and deliver the electrons to transport systems. But FAD drops them off at a *lower* step in a transport system "staircase," so its deliveries produce less ATP.

One final point. Glucose, recall, has more energy (stored in more covalent bonds) than the products of its full breakdown (carbon dioxide and water). About 686 kilocalories of energy are released when each mole of

Figure 7.7 Electron transport phosphorylation, the third and final stage of aerobic respiration. The reactions occur at electron transport systems and at ATP synthases, a type of transport protein, in the inner mitochondrial membrane. Each electron transport system consists of specific enzymes, cytochromes, and other proteins. The membrane divides the mitochondrion's interior into two compartments.

NADH and FADH$_2$ give up electrons and hydrogen to the transport systems. As electrons are transferred *through* the systems, the unbound hydrogen (H^+) is shuttled across the membrane, to the outer compartment:

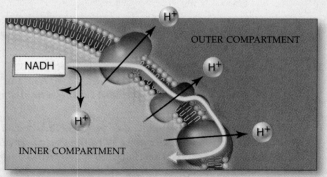

The H^+ concentration is now greater in the outer compartment. Concentration and electric gradients across the membrane have been set up. H^+ follows these gradients, through the interior of ATP synthases. Energy associated with the flow drives the formation of ATP from ADP and unbound phosphate (P_i). Hence the name, electron transport *phosphorylation*:

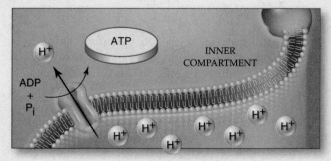

Do these events sound familiar? They should. ATP forms in much the same way in chloroplasts. By the *chemiosmotic* model, H^+ concentration and electric gradients across a cell membrane drive ATP formation. In this case, H^+ flows in the opposite direction compared to the flow in chloroplasts.

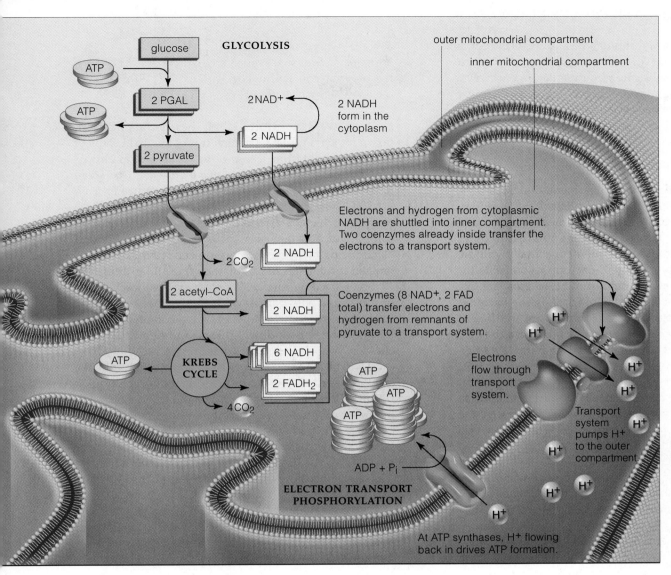

GLYCOLYSIS

glucose

ATP

ATP

2 PGAL

2 pyruvate

2 NAD$^+$

2 NADH form in the cytoplasm

2 NADH

outer mitochondrial compartment

inner mitochondrial compartment

Electrons and hydrogen from cytoplasmic NADH are shuttled into inner compartment. Two coenzymes already inside transfer the electrons to a transport system.

2 CO$_2$

2 NADH

2 acetyl–CoA

2 NADH

Coenzymes (8 NAD$^+$, 2 FAD total) transfer electrons and hydrogen from remnants of pyruvate to a transport system.

ATP

KREBS CYCLE

6 NADH

2 FADH$_2$

4 CO$_2$

Electrons flow through transport system.

H$^+$

H$^+$

ATP

ATP

ATP

ATP

Transport system pumps H$^+$ to the outer compartment

H$^+$

H$^+$

H$^+$

ADP + P$_i$

ELECTRON TRANSPORT PHOSPHORYLATION

H$^+$

H$^+$

H$^+$

At ATP synthases, H$^+$ flowing back in drives ATP formation.

2 ATP

a In glycolysis, 2 ATP used; 4 ATP form by *substrate-level phosphorylation*. So *net* yield is 2 ATP.

2 ATP

b In Krebs cycle of second stage, 2 ATP form by *substrate-level phosphorylation*.

4 ATP

c In third stage, NADH from glycolysis used to form 4 ATP by *electron transport phosphorylation*.

28 ATP

d In third stage, NADH and FADH$_2$ from second stage used to make 28 ATP by *electron transport phosphorylation*.

TYPICAL NET ENERGY YIELD: **36 ATP**

Figure 7.8 Summary of the harvest from the energy-releasing pathway of aerobic respiration. Commonly, thirty-six ATP form for each glucose molecule. The net yield varies according to shifting concentrations of reactants, intermediates, and end products. It also varies among different types of cells.

Cells differ in how they use the NADH from glycolysis, which can't enter mitochondria. These NADH give up electrons and hydrogen to transport proteins in the outer mitochondrial membrane, which shuttle them across. NAD$^+$ or FAD already inside accept them, forming NADH or FADH$_2$.

Any NADH inside a mitochondrion delivers electrons to the highest entry point into a transport system. When it does, enough H$^+$ is pumped across the inner membrane to make *three* ATP. FADH$_2$ delivers them to a lower entry point. Fewer hydrogen ions are pumped, so only *two* ATP can form.

In liver, heart, and kidney cells, for example, electrons and hydrogen from glycolysis enter the highest entry point of transport systems, so the energy harvest is thirty-eight ATP. More commonly, as in skeletal muscle and brain cells, they are transferred to FAD, so the harvest is thirty-six ATP.

glucose is degraded to these more stable end products. (Section 2.7, *Critical Thinking* question 7, gives a simple definition of molar weight.) Much of the freed energy escapes as heat, but about 39 percent is stored in ATP.

In the final stage of the aerobic pathway, many coenzymes deliver electrons to transport systems in a mitochondrion's inner membrane. The membrane forms two compartments.

Coenzymes give up hydrogen and electrons to the transport systems. Electron transfers through these systems cause the unbound hydrogen (H$^+$) to be shuttled into the inner compartment. The resulting H$^+$ concentration and electric gradients drive ATP formation as H$^+$ flows back across the membrane. Oxygen is the final electron acceptor.

Again, from start (glycolysis in the cytoplasm) to finish (in mitochondria), the pathway commonly has a net yield of thirty-six ATP for every glucose molecule metabolized.

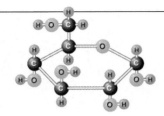

7.5

ANAEROBIC ROUTES OF ATP FORMATION

So far, we have tracked the fate of a glucose molecule through the pathway of aerobic respiration. We turn now to its use as a substrate for fermentation pathways. Remember, these are anaerobic pathways; they do *not* use oxygen as the final acceptor of the electrons that ultimately drive the ATP-forming machinery.

Fermentation Pathways

Diverse kinds of organisms use fermentation pathways. Many are prokaryotic cells and protistans that reside in marshes, bogs, mud, deep-sea sediments, the animal gut, canned foods, sewage treatment ponds, and other oxygen-free settings. Some fermenters actually will die if they are exposed to oxygen. Bacteria responsible for many diseases, including botulism and tetanus, are like this. Other kinds of fermenters, including the bacterial "employees" of yogurt manufacturers, are indifferent to the presence of oxygen. Still others can use oxygen, but they also might use a fermentation pathway when oxygen becomes scarce. Even your muscle cells do this.

Glycolysis serves as the first stage of fermentation pathways, just as it does for aerobic respiration. In these pathways also, enzymes split glucose and rearrange the fragments into two pyruvate molecules. Here again, two NADH form, and the net energy yield is two ATP.

However, fermentation reactions do not completely break down glucose to carbon dioxide and water. Also, they produce no more ATP beyond the tiny yield from glycolysis. *Fermentation's final steps only regenerate NAD⁺, the coenzyme with a key role in the breakdown reactions.*

Fermentation yields enough energy to sustain many single-celled anaerobic organisms. It even helps some aerobic cells through times of stress. But it isn't enough to sustain large, multicelled organisms. And that is one reason why you will never see an anaerobic elephant.

LACTATE FERMENTATION With these points in mind, take a look at Figure 7.9, which tracks the main steps of **lactate fermentation**. During this anaerobic pathway, the *pyruvate* molecules from glycolysis—the first stage of its reactions—accept hydrogen and electrons from NADH. This transfer regenerates NAD⁺. At the same time, it converts each pyruvate to lactate, a three-carbon compound. You may hear people call this compound "lactic acid," which is the non-ionized form. Lactate, its ionized form, is far more common in cellular fluids.

Some bacteria, such as *Lactobacillus*, rely exclusively on this anaerobic pathway. Their activities often spoil food. Even so, some fermenters have commercial uses, as when they break down glucose in huge vats where cheeses, yogurt, and sauerkraut are produced.

Also, in humans, rabbits, and many other animals, some types of cells use lactate fermentation for a quick fix of ATP. When your own demands for energy are intense but brief—say, during a short race—muscle cells use this pathway. They can't do so for long; they would throw away too much of glucose's energy for too little ATP. Muscles fatigue and lose their ability to contract when these cells deplete their glucose supply.

ALCOHOLIC FERMENTATION In the anaerobic route called **alcoholic fermentation**, enzymes convert each pyruvate molecule that formed during glycolysis to an intermediate form: acetaldehyde. The NADH transfers electrons and hydrogen to this form and thus converts it to an alcoholic end product—ethanol (Figure 7.10).

Some single-celled fungi called yeasts are famous for their use of this pathway. One type, *Saccharomyces cerevisiae*, makes bread dough rise. Bakers mix the yeast with sugar, then blend both into dough. Fermenting yeast cells release carbon dioxide. Bubbles of this gas expand the dough (make it rise). Oven heat forces the gas out of the dough, and a porous product remains.

Beer and wine producers use yeasts on a large scale. Vintners use wild yeasts living on grapes and cultivated strains of *S. ellipsoideus*, which remain active until the alcohol concentration in wine vats exceeds 14 percent. (Wild yeast cells die when the concentration exceeds 4 percent.) That is why some birds get drunk on naturally fermenting berries. That is why landscapers don't plant prodigious berry-producing shrubbery near highways;

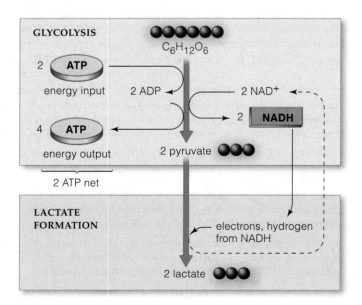

Figure 7.9 Lactate fermentation. In this anaerobic pathway, electrons end up in lactate, the reaction product.

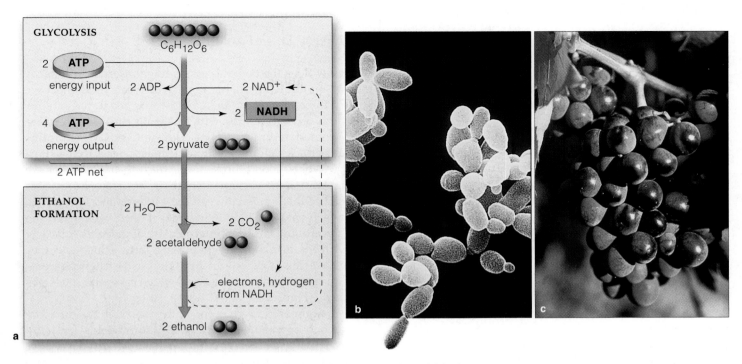

GLYCOLYSIS

$C_6H_{12}O_6$

2 **ATP**
energy input

2 ADP

2 NAD^+

2 **NADH**

4 **ATP**
energy output

2 pyruvate

2 ATP net

ETHANOL
FORMATION

2 H_2O

2 CO_2

2 acetaldehyde

electrons, hydrogen
from NADH

2 ethanol

a

b

c

Figure 7.10 (**a**) Alcoholic fermentation, one of the anaerobic pathways. Acetaldehyde, an intermediate of the reactions, is the final acceptor of electrons. Ethanol is the end product. Yeasts, single-celled organisms, use this pathway. (**b**) Activity of one species of *Saccharomyces* makes bread dough rise. (**c**) Another species lives on sugar-rich tissues of ripe grapes.

drunk birds doodle into windshields (Figure 7.11). Wild turkeys similarly have been known to get tipsy when they gobble fermenting apples in untended orchards.

Anaerobic Electron Transport

Especially among prokaryotic cells, we see less common energy-releasing pathways, some of which are topics of later chapters in the book. For example, many species have key roles in the global cycling of sulfur, nitrogen, and other crucial elements. Collectively, their metabolic

Figure 7.11 Robin feasting on the fermented berries of a pyracantha bush.

activities influence nutrient availability for organisms in ecosystems everywhere.

For example, certain archaebacteria and eubacteria use **anaerobic electron transport**. Electrons from some organic compound flow through transport systems in the plasma membrane. Often an inorganic compound in the environment serves as the final electron acceptor. The net energy yield varies, but it is always small.

Even as you read this, some anaerobic species that live in waterlogged soil are stripping electrons from a variety of compounds. They dump electrons on sulfate. Hydrogen sulfide, a putrid-smelling gas, is the result. Sulfate-reducing species also thrive in aquatic habitats that are enriched with decomposed organic material. They even live on the ocean floor, near hydrothermal vents. As described in Section 44.11, they form the food production base for unique communities.

In fermentation pathways, an organic substance that forms during the reactions serves as the final acceptor of electrons from glycolysis. The reactions regenerate NAD^+, which is required to keep the pathway operational.

In anaerobic electron transport, an inorganic substance (but not oxygen) usually serves as the final electron acceptor.

For each glucose molecule metabolized, the anaerobic pathways typically have a net energy yield of two ATP molecules, which only form during glycolysis.

ALTERNATIVE ENERGY SOURCES IN THE HUMAN BODY

So far, you have looked at what happens after a lone glucose molecule enters an energy-releasing pathway. Now you can start thinking about what cells do when they have too many or too few glucose molecules.

Carbohydrate Breakdown in Perspective

THE FATE OF GLUCOSE AT MEALTIME While you or any other mammal is eating, glucose and other small organic molecules are being absorbed across the gut lining, then blood transports them through the body. A rise in the glucose level in blood prompts the pancreas to release insulin. This hormone stimulates cells to take up glucose faster. Cells convert the incoming glucose to glucose–6–phosphate and "trap" it in the cytoplasm. (When phosphorylated, glucose cannot be transported back out, across the plasma membrane.) Look again at Figure 7.4, and you see that glucose–6–phosphate is the first activated intermediate of glycolysis.

If your glucose intake exceeds cellular demands for energy, ATP-producing machinery goes into high gear. Unless a cell is rapidly using its ATP, the cytoplasmic concentration of ATP can rise to high levels. When that happens, the glucose–6–phosphate gets diverted into a biosynthesis pathway. This pathway assembles glucose units into glycogen, a storage polysaccharide (Section 3.3). The pathway is especially favored in muscle and liver cells, which maintain the largest glycogen stores.

THE FATE OF GLUCOSE BETWEEN MEALS When you are not eating, glucose is not entering the bloodstream, and its level in the blood declines. If the decline were not countered, that would be bad news for the brain, your body's glucose hog. At any time, your brain is taking up more than two-thirds of the freely circulating glucose, because its many hundreds of millions of cells use this sugar alone as their preferred energy source.

The pancreas responds to the decline by secreting glucagon. This hormone prompts liver cells to convert glycogen back to glucose and send it back to the blood. Only liver cells do this; muscle cells won't give it up. The blood glucose level rises, and brain cells keep on functioning. Thus, *hormones control whether your body's cells use free glucose as an energy source or tuck it away.*

A word of caution: Don't let the preceding examples lead you to believe cells squirrel away large amounts of glycogen. In adult humans, glycogen makes up merely 1 percent or so of the body's total energy reserves, the energy equivalent of two cups of cooked pasta. Unless you eat on a regular basis, you will deplete the liver's small glycogen stores in less than twelve hours.

Of the total energy reserves in, say, a typical adult who eats well, 78 percent (about 10,000 kilocalories) is concentrated in body fat and 21 percent in proteins.

Energy From Fats

How does the body access its huge reservoir of fats? A fat molecule, recall, has a glycerol "head" and one, two, or three fatty acid "tails." Most fats get stored in your body as triglycerides, with three tails each. Triglycerides accumulate in fat cells of adipose tissue, which forms at buttocks and other strategic places beneath the skin.

When blood glucose levels decline, triglycerides are tapped as an energy alternative. Enzymes in fat cells cleave the bonds between the glycerol and fatty acids, which enter the blood. Enzymes in the liver convert the glycerol to PGAL, an intermediate of glycolysis. Nearly all cells take up the circulating fatty acids. Enzymes cleave the backbone of the fatty acids. The fragments are converted to acetyl–CoA, which can enter the Krebs cycle (Figures 7.6 and 7.12).

Compared to glucose, a fatty acid tail has far more carbon-bound hydrogen atoms, so it yields far more ATP. Between meals or during sustained exercise, fatty acid conversions supply about half of the ATP that muscle, liver, and kidney cells require.

What happens if you eat too many carbohydrates? Exceed the glycogen-storing capacity of your liver and muscle cells, and the excess gets converted to fats. *Too much glucose ends up as excess fat.* In the United States, 25 percent of the people have a combination of genes that lets them eat as much as they like without gaining weight. A diet too rich in carbohydrates keeps the other 75 percent fat. Their insulin levels stay elevated, which "tells" the body to store fat, not use it for energy. We return to this topic in Section 32.7 and Chapter 37.

Energy From Proteins

Eat more proteins than your body requires to grow and maintain itself, and its cells won't store them. Enzymes split dietary proteins into amino acid units. Then they remove the amino group ($-NH_3^+$) from each unit, and ammonia (NH_3) forms. What happens to the leftover carbon backbones? The outcome varies, depending on conditions of the moment. Maybe they'll get converted to fats or carbohydrates. Or maybe they'll enter the Krebs cycle (Figure 7.12), where coenzymes pick up the hydrogen and electrons stripped from carbon atoms. The ammonia that forms undergoes conversions and becomes urea. This nitrogen-containing waste product would be toxic if it accumulated to high concentrations. Normally your body excretes urea, in urine.

As this brief discussion makes clear, maintaining and accessing the body's energy reserves is complicated business. Hormonal controls over the disposition of glucose are special only because glucose is the fuel of choice for the all-important brain. However, as you will

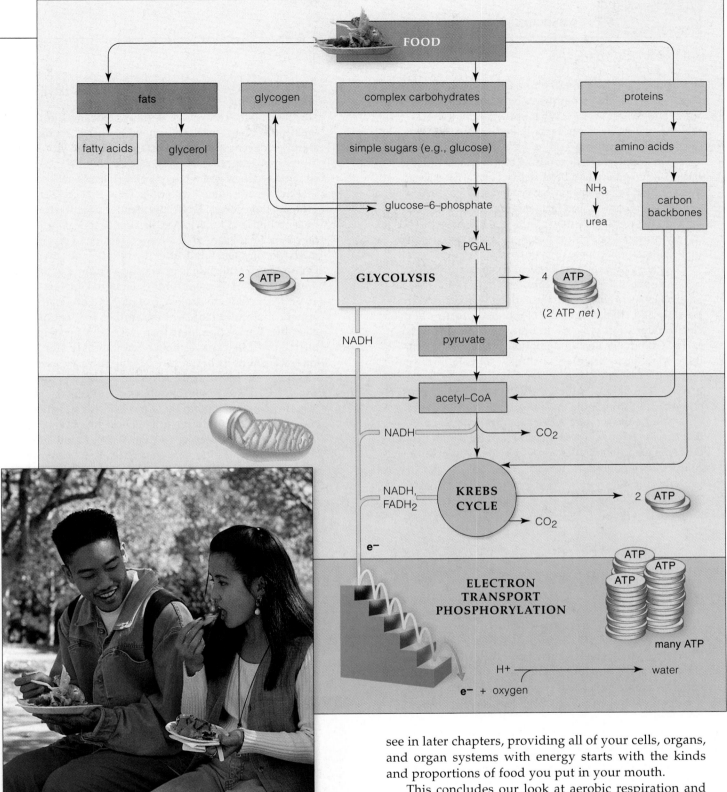

Figure 7.12 Alternative energy sources in the human body. The diagram shows reaction sites where a variety of organic compounds can enter the stages of aerobic respiration.

Complex carbohydrates, fats, and proteins cannot enter the aerobic pathway directly. In humans and other mammals, the digestive system, and individual cells, must first break apart these molecules into simpler, degradable subunits.

see in later chapters, providing all of your cells, organs, and organ systems with energy starts with the kinds and proportions of food you put in your mouth.

This concludes our look at aerobic respiration and other energy-releasing pathways. The section to follow may help you get a sense of how they fit into the larger picture of life's evolution and interconnectedness.

In humans and other mammals, the entrance of glucose or other organic compounds into an energy-releasing pathway depends on the kinds and proportions of carbohydrates, fats, and proteins in the diet as well as on the type of cell.

Perspective on Life

In this unit, you read about photosynthesis and aerobic respiration—the main pathways by which cells trap, store, and release energy. What you may not know is that the two pathways became linked, on a grand scale, over evolutionary time.

When life originated, some 3.8 billion years ago, the Earth's atmosphere had little free oxygen. The earliest single-celled organisms probably made ATP by reactions similar to glycolysis, and fermentation pathways must have dominated. More than a billion years passed. Then oxygen-producing photosynthetic cells appeared—and they changed the course of evolution.

Oxygen, a by-product of the noncyclic pathway of photosynthesis, began to accumulate in the atmosphere. Some cells were able to use the oxygen to accept electrons, possibly as a chance outcome of mutated proteins in their electron transport systems. In time, descendants of those early aerobic cells abandoned photosynthesis. Among them were the forerunners of animals and all other organisms that engage in aerobic respiration.

With aerobic respiration, the flow of carbon, hydrogen, and oxygen through the metabolic pathways of living organisms came full circle. For the final products of this aerobic pathway—carbon dioxide and water—are the same materials that are necessary to build organic compounds in photosynthesis:

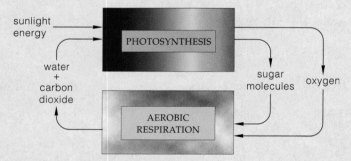

Perhaps you have difficulty fathoming the connection between yourself—a highly intelligent being—and such remote-sounding events as energy flow and the cycling of carbon, hydrogen, and oxygen. Is this really the stuff of humanity?

Think back, for a moment, on the structure of a water molecule. Two hydrogen atoms sharing electrons with an oxygen atom may not seem close to your daily life. Yet, through that sharing, water molecules show polarity—and they hydrogen-bond with one another. Their chemical behavior is a beginning for the organization of lifeless matter that leads to the organization of all living things.

For now you can imagine other molecules interspersed through water. The nonpolar kinds resist interaction with water, and the polar kinds dissolve in it. On their own, the phospholipids among them assemble into a two-layered film. Such lipid bilayers, remember, serve as the framework of all cell membranes, hence all cells. From the beginning, the cell has been the basic *living* unit.

The essence of life is not some mysterious force. It is metabolic control. With a cell membrane to contain them, reactions *can* be controlled. With mechanisms built into their membranes, cells can respond to energy changes and shifting concentrations of substances in the environment. The response mechanisms operate by "telling" proteins—enzymes—when and what to build or tear down.

And it is not some mysterious force that creates the proteins themselves. DNA, the slender double-stranded treasurehouse of inheritance, has the chemical structure—*the chemical message*—that allows molecule to reproduce molecule, one generation after the next. In your own body, DNA strands tell trillions of cells how countless molecules must be built or torn apart for their stored energy.

So yes, carbon, hydrogen, oxygen, and other atoms of organic molecules represent the stuff of you, and us, and all of life. But it takes more than molecules to complete the picture. Life exists as long as an unbroken flow of energy sustains its organization. Molecules are assembled into cells, cells into organisms, organisms into communities, and so on up through the biosphere. It takes energy inputs from the sun to maintain these levels of organization. And energy flows through time in one direction—from organized to less organized forms. Only as long as energy continues to flow into the web of life can life continue in all its rich expressions.

In short, life is no more *and no less* than a marvelously complex system of prolonging order. Sustained by energy transfusions from the sun, life continues through its capacity for self-reproduction. For with the hereditary instructions contained in DNA, energy and materials can be organized, generation after generation. Even with the death of individuals, life elsewhere is prolonged. With each death, molecules are released and may be cycled once more, as raw materials for new generations.

In this flow of energy and cycling of material through time, each birth is affirmation of our ongoing capacity for organization, each death a renewal.

SUMMARY

1. Phosphate-group transfers from ATP are central to metabolism. Autotrophic cells alone tap energy from the environment to make ATP for building carbohydrates. *All* cells make ATP by pathways that release chemical energy from organic compounds, such as glucose. *7.1*

2. After glucose enters an energy-releasing pathway, enzymes strip electrons and hydrogen from reaction intermediates formed along the way. Coenzymes pick these up and deliver them to other reaction sites, where the pathway ends. NAD^+ is the main coenzyme; FAD is also used in the aerobic route. Reduced forms of these coenzymes are designated NADH and $FADH_2$. *7.1–7.3*

3. All of the main energy-releasing pathways start with glycolysis. This stage of reactions begins and ends in the cytoplasm. It can be completed either in the presence of oxygen or in its absence. *7.1, 7.2*

   a. In glycolysis, enzymes break down glucose to two pyruvate molecules. Two NADH and four ATP form.

   b. The net energy yield is two ATP (because two ATP had to be invested up front to get the reactions going).

4. Aerobic respiration continues on through two more stages: (1) the Krebs cycle and a few steps preceding it, and (2) electron transport phosphorylation. The stages occur only in mitochondria of eukaryotic cells. *7.1, 7.3*

5. The second stage of the aerobic pathway starts when an enzyme strips a carbon atom from each pyruvate. Coenzyme A binds the two-carbon fragment to form acetyl–CoA, and transfers this to oxaloacetate, the entry point of the Krebs cycle. These cyclic reactions and a few steps before them load ten coenzymes with electrons and hydrogen (eight NADH, plus two $FADH_2$). Two ATP form. Three carbon dioxide molecules are released for each pyruvate that entered this second stage. *7.3*

6. The third stage of the aerobic pathway proceeds at a membrane dividing the mitochondrion's interior into two compartments. Electron transport systems and ATP synthases are embedded in this inner membrane. *7.4*

   a. Coenzymes deliver electrons and hydrogen from the first two stages to transport systems. These systems function to establish an H^+ concentration gradient and electric gradient across the membrane.

   b. H^+ follows the gradients and flows from the outer to the inner compartment. It does so through the interior of ATP synthases. Energy released by the ion flow drives the formation of ATP from ADP and unbound phosphate.

   c. Oxygen picks up spent electrons at the end of the transport system and combines with hydrogen ions to form water. That is, oxygen is the final acceptor of the electrons that initially resided in glucose.

7. Aerobic respiration has a typical net energy yield of thirty-six ATP for each glucose molecule metabolized. Yields vary by cell type and cellular conditions. *7.4*

8. The fermentation pathways and anaerobic electron transport also start with glycolysis, but they do not use oxygen. They are anaerobic, start to finish. *7.5*

   a. Lactate fermentation has a net energy yield of two ATP, which form in glycolysis. The remaining reaction regenerates NAD^+. The two NADH from glycolysis transfer electrons and hydrogen to two pyruvate from glycolysis. Two lactate molecules are the end products.

   b. Alcoholic fermentation has a net energy yield of two ATP from glycolysis, and its remaining reactions serve to regenerate NAD^+. Enzymes convert pyruvate from glycolysis to acetaldehyde, and carbon dioxide is released. The NADH from glycolysis transfer electrons and hydrogen to the two acetaldehyde molecules, thus forming two ethanol molecules, the end products.

   c. Certain bacteria use anaerobic electron transport. Electrons are stripped from various organic compounds and travel through transport systems in the bacterial cell's plasma membrane. An inorganic compound in the environment often serves as the final electron acceptor.

9. In humans and other mammals, simple sugars such as glucose from carbohydrates, glycerol and fatty acids from fats, and carbon backbones of amino acids from proteins can enter ATP-producing pathways. *7.6*

Review Questions

1. Is this true or false: Aerobic respiration occurs in animals but not plants, which make ATP only by photosynthesis. *7.1*

2. Using the diagram below of the aerobic pathway, fill in the blanks with the number of molecules of pyruvate, coenzymes, and end products. Write in the net ATP formed in each stage, then the net ATP formed from start (glycolysis) to finish. *7.1*

3. Is glycolysis energy-*requiring* or energy-*releasing*? Or do both kinds of reactions occur during glycolysis? *7.2*

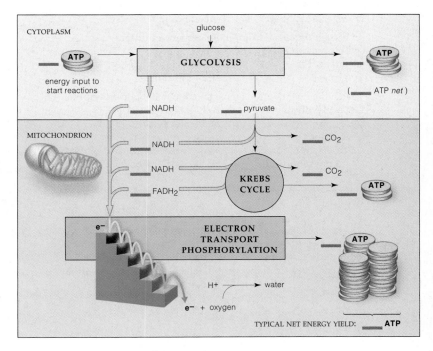

4. In what respect does *electron transport* phosphorylation differ from *substrate-level* phosphorylation? *7.2, Figure 7.7*

5. Sketch the double-membrane system of the mitochondrion and show where electron transport systems and ATP synthases are located. *7.3*

6. Name the compound that is the entry point for the Krebs cycle, and state whether it directly accepts the pyruvate from glycolysis. For each glucose molecule, how many carbon atoms enter the Krebs cycle? How many depart from it, and in what form? *7.3*

7. Is this statement true or false: Muscle cells cannot contract at all when deprived of oxygen. If true, explain why. If false, name the alternative(s) available to them. *7.5*

Self-Quiz   ANSWERS IN APPENDIX III

1. Glycolysis starts and ends in the _____ .
   a. nucleus   c. plasma membrane
   b. mitochondrion d. cytoplasm

2. Which of the following does *not* form during glycolysis?
   a. NADH   c. $FADH_2$
   b. pyruvate   d. ATP

3. Aerobic respiration is completed in the _____ .
   a. nucleus   c. plasma membrane
   b. mitochondrion d. cytoplasm

4. In the last stage of aerobic respiration, _____ is the final acceptor of electrons that originally resided in glucose.
   a. water   c. oxygen
   b. hydrogen   d. NADH

5. _____ engage in lactate fermentation.
   a. *Lactobacillus* cells   c. Sulfate-reducing bacteria
   b. Muscle cells   d. a and b

6. In alcoholic fermentation, _____ is the final acceptor of the electrons stripped from glucose.
   a. oxygen   c. acetaldehyde
   b. pyruvate   d. sulfate

7. The fermentation pathways produce no more ATP beyond the small yield from glycolysis, but the remaining reactions _____ .
   a. regenerate ADP   c. dump electrons on an inorganic
   b. regenerate NAD^+ substance (not oxygen)

8. In certain organisms and under certain conditions, _____ can be used as an energy alternative to glucose.
   a. fatty acids   c. amino acids
   b. glycerol   d. all of the above

9. Match the event with its most suitable metabolic description.
   ____ glycolysis   a. ATP, NADH, $FADH_2$, CO_2,
   ____ fermentation   and water form
   ____ Krebs cycle   b. glucose to two pyruvate
   ____ electron transport c. NAD^+ regenerated, two ATP net
       phosphorylation   d. H^+ flows through ATP synthases

Critical Thinking

1. Diana suspects that a visit to her family doctor is in order. After eating carbohydrate-rich food, she always experiences sensations of being intoxicated and becomes nearly incapacitated, as if she had been drinking alcohol. She even wakes up with a hangover the next day. Having completed a course in freshman biology, Diana has an idea that something is affecting the way her body is metabolizing glucose. Explain why.

2. The cells of your body absolutely do not use nucleic acids as alternative energy sources. Suggest why.

3. The human body's energy needs and its programs for growth depend on balancing the levels of amino acids in blood with proteins in cells. Cells of the liver, kidneys, and intestinal lining are especially important in this balancing act. When the levels of amino acids in blood decline, lysozymes in cells can rapidly digest some of their proteins (structural and contractile proteins are spared, except in cases of malnutrition). The amino acids released this way enter the blood and thereby help maintain the required levels.

   Suppose you embark on a body-building program. You already eat plenty of carbohydrates. However, a nutritionist recommends that you follow a protein-rich diet that includes protein supplements. Speculate on how extra dietary proteins will be put to use, and in which tissues.

4. Each year, Canada geese lift off in precise formation from their northern breeding grounds. They head south to spend the winter months in warmer climates, then make the return trip in spring. As is the case for other migratory birds, their flight muscle cells are efficient at using fatty acids as an energy source. (Remember, the carbon backbone of fatty acids can be cleaved into fragments that can be converted to acetyl–CoA for entry into the Krebs cycle.)

   Suppose a lesser Canada goose from Alaska's Point Barrow has been steadily flapping along for three thousand kilometers and is nearing Klamath Falls, Oregon. It looks down and notices a rabbit sprinting like the wind from a coyote with a taste for rabbit. With a stunning burst of speed, the rabbit reaches the safety of its burrow.

   Which energy-releasing pathway predominated in muscle cells in the rabbit's legs? Why was the Canada goose relying on a different pathway for most of its journey? And why wouldn't the pathway of choice in goose flight muscle cells be much good for a rabbit making a mad dash from its enemy?

5. Reflect on this chapter's introduction, then question 4. Now speculate on which energy-releasing pathway is predominating in agitated Africanized bees that are chasing a farmer through a cornfield.

Selected Key Terms

acetyl–CoA 7.3	Krebs cycle 7.1
aerobic respiration 7.1	lactate fermentation 7.5
alcoholic fermentation 7.5	mitochondrion 7.3
anaerobic electron transport 7.5	NAD^+ 7.1
electron transport	oxaloacetate 7.3
phosphorylation 7.1	pyruvate 7.1
FAD 7.1	substrate-level
glycolysis 7.1	phosphorylation 7.2

Readings

DeMauro, S. October 1998. "Hooked on Mitochondria." *Quest.* Volume 5, Number 5. A brief look at mitochondrial disorders.

Wolfe, S. 1995. *An Introduction to Molecular and Cellular Biology.* Belmont, California: Wadsworth. Exceptional reference text.

On-Line readings at Student Guide for InfoTrac: www.brookscole.com/biology

II Principles of Inheritance

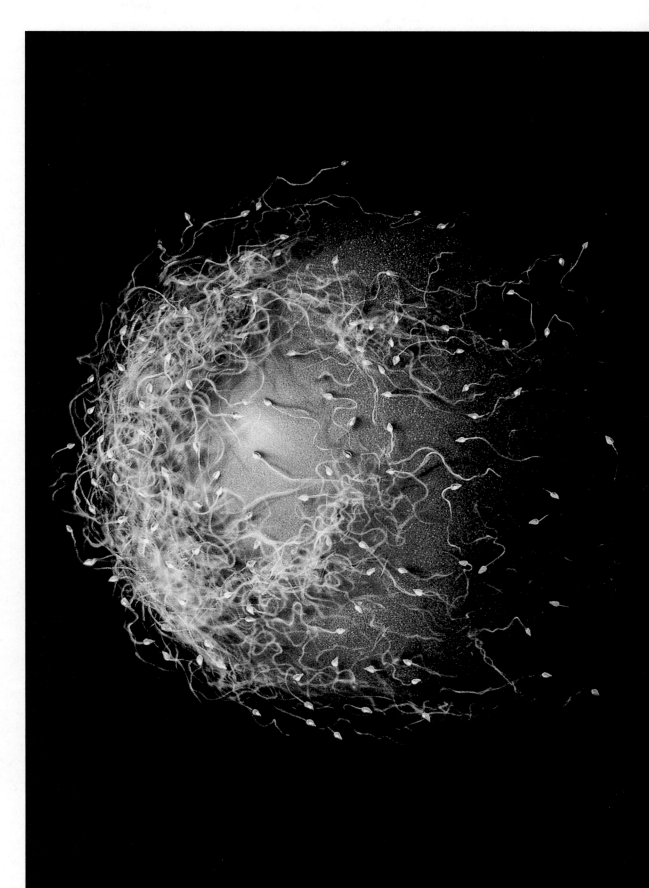

Human sperm, one of
which will penetrate
this mature egg and
so set the stage for the
development of a new
individual in the image
of its parents. This
exquisite art is based
on a scanning electron
micrograph.

8

CELL DIVISION AND MITOSIS

Silver In the Stream of Time

Five o'clock, and the first rays from the sun dance over the wild Alagnak River of the Alaskan tundra. It is September, and life is ending and beginning in the clear, cold waters. By the thousands, mature silver salmon have returned from the open ocean to spawn in their shallow home waters. The females are tinged with red, the color of spawners, and they are dying.

This morning you watch a female salmon releasing translucent eggs into a shallow depression, which her fins made in the gravel riverbed (Figure 8.1). Moments later a male sheds a cloud of sperm, and the eggs are fertilized. Trout and other predators eat most of them. But wait around and you'll see that some eggs survive and give rise to a new generation.

Within three years the pea-sized eggs have become streamlined salmon, fashioned from billions of cells. A few of their cells will develop into eggs or sperm. In time, on some September morning, they will take part in an ongoing story of birth, growth, death, and rebirth.

Your body, too, emerged through *cell divisions*, as it did for salmon and all other multicelled species. Inside your mother, a fertilized egg divided in two, then the

two became four, and so on until billions of cells were growing, developing in specialized ways, and dividing at different times to form different parts. Your body now has about 65 trillion living cells. Many are still dividing. Every five days, for instance, some divisions replace the tissue that lines your small intestine.

Understanding cell division—and, ultimately, how new individuals are put together in the image of their parents—begins with answers to three questions. *First,* what instructions are necessary for inheritance? *Second,* how are those instructions duplicated for distribution into daughter cells? *Third,* by what mechanisms are the duplicated instructions parceled out to daughter cells? We'll need more than one chapter to consider the nature of cell reproduction and other mechanisms of inheritance. Even so, the points made early in this chapter can help you keep the overall picture in focus.

Begin with the word **reproduction**. In biology, this means that parents produce a new generation of cells or multicelled individuals like themselves. The process starts in cells that are programmed to divide. And the ground rule for division is this: *Parent cells must provide*

sexually mature
female salmon

Figure 8.1 The last of one generation and the first of the next in Alaska's Alagnak River.

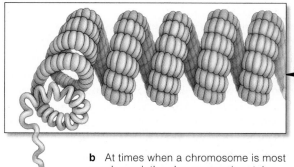

b At times when a chromosome is most condensed, the chromosomal proteins interact, and package loops of already coiled DNA into a "supercoiled" array.

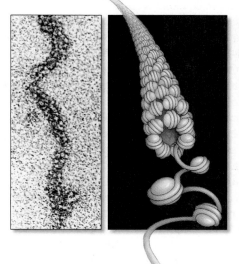

c At a deeper level of structural organization, the chromosomal proteins and DNA are organized as a cylindrical fiber.

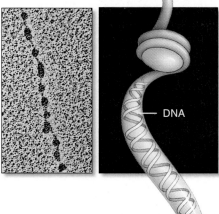

d Immerse a chromosome in saltwater and it loosens up to a beads-on-a-string organization. The "string" is one DNA molecule. Each "bead" is a nucleosome.

— DNA

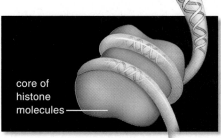

e A nucleosome consists of part of a DNA molecule looped twice around a core of histones.

core of histone molecules —

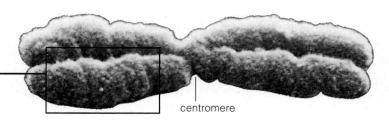

centromere

a A duplicated human chromosome at metaphase, at its most condensed.

Figure 8.3 One model of levels of organization in a human chromosome at its most condensed form.

Actually, your 46 chromosomes are like volumes of two sets of books. Each set is numbered, from 1 to 23. For example, you have two "volumes" of chromosome 22—that is, *a pair of them*. Except for certain pairings of sex chromosomes, both have the same length and shape. They carry the same portion of hereditary instructions for the same traits. Think of them as two sets of books on how to build a house. Your father gave you one set. Your mother had her own ideas about wiring, storage, plumbing, and so on; she gave you an alternate edition. Her set covers the same topics, but her instructions are slightly different for many of them.

We say the chromosome number is **diploid**, or $2n$, if a cell has two of each type of chromosome characteristic of the species. The body cells of humans, gorillas, pea plants, and a great many other organisms are like this. By contrast, as Chapter 9 describes, eggs and sperm of such organisms have a *haploid* chromosome number (n). This means that they contain only one of each type of chromosome characteristic of the species.

With mitosis, a diploid parent cell can produce two diploid daughter cells. This doesn't mean each merely gets forty-six or forty-eight or fourteen chromosomes. If only the total mattered, one cell might get, say, two pairs of chromosome 22 and no pairs whatsoever of chromosome 9. Neither would be able to function like the parent cell *without two of each type of chromosome*.

A eukaryotic chromosome consists of one DNA molecule and many proteins that structurally organize the DNA.

A chromosome number is the sum total of chromosomes in a body cell or germ cell of a particular organism. Cells with a diploid chromosome number have two of each type of chromosome, usually from two parents.

Mitosis is a nuclear division mechanism that sorts out a required number of chromosomes for each daughter cell. A separate mechanism divides the cytoplasm.

Mitosis keeps the chromosome number constant, division after division, from one cell generation to the next. Thus, if a parent cell is diploid, its daughter cells will be diploid.

THE CELL CYCLE

Let's start thinking about cell reproduction in terms of an orderly sequence of events called a **cell cycle**. It starts every time a new daughter cell forms by mitosis and cytoplasmic division and ends when that cell finishes its own division (Figure 8.4). Thus, *mitosis, cytoplasmic division, and interphase constitute one turn of the cell cycle.*

The Wonder of Interphase

Interphase is the portion of the cell cycle when a cell increases in mass, roughly doubles the number of its cytoplasmic components, and duplicates its DNA. For most types of cells, interphase is the longest portion of the cycle. Biologists divide it into three parts:

G1 Interval ("*Gap*") of cell growth and functioning before the onset of DNA replication

S Time of "*Synthesis*" (DNA replication)

G2 Second interval (*Gap*), after DNA replication, when the cell prepares for division

G1, S, and G2 are no more than code names for some amazing events. Consider what cells do with their DNA. If you could coax the DNA molecules from just one of your somatic cells to stretch in a single line, one after another, that line would extend past the fingertips of your outstretched arms. If you could do the same with salamander DNA, a single line of it would extend ten meters! The wonder is, enzymes and other proteins in cells can selectively access, activate, and silence DNA's instructions. They also can make base-by-base copies of each DNA molecule. They do all of this in interphase.

G1, S, and G2 of interphase have distinct patterns of biosynthesis. Most of your cells remain in G1, when they build nearly all of the proteins, carbohydrates, and lipids they use or export. Cells destined to divide enter S, when they copy their DNA as well as the histones and other proteins associated with it. During G2, these cells produce proteins that will drive mitosis to completion.

Once S begins, events normally proceed at about the same rate in all cells of a species and continue through mitosis. Given this observation, you may well assume the cycle has built-in molecular brakes. Apply brakes meant to work in G1, and the cycle stalls in G1. Lift the brakes, and the cycle runs to completion. Said another way, *control mechanisms govern the rate of cell division.*

Imagine a car losing its brakes just as it starts down a steep mountain road. As you will read later on in the book, that is how cancer starts. Controls over division are lost, and the cell cycle cannot stop turning.

The cell cycle lasts about the same length of time for cells of the same type. Its duration differs among cells of different types. As examples, all neurons (nerve cells) in your brain are arrested at interphase, and usually will not divide again. All red blood cells form (and replace your worn-out ones) at an average rate of 2 million to 3 million each second. The number of cells doubles every two hours early in the development of a sea urchin embryo.

Adverse conditions often disrupt a cell cycle. When deprived of a vital nutrient, for instance, the free-living cells called amoebas do not leave interphase. Even so, if any cell proceeds past a certain point in interphase, the cycle normally will continue regardless of outside conditions because of built-in controls over its duration.

Mitosis Proceeds Through Four Stages

A cell that's making the transition from interphase to mitosis has stopped constructing new cell parts. Its DNA has already been replicated. Major changes now proceed smoothly through four stages: **prophase**, **metaphase**, **anaphase**, and **telophase**.

Figure 8.5 shows these stages in a plant cell. Notice how all the chromosomes are changing positions. They aren't doing so on their own. Moving them is a **spindle**

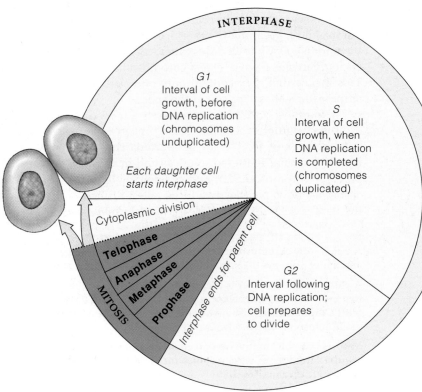

INTERPHASE

G1
Interval of cell growth, before DNA replication (chromosomes unduplicated)

S
Interval of cell growth, when DNA replication is completed (chromosomes duplicated)

Each daughter cell starts interphase

Cytoplasmic division

Telophase
Anaphase
Metaphase
Prophase
MITOSIS

Interphase ends for parent cell

G2
Interval following DNA replication; cell prepares to divide

Figure 8.4 Eukaryotic cell cycle, generalized. The length of each interval differs among different cell types.

A CELL AT
INTERPHASE:

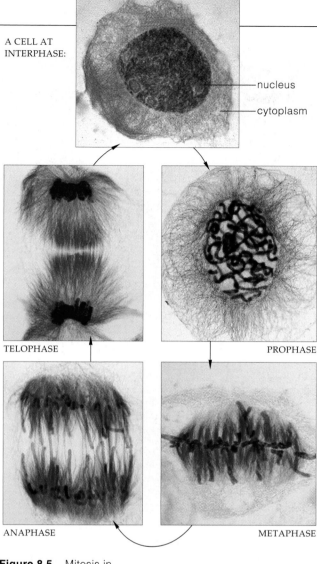

nucleus

cytoplasm

TELOPHASE

PROPHASE

ANAPHASE

METAPHASE

Figure 8.5 Mitosis in a cell from the African blood lily, *Haemanthus*. The chromosomes are stained *blue* and the many microtubules, *red*. Before reading further, take a moment to become familiar with the labels on the micrographs.

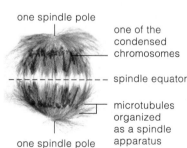

one spindle pole

one of the condensed chromosomes

spindle equator

microtubules organized as a spindle apparatus

one spindle pole

apparatus, of two sets of microtubules. Each set extends from one of the spindle's two poles (its end points), and each overlaps the other at the spindle equator, midway between the poles. This "bipolar" spindle will establish the final chromosome destinations before mitosis ends and the cell divides in two.

How important are spindle microtubules? One clue comes from plants of the genus *Colchicum*. These plants make colchicine, a poison that evolved against browsing animals. It blocks microtubule assembly and promotes their disassembly. This poison is a favorite of researchers who study cancer and other expressions of cell division.

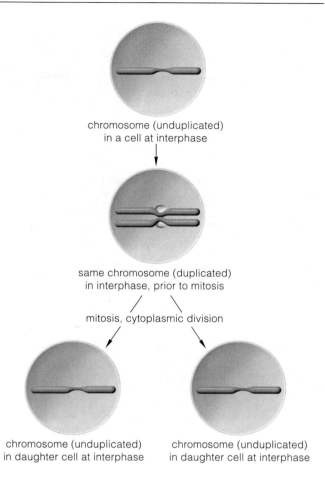

chromosome (unduplicated)
in a cell at interphase

same chromosome (duplicated)
in interphase, prior to mitosis

mitosis, cytoplasmic division

chromosome (unduplicated)
in daughter cell at interphase

chromosome (unduplicated)
in daughter cell at interphase

Figure 8.6 Simple way to think about how mitosis maintains the chromosome number, one generation after the next. For clarity, we track only one chromosome.

Another is mentioned in Section 8.6, *Critical Thinking* question 3. Spindles in cells fall apart within seconds or minutes after exposure to such microtubule poisons.

Before turning the page to consider the mechanism of mitosis, hold onto these thoughts: Chromosomes are duplicated *before* mitosis. A spindle apparatus moves the sister chromatids of duplicated chromosomes apart *during* mitosis. As Figure 8.6 shows, that is how mitosis maintains the chromosome number through turn after turn of the cell cycle.

Interphase, mitosis, and cytoplasmic division constitute one turn of the cell cycle.

In interphase, a new cell increases its mass, roughly doubles the number of its cytoplasmic components, and duplicates its chromosomes. The cycle ends after the cell undergoes mitosis and then divides. Molecular mechanisms control the rate of cell division.

Mitosis proceeds through four stages: prophase, metaphase, anaphase, and telophase. Its microtubular spindle moves chromosomes that were duplicated earlier, in interphase.

MITOSIS

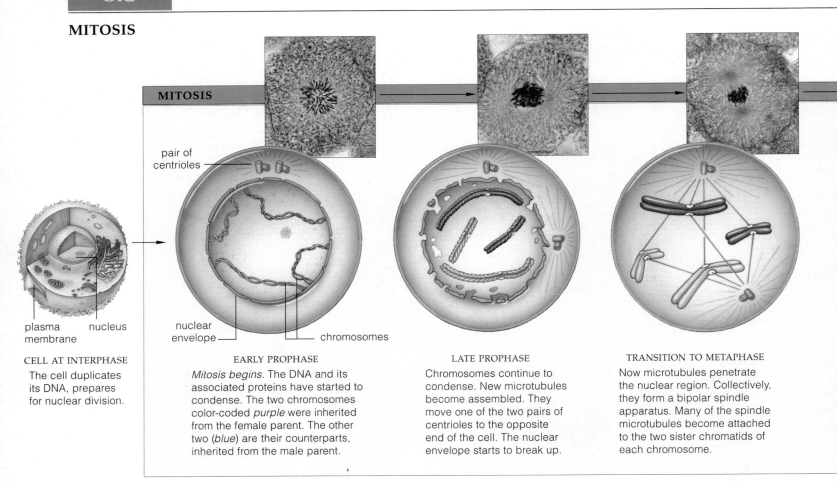

MITOSIS

CELL AT INTERPHASE
The cell duplicates its DNA, prepares for nuclear division.

plasma membrane

nucleus

pair of centrioles

nuclear envelope

chromosomes

EARLY PROPHASE
Mitosis begins. The DNA and its associated proteins have started to condense. The two chromosomes color-coded *purple* were inherited from the female parent. The other two (*blue*) are their counterparts, inherited from the male parent.

LATE PROPHASE
Chromosomes continue to condense. New microtubules become assembled. They move one of the two pairs of centrioles to the opposite end of the cell. The nuclear envelope starts to break up.

TRANSITION TO METAPHASE
Now microtubules penetrate the nuclear region. Collectively, they form a bipolar spindle apparatus. Many of the spindle microtubules become attached to the two sister chromatids of each chromosome.

Prophase: Mitosis Begins

We know a cell is in prophase, the first stage of mitosis, when its chromosomes are visible in light microscopes as threadlike forms. ("Mitosis" is after the Greek *mitos*, for thread.) Each chromosome was duplicated earlier, in interphase; each is two sister chromatids joined at the centromere. In early prophase, all of these chromatids twist and fold. By late prophase, they will be condensed into thicker, compact, rod-shaped forms.

Meanwhile, in the cytoplasm, most microtubules of the cytoskeleton are breaking down to tubulin subunits (Section 4.10). The subunits reassemble near the nucleus as microtubules of the spindle. While new microtubules are assembling, the nuclear envelope physically prevents them from interacting with the chromosomes inside the nucleus. However, the nuclear envelope starts to break up as prophase draws to a close (Figure 8.7).

Many cells have two barrel-shaped **centrioles**. Each centriole started duplicating itself during interphase, so there are two pairs of them when prophase is under way. Microtubules start moving one pair to the opposite pole of the newly forming spindle. Centrioles, recall, give rise to flagella or cilia. If you observe them in cells of an organism, you can bet that flagellated cells (such as sperm) or ciliated cells develop during its life cycle.

Figure 8.7 Mitosis in a generalized animal cell. By this nuclear division mechanism, each daughter cell ends up with the same chromosome number as the parent cell. For clarity, we track only two pairs of chromosomes from a diploid (2*n*) cell. The picture is almost always more complicated, as you may sense from the above micrographs of mitosis in a whitefish cell.

Transition to Metaphase

So much happens between prophase and metaphase that we often call this transitional time "prometaphase." The nuclear envelope breaks up completely into many small, flattened vesicles, so the chromosomes are now free to interact with microtubules. Some microtubules that have been lengthening from the two spindle poles harness each chromosome and pull on it. The two-way pulling orients it with respect to the poles. However, other microtubules form two sets that extend from the two poles and overlap at the spindle's midpoint. They push the poles apart by interacting and ratcheting past each other. The push–pull forces become balanced when the chromosomes reach the spindle's midpoint.

When all of the duplicated chromosomes are aligned midway between the poles of a completed spindle, we call this metaphase (*meta–* means "midway between"). The alignment is crucial for the next stage of mitosis.

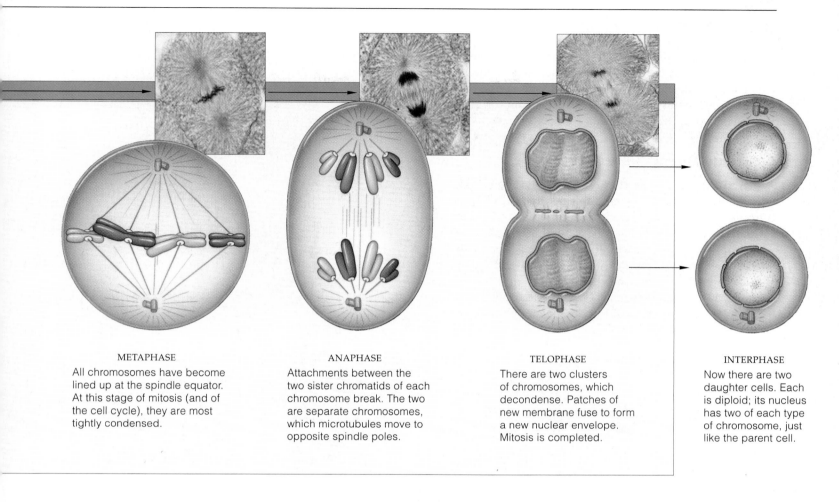

METAPHASE

All chromosomes have become lined up at the spindle equator. At this stage of mitosis (and of the cell cycle), they are most tightly condensed.

ANAPHASE

Attachments between the two sister chromatids of each chromosome break. The two are separate chromosomes, which microtubules move to opposite spindle poles.

TELOPHASE

There are two clusters of chromosomes, which decondense. Patches of new membrane fuse to form a new nuclear envelope. Mitosis is completed.

INTERPHASE

Now there are two daughter cells. Each is diploid; its nucleus has two of each type of chromosome, just like the parent cell.

From Anaphase Through Telophase

At anaphase, the sister chromatids of each chromosome separate from each other and move to opposite spindle poles. Two mechanisms bring this about.

First, microtubules that latched onto the centromere region of each chromosome shorten (they disassemble) and *pull* the chromosome to a spindle pole. Think of the centromere as a train chugging along a railroad track, except the track falls apart after the train passes over it. Activated motor proteins are the engine. They project in orderly arrays from the microtubules and drive the chromosome onward by sliding motions (Section 4.10).

Second, the spindle itself elongates as overlapping microtubules continue to ratchet past one another and *push* the spindle's two poles even farther apart. These microtubules, too, incorporate motor proteins, and they actively slide past one another where they overlap.

Once each chromatid is separated from its sister, we recognize it as a separate chromosome.

Telophase gets under way as soon as each of the two clusters of chromosomes arrives at a spindle pole. The chromosomes, no longer harnessed to the microtubules, return to threadlike form. Vesicles derived from the old nuclear envelope fuse and form patches of membrane around the chromosomes. Patch joins with patch, and soon a new nuclear envelope separates both clusters of chromosomes from the cytoplasm. If the parent cell was diploid, each cluster contains two chromosomes of each type. With mitosis, remember, each new nucleus has the same chromosome number as the parent nucleus. Once two nuclei form, telophase is over—and so is mitosis.

Prior to mitosis, each chromosome in a cell's nucleus is duplicated, so that it consists of two sister chromatids.

In prophase, microtubules assemble outside the nucleus and start to form a bipolar spindle. The nuclear envelope breaks down, so the microtubules are now free to harness and move the duplicated chromosomes.

At metaphase, all chromosomes are aligned and oriented, with respect to the poles, at the spindle equator.

At anaphase, microtubules move sister chromatids of each chromosome apart, to opposite spindle poles. Some pull on them; others push the spindle poles apart, increasing the distance between the former sister chromatids.

At telophase, there are two clusters of chromosomes, and a new nuclear envelope forms around each cluster.

Thus mitosis forms two daughter nuclei. Each has the same chromosome number as the parent cell's nucleus.

DIVISION OF THE CYTOPLASM

The cytoplasm usually divides at some time between late anaphase and the end of telophase. As you might well conclude by comparing Figure 8.8 with Figure 8.9, the actual mechanism of **cytoplasmic division** (or, as it is often called, cytokinesis) differs among organisms.

Cell Plate Formation in Plants

As described in Section 4.11, most plant cells are walled, which means their cytoplasm can't be pinched in two. Cytoplasmic division of such cells involves **cell plate formation**, as shown in Figure 8.8. By this mechanism, vesicles packed with wall-building materials fuse with one another and with remnants from the microtubular spindle. Together, they form a disklike structure—a cell plate. At this location, deposits of cellulose accumulate. In time, the cellulose deposits are thick enough to form

light micrograph of a cell plate forming in a dividing plant cell

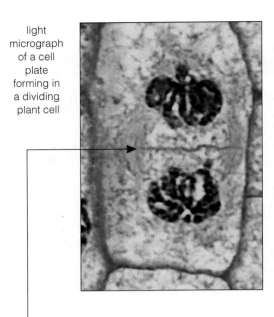

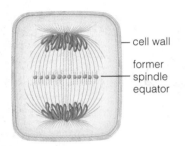 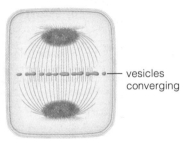 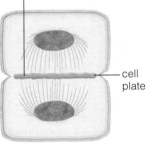 

cell wall
former spindle equator

vesicles converging

cell plate

a As mitosis ends, vesicles converge at the spindle equator. They contain cementing materials and structural materials for a new primary cell wall.

b A cell plate starts forming as membranes of the vesicles fuse. Materials inside the vesicles get sandwiched between two new membranes that elongate along the plane of the cell plate.

c Cellulose is deposited on the inside of the "sandwich." (In time, deposits will form two cell walls. Other deposits will form a middle lamella and cement the walls together; refer to Section 4.11.)

d A cell plate grows at its margins until it fuses with the parent cell's plasma membrane. During growth, new plant cells expand. Their primary wall is still thin, and new material gets deposited on it.

Figure 8.8 Cytoplasmic division of a plant cell, as brought about by cell plate formation.

 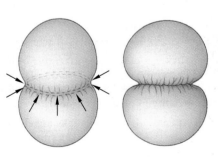  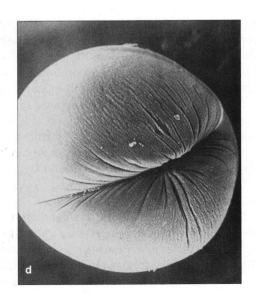

a Mitosis is over, and the spindle is now disassembling.

b At the former spindle equator, a ring of microfilaments attached to the plasma membrane contracts. As its diameter shrinks, it pulls the cell surface inward.

c Contractions continue until the ring cuts the cell in two.

Figure 8.9 (**a–c**) Cytoplasmic division of an animal cell. (**d**) Scanning electron micrograph of the cleavage furrow at the plane of the former spindle's equator. Beneath it, a band of microfilaments attached to the plasma membrane contracts and pulls the surface inward. The furrow deepens until the cell is cut in two.

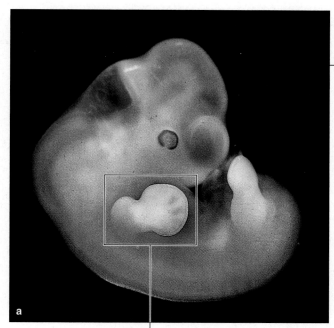

future arm and hand of embryo, five weeks old

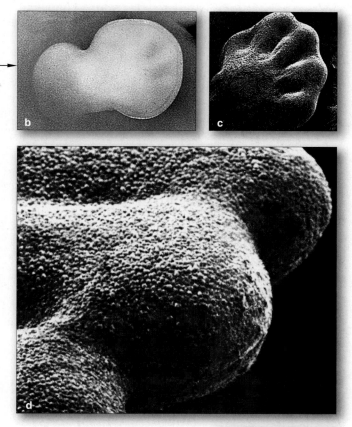

a crosswall. That new crosswall bridges the cytoplasm and divides the parent cell into two daughter cells.

Cytoplasmic Division of Animal Cells

Unlike plant cells, an animal cell isn't confined within a cell wall, and its cytoplasm typically "pinches in two." Look at Figure 8.9, a surface view of a newly fertilized animal egg. An indentation is forming about midway between the egg's two poles ("ends"). The indentation in its plasma membrane is a **cleavage furrow**. It's the first visible sign that the cytoplasm of an animal cell is dividing. The furrow will extend all around the cell and continue to deepen along the plane of the former spindle's midpoint until the cell is cut in two.

A band of microfilaments beneath the cell's plasma membrane generates force for a cut. These cytoskeletal elements are organized to interact and slide past one another (Section 4.10). As they do, they pull the plasma membrane inward until there are two daughter cells. Each has a nucleus, cytoplasm, and plasma membrane.

Perspective on Mitotic Cell Division

This concludes our introduction to mitotic cell division. Look now at your hands and try to visualize the cells making up your palms, thumbs, and fingers. Imagine the divisions that produced all the generations of cells that preceded them as you were developing early on, inside your mother (Figure 8.10). And be grateful for the astonishing precision of the mechanisms that led to their formation at certain times, in certain numbers, for the alternatives can be terrible indeed. Why? Good

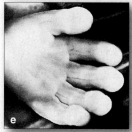

Figure 8.10 Transformation of a paddlelike structure into a human hand by mitosis, cytoplasmic divisions, and other processes that are necessary for embryonic development. (**d**) You can see many individual cells at this high magnification. (**e**) Fingers at a later stage of development.

health—and survival itself—depends absolutely on the proper timing and the completion of cell cycle events, including mitosis. Certain genetic disorders arise from mistakes in the duplication or distribution of even one chromosome. Also, when normal controls that prevent cells from dividing are lost, unchecked cell divisions may destroy surrounding tissues and, ultimately, the organism. Section 8.5 describes a landmark case of such losses, which we will explore further in Section 14.4.

After mitosis, a separate mechanism cuts the cytoplasm into two daughter cells, each with a daughter nucleus.

Cytoplasmic division in plants often involves the formation of a cell plate and a crosswall between the adjoining, new plasma membranes of daughter cells.

Cleavage is a form of cytoplasmic division in animals. Rings of microfilaments around a parent cell's midsection slide past one another in a way that pinches the cytoplasm in two.

Henrietta's Immortal Cells

Each human starts out as a single fertilized egg. By the time of birth, mitotic cell divisions and other processes have resulted in a human body of about a trillion cells. Even in adults, billions of cells still divide. For example, cells of the stomach's lining divide every day. Liver cells usually do not divide, but if part of the liver becomes injured or diseased, repeated cell divisions will yield more new cells until the damaged part is finally replaced.

In 1951, George and Margaret Gey of Johns Hopkins University were trying to develop a way to keep human cells dividing *outside* the body. With such isolated cells, these researchers and others could investigate basic life processes. They also could conduct studies of cancer and other diseases without having to experiment directly on patients and gamble with human lives. The Geys used normal and diseased human cells, which local physicians had sent them. But they couldn't stop the descendants of those precious cells from dying out within a few weeks.

Mary Kubicek, a laboratory assistant, worked with the Geys in their efforts to start a self-perpetuating lineage of cultured human cells. She was about to give up after dozens of failed attempts. Even so, in 1951 she decided to prepare one more sample of cancer cells for culture. She gave the sample the code name **HeLa cells**, for the first two letters of the patient's first and last names.

The HeLa cells began to divide. And divide. And divide again! By the fourth day there were so many cells that Kubicek subdivided them into more culture tubes. Unfortunately, tumor cells in the patient were just as vigorous. Six months after she was diagnosed as having cancer, tumor cells had spread to tissues throughout her body. Only eight months after the diagnosis, Henrietta Lacks, a young woman from Baltimore, was dead.

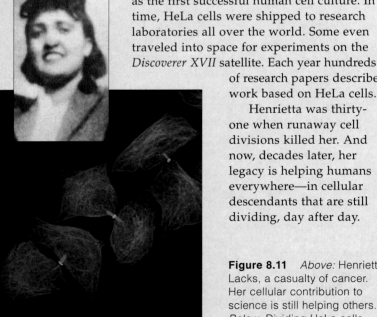

Although Henrietta passed away, some of her cells lived on in the Geys' laboratory as the first successful human cell culture. In time, HeLa cells were shipped to research laboratories all over the world. Some even traveled into space for experiments on the *Discoverer XVII* satellite. Each year hundreds of research papers describe work based on HeLa cells.

Henrietta was thirty-one when runaway cell divisions killed her. And now, decades later, her legacy is helping humans everywhere—in cellular descendants that are still dividing, day after day.

Figure 8.11 *Above:* Henrietta Lacks, a casualty of cancer. Her cellular contribution to science is still helping others. *Below:* Dividing HeLa cells.

SUMMARY
Gold indicates text section

1. Through specific division mechanisms, a parent cell provides each of its daughter cells with the hereditary instructions (DNA) and cytoplasmic machinery required to start up its own operation. *CI, 8.1*

   a. In eukaryotic cells, the nucleus divides by mitosis or meiosis. Cytoplasmic division typically follows.

   b. Prokaryotic cells divide by prokaryotic fission.

2. Each eukaryotic chromosome is one DNA molecule with numerous proteins attached. The chromosomes in a given cell differ in length, shape, and which portion of the hereditary instructions they carry. *8.1*

   a. The "chromosome number" is to the sum total of chromosomes in cells of a given type. Cells having a diploid chromosome number ($2n$) contain two of each kind of chromosome.

   b. Mitosis divides the nucleus into two equivalent nuclei, each with the same chromosome number as the parent cell. It maintains the chromosome number from one cell generation to the next.

   c. Mitosis is the basis of growth, tissue repair, and cell replacements among multicelled eukaryotes. It is the basis of asexual reproduction in many single-celled eukaryotes. (Meiosis occurs only in germ cells.)

3. A duplicated chromosome has two DNA molecules attached at the centromere. For as long as the two stay connected to each other, they are sister chromatids. *8.1*

4. A cell cycle starts when a new cell forms. It proceeds through interphase and ends when the cell reproduces by mitosis and cytoplasmic division. In interphase, a cell carries out its functions. When it is to divide again, the cell increases in mass, roughly doubles the number of its cytoplasmic components, then duplicates each of its chromosomes in preparation for division. *8.2*

5. Mitosis has four continuous stages: *8.3*

   a. Prophase. The duplicated, threadlike chromosomes start to condense. A spindle starts to form. The nuclear envelope starts to break up; its remnants form vesicles during the *transition* to metaphase (or prometaphase). Some microtubules from both poles of the developing spindle push the poles apart. Others directly attach to one of two sister chromatids of each chromosome.

   b. Metaphase. *At* metaphase, all chromosomes have become aligned at the spindle equator.

   c. Anaphase. Microtubules pull sister chromatids of each chromosome away from each other, to opposite spindle poles. Now each type of parental chromosome is represented by a daughter chromosome at both poles.

   d. Telophase. Chromosomes decondense to threadlike form. A new nuclear envelope forms around them. Each nucleus has the parental chromosome number.

6. Separate mechanisms divide the cytoplasm near the end of nuclear division or afterward (in plants, by cell plate formation; in animals, by cleavage). *8.4*

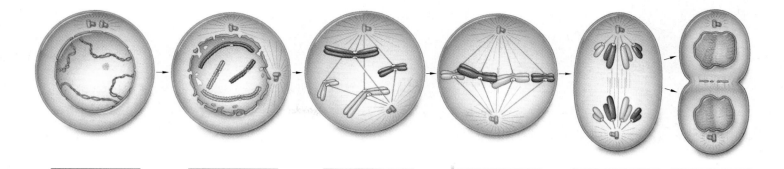

_____   _____   _____   _____   _____   _____

Review Questions

1. Define mitosis and meiosis, two mechanisms that operate in eukaryotic cells. Does either one divide the cytoplasm? *8.1*

2. Define somatic cell and germ cell. *8.1*

3. What are chromosomal proteins? Give an example, and briefly state how it interacts with a DNA molecule. *8.1*

4. What is a chromosome called when it is in the unduplicated state? In the duplicated state (with two sister chromatids)? *8.1*

5. Describe the microtubular spindle and its functions. What role do motor proteins play in its operation? *8.3*

6. Using the diagram above as a guide, name and describe the key features of the stages of mitosis. *8.3*

7. Briefly explain how cytoplasmic division differs between a typical plant cell and typical animal cell. *8.4*

Self-Quiz   ANSWERS IN APPENDIX III

1. Mitosis and cytoplasmic division function in _____ .
   a. asexual reproduction of single-celled eukaryotes
   b. growth, tissue repair, and sometimes asexual reproduction in many multicelled eukaryotes
   c. gamete formation in prokaryotes
   d. both a and b

2. A duplicated chromosome has _____ chromatid(s).
   a. one   b. two   c. three   d. four

3. In a chromosome, a _____ is a constricted region with attachment sites for microtubules.
   a. chromatid   b. cell plate   c. centromere   d. cleavage

4. A somatic cell having two of each type of chromosome has a(n) _____ chromosome number.
   a. diploid   b. haploid   c. tetraploid   d. abnormal

5. Interphase is the part of the cell cycle when _____ .
   a. a cell ceases to function
   b. a germ cell forms its spindle apparatus
   c. a cell grows and duplicates its DNA
   d. mitosis proceeds

6. After mitosis, the chromosome number of a daughter cell is _____ the parent cell's.
   a. the same as   c. rearranged compared to
   b. one-half   d. doubled compared to

7. Only _____ is not a stage of mitosis.
   a. prophase   b. interphase   c. metaphase   d. anaphase

8. Match each stage with the events listed.
   _____ metaphase   a. sister chromatids move apart
   _____ prophase   b. chromosomes start to condense
   _____ telophase   c. chromosomes decondense and
   _____ anaphase   daughter nuclei form
                     d. all duplicated chromosomes are
                       aligned at the spindle equator

Critical Thinking

1. Suppose you have a way to measure the amount of DNA in a single cell during the cell cycle. You first measure the amount at the G1 phase. At what points during the rest of the cell cycle would you predict changes in the amount of DNA per cell?

2. The cervix is part of the uterus, a chamber in which embryos develop. The *Pap smear* is a screening procedure that can detect *cervical cancer* in its earliest stages. Treatments range from freezing precancerous cells or killing them with a laser beam to removal of the uterus (a hysterectomy). Treatment is 90+ percent if this cancer is detected early. Survival chances plummet to less than 9 percent after the cancer spreads.

     Most cervical cancers develop slowly. Unsafe sex increases the risk. A key risk factor is infection by human papillomaviruses that cause genital warts (Section 39.19). In 93 percent of all cases, viral genes coding for the tumor-inducing proteins had become inserted into the DNA of previously normal cervical cells.

     Not all women request Pap smears. Many wrongly believe the procedure is costly. Many don't recognize the importance of abstinence or "safe" sex. Others simply don't want to think about whether they have cancer. Knowing what you've learned so far about the cell cycle and cancer, what would you say to a woman who falls into one or more of these groups?

3. Pacific yews (*Taxus brevifolius*) face extinction. People started stripping its bark and killing trees when they heard that *taxol*, a chemical extract from the bark, may fight breast and ovarian cancer. (Synthesizing taxol in laboratories may save the species.) Taxol is a poison that prevents microtubule disassembly. What does this tell you about its potential as an anticancer drug?

4. X rays and gamma rays emitted from some radioisotopes chemically damage DNA, especially in cells engaged in DNA replication. High-level exposure can result in *radiation poisoning*. Hair loss and damage to the lining of the stomach and intestines are two early symptoms. Speculate why. Also speculate on why highly focused radiation therapy is used against some cancers.

Selected Key Terms

anaphase *8.2*	cytoplasmic	mitosis *8.1*
cell cycle *8.2*	division *8.4*	nucleosome *8.1*
cell plate	diploid (chromosome	prophase *8.2*
formation *8.4*	number) *8.1*	reproduction *CI*
centriole *8.3*	germ cell *8.1*	sister
centromere *8.1*	HeLa cell *8.5*	chromatid *8.1*
chromosome *8.1*	histone *8.1*	somatic cell *8.1*
chromosome	interphase *8.2*	spindle
number *8.1*	meiosis *8.1*	apparatus *8.2*
cleavage furrow *8.4*	metaphase *8.2*	telophase *8.2*

Readings

Murray, A., and M. Kirschner. March 1991. "What Controls the Cell Cycle?" *Scientific American* 264(3): 56–63.

On-Line readings at Student Guide for InfoTrac:
www.brookscole.com/biology

MEIOSIS

Octopus Sex and Other Stories

The couple clearly are interested in each other. First he caresses her with one tentacle, then with another—and another and another. She reciprocates with a hug here, a squeeze there. This goes on for hours. Finally the male reaches under his mantle, a fold of tissue that drapes around most of his body. He removes a packet of sperm from a reproductive organ and inserts it into an egg chamber beneath the female's mantle. For every sperm that fertilizes an egg, a new octopus may develop.

Unlike the one-to-one coupling between a male and a female octopus, sex for the slipper limpet is a group enterprise. Slipper limpets are marine animals, relatives of the familiar land snails. Before becoming transformed into a sexually mature adult, a slipper limpet must go through a free-living stage of development called a larva. When a limpet larva is about to undergo its programmed transformation, it settles on a rock or pebble or shell. If it settles down all by itself, it will become a female. If another larva settles on the first limpet and proceeds to develop, that second limpet will function right off as a male. However, the second one will switch gears and develop into a female if a third limpet develops into a male on top of *it*. Later, that third limpet will also become a female if still another limpet develops into a male on top of it—and so on amongst ten or more limpets!

Slipper limpets typically live in such piles, with the bottom one always being the oldest female and the uppermost one being the youngest male (Figure 9.1a). Until they switch their sex, male limpets release sperm, which

Figure 9.1 Examples of variations in reproductive modes of eukaryotic organisms. (**a**) Slipper limpets, busily perpetuating the species by group participation in sexual reproduction. The tiny crab in the foreground is merely a passerby. (**b**) Live birth of an aphid, a type of insect that reproduces sexually in autumn but can switch to an asexual mode in summer.

fertilize a female's eggs, which grow up to become males but then, most likely, females—and so it goes, from one sexually flexible generation to the next.

Limpets are not alone in having unusual variations in their mode of reproduction. For example, sexual reproduction is common in many life cycles, but so are asexual episodes based on mitotic cell divisions. Orchids, dandelions, and many other plants reproduce very well with or without sex. Aquatic animals called flatworms can engage in sex or split their small body into two roughly equivalent parts, which each grow into a new flatworm.

And what about those aphids! In summer, nearly all aphids are females, which produce more females from *unfertilized* egg cells (Figure 9.1b). Only when autumn approaches do male aphids develop and do their part in the sexual phase of the life cycle. Even then, females that manage to survive over the winter can do without the opposite sex. Come summer, they begin another round of producing offspring all by themselves.

These examples only hint at the immense variation in reproductive modes among eukaryotic organisms. And yet, despite the variation, *sexual* reproduction dominates nearly all of the life cycles. And it always involves the same events. Briefly, cells set aside for sexual reproduction duplicate their chromosomes before they divide. **Germ cells**, immature reproductive cells that develop in male and female animals, are a fine example. They undergo meiosis and cytoplasmic division. Cellular descendants of germ cells mature and become **gametes**, or sex cells. When a male and female gamete manage to get together, they form the first cell of a new individual by way of fertilization.

With this chapter, we turn to the kinds of cells that serve as the bridge between generations of organisms. Specialized phases of reproduction and development, including asexual episodes, loop out from the basic life cycle of many eukaryotic species. Regardless of the specialized details, all of the life cycles turn on three events: *meiosis, the formation of gametes, and fertilization*. These three interconnected events are the hallmarks of sexual reproduction. As you will see in many chapters throughout the book, they have contributed to the diversity of life.

Key Concepts

1. Sexual reproduction proceeds through three events: meiosis, gamete formation, and fertilization. Sperm and eggs are familiar gametes.

2. Meiosis, a nuclear division mechanism, occurs only in cells that are set aside for sexual reproduction. The immature germ cells of male and female animals are examples. Meiosis sorts out a germ cell's chromosomes into four new nuclei. After meiosis, gametes form by way of cytoplasmic division and other events.

3. Cells with a diploid chromosome number contain two of each type of chromosome characteristic of the species. The two function as a pair during meiosis. Commonly, one chromosome of the pair is maternal, with hereditary instructions from a female parent. The other is paternal, with the same categories of hereditary instructions from a male parent.

4. Meiosis divides the chromosome number by half for each forthcoming gamete. Thus, if both parents have a diploid chromosome number ($2n$), the gametes that form will be haploid (n). Later, a union of two gametes at fertilization will restore the diploid number in the new individual ($n + n = 2n$).

5. Each pair of chromosomes typically swap segments during meiosis and so exchange hereditary information about certain traits. Also, meiosis randomly assigns one of each pair of chromosomes to a forthcoming gamete; *which* one of the pair ends up in a given gamete is a matter of chance. Hereditary instructions are further shuffled at fertilization. All three reproductive events lead to variations in traits among offspring.

6. In most plants, spore formation and other events intervene between meiosis and gamete formation.

COMPARING SEXUAL WITH ASEXUAL REPRODUCTION

When an orchid, flatworm, or aphid reproduces all by itself, what sort of offspring does it get? By the process of **asexual reproduction**, one parent alone produces offspring, and each offspring inherits the same number and kinds of genes as its parent. **Genes** are particular stretches of chromosomes—that is, of DNA molecules. Taken together, the genes for each species contain all the heritable bits of information necessary to make new individuals. Rare mutations aside, this means asexually produced individuals can only be *clones*, or genetically identical copies of the parent.

Inheritance gets much more interesting with **sexual reproduction**. This process involves meiosis, formation of gametes, and fertilization (union of the nuclei of two gametes). In most sexual reproducers, such as humans, the first cell of a new individual contains *pairs of genes* on pairs of homologous chromosomes. Typically, one of each pair is maternal and the other paternal in origin.

If instructions encoded in every pair of genes were identical down to the last detail, sexual reproduction would produce clones, also. Just imagine—you, every single person you know, the entire human population might be a clone, with everybody looking alike. But the two genes of a pair might *not* be identical. Why not? A gene's molecular structure can change; that is what we mean by mutation. Depending on their structure, two genes that happen to be paired in a person's cells may "say" slightly different things about a trait. Each unique molecular form of the same gene is called an **allele**.

Such tiny differences affect thousands of traits. For example, whether your chin has a dimple depends on which pair of alleles you inherited at one chromosome location. One kind of allele at that location says "put a dimple in the chin." Another kind says "no dimple." This leads us to a key reason why members of sexually reproducing species don't all look alike. *Through sexual reproduction, offspring inherit new combinations of alleles, which lead to variations in the details of their traits.*

This chapter gets into the cellular basis of sexual reproduction. More importantly, it starts you thinking about far-reaching effects of gene shufflings at different stages of the process. The process introduces variations in traits among offspring that are typically acted upon by agents of natural selection. Thus, *variation in traits is a foundation for evolutionary change.*

Asexual reproduction produces genetically identical copies of the parent. Sexual reproduction introduces variations in the details of traits among offspring.

Sexual reproduction dominates the life cycles of eukaryotic species. Meiosis, formation of gametes, and fertilization are the basic events of this process.

HOW MEIOSIS HALVES THE CHROMOSOME NUMBER

Think "Homologues"

Think back to the preceding chapter and its focus on mitotic cell division. Unlike mitosis, **meiosis** divides chromosomes into separate parcels not once but *twice* prior to cell division. Unlike mitosis, it is the first step leading to the formation of gametes.

Gametes, recall, are sex cells such as sperm or eggs. In nearly all multicelled eukaryotic organisms, gametes develop from cells that arise in specialized reproductive structures or organs. Figure 9.2 shows a few examples of where gametes form.

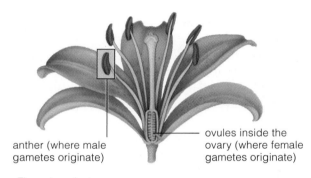

anther (where male gametes originate)

ovules inside the ovary (where female gametes originate)

a Flowering plant

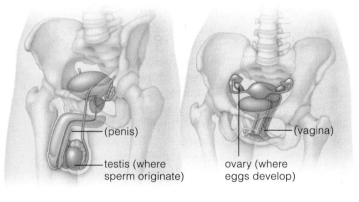

(penis)

(vagina)

testis (where sperm originate)

ovary (where eggs develop)

b Human male   **c** Human female

Figure 9.2 Examples of gamete-producing structures.

As you know, the **chromosome number** is the sum total of chromosomes in cells of a given type (Section 8.1). Germ cells start out with the same chromosome number as somatic cells (the rest of the body's cells). If a cell has a **diploid number** (2n), it has a *pair* of each type of chromosome, often from two parents. Except for a pairing of nonidentical sex chromosomes, each pair has the same length, shape, and assortment of genes. And they line up with each other at meiosis. We call them **homologous chromosomes** (*hom–* means alike).

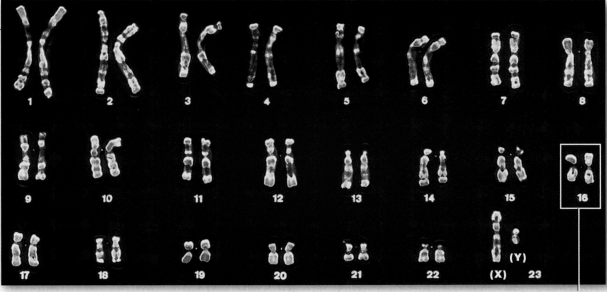

Figure 9.3 From a diploid cell of a human male, twenty-three pairs of homologous chromosomes. The last two are a pair of sex chromosomes (XY or XX). An XY pair does differ in length, shape, and which genes they carry, but the two still pair up as homologues during meiosis.

one pair of homologous chromosomes

As you can probably deduce from Figure 9.3, your own germ cells have 23 + 23 homologous chromosomes. After meiosis, 23 chromosomes—one of each type—end up in gametes. That is, meiosis halves the chromosome number, so gametes have a **haploid number** (n).

Two Divisions, Not One

Meiosis is like mitosis in some ways, but the outcome is different. As in mitosis, a germ cell duplicates its DNA in interphase. The two DNA molecules and their associated proteins remain attached at the centromere, the notably constricted region along their length. For as long as they remain attached, we call them **sister chromatids**:

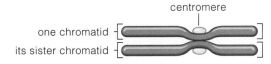

one chromosome in the duplicated state

As in mitosis, the microtubules of a spindle apparatus move the chromosomes in prescribed directions.

With meiosis alone, however, *chromosomes go through two consecutive divisions that end with the formation of four haploid nuclei*. There is no interphase between the nuclear divisions, which we call meiosis I and meiosis II:

	MEIOSIS I		MEIOSIS II
interphase (*DNA replication before meiosis I*)	PROPHASE I METAPHASE I ANAPHASE I TELOPHASE I	no interphase (*no DNA replication before meiosis II*)	PROPHASE II METAPHASE II ANAPHASE II TELOPHASE II

Each duplicated chromosome becomes aligned with its partner—*homologue to homologue*—during meiosis I.

After the two chromosomes of each pair have lined up with each other, they are moved apart:

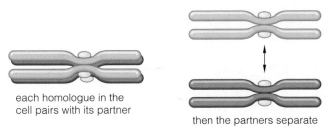

each homologue in the cell pairs with its partner

then the partners separate

The cytoplasm typically starts to divide at some point after each homologue is separated from its partner. The two daughter cells formed this way are haploid; each has *one* of each type of chromosome. But don't forget, the chromosomes are still in the duplicated state.

Next, during meiosis II, *the two sister chromatids of each chromosome are separated from each other*:

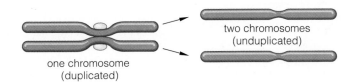

one chromosome (duplicated)

two chromosomes (unduplicated)

Each sister chromatid is now a chromosome in its own right. Four nuclei now form. Often the cytoplasm divides once more. The final outcome is four haploid cells.

Figure 9.4, on the next two pages, gives you a closer look at key events of meiosis and their consequences.

Meiosis is a type of nuclear division mechanism. It reduces the chromosome number of a parental cell by half—to the haploid number (n)—in daughter cells.

Meiosis proceeds only in immature cells set aside for sexual reproduction, such as animal germ cells. It is the first step leading to the formation of gametes.

VISUAL TOUR OF THE STAGES OF MEIOSIS

MEIOSIS I

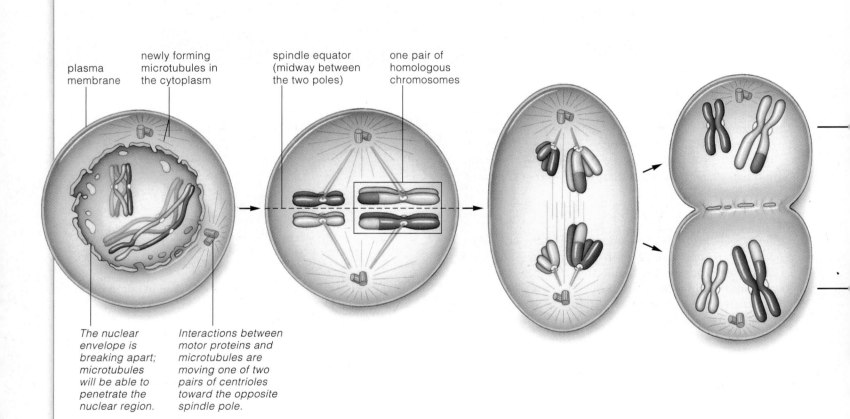

plasma membrane

newly forming microtubules in the cytoplasm

spindle equator (midway between the two poles)

one pair of homologous chromosomes

The nuclear envelope is breaking apart; microtubules will be able to penetrate the nuclear region.

Interactions between motor proteins and microtubules are moving one of two pairs of centrioles toward the opposite spindle pole.

PROPHASE I

Each duplicated chromosome is in threadlike form, but now starts to condense. It pairs with its homologue, and the two typically swap segments. The swapping, called crossing over, is indicated by the break in color on the pair of larger chromosomes. Some of the microtubules of a newly forming spindle become attached to each chromosome's centromere.

METAPHASE I

As in mitosis, motor proteins attached to microtubules move the chromosomes and move the spindle poles apart. They tug the chromosomes into position midway between the spindle poles. Thus the spindle becomes fully formed, owing to dynamic interactions among the motor proteins, microtubules, and the chromosomes themselves.

ANAPHASE I

Microtubules extending from the poles and overlapping at the spindle equator *lengthen* and push the poles apart. Other microtubules extending from the poles to the chromosomes *shorten*, thereby pulling each chromosome away from its homologous partner. These motions move the homologous partners to opposite poles.

TELOPHASE I

The cytoplasm of the germ cell divides at some point. There are now two haploid (*n*) cells. Each cell has one of each type of chromosome that was present in the parent (2*n*) cell. *However, all chromosomes are still in the duplicated state.*

Figure 9.4 Sketches of meiosis in a generalized animal cell. This nuclear division mechanism reduces the chromosome number in immature reproductive cells by half (to the haploid number) for forthcoming gametes. To keep things simple, we track only two pairs of homologous chromosomes. Maternal chromosomes are shaded *purple* and paternal chromosomes *blue*. The light micrographs above show corresponding stages in the formation of pollen grains in a lily (*Lilium regale*).

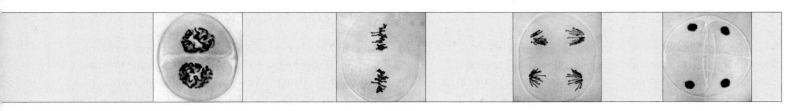

MEIOSIS II

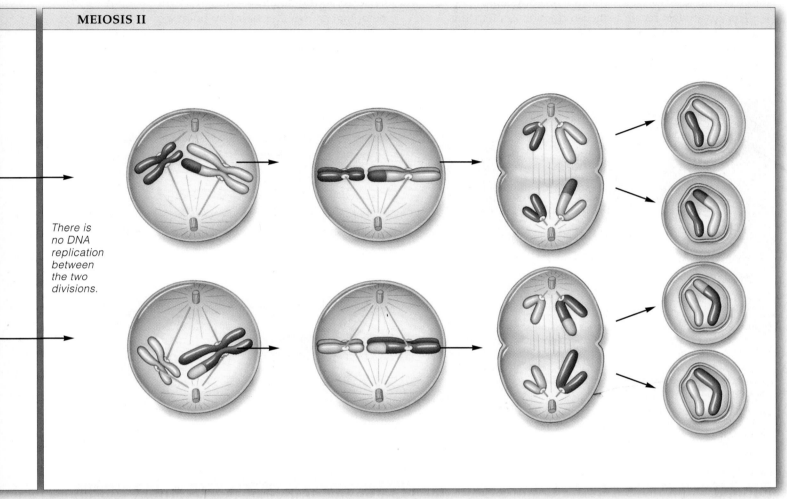

There is no DNA replication between the two divisions.

PROPHASE II

Microtubules have already moved one member of the centriole pair to the opposite pole of the spindle in each of the two daughter cells. Now, during prophase II, microtubules attach to the chromosomes, and motor proteins drive the movement of chromosomes toward the spindle's equator.

METAPHASE II

In each daughter cell, interactions among motor proteins, spindle microtubules, and each duplicated chromosome have moved all of the chromosomes so that they are positioned at the spindle equator, midway between the two poles.

ANAPHASE II

The attachment between the two chromatids of each chromosome breaks. Each of the former "sister chromatids" is now a chromosome in its own right. Motor proteins drive the movement of the newly separated chromosomes to opposite poles of the spindle.

TELOPHASE II

By the time telophase II is over, there will be four daughter nuclei. When cytoplasmic division is completed, each new, daughter cell will have a haploid chromosome number (*n*). All of the chromosomes will now be in the unduplicated state.

Of the four haploid cells that form by way of meiosis and cytoplasmic divisions, one or all may develop into gametes and function in sexual reproduction. In plants, cells that form after meiosis is over may develop into spores, which take part in a stage of the life cycle that precedes gamete formation. (The telophase II micrograph shows spores that will develop into pollen grains.)

A CLOSER LOOK AT KEY EVENTS OF MEIOSIS I

Study the overview in Sections 9.2 and 9.3, and you can sense the basic function of meiosis: *the reduction of the chromosome number by half for the forthcoming gametes.* However, two other major events occur during meiosis: crossing over at prophase I and the random alignment of homologues at metaphase I. Both contribute greatly to the adaptive advantage of sexual reproduction.

The advantage, recall, is the production of offspring with new combinations of alleles. Those combinations are translated into a new generation of individuals that differ in the details of some number of traits.

Crossing Over in Prophase I

Prophase I of meiosis is a time of major gene shufflings. Reflect on Figure 9.5a, which shows two chromosomes condensed to threadlike form. All chromosomes in a germ cell condense this way. As they do, each is drawn close to its homologue. Molecular interactions stitch homologues together point by point along their length, with little space between. The intimate, parallel array favors **crossing over**, a molecular interaction between two of the *non*sister chromatids of a pair of homologous

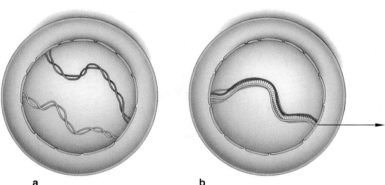

a   b

Figure 9.5   Key events of prophase I, the first stage of meiosis. For clarity, this diagram shows only one pair of homologous chromosomes and one crossover event. (Typically, more than one crossover occurs.) *Blue* signifies the paternal chromosome; *purple* signifies its maternal homologue.

(**a**) Both chromosomes were duplicated earlier, in interphase. Early in prophase I, the sister chromatids of each duplicated chromosome are in thin, threadlike form. They are positioned so closely together, they look like a single thread.

(**b**) Each chromosome becomes zippered to its homologue, so all four chromatids are intimately aligned. When the two sex chromosomes have different forms (such as X paired with Y) they still get zippered together, although only in a small region along their length.

(**c,d**) We show the pair of chromosomes as if they were already condensed, then teased apart to give you a sense of what goes on. Bear in mind, their double-stranded DNA molecules are still tightly aligned at this stage. The intimate contact allows one crossover (and usually more) to happen at intervals along the length of nonsister chromatids.

(**e**) Nonsister chromatids exchange segments. As prophase I ends, the chromosomes continue to condense into thicker, rodlike forms. Then they unzipper from each other except at places where they physically cross each other. Such places are called chiasmata (singular, chiasma, meaning "cross"). Each chiasma does not last long, but it's indirect evidence of a crossover at some place in the chromosomes.

(**f**) What is the function of crossing over? It breaks up old combinations of alleles and puts new ones together in pairs of homologous chromosomes.

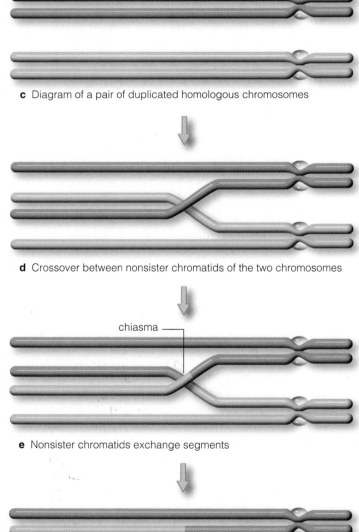

c Diagram of a pair of duplicated homologous chromosomes

d Crossover between nonsister chromatids of the two chromosomes

chiasma

e Nonsister chromatids exchange segments

f Homologues have new combinations of alleles

chromosomes. Nonsister chromatids break at the same places along their length. At these break points, they exchange corresponding segments—that is, genes.

Gene swapping would be pointless if each type of gene never varied. But remember, a gene can come in slightly different forms: alleles. You can bet that some number of the alleles on one chromosome will *not* be identical to their partner alleles on the homologue. Each crossover is a chance to swap slightly different versions of hereditary instructions for particular traits.

We will look at the mechanism of crossing over in later chapters. For now, it is enough to remember this: *Crossing over leads to genetic recombination, which in turn leads to variation in traits among offspring.*

Metaphase I Alignments

Major shufflings of whole chromosomes begin during the transition from prophase I to metaphase I, the second stage of meiosis. Suppose the shufflings are happening right now in one of your germ cells. Crossovers have already made genetic mosaics of the chromosomes, but put this aside in order to simplify tracking. Just call the twenty-three chromosomes you inherited from your mother the *maternal* chromosomes and the twenty-three homologues from your father the *paternal* chromosomes.

Microtubules already oriented one chromosome of each pair toward one spindle pole and its homologue toward the other pole (refer to Section 8.3). They are busily moving all of the chromosomes, which soon will become positioned at the spindle's equator.

Will all maternal chromosomes be directed to one spindle pole and all paternal chromosomes directed to the other? Maybe, but probably not. Remember, initial contacts between microtubules and chromosomes are random. Because of the random grabs, the positioning of maternal or paternal chromosomes at the spindle's equator at metaphase I follows no particular pattern. Carry this thought one step further. *Either one* of each pair of homologous chromosomes can end up at either spindle pole after they move apart at anaphase I.

Think about the possibilities when you are tracking merely three pairs of homologues. As you can see from Figure 9.6, by metaphase I, three pairs of homologues may be arranged in any one of four possible positions. Here, eight combinations (2^3) of maternal and paternal chromosomes are possible for forthcoming gametes.

Of course, a human germ cell has twenty-three pairs of homologous chromosomes, not just three. So every time a human germ cell gives rise to sperm or eggs, we can expect a grand total of *8,388,608* (or 2^{23}) possible combinations of maternal and paternal chromosomes!

Moreover, in each sperm or egg, many hundreds of alleles inherited from the mother might not "say" the

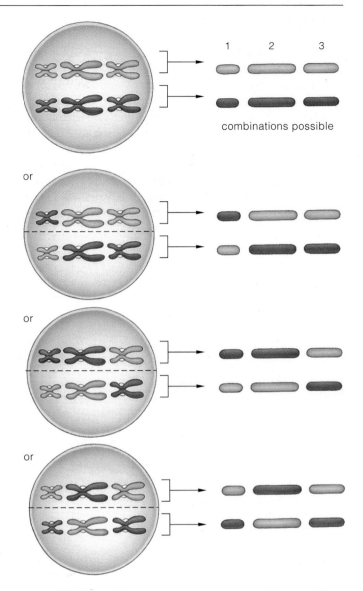

combinations possible

or

or

or

Figure 9.6 Possible outcomes of the random alignment of three pairs of homologous chromosomes at metaphase I. We label three types of chromosomes 1, 2, and 3. The maternal chromosomes are *purple*; paternal ones are *blue*. With merely four possible alignments, eight combinations of maternal and paternal chromosomes are possible in gametes.

exact same thing about hundreds of different traits as the alleles inherited from the father. Are you beginning to get an idea of why such fascinating combinations of traits show up even in the same family?

Crossing over is an interaction between a pair of homologous chromosomes. It breaks up old combinations of alleles and puts new ones together during prophase I of meiosis.

The random attachment and subsequent positioning of each pair of maternal and paternal chromosomes during metaphase I lead to different combinations of maternal and paternal traits in each generation of offspring.

FROM GAMETES TO OFFSPRING

The gametes that form following meiosis are not all the same in their details. For example, human sperm have one tail, opossum sperm have two, and roundworm sperm have none. Crayfish sperm look like pinwheels. Most eggs are microscopic in size, yet an ostrich egg tucked inside its shell is as large as a baseball. From its appearance alone, you might not believe that a plant's gamete is even remotely like an animal's.

Later chapters explain how gametes form in the life cycles of specific organisms, including humans. Figure 9.7 and the rest of this section may help you keep the details in perspective.

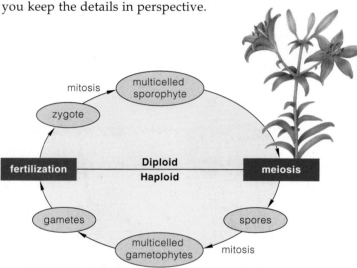

a Generalized life cycle for most kinds of plants

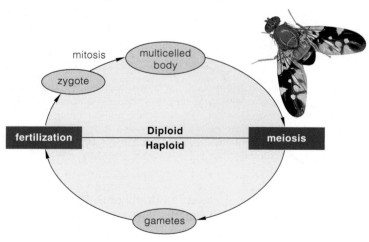

b Generalized life cycle for animals

Figure 9.7 Generalized life cycles for (**a**) most plants and (**b**) animals. The zygote is the first cell that forms when the nuclei of two gametes fuse at fertilization.

For plants, a sporophyte (spore-producing body) develops, by way of mitotic cell divisions, from the zygote. After meiosis, gametophytes (gamete-producing bodies) form. A lily plant is a sporophyte. Gametophytes form in parts of its flowers.

Chapters 20 through 24, 28, and 39 have specific examples of life cycles for representative organisms.

Gamete Formation in Plants

For pine trees, apple trees, roses, dandelions, corn, and nearly all other familiar plants, certain events intervene between meiosis and the time that gametes develop and mature. Among other things, spores form.

Spores are haploid resting cells, often walled, that are good at resisting drought, cold, and other adverse environmental conditions. When favorable conditions return, spores germinate (resume growth) and develop into a haploid body or structure that produces gametes. So *gamete*-producing bodies and *spore*-producing bodies develop during the life cycle of most kinds of plants. Figure 9.7a is a generalized diagram of these events.

Gamete Formation in Animals

In male animals, gametes form by a process known as spermatogenesis. A diploid germ cell grows in size in the male's reproductive system. It becomes a primary spermatocyte. This large immature cell enters meiosis and cytoplasmic divisions. Four haploid cells result and develop into spermatids (Figure 9.8). These immature cells change in form and develop a tail. Each becomes a **sperm**, a common type of mature male gamete.

In female animals, gametes form by a process called oogenesis. In human females, for instance, a diploid germ becomes an **oocyte**, or immature egg. Unlike sperm, an oocyte stockpiles many cytoplasmic components, and its daughter cells differ in size and function (Figure 9.9). When an oocyte divides after meiosis I, one daughter cell (a secondary oocyte) gets nearly all the cytoplasm. The other cell, the first polar body, is small. Later, both cells undergo meiosis II and cytoplasmic division. One daughter cell of the secondary oocyte develops into a second polar body. The other gets most of the cytoplasm and develops into a gamete. The mature female gamete is called an ovum (plural, ova) or, more often, an **egg**.

And so we have one egg and three polar bodies. The polar bodies don't function as gametes and aren't rich in nutrients or cytoplasm. In time they degenerate. But the very fact that they formed means the egg has a suitable (haploid) chromosome number. Also, by getting most of the cytoplasm, the egg has enough start-up machinery to support the new individual right after fertilization.

More Shufflings at Fertilization

The chromosome number characteristic of the parents is restored at **fertilization**, a time when a female and male gamete unite and their haploid nuclei fuse. Fertilization would double the chromosome number for each new generation if meiosis didn't precede it. Such doublings would disrupt the hereditary instructions, usually for

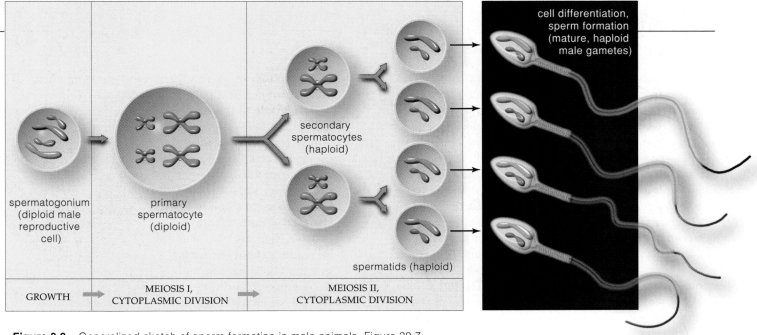

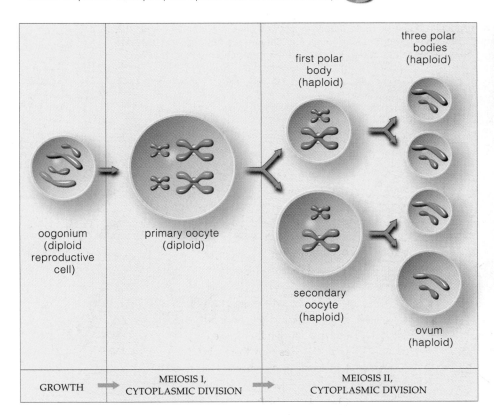

Figure 9.8 Generalized sketch of sperm formation in male animals. Figure 39.7 shows a specific example (how sperm form in human males).

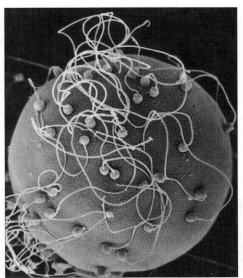

Figure 9.9 Egg formation in female animals. The eggs are far larger than sperm, as the micrograph of sea urchin gametes (*above*) suggests. Also, the three polar bodies are far smaller than the egg. Figure 39.9 gives a specific example (how human eggs form).

the worse. Why? The instructions work as an intricate, fine-tuned package in each individual.

Fertilization contributes to variation in offspring. Reflect on the possibilities for humans alone. During prophase I, an average of two or three crossovers takes place in each human chromosome. Even without these crossovers, the random positioning of pairs of paternal and maternal chromosomes at metaphase I results in one of millions of possible chromosome combinations

in each gamete. And of all the male and female gametes that are produced, *which* two actually get together is a matter of chance. The sheer number of combinations that can exist at fertilization is staggering!

Cumulatively, crossing over, the distribution of random mixes of homologous chromosomes into gametes, and fertilization contribute to variation in the traits of offspring.

MEIOSIS AND MITOSIS COMPARED

In this unit our focus has been on two nuclear division mechanisms. Single-celled eukaryotic species reproduce asexually by way of mitosis, followed by cytoplasmic division. Many multicelled eukaryotic species depend on mitosis and cytoplasmic division during episodes of asexual reproduction in their life cycle. All depend on it for growth and tissue repair. By contrast, meiosis occurs only in reproductive cells, such as germ cells that give rise to the gametes used in sexual reproduction. Figure 9.10 summarizes the main similarities and differences between the two nuclear division mechanisms.

The end results of the two mechanisms differ in a crucial way. *Mitotic cell division only produces clones—genetically identical copies of a parent cell. But meiotic cell division, in conjunction with fertilization, promotes variation in traits among offspring.* First, crossing over at prophase I of meiosis puts new combinations of alleles in chromosomes. Second, the random assignment of either member of a pair of homologous chromosomes to either pole of the spindle at metaphase I affects gametes, which end up with mixes of maternal and paternal alleles. And third, different combinations of alleles are brought together simply by chance during fertilization. In later chapters, you will be reading about the ways in which both meiosis and fertilization contribute to the truly stunning diversity and evolution of sexually reproducing organisms.

A *somatic cell* with a diploid chromosome number (2n) is at interphase. Before mitotic division begins, its DNA is replicated (all chromosomes are duplicated).

Figure 9.10 Summary of mitosis and meiosis. Both diagrams use a diploid (2n) animal cell as the example. They are arranged to help you compare similarities and differences between the division mechanisms. The maternal chromosomes are coded *purple* and the paternal chromosomes *blue*.

MEIOSIS I

A *germ cell* with a diploid chromosome number (2n) is at interphase. Before mitotic division begins, its DNA is replicated (all chromosomes are duplicated).

PROPHASE I

Each duplicated chromosome (consisting of two sister chromatids) condenses to threadlike form, then rodlike form. *Crossing over* occurs. Each chromosome unzips from its homologue. Each gets attached to the spindle in transition to metaphase.

METAPHASE I

All chromosomes are now positioned at the spindle's equator.

ANAPHASE I

Each chromosome is separated from its homologue. They are moved to opposite poles of the spindle.

TELOPHASE I

When the cytoplasm divides, there are two cells. Each has a haploid (n) number of chromosomes, but these are still in the duplicated state.

MITOSIS

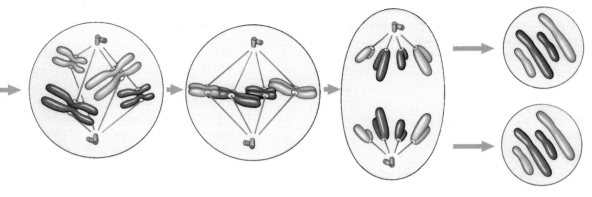

PROPHASE

Each duplicated chromosome (consisting of two sister chromatids) condenses from threadlike form to rodlike form. Each gets attached to the spindle during the transition to metaphase.

METAPHASE

All chromosomes are now positioned at the spindle's equator.

ANAPHASE

Sister chromatids of each chromosome are separated from each other. These new, daughter chromosomes are moved to opposite poles of the spindle.

TELOPHASE

When the cytoplasm divides, there are two cells. Each is diploid (2n)—*it has the same chromosome number as the parent cell.*

MEIOSIS II

There is no DNA replication between the two divisions.

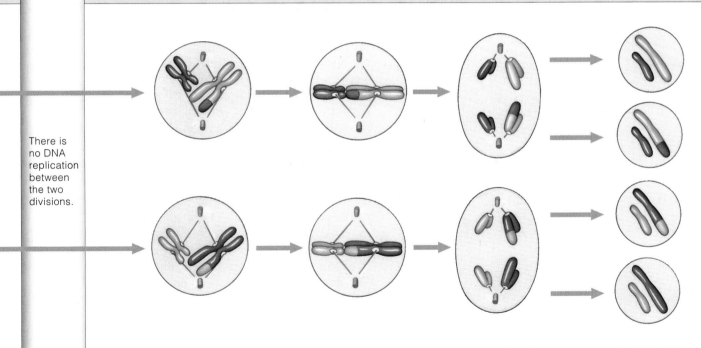

PROPHASE II

Before prophase II, the two centrioles in each new cell were moved apart and a new spindle formed. Now, each chromosome becomes attached to the spindle and starts moving toward its equator.

METAPHASE II

All chromosomes are now positioned at the spindle's equator.

ANAPHASE II

Sister chromatids of each chromosome are separated from each other. These new, daughter chromosomes are moved to opposite poles of the spindle.

TELOPHASE II

Four daughter nuclei form. When the cytoplasm divides, each new cell is haploid (n). *The original chromosome number has been reduced by half.* One or all of these cells may become gametes.

SUMMARY

Gold indicates text section

1. Eukaryotic life cycles often have asexual as well as sexual phases. Asexual reproduction results in clones. Sexual reproduction (by meiosis, gamete formation, and fertilization) leads to variation in traits. *CI, 9.1, 9.2*

   a. Meiosis, a nuclear division mechanism, reduces the chromosome number of a parent germ cell by half. It precedes the formation of haploid gametes, such as sperm in males and eggs in females.

   b. At fertilization, a sperm and egg nuclei fuse. This event restores the chromosome number (Figure 9.11).

2. A germ cell with a diploid chromosome number (2*n*) has *two* of each type of chromosome characteristic of its species. Commonly, one of each pair of chromosomes is maternal and the other is paternal. *9.2*

3. Each pair of maternal and paternal chromosomes has homology; the two are alike. Except for a pairing of nonidentical sex chromosomes (e.g., X with Y), the two have the same length, shape, and gene sequence along their length. They interact during meiosis. *9.2*

4. Chromosomes become duplicated during interphase. Each consists of two DNA molecules that stay attached (as sister chromatids) during mitosis. *9.2*

5. Meiosis consists of two consecutive divisions, which both require a microtubular spindle apparatus. *9.1–9.4*

   a. During meiosis I, spindle microtubules attach to the centromere region of each duplicated chromosome. Motor proteins attached to the microtubules separate it from its partner, the homologous chromosome.

   b. In meiosis II, similar interactions move the sister chromatids of each chromosome away from each other.

6. Meiosis I, the first nuclear division, is characterized by the following events and outcomes: *9.2–9.4*

   a. Crossing over occurs in prophase I. Two nonsister chromatids of each pair of homologous chromosomes break at corresponding sites and exchange segments. This puts new combinations of alleles together. Alleles (slightly different molecular forms of the same gene) specify different versions of the same trait.

   b. Different combinations of alleles lead to variation in the details of a given trait among offspring.

   c. Also in prophase I, a microtubular spindle forms outside the nucleus, and the nuclear envelope starts to break up. In cells with duplicated pairs of centrioles, one pair starts moving to the opposite spindle pole.

   d. All of the pairs of homologous chromosomes have become positioned at the spindle equator at metaphase I. Both the maternal chromosome and its homologue have become oriented at random, toward either pole.

   e. In anaphase I, spindle microtubules interact with each duplicated chromosome and move it away from its homologue, to opposite spindle poles.

7. Meiosis II (second nuclear division) is characterized by these events and outcomes: *9.2–9.4*

   a. At metaphase II, all the duplicated chromosomes are positioned at the spindle equator.

   b. Sister chromatids are moved apart in anaphase II. Each is now a separate, unduplicated chromosome.

   c. By the end of telophase II, four nuclei—each with a haploid chromosome number (*n*)—have been formed.

8. When the cytoplasm divides, there are four haploid cells. One or all may serve as gametes (or as plant spores that give rise to gamete-producing bodies). *9.4, 9.5*

9. Crossing over, the chance allocation of different mixes of pairs of maternal and paternal chromosomes to different gametes, and the chance of any two gametes meeting at fertilization all contribute to the immense variation in details of traits among offspring. *9.6*

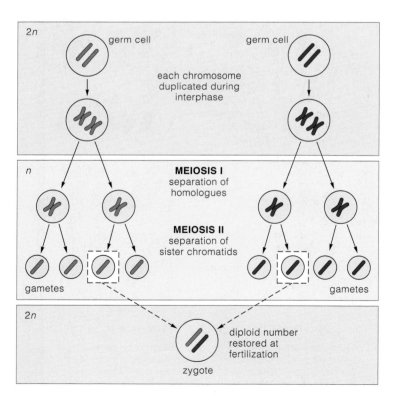

Figure 9.11 Summary of changes in chromosome number at different stages of sexual reproduction, using two diploid (2*n*) germ cells as the example. Meiosis reduces the chromosome number by half (*n*). Union of haploid nuclei of two gametes at fertilization restores the diploid number.

Review Questions

1. The diploid chromosome numbers for the somatic cells of a few organisms are listed at *right*. How many chromosomes will end up in the gametes of each organism? *9.2*

Fruit fly, *Drosophila melanogaster*	8
Garden pea, *Pisum sativum*	14
Corn, *Zea mays*	20
Frog, *Rana pipiens*	26
Earthworm, *Lumbricus terrestris*	36
Human, *Homo sapiens*	46
Chimpanzee, *Pan troglodytes*	48
Amoeba, *Amoeba*	50
Horsetail, *Equisetum*	216

2. A diploid germ cell has four pairs of homologous chromosomes, designated AA, BB, CC, and DD. How would the chromosomes of the gametes be designated? 9.2

The cell is at anaphase ____ rather than anaphase ____ . I know this because:

_____
_____
_____

3. Look at the chromosomes in the germ cell in the diagram at *right*. Is this cell at anaphase I or anaphase II? 9.3, 9.6

4. Define meiosis and describe its stages. In what respects is meiosis *not* like mitosis? 9.2, 9.3, 9.6

5. Actor Michael Douglas (Figure 9.12*a*) inherited a gene from each parent that influences the chin dimple trait. One form of the gene called for a dimple and the other didn't, but one is all it takes for this particular trait. Figure 9.12*b* shows what the chin of Mr. Douglas might have looked like if he had inherited two ordinary forms of the gene instead. What is the name for the alternative forms of the same gene? 9.1, 9.4

6. Outline the main steps by which gametes form in plants. Do the same for gamete formation in animals. 9.5

7. Genetically speaking, what is the key difference between the outcomes of sexual and asexual reproduction? 9.1, 9.6

Figure 9.12 Example of the chin dimple trait (actually a fissure in the chin surface).

Self-Quiz   ANSWERS IN APPENDIX III

1. Sexual reproduction requires _____ .
   a. meiosis   c. fertilization
   b. gamete formation   d. all of the above

2. An animal cell having two rather than one of each type of chromosome has a _____ chromosome number.
   a. diploid   c. normal gamete
   b. haploid   d. both b and c

3. Generally, a pair of homologous chromosomes _____ .
   a. carry the same genes   c. interact at meiosis
   b. are the same length, shape   d. all of the above

4. Meiosis _____ the parental chromosome number.
   a. doubles   c. maintains
   b. reduces   d. corrupts

5. Meiosis is a division mechanism that produces _____ .
   a. two cells   c. eight cells
   b. two nuclei   d. four nuclei

6. Before the onset of meiosis, all chromosomes are _____ .
   a. condensed   c. duplicated
   b. released from protein d. both b and c

7. Duplicated chromosomes move away from their homologue and end up at the opposite spindle pole during _____ .
   a. prophase I   c. anaphase I
   b. prophase II   d. anaphase II

8. Sister chromatids of each duplicated chromosome move apart and end up at opposite spindle poles during _____ .
   a. prophase I   c. anaphase I
   b. prophase II   d. anaphase II

9. Match each term with its description.
   ____ chromosome number   a. different molecular forms of the same gene
   ____ alleles   b. none between meiosis I and II
   ____ metaphase I   c. pairs of homologous chromosomes are now aligned at the spindle equator
   ____ interphase   d. sum total of chromosomes in all cells of a given type

Critical Thinking

1. Assume you can measure the amount of DNA in a primary oocyte, then in a primary spermatocyte, which gives you a mass *m*. What mass of DNA would you expect to find in each mature gamete (egg and sperm) that forms after meiosis? What mass of DNA would you expect to find (1) in an egg fertilized by one of the sperm and (2) in that egg after the first DNA duplication?

2. Adam has a pair of alleles that influence whether a person is right- or left-handed. One allele says "left," and its partner says "right." Visualize one of his germ cells, in which chromosomes are being duplicated prior to meiosis. Visualize what happens to the chromosomes during anaphase I and II. (It might help to use index cards as models of the sister chromatids of each chromosome.) What fraction of Adam's sperm will carry the gene for right-handedness? For left-handedness?

3. Adam also has one allele for long eyelashes, and a partner allele (on the homologous chromosome) for short eyelashes. What fraction of his sperm will have these gene combinations:
   right-handed, long eyelashes   left-handed, long eyelashes
   right-handed, short eyelashes   left-handed, short eyelashes

Selected Key Terms

allele *9.1*
asexual reproduction *9.1*
chromosome number *9.2*
crossing over *9.4*
diploid number *9.2*
egg (ovum) *9.5*
fertilization *9.5*
gamete *CI*
gene *9.1*

germ cell *CI*
haploid number *9.2*
homologous chromosome *9.2*
meiosis *9.2*
oocyte *9.5*
sexual reproduction *9.1*
sister chromatid *9.2*
sperm *9.5*
spore *9.5*

Readings

Klug, W., and M. Cummings. 1994. *Concepts of Genetics.* Fourth edition. New York: Macmillan.

Wolfe, S. 1995. *Introduction to Molecular and Cellular Biology.* Belmont, California: Wadsworth.

On-Line readings at Student Guide for InfoTrac: www.brookscole.com/biology

10

OBSERVABLE PATTERNS OF INHERITANCE

A Smorgasbord of Ears and Other Traits

Basketball ace Charles Barkley has them. So does actor Tom Cruise. Actress Joan Chen doesn't, and neither did a monk named Gregor Mendel. To see how you fit in with these folks, use a mirror to check out your ears. Is the fleshy lobe at the base of each ear attached to the side of your head? If so, you and Barkley and Cruise have something in common. Or is the fleshy lobe not attached, so that you can flap it back and forth? If so, you are like Chen and Mendel (Figure 10.1).

Charles Barkley

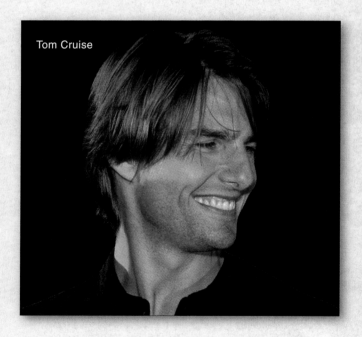
Tom Cruise

Whether a person is born with attached or detached earlobes depends on a single kind of gene. That gene comes in slightly different molecular forms—alleles. Only one form has information about detached lobes. The information is put to use while a human body is developing inside the mother. It calls for a death signal, which is sent to all cells positioned between the newly forming lobes and the head. Without that signal, the cells don't die and earlobes don't detach.

We all have genes for thousands of traits, including earlobes, cheeks, lashes, and eyeballs. Most traits vary in their details from one person to the next. Remember, we inherit pairs of genes on pairs of chromosomes. In some pairs, one allele has strong effects and overwhelms the other allele's contribution. The outgunned allele is said to be recessive to the dominant one. If you have detached earlobes, dimpled cheeks, long lashes, or large

Figure 10.1 Attached and detached earlobes of a few representative humans. This sampling provides observable evidence of a trait governed by a certain gene, which exists in different molecular forms in the human population. Do you have one or the other version of the trait? It depends on which molecular forms of the gene you inherited from your mother and father. As Gregor Mendel perceived, such easily observable traits can be used to identify patterns of inheritance from one generation to the next.

eyeballs, you have at least one and quite possibly two dominant alleles that affect the trait in a predictable way.

When both alleles of a pair are recessive, nothing masks their effect on a trait. You get *attached* earlobes with one pair of recessive alleles, *flat* feet with another pair, a *straight* nose with another pair, and so on.

How did we discover such remarkable things about our genes? It started with Mendel. By analyzing garden

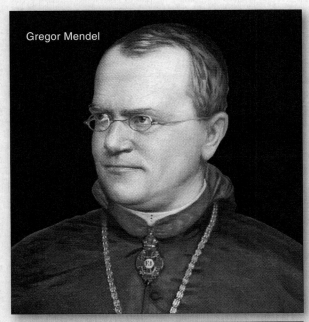

Gregor Mendel

Joan Chen

Key Concepts

1. Genes are units of information about heritable traits. Alleles, which are slightly different molecular forms of a gene, specify different versions of the same trait.

2. Genes have specific locations on the chromosomes of a species. Humans, pea plants, and other organisms with a diploid chromosome number inherit *pairs* of genes, at equivalent locations on pairs of homologous chromosomes.

3. When the two members of each pair of homologous chromosomes are moved apart from each other during meiosis, their pairs of genes are moved apart, also, and end up in different gametes. Gregor Mendel found indirect evidence of this gene segregation when he crossbred pea plants showing different versions of the same trait, such as purple versus white flowers.

4. Each pair of homologous chromosomes in a germ cell is sorted out for distribution into one gamete or another independently of how the other pairs of homologous chromosomes are assorted. Mendel discovered indirect evidence of this when he tracked many plants having observable differences in two traits, such as flower color and plant height.

5. The contrasting forms of traits that Mendel happened to study were specified by nonidentical alleles. One allele was dominant, in that its effect on a trait masked the effect of a recessive allele paired with it.

6. Not all traits have such clearly dominant or recessive forms. One allele of a pair may be fully or partially dominant over its partner or codominant with it. Two or more gene pairs often influence the same trait, and some single genes influence many traits.

7. The environment introduces variation in traits.

pea plants generation after generation, Mendel found indirect but *observable* evidence of how parents bestow units of hereditary information—genes—on offspring. This chapter focuses on both the methods and the results of Mendel's experiments. They remain a classic example of how a scientific approach can pry open important secrets about the natural world. And to this day, they serve as the foundation for modern genetics.

MENDEL'S INSIGHT INTO INHERITANCE PATTERNS

More than a century ago, people wondered about the basis of inheritance. It was common knowledge that sperm and eggs both transmit information about traits to offspring. But few suspected that the information is organized in units (genes). Instead, the idea was that a father's blob of information "blended" with the mother's blob, like cream into coffee, at fertilization.

However, carried to its logical conclusion, blending would slowly dilute a population's shared pool of hereditary information until there was only a single version left of each trait. If that were so, then why did, say, freckles keep showing up among the children of nonfreckled parents through the generations? Why weren't all the descendants of a herd of white stallions and black mares gray? The blending theory could scarcely explain the obvious variation in traits that people could observe with their own eyes. Nevertheless, few disputed the theory.

Charles Darwin was among the scholarly dissidents. According to the key premise of his theory of natural selection, individuals of a population show variation in heritable traits. Through the generations, the variations that improve the chance of surviving and reproducing show up with greater frequency than those that do not. Less advantageous variations may persist among fewer individuals or they may even disappear. It is not that separate versions of a given trait are "blended out" of the population. Rather, *each version of a trait may persist in a population, at frequencies that can change over time.*

Just before Darwin presented his theory, someone was gathering evidence that eventually would support his key premise. A monk, Gregor Mendel, had already guessed that sperm and eggs carry distinct "units" of information about heritable traits. By carefully analyzing traits of pea plants generation after generation, Mendel found indirect but *observable* evidence of how parents transmit genes to offspring.

Mendel's Experimental Approach

Mendel spent most of his adult life in a monastery in Brno, a city near Vienna that has since become part of the Czech Republic. However, Mendel was not a man of narrow interests who accidentally stumbled onto principles of great import. The monastery of St. Thomas was close to European capitals that were the centers of scientific inquiry.

Having been raised on a farm, Mendel was aware of agricultural principles and their applications. He kept abreast of the breeding experiments and developments described in the available literature. He was a member of the regional agricultural society. He won awards for developing improved varieties of vegetables and fruits. Shortly after entering the monastery, he took courses in mathematics, physics, and botany at the University of Vienna. Few scholars of his time showed interest in both plant breeding and mathematics.

carpel · stamen

Figure 10.2 Garden pea plant (*Pisum sativum*), the organism Mendel chose for experimental tests of his ideas about inheritance.

a Garden pea flower. The section shows the location of its stamens and carpel. Sperm-producing pollen grains form in stamens. Eggs develop, fertilization takes place, and seeds mature inside carpels.

b Pollen from a garden pea plant that breeds true for purple flowers is brushed onto a floral bud of a plant that breeds true for white flowers. The white flower had its stamens snipped off. This is one way to assure cross-fertilization of plants.

c Later, the cross-fertilized plant produces seeds (in pea pods). Each seed is allowed to grow and develop into a new plant.

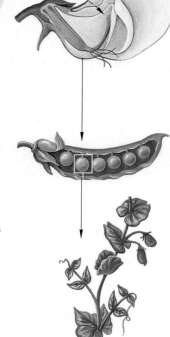

d The flower color of each new plant can be used as visible evidence of patterns in how hereditary material might be transmitted to it from each parent plant.

Shortly after his university training, Mendel began experimenting with the garden pea plant, *Pisum sativum* (Figure 10.2). This plant is self-fertilizing. Its male and female gametes (call them sperm and eggs) develop in different parts of the same flower, where fertilization occurs. Individual pea plants generally breed true for certain traits. In other words, successive generations are just like the parents in one or more traits, as when all offspring grown from seeds of self-fertilized, white-flowered parent plants have white flowers.

Pea plants may be cross-fertilized by transferring pollen from one plant's flower to the flower of another plant. Mendel knew he could open the flower buds of a plant that bred true for a trait, such as white flowers, and snip out its stamens. (Stamens bear pollen grains in which sperm develop.) He could brush those buds with pollen from a plant that bred true for a *different* version of the same trait—say, purple flowers. As he hypothesized, such clearly observable differences would help him track a given trait through many generations. If there were patterns to the trait's inheritance, *then the patterns might tell him something about heredity itself.*

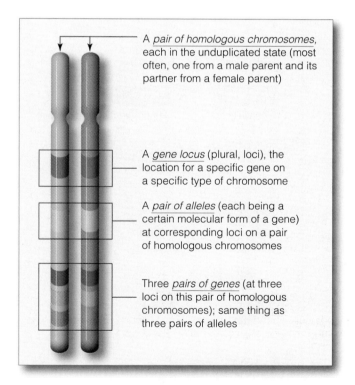

Figure 10.3 A few genetic terms illustrated. Diploid organisms have pairs of genes, on pairs of homologous chromosomes. For example, you inherited one chromosome of each pair from your mother and the other, homologous chromosome from your father.

Most genes come in slightly different molecular forms, called alleles. Different alleles specify different versions of the same trait. An allele at one location on a chromosome may or may not be identical to its partner on the homologous chromosome.

Some Terms Used in Genetics

Having read the chapter on meiosis, you already have insight into the mechanisms of sexual reproduction, which is more than Mendel had. Neither he nor anyone else of his era knew about chromosomes. So he could not have known that a chromosome number is reduced by half in gametes, then restored when gametes meet at fertilization. Even so, Mendel sensed what was going on. As we follow his thinking, let's simplify the story by substituting a few of the modern terms used in studies of inheritance (see also Figure 10.3):

1. **Genes** are units of information about specific traits, and they are passed from parents to offspring. Each gene has a specific location (locus) on a chromosome.

2. Cells with a diploid chromosome number ($2n$) have pairs of genes, on pairs of homologous chromosomes.

3. Mutation can alter a gene's molecular structure. The alteration may change the gene's information about a trait (as when the gene for flower color specifies purple and a mutated version specifies white). All the different molecular forms of the same gene are called **alleles**.

4. When offspring of genetic crosses inherit a pair of *identical* alleles for a trait, generation after generation, they are a **true-breeding lineage**. By contrast, when offspring of a genetic cross inherit a pair of *nonidentical* alleles for a trait, they are **hybrid offspring**.

5. When both alleles of a pair are identical, this is a *homozygous* condition. When the two are not identical, this is a *heterozygous* condition.

6. An allele is said to be *dominant* when its effect on a trait masks that of any *recessive* allele paired with it. We use capital letters for dominant alleles and lowercase letters for recessive ones. *A* and *a* are examples.

7. Putting this all together, a **homozygous dominant** individual has a pair of dominant alleles (*AA*) for the trait being studied. A **homozygous recessive** individual has a pair of recessive alleles (*aa*). And a **heterozygous** individual has a pair of nonidentical alleles (*Aa*).

8. Two terms help keep the distinction clear between genes and the traits they specify. **Genotype** refers to the particular alleles an individual carries. **Phenotype** refers to an individual's observable traits.

9. When tracking the inheritance of traits through generations of offspring, these abbreviations apply:

P	parental generation
F_1	first-generation offspring
F_2	second-generation offspring

MENDEL'S THEORY OF SEGREGATION

Mendel had an idea. In every generation, a plant might inherits two "units" (genes) of information about a trait, one from each parent. He used **monohybrid crosses** to test his idea. For such crosses, two parents that breed true for different forms of a trait produce F_1 offspring that are all heterozygous ($AA \times aa = Aa$). The actual experiment is an intercross between two of the identical F_1 heterozygotes—the "monohybrids" ($Aa \times Aa$).

Predicting Outcomes of Monohybrid Crosses

Mendel tracked many traits over two generations. For one set of experiments, he crossed true-breeding purple-flowered plants and true-breeding white-flowered ones. All plants of the next generation had purple flowers. Mendel let those F_1 plants self-fertilize. Later on, some of the F_2 offspring produced white flowers!

If Mendel's hypothesis were correct—if each plant inherited two units of information about flower color—then the "purple" unit had to be dominant. Why? It masked the unit for "white" in the F_1 plants.

Let's rephrase his thinking. Germ cells of pea plants are diploid, with pairs of homologous chromosomes. Assume one parent is homozygous dominant (AA) and the other is homozygous recessive (aa) for flower color. Following meiosis, a sperm or egg carries one allele for flower color (Figure 10.4). So when a sperm fertilizes an egg, only one outcome is possible: $A + a = Aa$.

Mendel knew about sampling error (Section 1.5). He crossed hundreds of plants and tracked thousands of offspring. He also counted and recorded the number of dominant and recessive plants. On average, three of every four F_2 plants had the dominant phenotype, and one had the recessive phenotype (Figure 10.5).

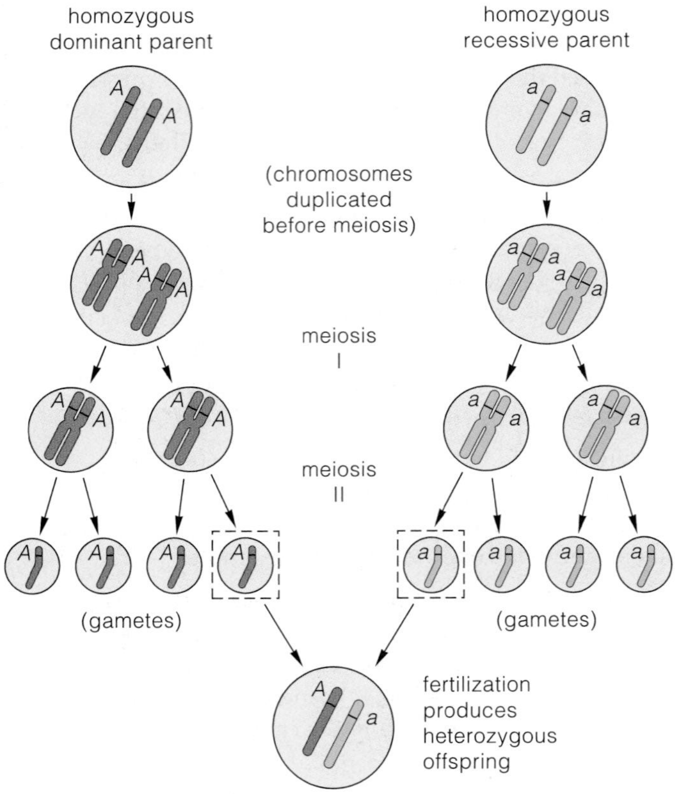

Figure 10.4 Monohybrid cross, showing how one gene of a pair segregates from the other gene. Two parents that breed true for two versions of a trait produce only heterozygous offspring.

Figure 10.5 *Right:* Numerical results from Mendel's monohybrid cross experiments with the garden pea plant (*P. sativum*). The numbers are his counts of the F_2 plants that carried dominant or recessive hereditary "units" (alleles) for the trait. On average, the dominant-to-recessive ratio was 3:1.

Trait Studied	Dominant Form	Recessive Form	F_2 Dominant-to-Recessive Ratio
SEED SHAPE	5,474 round	1,850 wrinkled	2.96:1
SEED COLOR	6,022 yellow	2,001 green	3.01:1
POD SHAPE	882 inflated	299 wrinkled	2.95:1
POD COLOR	428 green	152 yellow	2.82:1
FLOWER COLOR	705 purple	224 white	3.15:1
FLOWER POSITION	651 along stem	207 at tip	3.14:1
STEM LENGTH	787 tall	277 dwarf	2.84:1

Average ratio for all traits studied: **3:1**

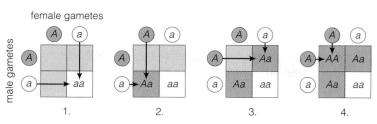

female gametes

male gametes

1.   2.   3.   4.

Figure 10.6 Punnett-square method of predicting the probable outcome of a genetic cross. Circles represent gametes. *Italic* letters on gametes represent dominant or recessive alleles. In the squares are the different genotypes possible among offspring. In this case, gametes are from a self-fertilizing heterozygous (*Aa*) plant.

Figure 10.7 *Right*: Results from one of Mendel's monohybrid crosses. On average, the dominant-to-recessive ratio among the second-generation (F_2) plants was 3:1.

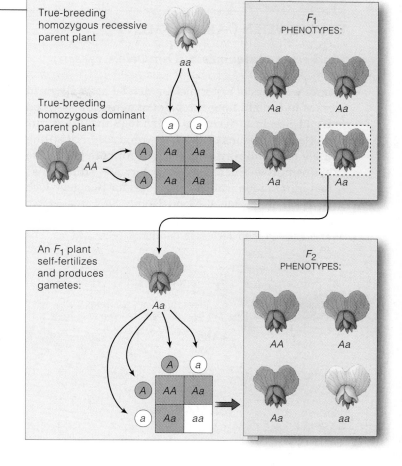

To Mendel, the ratio suggested that fertilization is a chance event with a number of possible outcomes. And he had an understanding of probability, which applies to chance events *and therefore could help him predict the possible outcomes of genetic crosses*. **Probability** simply means this: The chance that each outcome of a given event will occur is proportional to the number of ways in which the event can be reached.

The **Punnett-square method**, explained in Figure 10.6 and applied in Figure 10.7, may help you visualize the possibilities. As you can see, if half of a plant's sperm (or eggs) were *a* and half were *A*, then four outcomes would be possible each time a sperm fertilized an egg:

POSSIBLE EVENT:	PROBABLE OUTCOME:
sperm *A* meets egg *A*	1/4 *AA* offspring
sperm *A* meets egg *a*	1/4 *Aa*
sperm *a* meets egg *A*	1/4 *Aa*
sperm *a* meets egg *a*	1/4 *aa*

y this prediction, an F_2 plant has three chances in four of getting at least one dominant allele (purple flowers). It has one chance in four of getting two recessive alleles (white flowers). That is a probable phenotypic ratio of three purple to one white, or 3:1.

Mendel's observed ratios were not *exactly* 3:1. You can see this for yourself by looking at the numerical results listed in Figure 10.5. Why did Mendel put aside the deviations? To understand why, flip a coin several times. As we all know, a coin is just as likely to end up heads as tails. But often a coin ends up heads, or tails, several times in a row. So if you flip the coin only a few times, the observed ratio might differ greatly from the predicted ratio of 1:1. Flip the coin many, many times, and you are more likely to come close to the predicted ratio. Mendel understood the rules of probability—and performed a large number of crosses. Almost certainly, this kept him from being confused by minor deviations from the predicted results of the experimental crosses.

Testcrosses

By running **testcrosses**, Mendel gained support for his prediction. In this type of experimental test, an organism shows dominance for a specified trait but its genotype is unknown, so it is crossed to a known homozygous recessive individual. Test results may reveal whether the organism is homozygous dominant or heterozygous.

Regarding the monohybrid crosses just described, Mendel tested his prediction that the purple-flowered F_1 offspring were heterozygous by crossing them with true-breeding, white-flowered plants. If they were all homozygous dominant, then all the F_2 offspring would show the dominant form of the trait. If heterozygous, there would be about as many dominant as recessive plants. Sure enough, when old enough to flower, about half the F_2 plants had purple flowers (*Aa*) and half had white (*aa*). Can you construct two Punnett squares that show the possible outcomes of this testcross?

The results from Mendel's monohybrid crosses and testcrosses became the basis of a theory of **segregation**, which we state here in modern terms:

MENDEL'S THEORY OF SEGREGATION Diploid cells have pairs of genes, on pairs of homologous chromosomes. The two genes of each pair are separated from each other during meiosis, so they end up in different gametes.

INDEPENDENT ASSORTMENT

Predicting Outcomes of Dihybrid Crosses

In another series of experiments, Mendel used **dihybrid crosses** to explain how *two* pairs of genes are assorted into gametes. In such crosses, individuals that breed true for different versions of *two* traits produce F_1 offspring that are all identically heterozygous for both traits. The experiment is an intercross between two F_1 "dihybrids" —that is, two identical heterozygotes for two gene loci.

Let's diagram one of Mendel's dihybrid crosses. We can use A for flower color and B for height as dominant alleles and, as their recessive counterparts, a and b:

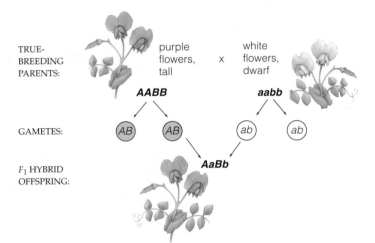

TRUE-BREEDING PARENTS:

purple flowers, tall   x   white flowers, dwarf

AABB   **aabb**

GAMETES: (AB) (AB)   (ab) (ab)

F_1 HYBRID OFFSPRING: **AaBb**

As Mendel would have predicted, the F_1 offspring from this cross are all purple-flowered and tall ($AaBb$).

When those F_1 plants reproduce, how will the two gene pairs be assorted into gametes? The answer partly depends on which chromosomes carry the pairs. Assume that one pair of homologous chromosomes carry the Aa alleles and a different pair carry the Bb alleles. Now think of how all chromosomes become positioned at the spindle equator during metaphase I of meiosis (Figures 9.4 and 10.8). The chromosome with the A allele might be positioned to move to either spindle pole (then into one of four gametes). Its homologue with the a allele might also move to either spindle pole. And the same can happen to the homologous chromosomes that carry the B and b alleles. After meiosis and gamete formation, then, four combinations of alleles are possible in sperm or eggs: $1/4\,AB$, $1/4\,Ab$, $1/4\,aB$, and $1/4\,ab$.

Given the alternative alignments of chromosomes at metaphase I, several allelic combinations are possible at fertilization. Simple multiplication (four kinds of sperm times four kinds of eggs) tells us sixteen combinations of gametes are possible in the F_2 offspring of a dihybrid cross. Use the Punnett-square method to diagram the probabilities (Figure 10.9). Now add up all the possible phenotypes and you get 9/16 tall purple-flowered, 3/16 dwarf purple-flowered, 3/16 tall white-flowered, and 1/16 dwarf white-flowered plants. That is a probable phenotypic ratio of 9:3:3:1. Results from one dihybrid cross that Mendel described were close to this ratio.

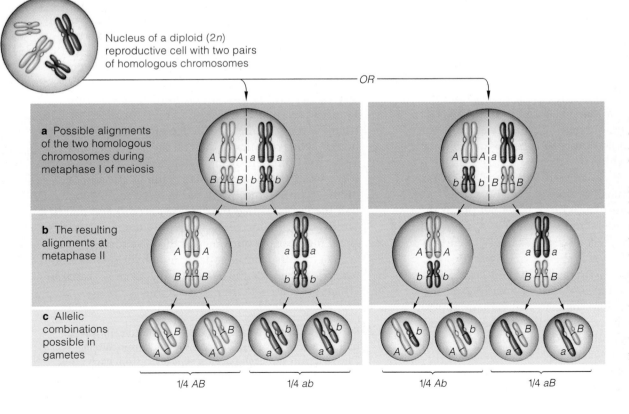

Nucleus of a diploid ($2n$) reproductive cell with two pairs of homologous chromosomes

OR

a Possible alignments of the two homologous chromosomes during metaphase I of meiosis

b The resulting alignments at metaphase II

c Allelic combinations possible in gametes

1/4 AB   1/4 ab   1/4 Ab   1/4 aB

Figure 10.8 One case of independent assortment. This example tracks two pairs of homologous chromosomes. An allele at a given locus on a chromosome may or may not be identical with its partner allele on the homologous chromosome. Either chromosome of a pair may become attached to either pole of the spindle during meiosis. As you can see, when just two pairs are being tracked, two different metaphase I lineups are possible.

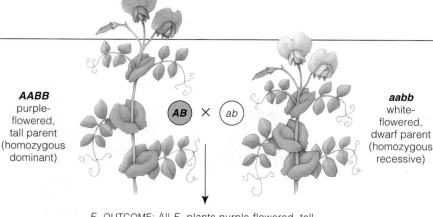

AABB
purple-
flowered,
tall parent
(homozygous
dominant)

AB × ab

aabb
white-
flowered,
dwarf parent
(homozygous
recessive)

Figure 10.9 Results from Mendel's dihybrid cross between parent plants that bred true for different versions of two traits: flower color and plant height. *A* and *a* represent the dominant and recessive alleles for flower color. *B* and *b* signify the dominant and recessive alleles for plant height. As the Punnett square indicates, the probabilities of certain combinations of phenotypes among the F_2 offspring occur in a 9:3:3:1 ratio, on average.

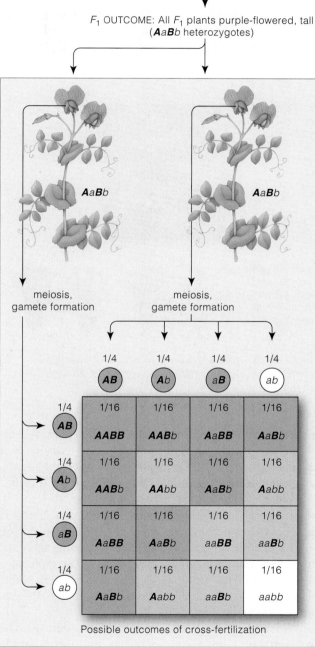

F_1 OUTCOME: All F_1 plants purple-flowered, tall (**A**a**B**b heterozygotes)

Possible outcomes of cross-fertilization

ADDING UP THE F_2 COMBINATIONS POSSIBLE:

- 9/16 or 9 purple-flowered, tall
- 3/16 or 3 purple-flowered, dwarf
- 3/16 or 3 white-flowered, tall
- 1/16 or 1 white-flowered, dwarf

The Theory in Modern Form

Mendel could do no more than analyze the numerical results from his dihybrid crosses, because he didn't know that seven pairs of homologous chromosomes carry the pea plant's "units" of inheritance. It just seemed to him that the two units for the first trait he was tracking had been assorted into gametes independently of the two units for the other trait. In time his interpretation became known as the theory of **independent assortment**, which we state here in modern terms: By the end of meiosis, each pair of homologous chromosomes—and the genes they carry—has been sorted for shipment into gametes independently of how the other pairs were sorted out.

Independent assortment and hybrid intercrosses lead to great genetic variation. In a monohybrid cross that involves only one gene pair, three genotypes are possible: *AA*, *Aa*, and *aa*. We can represent this as 3^n, where *n* is the number of gene pairs. With more pairs of genes, the number of possible combinations increases dramatically. When the parents differ in ten gene pairs, almost 60,000 genotypes are possible. When they differ in twenty gene pairs, the number approaches 3.5 billion!

In 1865, Mendel presented his ideas to the Brünn Natural History Society. His ideas had little impact. The next year he published a paper, and apparently it was read by few and understood by no one. In 1871 he became abbot of the monastery, and his experiments gradually gave way to administrative tasks. He died in 1884, never to know his experiments would become the starting point for the development of modern genetics.

Today, Mendel's theory of segregation still stands. Hereditary material is indeed organized in units (genes) that retain their identity and are segregated from each other for distribution into different gametes. But the theory of independent assortment does not apply to all gene combinations, as you will see in the next chapter.

MENDEL'S THEORY OF INDEPENDENT ASSORTMENT **By the end of meiosis, genes on pairs of homologous chromosomes have been sorted out for distribution into one gamete or another independently of gene pairs of other chromosomes.**

DOMINANCE RELATIONS

For the most part, Mendel studied traits having clearly dominant or recessive forms. As the remaining sections of this chapter will make clear, however, the expression of other traits is not as straightforward.

Incomplete Dominance

In **incomplete dominance**, one allele of a pair isn't fully dominant over its partner, so a heterozygous phenotype *somewhere in between* the two homozygous phenotypes emerges. Cross a true-breeding red snapdragon and a true-breeding white one. All F_1 offspring will have pink flowers. Cross two F_1 plants and expect red, pink, and white snapdragons in a predictable ratio (Figure 10.10). What causes this inheritance pattern? Red snapdragons have two alleles that allow them to make an abundance of red pigment molecules. White snapdragons have two mutant alleles that make them pigment-free. The pink ones are heterozygous; the red allele they carry specifies enough pigment to make flowers pink, but not red.

ABO Blood Types: A Case of Codominance

In **codominance**, a pair of nonidentical alleles specify two phenotypes, which are both expressed at the same time in heterozygotes. Example: One of the glycolipids at the plasma membrane of your red blood cells helps give these cells a unique identity. However, it comes in slightly different molecular forms. An analytical method, *ABO blood typing*, reveals which form a person has.

An enzyme dictates the glycolipid's final structure. In humans, the gene for that enzyme has three alleles. Two alleles, I^A and I^B, are codominant when paired. The third, i, is recessive; a pairing with either I^A or I^B masks its effect. Together, they are a **multiple allele system**, which we define as the presence of three or more alleles of a gene among individuals of a population.

Before each glycolipid became positioned at the cell surface, it was modified in the cytomembrane system (Section 4.7). An oligosaccharide chain was attached to

homozygous parent × homozygous parent

All F_1 offspring are heterozygous for flower color:

Cross two of the F_1 plants, and the F_2 offspring will show three phenotypes in a 1:2:1 ratio:

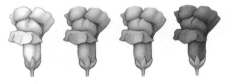

Figure 10.10 Visible evidence of incomplete dominance in heterozygous (pink) snapdragons, in which an allele for red pigment is paired with a "white" allele.

a lipid, then a sugar was attached to the chain. Alleles I^A and I^B specify different versions of the enzyme that attaches the sugar. The enzymes attach *different* sugars, which gives the glycolipid different identities: A or B.

Which alleles do you have? With either $I^A I^A$ or $I^A i$, you have type A blood. With $I^B I^B$ or $I^B i$, your blood is type B. With codominant alleles $I^A I^B$, it's AB—meaning you have both versions of the sugar-attaching enzyme. But if you are homozygous recessive (ii), the molecules never did get the final sugar attached to them. Your blood type isn't A or B; that is what type "O" means. Figure 10.11 summarizes the possibilities.

With *transfusions*, the blood of two people mixes. The recipient's immune system perceives any incompatible red blood cells as "nonself." In such cases, it acts against them and may cause death (Section 34.4).

One allele may be fully dominant, incompletely dominant, or codominant with its partner on the homologous chromosome.

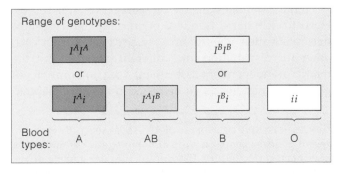

Range of genotypes:

$I^A I^A$		$I^B I^B$	
or		or	
$I^A i$	$I^A I^B$	$I^B i$	ii

Blood types: A — AB — B — O

Figure 10.11 Allelic combinations for ABO blood typing.

MULTIPLE EFFECTS OF SINGLE GENES

Expression of the alleles at just a single location on a chromosome may have positive or negative effects on two or more traits. This phenotypic outcome of a single gene's activity is known as **pleiotropy** (after the Greek *pleio–*, meaning more, and *–tropic*, meaning to change).

A genetic disorder called *sickle-cell anemia* is a classic example of how alleles at one locus can have extended effects. The disorder arises from a mutated gene for one of two kinds of polypeptide chains in hemoglobin. Hemoglobin, recall, is an oxygen-transporting protein in red blood cells. We designate the mutant allele Hb^S instead of Hb^A. Heterozygotes (Hb^A/Hb^S) usually show few symptoms of the disorder. Their red blood cells are able to produce enough normal hemoglobin molecules to compensate for the abnormal ones. In homozygotes

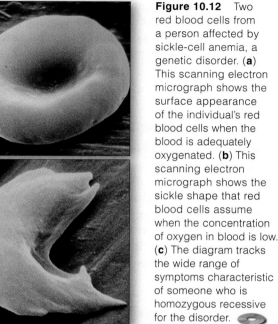

Figure 10.12 Two red blood cells from a person affected by sickle-cell anemia, a genetic disorder. (**a**) This scanning electron micrograph shows the surface appearance of the individual's red blood cells when the blood is adequately oxygenated. (**b**) This scanning electron micrograph shows the sickle shape that red blood cells assume when the concentration of oxygen in blood is low. (**c**) The diagram tracks the wide range of symptoms characteristic of someone who is homozygous recessive for the disorder.

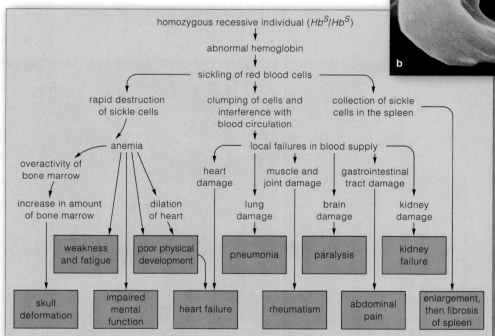

c

Inside the red blood cells of people who carry the mutant gene, abnormal hemoglobin molecules stick together into rod-shaped clumps. The rods distort the cells into a sickle shape, as in Figure 10.12*b*. (A sickle is a farm tool that has a long, crescent-shaped blade.)

The distorted cells rupture easily, and then the remnants clog and rupture capillaries. Rapid destruction of oxygen-transporting molecules leads to oxygen-starved cells in the affected tissues. Clumping also causes local failures in the circulatory system's capacity to deliver oxygen and carry away carbon dioxide and other metabolic wastes.

(Hb^S/Hb^S), red blood cells can only produce abnormal hemoglobin. This single abnormality may have severe repercussions throughout the body, for it disrupts the concentration of oxygen in the bloodstream.

Humans, like most organisms, depend on the intake of oxygen for aerobic respiration. Oxygen flows into our lungs, then diffuses into blood. There it binds to hemoglobin, which travels through arteries, arterioles, and then the small-diameter, thin-walled capillaries that thread past all living cells in the body. A steep oxygen concentration gradient exists between blood and the fluid in the surrounding tissues, so most of the oxygen diffuses into tissues, then into cells. The cellular uptake lowers its concentration in blood. Its decline is greatest at high altitudes and during strenuous activity.

Over time, ongoing expression of the mutant gene may damage tissues and organs throughout the body. Figure 10.12*c* tracks how the successive alterations in phenotype may proceed. In chapters to come, you will read about other aspects of this genetic disorder.

The alleles at a single gene location may have positive or negative effects on two or more traits.

The effects may not be simultaneous. Rather, they may have repercussions over time. The gene may lead to an alteration in one trait. That change may alter another trait, and so on.

INTERACTIONS BETWEEN GENE PAIRS

Often a trait results from interactions among products of two or more gene pairs. For example, two alleles of a gene may mask expression of another gene's alleles, so some expected phenotypes may not appear at all. Such interactions between the product of pairs of genes are called **epistasis** (meaning the act of stopping).

Hair Color in Mammals

Epistasis is common among the gene pairs responsible for skin or fur color in mammals. Consider the black, brown, or yellow fur of Labrador retrievers (Figure 10.13). The different colors arise from variations in the amount and distribution of melanin, a brownish-black pigment. A variety of enzymes and other products of many gene pairs affect different steps in the production of melanin and its deposition in certain body regions.

The alleles of one gene specify an enzyme required to produce melanin. Expression of allele *B* (black) has a more pronounced effect and is dominant to *b* (brown). Alleles of a different gene control the extent to which molecules of melanin will be deposited in a retriever's hairs. Allele *E* permits full deposition. Two recessive alleles (*ee*) reduce deposition, and fur will be yellow.

In some individuals, those two gene pairs are not able to interact, owing to a certain allelic combination at still another gene locus. There, a gene (*C*) calls for tyrosinase, the first of several enzymes in a melanin-producing pathway. An individual bearing one or two dominant alleles (*CC* or *Cc*) can make the functional enzyme. An individual bearing two recessive alleles (*cc*) cannot. When the biosynthetic pathway for melanin production gets blocked, then *albinism*—the absence of melanin—is the resulting phenotype (Figure 10.14).

a BLACK LABRADOR   **b** YELLOW LABRADOR   **c** CHOCOLATE LABRADOR

Figure 10.13 The heritable basis of coat color among Labrador retrievers. The trait arises through interactions among the alleles of two pairs of genes.

One kind of gene is involved in melanin production. Allele *B* (black) of this gene is dominant to allele *b* (brown). A different kind of gene influences the deposition of melanin pigment in individual hairs. Allele *E* of this gene promotes melanin deposition, but a pairing of recessive alleles (*ee*) of the gene blocks deposition, and a yellow coat results.

F_1 offspring of a dihybrid cross produce F_2 offspring in a 9:3:4 ratio, as the Punnett-square diagram at *right* indicates.

The yellow Labrador in (**b**) probably has genotype *BBee*, because it can produce melanin but cannot deposit pigment in hairs. After thinking about this photograph, can you say why?

HOMOZYGOUS PARENTS: $BBEE \times bbee$

            ↓

F_1 PUPPIES: *BbEe*

            ↓

ALLELIC COMBINATIONS POSSIBLE AMONG F_2 PUPPIES:

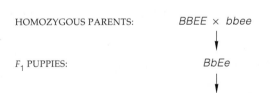

	BE	Be	bE	be
BE	BBEE	BBEe	BbEE	BbEe
Be	BBEe	BBee	BbEe	Bbee
bE	BbEE	BbEe	bbEE	bbEe
be	BbEe	Bbee	bbEe	bbee

RESULTING PHENOTYPES:

- ☐ 9/16 or 9 black
- ☐ 3/16 or 3 brown
- ☐ 4/16 or 4 yellow

Figure 10.14 A rare albino rattlesnake. Like other animals that can't produce melanin, its body surface is white, overall, and its eyes are pink. In birds and mammals, surface coloration arises largely from pigments in feathers, fur, or skin. In fishes, amphibians, and reptiles, it depends on color-bearing cells. Some of these cells contain melanin or yellow-to-red pigments. Others contain crystals that reflect light and thus alter the surface coloration.

The mutation affecting melanin production in the snake shown here had no effect on the production of its yellow-to-red pigments and light-reflecting crystals. That's why the snake's skin appears to be iridescent yellow as well as white. Its eyes look pink because melanin is absent from a tissue layer in each eyeball. Without melanin to absorb it, red light is reflected from blood vessels in the eyes.

Figure 10.15 Interaction between two genes that affect the same trait in domestic chicken breeds. The initial cross is between a Wyandotte (with a rose comb, **b**, on the crest of its head) and brahma (pea comb, **c**). With complete dominance at the locus for pea comb and at the locus for rose comb, products of the two gene pairs interact and give rise to a walnut comb (**a**). With full recessiveness at both gene loci, products of the genes interact and give rise to a single comb (**d**).

Comb Shape in Poultry

In some cases, interaction between two gene pairs results in a phenotype that neither pair can produce by itself. Geneticists W. Bateson and R. Punnett identified two interacting gene pairs (*R* and *P*) that affect comb shape in chickens. Allelic combinations of *rr* at one gene locus and *pp* at the other locus result in the least common phenotype, the single comb. The presence of dominant alleles (*R*, *P*, or both) results in varied phenotypes.

Take a look at Figure 10.15. The diagram shows the combinations of alleles that interact to specify the rose, pea, and walnut combs shown in the photographs.

Genes often interact, as when alleles of one gene mask the expression of another gene, and when some expected phenotypes may not appear at all.

HOW CAN WE EXPLAIN LESS PREDICTABLE VARIATIONS?

Regarding the Unexpected Phenotype

As Mendel demonstrated, the phenotypic effects of one or two pairs of certain genes show up in predictable ratios when you track them from one generation to the next. Besides this, interactions among two or more gene pairs also can produce phenotypes in predictable ratios, as the example of Labrador coat color demonstrated in Section 10.6.

However, even if you were to track a single gene over the generations, you might find that the resulting phenotypes were not quite what you had expected.

Consider *camptodactyly*, a rare genetic abnormality that affects the shape and the movement of fingers. People who carry the mutant allele for this heritable trait have immobile, bent fingers on both hands. Other people have immobile, bent fingers on the left or right hand only. The fingers of others who carry the mutant allele aren't affected in any obvious way.

What causes such odd variation? Remember, most organic compounds are synthesized by a sequence of metabolic steps. *And different enzymes, each the product of a gene, regulate different steps.* Maybe one gene mutated in one of a number of possible ways. Maybe the gene product blocks the pathway or causes it to run nonstop or not long enough. Maybe poor nutrition or another factor that is variable in the individual's environment affects a crucial enzyme in the pathway. These are the kinds of variable factors that often introduce far less predictable variations in the phenotypes resulting from gene expression.

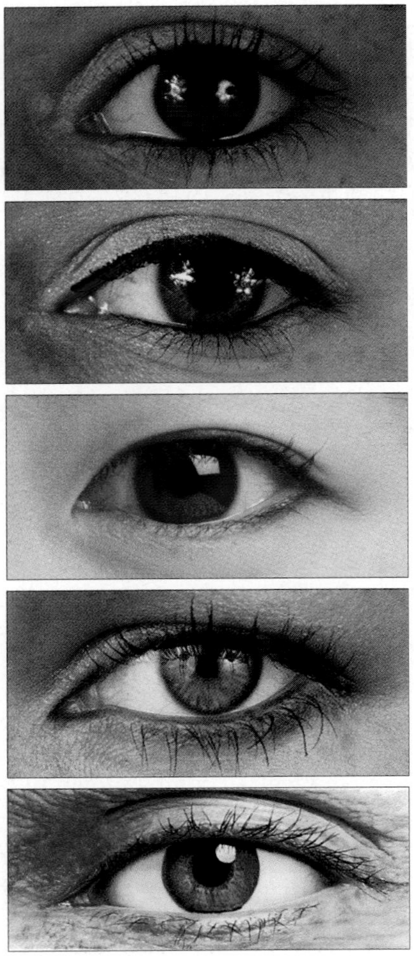

Continuous Variation in Populations

Generally, individuals of a population display a range of small differences in most traits. This characteristic of populations is called **continuous variation**. It's mainly an outcome of the number of genes affecting a trait and the number of environmental factors influencing their expression. Usually, the greater the number of genes and environmental factors, the more continuous will be the expected distribution of all the versions of that trait.

Look in a mirror at your eye color. The colored part is the iris, a doughnut-shaped, pigmented structure just beneath the cornea. Its color is the cumulative outcome of a number of gene products. Some products take part in the stepwise production and distribution of melanin, the same light-absorbing pigment that influences coat color in mammals. Dark eyes that seem to be almost black have heavy deposits of melanin molecules inside the iris. These molecules absorb most of the incoming light. Melanin deposits aren't as extensive in brown eyes, and some unabsorbed light is being reflected out. Light brown or hazel eyes have even less (Figure 10.16).

Green, gray, or blue eyes don't have green, gray, or blue pigments. Their iris incorporates some amount of melanin, but not much. As a result, many or most of the blue wavelengths of light that do enter the eyeball are reflected out.

How might you describe the continuous variation of some trait within a group, such as the students in Figure 10.17*a*? The students range from very short to very tall, with average heights much more common than either extreme. You might start out by dividing the full range of different phenotypes into measurable categories. Next, you can count all of the individual students in each category. Doing so will give you the relative frequencies of all the phenotypes, distributed across the range of measurable values.

The bar chart in Figure 10.17*c* plots the proportion of students in each category against the range of the measured phenotypes. Here, the shortest vertical bars represent categories with the least number of students. The tallest bar represents the category with the greatest number of students. Finally, draw a graph line around

Figure 10.16
A small sampling from the range of continuous variation in the color of human eyes. Products of different pairs of genes interact to produce and then deposit the pigment melanin. Among other things, melanin helps color the eye's iris. Different combinations of alleles result in small differences in eye color. So the frequency distribution for the eye-color trait appears to be continuous over a range from black to light blue.

1	4	8	10	16	16	15	15	14	13	13	11	9	8	8	5	1	2

Number of individuals

60	61	62	63	64	65	66	67	68	69	70	71	72	73	74	75	76	77

Height (inches)

a Students organized according to height, as an example of continuous variation

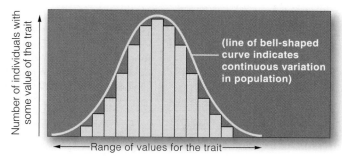

(line of bell-shaped curve indicates continuous variation in population)

Number of individuals with some value of the trait

Range of values for the trait

b Idealized bell-shaped curve for a population that displays continuous variation in some trait

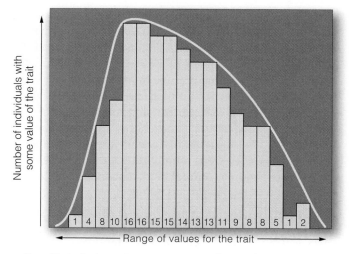

Number of individuals with some value of the trait

1	4	8	10	16	16	15	15	14	13	13	11	9	8	8	5	1	2

Range of values for the trait

c Specific bell-shaped curve corresponding to the distribution of the trait (height) illustrated by the photograph in (**a**)

Figure 10.17 Continuous variation in body height, a trait that is one of the characteristics of the human population.

(**a**) Suppose you wish to know the frequency distribution for height in a group of 169 biology students at Brigham Young University. You decide on how finely the range of possible heights should be divided. Then you measure each student and assign her or him to the appropriate category. Finally, you divide the number in each category by the total number of all students in all categories.

(**b**) Often, a bar graph is used to depict continuous variation in a population. In such graphs, the proportion of individuals in each category is plotted against the range of measured phenotypes. The curved line above this particular set of bars is an idealized example of the kind of "bell-shaped" curve that emerges for populations showing continuous variation in a trait. The bell-shaped curve in (**c**) is a specific example of this type of diagram.

all the bars and you end up with a bell-shaped curve. Such "bell curves" are typical of populations that show continuous variation in a trait.

Enzymes and other products of genes regulate each step of most metabolic pathways. Mutations, gene interactions, and environmental conditions may affect one or more of the steps. The outcome is variation in phenotypes.

For most traits, individuals of a population show continuous variation—that is, a range of small differences.

The greater the number of genes and environmental factors that can influence a trait, the more continuous will be the expected distribution of all versions of that trait.

ENVIRONMENTAL EFFECTS ON PHENOTYPE

We have mentioned, in passing, that the environment often contributes to variable gene expression among a population's individuals. Before leaving this chapter, consider a few examples of phenotypic variations that arise from environmental effects on genotype.

Possibly you've seen a Himalayan rabbit or Siamese cat. Both furry mammals carry an allele that specifies a heat-sensitive form of one of the enzymes for melanin production. The enzyme is less active at the surface of warm body parts. There, the fur color is lighter than in cooler parts, such as ears and limbs projecting from the main body mass. Figure 10.18 gives an experiment that demonstrated an environmental effect on this allele.

We also can identify environmental effects on genes that govern plant phenotypes. Plant a hydrangea in a garden and you may get pink instead of the expected blue blossoms (Figure 10.19). Genes for its floral color are affected by soil acidity.

Also consider an experiment with yarrow plants. Yarrow can grow from cuttings, which makes it a fine experimental organism; cuttings from the same plant all have the same genotype. Three cuttings were grown at three elevations. The two plants grown at the lowest and highest elevation fared best; the one at the medium elevation grew poorly, as Figure 10.20 shows.

Figure 10.19 Environmental effect on gene expression in *Hydrangea macrophylla*, a favorite garden plant. Even different plants that carry the same alleles have floral colors ranging from pink to blue. The color variation arises because of differences in the acidity of soil in which they happen to be growing.

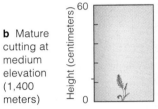

a Mature cutting at high elevation (3,050 meters)

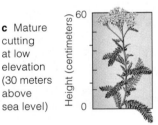

b Mature cutting at medium elevation (1,400 meters)

c Mature cutting at low elevation (30 meters above sea level)

Figure 10.20 Experiment demonstrating the effect of environmental conditions on gene expression in a yarrow plant (*Achillea millefolium*). Cuttings from the same parent plant were grown in the same soil medium but at three different elevations.

Icepack is strapped onto a hair-free patch.

New hair growing in patch exposed to cold is black.

Figure 10.18 Observable effect of environmental conditions on gene expression in animals. A Himalayan rabbit normally has black hair only on its long ears, nose, tail, and lower leg limbs. For one experiment, a patch of a rabbit's white fur was plucked clean, then an icepack was secured over the hairless patch. Where the colder temperature had been maintained, the hairs that grew back were black.

Himalayan rabbits are homozygous for the *ch* allele of the gene for tyrosinase, an enzyme required to make melanin. This allele specifies a heat-sensitive form of the enzyme, which functions only when the air temperature is below about 33°C. When cells that give rise to hairs grow under warmer conditions, they cannot make melanin, so hairs appear light. This happens in body regions massive enough to conserve a fair amount of metabolic heat. Ears and other slender extremities are cooler because they tend to lose metabolic heat faster.

But remember the sampling error trap (Chapter 1)? The researchers did the same growth experiments for *many* yarrow plants. Phenotypic variation was so great, they couldn't see a consistent pattern in it. For instance, a cutting from one plant developed best at the medium elevation. The conclusion? Different yarrow genotypes react differently across a range of environments.

And so we conclude this chapter, which has dealt with heritable and environmental factors that give rise to variations in phenotype. What is the take-home lesson? Simply this: An individual's phenotype is an outcome of complex interactions among its genes, enzymes and other gene products, and of environmental factors.

Individuals of most populations or species show complex variation for many traits. The variation arises not only from gene mutations and cumulative gene interactions. It arises also as responses to variations in environmental conditions.

1. A gene is a unit of information about a heritable trait. Alleles of a gene are different molecular versions of that information. Through experimental crosses with pea plants, Mendel gathered indirect evidence that diploid organisms have two genes for each trait and that genes keep their identity when transmitted to offspring. *10.1*

2. A homozygous dominant individual has inherited two dominant alleles (*AA*) for the trait being studied. A homozygous recessive has two recessive alleles (*aa*). A heterozygote has two nonidentical alleles (*Aa*). *10.1*

3. For monohybrid crosses, two individuals that breed true for different versions of the same trait produce F_1 offspring that are all heterozygous (*Aa*) for the trait. The experiment is an intercross between two of these F_1 heterozygotes—that is, the monohybrids (*Aa × Aa*). Mendel's monohybrid crosses with garden pea plants provided indirect evidence that some forms of a gene may be dominant over other, recessive forms. *10.2*

4. All F_1 offspring of the parental cross *AA × aa* were *Aa*. Mendel's crosses between F_1 monohybrids resulted in these combinations of alleles in F_2 offspring: *10.2*

	A	*a*
A	*AA*	*Aa*
a	*Aa*	*aa*

AA (dominant)
Aa (dominant)
Aa (dominant)
aa (recessive)
} the expected phenotypic ratio of 3:1

5. Results from Mendel's monohybrid crosses led to the formulation of a theory of segregation. In modern terms, diploid organisms have pairs of genes, on pairs of homologous chromosomes. The genes of each pair segregate from each other at meiosis, so each gamete formed ends up with one or the other gene. *10.2*

6. For dihybrid crosses, individuals that breed true for different versions of *two* traits produce F_1 offspring that are all identical heterozygotes for both traits. The experiment is an intercross between two F_1 dihybrids. Phenotypes of the F_2 offspring from Mendel's dihybrid crosses were close to a 9:3:3:1 ratio: *10.3*

9 dominant for both traits
3 dominant for *A*, recessive for *b*
3 dominant for *B*, recessive for *a*
1 recessive for both traits

7. Mendel's dihybrid crosses led to the formulation of a theory of independent assortment. In modern terms, by the end of meiosis, the gene pairs of two homologous chromosomes have been sorted out for distribution into one gamete or another, independently of how the gene pairs of other chromosomes were sorted out. *10.3*

8. Certain factors influence gene expression. *10.4–10.8*
    a. In some cases, one allele of a pair is incompletely dominant or codominant. *10.4*

    b. Products of pairs of genes may interact in ways that influence the same trait. *10.4, 10.6, 10.7*
    c. One gene may have positive or negative effects on two or more traits, a condition called pleiotropy. *10.5*
    d. Environmental conditions to which an individual is subjected may affect gene expression. *10.8*

Review Questions

1. Distinguish between these terms: *10.1*
    a. gene and allele
    b. dominant allele and recessive allele
    c. homozygote and heterozygote
    d. genotype and phenotype

2. Define a true-breeding lineage. What is a hybrid? *10.1*

3. Distinguish between monohybrid cross, dihybrid cross, and testcross. *10.2, 10.3*

4. Do segregation and independent assortment occur in mitosis, meiosis, or both? *10.2, 10.3*

5. What do the vertical and horizontal arrows of the diagram at right represent? What do the bars and the curved line represent? *10.7*

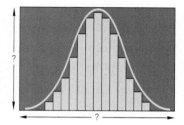

Self-Quiz   ANSWERS IN APPENDIX III

1. Alleles are _____ .
    a. different molecular forms of a gene
    b. different phenotypes
    c. self-fertilizing, true-breeding homozygotes

2. A heterozygote has a _____ for the trait being studied.
    a. pair of identical alleles
    b. pair of nonidentical alleles
    c. haploid condition, in genetic terms
    d. a and c

3. The observable traits of an organism are its _____ .
    a. phenotype   c. genotype
    b. sociobiology   d. pedigree

4. Second-generation offspring from a cross are the _____ .
    a. F_1 generation   c. hybrid generation
    b. F_2 generation   d. none of the above

5. F_1 offspring of the monohybrid cross *AA × aa* are _____ .
    a. all *AA*   c. all *Aa*
    b. all *aa*   d. 1/2 *AA* and 1/2 *aa*

6. Assuming complete dominance, the F_2 generation following the cross *Aa × Aa* will show a phenotypic ratio of _____ .
    a. 3:1   c. 1:2:1
    b. 9:1   d. 9:3:3:1

7. Crosses between F_1 pea plants resulting from the cross *AABB × aabb* lead to F_2 phenotypic ratios close to _____ .
    a. 1:2:1   c. 1:1:1:1
    b. 3:1   d. 9:3:3:1

8. Match each example with the most suitable description.
    ____ dihybrid cross   a. *bb*
    ____ monohybrid cross   b. *AaBb × AaBb*
    ____ homozygous condition   c. *Aa*
    ____ heterozygous condition   d. *Aa × Aa*

Critical Thinking—Genetics Problems

ANSWERS IN APPENDIX IV

1. One gene has alleles A and a. Another has alleles B and b. For each genotype listed, what type(s) of gametes will be produced? Assume independent assortment occurs before gametes form.
   a. *AABB*   c. *Aabb*
   b. *AaBB*   d. *AaBb*

2. Refer to Problem 1. What will be the genotypes of offspring from the following matings? Indicate the frequencies of each genotype among them.
   a. *AABB* × *aaBB*   c. *AaBb* × *aabb*
   b. *AaBB* × *AABb*   d. *AaBb* × *AaBb*

3. In one experiment, Mendel crossed a pea plant that bred true for green pods with one that bred true for yellow pods. All the F_1 plants had green pods. Which form of the trait (green or yellow pods) is recessive? Explain how you arrived at your conclusion.

4. Return to Problem 1. Assume you now study a third gene having alleles C and c. For each genotype listed, what type(s) of gametes will be produced?
   a. *AABB CC*   c. *Aa BB Cc*
   b. *Aa BB cc*   d. *Aa Bb Cc*

5. Mendel crossed a true-breeding tall, purple-flowered pea plant with a true-breeding dwarf, white-flowered plant. All F_1 plants were tall and had purple flowers. If an F_1 plant self-fertilizes, then what is the probability that a randomly selected F_2 offspring will be heterozygous for the genes specifying height and flower color?

6. At a gene location on a human chromosome, a dominant allele controls *tongue rolling*, an ability to curl up the two sides of the tongue (Figure 10.21). People who are homozygous for a recessive allele at that locus can't roll their tongue. At another gene locus, a dominant allele controls whether the earlobes will be attached or detached (refer to Figure 10.1). These two pairs of genes assort independently. Suppose a tongue-rolling, detached-earlobe woman marries a man who has attached earlobes and can't roll his tongue. Their first child has the phenotype of the father. Given this outcome,
   a. What are the genotypes of the mother, father, and child?
   b. What is the probability that a second child of theirs will have detached earlobes and won't be a tongue roller?

7. Bill and his wife, Marie, are hoping to have children. Both have notably flat feet and long eyelashes, and they tend to sneeze a lot (hence the name of the *achoo syndrome*). Dominant alleles give rise to these traits: A (foot arch), E (eyelash length), and S (chronic sneezing). Bill is heterozygous for all three dominant alleles and Marie is homozygous.
   a. What is Bill's genotype? What is Marie's genotype?
   b. Marie becomes pregnant four times. What is the probability each child will have the dominant phenotypes for all three traits?
   c. What is the probability that each child will have short lashes, high arches, and no chronic tendency to sneeze?

8. *DNA fingerprinting* is a method of identifying individuals by locating unique base sequences in their DNA molecules (Section 15.3). Before researchers refined the method, attorneys often relied on the ABO blood-typing system to settle disputes over paternity. Suppose that you, as a geneticist, are asked to testify during a paternity case in which the mother has type A blood, the child has type O blood, and the alleged father has type B blood. How would you respond to the following statements?
   a. Attorney of the alleged father: "The mother's blood is type A, so the child's type O blood must have come from the father. My client has type B blood; he could not be the father."
   b. Mother's attorney: "Because further tests prove this man is heterozygous, he must be the father."

9. Suppose you identify a new gene in mice. One of its alleles specifies white fur. A second allele specifies brown fur. You want to determine whether the relationship between the two alleles is one of simple dominance or incomplete dominance. What sorts of genetic crosses would give you the answer? On what types of observations would you base your conclusions?

10. Your sister moves away and gives you her purebred Labrador retriever, a female named Dandelion. Suppose you decide to breed Dandelion and sell puppies to help pay for your college tuition. Then you discover that two of her four brothers and sisters show *hip dysplasia*, a heritable disorder arising from a number of gene interactions. If Dandelion mates with a male Labrador known to be free of the harmful genes, can you guarantee to a buyer that puppies will not develop the disorder? Explain your answer.

11. A dominant allele W confers black fur on guinea pigs. A guinea pig that is homozygous recessive (ww) has white fur. Fred would like to know whether his pet black-furred guinea pig is homozygous dominant (WW) or heterozygous (Ww). How might he determine his pet's genotype?

12. Red-flowering snapdragons are homozygous for allele R^1. White-flowering snapdragons are homozygous for a different allele (R^2). Heterozygous plants (R^1R^2) bear pink flowers. What phenotypes should appear among F_1 offspring of the crosses listed? What are the expected proportions for each phenotype?
   a. $R^1R^1 \times R^1R^2$ c. $R^1R^2 \times R^1R^2$
   b. $R^1R^1 \times R^2R^2$ d. $R^1R^2 \times R^2R^2$

Notice, in Problem 12, that in cases of incomplete dominance it is inappropriate to refer to either allele of a pair as dominant or recessive. When the phenotype of a heterozygous individual is halfway between those of the two homozygotes, then there is no dominance. Such alleles are usually designated by superscript numerals, as shown here, rather than by uppercase letters for dominance and lowercase letters for recessiveness.

Figure 10.21 A student at San Diego State University exhibiting the tongue-rolling trait for the benefit of a tongue-roll-challenged student.

13. Two pairs of genes affect comb type in chickens (Figure 10.15). When both genes are recessive, a chicken has a single comb. A dominant allele of one gene, *P*, gives rise to a pea comb. Yet a dominant allele of the other (*R*) gives rise to a rose comb. An epistatic interaction occurs when a chicken has at least one of both dominants, *P— R —*, which gives rise to a walnut comb. Predict the F_1 ratios resulting from a cross between two walnut-combed chickens that are heterozygous for both genes (*PpRr*).

14. As described in Section 10.5, a single mutant allele gives rise to an abnormal form of hemoglobin (Hb^S instead of Hb^A). Homozygotes (Hb^SHb^S) develop sickle-cell anemia. But the heterozygotes (Hb^AHb^S) show few outward symptoms.

Suppose a woman's mother is homozygous for the Hb^A allele and her father is homozygous for the Hb^S allele. She marries a male who is heterozygous for the allele, and they plan to have children. For *each* of her pregnancies, state the probability that this couple will have a child who is:

    a. homozygous for the Hb^S allele

    b. homozygous for the Hb^A allele

    c. heterozygous Hb^AHb^S

15. Certain dominant alleles are so vital for normal development that an individual who is homozygous recessive for a mutant recessive form of the allele is unable to survive. Such recessive, *lethal alleles* can be perpetuated by heterozygotes.

Consider the Manx allele (M^L) in cats. Homozygous cats (M^LM^L) die when they are still embryos inside the mother cat. In heterozygotes (M^LM), the spine develops abnormally, and the cats end up with no tail whatsoever (Figure 10.22).

Two M^LM cats mate. Among their *surviving* progeny, what is the probability that any one kitten will be heterozygous?

16. A recessive allele *a* is responsible for *albinism*, an inability to produce or deposit melanin in tissues. Humans and some other organisms can have this phenotype (Figure 10.23). In each of the following cases, what are the possible genotypes of the father, of the mother, and of their children?

    a. Both parents have normal phenotypes; some of their children are albino and others are unaffected.

    b. Both parents are albino and have only albino children.

    c. The woman is unaffected, the man is albino, and they have one albino child and three unaffected children.

Figure 10.23 An albino male in India.

17. Kernel color in wheat plants is determined by two pairs of genes. Alleles of one pair show incomplete dominance over alleles of the other pair.

For the gene pair at one locus on the chromosome, allele A^1 imparts one dose of red color to the kernel, whereas allele A^2 does not. At the second locus, allele B^1 gives one dose of red color to the kernel, whereas allele B^2 does not. One kernel with genotype $A^1A^1B^1B^1$ is dark red. A different kernel with genotype $A^2A^2B^2B^2$ is white. All other genotypes have kernel colors in between the two extremes.

    a. Suppose you cross a plant grown from a dark red kernel with a plant grown from a white kernel. What genotypes and what phenotypes would you expect among the offspring?

    b. If a plant with genotype $A^1A^1B^1B^2$ self-fertilizes, what genotypes and what phenotypes would be expected among the offspring? In what proportions?

Selected Key Terms

allele *10.1*

codominance *10.4*

continuous variation *10.7*

dihybrid cross *10.3*

epistasis *10.6*

F_1 *10.1*

F_2 *10.1*

gene *10.1*

genotype *10.1*

heterozygous *10.1*

homozygous dominant *10.1*

homozygous recessive *10.1*

hybrid offspring *10.1*

incomplete dominance *10.4*

independent assortment *10.3*

monohybrid cross *10.2*

multiple allele system *10.4*

phenotype *10.1*

pleiotropy *10.5*

probability *10.2*

Punnett-square method *10.2*

segregation *10.2*

testcross *10.2*

true-breeding lineage *10.1*

Figure 10.22 Manx cat.

Readings

Fairbanks, D. J., and W. R. Andersen. 1999. *Genetics: The Continuity of Life.* Monterey, California: Brooks-Cole.

Orel, V. 1996. *Gregor Mendel: The First Geneticist.* New York: Oxford University Press.

11

CHROMOSOMES AND HUMAN GENETICS

The Philadelphia Story

Positioned at strategic locations in chromosomes are genes concerned with cell growth and division. Some specify enzymes and other proteins that perform these tasks. Neighboring genes control when, whether, and how fast the tasks proceed. If something disturbs the neighborhood, cell growth and division can spiral out of control and lead to cancer.

The first abnormal chromosome tied to cancer was named the *Philadelphia chromosome* after the city where someone discovered it. This chromosome shows up in cells of humans affected by a type of leukemia.

The disorder starts with stem cells in bone marrow. All stem cells are unspecialized and retain the capacity for mitotic cell division. A portion of their descendants divide and become specialized. With leukemias, there are too many descendants—far too many of the white blood cells in charge of the body's housekeeping and defense. Leukemic cells may crowd out the stem cells that give rise to red blood cells and platelets. Anemia and internal bleeding follow. Leukemic cells also enter blood and the liver, lymph nodes, spleen, and other organs, where they disrupt basic functions. Left to itself, the cancerous transformation kills the patient.

No one knew about the Philadelphia chromosome until microscopists learned how to identify its physical appearance. Chromosomes, recall, are most condensed at metaphase of mitosis. Then, their size, length, and centromere location are easiest to identify. A **karyotype** is a preparation of metaphase chromosomes based on their defining features. In this chapter, you will learn how to make a karyotype diagram from a photograph of metaphase chromosomes, just like microscopists do. The idea is to line up all of a cell's chromosomes in a certain way and to compare them with a standard karyotype for the species.

It hasn't been easy to identify the Philadelphia chromosome in karyotypes. But *spectral* karyotyping changed things. This newer research and diagnostic tool artificially colors chromosomes, and the colors are clues to their structure (Figure 11.1).

As it turns out, the Philadelphia chromosome is physically longer than its normal counterpart, human chromosome 9. The extra length is actually a piece of chromosome 22! What happened? By chance, both chromosomes broke inside a stem cell. Each broken piece was reattached, but on the wrong chromosome. At the end of chromosome 9, a gene with a role in cell division fused with a gene in chromosome 22. The altered gene region specifies an abnormal protein, and uncontrolled divisions of white blood cells follow.

The Philadelphia story is a glimpse into the world of modern genetics research. Reading it invites you to think about how far you have come in this unit of the book. You first looked at cell division, the starting point of inheritance. You looked at how chromosomes and the genes they carry are shuffled during meiosis, then at fertilization. You mulled over Mendel's insights into inheritance and some exceptions to his conclusions. Here, you have a current example of what we know about the chromosomal basis of inheritance. *How did we get to that understanding?* The answer requires a bit of history, which picks up from where we left Mendel.

In 1884 Mendel had just passed away, and his paper on pea plants had been gathering dust in a hundred

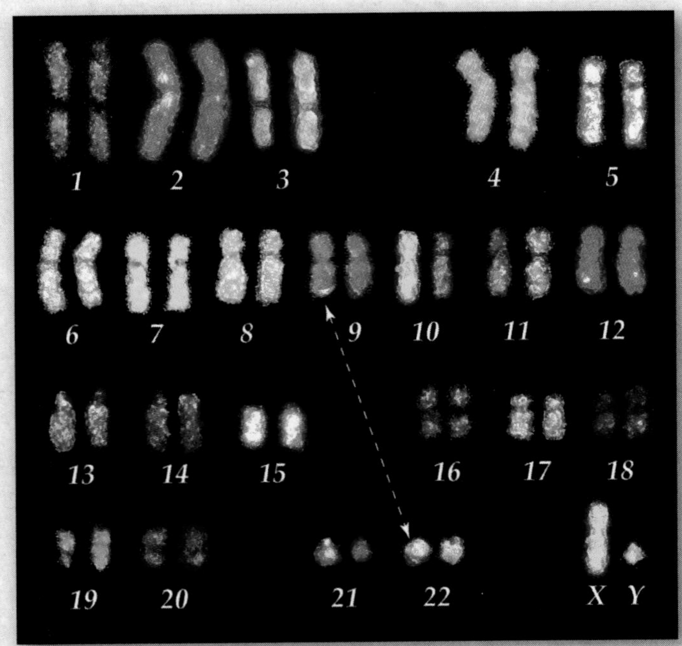

Figure 11.1 Revealing the origin of a killer—how reciprocal translocation between human chromosome 9 and chromosome 22 might appear with the help of spectral karyotyping, a new imaging technique. This particular translocation, the Philadelphia chromosome, results in a gene abnormality that gives rise to chronic myelogenous leukemia (CML), a type of cancer. Gleevec, a new oral drug, inactivates the abnormal protein product of the gene, thus arresting the dangerous cell proliferation. In preliminary tests, the drug put fifty of fifty-three CML patients into remission. So far, its side effects have been mild, compared to severe side effects of chemotherapy.

Figure 11.2 The DNA icon for an Internet site, *Rare Genetic Diseases in Children*, which allows interested individuals to seek information about specific genetic diseases and disorders. At this writing, the Philadelphia chromosome is on the site's topic board.

libraries for nearly two decades. Then the resolving power of microscopes improved and rekindled interest in the hereditary material. Walther Flemming had seen threadlike bodies—chromosomes—in dividing cells. Could chromosomes be the hereditary material?

Microscopists soon realized that each gamete has half the number of chromosomes of a fertilized egg. In 1887, August Weismann hypothesized that a special division process must reduce the chromosome number by half before gametes form. Sure enough, in that same year meiosis was discovered. Weismann promoted another hypothesis: If the chromosome number is halved at meiosis and restored at fertilization, then half of the chromosomes in our cells must come from the father and half from the mother. This view was hotly debated. It prompted a flurry of experimental crosses—just like the ones Mendel had carried out.

Finally, in 1900, researchers came across Mendel's paper while checking literature related to their own genetic crosses. To their surprise, their experimental results confirmed what Mendel's results had already suggested: Diploid cells have two units (genes) for each heritable trait, and the two units are segregated from each other before gametes form.

In the decades to follow, researchers learned more about chromosomes. You'll come across a few high points of their work as we continue with our look at inheritance. As the Philadelphia story tells you, the methods of analysis are not remote from your interests. An inherited collection of information in chromosomal DNA gives rise to traits that, for better or worse, define each organism, young and old alike (Figure 11.2).

Key Concepts

1. Cells of humans and many other sexually reproducing species contain pairs of homologous chromosomes that interact during meiosis. Typically, one chromosome of each pair is maternal in origin, and its homologue is paternal in origin.

2. Each kind of gene has its own specific location, or locus, in a particular type of chromosome.

3. The molecular form of a gene occupying a given locus may be slightly different from one chromosome to the next. All of the different molecular forms of a gene are called alleles.

4. The combination of alleles along the length of a chromosome does not necessarily remain intact through meiosis and gamete formation. By the event called crossing over, some alleles along its length swap places with their partner on the homologous chromosome. Alleles that swap places may or may not be identical.

5. Allelic recombinations contribute to variations in the phenotypes of offspring.

6. A chromosome may change structurally, as when a segment of it is deleted, duplicated, inverted, or moved to a new location. Also, the chromosome number of an individual's cells may change as a result of an improper separation of duplicated chromosomes during meiosis or mitosis.

7. Chromosome structure and the parental chromosome number rarely change. When changes do occur, they may result in genetic abnormalities or genetic disorders.

CHROMOSOMES AND INHERITANCE

Genes and Their Chromosome Locations

Earlier chapters described the structure of chromosomes and what happens to them during meiosis. To refresh your memory and get a general sense of where you are going from here, take a moment to read this list:

1. **Genes** are units of information about heritable traits. The genes of eukaryotic cells are distributed among a number of chromosomes. Each gene has its own location—a gene locus—in one type of chromosome.

2. Any cell with a diploid chromosome number ($2n$) has inherited pairs of **homologous chromosomes**. All but one pair are identical in length, shape, and gene sequence. The single exception is a pairing of nonidentical sex chromosomes, such as X with Y. The two members of a pair of homologous chromosomes interact and segregate from each other during meiosis.

3. A gene at one locus may have the same form or a slightly different one compared to its partner gene on the homologous chromosome. When considering a population as a whole, *which* forms are inherited usually varies from one individual to the next.

4. All the different molecular forms of a gene that are possible at a given locus are called **alleles**. New alleles arise only through mutation.

5. A *wild-type* allele is the most common form of a gene, either in a natural population or in a standard, laboratory-bred strain of a species. Any one of the less common forms of a gene is a *mutant* allele.

6. Genes on the same chromosome are physically linked together. The farther apart two linked genes are, the more vulnerable they are to **crossing over**. By this event, homologous chromosomes exchange corresponding segments (Figure 11.3a,b).

7. Crossing over results in **genetic recombination**. The term refers to nonparental combinations of alleles in gametes, then in offspring (Figure 11.3c).

8. **Independent assortment** refers to the random alignment of each pair of homologous chromosomes at metaphase I of meiosis. It results in nonparental combinations of alleles in gametes and offspring.

9. On rare occasions, the structure of chromosomes changes abnormally during mitosis or meiosis. So does the parental chromosome number.

Autosomes and Sex Chromosomes

In all but one case, a pair of homologous chromosomes are exactly alike in length, shape, and gene sequence. Microscopists discovered the exception in the late 1800s, using analytical methods such as karyotyping (Section 11.2). A distinctive chromosome is present in females *or* males of many species, but not in both. As an example,

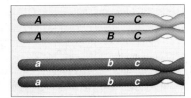

a A pair of duplicated homologous chromosomes (two sister chromatids each). This example has nonidentical alleles at three gene loci (*A* with *a*, *B* with *b*, and *C* with *c*).

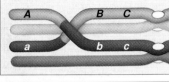

b In prophase I of meiosis, a crossover event occurs: two nonsister chromatids exchange corresponding segments.

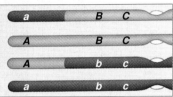

c This is the outcome of the crossover: genetic recombination between nonsister chromatids (which are shown after meiosis as unduplicated, separate chromosomes).

Figure 11.3   Review of crossing over. As shown earlier in Section 9.4, this event occurs in prophase I of meiosis.

a human male diploid cell has one **X chromosome** and one **Y chromosome** (XY). A human female diploid cell has two X chromosomes (XX). This inheritance pattern is typical of many species, including all mammals and fruit flies. We see a different pattern in butterflies, moths, some birds, and certain fishes. For them, inheriting two identical sex chromosomes results in a male. Inheriting two nonidentical sex chromosomes results in a female.

The human X and Y chromosomes differ physically; one is shorter than the other (Figure 11.4f). They also differ in which genes they carry. Even so, they are able to synapse (become zippered together briefly) in a small region along their length. The bit of zippering lets them function as homologues during meiosis.

The human X and Y chromosomes fall into the more general category of **sex chromosomes**. The term refers to types of chromosomes that, in certain combinations, determine a new individual's sex—whether a male or a female will develop. All other chromosomes in cells are the same in both sexes; they are **autosomes**.

Diploid cells have pairs of genes, on pairs of homologous chromosomes. At each gene locus, the alleles (alternative forms of a gene) may be identical or nonidentical.

Crossing over and other events during meiosis give offspring new combinations of alleles and parental chromosomes.

Abnormal events during meiosis or mitosis can change the structure and number of chromosomes.

Autosomes are the pairs of chromosomes that are the same in males and females of a species. One other pair, the sex chromosomes, govern the sex of a new individual.

Karyotyping Made Easy

Karyotype diagrams help answer questions about an individual's chromosomes. Chromosomes are the most condensed and easiest to identify in cells going through metaphase of mitosis. Technicians don't count on finding a cell that happens to be dividing in the body when they go looking for it. They culture cells in vitro (literally, "in glass"). They put a sample of cells, usually from blood, into a glass container. The container holds a solution that stimulates cell growth and mitotic cell divisions.

Adding colchicine to a culture medium arrests cell division at metaphase. This extract of *Colchicum* plants also blocks spindle formation (Section 4.10). If a spindle can't form, sister chromatids of duplicated chromosomes won't be able to separate at anaphase. By using suitable colchicine concentrations and exposure times, technicians can stockpile metaphase cells. This improves the odds of finding candidates for karyotype diagrams.

Following colchicine treatment, the culture medium is transferred to tubes of a centrifuge (Figure 11.4a). Cells have greater mass and density than the solution bathing them. The spinning force moves them farthest from the center of rotation, to the bottom of the attached tubes.

The cells are transferred to a saline solution. When immersed in this hypotonic fluid, they swell (by osmosis) and move apart. The metaphase chromosomes also move apart. The cells are ready to be mounted on a microscope slide, fixed as by air-drying, and stained.

Chromosomes take up some stains uniformly along their length, which allows identification of chromosome size and shape. By other staining procedures, horizontal bands show up along the length of the chromosomes of certain species. If researchers direct a ray of ultraviolet light at the chromosomes, the bands will fluoresce. You can see an example of this in Figure 9.3.

The chromosomes are photographed through the microscope, and the image is enlarged. The photograph is cut apart, one chromosome at a time. Individual cutouts are arranged according to size, shape, and length of the arms. All pairs of homologous chromosomes are horizontally aligned by their centromeres. Figure 11.4f shows a karyotype diagram prepared this way.

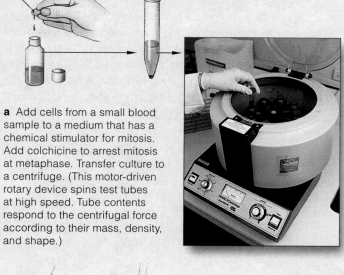

a Add cells from a small blood sample to a medium that has a chemical stimulator for mitosis. Add colchicine to arrest mitosis at metaphase. Transfer culture to a centrifuge. (This motor-driven rotary device spins test tubes at high speed. Tube contents respond to the centrifugal force according to their mass, density, and shape.)

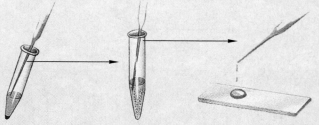

b Centrifugation forces cells to bottom of tube. Draw off culture medium. Add a dilute saline solution to tube. Add a fixative.

c Prepare and stain cells for microscopy.

d Put cells on a microscope slide. Observe.

e Photograph one cell through microscope. Enlarge image of its chromosomes. Cut the image apart. Arrange chromosomes as a set.

Figure 11.4 (a–e) How to prepare karyotypes. (f) Human karyotype. Human somatic cells hold 22 pairs of autosomes and 1 pair of sex chromosomes (XX or XY). That is a diploid number of 46. These are metaphase chromosomes; each is in the duplicated state, consisting of two sister chromatids joined at the centromere.

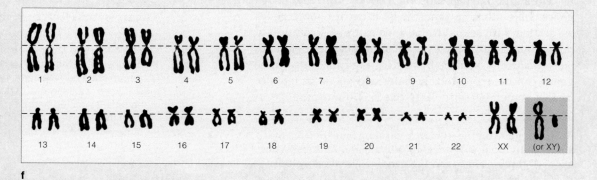

f

SEX DETERMINATION IN HUMANS

Analyzing human cells, as by karyotyping, has yielded evidence that every normal egg produced by a female has one X chromosome. Half the sperm cells produced by a male carry an X chromosome and half carry a Y.

If an X-bearing sperm fertilizes an X-bearing egg, the new individual will develop into a female. If the sperm happens to carry a Y chromosome, the individual will develop into a male (Figure 11.5).

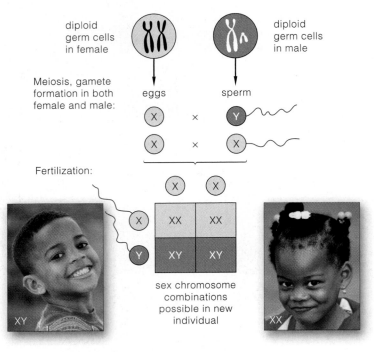

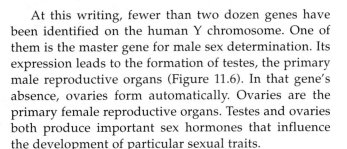

Figure 11.5 Pattern of sex determination in humans.

At this writing, fewer than two dozen genes have been identified on the human Y chromosome. One of them is the master gene for male sex determination. Its expression leads to the formation of testes, the primary male reproductive organs (Figure 11.6). In that gene's absence, ovaries form automatically. Ovaries are the primary female reproductive organs. Testes and ovaries both produce important sex hormones that influence the development of particular sexual traits.

The human X chromosome carries more than 2,300 genes. Like other chromosomes, it carries some genes associated with sexual traits, such as the distribution of body fat and hair. However, most of its genes deal with *nonsexual* traits, such as blood-clotting functions. These genes can be expressed in males as well as in females (males, remember, also carry one X chromosome).

A certain gene on the human Y chromosome dictates that a new individual will develop into a male. In the absence of the Y chromosome (and the gene), a female develops.

---

umbilical cord (lifeline between the embryo and the mother's tissues)

amnion (a protective, fluid-filled sac surrounding and cushioning the embryo)

a A human embryo, eight weeks old and about an inch long. The mass at left is part of the placenta, an organ that forms from maternal and embryonic tissues.

Figure 11.6 Boys, girls, and the Y chromosome.

For about the first four weeks of its existence, a human embryo has neither male nor female traits, regardless of whether it is XY or XX. Then ducts and other internal structures that can develop either way start forming.

(**a–c**) In an XX embryo, ovaries (the primary female reproductive organs) start to form automatically—*in the absence of a Y chromosome*. By contrast, in an XY embryo, testes (the primary male reproductive organs) start to form during the next four to six weeks. Apparently, a gene region on the Y chromosome governs a fork in the developmental road that can lead to maleness.

The newly forming testes start to produce testosterone and other sex hormones. These hormones are crucial for the development of a male reproductive system. By contrast, in an XX embryo, newly forming ovaries start to produce different kinds of sex hormones, particularly estrogens. Estrogens are crucial for the development of a female reproductive system.

The master gene for male sex determination is named *SRY* (short for the *Sex-determining Region* of the *Y* chromosome). The same gene has been identified in DNA from male humans, chimpanzees, mice, rabbits, pigs, horses, cattle, and tigers, among others. None of the females tested had the gene. Tests with mice indicate that the gene region becomes active about the time that testes are starting to develop.

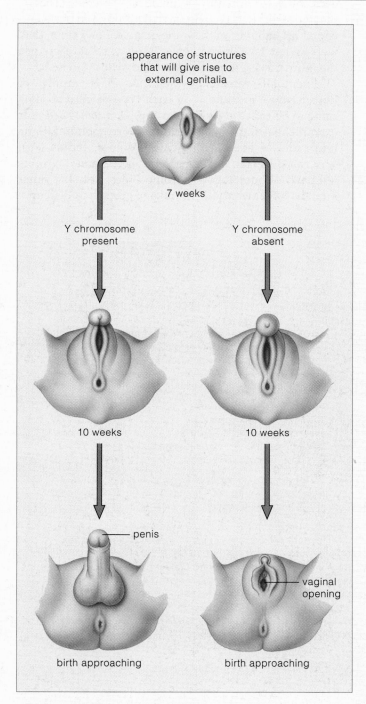

appearance of structures
that will give rise to
external genitalia

7 weeks

Y chromosome
present

Y chromosome
absent

10 weeks

10 weeks

penis

vaginal
opening

birth approaching

birth approaching

b External appearance of reproductive organs
forming in human embryos.

The *SRY* gene resembles DNA regions that specify regulatory
proteins. As described in Chapter 14, regulatory proteins
influence the activity of genes and their products. The *SRY*
gene product apparently regulates a cascade of reactions
that are necessary for male sex determination.

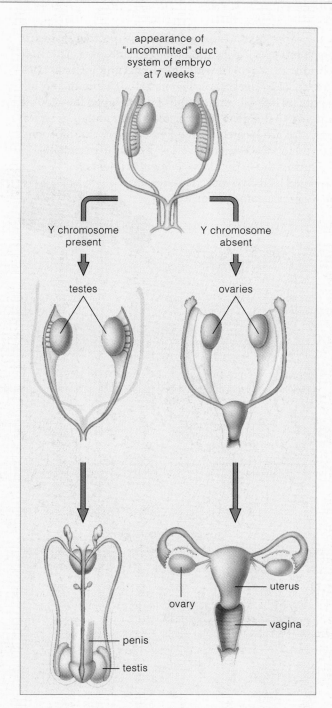

appearance of
"uncommitted" duct
system of embryo
at 7 weeks

Y chromosome
present

Y chromosome
absent

testes

ovaries

uterus

ovary

vagina

penis

testis

c A duct system that forms early in the human embryo. It
may develop into male *or* female primary reproductive organs,
depending on whether the *SRY* gene is present or absent.

EARLY QUESTIONS ABOUT GENE LOCATIONS

Linked Genes—Clues to Inheritance Patterns

By the early 1900s, researchers were suspecting that each gene has a specific location on a chromosome. Through hybridization experiments involving mutant fruit flies (*Drosophila melanogaster*), Thomas Hunt Morgan and his coworkers helped confirm this. For example, they found evidence that a gene for eye color and another gene for wing size are located on the *Drosophila* X chromosome. Figure 11.7 describes one series of their experiments.

During the early *Drosophila* experiments, it seemed as if two mutant genes on the X chromosome (*w* for white

eyes and *m* for miniature wings) were "linked." That is, they apparently traveled together through meiosis and ended up together in the same gamete. For a time, they were called "sex-linked genes." Today, researchers use the more precise terms *X-linked* and *Y-linked* genes.

Eventually, researchers identified a large number of linked genes, and those on each type of chromosome came to be called a **linkage group**. *D. melanogaster*, for example, has four linkage groups, corresponding to its four pairs of homologous chromosomes. Indian corn (*Zea mays*) has ten linkage groups, corresponding to its ten pairs of homologous chromosomes. You and other humans have twenty-three linkage groups, and so on.

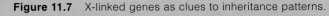

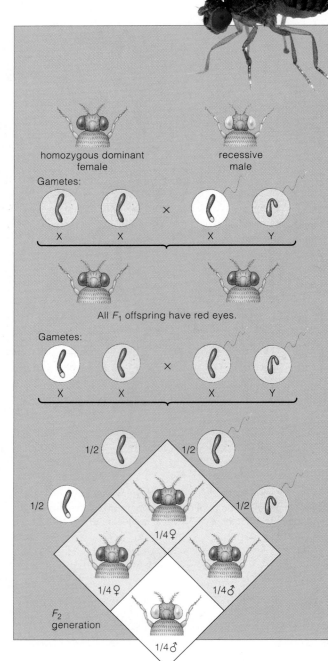

Figure 11.7 X-linked genes as clues to inheritance patterns.

In the early 1900s Thomas Morgan, an embryologist, was studying patterns of inheritance. He and his coworkers discovered a genetic basis for a relationship between sex determination and some nonsexual traits. For example, human males and females both have blood-clotting factors. Yet blood-clotting disorders (hemophilias) show up most often in males of a family lineage. This sex-specific outcome wasn't like anything Mendel saw; in his hybrid crosses of pea plants, it made no difference which parent carried a recessive allele.

Morgan studied eye color and other nonsexual traits of *Drosophila melanogaster*. This fruit fly can live in small bottles on agar, cornmeal, molasses, and yeast. A female lays hundreds of eggs in a few days and offspring reproduce in less than two weeks. In a single year, Morgan could track traits through nearly thirty generations of thousands of flies.

At first, all flies were wild type for eye color; they had brick-red eyes. Then Morgan got lucky; a mutation in some gene controlling eye color presented him with a white-eyed male.

Morgan established true-breeding strains of white-eyed males and females for **reciprocal crosses**. (In the first of such paired crosses, one parent displays the trait of interest. In the second cross, the other parent displays it.) First he let white-eyed males mate with homozygous red-eyed females. All F_1 offspring had red eyes, and some F_2 males had white eyes (*see diagram*). Then he let true-breeding red-eyed males mate with white-eyed females. The results gave Morgan an eyeful, so to speak. Half the F_1 offspring were red-eyed females, and half were white-eyed males. And of the F_2 offspring, 1/4 were red-eyed females, 1/4 white-eyed females, 1/4 red-eyed males, and 1/4 white-eyed males.

The odd results implied a relationship between an eye-color gene and sex determination. Probably the gene locus was on a sex chromosome. Which one? Because females (XX) could be white-eyed, the recessive allele had to be on one of their X chromosomes. What if the white-eyed males (XY) had the recessive allele on their X chromosome and their Y chromosome had no corresponding eye-color allele? In that case, they had white eyes because they had no dominant allele to mask the effect of the recessive one, as the diagram shows.

And so Morgan's idea of an X-linked gene dovetailed with Mendel's concept of segregation. By proposing that a specific gene is located on an X chromosome but not on the Y, Morgan explained his reciprocal crosses. His experimental results matched predicted outcomes.

homozygous dominant female

recessive male

Gametes:

X X × X Y

All F_1 offspring have red eyes.

Gametes:

X X × X Y

1/2 1/2

1/2

1/2

1/4 ♀

1/4 ♀

1/4 ♂

1/4 ♂

F_2 generation

Crossing Over and Genetic Recombination

If linked genes always stayed together through meiosis, then there would be no exceptions to Mendel's view of independent assortment. You could always predict the ratio of possible combinations of phenotypes among, say, the F_2 offspring of dihybrid crosses. (Here you may wish to review Section 10.3 and Figure 11.3.) Yet some results from the *Drosophila* experiments did not match expectations. Some genes weren't "tightly linked."

Example: In one set of experiments, true-breeding mutant females (white eyes, yellow body) were crossed with wild-type males (red eyes, gray body). As the researchers expected, 50 percent of the F_1 offspring had one or the other parental phenotype. However, of 2,205 F_2 offspring, 129 were recombinants. That's 1.3 percent.

As it turns out, some alleles tend to remain together more often than others through meiosis. They are closer together on the same chromosome and less vulnerable to crossovers! Imagine any two genes at two different locations on the same chromosome. *The probability that a crossover will disrupt their linkage is proportional to the distance that separates the two loci.* Suppose genes *A* and *B* are twice as far apart as two other genes, *C* and *D*:

We would expect crossing over to disrupt the linkage between *A* and *B* much more often.

Two genes are very closely linked when the distance between them is small. Their allelic combinations nearly always end up inside the same gamete. Linkage is more vulnerable to crossover when the distance between them is greater (Figure 11.8). When two gene loci are very far apart, crossing over disrupts linkage so often that those genes assort independently of each other into gametes.

Unlike fruit flies, humans do not lend themselves to experimental crosses. Even so, some tight gene linkages have been identified by tracking phenotypes in certain families through the generations.

For example, recessive alleles at two gene loci on the X chromosome cause *color blindness* and *hemophilia* (to be described shortly). One female carried both alleles, but she was symptom-free. Her father was, too, so she must have inherited a normal X chromosome from him (remember, males have only one X and one Y). The X chromosome she inherited from her mother must have carried both mutant alleles. She gave birth to six sons. Three developed color blindness and hemophilia; two were unaffected. Here was phenotypic evidence of no recombination. However, her sixth son started life as a fertilized, recombinant egg; he was color-blind only. For two mutant alleles in this family, the recombination frequency was 1/6 (or 0.167 percent).

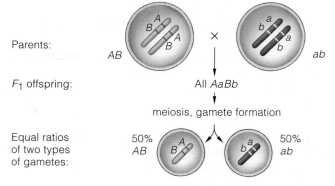

a Full linkage between two genes (no crossovers): half of the gametes have one parental genotype and half have the other.

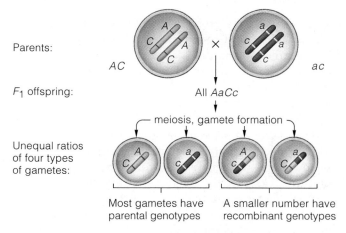

b Incomplete linkage; crossing over affected the outcome.

Figure 11.8 Examples of gene linkages that affect a dihybrid cross. In (**a**), linkage is complete; in (**b**), it is incomplete.

Tracking many families affected by these disorders and using biotechnologies (Chapter 15) can help show how closely these two genes are linked. Generally, the recombination frequencies are high because the two are very closely linked at one end of the X chromosome.

A final, important point: Don't think that crossing over is a rare event. In fact, for humans and most other eukaryotic species, meiosis cannot even be completed properly unless each pair of homologous chromosomes takes part in at least one crossover.

Genes at different loci on the same chromosome belong to the same linkage group and do not assort independently during meiosis.

Crossing over between homologous chromosomes disrupts gene linkages and results in nonparental combinations of genes in chromosomes.

The farther apart two genes are on the same chromosome, the greater will be the frequency of crossing over and genetic recombination between them.

HUMAN GENETIC ANALYSIS

Some organisms, including pea plants and fruit flies, are ideal for genetic analysis. They grow and reproduce rapidly in small spaces, under controlled conditions. It does not take very long to track a trait through many generations. Humans are another story. We live under variable conditions in diverse environments. We select our own mates and reproduce if and when we want to. Humans live as long as the geneticists who study them, so tracking traits through generations is tedious. Most human families are not large, so there are not enough offspring for easy inferences about inheritance.

Constructing Pedigrees

To get around the problems associated with analyzing human inheritance, geneticists put together pedigrees. A **pedigree** is a chart of the genetic connections among individuals. Standardized methods, definitions, and symbols that represent individuals are used when constructing it (Figure 11.9a).

When geneticists analyze pedigrees, they rely on their knowledge of probability and Mendelian inheritance patterns, which may yield clues to the genetic basis for a trait. For example, clues might suggest that an allele responsible for a certain disorder is dominant or recessive or that it is located on a certain autosome or sex chromosome.

Gathering a great many family pedigrees increases the numerical base for analysis. When any trait follows a simple Mendelian inheritance pattern, a geneticist has greater confidence for predicting the probability of its occurrence among children of prospective parents. We will return to this topic later.

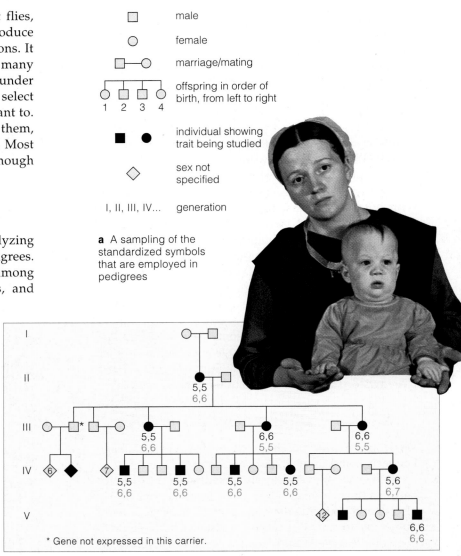

a A sampling of the standardized symbols that are employed in pedigrees

b Pedigree for a family in which polydactyly recurs as one symptom of Ellis–van Creveld syndrome (Section 16.11)

* Gene not expressed in this carrier.

Figure 11.9 (**a**) Some standardized symbols used in constructing pedigree diagrams. (**b**) Example of a pedigree for *polydactyly*. With this condition, an individual has extra fingers, extra toes, or both. Expression of the gene for this trait varies among individuals. *Black* numerals signify the known number of fingers on each hand; *blue* numerals signify the number of toes on each foot.

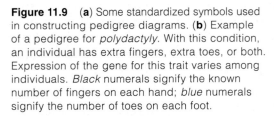

(**c**) From human genetic researcher Nancy Wexler, a pedigree for *Huntington disorder*, by which the human nervous system progressively degenerates. Wexler and her team pieced together an extended family tree for nearly 10,000 Venezuelans. Analysis of affected and unaffected individuals revealed that a dominant allele on human chromosome 4 is the genetic culprit. Wexler has a special interest in Huntington disorder; she herself has a 50 percent chance of developing it.

c Nancy Wexler and a pedigree she constructed to track Huntington disorder

Table 11.1 Examples of Human Genetic Disorders and Genetic Abnormalities

Disorder or Abnormality*	Main Consequences	Disorder or Abnormality*	Main Consequences
AUTOSOMAL RECESSIVE INHERITANCE		**X-LINKED RECESSIVE INHERITANCE**	
Albinism *10.6; 10.9 CT*	Absence of pigmentation	Color blindness *11.4, 11.6*	Inability to distinguish among some or all colors
Blue offspring *16.12 CT*	Bright blue skin coloration	Fragile X syndrome *11.6*	Mental retardation
Cystic fibrosis *4.12 CT*	Excessive glandular secretions leading to tissue, organ damage	Hemophilia *11.4, 11.6, 11.12 CI*	Impaired blood-clotting ability
Ellis–van Creveld syndrome *16.11*	Extra fingers, toes, short limbs	Muscular dystrophies *11.11 CT, 14.5 CT*	Progressive loss of muscle function
Fanconi anemia *11.10*	Physical abnormalities, bone marrow failure	Testicular feminization syndrome *32.2*	XY individual but having some female traits; sterility
Galactosemia *11.6*	Brain, liver, eye damage	X-linked anhidrotic dysplasia *14.3*	Mosaic skin (patches with or without sweat glands); other effects
Phenylketonuria (PKU) *11.10*	Mental retardation		
Sickle-cell anemia *10.5, 13.4, 16.9, 34.3*	Adverse pleiotropic effects on organs throughout body		
AUTOSOMAL DOMINANT INHERITANCE		**CHANGES IN CHROMOSOME NUMBER**	
Achondroplasia *11.6*	One form of dwarfism	Down syndrome *11.9; 11.11 CT, 39.9*	Mental retardation; heart defects
Achoo syndrome *10.9 CT*	Chronic sneezing	Klinefelter syndrome *11.9*	Sterility; retardation
Camptodactyly *10.7*	Rigid, bent little fingers	Turner syndrome *11.9*	Sterility; abnormal ovaries, abnormal sexual traits
Familial cholesterolemia *15 CI*	High cholesterol levels in blood; eventually clogged arteries	XYY condition *11.9*	Mild retardation or free of symptoms
Huntington disorder *11.5, 11.7*	Nervous system degenerates progressively, irreversibly	**CHANGES IN CHROMOSOME STRUCTURE**	
Marfan syndrome *11.11 CT*	Abnormal or no connective tissue	Chronic myelogenous leukemia *12 CI*	Overproduction of white blood cells in bone marrow; organ malfunctions
Polydactyly *11.5*	Extra fingers, toes, or both	Cri-du-chat syndrome *11.8*	Mental retardation; abnormally shaped larynx
Progeria *11.7*	Drastic premature aging		
Neurofibromatosis *13.5*	Tumors of nervous system, skin		

*Italic numbers indicate sections in which a disorder is described. *CI* signifies Chapter Introduction. *CT* signifies an end-of-chapter Critical Thinking question.

Regarding Human Genetic Disorders

Table 11.1 is a list of some heritable traits that have been studied in detail. A few are abnormalities, or deviations from the average condition. Said another way, a **genetic abnormality** is nothing more than a rare, uncommon version of a trait, as when a person is born with six toes on each foot instead of five. Whether an individual or society at large views an abnormal trait as disfiguring or merely interesting is subjective. As the classic novel *The Hunchback of Notre Dame* suggests, there is nothing inherently life-threatening or even ugly about it.

By comparison, a **genetic disorder** is an inherited condition that sooner or later will cause mild to severe medical problems. A **syndrome** is a recognized set of symptoms that characterize a given disorder.

Because alleles underlying severe genetic disorders put people at great risk, they are rare in populations. Why, then, don't they disappear entirely? First, we can expect rare mutations to introduce new copies of the alleles into the population. Second, in heterozygotes, a harmful allele is paired with a normal one that may cover its functions, so it still can be passed to offspring.

You may hear someone refer to a genetic disorder as a disease, but the terms aren't always interchangeable. A disease also is an abnormal alteration in the way the body functions, and it, too, is characterized by a set of symptoms. But **disease** is illness caused by infectious, dietary, or environmental factors, not by inheritance of mutant genes. That said, it may be appropriate to call an illness a *genetic* disease if previously workable genes become altered in a way that disrupts body functions.

With these qualifications in mind, we turn next to examples of inheritance in the human population. As you'll see, some show simple Mendelian patterns. Many traits have been traced to a dominant or recessive allele on an autosome or X chromosome. Others arise from changes in the structure or number of chromosomes.

For many genes, pedigree analysis might reveal simple Mendelian inheritance patterns that will allow inferences about the probability of their transmission to children.

A genetic abnormality is a rare or less common version of an inherited trait. A genetic disorder is an inherited condition that results in mild to severe medical problems.

EXAMPLES OF INHERITANCE PATTERNS

Autosomal Recessive Inheritance

For some traits, inheritance patterns reveal two clues that point to a recessive allele on an autosome. First, if both parents are heterozygous, any child of theirs will have a 50 percent chance of being heterozygous and a 25 percent chance of being homozygous recessive, as Figure 11.10 indicates. Second, if the parents are both homozygous recessive, any child of theirs will be, also.

On average, 1 in 100,000 newborns is homozygous for a recessive allele that causes *galactosemia*. It cannot make functional molecules of an enzyme that prevents a product of lactose breakdown from accumulating to toxic levels. Lactose normally is converted to glucose and galactose, then to glucose–1–phosphate (which is broken down by glycolysis or converted to glycogen). The full conversion is blocked in galactosemics:

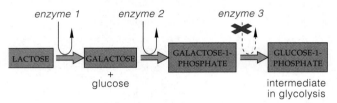

Galactose in blood rises to levels that can be detected in urine. Excess amounts damage the eyes, liver, and brain, and cause malnutrition, diarrhea, and vomiting. Untreated galactosemics often die in childhood. When they are quickly put on a restricted diet that excludes dairy products, they can grow up symptom-free.

Autosomal Dominant Inheritance

Two clues point to an autosomal dominant allele for a trait. First, the trait typically appears each generation; usually the allele is expressed, even in heterozygotes. Second, if one parent is heterozygous and the other is homozygous recessive, any child of theirs will have a 50 percent chance of being heterozygous (Figure 11.11*a*).

A few dominant alleles persist in populations even though they cause severe genetic disorders. Some persist by spontaneous mutations. For others, expression of the dominant allele may not interfere with reproduction or affected people reproduce before symptoms are severe.

For example, *Huntington disorder* is characterized by progressive involuntary movements and deterioration of the nervous system; death is inevitable (Figure 11.9). Symptoms may not start to show up until an affected individual is past age thirty. Most people have already reproduced by then. Affected individuals usually die in their forties or fifties, sometimes before they realize they transmitted the mutant allele to their children.

As another example, *achondroplasia* affects about 1 in 10,000 people. Commonly, the homozygous dominant

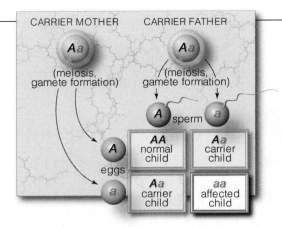

Figure 11.10 A pattern for autosomal recessive inheritance. In this case, both parents are heterozygous carriers of the recessive allele (coded *red*).

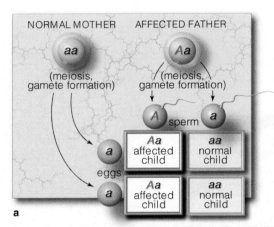

Figure 11.11 (**a**) A pattern for autosomal dominant inheritance. This dominant allele (*red*) is fully expressed in the carriers. (**b**) Infanta Margarita Teresa of the Spanish court and her maids, including the achondroplasic woman at the far right.

condition leads to stillbirth, yet heterozygotes are able to reproduce. While heterozygotes are young, the cartilage of the skeleton forms improperly. At maturity, affected people have abnormally short arms and legs relative to other body parts (Figure 11.11*b*). Adult achondroplasics are less than 4 feet, 4 inches tall. Often the dominant allele has no other phenotypic effects.

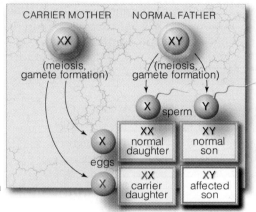

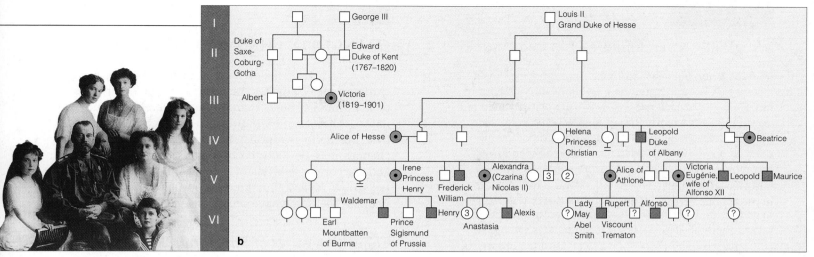

b

X-Linked Recessive Inheritance

Distinctive clues often show up when a recessive allele on an X chromosome causes a genetic disorder. First, males show the recessive phenotype more often than females. A dominant allele on the other X chromosome can mask the allele in females. The allele is not masked in males, who have one X chromosome (Figure 11.12*a*). Second, a son can't inherit the recessive allele from his father. A daughter can. If she is heterozygous, there is a 50 percent chance each son of hers will inherit the allele.

Color blindness, an inability to distinguish between some or all colors, is a common X-linked recessive trait. For instance, in red–green color blindness, an individual lacks some or all of the sensory receptors that normally respond to visible light of red and green wavelengths.

Hemophilia A, a blood-clotting disorder, is a case of X-linked recessive inheritance. Normally a blood-clotting mechanism quickly stops bleeding from minor injuries. Clotting requires several gene products, some on the X chromosome. If one of the X-linked genes is mutated in a human male, bleeding will be prolonged. About 1 in 7,000 males is affected by hemophilia A. Clotting time is close to normal in heterozygous females.

The frequency of hemophilia A was high in royal families of nineteenth-century Europe, in which close relatives often married. Queen Victoria of England was a

notable carrier (Figure 11.12*b*). At one time eighteen of her sixty-nine descendants housed the recessive allele.

Fragile X syndrome, an X-linked recessive disorder that causes mental retardation, affects 1 in 1,500 males in the United States. In cultured cells, an X chromosome carrying the mutant allele is constricted near the end of its long arm (Figure 11.13). This constriction is called a fragile site because the end of the chromosome tends to break away. The term is misleading, for the end breaks away only in cultured cells, not in cells in the body.

A mutant gene, not breakage, causes the syndrome. It specifies a protein required for normal development of brain cells. Within that gene, a segment of DNA is repeated several times. Certain mutations result in the addition of many more repeats in the DNA; hence their name, expansion mutations. The outcome is a mutant allele that cannot function properly. Because that allele is recessive, males who inherit it and females who are homozygous for it have fragile X syndrome. Expansion mutations are now known to be the cause of Huntington disorder and some other genetic disorders.

Genetic analyses of family pedigrees have revealed simple Mendelian inheritance patterns for certain traits, as well as for many genetic disorders that arise from expression of specific alleles on an autosome or X chromosome.

Figure 11.12 (**a**) One pattern for X-linked inheritance. In this example, assume the mother carries the recessive allele on one of her X chromosomes (*red*).

(**b**) *Above:* Partial pedigree for Queen Victoria's descendants, showing carriers and affected males who carried the X allele for hemophilia A. Of the Russian royal family members in the photograph, the mother was a carrier and Crown Prince Alexis was hemophilic. He was a focus of political intrigue that helped trigger the Russian Revolution of 1917.

Figure 11.13 Scanning electron micrograph of the fragile X chromosome from a cultured cell. The arrow points to the fragile site.

Progeria—Too Young to Be Old

Imagine being ten years old with a mind trapped inside a body that is rapidly getting a bit more shriveled, more frail—old—with each passing day. You are just barely tall enough to peer over the top of the kitchen counter, and you weigh less than thirty-five pounds. Already you are bald and have a crinkled nose. Possibly you have a few more years to live. Would you, like Mickey Hayes and Fransie Geringer, still laugh with your friends?

Of every 8 million newborn humans, one is destined to grow old far too soon. On one of its autosomes, that rare individual carries a mutant gene that gives rise to *Hutchinson–Gilford progeria syndrome*. Through hundreds, thousands, then many billions of DNA replications and mitotic cell divisions, terrible information encoded in that gene was systematically distributed to every cell in the growing embryo, and later in the newborn. Its legacy will be accelerated aging and a greatly reduced life span. The photograph of Mickey and Fransie in Figure 11.14 shows some of the symptoms.

The mutation causes gross disruptions in interactions among genes that bring about the body's growth and development. Observable symptoms start before age two. Skin that should be plump and resilient starts to thin. Skeletal muscles weaken. Tissues in limb bones that should lengthen and grow stronger start to soften. Hair loss is pronounced; extremely premature baldness is inevitable. There are no documented cases of progeria running in families, so a gene must undergo random, spontaneous mutation. Probably the gene is dominant over a normal partner on the homologous chromosome.

Most progeriacs can expect to die in their early teens as a result of strokes or heart attacks. These final insults are brought on by a hardening of the walls of arteries, a condition typical of advanced age. When Mickey turned eighteen, he was the oldest living progeriac. Fransie was seventeen when he died.

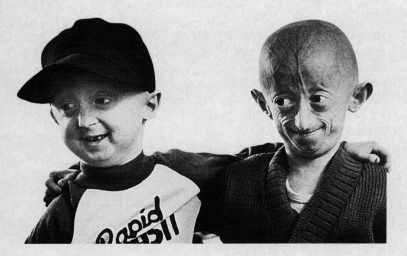

Figure 11.14 Two boys who met at a gathering of progeriacs at Disneyland, California, when they were not yet ten years old.

CHANGES IN CHROMOSOME STRUCTURE

On rare occasions, the physical structure of one or more chromosomes changes. The result is a genetic disorder or abnormality. Such changes spontaneously occur in nature. They also are induced in research laboratories by exposure to chemicals or by irradiation. Either way, changes may be detected by microscopic examination and karyotype analysis of cells at mitosis or meiosis. Let's now review four kinds of structural change. As you will see, some have severe or lethal consequences.

Major Categories of Structural Change

DUPLICATION Even normal chromosomes have gene sequences that are repeated several to many hundreds or thousands of times. These are **duplications**:

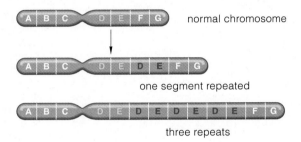

INVERSION With an **inversion**, a linear stretch of DNA within the chromosome becomes oriented in the reverse direction, with no molecular loss:

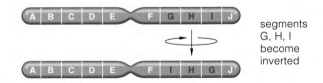

TRANSLOCATION As the chapter introduction showed, the Philadelphia chromosome that has been linked to a form of leukemia results from **translocation**, whereby a broken part of a chromosome becomes attached to a *non*homologous chromosome. Most translocations are reciprocal (both chromosomes exchange broken parts):

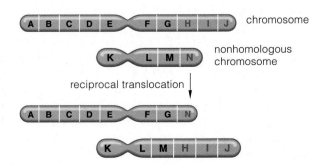

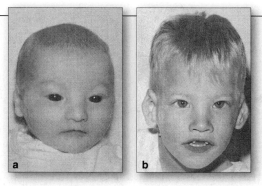

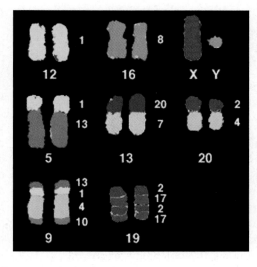

Gibbon chromosomes 12, 16, X, and Y. The three are structurally identical to human chromosomes 12, 16, X, and Y (compare Figure 11.1).

Gibbon chromosomes 5, 13, and 20 show translocations that correspond to human chromosomes 1/13, 20/7, and 2/4.

Gibbon chromosome 9 corresponds to parts of human chromosomes. Gibbon chromosome 19 duplications correspond to human chromosomes 2 and 17.

Figure 11.15 (**a**) A male infant who later developed cri-du-chat syndrome. The ears are positioned low on the side of the head relative to the eyes. (**b**) The same boy, four years later. By this age, affected humans no longer make mewing sounds typical of the syndrome.

Figure 11.16 Spectral karyotype of duplicated chromosomes from an ape. Colors identify which parts are structurally identical in human chromosomes.

DELETION Viral attacks, irradiation (ionizing radiation especially), chemical assaults, or other environmental factors may trigger a **deletion**, the loss of some segment of a chromosome:

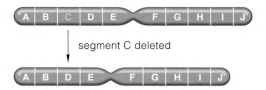

segment C deleted

Don't species with diploid cells have an advantage after a deletion? After all, wouldn't the genes on the homologous chromosome cover the loss? Maybe, but the sword of chance cuts both ways. If the remaining segment happens to carry a harmful recessive allele, nothing will mask or compensate for *its* effects.

Most deletions are lethal or cause serious disorders in mammals, for they disrupt normal gene interactions that underlie a program of growth, development, and maintenance activities. For example, one deletion from human chromosome 5 results in mental retardation and the development of an abnormally shaped larynx. When affected infants cry, they produce sounds rather like a cat's meow. Hence *cri-du-chat* (cat-cry), the name of this disorder. Figure 11.15 shows an affected child.

Does Chromosome Structure Evolve?

Changes in chromosome structure tend to be selected against rather than conserved over evolutionary time. Even so, for many species, we have interesting signs of past changes. For example, duplications are relatively rare. Many have not harmed their bearers. Over billions of years, the neutral ones have been accumulating and are now built into the DNA of all species.

Duplicates of gene sequences with neutral effects might have an adaptive advantage. Having two of the same gene could free up one of them for potentially beneficial mutations. The other gene would still issue the required product. The duplicated gene region could become slightly modified, then the products of the two genes could function in slightly different or novel ways.

Apparently, several kinds of duplicated, modified genes were pivotal in evolution. Think back on those gene regions for the polypeptide chains of hemoglobin (Section 3.5). In humans and other primates, the regions have multiple nucleotide sequences that are strikingly similar. The sequences specify whole families of chains, each with a slight structural difference. The structural differences translate into slightly different efficiencies in hemoglobin's capacity to bind and transport oxygen under a range of cellular conditions.

It also appears that certain duplications, inversions, and translocations helped put the primate ancestors of humans on a unique evolutionary road. They seem to have contributed to divergences that led to the modern apes and humans. Of the twenty-three pairs of human chromosomes, eighteen are nearly identical with their counterparts in chimpanzees and gorillas. The other five pairs differ at inverted and translocated regions. You can observe the dramatic similarities for yourself by comparing chromosomes of a human with those of a gibbon, one of the apes (Figure 11.16).

On rare occasions, a segment of a chromosome may get lost, inverted, moved to a new location, or duplicated.

Most chromosome changes are harmful or lethal when they alter gene interactions that underlie growth, development, and maintenance activities.

Over evolutionary time, many changes have been conserved; they confer adaptive advantages or have had neutral effects.

CHANGES IN CHROMOSOME NUMBER

Occasionally, abnormal events occur before or during cell division, so gametes and new individuals end up with the wrong chromosome number. The consequences range from minor to lethal physical changes.

Categories and Mechanisms of Change

With **aneuploidy**, individuals usually have one extra or one less chromosome. This condition is a major cause of human reproductive failure. Possibly it affects about half of all fertilized eggs. Aneuploids are implicated in most *miscarriages,* the spontaneous aborting of embryos before pregnancy reaches full term.

With **polyploidy**, individuals have three or more of each type of chromosome. About half of all flowering plant species are polyploids; also, colchicine exposure can induce the condition in undifferentiated plant cells (Section 11.2). Some insects, fishes, and other animals are polyploids. Polyploidy is lethal for humans. All but 1 percent or so of human polyploid embryos die before birth. The rare newborns die soon after birth. We will take a closer look at polyploidy in Section 17.3.

Chromosome numbers can change during mitotic or meiotic cell divisions. Suppose a cell cycle proceeds through DNA duplication and mitosis but is arrested before the cytoplasm divides. The cell is now *tetra*ploid, with four of each type of chromosome. Suppose one or more pairs of chromosomes fail to separate in mitosis or meiosis, an event called **nondisjunction**. Some or all of the forthcoming cells will have too many or too few chromosomes. Figure 11.17 gives an example.

The chromosome number also may change during fertilization. Visualize a normal gamete that unites by chance with an $n + 1$ gamete (one extra chromosome). The new individual will be "trisomic" ($2n + 1$); it will have three of one type of chromosome and two of every other type. What if an $n - 1$ gamete unites with a normal n gamete? Then, the new individual will turn out to be "monosomic" ($2n - 1$).

Change in the Number of Autosomes

Nearly all changes in the number of autosomes arise by nondisjunction in reproductive cells that give rise to gametes. Let's look at one of the most common of the resulting disorders. A trisomic 21 newborn, with three chromosomes 21, will develop *Down syndrome.* The symptoms vary, but most affected individuals do show moderate to severe mental impairment. Heart defects are common. The skeleton also develops abnormally, so older children have shortened body parts, loose joints, and poorly aligned bones of the hips, fingers, and toes. Muscles and muscle reflexes are weaker than normal, and speech and other motor skills develop slowly. With special training, trisomic 21 individuals often take part in normal activities; Figure 11.18 gives examples. As a group, they tend to be cheerful and affectionate, and they derive great pleasure from socializing.

Down syndrome is one of many disorders that can be detected before birth. Before detection procedures were widespread, about 1 in 700 newborns of all ethnic groups was trisomic 21. The number is now closer to 1 in 1,100, for some women elect to terminate pregnancy when the condition is detected (Section 11.10). The risk is greater if pregnant women are more than thirty-five years old, as indicated in Figure 11.18*b*.

Change in the Number of Sex Chromosomes

Most sex chromosome abnormalities arise as a result of nondisjunction during meiosis and gamete formation. Let's look at a few phenotypic outcomes.

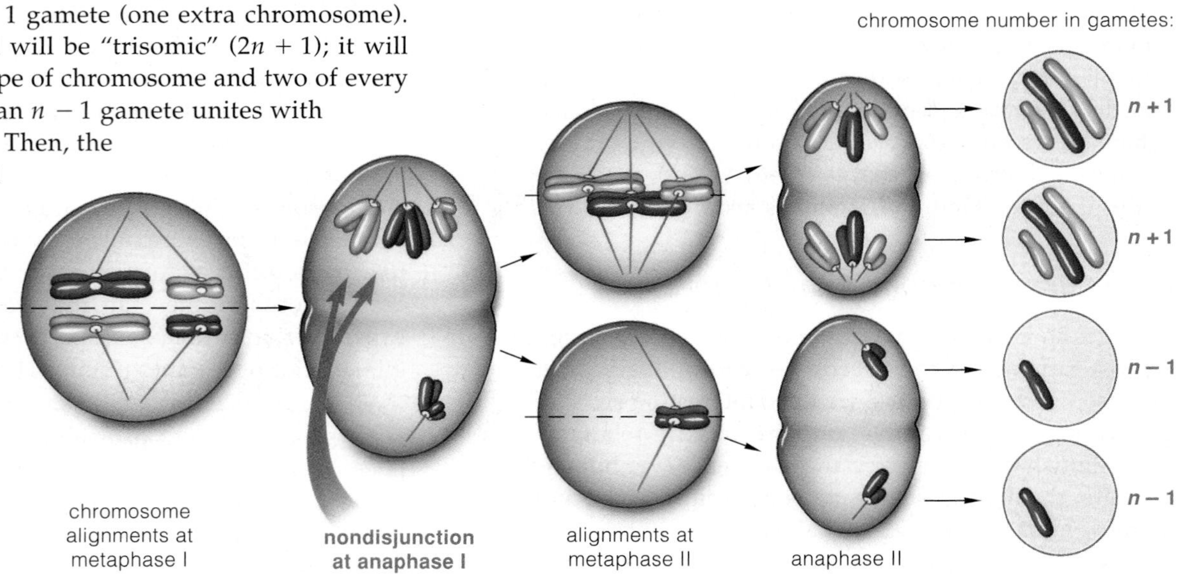

chromosome number in gametes:

| chromosome alignments at metaphase I | nondisjunction at anaphase I | alignments at metaphase II | anaphase II |

Figure 11.17 Example of nondisjunction. Of two pairs of homologous chromosomes shown, one pair fails to separate at anaphase I of meiosis. The chromosome number changes in the gametes. (Make a sketch of nondisjunction at anaphase II. What will the chromosome numbers be in gametes?)

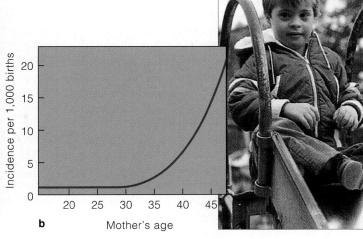

Figure 11.18 Down syndrome. (**a**) Karyotype revealing a trisomic 21 condition (*arrows*). (**b**) Relationship between the frequency of Down syndrome and mother's age at the time of childbirth. Results are from a study of 1,119 affected children who were born in Victoria, Australia, between 1942 and 1957. The young lady above was a lively participant in the Special Olympics, held annually in San Mateo, California.

TURNER SYNDROME Inheriting one X chromosome with no corresponding X or Y chromosome gives rise to *Turner syndrome*, which affects 1 in 2,500 to 10,000 or so newborn girls. Nondisjunction affecting sperm accounts for 75 percent of the cases. Compared to people having other sex chromosome abnormalities, there aren't many who have Turner syndrome. At least 98 percent of all X0 zygotes spontaneously abort early in pregnancy. About 20 percent of all spontaneously aborted embryos with detected chromosome abnormalities have been X0.

Despite the near-lethality, X0 survivors are not as disadvantaged as other aneuploids. They grow up well proportioned but only 4 feet, 8 inches tall, on average. Generally, their behavior is normal during childhood. But most Turner females are infertile. They do not have functional ovaries and so cannot produce eggs or sex hormones. Without the sex hormones, the development of secondary sexual traits, including breasts, is reduced. Possibly as a result of the arrested sexual development and small size, X0 females often are passive and easily intimidated by peers in their teens. Some benefit from hormone therapy and corrective surgery.

KLINEFELTER SYNDROME Of every 500 to 2,000 liveborn males, one inherits one Y and *two* X chromosomes. This XXY condition results mainly from nondisjunction in mothers (about 67 percent of the time, compared to 33 percent in fathers). Symptoms of the resulting *Klinefelter syndrome* develop after the onset of puberty. XXY males are taller than average and sterile, or nearly so. Testes usually are much smaller than average; the penis and

scrotum are not. Often facial hair is sparse, and breasts may be somewhat enlarged. Injections of testosterone, a male sex hormone, can reverse the feminized traits but not the low fertility. Some XXY males display mild mental impairment, but many fall in the normal range for intelligence. Except for low fertility, many affected individuals show no outward symptoms at all.

XYY CONDITION About 1 in 1,000 males has one X and two Y chromosomes. *XYY* males tend to be taller than average. Some may be mildly retarded, but most are phenotypically normal. At one time, all XYY males were thought to be genetically predisposed to become criminals. The erroneous conclusion was based on small numbers of cases in narrowly selected groups, such as prison inmates. Investigators often knew who the XYY males were, and this may have biased their evaluations. There were no **double-blind studies**, by which different investigators gather data independently of one another and then match them up only after both sets of data are completed. In this case, the same investigators gathered the karyotypes and the personal histories. Fanning the stereotype was a sensationalized report in 1968 that a mass-murderer of young nurses was XYY. He wasn't.

In 1976, one Danish geneticist issued a report on a large-scale study based on records of 4,139 tall males, twenty-six years old, who had reported to their draft board. Besides giving results of physical examinations and intelligence testing, the records provided clues to social and economic status, educational history, and any criminal convictions. Only twelve of the males were XYY, which left more than 4,000 for the control group. The only finding of significance was that tall, mentally impaired males who engage in criminal activity are just more likely to get caught—irrespective of karyotype.

Most changes in chromosome number arise as an outcome of nondisjunction during meiosis and gamete formation.

Prospects in Human Genetics

With the arrival of their newborn, parents typically ask, "Is our baby normal?" Quite naturally, they want their baby to be free of genetic disorders, and most of the time it is. But what are the options when it is not?

We do not approach heritable disorders and diseases the same way. We attack diseases with antibiotics, surgery, and other weapons. But how do we attack a heritable "enemy" that can be transmitted to offspring? Should we institute regional, national, or global programs to identify people who might be carrying harmful alleles? Do we tell them they are "defective" and run a risk of bestowing a disorder on their children? Who decides which alleles are harmful? Should society bear the cost of treating genetic disorders before and after birth? If so, should society also have a say in whether an affected embryo will be born at all, or whether it should be aborted? An **abortion** is the expulsion of a pre-term embryo or fetus from the uterus.

Such questions are only the tip of an ethical iceberg. We don't have answers that are universally accepted.

PHENOTYPIC TREATMENTS Often, symptoms of genetic disorders can be minimized or suppressed by dietary controls, adjustments to environmental conditions, and surgical intervention or hormone replacement therapy.

For example, dietary control works in *phenylketonuria*, or PKU. A certain gene specifies an enzyme that converts one amino acid to another—phenylalanine to tyrosine. If an individual is homozygous recessive for a mutated form of the gene, the first of these amino acids accumulates inside the body. If excess amounts are diverted into other pathways, then phenylpyruvate and other compounds may form. High levels of phenylpyruvate in the blood can impair the functioning of the brain. When affected people restrict their intake of phenylalanine, they are not required to dispose of excess amounts, so they can lead normal lives. Among other things, they can avoid soft drinks and other food products that are sweetened with aspartame, a compound that contains phenylalanine.

Environmental adjustments can counter or minimize symptoms of some disorders, as when albinos avoid direct sunlight. Surgery can repair conditions such as a *cleft lip*, an opening in the upper lip that did not seal, as it should have, during embryonic development. This condition usually arises by interactions among multiple genes from both parents and environmental factors.

GENETIC SCREENING Through large-scale screening programs in the general population, affected persons or carriers of a harmful allele often can be detected early enough to start preventive measures before symptoms develop. For example, most hospitals in the United States routinely screen newborns for PKU, so today it is less common to see people with symptoms of this disorder.

GENETIC COUNSELING If a first child or close relative has a severe heritable problem, prospective parents may worry about their next child. They may request help in evaluating their options from a qualified professional counseler. *Genetic counseling* often includes diagnosis of parental genotypes, detailed pedigrees, and genetic testing for hundreds of known metabolic disorders. Geneticists, too, may be contacted to help predict risks for genetic disorders. During genetic counseling, the prospective parents must be reminded that the same risk usually applies to each pregnancy.

PRENATAL DIAGNOSIS Methods of *prenatal diagnosis* can be used to determine the sex of embryos or fetuses and more than a hundred genetic conditions. (Prenatal means before birth. The term embryo applies until eight weeks after fertilization, after which the term fetus is appropriate.)

Suppose a woman who is forty-five years old becomes pregnant and worries about Down syndrome. She might request prenatal diagnosis by *amniocentesis* (Figure 11.19). With this diagnostic procedure, a clinician withdraws a tiny sample of the fluid inside the amnion, a membranous sac that surrounds the fetus. Some cells that the fetus has sloughed off are suspended in the sample. The cells are cultured and analyzed.

Chorionic villi sampling (CVS) is a different diagnostic procedure. A clinician withdraws cells from the chorion, a fluid-filled, membranous sac that surrounds the amnion.

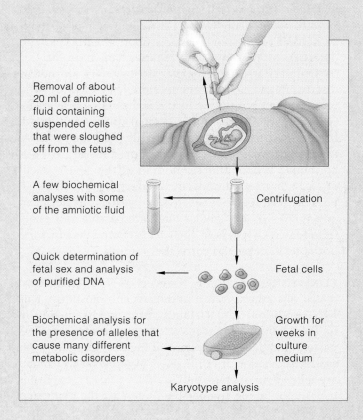

Removal of about 20 ml of amniotic fluid containing suspended cells that were sloughed off from the fetus

A few biochemical analyses with some of the amniotic fluid

Quick determination of fetal sex and analysis of purified DNA

Biochemical analysis for the presence of alleles that cause many different metabolic disorders

Centrifugation

Fetal cells

Growth for weeks in culture medium

Karyotype analysis

Figure 11.19 Amniocentesis, a prenatal diagnostic tool.

CVS can be perfomed weeks before amniocentesis. It can yield results as early as the ninth week of pregnancy.

Direct visualization of a developing fetus is possible with *fetoscopy*. An endoscope, a fiber-optic device, uses pulsed sound waves to scan the uterus and visually locate particular parts of the fetus, umbilical cord, or placenta (Figure 11.20). Fetoscopy has been used to diagnose blood cell disorders, such as sickle-cell anemia and hemophilia.

All three procedures may accidentally cause infection or puncture the fetus. Occasionally the punctured amnion does not reseal itself quickly, and an excessive amount of amniotic fluid may leak out and cause problems for the fetus. A mother-to-be who requests amniocentesis runs a 1 to 2 percent greater risk of miscarriage. For CVS, she runs a 0.3 percent risk that her future child will have missing or underdeveloped fingers and toes. Fetoscopy increases the risk of a miscarriage by 2 to 10 percent.

Parents-to-be probably should seek counseling from a doctor to help them weigh the risks and benefits of such procedures in terms of their own circumstances. They may wish to ask about the small overall risk of 3 percent that any child will have some kind of birth defect. They may ask about the severity of a genetic disorder that a child might be at risk of developing. And they might consider how old the woman is at the time of pregnancy.

REGARDING ABORTION   What happens when prenatal diagnosis does reveal a serious problem? Do prospective parents opt for induced abortion? We can only say here

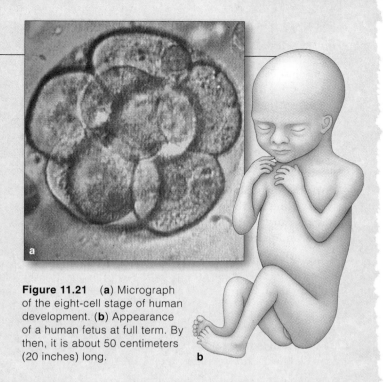

Figure 11.21   (**a**) Micrograph of the eight-cell stage of human development. (**b**) Appearance of a human fetus at full term. By then, it is about 50 centimeters (20 inches) long.

that they must weigh their awareness of the severity of the genetic disorder against ethical and religious beliefs. Worse, they must play out their personal tragedy on a larger stage, dominated by a nationwide battle between fiercely vocal "pro-life" and "pro-choice" factions. We return to this volatile issue in Sections 39.18 and 39.20.

PREIMPLANTATION DIAGNOSIS   Another procedure, *preimplantation diagnosis*, relies on **in-vitro fertilization**. In vitro, recall, means "in glass." Sperm and eggs from prospective parents are placed in an enriched medium in a glass or plastic petri dish. One or more eggs may get fertilized. In two days, mitotic cell divisions may convert one of these into a ball of eight cells (Figure 11.21*a*).

According to one view, the tiny, free-floating ball is a *pre*-pregnancy stage. Like the unfertilized eggs discarded monthly from a woman, it isn't attached to the uterus. Its cells all have the same genes and are not yet committed to giving rise to specialized cells of a heart, lungs, and other organs. Doctors take one of the undifferentiated cells and analyze its genes. If that cell has no detectable genetic defects, the ball is inserted into the uterus.

Some couples who are at risk of passing on muscular dystrophy, cystic fibrosis, and other disorders have opted for the procedure. Many *"test-tube" babies* have been born in good health and are free of the mutant alleles.

In 2000, genetic screening of a dozen fertilized eggs revealed an embryo free of a certain mutated gene. Both prospective parents carried the gene, and their first-born child, Molly, had developed *Fanconi anemia*. She had no thumbs, deformed arms, an incomplete brain, and the prospect of dying soon from leukemia. When her new brother was born, blood stem cells from the umbilical cord were transfused into her. The cells took hold. At this time Molly's bone marrow is producing red blood cells and platelets that are countering the anemia. The first embryo genetically selected to save a life is doing just that.

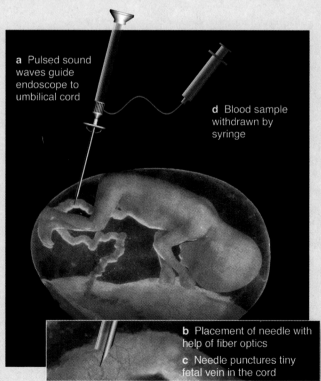

a Pulsed sound waves guide endoscope to umbilical cord

d Blood sample withdrawn by syringe

b Placement of needle with help of fiber optics

c Needle punctures tiny fetal vein in the cord

Figure 11.20   Fetoscopy for prenatal diagnosis.

SUMMARY

Gold indicates text section

1. Genes, the units of instruction for heritable traits, are arranged one after the other along chromosomes. Each gene has its own specific location, or locus, on one type of chromosome. Different molecular forms of that gene, called alleles, may occupy the locus. *11.1*

2. Human somatic cells are diploid (2n). They contain twenty-three pairs of homologous chromosomes that interact during meiosis. Each pair has the same length, shape, and genes (except for an XY pairing). *11.1*

3. An allele on one chromosome may or may not be identical to the allele at the equivalent locus on the homologous chromosome. *11.1*

4. Human females have two X chromosomes. Males have one X paired with one Y. All other chromosomes are autosomes (the same in both females and males). A gene on the Y chromosome determines sex. *11.1, 11.3*

5. Karyotypes, used in genetic analysis, are constructed to compare an individual's chromosomes (based upon their defining structural features) against standardized preparations of metaphase chromosomes. *CI, 11.2*

6. Genes on the same chromosome represent a linkage group. However, crossing over (breakage and exchange of segments between homologues) disrupts linkages. The farther apart two gene loci are along the length of a single chromosome, the greater will be the frequency of crossovers between them. *11.4*

7. Pedigrees are charts of genetic connections through lines of descent. Pedigrees provide clues to inheritance of a trait. *11.5*

8. Mendelian patterns of inheritance are characteristic of certain dominant or recessive alleles on autosomes or on the X chromosome. *11.5–11.7*

9. A chromosome's structure may be altered on rare occasions. A segment may be deleted, inverted, moved to a new location (translocated), or duplicated. *11.8*

10. The chromosome number can change. Gametes and offspring may get one more or one less chromosome than the parents (aneuploidy). They may get three or more of each type of chromosome (polyploidy). Nondisjunction during meiosis accounts for most of these changes in chromosome number. *11.9*

11. Changes in chromosome structure or number often result in genetic abnormalities or disorders. Phenotypic treatments, genetic screening, genetic counseling, and prenatal diagnosis are social responses. *11.8–11.10*

12. Crossing over adds to potentially adaptive variation in traits in a population. Most but not all changes in chromosome number or structure are harmful or lethal. Over evolutionary time, some have become established in chromosomes of all species. *11.1, 11.8*

Review Questions

1. What is a gene? What are alleles? *11.1*

2. Distinguish between: *CI, 11.1*
   a. homologous and nonhomologous chromosomes
   b. sex chromosomes and autosomes
   c. karyotype and karyotype diagram

3. Define genetic recombination, and describe how crossing over can bring it about. *11.1, 11.4*

4. Define pedigree. Explain the difference between genetic abnormality and genetic disorder, using examples. *11.5*

5. Contrast a typical pattern of autosomal recessive inheritance with that of autosomal dominant inheritance. *11.6*

6. Describe two clues that often show up when a recessive allele on an X chromosome causes a genetic disorder. *11.6*

7. Distinguish among a chromosomal deletion, duplication, inversion, and translocation. *11.8*

8. Define aneuploidy and polyploidy. Make a simple sketch of an example of nondisjunction. *11.9*

Self-Quiz   ANSWERS IN APPENDIX III

1. _____ segregate during _____ .
   a. Homologues; mitosis
   b. Genes on nonhomologous chromosomes; meiosis
   c. Homologues; meiosis
   d. Genes on one chromosome; mitosis

2. The probability of a crossover occurring between two genes on the same chromosome is _____ .
   a. unrelated to the distance between them
   b. increased if they are closer together on the chromosome
   c. increased if they are farther apart on the chromosome

3. Genetic disorders are caused by _____ .
   a. altered chromosome number   c. mutation
   b. altered chromosome structure   d. all of the above

4. A recognized set of symptoms that characterize a specific disorder is a _____ .
   a. syndrome   b. disease   c. pedigree

5. Chromosome structure can be altered by a _____ .
   a. deletion   c. inversion   e. all of the above
   b. duplication   d. translocation

6. Nondisjunction can be caused by _____ .
   a. crossing over in mitosis
   b. segregation in meiosis
   c. failure of chromosomes to separate during meiosis
   d. multiple independent assortments

7. A gamete affected by nondisjunction would have _____ .
   a. a change from the normal chromosome number
   b. one extra or one missing chromosome
   c. the potential for a genetic disorder
   d. all of the above

8. Match the chromosome terms appropriately.
   ____ crossing over   a. number and defining features of an
   ____ deletion   individual's metaphase chromosomes
   ____ nondisjunction   b. chromosome segment moves to a
   ____ translocation   nonhomologous chromosome
   ____ karyotype   c. disrupts gene linkages at meiosis
   ____ linkage group   d. causes gametes to have abnormal
                              chromosome numbers
                           e. loss of a chromosome segment
                           f. all genes on a given chromosome

Figure 11.22 Mutant fruit fly with vestigial wings.

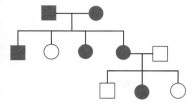

Figure 11.23 Go ahead, identify the mystery pedigree.

Figure 11.24 Two of your linked genes, on a pair of homologous chromosomes.

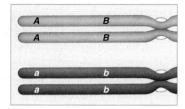

Critical Thinking—Genetics Problems

ANSWERS IN APPENDIX IV

1. Human females are XX and males are XY.
   a. Does a male inherit the X from his mother or father?
   b. With respect to X-linked alleles, how many different types of gametes can a male produce?
   c. If a female is homozygous for an X-linked allele, how many types of gametes can she produce with respect to that allele?
   d. If a female is heterozygous for an X-linked allele, how many types of gametes can she produce with respect to that allele?

2. Expression of a dominant allele at a gene locus governing wing length in *D. melanogaster* results in long wings. Figure 11.22 shows how homozygosity for a recessive allele results in formation of vestigial (short) wings. You cross a homozygous dominant, long-winged fly with a homozygous recessive, vestigial-winged fly. Then you ask a technician to expose the fertilized eggs to a level of x-rays known to induce mutation and deletions. Later, when the irradiated eggs develop into adults, most flies are heterozygous and have long wings. A few have vestigial wings. What might explain these results?

3. *Marfan syndrome* is a genetic disorder with pleiotropic effects. Expression of a mutant allele leads to the absence of a connective tissue (fibillin) or its abnormal formation in many organs. Affected people are often tall and thin, with double-jointed fingers. The spine curves abnormally. Arms, fingers, and lower limbs are disproportionately long. Eye lenses dislocate easily. Thin tissue flaps in the heart that act like valves to direct blood flow may billow the wrong way when the heart contracts. The heart beats irregularly. The main artery from the heart is fragile. Its diameter widens; its wall may tear.
   An estimated 40,000 people in the United States are affected, including a few prominent athletes. Spontaneous mutation gives rise to the allele. Genetic analysis shows it follows a pattern of autosomal dominant inheritance. What is the chance any child will inherit the allele if one parent is heterozygous for it?

4. In the Figure 11.23 pedigree, does the phenotype indicated by *red* circles and squares follow a Mendelian inheritance pattern that is autosomal dominant, autosomal recessive, or X-linked?

5. One kind of *muscular dystrophy*, a genetic disorder, is due to a recessive X-linked allele. Usually, symptoms start in childhood. Over time, a slowly progressing loss of muscle function leads to death, usually by age twenty or so. Unlike color blindness, this disorder is nearly always restricted to males. Suggest why.

6. Suppose you carry two linked genes with alleles *Aa* and *Bb*, respectively, as in Figure 11.24. If the crossover frequency between these two genes is zero, what genotypes would be expected among the gametes you produce, and with what frequencies?

7. Say you have alleles for left-handedness and straight hair on one chromosome and alleles for right-handedness and curly hair on the homologous chromosome. If the two loci for the alleles are very close together along the chromosome, how likely is it that crossing over will occur between them? If they are distant from each other, is a crossover more or less likely to occur?

8. Individuals showing *Down syndrome* usually have an extra chromosome 21, so their body cells contain 47 chromosomes.
   a. At which stages of meiosis I and II could a mistake occur that could result in the altered chromosome number?
   b. In a few cases, 46 chromosomes are present, including two normal-appearing chromosomes 21 and a longer-than-normal chromosome 14. Explain how this situation can arise.

9. In the human population, mutation of two different genes on the X chromosome causes two types of X-linked *hemophilia* (types A and B). In a few known cases, a woman is heterozygous for both mutant alleles (one on each of her two X chromosomes). All her sons should have either hemophilia A or B. Yet, on very rare occasions, such a woman gives birth to a son who does not have hemophilia, and his one X chromosome does not have either mutant allele. Explain how such an X chromosome could arise.

10. Think back on the chapter definitions of genetic disorder and genetic abnormality. Then think about how subjective we are in terms of our acceptance of a condition that is out of the ordinary. Example: Would you find a cleft lip as acceptable, even attractive, on someone not as gorgeous as Joaquin Phoenix, shown at right? You will not find an answer to this question in Appendix IV. We present it simply as a way to invite critical thinking about what we as a society consider to be "ideal" phenotypes.

Selected Key Terms

abortion *11.10*	genetic abnormality *11.5*	karyotype *CI*
allele *11.1*	genetic disorder *11.5*	linkage group *11.4*
aneuploidy *11.9*	genetic	nondisjunction *11.9*
autosome *11.1*	recombination *11.1*	pedigree *11.5*
crossing over *11.1*	homologous	polyploidy *11.9*
deletion *11.8*	chromosome *11.1*	reciprocal cross *11.4*
disease *11.5*	independent	sex chromosome *11.1*
double-blind	assortment *11.1*	syndrome *11.5*
study *11.9*	inversion *11.8*	translocation *11.8*
duplication *11.8*	in-vitro	X chromosome *11.1*
gene *11.1*	fertilization *11.10*	Y chromosome *11.1*

Readings

Fairbanks, D., and W. R. Andersen. 1999. *Genetics: The Continuity of Life.* Monterey, California: Brooks-Cole.

On-Line readings at Student Guide for InfoTrac:
www.brookscole.com/biology

12

DNA STRUCTURE AND FUNCTION

Cardboard Atoms and Bent Wire Bonds

One might have wondered, in the spring of 1868, why Johann Friedrich Miescher was collecting cells from the pus of open wounds and, later, from the sperm of a fish. Miescher, a physician, wanted to identify the chemical composition of the nucleus. These particular cells have very little cytoplasm, which makes it easier to isolate the nuclear material for analysis.

Miescher finally succeeded in isolating an organic compound having the properties of an acid. Unlike other substances in cells, it incorporated a notable amount of phosphorus. Miescher called the substance nuclein. He had discovered what came to be known many years later as **deoxyribonucleic acid**, or **DNA**.

The discovery did not cause even a ripple through the scientific community. At the time, no one really knew much about the physical basis of inheritance—that is, *which chemical substance encodes the instructions for reproducing parental traits in offspring*. Few even suspected that the cell nucleus might hold the answer. For a time, researchers generally believed hereditary instructions had to be encoded in the structure of some unknown class of proteins. After all, heritable traits are spectacularly diverse. Surely the molecules encoding information about those traits were structurally diverse also. Proteins are put together from potentially limitless combinations of twenty different amino acids, so the thinking was that they could function as the sentences (genes) in each cell's book of inheritance.

By the early 1950s, however, the results of many ingenious experiments clearly indicated that DNA is the substance of inheritance. Moreover, in 1951, Linus Pauling did something no one had done before. Through his training in biochemistry, a talent for model building, and a few great educated guesses, Pauling deduced the three-dimensional structure of the protein collagen. His discovery was truly electrifying. If someone could pry open the secrets of proteins, then why not assume the same might be done for DNA? And if the structural details of the DNA molecule were worked out, then wouldn't they provide clues to its biological functions? *Who would go down in history as having discovered the very secrets of inheritance?*

Scientists around the world started scrambling after that ultimate prize. Among them were James Watson, a young postdoctoral student from Indiana University, and Francis Crick, an unflappably exuberant researcher at Cambridge University. Exactly how could DNA, a molecule that consists of only four kinds of subunits, hold genetic information? Watson and Crick spent long hours arguing over everything they had read about the size, shape, and bonding requirements of the subunits of DNA. They fiddled with cardboard cutouts of the subunits. They badgered chemists to help them identify any potential bonds they might have overlooked. Then they assembled models from bits of metal connected by wire "bonds" bent at seemingly suitable angles.

In 1953, Watson and Crick put together a model that fit all the pertinent biochemical rules and all the facts about DNA they had gleaned from other sources. They had discovered the structure of DNA (Figures 12.1 and 12.2). And the breathtaking simplicity of that structure enabled them to solve another long-standing riddle—*how the world of life can show such unity at the molecular level and yet show such spectacular diversity at the level of whole organisms.*

Figure 12.1 James Watson and Francis Crick posing in 1953 by their newly unveiled structural model of DNA.

Figure 12.2 A recent computer-generated model of DNA, corresponding to the prototype that Watson and Crick put together decades ago.

Key Concepts

1. In all living cells, DNA molecules are the storehouses of information about heritable traits.

2. In a DNA molecule, two strands of nucleotides twist together, like a spiral stairway. Each strand consists of four kinds of nucleotides that are the same except for one component—a nitrogen-containing base. The four bases are adenine, guanine, thymine, and cytosine.

3. Great numbers of nucleotides are arranged one after another in each strand of the DNA molecule. In at least some regions, the order in which one kind of nucleotide follows another is unique for each species. Hereditary information is encoded in the particular sequence of nucleotide bases.

4. Hydrogen bonds connect the bases of one strand of the DNA molecule to bases of the other strand. As a rule, adenine pairs (hydrogen-bonds) with thymine, and guanine with cytosine.

5. Before a cell divides, its DNA is replicated with the assistance of enzymes and other proteins. Each double-stranded DNA molecule starts unwinding. As it does so, a new, complementary strand is assembled bit by bit on the exposed bases of each parent strand according to the base-pairing rule stated above.

With this chapter, we turn to investigations that led to our current understanding of DNA. The story is more than a march through details of its structure and function. *It also is revealing of how ideas are generated in science.* On the one hand, having a shot at fame and fortune quickens the pulse of men and women in any profession, and scientists are no exception. On the other hand, science proceeds as a community effort, with individuals sharing not only what they can explain but also what they do not understand. Even when an experiment fails to produce the anticipated results, it may turn up information that others can use or lead to questions that others can answer. Unexpected results, too, may be clues to something important about the natural world.

DISCOVERY OF DNA FUNCTION

Early and Puzzling Clues

The year was 1928. Frederick Griffith, an army medical officer, was attempting to develop a vaccine against *Streptococcus pneumoniae*, a bacterium that is one cause of the lung disease pneumonia. (When introduced into the body, vaccines mobilize internal defenses against a real attack. Many vaccines are preparations of killed or weakened bacterial cells.) Griffith never did develop a vaccine. But his work unexpectedly opened a door to the molecular world of heredity.

Griffith isolated and cultured two different strains of the bacterium. He noticed that colonies of one strain had a rough surface appearance, but those of the other strain appeared smooth. He designated the two strains R and S and used them in a series of four experiments:

1. Laboratory mice were injected with live R cells. The mice did not develop pneumonia, as Figure 12.3 shows. *The R strain was harmless.*

2. Other mice were injected with live S cells. The mice died. Blood samples taken from them teemed with live S cells. *The S strain was pathogenic* (disease-causing).

3. S cells were killed by exposure to high temperature. Mice injected with these cells did not die.

4. Live R cells were mixed with heat-killed S cells and injected into mice. The mice died—and blood samples from them teemed with *live* S cells!

What was going on in the fourth experiment? Maybe heat-killed S cells in the mixture weren't really dead. But if that were true, then mice injected with heat-killed S cells alone (experiment 3) would have died. Maybe harmless R cells in the mixture had mutated into a killer form. But if that were true, then mice injected with the R cells alone (experiment 1) would have died.

The simplest explanation was as follows: *Heat killed the S cells but did not destroy their hereditary material—including the part that specified "how to cause infection."* Somehow, that material had been transferred from the dead S cells to living R cells, which put it to use.

Further experiments showed the harmless cells had indeed picked up information on causing infections and were permanently transformed into pathogens. After a few hundreds of generations, descendants of those transformed bacterial cells were still infectious!

The unexpected results of Griffith's experiments intrigued Oswald Avery and his fellow biochemists. Later, they also transformed harmless bacterial cells with *extracts* of killed pathogenic cells. Finally in 1944, after rigorous chemical analyses, they felt confident in reporting that the hereditary substance in their extracts probably was DNA—not proteins, as was then widely believed. To give experimental evidence for their conclusion, they reported that they had added certain protein-digesting enzymes to some extracts, but cells exposed to those extracts were transformed anyway. To other extracts, they had added an enzyme that digests DNA but not proteins. Doing so blocked hereditary transformation.

Despite these impressive experimental results, many biochemists refused to give up on the proteins. Avery's findings, they said, probably applied only to bacteria.

Confirmation of DNA Function

By the early 1950s molecular detectives, including Max Delbrück, Alfred Hershey, Martha Chase, and Salvador Luria, were using viruses as experimental subjects. The viruses they had selected, called **bacteriophages**, infect *Escherichia coli* and other bacteria.

Viruses are biochemically simple infectious agents. They aren't alive, but they hold hereditary information about building more new virus particles. At some point after a virus infects a host cell, viral enzymes take over the cell's metabolic machinery, which starts churning out substances necessary to make new virus particles.

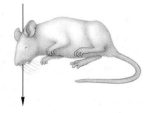

genetic material
viral coat
sheath
base plate
tail fiber

a

1 Mice injected with live cells of harmless strain R.

2 Mice injected with live cells of killer strain S.

3 Mice injected with heat-killed S cells.

4 Mice injected with live R cells *plus* heat-killed S cells.

Figure 12.3
Results of Griffith's experiments with a harmless and a pathogenic strain of *Streptococcus pneumoniae*, as described in the text above.

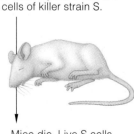
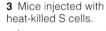
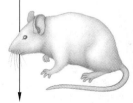

Mice do not die. No live R cells in their blood.

Mice die. Live S cells in their blood.

Mice do not die. No live S cells in their blood.

Mice die. Live S cells in their blood.

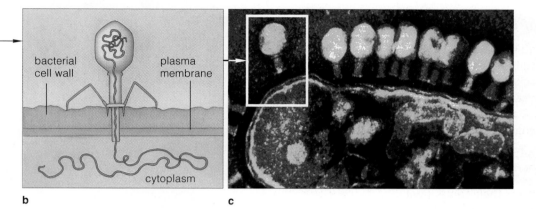

bacterial cell wall

plasma membrane

cytoplasm

b

c

Figure 12.4 (**a,b**) *Far left:* Structural organization of a T4 bacteriophage. The diagram shows the genetic material of this type of virus being injected into the cytoplasm of a host cell. The genetic material of this bacteriophage is DNA (the *blue*, threadlike strand). (**c**) Micrograph of T4 virus particles that infected the bacterium *Escherichia coli*, which thereby became an unwilling host.

Figure 12.5 Examples of the landmark experiments that pointed to DNA as the hereditary substance. In the 1940s, Alfred Hershey and Martha Chase were studying the biochemical basis of inheritance. They knew that certain bacteriophages consist of proteins and DNA. *Did the proteins, DNA, or both contain viral genetic information?*

To find a possible answer, Hershey and Chase designed two experiments. They started with two known biochemical facts: First, bacteriophage proteins incorporate sulfur (S) but not phosphorus (P). Second, by contrast, bacteriophage DNA incorporates phosphorus but not sulfur.

(**a**) In one experiment, some bacterial cells were grown on a culture medium that included a radioisotope of sulfur, ^{35}S. When bacterial cells synthesized proteins, they had to take up the radioisotope—which the researchers used as a tracer. (Here you may wish to review Section 2.2.)

After the cells were labeled with the tracer, bacteriophages were allowed to infect them. As the infection ran its course, the host cells synthesized viral proteins. These proteins also became labeled with ^{35}S. So did the new generation of virus particles.

Labeled bacteriophages were allowed to infect a new batch of unlabeled bacteria that were suspended in a fluid culture medium. Afterward, Hershey and Chase whirred the fluid in a kitchen blender. Whirring dislodged the viral protein coats from the cells, so the particles became suspended in the fluid medium. Chemical analysis revealed the presence of labeled protein in the fluid. There was no evidence of labeled protein *inside* the bacterial cells.

(**b**) For the second experiment, Hershey and Chase cultured more bacterial cells. The phosphorus available to the cells for synthesizing DNA included the radioisotope ^{32}P. Later, bacteriophages were allowed to infect the cells.

As predicted, viral DNA synthesized inside infected cells became labeled. So did a new generation of virus particles. The labeled particles were allowed to infect bacteria that were suspended in a fluid medium. They were dislodged from the host cells.

Analysis showed the labeled viral DNA was not in the fluid. DNA stayed *inside* host cells, where its hereditary instructions had to be used to make more virus particles. Here was evidence that DNA is the genetic material of this type of virus.

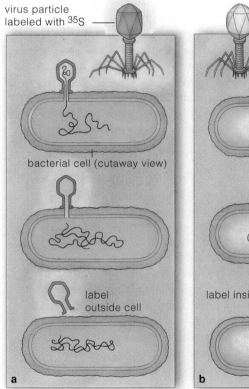
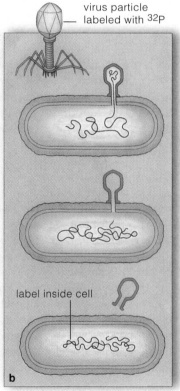

virus particle labeled with ^{35}S

virus particle labeled with ^{32}P

bacterial cell (cutaway view)

label outside cell

label inside cell

a

b

By 1952, researchers knew that some bacteriophages consist only of DNA and a protein coat. Also, electron micrographs revealed that the main part of the viruses remains *outside* the cells they are infecting (Figure 12.4). Possibly, such viruses were injecting genetic material alone *into* host cells. If that were true, was the material DNA, protein, or both? Through many experiments, researchers accumulated strong evidence that DNA, not proteins, serves as the molecule of inheritance. Figure 12.5 describes two of these landmark experiments.

Information for producing the heritable traits of single-celled and multicelled organisms is encoded in DNA.

DNA STRUCTURE

What Are the Components of DNA?

Long before the bacteriophage studies were under way, biochemists knew that DNA contains only four types of nucleotides—the building blocks of nucleic acids. Each **nucleotide** consists of a five-carbon sugar (which, in DNA, is deoxyribose), a phosphate group, and one of the following nitrogen-containing bases:

adenine	guanine	thymine	cytosine
A	G	T	C

All four types of nucleotides have their component parts organized the same way (Figure 12.6). But **T** and **C** are pyrimidines, which are single-ring structures. **A** and **G** are purines, which are larger, bulkier molecules; they have double-ring structures.

By 1949, Erwin Chargaff, a biochemist, had shared with the scientific community two crucial insights into the composition of DNA. First, the amount of adenine relative to guanine differs from one species to the next. Second, the amount of adenine in DNA always equals that of thymine, whereas the amount of guanine always equals that of cytosine. We may show this as:

$$A = T \quad \text{and} \quad G = C$$

The proportions of those four kinds of nucleotides relative to one another were tantalizing clues. In some way, the proportions almost certainly were related to the arrangement of the nucleotides in a DNA molecule.

The first convincing evidence of that arrangement emerged from Maurice Wilkins's research laboratory in England. Rosalind Franklin, one of Wilkins's colleagues, had obtained especially good **x-ray diffraction images** of DNA fibers. (Maybe for the reasons sketched out in Section 12.3, Franklin's contribution has only recently been acknowledged.) X-ray diffraction images can be made by directing a beam of x-rays at a molecule. The molecule scatters the beam in patterns that are captured on film. The pattern consists only of dots and streaks; it alone does not reveal molecular structure. However, researchers can use photographic images of the patterns to calculate the positions of the molecule's atoms.

DNA does not readily lend itself to x-ray diffraction. But researchers can rapidly spin a suspension of DNA

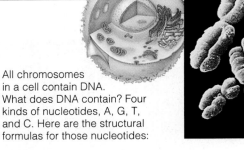

All chromosomes in a cell contain DNA. What does DNA contain? Four kinds of nucleotides, A, G, T, and C. Here are the structural formulas for those nucleotides:

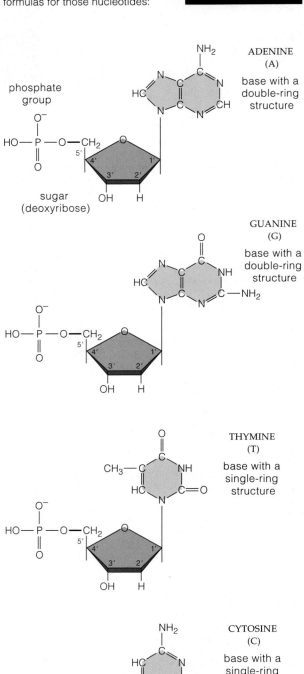

Figure 12.6 Four kinds of nucleotides used as building blocks for DNA. Small numerals on the structural formulas identify carbon atoms to which other parts of the molecule are attached.

Each nucleotide in a DNA molecule has a five-carbon sugar (coded *red*) and a phosphate group attached to the fifth carbon atom of its carbon ring structure. Each nucleotide also has one of four kinds of nitrogen-containing bases (*blue*) attached to the first carbon atom. The four kinds of nucleotides in DNA differ only in which base they have: adenine, guanine, thymine, or cytosine.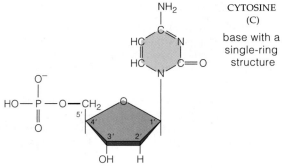

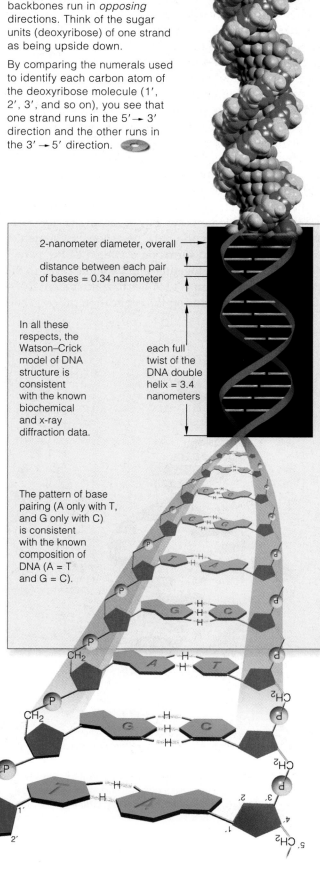

Figure 12.7 Composite of different models for a DNA double helix. The two sugar–phosphate backbones run in *opposing* directions. Think of the sugar units (deoxyribose) of one strand as being upside down.

By comparing the numerals used to identify each carbon atom of the deoxyribose molecule (1′, 2′, 3′, and so on), you see that one strand runs in the 5′→ 3′ direction and the other runs in the 3′ → 5′ direction.

2-nanometer diameter, overall

distance between each pair of bases = 0.34 nanometer

In all these respects, the Watson–Crick model of DNA structure is consistent with the known biochemical and x-ray diffraction data.

each full twist of the DNA double helix = 3.4 nanometers

The pattern of base pairing (A only with T, and G only with C) is consistent with the known composition of DNA (A = T and G = C).

molecules, spool them onto a rod, and gently pull them into gossamer fibers, like cotton candy. If the atoms in DNA were arranged in a regular order, x-rays directed at a fiber should scatter in a regular pattern that could be captured on film. As calculations based on Franklin's images strongly indicated, a DNA molecule is long and thin, with a 2-nanometer diameter. Some molecular configuration repeats every 0.34 nanometer along its length, and another every 3.4 nanometers.

Could the sequence of nucleotide bases be twisting, like a circular stairway? Certainly Pauling thought so. After all, he had discovered collagen's helical shape. He and everybody else—including Wilkins, Watson, and Crick—were thinking "helix." Watson later wrote, "We thought, why not try it on DNA? We were worried that *Pauling* would say, why not try it on DNA? Certainly he was a very clever man. He was a hero of mine. But we beat him at his own game. I still can't figure out why."

Pauling, it turned out, made a big chemical mistake. His model had hydrogen bonds at phosphate groups holding DNA's structure together. That does happen in highly acidic solutions. It doesn't happen in cells.

Patterns of Base Pairing

As Watson and Crick perceived, DNA consists of *two* strands of nucleotides, held together at their bases by hydrogen bonds (Figure 12.7). These bonds form when the two strands run in opposing directions and twist to form a double helix. Two kinds of base pairings form along the length of the molecule: A—T and G—C. This bonding pattern allows variation in the order of bases. For example, even a tiny stretch of DNA from a rose, a gorilla, a human, or any other organism might be:

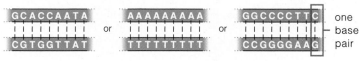

GCACCAATA or AAAAAAAAA or GGCCCCTTC } one
CGTGGTTAT   TTTTTTTTT   CCGGGGAAG } base pair

All DNA molecules show the same bonding pattern, but each species has unique base sequences in its DNA. *This molecular constancy and variation among species is the foundation for the unity and diversity of life.*

Intriguingly, recent computer simulations show that if you want to pack a string into the least space, coil it into a helix. Could this space-saving advantage also be a factor in nature? Did natural selection favor it in the molecular evolution of the DNA double helix? Maybe.

The pattern of base pairing between the two strands in DNA is constant for all species—A with T, and G with C. Each species also has some unique differences in *which* base pair follows the next along the length of its DNA molecules.

Rosalind's Story

In 1951, Rosalind Franklin arrived at King's Laboratory of Cambridge University with impressive credentials. Earlier, in Paris, she had refined existing procedures for x-ray diffraction while studying the structure of coal. She also had devised a new mathematical approach to interpreting x-ray diffraction images and had built three-dimensional models of molecules, as Pauling had done. Now she had been asked to run an x-ray crystallography laboratory, which she would create with state-of-the-art equipment. Her assignment? Investigate the structure of DNA.

No one bothered to tell her that, down the hall, Maurice Wilkins was already working on the puzzle. Even the graduate student assigned to assist her failed to mention it. And no one bothered to tell Wilkins about Franklin's assignment, so he assumed she was a technician hired to do his x-ray crystallography work because he could not

Figure 12.8 Rosalind Franklin.

do it himself. And so began a poisonous clash. To Franklin, Wilkins seemed inexplicably prickly; to Wilkins, Franklin displayed an appalling lack of deference that technicians usually show to researchers.

Wilkins had a prized cache of crystalline fibers of DNA, each having parallel arrays of hundreds of millions of DNA molecules—and these he gave to his "technician."

Five months later, Franklin gave a talk on what she had learned so far. DNA, she said, might have two, three, or four parallel chains twisted into a helix, with phosphate groups projecting outward. She had measured DNA's density and assigned DNA fibers to 1 of 230 categories of crystals, based on the symmetry of their parallel chains.

With his background in crystallography, Crick would have recognized the significance of that symmetry *if* he had been present. (To wit, *paired* chains running in opposite directions would look the same even if flipped 180°. Two paired chains? No. DNA's density ruled that out. But *one pair* of chains? Yes!) Watson was in the audience, but he didn't have a clue to what Franklin was talking about.

Later, Franklin created an outstanding x-ray diffraction image of wet DNA fibers that fairly screamed *Helix!* She also worked out DNA's length and diameter. But she had been working with dry fibers for so long she didn't dwell on her new data. Wilkins, however, did. In 1953, he let Watson see Franklin's exceptional x-ray diffraction image and reminded him of what she had reported fourteen months earlier. And when Watson and Crick finally did focus on her data, they had the final bits of information necessary to start building a DNA model—one with two helically twisted chains running in opposing directions.

Not until ten years after Franklin's untimely death in 1958 did Watson acknowledge her pivotal discoveries.

12.4

DNA REPLICATION AND REPAIR

How Is a DNA Molecule Duplicated?

The discovery of DNA structure was a turning point in studies of inheritance. Until then, no one could explain **DNA replication**, or how the molecule of inheritance is duplicated before the cell divides. Once Watson and Crick had assembled their model, Crick understood at once how this might be done.

As he knew, enzymes can easily break the hydrogen bonds between the two nucleotide strands of a DNA molecule. When these enzymes and other proteins act on the molecule, one strand can unwind from the other, thereby exposing stretches of nucleotide bases. Cells have stockpiles of free nucleotides, and these can pair with the exposed bases.

Each parent strand remains intact, and a companion strand is assembled on each one according to this base-pairing rule: A to T, and G to C. As soon as a stretch of a new, partner strand forms on a stretch of the parent strand, the two twist together into a double helix, in the manner shown in Figure 12.9. Because the parent DNA strand is conserved during the replication process, half

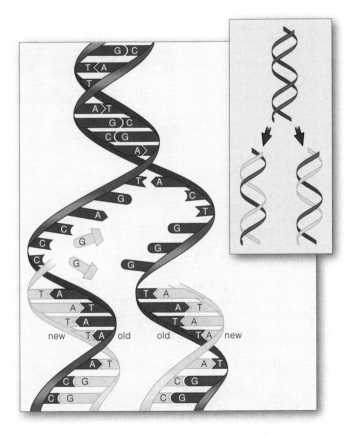

Figure 12.9 Overview of the semiconservative nature of DNA replication. The original two-stranded DNA molecule is shown in *blue*. Each parent strand remains intact. A new strand (*yellow*) is assembled on each one.

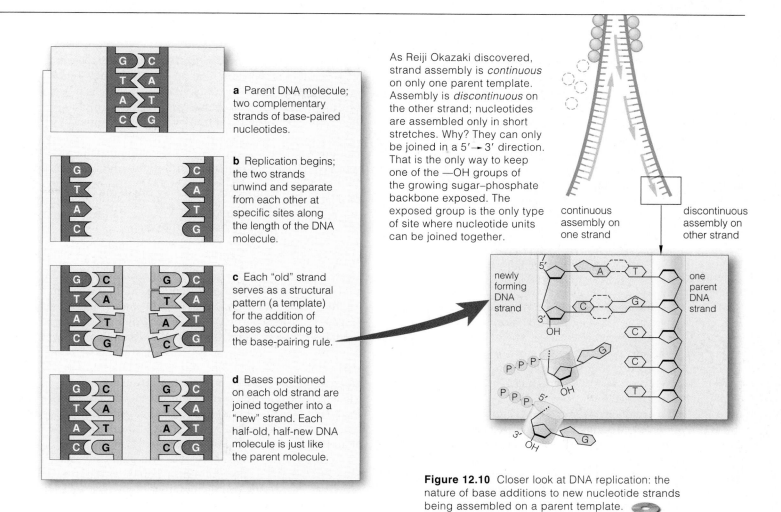

a Parent DNA molecule; two complementary strands of base-paired nucleotides.

b Replication begins; the two strands unwind and separate from each other at specific sites along the length of the DNA molecule.

c Each "old" strand serves as a structural pattern (a template) for the addition of bases according to the base-pairing rule.

d Bases positioned on each old strand are joined together into a "new" strand. Each half-old, half-new DNA molecule is just like the parent molecule.

As Reiji Okazaki discovered, strand assembly is *continuous* on only one parent template. Assembly is *discontinuous* on the other strand; nucleotides are assembled only in short stretches. Why? They can only be joined in a 5′→3′ direction. That is the only way to keep one of the —OH groups of the growing sugar–phosphate backbone exposed. The exposed group is the only type of site where nucleotide units can be joined together.

continuous assembly on one strand

discontinuous assembly on other strand

newly forming DNA strand

one parent DNA strand

Figure 12.10 Closer look at DNA replication: the nature of base additions to new nucleotide strands being assembled on a parent template.

of every double-stranded DNA molecule is "old" and half is "new" (Figure 12.9). That is why biologists refer to the process as *semiconservative* replication.

DNA replication uses a team of molecular workers. In response to cellular signals, the replication enzymes become active along the length of the DNA molecule. Together with other proteins, some enzymes unwind the strands in both directions and prevent them from rewinding. Enzyme action jump-starts the unwinding but isn't necessary to unzip hydrogen bonds between the strands; hydrogen bonds are individually weak.

Now enzymes called **DNA polymerases** attach short stretches of free nucleotides to the unwound portions of a parent template (Figure 12.10). The free nucleotides themselves actually drive the strand assembly. Each has three phosphate groups. DNA polymerase splits off two of them, releasing energy that drives the attachments.

DNA ligases fill in tiny gaps between the new short stretches to form a continuous strand. Then enzymes wind the template strand and complementary strand together to form a DNA double helix.

As you will read in Section 15.1, some replication enzymes have uses in recombinant DNA technology.

Monitoring and Fixing the DNA

DNA polymerases, DNA ligases, and other enzymes also engage in **DNA repair**. By this process, enzymes excise and fix altered parts of the base sequence in one strand of the double helix. Suppose a few bases are lost from a strand during replication. DNA polymerases "read" the complementary sequence on the other strand. They and other repair enzymes restore the original sequence. Also, remember crossing over, in which homologous chromosomes exchange segments? The same enzymes of that recombination event repair double-strand breaks. They initiate strand exchange with homologous DNA. Section 13.4 shows what happens when mutation or some other factor compromises the excision–repair function.

DNA is replicated prior to cell division. Enzymes unwind its two strands. Each strand remains intact throughout the process—it is conserved—and enzymes assemble a new, complementary strand on each one.

Enzymes involved in replication also repair the DNA where base-pairing errors have crept into the nucleotide sequence.

Cloning Mammals—A Question of Reprogramming DNA

Imagine the possibility of **cloning**—making a genetically identical copy—of yourself. Is the image that far-fetched? Consider this: Researchers have been cloning complex animals for more than a decade. For example, some use in vitro fertilization methods to grow cattle embryos in petri dishes. After a fertilized egg first starts dividing, they split the early cluster of cells. The two clusters develop as two identical-twin cattle embryos, get implanted in surrogate mothers, and are born as cloned calves (Figure 12.11a).

A researcher who clones farm animals derived from embryonic cells has to wait for the clones to grow up to see if they display a desired trait. Using a differentiated cell from an adult would be faster, for a prized genotype would already be known. But at one time, tricking a differentiated cell into reprogramming its DNA to direct the development of a whole embryo seemed impossible.

What does "differentiated" mean? When an embryo first grows from a fertilized egg, all of its cells have the same DNA and are pretty much alike. Then different embryonic cells start using different parts of their DNA. Their unique selections commit them to being liver cells, heart cells, brain cells, and other specialists in structure, composition, and function (Section 15.3).

In 1997 in Scotland, Ian Wilmut coaxed a differentiated sheep cell to become the "uncommitted" first cell of an embryo. His group had slipped nuclei from differentiated cells into unfertilized eggs from which the nucleus had been surgically removed. Of hundreds of modifed eggs, one developed into a whole animal. The cloned lamb, named Dolly, grew into a healthy adult and gave birth to a lamb of her own (Figure 12.11b).

In science, extraordinary claims call for extraordinary proof—in this case, successful repeats of the experiment at a time when no one thought mammals could ever be cloned. Researchers around the world have since cloned sheep and mice, cows, pigs, and goats. Some of the mice have been cloned through six generations. Researchers also cloned an endangered species, a guar. (The guar, a

wild ox, died two days after birth owing to a bacterial infection that is typically fatal for young calves.) Pigs are physiologically very much like us, so there is talk of using cloned pigs as sources of a variety of organs that can be harvested and transplanted into humans.

All of which raises some far-out questions. Is human cloning next? For example, will a human female be able to reproduce copies of herself without involving men? Will men be able to pay to have their DNA inserted into a cell stripped of its nucleus, have the cell implanted in a surrogate mother, and make a little repeat of themselves?

Aside from the ethical issues, of all cloning efforts, fewer than 3 percent result in healthy animals. And many of the "successful" clones develop severe medical problems, such as heart and lung abnormalities, sudden and gross obesity, and a compromised immune system.

It now appears that cloning invites random mistakes in gene expression. The cytoplasm of a mammalian egg contains proteins, mRNAs, and other components that have roles in guiding gene expression, starting with the early embryo (Sections 39.3 and 39.4). It takes months or years for an egg that is maturing inside an ovary to stockpile the required components in specific locations in the cytoplasm. Cloning processes use differentiated eggs that must reprogram new DNA within minutes or hours after it is inserted into them.

Even so, Lee Silver, a biologist at Princeton University, suggested that anyone who thinks cloning will move slowly is naive. One couple has already pledged up to 500 million dollars to a controversial company to clone their dead infant. The company says that it has lots of potential customers and has surrogate mothers lined up in various countries.

At this writing, several countries have banned human cloning, but the bans leave room for cloning technology with uses in basic research. Should human cloning be done at all? Both nonscientists and scientists alike are actively debating the issue.

Figure 12.11 (a) Clone of Holsteins, prized milk producers, that resulted from in vitro fertilization. (b) Dolly, a cloned sheep. She started her life as a differentiated cell that was extracted from an adult ewe, then induced to undergo mitotic cell divisions. She is shown with her first lamb. She clearly is able to breed normally and reproduce the old-fashioned way.

SUMMARY

Gold indicates text section

1. For all living cells, hereditary information is encoded in DNA (deoxyribonucleic acid). *CI, 12.1*

2. DNA consists of nucleotide subunits. Each of these has a five-carbon sugar (deoxyribose), one phosphate group, and one of four kinds of nitrogen-containing bases (adenine, thymine, guanine, or cytosine). *12.2*

3. A DNA molecule consists of two nucleotide strands twisted together as a double helix. Bases of one strand pair (hydrogen-bond) with bases of the other. *12.2*

4. The bases of the two strands in a DNA double helix pair in constant fashion. Adenine pairs with thymine (**A** to **T**), and guanine with cytosine (**G** to **C**). *Which* base pair follows the next (**A–T, T–A, G–C,** or **C–G**) varies along the length of the strands. *12.2*

5. Overall, the DNA of one species includes a number of unique stretches of base pairs that set it apart from the DNA of all other species. *12.2*

6. During DNA replication, enzymes unwind the two strands of a double helix and assemble a new strand of complementary sequence on each parent strand. Two double-stranded molecules result. One strand of each molecule is old (is conserved); the other is new. *12.4*

7. Some of the enzymes involved in DNA replication also repair DNA where base-pairing errors have been introduced into the nucleotide sequence. *12.4*

Review Questions

1. Name the three molecular parts of a nucleotide in DNA. Also name the four different bases in these nucleotides. *12.2*

2. What kind of bond joins two DNA strands in a double helix? Which nucleotide base-pairs with adenine? With guanine? *12.2*

3. Explain how DNA molecules can show both constancy and variation from one species to the next. *12.2*

Self-Quiz   ANSWERS IN APPENDIX III

1. Which is *not* a nucleotide base in DNA?
   a. adenine   c. uracil   e. cytosine
   b. guanine   d. thymine

2. What are the base-pairing rules for DNA?
   a. A–G, T–C   c. A–U, C–G
   b. A–C, T–G   d. A–T, G–C

3. A DNA strand having the sequence C–G–A–T–T–G would be complementary to the sequence _____ .
   a. C–G–A–T–T–G   c. T–A–G–C–C–T
   b. G–C–T–A–A–G   d. G–C–T–A–A–C

4. One species' DNA differs from others in its _____ .
   a. sugars   b. phosphates   c. base sequence

5. When DNA replication begins, _____ .
   a. the two DNA strands unwind from each other
   b. the two DNA strands condense for base transfers
   c. two DNA molecules bond
   d. old strands move to find new strands

6. DNA replication requires _____ .
   a. free nucleotides   c. many enzymes
   b. new hydrogen bonds   d. all of the above

7. Match the DNA terms appropriately.
   _____ DNA polymerase   a. two nucleotide strands that
   _____ constancy in   are twisted together
        base pairing   b. A with T, G with C
   _____ replication   c. hereditary material duplicated
   _____ DNA double helix   d. replication enzyme

Critical Thinking

1. Chargaff's data suggested that adenine pairs with thymine, and guanine pairs with cytosine. What other data available to Watson and Crick suggested that adenine–guanine and cytosine–thymine pairs normally do not form?

2. One of Matthew Meselson and Frank Stahl's experiments supported the semiconservative model of DNA replication. The researchers made "heavy" DNA by growing *Escherichia coli* in a medium enriched with ^{15}N, a heavy isotope of nitrogen. They prepared "light" DNA by growing *E. coli* in the presence of ^{14}N, the more common isotope. An available technique helped them identify which replicated molecules were heavy, light, or hybrid (one heavy strand, one light). Use two pencils of different colors, one for heavy strands and one for light. Starting with a DNA molecule having two heavy strands, sketch daughter molecules that would form after replication in a ^{14}N-containing medium. Sketch the four DNA molecules that would form if the daughter molecules were replicated a second time in the ^{14}N medium.

3. Mutations (permanent changes in base sequences of genes) are the original source of genetic variation. This variation is the raw material of evolution. Yet how can both statements be true, given that cells have efficient mechanisms to repair DNA before mutations can become established?

4. As Section 4.12 indicates, a pathogenic strain of *E. coli* has acquired an ability to produce a dangerous toxin that causes medical problems and fatalities. This is especially the case for young children who have ingested undercooked, contaminated beef. Develop hypotheses to explain how a normally harmless bacterium such as *E. coli* can become a pathogen.

5. In 1999, scientists discovered a woolly mammoth that had been frozen in glacial ice for the past 20,000 years. They thawed it very carefully so they could use its DNA to clone a woolly mammoth. It turns out there wasn't enough material to work with. But they plan to try again the next time a frozen woolly mammoth comes along. Consider Section 12.5, then speculate on the pros and cons of cloning an extinct animal.

Selected Key Terms

adenine (A) *12.2*
bacteriophage *12.1*
cloning *12.5*
cytosine (C) *12.2*
deoxyribonucleic acid (DNA) *CI*
DNA ligase *12.4*
DNA polymerase *12.4*

DNA repair *12.4*
DNA replication *12.4*
guanine (G) *12.2*
nucleotide *12.2*
thymine (T) *12.2*
x-ray diffraction
   image *12.2*

Readings

Watson, J. 1978. *The Double Helix.* New York: Atheneum. Highly personal view of scientists and their methods, interwoven into an account of how DNA structure was discovered.

13

FROM DNA TO PROTEINS

Beyond Byssus

Picture a mussel, of the sort shown in Figure 13.1. Hard-shelled but soft of body, it is using its muscular foot to probe a wave-scoured rock. At any moment, pounding waves can whack the mussel into the water, hurl it repeatedly against the rock with shell-shattering force, and so offer up a gooey lunch for gulls.

By chance, the mussel's foot comes across a crevice in the rock. The foot moves, broomlike, and sweeps the crevice clean. It presses down, forcing air out from underneath it, then arches up. The result is a vacuum-sealed chamber, much like the one that forms when a plumber's rubber plunger is being squished down and up to unclog a drain. Into this vacuum chamber the mussel spews a fluid that's made of keratin and other proteins. The fluid bubbles into a sticky foam. Now, by curling its foot into a tubelike shape and pumping the foam through it, the mussel produces sticky threads about as wide as a human whisker. It varnishes these threads with another type of protein and ends up with an adhesive. With this adhesive, which we call byssus, the mussel anchors itself to the rock.

Byssus is the world's premier underwater adhesive. Nothing humans have manufactured comes close to it; water degrades or deforms synthetic adhesives. Byssus fascinates biochemists, dentists, and surgeons looking for better ways to do tissue grafts and rejoin severed nerves. Genetic engineers insert mussel DNA into yeast cells. These cells, which reproduce in huge numbers, are

Figure 13.1 Mussels (*Mytilus californianus*) busily demonstrating the importance of proteins for survival. When mussels come across a suitable anchoring site, they use their muscular foot like a plumber's plunger and create a vacuum chamber. In this chamber they manufacture the world's best underwater adhesive from a mix of proteins. The adhesive anchors them to rocks in their wave-swept habitat.

"factories" for translating mussel genes into useful quantities of proteins. This exciting work, like the mussel's own byssus-building efforts, starts with one of life's universal precepts: *Every protein is synthesized in accordance with instructions in DNA.*

You are about to trace the steps leading from DNA to proteins. Many enzymes are players in this pathway. So is another kind of nucleic acid besides DNA. The same steps produce *all* proteins, from mussel-inspired adhesives to the keratin in your hair and fingernails to the insect-digesting enzymes of a Venus flytrap.

Start out by thinking of each cell's DNA as a book of protein-building instructions. The alphabet used to create the book is simple enough: A, T, G, and C (for the nucleotide bases adenine, thymine, guanine, and cytosine). How do you get from that alphabet to a protein? The answer starts with DNA's structure.

DNA, recall, is a double-stranded molecule. Which kind of nucleotide base follows the next along the length of a strand—the **base sequence**—differs from one kind of organism to the next. The two strands unwind entirely from each other when DNA is being replicated. However, at other times in a cell's life, the two strands unwind only in certain regions to expose certain base sequences—genes. Most genes contain instructions for building proteins.

It takes two steps, **transcription** and **translation**, to carry out a gene's protein-building instructions. In eukaryotic cells, transcription proceeds in the nucleus. A newly exposed base sequence in DNA serves as a structural pattern—a template—for assembling a strand of **ribonucleic acid** (RNA) from the cell's pool of free nucleotides. Sooner or later the RNA moves into the cytoplasm, where translation proceeds. At this second step, RNA directs the assembly of amino acids into polypeptide chains. The newly formed chains become folded into the three-dimensional shapes of proteins.

In short, DNA guides the synthesis of RNA, then RNA guides the synthesis of proteins:

$$\text{DNA} \xrightarrow{\textit{transcription}} \text{RNA} \xrightarrow{\textit{translation}} \text{PROTEIN}$$

The newly synthesized proteins will play structural and functional roles in cells. Some even will have roles in synthesizing more DNA, RNA, and proteins.

Key Concepts

1. Life cannot exist without enzymes and other proteins. Proteins consist of polypeptide chains, which consist of amino acids. The sequence of amino acids corresponds to a gene, which is a sequence of nucleotide bases in a DNA molecule.

2. The path leading from genes to proteins consists of two steps, called transcription and translation.

3. In transcription, the double-stranded DNA molecule is unwound at a gene region, then an RNA molecule is assembled on the exposed bases of one of the strands.

4. Translation requires three classes of RNA molecules: messenger RNA, transfer RNA, and ribosomal RNA.

5. In translation, amino acids are joined together to form a polypeptide chain, in a sequence specified by messenger RNA. Transfer RNA delivers the amino acids one at a time to the construction site. Ribosomal RNA catalyzes the chain-building reaction.

6. With few exceptions, the genetic "code words" by which DNA's instructions are translated into proteins are the same in all species of organisms.

7. A mutation is a permanent alteration in a gene's base sequence. Mutations are the original source of genetic variation in populations.

8. Mutations cause changes in protein structure, protein function, or both. The changes may lead to small or large differences in traits among individuals of a population.

HOW IS DNA TRANSCRIBED INTO RNA?

The Three Classes of RNA

Before turning to the details of protein synthesis, be clear on one point. The chapter introduction may have left you with the impression that synthesis of proteins requires only one class of RNA molecules. Actually, it requires three. Transcription of most genes produces **messenger RNA**, or **mRNA**—the only class of RNA that carries *protein-building* instructions. Transcription of some other genes produces **ribosomal RNA**, or **rRNA**, a major component of ribosomes. Ribosomes, recall, are the structural units upon which polypeptide chains are assembled. Transcription of still other genes produces **transfer RNA**, or **tRNA**, which delivers amino acids one by one to a ribosome in the order specified by mRNA.

The Nature of Transcription

An RNA molecule is almost but not quite like a single strand of DNA. RNA, too, consists of only four types of nucleotides. Each nucleotide has a five-carbon sugar, ribose (not DNA's deoxyribose), a phosphate group, and a base. Three types of bases—adenine, cytosine, and guanine—are the same in RNA and DNA. But in RNA, the fourth type of base is **uracil**, not thymine (Figure 13.2). Like thymine, uracil can pair with adenine. This means a new RNA strand can be put together on a DNA region according to base-pairing rules (Figure 13.3).

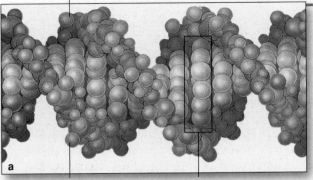

sugar-phosphate backbone of one strand of nucleotides in a DNA double helix

a

sugar-phosphate backbone of the other strand of nucleotides

part of the sequence of base pairs in DNA

Figure 13.4 The process of gene transcription, by which an RNA molecule is assembled on a DNA template. The sketch in (**a**) shows a gene region in part of a DNA double helix. In this region, the base sequence of one of the two nucleotide strands (not both) is about to be transcribed into an RNA molecule, in the manner shown in (**b**) through (**e**).

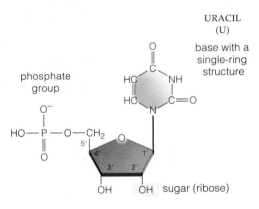

URACIL
(U)

base with a single-ring structure

phosphate group

sugar (ribose)

Figure 13.2 Structural formula for one of the four types of RNA nucleotides. The three others have a different base (adenine, guanine, or cytosine instead of the uracil shown here). Compare Section 12.2, which shows DNA's four nucleotides. Notice how the sugars of DNA and RNA differ at one group only (*yellow*).

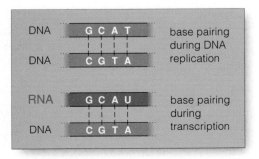

DNA	G C A T	base pairing during DNA replication
DNA	C G T A	
RNA	G C A U	base pairing during transcription
DNA	C G T A	

Figure 13.3 An example of base pairing of RNA with DNA during transcription, as compared to base pairing during DNA replication.

Transcription is like DNA replication in another way. Enzymes add nucleotides to a growing RNA strand one at a time, in the $5' \rightarrow 3'$ direction. Section 12.4 has a simple explanation of strand assembly.

But transcription *differs* from DNA replication in three key respects. First, only a selected stretch of one DNA strand, not the whole molecule, is used as the template. Second, instead of DNA polymerase, the enzyme **RNA polymerase** catalyzes nucleotide additions to a growing RNA strand. Third, at the end of transcription there is a single, free strand of RNA nucleotides, not a double helix.

Transcription is initiated at a **promoter**, a base sequence in DNA that signals the start of a gene. Proteins position an RNA polymerase on DNA and help it bind to the promoter. The enzyme moves along the DNA strand, joining one nucleotide after another (Figure 13.4). When it reaches a certain base, the new molecule is released as a free transcript.

Finishing Touches on mRNA Transcripts

In eukaryotic cells alone, a new mRNA molecule is unfinished. This "pre-mRNA" must be modified before its protein-building instructions can be put to use. Just as a dressmaker might snip off some threads or add bows on a dress before it leaves the shop, so do eukaryotic cells tailor their pre-mRNA.

For example, enzymes attach a cap to the 5' end of a pre-mRNA molecule. The cap is a nucleotide

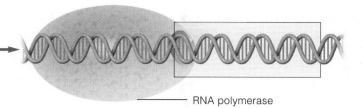

— RNA polymerase

b A molecule of RNA polymerase binds with a promoter region in the DNA. It will recognize the base sequence located downstream from that site as a template for linking together the nucleotides adenine, cytosine, guanine, and uracil into a strand of RNA.

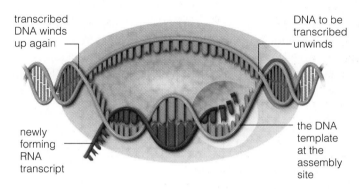

transcribed DNA winds up again

DNA to be transcribed unwinds

newly forming RNA transcript

the DNA template at the assembly site

c All through transcription, the DNA double helix becomes unwound in front of the RNA polymerase. Short lengths of the newly forming RNA strand briefly wind up with its DNA template strand. New stretches of RNA unwind from the template (and the two DNA strands wind up again).

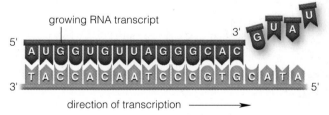

growing RNA transcript

GUAU

5' AUGGUGUUAGGGCAC
3' TACCACAATCCCGTGCATA 5'

direction of transcription →

d What happened at the assembly site? RNA polymerase catalyzed the base-pairing of RNA nucleotides, one after another, with exposed bases on the DNA template strand.

5' AUGGUGUUAGGGCACGUAU 3'

e At the end of the gene region, the last stretch of the new mRNA transcript is unwound and released from the DNA.

that has a methyl group and phosphate groups bonded to it. Also, enzymes attach a tail of about 100 to 200 nucleotides to pre-mRNA transcripts. The tail becomes wound up with proteins. Later, in the cytoplasm, the cap helps bind the mRNA to a ribosome. Also, enzymes will slowly destroy the wound-up tail from the tip on back. Such tails "pace" the access of enzymes to mRNA.

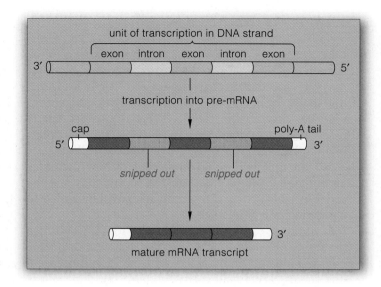

unit of transcription in DNA strand

exon intron exon intron exon

3'   5'

transcription into pre-mRNA

cap   poly-A tail

5'   3'

snipped out   *snipped out*

3'

mature mRNA transcript

Figure 13.5 Transcription and modification of new mRNA in the nucleus of eukaryotic cells. Its cap is a nucleotide with functional groups attached. Its tail is a string of adenine nucleotides (hence the name, poly-A tail).

That controlled access determines how long an mRNA molecule will last. It helps keeps the protein-building messages intact for as long as the cell requires them.

Besides these alterations, the pre-mRNA itself gets modified. Most eukaryotic genes contain one or more **introns**, base sequences that must be removed before a pre-mRNA molecule can be translated. These introns intervene between **exons**, the parts that are still in the mRNA when it's translated into protein. As Figure 13.5 shows, introns are transcribed along with the exons but are snipped out before the mRNA leaves the nucleus in mature form. (Think of it this way: *Ex*ons are *ex*ported from the nucleus and *in*trons stay *in* the nucleus.)

It could be that some introns are evolutionary junk, the leftovers of past mutations that led nowhere. Yet many are sites where instructions for building one type of protein can be snipped apart and spliced together in various ways. Alternative splicing allows different cells in your body to use the same gene to make different versions of a pre-mRNA transcript. As a result, the cell has different versions of the protein at its disposal. We return to this topic in Section 14.3.

During gene transcription, a sequence of exposed bases in one of the two strands of a DNA molecule serves as the template for assembling a single strand of RNA. The assembly follows these base-pairing rules: adenine only with uracil, and cytosine only with guanine.

Before leaving the nucleus, each new mRNA transcript, or pre-mRNA, undergoes modification into final form.

DECIPHERING THE mRNA TRANSCRIPTS

What Is the Genetic Code?

Like a strand of DNA, an mRNA molecule is a linear sequence of nucleotides. What are the protein-building "words" encoded in that sequence? Gobind Khorana, Marshall Nirenberg, and other investigators came up with the answer. They showed that ribosomes "read" nucleotide bases *three at a time*, as triplets. Base triplets in an mRNA strand were given this name: **codons**.

Figure 13.6 will give you an idea of how the order of different codons in an mRNA strand dictates the order in which particular amino acids will be added to a growing polypeptide chain.

Count the codons listed in Figure 13.7, and you see that there are sixty-four kinds. Notice how most of the twenty kinds of amino acids correspond to more than one codon. Glutamate corresponds to the code words GAA *and* GAG, for example. AUG codes for the amino acid methionine and is the start point for translation of all mRNA transcripts. Said another way, the "three-bases-at-a-time" selections start at a particular AUG in the transcript's nucleotide sequence. Methionine is the first amino acid in all new polypeptide chains. Codons UAA, UAG,

and UGA do not correspond to amino acids. They are STOP signals that prevent further additions of amino acids to a new polypeptide chain.

The set of sixty-four different codons is the **genetic code**. It is the basis of protein synthesis in all organisms.

Structure and Function of tRNA and rRNA

In a cell's cytoplasm are pools of free amino acids and free tRNA molecules. The tRNAs each have a molecular "hook," an attachment site for amino acids. They also have an **anticodon**, a nucleotide triplet that can base-pair with a codon (Figure 13.8). When tRNAs bind to the codons, they automatically position their attached amino acids in the order specified by mRNA.

A cell has a cytoplasmic pool of sixty-four kinds of codons, but it is able to utilize fewer kinds of tRNAs.

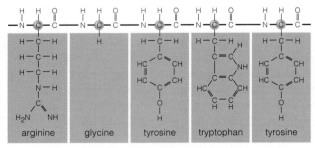

a Part of the amino acid sequence that was put together when mRNA was translated into a polypeptide chain

b Part of the mRNA strand that was transcribed from DNA

c The base sequence of a gene region in DNA

Figure 13.6 The correspondence between genes and proteins. (**a**) A sequence of amino acids, part of a protein's polypeptide chain. (**b**) An mRNA transcript. Every three nucleotide bases, equaling one codon, called for one of the chain's amino acids. (**c**) Exposed bases on one strand of a DNA double helix that was unwound during transcription. It was the template for assembling the mRNA strand. Referring to Figure 13.7, can you fill in the blank codon for tryptophan in the mRNA strand?

first base	second base				third base
	U	C	A	G	
U	phenylalanine	serine	tyrosine	cysteine	U
	phenylalanine	serine	tyrosine	cysteine	C
	leucine	serine	STOP	STOP	A
	leucine	serine	STOP	tryptophan	G
C	leucine	proline	histidine	arginine	U
	leucine	proline	histidine	arginine	C
	leucine	proline	glutamine	arginine	A
	leucine	proline	glutamine	arginine	G
A	isoleucine	threonine	asparagine	serine	U
	isoleucine	threonine	asparagine	serine	C
	isoleucine	threonine	lysine	arginine	A
	methionine (or START)	threonine	lysine	arginine	G
G	valine	alanine	aspartate	glycine	U
	valine	alanine	aspartate	glycine	C
	valine	alanine	glutamate	glycine	A
	valine	alanine	glutamate	glycine	G

Figure 13.7 The genetic code. Codons in mRNA are nucleotide bases "read" in blocks of three. Sixty-one of the base triplets correspond to specific amino acids. Three others are signals that stop translation. The vertical column at *left* lists choices for the first of the three nucleotides in mRNA codons. The top horizontal row lists choices for the second codon. The vertical column at *right* lists choices for the third. For example, reading from left to right, the triplet U G G corresponds to tryptophan. Both U U U and U U C correspond to phenylalanine.

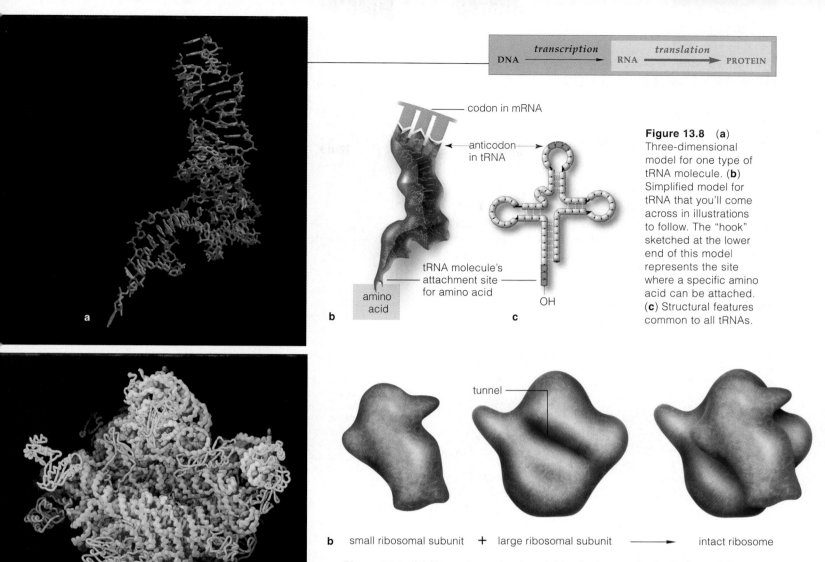

codon in mRNA

anticodon in tRNA

tRNA molecule's attachment site for amino acid

amino acid

OH

b

c

Figure 13.8 (**a**) Three-dimensional model for one type of tRNA molecule. (**b**) Simplified model for tRNA that you'll come across in illustrations to follow. The "hook" sketched at the lower end of this model represents the site where a specific amino acid can be attached. (**c**) Structural features common to all tRNAs.

a

tunnel

b small ribosomal subunit **+** large ribosomal subunit ⟶ intact ribosome

Figure 13.9 (**a**) Three-dimensional model for the large subunit of a bacterial ribosome. It consists of two rRNA molecules (*light gray*) and thirty-one structural proteins (*gold*), which stabilize the structure. At one end of a tunnel inside this subunit, rRNA catalyzes polypeptide chain assembly. This is an ancient, highly conserved molecular structure. Its role in life is so vital that the corresponding large subunit of eukaryotic ribosomes, which is bigger, is probably very similar in structure and function. (**b**) Model for the small and large subunits of a ribosome.

How do tRNAs match up with more than one type of codon? According to base-pairing rules, adenine must pair with uracil, and cytosine with guanine. For codon–anticodon interactions, however, the rules loosen up for the first and third bases. To give one example, AUU, AUC, and AUA specify isoleucine. All three of these codons can pair with a single type of tRNA that carries isoleucine. Such freedom in codon–anticodon pairing at a base is known as the "wobble effect."

Before anticodons interact with codons of an mRNA strand, that strand must bind to a ribosome. As shown in Figure 13.9, each ribosome has two subunits. These are assembled in the nucleus from rRNA and structural proteins, which stabilize the ribosome structure. rRNA's enzyme action drives protein synthesis.

At some point the subunits are shipped separately to the cytoplasm. There, intact, functional ribosomes are put together, each from two subunits—but only when messages encoded in mRNA are to be translated.

The nucleotide sequence of both DNA and mRNA encodes protein-building instructions. The genetic code is a set of sixty-four base triplets, which are nucleotide bases read in blocks of three. A codon is a base triplet in mRNA.

Different combinations of codons specify the amino acid sequence of different polypeptide chains, start to finish.

mRNAs are the only molecules that carry protein-building instructions from DNA into the cytoplasm.

tRNAs deliver amino acids to ribosomes, where they base-pair with codons in the order specified by mRNA.

Ribosomes, composed of rRNA and proteins, are structures upon which amino acids are assembled into polypeptide chains. An rRNA catalyzes chain assembly.

HOW IS mRNA TRANSLATED?

Stages of Translation

The protein-building code built into mRNA transcripts of DNA becomes translated at intact ribosomes in the cytoplasm. Translation proceeds through three stages: initiation, elongation, and termination.

During the stage called *initiation*, an initiator tRNA (the only one that can start transcription) and an mRNA transcript are both loaded onto a ribosome. First, the initiator tRNA binds with the small ribosomal subunit.

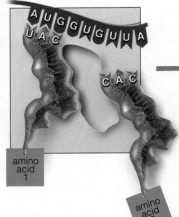

binding site for mRNA

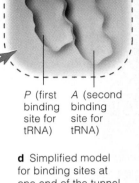

ELONGATION

c As the final step of the initiation stage, a large ribosomal subunit joins with the small one. Once this initiation complex has formed, chain *elongation*—the second stage of translation—can get under way.

intact ribosome

P (first binding site for tRNA) *A* (second binding site for tRNA)

d Simplified model for binding sites at one end of the tunnel through the large ribosomal subunit, as shown in Figure 13.9. One site is for an mRNA transcript. Two others are for tRNAs that deliver amino acids to the intact ribosome.

amino acid 1

amino acid 2

e The initiator tRNA has become positioned in the first tRNA binding site (called *P*) on the ribosomal platform. Its anticodon matches up with the START codon (AUG) of the mRNA, which also has become positioned in *its* binding site. Another tRNA is about to move into the platform's second tRNA binding site (called *A*). It is one that can bind with the codon following the START codon.

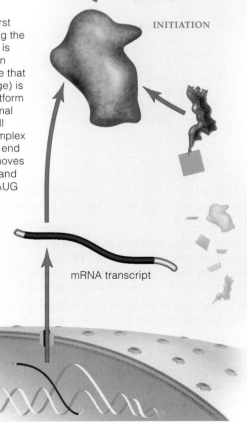

INITIATION

b *Initiation*, the first stage of translating the mRNA transcript, is about to begin. An initiator tRNA (one that can start this stage) is loaded onto a platform of a small ribosomal subunit. The small subunit/tRNA complex attaches to the 5' end of the mRNA. It moves along the mRNA and "scans" it for an AUG START codon.

mRNA transcript

a A mature mRNA transcript leaves the nucleus by passing through pores across the nuclear envelope. Thus it enters the cytoplasm, which contains pools of many free amino acids, tRNAs, and ribosomal subunits.

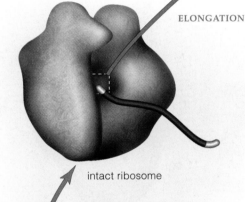

Figure 13.10 Translation, the second step of protein synthesis.

The START codon for the transcript, AUG, matches up with that tRNA's anticodon. At the same time, the AUG binds with the small subunit. Second, a large ribosomal subunit binds with the small subunit. When joined this way, the ribosome, mRNA, and tRNA are an initiation complex (Figure 13.10*b*). The next stage can begin.

In the *elongation* stage of translation, a polypeptide chain is assembled as the mRNA passes between the two ribosomal subunits, a bit like a thread being moved through the eye of a needle. Part of the rRNA molecule located at the center of the large ribosomal subunit has unusual acidity. This region functions as an enzyme. It catalyzes the joining of individual amino acids in the sequence dictated by the codon sequence in the mRNA molecule. Figure 13.10*f–i* shows how a peptide bond forms between the most recently attached amino acid and the next one delivered to the intact ribosome while a polypeptide chain is growing. Here, you might wish to refer to the description and sketch of peptide bond formation in Section 3.4 (Figure 3.15).

During the last stage of translation, *termination*, a STOP codon in the mRNA moves onto the platform. No tRNA has a corresponding anticodon. Proteins known as release factors bind to the ribosome. They trigger enzyme activity that detaches the mRNA *and* the chain from the ribosome (Figure 13.10*j–l*).

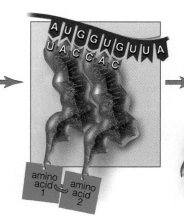

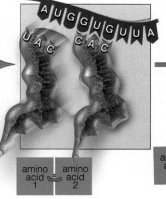

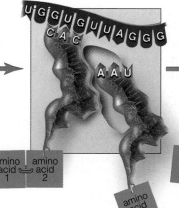

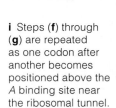

f Enzyme action breaks the bond between the initiator tRNA and the amino acid hooked to it. At the same time, enzyme action also catalyzes the formation of a peptide bond between that amino acid and the one hooked to the second tRNA. Then the initiator tRNA is released from the ribosome.

g Now the first amino acid is attached only to the second one—which is still hooked to the second tRNA. The ribosome is about to move this tRNA into the *P* site and slide the mRNA along with it by one codon. This will align the third codon in the *A* site.

h A third tRNA is about to move into the vacated *A* site. Its anticodon can base-pair with the third codon of the mRNA transcript. Now the ribosome will catalyze the formation of a peptide bond between amino acids 2 and 3.

i Steps (**f**) through (**g**) are repeated as one codon after another becomes positioned above the *A* binding site near the ribosomal tunnel.

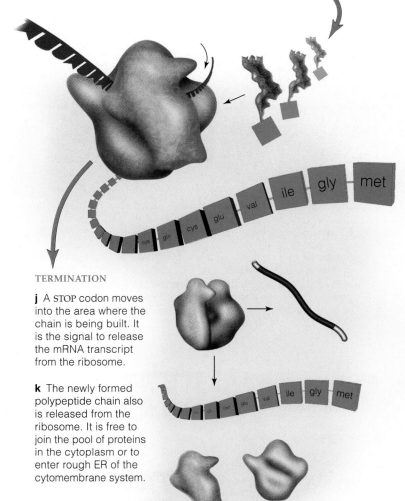

What Happens to the New Polypeptides?

Unfertilized eggs and other cells that will be called upon to rapidly synthesize many copies of different proteins usually stockpile mRNA transcripts in their cytoplasm. In cells that are rapidly using or secreting proteins, you often observe polysomes. Each polysome is a cluster of many ribosomes translating one mRNA transcript at the same time. The transcript threads through all of them, one after another, like the thread of a pearl necklace.

After new polypeptide chains are synthesized, many join the cytoplasmic pool of free proteins. Many others enter the ribosome-studded, flattened sacs of rough ER, which is part of the cytomembrane system (Section 4.7). There they take on final form before they are shipped to their ultimate destinations inside or outside the cell.

TERMINATION

j A STOP codon moves into the area where the chain is being built. It is the signal to release the mRNA transcript from the ribosome.

k The newly formed polypeptide chain also is released from the ribosome. It is free to join the pool of proteins in the cytoplasm or to enter rough ER of the cytomembrane system.

l The two ribosomal subunits separate.

Translation is initiated when a small ribosomal subunit and an initiator tRNA arrive at an mRNA's START codon and then a large ribosomal subunit binds to them.

tRNAs deliver amino acids to the ribosome in the order dictated by the sequence of mRNA codons, to which the tRNA anticodons base-pair. A polypeptide chain lengthens as peptide bonds form between the amino acids.

Translation ends when a STOP codon triggers events that cause the chain and mRNA to detach from the ribosome.

13.4

DO MUTATIONS AFFECT PROTEIN SYNTHESIS?

Whenever a cell puts its genetic code into action, it is making precisely those proteins that form its structure and carry out its functions. If something changes a gene's code words, the resulting protein may change, also. If the protein is central to cell architecture or metabolism, we can expect the outcome to be an abnormal cell.

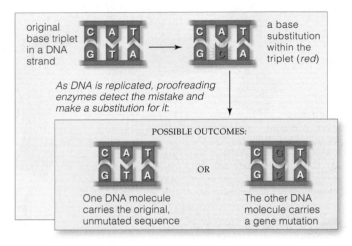

a Example of a base-pair substitution

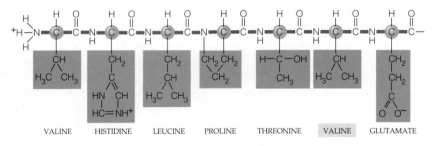

VALINE HISTIDINE LEUCINE PROLINE THREONINE VALINE GLUTAMATE

b Outcome of the base-pair substitution

Figure 13.11 Common types of mutations. (**a**) One example of a base-pair substitution. (**b**) This base-pair substitution is a type of molecular change that caused a single amino acid to be replaced in the beta chains of hemoglobin (valine instead of glutamate). Sickle-cell anemia is the result.

Gene sequences do change. Sometimes one base gets substituted for another in the nucleotide sequence. At other times, an extra base is inserted into the sequence or a base is lost from it. Such small-scale changes in the nucleotide sequence of genes in the DNA molecule are **gene mutations**. There is some leeway here; remember, more than one codon may specify the same amino acid. For instance, if UCU were changed to UCC, it probably wouldn't have dire effects because both codons specify serine. More often, however, gene mutations give rise to proteins with altered or lost functions.

Common Gene Mutations and Their Sources

Check out Figure 13.11, which shows a common gene mutation. One base (adenine) was wrongly paired with another (cytosine) as DNA was replicated. Proofreading enzymes usually detect such errors in newly replicated DNA strands and fix them. If they do not, a mutation will become established in a DNA molecule in the next round of replication. Because of this type of mutation— a **base-pair substitution**—one amino acid may replace another during protein synthesis. That's what happened in people who carry Hb^S, the mutant allele that causes sickle-cell anemia (Section 10.5).

Figure 13.12 shows a different mutation. One *extra* base was inserted into a gene region. Remember, DNA polymerases read nucleotide sequences in blocks of three. The insertion shifted the "three-bases-at-a-time" reading frame by one base; hence the term *frameshift* mutation. The altered gene has a different message, so an altered version of the protein will be synthesized. Frameshift mutations fall in broader categories of mutation called **insertions** and **deletions**. In such cases, one to several base pairs are inserted into DNA or deleted from it.

As another example, **transposons**—or transposable elements—may bring about mutation when they jump around in the genome. Barbara McClintock discovered that these segments of DNA move spontaneously from

mRNA transcribed from the DNA

PART OF PARENTAL DNA TEMPLATE

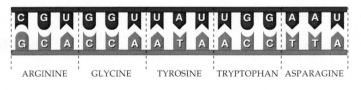

resulting amino acid sequence   ARGININE   GLYCINE   TYROSINE   TRYPTOPHAN   ASPARAGINE

altered message in mRNA

A BASE INSERTION *(RED)* IN DNA

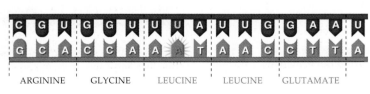

the altered amino acid sequence   ARGININE   GLYCINE   LEUCINE   LEUCINE   GLUTAMATE

Figure 13.12 Example of an insertion, a mutation in which an extra base gets inserted into a gene region of DNA. This insertion has caused a *frameshift*; it has changed the reading frame for base triplets in the DNA and in the mRNA transcript of that region. As a result, the wrong amino acids will be called up when the mRNA transcript becomes translated into protein.

Figure 13.13 Barbara McClintock, who won a Nobel Prize for her research that showed some DNA segments slip into and out of different locations in DNA molecules. The segments are transposons. In her hands is an ear of Indian corn (*Zea mays*). The curiously nonuniform coloration of its kernels sent her on the road to discovery.

Each corn kernel is a seed that can grow into a new corn plant. All of its cells have the same pigment-coding genes. Yet some kernels are colorless or spottily colored. In the ancestor of the plant from which this ear of corn was plucked, a gene in a germ cell left its position in one DNA molecule, invaded a different DNA molecule, and shut down a pigment-encoding gene.

The plant inherited the mutation. As cell divisions proceeded in the growing plant, none of the mutated cell's descendants could synthesize pigment molecules. Wherever they were located, the kernel tissue was colorless. Later, in some cells, the movable DNA slipped out of the pigment-encoding gene. All descendants of *those* cells produced pigment—and colored kernel tissue.

one location to another in the same DNA molecule or to a different one. Often they inactivate genes into which they become inserted. Their unpredictability can cause interesting variations in traits. You can read about two examples in Figure 13.13 and *Critical Thinking* question 3 at this chapter's end.

Causes of Gene Mutations

Many mutations arise spontaneously when DNA is being replicated. This shouldn't be surprising, given the swift pace of replication and the huge pools of free nucleotides concentrated near a growing DNA strand. Proofreading and repair enzymes detect and fix most mistakes, but a small number slip past with predictable frequency.

Each gene has a characteristic **mutation rate**. This is the probability that it will mutate spontaneously during a specified interval, such as each DNA replication cycle. (The *rate* isn't the same as the *frequency* of a mutation— the number of times it occurred in some population, as in 1 million gametes that gave rise to 500,000 people.)

Not all mutations are spontaneous. Many result after exposure to mutagens, or mutation-causing agents in the environment. Two classes of radiation are mutagenic. High-energy wavelengths of **ionizing radiation**, such as x rays, can damage DNA directly. They also damage it indirectly by the action of free radicals that form when they ionize water and other molecules. Repair enzymes may not restore the altered base sequences. Ionizing radiation that deeply penetrates living tissue leaves a trail of free radicals. X ray doses used for dental work and internal medicine diagnoses are extremely low to minimize mutation. However, x rays have a cumulative effect, so repeated exposure even to low levels over the years can cause problems.

Nonionizing radiation simply boosts electrons to a higher energy level. But DNA easily absorbs one form, ultraviolet (UV) light. Two of its nucleotides, cytosine

and thymine, are highly vulnerable to excitation that can alter their base-pairing properties. The next chapter's introduction describes one possible mutation.

Natural and synthetic chemicals accelerate the rate of spontaneous mutations. For instance, substances called **alkylating agents** transfer methyl and ethyl groups to reactive sites in the bases or phosphate groups of DNA. At the alkylated sites, DNA is more susceptible to base-pair disruptions that invite mutation. Many **carcinogens**, or cancer-causing agents, operate by alkylating DNA.

The Proof Is in the Protein

Spontaneous mutations are rare in terms of a human life. The rate for eukaryotes in general ranges between 10^{-4} and 10^{-6} per gene per generation. If one arises in a somatic cell, its good or bad effects won't endure, for it cannot be passed on to offspring. If the mutation arises in a germ cell or gamete, however, it may well enter the evolutionary arena. The same can happen if a mutation arises in an asexually reproducing organism or cell.

In all such cases, nature's test is this: *A protein that is specified by a heritable mutation may have harmful, neutral, or beneficial effects on the individual's ability to function in the prevailing environment.* Also, gene mutations have had powerful evolutionary consequences—and that will be a major theme of the next unit of the book.

A gene mutation is an alteration in one to several bases in the nucleotide sequence of DNA. The most common are base-pair substitutions, base insertions, and base deletions.

Each gene has a spontaneous and characteristic mutation rate, which may be accelerated by exposure to harmful radiation and certain chemicals in the environment.

A protein specified by a mutated gene may have harmful, neutral, or beneficial effects on the ability of an individual to function in the prevailing environment.

Gold indicates text section

1. Single cells and multicelled organisms cannot stay alive unless they build enzymes and other proteins. A protein consists of one or more polypeptide chains. Each chain is a linear sequence of amino acids. *CI, 13.1*

   a. The amino acid sequence of a polypeptide chain corresponds to a gene region in one strand of the DNA double helix. That region is a sequence of nucleotide bases. DNA's bases are adenine, thymine, guanine, and cytosine (A, T, G, and C).

   b. Transcription and translation are two steps in the path from genes to proteins (Figure 13.14*a*):

$$\text{DNA} \xrightarrow{\text{transcription}} \text{RNA} \xrightarrow{\text{translation}} \textbf{PROTEIN}$$

2. The path requires three classes of RNA molecules, or ribonucleic acids: *CI*

   a. Messenger RNA (mRNA) is the only class of RNA that carries a protein-building message. *13.1*

   b. Ribosomal RNA (rRNA) and structural proteins that stabilize it are the components of ribosomes. All polypeptide chains are assembled on ribosomes. *13.2*

   c. Transfer RNA (tRNA) binds free amino acids in the cytoplasm and gives them up at a ribosome, in the sequence dictated by a sequential message in mRNA. Different kinds bind different amino acids. *13.2*

3. In transcription, DNA is unwound at a gene region. Exposed bases on one strand function as a template for assembling an RNA strand from the cell's pool of free nucleotides. The RNA-to-DNA rules for base-pairing are that guanine pairs with cytosine, and *uracil*—not thymine—pairs with adenine: *CI, 13.1*

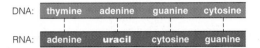

DNA:	thymine	adenine	guanine	cytosine
RNA:	adenine	**uracil**	cytosine	guanine

   a. Different RNAs are assembled on different genes.

   b. In eukaryotic cells, the mRNA transcripts become modified into final form before being shipped from the nucleus. We call this transcript processing.

4. In translation, mRNA, tRNAs, and rRNA interact to build polypeptide chains. Afterward, the chains twist, fold, and may be additionally modified into a protein's final, three-dimensional shape. *CI, 13.2, 13.3*

   a. Translation follows a genetic code. The code is a set of sixty-four base triplets; each is a series of three nucleotide bases. *Triplet* refers to the way the bases are "read" three at a time during translation at a ribosome.

   b. A base triplet in mRNA is a codon. An anticodon is a complementary triplet in a tRNA molecule. Some combination of codons specifies what the amino acid sequence of a polypeptide chain will be, start to finish.

5. Translation proceeds through three stages: *13.3*

   a. Initiation. One small ribosomal subunit and one initiator tRNA bind with the mRNA and move along it until they encounter an AUG START codon. The small subunit binds with a large ribosomal subunit.

   b. Chain elongation. tRNAs deliver amino acids to an intact ribosome. Their anticodons base-pair with the mRNA codons. Part of the rRNA of the large ribosomal subunit catalyzes peptide bond formation between every two amino acids, forming a new polypeptide chain.

   c. Chain termination. An mRNA STOP codon moves onto the ribosomal platform, making the polypeptide chain and the mRNA detach from the ribosome.

6. Gene mutations are heritable, small-scale changes in the base sequence of DNA. Many arise spontaneously while DNA is being replicated or after it is exposed to ultraviolet or ionizing radiation, alkylating agents, or some other mutagen in the environment. *13.4*

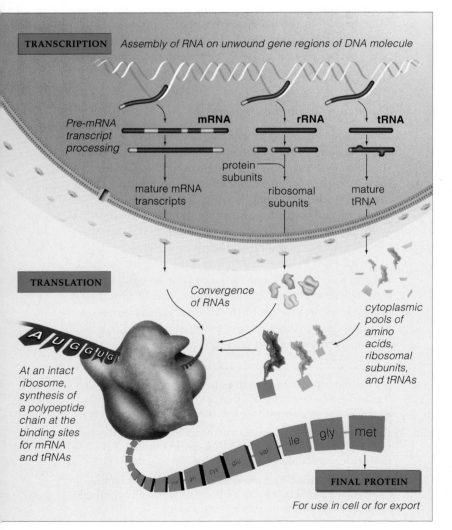

TRANSCRIPTION *Assembly of RNA on unwound gene regions of DNA molecule*

Pre-mRNA transcript processing

mRNA   **rRNA**   **tRNA**

protein subunits

mature mRNA transcripts   ribosomal subunits   mature tRNA

TRANSLATION

Convergence of RNAs

cytoplasmic pools of amino acids, ribosomal subunits, and tRNAs

At an intact ribosome, synthesis of a polypeptide chain at the binding sites for mRNA and tRNAs

AUG GUG

...cys glu val ile gly met

FINAL PROTEIN

For use in cell or for export

Figure 13.14 (**a**) Summary of the flow of genetic information from DNA to proteins in eukaryotic cells. DNA is transcribed into RNA in the nucleus; RNA is translated in the cytoplasm. Prokaryotic cells don't have a nucleus; both transcription and translation proceeds in their cytoplasm.

Review Questions

1. Are the polypeptide chains of proteins assembled on DNA? If so, state how. If not, state how they are assembled, and on which molecules. *CI, 13.1*

2. Name the three classes of RNA and state the function of each class in protein synthesis. *13.1, 13.2*

3. The pre-mRNA transcripts of eukaryotic cells contain both introns and exons. Are the introns or exons snipped out before the transcript leaves the nucleus? *13.1*

4. Distinguish between codon and anticodon. *13.2*

5. Name the three stages of translation. Briefly describe the key events of each stage. *13.3*

6. Define gene mutation. Give three examples of agents that cause mutations. *13.4*

7. Do all mutations arise spontaneously? Are environmental agents always the trigger for mutation? *13.4*

8. Define and then state the possible outcomes of the following types of mutation: base-pair substitution, base insertion, and insertion of a transposon at a new location in the DNA. *13.4*

Self-Quiz   ANSWERS IN APPENDIX III

1. DNA contains many different genes that are transcribed into different _____ .
   a. proteins
   b. mRNAs only
   c. mRNAs, tRNAs, and rRNAs
   d. all are correct

2. An RNA molecule is _____ .
   a. a double helix
   b. usually single-stranded
   c. always double-stranded
   d. usually double-stranded

3. An mRNA molecule is produced by _____ .
   a. replication
   b. duplication
   c. transcription
   d. translation

4. Each codon calls for a specific _____ .
   a. protein
   b. polypeptide
   c. amino acid
   d. carbohydrate

5. Referring to Figure 13.7, use the genetic code to translate the mRNA sequence UAUCGCACCUCAGGAGACUAG. Notice that the first codon in the frame is UAU. Which amino acid sequence is being specified?

   a. TYR—ARG—THR—SER—GLY—ASP—

   b. TYR—ARG—THR—SER—GLY

   c. TYR—ARG—TYR—SER—GLY—ASP—

6. Anticodons pair with _____ .
   a. mRNA codons
   b. DNA codons
   c. tRNA anticodons
   d. amino acids

7. Match the terms with the most suitable description.
   ____ alkylating agent
   ____ chain elongation
   ____ exons
   ____ genetic code
   ____ anticodon
   ____ intron
   ____ codon

   a. parts of mature mRNA transcript
   b. base triplet coding for amino acid
   c. second stage of translation
   d. base triplet that pairs with codon
   e. one environmental agent that induces mutation in DNA
   f. set of sixty-four codons for mRNA
   g. parts removed from a pre-mRNA transcript

Figure 13.15
Soft skin tumors on a person with neurofibromatosis, an autosomal dominant disorder.

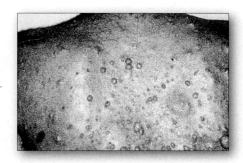

Critical Thinking

1. Sandra discovered a tRNA with a mutation in DNA that encodes the anticodon 3'-AAU instead of 3'-AUU. In cells with the mutated tRNA, what will be the effect on protein synthesis?

2. A DNA polymerase made an error during the replication of an important gene region of DNA. None of the DNA repair enzymes detected or repaired the damage. A portion of the DNA strand with the error is shown here:

   ..AATTCCGACTCCTATGG
   ..TTAAGGTTGAGGATACC

After the DNA molecule is replicated and two daughter cells have formed, one cell is carrying a mutation and the other cell is normal. Develop a hypothesis to explain this observation.

3. *Neurofibromatosis* is a human autosomal dominant disorder caused by mutations in the *NF1* gene. It is characterized by soft, fibrous tumors in the peripheral nervous system and skin as well as abnormalities in muscles, bones, and internal organs (Figure 13.15).

Because the gene is dominant, an affected child usually has an affected parent. Yet in 1991, scientists reported on a boy who had neurofibromatosis yet whose parents did not. When they examined both copies of his *NF1* gene, they found the copy he had inherited from his father contained a transposon. Neither the father nor the mother had a transposon in any of the copies of their own *NF1* genes. Explain the cause of neurofibromatosis in the boy and how it arose.

Selected Key Terms

alkylating agent *13.4*	mRNA (messenger RNA) *13.1*
anticodon *13.2*	mutation rate *13.4*
base sequence *CI*	nonionizing radiation *13.4*
base-pair substitution *13.4*	promoter *13.1*
carcinogen *13.4*	ribonucleic acid (RNA) *CI*
codon *13.2*	RNA polymerase *13.1*
deletion (of base) *13.4*	rRNA (ribosomal RNA) *13.1*
exon *13.1*	transcription *CI*
gene mutation *13.4*	translation *CI*
genetic code *13.2*	transposon *13.4*
insertion (of base) *13.4*	tRNA (transfer RNA) *13.1*
intron *13.1*	uracil *13.1*
ionizing radiation *13.4*	

Readings

Crick, F. 1988. *What Mad Pursuit: A Personal View of Scientific Discovery.* New York: HarperCollins. Crick's autobiography.

Fairbanks, F., and W. R. Anderson. 1999. *Genetics: The Continuity of Life.* Monterey, California: Brooks-Cole.

On-Line readings at Student Guide for InfoTrac:

14

CONTROLS OVER GENES

When DNA Can't Be Fixed

Not long ago, Laurie Campbell turned eighteen and happened to notice a black mole on her skin. It was an odd, encrusted lump with a ragged border. No fool, she quickly made an appointment with her family doctor, who ordered a biopsy. The mole turned out to be a *malignant melanoma*, the deadliest form of skin cancer.

Moles and other tumors are tissue masses that arise after mutations in genes that control cell growth and division. Their cells don't respond to normal controls; they go on dividing as long as conditions for growth remain favorable. All **cancers** are malignant forms of tumors. Laurie was lucky. She detected a cancer in its earliest stage before it could spread through her body. Now she regularly checks out other skin moles. She is aware of having become a statistic. In 2000, there were more than a million cases of skin cancer in the United States and 7,700 deaths from malignant melanoma.

Laurie is smart. She plotted the position of each mole on her body. Once a month, she uses that body map as a guide for a quick, thorough self-examination. Figure 14.1 shows examples of what she looks for. Laurie also schedules a medical examination every six months.

Ultraviolet wavelengths in the sun's rays, tanning lamps, and other sources of nonionizing radiation can cause skin cancer. For instance, they promote covalent bonding between two neighboring thymine bases in a DNA strand. The result is an abnormal, bulky structure in a DNA molecule—a thymine dimer.

Normally, at least seven gene products interact as a DNA repair mechanism to remove the bulky lesion. But mutation in one or more of those genes can disrupt the mechanism. When that happens, thymine dimers can accumulate in skin cells. They may trigger cancers and other lesions.

thymine dimer

You are at risk if you habitually irritate moles, as by shaving or wearing abrasive clothing. You are at risk if your skin is chronically chapped, cracked, or sore. You are at risk if there is a history of cancer in your family or if you have endured radiation therapy. Like Laurie, you are at risk simply if you burn easily. Tanning or otherwise staying out in the sun without protection is ill advised; damaged DNA is a reality for all of us.

Why start a chapter with such awful prospects? Doing so might help focus your attention on how lucky individuals are when gene controls operate as they should. **Gene controls** are molecular mechanisms that govern when and how fast specific genes will be transcribed and translated, and whether gene products will be switched on or silenced.

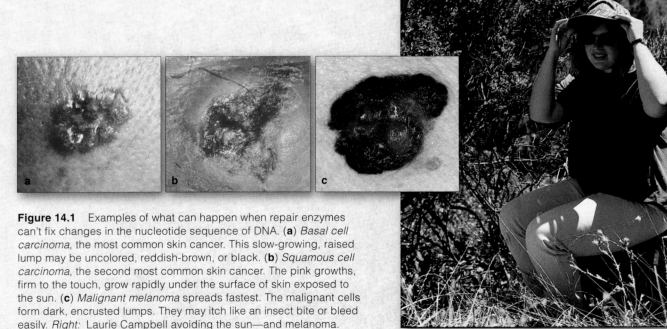

Figure 14.1 Examples of what can happen when repair enzymes can't fix changes in the nucleotide sequence of DNA. (**a**) *Basal cell carcinoma*, the most common skin cancer. This slow-growing, raised lump may be uncolored, reddish-brown, or black. (**b**) *Squamous cell carcinoma*, the second most common skin cancer. The pink growths, firm to the touch, grow rapidly under the surface of skin exposed to the sun. (**c**) *Malignant melanoma* spreads fastest. The malignant cells form dark, encrusted lumps. They may itch like an insect bite or bleed easily. *Right:* Laurie Campbell avoiding the sun—and melanoma.

In all prokaryotic and eukaryotic cells, many of the controls make transcription rates rise or fall in response to concentrations of nutrients and other substances. For example, genes that specify the enzymes required to metabolize sugars aren't turned on and off; they are transcribed on an ongoing basis. But controls adjust the actual rates of transcription according to how much sugar is available. Controls activate, inactivate, and degrade gene products, including the sugar-digesting enzyme molecules that have finished their task.

In eukaryotic cells only, other controls guide mRNA transcript processing, transport of mature RNAs from the nucleus, and rates of translation in the cytoplasm. (Life is generally short for prokaryotic cells, and their mRNA gets translated as soon as it starts peeling off the DNA.) Other controls guide the modification of new polypeptide chains in the cytomembrane system.

Controls also guide the contribution that eukaryotic cells will make to a long-term program of growth and development. This is especially so for large, complex, multicelled species. As part of the program, different cell lineages activate and suppress fractions of their genes in unique ways. Certain genes are expressed once, some of the time, or not at all. The result? Most cells become specialized in structure, composition, and function. We call this process **cell differentiation.**

For example, all of your cells inherited the same genes, because every one of them descended from a single fertilized egg. They all use genes that specify the enzymes of glycolysis on an ongoing basis. Yet only the immature red blood cells use the genes that specify hemoglobin. Your liver cells access the genes required to synthesize enzymes that neutralize certain toxins, but they are the only ones that do. When your eyes first formed, only certain cells accessed the genes necessary to synthesize crystallin. No other cells used the genes for this type of protein, which helped make transparent fibers of the lens in each eye.

Explaining control of gene activity is like explaining a full symphony orchestra to someone who has never seen one or heard it perform. You have to identify its many separate parts before their interactions start to make sense! Gene controls weave through the story line of many chapters to follow. For that reason, take some time now to become acquainted with just a few of the molecular players and their amazing functions.

Key Concepts

1. In cells, a variety of controls govern when, how, and to what extent genes are expressed. The control elements operate in response to changing chemical conditions and to reception of external signals.

2. Control is exerted by way of regulatory proteins and other molecules that operate before, during, or after gene transcription. The control elements interact with DNA, with RNA that has been transcribed from DNA, or with gene products—that is, with the resulting polypeptide chains or final proteins.

3. Prokaryotic cells depend on rapid control over short-term shifts in nutrient availability and other aspects of their surrounding environment. They commonly rely on regulatory proteins that help make quick adjustments in rates of gene transcription.

4. Eukaryotic cells also rely on controls over short-term shifts in diet and levels of activity. In complex multicelled species, they rely as well on long-term controls over an intricate program of growth and development.

5. Controls over eukaryotic cells come into play when new cells contact one another in developing tissues. They also come into play when these cells start interacting with their neighbors by way of hormones and other signaling molecules.

6. Although all cells of a multicelled organism inherit the same genes, different cell types activate or suppress many of the genes in different ways. Their controlled, selective use of their store of genes leads to the synthesis of the proteins that give each type of cell its distinctive structure, function, and products.

TYPES OF CONTROL MECHANISMS

Different mechanisms control gene expression through interactions with DNA, RNA, and gene products—that is, with new polypeptide chains or final proteins. Some respond to rising or falling concentrations of a nutrient or some other substance. Other mechanisms respond to external signals for change.

The control agents include **regulatory proteins** that intervene before, during, or after gene transcription or translation. They also include signaling molecules such as **hormones**, which initiate change in some cell activity when they dock at suitable receptors.

With **negative control**, certain regulatory proteins trigger a slowdown in some activity. Hence their name, *repressor* proteins. With **positive control**, the regulatory proteins promote or enhance some activity. Hence their name, *activator* proteins. With respect to transcription, control mechanisms involve noncoding base sequences in DNA (which do not specify proteins). Examples are **promoters**, which signal the beginning of a gene; and **enhancers**, which are binding sites for activator proteins.

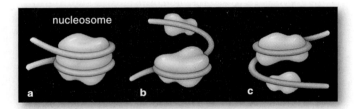

Figure 14.2 How DNA–histone packing in nucleosomes may be loosened to make gene regions available for transcription. Attaching an acetyl group to a histone makes it loosen its grip on DNA wound around it. Enzymes that attach or detach acetyl groups are known to be associated with transcription.

In eukaryotes, control also is exerted by chemical modification. For instance, regions of newly replicated DNA can be shut down by **methylation**: attachment of methyl groups ($-CH_3$) to nucleotide bases. Inactivated genes are often heavily methylated. And certain genes are activated by demethylation. Also, access to genes is partly controlled through **acetylation**: attaching acetyl groups from the histones that structurally organize the DNA (Section 8.1 and Figure 14.2).

When, how, and to what extent a gene is expressed depends on the type of cell and its functions, on the cell's chemical environment, and on signals for change.

Gene expression is controlled by regulatory proteins that interact with one another, with control elements built into the DNA, with RNA, and with gene products.

Control also is exerted through chemical modifications that inactivate or activate specific gene regions of DNA.

BACTERIAL CONTROL OF TRANSCRIPTION

Think about the dot of the letter "i." About a thousand bacterial cells would stretch side by side across the dot. Just imagine, each of those microscopic specks depends as much on gene controls as you do! When nutrients are plentiful and other environmental conditions also favor growth, the cells swiftly grow and divide. Gene controls promote the rapid synthesis of enzymes that catalyze nutrient digestion and other growth-related activities. Translation starts even before the RNA transcripts have been completed. Remember, bacteria have no nucleus to keep DNA away from ribosomes in the cytoplasm.

Often, prokaryotic genes for enzymes of a metabolic pathway are clustered together in the same sequence as the reaction steps. All genes in such a sequence may be transcribed as one continuous mRNA transcript.

With this bit of background, consider how one kind of prokaryote adjusts transcription rates downward or upward, depending on the availability of nutrients.

Negative Control of the Lactose Operon

Escherichia coli is an enteric bacterium (gut dweller) in all mammals. It survives on glucose, milk sugars such as lactose, and other ingested nutrients. Like other adult mammals, you probably don't drink milk around the clock. When you do, *E. coli* cells in your large intestine (colon) swiftly transcribe three genes for enzymes with roles in sugar-digesting reactions, starting with lactose.

A promoter precedes the three genes, which are next to one another. A different control element intervenes between the promoter and the bacterial genes. It is an **operator**, a binding site for a repressor that can block transcription (Figure 14.3). An arrangement in which a promoter and operator service more than one bacterial gene is an **operon**. Elsewhere in *E. coli* DNA, a different gene codes for this particular repressor protein, which can bind with the operator *or* with a lactose molecule.

When lactose concentration is low, a repressor binds with the operator (Figure 14.4). Being a large molecule, it extends over the promoter and blocks transcription. *So lactose-degrading enzymes aren't built when they aren't required.* When many lactose molecules are moving past in the gut, the chances are greater that one will enter a

Figure 14.3 Computer model for one type of repressor protein. This repressor is bound to an operator in a prokaryotic DNA molecule.

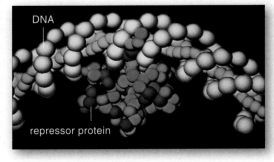

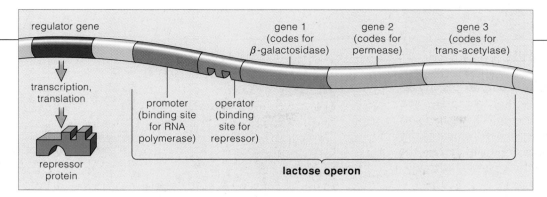

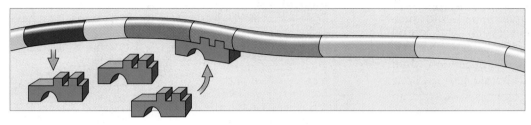

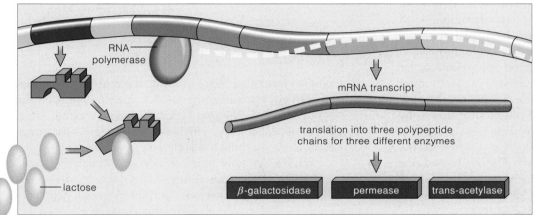

regulator gene

gene 1
(codes for
β-galactosidase)

gene 2
(codes for
permease)

gene 3
(codes for
trans-acetylase)

transcription,
translation

promoter
(binding site
for RNA
polymerase)

operator
(binding
site for
repressor)

repressor
protein

lactose operon

a A repressor protein exerts negative control over three genes of the lactose operon by binding to the operator and inhibiting transcription.

b When the concentration of lactose is low, the repressor is free to block transcription. Being bulky, it overlaps the promoter and prevents binding by RNA polymerase. The enzymes (not needed) are not produced.

RNA polymerase

mRNA transcript

translation into three polypeptide chains for three different enzymes

β-galactosidase

permease

trans-acetylase

lactose

c At high concentration, lactose is an inducer of transcription. It binds to and distorts the shape of the repressor—which now cannot bind to the operator. The promoter is exposed and the genes can be transcribed.

Figure 14.4 Negative control of the lactose operon. The operon's first gene codes for an enzyme that splits lactose (a disaccharide) into glucose and galactose. The second enzyme codes for an enzyme that transports lactose into cells. The third helps metabolize certain sugars.

bacterium and bind with the repressor. Binding alters the repressor's shape. The alteration prevents it from binding with the operator, so nothing is stopping RNA polymerase from transcribing the genes. Thus, *lactose-degrading enzymes are synthesized only when required.*

Positive Control of the Lactose Operon

E. coli cells pay far more attention to glucose than to lactose. They transcribe genes for glucose breakdown faster, and continuously. Even if lactose is in the gut, the lactose operon isn't used much *unless there is no glucose.* With no glucose, an activator protein called CAP goes to work. It happens that the lactose operon's promoter is not good at binding RNA polymerase. It does better when CAP adheres to it first. But CAP won't bind to the promoter without first binding to a chemical messenger, cAMP (cyclic adenosine monophosphate, which is like ATP with two phosphate groups removed). When the two adhere to a promoter, they are like an invitation to RNA polymerase to join up and start transcription.

When glucose is plentiful, glycolysis runs full bore, ATP forms, and an enzyme required for cAMP formation is inhibited. Transcription of the lactose operon genes slows almost to a standstill. When glucose is scarce and lactose is available, though, the enzyme makes cAMP. Molecules of cAMP accumulate, CAP–cAMP complexes form, and lactose operon genes are transcribed fast. So the cells use lactose as an alternative energy source.

We humans are born with lactase, a lactose-digesting enzyme in cells of the intestinal lining. Before three years have passed, lactase levels start to decline in genetically predisposed individuals, and *lactose intolerance* follows. Lactose can't be degraded. It moves into the colon and fans the growth of resident bacterial populations. One gaseous by-product of their metabolism accumulates and distends the colon, causing pain. Short fatty acid chains released by the reactions lead to diarrhea, which is often severe. People avoid the symptoms by drinking milk that incorporates predigested lactose or by taking lactase pills when drinking or eating dairy products.

Transcription rates of bacterial genes for nutrient-digesting enzymes are quickly adjusted downward and upward by control systems that respond to nutrient availability.

CONTROLS IN EUKARYOTIC CELLS

By some estimates, the cells of all complex organisms rarely use more than 5 to 10 percent of their genes at a given time. One way or another, control mechanisms are keeping most of the genes repressed. Which genes are active depends on the type of organism and on the stage of growth and development it's passing through.

Case Study: X Chromosome Inactivation

We introduce a classic case of eukaryotic gene control with this question: At the cellular level, what do female humans and female calico cats have in common? Both have two X chromosomes in their diploid cells, and one is in its threadlike form, but the other is scrunched up even at interphase. The scrunching isn't a chromosome abnormality. It is a programmed shutdown of all but about three dozen genes on *one* of two homologous X chromosomes. This **X chromosome inactivation** occurs in the diploid cells of all female placental mammals.

One X chromosome is inactivated when the females are early embryos, no more than a tiny ball of dividing cells. The outcome is random; *either* chromosome may be condensed this way. One cell might shut down the maternal X chromosome, another cell right next to it might shut down the paternal X chromosome or the maternal X chromosome, and so on. The condensed one looks like a dense spot in an interphase nucleus (Figure 14.5a). Researchers named a condensed X chromosome a **Barr body** after Murray Barr, its discoverer.

Once that first random selection is made in a cell, however, all the descendant cells make the exact same selection as they go on dividing to form tissues. When fully developed, then, *each female mammal bears patches of tissue where genes of the maternal X chromosome are being expressed—and patches where genes of the paternal X chromosome are being expressed.* She is a "mosaic" for the X chromosomes! As you know, a pair of alleles on two homologous chromosomes might or might not be identical. When they are not, skin and other tissues of female mammals may have different features from one patch to the next. Mary Lyon, a geneticist, was the discoverer of this **mosaic tissue effect** of random X chromosome inactivation.

We see the effect in human females who are heterozygous for a recessive allele that blocks sweat gland formation. Their skin is a mosaic of tissues with and without sweat glands. This is one symptom of *anhidrotic ectodermal dysplasia.*

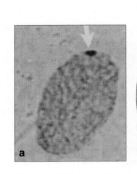

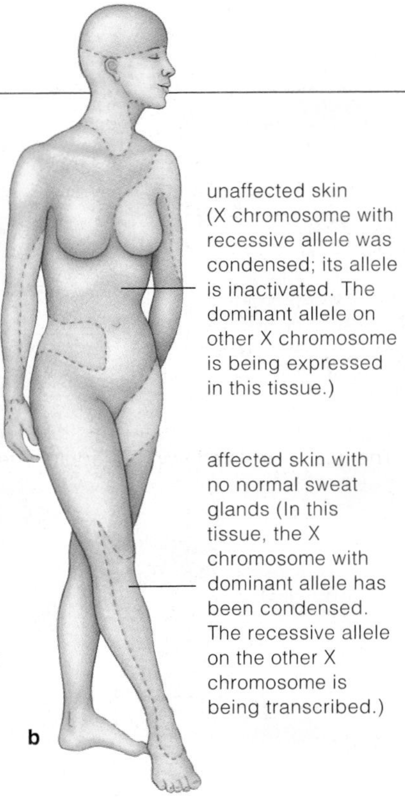

Figure 14.5 (**a**) Light micrograph revealing a Barr body (*arrow*) in a cell's nucleus. A Barr body is an inactivated X chromosome. (**b**) A mosaic tissue effect that shows up in a human female with anhidrotic ectodermal dysplasia.

unaffected skin (X chromosome with recessive allele was condensed; its allele is inactivated. The dominant allele on other X chromosome is being expressed in this tissue.)

affected skin with no normal sweat glands (In this tissue, the X chromosome with dominant allele has been condensed. The recessive allele on the other X chromosome is being transcribed.)

Where sweat glands are missing, the mutant allele is on an active X chromosome; the dominant allele is on the condensed one (Figure 14.5b). We see the same effect in female calico cats, which are heterozygous for a coat color allele on their X chromosomes (Figure 14.6).

What's the point of the shutdown? Remember, male and female mammals differ in their sex chromosomes (XY versus XX). However, a subset of genes, primarily on the X chromosome's long arm, must be expressed at the *same* levels in males and females. Otherwise, early embryos will not develop properly. Inactivating one X chromosome in XX embryos is a **dosage compensation** mechanism; it balances gene expression between sexes during critical early stages of development. The control mechanism works by activating the *Xist* gene on one X chromosome. The gene product interacts with control elements in DNA to condense only that chromosome.

Case Study: Hormones as Control Agents

Hormones, a type of signaling molecule, promote or inhibit gene expression in target cells. Any cell that has receptors for a given hormone is its target. In animals,

Figure 14.6 Why is this female calico cat "calico"? She is heterozygous for an allele on the X chromosome that deals with melanin, a dark pigment. Her coat color depends on which allele was randomly inactivated in the cells making up a given tissue region. In regions where the *active* X chromosome bears the dominant allele, you see dark fur. In regions where the *condensed* X chromosome bears the dominant allele, nothing is masking the recessive allele, so you see yellow fur. The white patches have a different basis. They arise from a gene interaction involving the spotting gene, which blocks melanin synthesis entirely.

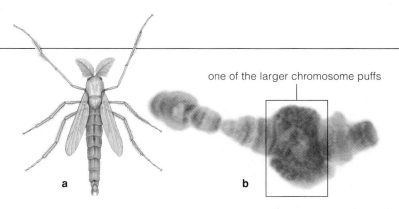

one of the larger chromosome puffs

a b

Figure 14.7 (**a**) Larva of a midge (*Chironomus*). Midges are flies. The tiny, short-lived, winged adults often swarm together, which helps them find mates fast. They look like mosquitoes but don't bite. Most midge larvae develop and feed rapidly. They continuously transcribe genes for saliva's protein components. Ecdysone is a hormone that binds to a transcription-promoting regulatory protein. The chromosomes loosen and puff out in regions being transcribed. (**b**) Puffs are largest and most diffuse where transcription is most intense. Staining techniques reveal the banding patterns.

hormone molecules secreted from cells enter the blood, then are transported to cells that may be quite distant.

Some hormones bind to membrane receptors at the surface of target cells. Others enter cells and bind to a regulatory protein involved in transcription, as when one type activates an enhancer. RNA polymerase will bind avidly to promoter–enhancer complexes.

Consider **ecdysone**, a hormone with roles in many insect life cycles. Insect larvae grow rapidly during the cycles. They feed voraciously on plant parts and other organic matter, and use up lots of saliva to prepare the food for digestion. Their salivary gland cells have made many copies of DNA. These copies remain together in parallel to form a **polytene chromosome**. The cells have ecdysone receptors. When ecdysone docks at a receptor, it triggers rapid transcription of genes in the multiple copies of DNA. Gene regions affected by the hormone puff out during transcription (Figure 14.7). After this, translation of the mRNA transcripts results in protein components of saliva.

Some vertebrate hormones have widespread effects on genes because many types of cells have receptors for them. For instance, somatotropin (growth hormone) promotes the synthesis of the proteins required for cell division, hence for the body's growth. Most cells have somatotropin receptors. Other hormones signal certain cells at certain times. Prolactin is like this. A few days after a female mammal gives birth, prolactin meets up with mammary gland cells and stimulates genes with roles in milk production. Liver and heart cells have the same genes, but they don't have prolactin receptors.

Case Study: Sunlight and Phytochrome

Plant a few seeds from a corn plant in a pot filled with moist, nutrient-rich soil. Let the seeds germinate in full darkness. Eight days later, you will have pale, spindly

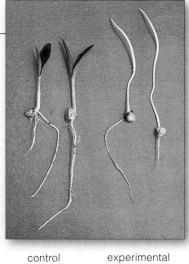

Figure 14.8 Experiment that reveals phytochrome's effect on corn seedlings. Two seedlings (the control group) grew in a greenhouse, in sunlight. Two seedlings (the experimental group) grew in the dark for eight days. They could not activate phytochrome, which helps control transcription of genes for the proteins that activate chlorophyll and make rubisco. Leaves of the experimental group never did green up.

control group experimental group

seedlings that have no chlorophyll (Figure 14.8). If you then expose the seedlings to a single burst of dim light from a flashlight, in less than ten minutes they will be producing functional chlorophyll molecules.

Phytochrome, a blue-green pigment, is a regulatory protein that influences gene transcription in ways that help plants adapt to changes in light conditions.

Far-red wavelengths inactivate phytochrome—and red wavelengths activate it. Far-red prevails at sunset, at night, or in the shade; red prevails at sunrise. The presence of these wavelengths varies from day to night and as the seasons change. By controlling phytochrome activity, the variation controls transcription at certain times of day and year. The genes under control specify enzymes and other proteins that help seeds germinate, stems lengthen and branch, and leaves grow. The gene products also help flowers, fruits, and seeds form.

Experiments by Elaine Tobin and her coworkers at the University of California, Los Angeles, give evidence of phytochrome control. Duckweed (*Lemna*) seedlings grown in the dark markedly increased the number of certain mRNA transcripts after a one-minute exposure to red light. What was transcribed? Genes that specify chlorophyll-binding proteins and rubisco, an enzyme that mediates carbon fixation (Sections 6.5 and 6.6). In the absence of these proteins, chloroplasts don't develop properly and they don't turn green.

As multicelled eukaryotic organisms develop, new cells inherit the same genes, but each type selectively activates and suppresses them. Selective gene expression leads to cell differentiation, whereby cells become specialized in composition, structure, and function.

X chromosome inactivation in XX mammalian embryos is a control mechanism that balances gene expression between sexes. It involves interaction between a gene product and control elements built into an X chromosome.

Hormones and other signals from the external environment help control selective gene expression in eukaryotic cells.

Lost Controls and Cancer

THE CELL CYCLE REVISITED   Every second, millions of cells in your skin, bone marrow, gut lining, liver, and elsewhere divide and replace worn-out, dead, and dying predecessors. They don't divide willy-nilly. Mechanisms control the expression of genes that specify enzymes and other proteins required for cell growth, DNA replication, chromosome movements, and cytoplasmic division. And they control when the division machinery is put to rest.

The cell cycle has built-in checkpoints, where proteins monitor chromosome structure and other aspects of the preceding phase of the cycle. They also sense whether or not conditions favor division. The proteins are products of **checkpoint genes**, which are the basis of mechanisms that advance, delay, or block the cell cycle.

For example, the protein products called **kinases** are a class of enzymes that phosphorylate molecules. Some signal that DNA replication is finished and call on the cell to make the transition from interphase to mitosis (Section 8.2). The products called **growth factors** invite transcription of genes with roles in the body's growth. Example: one of these, epidermal growth factor (EGF), activates tyrosine kinase when it binds to target cells in epithelial tissues. Binding is a signal to start mitosis. The product of checkpoint gene *p53* can put the brakes on division when chromosomes are damaged.

ONCOGENES   Many cancers are known to arise through mutations in one or more checkpoint genes. A mutated gene that has the potential to induce cancer falls into the broader category of **oncogenes**.

Oncogenes can form following insertion of viral DNA into a cell's DNA. They also can form after mutagens alter the structure of DNA. As you read in Section 13.4, ultraviolet radiation and nonionizing radiation, such as x-rays, are mutagens. So are many natural and synthetic

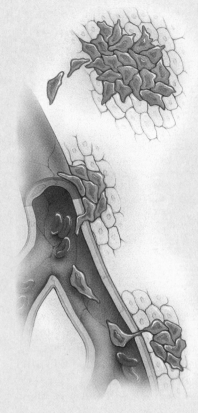

a Cancer cells break away from their home tissue.

b The metastasizing cells attach to the wall of a blood vessel or lymph vessel and secrete digestive enzymes onto it. They cross the breached wall.

c Cancer cells creep or tumble along inside blood vessels, then leave the bloodstream the same way they got in. They start new tumors in new tissues.

Figure 14.10   Steps in metastasis.

compounds, including asbestos and certain substances in tobacco smoke.

And remember those chromosome alterations and gene mutations described in Sections 11.8 and 13.4? Some cancers arise when base substitutions or deletions alter a gene or one of the control elements that deal with its transcription. Other cancers may arise by translocations or by destabilizing transposons.

CHARACTERISTICS OF CANCER   What happens to a cell when it undergoes cancerous transformation? At the very least, all cancer cells display four characteristics. First, *the plasma membrane and cytoplasm change profoundly.* That membrane becomes more permeable, its proteins become lost or altered, and abnormal proteins form. The cytoskeleton shrinks drastically, becomes disorganized, or both (Figure 14.9). Enzyme action shifts, as in amplified reliance on glycolysis. Second, *cancer cells grow and divide abnormally.* Controls against overcrowding in tissues are lost; cell populations reach high densities. The number of small blood vessels that service the growing cell mass increases abnormally. Third, *cancer cells have a weakened capacity for adhesion.* Recognition proteins become lost or altered, so cells can't stay anchored in proper tissues. They break away and may establish colonies in other, distant tissues. **Metastasis** (meh-TA-stuh-SIS) is the name for the process of abnormal cell migration and tissue invasion (Figure 14.10). Fourth, *cancer cells are lethal.*

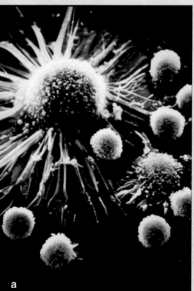

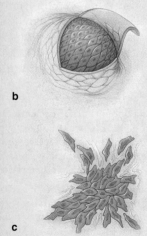

Figure 14.9   (**a**) A patrol of white blood cells surrounding a cancer cell. (**b**) Benign tumor. (**c**) Malignant tumor.

Signal to die docks at receptor.

Signal causes activation of ICE-like proteases.

Figure 14.11 Artist's representation of weapons of cell death being unleashed inside a cell. Normally, cells self-destruct when they finish their functions or become altered in ways that could threaten the body as a whole. When controls over programmed cell death are lost, an altered cell may start dividing abnormally and give rise to cancer.

Unless cancer cells are eradicated, their uncontrollable divisions put the individual on a painful road to death.

Cells of common skin moles and other noncancerous, *benign* tumors grow abnormally but slowly, and they retain the surface recognition proteins that hold them in their home tissue. Unless benign tumors become too large or irritating, doctors usually leave them alone.

Abnormal cells of a *malignant* tumor grow and divide rapidly, with destructive physical and metabolic effects on the surrounding tissues. These are grossly disfigured, metastasizing cancer cells (Figure 14.9c). They break loose from their home tissue, enter lymph or blood vessels, then slip out and invade other tissues where they do not belong. There, they may start growing as new tumors.

Each year in the developed countries alone, 15 to 20 percent of all deaths result from cancer. Cancer isn't just a human problem. Researchers have observed cancers in most of the animal species they have studied to date.

HERE'S TO SUICIDAL CELLS! We conclude with a case study of a gene and its cancerous transformation.

The first cell of a new multicelled individual contains marching orders that will guide its descendants along a program of growth, development, and reproduction, then on to death. As part of that program, many cells heed calls to self-destruct when they finish their prescribed function. If they become altered in ways that might pose a threat to the body as a whole, as by infection or cancer, they can execute themselves. **Apoptosis** (app-uh-TOE-sis) is the name for this form of cell death. It starts with molecular signals that activate and unleash lethal weapons of self-destruction, which were stockpiled earlier in the cell.

Protein-cleaving enzymes called **ICE-like proteases** are such weapons. Think of them as folded pocketknives or lethal Ninja weapons. When popped open, they chop apart structural proteins, including the building blocks of cytoskeletal elements and nucleosomes that organize the DNA (Figure 14.11).

A body cell in the act of suicide shrinks away from its neighbors. Its cytoplasm seems to roil, and its surface repeatedly bubbles outward and inward. No longer are its chromosomes extended through the nucleoplasm; they bunch up near the nuclear envelope. The nucleus, then the cell, breaks apart. Suicidal cells or remnants of them are swiftly engulfed by phagocytic white blood cells, which patrol and protect the body's tissues.

The timing of cell death is predictable in some cells, such as pigment-packed keratinocytes. These form the densely packed sheets of dead cells that are continually sloughed off and replaced at the skin surface. They have a three-week life span, more or less. Keratinocytes and other body cells, even kinds that are supposed to last for a lifetime, can be induced to die ahead of schedule. All it takes is sensitivity to signals that can activate ICE-like proteases and other enzymes of death.

Control genes suppress or trigger programmed cell death. One of these, *bcl-2*, helps keep normal body cells from dying before their time. The knives stay sheathed in cancer cells, which are supposed to—but don't—commit suicide on cue. Maybe they lost normal receptors that would permit contact with their signaling neighbors. Maybe they are receiving abnormal signals about when to grow, divide, or cease dividing. In some cancers, for instance, the suppressor gene *bcl-2* has been tampered with or shut down. Apoptosis is no longer in the cards.

Cancer is a multistep process involving more than one oncogene. But researchers have already identified many of the mutated forms of checkpoint genes that contribute to it. Remember how one form of leukemia, described in Chapter 11's introduction, is responding to the molecular targeting of one gene's product? More anticancer drugs and gene therapies should soon follow.

14.5

SUMMARY   *Gold* indicates text section

1. Cells are equipped with controls that govern gene expression; that is, which gene products appear, when, and in what amounts. When control mechanisms come into play depends on cell type, on prevailing chemical conditions, and on signals from other cell types that can change a target cell's activities. *CI, 14.1*

   a. In all prokaryotic and eukaryotic cells, many of the controls adjust transcription rates in response to concentrations of nutrients and other substances.

   b. In eukaryotic cells of complex organisms, controls contribute to a program of growth and development. They allow a fraction of the same genes to be expressed selectively in different lineages of cells. Selective gene expression is the beginning of diverse specializations in cell structure, composition, and function. We call this process cell differentiation.

2. Regulatory proteins (e.g., activators and inhibitors of transcription), hormones, and some DNA sequences are typical control agents. *14.1–14.3*

   a. Diverse control agents interact with one another, with operators and other control elements built into DNA, with RNA, and with gene products.

   b. Gene expression also is controlled with chemical modifications that inactivate and activate gene regions. X chromosome inactivation is an example.

   c. Control elements called promoters are the binding sites in DNA that signal the start of a gene. Enhancers are binding sites in DNA for some activator proteins.

3. In all cell types, two of the most common types of control systems inhibit or enhance gene transcription. In negative control systems, a regulatory protein binds at a specific DNA sequence to block transcription of one or more genes. In positive control systems, a regulatory protein binds to DNA to promote transcription. *14.1*

4. Unlike eukaryotic cells, most prokaryotic cells do not require many genes to live, grow, and reproduce. Most of their control systems adjust transcription rates in response to nutrient availability. Control of operons (groupings of functionally related prokaryotic genes and control elements) is an example. *14.2*

5. Compared to prokaryotic cells, eukaryotic cells use more complex gene controls. Gene expression changes in response to external conditions and also is subject to long-term controls over growth and development. *14.3*

   a. X chromosome inactivation in XX embryos of all female mammals is an outcome of interactions between a product of a gene (*Xist*) and control elements in one X chromosome. This interaction is an example of dosage compensation, a control mechanism that maintains a crucial balance in gene expression between the sexes.

   b. Hormones and phytochromes are two examples of signaling molecules that have roles in selective gene expression. Both can affect transcription.

6. Checkpoint gene products advance, delay, or block the cell cycle. They include growth factors and protein kinases. Some checkpoint gene mutations lead to cancer. A mutated form of such a gene is an oncogene. *14.4*

7. Cancer involves loss of normal controls over the cell cycle and mechanisms of programmed cell death. In all cancer cells, the structure and function of the plasma membrane and cytoplasm are severely compromised. Cancer cells grow and divide abnormally. Mechanisms for self-destruction are suppressed. Cancer cells have a weakened capacity for adhesion to their home tissue. Unless eradicated, they are lethal. *14.4*

Review Questions

1. A plant, fungus, and animal consist of diverse cell types. How might the diversity arise, given that body cells in each of these organisms inherit the same set of genetic instructions? As part of your answer, define cell differentiation and the general way that selective gene expression brings it about. *CI, 14.3*

2. In what fundamental way do negative and positive controls of transcription differ? Is the effect of one or the other form of control (or both) reversible? *14.1*

3. Distinguish between: *14.1, 14.2*
   a. repressor protein and activator protein
   b. promoter and operator
   c. repressor and enhancer

4. Describe one type of control of transcription for the lactose operon in *E. coli*, a prokaryotic cell. *14.2*

5. Using the sketch on the facing page, define three types of gene controls and note the levels at which they take effect. Does the sketch work for both prokaryotic and eukaryotic cells? Why or why not? *14.1–14.3*

6. What is a Barr body? Does it appear in the cells of human males, human females, or both? Explain your answer. *14.3*

7. Briefly describe the general characteristics of normal cells. Then explain the difference between a benign tumor and a malignant tumor. *CI, 14.4*

Self-Quiz   ANSWERS IN APPENDIX III

1. Cell differentiation _____ .
   a. occurs in all complex multicelled organisms
   b. requires different genes in different cells
   c. involves selective gene expression
   d. both a and c
   e. all of the above

2. The expression of a given gene depends on the _____ .
   a. type of cell and its functions   c. environmental signals
   b. chemical conditions   d. all of the above

3. Regulatory proteins interact with _____ .
   a. DNA   c. gene products
   b. RNA   d. all of the above

4. A base sequence signaling the start of a gene is a(n) _____ .
   a. promoter   c. enhancer
   b. operator   d. activator protein

5. In prokaryotic cells but not eukaryotic cells, a(n) _____ precedes the genes of an operon.
   a. lactose molecule   c. operator
   b. promoter   d. both b and c

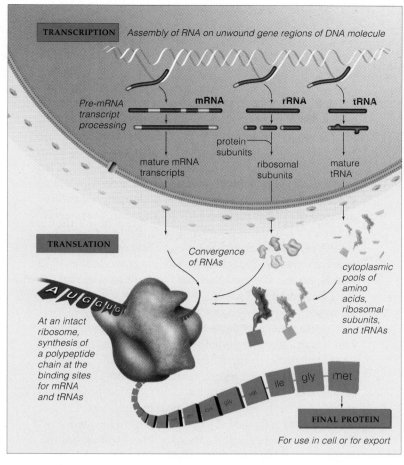

TRANSCRIPTION *Assembly of RNA on unwound gene regions of DNA molecule*

Pre-mRNA transcript processing

mRNA **rRNA** **tRNA**

mature mRNA transcripts

protein subunits

ribosomal subunits

mature tRNA

TRANSLATION

Convergence of RNAs

cytoplasmic pools of amino acids, ribosomal subunits, and tRNAs

A U G G U G

At an intact ribosome, synthesis of a polypeptide chain at the binding sites for mRNA and tRNAs

ile gly met

FINAL PROTEIN

For use in cell or for export

6. An operon most typically governs _____ .
   a. bacterial genes   c. genes of all types
   b. a eukaryotic gene   d. DNA replication

7. Eukaryotic genes guide _____ .
   a. fast short-term activities   c. development
   b. overall growth   d. all of the above

8. X chromosome inactivation is an example of _____ .
   a. a chromosome abnormality   c. dysfunctional calico cats
   b. dosage compensation   d. dysfunctional women

9. Hormones may _____ gene transcription in target cells.
   a. promote   c. participate in
   b. inhibit   d. both a and b

10. Apoptosis is _____ .
    a. cell division after severe tissue damage
    b. programmed cell death by suicide
    c. a popping sound in mutated toes

11. ICE-like proteases are _____ .
    a. structural proteins   c. environmental signals
    b. lethal weapons   d. low-temperature enzymes

12. Match the terms with their most suitable descriptions.
    _____ phytochrome   a. inhibits gene transcription
    _____ Barr body   b. mutated form of any gene that
    _____ oncogene   can induce cancer
    _____ repressor   c. hormone with key role in
    _____ ecdysone   insect life cycles
                          d. helps plants adapt to daily
                             and seasonal changes in light
                          e. inactivated X chromosome

Critical Thinking

1. Geraldo isolated a strain of *E. coli* in which a mutation has affected the capacity of CAP to bind to a region of the lactose operon, as it would do normally. State how the mutation affects transcription of the lactose operon when cells of this bacterial strain are subjected to these conditions:
   a. Lactose and glucose are available.
   b. Lactose is available but glucose is not.
   c. Both lactose and glucose are absent.

2. *Duchenne muscular dystrophy*, a genetic disorder, affects boys almost exclusively. Early in childhood, muscles begin to atrophy (waste away) in affected individuals, who typically die in their teens or early twenties as a result of respiratory failure. Muscle biopsies of women who carry an allele associated with the disorder reveal some body regions with atrophied muscle tissue. Yet muscles adjacent to a region of atrophy is normal or even larger and more chemically active, as if to compensate for the weakness of the adjoining region. Give a brief explanation of these observations.

3. Unlike most rodents, guinea pigs are well developed at the time of birth. Within a few days, they can eat grass, vegetables, and other plant material. Suppose a breeder decides to separate the baby guinea pigs from their mothers after three weeks. He wants to keep the males and females in different cages. But it is difficult to identify the sex of guinea pigs when they are so young. Suggest a simple test that the breeder can perform to identify their sex.

4. Individuals affected by *pituitary dwarfism* cannot synthesize somatotropin (also called growth hormone). Children with this genetic abnormality will be below the range of normal heights. Develop a hypothesis to explain why therapy that is based on somatotropin injections may be effective.

5. The closer a mammalian species is to humans in its genetic makeup, the more useful information it may yield in laboratory studies of the mechanisms of cancer. Do you support the use of any mammal for cancer research? Why or why not?

Selected Key Terms

acetylation *14.1*	metastasis *14.4*
apoptosis *14.4*	methylation *14.1*
Barr body *14.3*	mosaic tissue effect *14.3*
cancer *CI*	negative control system *14.1*
cell differentiation *CI*	oncogene *14.4*
checkpoint gene *14.4*	operator *14.2*
dosage compensation *14.3*	operon *14.2*
ecdysone *14.3*	phytochrome *14.3*
enhancer *14.1*	polytene chromosome *14.3*
gene control *CI*	positive control system *14.1*
growth factor *14.4*	promoter *14.1*
hormone *14.1*	regulatory protein *14.1*
ICE-like protease *14.4*	X chromosome
kinase *14.4*	inactivation *14.3*

Readings

Duke, R., D. Ojcius, and J. Ding-E Young. December 1996. "Cell Suicide in Health and Disease." *Scientific American*, 80–87.

Murray, A., and M. Kirschner. March 1991. "What Controls the Cell Cycle?" *Scientific American*, 264(3): 56–63.

Tijan, R. February 1995. "Molecular Machines That Control Genes." *Scientific American*, 54–61.

Travis, J. 5 August 2000. "Silence of the Xs." *Science News* 158: 92–94. Survey of recent investigations into the mechanism of X chromosome inactivation. Raises interesting questions about what has been called "junk DNA" (base sequences that have accumulated in DNA yet have no known function).

15

RECOMBINANT DNA AND GENETIC ENGINEERING

Mom, Dad, and Clogged Arteries

Butter! Bacon! Eggs! Ice cream! Cheesecake! Possibly you think of such foods as enticing, off-limits, or both. After all, who among us doesn't know about animal fats and the dreaded cholesterol?

Soon after you feast on these fatty foods, cholesterol enters the bloodstream. Cholesterol is important. It is a structural component of animal cell membranes, and without membranes, there would be no cells. Cells also remodel cholesterol into a variety of molecules, such as the vitamin D necessary for the development of good bones and teeth. Normally, however, the liver itself synthesizes enough cholesterol for your cells.

Some proteins circulating in blood combine with cholesterol and other substances to form lipoprotein particles. *High*-density lipoproteins, or *HDLs*, transport cholesterol to the liver, where it's metabolized. Usually, *low*-density lipoproteins, or *LDLs*, end up inside cells that store or use cholesterol. But sometimes too many LDLs form. The excess infiltrates the elastic walls of arteries and helps form atherosclerotic plaques (Figure 15.1). These abnormal masses interfere with blood flow. If they clog one of the tiny coronary arteries that deliver blood to the heart, a heart attack may result.

How well you handle dietary cholesterol depends on the genes you got from your parents. One of the genes specifies a protein receptor for LDLs. Inherit two "good" alleles and your blood level of cholesterol will stay low. Your arteries may never clog up, even with a high-fat diet. Inherit two copies of a certain mutated allele, and

you will develop *familial cholesterolemia*, a rare genetic disorder. Cholesterol levels get so high, many of those affected die of heart attacks in childhood or their teens.

In 1992 a woman from Quebec, Canada, became a milestone in the history of genetics. She was thirty years old. Like two brothers felled in their early twenties by heart attacks, she had inherited the mutant allele for the LDL receptor. She survived a heart attack at sixteen. At twenty-six, she had coronary bypass surgery.

At the time, people were hotly debating the risks and the promises of **gene therapy**—the transfer of one or more normal or modified genes into an individual's body cells to correct a genetic defect or boost resistance to disease. The woman opted for an untried procedure to give her body working copies of the good gene.

Surgeons removed about 15 percent of her liver. Researchers put cells from it in a nutrient-rich medium to promote cell growth and division. *And they spliced the functional allele for the LDL receptor into the genetic material of a harmless virus.* The modified virus infected the cultured liver cells, thereby inserting copies of the good gene into them. When the liver cells went on to reproduce, they made copies of the good gene, also.

About a billion modified cells were infused into the woman's portal vein, a blood vessel that leads directly to the liver. Some cells took up residence in the liver and started to make the missing cholesterol receptor. Two years later, a fraction of the woman's liver cells were behaving normally and sponging up cholesterol from blood. Her LDL blood levels declined nearly 20 percent. There was no sign of the arterial clogging that nearly killed her. Her cholesterol levels are still higher than normal. It's too soon to know whether her life will be prolonged. Yet the intervention gives strong evidence that gene therapy may be able to help some patients.

As you might gather from this pioneering clinical work, recombinant DNA technology has huge potential for medicine, agriculture, and industry. Think of it! For thousands of years, we humans have been changing

Figure 15.1 Life-threatening plaques (*bright yellow, white*) in one of the coronary arteries. These small arteries deliver blood to the heart. Abnormally high levels of cholesterol contribute to the formation of plaques that may clog them. Gene therapies based on recombinant DNA technology have the potential to counter many such threats to health.

Figure 15.2 One outcome of genetic manipulation by way of artificial selection practices: a large kernel from a modern strain of corn next to the tiny kernels of an ancestral species, which was recovered from a prehistoric cave in Mexico.

genetically based traits. By artificial selection practices, we produced new plants and new breeds of cattle, cats, dogs, and birds from wild ancestral stocks. We were selective agents for meatier turkeys, sweeter oranges, seedless watermelons, spectacular ornamental roses, and big juicy corn kernels (Figure 15.2). We conjured up hybrids such as the mule (horse × donkey) and plants that bear tangelos (tangerine × grapefruit).

Of course, we have to remember we're newcomers on the evolutionary stage. During the 3.8 billion years before we even made our entrance, nature conducted uncountable numbers of genetic experiments. Nature's tools have included mutation, crossing over, and other events that introduce changes in genetic messages. The countless changes gave rise to life's rich diversity.

But the striking thing about human-directed change is that the pace has picked up. Researchers analyze genes with **recombinant DNA technology**. They cut and recombine DNA from different species and insert it into bacterial, yeast, or mammalian cells. The cells replicate their DNA and divide rapidly. They copy the foreign DNA as if it were their own and churn out useful quantities of recombinant DNA molecules. The technology also is the basis of **genetic engineering**. By this process, genes are isolated, modified, and inserted into the same organism or into a different one. Protein products of the modified genes may cover the function of their missing or malfunctioning counterparts.

The new technology does not come without risks. With this chapter, we consider its basic aspects. At the chapter's end, we also address some ecological, social, and ethical questions related to its application.

Key Concepts

1. Genetic experiments have been proceeding for billions of years, through gene mutations, crossing over, genetic recombination, and other natural events.

2. Humans are now purposefully bringing about genetic changes by way of recombinant DNA technology. Such manipulations are called genetic engineering.

3. With this technology, researchers isolate, cut, and splice together gene regions from different species. Then they greatly amplify the number of copies of genes that interest them. The genes, and sometimes the proteins they specify, are produced in quantities that are large enough to use for research and for practical applications.

4. Three activities are at the heart of recombinant DNA technology. First, procedures based on specific types of enzymes are used to cut DNA molecules into fragments. Second, the fragments are inserted into cloning tools, such as plasmids. Third, the fragments containing the genes of interest are identified, then copied rapidly and repeatedly.

5. Genetic engineering involves isolating, modifying, and inserting genes back into the same organism or into a different one. The goal is to beneficially modify traits that the genes influence. Human gene therapy, which focuses on controlling or curing genetic disorders, is an example.

6. The new technology raises social, legal, ecological, and ethical questions regarding its benefits and risks.

A TOOLKIT FOR MAKING RECOMBINANT DNA

Restriction Enzymes

In the 1950s, the scientific community was agog over the discovery of DNA's structure. The excitement gave way to frustration. No one could figure out the sequence of nucleotides, the order of genes and gene regions along a chromosome. Robert Holley and his colleagues did sequence a small tRNA. They used digestive enzymes that broke the molecule into fragments small enough to be chemically characterized. But DNA molecules? These are far longer than RNAs. No one knew how to cut one into fragments long enough to have unique, and thus analyzable, sequences. Digestive enzymes can cut DNA, but not in any particular order. There was no telling how the fragments had been arranged in the molecule.

Then, by accident, Hamilton Smith discovered that *Haemophilus influenzae* chops up foreign DNA inserted into it by a bacteriophage. Extracts from *H. influenzae* cells included an enzyme that restricts itself to a specific kind of site in DNA. It was named a **restriction enzyme**.

In time, several hundred strains of bacteria offered up a toolkit of restriction enzymes that recognize and cut specific sequences of four to eight bases in DNA.

Table 15.1 lists a few. They are among many that make *staggered* cuts, which leave single-stranded "tails" on the end of a DNA fragment. Tails made by *Taq*I are two bases long (CG). *Eco*RI makes tails four bases long (AATT).

How many cuts do restriction enzymes make? That depends in part on the molecule. Example: The eight-base sequence recognized by *Not*I is rare in the DNA of mammals, so most fragments are tens of thousands of base pairs long. That's long enough for studying the organization of a **genome**—all of the DNA in a haploid number of chromosomes for a species. For the human species, the genome is about 3.2 billion base pairs long.

Modification Enzymes

DNA fragments with staggered cuts have sticky ends. "Sticky" means a restriction fragment's single-stranded tail can base-pair with a complementary tail of any other DNA fragment or molecule cut by the same restriction enzyme. *Mix DNA fragments cut by the same restriction enzyme, and the sticky ends of any two fragments having complementary base sequences will base-pair and form a recombinant DNA molecule:*

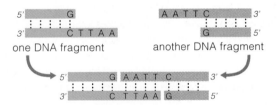

Notice the nicks where such fragments base-pair. **DNA ligase**, a modification enzyme, can seal these nicks:

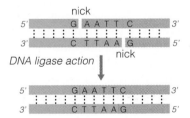

Cloning Vectors for Amplifying DNA

Restriction and modification enzymes make it possible to insert foreign DNA into bacterial cells. As you know, each bacterial cell has only one chromosome, a circular DNA molecule. Many also inherit plasmids. A **plasmid** is a very small circle of extra prokaryotic DNA that has just a few genes and that gets replicated along with the bacterial chromosome (Figure 15.3). Bacteria usually can survive without plasmids, but some of the genes offer benefits, as when they confer resistance to antibiotics.

Table 15.1	Examples of Restriction Enzymes	
Bacterial Source	Enzyme's Abbreviation	Its Specific Cut
Thermus aquaticus	*Taq*I	5′ T C G A 3′ 3′ A G C T 5′
Escherichia coli	*Eco*RI	5′ G A A T T C 3′ 3′ C T T A A G 5′
Nocardia otitidus caviarum	*Not*I	5′ G C G G C C G C 3′ 3′ C G C C G G C G 5′

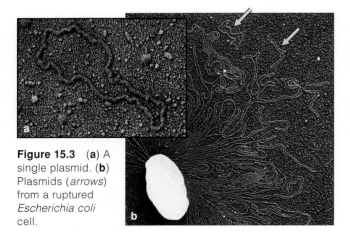

Figure 15.3 (a) A single plasmid. (b) Plasmids (*arrows*) from a ruptured *Escherichia coli* cell.

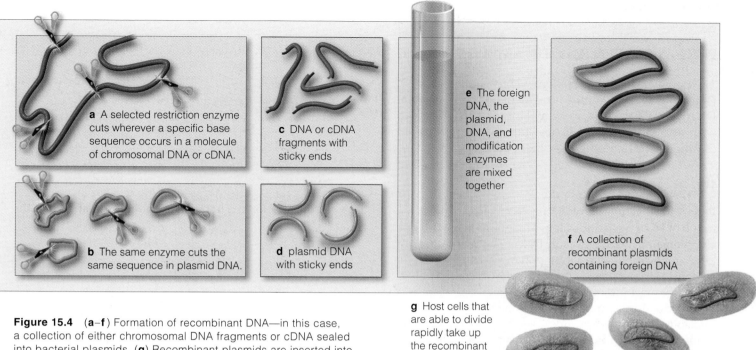

Figure 15.4 (**a–f**) Formation of recombinant DNA—in this case, a collection of either chromosomal DNA fragments or cDNA sealed into bacterial plasmids. (**g**) Recombinant plasmids are inserted into host cells that can rapidly amplify the foreign DNA of interest.

a A selected restriction enzyme cuts wherever a specific base sequence occurs in a molecule of chromosomal DNA or cDNA.

b The same enzyme cuts the same sequence in plasmid DNA.

c DNA or cDNA fragments with sticky ends

d plasmid DNA with sticky ends

e The foreign DNA, the plasmid, DNA, and modification enzymes are mixed together

f A collection of recombinant plasmids containing foreign DNA

g Host cells that are able to divide rapidly take up the recombinant plasmids

Under favorable conditions, bacteria divide rapidly and often—every thirty minutes, for some species—so that huge populations of genetically identical cells form. Before cell division, replication enzymes duplicate the bacterial chromosome. They also replicate the plasmid, sometimes repeatedly, so a cell can hold many identical copies of foreign DNA. In research laboratories, foreign DNA typically is inserted into a plasmid for replication. The outcome is called a **DNA clone**, because bacterial cells have made many identical, "cloned" copies of it.

A modified plasmid that can accept foreign DNA is a **cloning vector**. It can function as a taxi for delivering foreign DNA into a bacterium, yeast, or some other cell that can start a "cloning factory." This is a population of rapidly dividing descendant cells, all with identical copies of the foreign DNA (Figure 15.4).

Reverse Transcriptase To Make cDNA

Researchers analyze genes to unlock secrets about gene products and how they are put to use. Yet even when a host cell takes up a gene, it may not be able to make the protein. For instance, most eukaryotic genes have introns (noncoding sequences). mRNA transcripts of the genes can't be translated until the introns are snipped out and coding regions (exons) spliced together. Bacterial cells can't snip out the introns, so they often can't translate human genes into proteins.

Researchers get around this problem with **cDNA**, a DNA strand "copied" from a mature mRNA transcript. **Reverse transcriptase** catalyzes transcription in reverse. This enzyme from RNA viruses builds a complementary DNA strand on an mRNA transcript (Figure 15.5). Then

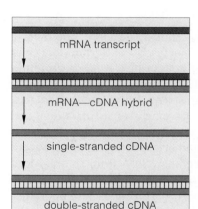

mRNA transcript

mRNA—cDNA hybrid

single-stranded cDNA

double-stranded cDNA

a Reverse transcriptase catalyzes the assembly of a single DNA strand on a mature mRNA transcript. The result is an mRNA–cDNA hybrid molecule.

b Enzymes remove the mRNA and use the cDNA strand as a template to assemble another DNA strand. The result is double-stranded cDNA, "copied" from an mRNA template.

Figure 15.5 Formation of cDNA from an mRNA transcript.

other enzymes remove the RNA from the mRNA–cDNA molecule and substitute a complementary DNA strand. The outcome is double-stranded cDNA.

Double-stranded cDNA can be further modified, as by attaching signals for transcription and translation. The modified cDNA can be inserted into a plasmid for amplification. Bacterial cells exposed to the recombinant plasmids often take them up and then use the cDNA instructions for synthesizing a protein of interest.

Restriction enzymes and modification enzymes cut apart chromosomal DNA or cDNA and splice it into plasmids and other cloning vectors. Recombinant plasmids can be taken up by bacteria or other cells that divide rapidly, thus making multiple, identical quantities of the foreign DNA.

PCR—A FASTER WAY TO AMPLIFY DNA

The polymerase chain reaction, widely known as **PCR**, is another way to amplify fragments of chromosomal DNA or cDNA. These copy-making reactions proceed with astounding speed in test tubes, not in bacterial cloning factories. Primers get them going.

What Are Primers?

Primers are synthetic, short nucleotide sequences, ten to thirty or so nucleotides long, that base-pair with any complementary sequences in DNA. The workhorses of DNA replication—the DNA polymerases—chemically recognize primers as START tags. Following a computer program, machines synthesize a primer one step at a time. How do researchers decide on the order of these nucleotides? First, they must identify short nucleotide sequences located just before and just after the DNA region from a cell that interests them. Then they build primers that have the complementary sequences.

What Are the Reaction Steps?

For PCR, the enzyme of choice is a DNA polymerase extracted from *Thermus aquaticus*, a bacterium that lives in superheated water of hot springs. It has been found also in water heaters. This enzyme is not destroyed at the elevated temperatures required to unwind a DNA double helix. Most DNA polymerases are denatured and permanently lose their activity at such temperatures.

Researchers mix together primers, the molecules of DNA polymerase and cellular DNA from an organism, and free nucleotides. Next, they expose the mixture to precise cycles of temperature. During each temperature cycle, the two strands of all the DNA molecules in the mixture unwind from each other.

Primers become positioned on exposed nucleotides at targeted sites according to base-pairing rules (Figure 15.6). Each round of reactions doubles the number of DNA molecules amplified from the targeted site. Thus, if there are 10 such molecules in a test tube, there soon will be 20, then 40, 80, 160, 320, 640, 1,280, and so on. Any targeted region from a single DNA molecule can be rapidly amplified into billions of molecules.

In short, *PCR amplifies samples that contain even tiny amounts of DNA.* At this writing, it is the amplification procedure of choice in thousands of laboratories all over the world. As you'll see in the next section, such samples can be obtained even from a single hair follicle or drop of blood left at the scene of a crime.

PCR is a method of amplifying chromosomal DNA or cDNA inside test tubes. Compared to cloning methods, PCR is far more rapid.

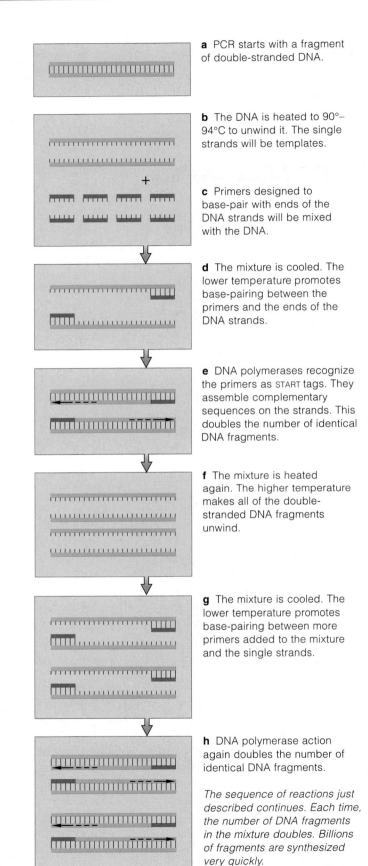

a PCR starts with a fragment of double-stranded DNA.

b The DNA is heated to 90°–94°C to unwind it. The single strands will be templates.

c Primers designed to base-pair with ends of the DNA strands will be mixed with the DNA.

d The mixture is cooled. The lower temperature promotes base-pairing between the primers and the ends of the DNA strands.

e DNA polymerases recognize the primers as START tags. They assemble complementary sequences on the strands. This doubles the number of identical DNA fragments.

f The mixture is heated again. The higher temperature makes all of the double-stranded DNA fragments unwind.

g The mixture is cooled. The lower temperature promotes base-pairing between more primers added to the mixture and the single strands.

h DNA polymerase action again doubles the number of identical DNA fragments.

The sequence of reactions just described continues. Each time, the number of DNA fragments in the mixture doubles. Billions of fragments are synthesized very quickly.

Figure 15.6 The polymerase chain reaction (PCR).

DNA Fingerprints

Except for identical twins, no two people have exactly the same base sequence in their DNA. By detecting the differences in sequences, scientists can distinguish one person from another. As you know, each human has a unique set of fingerprints, a marker of identity. Like all other sexually reproducing species, each human also has a **DNA fingerprint**, a unique array of DNA sequences that were inherited from parents in a Mendelian pattern. DNA fingerprints are so accurate that they can reveal differences even between full siblings.

More than 99 percent of the DNA in all humans is exactly the same. However, DNA fingerprinting focuses only on the portion that tends to be highly variable from one person to the next. Throughout the human genome are **tandem repeats**—short DNA sequences present in many copies, one after the other, in a chromosome.

For example, one individual's DNA might have the five bases TTTTC repeated four times in one location, yet another individual might have them repeated fifteen times in the same location. One might have five repeats of the bases CGG, and another might have fifty of them.

Repetitive DNA sequences lengthened through many generations. They slipped into the DNA as spontaneous mutations or as mistakes that proofreading and repair enzymes missed while the DNA was being replicated. The mutation rate is high in these regions (Section 13.4).

Researchers detect such differences at tandem-repeat sites with **gel electrophoresis**. This laboratory technique uses an electric field to force molecules through a viscous gel. In this case, it separates DNA fragments according to length. Size alone dictates how far each fragment will move through the gel. Tandem repeats of different sizes migrate at different rates.

A gel is immersed in a buffered solution, then DNA fragments from individuals are added to the gel. When an electric current is applied to the solution, one end of the gel takes on a negative charge, and the other end a positive charge. With their negatively charged phosphate groups, DNA fragments migrate through the gel toward the positively charged pole. They do so at different rates and separate into bands according to length. The smaller the fragment, the farther it migrates. After a set time, researchers identify fragments of different lengths by staining the gel or specifically highlighting fragments that contain tandem repeats.

DNA fingerprints help forensic scientists identify criminals, victims, and innocent suspects. A few drops of blood, semen, or cells from a hair follicle at a crime scene or on a suspect's clothing often yield enough to do the trick. For example, Figure 15.7 shows some tandem-repeat DNA fragments separated by gel electrophoresis. They are DNA fingerprints from seven people and from blood collected at a crime scene. Notice how much all but two of the DNA fingerprints differ. Can you identify which pattern exactly matches the pattern from the crime scene?

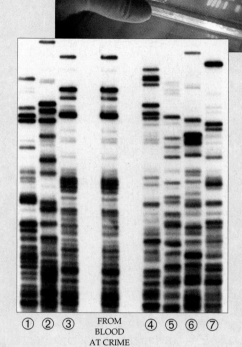

① ② ③ FROM BLOOD AT CRIME SCENE ④ ⑤ ⑥ ⑦

Figure 15.7 *Left:* Comparison of DNA fingerprints from a bloodstain discovered at a crime scene and from blood samples of seven suspects (the circled numbers). Given what you have learned about gel electrophoresis, point out which of the seven is a match. *Above:* The gel shown in this photograph had been stained earlier with a substance that can make DNA fragments fluoresce when placed under ultraviolet light.

DNA fingerprint analysis even confirmed that human bones exhumed from a shallow pit in Siberia belonged to five members of the Russian imperial family, all shot to death in secrecy in 1918.

When DNA fingerprinting was first used as evidence in court, attorneys challenged conclusions based on it. But DNA fingerprinting is now firmly established as an accurate, unambiguous procedure. Today it is routinely submitted as evidence in disputes over paternity. It is widely used to convict the guilty and exonerate innocent suspects. To give an early example, in 1998, an eleven-year-old girl was raped, stabbed, and strangled to death in Elisabethfehn, Germany. In the largest mass screening yet conducted, the police requested blood samples from 16,400 males in and around the vicinity. A single DNA fingerprint from a thirty-year-old mechanic matched the evidence, and he confessed to the crime.

The variation in tandem repeats also can be detected as restriction fragment length polymorphisms, or RFLPs (DNA fragments of different sizes that had been cleaved by restriction enzymes). In the case of tandem repeats, a restriction enzyme cleaves the DNA flanking the repeat. Alternatively, researchers might use PCR to amplify the tandem-repeat region. Either way, differences in the size of the fragments, which reveal genetic differences, can be detected with electrophoresis.

HOW IS DNA SEQUENCED?

In 1995, researchers accomplished what was little more than a dream a few decades ago. They determined the full DNA sequence for a species: *Haemophilus influenzae*, a bacterium that causes human respiratory infections. Since then, the genomes of other species have been fully sequenced. A draft sequence of coding regions of the human genome has been completed.

The molecular sleuths are using **automated DNA sequencing**. This laboratory method gives the sequence of cloned DNA or PCR-amplified DNA in a few hours.

Researchers use the four standard nucleotides (T, C, A, and G) for automated DNA sequencing. They also use four modified versions, which we can represent as T*, C*, A*, and G*. Each modified version is labeled with a molecule that fluoresces a certain color when it passes through a laser beam. Every time one of these modified nucleotides gets incorporated into a growing DNA strand, it arrests DNA synthesis.

Before the reactions, researchers mix the eight kinds of nucleotides together. They add millions of copies of the DNA to be sequenced along with a type of primer and molecules of DNA polymerase. Then they separate the DNA into single strands, and the reactions begin.

The primer binds with its complementary sequence on one of the strands. DNA polymerase synthesizes a new DNA strand right behind the primer. One by one, it adds nucleotides in the order dictated by the exposed sequence. Each time, one of the standard nucleotides *or* one of the modified versions may be attached.

Suppose that DNA polymerase encounters a T in a DNA template strand. It will catalyze the base-pairing of either A or A* to it. If A is added to the new strand, replication will continue. But if A* is added, replication will stop; the modified nucleotide will *block* addition of any more nucleotides to that strand. The same thing happens at each nucleotide in a template strand. When a standard nucleotide is attached, replication proceeds; when a modified version is attached, replication stops.

Remember, the starting mixture contained millions of identical copies of the DNA sequence. Because either a standard or a modified nucleotide could be added at every exposed base, the new strands end at different locations in the sequence. The mixture now contains millions of copies of tagged fragments having different lengths. These can now be separated by length into *sets* of fragments. And each set corresponds to only one of the nucleotides in the entire base sequence.

The automated DNA sequencer is a machine that separates the sets of fragments by gel electrophoresis. The set having the shortest fragments migrates fastest through the gel and peels away from it first. The last set to peel away has the longest fragments. Because its fragments have a modified nucleotide at the 3′ end, the set fluoresces a particular color as it passes through a

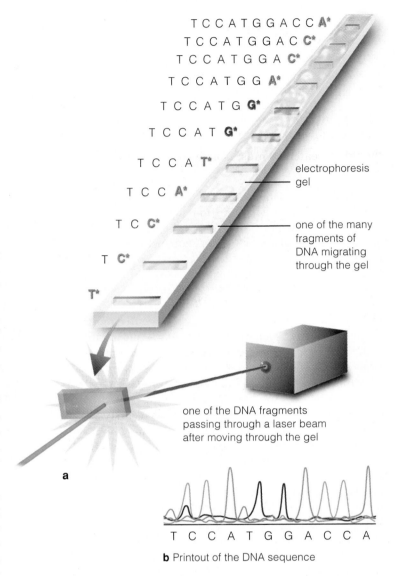

electrophoresis gel

one of the many fragments of DNA migrating through the gel

one of the DNA fragments passing through a laser beam after moving through the gel

a

T C C A T G G A C C A

b Printout of the DNA sequence

Figure 15.8 Automated DNA sequencing. (**a**) DNA fragments from an organism's genome become labeled at their end with a modified nucleotide that fluoresces a certain color. (**b**) Printout of the DNA sequence used in this example. Each peak indicates absorbance by a particular labeled nucleotide.

laser beam (Figure 15.8*a*). The automated sequencer detects the color and indicates which nucleotide is on the end of the fragments in each set. It assembles the information from all nucleotides in the sample, and in this way it reveals the entire DNA sequence.

Figure 15.8*b* shows the printout from an automated DNA sequencer. Each peak in the tracing represents the detection of a particular color as the sets of fragments reached the end of the gel.

With automated DNA sequencing, the order of nucleotides in a cloned or amplified DNA fragment can be determined.

FROM HAYSTACKS TO NEEDLES—ISOLATING GENES OF INTEREST

Any genome consists of thousands of genes. *E. coli* has about 4,000, for instance; humans apparently have about 30,000. What if you wanted to learn about or modify the structure of any one of those genes? First you'd have to isolate that gene from all others in the genome.

There are several ways to do this. If part of the gene sequence is already known, you can design primers to amplify the entire gene or part of it by PCR. Often, though, researchers must isolate and clone a gene. First they make a **gene library**. This is a mixed collection of bacteria that contain different cloned DNA fragments. One is the gene of interest. A *genomic* library contains cloned DNA fragments from an entire genome. A *cDNA* library contains DNA derived from mRNA. Often it is the most useful because it is free of introns. But the gene is still hidden like a needle in a haystack. How can you isolate it? One way is to use a nucleic acid probe.

What Are Probes?

A **probe** is a very short stretch of DNA labeled with a radioisotope so that it can be distinguished from other DNA molecules in a given sample. Part of the probe must be able to base-pair with some portion of the gene of interest. Any base-pairing that takes place between sequences of DNA (or RNA) from different sources is called **nucleic acid hybridization**.

How do you acquire a suitable probe? Sometimes part of the gene or a closely related gene has already been cloned, in which case it can be used as the probe. If the gene's structure is a mystery, you still may be able to work backward from the amino acid sequence of its protein product, assuming that protein is already available. By using the genetic code as a guide (Section 13.2), you could build a DNA probe that is more or less similar to the gene of interest.

Screening for Genes

Once you have a gene library and a suitable probe, you are ready to hunt down the gene. Figure 15.9 shows the steps of one isolation method. The first step is to take bacterial cells of the library and spread them apart on the surface of a gelled growth medium in a petri plate. When spread out sufficiently, individual cells undergo division. Each cell starts a colony of genetically identical cells. The bacterial colonies appear as hundreds of tiny white spots on the surface of a culture medium.

After colonies appear, you lay a nylon filter on top of the colonies. Some cells stick to the filter at locations that mirror the locations of the original colonies. You use solutions to rupture the cells, and the DNA so released sticks to the filter. You make the DNA unwind to single strands and then add the probes. The probes hybridize

a Bacterial colonies, each derived from a single cell, grow on a culture plate. Each colony is about 1 millimeter across.

b A nitrocellulose or nylon filter is placed on the plate. Some cells of each colony adhere to it. The filter mirrors how the colonies are distributed on the culture plate.

c The filter is lifted off and put into a solution. The cells stuck to it lyse, but the cellular DNA sticks to the filter.

d Also, the DNA is denatured to single strands at each site. A radioactively labeled probe is added to the filter. It binds to DNA fragments that have a complementary base sequence.

e The probe's location is identified by exposing the filter to x-ray film. The image that forms on the film reveals the colony that has the gene of interest.

Figure 15.9 Use of a probe to identify bacterial colonies that have taken up a gene library.

only with DNA from the colony that took up the gene of interest. If you expose the probe-hybridized DNA to x-ray film, the pattern formed by the radioactivity will identify that colony. With this information you can now culture cells from that colony alone, knowing that it will be the only one that can replicate the cloned gene.

Probes may be used to identify one particular gene among many in gene libraries. Bacterial colonies that have taken up the library can be cultured to isolate the gene.

USING THE GENETIC SCRIPTS

As researchers decoded the genetic scripts of species, they opened the door to astonishing possibilities. For example, genetically engineered bacteria now produce medically valued proteins. Huge bacterial populations produce useful quantities of the desired gene products in stainless steel vats. *Diabetics* who must receive insulin injections every day are among the beneficiaries. At one time, medical supplies of this pancreatic hormone were extracted only from pigs and cattle. Later on, synthetic genes for human insulin were transferred into *E. coli* cells. The cells were the start of populations that became the first large-scale, cost-effective bacterial factory for proteins. Besides insulin, human somatotropin, blood-clotting factors, hemoglobin, interferon, and a variety of drugs and vaccines are also manufactured with the help of genetically engineered bacteria.

Other kinds of modified bacteria hold potential for industry and for cleaning up environmental messes. For instance, as you know, many microorganisms can break down organic wastes and help cycle nutrients through ecosystems. Certain modified bacteria can break down crude oil into less harmful compounds. When sprayed onto oil spills, as from a shipwrecked supertanker, they might help avert an environmental disaster. Others are genetically engineered to sponge up excess phosphates or heavy metals from the environment.

In addition, bacterial species that contain plasmids offer benefits for basic research, agriculture, and gene therapy. For instance, deciphering the messages encoded in bacterial genes helps us reconstruct the evolutionary history of life. Sections 18.6 and 18.7 will give you an idea of how comparisons of DNA or RNA from different organisms can reveal evolutionary secrets. The next two sections give a few more examples of benefits.

What about the "bad" bunch—the pathogenic fungi, bacteria, and viruses? Natural selection favors mutated genes that improve a pathogen's chances of evading a host organism's natural defenses. Mutation is frequent among pathogens that reproduce rapidly. So designing new antibiotics and other defenses against new gene products is a constant challenge. But if we learn about the genes, we may have advance warning of the "plan of attack." The story of HIV, a rapidly mutating virus that causes AIDS, gives insight into the magnitude of the problem (Section 35.10).

Knowing about genes allows us to genetically engineer beneficial microorganisms for uses in medicine, industry, agriculture, and environmental remediation.

Knowing about genes can help us reconstruct evolutionary histories of species.

Knowing about genes may help us design effective, timely counterattacks against rapidly mutating pathogens.

DESIGNER PLANTS

Regenerating Plants From Cultured Cells

Many years ago, Frederick Steward and his coworkers cultured cells from carrot plants. They induced some of them to develop into small embryos. As you will read in Section 28.5, some embryos grew into whole plants. Today, researchers routinely regenerate crop plants and many other plant species from cultured cells. They use ingenious methods to pinpoint a gene in a culture that may contain even millions of cells. Suppose they add a pathogen's toxin to a culture medium. Suppose a few cells happen to carry a gene that confers resistance to the toxin. Those cells will be the only ones to survive.

Whole plants regenerated from preselected cultured cells can be hybridized with other varieties of plants. In this way, desired genes might be transferred to a plant lineage to improve herbicide resistance, pest resistance, and other traits that can benefit crops and gardens.

Consider this larger aspect of the research: The food supply for most of the human population is vulnerable. Generally, farmers prefer genetically similar varieties of high-yield plants and have discarded the more diverse, older varieties. But genetic uniformity makes food crops vulnerable to many kinds of pathogenic fungi, viruses, and bacteria. That is why botanists comb the world for seeds of the wild ancestors of potatoes, corn, and other crop plants. They send their prizes—seeds with genes of a plant's lineage—to **seed banks**. These safe storage facilities are designed to preserve genetic diversity.

The problem is huge. Example: In 1970 a new fungal strain of *Southern corn leaf blight* destroyed much of the United States corn crop. All of the plants carried the same gene that conferred susceptibility to the disease. Ever since that devastating epidemic, seed companies have been much more attentive to offering genetically diverse corn seeds. And where does the diversity come from? Plant breeders call on the seed banks.

How Are Genes Transferred Into Plants?

The Ti (Tumor-inducing) plasmid from *Agrobacterium tumefaciens* is one vehicle for inserting new or modified genes into plants. *A. tumefaciens* infects many kinds of flowering plants (Figure 15.10). Some Ti plasmid genes

crown gall tumor

Figure 15.10 Crown gall tumor on a woody plant, an abnormal tissue growth triggered by a gene in the Ti plasmid.

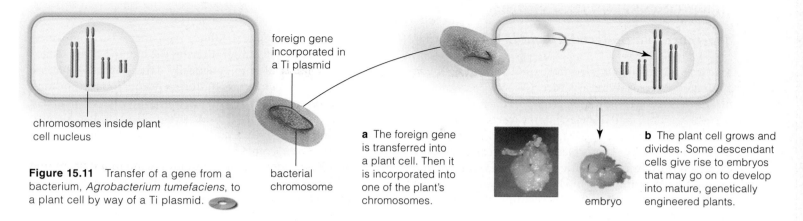

Figure 15.11 Transfer of a gene from a bacterium, *Agrobacterium tumefaciens*, to a plant cell by way of a Ti plasmid.

chromosomes inside plant cell nucleus

foreign gene incorporated in a Ti plasmid

bacterial chromosome

a The foreign gene is transferred into a plant cell. Then it is incorporated into one of the plant's chromosomes.

b The plant cell grows and divides. Some descendant cells give rise to embryos that may go on to develop into mature, genetically engineered plants.

embryo

invade a plant's DNA, then induce the formation of abnormal tissue masses. We call the masses crown gall tumors. Before introducing the plasmid into plant cells, researchers remove the tumor-inducing genes and then insert a desired gene into it. They place plant cells in a culture of the modified bacteria. When some cells take up the gene, whole plants may be regenerated from the cellular descendants, as in Figure 15.11. The expression of foreign genes in plants can sometimes be dramatic, as in Figure 15.12.

In nature, *A. tumefaciens* infects plants called dicots, which include beans, peas, potatoes and other vital crops. Geneticists are working to modify the bacterium so that it might also deliver genes into monocots that are vital food crops, including wheat, corn, and rice. Some use electric shocks or chemicals to deliver modified genes into plant cells. Some also blast microscopic particles coated with DNA into them.

Despite many obstacles, improved varieties of crop plants have been engineered or are in the works. For example, genetically engineered cotton plants display resistance to a herbicide (Figure 15.13). Farmers spray the herbicide in fields of modified cotton plants to kill weeds. The plants are not affected.

Also on the horizon are engineered plants that can be factories for pharmaceuticals. A few years ago, for example, tobacco plants that were engineered to produce hemoglobin and some other proteins were planted in a test field in North Carolina. Afterward, ecologists found

Figure 15.12 A modified plant that glows in the dark. A firefly gene was inserted into the plant's DNA and is being expressed. (Refer to Section 6.9, *Critical Thinking* question 4). The gene's product is luciferase, an enzyme with a role in bioluminescence.

Figure 15.13 (**a**) Control plant (*left*) and three genetically engineered aspen seedlings. Vincent Chiang and coworkers suppressed a regulatory gene involved in a lignin biosynthetic pathway. The modified plants synthesized normal lignin, but not as much. Lignin production decreased by as much as 45 percent—yet cellulose production increased 15 percent. Root, stem, and leaf growth were greatly enhanced. Plant structure did not suffer. Wood harvested from such trees might make it easier to manufacture paper and some clean-burning fuels, such as ethanol. (Lignin, a tough polymer, strengthens the secondary cell walls of plants. Before paper can be made from wood, the lignin must be chemically extracted.)

(**b**) *Left:* Cotton plant, used as the control. *Right:* Genetically engineered cotton plant with a gene for herbicide resistance. Both plants were sprayed with a weedkiller that now has widespread application in cotton fields.

no trace of foreign genes or proteins in the soil or in other plants or animals in the neighborhood. Another example: In a Stanford University laboratory, mustard plants synthesized biodegradable plastic beads that are suitable for use in manufacturing plastics.

Genetically diverse plant species are essential to protecting our vulnerable food supply. Genetic engineers can design plants with new beneficial traits.

GENE TRANSFERS IN ANIMALS

Supermice and Biotech Barnyards

The first mammals enlisted for experiments in genetic engineering were laboratory mice. Consider an example of this work. R. Hammer, R. Palmiter, and R. Brinster corrected a hormone deficiency that leads to dwarfism in mice. Insufficient levels of somatotropin (or growth hormone) cause the abnormality. The researchers used a microneedle to inject the gene for rat somatotropin into fertilized mouse eggs, which they implanted in an adult female. The gene was successfully integrated into mouse DNA, and the eggs developed into mice. Young mice in which that foreign gene was expressed grew 1–1/2 times larger than their dwarf littermates. In other experiments, researchers transferred the gene for human somatotropin into a mouse embryo. The gene became integrated into mouse DNA, and the modified embryo grew up to be a "supermouse" (Figure 15.14).

Today, human genes are being inserted into mouse embryos as part of ongoing research into the molecular basis of genetic disorders, such as Alzheimer's disease. In addition to microneedles, microscopic laser beams are used to penetrate cultured cells.

Animals of "biotech barnyards" are competing with bacterial factories as genetically engineered sources of proteins. For example, goats produce CFTR protein (for treating cystic fibrosis) and TPA (which may lessen the severity of a heart attack). Cattle may produce human collagen for repairing cartilage, bone, and skin.

In 1997 researchers reported the first case of cloning a mammal from an adult cell. They extracted a nucleus from a ewe's mammary gland cell and inserted it into an egg (from another ewe) from which the nucleus had been removed. Signals from the egg cytoplasm sparked the development of an embryo, which was implanted into a surrogate mother. As Section 12.5 describes, this resulted in *Dolly*, a clone that was herself cloned later on.

Researchers had been cloning animals from embryonic tissue for years, but not adults that already displayed some sought-after trait. Now, if they could refine cloning processes, they might be able to maintain a genotype indefinitely.

Not long after the experiment that resulted in Dolly, genetically engineered clones of mice, cattle, and other animals were being produced. As an example, Steve Stice and his colleagues employed similar cloning methods to make designer cattle. They started with a culture of cells from cattle. They induced the development of six genetically identical calves. And they also engineered specific, targeted alterations in the cloning cell lineage.

Stice's goal is to genetically engineer cattle that are resistant to bovine spongiform encephalopathy, or mad cow disease (Section 20.5). People who eat infected meat also are vulnerable to the disease. Stice's technique also is used to put a human serum albumin gene into dairy cow chromosomes. We use albumin to control blood pressure. At present, this protein must be separated from large quantities of donated human blood. Imagine how much easier it would be to obtain large quantities of the protein from bountiful supplies of milk.

Given how rapidly the technologies are developing, is human cloning not far behind? And what about the possibility of genetically engineering the clones? Here you may wish to reflect once more on Section 12.5.

Mapping and Using the Human Genome

In the early 1980s, biologists were in an uproar over a costly proposal before the National Institutes of Health (NIH) to map the entire human genome. Many argued that benefits would be incalculable—not only for pure research, but also for efforts to cure genetic disorders and combat diseases. Others saw it as a boondoggle; they said such mapping couldn't be done and would divert NIH funding from worthy endeavors. But then Leroy Hood, a molecular biologist, invented automated DNA sequencing (Section 15.4), and the race was on.

By 1990, the Human Genome Initiative was under way. The international effort came with a projected price tag of 3 billion dollars—about 92 cents for each base pair in the heritable script for human life. For the first phase, biologists were to pinpoint the location of the stretches of nucleotides that were already sequenced. Once those "reference markers" were assigned to the chromosomes, biologists could fill in the long spans between them.

By 1991, not even 2,000 genes had been sequenced. But the pace picked up after Craig Venter realized that even a tiny bit of cDNA could be used as a molecular hook to drag the whole sequence of its parent gene out of a

Figure 15.14 Evidence of a successful gene transfer. Two ten-week-old mouse littermates. *Left:* This one weighs 29 grams. *Right:* This one weighs 44 grams. It grew from a fertilized egg into which a gene for human somatotropin had been inserted.

cDNA library. He named the hooks ESTs (for *Expressed Sequence Tags*). By tapping commercial sources, Ventner and his colleague, Mark Adams, put together a cDNA library that was produced from various neurons, the communication cells of nervous systems. Within a few months, the group had sequenced 2,000 more genes. By 1994, many thousands of genes had been decoded.

Later, Ventner and Hamilton Smith used a software program, the TIGR Assembler, to work out the first full genome of an organism—the bacterium *H. influenzae* you read about in Section 15.1. Through application of advanced technologies, a rough draft of the human genome was completed ahead of schedule. Noncoding portions, which are extensive, are still being analyzed.

Even when the whole human genome is sequenced, it will not be easy to manipulate it to advantage. Put yourself in the position of some genetic researcher. You have modified a human gene and you have to figure out how to get it into a host cell of a particular tissue. That's not all. You have to be sure the modified gene gets inserted at a suitable site—say, at or very near the locus for its abnormal counterpart in a chromosome. You must also make sure the cell will synthesize the specified protein at suitable times, in suitable amounts.

Today, most experimenters employ stripped-down viruses as vectors to put genes into cultured human cells. They know that cells can incorporate foreign genes into their DNA. But the viral genetic material can undergo rearrangements, deletions, and other alterations, which can shut down or disrupt gene expression. What about developing synthetic, streamlined versions of human chromosomes? Maybe the cellular machinery for DNA replication and protein synthesis will work on them.

Certain gene therapies simply deliver genetically modified cells into a tissue. This may help, even if the cells only make 10 to 15 percent of a required protein. The chapter introduction is a case in point. But no one can yet predict where the new genes will end up. The danger is that a gene insertion will disrupt the function of some other gene, including the genes controlling cell growth and cell division. One-for-one gene swaps by a process called homologous recombination are possible. Oliver Smithies, one of the best at this, currently puts genes right where they are supposed to go, but only once every 100,000 or so tries. The body may be a bit forgiving of a small load of cells bearing poor swaps. But mistakes in germ cells—which give rise to all cells of a new individual—probably would have unexpected and disastrous effects on the body's development.

Although the technical details are still being worked out, modified genes are being transferred into cells of humans and other mammals during experimental and clinical trials.

Who Gets Enhanced?

This chapter opened with a historic, inspiring case, the first proof that gene therapy may help save human lives. It closes with questions that invite you to consider some social and ethical issues surrounding the application of recombinant DNA technology to the human genome.

To most of us, human gene therapy to correct genetic abnormalities seems like a socially acceptable goal. Let's take this idea one step further. Is it also socially desirable or acceptable to change certain genes of a normal human individual (or sperm or egg) to alter or enhance traits?

The idea of selecting desirable human traits is called *eugenic engineering*. Yet who decides which forms of a trait are most "desirable"? For example, would it be okay to engineer taller or blue-eyed or fair-skinned boys and girls? Would it be okay to engineer "superhumans" with amazing strength or intelligence? Are there some people narcissistic enough to commission a clone of themselves, one with a few genetically engineered "enhancements"?

Recently, more than 40 percent of Americans surveyed say gene therapy would be okay to make their offspring smarter or better looking. A poll of British parents found 18 percent willing to use genetic enhancement to prevent children from being aggressive and 10 percent to keep them from growing up to be homosexual.

There are those who say the DNA of any organism must never be altered. Put aside the fact that nature itself alters DNA much of the time and has done so for nearly all of life's history. The concern is that *we* do not have the wisdom to bring about beneficial changes without causing great harm to ourselves or to the environment. To be sure, when it comes to altering human genes, one is reminded of our very human tendency to leap before we look. And yet, when it comes to restricting genetic modifications of any sort, one also is reminded of another old saying: "If God had wanted us to fly, he would have given us wings." Something about the human experience did give us a capacity to imagine wings of our own making— and that capacity carried us to the frontiers of space.

Where are we going from here? To gain perspective on the question, spend some time reading the history of the human species. It is a history of survival in the face of challenges, threats, bumblings, and sometimes disasters on a grand scale. It is also a story of our connectedness with the environment and with one another.

The basic questions confronting you are these: Should we be more cautious, believing the risk takers may go too far? And what do we as a species stand to lose if risks are not taken? At this writing, a young man with a genetic disorder died after gene therapy that used an adenovirus as a vector. Yet two *SCID-X1* infants with a mutant gene that disabled their immune system may be cured. A viral vector put copies of the nonmutated gene into stem cells taken from their bone marrow. Months after the modified cells were reinfused into their bone marrow, both "bubble boys" left their isolation tents. Their immune system is working normally; they are healthy and living at home.

SAFETY ISSUES

Many years have passed since foreign DNA was first transferred into a plasmid. That gene transfer ignited a debate that will continue well into the next century. The issue is this: *Do potential benefits of gene modifications and gene transfers outweigh potential dangers?*

Genetically engineered bacteria are "designed" so they cannot survive except in the laboratory. As added precautions, "fail-safe" genes are built into the foreign DNA in case they escape. These genes are silent unless the captives are exposed to environmental conditions— whereupon the genes get activated, with lethal results for their owner. Say the package includes a *hok* gene next to a promoter of the lactose operon (Section 14.2). Sugars are plentiful in the environment. If they were to activate the *hok* gene, the protein product of that gene would destroy membrane function and the wayward cell.

What about a worst-case scenario? Remember how retroviruses are used to insert genes into cultured cells? If they escape from laboratory isolation, what might be the consequences? Example: Check out Section 42.8.

And what about genetically engineered plants and animals released into the environment? For example, Steven Lindlow thought about how frost destroys many crops. Knowing that a surface protein of a bacterium promotes formation of ice crystals, he excised the "ice-forming" gene from bacterial cells. As he hypothesized, spraying "ice-minus bacteria" on strawberry plants in an isolated field prior to a frost would help plants resist freezing. He actually had deleted a *harmful* gene from a species, yet a bitter legal battle ensued. The courts ruled in his favor. His coworkers sprayed a strawberry patch; nothing bad happened.

Then there was the potato plant designed to kill the insects that attack it. It also was too toxic for people to eat. Or think of how crop plants compete poorly with weeds for nutrients. Many have been designed to resist weedkillers so farmers can spray for weeds and not worry about killing their crops. Herbicides do have toxic effects on more than their targets (Section 3.7). If crop plants offer herbicide resistance, will farmers be less or more apt to spread them about?

And what if engineered plants or animals transfer modified genes to organisms in the wild? Think of how the advantage in acquiring resources would tilt toward vigorous weeds blessed with herbicide-resistant genes. Such possibilities are why standards for rigorous and extended safety tests must be in place *before* modified organisms enter the environment.

For more on safety issues, turn to *Critical Thinking* question 7 at the chapter's end.

Rigorous safety tests are carried out before genetically modified organisms are released into the environment.

SUMMARY — Gold indicates text section

1. Uncountable numbers of gene mutations and other forms of genetic "experiments" have been proceeding in nature for at least 3 billion years. *CI*

2. Through artificial selection practices, humans have manipulated the genetic character of many species for many thousands of years. Currently, recombinant DNA technology has enormously expanded our capacity to genetically modify organisms. *CI*

3. In genetic engineering, specific genes are modified and inserted into the same organism or a different one. In gene therapy, copies of normal or modified genes are inserted into individuals in order to correct a genetic defect or boost resistance to disease. *CI*

4. A genome is all the DNA in the haploid chromosome number for a species. By recombinant DNA technology, a genome can be cut into fragments, then the fragments can be amplified (copied over and over again) to make useful quantities that permit analysis of the nucleotide sequence of the genome or a specific portion of it. *15.1*

5. Researchers work with chromosomal DNA or with cDNA (a DNA strand transcribed by the enzyme reverse transcriptase from a mature mRNA transcript). The first choice is better for questions about DNA regions that control gene expression or that contain introns. cDNA is a better choice for questions about the amino acid sequence of a protein. Either way, researchers use restriction enzymes to cut DNA into fragments. *15.1*

6. The use of plasmids or other cloning vectors is one way to amplify DNA fragments. Many bacteria contain plasmids: small circles of DNA with a few genes in addition to those of the bacterial chromosome. *15.1*

a. Certain restriction enzymes make staggered cuts that leave the fragments with single-stranded tails. Such tails base-pair with complementary tails of any other DNA cut by the same enzyme, such as plasmid DNA.

b. DNA ligase, a modification enzyme, seals base-pairing sites between plasmid DNA and foreign DNA. A plasmid modified to accept foreign DNA is a cloning vector; it can deliver foreign DNA into a bacterium or some other cell that can start a population of rapidly dividing descendant cells. All of the descendants have identical copies of the foreign DNA. Collectively, all of the identical copies are a DNA clone.

7. Currently, the polymerase chain reaction (PCR) is the fastest way to amplify fragments of chromosomal DNA or cDNA. Bacterial factories are not needed; the reactions occur in test tubes. The reactions use a supply of nucleotide building blocks and primers: synthetic, short nucleotide sequences that will base-pair with any complementary DNA sequence and that enzymes (DNA polymerases) recognize to start replication. Each round of replication doubles the number of fragments. *15.2*

8. For sexually reproducing species, no two individuals have exactly the same DNA base sequence (except for identical twins). When restriction enzymes are used to cut an individual's DNA, the result is a unique array of restriction fragments called a DNA fingerprint. *15.3*

a. The DNA in all humans is more than 99 percent identical except at tandem repeats (short stretches of repeated base sequences, such as TTTC). The number and combination of tandem repeats is unique in each individual and can be detected by gel electrophoresis.

b. DNA fingerprinting has uses in forensic science and in resolving paternity suits.

9. Automated DNA sequencing can rapidly reveal the base sequence of cloned DNA or PCR-amplified DNA fragments. It labels fragments with one of four modifed nucleotides, then separates them according to length by gel electrophoresis. Each label fluoresces a certain color under a laser beam. A machine reads each as it peels off the gel and assembles the whole sequence. *15.4*

10. A gene library is a mixed collection of bacterial cells that took up different cloned DNA or cDNA fragments. A gene may be isolated from the library with use of a probe, a very short stretch of radioactively labeled DNA that is known or suspected to be similar to or identical with part of that gene and can base-pair with it. A base pairing between nucleotide sequences from different sources is called nucleic acid hybridization. *15.5*

11. Generally, recombinant DNA technology and genetic engineering have potential for research and applications in medicine and agriculture, in the home and industry. As with any new technology, the potential benefits must be weighed against potential risks, including ecological and social repercussions. *CI, 15.6–15.10*

Review Questions

1. Distinguish between recombinant DNA technology and genetic engineering. *CI*

2. Distinguish these terms from one another: *15.1*
   a. chromosomal (genomic) DNA and cDNA
   b. cloning vector and DNA clone

3. Define PCR. Can fragments of chromosomal DNA, cDNA, or both be amplified by PCR? *15.2*

4. Define DNA fingerprinting. Briefly describe which portions of the DNA are used in DNA fingerprinting. *15.3*

5. Outline the steps of automated DNA sequencing. *15.4*

6. Define cDNA library, then briefly explain how a gene can be isolated from it. Define probe and nucleic acid hybridization as part of your answer. *15.5*

7. Give three examples of applications that can be derived from knowledge of an organism's genome. *CI, 15.6–15.8*

8. Name one of the ways in which modified genes have been inserted into mammalian cells. *15.8*

9. Define gene therapy. Once the human genome has been fully sequenced, why will it be difficult to manipulate its genes to advantage? *CI, 15.8*

Self-Quiz ANSWERS IN APPENDIX III

1. _____ is the transfer of normal genes into body cells to correct a genetic defect.
   a. Reverse transcription c. Gene mutation
   b. Nucleic acid hybridization d. Gene therapy

2. DNA fragments result when _____ cut DNA molecules at specific sites.
   a. DNA polymerases c. restriction enzymes
   b. DNA probes d. RFLPs

3. _____ are small circles of bacterial DNA that are separate from the circular bacterial chromosome.

4. Foreign DNA that was inserted into a plasmid and then replicated many times in a population of bacteria is a _____ .
   a. DNA clone c. DNA probe
   b. gene library d. gene map

5. By reverse transcription, _____ is assembled on _____ .
   a. mRNA; DNA c. DNA; enzymes
   b. cDNA; mRNA d. DNA; agar

6. PCR stands for _____ .
   a. polymerase chain reaction
   b. polyploid chromosome restrictions
   c. polygraphed criminal rating
   d. politically correct research

7. By gel electrophoresis, fragments of a gene library can be separated according to _____ .
   a. shape b. length c. species

8. Automated DNA sequencing relies on _____ .
   a. supplies of standard and labeled nucleotides
   b. primers and DNA polymerases
   c. gel electrophoresis and a laser beam
   d. all of the above

9. Match the terms with the most suitable description.
   ____ DNA fingerprint
   ____ Ti plasmid
   ____ nature's genetic experiments
   ____ nucleic acid hybridization
   ____ Human Genome Initiative
   ____ eugenic engineering

   a. selecting "desirable" traits
   b. deciphering 3.2 billion base pairs of 23 human chromosomes
   c. used in some gene transfers
   d. unique array of DNA fragments inherited in Mendelian pattern from each of two parents
   e. base pairing of nucleotide sequences from different DNA or RNA sources
   f. mutations, crossovers

Critical Thinking

1. In the following diagram, which restriction enzyme made the cuts (indicated in *red*) in part of a DNA molecule from two different organisms?

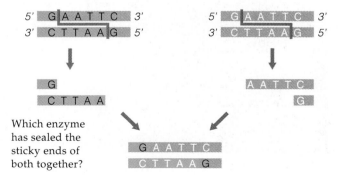

Which enzyme has sealed the sticky ends of both together?

2. Ryan, a forensic scientist, obtained a very small bit of DNA from material at a crime scene. In order to examine the sample by DNA fingerprinting, he must amplify the sample by PCR, the polymerase chain reaction. By his estimate, there are 50,000 copies of the DNA in his sample. Calculate the number of copies Ryan will have after fifteen cycles of PCR.

3. A game warden in Africa confiscated eight ivory tusks from elephants. Some tissue is still attached to the tusks. Now she must determine whether the tusks were taken illegally from northern populations of endangered elephants or from other populations of elephants to the south that can be hunted legally. How can she use DNA fingerprinting to find the answer?

4. The Human Genome Initiative is undergoing completion, and knowledge about a number of the newly discovered genes is already being used to detect genetic disorders. Ask yourself: What will be done with genetic information about individuals? Will insurance companies and potential employers request it? At this writing, many women have already refused to take advantage of genetic screening for a gene associated with the development of breast cancer. Should medical records about people participating in genetic research and genetic clinical services be made available to other individuals? If not, how could such information be protected?

5. Lunardi's Market put out a bin of tomatoes having splendid vine-ripened redness, flavor, and texture. The sign posted above the bin identified them as genetically engineered produce. Most shoppers selected unmodified tomatoes in the adjacent bin even though those tomatoes were pale pink, mealy-textured, and tasteless. Which ones would you pick? Why?

6. Avoiding genetically engineered food is probably impossible in the United States. At least 45 percent of cotton, 38 percent of soybean, and 25 percent of corn crops have been engineered to withstand weedkillers or to make their own pesticides. For years, modified corn and soybeans have found their way into breakfast cereals, tofu, soy sauce, vegetable oils, beer, soft drinks, and other food products. They are fed to farm animals.

By contrast, public resistance to genetically engineered food is high in Europe, especially Britain. Many people, including the Prince of Wales, speak out against what the tabloids call "Frankenfood." Protesters routinely vandalize crops (Figure 15.15). Worries abound that such foods may be more toxic, have lower nutritional value, and promote natural selection for antibiotic resistance. Some people worry that designer plants will cross-pollinate with wild plants to produce "superweeds."

Biotechnologists envision a new Green Revolution. They argue that designer plants can hold down food production costs, reduce dependence on pesticides and herbicides, enhance crop yields, and offer improved flavor, nutritional value, even salt tolerance and drought tolerance. Yet the chorus of critics in Europe may provoke a trade war with the United States. The issue is not small potatoes, so to speak. In 1998, the value of

agricultural exports reached about 50 billion dollars. Flattery, threats, and bullying are rampant on both sides of the Atlantic. Restrictions on genetic engineering will profoundly impact United States agriculture, and inevitably the impact will trickle down to what you eat and how much you pay for it.

All of which invites you to read up on scientific research related to this issue and form your own opinions. The alternatives are to be swayed either by media hype (the term Frankenfood, for instance) or by biased reports from groups (such as chemical manufacturers) with their own agendas.

Possibly start with Christopher Bond's article "Politics, Misinformation, and Biotechnology" in the 18 February 2000 *Science* (287:1201). Bond argues that biotechnology attempts to solve real-world problems of sickness, hunger, and dwindling resources; and that "hysteria and unworkable propositions advanced by those who can afford to take their next meal for granted have little currency among those who are hungry."

7. What if it were possible to create life in test tubes? This is the question behind attempts to model and eventually create *minimal organisms*, which we define as living cells having the smallest set of genes necessary to survive and reproduce.

As recent experiments by Craig Ventner and Claire Fraser revealed, *Mycoplasma genitalium*, a bacterium with 517 genes (and 2,209 transposons) might be a candidate. By disabling its genes one at a time in the laboratory, they discovered that it may have only 265–350 essential protein-coding genes.

What if those genes were to be synthesized one at a time and inserted into an engineered cell consisting only of a plasma membrane and cytoplasm? Would the cell come to life? The possibility that it might prompted Ventner and Fraser to seek advice from a panel of bioethicists and theologians. As Arthur Caplan, a bioethicist at the University of Pennsylvania reported, no one on the panel objected to synthetic life research. They felt that much good might come of it, provided scientists didn't claim to have found "the secret of life." The 10 December 1999 issue of *Science* includes an essay from the panel and an article on *M. genitalium* research. Read both, then write down your thoughts about creating life in a test tube.

Selected Key Terms

automated DNA sequencing *15.4*	genome *15.1*
cDNA *15.1*	nucleic acid hybridization *15.5*
cloning vector *15.1*	PCR *15.2*
DNA clone *15.1*	plasmid *15.1*
DNA fingerprint *15.3*	primer *15.2*
DNA ligase *15.1*	probe (nucleic acid) *15.5*
gel electrophoresis *15.3*	recombinant DNA technology *CI*
gene library *15.5*	restriction enzyme *15.1*
gene therapy *CI*	reverse transcriptase *15.1*
genetic engineering *CI*	seed bank *15.7*
	tandem repeats *15.3*

Readings

Cho, M., et al. 10 December 1999. "Ethical Considerations in Synthesizing a Minimal Genome." *Science* 286: 2087–2090.

Pennisi, E. June 2000. "Finally, The Book of Life and Instructions for Navigating It." *Science* 288: 5475.

Watson, J. D., et al. 1992. *Recombinant DNA*. Second edition. New York: Scientific American Books.

"The Human Genome." *Science* 291: 5507. The entire 16 February 2001 issue consolidates current understandings and questions.

Figure 15.15 Activists ripping some genetically modified crop plants from a field in Great Britain.

On-Line readings at Student Guide for InfoTrac: www.brookscolecom/biology

III Principles of Evolution

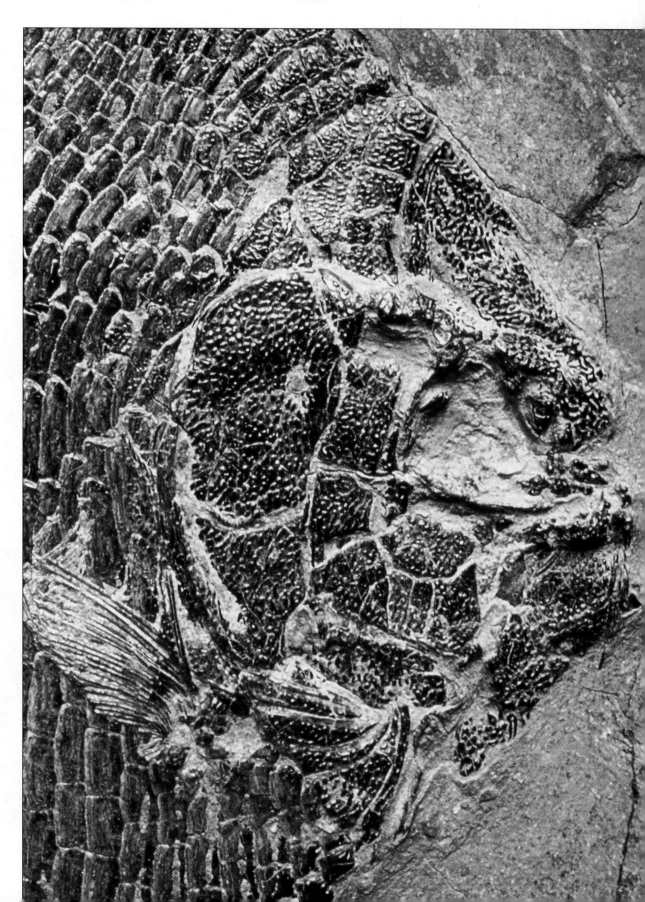

Millions of years ago, a bony fish died, and sediments gradually buried it. Today its fossilized remains are studied as one more piece of the evolutionary puzzle.

MICROEVOLUTION

Designer Dogs

We humans have tinkered rather ruthlessly with the modern descendants of a long and distinguished lineage. The lineage originated some 40 million years ago with the appearance of tree-dwelling carnivores that looked rather like small weasels. Their descendants evolved along separate branchings that now include weasels, badgers, otters, skunks, bears, pandas, raccoons —and dogs.

About 50,000 years ago, we began domesticating wild dogs. No doubt the advantages of doing so were important. Times were tough in the days before police protection and supermarkets. Dogs welcomed to the hearth could guard people and their possessions. They could corner, kill, eat, and thus dispose of big rats and other unwelcome vermin.

By 14,000 years ago, we started to develop different varieties (breeds) through artificial selection. Individual dogs having desired forms of traits were selected from each new litter and, later, encouraged to breed. Those having undesired forms of traits were passed over.

After favoring the pick of the litter over hundreds or thousands of generations, we ended up with sheep-herding border collies, badger-hunting dachshunds, wily retrievers, and sled-pulling huskies. And at some point we began to delight in the odd, extraordinary dog. Imagine! In practically no time at all, evolutionarily speaking, we picked our way through the pool of variant dog alleles and came up with such extremes as Great Danes and chihuahuas (Figure 16.1).

Sometimes our canine designs exceeded the limits of biological common sense. For example, how long would a tiny, finicky-eating, nearly hairless, nearly defenseless chihuahua last in the wild? Not long. What about the English bulldog, bred for a very short snout and a compressed face? Long ago, breeders thought these particular traits would allow the dogs to get a better grip on the nose of a bull. (Why they wanted dogs to bite bulls is a story in itself.) So now the roof of the bulldog mouth is ridiculously wide and often flabby, so bulldogs have trouble breathing. Sometimes they get so short of air they pass out.

Through our centuries-old fascination with artificial selection, we produced thousands of varieties of crop plants, ornamental plants, cats, cattle, and birds as well as dogs. With the currently available technologies of genetic engineering, we are now mixing the genes of many different species and producing incredible new varieties, including tobacco plants that produce useful quantities of hemoglobin, and mustard plants that produce plastic.

So, when you hear someone wonder about whether "evolution" takes place, remind yourself that **evolution** simply means genetic change in a line of descent over the generations. Selective breeding practices provide abundant, tangible evidence that heritable changes do, indeed, occur. The actual mechanisms that bring about change are the focus of this chapter. Later chapters will explain their role in the evolution of new species from parent species.

Figure 16.1 Two designer dogs. About 50,000 years ago, humans began domesticating wild dogs. From that ancestral stock, artificial selection resulted in startling diversity among rather closely related breeds, such as the Great Dane (*legs, left*) and the chihuahua (*possibly fearful of being stepped on, right*).

Key Concepts

1. From the fifteenth century onward, explorers found puzzling differences in the world distribution of species. Anatomists identified similarities and differences in the body structure and patterning among the embryos and adult forms of major groups of animals. In sequences of sedimentary rocks, geologists found sequences of fossils. The findings were suggestive of biological evolution, or heritable changes in lines of descent over time.

2. Charles Darwin came of age at a time when scholars were attempting to reconcile the new findings from global explorations, comparative morphology, and geology with prevailing cultural beliefs. He, and Alfred Wallace, came up with a theory of evolution by natural selection to explain the findings. Their evolutionary theory and the theories of others start with variation in traits.

3. All individuals of a population generally have the same number and kinds of genes. Those genes give rise to an array of traits that characterize the population.

4. Mutations may result in two or more slightly different molecular forms of a gene—alleles—that influence a trait in different ways. Individuals of a population vary in the details of a trait when they inherit different combinations of alleles.

5. Any allele may become more or less common in the population relative to other kinds at a gene locus, or it may disappear. *Microevolution* refers to changes in the allele frequencies of a population over time.

6. Allele frequencies change as an outcome of mutation, gene flow, genetic drift, and natural selection. Mutation alone produces new alleles. Gene flow, genetic drift, and natural selection shuffle existing alleles into, through, or out of populations.

EARLY BELIEFS, CONFOUNDING DISCOVERIES

The Great Chain of Being

Our story begins more than two thousand years ago, when the seeds of biological inquiry were starting to take hold among the ancient Greeks. At the time, popular belief held that supernatural beings intervened directly and often in human affairs. As an example, people "knew" that angry gods inflicted a common ailment known as the sacred disease. And yet, from one of the physicians of the school of Hippocrates, these thoughts come down to us:

It seems to me the disease called sacred . . . has a natural cause, as other diseases have. Men think it divine merely because they do not understand it. But if they called everything divine that they did not understand, there would be no end of divine things! . . . If you watch these fellows treating the disease, you see them use all kinds of incantations and magic—but they are also careful in regulating diet. Now if food makes the disease better or worse, how can they say it is the gods who do this? . . . It does not really matter whether you call such things divine or not. In Nature, all things are alike in this, in that they can be traced to preceding causes.

— On The Sacred Disease (400 B.C.)

He perceived that all aspects of nature have underlying causes—a key premise of modern science (Section 1.5).

Aristotle was foremost among the early naturalists, and he described the world around him in great detail. He had no reference books or instruments to guide him, for biological science in the Western world began with the great thinkers of this age. Yet here was a man who was no mere collector of random tidbits of information. In his thoughtful descriptions of nature we see evidence of a mind perceiving connections between observations and attempting to explain the order of things.

Aristotle believed (as did others) that each kind of organism was distinct from all the rest. Nevertheless, he wondered about organisms with traits that seemed to blur such distinctions. For example, some sponges look very much like plants but do not make their own food, as plants do. They capture and digest it, as animals do. In time, Aristotle came to view nature as a continuum of organization, from lifeless matter through complex forms of plant and animal life.

By the fourteenth century, Aristotle's idea had been transformed into a rigid view of life. A Chain of Being was seen to extend from the "lowest" forms of life to humans, then on up to spiritual beings. Each kind of being, or "species" as it was called, was a separate link in the great chain. All links were designed and forged at the same time, at the same center of creation, and had not changed since then. Scholars thought that once they discovered, named, and described all of the links, the meaning of life would be revealed to them.

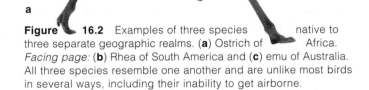

a

Figure 16.2 Examples of three species native to three separate geographic realms. (**a**) Ostrich of Africa. *Facing page:* (**b**) Rhea of South America and (**c**) emu of Australia. All three species resemble one another and are unlike most birds in several ways, including their inability to get airborne.

Questions From Biogeography

Until the fifteenth century, naturalists were not aware that the world is a great deal bigger than Europe, so the task of locating and describing all species seemed manageable. Then globe-spanning explorations began. Naturalists were soon overwhelmed by descriptions of tens of thousands of plants and animals that explorers were discovering in Asia, Africa, the Pacific Islands, and the New World.

In 1590 the naturalist Thomas Moufet attempted to sort through the bewildering array. He simply gave up and wrote such gems as this description of locusts and grasshoppers: "Some are green, some black, some blue. Some fly with one pair of wings; others with more; those that have no wings they leap; those that cannot fly or leap they walk . . . Some there are that sing, others are silent." It was not a work of subtle distinctions.

Even so, a few scholars began to examine the world distribution of organisms, a discipline now known as **biogeography** (Section 18.3 and Chapter 44). They soon realized that many plants and animals are unique to isolated places, such as remote oceanic islands. They also were aware that certain species separated by great distances resemble one another (Figure 16.2).

How did so many species get from one center of creation to oceanic islands and other isolated, remote locations? And what do the similarities and differences among them mean?

Questions From Comparative Morphology

By the eighteenth century, many scholars were engaged in **comparative morphology**, the systematic study of similarities and differences in the body plans between major groups, such as different kinds of vertebrates. Think of the bones of a human arm, whale flipper, and

bat wing. They differ in size, shape, and function. Yet all have similar locations in the body. They consist of the same tissues, arranged in the same overall patterns. They develop in similar ways in embryos. Comparative morphologists who deduced all of this wondered: Why are some animals that are so different in some features so much alike in others?

By one hypothesis, basic body plans were so perfect there was no need to come up with a new one for each organism at the time of their creation. Yet if that were so, then how could there be parts with no function? For example, some snakes have bones that correspond to a pelvic girdle—a set of bones to which hind legs attach (Figure 16.3). Snakes don't have legs. Why the bones? Humans have bones that correspond to a few tailbones of many other mammals, but they don't have a tail. Why do they have parts of one? (After reading Chapter 18, you may have a good idea.)

Questions About Fossils

From the late 1600s on, geologists added to the growing confusion. They began mapping layers of rocks at sites where erosion or quarrying had cut deep into the earth. They found similar layers around the world. (Such beds consist of distinct, multiple layers of sedimentary rock. Section 18.3 shows an example.) Most agreed that sand and other sediments had been deposited at different times and had slowly compacted into rock layers. They realized that certain layers contained distinctive **fossils**, which were thought to be stone-hard evidence of life in ancient times.

For example, deep layers contained fossils of simple marine organisms. Some fossils in the overlying layers were similar but structurally more complex. The fossils in layers above these closely resembled living marine organisms. What did increasing structural complexity among fossils of a given type represent? Were these *sequences* of fossils, separated in time? Another puzzle: many fossils were unique in some traits yet similar to

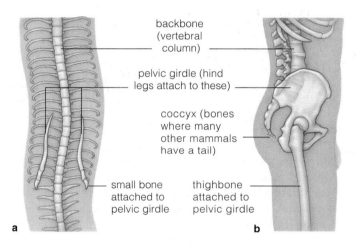

Figure 16.3 (**a**) Python bones corresponding to the pelvic girdle of other vertebrates, including humans (**b**). Small "hind limbs" protrude through the skin on the underside of the snake.

certain existing species in other traits! Could species so similar yet so far apart in time be *related* in some way?

Taken as a whole, the findings from biogeography, comparative morphology, and geology did not fit with prevailing beliefs. Georges-Louis Leclerc de Buffon and a few others started to formulate novel hypotheses. If dispersal of all species from a center of creation was not possible, given the vast oceans and other barriers, *then perhaps species had originated in more than one place.* And if they were not created in a perfect state—and fossil sequences and the presence of "useless" body parts in certain organisms suggested they were not—*then perhaps species had been modified over time.* Awareness of change in lines of descent—evolution—was in the wind.

Awareness of biological evolution emerged over centuries, through the cumulative observations of many naturalists, biogeographers, comparative anatomists, and geologists.

A FLURRY OF NEW THEORIES

Squeezing New Evidence Into Old Beliefs

In the nineteenth century, naturalists tried to reconcile the growing evidence of change in lines of descent with a traditional conceptual framework that did not allow for change. Foremost among them was Georges Cuvier, a respected anatomist. For years he had compared body plans of fossils and living organisms. He acknowledged the abrupt changes in the fossil record, corresponding to discontinuities between some layers of sedimentary beds. Was the record evidence of changing populations of organisms that lived in those ancient environments? Cuvier thought so. And he was right, as you will see from the evolutionary story in Chapters 18 and 19.

Based on that inference, Cuvier constructed his own theory of catastrophism. There was one time of creation, he said, that populated the world with every species. A global catastrophe destroyed many of them. Survivors repopulated the world. It was not that survivors were *new* species; naturalists simply hadn't yet found fossils of them that would date to the time of creation. Later catastrophes wiped out more species, and repopulation by the survivors followed, as recorded by fossils.

Many scholars accepted his theory, but others kept at the puzzle. One hypothesis—inheritance of *acquired* characteristics—was pushed by Jean Lamarck. During each individual's life, thought Lamarck, environmental pressures and internal "needs" bring about permanent changes in body form and functioning, then offspring inherit the necessary changes. And so life, created long ago in a simple state, gradually improved. The force for change was a drive toward perfection, up the Chain of Being. The drive was centered in nerves that directed an unknown "fluida" to body parts in need of change.

Apply his hypothesis to modern giraffes. Say they had a short-necked ancestor. Pressed by a need to find food, it kept stretching its neck to browse upon leaves beyond the reach of other animals. Stretching directed fluida to its neck, which lengthened permanently. The longer neck was inherited by offspring, which stretched their necks, also. So generations of animals desiring to reach ever loftier leaves led to the modern giraffe.

As Lamarck correctly inferred, the environment *is* a factor in evolution. His hypothesis, however, like others proposed at the time, has not been supported by tests carried out since then. There is no evidence that the environment modifies traits of all existing individuals in a way that can be passed on to offspring.

Voyage of the Beagle

In 1831, in the midst of this confusion, Charles Darwin was twenty-two years old and wondering what to do with his life. Ever since he was eight, he had wanted to

Figure 16.4 (**a**) Charles Darwin and (**b**) a blue-footed booby, one of many species he observed during his five-year voyage around the world on the *Beagle*. (**c**) A replica of the ship, sailing off the coast of South America. On this trip, Darwin also ventured into the Andes. He observed fossils of marine organisms in rock layers 3.6 kilometers above sea level. (**d**,**e**) The Galápagos Islands are isolated in the ocean far to the west of Ecuador. They arose through volcanic action about 5 million years ago, so species could not have originated there. Winds or ocean currents must have carried them to the new islands.

hunt, fish, collect shells, or simply watch insects and birds—anything but sit in school. Later, at his father's insistence, he did try to study medicine in college. The crude, painful procedures used on patients at that time sickened him. His by-now exasperated father urged him to become a clergyman, so he packed for Cambridge. His grades were good enough to earn him a degree in theology. But he spent most of his time among faculty members with leanings toward natural history.

John Henslow, a botanist, perceived Darwin's real interests. He arranged for Darwin to function as ship's naturalist aboard H.M.S. *Beagle*. The *Beagle* was about to embark on a five-year voyage that would take Darwin around the world (Figure 16.4). Abruptly, the young man who hated schoolwork and had no formal training as a naturalist started to work with enthusiasm.

The *Beagle* sailed first to South America to complete work on mapping the long coastline. During the Atlantic crossing, Darwin collected and studied marine life. He read Henslow's parting gift, the first volume of Charles Lyell's *Principles of Geology*. During stops at the coast and at islands, he saw diverse species in environments

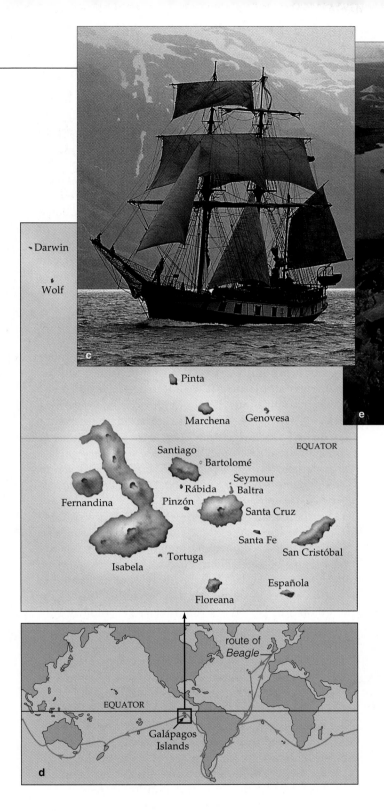

Darwin

Wolf

Pinta

Marchena   Genovesa

EQUATOR

Santiago

Bartolomé

Seymour
Rábida   Baltra

Fernandina   Pinzón

Santa Cruz

Santa Fe

Tortuga   San Cristóbal

Isabela

Española

Floreana

route of
Beagle

EQUATOR

Galápagos
Islands

d

ranging from sandy shores to high mountains. And he started circling the question of evolving life, which was now on the minds of many respected individuals.

Darwin started mulling over a rather radical theory that Lyell was advancing in his book. Lyell and other geologists were arguing that catastrophes had no more and possibly less effect on Earth history than did subtle processes of change. They had thought about how long it takes for rains, the pounding surf, and other forces of nature to sculpt the landscape. For years geologists had

chipped away at layers of sandstones, limestones, and other rocks, which form after sediments erode from the land and accumulate in the beds of rivers and seas. They thought about how sedimentary beds often consist of a number of stacked layers. If, they hypothesized, the deposition took place as gradually in the past as it did in their own era, then surely it took many millions of years—not a few thousand—for such thick stacks to form. They even managed to incorporate earthquakes and other infrequent events into their view of Earth history. After all, major floods, more than a hundred great earthquakes, and twenty or so volcanic eruptions typically occur each year, so catastrophes aren't unusual.

Their view of gradual, uniformly repetitive change became the **theory of uniformity**. It directly challenged prevailing views of the age of the Earth.

The theory bothered scholars who firmly believed the Earth was only about 6,000 years old. They thought people had recorded everything that happened during those thousands of years, and in all that time no one ever mentioned seeing a species evolve. Yet, by Lyell's calculations, it must have taken millions of years to mold the present landscape. *Wasn't that enough time for species to evolve in diverse ways?* Later, Darwin thought so. But exactly *how* did they evolve? He would end up devoting the rest of his life to that burning question.

Prevailing beliefs can influence how we interpret clues to natural processes and their observable outcomes.

Darwin's observations during a global voyage helped him think about natural processes in a novel way.

DARWIN'S THEORY TAKES FORM

Old Bones and Armadillos

After Darwin returned to England in 1836, he talked with other naturalists about possible evidence that life evolves. By carefully studying all of the notes from his journey, he came up with some possibilities.

In Argentina, for example, he had observed fossils of glyptodonts, which are now extinct. Of all animals on Earth, only living armadillos are like glyptodonts (Figure 16.5). And of all places on Earth, armadillos live only in the same places where glyptodonts once lived. If the two kinds of animals had been created at the same time, lived in the same place, and were so much alike, why is only one still alive? Wouldn't it be reasonable to assume glyptodonts were early relatives of armadillos? Many of their shared traits might have been retained through countless generations. Other traits might have been modified in the armadillo branch of a family tree. Descent with modification—it seemed possible. What, then, could be the driving force for evolution?

A Key Insight—Variation in Traits

While Darwin assessed his notes, an influential essay by Thomas Malthus, a clergyman and economist, made him look closely at a topic of social interest. Malthus had correlated population size with famine, disease, and war. Humans, he claimed, run out of food, living space, and

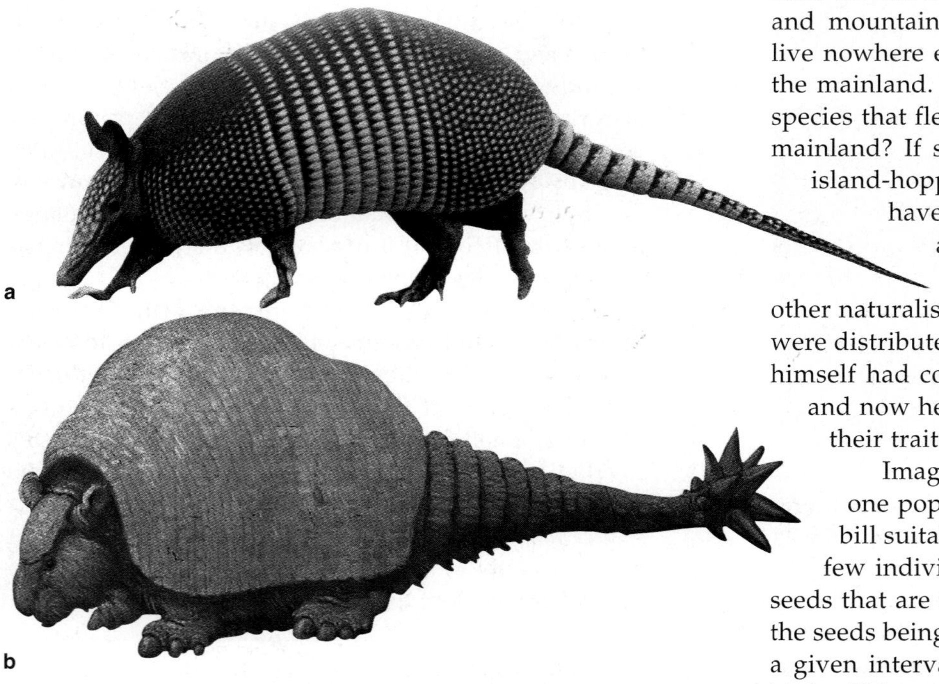

Figure 16.5 (**a**) An armadillo. (**b**) A glyptodont, long extinct. Although separated in time, these animals share unusual traits and a restricted distribution. To Darwin, they were a clue that helped him develop a theory of evolution by natural selection.

other resources because they reproduce too much. The larger a population gets, the more people there are to reproduce in each generation. Population size burgeons, resources dwindle, and the struggle to live intensifies. Many people starve, get sick, and engage in war and other forms of competition for remaining resources.

After Darwin reflected on his personal observations, he suspected that any population has the capacity to produce more individuals than the environment is able to support. To give one example, a single sea star can release 2,500,000 eggs per year, but the seas obviously do not fill up with sea stars. Nearly all of the eggs and larvae of each generation end up inside the bellies of predators. Many of the survivors starve or succumb to disease or some other environmental assault.

Assume that the environment restricts the number of reproducing individuals. Which individuals will be the winners and losers? What influences the outcome? Darwin thought about the populations he had observed during his voyage. As he recalled, individuals were not alike down to the last detail. They showed variations in size, coloration, and other traits. *And it dawned on him that variations in traits might affect an individual's ability to secure resources—and therefore to survive and reproduce in particular environments.*

Did the Galápagos Islands show evidence of this? Between these volcanic islands and the South American coastline are 900 kilometers of open ocean. The islands offer diverse habitats along their rocky shores, deserts, and mountain flanks. Nearly all of their inhabitants live nowhere else—although they resemble species on the mainland. Were these remote islands colonized by species that flew, floated, or were blown over from the mainland? If so, then in the different island habitats, island-hopping descendants of the colonizers might have undergone modifications, over time, as adaptive responses to local conditions.

As Darwin eventually learned from other naturalists in England, thirteen species of finches were distributed throughout the Galápagos Islands. He himself had collected specimens of some of the birds, and now he attempted to correlate the variations in their traits with environmental challenges.

Imagine yourself in his place. The finches of one population, you notice, have a large, strong bill suitable for cracking seeds (Figure 16.6). Yet a few individuals with a slightly stronger bill crack seeds that are too tough for their neighbors. If most of the seeds being produced in a particular habitat during a given interval have hard coats, then a strong-billed bird will have a competitive edge. It will have a better chance than the other birds of surviving and producing offspring. Assuming the trait is heritable, the same will be true of that bird's strong-billed descendants.

Figure 16.6 Four species of finches that live on the Galápagos Islands. (**a**) *Geospiza conirostris* and (**b**) *G. scandens*, both with a bill adapted for eating cactus flowers and fruits. Other finches have thick, strong bills that crush cactus seeds. (**c**) *Certhidea olivacea*, a tree-dwelling finch, resembles warblers in song and behavior. It uses its slender beak to probe for insects. (**d**) *Camarhynchus pallidus* feeds on wood-boring insects such as termites. It has learned to break cactus spines and twigs to suitable lengths, and then hold the "tools" and use them to probe bark for insects hidden from its view.

Figure 16.7 Alfred Wallace, who studied in Malaysia. Wallace worked out a theory of natural selection long after Darwin did but was first to report it. He quickly circulated a brief letter that described his views about the process, making it "his" theory.

Take these thoughts one step further. If factors in the environment continue to "select" the most adaptive version of a trait, then the population will become one of mostly strong-billed birds. *And a population is evolving if forms of heritable traits are changing over the generations.*

Recall, from Section 1.4, that Darwin offered pigeon breeding and other *artificial* selection practices as a way to explain *natural* selection of traits in the wild. When breeders favor pigeons with, say, black-feathered tails,

they encourage black-tailed pigeons of each generation to mate but not white-tailed ones. It was an easy way to show how selection could lead to an increase in the frequency of one form of a trait in a captive population.

Anticipating that his view would be controversial, Darwin waited to announce it and searched for flaws in his reasoning. He waited too long. More than a decade after he wrote but did not publish his theory, another respected naturalist—Alfred Wallace—sent him a letter (Figure 16.7). Wallace had developed the same theory but was the first to circulate a short letter about it! Like Wallace, most scholars thought Darwin should get most of the credit and prevailed on him to formally present a paper at the same time Wallace presented his. The next year, in 1859, Darwin published his detailed evidence in support of the theory, *On the Origin of Species*.

Although you may have heard that Darwin's book fanned an intellectual firestorm, the idea that diversity is the product of evolution was accepted almost at once by most naturalists. But Darwin's specific explanation, of gradual evolution by natural selection, was fiercely debated. Nearly seventy years passed before advances in a new field, genetics, led to widespread acceptance of his explanation. Until that happened, people generally associated Darwin's name mainly with the premise that life evolves—something others had suggested before him.

DARWIN'S THEORY OF EVOLUTION BY NATURAL SELECTION. A population can evolve (change over time) when individuals differ in one or more heritable traits that are responsible for differences in the ability to survive and reproduce.

INDIVIDUALS DON'T EVOLVE—POPULATIONS DO

Examples of Variation in Populations

As Charles Darwin perceived, *individuals do not evolve; populations do*. By definition, a **population** is a group of individuals of the same species that are occupying a given area. To understand how a population evolves, start with variation in the features of its individuals.

Certain features characterize every population. All members share the same body plan, as when jays have two wings, feathers, three toes forward and one toe back, and so on. These are *morphological* traits (*morpho–*, meaning form). Cells and body parts of all members work much the same way during metabolism, growth, and reproduction. These are *physiological* traits, relating to how the body functions in its environment. And the members respond the same way to basic stimuli, as when babies instinctively imitate adult facial expressions. These are *behavioral* traits.

Especially in sexually reproducing species, details of most traits vary among individuals. Pigeon feathers or snail shells vary in patterns or colors within a population (Section 1.4 and Figure 16.8). Some individuals of a frog population may be more sensitive to winter cold or better at attracting a mate than others. Humans differ in the distribution, color, texture, and amount of hair. These examples merely hint at the stunning variation in populations; almost every trait of every species is variable.

Many traits, such as those of Gregor Mendel's pea plants, vary in *qualitatively different* ways. They come in two or more distinct forms (or morphs) in a population, as when feathers are yellow or white. Such qualitative variation is known as **polymorphism**. Other traits, such as eye color and height, show continuous, *quantitatively different* variation. That is, individuals of a population show small, incremental differences in traits that may be quantified, as Section 10.7 describes.

The "Gene Pool"

Information about heritable traits occurs in genes, which are specific regions of DNA molecules. In general, all individuals of a population have the same number and kinds of genes. We say "in general" because males and females of sexually reproducing populations differ in a number of genes on the sex chromosomes.

Think of all the genes in the entire population as a **gene pool**—a pool of genetic resources that, in theory at least, is shared by all members of a population and then their offspring, the next generation. Each kind of gene in that pool is most often present in two or more slightly different molecular forms, called **alleles**.

Figure 16.8 *Facing page:* From different Caribbean islands, variation in shell color and banding patterns in populations of the same snail species. Different individuals carry different alleles for the genes that specify most traits. The shells differ because their owners had different combinations of alleles at particular gene locations along their chromosomes.

Individuals have different combinations of alleles. This leads to variations in phenotype—differences in the details of traits. For example, whether your hair is black, brown, red, or blond depends on which alleles of certain genes you inherited from your two parents. And don't forget: *Offspring inherit genes, not phenotypes.* Environmental conditions often modify expression of a gene (Section 10.8). But phenotypic variation resulting from their effects last no longer than the individual.

Which alleles end up in a given gamete and later in the new individual? Five events shape the outcome, as described in earlier chapters and summarized here:

1. Gene mutation (produces new alleles)

2. Crossing over at meiosis I (results in novel combinations of alleles in chromosomes)

3. Independent assortment at meiosis I (puts mixes of maternal and paternal chromosomes in gametes)

4. Fertilization (combines alleles from two parents)

5. Change in chromosome number or structure (leads to the loss, duplication, or repositioning of genes)

Of all the events just listed, only mutation *creates* new alleles. The other four only shuffle *existing* alleles into different combinations—but what a shuffle! Consider this: Each human gamete inherits one of 10^{600} possible combinations of alleles. Not even 10^{10} humans are alive today. Unless you are an identical twin, it is extremely unlikely that any other person with your precise genetic makeup has ever lived, or ever will.

Stability and Change in Allele Frequencies

Imagine yourself in a big garden in summer, watching butterflies flitting about. They look the same except in wing color. A few have white wings; more have blue. Perhaps, you muse, the "blue" allele is more common than the "white" allele. By using genetic analysis, you could identify the **allele frequencies**, the abundance of each kind of allele in the population. You could track the rate of genetic change over the generations.

Suppose you start with the "Hardy–Weinberg rule," as given in Section 16.5, as a theoretical reference point for measuring patterns of change. At that point, called **genetic equilibrium**, frequencies of alleles at a given gene locus remain stable one generation after the

you know, gene interactions underlie the phenotypes of complex organisms. A mutation that has severe effects on phenotype typically leads to death; it is a **lethal mutation**.

By comparison, a **neutral mutation** does not help *or* harm an individual. Natural selection can neither increase nor decrease the frequency of neutral mutations in the population, for these do not have any discernible effect on the individual's chances of surviving a n d reproducing. For example, if you carry a mutant gene that resulted in attached earlobes instead of detached ones, this alone should not stop you from surviving and reproducing just as well as anybody else.

next. The population is *not* evolving in terms of that gene, for five conditions are being met. First, no gene mutations have occurred. Second, that population is very large. Third, it is isolated from other populations of the species. Fourth, the gene has no effect at all on survival or reproduction. Fifth, all mating is random.

Rarely, if ever, do all five conditions prevail at the same time in nature. Gene mutation is infrequent but inevitable. Three processes—*natural selection, gene flow,* and *genetic drift*—may drive a population away from genetic equilibrium within even a few generations. The term **microevolution** refers to small-scale changes in allele frequencies brought about by mutation, natural selection, gene flow, and genetic drift.

Every so often, a mutation bestows an advantage. For instance, a product of a mutant gene that affects growth might make a corn plant grow larger or faster and so give it the best access to sunlight and nutrients. Or maybe a neutral mutation turns out to be beneficial after conditions in the environment change. Even if the advantage is small, chance events or natural selection might preserve the mutant gene in the DNA and favor its representation in the next generation.

Mutations are so rare, they usually have little or no immediate effect on the population's allele frequencies. But beneficial mutations, and neutral ones, have been accumulating in lineages for billions of years. Through all that time, they have represented the raw material for evolutionary change—the basis for the staggering range of biological diversity, past and present.

In the evolutionary view, then, the reason you don't look like a bacterium or an avocado or an earthworm or even neighbors down the street began with different mutations that originated at different times in the past, in different lines of descent.

Mutations Revisited

Mutations, remember, are heritable changes in DNA that usually give rise to altered gene products. They are the only source of new alleles. We cannot predict exactly when or in which individual they will appear. Yet each gene has its own **mutation rate**, which is the probability of its mutating during or in between DNA replications. On average, the rate is between 10^{-5} and 10^{-6} per gene locus per gamete, each generation. In a single reproductive season, about 1 gamete in 100,000 to 1,000,000 has a new mutation at any given locus.

Mutations may give rise to structural, functional, or behavioral alterations that can diminish an individual's chances of surviving and reproducing. Even one small biochemical change can have devastating effects.

For instance, skin, bones, tendons, and many other vertebrate organs cannot develop without collagen, a structural protein. If the gene specifying the molecular form of collagen mutates, drastic changes in the lungs, arteries, skeleton, and other body parts may ensue. As

Certain morphological, physiological, and behavioral traits characterize a population. The traits differ in their details from one individual to the next.

Differences in the combinations of alleles that individuals carry give rise to variations in phenotype. By phenotypic variation, we mean differences in the details of structural, functional, and behavioral traits that the individuals of a population have in common.

In sexually reproducing species, a population's individuals share a pool of genetic resources—that is, a gene pool.

Mutation alone creates *new* alleles. Natural selection, gene flow, and genetic drift change *allele frequencies* in a gene pool. The evolutionary story starts with these changes.

When Is a Population *Not* Evolving?

Look back on Figure 10.1, which shows the earlobes of Tom Cruise, Gregor Mendel, and other celebrated people. Now ask this: Is the number of individuals with *detached* earlobes staying the same, one generation after the next, in the whole human population? What about the number of individuals who have *attached* earlobes? Stated more broadly, how do we know whether or not a population is evolving with respect to earlobes or any other trait?

Almost a hundred years ago, a mathematician and a doctor came up with a way to find the answer. Working independently, they thought about what it would take to maintain an idealized population's allele frequencies. They came up with what we now call the **Hardy–Weinberg rule**, in their honor. They started with a formula that represents this idea: In any population at genetic equilibrium, the proportions of genotypes at a gene locus for which there are two kinds of alleles are

$$p^2\ AA + 2pq\ Aa + q^2\ aa = 1$$

where *p* is the frequency of allele *A*, and *q* is the frequency of allele *a*. They applied this rule to the formula: *The allele frequencies will not change through successive generations if there is no mutation, if the population is infinitely large and is isolated from other populations of the same species, if mating is random with respect to the alleles, and if all individuals survive and reproduce equally.* Stated differently, the rule presents a hypothesis for determining whether any of the listed assumptions currently apply to a population.

Suppose you want to test whether a population's allele frequencies stay the same from one generation to the next in the absence of evolutionary forces. You decide to track a pair of alleles through a population of butterflies. These sexually reproducing organisms have pairs of genes, on pairs of homologous chromosomes. You start by assuming the population is presently at genetic equilibrium. The pair of alleles you are interested in affect wing color. Allele *A* specifies dark-blue wings. Allele *a* is associated with white wings. And the heterozygous (*Aa*) condition results in medium-blue wings.

In the population as a whole, the frequencies of *A* and *a* must add up to 1. For example, if *A* occupies half of all the loci for this gene in the population, then *a* must occupy the other half (0.5 + 0.5 = 1). If *A* occupies 90 percent of all the loci, then *a* must occupy 10 percent (0.9 + 0.1 = 1). No matter what the proportions of the two kinds of alleles,

$$p + q = 1$$

During meiosis in germ cells, each allele segregates from its partner, and the two end up in separate gametes. Therefore, *p* is also the proportion of gametes carrying the *A* allele, and *q* is the proportion with the *a* allele. To find the expected frequencies of the three genotypes (*AA*, *Aa*, and *aa*) that are possible in the next generation, you construct a Punnett square:

The frequencies of the genotypes add up to 1:

$$p^2 + 2pq + q^2 = 1$$

To see whether the allele frequencies and genotypic frequencies will stay the same through the generations, you work through an example. You find the population has 1,000 butterflies, each of which produces two gametes:

490 *AA* individuals produce 980 *A* gametes
420 *Aa* individuals produce 420 *A* and 420 *a* gametes
 90 *aa* individuals produce 180 *a* gametes

You notice the frequency of *A* among the 2,000 gametes is

$$(980 + 420)/2,000 = 0.7$$

Also,

$$q = (420 + 180)/2,000 = 0.3$$

At fertilization, the gametes combine at random and give rise to the next generation, as given in the Punnett square. Assuming the population remains constant at 1,000 individuals, you now have

$$p^2\ AA = 0.7 \times 0.7 = 0.49 \qquad (490\ AA\ \text{individuals})$$
$$2pq\ Aa = 2 \times 0.7 \times 0.3 = 0.42 \quad (420\ Aa\ \text{individuals})$$
$$q^2\ aa = 0.3 \times 0.3 = 0.09 \qquad (90\ aa\ \text{individuals})$$

and

$$p^2 + 2pq + q^2 = 0.49 + 0.42 + 0.09 = 1$$

The allele frequencies have not changed:

$$A = \frac{2 \times 490 + 420}{2,000\ \text{alleles}} = \frac{1,400}{2,000} = 0.7 = p$$

$$a = \frac{2 \times 90 + 420}{2,000\ \text{alleles}} = \frac{600}{2,000} = 0.3 = q$$

You notice that the genotype frequencies haven't changed, either. And as long as the five assumptions of the Hardy–Weinberg rule are being met, frequencies should stay the same through the generations. To test this, you calculate allele frequencies in the gametes of the *next* generation:

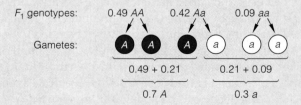

which is back where you started from. Because the allele frequencies for dark-blue, medium-blue, and white wings are the same as they were in the original gametes, they will yield the same phenotypic frequencies you observed in the second generation.

You could do similar calculations for other wing colors. You could go on with your calculations until you run out of paper (or patience). As long as the five assumptions continue to hold, however, the allele frequencies and the range of values for the wing-color trait will not change, as you can see from Figure 16.9.

Therefore, when genotypes and phenotypes do *not* show up in the proportions you predicted on the basis of the Hardy–Weinberg rule, this tells you that one or more conditions of the rule are being violated. And the hunt can begin for the specific evolutionary force, or forces, driving the change.

STARTING POPULATION

490 *AA* butterflies
Dark-blue wings

420 *Aa* butterflies
Medium-blue wings

90 *aa* butterflies
White wings

THE NEXT GENERATION

490 *AA* butterflies

420 *Aa* butterflies

90 *aa* butterflies

NO CHANGE

THE NEXT GENERATION

490 *AA* butterflies

420 *Aa* butterflies

90 *aa* butterflies

NO CHANGE

Figure 16.9 A hypothetical population of butterflies at genetic equilibrium.

16.6

NATURAL SELECTION REVISITED

We now turn from our idealized population that never changes to the real-world processes of change. Of these processes, natural selection might account for most of the morphological and physiological changes that have accrued throughout the history of life.

Darwin, recall, was able to explain natural selection after correlating his understanding of inheritance with certain features of populations and their environments. Before we consider the modes of natural selection, let's restate his correlations in modern terms:

1. *Observation:* All populations in nature have the reproductive capacity to increase in numbers through the generations.

2. *Observation:* No population is able to increase indefinitely, for its individuals will run out of food, living space, and other resources that sustain it.

3. *Inference:* Sooner or later, the individuals of a population will end up competing for resources.

4. *Observation:* All of the individuals generally have the same genes, which specify the same assortment of traits. Collectively, their genes represent a pool of heritable information.

5. *Observation:* Most, if not all, kinds of genes occur in slightly different molecular forms (alleles), which give rise to differences in phenotypic details.

6. *Inferences:* Some phenotypes are better than others at helping the individual compete for resources, hence to survive and reproduce. Alleles for those phenotypes therefore increase in the population, and other alleles decrease. In time the genetic change leads to increased **fitness**—an increase in adaptation to the environment.

7. *Conclusions:* **Natural selection** is the outcome of variations in shared traits that affect which individuals of a population survive and reproduce each generation. This microevolutionary process results in adaptation to the environment.

Evolutionary biologists have been documenting the results of natural selection in thousands of field studies of populations of all kinds of organisms. They find that this microevolutionary process has different results. As you will see in sections to follow, sometimes the result is a shift in the range of values for a given trait in some direction. At other times, the result may be stabilization or disruption of an existing range of values.

As Darwin perceived, natural selection is the outcome of variations in shared traits that influence which individuals of a population survive and reproduce in each generation. Natural selection can lead to increased fitness—that is, to an increase in adaptation to the environment.

DIRECTIONAL CHANGE IN THE RANGE OF VARIATION

What Is Directional Selection?

In cases of **directional selection**, allele frequencies that give rise to a range of variation in phenotype tend to shift in a consistent direction. The shift is a response to directional change in the environment or to one or more novel environmental conditions. A new mutation that proves adaptive also sets it in motion. Either way, the forms at one end of the phenotypic range become more common than the midrange forms (Figure 16.10). Let's look at a few documented cases of this outcome.

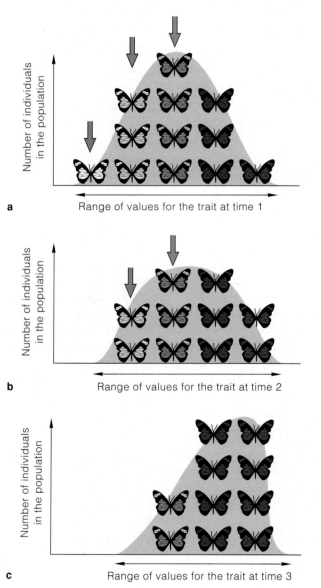

a — Range of values for the trait at time 1

b — Range of values for the trait at time 2

c — Range of values for the trait at time 3

Figure 16.10 Directional selection, using phenotypic variation within a population of butterflies as the example. A bell-shaped curve (*green*) represents the range of continuous variation in wing color. The most common form (*medium blue*) is between extreme forms of the trait (*white* at one end of the curve, *deep purple* at the other). *Orange* arrows signify which forms are being selected against over time.

The Case of the Peppered Moths

In England, biologists tracked directional selection in about a hundred moth species, including the peppered moth (*Biston betularia*). Peppered moths feed and mate at night. During the day, they rest motionless on birches and other trees. Their wings and body have a mottled pattern in shades that range from light gray to nearly black. Their behavior, coloration, and wing patterning camouflage them from moth-eating birds. Like all birds, the ones that prey on moths hunt during the day.

The industrial revolution started in England in the 1850s. Outpourings of sooty smoke altered conditions in many parts of the surrounding countryside. Before then, light moths were the most common form, and a dark form was rare. Also before conditions changed, light-gray speckled lichens grew thickly on tree trunks. Lichens can camouflage light moths that rest on them, but not dark ones (Figure 16.11*a*).

Lichens are sensitive to air pollution. Between 1848 and 1898, soot and other factory pollutants started to kill the lichens and darken tree trunks. In the modified environment, the less common moth form was better camouflaged (Figure 16.11*b*). Moth collectors came up with a hypothesis: If original conditions favored light moths, then the *changed* conditions favored dark ones.

In the 1950s, H. B. Kettlewell used a *mark–release–recapture method* to test that prediction. He bred both moth forms in captivity and marked hundreds of them so they could be easily identified after being released in the wild. He released them near highly industrialized areas around Birmingham and in an unpolluted area of Dorset. Later on, more dark moths were recaptured in the polluted area and more light ones in the pollution-free area (Table 16.1). Observers also were stationed in blinds near groups of moths that had been tethered to trees. They directly observed birds capturing more of the light moths around Birmingham and more of the dark moths around Dorset. Here was strong evidence that directional selection was operating.

Strict pollution controls went into effect in 1952. Lichens made a comeback. Tree trunks became free of soot, for the most part. As you might have predicted, phenotypes shifted in the reverse direction as a result of natural selection. Where pollution has decreased, the frequency of dark moths has been decreasing as well.

Pesticide Resistance

Widespread use of chemical pesticides in agriculture also results in directional selection. Initial applications kill most of the insects, worms, or other pests, but some individuals usually manage to survive. Some aspect of their structure, physiology, or behavior helps them resist

Figure 16.11 Peppered moths (*Biston betularia*). The light form of the moth is less visible to birds on lichen-covered tree trunks (**a**). The dark form is less visible after pollution kills the lichens and darkens the trunks (**b**).

Table 16.1 *Marked Peppered Moths Recaptured in Polluted and Nonpolluted Areas*

	Near Birmingham (pollution high)	Near Dorset (pollution low)
LIGHT-GRAY MOTHS		
Released	64	393
Recaptured	16 (25%)	54 (13.7%)
DARK-GRAY MOTHS		
Released	154	406
Recaptured	82 (53%)	19 (4.7%)

Data after H. B. Kettlewell.

the chemical effects. When the resistance has a heritable basis, it becomes more common in each new generation. These chemicals are agents of selection that favor the most resistant forms! Today, 450 different species resist one or more pesticides. Worse, the pesticides also kill natural predators of the pests. When freed from natural constraints, the populations of resistant pests burgeon, and crop damage is greater than ever. This outcome of directional selection is called *pest resurgence*.

Maybe pesticide use will decline in fields of plants that are genetically engineered to resist pests. Even modified plants will not win the coevolutionary arms race, but they might help keep food supplies one step ahead of the pests. This is a contentious issue; many consumers are leery of genetically engineered food, as you read in Section 15.11 (*Critical Thinking* question 5).

What about using *biological controls*? By this practice, natural enemies of pests, including parasitic wasps and predatory beetles, are raised in commercial insectaries. Large populations are released at preselected sites. The practice has an advantage in that a control species can coevolve with the pests. Farmers must replace the ones that migrate from the fields or are destroyed at harvest time. They pass on the cost to consumers.

Antibiotic Resistance

When your grandparents were young, tuberculosis, pneumonia, and scarlet fever caused one-fourth of the annual deaths in the United States. Since the 1940s, we have been relying on natural and synthetic antibiotics to fight such bacterial diseases. **Antibiotics**, remember, are toxins that some microorganisms in soil release to destroy bacterial competitors for nutrients (Section 1.4). Streptomycins, for example, prevent protein synthesis in target cells. Penicillins disrupt formation of covalent bonds that hold bacterial cell walls together. Penicillin derivatives weaken the wall until it ruptures.

Antibiotics must be prescribed with restraint and care. Why? Besides executing their intended function, they often disrupt resident bacterial populations that compete for nutrients in the intestines. They can disrupt bacterial and yeast populations in the vagina, as well. Imbalances follow and lead to secondary infections.

Also, antibiotics have been overprescribed in the human population. All too frequently, people demand them for simple infections that they often can fight on their own. As a consequence, antibiotics have lost their punch. Over time, they killed the most susceptible cells of target populations. However, they also favored their replacement by much more resistant cells. Each year, millions of people around the world die from cholera, tuberculosis, and other bacterial diseases. Vancomycin, once held in reserve as "the antibiotic of last resort," is no longer effective against some pathogenic strains of enteric (gut-dwelling) bacteria. In 1996, the World Health Organization announced that, in the dangerous race for supremacy, the pathogens are sprinting ahead.

With directional selection, allele frequencies underlying a range of variation tend to shift in a consistent direction in response to directional change in the environment.

SELECTION AGAINST OR IN FAVOR OF EXTREME PHENOTYPES

As you have seen, natural selection can bring about a directional shift in a population's range of phenotypic variation. Depending on the prevailing environmental conditions, the process also may favor either the most common or the most extreme phenotypes in that range.

Stabilizing Selection

In **stabilizing selection**, intermediate forms of a trait in a population are favored, so alleles for extreme forms are eliminated (Figure 16.12). This mode of selection tends to counter mutation, gene flow, and genetic drift, and it preserves the most common phenotypes.

The gallmaking fly *Eurosta solidaginis* is evidence of stabilizing selection. Its larvae bore into stems of a tall goldenrod (*Solidago altissima*) and feed on tissues until they metamorphose into adults. In response, the tissue cells multiply rapidly and encase the invaders in a gall, a tumorous mass. We know from genetic analyses that flies with different phenotypes cause galls of different sizes—large, small, and in between—to form.

A wasp parasitizes the larvae (Figure 16.13). It can penetrate small galls, so flies that cause small galls to form are at risk. Also, the downy woodpecker and some other birds eat the larvae. They can peck through large galls, so the flies that cause large galls to form are at risk. Think it through, and you see that the flies that cause *intermediate-sized* galls to form are favored. Their larvae are less vulnerable to wasps *and* to birds.

And so parasitic wasps work against one extreme phenotype and predatory birds work against the other. In this fly population, the intermediate phenotype has the highest survival rate and fitness.

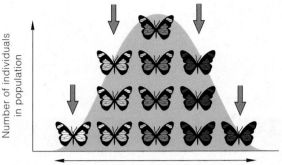

Range of values for wing-color trait at time 1

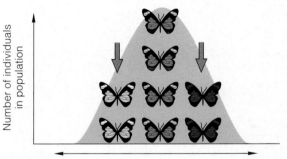

Range of values for wing-color trait at time 2

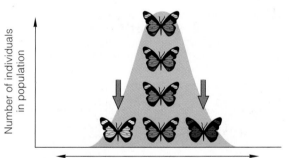

Range of values for wing-color trait at time 3

Figure 16.12 Stabilizing selection, using phenotypic variation within a population of butterflies as the example.

c Parasitic wasp (an agent of selection)

a Gall-making fly (producer of tasty larvae)

b Gall on a goldenrod stem that houses fly larvae

d Downy woodpecker (an agent of selection)

Figure 16.13 Example of stabilizing selection. Larvae of the fly *Eurosta solidaginis* (**a**) induce formation of a type of tumor called a gall (**b**) on goldenrod stems. (**c**) A parasitic wasp (*Eurytoma gigantea*) has an egg-laying device that penetrates only the thin wall of small galls. Its eggs develop into larvae, then the wasp larvae eat fly larvae. (**d**) Downy woodpeckers (*Dendrocopus pubescens*) and other birds can use their bill to chisel into large-size galls. They, too, eat the larvae.

Warren Abrahamson and his coworkers monitored twenty *Eurosta* populations in Pennsylvania. They found that larvae inside small and large galls have low relative fitnesses. They also found that larvae in intermediate-size galls have relatively high fitnesses. The fly's natural enemies target larvae in small and large galls; they stabilize the fly's range of phenotypic variation.

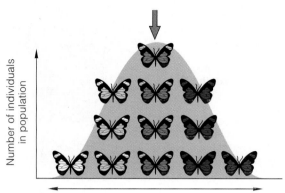

a   Range of values for wing-color trait at time 1

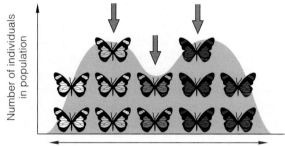

b   Range of values for wing-color trait at time 2

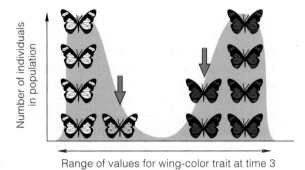

c   Range of values for wing-color trait at time 3

Figure 16.14 Disruptive selection, using phenotypic variation within a population of butterflies as the example.

Disruptive Selection

In **disruptive selection**, forms at both ends of a range of variation are favored and intermediate forms are selected against (Figure 16.14). Thomas Smith found an example of this in a rain forest in Cameroon, West Africa. Smith had read about unusual variation in populations of the black-bellied seedcracker (*Pyrenestes ostrinus*). These African finches have large or small bills—but no sizes in between. The pattern holds for females and males through the geographic range. (This is just remarkable; imagine every person in Texas being four feet *or* six feet tall, with no intermediates.) If the bill pattern is unrelated to gender or geography, what causes it?

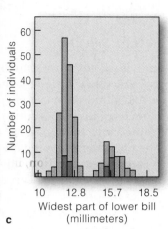

Figure 16.15 Disruptive selection among African finches. Feeding trials showed that birds with large bills are better users of hard seeds. Small-billed birds are better at using soft seeds, not hard ones. (**a**,**b**) Two individuals displaying small and large bill sizes.

(**c**) Graph of survival of juvenile birds during the dry season, when competition for resources is most intense. Individuals with very small, very large, or intermediate-size bills can't feed efficiently on either type of seed; they survive poorly. For this graph, *tan* bars indicate the number of nestlings; *orange* bars indicate the survivors among them. The findings are based on measurements of 2,700 netted individuals.

Smith hypothesized: Seed-cracking ability directly affects finch survival. If only two bill sizes persist, then disruptive selection may be eliminating birds that have intermediate-size bills. What selection pressures could be at work on feeding performance? Cameroon's swamp forests flood during the wet season; lightning-sparked fires burn during the dry season. Two species of sedge (fire-resistant, grasslike plants) dominate these forests. One sedge has hard seeds and the other has soft. When finches reproduce, hard *and* soft seeds are abundant.

All birds prefer soft seeds for as long as they can get them. However, the birds with small bills are better at cracking soft seeds; birds with large bills are better at cracking hard ones. Soft seeds and other food supplies dwindle as the dry season peaks. That is when small-billed birds, and the youngest ones, are at a competitive disadvantage. Many don't survive (Figure 16.15).

Smith also performed experimental crosses between large- and small-billed birds. All offspring had large *or* small bills. Along with other data, the crosses suggest that two alleles at one autosomal gene locus control bill size and feeding performance.

With stabilizing selection, intermediate phenotypes are favored and extreme phenotypes at both ends of the range of variation are eliminated.

With disruptive selection, intermediate forms are selected against; extreme forms in the range of variation are favored.

SPECIAL TYPES OF SELECTION

Sexual Selection

Individuals of most sexually reproducing species have a distinctively male or female phenotype. We call this sexual dimorphism (*dimorphos*, "having two forms"). How does this condition arise, and what maintains it? Natural selection is taking the form of **sexual selection**. The traits being favored are advantageous, with respect to survival and reproduction, simply because males or females prefer them. Through nonrandom mating, the alleles for preferred traits prevail over the generations.

Sexual dimorphism is particularly striking among many mammals and birds, as in Figure 16.16. The males of many species are larger and have flashier coloration and patterning. They are often more aggressive than the females. Remember those male bighorn sheep butting

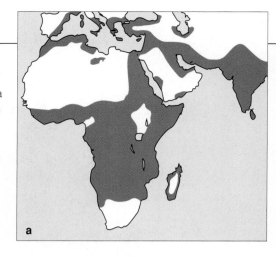

Figure 16.17 (**a**) Distribution of malaria cases in Africa, Asia, and the Middle East in the 1920s, before the start of mosquito control programs. (**b**) The distribution and frequency of people with the sickle-cell trait. Note the close correlation between the maps.

heads (Section 1.3)? Fighting wastes time and energy, and it may cause serious injuries. Why, then, do alleles that contribute to aggressive behavior still persist in a population? An increased chance of mating may offset the costs. Male bighorn sheep fight only to control areas where receptive females gather during a winter rutting season. Winners mate often, with a number of females. Losers won't challenge a stronger, larger male.

Often, females are the agents of selection. They exert direct control over reproductive success by choosing their mates. We return to this topic in Chapter 41.

Maintaining Two or More Alleles

Balancing selection refers to all forms of selection that maintain two or more alleles for a trait in a population. When this type of genetic variation persists over time, we call it **balanced polymorphism** (after *polymorphos*, "having many forms"). A population shows balanced polymorphism when nonidentical alleles for some trait are being maintained at frequencies above 1 percent. Allele frequencies might shift slightly, but over time, they often return to the same values. Smith's African finches are a good example of the effect of balancing selection through the generations.

Sickle-Cell Anemia—Lesser of Two Evils?

Balanced polymorphism may emerge when conditions favor heterozygotes, which carry nonidentical alleles for the trait. Compared to homozygotes, which carry identical alleles for the trait, they have higher fitness.

Let's look at the environmental pressures that favor an Hb^A/Hb^S pairing in humans. Hb^S specifies a mutant form of hemoglobin, an oxygen-transporting protein in blood. Homozygotes (Hb^S/Hb^S) develop *sickle-cell anemia*, a genetic disorder that has serious pleiotropic effects on phenotype (Section 10.8). The frequency of Hb^S is high in tropical and subtropical regions of Asia and Africa. Often, Hb^S/Hb^S homozygotes die in their early teens or early twenties. Yet in those same regions, the heterozygotes (Hb^A/Hb^S) make up nearly one-third

Figure 16.16 An outcome of sexual selection. This male bird of paradise (*Paradisaea raggiana*) is engaged in a flashy courtship display. He caught the eye (and, perhaps, the sexual interest) of the smaller, less colorful female. Males of this species compete fiercely for females, the selective agents. (Why do you suppose drab-colored females have been favored?)

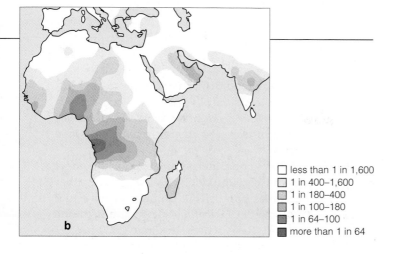

☐	less than 1 in 1,600
☐	1 in 400–1,600
☐	1 in 180–400
☐	1 in 100–180
☐	1 in 64–100
■	more than 1 in 64

b

of the human population! Why is this combination of alleles maintained at such high frequency?

The balancing act, an outcome of natural selection, is most pronounced in areas with the highest incidence of *malaria* (Figure 16.17 and Section 20.9). A mosquito transmits *Plasmodium*, the parasitic agent of disease, to humans. The parasite multiplies in the liver, and later in red blood cells. The cells rupture and so release new parasites during severe, recurring bouts of infection.

Who is more likely to survive recurring infections? Hb^A/Hb^S heterozygotes. The allelic combination gives them two forms of hemoglobin, with interesting results. They produce enough *nonmutated* molecules to support body functions—and the *altered* molecules distort red blood cells in a way that slows down circulation. The slowdown in blood flow is a factor in getting through recurrences because it hampers the parasite's ability to travel and rapidly infect new cells. The distorted cells are later disposed of as they trickle through the spleen.

So the persistence of the "harmful" Hb^S allele is a matter of relative evils. Natural selection has favored one allelic combination, Hb^A/Hb^S, because its bearers show greater fitness in places where malaria is most prevalent. In these environments, the combination has a higher fitness than either Hb^S/Hb^S or Hb^A/Hb^A.

Malaria has been a selective force for over 2,000 years in tropical and subtropical habitats of Asia and the Middle East. Although sickle-cell anemia occurs at high frequencies in all these regions, its symptoms are far less severe than they are in Central Africa—where the Hb^S allele became established much later in time. It seems likely that other gene products are mediating some of the Hb^S allele's widespread effects; in some way they minimize symptoms of the disorder.

With sexual selection, some version of a trait simply gives the individual an advantage in reproductive success. Sexual dimorphism is one outcome of sexual selection.

Balanced polymorphism is a state in which natural selection is maintaining two or more alleles through the generations at frequencies greater than 1 percent.

GENE FLOW

Individuals of the same species don't necessarily stay put. A population loses alleles whenever an individual permanently leaves it, an act called *emigration*. It gains alleles when new individuals permanently move in, an act called *immigration*. This **gene flow**, a physical flow of alleles between populations, tends to counter genetic differences that we expect to arise through mutation, natural selection, and genetic drift. And it helps keep separated populations genetically similar.

Think of the acorns that blue jays disperse when they store nuts for the winter. Each fall the jays may make hundreds of round trips from acorn-bearing oak trees to bury acorns in the soil of their home territories, which may be up to a mile away (Figure 16.18). The alleles flowing in with "immigrant acorns" help reduce genetic differences that might otherwise arise among neighboring stands of oak trees. Gene flow apparently is operating among peppered moths, also, for the form that doesn't match the prevailing background color is being maintained at frequencies higher than expected.

Figure 16.18 Gene flow among oak populations, courtesy of feathered travel agents. Blue jays hoard acorns in their home territory, but they might shop at nut-bearing trees up to a mile away. Some acorns contribute to the allele pool of an oak population some distance away from the parent tree.

Or think of the millions of people from politically explosive, economically bankrupt countries who seek a more stable homeland. The scale of their movements is unprecedented, but not unique. Throughout human history, immigrations may have minimized many of the genetic differences that otherwise would have built up among geographically separated groups of people.

Gene flow is the physical movement of alleles into and out of a population, through immigration and emigration.

GENETIC DRIFT

Chance Events and Population Size

Genetic drift is a random change in allele frequencies over the generations, brought about by chance alone. The magnitude of its effect on genetic diversity and on the range of phenotypes relates to population size. Its impact tends to be minor or insignificant in very large populations but significant in small ones.

Sampling error, a rule of probability, helps explain the difference. By this rule, the fewer times a chance event occurs, the greater will be the variance from the expected outcome of that occurrence. You saw a simple demonstration of this in Section 1.5.

Also think back on the coin-flipping example given earlier, in Section 10.2. Each time you flip a coin, there is a 50 percent chance it will turn up heads or tails. With, say, only ten flips, the odds are great that the coin will turn up heads 7 times and tails 3 times. But with a thousand flips, the odds it will turn up heads 700 times and tails 300 times are almost nil. Similarly, sampling error applies each time random mating and fertilization take place in a population.

Bear in mind, genetic drift has nothing to do with how a population got small in the first place. *Genetic drift simply increases the chance that an allele will become more or less prevalent when the number of individuals in a population is small.*

Figure 16.19 is a computer simulation of the effect of genetic drift in two populations: one large, the other small. The outcomes parallel the findings of an actual experiment with beetles (*Tribolium*) that mate with one another at random. Researchers grouped 1,320 beetles into twelve populations of 10 beetles and twelve of 100. Which beetles ended up in which group was a matter of chance. At the outset, the frequency of a wild-type allele (call it *A*) was 0.5. The researchers tracked allele *A* for twenty generations. Every time, they randomly removed some offspring to maintain each population's original size. At the end of the experiment, *A* was not the only allele left in the large groups, but it was fixed in seven of the small groups. **Fixation** means that only one kind of allele remains at a particular locus in the population; all individuals are homozygous for it.

Thus, *in the absence of other forces, random change in allele frequencies leads to the homozygous condition and a loss of genetic diversity over the generations.* This happens in all populations; it just happens faster in small ones. Once alleles inherited from an original population are fixed, their frequencies will not change again unless mutation or gene flow introduce new alleles.

Bottlenecks and the Founder Effect

Genetic drift is pronounced when very few individuals rebuild a population or found a new one. This happens after a **bottleneck**, a severe reduction in population size brought about by intense selection pressure or some calamity. Say contagious disease, habitat loss, hunting, or an erupting volcano nearly wipes out a population. Even if a moderate number do survive the bottleneck, allele frequencies will have been altered at random.

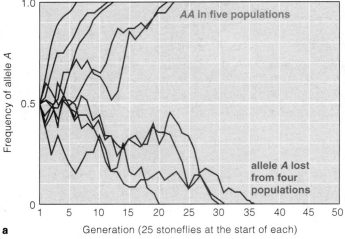

a

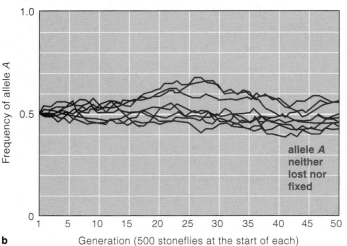

b

Figure 16.19 Computer simulation of the effect of genetic drift on one allele's frequency in small populations and in large populations. Equal fitness is assumed for three simulations (*AA* = 1, *Aa* = 1, and *aa* = 1).

(**a**) The size of nine populations of a species (stoneflies, in this case) was maintained at 25 breeding individuals in each generation, through fifty generations. The five graph lines reaching the top of the diagram tell you that allele *A* became fixed in five of the small populations. The four lines plummeting off the bottom of the diagram tell you it was lost from four of them. As you can see, *alleles can be fixed or lost even in the absence of selection.*

(**b**) The size of nine other populations was maintained at 500 individuals in each generation, through fifty generations. Allele *A* did not become fixed in any of these large populations. The magnitude of genetic drift was much less in every generation than in the small populations.

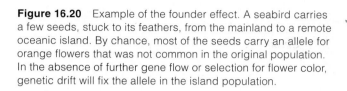

Figure 16.20 Example of the founder effect. A seabird carries a few seeds, stuck to its feathers, from the mainland to a remote oceanic island. By chance, most of the seeds carry an allele for orange flowers that was not common in the original population. In the absence of further gene flow or selection for flower color, genetic drift will fix the allele in the island population.

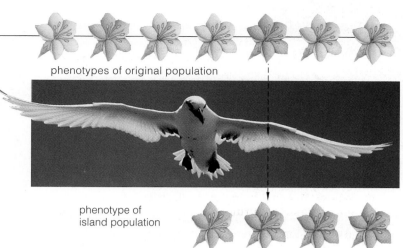

phenotype of island population

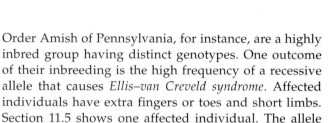

Example: In the 1890s hunters killed all but twenty of a large population of northern elephant seals. The population eventually recovered. After its size reached 30,000, electrophoretic analysis of a sample of twenty-four genes revealed no variation in the population. A number of alleles had been lost during the bottleneck.

The genetic outcome can be similarly dicey after a few individuals leave a population and establish a new one elsewhere. This form of bottlenecking is called a **founder effect**. By chance, the allele frequencies of the founders may not be the same as those of the original population. In the absence of further gene flow, natural selection will influence those frequencies in drastically different ways because of its interaction with genetic drift. As you might deduce, the founder effect is quite pronounced on isolated islands (Figure 16.20).

Genetic Drift and Inbred Populations

Inbreeding refers to nonrandom mating among closely related individuals, which have many identical alleles in common. Inbreeding is a form of genetic drift in a small population—that is, within the group of relatives that are preferentially interbreeding. Like genetic drift, inbreeding leads to the homozygous condition. It also can lower fitness when the alleles that are increasing in frequency are recessive and have harmful effects.

Most human societies forbid or discourage incest (inbreeding between parents and children or between siblings). But inbreeding among other close relatives is common in small communities that are geographically or culturally isolated from a larger population. The Old

Order Amish of Pennsylvania, for instance, are a highly inbred group having distinct genotypes. One outcome of their inbreeding is the high frequency of a recessive allele that causes *Ellis–van Creveld syndrome*. Affected individuals have extra fingers or toes and short limbs. Section 11.5 shows one affected individual. The allele probably was rare when the small number of founders immigrated to Pennsylvania. Today, though, about 1 in 8 are heterozygous and 1 in 200 are homozygous for it.

Bottlenecks and inbreeding are a bad combination for endangered species, the populations of which have become smaller and vulnerable to extinction. Cheetahs, for instance, apparently survived a drastic bottleneck in the nineteenth century. Survivors mated with their own offspring when no other options were available.

Inbreeding among survivors and their descendants resulted in strikingly similar alleles among the 20,000 existing cheetahs (Figure 16.21). One, a mutated allele, affects fertility. Most male cheetahs have a low sperm count, and 70 percent of the sperm are abnormal. Other shared alleles result in far lower resistance to disease. Therefore, infections that are seldom life-threatening to other cat species can be devastating to cheetahs. In one outbreak of *feline infectious peritonitis* in a wild animal park, the viral pathogen had very little effect on captive lions, but it killed many cheetahs. For them, infection triggered uncontrollable inflammation. Fluid filled the body cavity that houses the heart and other organs, and the cats died in agony. There is no vaccine.

The Florida panther also is highly inbred. Pressure from hunting and urban sprawl has helped put this cat on the endangered species list. Only seventy remain.

Figure 16.21 A few of the remaining cheetahs, which have some bad alleles that made it through a severe bottleneck.

Genetic drift is the random change in allele frequencies over the generations, brought about by chance alone. The magnitude of its effect is greatest in small populations, such as the ones that make it through a bottleneck.

Barring mutation, selection, and gene flow, the chance losses and increases of the various alleles at a given locus lead to the homozygous condition and a loss of genetic diversity.

16.12

SUMMARY   *Gold* indicates text section

1. Awareness of evolution—changes in lines of descent over time—emerged through comparisons of the body structure and patterning for major groups of animals (comparative morphology), questions about the world distribution of plants and animals (biogeography), and observations of fossils in sedimentary rock layers. As first proposed by Charles Darwin and Alfred Wallace, evolution can occur by natural selection. *16.1–16.3*

2. Individuals of a population generally have the same number and kind of genes. But genes come in different allelic forms, and this leads to variations in traits. *16.4*

3. A population is evolving when some forms of a trait (and the underlying alleles) are becoming more or less common relative to the other kinds one generation after the next. This happens through mutation, gene flow, genetic drift, and natural selection (Table 16.2). *16.4*

4. Gene mutations (heritable changes in DNA) are the only source of *new* alleles. New combinations of existing alleles arise by crossing over, independent assortment at meiosis, and mixing of alleles at fertilization. *16.4*

5. Genetic equilibrium, a state in which a population is not evolving, is used as a baseline to measure change. The Hardy–Weinberg rule is this: Genetic equilibrium occurs only if there is no mutation, the population is very large and isolated from other populations of the same species, there is no selection, mating is random, and all members survive and reproduce equally. *16.5*

6. Natural selection is the difference in survival and reproduction among individuals of a population that differ in details of heritable traits. It leads to increased fitness (increased adaptation to the environment). *16.6*

   a. Pressures may shift a range of variation for a trait in one direction (*directional* selection), favor extremes (*disruptive* selection), or eliminate extremes and favor intermediate forms (*stabilizing* selection). *16.7, 16.8*

   b. Selection can result in balanced polymorphism. A population is in this state when nonidentical alleles for a given trait are being maintained over the generations at frequencies greater than 1 percent. *16.9*

   c. Sexual selection, by females or males, can lead to forms of traits that favor reproductive success. Sexual dimorphism (persistence in the phenotypic differences between males and females) is one outcome. *16.9*

7. Gene flow is a change in allele frequencies brought about by the physical movement of alleles into and out of a population (by immigration and emigration). *16.10*

8. Genetic drift is a change in allele frequencies over the generations due to chance alone. Its effect is greater in small populations than in large ones. *16.11*

9. Drift has greatest impact following a bottleneck (a severe reduction, then a recovery, in population size). Alleles making it through a bottleneck may give rise to

Table 16.2 *Summary of Microevolutionary Processes*

MUTATION	A heritable change in DNA
NATURAL SELECTION	Change or stabilization of allele frequencies owing to differences in survival and reproduction among variant individuals of a population
GENETIC DRIFT	Random fluctuation in allele frequencies over time due to chance occurrences alone
GENE FLOW	Change in allele frequencies as individuals leave or enter a population

differences in the range of phenotypes compared to the original population. In bottlenecking, one type of founder effect, a few emigrants from one population establish a small subpopulation in a new environment. *16.11*

Review Questions

1. Define biogeography and comparative anatomy. How did studies in both disciplines contradict the idea that species have remained unchanged since the time of their creation? *16.1*

2. Define evolution. Define evolution by natural selection. Can an individual evolve? *16.1, 16.4, 16.5*

3. Name three broad categories of heritable traits that help characterize a population. *16.4*

4. Explain the difference between continuous variation and polymorphism. *16.4*

5. How do lethal mutations and neutral mutations differ? *16.4*

6. Define genetic equilibrium. Which occurrences can drive allele frequencies away from genetic equilibrium? *16.4, 16.5*

7. Define fitness, with respect to phenotype. *16.6*

8. Identify the mode of selection (stabilizing, directional, or disruptive) for the diagrams on the facing page. *16.7, 16.8*

9. Define bottleneck and the founder effect. Are these cases of genetic drift, or do they merely set the stage for it? *16.11*

Self-Quiz   ANSWERS IN APPENDIX III

1. Biologists define evolution as _____ .
   a. the origin of a species
   b. heritable change in a line of descent over generations
   c. inheritance of characteristics acquired by the individual

2. Darwin saw that populations of Galápagos finches _____ .
   a. show variation in traits
   b. resemble birds in South America
   c. are adapted to different island habitats
   d. all of the above

3. Individuals don't evolve; _____ do.

4. Genetic variation gives rise to variation in _____ traits.
   a. morphological   c. behavioral
   b. physiological   d. all of the above

5. Sickle-cell anemia first appeared in Asia, the Middle East, and Africa. The causative allele entered the U.S. population when people were forcibly brought over from Africa prior to the Civil War. In microevolutionary terms, this is a case of _____ .
   a. mutation   c. gene flow
   b. genetic drift   d. natural selection

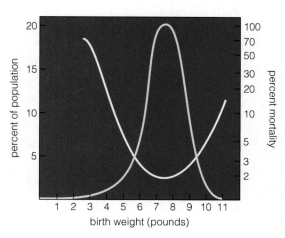

Figure 16.22 The weight distribution for 13,730 human newborns (the *yellow* curve) correlated with mortality rate (*white* curve).

6. Natural selection may occur when there are _____ .
   a. differences in the adaptiveness of forms of traits to prevailing environmental conditions
   b. differences in survival and reproduction among individuals that differ in one or more traits
   c. both a and b

7. Directional selection _____ .
   a. eliminates uncommon forms of alleles
   b. shifts allele frequencies in a steady, consistent direction
   c. favors intermediate forms of a trait
   d. works against adaptive traits

8. Disruptive selection _____ .
   a. eliminates uncommon forms of alleles
   b. shifts allele frequencies in a steady, consistent direction
   c. doesn't favor intermediate forms of a trait
   d. both b and c

9. Match the evolution concepts.
   ____ gene flow   a. source of new alleles
   ____ natural   b. changes in a population's allele
        selection   frequencies due to chance alone
   ____ mutation   c. allele frequencies change owing to
   ____ genetic   immigration, emigration, or both
        drift   d. differences in survival and reproduction
                           among variant individuals

Critical Thinking

1. Prospects for human newborns of very high or very low birth weight are not good. Being born too small increases the risk of stillbirth or early infant death (Figure 16.22). Also, pre-term instead of full-term pregnancies increase the risks. Are these outcomes of directional, disruptive, or stabilizing selection?

2. For centuries *tuberculosis,* or TB, killed people throughout the world. A bacterium (*Mycobacterium tuberculosis*) causes this contagious lung disease. TB declined steadily in the United States over the years, but caseloads are now rising. The AIDS epidemic, an influx of immigrants from regions where TB is still common, and severe overcrowding in tenements and homeless shelters contribute to the increase. Antibiotics once cured TB in six to nine months. But antibiotic-resistant strains of *M. tuberculosis* have evolved. Two or even three different antibiotics are used simultaneously to block different metabolic pathways of the bacterium. How might this more aggressive approach help get around the problem of antibiotic resistance?

3. A few families in a remote region of Kentucky show a high frequency of *blue offspring,* an autosomal recessive disorder.

Skin of affected individuals appears bright blue. Homozygous recessives lack the enzyme diaphorase. The enzyme catalyzes reactions that maintain hemoglobin in its normal molecular form. Without it, a blue form of hemoglobin accumulates in blood. Skin and its blood capillaries are transparent, so the pigment colors in blood show through and contribute to skin coloration. This gives the skin of nonmutated individuals a pinkish cast—and blue skin its color.

Formulate a hypothesis to explain why the blue-offspring trait is rather common among a cluster of families but rare in the human population at large.

4. In 1996, after reflecting on evidence of evolution that has been accumulating for more than a hundred years, Pope John Paul II acknowledged that the theory of evolution "has been progressively accepted by researchers, following a series of discoveries in various fields of knowledge. The convergence, neither sought nor fabricated, of the results of work that was conducted independently is in itself a significant argument in favor of this theory." His words infuriated individuals who believe in a strict, literal interpretation of the biblical account of creation. Some of these individuals have been demanding that biology instructors treat evolution as "just a theory" and give equal time to the biblical interpretation.

Refer to Section 1.5 on the nature of scientific inquiry. Do you equate scientific and religious explanations of life's origin and history as alternative theories? Why or why not?

Selected Key Terms

allele *16.4*	genetic drift *16.11*
allele frequency *16.4*	genetic equilibrium *16.4*
antibiotic *16.7*	Hardy–Weinberg rule *16.5*
balanced polymorphism *16.9*	inbreeding *16.11*
biogeography *16.1*	lethal mutation *16.4*
bottleneck *16.11*	microevolution *16.4*
comparative morphology *16.1*	mutation rate *16.4*
directional selection *16.7*	natural selection *16.6*
disruptive selection *16.8*	neutral mutation *16.4*
evolution *CI*	polymorphism *16.4*
fitness *16.6*	population *16.4*
fixation *16.11*	sampling error *16.11*
fossil *16.1*	sexual selection *16.9*
founder effect *16.11*	stabilizing selection *16.8*
gene flow *16.10*	uniformity (theory of) *16.2*
gene pool *16.4*	

Readings

Darwin, C. 1957. *Voyage of the Beagle.* New York: Dutton.

Smith, T. B. January 1991. "A Double-Billed Dilemma." *Natural History,* 14–21.

SPECIATION

The Case of the Road-Killed Snails

If you happen to be a snail living in a garden in Bryan, Texas, it doesn't take much to keep your genes away from snails in a backyard across the street. By day, the sunbaked asphalt would be about as inviting to a snail as a desert would be to a catfish. Besides, day or night, a street-traversing snail is vulnerable to cars, trucks, skateboards, and bicycles (Figure 17.1*a*). That strip of asphalt is a formidable barrier to the flow of genes between populations. For snails, that is.

Whether any physical barrier deters gene flow depends in large part on the organism's mode of locomotion or dispersal. It also depends on how fast and how long an organism *can* move in response to environmental factors or its own hormonal signals.

A snail is not swift, and it does not roam very far from its home population. Compare it with the wild black duck, leg-banded in Virginia in 1969, that turned up eight years later in Korea. Compare it to the wandering albatross, one of the supreme barrier busters. After lifting off from Kerguelen Island in the Indian Ocean, one of these birds soared westward past the southern tip of Africa, across the Atlantic, and around South America's Cape Horn. After traveling thirteen thousand kilometers, it landed in Chile. With a wingspan of 3.65 meters (12 feet) and a lightweight body, that bird was able to exploit the great prevailing winds of the Southern Hemisphere.

And yet, in 1859, some snails did cross an ocean. Humans had transported garden-variety snails (*Helix aspersa*) from France, then released them in California. The idea was that the snails would multiply and meet culinary demands for that French delicacy *escargots aux fines herbes*. It was bad enough that the importers brought over a less tasty species by mistake. Worse, the exotic species exceeded expectations and became an absolute nuisance in

Figure 17.1 (**a**) A snail (*Helix aspersa*) encountering a major barrier to gene flow. It only appears to be reading the warning label. (**b**) Results from a study of neighboring populations of snails, all descended from founders that ended up in a small town in Texas in the 1930s. For each population, a circle represents relative abundances of three alleles (color-coded *green*, *orange*, and *blue*) for an enzyme, leucine aminopeptidase. Genetic variation is greater between populations living on opposite sides of Twenty-Second Street. For example, notice the higher frequencies of the allele coded *orange* in the block to the west of the street.

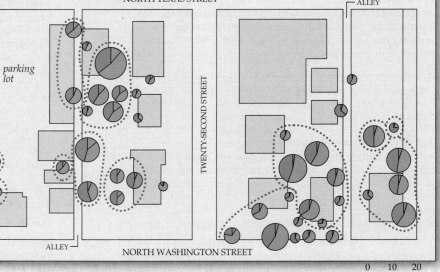

gardens and commercial nurseries through much of the southwestern United States.

By the 1930s, *H. aspersa* had hitched rides to Bryan, Texas, possibly as eggs in the soil of plant containers. They founded small colonies in patches of vegetation. Forty years later Robert Selander, now of Pennsylvania State University, was on his hands and knees with a few graduate students, scouring the patches on two adjacent city blocks. Why? He was interested in the effect of genetic drift on introduced populations. He and his students collected every snail—2,218 of them—from fourteen local colonies. For each colony, they figured out the allele frequencies for five different genes.

The analysis revealed some genetic variation among colonies on the same block—and major differences in allele frequencies *between* different blocks. Figure 17.1*b* shows an example for one of the genes studied.

What if the genetic differences between populations continue to increase, as by natural selection or genetic drift? Will the time come when snails from opposite sides of a street can no longer interbreed successfully even if they manage to get together? In other words, *will they become members of separate species?* Or will stabilizing selection work against significant genetic divergence by eliminating extreme phenotypes from neighboring populations? After all, how far can a workable package of *H. aspersa* genes evolve in such close-together colonies that encounter such similar environmental pressures in a small town in Texas?

And with these questions, we arrive at the topic of **speciation**—of changes in allele frequencies that are significant enough to mark the formation of daughter species from a parent species. Charles Darwin proposed long ago that natural selection might lead to speciation. But he wasn't sure how this actually happens. No one was around to watch species originate in the distant past. No one lives long enough to know whether many current populations are at intermediate stages leading to speciation. Even so, evidence is accumulating that supports certain speciation models.

As you read through this chapter, it is important for you to recognize that speciation is not the same thing as natural selection. Evolutionary biologists now view speciation as a *potential consequence* of natural selection and of other microevolutionary processes that may work in concert with it.

Key Concepts

1. A species consists of one or more populations of individuals that can interbreed under natural conditions and produce fertile offspring and that are reproductively isolated from other such populations. This definition is restricted to sexually reproducing species.

2. Populations of the same species have a shared genetic history, they are maintaining genetic contact over time, and they are evolving independently of other species.

3. Speciation is the process by which daughter species evolve from a parent species.

4. By one model, speciation starts when a geographic barrier arises between populations or subpopulations of a species. Thereafter, mutation, natural selection, and genetic drift operate independently in each population and lead to irreversible genetic divergence of one from the other.

5. Such populations may come to differ in certain alleles that affect morphological, physiological, or behavioral traits associated with reproduction. When the genetic differences affecting reproduction become great enough, the populations can no longer interbreed even if they end up coexisting in the same area later on. Speciation is completed.

6. The model of speciation just described, which applies to geographically separated populations, is known as allopatric speciation. This might be the way most species originate in nature. Sympatric speciation and parapatric speciation might be less prevalent routes.

7. By sympatric speciation, species form within the home range of their parent species. By parapatric speciation, adjacent populations become distinct species while still maintaining contact along a common border between their home ranges.

8. The timing, rate, and direction of speciation differ among the branches of a given lineage. They also differ from one lineage to the next. The extinction of some number of species is inevitable for all lineages.

ON THE ROAD TO SPECIATION

What Is a Species?

"If it looks like a duck, walks like a duck, and quacks like a duck, probably it's a duck." Let's use this familiar saying as a starting point for defining what is and what is not a species. **Species** is a Latin word. It simply means "kind," as in "a particular kind of duck."

Early naturalists defined species mostly in terms of morphological traits. But common sense told them to consider other factors. Example: Sometimes individuals of a species look very different only because they grew and developed under different environmental conditions. The two plants in Figure 17.2 are a case in point.

In fact, individuals of most species vary enormously in morphological details. So perhaps we should look also for a basic function that unites the populations of a species and isolates them from all other species.

Reproduction is such a basic, defining function. It is actually central to the **biological species concept**. Ernst Mayr, an evolutionary biologist, phrased the concept this way: Species are groups of interbreeding natural populations that are reproductively isolated from other such groups. In his view, it doesn't matter how much phenotypes vary. The populations belong to the same species as long as their members have the same form, physiology, and behaviors that let them interbreed and produce fertile offspring. Being able to contribute to a shared gene pool is qualification for membership.

Mayr's species concept does not apply to asexually reproducing organisms, such as bacteria. However, it's a useful guide for research into factors that define the vast majority of species, which do reproduce sexually.

If we subscribe to Mayr's concept, then speciation is attainment of reproductive isolation. Bear in mind, this does not mean that reproductive isolation evolves purposefully to promote the formation of a species or to maintain its separate identity. Rather, *any structural, functional, or behavioral difference that favors reproductive isolation is simply a by-product of genetic change.*

Gene flow alone counters genetic changes between populations of the same species. **Gene flow**, recall, is a movement of alleles into and out of populations by immigration and emigration. It exerts a homogenizing effect even when two populations are geographically separate as long as some genes are exchanged between them. Thus, gene flow is a microevolutionary process that helps maintain a common reservoir of alleles.

What if some geographic barrier arises and blocks intermingling of genes between some populations or subpopulations of a species? Genetic divergence will follow. Through **genetic divergence**, the gene pools of genetically separated populations accumulate distinct differences, for mutation, natural selection, and genetic drift are modifying each one independently of others.

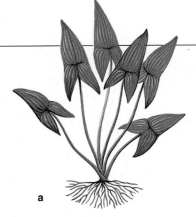

Figure 17.2 Pronounced morphological differences between two plants of the same species. Shown are mature leaves of arrowheads (*Sagittaria sagittifolia*) growing (**a**) on land and (**b**) in water. The difference is attributable to adaptive responses to different environmental conditions, not to genetic differences.

populations of one species (*gold*)

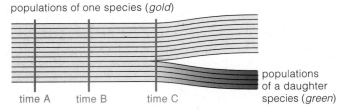

time A   time B   time C

populations of a daughter species (*green*)

Figure 17.3 Simplified diagram of genetic divergence. Each horizontal line represents a different population.

Figure 17.3 is a simple way to think about the process. In this sketch, each horizontal line represents a single population. At times A and B, gene flow was keeping them in genetic contact, so they belonged to the same species. However, a geographic barrier isolated some populations at time C and triggered their divergence.

Most species arose by gradual genetic divergence. Yet, as you will see, many flowering plants and other species arose far more rapidly. Said another way, *the process of speciation can vary in its duration.*

Depending on how populations interact and their patterns of distribution, *speciation also may vary in its details.* The preceding paragraphs started you thinking about what is probably the main route, which is called allopatric speciation. Later in the chapter, you will be taking a closer look at this route and two others.

Reproductive Isolating Mechanisms

Let's now define **reproductive isolating mechanisms** as any heritable feature of body form, functioning, or behavior that prevents interbreeding between one or more genetically divergent populations. Some prevent successful mating or pollination between individuals of the divergent populations, so hybrid zygotes can't form. Let's look first at some pre-zygotic mechanisms.

Figure 17.4 (**a**) Behavioral isolation, represented by a sampling of courtship displays that precede sex between a male and female albatross. Individuals of their species recognize tactile, visual, and acoustical components of the displays.

(**b**) A case of temporal isolation. Adult *Magicicada septendecim*. This periodical cicada matures underground and emerges to reproduce every 17 years. Often its populations overlap the habitats of a sibling species (*M. tredecim*), which reproduces every 13 years. Adults live only a few weeks. Birds consider them a taste thrill and may gorge frenziedly on them to the point of vomiting; which makes one suspect why periodical cicadas stay underground for so long.

(**c**) Mechanical isolation, as demonstrated by the precise fit between a zebra orchid and a wasp species, one of its few pollinators.

Behavioral isolation: Behavioral differences bar gene flow between related species in the same territory. For instance, before male and female birds copulate, they often participate in courtship rituals (Figure 17.4a). A female is genetically prewired to recognize distinctive singing, head bobbing, wing spreading, or prancing of a male of her species as an overture to sex. Females of different species usually ignore his behavior.

Temporal isolation: What if individuals of diverging populations retain the potential to interbreed? If they reproduce at different times, this makes no difference. Most animals mate and most plants get pollinated fast, often in less than a day, so there is not much chance of overlap with others. Periodical cicadas (*Magicicada*) of the eastern United States are separated in time to an extreme. These winged insects mature underground, where they feed on juicy roots. Three species differ in size, color, and song. Every 17 years they emerge and reproduce (Figure 17.4b). Each has a "sibling species," which is about the same, morphologically. Its sibling species emerges every 13 years. Only once every 221 years do the two release gametes at the same time!

Mechanical isolation: Incompatibility between body parts is a way to keep potential mates or pollinators mechanically isolated (Figure 17.4c). Two sage species keep their pollen to themselves by attracting different insect pollinators. The pollen-bearing stamens of one extend from a nectar cup, above petals that form a big platform for big pollinators. Small bees that land on it typically do not brush against the stamens and pick up pollen. By contrast, the landing platform of the other species is too small to hold big pollinators.

Ecological isolation: Populations adapted to different microenvironments in a single habitat may be isolated ecologically. In seasonally dry foothills of the Sierra Nevada are two species of manzanita. We find one in open forests of conifers at elevations between 600 and 1,850 meters, and the other at elevations between 750 and 3,350 meters. Where their ranges overlap, the two rarely hybridize. Like other manzanitas, both species have built-in, physiological mechanisms that enhance water conservation during the dry season. But one is adapted to far more sheltered sites, where water stress is not intense. The other species is adapted to drier, more exposed sites on rocky hillsides. These ecological differences make cross-pollination unlikely.

Gametic mortality: Gametes of different species may have evolved incompatibilities at the molecular level. For example, when the pollen of one species lands on a plant of a different species, it usually doesn't recognize the molecular signals that are supposed to trigger its growth down through the plant's tissues, to the egg.

What about isolation that occurs *after* zygotes form at fertilization? Post-zygotic mechanisms may kick in while an embryo is developing. Unsuitable interactions among some of its genes or gene products cause early death, sterility, or hybrids of low fitness. The hybrid offspring are often weak; survival rates are low. Some hybrids are sturdy but sterile. Mules, the offspring of a female horse and male donkey, are like this.

THE BIOLOGICAL SPECIES CONCEPT. **A species is one or more populations of individuals that (1) are interbreeding under natural conditions and producing fertile offspring, and (2) are reproductively isolated from other such populations.**

Speciation is the process by which a daughter species forms from a population or subpopulation of a parent species. The process varies in its details and in the length of time it takes to complete reproductive isolation.

By what may be the most common speciation mechanism, a geographic barrier separates populations of a species. Mutation, natural selection, and genetic drift independently operate in each population, so differences build up in the gene pools. This outcome is called genetic divergence.

Reproductive isolating mechanisms may evolve simply as by-products of the genetic changes. They are heritable traits that, one way or another, prevent interbreeding.

SPECIATION IN GEOGRAPHICALLY ISOLATED POPULATIONS

Allopatric Speciation Defined

If we assume *physical separation* between populations promotes the genetic changes necessary for a species to originate, then allopatry may be the main speciation route. By the model for **allopatric speciation**, some type of physical barrier arises and shuts down all gene flow between populations or subpopulations of a species. (*Allo–* means different, and *patria* can be taken to mean homeland.) Reproductive isolating mechanisms evolve in the genetically diverging populations. The process of speciation is completed when individuals of the two populations no longer will interbreed even if changing circumstances put them back together in the same area.

Whether a geographic barrier proves to be effective at blocking gene flow between populations depends on an organism's means of travel, how fast it can travel, and whether it is compelled or behaviorally inclined to disperse. You have only to reflect on the garden snails and the wandering albatross described at the start of this chapter to conclude that this is so.

The Pace of Geographic Isolation

Some measurable distance separates the populations of most species, and gene flow among them is more of an intermittent trickle than a steady stream. Also, barriers can rapidly arise and shut off the trickles. For example, in the 1800s, a monstrous earthquake buckled part of the Midwest and altered the course of the Mississippi River. The new course cut right through the habitats of certain insects that could not swim or fly, so it cut off gene flow between populations on opposite shores.

Geographic isolation also occurs over great spans of time. We see this in parts of the fossil record that reflect the breakup of continuous environments.

This happened after huge glaciers advanced down into North America and Europe during past ice ages and cut off populations from one another. In time the glaciers retreated, and the descendants of the separated populations met up. Although related by descent, some were no longer reproductively compatible; they had evolved into new species. Genetic divergences did not proceed as far between other separated populations; their descendants are still interbreeding. Reproductive isolation was incomplete, so speciation did not occur.

Another example: As you might know, the Earth's crust is fractured into gigantic plates, somewhat like a cracked eggshell. In the past, imperceptibly slow but colossal movements of the plates caused land masses to collide. This happened at what is now called Central America. Collisions uplifted part of an ancient ocean basin and formed the Isthmus of Panama. You will read about this in Section 18.3. For now, it's enough to know that this geographic barrier divided marine populations and opened the door to allopatric speciation.

In the 1980s, John Graves compared four enzymes from the muscle cells of two related species of isthmus fishes (Figure 17.5). Both species are strong swimmers. As Graves knew, temperature has different effects on different enzymes. Seawater on the Pacific side of the isthmus is cooler by about 2°–3°C than on the Atlantic side. It also varies seasonally. Graves discovered that the four "Pacific" enzymes he studied function better at lower temperatures than the equivalent "Atlantic" enzymes. He then subjected the four pairs of enzyme molecules to gel electrophoresis (Section 15.3). In this case, the laboratory technique revealed slight differences in electric charge between two pairs, which differ slightly in amino acid sequences.

Graves drew two tentative conclusions: First, selection pressures imposed by small differences in environmental conditions may have triggered divergences in the molecular structure of certain enzymes. Second, it appears that closely related fish populations on opposite sides of the isthmus are diverging from each other. Why? Alternative molecular forms of the enzymes being studied are already displaying detectable differences in catalytic activity.

How can we interpret results from this study of geographical isolation? For the populations on each side of the isthmus, a gradual accumulation of neutral or adaptive gene mutations might be a microevolutionary "foot in the door." The genetic divergences among the related fish populations might be evidence of speciation in progress.

ISTHMUS OF PANAMA

Figure 17.5 (**a**) Blue-headed wrasse (*Thalassoma bifasciatum*) from the Atlantic Ocean near the Isthmus of Panama. (**b**) Cortez rainbow wrasse (*T. lucasanum*) from the Pacific Ocean near the isthmus. The species may be related by descent from a shared ancestral population that split when geologic forces created the isthmus. You might see individuals that differ in body coloration and patterning from these specimens. Such variation is common among reef fishes.

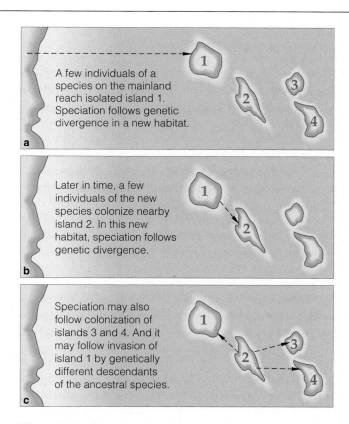

Figure 17.6 Sketch of allopatric speciation on an isolated archipelago. (Can you envision other possibilities?)

In panel a:

A few individuals of a species on the mainland reach isolated island 1. Speciation follows genetic divergence in a new habitat.

In panel b:

Later in time, a few individuals of the new species colonize nearby island 2. In this new habitat, speciation follows genetic divergence.

In panel c:

Speciation may also follow colonization of islands 3 and 4. And it may follow invasion of island 1 by genetically different descendants of the ancestral species.

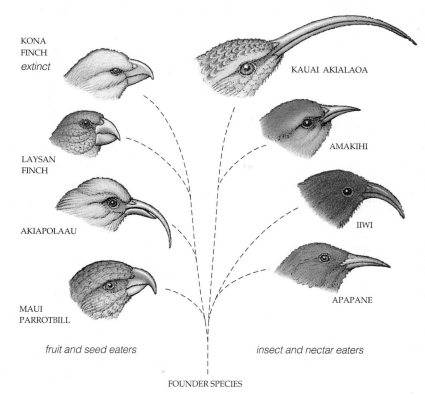

KONA FINCH *extinct*

KAUAI AKIALAOA

LAYSAN FINCH

AMAKIHI

AKIAPOLAAU

IIWI

MAUI PARROTBILL

APAPANE

fruit and seed eaters

insect and nectar eaters

FOUNDER SPECIES

Figure 17.7 A few Hawaiian honeycreepers, a fine example of how a new arrival in species-poor habitats on an isolated archipelago can be the start of a flurry of allopatric speciation.

Allopatric Speciation on Archipelagos

An **archipelago** is an island chain some distance away from a continent. Some archipelagos are so close to the mainland that gene flow is more or less unimpeded, and there is little if any speciation. The Florida Keys are like this. Other archipelagos are isolated enough to serve as laboratories for the study of evolution (Figure 17.6). Foremost among these are the Hawaiian Archipelago, nearly 4,000 kilometers from the California coast, and the Galápagos Islands, about 900 kilometers from the Ecuadorian coast. The islands of both chains are merely the tops of great volcanoes, some still active, that rise from the deep seafloor. When each volcano first broke the surface of the sea, its fiery surface was devoid of life.

Remember those Galápagos finches (Section 16.3)? A few finches from the mainland apparently evolved in isolation on one Galápagos island. Later, some of their descendants reached other islands in the chain. In those new and unoccupied habitats—even in different parts of the same habitats—conditions varied, and the island hoppers were subject to a variety of selection pressures. In time, genetic divergences within and between the islands paved the way for new episodes of allopatric speciation. Later on, some island-hopping new species even invaded the island of their ancestors. The distances between islands are enough to foster divergences, but not enough to stop occasional invasions.

Bursts of speciation in the Hawaiian Archipelago have been quite dramatic. They are a premier example of adaptive radiation in a species-poor environment, as described in Section 17.4. The youngest island, Hawaii, formed less than a million years ago. Here alone we see diverse habitats, ranging from lava beds to rain forests to alpine grasslands and high, snow-capped volcanoes. When the ancestors of Hawaiian honeycreepers arrived, they found a veritable buffet of fruits, seeds, nectars, and tasty insects—and not many competitors for them. The near absence of competition from other bird species fanned allopatric speciations. Figure 17.7 only hints at the resulting variation that arose among the different species of Hawaiian honeycreepers. Today, these and thousands of other species of animals and plants that evolved in the archipelago are found nowhere else. As another example of the potential for speciation, the Hawaiian Islands make up less than 2 percent of the world's land mass. Yet they are home to 40 percent of all species of fruit flies (*Drosophila*).

ALLOPATRIC SPECIATION. **Some type of physical barrier intervenes between populations or subpopulations of a species and prevents gene flow among them. By doing so, it favors genetic divergence and speciation.**

MODELS FOR OTHER SPECIATION ROUTES

Geographic isolation may not always be required for reproductive isolation. In the models for sympatric and parapatric speciation, populations diverge genetically even while they are in contact with one another.

Sympatric Speciation

By the model for **sympatric speciation**, a species may form *within* the home range of an existing species, in the absence of a physical barrier. (*Sym–* means together with, as in "together with others in the homeland.")

EVIDENCE FROM CICHLIDS IN AFRICA   In 1994, Ulrich Schliewen and his coworkers found evidence for the sympatric speciation model. They studied fishes called cichlids that live together in two lakes in Cameroon, West Africa. Each lake basin is the collapsed cone of a small volcano (Figure 17.8). Yet eleven cichlid species coexist in one lake. Nine coexist in the other.

The researchers analyzed mitochondrial DNA from all species in one lake. They did the same for all species in the other lake as the basis for comparison. In their nucleotide sequences, the species of each lake are like each other and *not* like related species in nearby lakes and rivers. For example, nine species in one lake are the only cichlids having a unique base-pair substitution in the gene that specifies the protein cytochrome *b*.

Physical and chemical conditions are too uniform in the lakes to foster geographic separation. For instance, shorelines are so uniform that even small topographic barriers are absent. Thus, allopatric populations could not have evolved, even on a small scale. The lakes are isolated except for tiny creeks trickling in from higher elevations; there has been no gene flow from outside. The lakes must have been colonized before connections with a nearby river system were severed. Also, cichlids are not sluggish swimmers. Besides being very mobile, individuals of different species often make contact in the small lakes; they must be living in sympatry.

Because clusters of species are more closely related to one another than to species anywhere else, we may assume they share a common ancestor. Because of their restricted distribution, we may assume they formed in the same lake. And yet nothing about the lakes isolates different populations. Therefore, the ancestors of each cichlid species were never isolated from one another while they were evolving in separate directions.

Within each lake, species do show small degrees of *ecological* separation; differences in feeding preferences put them in different places. Some cichlid species feed in open waters and others at the lake bottom. Even so, they all *breed* close to the bottom, in sympatry. It may be that small-scale ecological separation was enough to influence sexual selection among potential mates. Over the generations, it may have afforded the reproductive isolation that can lead to speciation.

SPECIATION BY WAY OF POLYPLOIDY   Quite possibly, sympatric speciation has been a prevalent evolutionary event among flowering plants. Consider that about half of all known species of flowering plants are polyploid. **Polyploidy**, remember, is a change in the chromosome number; offspring inherit three or more of each type of chromosome characteristic of the parental stock. Such changes arise when chromosomes separate improperly during meiosis or mitosis. They also arise when a germ cell replicates its DNA but fails to divide, then goes on to function as a gamete (Section 11.9).

Speciation may have been rapid for many flowering plants that engage in self-fertilization or some asexual reproductive mode. Suppose they produced polyploid offspring. If the extra chromosomes paired with each other at meiosis, maybe the extra set of genes did no harm. Common bread wheat is one of the species that might have arisen by polyploidy. In this case, though, cross-fertilization also was involved (Figure 17.9).

Allen Orr has evidence that polyploid animals are rare because of a failed **dosage compensation**. By this normal event, genes on sex chromosomes are expressed at the same levels in females *and* males. For instance, X chromosome inactivation in female mammals means the cells of both males and females have a single active X chromosome (Section 14.3). Polyploidy skews dosage compensation, with bad or fatal effects. The mechanism is absent in plants, so their chromosome doublings are not as problematic as they are in animals.

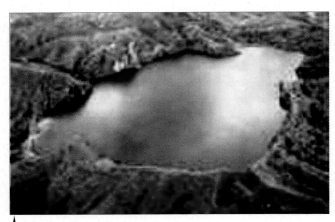

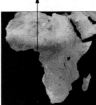

Figure 17.8   Lake Barombi Mbo in Cameroon, West Africa. Nine kinds of cichlids apparently evolved by way of sympatric speciation in this small, isolated crater lake. There is not even microgeographic isolation. Cichlid populations separate by feeding preferences, but they all breed near the lake bottom, in sympatry.

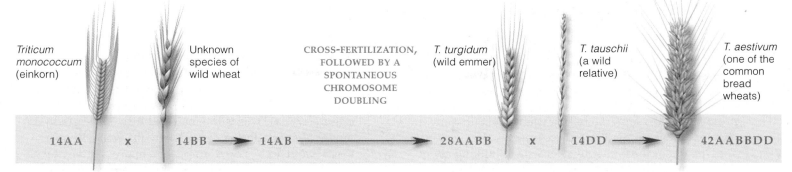

| 14AA | x | 14BB → 14AB | 28AABB | x | 14DD → | 42AABBDD |

Triticum monococcum (einkorn)

Unknown species of wild wheat

CROSS-FERTILIZATION, FOLLOWED BY A SPONTANEOUS CHROMOSOME DOUBLING

T. turgidum (wild emmer)

T. tauschii (a wild relative)

T. aestivum (one of the common bread wheats)

a By 11,000 years ago, humans had started to cultivate wild wheats. Einkorn (*Triticum monococcum*) is still around. It has a diploid chromosome number of 14 (two sets of 7 chromosomes, shown above as 14AA). Long ago, einkorn wheat probably hybridized with another wild wheat species having the same chromosome number.

b If the AB hybrid offspring were sterile but self-fertilizing, an interbreeding population of AB plants could have arisen by asexual reproduction. About 8,000 years ago, polyploidy must have arisen in such a population. Wild emmer (*T. turgidum*) plants are tetraploid (AABB), with a chromosome number of 28 (two sets of 14). They are fertile; at meiosis, the A chromosomes pair with each other, and the B chromosomes pair with each other.

c Later, an AABB plant probably hybridized with *T. tauschii*, a wild relative of the wild emmer. Its diploid chromosome number must have been 14 (two sets of 7 DD). Populations of the hybrid descendants now include common bread wheats. One of these, *T. aestivum*, has a chromosome number of 42 (six sets of 7 AABBDD).

Figure 17.9 Presumed sympatric speciation in wheat by polyploidy and hybridizations. Wheat grains 11,000 years old have been found in the Near East. Diploid wild wheats still grow there.

BULLOCK'S ORIOLE

BALTIMORE ORIOLE

BULLOCK'S ORIOLE   BALTIMORE ORIOLE

HYBRID ZONE

Figure 17.10 A possible setting for parapatric speciation? Hybrids arise along a common border of the geographic ranges of two species of orioles. To the east are Baltimore orioles; to the west are Bullock's orioles. Hybridization was once so common that these two birds were considered subspecies. In 1997, the American Ornithologist Union decided they are separate species.

Parapatric Speciation

By the model for **parapatric speciation**, neighboring populations become distinct species while maintaining contact along a common border. (*Para–* means near, as in "near another homeland.") Interbreeding individuals produce hybrid offspring in the region, which is called a **hybrid zone**. Evidence that might support parapatric speciation is sketchy. Why? It's often difficult to know whether or not geographically separated populations are only subspecies. *Subspecies* refers to geographically distinct populations of the same species.

For example, Bullock's orioles differ from Baltimore orioles in the color patterning of most of their feathers.

Male orioles have different territorial songs and mate with their own kind, usually. As K. Corbin determined, the birds also differ in allele frequencies for several enzymes. Their geographic ranges differ but overlap in the American Midwest (Figure 17.10). Interbreeding was once common in the hybrid zone but is becoming less frequent. Is it an example of parapatric speciation in progress? Maybe. But it might also be an example of "secondary contact." Previously isolated subspecies that diverged recently from a common ancestor may have been getting together again.

SYMPATRIC SPECIATION. **Daughter species arise from a group of individuals within an existing population. This seems to have been a common speciation event among the polyploid species of flowering plants.**

PARAPATRIC SPECIATION. **Adjacent populations evolve into distinct species even while maintaining contact along their common border.**

PATTERNS OF SPECIATION

Branching and Unbranched Evolution

All species—past and present—are related by descent. They share genetic connections through lineages that extend back in time to the molecular origin of the first prototypic cells, some 3.8 billion years ago. Subsequent chapters focus on evidence that supports this view. In anticipation of those chapters, let's start thinking about ways to interpret the large-scale histories of species.

The fossil record provides evidence of two patterns of evolutionary change in lineages. One is branched, the other unbranched. The first is called **cladogenesis** (from the Greek *klados*, meaning branch; and *genesis*, meaning origin). This is the pattern by which a lineage splits, with populations becoming genetically isolated and then diverging in different evolutionary directions. This is the pattern of speciation described earlier.

By the second pattern, **anagenesis**, changes in allele frequencies and in morphology accumulate within an unbranched line of descent. (In this context, *ana*– means renewed.) Directional changes are confined within one lineage, and gene flow never does cease among all of its populations. At some point in time, allele frequencies and morphology have changed so much that we assign a separate name to the descendants. Such changes are occurring in the moth populations you read about earlier.

Evolutionary Trees and Rates of Change

Evolutionary trees summarize information about the continuity of relationship among species. Figure 17.11 is a simple way to start thinking about how tree diagrams are constructed. Each *branch* represents a single line of descent from a common ancestor. And each *branch point* represents a time of genetic divergence and speciation, as brought about by microevolutionary processes.

Figure 17.11 also shows how an evolutionary tree diagram conveys *rates of change*, the estimated length of time between speciation events. Branches having slight angles imply that the species emerged through many small morphological changes over long spans of time (Figure 17.11*a*). This is the main premise of the **gradual model of speciation**, which actually fits well with many fossil sequences. For example, in many sedimentary rock layers we find sequences of intricately perforated shells of foraminiferans, a common type of protistan. The observable characteristics of foraminiferan fossils in the sequences offer evidence of slow change.

Alternatively, evolutionary tree diagrams for some lineages are constructed with short horizontal branches that make an abrupt 90-degree turn, as in Figure 17.11*b*. The **punctuation model of speciation** is consistent with such diagrams. By this model, most of the changes in morphology are compressed into a brief period when populations are starting to diverge—say, within only hundreds or thousands of years. Bottlenecks, founder effects, strong directional selection, or a combination of these foster rapid speciation. The daughter species recover fast from the adaptive wrenching, then change little during the next 2 million to 6 million years or so. The proponents of this model argue that reproductive cohesion has indeed prevailed for about 99 percent of the history of most lineages. They cite strong evidence of abrupt change in many parts of the fossil record.

Changes in lines of descent apparently have been gradual, abrupt, or both. Species originated at different times and they differ in how long they have persisted. Remember, some lineages have endured without much change, producing a species here and losing a species there, often over millions of years. Other lineages have branched bushily and sometimes spectacularly during times of adaptive radiation.

Adaptive Radiations

An **adaptive radiation** is a burst of divergences from a single lineage that give rise to many new species, each adapted to an unoccupied or a new habitat or to using a novel resource. Figure 17.12 provides an example. In the past, lineages often diversified in such a way when the member species found themselves in vacant **adaptive zones**. Think of

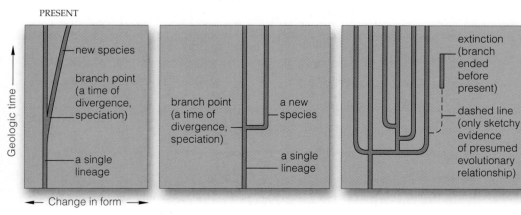

a Branching at a slight angle; a new species formed through gradual changes in traits over geologic time.

b Horizontal branching; very rapid change in traits around the time of speciation. A long vertical branch means the traits of the new species did not change much thereafter.

c Many branchings of the same lineage at or near the same point in geologic time; an adaptive radiation has taken place.

Figure 17.11 How to read evolutionary tree diagrams.

Figure 17.12 Example of adaptive radiation, as shown by an evolutionary tree diagram. This one started about 65 million years ago, at the start of a geologic era called the Cenozoic, and it led to mammals as different as the opossum and walrus. Variations in the branch widths correspond to the number of groups in the lineage, as represented at different points in time. The wider the branch, the greater the species diversity.

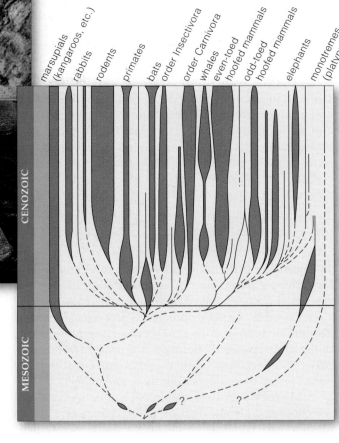

adaptive zones as ways of life, such as "burrowing in seafloor sediments" or "catching winged insects in the air at night." A lineage may radiate into such zones if it has physical, evolutionary, or ecological access to them.

Physical access means a lineage happens to be there when adaptive zones open up. For example, mammals were once distributed through uniform tropical regions of a huge continent. The continent broke up into several land masses that gradually drifted apart. Habitats and resources changed in many different ways on different masses and set the stage for independent radiations.

Evolutionary access means that a modification to an existing structure or function will permit a lineage to exploit the environment in new or more efficient ways. Such modifications are **key innovations**. Example: The forelimbs of certain five-toed vertebrates evolved into wings. And that innovation opened up novel adaptive zones for the ancestors of birds and bats (Section 18.4).

Ecological access means that the lineage can enter vacated adaptive zones or displace the resident species. Sections 25.2, 42.7, and 42.8 offer examples.

Extinctions—End of the Line

Extinction is the irrevocable loss of a species. The fossil record shows twenty or more **mass extinctions**—large, catastrophic events in which entire families and other major groups were abruptly wiped out. For example, 95 percent of all known species disappeared 250 million years ago. In other times, only certain groups vanished.

By studying the statistical distribution of extinctions, David Raup found that most were clustered in time even though they differed in size. Like George Simpson before him, he realized that times of reduced diversity follow extinction events, then new species arise and fill vacant adaptive zones. Apparently, luck has a lot to do

with it. For example, in the past, asteroids repeatedly hit the Earth. Some survivors of one impact radiated into the adaptive zones that dinosaurs had occupied. Among them were mammals—some of which gave rise to your early ancestors. As another example, species not widely distributed tend to be hit hard by pronounced declines in global temperature. This is usually the case for species adapted to warm regions. Unlike species in cold regions, they have nowhere else to go. Asteroids, drifting continents, climatic change—as you will see in later chapters, these are some of the major factors that have contributed to a general pattern of extinctions. In Chapter 25, you also will read about a mass extinction that is currently under way, thanks largely to us.

Lineages have changed gradually, abruptly, or both. Their member species originated at different times and have differed in how long they have persisted.

Adaptive radiations are bursts of divergences from a single lineage that gave rise to many new species, each adapted to a vacant or new habitat or to using a novel resource.

Repeated and often large extinctions occurred in the past. After times of reduced diversity, new species formed and occupied new or vacated adaptive zones.

Taken together, the persistence, branchings, and extinctions of species account for the full range of biological diversity at any point in geologic time.

---

1. A species is a single kind of organism, recognized partly in terms of its morphology. *17.1*

2. By the biological species concept, a species is one or more populations of individuals that are interbreeding and producing fertile offspring under natural conditions and that are reproductively isolated from other such populations. The concept identifies a species mainly in terms of the portion of alleles that promote or maintain reproductive isolation. Also, it applies only to sexually reproducing organisms. *17.1*

3. The populations of a species have a shared genetic history, are maintaining genetic contact over time, and are evolving independently of other species. *17.1*

4. Speciation is the process by which daughter species form from a population or subpopulation of a parent species; the attainment of reproductive isolation. *17.1*

a. Example: Genetic divergence proceeds after some geographic barrier stops gene flow among populations of the same species. Differences accumulate in the gene pools of the isolated populations as microevolutionary processes operate independently in each one.

b. Microevolutionary processes—mutation, natural selection, and genetic drift—may by chance give rise to reproductive isolating mechanisms. Such mechanisms prevent interbreeding between populations and thus promote irreversible genetic differences between them.

c. Speciation also can occur instantaneously, as by polyploidy or other changes in chromosome number.

5. Prezygotic isolating mechanisms prevent mating or pollination between individuals of populations. They include differences in reproductive timing or behavior, incompatibilities in reproductive structures or gametes, and occupation of different microenvironments in the same area. Postzygotic mechanisms lead to early death, sterility, or unfit hybrid offspring. They take effect after fertilization, while the embryo develops. *17.1*

6. Here are three current models for speciation:

a. Allopatric speciation. Geographic barriers prevent gene flow between the populations of a species, which genetically diverge in such a way that interbreeding will not occur in nature even if their individuals make contact with each other later on. Allopatric speciation might be the most prevalent speciation route. *17.2*

b. Sympatric speciation. Reproductive isolation of individuals in the same home range leads to speciation. Speciation by polyploidy is an example. *17.3*

c. Parapatric speciation. Adjacent populations give rise to a new species while maintaining contact along a border between their home ranges. *17.3*

7. Lineages differ in their time, rates, and direction of speciation. Speciation can proceed gradually, rapidly, or both. Extensive branching of a single lineage (a burst of divergences from it) in the same geologic time span is an adaptive radiation. The lineage may be physically in place when new adaptive zones (ways of life) open up. It may access zones by key innovations. It may enter vacated zones or displace resident species. *17.4*

8. Evolutionary tree diagrams indicate the relationships among groups of species. Each tree branch represents a line of descent (lineage). Branch points are speciation events, as brought about by natural selection, genetic drift, and other microevolutionary processes. *17.4*

9. Extinction is the irrevocable loss of a species. The fossil record shows twenty or more mass extinctions: large, catastrophic events in which entire families or other major groups were abruptly lost. *17.4*

10. Persistences, branchings, and extinctions of species account for the range of biodiversity (Table 17.1). *17.4*

Table 17.1 *Summary of Processes and Patterns of Evolution*

MICROEVOLUTIONARY PROCESSES

Mutation	Original source of alleles
Gene flow	Preserves species cohesion
Genetic drift	Erodes species cohesion
Natural selection	Preserves or erodes species cohesion, depending on environmental pressures

Stability or change in a species is the outcome of balances or imbalances among all of these processes, the effects of which are influenced by population size and by the prevailing environmental conditions.

MACROEVOLUTIONARY PROCESSES

Genetic persistence	Basis of the unity of life. The biochemical and molecular basis of inheritance extends from the origin of first cells through all subsequent lines of descent.
Genetic divergence	Basis of life's diversity, as brought about by adaptive shifts, branchings, and radiations. Rates and times of change varied within and between lineages.
Genetic disconnect	Extinction. End of the line for a species. Mass extinctions are catastrophic events in which major groups are lost abruptly and simultaneously.

Review Questions

1. How does the biological species concept differ from a definition of species based on morphological traits alone? *17.1*

2. Define speciation and describe a speciation model. *CI, 17.2, 17.3*

3. Give examples of reproductive isolating mechanisms. *17.1*

4. With respect to evolutionary tree diagrams, describe what each of the following features represents: *17.4*
   a. a single line
   b. soft-angled branching of line
   c. horizontal branching
   d. vertical continuation of branch
   e. many branchings of a line
   f. dashed line
   g. branch ending before present

5. Define and give an example of an adaptive radiation. *17.4*

6. Define extinction. What kinds of events might bring about extinctions? *17.4*

Self-Quiz   ANSWERS IN APPENDIX III

1. Sexually reproducing individuals of a species _____ .
   a. can interbreed under natural conditions
   b. can produce fertile offspring
   c. have a shared genetic history
   d. all of the above

2. Reproductive isolating mechanisms _____ .
   a. prevent interbreeding   c. reinforce genetic divergence
   b. prevent gene flow   d. all of the above

3. When potential mates occupy overlapping ranges but reproduce at different times, this is a case of _____ isolation.
   a. postzygotic   c. temporal
   b. mechanical   d. gametic

4. In an evolutionary tree diagram, a branch point represents a _____ , and a branch that ends represents _____ .
   a. single species; incomplete data on lineage
   b. single species; extinction
   c. time of divergence; extinction
   d. time of divergence; speciation complete

5. An evolutionary tree diagram with horizontal branches that abruptly become vertical is consistent with the _____ .
   a. gradual model of speciation
   b. punctuation model of speciation
   c. idea of small changes in form over long spans of time
   d. both a and c

6. Match each term with its most suitable description.
   ____ cladogenesis   a. burst of genetic divergences
   ____ anagenesis   from a single lineage
   ____ adaptive   b. catastrophic disappearance
   ____ radiation   of major groups of organisms
   ____ extinction event   c. branching lineages
   ____ mass extinction   d. a species lost from a lineage
                            e. genetic and morphological
                               change in unbranched lineage

Critical Thinking

1. You notice several duck species in the same lake habitat. All the females of the various species look quite similar to one another. But the males differ in the patterns and colors of their feathers. Speculate on which forms of reproductive isolation may be keeping each species distinct. How does the appearance of the male ducks provide a clue to the answer?

2. One family of mammals includes true horses (*Equus*), zebras (*Hippotigris* and *Dolichohippus*), and donkeys and asses (*Asinus*). Zebroids are hybrid offspring of wild zebras and domesticated horses that were confined to the same pasture (Figure 17.13). The unnatural confinement breached the reproductive barriers between the two lineages. The barriers have been in place since a genetic divergence got under way more than 3 million years ago. What does the breach suggest about the genetic changes required to attain irreversible reproductive isolation in nature?

3. Suppose a small population that was founded by only a few individuals gets cut off from gene exchange with the main population of their species. By Mayr's view of the *founder effect*, allele frequencies for a few gene products change as a result of genetic drift; then the change triggers a number of changes in other genes that are affected by those alleles. If that is so, then

Figure 17.13   A mixed herd of zebroids and horses.

genetic changes in newly founded populations may be so great that speciation is rapid—so rapid that we will seldom be able to document it through the fossil record. Does this view correspond to the gradual model or the punctuated model of speciation?

4. A key innovation, recall, is some modification in structure or function that permits a species to exploit the environment in a more efficient or novel way, compared to the ancestral species. Identify a key innovation of an existing species, such as humans. Then describe how that innovation might be the basis of an adaptive radiation in environments of the distant future.

5. Richard Lenski uses bacterial populations in culture tubes to develop model systems for studying evolution. He is fond of saying that such a population is the equivalent of the entire human population, that he can replicate it several times over, that bacteria produce several generations in a day, and that he can store them in the deep freeze, then bring them back to active form, unaltered, to directly compare ancestors and descendants. Are bacterial models relevant to evolutionary studies of sexually reproducing organisms? Before you answer, read a short article by P. Raine and M. Travisano entitled "Adaptive Radiation in a Heterogenous Environment" (*Nature*, 2 July 1998, 69–72).

Selected Key Terms

adaptive radiation *17.4*	hybrid zone *17.3*
adaptive zone *17.4*	key innovation *17.4*
allopatric speciation *17.2*	mass extinction *17.4*
anagenesis *17.4*	parapatric speciation *17.3*
archipelago *17.2*	polyploidy *17.3*
biological species concept *17.1*	punctuation model,
cladogenesis *17.4*	speciation *17.4*
dosage compensation *17.3*	reproductive isolating
evolutionary tree *17.4*	mechanism *17.1*
extinction *17.4*	speciation *CI*
gene flow *17.1*	species *17.1*
genetic divergence *17.1*	sympatric speciation *17.3*
gradual model, speciation *17.4*	

Readings

Futuyma, D. 1998. *Evolutionary Biology*. Third edition. Sunderland, Massachusetts: Sinauer.

Mayr, E. 1976. *Evolution and the Diversity of Life*. Cambridge, Massachusetts: Belknap Press of Harvard University Press.

On-Line readings at Student Guide for InfoTrac:
www.brookscole.com/biology

18

THE MACROEVOLUTIONARY PUZZLE

Legends of the Flood

So far in this unit, you have been considering some evolutionary theories that biologists routinely use to interpret the past and present, even to predict possible futures for the natural world. Today the theories are widely accepted in society at large. This was not the case when early evolutionists in Europe started to work out the details. They rankled more than a few people with their astounding theories, which revealed connections between the geologic record, the fossil record, and the sweep of biological diversity.

As evidence accumulated in support of the theories, many zoologists, geologists, and others came to accept them as guides for research. Others couldn't reconcile them with the Bible, the premier book of the Western world. That book had helped many generations cope with the uncertainties of life, with death, and with natural disasters, including a stupendous flood. Such a flood did indeed occur, but it was assigned meaning at a time when people didn't know much about geologic processes.

For example, William Ryan and Walter Pitman drew from marine geology, genetics, archaeology, botany, and linguistics to discover the location of the biblical flood. They zeroed in on the Black Sea. This landlocked sea is 1,800 meters (6,000 feet) deep. Rainfall and large rivers freshen it, and excess water flows through the Bosporus and Dardanelles straits to the Aegean Sea (Figure 18.1).

What Ryan and Pitman discovered coincides with the biblical account of a great flood. It coincides with stone tablets inscribed in Babylon long before the Bible was written. It coincides with the epic of Gilgamesh and other myths from Greece, Mesopotamia, and the Levant. What follows is the picture they pieced together.

Starting about 120,000 years ago, the climate cooled and evaporation from the sea increased dramatically. So much seawater became airborne, then accumulated on land as snow, that the sea level fell 120 meters (400 feet) below what it is today. Snow compacted into thick ice sheets that advanced over much of Europe, North America, and Asia. Then, about 20,000 years ago, the global climate warmed and a colossal meltdown began. Meltwater drained mainly to the sea, but some collected on land as immense freshwater lakes. One lake was the start of the Black Sea. It rose high enough to connect with a narrow outlet to the Aegean Sea (Figure 18.2a).

Glacial conditions returned, then the climate warmed again. This time nearly all meltwater was diverted to the sea. Little rain fell on land, and evaporation gradually lowered the lake's surface. By 7,600 years ago, it was 150 meters (500 feet) below its outlet to the sea.

When vast ice sheets prevailed, cool, dry deserts and grasslands formed to the south. Modern humans (*Homo sapiens*) witnessed the change. Hunter–gatherers moved into fertile valleys and river deltas in Europe and Asia. Where they encountered water, wild game, and edible plants, they founded settlements—only to abandon them when the climate changed and resources vanished. The lake beckoned the displaced more than once. Being so far below sea level, the lake edge was warmer than the surrounding arid mountains and plateaus. It welcomed the thirsty when springs, ponds, and other lakes dried up. In the moist, fertile soil of that Edenlike oasis, wild wheat and barley took root. So did trading cultures and early attempts at farming.

All the while, the sea was rising. By 7,600 years ago it was high enough to lap over the crest of the Bosporus valley, which plunged to the lake below. The first trickles cut into soft earth. But the cut deepened and water thundered through at 200 times the flow volume of Niagara Falls. For the next ten months, rapidly advancing water submerged valleys and villages, driving humans and wildlife before it.

Figure 18.1 The geologic region where global climate changes led to a catastrophe that altered the course of human history.

a Direction of meltwater flows from Eurasian ice sheet that started about 20,000 years ago

b Freshwater lake (future Black Sea) connects with Mediterranean 14,500 years ago

c The flow of seawater in the same region during flood 7,600 years ago

d Present-day exchange of seawater and freshwater between the Mediterranean and Black seas

Figure 18.2 Historical connection between the Mediterranean and the ice age freshwater lake that became the Black Sea.

These were farmers and fishermen, not nomads who would simply pack up and move on. They built houses and boats; they traded obsidian, leather, herbs, and finely incised pottery. They could not comprehend the force behind their mass dispersal. In time, small bands settled in the Levant, Mesopotamia, Egypt, the Persian Gulf, central Asia, and western Europe. They brought with them precious seeds, farm animals, and fragments of languages, ideas, and technologies. And they carried with them a terrifying story of a monstrous flood.

The story endured. It influenced investigations into processes of nature, as when sixteenth-century biblical scholars argued that the Earth was about 6,000 years old. To them, that was plenty of time for man's creation, fall from grace, and salvation after a punishing flood. The geologic changes that made it into oral and written traditions were such extraordinary catastrophes that people assumed they were divinely invoked. Otherwise the Earth was just an unchanging stage for the human drama. This was a core hypothesis of what came to be known as theories of **catastrophism**.

Biological science emerged within the framework of such influential beliefs, and so did awareness of changes in the Earth and its creatures. Understand this, and you will gain insight into why acceptance of the very idea of biological evolution was so long in coming.

Key Concepts

1. In the evolutionary view, all species that have ever lived are related by descent—some closely and others remotely so.

2. Macroevolution is the name for the major patterns, trends, and rates of change among lines of descent, or lineages, during the history of life. The fossil record, the geologic record, and radiometric dating of rocks yield evidence of macroevolution. So do morphological comparisons of existing groups of species.

3. The fossil record has enough detail to reveal major patterns of descent with modification. The record will never be complete in all of its details, but gaps are to be expected. We know that parts were lost owing to major movements of the Earth's crust and other geologic events. We also know there is a low probability of fossilization in the first place, and of even finding a fossil.

4. Biogeography is the study of the distribution of species through time and through the environment. A theory about crustal movements helps explain some puzzling aspects of the global distribution of life in the past.

5. Comparisons of body forms among major groups may reveal evolutionary relationship. Adults and embryos of different lineages often show similarities in one or more body parts that reflect descent from a common ancestor.

6. Biochemical comparisons within and between major lineages provide strong evidence of macroevolution.

7. Biologists use taxonomy, phylogenetic reconstruction, and classification to study patterns in the diversity of life. Taxonomy is concerned with identifying and naming new species. Phylogenetic reconstruction infers evolutionary connections by certain analytical methods. Classification is concerned with organizing information about species into retrieval systems.

FOSSILS—EVIDENCE OF ANCIENT LIFE

About 500 years ago, Leonardo da Vinci was brooding about seashells entombed in rocks of northern Italy's high mountains, hundreds of kilometers from the sea. How did they get there? By the biblical explanation, floodwaters had deposited the shells in the mountains (Figure 18.3). But many shells were thin, fragile, and intact. Surely they would have been battered to bits if they had been swept across such distances and all the way up into the mountains.

Leonardo also brooded about the rocks. They were stacked like cake layers, and some contained shells but others had none. Then he remembered how large rivers swell with spring floodwaters and deposit silt in the sea. Were the layers deposited long ago, in separate intervals? If so, then shells in the mountains could be evidence of layered communities of organisms that once lived in the seas! Leonardo didn't announce his novel idea. Perhaps he suspected it would be considered heresy and invite deafening silence, imprisonment, or worse.

By the 1700s, **fossils** were accepted as the buried remains and impressions of organisms that had lived in the past. (*Fossil* comes from a Latin word for something that was "dug up.") They were still being interpreted through the prism of prevailing cultural beliefs, as when a Swiss naturalist excitedly unveiled the remains of a giant salamander and announced that they were the skeleton of a man who had drowned in the great flood. By midcentury, however, scholars began to question the interpretations.

Extensive mining, quarrying, and canal excavations were under way. The diggers were finding similar rock layers and similar fossil sequences in distant places, such as the cliffs on both sides of the English Channel. The layers appeared to record the passing of geologic time. Thus, the ordered array of fossils within the rock layers might be a historical record of life—*a fossil record.*

Fossilization

Figure 18.4 shows a few fossils. Most fossils discovered so far are bones, teeth, shells, seeds, spores, and other hard parts. Unless organisms are buried quickly, their soft parts decompose or scavengers make short work of them. Fossilized feces (coprolites) hold residues of species that were eaten in ancient environments. Also, imprints of leaves, stems, tracks, burrows, and other *trace* fossils give indirect evidence of past life.

Fossilization is a very slow process that starts when an organism, or traces of it, becomes buried in volcanic ash or sediments. Sooner or later, water infiltrates the organic remains, which become infused with dissolved metal ions and other inorganic compounds. More and more sediments gradually accumulate above the burial site. They exert ever increasing pressure on the remains. Over great spans of time, the pressure and the chemical changes transform those remains to stony hardness.

Preservation is favored when organisms are buried rapidly in the absence of oxygen. Gentle entombment by volcanic ash or anaerobic mud is best. Preservation also is favored when a burial site remains undisturbed. Most often, however, erosion and other geologic insults crush, deform, break, or scatter the fossils. This is one reason why fossils are relatively rare.

Fossils in Sedimentary Rock Layers

Stratified layers of sedimentary rock, stacked one on top of the other, are a rich source of fossils. Long ago, each layer formed by the gradual deposition of volcanic ash, silt, and other materials. Example: Sand and silt piled up when ancient rivers transported them from land to the sea, just as Leonardo thought. Over time, the sand was compressed into sandstone, the silt into shale. Deposits were not continuous. For instance, as you read in the introduction, sea levels changed when a long-term cooling trend ushered in ice ages. Deposition ceased in some places but resumed after the climate warmed and the sea level rose again.

We call the formation of sedimentary rock layers **stratification**. We can expect that the deepest layers were the first to form, and layers closest to the surface, the last. We can expect that most of the layers formed horizontally, because particles tend to settle in response to gravity. You may

Figure 18.3 From the Sistine Chapel in Italy, Michelangelo's painting of the onset of the biblical flood.

Figure 18.4 (**a**) A fossil hunter's dream: a complete skeleton of a bat that lived 50 million years ago. Erosion and other forces left few ancient burial sites undisturbed, so intact fossils are rare. Even the jumbled parts of the ducklike birds in (**b**) are a good find. It will take hours of preparation and analysis to identify the species. (**c**) Fossilized parts of the oldest known land plant (*Cooksonia*). Its stems were not even seven centimeters tall. (**d**) The skeletal remains of an ichthyosaur. This marine reptile lived 200 million years ago. Section 19.6 includes art that shows what it looked like.

come across tilted or ruptured layers, as along a road cut into the side of a mountain. They usually are a sign of geologic disturbance after the time of stratification.

Understand how rock layers form, and you realize that fossils in a particular layer are from a similar age in Earth history. Specifically, *the older the layer, the older the fossils*. Given that rock layers formed in sequence, then their fossil assemblages are unique to sequential ages. That is why fossils can be used to assign relative dates to the record of the rocks. The next section will show how the fossil layers have been used as reference points in constructing the geologic time scale.

Interpreting the Fossil Record

Fossils that have been recovered for more than 250,000 known species are clues to life's history. Judging from the current range of diversity, there must have been many, many millions more species than that. We never will be able to recover fossils for most of them, so the record of past life is incomplete. Why is this so?

Think about the odds against finding evidence of an extinct, ancient species. At least one of its individuals had to die and be gently buried before something else ate it or decomposed it. The burial site had to escape obliteration by erosion, lava flows, and other geologic events. And the fossil had to end up in a place where someone could actually find it, as in a sedimentary layer exposed by a river cutting down through a canyon.

Also, most species of ancient communities did not lend themselves to preservation. For example, unlike bony fishes and hard-shelled mollusks, jellyfishes and soft-bodied worms don't show up as much in the fossil record. Probably they were just as common, or more so.

Also think about population density and body size. A population of plants might release millions of spores in a growing season. The earliest humans lived in small groups and produced few offspring. What are the odds of finding even one fossilized human bone compared to finding spores of plants that lived at the same time?

Finally, think about a line of descent, or **lineage**, on a remote volcanic island that ended up sinking into the sea. Or think about one lineage that lasted only briefly and another that endured for billions of years. Which is more likely to be represented in the fossil record?

Fossils are physical evidence of organisms that lived in the distant past. Stratified layers of sedimentary rock that are rich in fossils are a historical record of life. The deepest layers generally hold the oldest fossils.

The fossil record is an incomplete record of life in the past. Major geologic changes have obliterated much of it. Also, our sampling of the record is slanted toward species that had large bodies, hard parts, dense populations, broad distribution, and persistence through time.

Even so, the fossil record is now substantial enough to reconstruct the sequence of changes in the history of life.

Dating Pieces of the Macroevolutionary Puzzle

ORIGIN OF THE GEOLOGIC TIME SCALE   Probably for as long as they have been digging up rocks, people have been finding fossilized leaves, shells, and other evidence of past life. *How do we know how old the fossils are?*

We didn't have a clue until long after early geologists thought about fossils in stratified rock layers. As they hypothesized, if newly formed rocks accumulate on top of older ones, then fossils in the upper layers must be younger than those in layers deposited on top. With this in mind, they constructed a chronology of Earth history—a **geologic time scale**—by counting backwards through layer upon layer of sedimentary rock.

Through comparative studies, they discovered four abrupt transitions in fossil sequences from around the world. They decided to use the transitions as boundaries between four great intervals in time. The first interval,

corresponding to the oldest known fossils, was named the Proterozoic. They named later intervals the Paleozoic, the Mesozoic, and the "modern" era, the Cenozoic.

Figure 18.5 is a modern version of the geologic time scale. We correlate it with **macroevolution**—with major patterns, trends, and rates of change among lineages. For example, all modern animal phyla emerged abruptly through adaptive radiations at the Proterozoic–Paleozoic boundary (the "Cambrian explosion"). The Paleozoic–Mesozoic boundary marks a mass extinction of nearly all families on land and in the seas. The Mesozoic–Cenozoic boundary marks a mass extinction of dinosaurs and other reptiles; an adaptive radiation of mammals followed.

The Proterozoic turned out to be an immense span. It, too, was divided into great intervals. Life originated in one of those intervals, the Archean eon.

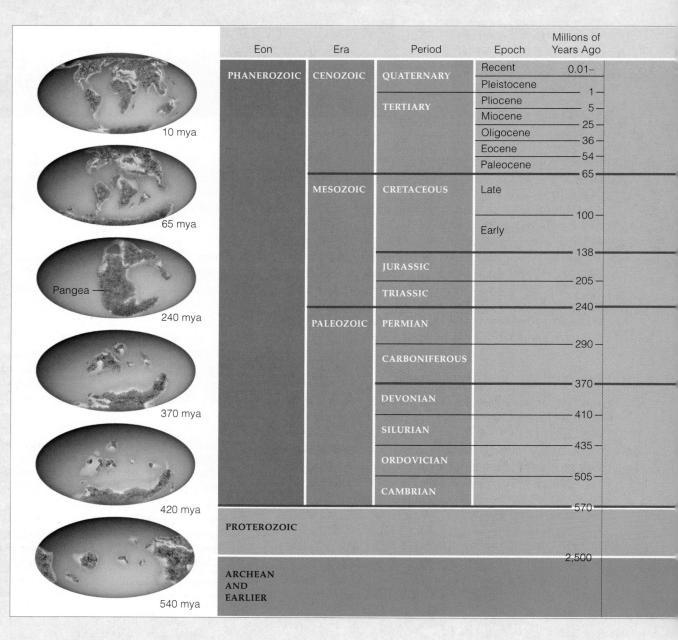

Figure 18.5   Geologic time scale. The major boundaries mark times of the greatest mass extinctions. Radiometric dating methods allowed researchers to assign absolute dates. Life originated during the Archean. The maps are an overview of changes in the distribution of land masses and oceans over time. The next chapter correlates the geologic changes with the history of life, as previewed in the far-right column.

The time spans shown here are not to scale. If they were, the Archean and Proterozoic portions would run off the page. Think of the spans as minutes on a clock that runs from midnight to noon. If we say that life originated at midnight, the Paleozoic began at 10:04 A.M., the Mesozoic at 11:09 A.M., and the Cenozoic at 11:47 A.M. The Recent epoch of the Cenozoic started in the last 0.1 second before noon.

Eon	Era	Period	Epoch	Millions of Years Ago
PHANEROZOIC	CENOZOIC	QUATERNARY	Recent	0.01–
			Pleistocene	1–
		TERTIARY	Pliocene	5–
			Miocene	25–
			Oligocene	36–
			Eocene	54–
			Paleocene	65–
	MESOZOIC	CRETACEOUS	Late	
				100–
			Early	
				138–
		JURASSIC		205–
		TRIASSIC		240–
	PALEOZOIC	PERMIAN		290–
		CARBONIFEROUS		370–
		DEVONIAN		410–
		SILURIAN		435–
		ORDOVICIAN		505–
		CAMBRIAN		570–
PROTEROZOIC				2,500
ARCHEAN AND EARLIER				

10 mya

65 mya

Pangea — 240 mya

370 mya

420 mya

540 mya

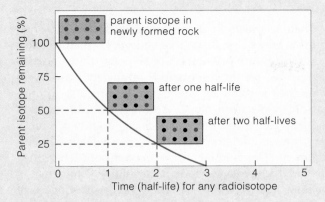

Figure 18.6 Decay of radioisotopes at a fixed rate to more stable form. The half-life of each kind is the time it takes for 50 percent of a sample to decay. After two half-lives, 25 percent of the sample has decayed, and so on.

a A living mollusk takes up trace amounts of ^{14}C carbon and a lot more ^{12}C, the more stable and most common carbon isotope.

b The mollusk dies, gets buried in sediments, and later is fossilized. As soon as it died, the proportion of ^{14}C to ^{12}C started declining owing to radioactive decay. ^{14}C has a half-life of 5,370 years. Half will be gone from the fossil in 5,370 years, half of what remains will be gone in another 5,370 years, and so on.

c Much later, someone unearths the fossil. Measuring the ^{14}C/^{12}C ratio reveals the half-life reductions since death. In this case, the ratio is one-eighth of the ^{14}C/^{12}C ratio in living organisms. The mollusk lived about 16,000 years ago.

Figure 18.7 Dating a fossil with the help of carbon 14 (^{14}C). This radioisotope forms in the atmosphere, where it combines with oxygen to form carbon dioxide. Along with far greater quantities of the more stable carbon isotopes, trace amounts of carbon 14 enter the web of life through photosynthesis. All organisms incorporate it.

**Major Geologic and Biological Events
That Occurred Millions of Years Ago (mya)**

1.65 mya to present. Major glaciations. Modern humans evolve. The most recent *extinction crisis* is under way.

65–1.65 mya. Major crustal movements, collisions, mountain building. Tropics, subtropics extend poleward. When climate cools, dry woodlands, grasslands emerge. *Adaptive radiations* of flowering plants, insects, birds, mammals.

65 mya. Asteroid impact; *mass extinction* of all dinosaurs and many marine organisms.

135–65 mya. Pangea breakup continues, inland seas form. *Adaptive radiations* of marine invertebrates, fishes, insects, and dinosaurs. Origin of angiosperms (flowering plants).

181–135 mya. Pangea starts to break up. Marine communities flourish. *Adaptive radiations* of dinosaurs.

205 mya. Asteroid impact? *Mass extinction* of many species in seas, some on land. Mammals, some dinosaurs survive.

240–205 mya. *Adaptive radiations* of marine invertebrates, fishes, dinosaurs. Gymnosperms dominant land plants. Origin of mammals.

Mass extinction. Ninety percent of all known families lost.

280–240 mya. Supercontinent Pangea and world ocean form. On land, *adaptive radiations* of reptiles and gymnosperms.

370–280 mya. Recurring ice ages. On land, *adaptive radiations* of insects, amphibians. Spore-bearing plants dominate; cone-bearing gymnosperms present. Origin of reptiles.

370 mya. *Mass extinction* of many marine invertebrates, most fishes.

435–370 mya. Major crustal movements. Ice ages. *Mass extinction* of many marine species. Vast swamps form. Origin of vascular plants. *Adaptive radiation* of fishes continues. Origin of amphibians.

500–435 mya. Major crustal movements. *Adaptive radiations* of marine invertebrates, early fishes.

550–500 mya. All land masses near equator. Simple marine communities flourish until origin of animals with hard parts.

700–550 mya. Supercontinent breaks up. Ice age. *Mass extinction.*

2,500–570 mya. Oxygen accumulates in atmosphere. Origin of aerobic metabolism. Origin of eukaryotic cells. Divergences lead to eukaryotic cells, then protistans, fungi, plants, animals.

3,800–2,500 mya. Origin of photosynthetic prokaryotic cells.

4,600–3,800 mya. Origin of Earth's crust, first atmosphere, first seas. Chemical, molecular evolution leads to origin of life (from proto-cells to anaerobic prokaryotic cells).

RADIOMETRIC DATING The geologic time scale only had relative ages until the discovery of radioactive decay. Absolute dates have been assigned to it by **radiometric dating**. This method is used to measure the proportions of (1) a radioisotope in a mineral trapped long ago in a new rock and (2) a daughter isotope that formed from it in the same rock. Remember, the number of protons and neutrons in the atomic nucleus of a radioactive element is *unstable* (Section 2.1). The nucleus spontaneously decays —it gives up energy and one or more particles of itself— until it reaches a more stable configuration.

Half-life is the time it takes for half of a given quantity of a radioisotope to decay into a less unstable, daughter isotope (Figure 18.6). The clock-like decay rate is constant; changes in pressure, temperature, or chemical state can't alter it. For example, a mineral in most volcanic rocks contains uranium 238, which decays into thorium 234. The still-unstable daughter isotope decays into something else, and so on through a series of intermediate daughter isotopes to the final, most stable configuration for this series: lead 206. Uranium 238 has a half-life of 4.5 billion years. Researchers using it realized that the Earth formed more than 4.6 billion years ago. As another example, with its half-life of 5,370 years, carbon 14 can be used to date more recent fossils (Figure 18.7).

Radiometric dating works for volcanic rocks or ashes, but most fossils are in sedimentary rocks. The only way to date most fossil-containing rocks is to determine their position relative to volcanic rocks in the same area. The dating method has an error factor of less than 10 percent.

EVIDENCE FROM BIOGEOGRAPHY

While early geologists were busily discovering fossils and mapping the record of the rocks, Charles Darwin was so taken with one of their theories of Earth history that it helped shape his view of life. By the **theory of uniformity**, recall, mountain building and erosion had repeatedly worked over the Earth's surface in precisely the same ways through time (Section 16.2). Great stacks of sedimentary layers, as in Figure 18.8, were evidence of this. Yet the more geologists clinked their hammers against the rocks, the more they realized that repetitive change was only part of the picture. Like life, the Earth had changed irreversibly; it evolved.

An Outrageous Hypothesis

For instance, the Atlantic coasts of South America and Africa seemed to "fit" like jigsaw puzzle pieces. Were all continents once joined as a "supercontinent" that later broke up? Did the fragments drift apart? This idea came to be known as continental drift. Geologists also wondered if mountain ranges that parallel many coasts had formed when some of the huge fragments collided.

Models for the proposed supercontinent abounded. Alfred Wegener gave the name **Pangea** to his version. In constructing his model, he used glacial deposits as clues to infer past climate zones. Other clues came from the world distribution of fossils and existing species.

Most scholars could not accept the hypothesis. They knew the continents are thinner and not as dense as the mantle, a rocky region below them. To them, continents plowing on their own across the mantle seemed about as plausible as a Kleenex tissue plowing across water in a bathtub. The theory of uniformity had become so entrenched, it was shutting out alternative ideas.

Then scientists made an amazing discovery. When iron-rich rocks first form, their internal structure takes on a north–south orientation in response to the Earth's magnetism. Yet the orientations in ancient rocks were *not* aligned with the magnetic poles. Using 200-million-year-old rocks from North America and Europe as a magnetic guide, scientists made sketches in which rocks were rotated into their inferred, original north–south alignment. The alignments dovetailed only when they moved North America and western Europe together.

More puzzles! Deep-sea probes revealed evidence of seafloor spreading. Molten rock erupts at mid-oceanic ridges in the seafloor. It flows laterally in both directions from the ridges, then hardens to form new crust. The spreading new crust forces the older crust down into deep trenches elsewhere in the seafloor. The ridges and trenches are the edges of huge, relatively thin plates, somewhat like the pieces of a cracked eggshell (Figure 18.9). The crustal plates move exceedingly slowly. But as they do, they raft continents to new positions.

What did the findings mean? Continental drift had to be just one part of a broader explanation of crustal movements, a **plate tectonics theory**. Researchers soon found ways to demonstrate the predictive power of the new theory. For instance, in Africa, India, Australia, and South America, the same succession of coal seams, glacial deposits, and basalt held fossils of a seed fern (*Glossopteris*, Figure 18.10) and a mammal-like reptile (the therapsid *Lystrosaurus*, Section 19.6). The plant's seeds were too heavy and the animal was too small for geographic dispersal across the open ocean. But what if both had originated on **Gondwana**—another proposed supercontinent—which was even older than Pangea?

A geologist predicted that fossils of *Glossopteris* and *Lystrosaurus* would also be discovered in Antarctica in a series of glacial deposits, coal seams, and basalt, like that in the southern continents. Sure enough, Antarctic explorers found the same series and fossils—evidence favoring the prediction and the plate tectonics theory.

Drifting Continents, Changing Seas

Take a look at Figure 18.9. In the remote past, crustal movements put huge land masses on collision courses. In time the masses converged to form supercontinents.

Figure 18.8 A splendid slice through time: the Grand Canyon of the American Southwest, once part of an ocean basin. Its sedimentary rock layers formed slowly, over hundreds of millions of years. Later, geologic forces lifted the ancient stacked layers above sea level. Later still, the erosive force of rivers carved the deep canyon walls and exposed the layers.

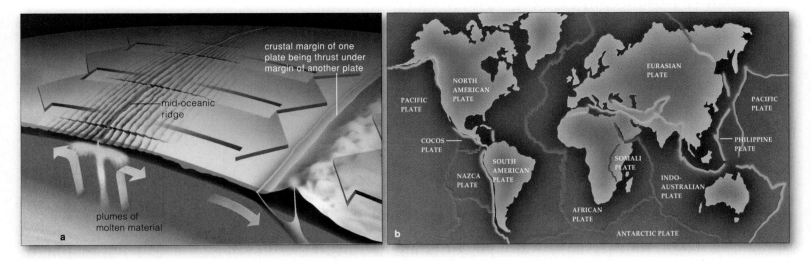

Figure 18.9 Some forces of geologic change. (**a**) The Earth's crust is fractured into rigid plates that slowly split apart, drift, and collide. Driving the movements are huge plumes of molten material that well up from the interior, spread laterally beneath the crust, and rupture the crust at ridges on the ocean floor. Plumes also cause deep rifting and splitting within continents. Such rifting is happening today in Missouri, at Lake Baikal in Russia, and in eastern Africa. Long ago, superplumes violently ruptured the crust. Volcanoes still form at these "hot spots," as they do today in the Hawaiian Archipelago and elsewhere.

At mid-oceanic ridges, molten material seeps out, cools, and slowly forces seafloor away from the rupture. Seafloor spreading displaces plates away from these ridges. Often, the displacement forces other parts of one plate under an adjoining plate and uplifts it. The Cascades, the Andes, and other mountain ranges paralleling the coasts of continents formed this way. (**b**) Present configuration of the crustal plates.

Glossopteris fossil

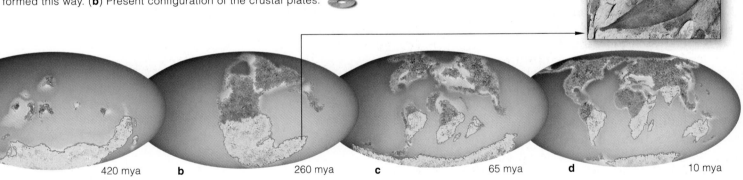

a   420 mya   b   260 mya   c   65 mya   d   10 mya

Later, these split apart at deep rifts and formed new ocean basins. Gondwana was an early supercontinent. It drifted south from the tropics, across the south polar region, and then north until it crunched into other land masses to form a single world continent—Pangea. That supercontinent extended from pole to pole; one world ocean lapped its coastlines. All the while, slow erosive forces of water and wind resculpted the surface of the land. As if this weren't enough, asteroids and meteorites bombarded the crust. The impacts and their aftermath had long-term effects on global temperature and climate.

All of those changes in the land, oceans, and atmosphere profoundly influenced the evolution of life. Imagine early life flourishing in the warm, shallow waters along the shorelines of continents. When continents collided, the shorelines vanished. Such changes were devastating for

Figure 18.10 Reconstructions of drifting continents. (**a**) Gondwana (*yellow*), 420 million years ago. (**b**) All land masses collided to form Pangea. About 260 million years ago, certain seed ferns (*Glossopteris*) and other plants lived on the Gondwana portion of this supercontinent but nowhere else. (**c**) Position of fragments from Pangea's breakup 65 million years ago, when the last dinosaurs disappeared. (**d**) The fragments 10 million years ago, closing in on their present positions.

many lineages. Yet, even as old habitats vanished, new ones opened for the survivors—and evolution took off in new directions. As you will see in the unit to follow, the histories of the Earth and life are inseparable.

Over the past 3.8 billion years, changes in the Earth's crust, the atmosphere, and the oceans profoundly affected the evolution of life.

EVIDENCE FROM COMPARATIVE MORPHOLOGY

Evolution, remember, simply means heritable changes in lines of descent. **Comparative morphology** provides good evidence of descent with modification. This field of inquiry focuses on the body form and structures of groups of organisms, such as vertebrates and flowering plants. Sometimes the studies reveal similarity in one or more body parts that we may attribute to descent from a common ancestor. Such body parts are known as **homologous structures**. (*Homo–* means the same.) The similarities are evident even when different kinds of organisms use the body parts for different functions.

Morphological Divergence

Recall, from Chapter 17, that populations of the same species diverge genetically when gene flow ends among them. Over long time spans, the populations typically diverge in morphological traits that help characterize their species. Change from the body form of a common ancestor is a major pattern of macroevolution. We call this **morphological divergence**. (*Morpho–* is Greek for body form.)

Even when two related species diverge considerably with respect to morphology, they remain alike in other ways. Remember, they started out with the same body form and the same structures, then they evolved in different directions by different modifications to the shared plan. Look carefully, and you should be able to identify underlying similarities.

An example: All vertebrates on land are descended from the first amphibians. Divergences led to reptiles, then to birds and mammals. We know about the "stem" reptile (the one ancestral to all other reptiles and to birds and mammals). We have the fossilized, five-toed limb bones of a species that crouched quite low to the ground (Figure 18.11*a*). Its descendants expanded into many new habitats on land; a few even returned to the seas when the environment changed (Figure 18.11*b–g*).

That five-toed limb was a key innovation for the pterosaurs, birds, and bats. It was evolutionary clay that became molded into different kinds of limbs with different functions. In lineages leading to penguins and porpoises, it was modified into flippers that assisted in swimming. In a lineage leading to the modern horse, it

Figure 18.11 Morphological divergence among vertebrate forelimbs, starting with the stem reptile. Similarities in the number and position of skeletal elements were preserved as diverse forms evolved. Some of the bony elements were lost during the course of evolution (compare numbers 1 through 5). The drawings are not to the same scale.

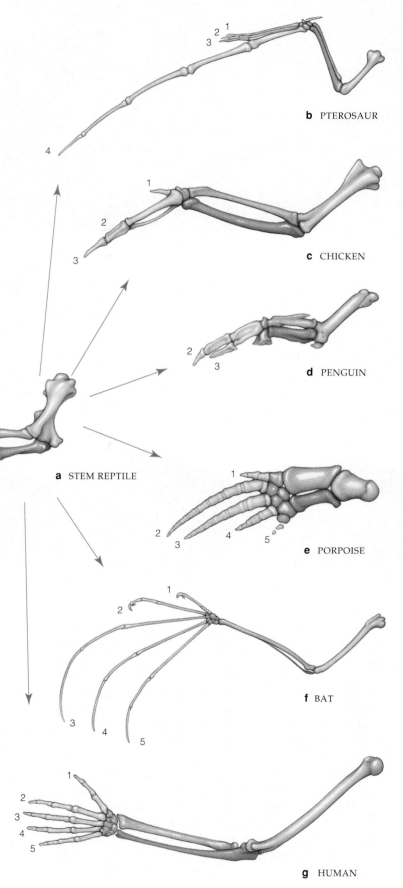

b PTEROSAUR

c CHICKEN

d PENGUIN

a STEM REPTILE

e PORPOISE

f BAT

g HUMAN

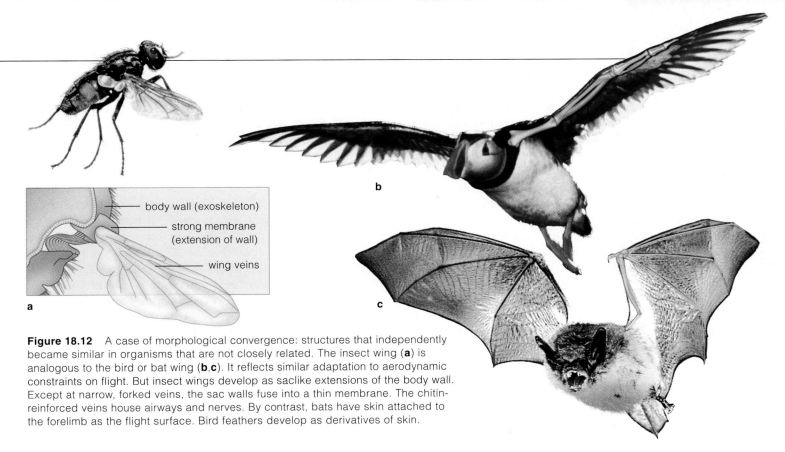

Figure 18.12 A case of morphological convergence: structures that independently became similar in organisms that are not closely related. The insect wing (**a**) is analogous to the bird or bat wing (**b,c**). It reflects similar adaptation to aerodynamic constraints on flight. But insect wings develop as saclike extensions of the body wall. Except at narrow, forked veins, the sac walls fuse into a thin membrane. The chitin-reinforced veins house airways and nerves. By contrast, bats have skin attached to the forelimb as the flight surface. Bird feathers develop as derivatives of skin.

was modified into long, one-toed limbs suitable for fast running. Among moles, it became a stubby shape that is good for burrowing into the earth. Among elephants, it became strong and pillarlike, suitable for supporting great weight. The five-toed limb became modified into the human arm and hand. The human hand also has a thumb positioned in opposition to four fingers; it is the basis of grasping and precision movements.

Even though the forelimbs of vertebrates are not all the same in size, shape, or function from one group to the next, there is clear resemblance in the structure and positioning of the bony elements. There is resemblance, too, in the nerves, blood vessels, and muscles present in the forelimbs. Comparisons of vertebrate embryos also reveal strong resemblance in how these structures develop (Chapter 39). Such similarities in body parts point to common ancestry.

Morphological Convergence

Body parts with similar forms or functions in different lineages aren't always homologous structures. They may have evolved *independently* in lineages that aren't closely related but that faced similar environmental pressures. When comparable body parts were used in much the same way in response, they sometimes became altered in similar ways—and ended up resembling each other. **Morphological convergence** refers to the independent evolution of body structures that did become similar in organisms that are not closely related.

For instance, you just saw how the bird wing and bat wing are homologous structures. In birds, however,

the flight surface evolved as a sweep of feathers derived from skin, which the forelimb structurally supports. In bats, the flight surface is a thin membrane, an extension of the skin itself, that is attached to and reinforced by bony elements of the forelimb (Figure 18.12).

The insect wing, too, resembles bird and bat wings in its function. But it shows no underlying homology. The insect wing develops as an extension of an outer body wall reinforced with chitin. It has no underlying bony elements to support it, as Figure 18.12 shows.

The convergent resemblance is evidence that birds, bats, and insects—three separate lineages—adapted independently to the aerodynamic constraints on flight. Their wings are **analogous structures**. They were not derived from comparable body parts; the convergent resemblance is only indicative of similar adaptations. (The Greek *analogos* means similar to one another.)

With morphological *divergence*, comparable body parts became modified in different ways in different lines of descent from a common ancestor.

Such divergences resulted in homologous structures. Even if they differ in size, shape, or function, these body parts have an underlying similarity owing to shared ancestry.

In morphological *convergence*, dissimilar body parts evolved independently in lineages that are not closely related but that responded to similar environmental pressures.

Analogous structures are the outcome of morphological convergence. Such body parts resemble one another because they are adaptations to similar environmental constraints, not because of close shared ancestry.

EVIDENCE FROM PATTERNS OF DEVELOPMENT

Comparing the patterns by which groups of plants or animals undergo growth and development yields good evidence of evolution. Remember, the body of each new plant and animal follows a long-term, inherited pattern of development (Sections 15.1 and 15.3). Later chapters will show how the body develops in stages. For now, simply note that it cannot develop properly unless each stage is successfully completed before the next begins.

Many constraints on evolution come into play as an embryo is developing. If gene mutations or changes in chromosome number or structure arise, they tend to be selected against when they disrupt an essential stage in the embryo's development. However, every so often, a change has a neutral or beneficial effect. Let's look at examples of how it might shift a developmental step in a way that natural selection favors.

Developmental Program of Larkspurs

Flowering plants offer evidence of evolution through changes in rates of development. Consider *Delphinium decorum*, a larkspur. A ring of petals guides honeybees to the nectar-storing tube of its flowers (Figure 18.13). At the flower's center, outward-bulging reproductive structures give bees something to hang on to as they gather nectar and pollinate the flower. *D. nudicaule*, a larkspur of more recent origin, has compact flowers and no landing platform. This is not a problem for its pollinators—hummingbirds—which can hover in front of the flower as they gather nectar from it.

At maturity, *D. nudicaule* flowers strongly resemble buds of *D. decorum* (Figure 18.13). That resemblance is probably a result of some change in a pattern of gene expression; something slowed down the rate of floral development in *D. nudicaule*. Think about the plant in which the first slowdown appeared. It could produce only compact flowers that discouraged bees—but not hummingbirds. Now dependent on novel pollinators, the plant was reproductively isolated from *D. decorum*. In time it gave rise to *D. nudicaule*, a daughter species.

Developmental Program of Vertebrates

Now consider the diverse vertebrates, which range from fishes to amphibians, reptiles, birds, and mammals. By comparing the ways in which their embryos develop, we can find compelling evidence of their evolutionary connection with one another.

The life cycles for all vertebrate lineages proceed through stages of embryonic development. The early stages are strikingly similar. Think about Figure 18.14*a*. Without the labels, would you know which embryo is an early stage of a fish, lizard, chicken, or human?

The early embryos of vertebrates strongly resemble one another because they inherited the same ancient plan for development. Tissues form as cells divide and start to interact in prescribed ways. Then the gut, heart, bones, skeletal muscles, and other parts grow and develop in ordered spatial patterns that depend on the successful completion of each developmental step (Chapter 39).

During the evolution of all vertebrates, most DNA alterations that disrupted pivotal steps of development probably had lethal effects on later stages. At least the similarity among early embryos of the different groups suggests that such disruptions were selected against.

How, then, did adults of different groups get to be so different? Some differences probably came about by heritable changes in the onset, rate, or completion time

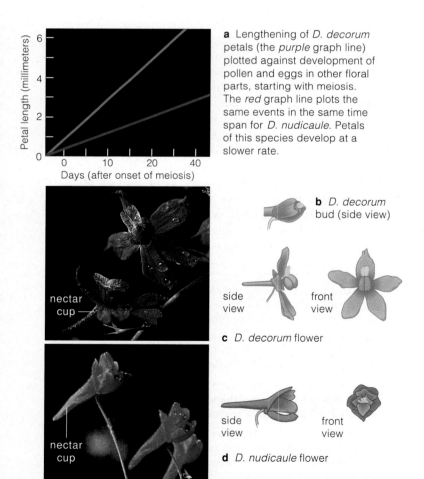

a Lengthening of *D. decorum* petals (the *purple* graph line) plotted against development of pollen and eggs in other floral parts, starting with meiosis. The *red* graph line plots the same events in the same time span for *D. nudicaule*. Petals of this species develop at a slower rate.

Petal length (millimeters) — Days (after onset of meiosis)

b *D. decorum* bud (side view)

c *D. decorum* flower — side view, front view

nectar cup

d *D. nudicaule* flower — side view, front view

nectar cup

Figure 18.13 From comparative morphology, evidence of an evolutionary relationship between two larkspurs (*Delphinium*). The time required for flowers to develop and mature is similar in both species. The *rate* of development is slower for most floral structures of *D. nudicaule*. Changes in petal shape are not nearly as great as they are for the other species, *D. decorum*.

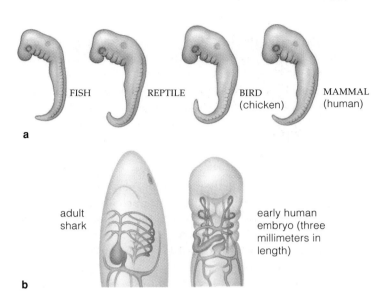

FISH   REPTILE   BIRD   MAMMAL
                  (chicken) (human)

a

adult shark   early human embryo (three millimeters in length)

b

Figure 18.14 From comparative embryology, some evidence of evolutionary relationship among vertebrates.

(**a**) Adult vertebrates show great diversity, yet very early embryos retain striking similarities. This is evidence of change in a shared program of development. (**b**) Fishlike structures still form in early embryos of reptiles, birds, and mammals. In fish embryos, for example, a two-chambered heart (*orange*), certain veins (*blue*), and parts of arteries called aortic arches (*red*) develop and persist in adult fishes. The same structures form in an early human embryo but do not persist as such in adults.

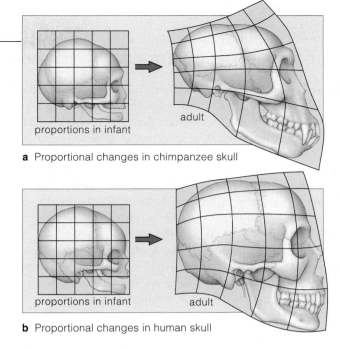

proportions in infant   adult

a Proportional changes in chimpanzee skull

proportions in infant   adult

b Proportional changes in human skull

Figure 18.15 Some pronounced morphological differences between two lines of descent, thought to be an outcome of changes in the timing of developmental steps. This example compares proportional changes in the skull bones of a human and a chimpanzee. The skulls are quite similar in infants.

Think of the representations of infant skulls as paintings on a blue rubber sheet divided into a grid. Stretching the sheet deforms the grid's squares. For the adult skulls, differences in size and shape within corresponding grid sections reflect differences in growth patterns.

of developmental steps. Such changes could have made the relative sizes of tissues or organs larger or smaller, and so the form or structure of body parts would have changed as a result. They could also have led to adult forms that retain some juvenile features.

The "rubber grid" in Figure 18.15 indicates how a change in growth rates might underlie the pronounced proportional differences in skull bones of two primates: chimpanzees and humans. From infants to adults, the grid for humans stays fairly intact in the cranial region that encloses the brain. For chimps, facial bone growth outstrips the growth of cranial bones; the grid becomes greatly modified. Adult skulls for the two lineages end up with significant proportional differences.

Roy Britten, Wanda Reynolds, and other molecular biologists hypothesize that transposons, not single gene mutations, gave rise to most of the variation among lineages. As you read in Chapter 13.4, **transposons** are DNA segments that can move spontaneously to new locations in a genome. And remember enhancers and promoters? These short DNA sequences help switch genes on and off (Sections 14.1 and 14.3). Transposons have their *own* enhancers and promoters. Thus, when they insert themselves next to a gene at a new location, they may have powerful regulatory effects on it.

Only primates carry the *Alu* transposon. They have for at least 30 million years. The Alu enhancer binds

estrogen, thyroid hormones, and other hormones that control the timing of many developmental events. Was such a transposon pivotal in primate evolution?

Think about this: Between 6 million and 4 million years ago, the forerunners of humans and chimpanzees diverged from the same ancestral stock. More than 98 percent of human DNA is still identical to chimpanzee DNA. How can the remainder account for the large morphological differences between the two primates? Humans happen to carry about a million copies of the Alu transposon, which make up more than 5 percent of our genome. At least some of those copies might have fanned or restricted gene expression in certain tissues. Reflect again on the proportional changes in developing skull bones between humans and chimpanzees. Among early ancestors of humans, something did change the duration of bone growth in a heritable way. Ever since, the rapid growth characteristic of the chimpanzee skull bones has been reduced in humans.

Similarities in patterns of development may be clues to evolutionary relationship among plant and animal lineages.

Heritable changes that alter key steps in a developmental program may be enough to bring about major differences in the adult forms of related lineages. Transposons as well as single gene mutations may bring about such changes.

EVIDENCE FROM COMPARATIVE BIOCHEMISTRY

All species have a mix of ancestral and novel traits. The kinds and numbers of traits they do or don't share are clues to how closely they are related. This is also the case for biochemical traits. Each species, recall, inherits a DNA base sequence, which encodes instructions for making the RNAs and proteins required to produce an individual. But mutations accumulated in the diverging lineages that led to new species. Thus, *we can expect the DNA, RNA, and proteins of closely related species to show more similarities than those of distantly related species.*

For instance, from morphological studies alone, you might already have an idea that monkeys, humans, chimpanzees, and other primates are related. You can test this idea by searching for differences in the amino acid sequence of a protein such as hemoglobin, which all three primates synthesize. Similarly, you could test it by determining how closely the DNA base sequences match up. Research tools used for such comparisons are quantitative measures of relatedness even between species that cannot be distinguished from each other on the basis of form alone. Example: Automated DNA sequencing can reveal the sequence of cloned DNA or PCR-amplified DNA in a few hours (Section 15.4). For many species, the DNA base sequences and the amino acid sequences for a number of proteins already have been worked out and are available on the Internet.

Protein Comparisons

Suppose the amino acid sequences of a gene's product are the same or nearly so in two species. The absence of mutation implies the species are close relatives. If the sequences differ greatly, many neutral mutations must have accumulated in them. A very long time must have passed since the species shared a common ancestor.

One highly conserved gene specifies cytochrome *c*, a protein component of the electron transport chains in species ranging from aerobic bacteria to humans. The protein's primary structure in humans consists of 104 amino acids. Figure 18.16 shows how the amino acid sequences for cytochrome *c* from a fungus, a plant, and an animal show striking similarity. Now think about this:

The *entire* cytochrome *c* sequence is identical in humans and chimpanzees. It differs by 1 amino acid in rhesus monkeys, 18 in chickens, 19 in turtles, and 56 in yeasts. On the basis of this biochemical information, would you assume that humans are more closely related to a chimpanzee or a rhesus monkey? A chicken or a turtle?

Nucleic Acid Comparisons

Typically, structural alterations that resulted from gene mutations are dispersed through nucleotide sequences of DNA and RNA molecules. Some number of unique alterations have accumulated in each lineage.

Nucleic acid hybridization refers to a base-pairing between DNA or RNA sequences from two different sources. Section 15.5 describes how researchers induce a DNA molecule from different species to unwind, then to recombine as hybrid molecules. The amount of heat required to pull a hybrid molecule apart is a measure of the similarity between its two strands. It takes more heat energy to disrupt the hybridized DNA of closely related species. And so DNA–DNA hybridization yields rough measures of evolutionary distance. Figure 18.17 describes how comparisons based on this method were used to build a family tree for raccoons and pandas.

Comparisons of nuclear DNA, mitochondrial DNA (mDNA), and ribosomal RNA (rRNA) are faster and more accurate with automated DNA sequencing. The restriction fragments cleaved from DNA are compared after they are separated by gel electrophoresis (Section 15.4). mDNA lends itself to comparisons of eukaryotic species. Its base sequence is rather short, and it mutates ten times faster than nuclear DNA. rRNA lends itself to comparisons of prokaryotic species. Genes for rRNA are so vital to all cells, they haven't mutated much, even in remotely related species. As you will see, they have been used to work out early divergences in life's history.

Computer programs can construct family trees based on DNA sequencing. It helps when the trees correlate with morphological comparisons and the fossil record. Why? Gene swapping can slant the interpretations. And it now appears that gene swapping has been rampant, particularly among the earliest prokaryotic species.

Molecular Clocks

By making certain assumptions about the constancy of evolutionary change at the molecular level, researchers also use molecular comparisons to identify the timing of divergences.

Figure 18.16 Primary structure of three versions of cytochrome *c*, a major protein component of electron transport systems in cells. This amino acid sequence is highly conserved, even in evolutionarily distant lineages. The three sequences are from a representative yeast (*top row*), wheat (*middle row*), and primate (*bottom row*). The gold sequences highlight amino acids that are identical in all three species. The probability that this pronounced molecular resemblance resulted by chance alone is extremely low.

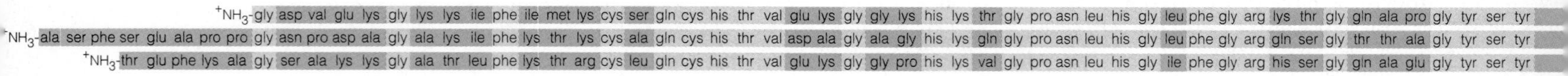

$^+NH_3$-gly asp val glu lys gly lys lys ile phe ile met lys cys ser gln cys his thr val glu lys gly gly lys his lys thr gly pro asn leu his gly leu phe gly arg lys thr gly gln ala pro gly tyr ser tyr

$^+NH_3$-ala ser phe ser glu ala pro pro gly asn pro asp ala gly ala lys ile phe lys thr lys cys ala gln cys his thr val asp ala gly ala gly his lys gln gly pro asn leu his gly leu phe gly arg gln ser gly thr thr ala gly tyr ser tyr

$^+NH_3$-thr glu phe lys ala gly ser ala lys lys gly ala thr leu phe lys thr arg cys leu gln cys his thr val glu lys gly gly pro his lys val gly pro asn leu his gly ile phe gly arg his ser gly gln ala glu gly tyr ser tyr

RACCOON   RED PANDA

GIANT PANDA

SPECTACLED   SLOTH   SUN   BLACK   POLAR   BROWN
BEAR   BEAR   BEAR   BEAR   BEAR   BEAR

DIVERGENCE
15–20 million years ago

DIVERGENCE
approximately
40 million years ago

Figure 18.17 How the red panda, giant panda, and various bears are related according to DNA–DNA hybridization studies.

A giant panda is a plant-eating mammal with the gut of a meat-eating animal. It can't digest the tough cellulose fibers of plants as efficiently as hoofed plant-eating mammals do. Such mammals pummel plants inside multiple stomach chambers, where huge populations of bacteria produce cellulose-digesting enzymes (Section 37.1). A panda's diet consists of bamboo which, pound for pound, is not nearly as nutrient-packed as meat. Also, a panda must grasp bamboo and eat large quantities of it to get enough proteins, fats, and carbohydrates. Its rounded, short-toed paws aren't good at holding and stripping leaves from bamboo stalks. But each front paw has a thumblike bony digit. When a panda uses a "thumb" in opposition to the five toes on a paw, it can grip a bamboo meal.

The giant panda looks like a bear but resembles the red panda, which lives in the same general area in China. The red panda eats bamboo. Then again, bears and the red panda don't have thumbs. Is the giant panda more closely related to red pandas or bears?

Researchers studied the extent to which single-stranded DNA from a giant panda hybridizes with DNA from a red panda and from bears. The more mismatched nucleotide bases, the greater the evolutionary distance between these mammals. The results support an earlier hypothesis. The three are related by descent from a shared ancestor that lived more than 40 million years ago. A divergence took place among the descendants. One branching led to modern raccoons and to the red panda. Another led to bears. Between 20 million and 15 million years ago, a split from the bear lineage gave rise to ancestors of the giant panda. Therefore, the giant panda appears to be much more closely related to bears than to the red panda.

Think about our own species. When compared to our close primate relatives, we have some unique traits, the result of mutations that accumulated after divergence from the ancestral primate stock. We also share many genes with other species. For example, the last shared ancestor of certain yeasts and humans lived hundreds of millions of years ago. Less than 10 percent of the known yeast gene products have changed significantly in overall structure; the rest are highly conserved.

Even so, neutral mutations have introduced slight structural differences in the highly conserved genes of different lineages. **Neutral mutations**, recall, have little or no effect on survival or reproduction. They can slip past agents of selection and accumulate in the DNA.

By some calculations, neutral mutations in highly conserved genes accumulated at a regular rate. Think of the accumulation of neutral mutations in a lineage as a series of predictable ticks of a **molecular clock**. Then turn the hands of the clock back, so that the total number of ticks will "unwind" down through the great geologic intervals of the past. Where the last tick stops, that is roughly the time of origin for the lineage.

How are molecular clocks calibrated in time? The number of differences between the base sequences or amino acid sequences between species is plotted against a series of branch points inferred from the fossil record. Such a graph reflects relative divergence times among species, even among phyla and kingdoms.

Biochemical similarity is greatest among the most closely related species and weakest among the most remote.

ala ala asn lys asn lys gly ile ile trp gly glu asp thr leu met glu tyr leu glu asn pro lys lys tyr ile pro gly thr lys met ile phe val gly ile lys lys lys glu glu arg ala asp leu ile ala tyr leu lys lys ala thr asn glu-COO⁻
ala ala asn lys asn lys ala val glu trp glu glu asn thr leu tyr asp tyr leu leu asn pro lys lys tyr ile pro gly thr lys met val phe pro gly leu lys lys pro gln asp arg ala asp leu ile ala tyr leu lys lys ala thr ser ser-COO⁻
asp ala asn ile lys lys asn val leu trp asp glu asn asn met ser glu tyr leu thr asn pro lys lys tyr ile pro gly thr lys met ala phe gly gly leu lys lys glu lys asp arg asn asp leu ile thr tyr leu lys lys ala cys glu-COO⁻

HOW DO WE INTERPRET THE EVIDENCE?

Identifying, Naming, and Classifying Species

Taxonomy is a field of biology that attempts to identify, name, and classify species. Taxonomists face a very big hurdle, for available information about species can be interpreted differently and arranged in different ways.

Let's start with the system of naming species, which all biologists agree upon. Corn plants, vanilla orchids, houseflies, humans—these and all other species have a scientific name that gives people throughout the world a way to know they're talking about the same organism (Figure 18.18). Every taxonomist assigns each name on the basis of a system developed by Carolus Linnaeus, an early naturalist. His **binomial system** requires that each species be assigned a two-part Latin name.

The first part of the species name is generic; it is descriptive of similar species that are thought to be the same type of organism and are grouped together. Such a grouping is a **genus** (plural, genera). The second part is the **specific name**. Together with the generic name, it refers to one kind of organism only. For example, there is only one *Ursus maritimus*, or polar bear. Other bears include *Ursus arctos* (brown bear) and *Ursus americanus* (black bear). Notice how the first letter of the generic name is capitalized but the second (specific) name is not. The specific name is never used without the full or abbreviated generic name preceding it, because it also can be the second name of a species in a different group. For example, *U. americanus* means black bear, *Homarus americanus* means Atlantic lobster, and *Bufo americanus* means American toad. Hence, one would not order *americanus* for dinner unless one is willing to take what one gets.

Groupings of Species—The Higher Taxa

Classification schemes are organized ways to retrieve information about species. There was a time when just about everything in nature was classified as animal, vegetable, or mineral. Linneaus devised a two-kingdom scheme, with all of life divided into animals or plants according to similarities and differences in traits. His scheme lasted for more than two centuries, although it had nowhere to put the tens of thousands of bacterial species. Biologists later came up with more inclusive groupings to reflect relationships. Groupings of species are called **higher taxa** (singular, taxon). Family, order, class, phylum, and kingdom are examples. Taxonomists now use **phylogeny** to assign species to higher taxa. The term refers to evolutionary relationships among species, from the most ancestral forms on through the genetic divergences that led to all descendant species.

Clues to evolutionary relationship are embedded in the fossil record, the geologic record, morphology (as through homologous structures), and biochemistry. Yet how should they be interpreted? *Classical* taxonomy uses degrees of morphological divergence to construct evolutionary tree diagrams. With *cladistic* taxonomy, the branch points in tree diagrams are the measure (*clad–* means branch). Only species that share derived traits are grouped past a branch point for the last common ancestor. A **derived trait** is a novel feature that evolved only once and is shared *only* by the descendants of the ancestral species in which it evolved.

The evolutionary tree diagrams called **cladograms** group taxa on the basis of derived traits to reflect how recently they shared a common ancestor. For example, sharks, crocodiles, birds, and mammals all have a heart. Also, crocodiles, birds, and mammals have two lungs, but sharks do not. Birds have feathers as well as a heart and lungs, but crocodiles and mammals do not. Only mammals have fur. Pulling this together into the form of a cladogram,

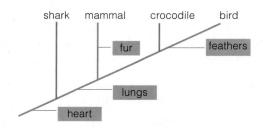

Bear in mind, cladograms do not directly convey "who came from whom." Groups closer together on the cladogram simply share a more recent common ancestor.

KINGDOM	Plantae	Plantae	Animalia	Animalia
PHYLUM	Anthophyta	Anthophyta	Arthropoda	Chordata
CLASS	Monocotyledonae	Monocotyledonae	Insecta	Mammalia
ORDER	Commelinales	Orchidales	Diptera	Primates
FAMILY	Poaceae	Orchidaceae	Muscidae	Hominidae
GENUS	*Zea*	*Vanilla*	*Musca*	*Homo*
SPECIES	*Z. mays*	*V. planifolia*	*M. domestica*	*H. sapiens*
COMMON NAME	corn	vanilla orchid	housefly	human

Figure 18.18 Taxonomic classification of four organisms. Each is assigned to ever more inclusive categories (higher taxa), from species to kingdom.

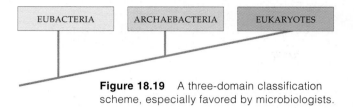

Figure 18.19 A three-domain classification scheme, especially favored by microbiologists.

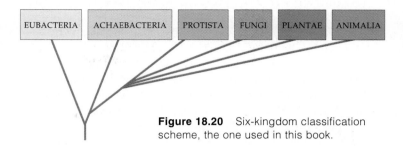

Figure 18.20 Six-kingdom classification scheme, the one used in this book.

Phylogenetic Classification Schemes

Robert Whittaker proposed one of the first phylogenetic schemes. He assigned species to one of five kingdoms and connected the evolutionary dots among them. He used morphological similarities and differences, modes of nutrition, cell structure, and developmental features as clues to the connections.

Whittaker's scheme places all the prokaryotic cells in kingdom **Monera**. Collectively, these single cells show more metabolic diversity than eukaryotes. They include many thousands of producers (photoautotrophs plus chemoautotrophs), and pathogens and decomposers (heterotrophs). Kingdom **Protista** has single-celled and multicelled eukaryotes with more internal complexity than prokaryotes, but not as much fungi, plants, and animals. Inside its boundaries are photoautotrophs and heterotrophs, such as diverse pathogens and parasites. Kingdom **Fungi** holds the multicelled, heterotrophic eukaryotes that depend on extracellular digestion and absorption. Different kinds are decomposers (nutrient cyclers), pathogens, and parasites. Kingdom **Plantae** holds the multicelled producers (photoautotrophs) with vascular tissues. We have named over 295,000 kinds. All of the species in kingdom **Animalia** are eukaryotic, multicelled, and heterotrophic. They range from the microscopic to giants. We know of more than a million kinds, including herbivores, carnivores, and parasites.

Whittaker's scheme enjoyed wide acceptance until recently. For years, biologists knew that the chemical composition, plasma membrane, and wall of single-celled prokaryotes called archaebacteria have unique features. The differences did not seem to be significant enough to pull them from kingdom Monera. Then the microbiologist Carl Woese and other researchers at the University of Illinois employed new DNA sequencing methods (Chapter 15). After comparing the rRNA base sequences from a variety of organisms, they proposed dividing up the prokaryotic cells into two major taxa. Archaebacteria, they said, are as different from true bacteria (eubacteria) as they are from eukaryotic cells.

Strong evidence favoring their conclusion came in 1996. Carol Bult and her colleagues sequenced all 1.7 million base pairs of DNA taken from the bacterium *Methanococcus jannaschii*. Then Woese and Bult worked to decipher the sequence. More than half the genes had never been seen before. Some are closer to the genes of humans and other eukaryotes than to eubacterial genes. Does the biochemical evidence point to three very early branchings in the history of life? Maybe.

A **THREE-DOMAIN SCHEME** Evidence of this primordial branching has given rise to a *three-domain* scheme, as in Figure 18.19. As you may suspect, microbiologists (who know the most about prokaryotic cells) are its fiercest advocates. Those showing the most reluctance to adopt it argue that it doesn't give suitable weight to the far greater biodiversity of eukaryotes. Besides, recent gene sequencing studies suggest that gene swapping was widespread among early prokaryotes. If that were the case, then the three-domain scheme has to go back to the drawing board for refinement as more data emerge.

A **SIX-KINGDOM SCHEME** At this point we don't have a universally accepted scheme. However, consensus is growing among biologists to replace the five-kingdom scheme with one that better reflects life's evolutionary history. In a **six-kingdom classification scheme**, the first great divergence after the origin of life gave rise to the ancestors of eubacteria and archaebacteria. Soon after, the road to archaebacteria forked, and the new road led to all single-celled eukaryotes. Early eukaryotic species later gave rise to all multicelled forms of life.

We adhere to the six-kingdom scheme in this book. **Archaebacteria** and **Eubacteria** are each accorded the rank of kingdom, just like protistans, fungi, plants, and animals (Figure 18.20).

Each species is assigned a two-part scientific name (genus and specific name). It is classified by ever more inclusive groupings of species, the higher taxa.

Classification schemes organize and simplify the retrieval of information about species. Phylogenetic schemes attempt to reflect evolutionary relationships among species.

Reconstructing the evolutionary history of a given lineage must be based on detailed understanding of the fossil record, morphology, life-styles, and habitats of its representatives, and on biochemical comparisons with other groups.

Recent evidence, especially from comparative biochemistry, favors a six-kingdom classification scheme: Archaebacteria, Eubacteria, Protista, Fungi, Plantae, and Animalia.

Case Study: Interpreting and Misinterpreting the Evidence

At the time Charles Darwin formally presented his theory of evolution by natural selection, it faced challenges. If scholars subscribed to the theory, then they would be tentatively accepting that all species are related by descent to ever more ancient species. Darwin saw with his own eyes that artificial selection practices by pigeon breeders and others could mold the traits of a population in no time at all. So he argued it was entirely possible for one species to evolve into a separate species, with one or more traits that were uniquely its own, over hundreds or thousands of generations. But if each species evolved from another, then where were the transitional forms, the so-called "missing links"? Where were the species with traits that bridged two major groups of organisms?

More than a century would pass before more fossil finds as well as molecular and genetic analysis yielded answers. In Darwin's time, though, the presumed absence of transitional forms was a major point of contention in discussions of the theory. Ironically, a fossil of just such a transitional form had already been unearthed at a limestone quarry near Solnhofen, in southern Germany.

At first the fossilized skeleton was labeled a small theropod (meat-eating) dinosaur. It was about the size of a pigeon. The specimen's shape said "dinosaur." Like the theropods, it had a long, bony tail, clawed fingers, and a heavy jaw with short, spiky teeth (Figure 18.21).

Later, diggers unearthed another fossil of the same type. Later still, someone noticed the feathers. If the fossils were of dinosaurs, what were they doing with *feathers*? Close examination revealed that the feathers were like those of modern birds. The specimen type was named *Archaeopteryx* (meaning "ancient winged one").

Between 1860 and 1988, six specimens of *Archaeopteryx* and a fossilized feather were found. Anti-evolutionists tried to dismiss them as forgeries. Someone, they asserted, had pressed bones and feathers of existing birds against wet plaster. After the imprinted plaster casts dried, they merely looked like fossils. But microscopic examination confirmed that the fossils are real. Further confirmation of their antiquity is found in the remains of clearly ancient species of worms, jellyfishes, and many other kinds of organisms preserved in the same limestone layers.

Through radiometric dating methods, we now know that *Archaeopteryx* lived 150 million years ago. When you examine the photograph in Figure 18.21, you might wonder: How could the remains of the winged creatures be so well preserved after so much time? As it happened, *Archaeopteryx* lived in tropical forests next to a large lagoon. The lagoon was warm, still, and stagnant; coral reefs barred inputs of oxygenated water from the sea. So it was not a favorable habitat for scavenging animals that might have feasted on—and obliterated the remains of—*Archaeopteryx* and other organisms that fell from the skies or drifted offshore. With each tropical storm surge, though, fine sediments swept over the reefs. They gently buried the carcasses littered at the bottom of the lagoon. Over time, the soft mud became compacted and hardened. It became an exceptionally fine limestone tomb for more than 600 species, including *Archaeopteryx*.

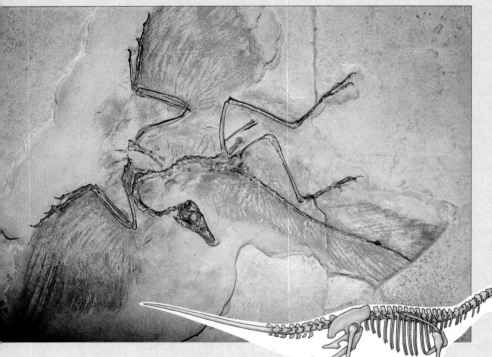

Figure 18.21 *Above:* A photograph of one of the *Archaeopteryx* fossils. *Right:* Comparison of the skeletons of a dinosaur (a type of reptile) and of *Archaeopteryx*, which was on or very near the evolutionary road leading from reptiles to birds. More recently discovered fossils show that dinosaurs closely related to *Dromaerosaurus* did indeed have feathers.

Dromaeosaurus, a small, two-legged dinosaur that had the most traits in common with birds. Like birds, it had long, slender hind legs, three toes forward and one toe pointing in reverse, and long forelimbs. Unlike birds, it had no feathers.

Archaeopteryx, which had feathers. Like dinosaurs, though, it also had a long, slender bony tail, a heavy jaw with serrated teeth, and three long fingers.

SUMMARY

1. The fossil record, the geologic record, comparative morphology, and comparative biochemistry have yielded extensive evidence of evolution. The evidence is based on similarities and differences in body form, functions, behavior, and biochemistry. *18.1–18.8*

2. Fossils are recognizable, physical evidence of life in the distant past. They start to form after an organism or traces of it become buried in volcanic ash or sediments. The organic remains become infused with minerals. Sediments slowly accumulate above the burial site; and the increasing pressure, along with chemical changes, transforms the remains to stony hardness. *18.1, 18.8*

   a. Fossils are present in layers of sedimentary rock. The deepest layers accumulate first; they are the oldest. Thus, the older the layers, the older the fossils.

   b. The completeness of the fossil record is variable in terms of the species represented, where they lived, and the stability of burial sites since fossilization occurred.

3. All species, past and present, are related by way of descent from their common ancestors, starting with the origin of the first living cells on Earth. Macroevolution refers to patterns, trends, and rates of change among groups of species over long spans of time. These clues to the past provide us with insight into the sweeping continuity of relationship in nature. *18.2*

4. Scientists use abrupt transitions in the sequence of fossil assemblages as the boundaries for intervals in the geologic time scale. In the modern version of the scale, life originated during the Archean eon. Oldest to most recent fossils extend from the Proterozoic eon through the Paleozoic, Mesozoic, and Cenozoic eras. *18.2*

5. Like life, the Earth has evolved. Some of its features change repetitively, as when mountains rise and erode slowly. Other features have changed irreversibly in ways that are explained by plate tectonics theory: *18.3*

   a. The Earth's crust is fractured into huge, thin, rigid plates that slowly split apart, drift, and collide with one another, rafting the land masses along with them.

   b. Great plumes of molten material welling up from Earth's interior drive the movements. At mid-oceanic ridges, material seeps out, cools, hardens, and laterally displaces older seafloor. Seafloor spreading has forced older crust down elsewhere, into deep trenches. Great mountain ranges that parallel coasts formed when one plate thrust under another, which uplifted as a result.

   c. Large-scale, long-term changes in the Earth's land masses have changed the ocean and atmosphere, and they have profoundly influenced the evolution of life.

6. Comparative morphology often reveals similarities that reflect evolutionary relationship. *18.4*

   a. In cases of morphological divergence, comparable body parts have become modified in different ways in different lines of descent from a common ancestor. The modifications resulted in homologous structures: body parts that have underlying similarities owing to shared ancestry, even if they differ in size, shape, or function.

   b. In cases of morphological convergence, dissimilar body parts evolve independently in different lineages but in response to similar environmental pressures. The modifications resulted in analogous structures: body parts that resemble one another through adaptation to similar environmental pressures, not by shared ancestry.

7. Similarities in patterns of development may be clues to evolutionary relationship. Changes in DNA that alter the onset, rate, or time of developmental steps probably brought about most morphological differences between related lineages. For example, transposons altered gene expression in humans, relative to chimpanzees. *18.5*

8. Comparative biochemistry reveals similarities and differences among species at the molecular level. *18.6*

   a. Nucleic acid hybridization, a base-pairing between DNA or RNA from two sources, is a rough measure of evolutionary distance between them. Automated gene sequencing is a faster way to compare nuclear DNA, mitochondrial DNA, and ribosomal RNA. Computer programs use the results for constructing family trees.

   b. Conserved genes have an accumulation of neutral mutations. Like predictable ticks of a molecular clock, the mutations are calibrated in time. The number of differences between the base sequences or amino acid sequences from different species is plotted against a series of branch points, inferred from the fossil record, to identify relative divergence times among groups.

9. Taxonomists attempt to identify, name, and classify species. By the Linnaean binomial system, each kind of organism is assigned a two-part Latin name. The first part (genus) identifies morphologically similar species derived from a common ancestor. In combination with the second part of the name (species epithet), it is the name of the particular species. *18.7*

10. All classification schemes are organized ways to retrieve information on species. They use higher taxa: ever more inclusive groupings from species to genera, families, orders, classes, phyla, and kingdoms. Those that are phylogenetic are constructed to reflect inferred evolutionary relatedness. *18.7*

11. This book uses a six-kingdom phylogenetic scheme: Archaebacteria, Eubacteria, Protista, Plantae, Fungi, and Animalia. *18.7*

Review Questions

1. Distinguish macroevolution from microevolution. *(Compare CI with Section 16.4)*

2. Name some of the reasons why biologists expect to find gaps in the fossil record of what apparently is the continuous evolution of life from the time of its origin. *18.1*

3. Name three eras of the geologic time scale. *18.2*

4. Did life originate in the Archean or Proterozoic eon? *18.2*

5. Define radiometric dating. What does half-life mean? *18.2*

6. Give an example of how immense crustal movements have influenced the evolution of life. *18.3*

7. Define and give examples of the difference between: *18.4*
   a. homologous and analogous structures
   b. morphological divergence and convergence

8. Comparative morphology refers to comparisons of body form and structures, embryonic and adult, for major lineages. Describe an example of such a comparison. *18.4, 18.5*

9. Name a protein specified by a gene that has been highly conserved in organisms ranging from bacteria to humans. *18.6*

10. Why do evolutionary biologists apply heat energy to hybrid molecules that contain DNA from two species? *18.6*

11. What type of mutation is the basis of a molecular clock? What does the last tick of a molecular clock signify? *18.6*

12. Name all the kingdoms in the six-kingdom classification scheme and give examples of representative organisms. *18.7*

Self-Quiz   ANSWERS IN APPENDIX III

1. Morphological convergences may lead to _____ .
   a. analogous structures   c. divergent structures
   b. homologous structures   d. both a and c

2. Heritable changes in DNA underlying morphological differences between lineages _____ .
   a. may have been caused mostly by transposons
   b. affected the onset, rate, and time of development steps
   c. both a and b

3. A classification system that is _____ is based on presumed evolutionary relationship.
   a. epigenetic   c. phylogenetic
   b. credited to Linnaeus   d. both b and c

4. *Pinus banksiana, Pinus strobus,* and *Pinus radiata* are _____ .
   a. three families of pine trees
   b. three different names for the same organism
   c. three species belonging to the same genus
   d. both a and c

5. Increasingly inclusive taxa range from _____ to _____ .
   a. kingdom; species   c. genera; kingdom
   b. kingdom; genera   d. species; kingdom

6. Match these terms suitably.
   ____ phylogeny   a. accumulation of neutral mutations
   ____ fossil   b. evidence of life in distant past
   ____ stratification   c. similar body parts in different
   ____ homologous   lineages owing to common descent
        structure   d. e.g., insect wing and bird wing
   ____ molecular clock   e. evolutionary relationship among
   ____ analogous   species, ancestors to descendants
        structure   f. layers of sedimentary rock

Critical Thinking

1. At the end of your backbone are several small, fused bones, called the coccyx (Section 16.1, Figure 16.3). Could the coccyx be a vestigial structure—all that's left of a tail that was a feature of the evolutionarily distant vertebrate (and primate) ancestors of humans? Or is it the start of a newly evolving structure with an as-yet-undetermined function? Make an educated guess, then describe some ways in which you might test whether your guess is plausible.

2. Comparative biochemistry helps us estimate evolutionary relationship and approximate times for divergences from ancestral stocks. Base sequence comparisons and amino acid comparisons yield good estimates. Reflect on the genetic code (Section 14.3), then suggest why it may be a useful measure of mutations, mutation rates, and biochemical relatedness.

3. Shannon thinks there are too many kingdoms and sees no good reason to make another one for something as small as archaebacteria. "Keep them with the other prokaryotes!" she says. Taxonomists would call her a "lumper." By contrast, Andrew is a "splitter." He sees no good reason to withhold kingdom status from archaebacteria simply because they are part of a microscopic world that not many people know about. Which may be the more useful: more or fewer boundaries between groups of organisms? Explain your answer.

4. When walking along a path cut into the side of a steep mountain, you pass rocky layers that are fractured and folded back on themselves in bizarre patterns. You look closely and discover a fossilized shell in the "highest" layer in the series. How would you use the plate tectonics theory to help you decide whether the fossil is of recent or ancient origin?

Selected Key Terms

analogous structure *18.4*	macroevolution *18.2*
Animalia *18.7*	molecular clock *18.6*
Archaebacteria *18.7*	Monera *18.7*
binomial system *18.7*	morphological convergence *18.4*
catastrophism *CI*	morphological divergence *18.4*
cladogram *18.7*	neutral mutation *18.6*
classification scheme *18.7*	nucleic acid hybridization *18.6*
comparative morphology *18.4*	Pangea *18.3*
derived trait *18.7*	phylogeny *18.7*
Eubacteria *18.7*	Plantae *18.7*
fossil *18.1*	plate tectonics theory *18.3*
fossilization *18.1*	Protista *18.7*
Fungi *18.7*	radiometric dating *18.2*
genus *18.7*	six-kingdom classification
geologic time scale *18.2*	scheme *18.7*
Gondwana *18.3*	specific name *18.7*
half-life *18.2*	stratification *18.1*
higher taxon (taxa) *18.7*	taxonomy *18.7*
homologous structure *18.4*	theory of uniformity *18.3*
lineage *18.1*	transposon *18.5*

Readings

Brooks, D. R., and D. A. McLennan. 1991. *Phylogeny, Ecology, and Behavior.* Chicago: University of Chicago Press.

Dott, R., Jr., and R. Batten. 1998. *Evolution of the Earth.* Fourth edition. New York: McGraw-Hill. Good historical perspective on correlations between biological and geologic evolution.

Ochert, A., December 1999. "Transposons." *Discover,* 59–66.

Ridley, M. 1999. *Genome: The Autobiography of a Species in 23 Chapters.* New York: HarperCollins.

On-Line readings at Student Guide for InfoTrac:
www.brookscole.com/biology

IV Evolution and Biodiversity

Patterns of diversity in nature, represented by plants of exquisite colors and by fungi that are intertwined with other organisms in structures called lichens. Some of the lichens are branching; flattened ones encrust this chunk of granite.

THE ORIGIN AND EVOLUTION OF LIFE

In the Beginning . . .

Some clear evening, watch the moon as it rises from the horizon and think of the 380,000 kilometers between it and you. *Five billion trillion times* farther away from you are galaxies—systems of stars—at the boundary of the known universe. Wavelengths traveling through space move faster than anything else—millions of meters per second—yet long wavelengths that originated from faraway galaxies many billions of years ago are only now reaching the Earth.

By all known measures, all of the near and distant galaxies in the vast space of the universe are moving away from one another, which means the universe must be expanding. One prevailing view of how the colossal expansion came about may account for every bit of matter in every living thing.

Think about how you rewind a videotape on a VCR. Then imagine "rewinding" the universe. As you do this, the galaxies start moving back together. After 12 to 15 billion years of rewinding, all galaxies, all matter, and all of space are compressed into a hot, dense volume about the size of the sun. You have arrived at time zero.

That incredibly hot, dense state lasted only for an instant. What happened next is known as the **big bang**,

Figure 19.1 Part of the great Eagle nebula, a hotbed of star formation 7,000 light-years from Earth, in the constellation Serpens. (The Latin *nebula* means mist.) Not shown in this image are a few huge, young stars above the pillars. For the past few million years, intense ultraviolet radiation from the stars has been eroding the less dense surface of the pillars. Immense globs of denser gases and dust that have resisted erosion are visible at the surface. Each pillar is wider than our solar system—more than 10 billion miles across! New stars are hatching from the protruding globs; some shine brightly on the tips of gaseous streamers.

a stupendous, nearly instantaneous distribution of matter and energy throughout the universe. About a minute later, temperatures dropped a billion degrees. Fusion reactions created most of the light elements, including helium, which are still the most abundant elements in the universe. Radio telescopes can detect cooled, diluted background radiation—a relic of the big bang—left over from the beginning of time.

Over the next billion years, uncountable numbers of gaseous particles collided and condensed under gravity's force to become the first stars. When the stars were massive enough, nuclear reactions ignited inside them and gave off tremendous light and heat. Massive stars continued to contract, and many became dense enough to promote the formation of heavier elements.

All stars have a life history, from birth to an often spectacularly explosive death. In what might be called the original stardust memories, the heavier elements released during the explosions became swept up in the gravitational contraction of new stars. They became the raw materials for the formation of even heavier elements. Even as you read this page, the Hubble space telescope is revealing astounding glimpses of ongoing star-forming activity, as in the dust clouds of Orion, Serpens, and other constellations (Figure 19.1).

Now imagine a time long ago, when explosions of dying stars ripped through our galaxy and left behind a dense cloud of dust and gas that extended trillions of kilometers in space. As the cloud cooled, countless bits of matter gravitated toward one another. By 4.6 billion years ago, the cloud had flattened into a slowly rotating disk. At the dense, hot center of that disk, the shining star of our solar system—the sun—was born.

The remainder of this chapter is a sweeping slice through time, one that cuts back to the formation of the Earth and life's chemical origins. It is the starting point for the next six chapters, which will take us along lines of descent that led to the present range of diversity.

The story is not complete. Even so, the available evidence from many avenues of research points to a principle that can help us organize bits of information about the past: *Life is a magnificent continuation of the physical and chemical evolution of the universe, of galaxies and stars, and of the planet Earth.*

Key Concepts

1. We have evidence that life originated about 3.8 billion years ago. The origin and subsequent evolution of life have been correlated with the physical and chemical evolution of the universe, the stars, and the planet Earth.

2. All inorganic and organic compounds necessary for self-replication, membrane assembly, and metabolism—for the structure and functions of living cells—could have formed spontaneously under conditions that apparently prevailed on the early Earth.

3. The history of life, from its chemical beginnings to the present, spans five intervals of geologic time. It extends through two great eons—the Archean and Proterozoic—and the Paleozoic, Mesozoic, and Cenozoic eras.

4. Not long after life originated, divergences led to two great prokaryotic lineages called the archaebacteria and eubacteria. Soon afterward, the ancestors of eukaryotes diverged from the branch leading to archaebacteria.

5. Archaebacteria and eubacteria were the dominant life forms of the Archean and Proterozoic eons. Eukaryotic cells originated late in the Proterozoic and soon became spectacularly diverse. A theory of endosymbiosis helps explain the profusion of specialized organelles that evolved in eukaryotic cells.

6. All six kingdoms of life have a history of persistences, extinctions, and radiations of different lineages.

7. Throughout the history of life, asteroid impacts, drifting and colliding continents, and other environmental assaults have had profound impact on the direction of evolution.

CONDITIONS ON THE EARLY EARTH

Origin of the Earth

Figure 19.1 shows part of one of the vast clouds in the universe. The clouds are mostly hydrogen gas, along with water, iron, silicates, hydrogen cyanide, ammonia, methane, formaldehyde, and other small inorganic and organic substances. The contracting cloud that became our solar system probably was similar in composition. The cloud's edges cooled between 4.6 and 4.5 billion years ago. Electrostatic attractions and gravity's pull caused mineral grains and ice orbiting the new sun to start clumping together (Figure 19.2). In time, larger, faster clumps collided and shattered. Some grew more massive by sweeping up asteroids, meteorites, and other rocky remnants of collisions. They evolved into planets.

While the Earth formed, heat generated by asteroid impacts, internal compression, and radioactive decay of minerals melted much of its rocky interior. Molten nickel, iron, and other heavy materials moved into the interior; lighter ones floated to the surface. The process resulted in a crust, mantle, and core. The **crust** became an outer zone of basalt, granite, and other low-density rocks. It rested on a zone of intermediate-density rocks, the **mantle**. In turn, the mantle enveloped an immense core of high-density, partially molten nickel and iron.

Four billion years ago, the Earth was a thin-crusted inferno (Figure 19.3*a*). Less than 200 million years later, life appeared on its surface! We have no record of the event. As far as we know, movements in the mantle and crust, volcanic activity, and erosion obliterated all traces of it. Still, we can put together a plausible explanation of how life originated by considering three questions:

First: Can we identify physical and chemical conditions that prevailed on the Earth when life originated?

Second: Do the known principles of physics, chemistry, and evolution support or disprove the hypothesis that large organic molecules formed spontaneously and evolved into molecular systems with the fundamental properties of life?

Third: Can we design experiments to test the hypothesis that living systems emerged through chemical evolution?

The First Atmosphere

Hot gases blanketed the Earth when the first patches of crust formed. We suspect that this first atmosphere was a mix of gaseous hydrogen (H_2), nitrogen (N_2), carbon monoxide (CO), and carbon dioxide (CO_2). Did it hold gaseous oxygen (O_2)? Probably not. Rocks subjected to intense heat, as happens during volcanic eruptions, do release oxygen, but not much. Also, free oxygen would have reacted at once with other elements. Remember, oxygen has an electron vacancy in its outermost shell, and it tends to bond with other atoms (Section 2.3).

If the early atmosphere had not been relatively free of oxygen, organic compounds necessary to assemble cells would not have been able to form on their own, spontaneously. Any oxygen would have attacked their structure and disrupted their functioning.

What about water? Dense clouds cloaked the early Earth, yet any water falling on the molten surface must have evaporated at once. But after the crust cooled and grew solid, the rains began. For millions of years, runoff eroded mineral salts from rocks. The salt-laden waters collected in crustal depressions, forming the first seas.

If liquid water had not accumulated, membranes—which take on their bilayer organization only in water—could not have formed. No membrane, no cell. Life at its most basic level *is* the cell, which has a capacity to survive and reproduce on its own.

Synthesis of Organic Compounds

A cell is made of proteins, complex carbohydrates and lipids, and nucleic acids. Existing cells assemble these molecules from smaller organic compounds: the simple sugars, fatty acids, amino acids, and nucleotides. Energy from the environment drives these synthesis reactions. Were small organic compounds present on the early Earth? Were there sources of energy that spontaneously drove their assembly into the large molecules of life?

Mars, meteorites, the Earth's moon, and the Earth formed at the same time, from the same cosmic cloud. Rocks from Mars, meteorites, and the moon all hold precursors of biological molecules, so these precursors must have been on the early Earth. *Energy from the sun, lightning, or heat vented from the crust could have been enough to drive their combination into organic molecules.*

Stanley Miller conducted the first experimental test of that prediction. He mixed water, methane, hydrogen, and ammonia in a reaction chamber (Figure 19.3*b*). He

Figure 19.2 Representation of the cloud of dust, gases, and clumps of rock and ice around the early sun.

Figure 19.3 (**a**) Representation of the Earth during its formation, when the moon's orbit was much closer than it is today. If the Earth had condensed into a planet of smaller diameter, its gravitational mass would not have been great enough to hold on to an atmosphere. If it had settled into an orbit closer to the sun, water would have evaporated from its hot surface. If the Earth's orbit had been more distant from the sun, its surface would have been far colder, and water would have been locked up as ice. And without liquid water, life as we know it never would have originated on Earth.

(**b**) Stanley Miller's experimental apparatus, used to study the synthesis of organic compounds under conditions that presumably existed on the early Earth. The condenser cools circulating steam so that water droplets form.

kept the mixture circulating and exposed it to a spark discharge to simulate lightning. Amino acids and other small organic compounds formed in less than a week.

In other experiments that simulated conditions on the early Earth, glucose, ribose, deoxyribose, and other sugars formed spontaneously from formaldehyde, and adenine from hydrogen cyanide. Ribose and adenine are part of ATP, NAD$^+$, and other vital nucleotides.

However, if *complex* organic compounds had formed directly in seawater, could they have lasted long? The spontaneous direction of the reactions would have been toward hydrolysis, not condensation, in water. So how did more lasting bonds form?

By one hypothesis, clay in the rhythmically drained muck of tidal flats and estuaries served as templates (structural patterns) for the spontaneous assembly of proteins and other complex compounds. Clay consists of thin, stacked layers of aluminosilicates, and its metal ions attract amino acids. Expose clay to amino acids, warm it with the sun's rays, then alternately moisten it and dry it out. Condensation reactions will occur and form proteins and other complex organic compounds.

By another hypothesis, complex organic compounds formed spontaneously near hydrothermal vents on the seafloor, where species of archaebacteria are thriving today (Section 6.7). Experimental tests by Sidney Fox and others show that when amino acids are heated and placed in water, they spontaneously order themselves into small protein-like molecules he calls proteinoids.

However the first proteins formed, their molecular structure dictated how they could interact with other compounds. Suppose some proteins functioned as weak enzymes and hastened bond formation between amino acids. An enzyme-directed synthesis of proteins would have had selective advantage. Protein configurations that promoted such reactions would win the chemical competition for available amino acids. They would have been favored in another way—for proteins have the capacity to bind metal ions and other metabolic agents.

As you will see, the evolution of metabolism was based on such chemical modifications. For now, simply reflect on the possibility that selection was at work before cells appeared, favoring the chemical evolution of enzymes and other complex organic compounds.

Many different experiments offer indirect evidence that the complex organic molecules characteristic of life could have formed under the physical and chemical conditions that prevailed on the early Earth.

EMERGENCE OF THE FIRST LIVING CELLS

Origin of Agents of Metabolism

A defining characteristic of life is *metabolism*. The word refers to all the reactions by which cells harness energy and use it to drive their activities, such as biosynthesis. During the first 600 million years or so of Earth history, enzymes, ATP, and other organic compounds may have assembled spontaneously, perhaps in the same places. If so, the close association naturally promoted chemical interactions and the start of metabolic pathways.

Imagine an ancient estuary rich in clay deposits. It is a coastal region where seawater mixes with mineral-rich water being drained from the land. There, beneath the sun's rays, countless aggregations of organic molecules stick to the clay. At first there are quantities of an amino acid; call it *D*. All over the estuary, *D* molecules get incorporated into new proteins—until the supply of *D* dwindles. However, suppose a weakly catalytic protein is in the estuary. It can promote the formation of *D* by acting on a plentiful, simpler substance—*C*. By chance, clumps of organic molecules include that enzyme-like protein. The clumps have an advantage in the chemical competition for starting materials.

In time, *C* molecules become scarce. At that point, the advantage tilts to molecular clumps that promote formation of *C* from even simpler substances *B* and *A*. Assume *B* and *A* are carbon dioxide and water. As you know, the atmosphere and the seas have essentially unlimited amounts of carbon dioxide and water. Thus chemical selection has favored a synthetic pathway:

$$A + B \longrightarrow C \longrightarrow D$$

Finally, suppose some clumps are better than others at absorbing and using energy. Which molecular parts could bestow such an advantage? Think of the energy-trapping pathway that now dominates the world of life: photosynthesis. It starts at pigments called chlorophylls. The part of the chlorophyll molecule that absorbs light and gives up electrons is a porphyrin ring structure. Porphyrins also are present in cytochromes, which are part of electron transport systems in all photosynthetic and aerobically respiring cells. Porphyrin molecules are spontaneously assembled from formaldehyde—one of the molecular legacies of cosmic clouds (Figure 19.4). Was porphyrin a major electron transporter of some of the early metabolic pathways? Perhaps.

Origin of Self-Replicating Systems

Another defining characteristic of life is a capacity for reproduction, which now starts with protein-building instructions in DNA. Molecules of DNA are fairly stable and are easily replicated before each cell division. As you know from earlier chapters, molecules of RNA,

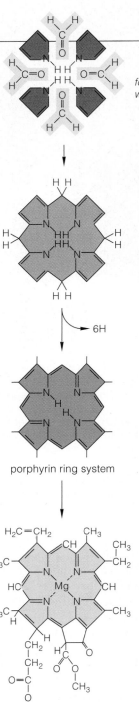

four formaldehyde molecules with four pyrrole rings

porphyrin ring system

chlorophyll *a*

Figure 19.4 One hypothetical sequence by which formaldehyde, an organic compound, underwent chemical evolution into porphyrin. Formaldehyde was present when the Earth formed. Porphyrin is the light-trapping and electron-donating component of chlorophyll molecules. It also is a component of cytochrome, which is a protein component of electron transport systems that are part of many metabolic pathways.

enzymes, and other factors are necessary to translate the DNA instructions into proteins.

Most existing enzymes get assistance from small organic molecules or metal ions called coenzymes. Intriguingly, some categories of coenzymes have a structure identical to that of RNA nucleotides. Another clue: Mix together and then heat up precursors of ribonucleotides and short chains of phosphate groups, and they self-assemble into single strands of RNA. On the early Earth, energy from the sun's rays or from geothermal events might have driven the spontaneous formation of RNA from such starting materials.

Very simple self-replicating systems of RNA, enzymes, and coenzymes have been produced in some laboratories. Did RNA become the information-storing templates on which simple proteins were synthesized? Maybe. We've learned that rRNA catalyzes the translation of mRNA into proteins (Section 13.2). A ribosome is like irregularly shaped pieces of a puzzle joined together. The pieces include many proteins that stabilize the rRNA just about everywhere except at an active site where the large ribosomal subunit contacts the smaller subunit. Ribosomes, remember, are highly conserved structures. The ones in eukaryotic cells are not that different from those in prokaryotic cells. We might

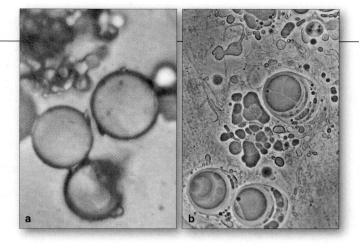

Figure 19.5 Microscopic spheres of (**a**) proteins and (**b**) lipids that self-assembled under abiotic conditions.

take this to mean that rRNA's catalytic behavior is a metabolic function that emerged early in Earth history.

It is possible that an **RNA world** in which RNA was a template for protein synthesis preceded the origin of DNA. Remember, DNA and RNA nucleotides are not that different. Their sugar component differs by a single functional group. Three of their four bases are identical, and RNA's uracil differs from DNA's thymine by only a single functional group.

Then what's the point of having DNA? Compared to RNA, this double-stranded molecule can form long nucleotide chains in far more stable fashion. Computer analyses also reveal that helical coiling of a strandlike molecule is the best way to pack the most subunits in the smallest space. DNA is just a better way to *package* more and more information about building proteins.

Until we identify the chemical ancestors of RNA and DNA, the story of life's origin will be incomplete. Filling in the details will require imaginative sleuthing. For instance, researchers ran a computer program that incorporated information about natural energy sources and simple inorganic compounds of the sort thought to have been present on the early Earth. They asked their advanced computer to subject the chosen compounds to random chemical competition and natural selection as might have occurred untold billions of times in the past. They ran their program again and again. Always the outcome was the same: *Simple precursors invariably evolved—and they spontaneously organized themselves as interacting systems of large, complex molecules.*

Origin of the First Plasma Membranes

Experimental tests are more revealing of the origin of the plasma membrane—the outermost membrane of all living cells. This cellular component consists of a lipid bilayer, studded with diverse proteins that carry out different functions. Its main role is to control which substances move into and out of the cell. Without such control, cells can neither exist nor reproduce.

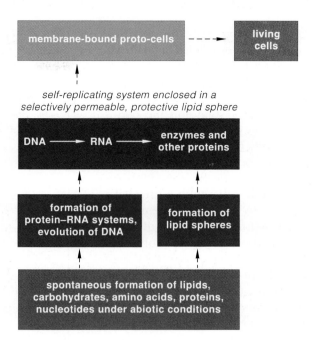

Figure 19.6 One possible sequence of events that led to the first self-replicating systems, then to the first living cells.

Probably, one avenue of molecular evolution led to **proto-cells**: simple membrane sacs that surrounded and protected information-storing templates and metabolic agents from the environment. We do know that simple membrane sacs can form spontaneously. For example, in one experiment, S. Fox heated amino acids until they formed protein-like chains, which he then placed in hot water. After cooling, the chains assembled into small, stable spheres (Figure 19.5*a*). Like membranes of cells, these proteinoid spheres were selectively permeable to various substances. In other experiments, the spheres picked up free lipid molecules from their surroundings, and a lipid–protein film formed at their surface.

In still other experiments, by David Deamer and his coworkers, fatty acids and glycerol combined to form long-tail lipid molecules under laboratory conditions that simulated conditions in evaporating tidepools. The lipids self-assembled into tiny, water-filled sacs, many of which were like cell membranes (Figure 19.5*b*).

In short, there are gaps in the story of life's origins. But there also is experimental evidence that chemical evolution led to the organic molecules and structures characteristic of life. Figure 19.6 summarizes likely steps in the chemical evolution that preceded the first cells.

Although the story is not yet complete, many laboratory experiments and computer simulations indirectly show that chemical and molecular evolution could have given rise to proto-cells.

19.3

ORIGIN OF PROKARYOTIC AND EUKARYOTIC CELLS

The first living cells originated in the **Archean** eon, which lasted from 3.9 billion to 2.5 billion years ago. Those cells emerged as molecular extensions of the evolving universe, of our solar system and the Earth. Maybe they originated at tidal flats or hydrothermal vents (Section 6.7). Fossils show they were like existing bacteria. Specifically, they all were **prokaryotic cells**, with no nucleus. Maybe they were only membrane-bound, self-replicating sacs of RNA, DNA, and other organic molecules. Free oxygen was absent, so they secured energy through anaerobic pathways—fermentation, probably. Energy was plentiful, for geologic processes had already enriched the seas with organic compounds. So "food" was available, there were no predators, and cell structures were free from oxygen attacks.

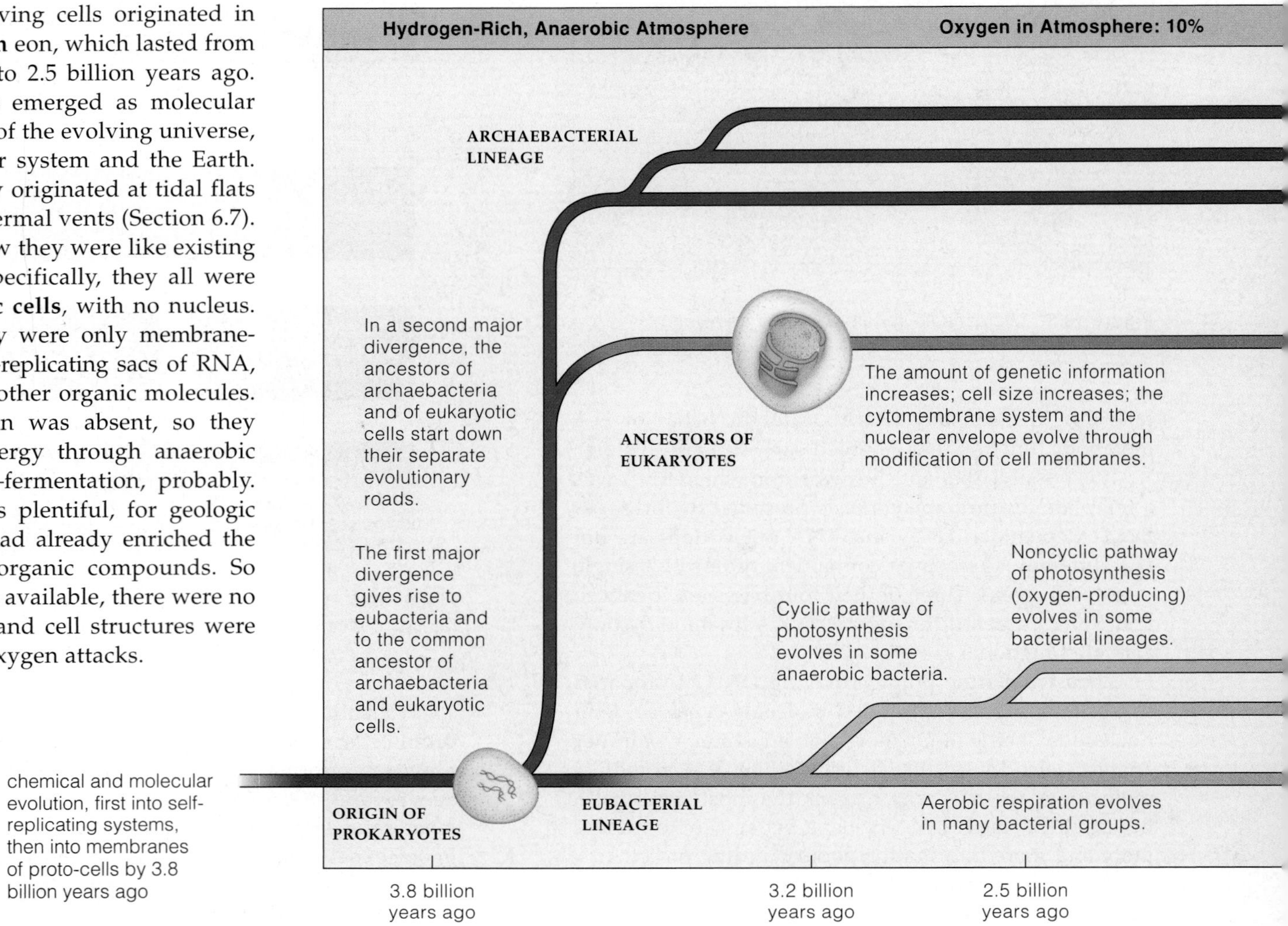

| Hydrogen-Rich, Anaerobic Atmosphere | Oxygen in Atmosphere: 10% |

ARCHAEBACTERIAL LINEAGE

In a second major divergence, the ancestors of archaebacteria and of eukaryotic cells start down their separate evolutionary roads.

ANCESTORS OF EUKARYOTES

The amount of genetic information increases; cell size increases; the cytomembrane system and the nuclear envelope evolve through modification of cell membranes.

The first major divergence gives rise to eubacteria and to the common ancestor of archaebacteria and eukaryotic cells.

Cyclic pathway of photosynthesis evolves in some anaerobic bacteria.

Noncyclic pathway of photosynthesis (oxygen-producing) evolves in some bacterial lineages.

chemical and molecular evolution, first into self-replicating systems, then into membranes of proto-cells by 3.8 billion years ago

ORIGIN OF PROKARYOTES

EUBACTERIAL LINEAGE

Aerobic respiration evolves in many bacterial groups.

3.8 billion years ago

3.2 billion years ago

2.5 billion years ago

Some populations of those first prokaryotic cells apparently diverged in two major directions shortly after the time of origin. One lineage gave rise to the **eubacteria**. The other lineage gave rise to the common ancestor of **archaebacteria** and **eukaryotic cells** (Figure 19.7).

Between 3.5 and 3.2 billion years ago, light-trapping pigments, electron transport systems, and other metabolic machinery evolved in some anaerobic eubacteria. With these innovations, the cells became the earliest practitioners of the cyclic pathway of photosynthesis. (You read about this ATP-forming pathway in Section 6.3). Sunlight, an unlimited source of energy, had been tapped. For nearly 2 billion years, photosynthetic descendants of those cells dominated the living world. Their tiny but numerous populations formed large mats in which sediments collected. The mats built up, one atop the other. Calcium deposits hardened and preserved the mats, which came to be called **stromatolites** (Figure 19.8).

Figure 19.7 Evolutionary tree of life showing possible connections among major lineages and the origins of some eukaryotic organelles.

Figure 19.8 Stromatolites exposed at low tide in western Australia's Shark Bay. These mounds started forming 2,000 years ago. They are structurally identical with stromatolites that formed more than 3 billion years ago.

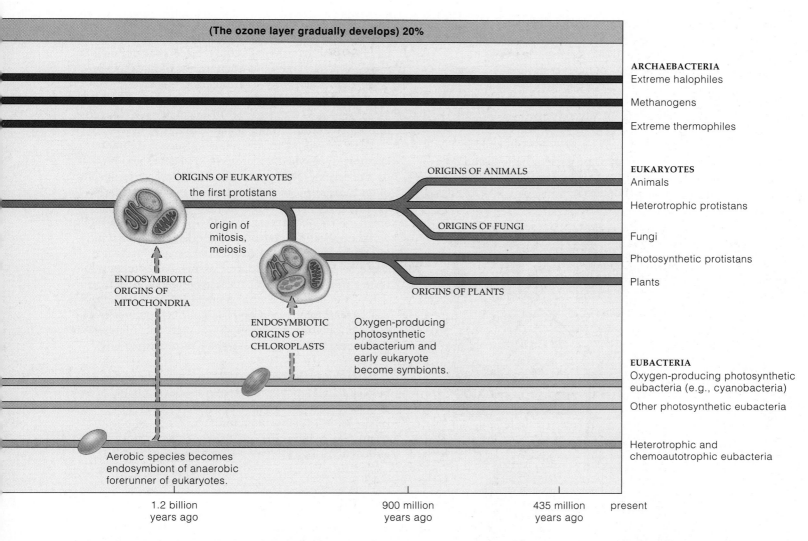

(The ozone layer gradually develops) 20%

ARCHAEBACTERIA
Extreme halophiles

Methanogens

Extreme thermophiles

ORIGINS OF EUKARYOTES
the first protistans

ORIGINS OF ANIMALS

EUKARYOTES
Animals

Heterotrophic protistans

origin of
mitosis,
meiosis

ORIGINS OF FUNGI

Fungi

Photosynthetic protistans

ENDOSYMBIOTIC
ORIGINS OF
MITOCHONDRIA

Plants

ORIGINS OF PLANTS

ENDOSYMBIOTIC
ORIGINS OF
CHLOROPLASTS

Oxygen-producing
photosynthetic
eubacterium and
early eukaryote
become symbionts.

EUBACTERIA
Oxygen-producing photosynthetic
eubacteria (e.g., cyanobacteria)

Other photosynthetic eubacteria

Heterotrophic and
chemoautotrophic eubacteria

Aerobic species becomes
endosymbiont of anaerobic
forerunner of eukaryotes.

1.2 billion
years ago

900 million
years ago

435 million
years ago

present

By the dawn of the **Proterozoic** eon, 2.5 billion years ago, photosynthetic machinery had become altered in some eubacterial species, and the noncyclic pathway of photosynthesis emerged. Oxygen, one of the pathway's by-products, started to accumulate. In time, this had two irreversible effects. First, *an oxygen-rich atmosphere stopped the further chemical origin of living cells*. Except in a few anaerobic habitats, complex organic compounds could no longer form spontaneously and resist attack. Second, *aerobic respiration became the dominant energy-releasing pathway*. In many prokaryotic lineages, selection favored metabolic equipment that neutralized oxygen by using it as an electron acceptor! This key innovation contributed to the rise of multicelled eukaryotes and their invasion of far-flung environments (Section 7.7).

Eukaryotic cells evolved in the Proterozoic, possibly before 1.2 billion years ago. We have fossils, 900 million years old, of complex algae, fungi, and plant spores. As you know, organelles are the premier defining features of eukaryotic cells. *Where did they come from?* The next section describes a few plausible hypotheses.

Rodinia, the first supercontinent, formed 1.1 billion years ago. By about 800 million years ago, stromatolites along its shorelines were declining dramatically. Were newly evolved, bacteria-eating *animals* using them as a concentrated source of food? Fossilized embryos—cell clusters no wider than a pin—give hints that the first animals were soft-bodied forerunners of the modern sponges and marine worms. Before 570 million years ago, in precambrian times, some of their descendants started the first adaptive radiation of animals.

The first living cells evolved by about 3.8 billion years ago, during the Archean eon. All were prokaryotic. Most, if not all, probably made ATP by fermentation routes.

Early on, ancestors of archaebacteria and eukaryotic cells diverged from the lineage that led to modern eubacteria.

Oxygen-releasing photosynthetic bacteria evolved. In time, the oxygen-enriched atmosphere put an end to the further spontaneous chemical evolution of life. That atmosphere was a key selection pressure in the evolution of eukaryotic cells.

Where Did Organelles Come From?

Thanks to Andrew Knoll, William Schopf, Jr., and other globe-hopping microfossil hunters, we have tantalizing evidence of early life. One fossil treasure is a strand of bacterial cells 3.5 billion years old; it was around not long after life originated. Others, from the Proterozoic, were eukaryotic cells with a few membrane-bound organelles in the cytoplasm (Figure 19.9). Most of their descendants have a profusion of organelles (Figure 19.10).

Where did eukaryotic organelles come from? Speculations abound. Some probably evolved by gene mutations and natural selection. For others, researchers make a case for evolution by way of endosymbiosis.

ORIGIN OF THE NUCLEUS AND ER Prokaryotic cells do not have an abundance of organelles. But some species have interesting infoldings of the plasma membrane that incorporate enzymes and other agents of metabolism (Figure 19.11). In early forerunners of eukaryotic cells, similar infoldings into the cytoplasm may have served as passageways to the surface. They may have evolved into ER channels and into an envelope around the DNA.

What would be the advantage of such membranous enclosures? Maybe they protected the genes and protein products from "foreigners." Remember, bacterial species often transfer plasmid DNA among themselves. So do very simple eukaryotic cells called yeasts. A nuclear envelope may have been favored because it got the cell's genes, replication enzymes, and transcription enzymes out of the cytoplasm. It meant vital genetic messages could be copied and read, free of metabolic competition from a potentially unmanageable hodgepodge of foreign genes. Similarly, ER channels might have protected the cell's vital proteins and other organic compounds from metabolically hungry "guests"—foreign cells that became permanent residents in a host cell, as described next.

A THEORY OF ENDOSYMBIOSIS Accidental partnerships between a variety of prokaryotic species probably arose countless times in lineages that gave rise to eukaryotic cells. Maybe some resulted in the origin of mitochondria, chloroplasts, and other organelles. This is a story of endosymbiosis, as developed in greatest detail by Lynn

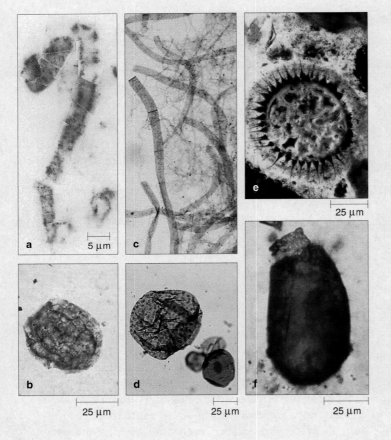

a   5 µm
b   25 µm
c
d   25 µm
e   25 µm
f   25 µm

Figure 19.9 From Australia, (**a**) a strand of walled prokaryotic cells 3.5 billion years old and (**b**) one of the oldest known eukaryotes— a protistan 1.4 billion years old. From Siberia, (**c**) a multicelled alga 900 million to 1 billion years old and (**d**) eukaryotic microplankton 850 million years old. (**e**) From China, a eukaryotic cell that lived about 560 million years ago. (**f**) From Spitsbergen, Norway, a protistan that was alive 50 million years before the dawn of the Cambrian.

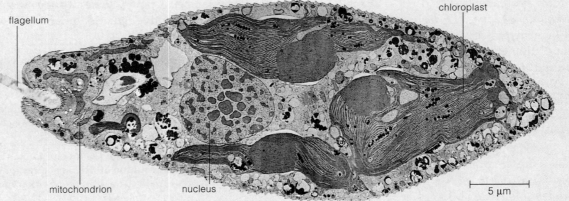

flagellum
chloroplast
mitochondrion
nucleus
5 µm

Figure 19.10
A fine example of the profusion of diverse organelles that are hallmarks of eukaryotic cells: *Euglena*, a single-celled protistan, sliced lengthwise for this micrograph. It also has a long flagellum, which could not fit in the image area at this magnification.

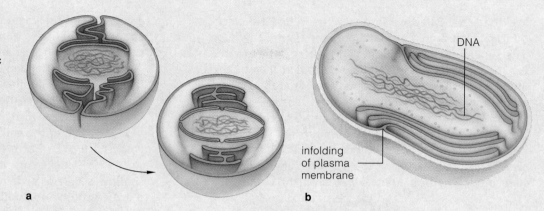

Figure 19.11 (a) Model for the origin of endoplasmic reticulum and the nuclear envelope. In the prokaryotic ancestors of eukaryotic cells, infoldings of the plasma membrane gave rise to both cell components. (b) Such infoldings are present in the cytoplasm of many kinds of existing bacteria, including *Nitrobacter*, sketched here in cutaway view.

a   b

DNA

infolding
of plasma
membrane

Margulis. *Endo–* means within; *symbiosis* means living together. In cases of **endosymbiosis**, one species (the guest) spends its entire life cycle *inside* another species (the host). The interaction benefits one or both.

According to one hypothesis, eukaryotic cells evolved by endosymbiosis long after the noncyclic pathway of photosynthesis emerged and oxygen had accumulated to significant levels in the atmosphere. In some groups of bacteria, one type of electron transport system in the plasma membrane had evolved and now incorporated "extra" cytochromes—which donated electrons to oxygen. With this key innovation, such bacteria started extracting energy from organic compounds by aerobic respiration.

Possibly before 1.2 billion years ago, the forerunners of eukaryotes were engulfing aerobic bacteria. Maybe they resembled existing soft-bodied, amoebalike cells that weakly tolerate free oxygen. Maybe they trapped food by sending out cytoplasmic extensions and used endocytic vesicles to get food into the cytoplasm for digestion.

A key point of the theory is that some aerobic bacteria resisted digestion and thrived in the protected, nutrient-rich environment. In time, they were releasing extra ATP. Their hosts came to depend on that ATP for their growth, increased activity, and assembly of hard parts and other structures. The guests benefited, too. They no longer had to seek food or build components for metabolic work the hosts did for them. In time the anaerobic and aerobic cells were incapable of independent life. The guests had evolved into mitochondria, supreme suppliers of ATP.

EVIDENCE OF ENDOSYMBIOSIS Strong evidence favors Margulis's theory. There are plenty of examples of nature continuing to tinker with endosymbionts, including the cell in Figure 19.12. Its mitochondria are like bacteria in size and structure. The inner mitochondrial membrane is like a bacterial plasma membrane. Each mitochondrion replicates its own DNA and divides independently of the host cell's division. A few genetic code words in its DNA and mRNA have unique meanings. They are translated into a few proteins required for special mitochondrial tasks. Compared to the genetic code of living cells, the "mitochondrial code" has a few distinct differences.

Chloroplasts, too, may be stripped-down descendants of oxygen-evolving, photosynthetic bacteria. Perhaps predatory aerobic bacteria engulfed photosynthetic cells. Maybe the cells escaped digestion, absorbed nutrients from the host's cytoplasm, and continued to function. By providing aerobically respiring hosts with oxygen, they would have promoted endosymbiosis.

In their metabolism and overall nucleic acid sequence, chloroplasts resemble some eubacteria. Their DNA self-replicates, and they divide independently of the cell's division. Chloroplasts vary in shape and light-absorbing pigments, just as photosynthetic eubacteria do. Did they originate a number of times in different lineages? Maybe.

Adding to the intrigue are ciliated protistans and marine slugs that "enslave" chloroplasts! The slugs eat algae but retain algal chloroplasts in their gut cells. The chloroplasts draw nutrients from host tissues. They photosynthesize and release oxygen for weeks.

However they arose, new kinds of cells did appear on the evolutionary stage. They had cytomembranes, a nucleus, and mitochondria, chloroplasts, or both. They were eukaryotic cells, the first **protistans**. And with their efficient metabolic strategies, early protistans underwent rapid divergences and adaptive radiations. In no time at all, evolutionarily speaking, some of their descendants gave rise to the great kingdoms of plants, fungi, and animals, as sketched out earlier in Figure 19.7.

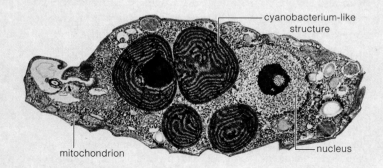

cyanobacterium-like
structure

mitochondrion   nucleus

Figure 19.12 *Cyanophora paradoxa*, a protistan. Its mitochondria resemble bacteria. Its photosynthetic structures look like spherical cyanobacteria (which are photosynthetic) without the cell wall.

LIFE IN THE PALEOZOIC ERA

We divide the **Paleozoic** into the Cambrian, Ordovician, Silurian, Devonian, Carboniferous, and Permian periods. The supercontinent Rodinia had split apart before this era. Some of its fragments now straddled the equator, and warm, shallow seas lapped their margins. Global conditions restricted any pronounced seasonal changes in winds, ocean currents, and the upward churning of nutrient deposits on the seafloor. Nutrient supplies in equatorial seas were stable but limited. Polar regions were inhospitably icy; and to the south, other fragments had collided to form a new supercontinent, Gondwana:

GONDWANA

Most major animal phyla had evolved earlier in the precambrian seas. Possibly some of their ancestors were among the **Ediacarans**, peculiar organisms shaped like fronds, disks, and blobs that nearly defy classification (Figure 19.13a,b). Like Ediacarans, the early Cambrian animals had highly flattened bodies with a fine surface-to-volume ratio for taking up nutrients (Figure 19.13c). Most thrived on or in seafloor sediments, where dead organisms and organic debris settled. They ranged from sponges to simple vertebrates, and they were diverse.

This was a time when new predators and prey with armorlike shells, spines, mouths, and amazing feeding structures rapidly evolved. We find fossilized animals with punctures, missing chunks, and healed wounds in sedimentary beds. Things were starting to get lively!

What caused the "Cambrian explosion" of diversity? Were asteroids an impetus? After tapering off about 3 billion years ago, the impact rate apparently increased abruptly around this time. Were genes governing early growth and development less intertwined than they are today? If so, selection against novel traits might have been less severe. What about the warm, shallow waters of the equatorial seas? They must have been adaptive zones, with novel opportunities for getting food.

Early in the Ordovician, widespread flooding of the land masses gave rise to broad, tropical seaways. The emergence of the vast marine environments promoted adaptive radiations. Many new reef organisms, such as the swift, shelled predators called nautiloids, evolved. Among their descendants are the chambered nautiluses (Section 25.3). Trilobites, one of the most common but unswift invertebrates, nearly vanished (Figure 19.13d).

Huge glaciers formed across the southern portion of Gondwana at the end of the Ordovician—one of the coldest times in Earth history. Enormous volumes of water were locked up in ice sheets, and this drained shallow seas everywhere. More than 70 percent of all marine groups—including the reef builders, trilobites, placoderms, and jawless fishes—became extinct at the Ordovician–Silurian boundary.

Gondwana drifted north during Silurian times. The sea level rose as great ice sheets melted, and adaptive radiations occurred in the seas. Large coral reefs formed, algae flourished, and the first jawed fishes appeared. Small plants took hold in wet lowlands (Figure 19.14b,c).

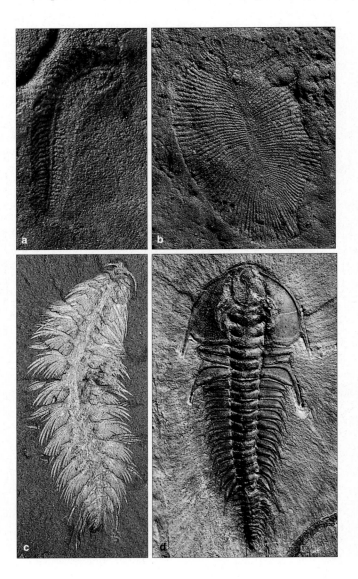

Figure 19.13 Representatives from the precambrian and Cambrian seas. Two Ediacarans, about 600 million years old: (**a**) *Spriggina* and (**b**) *Dickensonia*. The oldest known Ediacarans lived 610 million years ago and the most recent in Cambrian times, 510 million years ago. (**c**) From British Columbia's Burgess Shale, a fossilized marine worm. (**d**) A beautifully preserved fossil of one of the earliest trilobites.

Figure 19.14 (**a**) Life in the Silurian seas. Some of these shelled animals (nautiloids) were as long as a kayak. (**b**) At the close of the Silurian and on into the Devonian, forerunners of modern ferns and club mosses dominated swamps and marshes. (**c**) Fossils of a Devonian plant (*Psilophyton*) that may have been one of the earliest ancestors of the conifers and other seed-bearing plants. (**d**) Reptiles (*Dimetrodon*) of a hotter, drier time, the Permian. Some fossils of these carnivores have been found in Texas. Giant club mosses and horsetails had declined. Conifers, cycads, and other gymnosperms replaced them during the Permian.

├─── 5 cm

So did fungi and invertebrates, such as spiders, insects, and segmented worms. The Devonian was a pivotal time in vertebrate evolution. Armor-plated fishes with massive jaws diversified in the seas and in fresh water. Fishes that gave rise to amphibians invaded the land. They had lobed fins—the forerunners of legs and other limbs—and simple lungs. Lobed fins and lungs were key innovations that would prove to be advantageous for life out of water, in dry land habitats (Section 24.5).

Warm-water species especially were hit hard by a Devonian mass extinction. Global cooling, a decline in sea level, a meteorite impact, or some combination of these might have been the cause. Afterward, plants and insects embarked on adaptive radiations on land.

In the Carboniferous, land masses were submerged and drained many times. Organic debris accumulated, became compacted, and was converted to coal (Section 22.2). Insects, amphibians, and early reptiles flourished in vast swamp forests of the Permian (Figure 19.14*d*). The early ancestors of cycads, ginkgos, conifers, and other seed-producing plants dominated the forests.

Late in the Permian, 95 percent of all known species were lost in the greatest of all mass extinctions. New evidence suggests that "The Great Dying" followed a direct hit by comet. Whatever the trigger, there was a colossal outpouring of lava, a severe disruption of the global carbon cycle, and a major glaciation.

Pangea had already been forming long before then. By the time the Permian drew to a close, it extended from pole to pole. One world ocean, including the vast Tethys Sea, lapped its margins:

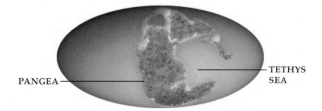

PANGEA ——   —— TETHYS SEA

And just as the Paleozoic mass extinctions had done, the new distribution of oceans, land masses, and land elevations redirected the course of life's evolution.

Early in the Paleozoic era, organisms of all six kingdoms were flourishing in the seas. By the end of the era, many lineages had successfully invaded the land, including the wet lowlands of the new supercontinent Pangea.

LIFE IN THE MESOZOIC ERA

Speciation on a Grand Scale

We divide the **Mesozoic** into the Triassic, Jurassic, and Cretaceous periods. It lasted about 175 million years. In the middle of the Jurassic, the supercontinent Pangea started to break up. Its fragments slowly drifted apart. The fossil record tells us those geographically isolated fragments favored divergences and speciation:

This was an era of spectacular expansion in the range of global diversity. Invertebrates and fishes underwent adaptive radiations in the seas. Conifers and other seed-bearing **gymnosperms,** insects, and reptiles became the visibly dominant lineages on land. Flowering plants—**angiosperms**—emerged during the late Jurassic or the early Cretaceous. In less than 40 million years, they would displace the conifers and related plants in most environments (Figure 19.15 and Section 22.8).

Rise of the Ruling Reptiles

Early in the Triassic, the first **dinosaurs** evolved from a reptilian lineage. They were not much bigger than a turkey. Possibly most species had high metabolic rates, and maybe they were warm-blooded. Many sprinted about on two legs. But the Triassic dinosaurs weren't the dominant land animals. The rulers were *Lystrosaurus* and other plant-eating, mammal-like reptiles that were too big to be bothered by most predators (Figure 9.16).

Adaptive zones opened up for the dinosaurs after a catastrophe that left a string of five craters in what is now France, Quebec, Manitoba, and North Dakota. One crater is as big as Rhode Island. About 214 million years ago, fragments from an asteroid or a comet fell in a row as the Earth spun beneath them, just as a string of huge fragments from comet Shoemaker–Levy 9 hit Jupiter in 1994. The blast waves, global firestorm, lava flows, and earthquakes must have been stupendous. Most animals lucky enough to survive this time of mass extinction were smaller, had higher rates of metabolism, and were less vulnerable than others to drastic changes in outside temperatures. These same attributes also characterized survivors of later mass extinctions.

Descendants of the surviving dinosaurs became the ruling reptiles; they endured for 140 million years. Some species reached monstrous proportions. Among them were the ultrasaurs, taller than a four-story building.

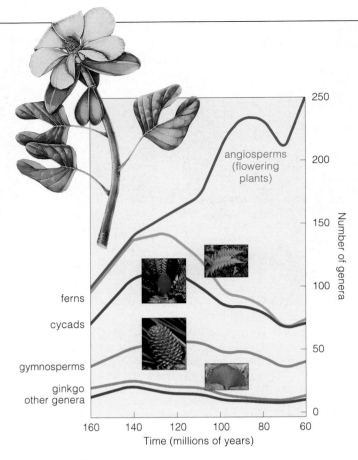

Figure 19.15 Range of diversity among vascular plants of the Jurassic and Cretaceous. Conifers and other gymnosperms were dominant. They started to decline even before flowering plants began a great adaptive radiation that continued into the Cenozoic. *Upper left:* From the Cretaceous, a floral shoot of *Archaeanthus linnenbergeri.* In many traits, this now-extinct flowering plant resembled living magnolias.

Many dinosaurs perished in another mass extinction at the end of the Jurassic, then in a pulse of extinctions during the Cretaceous. Huge outpourings of molten rock may have triggered changes in the global climate. Or maybe asteroids or comets inflicted the blows. Not all lineages survived. Yet some did recover, and new ones evolved. Duckbilled dinosaurs appeared in forests and swamps. Tanklike *Triceratops* and other plant eaters flourished in open habitats. They were prey for the agile and swift *Velociraptor* of motion picture fame.

About 120 million years ago, global temperatures skyrocketed 25 degrees. By one theory, plumes spread out beneath the crust and "greased" the crustal plates into moving twice as fast. A superplume or a rash of them broke through the crust. In what is now the South Atlantic, the crust opened like a zipper. Basalt and lava poured from fissures; volcanoes spewed nutrient-rich ashes. The plumes released great quantities of carbon dioxide, one of the key "greenhouse" gases that absorb some heat radiating from the Earth before it escapes to outer space. The nutrient-enriched planet warmed, and

Figure 19.16 Life in the Mesozoic. (**a**) *Lystrosaurus*, a therapsid (mammal-like reptile) of the Triassic. This tusked herbivore, about 1 meter (three feet) long, fed on fibrous plants in dry floodplains. As Section 18.3 describes, its fossils are distributed on widely separated continents and are evidence in favor of the plate tectonics theory.

(**b**) *Temnodontosaurus*, an ichthyosaur that hunted large squid, ammonites, and other prey in the warm, shallow seaways of the early Jurassic. Fossils meauring 9 meters (30 feet) long have been found in England and Germany. (**c**) From the Cretaceous, cycads, ferns, *A. linnenbergeri* and other flowering plants, and *Deinonychus*, an agile dinosaur with keen eyesight, serrated teeth, and powerful jaws.The short forelimbs of this bipedal carnivore had sharp claws, including a large one shaped like a sickle; hence its name, which means "terrible claw." Fossils 3 meters (nine feet) long have been found in Texas.

it remained warm for 20 million years. Photosynthetic organisms flourished on land and in the shallow seas. Their remains were slowly buried and converted into the world's vast oil reserves.

About 65 million years ago, the last dinosaurs and many marine organisms vanished in a mass extinction. As described in the next section, their disappearance apparently coincided with a direct hit by an asteroid the size of Mount Everest. Imperceptibly over time, the impact site slowly drifted into a position that we now call the northern Yucatán peninsula.

The Mesozoic was a time of major adaptive radiations and of a mass extinction in which the last dinosaurs and many marine organisms disappeared.

Horrendous End to Dominance

It's only been about 55,000 years since the first modern humans walked the Earth. Since then, how many times have people, puffed up with self-importance, set out to conquer neighbors, lands, and seas? Think about it. Then think about the dinosaurs. Were they good at reigning supreme? No question about it; their lineage dominated the land for 140 million years. In the end, did it matter? Not a bit. Sixty-five million years ago, at the Cretaceous–Tertiary (K–T) boundary, nearly all members of their most excellent lineage perished. Why? Bad luck.

A thin layer of iridium-rich rock distributed around the world dates precisely to the K–T boundary. Iridium is rare here but not in asteroids. **Asteroids** are rocky, metallic bodies, 1,000 kilometers to a few meters across, that are hurtling through space. When planets of our solar system were forming, their gravitational pull swept most asteroids from the sky. At least 6,000 still orbit the sun in a belt between Mars and Jupiter (Figure 19.17). The orbits of dozens of others take them across Earth's orbit, like Russian roulette on a cosmic scale.

By analyzing iridium levels in soils, gravity maps, and other evidence, Walter Alvarez and Luis Alvarez hypothesized that an asteroid impact caused the K–T mass extinction. Researchers later identified the impact site. Massive movements in the crust transported it to what is now the northern Yucatán peninsula of Mexico (Figure 19.18). The impact crater is 9.6 kilometers deep and 300 kilometers across—wider than Connecticut. This crater as well as other evidence strongly supports what is now called the **K–T asteroid impact theory**.

To make a crater that big, the asteroid had to hit the Earth at 160,000 kilometers per hour. It blasted at least 200,000 cubic kilometers of dense gases and debris into the sky. The crust heaved violently. Monstrous waves 120 meters high raced across the ocean. They obliterated life on islands, then slammed into coasts of continents. Researchers hypothesized that the atmospheric debris kept out sunlight for months. If so, plants and other

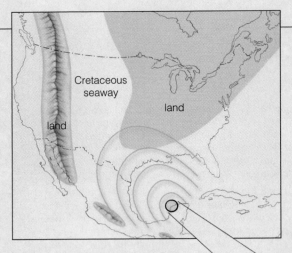

Figure 19.18 Artist's interpretation of what might have happened in the last few minutes of the Cretaceous.

producers would have withered and died. Animals and other consumers would have starved to death.

The hypothesis had problems. By some calculations, the volume of debris blasted aloft would not have been large enough to have had such consequences.

Then comet Shoemaker–Levy 9 slammed into Jupiter. Particles blasted into the Jovian atmosphere triggered an intense heating of an area larger than the Earth. That event supports a **global broiling hypothesis**, proposed first by H. J. Melosh and his colleagues. Briefly stated, energy released at the K–T impact site was equivalent to detonating 100 million nuclear bombs. Trillions of tons of vaporized debris rose in a colossal fireball, then rapidly condensed into particles the size of sand grains. Seconds later, a cooler fireball made of steam, carbon dioxide, and unmelted rock formed. When the debris fell to the Earth, it raised the atmosphere's temperature by thousands of degrees. The sky must have been ten times hotter than it gets above Death Valley in summer. In one horrific glowing hour, nearly all plants erupted in flames and every animal out in the open was broiled alive.

Things haven't settled down much since dinosaurs disappeared. For instance, about 2.3 million years ago, a huge object from outer space hit the Pacific Ocean. At about the same time, vast ice sheets started to form abruptly in the Northern Hemisphere. Long-term shifts in climate may have been ushering in that most recent ice age, but the global impact would have accelerated it. Water vaporized by the impact could have contributed to a global cloud cover that limited the sunlight reaching the surface. Ancestors of humans were around when this happened. The extreme shift in climate surely put their adaptability to the test.

Almost certainly, severe environmental tests await all existing lineages. When we become too smitten with our importance in the world of life, we might do well to step back, from time to time, and reflect on what is going on above and beneath the Earth's surface. Asteroids and superplumes do have a way of leveling the playing field, as they did for gigantic dinosaurs and tiny mammals.

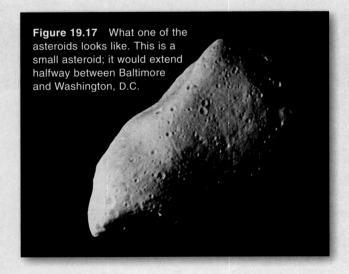

Figure 19.17 What one of the asteroids looks like. This is a small asteroid; it would extend halfway between Baltimore and Washington, D.C.

LIFE IN THE CENOZOIC ERA

The breakup of Pangea triggered events that continued into the present era, the **Cenozoic**. At the dawn of the Cenozoic, major land masses were on collision courses:

Coastlines fractured. The Cascades, Andes, Himalayas, and Alps rose through volcanic activity, uplifting, and other events at crustal rifts and plate boundaries. These geologic changes brought about major shifts in climate that influenced the further evolution of life.

During the Paleocene epoch, climates were warmer and wetter. Tropical and subtropical forests extended farther north and south than they do today. Woodlands and forests spread even into polar regions. Although their key traits evolved before dinosaurs left the scene, mammals now began a major adaptive radiation.

The climate warmed even more during the Eocene epoch. Subtropical forests extended north into polar regions. A variety of mammals, including primates, bats, rhinos, hippos, elephants, horses, and carnivores, emerged. By the late Eocene, the climate grew more cool and dry, and the seasons became pronounced. Woodlands and dry grasslands dominated vast land regions. The patterns of vegetational growth changed so greatly that many mammals became extinct.

From the Oligocene through the Pliocene, the woodlands and grasslands were home to many grazing and browsing animals. Among them were camels and the "giraffe rhinoceros," along with some fearsome carnivores that stalked them (Figure 19.19).

Currently, the distribution of land masses favors biodiversity. The richest ecosystems are tropical forests of South America, Madagascar, and Southeast Asia, and the tropical marine ecosystems of the Pacific archipelagos. Yet we are in the midst of a new mass extinction. About 50,000 years ago, nomadic humans followed the migrating herds of wild animals around the Northern Hemisphere. By a few thousand years later, major groups of mammals were extinct. The pace of extinction has since accelerated as humans hunt

Figure 19.19 Cenozoic species. (**a**) During the early Paleocene, in what is now Wyoming, mammals lived in dense forests of sequoia and other plants. On the ground, raccoonlike *Chriacus* faces a tree-climbing rodent (*Ptilodus*). Higher up is *Peradectes*, a marsupial. (**b**) *Indricotherium*, the "giraffe rhinoceros." At 15 tons and 5.5 meters tall at the shoulder, it was the largest mammal known. It browsed in late Eocene and early Miocene woodlands. (**c**) Saber-tooth cats (*Smilodon*) and small horses from the Pleistocene. Fossils have been recovered from pitch pools in Rancho LaBrea, California.

for food, fur, feathers, or fun, and as natural habitats are converted for farm animals and crops. We'll revisit the current extinction crisis in Chapters 25 and 45.

Crustal movements and mountain building during the Cenozoic triggered major shifts in climate. Conditions favored a great adaptive radiation of mammals, first in tropical forests, then in woodlands and grasslands.

SUMMARY

Cambrian. Explosion of diversity in equatorial seas; icy polar regions

Silurian. Sea level rises, diverse marine life; plants, invertebrates invade land

1. The evolutionary story of life begins with the "big bang," a model for the origin of the universe. By this model, all matter and space were once compressed in a fleeting state of incredible heat and density. Time began with the near-instantaneous distribution of matter and energy through the known universe, which has been expanding ever since. Helium and other light elements formed just after the big bang. Heavier elements were created by the formation, evolution, and death of stars. All elements of the solar system, the Earth, and life are products of this physical and chemical evolution. *CI*

2. Four billion years ago, the Earth had a high-density core, an intermediate-density mantle, and a thin, highly unstable crust of low-density rocks. Probably gaseous hydrogen, nitrogen, carbon monoxide, and carbon dioxide made up most of the first atmosphere. Under prevailing conditions, neither free oxygen nor liquid water could accumulate at the Earth's surface. *19.1*

3. For hundreds of millions of years after the crust cooled, runoff from rains put dissolved mineral salts and other compounds in crustal depressions, where the early seas formed. Life as we know it would not have originated without that salty liquid water. *19.1*

4. Many studies and experiments have yielded indirect evidence that life originated under the conditions that existed on the early Earth. *19.1, 19.2*

 a. The composition of cosmic clouds and rocks from planets and the Earth's moon indicate that precursors of complex molecules associated with life were present. In laboratory tests simulating the primordial conditions, including the absence of free oxygen, these precursors spontaneously assembled into sugars (such as glucose), amino acids, and other organic compounds. *19.1*

 b. Chemical principles and computer simulations suggest that metabolic pathways evolved by chemical competition for limited supplies of organic molecules that had accumulated by natural geologic processes in the seas. In the laboratory, self-replicating systems of RNA, enzymes, and coenzymes have been synthesized. Double-stranded, helically coiled DNA probably evolved as the most efficient way of packing the most protein-building instructions in the smallest space. Also, its long nucleotide chains are more stable than RNA. *19.2*

 c. In laboratory simulations of conditions thought to have prevailed on the early Earth, lipids as well as lipid–protein membranes having some of the properties of cell membranes have formed spontaneously. *19.2*

5. Since its origin about 3.8 billion years ago, life has been influenced by changes in the Earth's atmosphere, crust, and oceans. Forces of change included asteroid impacts, volcanism, and crustal plate movements. The forces altered the distribution of land masses and the seas (Figure 19.20) as well as global and local climates.

Another major force of change: activities of organisms, especially the oxygen-releasing photosynthesizers and, far more recently, the human species. *19.2–19.8*

6. The history of life spans the following great intervals (here you may wish to review Section 18.2): *19.2–19.8*
 a. Archean: 3.9 billion to 2.5 billion years ago
 b. Proterozoic: 2.5 billion to 570 million years ago
 c. Paleozoic: 570 to 240 million years ago
 d. Mesozoic: 240 to 65 million years ago
 e. Cenozoic: 65 million years ago to the present

7. The first living cells arose in the Archean and were prokaryotic. The first divergence led to eubacteria and to the ancestor of archaebacteria and eukaryotes. Some eubacteria used a cyclic photosynthesis pathway. *19.3*

8. A noncyclic photosynthetic pathway evolved in the Proterozoic in some eubacterial lineages. Oxygen, the pathway's by-product, accumulated in the atmosphere; it blocked the further spontaneous formation of organic molecules. The spontaneous origin of life was over. The abundance of free oxygen was a selective pressure that favored aerobic respiration, a key step in the origin of eukaryotes. It helped form an ozone layer. This shield against ultraviolet radiation allowed some lineages to move from the seas into low wetlands. *19.3, 19.5*

9. Mitochondria and chloroplasts probably evolved by endosymbiosis between aerobic bacteria and anaerobic bacteria that were the forerunners of eukaryotes. *19.4*

10. By the early Paleozoic, diverse organisms of all six lineages were established in the seas. Before it ended, land was invaded. Ever since, there have been pulses of mass extinctions and adaptive radiations. *19.5–19.8*

11. Dinosaurs were the dominant land animals for 140 million years. Their lineage ended at the Cretaceous–Tertiary (K–T) boundary when an asteroid hit the Earth. In the Cenozoic, mammals began a great radiation into adaptive zones vacated by the dinosaurs. *19.7–19.8*

Review Questions

1. Compare chemical and physical conditions that are thought to have prevailed on the Earth 4 billion years ago with current conditions. *19.1*

2. Describe some experimental evidence for the (a) spontaneous origin of large organic molecules, (b) self-assembly of proteins, and (c) formation of organic membranes and spheres. *19.1, 19.2*

3. Summarize the key points of the theory of endosymbiotic origins for mitochondria and chloroplasts. Cite evidence that favors this theory. *19.4*

Devonian. Jawed fishes evolve, diversify; ancestors of amphibians invade land

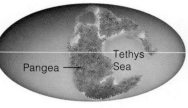
Permian into Triassic. Swamp forests (coal source); seed plants evolve

Cretaceous into Tertiary. Extinction of dinosaurs; rise of mammals

Middle Miocene, polar regions again iced over, as in Cambrian

4. Describe the prevailing conditions that probably favored the Cambrian "explosion" of diversity among marine animals, as evidenced by the fossil record. *19.5*

5. When did plants, fungi, and insects invade the land? What kind of vertebrates first invaded the land, and when? *19.5*

6. What were global conditions like when the gymnosperms and dinosaurs originated? *19.6*

7. Briefly explain how an asteroid impact and "global broiling" may have caused the mass extinction at the K–T boundary. *19.7*

8. Would you expect the Paleozoic, Mesozoic, or Cenozoic to be called "the age of mammals"? As part of your answer, explain the differences between global conditions in each era. *19.8*

Self-Quiz   ANSWERS IN APPENDIX III

1. Through study of the geologic record, we know that the evolution of life has been profoundly influenced by _____ .
   a. tectonic movements of the Earth's crust
   b. bombardment of the Earth by celestial objects
   c. profound shifts in land masses, shorelines, and oceans
   d. physical and chemical evolution of the Earth
   e. all of the above

2. _____ was the first to obtain indirect evidence that organic molecules could have been formed on the early Earth.
   a. Darwin   b. Miller   c. Fox   d. Margulis

3. An abundance of _____ was conspicuously absent from the Earth's atmosphere 4 billion years ago.
   a. hydrogen   c. carbon monoxide
   b. nitrogen   d. free oxygen

4. Life as we know it originated by _____ .
   a. 4.6 billion years ago   c. 3.8 billion years ago
   b. 2.8 million years ago   d. 3.8 million years ago

5. Which of the following statements is false?
   a. The first living cells were prokaryotic.
   b. The cyclic pathway of photosynthesis first appeared in some eubacterial species.
   c. Oxygen began accumulating in the atmosphere after the noncyclic pathway of photosynthesis evolved.
   d. In the Proterozoic, the rise in atmospheric oxygen enhanced the spontaneous formation of organic molecules.
   e. All are correct.

6. The first eukaryotic cells emerged during the _____ .
   a. Paleozoic   c. Archean   e. Cenozoic
   b. Mesozoic   d. Proterozoic

7. Match the geologic time interval with the events listed.
   ____ Archean   a. major radiations of dinosaurs, origin
   ____ Proterozoic   of flowering plants and mammals
   ____ Paleozoic   b. chemical evolution, origin of life
   ____ Mesozoic   c. radiations of flowering plants, insects,
   ____ Cenozoic   birds, mammals; humans emerge
                    d. free oxygen present; origin of aerobic
                       metabolism, protistans, fungi, animals
                    e. rise of early plants on land, origin
                       of amphibians, and origin of reptiles

Figure 19.20 Summary of changes in the distribution of land masses and the seas that influenced life's evolution.

Critical Thinking

1. Briefly explain, in terms of hydrophilic and hydrophobic interactions, how proto-cells might have formed in water from aggregations of lipids, proteins, and nucleic acids.

2. Reflect on Figure 19.20 and on this: The Atlantic Ocean is widening, and the Pacific Ocean and Indian Ocean are closing in on each other. Many millions of years from now, continents will collide and form a second Pangea. Write a short essay on what conditions might be like on that future supercontinent and what types of species might survive on it. (*Hint:* Check out the Paleomap Project animations on the Internet.)

3. According to one estimate, there is a chance that about 10^{20} planets have formed in the universe that are capable of sustaining life—but there is only one chance at intelligent life per planet. Given your knowledge of molecular biology and evolutionary processes, do you find this estimate plausible? If so, speculate on why the odds are so low.

4. We know of a number of large asteroids that may intersect Earth's orbit in the distant future. There probably are a number we don't know about. Would you use this as an excuse not to worry about polluting the environment, not to take care of your physical health (as by avoiding drugs), and not to care about our cultural evolution? Why or why not?

Selected Key Terms

angiosperm *19.6*	global broiling hypothesis *19.7*
archaebacterium *19.3*	gymnosperm *19.6*
Archean *19.3*	K–T asteroid impact theory *19.7*
asteroid *19.7*	mantle, of Earth *19.1*
big bang *CI*	Mesozoic *19.6*
Cenozoic *19.8*	Paleozoic *19.5*
crust, of Earth *19.1*	prokaryotic cell *19.3*
dinosaur *19.6*	Proterozoic *19.3*
Ediacaran *19.5*	protistan *19.4*
endosymbiosis theory *19.4*	proto-cell *19.2*
eubacterium *19.3*	RNA world *19.2*
eukaryotic cell *19.3*	stromatolite *19.3*

Readings

de Duve, C. April 1996. "The Birth of Complex Cells." *Scientific American* 274(4): 50–57.

Dott, R., Jr., and R. Batten. 1994. *Evolution of the Earth.* Fifth edition. New York: McGraw-Hill.

Impey, C., and W. Hartman. 2000. *The Universe Revealed.* Belmont, California: Wadsworth. Paperback.

Wright, K. March 1997. "When Life Was Odd." *Discover* 18(3): 52–61. Update on the puzzling Ediacarans.

On-Line readings at Student Guide for InfoTrac:
www.brookscole.com/biology

PROKARYOTES, VIRUSES, AND PROTISTANS

The Mostly Microscopic Multitudes

Did a friend ever mention that you are nearly 1/1,000 of a mile tall? Probably not. What would be the point of measuring people in units as big as miles? Even so, we think that way, in reverse, whenever we measure microorganisms. For the most part, **microorganisms** are prokaryotic cells and single-celled protistans too small to be seen without the aid of a microscope.

The bacterial cells in Figure 20.1 are a case in point. To measure them, you would have to divide one meter into a thousand units, or millimeters. Next, you would have to divide one of the millimeters into a thousand smaller units, or micrometers. To give you a sense of how small that is, a single millimeter would be about as small as the dot of this "i." And a *thousand* bacteria would fit side by side on top of the dot!

To be sure, viruses are smaller still, as you might deduce after looking at Figure 20.1. We measure them in nanometers (billionths of a meter). But viruses are not alive. We consider them in this chapter because one kind or another infects just about every living thing.

Prokaryotic cells are the smallest organisms, but they vastly outnumber the individuals in all other kingdoms combined. Their reproductive potential is staggering.

Under ideal conditions, cells of some species divide about every twenty minutes. If that reproduction rate were to hold constant, a single bacterium would have nearly a billion descendants in ten hours! Similarly, huge numbers of single-celled protistans reproduce rapidly in aquatic habitats. Among them are a variety of producers of food webs, helpful symbionts, herbivores and other predators, and parasites (Figure 20.2).

So why don't the unseen multitudes take over the world? Sooner or later, their burgeoning populations use up available nutrients and pollute the surroundings with their own metabolic wastes. They alter the very conditions that initially favored their reproduction. Besides, certain kinds of viruses attack just about every type of bacterium and help keep their population sizes in check. So do seasonal changes in living conditions.

Of course, you probably do not find much comfort in this when you serve as host for one of the pathogenic types. **Pathogens** are infectious, disease-causing agents that invade target organisms and multiply inside or on them. Disease follows when the metabolic activities of pathogenic cells damage body tissues and interfere with their normal functioning.

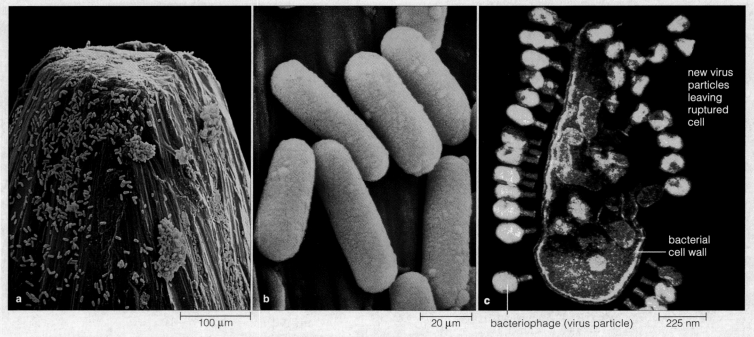

new virus particles leaving ruptured cell

bacterial cell wall

a | 100 µm

b | 20 µm

c | bacteriophage (virus particle) | 225 nm

Figure 20.1 (**a**,**b**) How small are prokaryotic cells? Shown here, *Bacillus* cells on the tip of a pin. (**c**) How small are viruses? Bacteriophage particles are one example. They attacked this bacterial cell, which ruptured open.

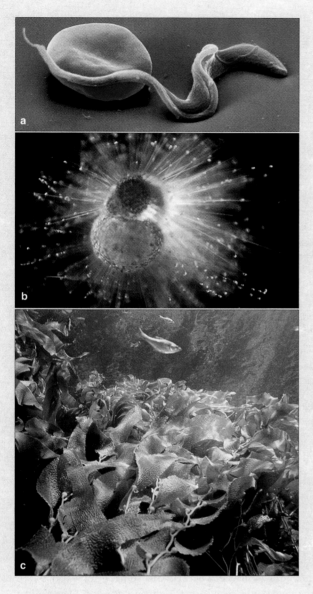

Figure 20.2 How diverse are protistans? Shown here, (**a**) *Trypanosoma brucei*, a parasitic, animal-like flagellate that causes African sleeping sickness and (**b**) a predatory foraminiferan that has golden algae living on its spines. Both are single-celled protistans. (**c**) Kelp (*Macrocystis*). Populations of this plantlike, multicelled species form underwater forests as tall as those on land.

Certain pathogens can indeed make you suffer, but they shouldn't give all prokaryotes and protistans a bad name. Think of how single-celled photoautotrophs in the seas provide food and oxygen for entire communities and even affect the global cycling of carbon (Section 6.8). Think of kelp forests, the basis of marine ecosystems (Figure 20.2*c*). Or think of bacterial species that feed on organic debris. Together with other decomposers, they help cycle the nutrients that sustain entire communities.

From the human perspective, bacteria and protistans are good or bad, or even dangerous. Basically, however, they are simply surviving and reproducing like the rest of us, in ways that are the topics of this chapter.

Key Concepts

1. In structural terms, the simplest forms of life are the prokaryotic cells called archaebacteria and eubacteria. Most are microscopically small. Viruses are smaller still.

2. Prokaryotic cells do not have a profusion of internal, membrane-bound organelles, as eukaryotic cells do. As a group, they show great metabolic diversity, and many kinds display complex behavior. Most reproduce by prokaryotic fission.

3. A virus, a noncellular infectious particle, consists of nucleic acid (either DNA or RNA), a protein coat, and sometimes an outer envelope. It cannot replicate without pirating the metabolic machinery of a specific host cell.

4. Nearly all viral multiplication cycles proceed through five steps: Attachment to a host cell, penetration of its plasma membrane, replication of viral DNA or RNA and synthesis of viral proteins, assembly of new viral particles, and then release from the infected cell.

5. Protistans are easily distinguishable from prokaryotes but difficult to classify with respect to other eukaryotes. Protistans differ enormously in size, morphology, and life-styles.

6. Fungus-like protistans include parasitic and predatory molds that produce spores. Animal-like protistans include nonphotosynthetic flagellated protozoans, sporozoans, and ciliates. Plant-like protistans include photosynthetic flagellates and the red, brown, and green algae.

7. Most of us tend to judge microorganisms through the prism of human interests. Yet their lineages are the most ancient, their adaptations are diverse, and they are simply surviving and reproducing like the rest of us.

CHARACTERISTICS OF PROKARYOTIC CELLS

Prokaryotic cells—bacteria—are the most far-flung and abundant organisms. Thousands of species live nearly everywhere, including deserts, hot springs, glaciers, and the seas. Some have persisted 2,780 meters underground for millions of years! A handful of rich soil may hold billions of prokaryotic cells. Trace any lineage back far enough and you will find prokaryotic ancestors. From *Escherichia coli* to redwoods and elephants, all organisms on Earth interconnect, regardless of size, numbers, and evolutionary distance.

Two Prokaryotic Kingdoms

Just a few decades ago, reconstructing the evolutionary history of prokaryotic cells seemed to be an impossible task. Except for the stromatolites, shown in Section 19.3, the oldest groups are not well represented in the fossil record. Most are not represented at all.

Given their elusive histories, the many thousands of known species are traditionally classified by **numerical taxonomy**. By this practice, the traits of an unidentified prokaryotic cell are compared with those of a known group. Such traits include cell shape, cell wall staining attributes, motility, nutritional requirements, metabolic patterns, and endospores, if any. The greater the total number of traits a cell has in common with the known group, the closer is their inferred relatedness.

Comparative biochemistry, especially new methods of gene sequencing, are yielding insight into prokaryotic phylogenies (Section 18.7 and Appendix I). Remember, rRNAs are essential for protein synthesis, so they can't undergo drastic changes in their overall base sequence without loss of function. The numerous small changes that accumulated in the rRNAs of different lineages can be directly measured. Surprisingly, the measurements are uniting some groups that did not even appear to be related on the basis of other tests. For example, it now seems a key divergence began shortly after prokaryotic cells first appeared on Earth (Section 19.3). One branch led to **eubacteria**, the most common prokaryotic cells. (Here, the *eu–* signifies "typical.") The other branch led to the common ancestor of both the **archaebacteria** and all eukaryotic cells. Table 20.1 and Figure 20.3 show the characteristics both kingdoms still share.

Splendid Metabolic Diversity

All organisms take in energy and carbon to meet their nutritional requirements. But prokaryotic cells display the most diversity in the way they secure resources.

Like plants, *photoautotrophic* species build organic compounds by photosynthesis; they are "self-feeders." They tap sunlight for energy and use carbon dioxide as their carbon source. Their plasma membrane contains

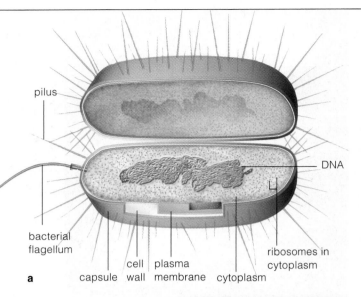

pilus

bacterial flagellum

capsule

cell wall

plasma membrane

cytoplasm

DNA

ribosomes in cytoplasm

a

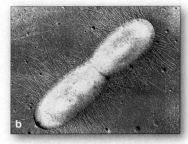

Figure 20.3 (**a**) General body plan of prokaryotic cells. (**b**) Example of pili, filamentous structures that project from the surface of many types of prokaryotic cells. This *Escherichia coli* cell is dividing in two.

b

the photosynthetic machinery. Like plants, the oxygen-releasing photosynthesizers get electrons and hydrogen from water molecules. Others are strict anaerobes; they die when exposed to oxygen. They obtain electrons and hydrogen from inorganic compounds, such as hydrogen sulfide and gaseous hydrogen, or H_2. *Chemoautotrophic* archaebacteria and eubacteria are self-feeders, too. Most types use carbon dioxide as a carbon source. Depending on the species, they strip electrons and hydrogen from certain organic compounds or inorganic substances in the environment. Gaseous hydrogen, sulfur, iron, and nitrogen compounds are their kinds of "food."

Photoheterotrophic species are not self-feeders. They use sunlight for energy but get carbon from fatty acids, complex carbohydrates, and other compounds already produced by other organisms. *Chemoheterotrophic* species

Table 20.1 *Characteristics of Prokaryotic Cells*

1. No membrane-bound nucleus.
2. Generally a single chromosome (a circular DNA molecule); many species also contain plasmids.
3. Cell wall present in most species.
4. Reproduction mainly by prokaryotic fission.
5. Great metabolic diversity among species.

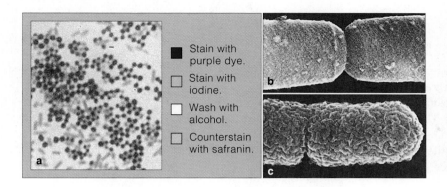

Figure 20.4 Prokaryotic cell walls. (**a**) Gram staining. Cocci and rods smeared on a slide are stained with a purple dye (such as crystal violet), washed off, then stained with iodine. All cells are purple. The slide is washed with alcohol to strip Gram-negative cells of color. The slide is counterstained (with safranin), washed, and dried. The Gram-positive cells (here, *Staphylococcus aureus*) remain purple. The counterstain colors Gram-negative cells (*E. coli*) pink. (**b**) Dividing *Bacillus subtilis* cell. Its wall has a smooth surface texture. (**c**) Dividing *E. coli* cell. Its wrinkled cell wall is characteristic of Gram-negative bacteria.

Figure 20.5 *Helicobacter pylori* cell, with its tuft of flagella. This pathogen can colonize the stomach lining. Untreated infection may lead to gastritis, peptic ulcers, and possibly stomach cancer. *H. pylori* can contaminate water and food, especially unpasteurized milk. Treatment that combines antibiotics with an antacid gets rid of the pathogen, after which ulcers heal.

aren't self-feeders, either. Parasitic types get glucose and other nutrients from living hosts. The saprobes get them from organic products or remains of other organisms.

Cell Sizes and Shapes

The chapter introduction started you thinking about the size of prokaryotic cells. Few species have a width or length greater than 10 micrometers. Three shapes are common among archaebacteria and eubacteria. A round shape is called a coccus (plural, cocci). A rod shape is a bacillus (plural, bacilli). A cell body that is twisted one or more times is a spirillum (plural, spirilla):

coccus   bacillus   spirillum

Don't let the simple categories fool you. Cocci may also be oval or flattened. Bacilli may be skinny like straws or tapered like cigars. Some cells have starlike surface extensions; some spiral types look like corkscrews or commas. Besides this, cells often remain stuck together after cell division, and together they form chains, sheets, and other aggregations.

Structural Features

Only archaebacteria and eubacteria are prokaryotic cells. They originated long before nucleated cells (*pro–*, before; *karyon*, nucleus). None has the profusion of organelles typical of eukaryotic cells. They have only one circular DNA molecule and maybe one or more plasmids. This doesn't mean prokaryotic cells are somehow inferior to eukaryotic cells. Being small-bodied, fast reproducers, they get along well without great internal complexity.

A cell wall usually surrounds the plasma membrane. Remember, this type of semirigid, permeable structure helps a cell maintain shape and resist rupturing when internal fluid pressure rises (Section 4.11). Eubacterial cell walls are made of peptidoglycan (many crosslinked strands of polysaccharides). Clinicians identify many species by wall composition and structure. Figure 20.4 describes a common staining reaction, the Gram stain.

A sticky mesh (glycocalyx) often encloses the wall. It consists of polysaccharides, polypeptides, or both. When highly organized and attached firmly to the wall, we call the mesh a capsule. When less organized and loosely attached to the wall, we call it a slime layer. The mesh helps a cell attach to teeth, mucous membranes, rocks in streams, and other interesting surfaces. It also helps some encapsulated types resist being engulfed by phagocytic, infection-fighting cells of host species.

Some prokaryotic cells have one or more bacterial flagella (singular, flagellum), as in Figure 20.5. These motile structures differ structurally from the eukaryotic flagellum and don't work the same way; they rotate like a propeller. Many types have pili (singular, pilus). These short, filamentous proteins project above the wall, as in Figure 20.3. Some help the cell adhere to surfaces or to one another as a prelude to conjugation (Section 20.3).

Archaebacteria and Eubacteria are kingdoms of prokaryotic cells. These cells have a circular bacterial chromosome and often extra DNA in the form of plasmids. Nearly all species have a cell wall around the plasma membrane. A capsule or slime layer often surrounds the wall.

Prokaryotic cells are classified by biochemical comparisons and numerical taxonomy (the total percentage of observable traits they have in common with a known group).

20.2

MAJOR BACTERIAL GROUPS

Archaebacteria

The major groups of archaebacteria are methanogens, extreme halophiles, and extreme thermophiles. All are unique in some aspects of their chemical composition, metabolism, structure, and nucleic acid sequences. They all differ as much from other prokaryotes as from eukaryotes. They are adapted to conditions as hostile as those confronting the first cells on Earth. Hence the name of the kingdom (*archae–* means beginning).

Methanogens ("methane makers") are at home in swamps, sewage, stockyards, animal guts, and other oxygen-free habitats (Figure 20.6*a,b*). Their anaerobic ATP-producing pathway ends in methane (CH_4). Each year, the methanogens release 2 billion tons of methane from termite guts, ruminant guts, wetlands, rice paddies, and landfills. As Section 42.3 describes, the tremendous quantities of this by-product influence carbon dioxide levels in the atmosphere and the global carbon cycle.

Extreme halophiles ("salt lovers") live in very salty water, as in brackish ponds and salt lakes, and near hydrothermal vents (Figure 20.6*c*). Halophiles also spoil salted fish, animal hides, and commercial sea salt. Most make ATP by aerobic pathways. Some also switch to a photosynthetic pathway when oxygen is low.

Extreme thermophiles ("heat lovers") live in highly acidic soils, hot springs, even coal mine wastes (Figure 20.6*d*). Some are the start of food webs at hydrothermal vents, where water reaches 110°C. They get electrons for ATP production from hydrogen sulfide. They are cited as evidence that life originated deep in the oceans.

Eubacteria

Nearly all of the 400 known genera of prokaryotes are eubacteria. Unlike archaebacteria, they have fatty acids in their plasma membrane. In most cases, their cell wall incorporates peptidoglycan. Until the phylogenies are worked out, we must classify them mainly by numerical taxonomy. Here, we focus on their modes of nutrition.

Photoautotrophs. Cyanobacteria, or blue-green algae, are common oxygen-releasing, photosynthetic types. You may have seen them in ponds and lakes, where mucus-sheathed chains of cells form slimy mats in nutrient-enriched water (Figure 20.7). *Anabaena* and other types can convert nitrogen gas to ammonia for biosynthesis. If nitrogen compounds dwindle, modified cells called **heterocysts** synthesize a nitrogen-fixing enzyme. They produce and share nitrogen compounds with other cells in the chains and get carbohydrates in return.

Anaerobic photosynthesizers such as green bacteria get electrons from hydrogen sulfide or hydrogen gas, not water. They might resemble anaerobic bacteria in which the cyclic pathway of photosynthesis evolved.

Figure 20.6 Archaebacteria. (**a**) *Methanococcus jannaschii*. Its DNA is fully sequenced (Section 18.7). (**b**) The stomachs of cattle and other ruminants house methanogens. Cattle release methane as they belch. (**c**) Commercial seawater evaporating ponds, San Francisco Bay, a home of *Halobacterium*. This extreme halophile thrives in places where salt levels greatly exceed that of seawater —for example, 300 versus 35 grams per liter. (**d**) Hot, sulfur-rich water spilling from Emerald Pool in Yellowstone National Park. The spring's yellow-orange rim is evidence of the abundance of carotenoids in extreme thermophiles. See also Figure 1.7a.

Chemoautotrophs. Although small, eubacteria in this category have mighty roles in the cycling of nitrogen and other nutrients. For instance, without nitrogen, there are no amino acids. Without amino acids, there is no life. Nitrifying bacteria get electrons from ammonia. Plants use the end product, nitrate, as a nitrogen source.

Chemoheterotrophs. Most eubacteria are in this group. Many have roles as decomposers and as human helpers. We use *Lactobacillus* to help make pickles, buttermilk, and yogurt; we use actinomycetes to make antibiotics. *E. coli* in our gut produces vitamin K and compounds that help us digest fat. It also helps newborns digest milk, and it prevents many food-borne pathogens from colonizing the human gut. Sugarcane and corn rely on the nitrogen-fixing spirochete *Azospirillum*. They take up some nitrogen fixed by this symbiont and give some sugars to it. Legumes such as beans, peas, alfalfa, and clover benefit from *Rhizobium*, a symbiont in their roots.

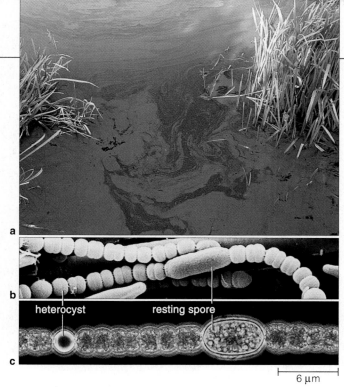

Figure 20.7 A common eubacterium. (**a**) Huge populations of cyanobacteria form dense mats near the surface of a nutrient-enriched pond. (**b,c**) Some cells become resting spores when conditions restrict growth. A nitrogen-fixing heterocyst is also shown. Section 4.12 gives other examples of eubacteria.

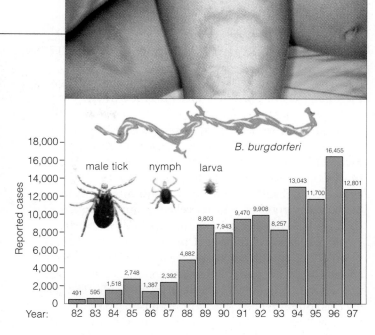

Figure 20.8 Deer ticks (*Ixodes*), the most common vector between all stages of the spirochete *Borrelia burgdorferi* and human hosts. The large photograph shows a rash that developed after infection by *B. burgdorferi*. It is a symptom of the most common tick-borne disease in the United States: Lyme disease. Tick bites deliver the spirochete to new hosts. The graph shows the number of Lyme disease cases in the United States between 1982 and 1997.

Most of the pathogenic prokaryotes are eubacteria. Examples: Pseudomonads are decomposers in soil, but they transfer plasmids with antibiotic-resistance genes. Some *E. coli* infections trigger diarrhea, a leading cause of infant deaths in developing countries. If *Clostridium botulinum* toxins taint fermented grain and improperly sterilized canned foods, you can get *botulism*. This form of poisoning disrupts breathing and can lead to death. *C. tetani*, a relative, causes *tetanus* (Section 30.5).

Like many other soil-dwelling bacteria, *C. tetani* can make endospores. This type of resting structure forms around a copy of the bacterial chromosome and some cytoplasm inside the cell. It forms when the supplies of nitrogen or some other nutrient dwindle and growth is arrested. An endospore is released as a spore when the plasma membrane ruptures. It resists heat, drying out, irradiation, acids, disinfectants, and boiling. It develops into a cell when favorable conditions return.

Many of the chemoautotrophs use insects to taxi to new hosts. Example: Bites from infected blood-sucking ticks transmit *Borrelia burgdorferi* from deer and some other wild animals to humans, and *Lyme disease* is the result. A "bull's-eye" rash, as in Figure 20.8, may form around the bite. Severe headaches, backaches, chills, and fatigue follow. Treatment must be prompt.

Regarding the "Simple" Bacteria

Bacteria are small. Their insides are not elaborate. *But bacteria aren't simple.* Their behavior reinforces this point.

Bacteria move into nutrient-rich regions. Aerobes move toward oxygen; anaerobes avoid it. Photosynthetic types move into light or away from intense light. Many will tumble away from toxins. Such behaviors often start with membrane receptors that change shape when they are stimulated by light or chemical compounds. When a bacterium moves in a different direction, the receptors are stimulated in a different way. This triggers a fleeting "memory," a change in some biochemical condition that can be compared against that of the immediate past.

Some species even show collective behavior, as when millions of *Myxococcus xanthus* cells form a "predatory" colony. The cells secrete enzymes that digest "prey"— cyanobacteria and other microorganisms that get stuck to the colony. Then they absorb breakdown products. What's more, these cells migrate, change direction, and move as a single unit toward what may be food!

Many myxobacteria produce fruiting bodies (spore-bearing structures). Under suitable conditions, cells in a colony differentiate and form a slime stalk, branches, and clusters of spores. The spores are dispersed when a cluster bursts open, and each may form a new colony. Certain eukaryotic species also form such structures.

Archaebacteria and Eubacteria are kingdoms of prokaryotic cells. Archaebacterial species are adapted to habitats that are anaerobic, extremely salty, or acidic. They are thought to resemble the first cells. Eubacterial species are found in most habitats and are the most common prokaryotic cells.

BACTERIAL GROWTH AND REPRODUCTION

Between divisions, prokaryotic cells grow by increasing their component parts. We measure growth of a large, multicelled organism in terms of increases in size, but doing so for a microscopically small bacterium is a bit pointless. Instead, we measure bacterial growth as an increase in the number of cells in a population. Under ideal conditions, each cell divides in two, the division of two cells results in four, four result in eight, and so forth. Many types divide every half hour. Such rates of increase lead to large population sizes in short order.

In nature, conditions that promote growth of some populations might not sustain growth of others. Most species could not survive extreme conditions, as in very hot deserts, Antarctic ice, or deep in the Earth. The few species found in such places reproduce only rarely.

A cell that has nearly doubled in size divides in two. Each daughter cell gets a bacterial chromosome: a circle of double-stranded DNA with very few proteins

Figure 20.10
Sketch of bacterial conjugation. Like many organisms in other kingdoms, many bacteria transfer genes. Recombinations occur between the same or different species. For this sketch, the plasmid's size is hugely exaggerated. The bacterial chromosome is far larger, as you saw in Figure 15.3.

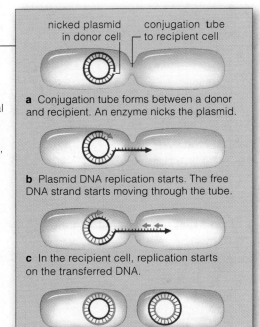

a Conjugation tube forms between a donor and recipient. An enzyme nicks the plasmid.

b Plasmid DNA replication starts. The free DNA strand starts moving through the tube.

c In the recipient cell, replication starts on the transferred DNA.

d The cells move apart and the plasmid in each forms a circle.

Figure 20.9
Prokaryotic fission. This cell division mechanism is used only by prokaryotic cells.

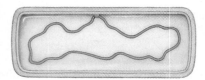

a The bacterial chromosome is attached to the plasma membrane before DNA replication.

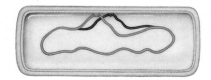

b Replication starts and proceeds in two directions from some point in the bacterial chromosome.

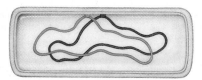

c The DNA copy is attached at a membrane site near the attachment site of the parent DNA molecule.

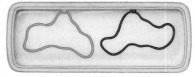

d The two DNA molecules are moved apart by membrane growth between the two attachment sites.

e New membrane and new wall material are added transversely, through the cell's midsection.

f The ongoing, orderly deposition of membrane and wall material at the midsection cuts the cell in two.

attached to it. In some species, a daughter cell buds off of its parent. Most often, however, bacteria reproduce by a division mechanism called **prokaryotic fission**.

As prokaryotic fission gets under way, a parent cell replicates its DNA (Figure 20.9). Both DNA molecules are attached to the plasma membrane at adjacent sites. The cell makes lipid and protein molecules, which are added to the plasma membrane between the two sites. The membrane additions make the two molecules of DNA move apart. Wall material is deposited above the growing membrane. In time, new membrane and new wall material at the cell midsection cut the cytoplasm. Two genetically equivalent daughter cells result.

Daughter cells often inherit one or more plasmids (Section 15.1). These are small circles of extra DNA, replicated independently of the bacterial chromosome.

Genes on the F (Fertility) plasmid confer the means for **bacterial conjugation**. By this mechanism, a donor cell transfers plasmid DNA to a recipient cell. *Salmonella*, *Streptococcus*, *E. coli*, and many other species do this. The F plasmid carries genetic instructions for making a sex pilus. Sex pili at the donor cell surface hook onto a recipient cell and pull it next to the donor. After the two cells make contact, a conjugation tube develops between them. Then the plasmid DNA is transferred through the tube in the manner shown in Figure 20.10.

Nearly all bacteria reproduce by prokaryotic fission, a cell division mechanism that follows DNA replication. Each daughter cell inherits a single bacterial chromosome (one DNA molecule). Many species also transfer plasmid DNA.

DEFINING CHARACTERISTICS OF VIRUSES

In ancient Rome, *virus* meant "poison" or "venomous secretion." In the late 1800s, this rather nasty word was bestowed on newly discovered pathogens, smaller than the bacteria being studied by Louis Pasteur and others. Many viruses deserve the name. They attack humans, cats, cattle, birds, insects, plants, fungi, protistans, and bacteria. You name it, there are viruses that can infect it.

Today we define a **virus** as a noncellular infectious agent that has two characteristics. First, a viral particle consists of a protein coat wrapped around a nucleic acid core—that is, around its genetic material. Second, the virus cannot reproduce itself. It can be reproduced only after its genetic material and its few enzymes enter a host cell and subvert the cell's biosynthetic machinery.

The genetic material of a virus is DNA or RNA. The coat consists of one or more types of protein subunits organized into a rodlike or a many-sided (polyhedral) shape, as in Figure 20.11. The coat protects the genetic material during the journey to a new host cell. It also incorporates proteins that can latch on to receptors of a host cell. An envelope, mostly of membrane remnants from a previously infected cell, encloses some viruses. Such envelopes bristle with glycoprotein spikes. The coat of a complex virus has accessory structures, such as a sheath and tail fibers, attached to it.

The vertebrate immune system detects certain viral proteins. The problem is, genes for many viral proteins mutate at high frequencies, so a virus often can elude the immune fighters. For instance, people susceptible to lung infections get new "flu shots" each year because envelope spikes on influenza viruses keep changing.

Each kind of virus multiplies only in certain hosts. It can't be studied easily unless an investigator cultures living host cells. This is why much of our knowledge of viruses comes from **bacteriophages**, a group of viruses that infect bacterial cells. Unlike cells of humans and other complex, multicelled species, bacterial cells can be cultured easily and rapidly. This also is why bacteria and bacteriophages were used in early experiments to determine the function of DNA (Section 12.1). They are still used as research tools in genetic engineering.

Many animal viruses cause diseases, such as AIDS, the common cold, some cancers, and influenza (Figure 20.12). Plant viruses breach cell walls to cause diseases. They typically hitch rides on the piercing or sucking devices of insects that feed on plant juices.

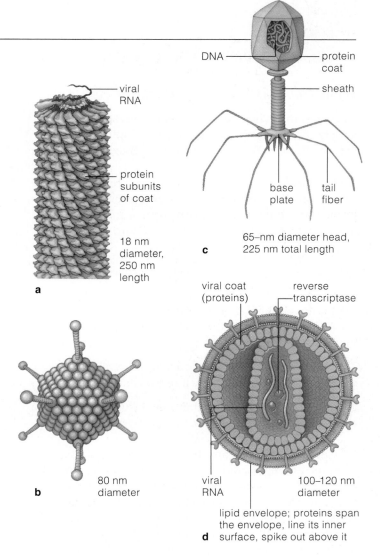

Figure 20.11 Body plans of viruses. (**a**) *Helical* viruses, such as this tobacco mosaic virus, have a rod-shaped coat of protein subunits coiled helically around nucleic acid. (**b**) This adenovirus and other *polyhedral* viruses have a many-sided coat. (**c**) T-even bacteriophages and other *complex* viruses have accessory parts attached to the coat. (**d**) Membrane encases the coat of *enveloped* viruses. HIV, shown here, is an example.

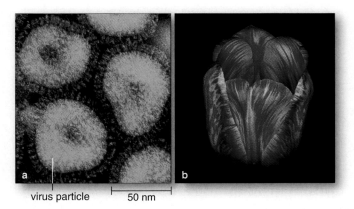

Figure 20.12 (**a**) Particles of an enveloped RNA virus that causes influenza in humans. Spikes project from its envelope. (**b**) Streaking of a tulip blossom. A harmless virus infected some (but not all) cells that normally produce red pigment.

A virus is a nonliving infectious particle that consists of nucleic acid enclosed in a protein coat and sometimes an outer envelope. It cannot be replicated without pirating the metabolic machinery of a specific type of host cell.

VIRAL MULTIPLICATION CYCLES

What Happens During Viral Infections?

Once they infect host cells, different viruses multiply in a variety of ways. Even so, nearly all multiplication cycles proceed through five basic steps, as outlined here:

1. *Attachment.* A virus attaches to a host cell. Any cell is a suitable host if molecular groups on a virus particle are able to chemically recognize and lock on to specific molecular groups at the cell surface.

2. *Penetration.* Either the whole virus or its genetic material alone penetrates the cell's cytoplasm.

3. *Replication and synthesis.* In an act of molecular piracy, the viral DNA or RNA directs the host cell into producing many copies of viral nucleic acids and viral proteins, including enzymes.

4. *Assembly.* The viral nucleic acids and viral proteins are put together to form new infectious particles.

5. *Release.* One way or another, new virus particles are released from the cell.

Consider the lytic and lysogenic pathways that are common among bacteriophage multiplication cycles. In a *lytic* pathway, steps 1 through 4 proceed rapidly, and new particles are released when a host cell undergoes **lysis.** Here, lysis means that damage to a cell's plasma membrane, wall, or both is allowing the cytoplasm to dribble out—and so the cell dies. Late into most lytic pathways, the host cell synthesizes a viral enzyme that causes the damage and swift destruction (Figure 20.13).

In a *lysogenic* pathway, a latent period extends the cycle. The virus does not kill its host outright. Instead, a viral enzyme cuts a host chromosome, then integrates viral genes into it. When an infected host cell prepares to divide, it replicates the recombinant molecule. As a result of that single instance of genetic recombination, miniature time bombs are passed on to all of the cell's descendants. Later on, a molecular signal or some other stimulus may reactivate the cycle.

Latency is part of the multiplication cycles of many viruses, not just bacteriophages. Type I *Herpes simplex*, the cause of *cold sores* (fever blisters), is a case in point. Almost everyone harbors this virus. It remains latent in facial tissues inside a ganglion (a cluster of neuron cell bodies). Sunburn and other stress factors can reactivate the virus. When this happens, the virus particles move down to the neuron endings, near the skin. There they infect epithelial cells and cause painful skin eruptions.

The multiplication cycle of the RNA viruses has an interesting twist to it. In a host cell's cytoplasm, their RNA serves as a template for synthesizing either DNA or mRNA. For example, HIV, a retrovirus, carts its own

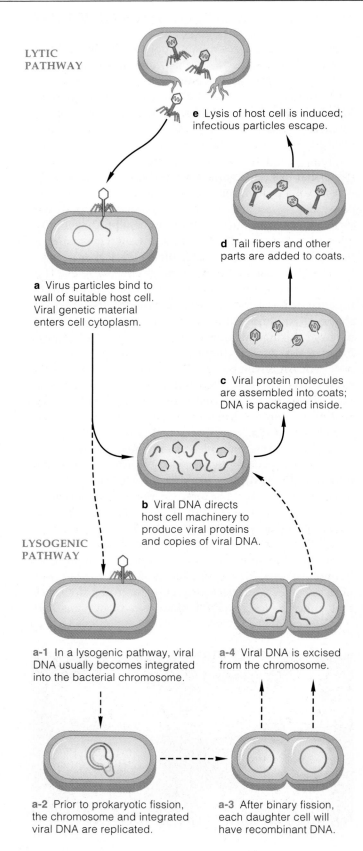

LYTIC PATHWAY

e Lysis of host cell is induced; infectious particles escape.

d Tail fibers and other parts are added to coats.

a Virus particles bind to wall of suitable host cell. Viral genetic material enters cell cytoplasm.

c Viral protein molecules are assembled into coats; DNA is packaged inside.

b Viral DNA directs host cell machinery to produce viral proteins and copies of viral DNA.

LYSOGENIC PATHWAY

a-1 In a lysogenic pathway, viral DNA usually becomes integrated into the bacterial chromosome.

a-4 Viral DNA is excised from the chromosome.

a-2 Prior to prokaryotic fission, the chromosome and integrated viral DNA are replicated.

a-3 After binary fission, each daughter cell will have recombinant DNA.

Figure 20.13 Generalized multiplication cycle for some of the bacteriophages. Virus particles may be produced and released by a lytic pathway. For certain viruses, the lytic pathway may expand to include a lysogenic pathway.

Figure 20.14 Multiplication cycle of one type of enveloped DNA virus, here infecting a generalized animal cell.

enzymes into cells. These assemble DNA on viral RNA by reverse transcription (Sections 15.1 and 35.10).

Like other enveloped viruses, the herpes viruses can enter a host cell by their own version of endocytosis, then leave it by budding from the plasma membrane. Figure 20.14 shows how they accomplish this.

What About Viroids and Prions?

Viroids, tightly folded strands or circles of RNA, are smaller than anything in viruses. They, too, can cause infections even though they don't have protein-coding genes. They have no protein coat, but the tight folding might help protect them from a host's enzymes. These bits of "naked" RNA cause many diseases in plants. Each year, they destroy many millions of dollars' worth of potatoes, citrus, and other crop plants.

Viroids actually resemble introns, those noncoding portions of eukaryotic DNA you read about in Section 13.1. They might be self-splicing introns that escaped from DNA molecules. Or they might be remnants left over from an ancient RNA world (Section 19.2).

Prions are largely—if not entirely—abnormal, less soluble forms of proteins necessary for the operation of neurons, the communication cells of nervous systems. Prions catalyze the conversion of the normal proteins into more prions. They coagulate as massive deposits inside the brain and cause fatal degenerative diseases. Have you heard of *kuru* and *Creutzfeldt–Jakob diseases* (CJD) in humans? Both diseases progressively destroy muscle coordination and brain function. Each year, 1 in 1 million people develops CJD. Symptoms include the loss of vision and speech, rapid mental deterioration, spastic paralysis, and inevitable death; there is no cure. Scrapie, a prion-linked disease of sheep, is so named because infected sheep rub against trees or posts until they scrape off most of their wool.

Prions also cause the fatal *mad cow disease*, or BSE (bovine spongiform encephalopathy). Infected animals stagger about, drool heavily, and show other symptoms. Spongy holes and amyloid deposits form in the brain. BSE causes a variant CJD in humans. Before 2000, about 750,000 infected cattle entered the human food chain. Ground-up tissues of sheep that died of scrapie had been used as a supplement in cattle feed; contaminated tissue as small as a peppercorn can lead to infection.

Scrapie has been around for two centuries in Britain but posed no risk to humans before now. Did BSE arise by some mutation that crossed the species barrier from sheep to cattle? Maybe. Today, comprehensive safeguard

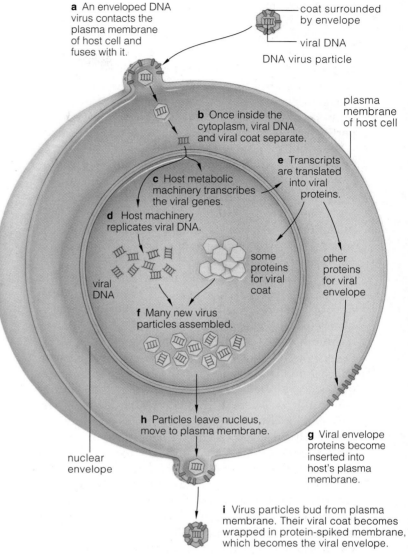

a An enveloped DNA virus contacts the plasma membrane of host cell and fuses with it.

coat surrounded by envelope

viral DNA

DNA virus particle

b Once inside the cytoplasm, viral DNA and viral coat separate.

plasma membrane of host cell

e Transcripts are translated into viral proteins.

c Host metabolic machinery transcribes the viral genes.

d Host machinery replicates viral DNA.

some proteins for viral coat

other proteins for viral envelope

viral DNA

f Many new virus particles assembled.

h Particles leave nucleus, move to plasma membrane.

g Viral envelope proteins become inserted into host's plasma membrane.

nuclear envelope

i Virus particles bud from plasma membrane. Their viral coat becomes wrapped in protein-spiked membrane, which becomes the viral envelope.

j The finished particle is equipped to infect a new potential host cell.

measures are in place, but consumer confidence has plummeted. Export markets dried up; the British cattle industry has not recovered. Depending on assumptions about the incubation period and other variables, the number of deaths resulting from this outbreak range from less than 100 to hundreds of thousands.

Viral multiplication cycles include five steps: Attachment to a suitable host cell, cell penetration, viral DNA or RNA replication and synthesis of viral proteins, assembly of new viral particles, and release from the infected cell.

Bacteriophage replication cycles commonly follow a rapid, lytic pathway and an extended, lysogenic pathway. The replication cycles of RNA viruses involve the use of viral RNA as a template for synthesizing DNA or mRNA.

KINGDOM AT THE CROSSROADS

We turn now to the protistans. Their story starts about 2.5 billion years ago. In tidal flats and sediments, huge populations of photosynthetic bacteria were releasing oxygen. At first the oxygen latched onto iron in rocks. When iron-rich deposits rusted out (oxidized), oxygen accumulated in water and the air. That environmental change triggered rampant competition for resources.

In the presence of so much free oxygen, energy-rich organic compounds could no longer build up through geochemical processes. Organic compounds produced by living cells were now the premier source of carbon and energy. In some lineages, modified cells had novel ways of getting and using organic compounds. Many prokaryotic species entered new kinds of partnerships, predations, and parasitic relationships. As outlined in Section 19.4, some of the evolutionary experiments gave rise to the first eukaryotic cells.

Of all existing organisms, **protistans** are most like the earliest, structurally simple eukaryotes. They differ from bacteria in several respects. At the least, they have a nucleus, large ribosomes, mitochondria, endoplasmic reticulum, and Golgi bodies. Their chromosomes are DNA molecules with many histones and other proteins attached to them. Protistans assemble microtubules for use in a cytoskeleton, in chromosome-moving spindles, and in the 9+2 core of flagella or cilia. Many species have chloroplasts. In this kingdom, mitosis and meiosis are the cell division mechanisms, not prokaryotic fission.

The photosynthetic types range from microscopic single cells to immense seaweeds. Other types resemble some bacteria and fungi. Some of the predators and parasites resemble animals. Although the vast majority of protistans are single-celled, nearly every lineage also includes multicelled species. Gene sequencing studies are revealing close evolutionary ties to other kingdoms.

You'll be surveying these major lineages: Chytrids, water molds, slime molds, protozoans, and sporozoans are all heterotrophs. The euglenoids, chrysophytes, and dinoflagellates include species that are photosynthetic, heterotrophic, or both. Most red, brown, and green algae are evolutionarily committed to photosynthesis.

Opinions differ on how to classify many of these diverse organisms. For example, many biologists view the multicelled algae as protistans, and about as many view them as plants. Don't worry about memorizing protistan classification schemes. Simply become familiar with the major groups. Soon enough, investigators will bring details of the evolutionary picture into focus.

Protistans are organisms most like the earliest eukaryotes. Structural and functional traits distinguish them from prokaryotic cells. Different protistan groups appear to have close evolutionary ties to plants, fungi, and animals.

PARASITIC AND PREDATORY MOLDS

Three groups of protistans, the **chytrids**, **water molds**, and **slime molds**, have species that resemble the fungi in some respects. Like fungi, they form spore-bearing structures and are heterotrophs. Like fungi, many are saprobic decomposers or parasites. "Saprobes" secrete digestive enzymes that break down organic compounds made by other organisms, then absorb the breakdown products. *Unlike* fungi, the slime molds are phagocytic predators. And *unlike* fungi, members of these groups produce motile cells during the life cycle.

Turn a microscope on a droplet of seawater from an estuary or fresh water from a pond and possibly you will see chytrids (Chytridiomycota). These cells absorb nutrients from dead tissues in plants and from plant debris. The 575 species are mostly saprobic decomposers; some are parasites. In damp places, their populations form fuzzy, pale masses that resemble fungal molds.

You may also see water molds (Oomycota), distant relatives of yellow-green and brown algae. Most of the 580 known types are saprobes in aquatic habitats; some are parasites. Those pale, cottony growths you may have seen on some aquarium fish are absorptive structures (mycelia) of a parasitic type, *Saprolegnia* (Figure 20.15a).

Some water molds are serious pathogens. *Plasmopara viticola* infects grapes and causes *downy mildew* (Figure 20.15b). *Phytophthora infestans* truly lives up to its name —plant destroyer. Over a century ago, Irish peasants grew and ate mostly potatoes. Between 1845 and 1860, growing seasons were cool and damp, year after year. Countless *P. infestans* spores moved unimpeded through thin films of water. *Late blight*, a rotting of plant parts, was rampant. Over a fifteen-year period, a third of the population of Ireland starved to death, died during an outbreak of typhoid fever (a secondary effect), or fled to the United States and elsewhere. Related *Phytophthora* species have attacked Australian eucalyptus forests and have caused a *sudden oak death* epidemic in California; they can destroy a healthy tree in one growing season.

Slime molds are free-living, amoebalike cells during part of the life cycle. The group includes cellular slime

Figure 20.15 Effects of two water molds. (**a**) *Saprolegnia*, a parasitic water mold, destroyed tissues of this aquarium fish. (**b**) *Plasmopara viticola* causes downy mildew in grapes. It is a recurring threat to vineyards in Europe and North America.

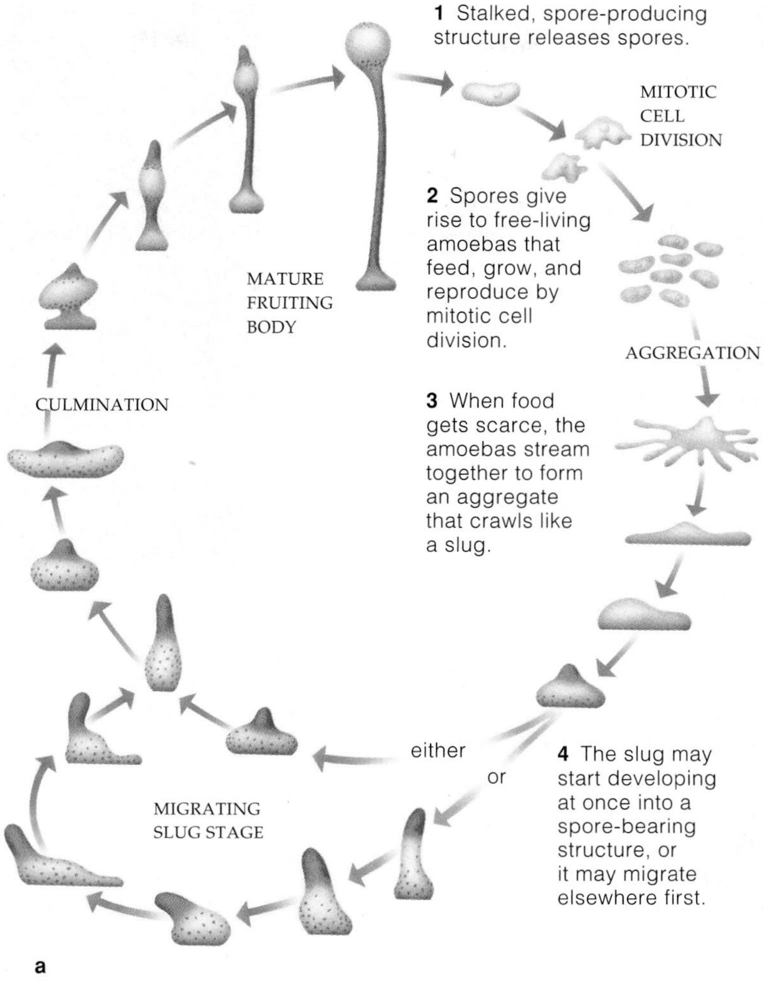

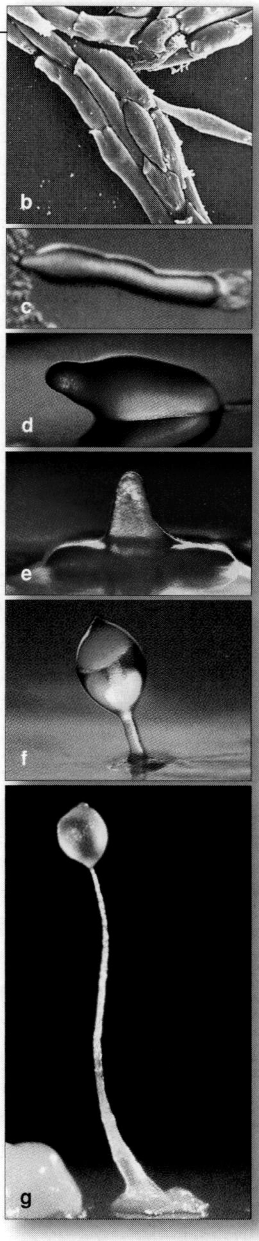

Figure 20.16 A cellular slime mold, *Dictyostelium discoideum*. (**a**) Its life cycle includes a spore-producing stage. Spores give rise to free-living amoebas, which grow and divide until food (soil bacteria) dwindles. Then, amoebas secrete cyclic AMP, a chemical signal that makes them stream toward one another. Their plasma membrane becomes sticky, so they adhere to one another. A cellulose sheath forms around the cell mass, which starts crawling like a slug (**b–d**). Some slugs may incorporate 100,000 amoebas.

In a migrating slug, the amoebas differentiate into prestalk (*red*), prespore (*white*), and anteriorlike cells (*brown dots*). Prestalk cells secrete ammonia in amounts that vary with temperature and light intensity. The slug moves fastest in response to intermediate levels of ammonia (not too little, not too much). Those levels correspond to warm, moist conditions (not too cold or hot, not soppy or dry). Prestalk and prespore cells differentiate and form a stalked, spore-bearing structure (**e–g**). Anteriorlike cells, which sort into two groups, may help elevate nonmotile spores for dispersal from the top of the spore-bearing structure.

1 Stalked, spore-producing structure releases spores.

MITOTIC CELL DIVISION

MATURE FRUITING BODY

2 Spores give rise to free-living amoebas that feed, grow, and reproduce by mitotic cell division.

AGGREGATION

CULMINATION

3 When food gets scarce, the amoebas stream together to form an aggregate that crawls like a slug.

MIGRATING SLUG STAGE

either

or

4 The slug may start developing at once into a spore-bearing structure, or it may migrate elsewhere first.

a

molds (Acrasiomycota) and plasmodial slime molds (Myxomycota). Like true amoebas, they are predatory. Their cells crawl on rotting plant parts and eat bacteria, yeasts, spores, and organic compounds. If nutrients are scarce, starving amoebas gather into a slimy mass that may migrate to a better place and form a spore-bearing structure (Figure 20.16*a*). Slime molds also reproduce sexually.

One of the 500 species of plasmodial slime molds is shown in Figure 20.17. The plasma membrane of these amoebalike cells breaks down when they get together. Cytoplasm can flow freely from cell to cell and thereby distributes nutrients and oxygen throughout the mass. A streaming mass of this sort (the plasmodium) often will occupy several square meters and migrate if food runs out. You might have seen one crossing a lawn or a road, or even climbing a tree.

Figure 20.17 *Physarum*, a plasmodial slime mold. This aggregation of cells is migrating along a rotting log.

Most chytrids and water molds are saprobic decomposers of aquatic habitats. Some are single cells. Like slime molds, many are free-living predators some of the time and simple experiments in multicellularity at other times.

During a slime mold life cycle, amoeboid cells aggregate to form a migrating mass. Cells in the mass differentiate, then form reproductive structures and spores or gametes.

THE ANIMAL-LIKE PROTISTANS

About 65,000 protistan species are informally called the **protozoans** ("first animals"). Why? They may resemble the single-celled, heterotrophic protistans that gave rise to animals. Among these amoeboid protozoans, ciliated protozoans, and animal-like flagellates are predators, parasites, and grazers that actively move about. You'll find free-living species in the seas, damp soil, ponds, lakes, and other aquatic habitats. The symbionts and parasites live in or on moist tissues of other organisms. Some protozoans are notorious pathogens.

Protozoans reproduce either asexually or sexually. Many alternate between reproductive modes when the environment changes. Most often, they reproduce by **binary fission**, an asexual process that divides the body in two parts. The division plane is random for amoebas, longitudinal for flagellates, and transverse for ciliates. Some species also bud from a parent. The life cycles of many parasitic types have encysted stages. Their **cysts** are resistant body coverings made of secreted material; they help the parasite survive stressful conditions.

Amoeboid Protozoans

Amoeboid protozoans (Sarcodina) are soft-bodied cells, some with hard parts. Naked amoebas, foraminiferans, heliozoans, and radiolarians are in this group. All use pseudopods for moving and for capturing prey (Figure 20.18*a*). Some are gut-dwelling species. Free-living types live in damp soil, fresh water, and saltwater. Most are phagocytes that engulf other protozoans and bacteria. A few parasitize certain invertebrates or vertebrates.

Entamoeba histolytica, a pathogen, completes its life cycle in human intestines. It causes *amoebic dysentery*, an infectious disease most prevalent in subtropical and tropical areas where sanitation is poor. Worldwide, this is a leading cause of death among infants and toddlers, whose immune systems are not yet well developed.

Unlike naked amoebas, foraminiferans have a highly perforated external skeleton of secretions hardened with calcium carbonate (Figures 20.2 and 20.18*c*). Thin, mucus-covered pseudopods extend through the perforations to trap prey. Like their ancestors, most species live on the seafloor. Of the species we know about, 99 percent are extinct. Over time, fossilized foraminiferan shells were compressed to form a distinctive type of sedimentary rock (Figure 20.18*d*). We mine the calcified remains for the manufacture of cement and blackboard chalk.

The radiolarians and heliozoans have many slender, reinforced pseudopods radiating from their body. They are well represented in the fossil record, for their silica-hardened parts resist breakdown (Figure 20.18*b*). Most species drift with ocean currents, as part of plankton. **Plankton** refers to any aquatic community of passive drifters and motile organisms, mostly microscopic.

Figure 20.18 Amoeboid protozoans. (**a**) *Amoeba proteus*. (**b**) Radiolarian shell. (**c**) Foraminiferan shell. (**d**) The white cliffs of Dover, England, formed more than 200 million years ago by fossilization and compression of countless foraminiferan shells.

Ciliated Protozoans

Ciliated protozoans (Ciliophora) typically have profuse arrays of cilia at their surface. You read about these fine motile structures in Section 4.10. Most species use them to swim in fresh water or seawater, where they prey on bacteria, algae, and one another. All of the cilia beat in a synchronized pattern over the body surface, calling to mind a soft wind through a field of tall grasses.

Paramecium is typical of the group (Figure 20.19). A mature cell is usually 150 to 200 micrometers long. Its outer body cover, a pellicle, consists of rigid or flexible membranes. A gullet starts as an oral depression at the body surface; there, cilia sweep food-laden water into the cell. Food gets enclosed in enzyme-filled vesicles and is digested. Like amoeboid protozoans, a ciliate's internal solute concentrations are greater than those of external fluids. The cell must counter water's tendency to diffuse inward by osmosis. Like amoebas, it depends on **contractile vacuoles**. Tubes extend from the center of these organelles and collect excess water that moves osmotically into the cell. A filled vacuole contracts and forces water through a pore to the outside (Section 5.10).

Like other protistan groups, ciliates show diversity in life-styles. About 65 percent of the known species are free-living, motile cells. Others attach to substrates, often by stalks. Some form colonies. Many interact with other organisms as symbionts or parasites.

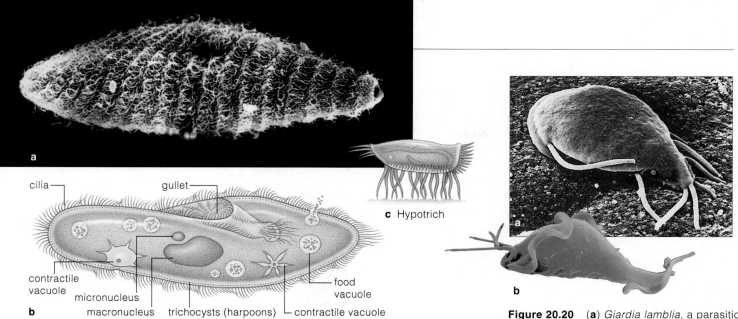

cilia —

gullet —

c Hypotrich

b

contractile vacuole

micronucleus

macronucleus   trichocysts (harpoons)   contractile vacuole

food vacuole

Figure 20.19   (**a**,**b**) Body plan and scanning electron micrograph of *Paramecium*, a ciliated protozoan. (**c**) Sketch of one of the most animal-like ciliates, which run around on leglike tufts of cilia. Some have a "head" end with modified sensory cilia.

a

b

Figure 20.20   (**a**) *Giardia lamblia*, a parasitic protozoan, adhering to human epithelium by a sucking disk. (**b**) *Trichomonas vaginalis*. This pathogenic species causes trichomoniasis, a highly prevalent, sexually transmitted disease.

Figure 20.21   Diagram of *Trypanosoma brucei*. This parasitic protozoan causes African sleeping sickness. Figure 20.2*a* is a micrograph of this animal-like flagellate.

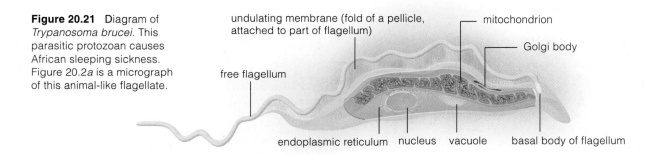

undulating membrane (fold of a pellicle, attached to part of flagellum)

mitochondrion

Golgi body

free flagellum

endoplasmic reticulum   nucleus   vacuole   basal body of flagellum

Ciliates reproduce sexually and asexually. Things get interesting because one cell often has two types of nuclei. When it reproduces asexually by binary fission, its small, diploid *micro*nucleus divides by mitosis. Its large *macro*nucleus splits in two a bit sloppily; some DNA may spill out. Most often, sexual reproduction is by a form of conjugation. Cells repeatedly divide their micronuclei, swap two of these with a partner, and let two micronuclei fuse and form a diploid macronucleus to replace the one that disappears. And you probably thought sex among single-celled critters was simple!

Animal-Like Flagellates

Animal-like flagellates (Mastigophora) are free-living predators and parasites bearing one to several flagella. Free-living types live in fresh water and seawater. The parasites live in moist tissues of plants and animals, including humans. *Giardia lamblia*, an internal parasite, causes diarrhea (Figure 20.20*a*). It contaminates surface waters of aquatic habitats and reservoirs. Cattle, wild animals, and people often harbor the parasite and its cysts, which leave the host in feces. The parasite infects people who drink water or eat food tainted with feces. Many trichomonads are parasitic flagellates. *Trichomonas vaginalis*, a worldwide nuisance, can infect new human hosts during sexual intercourse (Figure 20.20*b*). Unless a trichomonad infection is treated, it may damage the urinary and reproductive tracts.

Trypanosomes, too, are parasites. *Trypanosoma brucei* causes *African sleeping sickness* (Figures 20.2*a* and 20.21). Tsetse fly bites transmit it to hosts; it can damage the nervous system. In South America and Mexico, *T. cruzi* causes *Chagas disease*. Bugs pick it up when they feed on infected humans and other animals. *T. cruzi* multiplies in the insect gut and is excreted onto a host. Scratches in skin invite infection. The liver, spleen, eyelids, and the face swell. The infection severely damages the brain and heart. There is no cure for Chagas disease.

Protozoans are single-celled protistans. In some respects they are animal-like, as in the way they actively move about to secure food. They include predators, parasites, and grazers. Some pathogenic species cause dangerous diseases in humans and other animals.

THE NOTORIOUS SPOROZOANS

Sporozoan is an informal name for parasitic protistans that complete part of the life cycle inside specific cells of hosts. These parasites form a motile infective stage (sporozoites); some have encysted stages. One end of the body has a complex of structures to penetrate hosts.

Many sporozoans, such as *Toxoplasma*, cause human diseases (Figure 20.22). *Cryptosporidium* contaminates most freshwater supplies. One strain has put thousands of people in hospitals. Intestinal disorders and other symptoms usually subside after ten days. People with weakened immune systems suffer for months or years.

Four parasitic species of *Plasmodium* cause *malaria*. This long-lasting disease affects more than 100 million people and kills 2.7 million of them every year. Only female *Anopheles* mosquitoes transmit it to humans. A host's bloodstream transports a motile, infective stage

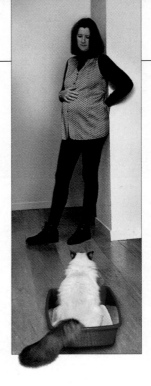

Figure 20.22 *Toxoplasma* is a sporozoan that uses domestic and wild cats as its definitive hosts. Its intermediate hosts are other animals, including humans. Cysts may be in undercooked or raw meat. Flies and cockroaches also move these cysts from cat feces onto food. The disease toxoplasmosis has flu-like symptoms. It isn't common, but it is dangerous for immune-compromised people and during pregnancies. If it infects an unborn child, miscarriage or birth defects result. Any cat, no matter how coddled, can pass cysts. Pregnant women should avoid stray cats and never empty litterboxes, clean up housecat "accidents," or clean out children's sandboxes.

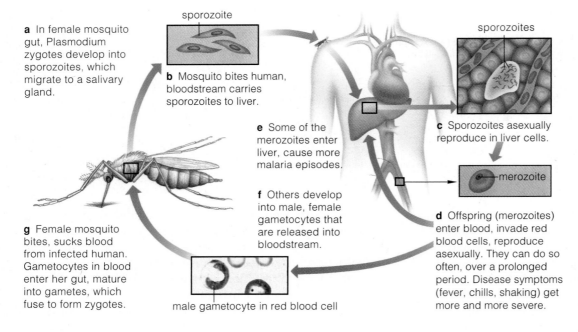

a In female mosquito gut, Plasmodium zygotes develop into sporozoites, which migrate to a salivary gland.

sporozoite

b Mosquito bites human, bloodstream carries sporozoites to liver.

e Some of the merozoites enter liver, cause more malaria episodes.

f Others develop into male, female gametocytes that are released into bloodstream.

g Female mosquito bites, sucks blood from infected human. Gametocytes in blood enter her gut, mature into gametes, which fuse to form zygotes.

male gametocyte in red blood cell

sporozoites

c Sporozoites asexually reproduce in liver cells.

merozoite

d Offspring (merozoites) enter blood, invade red blood cells, reproduce asexually. They can do so often, over a prolonged period. Disease symptoms (fever, chills, shaking) get more and more severe.

Figure 20.23 Life cycle of a sporozoan that causes malaria, a serious disease.

are warm-bodied and have little free oxygen in blood (hemoglobin binds most of it). That stage matures in mosquitoes, which have a lower body temperature and slurp oxygen from the air, along with infected blood.

Historically, malaria has been most prevalent in both the tropical and subtropical regions of Africa. Reported cases in North America and elsewhere are dramatically increasing. This is a result of unprecedented levels of human immigration and globe-hopping travels. Worse, older anti-malarial drugs no longer work against some *Plasmodium* strains. Developing vaccines (preparations that induce the body to build up resistance to a specific pathogen) is difficult. Experimental vaccines just aren't equally effective against all of the stages that develop during the *Plasmodium* life cycle. This is generally the case for developing vaccines against any parasite that has a complex life cycle.

to the liver, where the stage divides repeatedly. Some of the resulting cells (merozoites) penetrate and asexually reproduce inside red blood cells, which they destroy. Others will mature into gametes (Figure 20.23).

Disease symptoms start after infected cells rupture and release merozoites, metabolic wastes, and cellular debris into the blood. They include shaking, chills, a burning fever, and drenching sweat. After one episode, they subside for a few weeks or months, and infected people may feel okay. Relapses recur. In time, anemia may develop and the liver may be grossly enlarged.

The *Plasmodium* life cycle is sensitive to human and mosquito body temperatures and oxygen levels. The gamete-producing stage cannot mature in humans, who

Sporozoans, parasites with a complex of host-penetrating structures, complete their life cycle in specific host cells. Motile infective stages and often encysted stages form.

The Nature of Infectious Diseases

CATEGORIES OF DISEASES   Just by being human, you are a potential host for many pathogenic bacteria, viruses, fungi, protozoans, and parasitic worms. When a pathogen invades your body, it multiplies in cells and tissues; we call this an **infection**. Its outcome, **disease**, results when body defenses cannot be mobilized fast enough to keep the pathogen's activities from interfering with normal functions. With *contagious* diseases, the pathogen can be transmitted by direct contact with body secretions from infected individuals, as in explosive wet sneezes.

During an **epidemic**, a disease spreads rapidly through part of a population for a limited time, then subsides. If an epidemic breaks out in several countries at the same time, this is a pandemic. The AIDS pandemic is like this (Section 35.10). *Sporadic* diseases such as whooping cough break out irregularly and affect very few people. *Endemic* diseases pop up more or less continually but don't spread far in large populations. Tuberculosis is like this.

AN EVOLUTIONARY VIEW   Now look at disease in terms of the pathogen's prospects for survival. *A pathogen stays around only for as long as it has access to outside sources of energy and raw materials.* To a microscopic organism or virus, a human is a treasurehouse of both. With such bountiful resources, it can multiply or replicate itself to amazing population sizes. Evolutionarily speaking, the ones that leave the most descendants win.

Two barriers prevent world dominance by pathogens. First, any species with a history of being attacked by a specific pathogen has coevolved with it and has built-in defenses against it. A premier example is the vertebrate immune system, as described in Chapter 35. Second, if a pathogen is too good at rapidly killing an individual, it may vanish before having time to infect a new one. This is one reason why most have less-than-fatal effects on a host. After all, infected people who live longer can spread more germs and contribute to a pathogen's reproductive or replicative success. Usually, an individual dies only if it becomes host to overwhelming numbers of a pathogen, if it is a novel host (one that has no coevolved defenses against the pathogen), or if a mutant pathogenic strain emerges and can surmount the host's current defenses.

Being equipped with an evolutionary perspective, you probably can work out the numbers on your own. To wit: *The greater the population density of host individuals, the greater the kinds and frequencies of infectious diseases transmitted among them.* This brings us to the bad news.

DRUG-RESISTANT STRAINS   There is an old saying that, when you attack nature, it will come back at you with a pitchfork. At this point in the book, we have looked at antibiotic resistance in several contexts. Here we reinforce the point that infectious diseases are on the upsurge.

Example: Because of economic pressures, the number of working mothers has skyrocketed in the United States. So has the number of preschoolers enrolled in day-care centers. Each center is a crowded population of hosts

intestinal epithelium   *Cryptosporidium* cyst   emerging sporozoite

Table 20.2 *The Eight Deadliest Infectious Diseases*

Disease	Main Agents	Estimated New Cases per Year	Estimated Deaths per Year
Acute respiratory infections*	Bacteria, viruses	1 billion	4.7 million
Diarrheas**	Bacteria, viruses, protozoans	1.8 billion	3.1 million
Tuberculosis	Bacteria	9 million	3.1 million
Malaria	Sporozoans	110 million	2.5–2.7 million
AIDS	HIV	5.6 million	2.6 million
Measles	Viruses	200 million	1 million
Hepatitis B	Virus	200 million	1 million
Tetanus	Bacteria	1 million	500,000

* Includes pneumonia, influenza, and whooping cough.
** Includes amoebic dysentery, cryptosporidiosis, and gastroenteritis.

whose developing immune systems are vulnerable to disease agents, including *Streptococcus pneumoniae*. This bacterium causes pneumonia, meningitis, and middle-ear infections. In people of all ages, *S. pneumoniae* infections result in 40,000 to 50,000 deaths each year. As recently as 1988, drug-resistant strains were almost unheard of. Now they are becoming the rule, not the exception.

Most of our 6.1 billion selves live in crowded cities. During any specified interval, as many as 50 million of us are on the move within and between countries in search of a better life. Is it any wonder the pathogenic agents of cholera, tuberculosis, and other diseases are striking with a vengeance (Table 20.2)?

PATHOGENS IN FOOD SUPPLIES   Food poisoned by infectious agents hits more than 80 million people in the United States each year. Treating food poisoning costs 5 billion to 22 billion dollars. About 9,000 people a year die from it, mainly the very young, the very old, or those with compromised immune systems.

People often dismiss their misery as "the 24-hour flu," not infection by *Salmonella enteriditis* and pathogenic *E. coli* strains. The agents can lurk in undercooked beef or poultry, water, unpasteurized milk or cider, and vegetables grown in manure-fertilized fields. Example: *S. enteriditis* once found its way into some ingredients for a popular ice cream. Almost 225,000 ice cream eaters got stomach cramps, diarrhea, and other symptoms of *salmonella*.

Pathogens in the kitchen may cause at least half of all food poisonings. Carlos Enriquez and his colleagues at the University of Arizona sampled 75 dishrags and 325 sponges in several homes. They found *Salmonella, E. coli, Pseudomonas,* and *Staphylococcus*, which can live up to two weeks in a wet sponge. You can sanitize sponges simply by putting them through a dishwasher cycle.

20.11

THE (MOSTLY) SINGLE-CELLED ALGAE

The rest of this chapter surveys "the algae," the largely photosynthetic protistans. The majority of euglenoids, chrysophytes, and dinoflagellates are single cells, which isn't the case for red, brown, and green algae. Most are components of *phyto*plankton—aquatic communities of mainly microscopic photoautotrophs that are the start of nearly all food webs in ponds, lakes, streams, rivers, the seas, and all other water provinces. Under favorable conditions, certain species undergo an **algal bloom**, a stupendous increase in population size.

A Sampling of Single-Celled Diversity

Fresh water or stagnant ponds and lakes are home to **euglenoids** (Euglenophyta). Most of the 1,000 species of these flagellated, free-living cells are photosynthetic. But all are descended from phagocytic ancestors and still retain a heterotrophic feeding apparatus, now used to take in organic compounds, bacteria, or eukaryotic prey. *Euglena* affords hints of an endosymbiotic past with green algae; its chloroplasts, too, contain chlorophylls *a* and *b* plus carotenoids (Figure 20.24*a* and Section 19.4). Like some protozoans, *Euglena* has a pellicle, a flexible, translucent cover of spiral strips of a protein-rich material. It has a long flagellum and a contractile vacuole that squirts excess water out of the

long flagellum

contractile vacuole   chloroplast

eyespot shielding a light-sensitive receptor

a   nucleus   Golgi body   mitochondrion   pellicle   ER

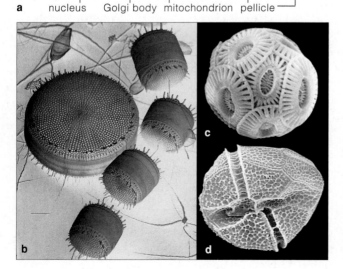

Figure 20.24 Single-celled algae. (**a**) Body plan for *Euglena*. Compare this with the electron micrograph in Section 19.4. (**b**) Diatom shells. (**c**) Coccolithophore shell. (**d**) Dinoflagellate shell.

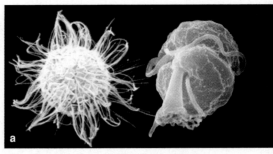

Figure 20.25 (**a**) Two of the more than twenty-four stages in the life cycle of *Pfiesteria piscicida*. (**b**) The kind of damage inflicted by this cell from hell.

encysted stage   flagellated stage

cell body. An eyespot made of pigments partly shields a light-sensitive receptor, which helps guide the cell to places where light is most suitable for its activities.

Chrysophytes (Chrysophyta) are mostly free-living, photosynthetic cells that contain a distinctive array of pigments: chlorophylls *a*, c_1, and c_2. The golden algae, yellow-green algae, diatoms, and coccolithophores are members of this large group.

Many of the 500 species of golden algae have silica scales or other hard parts and an abundance of a golden-brown pigment, fucoxanthin. Some types resemble true amoebas. We find yellow-green algae in many aquatic habitats. Most of these species are nonmotile, although they all produce flagellated gametes.

The 5,600 existing species of diatoms have a silica "shell." It consists of two perforated parts that overlap like a pillbox (Figure 20.24*b*). For 100 million years or so, finely crushed shells of at least 35,000 extinct species of these mostly photosynthetic cells accumulated at the bottom of lakes and seas. Many sediments contain the deposits; we quarry them for use in filters, insulation, and abrasives. Each year, for instance, 270,000+ metric tons are quarried near Lompoc, California.

Calcium carbonate plates protect coccolithophores (Figure 20.24*c*). Most of the 500 or so existing species are photosynthesizers of marine habitats, especially in the tropics. In the past, accumulations of the plates formed sea sediments, chalk deposits, and limestone deposits, such as Dover's white cliffs. During blooms, these algae secrete mucus that can clog fish gills and a substance so noxious it makes migrating fish change their routes.

Most **dinoflagellates** (Pyrrhophyta) are producers in marine phytoplankton; a few of the 1,200 species are symbionts with corals. Bioluminescent ones make warm seawater shimmer at night. They all have flagella and cellulose plates (Figure 20.24*d*). When they reproduce, the cell and its plates divide into two daughter cells.

Dinoflagellates are yellow-green, green, blue, brown, or red, depending on the pigment array. *Red tides* form near coasts when certain species undergo blooms that turn the water rust-red or brown. Some produce toxins that cause massive kills of fish, birds, and other animals.

Algal Blooms and the Cell From Hell

Since 1991, algal blooms have killed more than a billion fish near the coastlines of North Carolina, Virginia, and Maryland. The agent of death, *Pfiesteria piscicida,* is a dinoflagellate known as "the cell from hell."

The life cycle of *P. piscicida* requires at least twenty-four encysted, flagellated, and amoeboid stages (Figure 20.25). *P. piscicida* waits out harsh conditions as cysts in sediments of estuaries. Its flagellated stages attach to prey and suck out juices. When a large school of oily menhaden or other fish lingers to feed, their excretions incite the encysted cells to emerge and release toxins. The toxins stop fishes from escaping (by slowing them down), then eat away patches of fish skin. *P. piscicida* feeds on sloughed epidermis, blood, and tissue juices oozing from the open sores. And then the flagellates give rise to amoeboids, which feast on the dead fish. All of these changes may unfold within a matter of hours.

Nitrogen, phosphorus, and other nutrients in runoff from fields and in raw sewage fan algal blooms all over the world. Blooms usually develop in warm, shallow water that has become enriched with nutrients. JoAnn Burkholder, a North Carolina State University botanist, correlated *Pfiesteria* blooms with hundreds of millions of gallons of raw sewage from hog and chicken farms. More than 16 million hogs live in 3,500 industrial-scale hog farms in eastern North Carolina. Untreated wastes in holding lagoons often spill into rivers that drain into the poorly flushed Chesapeake Bay estuaries and into rivers in Alabama, Delaware, Florida, and Virginia. For instance, after one heavy rain, a spill from a hog farm exceeded by three times the volume of oil spilled from the tanker *Valdez* (Section 45.8). In 1997 alone, *Pfiesteria* inflicted a 60 million dollar loss on fishing and tourism industries. The magnitude of the loss finally prodded health, agricultural, and environmental agencies into coordinating research to address the problem.

Single-celled photosynthetic protistans include diverse species of euglenoids, chrysophytes, and dinoflagellates. Most species are components of phytoplankton, the food-producing base of most aquatic habitats.

RED ALGAE

Of the 4,100 known species of **red algae** (Rhodophyta), nearly all are marine; only 200 live in freshwater. The life cycle for most species includes multicelled stages, but these do not have tissues or organs. Reproductive modes are diverse, with complex asexual and sexual phases. Figure 20.26 shows one example of a life cycle.

Not all "red algae" are red. Diverse photosynthetic species are green, purple, or almost black. Phycobilins and other accessory pigments mask their chlorophyll *a* (Section 6.2). Parasitic species have little pigmentation.

Red algae have a flexible, slippery texture owing to mucous material in their cell walls. Agar is made from extracts of wall material of several species. This inert, gelatinous substance has uses as a moisture-preserving agent in baked goods and cosmetics, as a setting agent for jellies, and as a culture gel. Carrageenan, extracted from *Eucheuma,* is a stabilizer in paints, dairy products, and many other emulsions.

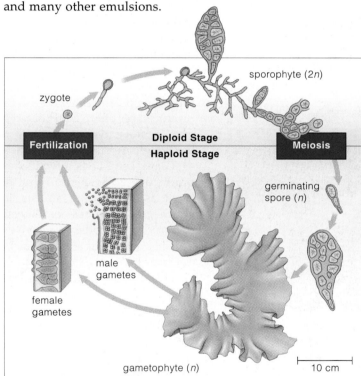

Figure 20.26 Life cycle of *Porphyra.*

Humans find different species of *Porphyra* tasty as well as nutritious. You may know it from sushi bars as *nori,* an algal wrapping for rice and fish.

Red algae are photosynthetic protistans with phycobilins and other accessory pigments that mask chlorophyll *a.* Multicelled stages usually form during their life cycles. Most species are aquatic and show great diversity in size, morphology, life-styles, reproductive modes, and habitats.

BROWN ALGAE

You find most of the 1,500 or so species of **brown algae** (Phaeophyta) living in cool or temperate marine waters, from the intertidal zone into the open ocean. Different kinds are brown, olive-green, or golden, depending on the array of pigments. Brown algae have carotenoids, such as fucoxanthin, and the chlorophylls a, c_1, and c_2. They range in size from the giant kelps—thirty meters long—to the microscopic, threadlike *Ectocarpus*.

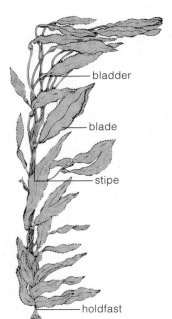

Macrocystis, *Laminaria*, and the other giant kelps are the largest, most complex of all protistans. The multicelled sporophytes have stipes (stem-like parts), blades (leaf-like parts), and holdfasts (anchoring structures). Gas-filled bladders on the stipes assist in buoyancy and in keeping blades upright (Figures 20.2c and 20.27a). Each stipe holds tubelike arrays of elongated cells. Through these tubes, sugars and other dissolved products of photosynthesis can quickly move to all living cells in the body.

Giant kelp beds function as productive ecosystems. Think of them as underwater forests in which great numbers of diverse bacteria and protistans, as well as fishes and other animals, carry out their lives.

Humans commercially harvest *Macrocystis* and other kinds of brown algae. We use extracts from them when making ice cream, jelly beans, puddings, salad dressings, canned foods, frozen foods, beer, cough syrups, toothpastes, floor polishes, a number of cosmetics, and paper. Alginic acid in the cell walls of some species is used to make algins, which are added to various products as thickening, emulsifying, and suspension agents. Many people, especially in the Far East, harvest kelps as sources of food and mineral salts, and as a fertilizer for crops.

Figure 20.27 Multicelled brown algae. (**a**) Body plan of *Macrocystis*. (**b**) *Postelsia palmaeformis*, which grows along rocky shorelines.

The brown algae range in size from the microscopic to the largest of all of the multicelled protistans. Nearly all species live in marine habitats, ranging from the intertidal zone to surface waters of the open ocean.

GREEN ALGAE

Of all protistans, the **green algae** (Chlorophyta) are structurally and biochemically most like the plants and are their closest relatives. Nearly all are photosynthetic. Like plants, they have chlorophylls a and b. They, too, temporarily store carbohydrates in starch grains inside their chloroplasts. The cell wall of some species consists of cellulose, pectins, and other polysaccharides typical of plants.

Figure 20.28 is a sampling of the more than 7,000 known species of green algae. They include single-celled, colonial, filamentous, sheetlike, and tubular forms. You won't be able to see many species without the aid of a microscope. Most, including the *Micrasterias* cell shown in Figure 20.28b, live in freshwater. *Micrasterias* is one of thousands of desmid species, which are important food producers in nutrient-poor ponds and peat bogs. Green algae also grow at the ocean surface, in marine sediments, just below the surface of soil, and on rocks,

Figure 20.28 Representative green algae. (**a**) *Volvox*, a colony of interdependent green algal cells with resemblances to free-living, flagellated cells of the genus *Chlamydomonas*. (**b**) A *Micrasterias* cell reproducing itself. (**c**) From a marine habitat, *Acetabularia*, fancifully called the mermaid's wineglass. Each individual in the cluster is a multinucleate cell mass with a rootlike structure, a stalk, and a cap in which gametes form. (**d**) Sea lettuce (*Ulva*) grows in estuaries and attaches to kelps in the seas. Reproductive cells form inside and are released from the margins of the sporophyte and gametophyte, both of which have the same form shown in the diagram.

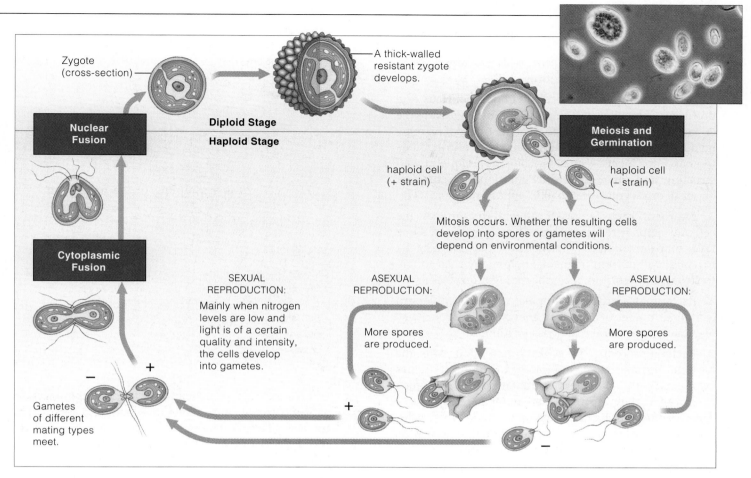

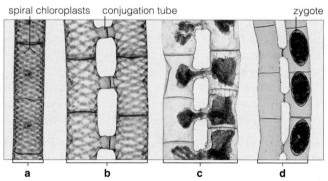

Figure 20.29 Life cycle of a species of *Chlamydomonas*, one of the most common green algae of freshwater habitats. This single-celled species reproduces asexually, most often. It also reproduces sexually under certain environmental conditions.

spiral chloroplasts   conjugation tube   zygote

a   b   c   d

Figure 20.30 One mode of sexual reproduction in *Spirogyra*, or watersilk. (**a**) This green alga has spiral, ribbonlike chloroplasts. (**b**) A conjugation tube forms between cells of adjacent haploid filaments of different mating strains. (**c**,**d**) The cellular contents of one strain pass through the tubes into cells of the other strain, where zygotes form. Zygotes develop thick walls. They undergo meiosis when they germinate and give rise to haploid filaments.

tree bark, other organisms, and snow. Some types are symbionts with fungi, protozoans, and a few marine animals. A colonial form (*Volvox*) is a hollow, whirling sphere of 500 to 60,000 flagellated cells. Those beautiful white, powdery beaches in tropical regions are largely the work of uncountable numbers of green algal cells (*Halimeda*) that formed calcified cell walls, then died and disintegrated. Some day, green algae might even accompany astronauts in space. They can grow in very small spaces, give off vital oxygen, and take up carbon dioxide exhaled by an aerobically respiring crew.

Green algae employ diverse modes of reproduction. *Chlamydomonas* provides a classic example. This single-celled alga is five to ten micrometers wide, on average. It is able to reproduce sexually, but most of the time it engages in asexual reproduction, with typically 16 to 32 daughter cells forming by mitotic cell division in the confines of the parent cell wall. Daughter cells may live at home for a while. But sooner or later they depart by secreting enzymes, which digest what's left of their parent. Figure 20.29 shows the life cycle of one species.

To give a final example, Figure 20.30 shows *Spirogyra*, a filamentous green alga, reproducing sexually.

The green algae show great diversity in size, morphology, life-styles, and habitats. The structure and biochemistry of some groups suggest that they have close evolutionary ties to the plant kingdom.

SUMMARY   *Gold* indicates text section

1. The microscopic world teems with prokaryotic cells, single-celled protistans, and viruses. *CI*

2. Bacteria are prokaryotic cells without a nucleus or a profusion of organelles. The most common types are the eubacteria. The archaebacteria (methanogens, extreme halophiles, and extreme thermophiles) live in habitats too harsh for most organisms. They are unique in wall structure and other features. Evolutionarily, they are closer to eukaryotic cells than to eubacteria. *20.1, 20.2*

3. Nearly all bacteria have a cell wall that protects the cell and helps it resist rupturing. A capsule or a slime layer may surround it. Some cells use bacterial flagella for motility. Pili (thin filaments) help many cells adhere to surfaces or assist in bacterial conjugation. *20.1*

4. Bacteria as a group show great metabolic diversity, as in their modes of acquiring energy and carbon. They also make behavioral responses to stimuli. *20.1, 20.2*

5. Bacteria reproduce by prokaryotic fission. After the bacterial chromosome is replicated, a parent cell divides into genetically equivalent daughter cells. If plasmids are present, they are replicated independently of the chromosome and may be transmitted to daughter cells. They also may be transferred to cells of the same species or a different species by bacterial conjugation. *20.3*

6. Viruses are nonliving, noncellular agents that infect particular species of nearly all organisms. *20.4*

   a. A virus particle has a core of DNA or RNA and a protein coat that may be enclosed in a lipid envelope from which viral proteins project. Coats of complex viruses have tail fibers and other accessory structures.

   b. A virus particle cannot reproduce on its own. Its genetic material must direct a host cell to synthesize materials necessary to produce new virus particles.

7. Nearly all viral multiplication cycles include five steps: Attachment to a suitable host cell, penetration of it, viral DNA or RNA replication and protein synthesis, assembly of new viral particles, and release. *20.5*

8. Unlike bacteria, protistans and other eukaryotic cells have a nucleus, mitochondria, endoplasmic reticulum with distinct ribosomes, microtubules, and at least two chromosomes (each with DNA and numerous proteins). Many contain chloroplasts. Like other eukaryotic cells, protistans engage in mitosis and meiosis. *20.6*

9. Many protistans are heterotrophs. They include water molds, chytrids, slime molds (cellular and plasmodial types), protozoans (amoebas, animal-like flagellates, and ciliated species), and sporozoans (Table 20.3). *20.6*

10. Like fungi, chytrids and water molds are decomposers and parasites. Their enzymes digest organic matter, and they absorb the breakdown products. Slime molds are phagocytes that form spore-producing structures. *20.7*

Table 20.3 *Summary of Major Protistan Groups*
HETEROTROPHS *Decomposers, predators, grazers, parasites*
Chytrids
Water molds
Cellular slime molds
Plasmodial slime molds
Protozoans: Amoeboid protozoans (naked amoebas, foraminiferans, heliozoans, radiolarians)
Ciliated protozoans
Animal-like flagellated protozoans
Sporozoans
AUTOTROPHS, HETEROTROPHS *Photosynthesizers, predators*
Euglenoids
Dinoflagellates
MOSTLY AUTOTROPHS *Photosynthesizers, some parasites*
Chrysophytes   Golden algae, yellow-green algae, coccolithophores, diatoms
Red algae
Brown algae
Green algae

11. Like animals, protozoans are predators, grazers, or parasites; some cause dangerous human diseases. The amoebas, foraminiferans, heliozoans, and radiolarians are amoeboid protozoans with or without hard parts. Ciliated protozoans use cilia for motility or for feeding. Animal-like protozoans live freely in aquatic habitats. Like sporozoans, some are parasites. *20.8, 20.9*

12. The euglenoids, chrysophytes, and dinoflagellates are mostly single celled (Table 20.3). Many are part of phytoplankton and undergo algal blooms. *20.11*

13. Red, brown, and green algae differ hugely in body plan, size, reproductive modes, and habitats. Most are multicelled photosynthesizers (Table 20.3). Some species of brown algae are the largest protistans. Plants may be descended from green algal ancestors. *20.12–20.14*

14. It has taken us more than one chapter to consider prokaryotes and protistans—the simplest eukaryotes. Table 20.4 summarizes the similarities and differences among the prokaryotes and eukaryotes.

Review Questions

1. Describe key metabolic and structural features of bacteria. Make sketches of the three basic shapes of bacterial cells. *20.1*

2. Describe one or two eubacteria that benefit humans. Name one or two that cause human diseases. *20.1, 20.2*

3. How is bacterial growth measured? *20.3*

4. Define and describe the characteristics of a virus. *20.4*

5. Review Table 20.4. Then cover it with a sheet of paper and list the major categories of protistans. *20.6, Table 20.4*

6. Select three protistan species, then briefly explain how they affect our affairs, such as human health. *20.8–20.11*

Table 20.4 Comparison of Prokaryotes With Eukaryotes

	PROKARYOTES	EUKARYOTES
Organism	Bacteria only	Protistans, fungi, plants, and animals
Level of organization	Single-celled	Single-celled (protistans, mostly), or multicelled, usually tissues, organs
Cell size	Small (1–10 µm)	Large (10–100 µm)
Cell wall	Nearly all are walled	Cellulose or chitin; none in animals
Organelles	Very rarely	Typically profuse
Metabolism	Anaerobic, aerobic	Aerobic modes predominate
Genetic material	Bacterial chromosome, sometimes plasmids	Complex chromosomes (DNA, many associated proteins) within a nucleus
Mode of division	Prokaryotic fission, mostly; also budding	Nuclear division (mitosis, meiosis, or both), and then cytoplasmic division

Self-Quiz   ANSWERS IN APPENDIX III

1. _____ are confined to extreme environments, probably much like those that prevailed on the early Earth.
   a. Cyanobacteria   c. Archaebacteria
   b. Eubacteria   d. Protozoans

2. Bacteria reproduce by _____ .
   a. mitosis   c. prokaryotic fission
   b. meiosis   d. longitudinal fission

3. A nondividing bacterial cell has _____ chromosome(s).
   a. one   b. two   c. four   d. many

4. Viruses have a _____ and a _____ .
   a. DNA core; carbohydrate coat
   b. DNA or RNA core; plasma membrane
   c. DNA or RNA core; protein coat

5. The multiplication cycle of all viruses requires _____ .
   a. penetration   b. budding   c. latency

6. During a _____ life cycle, amoeboid cells aggregate and form a migrating mass.
   a. slime mold   b. water mold   c. chytrid   d. sporozoan

7. Amoebas, foraminiferans, and radiolarians are _____ .
   a. protozoans   b. ciliates   c. algae   d. sporozoans

8. Trypanosomes cause which disease(s)?
   a. toxoplasmosis   d. African sleeping sickness
   b. Chagas disease   e. malaria
   c. amoebic dysentery   f. both b and d

9. Euglenoids and chrysophytes are mostly _____ .
   a. photoautotrophs   c. heterotrophs
   b. chemoautotrophs   d. omnivorous

10. Single-celled photosynthetic protistans, which include most euglenoids, chrysophytes, and dinoflagellates, are members of _____ , the "pastures" of most aquatic habitats.
   a. zooplankton   c. brown algae
   b. red algae   d. phytoplankton

11. Match the terms with the most suitable descriptions.
   ____ archaebacteria   a. bacterial cell division mechanism
   ____ eubacteria   b. nonliving infectious particle
   ____ virus   c. agent of malaria
   ____ *Plasmodium*   d. methanogens, extreme halophiles, and extreme thermophiles
   ____ prokaryotic fission   e. most common prokaryotic cells

Critical Thinking

1. Think about Figure 20.1. What does it tell you about the risks that drug abusers take when they share unsterilized needles? What does it tell you about medical uses of 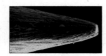 needles in developing countries where hygiene is marginal?

2. *Salmonella* bacteria cause food poisoning. Often they live in poultry and eggs. Newly hatched chicks acquire them by eating bacteria-laden feces of healthy adult chickens. Harmless bacteria ingested the same way colonize intestinal cells, leaving no place for *Salmonella* to take hold. Some farmers raise thousands of chicks in confined quarters with no adult chickens. Should they feed chicks a known mixture of bacteria from a lab or a mixture of unknown bacteria from healthy adult chickens? Devise an experiment to test which might stop *Salmonella* colonization.

3. Some strains of picornaviruses cause the highly contagious *foot and mouth disease*. Their hosts include cattle, water buffalo, sheep, and pigs. Symptoms include extensive lesions and major tissue erosion, lameness, weight loss, milk loss, and often death.

The viral strain causing a recent epidemic was first identified in Asia, then in Europe and elsewhere. It devastated farmers still recovering from an outbreak of BSE. There is no vaccine; at this writing, more than a million animals have been destroyed to stop the spread of the disease. How is the virus influencing travel and the global economy? *Hint:* Search for foot and mouth disease on the Internet, starting with http://www.cdc.gov/.

4. Suppose you vacation in a developing country. Sanitation and standards of personal hygiene are poor. Having read about parasitic protozoans in water and damp soil, what would you consider safe to drink? Which foods may be best to avoid or which food preparation methods might make them safe to eat?

5. As you read in this chapter, input from heavily fertilized cropland or raw sewage fans algal blooms. Think about it. Do you accept massive kills of aquatic species, birds, and other forms of wildlife as an unfortunate but necessary side effect of human life? If not, how would you stop the pollution? If you could stop it, what sorts of measures would you take to feed the enormous human population, which is dependent on high-yield (and very heavily fertilized) crops? How do you suggest we dispose of or prevent the accumulation of fecal matter and other wastes of more than 6.1 billion people?

Selected Key Terms

algal bloom 20.11
archaebacterium 20.1
bacterial
   conjugation 20.3
bacteriophage 20.4
binary fission 20.8
brown alga 20.13
chrysophyte 20.11
chytrid 20.7
contractile
   vacuole 20.8
cyst 20.8
dinoflagellate 20.11
disease 20.10
epidemic 20.10
eubacterium 20.1
euglenoid 20.11
extreme halophile 20.2
extreme
   thermophile 20.2
green alga 20.14
heterocyst 20.2
infection 20.10
lysis 20.5
methanogen 20.2
microorganism CI
numerical
   taxonomy 20.1
pathogen CI
plankton 20.8
prion 20.5
prokaryotic
   fission 20.3
protistan 20.6
protozoan 20.8
red alga 20.12
slime mold 20.7
sporozoan 20.9
viroid 20.5
virus 20.4
water mold 20.7

Readings

Margulis, L. 1993. *Symbiosis in Cell Evolution*. Second edition. New York: Freeman. Paperback.

Satchell, M. 28 July 1997. "The Cell From Hell." *U.S. News and World Report*, 26–28.

FUNGI

Ode to the Fungus Among Us

When push comes to shove, plants need certain fungi more than they need us. (Actually, they don't need us at all.) Fungi were there, as symbionts, when plants first invaded the land. **Symbiosis**, recall, refers to species that live together and closely interact. In cases of **mutualism**, the interaction benefits both partners or does one of them no harm. Lichens and mycorrhizae are like this.

A **lichen** is a vegetative body in which a fungus is intertwined with one or more photosynthetic organisms. From the Antarctic to the Arctic, lichens live in habitats that are just too hostile to support most organisms.

Lichens absorb minerals from substrates. They make antibiotics against bacteria that can decompose them. They make toxins against invertebrate larvae that graze upon them. Coincidentally, their metabolic products enrich soil or help form new soil from bedrock. This is what happens when lichens colonize barren sites, such as bedrock exposed by retreating glaciers. Conditions improve so much, other species move in and replace the pioneers. This probably happened when plants invaded the land. Cyanobacteria-containing lichens even help maintain ecosystems. They capture nitrogen and convert it to a form that plants use. *Lobaria* alone dishes up 20 percent of the nitrogen needed by old-growth forests in the Pacific Northwest (Figure 21.1).

Lichens also give early warning of deteriorating environmental conditions. How? They absorb toxins but can't get rid of them. From studies in New York City and in England, we know that when lichens die around human habitats, air pollution is getting bad.

a SULFUR SHELF FUNGUS *Polyporus*

Figure 21.2 Fungal species from southeastern Virginia. This sampling hints at the rich diversity within the kingdom Fungi.

Figure 21.1 *Lobaria oregana*, one of the premier lichens.

Some fungi and young tree roots also are mutualists. Their interaction is a **mycorrhiza** (plural, mycorrhizae), which means "fungus-root." Underground parts of the fungus grow through the soil and afford a huge surface area for absorption. The fungus swiftly takes up many ions of phosphorus and other minerals when these are abundant, and it releases ions to the plant when they are scarce. And what does the plant give the fungus in return? It gives up some sugars. The loss is a trade-off; many plants can't grow well without mycorrhizae.

Many fungi also help plants by being **decomposers**. Figure 21.2 shows just a few of these beneficial species. Like all other decomposers, they break down organic compounds in their surroundings. But few organisms besides fungi digest dinner while it's still on the table. As fungi grow into or on organic matter, their enzyme secretions digest it into bits that fungal cells absorb. We call this mode of nutrition **extracellular digestion and absorption**. Plants benefit when they take up some of the released nutrients. And plants, remember, are the primary producers of nearly all ecosystems.

Keep this global perspective in mind. Why? As you will see, some fungi do cause diseases in humans, pets

b PURPLE CORAL FUNGUS *Clavaria*

c RUBBER CUP FUNGUS *Sarcosoma*

d BIG LAUGHING MUSHROOM *Gymnophilus*

e TRUMPET CHANTERELLE *Craterellus*

f SCARLET HOOD *Hygrophorus*

and farm animals, ornamental plants, and crop plants. Some are notorious spoilers of food supplies. Others, however, help us manufacture substances ranging from antibiotics to cheeses. We show a tendency to assign "value" to fungi and other organisms in terms of their direct effect on our lives. There is nothing wrong with battling dangerous species and admiring the beneficial ones. Yet we should not lose sight of the greater roles of fungi or any other kind of organism in nature.

Key Concepts

1. Fungi are heterotrophs. Together with heterotrophic bacteria, they are the biosphere's decomposers. The saprobic types get nutrients from nonliving organic matter. Parasitic types get them from tissues of living hosts.

2. Fungi secrete enzymes that digest food outside their body, then fungal cells absorb breakdown products. Their metabolic activities release carbon dioxide to the atmosphere and return many nutrients to the soil, where they become available to plants and other producers.

3. Most fungi are multicelled. A mycelium, the food-absorbing portion of a fungal body, develops during their life cycle. Each mycelium is a mesh of hyphae. Hyphae are elongated filaments that grow by mitotic cell divisions.

4. In many species, some modified hyphae become interwoven into a reproductive structure, as typified by a mushroom. Fungal spores develop in or on such structures. After germinating, a spore may grow and develop into new mycelium.

5. Many fungi are symbionts. Lichens consist of fungi partnered with algae and other organisms. Mycorrhizae consist of fungal species locked in mutually beneficial relationships with young roots of land plants. Metabolic activities of the cells making up fungal hyphae provide the plants with nutrients. The plants provide the fungi with carbohydrates.

6. We tend to assign value to plants and fungi in terms of their direct effect on our lives. Our battles with the "bad" ones and reliance on the "good" ones should start from a solid understanding of their long-established roles in nature.

CHARACTERISTICS OF FUNGI

Mode of Nutrition

Fungi are heterotrophs. This means they require organic compounds that other organisms synthesize. Most are **saprobes**, which take nutrients from nonliving organic matter and cause its decay. Others are **parasites**, which extract nutrients from tissues of a living host. When the cells of any species grow in or on organic matter, they secrete digestive enzymes and then absorb breakdown products. This mode of extracellular digestion is good for plants, because plants readily absorb a portion of the released nutrients and carbon dioxide by-products. Without fungi and heterotrophic bacteria, communities would be slowly buried in their own garbage, nutrients wouldn't be cycled, and life would be over.

Major Groups

For many of us, "fungi" are mushrooms sold in grocery stores. But the commercial mushrooms are fungal body parts of only a few species. This kingdom is stunning in its diversity. Figure 21.2 shows only a few of the 56,000 fungal species we know about. There might be at least a million more we don't know about!

The fossil record suggests that fungi evolved before 900 million years ago. Some accompanied simple plants onto land 430 million years ago. Three major lineages were well established about 100 million years after that. We call them the **zygomycetes** (Zygomycota), **sac fungi** (Ascomycota), and **club fungi** (Basidiomycota). Other, puzzling kinds known as "imperfect fungi" are lumped together but aren't a formal taxonomic group. The vast majority of species in all these groups are multicelled.

Key Features of Fungal Life Cycles

Fungi reproduce asexually quite often. They reproduce sexually, too, when the opportunity presents itself. They form stupendous numbers of nonmotile spores. **Spores** are reproductive cells and multicelled structures, often walled, that germinate after dispersal from the parent. In multicelled species, a spore gives rise to a **mycelium** (plural, mycelia). This mesh of branched filaments has a good surface-to-volume ratio and grows quickly over or into organic matter. Each filament in a mycelium is a **hypha** (plural, hyphae). Hyphal cells commonly have chitin-reinforced walls. Their cytoplasm interconnects, so nutrients flow unimpeded through the mycelium.

Fungi are major decomposers that engage in extracellular digestion and absorption of organic matter. Most species are saprobic, and some are parasitic. The multicelled types form absorptive mycelia and spore-producing structures.

CONSIDER THE CLUB FUNGI

A Sampling of Spectacular Diversity

Fungal life cycles and life-styles show dizzying variety. The most we can do here is to sample a few species, starting with the club fungi. The 25,000 or so club fungi include mushrooms, shelf fungi, coral fungi, puffballs, and stinkhorns. Figures 21.2 through 21.4 and 21.6d show examples. Some of the saprobic species are important decomposers of litter on and in soil. Other species are symbionts that live in close association with the young roots of trees. Fungal rusts and smuts often destroy fields of wheat, corn, and other crop plants. Cultivation of the common mushroom (*Agaricus brunnescens*) is a multimillion-dollar business; this is the mushroom of grocery-store and pizza-topping fame. Yet some of its

Figure 21.3 Club fungi. Except for the rubber cup fungus, all the species in Figure 21.2 are also club fungi. (**a**) The red coral fungus *Ramaria*. (**b**) Fly agaric mushroom (*Amanita muscaria*). This species induces hallucinations. In ancient societies of Central America, Russia, and India, it was used in rituals to induce trances. (**c**) California's *A. ocreata*. Eat this one and you might die. In outward appearance, it resembles its relative *A. phalloides*, the death cap mushroom. Taste as little as five milligrams of the death cap's toxin, and you will start to vomit and suffer diarrhea eight to twenty-four hours later. Your liver and kidneys will degenerate. You may die within a few days.

Figure 21.4 Generalized life cycle that applies to many club fungi. When hyphal cells of two compatible mating strains make contact, their cytoplasm fuses but the nuclei do not. Cell divisions result in a dikaryotic mycelium in which each cell has two nuclei. When conditions are favorable, mushrooms form. Club-shaped, spore-bearing structures develop on gills, the mushroom cap's leaflike inner surface. Inside each structure, the two nuclei fuse to form one diploid zygote. With zygote formation, the cycle starts again.

The scanning electron micrograph below shows part of a mycelium, the underground portion of a club fungus that absorbs water and dissolved nutrients.

After nuclear fusion, the club-shaped structure (now 2*n*) will produce and bear haploid spores at the four tips of the cell.

nuclear fusion   Diploid Stage / Haploid Stage   **meiosis**

Club-shaped structures having two nuclei (*n* + *n*) form at the margin of each gill.

gills

spore (*n*)

Spores are released.

Each germinating spore gives rise to a hypha that grows and becomes a branching mycelium.

cap

stalk

After cytoplasmic fusion, a "dikaryotic" (*n* + *n*) mycelium gives rise to spore-bearing bodies (e.g., mushrooms).

cytoplasmic fusion

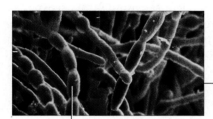

one hyphal cell among the many hyphae that make up a mycelium

relatives produce toxins that can kill you or any other organism that nibbles on them.

Have you ever wondered which organisms are the oldest and the largest? The honey mushroom, *Armillaria ostoyae*, might be the winner. In Oregon, one individual has been spreading through forest soil for 2,400 years. It now extends through 2,200 acres to an average depth of 3 feet. Imagine 1,665 football fields side by side and you get an idea of how big that is. So far, scientists have not identified any larger organism. *A. ostoyae* definitely is not one of the symbionts. Its hyphae penetrate and clog tree roots, and ultimately kill the tree.

Example of a Fungal Life Cycle

Probably you know what *A. brunnescens*, the common mushroom, looks like, so let's use it as an example of a fungal life cycle. Like most of the other club fungi, it produces **mushrooms**. These reproductive bodies are short-lived, aboveground parts; the living mycelium is buried in soil or decaying wood. Each mushroom has a stalk and a cap with fine tissue sheets (gills) that line its inner surface. Its spores, called **basidiospores**, form in club-shaped structures on the gills. When a spore that's been dispersed from a mushroom lands on a suitable site, it germinates and gives rise to a haploid mycelium.

If a hyphal cell of one strain of club fungus meets up with that of a compatible mating strain, they may undergo cytoplasmic fusion. But their nuclei won't fuse immediately. The fused portion may start a *dikaryotic* mycelium, in which hyphal cells have one nucleus of each mating type (Figure 21.4). A mycelium may grow extensively. Mushrooms arise from it when plentiful nutrients and moisture favor reproduction. Each spore-producing structure of a mushroom is dikaryotic at first, then its two nuclei fuse and form a short-lived zygote. The zygote undergoes meiosis, haploid spores develop on small stalks, then air currents disperse them.

When you notice mushrooms or any other fungus growing outdoors, think twice before nibbling on them. Unlike the common mushroom just described, many species are toxic (Figure 21.3). No one should eat any mushrooms that were gathered in the wild until they are accurately identified as edible. As the saying goes, there are old mushroom hunters and bold mushroom hunters, but no old, bold mushroom hunters.

The familiar mushrooms are reproductive structures of club fungi. A mushroom's gills display club-shaped, spore-bearing structures on their surface. Although some mushrooms are edible, others are toxic enough to kill.

A Look at the Unloved Few

You know you are a serious student of biology when you view organisms objectively in terms of their place in nature, not in terms of their impact on humans in general and you in particular. As a student you salute saprobic fungi as vital decomposers and praise parasitic fungi that help keep populations of harmful insects and weeds in check. The true test is when you open the fridge to get a bowl of high-priced raspberries and discover a fungus beat you to them. The true test is when a fungus starts feeding on warm, damp tissues between your toes and turns skin scaly, reddened, and cracked (Figure 21.5*a*).

Which home gardeners wax poetic about black spot or powdery mildew on roses? Which farmers happily hand over millions of dollars a year to sac fungi that attack corn, wheat, peaches, and apples (Figure 21.5*b*)? Who rejoices that a sac fungus, *Cryphonectria parasitica*, blitzed the chestnut trees in eastern North America?

Who willingly inhales airborne spores of *Ajellomyces capsulatus*? After landing on soil, these dimorphic beasties form mycelia. When they alight in moist lung tissues, they form yeastlike cells that cause *histoplasmosis*, a respiratory disease. The body's macrophages normally eliminate the threat, but debris from the battle results in calcified lung tissue. Heavy exposure to spores invites pneumonia.

And household molds! Thank them for sinus, ear, and lung infections, hearing losses, memory losses, and asthma attacks, boosted 300 percent in the past twenty years. The worst culprits are *Stachybotrys*, *Memnoliella*, *Cladosporium*, and certain *Aspergillus* and *Penicillium* species.

Some fungi have even tweaked human history. One notorious species, *Claviceps purpurea*, parasitizes rye and other cereal grains. Give it credit; we use some of its by-products (alkaloids) to treat migraines and to shrink the uterus after childbirth to prevent hemorrhaging. But the alkaloids are toxic when eaten in large amounts. Eat a lot of bread made with tainted rye flour and you end up with *ergotism*. The symptoms include vomiting, diarrhea, hallucinations, hysteria, and convulsions. Untreated, the disease turns limbs gangrenous and brings on death.

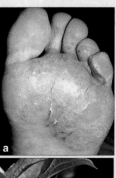

Ergotism epidemics were common in Europe during the Middle Ages, when rye was a major crop. They thwarted Peter the Great, the Russian czar who became obsessed with conquering ports along the Black Sea for his nearly landlocked empire. Soldiers laying siege to the ports ate mostly rye bread and fed rye to their horses. The soldiers went into convulsions and the horses into "blind staggers." It may also be that ergotism outbreaks were used as an excuse to launch witch-hunts in the early American colonies.

Figure 21.5 Love those fungi! (**a**) Athlete's foot, thanks to *Epidermophyton floccosum*. (**b**) Apple scab, trademark of *Venturia inaequalis*.

21.4

SPORES AND MORE SPORES

A fungus has a thing about spores. It produces sexual spores, asexual spores, or both, depending on contact with a suitable hypha, food availability, and how cool or damp conditions are. Its spores are usually small and dry, and air currents disperse them. Each spore that germinates can be the start of a hypha and a mycelium. Stalked reproductive structures may develop on many of the hyphae and produce asexual spores. After these spores germinate, each may be the start of still *another* extensive mycelium. In no time at all, that one fungus and staggering numbers of its descendants are busily decomposing organic stuff or pirating nutrients from a host! Just look at what happened to that slice of stale bread shown in Figure 21.6*a*.

Each fungal class has unique sexual spores. Club fungi form basidiospores, zygomycetes form spores by way of zygosporangia, and sac fungi form ascospores.

Consider the zygomycetes. Parasitic species feed on insects. Most saprobic types live in soil, decaying plant and animal material, and stored food. You just saw

Figure 21.6 (**a**) Moldy bread. (**b,c**) *Sarcoscypha coccinia*, the scarlet cup fungus. Saclike structures on the cup's inner surface produce sexual spores by meiosis. (**d**) One of the edible morels, *Morchella deliciosa*. (**e**) From *Eupenicillium*, chains of asexual spores of a type called conidiospores. The spores drift away from the chains, like dust, after even the slightest jiggling. "Conidia" means dust. (**f**) Cells of *Candida albicans*, an agent of "yeast" infections of the vagina, mouth, intestines, and skin.

Figure 21.7 Life cycle of the black bread mold *Rhizopus stolonifer.* Asexual phases are the most common. Mating strains (+ and −) also can reproduce sexually. Either way, haploid spores and mycelia form. A + hypha and − hypha fuse. Two gamete-producing structures form; each has a few haploid nuclei. Their nuclei fuse to form a zygote. The zygote develops a thick wall; it is now a zygospore. It may be dormant for months. Meiosis occurs when the zygospore germinates. Asexual spores form in spore sacs.

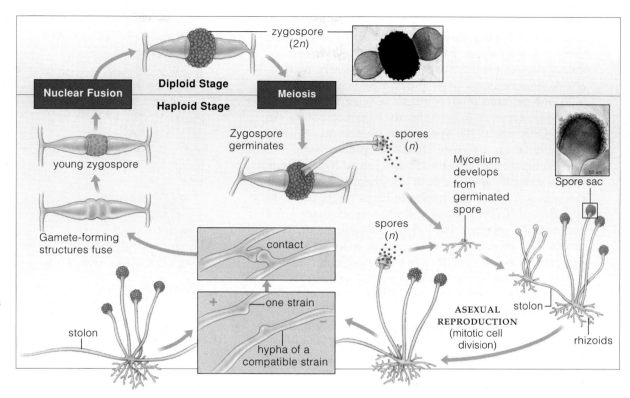

what *Rhizopus stolonifer*, the black bread mold, does to bread. A thick-walled sexual spore—a diploid zygote—forms when it reproduces sexually. A **zygosporangium**, a thin, clear covering, encloses the zygote (Figure 21.7). The zygote proceeds through meiosis. Then it gives rise to a specialized hypha, which bears a spore sac. Some number of spores form in the sporangium; each may give rise to a mycelium. Stalked hyphae grow out from such mycelia. Asexual spores form inside the spore sac on top of each stalk (Figure 21.7).

Or consider the 30,000+ kinds of sac fungi. The vast majority are multicelled. **Ascospores**, the sexual spores of sac fungi, form in sac-shaped cells. In multicelled species, reproductive structures of tightly intertwined hyphae enclose the cells, which we call asci. They look like flasks, globes, and cups (Figures 21.2*b* and 21.6).

The vast majority of sac fungi are multicelled. They include high-priced truffles and morels (Figure 21.6*d*). Truffles are symbionts with roots of hazelnut and oak trees. Pigs and dogs are trained to snuffle out truffles. Other multicelled sac fungi include certain species of *Penicillium* that "flavor" Camembert and Roquefort cheeses and make penicillins, used as antibiotics. We use *Aspergillus* to make citric acid for candies and soft drinks, and to ferment soybeans for soy sauce. Most red, bluish-green, and brown fungal molds that spoil stored foods also are multicelled. The salmon-colored *Neurospora sitophila* can run amok in bakeries and in research laboratories. It produces so many spores that it

is extremely difficult to eradicate. One of its relatives, *N. crassa*, is an important organism in genetic research.

Sac fungi also include about 500 species of single-celled yeasts (although still other yeasts are classified as club fungi). Yeasts reproduce sexually when two cells fuse and become a spore-producing sac. Some types live in the nectar of flowers and on fruits and leaves. Bakers and vintners put fermenting by-products of vast populations of yeasts to use. For example, the carbon dioxide by-products of *Saccharomyces cerevisiae* leaven bread. The commercial production of wine and beer depends on its ethanol end product. Many yeast strains with desired properties have been developed through artificial selection and genetic engineering. Then again, *Candida albicans*, a notorious relative of "good" yeasts, causes vexing infections in humans (Figure 21.6*f*).

What about imperfect fungi? Taxonomically, they are put in a holding station not because they are somehow defective but mainly because no one yet knows what kind of sexual spores they produce (if any). Previously orphaned *Aspergillus, Candida,* and *Penicillium* species were recently reunited with their kin—other sac fungi.

Through their rapid production of asexual and sexual spores, fungi take quick advantage of available organic matter, whether it has been discarded or is part of a living or dead organism. Their penchant for spore production is central to their success as decomposers and parasites.

THE SYMBIONTS REVISITED

Recall, from the introduction, that symbiosis refers to species that live together in close ecological association. Often one is a parasite's victim, not a partner. In cases of mutualism, interaction benefits both partners or does one of them no harm. In lichens, a fungus is intertwined with one or more photosynthetic species. Of 13,500 or so known types of lichens, nearly half incorporate sac fungi. Only about 100 species are photosynthetic, and most often these are green algae and cyanobacteria.

A lichen forms after the tip of a fungal hypha binds with a suitable host cell. Both lose their wall, and their cytoplasm fuses or the hypha induces the host cell to cup around it. The fungus and photosynthesizer grow and multiply together. The lichen commonly has distinct layers. The overall pattern of growth is leaflike, erect, pendulous, or flattened (Figures 21.1 and 21.8).

Lichens typically colonize sites that are hostile for most organisms, including sunbaked or frozen rocks, gravestones, even the tops of giant Douglas firs. Almost always, the fungus is the largest portion. Cyanobacteria typically reside in a separate structure inside or outside the main body. The fungus gets a long-term source of nutrients, which it absorbs from photosynthetic cells. Nutrient withdrawals affect the growth of the cells, but the lichen may help shelter them. When more than one fungus is present, it might be a mycobiont, a parasite, or even an opportunist using the lichen as a substrate.

What about those fungal mutualists with young tree roots? Plant growth improves with mycorrhizae, as you can see from Figure 21.9a. The hyphae of some species form a dense net around living cells in roots but do not penetrate them (Figure 21.9b). Others form a velvety

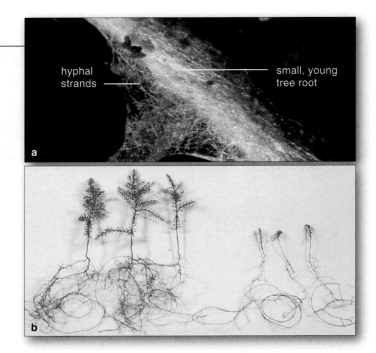

Figure 21.9 (**a**) Mycorrhiza of a hemlock tree. (**b**) Effects of the presence or absence of mycorrhizae on plant growth. The juniper seedlings at left are six months old. They were grown in sterilized, phosphorus-poor soil with a mycorrhizal fungus. The seedlings at right were grown without the fungus.

wrapping around tree roots as the mycelium radiates through soil. These are common in temperate forests. They help trees survive seasonal shifts in temperature and rainfall. About 5,000 fungal species, mostly club fungi, enter into such associations.

Different mycorrhizae develop in about 80 percent of all vascular plants. Fungal hyphae penetrate plant cells, as they do in lichens. Zygomycetes are the fungal partner. Their hyphae branch extensively, forming tree-shaped absorptive structures in plant cells. The hyphae also extend for several centimeters into the soil. Section 27.2 offers a closer look at these beneficial species.

A final note: Since the early 1900s, collectors have recorded data on wild mushroom populations in the forests of Europe. They find the number and kinds of fungi are severely declining. Mushroom gatherers can't be the cause, because inedible as well as edible species are vanishing. However, the decline does correlate with rising air pollution. Vehicle exhaust, smoke from coal burning, and emissions from nitrogen fertilizers pump ozone, nitrogen oxides, and sulfur oxides into the air. Normally, as a tree ages, one species of mycorrhizal fungus gives way to another, in predictable patterns. When fungi die, trees lose this vital support system, and they become vulnerable to severe frost and drought. Are the North American forests at risk, also? Conditions there are deteriorating in comparable ways.

dispersal fragment
(cells of fungus and of
photosynthetic species)

outer layer
of fungal cells

photosynthetic
species

inner layer of
loosely woven
hyphae

outer layer
of fungal cells

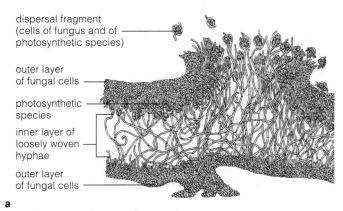

Figure 21.8 (**a**) Structure of one kind of lichen, cross-section. (**b**) Encrusting, flattened lichens growing on a rock.

Lichens and mycorrhizae are both symbiotic associations between fungi and other organisms, with mutual benefits.

SUMMARY

1. Fungi are heterotrophs and major decomposers. The saprobes feed on nonliving organic matter; parasites feed on tissues of living organisms. Some fungal species are symbiotic partners with other organisms. Cells of all species secrete digestive enzymes that break down food to small molecules, which the cells absorb. *21.1*

2. Nearly all fungi are multicelled. The food-absorbing part, the mycelium, is a mesh of filaments (hyphae). The aboveground reproductive parts (e.g., mushrooms) form from tightly interwoven hyphae. *21.1, 21.2*

3. Major groups are zygomycetes, ascomycetes (sac fungi), and basidiomycetes (club fungi). Each produces distinctive sexual and asexual spores. When a sexual phase can't be detected or is absent from the life cycle, a fungus is assigned to an informal category called the imperfect fungi. *21.2, 21.4*

4. Lichens are mutualistic associations of fungi with photosynthesizers (e.g., green algae and cyanobacteria). Mycorrhizae are mutualistic associations of a fungus and young roots. Fungal hyphae give up nutrients to the plant and get back carbohydrates. *CI, 21.5*

Review Questions

1. Describe the fungal mode of nutrition, and explain how the structure of mycelia facilitates this mode. *21.1*

2. How does a lichen differ from a mycorrhiza? *CI, 21.5*

3. List conditions that influence fungal spore formation. *21.4*

4. What makes fungi such successful decomposers? *21.3*

Self-Quiz   ANSWERS IN APPENDIX III

1. A mycorrhiza is a _____ .
   a. fungal disease of the foot   c. parasitic water mold
   b. fungus–plant relationship   d. fungus of barnyards

2. Parasitic fungi obtain nutrients from _____ .
   a. tissues of living hosts   c. only living animals
   b. nonliving organic matter   d. none of the above

3. Saprobic fungi derive nutrients from _____ .
   a. nonliving organic matter   c. living animals
   b. living plants   d. both b and c

4. New mycelia form after _____ germinate.
   a. hyphae   b. mycelia   c. spores   d. mushrooms

5. A mushroom is _____ .
   a. the food-absorbing part of a fungal body
   b. the part of the fungal body not constructed of hyphae
   c. a reproductive structure
   d. a nonessential part of the fungus

6. Match the terms appropriately.
   ____ zygomycetes   a. mushrooms, shelf fungi
   ____ conidia   b. type of sexual spore
   ____ hypha   c. *Penicillium*'s chains of asexual spores
   ____ club fungi   d. each filament in a mycelium
   ____ ascospore   e. black bread mold
   ____ sac fungi   f. truffles, morels, some yeasts

Figure 21.10 Reproductive structures of *Pilobolus*, a name from a Greek word for "hat-thrower." The "hats" actually are spore sacs.

Critical Thinking

1. *Pilobolus* is a type of fungus that commonly dines on horse dung. Each morning, stalked reproductive hyphae emerge from irregularly spaced piles of dung. By early afternoon, they have dispersed spores to sunlit grasses where horses feed. The spores pass through the horse gut unharmed and exit with their own pile of dung. At the tip of each stalked hypha is a dark-walled, spore-containing sac (Figure 21.10). Just below the sac, the stalk is differentiated into a vesicle, swollen with a fluid-filled central vacuole. At the base of the vesicle is a ring of light-sensitive, pigmented cytoplasm. The stalk bends as it grows until its wall is parallel with the sun's rays and light strikes all of the ring. When that happens, turgor pressure builds up inside the central vacuole until the vesicle ruptures. The forceful blast can propel spore sacs two meters away—which is amazing, considering that the stalk is less than ten millimeters tall. Reflect on the examples of fungi discussed in this chapter. Would you say *Pilobolus* is a zygomycete, club fungus, or sac fungus?

2. Renee sees in the laboratory that the fungus *Trichoderma* grows well in distilled water. It continues to do so even after she rigorously treats the water and glassware to remove all traces of organic carbon. This fungus is not a photoautotroph. Suggest a metabolic life-style that lets it grow under these conditions.

3. *Trichoderma* is being tested as a natural pest control agent. Laboratory experiments demonstrated that some strains of this fungus combat other fungi that cause plant diseases. Some even promote seed germination and plant growth. During one set of twenty trials, workers increased lettuce yields by 54 percent. What concerns must be addressed before *Trichoderma* can be released into the environment for commercial applications?

Selected Key Terms

ascospore *21.4*
basidiospore *21.2*
club fungus
   (basidiomycetes) *21.1*
decomposer *CI*
extracellular digestion
   and absorption *CI*
fungus *21.1*
hypha *21.1*
lichen *CI*
mushroom *21.2*
mutualism *CI*
mycelium *21.1*
mycorrhiza *CI*
parasite *21.1*
sac fungus
   (ascomycetes) *21.1*
saprobe *21.1*
spore (fungal) *21.1*
symbiosis *CI*
zygomycetes *21.1*
zygosporangium *21.4*

Readings

Moore-Landecker, E. 1996. *Fundamentals of the Fungi*. Fourth edition. Englewood Cliffs, New Jersey: Prentice-Hall.

On-Line readings at Student Guide for InfoTrac:
www.brookscole.com/biology

PLANTS

Pioneers in a New World

Seven hundred million years ago, no shorebirds stirred and noisily announced the dawn of a new day. There were no crabs to clack their claws together and skitter off to burrows. The only sounds were the rhythmic, muffled thuds of waves in the distance, at the outer limits of another low tide. More than 3 billion years before, life had originated in the waters of the Earth. Now, quietly, the invasion of the land was under way.

Why did it happen then? Astronomical numbers of photosynthetic cells had come and gone, and oxygen-producing types had changed the atmosphere. High above the Earth, the sun's energy had converted much of the oxygen into a dense ozone layer. That layer was a shield against lethal doses of ultraviolet radiation, which had kept life beneath the surface of water and mud.

Were cyanobacteria the first to adapt to intertidal zones, where mud dried out with each retreating tide? Were they the first in the shallow, freshwater streams flowing down to the coasts? Probably. Fossils suggest that later in time, some green algae and fungi made the journey together. Ancient species of green algae that lived near the water's edge or made it onto land gave rise to plants. And through millions of years thereafter, descendant plants became the basis of communities in coastal lowlands, near the snow line of high mountains, and in just about all places in between (Figure 22.1).

We have tantalizing fossils of the first pioneers. We are learning more about them through comparative studies of many modern species. Today, as in the late precambrian, cyanobacteria and green algae form mats in nearshore waters and along freshwater streams of the sort shown in Figure 22.1a. After a volcano erupts or a glacier retreats, cyanobacteria are the first to colonize the barren rocks. Symbiotic interactions between green algae and fungi follow. Their organic by-products and remains accumulate and create pockets of soil. Mosses and other plants soon become established in the newly forming soil and further enrich it.

With this new chapter we turn to the story of plants. Nearly all species are multicelled photoautotrophs that absorb energy from the sun, carbon dioxide from the air, and minerals dissolved in water to synthesize organic compounds. These metabolic wizards also split water molecules. In doing so, they get stupendous numbers of the electrons and hydrogen atoms required for growth into multicellular forms as tall as redwoods, as vast as an aspen forest that is actually one interconnected clone.

We know of at least 295,000 kinds of existing plants. Be glad their ancient ancestors left the water. Without those pioneers in a new world, we humans and other land-dwelling animals never would have made it onto the evolutionary stage.

Figure 22.1 (a) Filaments of a green alga massed in a shallow stream. More than 400 million years ago, green algal species that may have been ancestral to all plants lived in similar streams that meandered down to the shores of ancient continents. Land-dwelling descendants of ancestral forms: (b) Moss growing on rocks. (c) Ponderosa pine high above the floor of Yosemite Valley in the Sierra Nevada of California. (d) Flowers of one of the most highly prized of all flowering plants—orchids—growing on a branch of a living tree in a tropical rain forest. With this chapter, we turn to the beginning—and end of the line—of some ancient lineages.

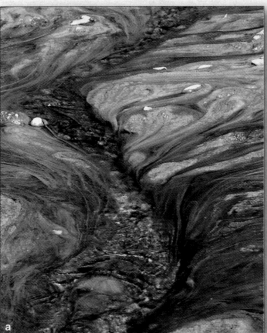

Key Concepts

1. With very few exceptions, the plant kingdom consists of multicelled photoautotrophs. From earlier chapters, you know that plants use chlorophylls *a* and *b* as their main photosynthetic pigments. In this respect they are like green algae, their closest relatives.

2. Unlike their algal ancestors, which were adapted to aquatic habitats, nearly all existing plants live on land.

3. In general, plants are structurally adapted to intercept sunlight, absorb water and mineral ions, and conserve water. Their lignin-reinforced tissues help them grow upright. Root systems mine the soil for water and ions, and internal tissues conduct water and solutes to all of the living cells in belowground and aboveground parts.

4. Land plants are reproductively adapted to withstand dry periods. During the life cycle, a sporophyte develops roots, stems, and leaves. It holds on to its developing gametes and supplies them with water and nutrients. And it disperses the new generation in ways that are responsive to prevailing conditions in its habitat.

5. Early divergences gave rise to the bryophytes, then seedless vascular plants, and then seed-bearing vascular plants. Of these categories, the seed producers were the most successful in radiating into drier environments.

6. The seed-bearing vascular plants called gymnosperms include the cycads, ginkgos, gnetophytes, and conifers. The angiosperms, another group of vascular plants, bear flowers as well as seeds. Two classes of flowering plants are known informally as the dicots and the monocots.

EVOLUTIONARY TRENDS AMONG PLANTS

Overview of the Plant Kingdom

At least 295,000 species of photoautotrophs and a few heterotrophs are grouped in the plant kingdom. Most are **vascular plants**, defined in part by internal tissues that conduct water and solutes through roots, stems, and leaves. Fewer than 19,000 species are *non*vascular plants called **bryophytes**. Plants, like photoautotrophic bacteria and protistans, are producers for ecosystems.

Liverworts, hornworts, and mosses are bryophytes. The whisk ferns, lycophytes, horsetails, and ferns are *seedless* vascular plants. Cycads, ginkgos, gnetophytes, and conifers belong to a group of *seed-bearing* vascular plants called **gymnosperms**. The **angiosperms**, another group of vascular plants, bear flowers and seeds. Dicots and monocots are two classes of flowering plants.

Ancestors of plants had evolved in the seas by 700 million years ago. About 265 million years later, simple stalked plants were growing along coasts and streams. Then the evolutionary pace picked up; it took only 60 million years or so for diverse plants to radiate through much of the land. Long-term changes in their structure and reproduction explain how they were able to do so.

Evolution of Roots, Stems, and Leaves

Simple underground structures started to form when plants colonized the land, and they evolved into root systems in vascular plant lineages. Most **root systems** have many underground absorptive structures which, taken together, have a large surface area. They rapidly take up soil water and dissolved mineral ions. In many species, the root system also anchors the plant.

Aboveground, **shoot systems** evolved. These have stems and leaves that absorb energy from the sun and carbon dioxide from the air. Stems grew and branched only after plants developed a biochemical capacity to synthesize and deposit **lignin**, an organic compound, in cell walls. Collectively, lignin-strengthened cell walls structurally support stems, which grow in patterns that increase the total light-intercepting surface of leaves.

In many plant lineages, cellular pipelines for water and solutes contributed to the evolution of roots, stems, and leaves. They evolved as components of xylem and phloem, two vascular tissues. **Xylem** distributes water and dissolved ions through the plant; **phloem** distributes dissolved sugars and other photosynthetic products.

Life on land also depended on water conservation, which had not been a problem in most aquatic habitats. Shoots became protected by a **cuticle**, a waxy coat that helps conserve water on hot, dry days. Also, **stomata** (singular, stoma), tiny openings across the surfaces of leaves and some stems, evolved. They control carbon dioxide absorption and restrict evaporative water loss. Later chapters describe these tissue specializations.

From Haploid to Diploid Dominance

Life cycles changed as early plants radiated into higher, drier places. Think of the gametes of algae, which get together only in liquid water. Remember **gametophytes**, the gamete-producing bodies described in Section 9.5? They dominate the *haploid* (n) phase of algal life cycles. The diploid ($2n$) phase simply is the zygote.

Now look at Figure 22.2. *The diploid phase dominates most plant life cycles.* After the diploid zygote forms at fertilization, mitotic cell divisions and cell enlargements convert it into a **sporophyte**, a multicelled, diploid plant body (for example, a pine tree). After some of its cells undergo meiosis, haploid resting cells—**spores**—form. (Sporophyte means spore-producing body.) Mitotic cell divisions of spores produce the gametophytes.

The shift to diploid dominance was an adaptation to land. As you probably know, most habitats on land show seasonal changes in the availability of free water and nutrients. It's likely that natural selection favored the further development of root systems. It must have mycorrhizae in which young roots interact with fungal symbionts; the interaction helps plants take up water and minerals even during dry seasons (Section 21.5).

Unlike algae and bryophytes, vascular plants have sporophytes that are structurally more complex and larger than their gametophytes. Especially among the

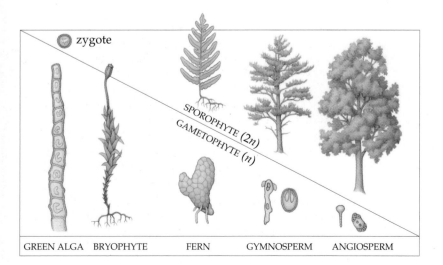

zygote

SPOROPHYTE (2n)
GAMETOPHYTE (n)

GREEN ALGA   BRYOPHYTE   FERN   GYMNOSPERM   ANGIOSPERM

Figure 22.2 Evolutionary trend among plants, from gametophyte (haploid) dominance to sporophyte (diploid) dominance in the life cycle. These representative species range from a green alga (*Ulothrix*) to a flowering plant. The trend occurred as early plants were colonizing habitats on land. See also Section 9.5.

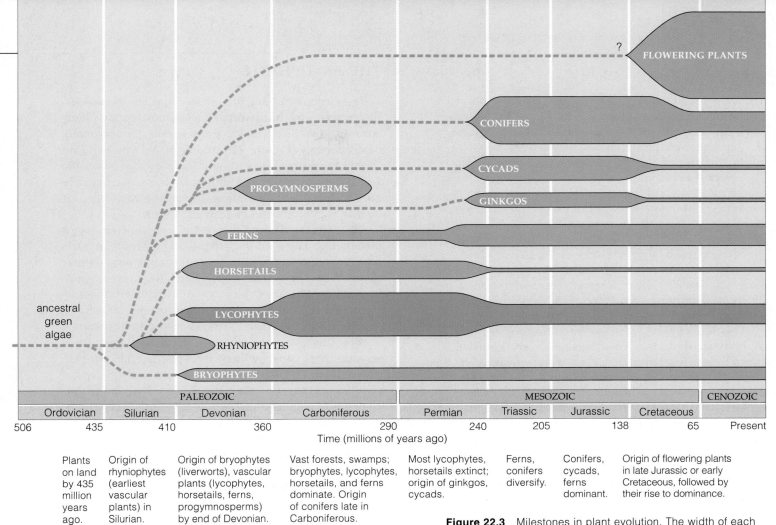

ancestral green algae							

PALEOZOIC — **MESOZOIC** — **CENOZOIC**

Ordovician	Silurian	Devonian	Carboniferous	Permian	Triassic	Jurassic	Cretaceous	
506	435	410	360	290	240	205	138	65	Present

Time (millions of years ago)

Plants on land by 435 million years ago.

Origin of rhyniophytes (earliest vascular plants) in Silurian.

Origin of bryophytes (liverworts), vascular plants (lycophytes, horsetails, ferns, progymnosperms) by end of Devonian.

Vast forests, swamps; bryophytes, lycophytes, horsetails, and ferns dominate. Origin of conifers late in Carboniferous.

Most lycophytes, horsetails extinct; origin of ginkgos, cycads.

Ferns, conifers diversify.

Conifers, cycads, ferns dominant.

Origin of flowering plants in late Jurassic or early Cretaceous, followed by their rise to dominance.

Figure 22.3 Milestones in plant evolution. The width of each lineage correlates with the range of diversity over time.

gymnosperms and, later, angiosperms, the sporophyte became the dominant phase of the life cycle. It could retain, nourish, and protect developing gametophytes and the young sporophytes. What was the advantage? *Fertilization and dispersal of each new generation could be timed with the arrival of suitable environmental conditions.*

Evolution of Pollen and Seeds

Like some seedless species, seed-bearing plants produce two types of spores. We call this condition *hetero*spory, as opposed to *homo*spory (one type of spore). In both gymnosperms and angiosperms, macrospores give rise to female gametophytes in which eggs form and get fertilized. Smaller, microspores are the start of **pollen grains**, cellular structures that become mature, sperm-bearing male gametophytes. Pollen grains hitch rides to the eggs on air currents, insects, birds, and so on; they don't require free-standing water to reach them. In this respect they differ enormously from algae. The evolution of pollen grains contributed to the successful radiation of seed-bearing plants into high and dry habitats.

In the drier habitats, seed production also proved adaptive. Female gametophytes of seed-bearing plants

form within sporophyte tissues. Each **seed** consists of an embryo sporophyte, nutritious tissues, and a protective coat. It helps the embryo sporophyte through dry and otherwise hostile conditions. It was no coincidence that seed plants rose to dominance during Permian times, when shifts in climate were extreme.

Before turning to the spectrum of diversity among plants, take a look at Figure 22.3. You may wish to use it as a map of the branching evolutionary roads.

The plant kingdom includes multicelled, photosynthetic species called bryophytes, seedless vascular plants, and seed-bearing vascular plants. Most of these live on land.

In most lineages, structural adaptations to life on land included a root system, shoot system, waxy cuticle, stomata, vascular tissues, and lignin-reinforced tissues.

Sporophytes with well-developed roots, stems, and leaves came to dominate the life cycle of most land plants. Parts of these complex sporophytes nourish and protect fertilized eggs and embryos until conditions favor their growth.

Some plants started to produce two types of spores, not one. This led to the evolution of male gametes well adapted for dispersal without liquid water and to the evolution of seeds.

Ancient Carbon Treasures

The simple evolutionary tree in Figure 22.3 only hints at past events that still influence our daily lives. Example: Three hundred million years ago, about halfway through the Carboniferous, mild climates prevailed and dense swamp forests carpeted the wet lowlands of continents. The absence of pronounced seasonal swings in temperature favored plant growth through much of the year. Plants having lignin-reinforced tissues and well-developed root and shoot systems had the competitive edge under the prevailing conditions. Some of them evolved into giants. Massive-stemmed lycophyte trees—giant club mosses—topped out at nearly forty meters. Horsetails, including species of *Calamites*, were close to twenty meters tall. Often, stems that grew fast from spreading underground rhizomes formed dense thickets.

This was a period when sea levels rose and fell fifty times. When the seas receded, swamp forests flourished. When the sea level rose, forest plants were submerged and became buried in sediments that protected them from decay. Gradually the sediments compressed the saturated, undecayed remains into what we now call peat. As more sediments accumulated, increased heat and pressure made the peat even more compact. It became **coal** (Figure 22.4).

Coal has a high percentage of carbon; it is energy-rich. We use it as a "fossil fuel." It took a fantastic amount of photosynthesis, burial, and compaction to form each large seam of coal. It has taken us only a few centuries to deplete many of the world's known coal deposits.

Often you will hear about annual "production rates" for coal or some other fossil fuel. But how much do we really produce each year? Nothing. We simply *extract* it from the Earth. Coal is a nonrenewable source of energy.

Figure 22.4   (a) Reconstruction of a Carboniferous swamp forest. (b) Part of a seam of coal.

THE BRYOPHYTES

Today, bryophytes include 18,600 species of **mosses**, **liverworts**, and **hornworts**. These nonvascular plants are mostly well adapted to grow in fully or seasonally moist habitats (Figures 22.1 and 22.5). Yet you will find certain mosses growing slowly in deserts, even on the windswept plateaus of Antarctica. Mosses particularly are sensitive to air pollution. In regions where the air quality is poor, mosses are few or absent.

Bryophytes are less than twenty centimeters (eight inches) tall. Their leaflike, stemlike, and rootlike parts have no xylem or phloem; hence the name nonvascular plants. Like some algae and lichens, they can dry out, then revive after absorbing water. Most have rhizoids: elongated cells or threadlike structures that attach the gametophytes to soil and serve as absorptive structures.

Bryophytes are the simplest plants to display three features that evolved in early land plants. *First*, a cuticle prevents water loss from aboveground parts. *Second*, a cellular jacket around the parts that produce sperm and eggs holds in moisture. *Third*, of all plants, bryophytes alone have large gametophytes that do not depend on sporophytes for nutrition. Instead, embryo sporophytes start to develop inside some gametophyte tissues. Even at maturity, the sporophytes are never dispersed. They remain *attached to* the gamete-producing body and still gain some nutritional support from it.

With 10,000 species, mosses are the most common bryophytes. The gametophytes of some species grow in clusters and form low, cushiony mounds (Figure 22.6b). Those of others commonly grow in branched, feathery patterns on tree trunks and branches in humid climates. Eggs and sperm develop in tiny, jacketed vessels at the shoot tips of gametophytes. Sperm reach the eggs by swimming through moisture clinging to plants. Zygotes give rise to sporophytes, each composed of a stalk and a jacketed structure in which spores will develop.

Figure 22.6 shows one of 350 kinds of peat mosses (*Sphagnum*). Whereas most bryophytes grow slowly, the peat mosses can grow fast enough to yield twelve metric tons of organic matter per hectare, which is about twice as much as corn plants yield. They soak up five times as much water as cotton does, owing to large, dead cells in their leaflike parts. The acids they produce inhibit the growth of bacterial and fungal decomposers. Because of their high absorbency and fine antiseptic properties, they were used as an emergency poultice on wounds of soldiers during World War I.

The remains of peat mosses slowly accumulate into compressed, exceedingly moist mats called **peat bogs**. In cold and temperate regions, peat bogs cover an area equal to one-half of the United States. Only the most acid-tolerant plants, including cranberries, blueberries, larch, and Venus flytraps, can grow in the bogs, which can be as acidic as vinegar.

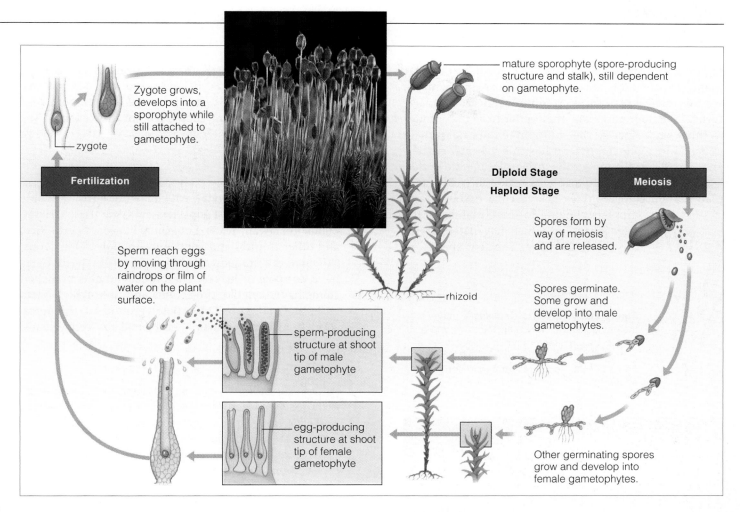

Zygote grows, develops into a sporophyte while still attached to gametophyte.

zygote

Fertilization

mature sporophyte (spore-producing structure and stalk), still dependent on gametophyte.

Diploid Stage

Haploid Stage

Meiosis

Sperm reach eggs by moving through raindrops or film of water on the plant surface.

rhizoid

Spores form by way of meiosis and are released.

Spores germinate. Some grow and develop into male gametophytes.

sperm-producing structure at shoot tip of male gametophyte

egg-producing structure at shoot tip of female gametophyte

Other germinating spores grow and develop into female gametophytes.

Figure 22.5 Life cycle of a moss (*Polytrichum*), a bryophyte. The moss sporophyte remains attached to the gametophyte and depends on it. The gametophyte provides it with nutrients and water.

Figure 22.6 (**a**) Peat bog in Ireland. This family is cutting out blocks of peat and stacking them to dry as a fuel source for their home. The Irish also harvest enough peat to generate electricity in peat-burning power plants. (**b**) Gametophyte of a peat moss (*Sphagnum*). A few sporophytes are attached to them. The brown, jacketed structures on white stalks are the sporophytes.

Nearly all peat that's harvested and dried in Ireland and elsewhere is burned to generate electricity in power plants. Compared to coal burning, peat fires generate fewer pollutants. Every so often, the peat harvesters come across exceptionally well-preserved bodies of humans who lived 2,000 to 3,000 years ago. The acidic bogs, which kept the bodies from decomposing, apparently were sites of ceremonial human sacrifices.

Bryophytes are nonvascular plants with flagellated sperm that require liquid water to reach and fertilize the eggs.

A sporophyte of these plants develops within gametophyte tissues. It remains attached to the gametophyte and receives some nutritional support from it.

EXISTING SEEDLESS VASCULAR PLANTS

Figure 22.7 shows one of the early seedless vascular plants. Descendants of certain lineages are still with us; we call them **whisk ferns**, **lycophytes**, **horsetails**, and **ferns**. Like their ancestors, they differ from bryophytes in three key respects. The sporophyte does not remain attached to a gametophyte, it has true vascular tissues, and it is the larger, longer lived phase of the life cycle.

Most seedless vascular plants reside in wet, humid places. Their gametophytes lack vascular tissues. Water droplets clinging to these plants are the only means by which flagellated sperm can reach the eggs. The few species in

Figure 22.7 Sketch of *Cooksonia*, one of the earliest known vascular plants, no more than a few centimeters tall. It probably grew in mud flats. Its upright, branching stems had a cuticle. Its spores formed in structures at stem tips. Compare Figure 18.4*c*.

dry habitats reproduce sexually during brief, seasonal pulses of heavy rains. In one sense, the whisk ferns, lycophytes, horsetails, and ferns are the "amphibians" of the plant kingdom. They have not fully escaped the aquatic habitats of their ancestors.

Whisk Ferns

Whisk ferns (Psilophyta), which are not ferns, resemble whisk brooms. Florist suppliers cultivate these striking plants in Hawaii, Texas, Louisiana, Florida, Puerto Rico, and other tropical and subtropical regions. One genus, *Psilotum*, is a unique vascular plant, for its sporophytes have no roots or leaves. The photosynthetic, branched stems have scalelike projections and, internally, xylem and phloem (Figure 22.8*a*). Belowground are **rhizomes**, short, branched, and mainly horizontal absorptive stems.

Lycophytes

About 350 million years ago, lycophytes (Lycophyta) included tree-sized members of swamp forests. About 1,100 far tinier species exist today. The most familiar are club mosses, members of communities in the Arctic, the tropics, and regions between. Many form mats on forest floors. One type, the resurrection plants, "come back to life" after being dried up for extended periods.

Sporophytes of most club mosses have leaves and a branching rhizome that gives rise to vascularized roots and stems. Some have nonphotosynthetic, cone-shaped leaf clusters with spore-producing structures (Figure 22.8*b*). Each cluster is a **strobilus** (plural, strobili). After spores disperse, they germinate and develop into small, free-living gametophytes. *Selaginella* is heterosporous; two kinds of spores develop in the same strobilus.

Horsetails

Tree-sized sphenophytes (Sphenophyta) flourished in ancient swamp forests. Twenty-five or so smaller species of one genus, *Equisetum*, made it to the present. These are the horsetails, and their body plan hasn't changed much over the past 300 million years.

Horsetails thrive in streambank muds, vacant lots, roadsides, and other disrupted habitats. Figure 22.8*c–e* shows the vegetative, photosynthetic stems and fertile

Figure 22.8 (**a**) Sporophytes of a whisk fern (*Psilotum*), a seedless vascular plant. Tips of stubby branchlets bear pumpkin-shaped spore-producing structures. (**b**) Sporophyte of a lycophyte (*Lycopodium*). (**c**) Vegetative stem of *Equisetum*; it loosely resembles a horsetail. (**d**) Nonphotosynthetic, fertile stems of *Equisetum*. At their tip is a strobilus, a cluster of spore-producing structures. (**e**) Closer look at a fertile stem. Each petal-shaped part holds many spores, formed by way of meiosis.

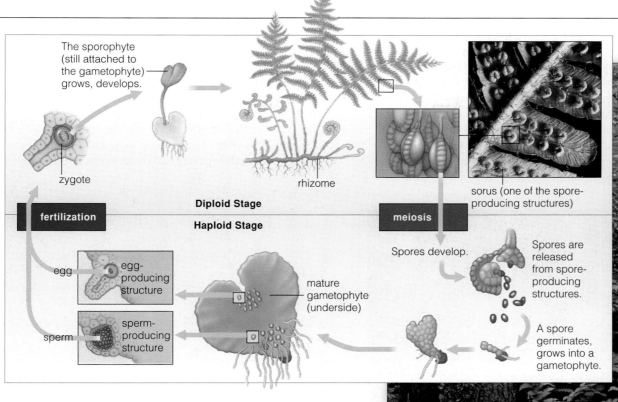

The sporophyte (still attached to the gametophyte) grows, develops.

zygote

rhizome

Diploid Stage

fertilization

Haploid Stage

meiosis

sorus (one of the spore-producing structures)

egg

egg-producing structure

sperm

sperm-producing structure

mature gametophyte (underside)

Spores develop.

Spores are released from spore-producing structures.

A spore germinates, grows into a gametophyte.

Figure 22.9 Life cycle of a fern. The photograph shows ferns growing in a moist habitat in Indiana. Ferns with finely divided fronds are in the foreground.

stems of one species. Its spores give rise to free-living gametophytes 1 millimeter to 1 centimeter across. The sporophytes of most horsetails have rhizomes, hollow photosynthetic stems, and scale-shaped leaves. Stems contain a ringlike array of xylem and phloem strands. Ribs reinforced with silica structurally support the stems and give them a texture like sandpaper. Pioneers of the American West, who did not have many places to store and wash towels, gathered horsetails on their westward journeys and used them as pot scrubbers.

Ferns

With 12,000 or so species, the ferns (Pterophyta) are the largest and the most diverse group of seedless vascular plants. All but about 380 are native to the tropics, but you find them in gardens and in homes throughout the world. Their size range is stunning. The leaves of some floating types are less than 1 centimeter across; some of the tropical tree ferns are 25 meters (82 feet) tall. One climbing fern has a modified leaf stalk 30 meters long.

Most ferns have vascularized rhizomes that give rise to roots and leaves. Exceptions include tropical tree ferns and epiphytes. (*Epiphyte* refers to any aerial plant that grows attached to tree trunks or branches.) While they develop, young fern leaves (fronds) are coiled in a way that resembles a fiddlehead. When mature, these fronds commonly are divided into leaflets.

You may have noticed rust-colored patches on the lower surface of many fern fronds. Each patch, a cluster of spore-producing structures, is a sorus (plural, sori). At dispersal time, the spore-producing structures snap open with such force that the released spores catapult through the air. Each spore that germinates develops into a small gametophyte. You can see an example, a green, heart-shaped gametophyte, in Figure 22.9.

Seedless vascular plants (whisk ferns, lycophytes, horsetails, and ferns) have sporophytes adapted to conditions on land. Yet they have not entirely escaped their aquatic ancestry. When reproducing sexually, they must have water clinging to them so their flagellated sperm can reach the eggs.

THE RISE OF THE SEED-BEARING PLANTS

About 360 million years ago, as the Devonian gave way to the Carboniferous, the first seed-bearing plants arose. In terms of diversity, numbers, and distribution, they would become the most successful groups of the plant kingdom. Seed ferns, gymnosperms, and (much later) angiosperms were the dominant groups. They differed from seedless vascular plants in three crucial respects.

First, compared with seedless vascular plants, the gymnosperms had thickened cuticles, stomata recessed below leaf surfaces, and other water-conserving traits. These traits gave gymnosperms the competitive edge after the Carboniferous gave way to the cooler, drier climates of the Permian. Swamps vanished; and cycads, conifers, and other gymnosperms rose to dominance.

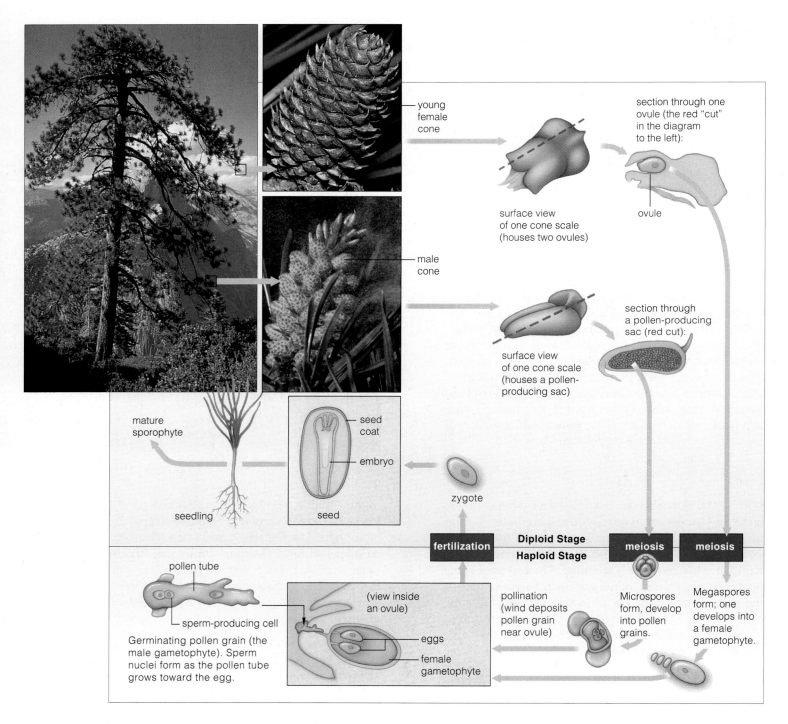

Figure 22.10 Life cycle of one of the conifers, the ponderosa pine (*Pinus ponderosa*).

Second, seed-bearing plants form **microspores** that give rise to pollen grains, a type of sperm-bearing male gametophyte. And they form **megaspores** that give rise to female gametophytes with egg cells.

A pollen grain is like a suitcase for transporting its contents—sperm—to eggs even in times of prolonged drought. The seedless vascular plants don't have the same survival advantage. Without predictable rainfall and moisture, their sperm cannot reach the eggs, and this lowers chances of their reproductive success. By contrast, pollen grains of gymnosperms simply drift on currents of air. Some angiosperm pollen grains do the same, but many others use insects, birds, bats, and other animals as taxis to eggs. **Pollination** is the name for the arrival of pollen grains on female reproductive structures. By this process, seed-bearing plants escaped dependence on free water for fertilization.

Third, embryo sporophytes are dispersed from the parent plant in seeds. A seed originates from a female reproductive structure on the sporophyte: an ovule. An **ovule** consists of a female gametophyte with its egg, nutrient-rich tissue, and a jacket of cell layers that will become the seed coat. A zygote will form in the ovule when a sperm fertilizes the egg. An embryo sporophyte will develop, and when the time comes to leave the parent plant, the seed coat will protect it on its journey.

As an example of this, think about what happens on the woody scales of pine cones. Wedged in between the scales of female cones are ovules in which megaspores give rise to female gametophytes. In male "cones" (very small strobili), microspores develop into pollen grains. Each spring, millions of pollen grains drift from a tree's male cones. Pollination is over when some land on the ovules of female cones. After a pollen grain germinates, cell divisions and cell enlargements transform it into a tubular structure, the sperm-bearing male gametophyte (Figure 22.10). Chemical gradients guide its growth through the female tissues. For pines, it takes sperm months or years after pollination to reach the egg.

As with other gymnosperms, seed formation begins at the ovule. An embryo sporophyte starts developing from the fertilized egg. The outer layers of the jacket around the female gametophyte and embryo become a hardened coat. Nutrients inside help see the embryo sporophyte through the critical time of germination, before its roots and shoots become fully functional.

Gymnosperms and other seed-bearing plants differ from seedless plants in three key respects. They have tissue adaptations to dry conditions. They produce pollen grains that deliver sperm to eggs even in the absence of liquid water. They produce seeds, a way to disperse nutrient-rich tissue and embryo sporophytes inside a protective coat.

Good-bye, Forests

Conifers dominated many habitats on land during the Mesozoic, but their slow reproductive pace put them at a competitive disadvantage when the flowering plants began their great adaptive radiation (Section 19.6). Coniferous forests still predominate in the far north, at higher elevations, and in some parts of the Southern Hemisphere. However, existing conifers face more than competition with flowering plants for resources. Now they are vulnerable to **deforestation**: the removal of all trees from large tracts, as by clear-cutting (Figure 22.11). Conifers just have the bad luck to be premier sources of lumber, paper, and other wood products demanded by human societies. Throughout the world, huge tracts of land that were deforested years ago still show no signs of recovery. We return to this topic in Chapter 45.

For now, keep this in mind: Where flowering plants flourish, the conifers are at a competitive disadvantage, partly because they take so long to reproduce. Rampant deforestation isn't helping them one bit, either.

Figure 22.11 Clear-cut peaks in Alaska. This isn't an isolated example. In the early 1980s, 400 million board feet of timber were cut in Washington's Olympic Peninsula each year. In Arkansas, about one-third of the Ouachita National Forest has been clear-cut. Its once-diverse forest communities have been replaced by "tree farms" of a single species of pine.

GYMNOSPERM DIVERSITY

With a bit of history behind us, we turn now to a survey of some existing gymnosperms. Unlike the seeds of flowering plants, which are enclosed in a reproductive chamber (an ovary), gymnosperm seeds develop in an exposed location: on top of a spore-producing structure. (*Gymnos* means naked; *sperma* is taken to mean seed.)

Conifers are woody trees and shrubs that produce scalelike or needlelike leaves and bear seeds exposed on scales of female cones. (Cones are clusters of modified leaves around spore-producing structures.) Most species shed some leaves all year but stay leafy, or *evergreen*. A few are *deciduous*; they shed leaves all at once in cold, dry seasons. The most abundant trees (pines), the tallest (redwoods), and oldest (bristlecone pine, as in Figure 22.12a) are conifers. One bristlecone pine sprouted when Egyptians built the Great Sphinx, 4,725 years ago. The group also includes firs, spruces, yews, and junipers.

Ginkgos were diverse in dinosaur times. The only surviving species is the maidenhair tree, *Ginkgo biloba* (Figure 22.12b). Like a few gymnosperms, these plants are deciduous. Several thousand years ago, they were planted around temples in China. Natural populations nearly became extinct even though these trees seem hardy. Perhaps too many were cut down for firewood. Today, male ginkgo trees are widely planted. They have attractive, fan-shaped leaves and are resistant to insects, disease, and air pollutants. Female trees are not

Figure 22.13 (**a**) A cycad's seed-bearing cone. (**b**) *Ephedra.* Its branches are popular with florists. (**c**) *Welwitschia.*

favored. Their thick, fleshy seeds, which are the size of small plums, give off quite a stench.

The 185 or so existing **cycad** species produce pollen-bearing cones on "male" plants and seed-bearing cones on "female" ones; they rely on insects and air currents to transfer their pollen. Cycad leaves resemble a palm tree's (Figure 22.13a), but palms are flowering plants. Most cycads live in tropical and subtropical regions. Two species (*Zamia*) grow wild in Florida and are widely planted as ornamentals. Cycad seeds and a flour made from the trunks are edible after their toxic alkaloids are removed. Many species are vulnerable to extinction.

At present, there are three genera of woody plants called **gnetophytes**. *Gnetum* trees and leathery leafed vines thrive in the humid tropics. The shrubby *Ephedra* lives in California deserts and some other arid regions (Figure 22.13b). Photosynthesis occurs in its green stems. *Welwitschia mirabilis* grows in hot deserts of south and west Africa. Its sporophyte is mainly a deep-reaching taproot. Its exposed part, a woody, disk-shaped stem, has cones and one or two strap-shaped leaves that split lengthwise repeatedly as the plant ages (Figure 22.13c).

Conifers, cycads, ginkgo, and gnetophytes are groups of gymnosperms. Like their ancestors, they produce their seeds on exposed surfaces of cones and other types of spore-producing structures.

Figure 22.12 (**a**) Bristlecone pine (*Pinus longaeva*) growing high in the Sierra Nevada. (**b**) Ginkgo and (**c**) its fleshy-coated seeds.

ANGIOSPERMS—THE FLOWERING, SEED-BEARING PLANTS

Only angiosperms produce specialized reproductive structures called **flowers** (Figure 22.14). *Angeion*, which means vessel, refers to the female reproductive parts at the center of a flower. The enlarged base of the "vessel" is the floral ovary, where ovules and seeds develop.

Most flowering plants coevolved with **pollinators** —insects, bats, birds, and other animals that withdraw nectar or pollen from a flower and, in so doing, transfer pollen to its female reproductive parts. The recruitment of animals as assistants in reproduction probably has contributed greatly to the success of flowering plants, which have dominated the land for 100 million years.

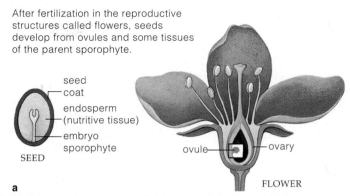

After fertilization in the reproductive structures called flowers, seeds develop from ovules and some tissues of the parent sporophyte.

seed
coat
endosperm
(nutritive tissue)
embryo
sporophyte
SEED

ovule — ovary
a
FLOWER

Figure 22.14 (**a**) Unique to angiosperms— the flower, a reproductive structure that functions in pollination and seed formation. Representing angiosperm diversity: (**b**) A passion flower giving up nectar to a pollinator, a hummingbird. (**c**) A parasitic plant, dwarf mistletoe (*Arceuthobium*). It hurts the growth of forest trees in the western United States. (**d**) Indian pipe (*Monotropa uniflora*). This rare nonphotosynthetic species withdraws nutrients from mycorrhizae on the roots of photosynthetic plants. (**e**) A water lily (*Nymphaea*), one of the few aquatic plants.

At least 260,000 species thrive in a variety of habitats. They range in size from tiny duckweeds (a millimeter or so long) to towering *Eucalyptus* trees (some are more than 100 meters tall). A few species, including mistletoes and Indian pipe, aren't even photosynthetic. They withdraw nutrients directly from other plants or from mycorrhizae.

There are two classes of flowering plants, called the **dicots** and **monocots** (more formally, the Dicotyledonae and Monocotyledonae). Among the 180,000 dicots are most herbaceous (nonwoody) plants, such as cabbages and daisies; most flowering shrubs and trees, such as maples and apple trees; water lilies; and cacti. Among 80,000 or so species of monocots are the orchids, palms, lilies, and grasses, including rye, sugarcane, corn, rice, and wheat, as well as many other highly valued crop plants (Appendix I).

The next unit deals with the structure and function of flowering plants. For now, simply start thinking about how a large sporophyte dominates the life cycles. It retains and nourishes gametophytes; its sperm are dispersed within pollen grains. Endosperm, a nutritive tissue, surrounds embryo sporophytes inside the seeds of flowering plants. As the seeds develop, the ovaries (along with other structures) mature into fruits. Fruits protect and help disperse embryos. Figure 22.15, in the next section, is an overview of these life cycle events.

Angiosperms are the most successful plants, in terms of diversity, numbers, and distribution. They alone produce flowers. Most species coevolved with animal pollinators.

VISUAL OVERVIEW OF FLOWERING PLANT LIFE CYCLES

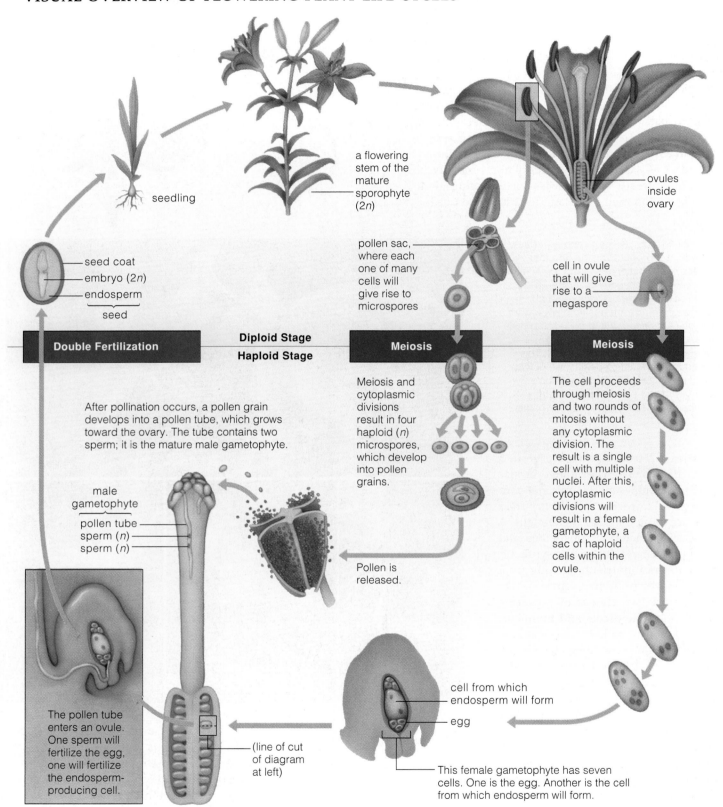

seedling

a flowering stem of the mature sporophyte (2n)

ovules inside ovary

seed coat
embryo (2n)
endosperm
seed

pollen sac, where each one of many cells will give rise to microspores

cell in ovule that will give rise to a megaspore

Double Fertilization

Diploid Stage

Meiosis

Meiosis

Haploid Stage

After pollination occurs, a pollen grain develops into a pollen tube, which grows toward the ovary. The tube contains two sperm; it is the mature male gametophyte.

Meiosis and cytoplasmic divisions result in four haploid (n) microspores, which develop into pollen grains.

The cell proceeds through meiosis and two rounds of mitosis without any cytoplasmic division. The result is a single cell with multiple nuclei. After this, cytoplasmic divisions will result in a female gametophyte, a sac of haploid cells within the ovule.

male gametophyte

pollen tube
sperm (n)
sperm (n)

Pollen is released.

The pollen tube enters an ovule. One sperm will fertilize the egg, one will fertilize the endosperm-producing cell.

(line of cut of diagram at left)

cell from which endosperm will form

egg

This female gametophyte has seven cells. One is the egg. Another is the cell from which endosperm will form.

Figure 22.15 Representative flowering plant life cycle. This example is for a lily (*Lilium*), one of the monocots. "Double" fertilization is a distinctive feature of flowering plant life cycles. A male gametophyte delivers two sperm to an ovule. One sperm fertilizes the egg, and the other fertilizes a cell that gives rise to endosperm, a tissue that will nourish the forthcoming embryo. Section 28.3 provides a closer look at flowering plant life cycles, using a dicot as the example.

Seed Plants and People

Which plants provide taste thrills and which kill? Starting with trials and errors of the earliest human species, we've acquired intimate knowledge of plants. By 300,000 years ago, *Homo erectus* clans in China were stashing pine nuts, walnuts, hazelnuts, and rose hips in caves and roasting seeds. At least by then, grown-ups were teaching their children which plants are edible and which are toxic. Through language, youngsters learned about the plants that had evolved in their parts of the world.

About 11,000 years ago, people started to domesticate wheat, barley, and other plants, this being a way to count on having reliable quantities of food. Of an estimated 3,000 species that different populations recognized as food, only about 200 became *the* major crops (Figure 22.16).

Plant lore still threads through our lives in other ways. We learned to use fast-growing, soft-wooded conifers for lumber and paper—and slow-growing, hard-wooded flowering plants, such as cherry, maple, and mahogany, for fine furniture. We make twine and rope from century plant leaves (*Agave*), cords and textiles from the leaf fibers of Manila hemp, and thatched roofs from grasses and palm fronds. Insecticides derived from Mexican cockroach plants kill cockroaches, fleas, lice, and flies. Extracts of neem tree leaves kill nematodes, insects, and mites but not the natural predators of those common pests.

Flowers grace our homes and customs (Figure 22.17). Their oils impart scents to perfumes; oils of eucalyptus and camphor have medicinal uses. Digitalin extracted from foxglove (*Digitalis purpurea*) stabilizes the heartbeat and blood circulation. Juices from *Aloe vera* leaves soothe sun-damaged skin. Alkaloids from periwinkle leaves can even slow the growth of some cancer cells.

And have we, in all this time, learned to abuse plants, also? You bet. Cocaine (from coca leaves), tobacco, and marijuana come to mind. Can you name other examples?

Figure 22.17 A bride tossing her bouquet to single women who attended her wedding, a ritual that supposedly reveals who will be married next. (A sixty-eight-year-old optimist caught this one.)

Figure 22.16 A few of the prized seed plants. (**a**) Indonesians gathering tender shoots of tea plants, evergreen shrubs related to garden camellias. Plants growing on moist, cool hills have leaves with the best flavors. Only the end bud and two or three of the youngest leaves are picked for fine teas. (**b**) From the American Midwest, mechanized harvesting of a field of common bread wheat, *Triticum*. (**c**) Triticale, a popular hybrid grain; parental stocks are wheat and rye (*Secale*). It has wheat's high yield and rye's tolerance of harsh conditions. (**d**) From Hawaii, sugarcane (*Saccharum officinarum*). Sap extracted from its cut stems is boiled down to make sucrose crystals (table sugar) or syrups. (**e**) Fruits of *Theobroma cacao*. Each one holds up to forty seeds, which are processed into cocoa butter or chocolate essences. The average American who buys 8–10 pounds of chocolate per year might not be happy that a pathogenic fungus (*Monilia*) may be driving *T. cacao* to extinction.

SUMMARY   *Gold* indicates text section

1. Ancient green algae apparently gave rise to plants. Plants invaded land by 435 million years ago. Nearly all species are multicelled photoautotrophs. Table 22.1 summarizes and compares major phyla. *CI, 22.1*

2. Table 22.2 summarizes key trends in plant evolution as identified by comparisons among lineages: *22.1*

a. Structural adaptations to dry habitats, including stomata, a cuticle, and vascular tissues (xylem, phloem).

b. A shift from haploid to diploid dominance during the life cycle. Complex sporophytes evolved; they hold on to, nourish, and protect spores and gametophytes.

c. A shift to two spore types, not one (homospory to heterospory). For gymnosperms and flowering plants, this led to the evolution of pollen grains and seeds.

3. Nearly all vascular plants live on land. Their cuticle and stomata conserve water. Root systems mine soil for nutrients. The upright and branched growth patterns

---

Table 22.1 *Comparison of Major Plant Groups*

Nonvascular land plants. Fertilization requires free water. Haploid dominance. Cuticle, stomata present in some.

BRYOPHYTES	18,600 species. Moist, humid habitats.

Seedless vascular plants. Fertilization requires free water. Diploid dominance. Cuticle, stomata present.

WHISK FERNS	7 species, sporophytes with no obvious roots or leaves. *Psilotum*.
LYCOPHYTES	1,100 species with simple leaves. Mostly wet or shady habitats.
HORSETAILS	25 species of single genus. Swamps, disturbed habitats.
FERNS	12,000 species. Wet, humid habitats in mostly tropical, temperate regions.

Gymnosperms—vascular plants with "naked seeds." Free water not required for fertilization. Diploid dominance. Cuticle, stomata present.

CONIFERS	550 species, mostly evergreen, woody trees and shrubs having pollen- and seed-bearing cones. Widespread distribution.
CYCADS	185 slow-growing tropical, subtropical species.
GINKGO	1 species, a tree with fleshy-coated seeds.
GNETOPHYTES	70 species. Limited to some deserts, tropics.

Angiosperms—vascular plants with flowers and protected seeds. Free water not required for fertilization. Diploid dominance. Cuticle, stomata present.

FLOWERING PLANTS

Monocots	80,000 species. Floral parts often arranged in threes or in multiples of three; one seed leaf; parallel leaf veins common.
Dicots	At least 180,000 species. Floral parts often arranged in fours, fives, or multiples of these; two seed leaves; net-veined leaves common.

Table 22.2 *Evolutionary Trends Among Plants*

Bryophytes	Ferns	Gymnosperms	Angiosperms

Nonvascular ⟶ Vascular ─────────────⟶

Haploid dominance ⟶ Diploid dominance ─────────⟶

Spores of one type ⟶ Spores of two types ─────⟶

Motile gametes ─────⟶ Nonmotile gametes* ⟶

Seedless ─────⟶ Seeds ─────────⟶

* Require pollination by wind, insects, animals, etc.

of shoot systems intercept sunlight and carbon dioxide. Tissues enclose and protect spores and gametes. *22.1*

4. Mosses, liverworts, and hornworts are bryophytes. These nonvascular plants have no complex xylem and phloem. They require free water for fertilization. *22.3*

5. Whisk ferns, lycophytes, horsetails, and ferns are all seedless vascular plants. Their flagellated sperm require ample water to swim to the eggs. *22.4*

6. Gymnosperms and flowering plants (angiosperms) are seed-bearing vascular plants. They produce pollen grains (mature male gametophytes) that develop from microspores, and female gametophytes (with egg cells) that develop from megaspores. *22.5, 22.8, 22.9*

a. Megaspores form in ovules: female reproductive structures consisting of a female gametophyte, nutritive tissue, and a jacket of cell layers. Part of the outer jacket develops into a seed coat. A seed is a mature ovule.

b. The evolution of pollen grains freed these plants from dependence on water for fertilization. Their seeds are efficient means of dispersing new generations, even during hostile conditions. Pollen grains and seeds were key adaptations in the move to high and dry habitats.

7. Only angiosperms produce flowers. Most coevolved with pollinators, such as insects, that move the pollen grains to female reproductive parts. Their seeds contain a nutritive tissue (endosperm). Most seeds are encased in fruit, which helps disperse them. *22.8, 22.9*

Review Questions

1. List a few structural and reproductive modifications that helped plants invade and diversify in habitats on land. *22.1*

2. Does the haploid phase or diploid phase dominate the life cycles of most plants? *22.1*

3. Name representatives of the following groups of plants and then compare their main characteristics: (*refer to Table 22.1*)
   a. Bryophytes and seedless vascular plants *22.3, 22.4*
   b. Gymnosperms and angiosperms *22.5, 22.7–22.9*

4. Distinguish between:
   a. Root system and shoot system *22.1*
   b. Xylem and phloem *22.1*
   c. Sporophyte and gametophyte *22.1*
   d. Ovule and seed *22.1, 22.5*
   e. Microspore and megaspore *22.5*

Figure 22.18 From forests to urban housing developments—where many conifers end up.

Self-Quiz   ANSWERS IN APPENDIX III

1. Which of the following statements is *not* true?
   a. Monocots and dicots are two classes of angiosperms.
   b. Bryophytes are nonvascular plants.
   c. Lycophytes and angiosperms are both vascular plants.
   d. Gymnosperms are the simplest vascular plants.

2. Which does *not* apply to gymnosperms and angiosperms?
   a. vascular tissues   c. single spore type
   b. diploid dominance   d. all of the above

3. Of all land plants, bryophytes alone have independent _____ and attached, dependent _____ .
   a. sporophytes;   c. rhizoids; zygotes
      gametophytes   d. rhizoids; stalked
   b. gametophytes;   sporangia
      sporophytes

4. Whisk ferns, lycophytes, horsetails, and ferns are classified as _____ plants.
   a. multicelled aquatic   c. seedless vascular
   b. nonvascular seed   d. seed-bearing vascular

5. A seed is _____ .
   a. a female gametophyte   c. a mature pollen tube
   b. a mature ovule   d. an immature embryo

6. Match the terms appropriately.
   ____ gymnosperm   a. gamete-producing body
   ____ sporophyte   b. help control water loss
   ____ lycophyte   c. "naked" seeds
   ____ ovule   d. only plant that produces
   ____ bryophyte   flowers
   ____ gametophyte   e. spore-producing body
   ____ stomata   f. nonvascular land plant
   ____ angiosperm   g. seedless vascular plant
                h. seed originates from it

Critical Thinking

1. Figure 22.18 shows a forest in the Nahmint Valley of British Columbia, before and after logging. It also shows wood frames of homes that are in the process of being built. Reflect on these photographs and Figure 22.11. To stop the loggers, would you chain yourself to a tree in an old-growth forest scheduled for clear-cutting? If your answer is yes, would you also give up the chance of owning a wood-frame home (as most homes are, in developed countries)? What about forest products, including newspapers, toilet tissue, and fireplace wood?

2. With respect to question 1, multiply each of your answers by 6.1 billion (there are more than that many people today) and describe what might happen when, inevitably, we run out of trees. Also describe what you might consider to be some of the pros and cons of tree farms—say, of a single species of pine.

3. Elliot Meyerowitz of the California Institute of Technology has studied the genetic basis of flower formation in *Arabidopsis thaliana*. By inducing mutations in seeds of this small weed, he discovered three genes (call them *A*, *B*, and *C*) that interact in different parts of a developing flower. Gene interactions lead to the formation of different structures—sepals, petals, stamens, and carpels—from the same mass of undifferentiated tissue. From what you know of gene control, suggest ways in which the *A*, *B*, and *C* genes may be controlling flower development.

4. Genes nearly identical to the *A*, *B*, and *C* genes of *A. thaliana* also have been isolated from snapdragons and other flowering plants. Ancestors of these plants evolved by 150 million years ago. They quickly rose to dominance in nearly all land habitats.
   Review the general introduction to adaptive radiation in Section 17.4. Then speculate on how the spectacular and rapid adaptive radiation of flowering plants came about.

Selected Key Terms

angiosperm 22.1	hornwort 22.3	pollinator 22.8
bryophyte 22.1	horsetail 22.4	rhizome 22.4
coal 22.2	lignin 22.1	root system 22.1
conifer 22.7	liverwort 22.3	seed 22.1
cuticle 22.1	lycophyte 22.4	shoot system 22.1
cycad 22.7	megaspore 22.5	spore 22.1
deforestation 22.6	microspore 22.5	sporophyte 22.1
dicot 22.8	monocot 22.8	stoma (stomata) 22.1
fern 22.4	moss 22.3	strobilus 22.4
flower 22.8	ovule 22.5	vascular plant 22.1
gametophyte 22.1	peat bog 22.3	whisk fern 22.4
ginkgo 22.7	phloem 22.1	xylem 22.1
gnetophyte 22.7	pollen grain 22.1	
gymnosperm 22.1	pollination 22.5	

Readings

Gray, J., and W. Shear. September–October 1992. "Early Life on Land." *American Scientist* 80:444–456.

Moore, R., W. D. Clark, and D. S. Vodopich. 1998. *Botany.* Second edition. New York: WCB/McGraw Hill.

23

ANIMALS: THE INVERTEBRATES

Madeleine's Limbs

In August of 1994, about 900 million years after the first animals appeared on Earth, Madeleine made *her* entrance. As they are wont to do, her grandmothers and aunts made a quick count on the sly—arms, legs, ears, and eyes, two of each; fully formed mouth and nose—just to be sure these were present and accounted for.

One grandmother, having been too long in the company of biologists, experienced an epiphany as she witnessed Madeleine's birth. In that profound instant she sensed ancestral connections emerging from the distant past and, through her, into the future.

Madeleine's body plan did not emerge out of thin air. Thirty-five thousand years ago, people just like us were having children just like Madeleine. If we are interpreting the fossil record correctly, then five million years ago the offspring of individuals on the road to modern humans resembled her in some respects but not others. Sixty million years ago, primate ancestors of those individuals were giving birth precariously, up in the trees. Two hundred fifty million years ago, mammalian ancestors of those primates were giving birth—and so on back in time to the very first animals, which had no limbs or eyes or noses at all.

We have few clues to what the very first animals looked like. Yet one thing is clear. By the dawn of the Cambrian, they had given rise to all major groups of invertebrates —animals without backbones—even to Madeleine's backboned but limbless ancestors.

And what stories those Cambrian animals tell! One community flourished 530 million years ago in a submerged basin between a reef and the coast of an early continent. Sediments were protected from ocean currents and had piled up against the steep reef. About 500 feet below the surface, the water was oxygenated and clear. Small, well-developed animals lived in, on, and above the dimly lit, muddy sediments (Figure 23.1).

Like castles built from wet sand along a seashore, their living quarters were unstable. The animals were wiped out when part of the bank above them slumped abruptly. The underwater avalanche buried them in sediments, so scavengers could not remove all traces of the dead. Over time, muddy silt rained down. Pressure and chemical change transformed the sediments into finely stratified shale. And soft parts of the flattened animals became shimmering, mineralized films.

Sixty-five million years ago, part of the seafloor was plowing under the North American plate, and western Canada's mountain ranges were rising. By 1909, the fossils had traveled high into the eastern mountains of British Columbia. In that year a fossil hunter tripped on a chunk of shale, which split into thin, fine layers—and the Burgess Shale story came to light.

In this chapter and the next, you will compare body plans of different groups of animals. Such comparisons give insight into evolutionary relatedness and help us construct family trees, as in Figure 23.2. Don't assume that structurally simple animals of the most ancient lineages are somehow evolutionarily stunted or "primitive." As you will see, they, too, are exquisitely adapted to their environment.

Figure 23.1 Reconstruction of a few Cambrian animals known from fossils of the Burgess Shale, British Columbia.

CHORDATES

ECHINODERMS

ARTHROPODS

ANNELIDS

MOLLUSKS

COELOMATE
ANCESTRY

ROTIFERS

ROUNDWORMS

BILATERAL
ANCESTRY

FLATWORMS

RADIAL
ANCESTRY

CNIDARIANS

MULTICELLED
ANCESTRY

SPONGES

SINGLE-CELLED, PROTISTAN-LIKE ANCESTORS

Figure 23.2 A simple evolutionary tree diagram that shows presumed relationships among major groups of animals. Take a moment to study this diagram. We will use it repeatedly as a road map through our discussions of each group, starting with the invertebrates and ending, in the chapter to follow, with chordates —including young Madeleine (*above*).

As you poke through the branches of the animal family tree, keep the greater evolutionary story in mind. At each branch point, microevolutionary processes gave rise to workable changes in body plans. Madeleine's uniquely human traits, and yours, emerged through modification of certain traits that had evolved earlier in countless generations of vertebrates and, before them, in ancient invertebrate forms.

Key Concepts

1. All animals are multicelled, aerobic heterotrophs that ingest or parasitize other organisms. Nearly all kinds have tissues, organs, and organ systems, and most are motile during at least part of their life cycle. Animals reproduce sexually and often asexually, and their embryos develop through a series of continuous stages.

2. Animals originated late in the precambrian. More than 2 million modern species have been identified. Of these, more than 1,950,000 are invertebrates—animals without a backbone. Fewer than 50,000 species are vertebrates—animals with a backbone.

3. Comparisons of the body plans of existing animals, in conjunction with the fossil record, reveal that there were several trends in the evolution of certain lineages. The most revealing aspects of an animal's body plan are its type of symmetry, gut, and cavity (if any) between the gut and body wall; whether it has a distinct head end; and whether it is divided into a series of segments.

4. The placozoans and sponges are structurally simple animals with no body symmetry. Both are at the cellular level of body construction. Cnidarians and comb jellies show radial symmetry, and they are at the tissue level of body construction.

5. Flatworms, roundworms, rotifers, and nearly all other animals that are more complex than the cnidarians show bilateral symmetry. Complex animals have tissues, organs, and organ systems.

6. Not long after the flatworms evolved, divergences began that were the start of two major lineages. One evolutionary branching gave rise to mollusks, annelids, and arthropods. The other gave rise to echinoderms and chordates.

7. By biological measures, including diversity, sheer numbers, and distribution, the arthropods—especially insects—have been the most successful animal group.

OVERVIEW OF THE ANIMAL KINGDOM

General Characteristics of Animals

What, exactly, are **animals**? We can only define them by a list of characteristics, not with a sentence or two. *First*, animals are multicelled. In most cases their body cells form tissues that become arranged as organs and organ systems. The body cells of nearly all species have a diploid chromosome number. *Second*, all animals are heterotrophs that get carbon and energy by ingesting other organisms or by absorbing nutrients from them. *Third*, animals require oxygen for aerobic respiration. *Fourth*, animals reproduce sexually and, in many cases, asexually. *Fifth*, most animals are motile during at least part of the life cycle. *Sixth*, the life cycle includes stages of embryonic development. Briefly, mitotic cell divisions transform an animal zygote into a multicelled embryo. The embryonic cells give rise to primary tissue layers: **ectoderm**, **endoderm**, and, in most species, **mesoderm**. The layers in turn give rise to all tissues and organs of the adult, as described in Section 29.6 and Chapter 39.

Diversity in Body Plans

Mammals, birds, reptiles, amphibians, and fishes are the most familiar animals. All are **vertebrates**, the only animals with a "backbone." And yet, of probably more than 2 million species of animals, fewer than 50,000 are vertebrates! What we call the **invertebrates** are animals with plenty of diverse features, but not a backbone.

Animals are grouped into more than thirty phyla. Table 23.1 lists the groups we describe in this book. The characteristics they share with one another arose early in time, before divergences from a common ancestor gave rise to separate lineages. Later, as morphological differences accumulated among them, the lineages took off in amazingly diverse directions. How might we get a conceptual handle on their modern-day descendants— on animals as different as flatworms, hummingbirds, spiders, toads, humans, and giraffes? We can compare their similarities and differences with respect to five basic features. These are body symmetry, cephalization, type of gut, type of body cavity, and segmentation.

BODY SYMMETRY AND CEPHALIZATION   With very few exceptions, animals are radial or bilateral. Those with **radial symmetry** have body parts arranged regularly around a central axis, like spokes of a bike wheel. Thus a cut down the center of a hydra (Figure 23.3*a*) divides it into equal halves; another cut at right angles to the first divides it into equal quarters. Radial animals live

Table 23.1   *Animal Phyla Described in This Book*

Phylum	Some Representatives	Existing Species
PLACOZOA *Trichoplax*	Simplest animal; plate-like but with layers of cells	1
PORIFERA Poriferans	Sponges	8,000
CNIDARIA Cnidarians	Hydrozoans, jellyfishes, corals, sea anemones	11,000
PLATYHELMINTHES Flatworms	Turbellarians, flukes, tapeworms	15,000
NEMATODA Roundworms	Pinworms, hookworms	20,000
ROTIFERA Rotifers	Tiny body with crown of cilia, great internal complexity	2,000
MOLLUSCA Mollusks	Snails, slugs, clams, squids, octopuses	110,000
ANNELIDA Segmented worms	Leeches, earthworms, polychaetes	15,000
ARTHROPODA Arthropods	Crustaceans, spiders, insects	1,000,000+
ECHINODERMATA Echinoderms	Sea stars, sea urchins	6,000
CHORDATA Chordates	Invertebrate chordates: Tunicates, lancelets	2,100
	Vertebrates: Fishes	21,000
	Amphibians	4,900
	Reptiles	7,000
	Birds	8,600
	Mammals	4,500

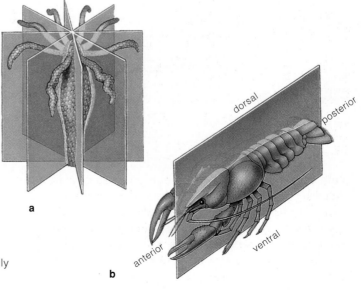

a

b

Figure 23.3   Examples of body symmetry. (**a**) A hydra, which is radially symmetrical, and (**b**) a bilaterally symmetrical crayfish.

Figure 23.4 Type of body cavity (if any) in animals.

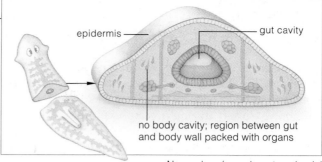

epidermis — gut cavity

no body cavity; region between gut and body wall packed with organs

a No coelom (*acoelomate* animals)

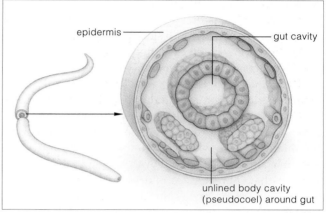

epidermis — gut cavity

unlined body cavity (pseudocoel) around gut

b Pseudocoel (*pseudocoelomate* animals)

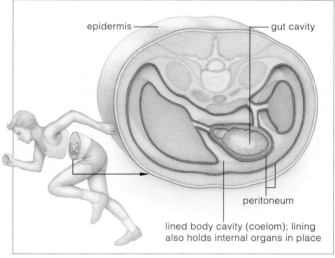

epidermis — gut cavity

peritoneum

lined body cavity (coelom); lining also holds internal organs in place

c Coelom (*coelomate* animals)

in water. Their body plan is adapted to intercepting food that is coming toward them from any direction.

Animals with **bilateral symmetry** have a right half and left half that are mirror images of each other. Most have an *anterior* end (head) and an opposite, *posterior* end. They have a *dorsal* surface (a back) and an opposite, *ventral* surface (Figure 23.3*b*). As fossils show, this body plan evolved among the first forward-creeping species. Their forward end encountered food and other stimuli first, so there was selection for **cephalization**. By this evolutionary process, sensory structures and nerve cells became more concentrated in the head of many animals. Bilateral body plans and cephalization evolved jointly, in that heritable changes in one affected changes in the other. Over time, pairs of muscles and pairs of sensory structures, nerves, and brain regions were the outcome.

TYPE OF GUT The **gut**, a sac projecting into the body or part of a tube through it, is where food is digested, then absorbed into the internal environment. A saclike gut has a single opening (mouth) for taking in food and disposing of residues. A tubular gut is part of a digestive system that has openings at two ends (mouth and anus). Such "complete" systems have specialized regions that break down, digest, absorb, store, and get rid of material. As more efficient digestive systems evolved, they helped make possible increases in body size and activity.

BODY CAVITIES In between the gut and the body wall of most bilateral animals is a body cavity (Figure 23.4). One type of cavity, a **coelom**, has a unique tissue lining called a peritoneum. This lining also encloses organs in the coelom and helps hold them in place. For example, your body has a coelom, and a sheetlike muscle called a diaphragm divides it into two smaller cavities. Your heart and lungs are positioned in the upper (thoracic) cavity, and your stomach, intestines, and other organs occupy the lower (abdominal) cavity.

The coelom turned out to be a key innovation in the evolution of larger, more complex animals from small ancestral forms. It favored increases in size and activity by cushioning and protecting internal organs.

Some of the less complex invertebrates don't even have a body cavity; tissues fill the region between their gut and body wall. Others have a pseudocoel ("false coelom")—a body cavity, but no peritoneum.

SEGMENTATION Segmented animals have a repeating series of body units that may or may not be similar to one another. The many segments of earthworms have a similar outward appearance. Insect segments are fused into three units (head, thorax, and abdomen) and differ greatly from one another. Especially among the insects, diverse head parts, legs, wings, and other appendages evolved from less specialized segments.

Animals are multicelled, aerobically respiring heterotrophs that ingest other organisms or absorb nutrients from them. Nearly all animals have tissues, organs, and organ systems. Most have a diploid chromosome number.

Animals reproduce sexually and, in many cases, asexually. They go through a period of embryonic development, and most are motile during at least part of the life cycle.

Animal body plans differ in body symmetry, cephalization, type of gut, type of body cavity, and segmentation.

PUZZLES ABOUT ORIGINS

Judging from recent genetic evidence and radiometric dating of fossilized tracks, burrows, and microscopic embryos, animals originated between 1.2 billion and 670 million years ago, during precambrian times. *Where did they come from?* They probably evolved from protistan lineages, but we don't know which ones (Section 19.4).

By one hypothesis, the forerunners of animals were ciliates, like *Paramecium*, and had multiple nuclei in a one-celled body. As they evolved, each nucleus became compartmentalized in individual cells of a multicelled body. However, we don't know of any existing animal that develops by compartmentalization.

By another hypothesis, multicelled animals arose from spherical colonies of a number of flagellated cells, maybe like the *Volvox* colony shown in Section 20.14. In time, as a result of mutation, some cells in the colony became modified in ways that enhanced reproduction and other specialized tasks. And so began the division of labor that characterizes multicellularity.

Suppose colonies became flattened over time and started creeping over the seafloor. A creeping life-style could have favored the evolution of layers of cells, such as those of *Trichoplax adhaerens*. This is the only known **placozoan** (Placozoa, after *plax*, meaning plate; and *zoon*, meaning animal). *T. adhaerens* is a soft-bodied marine animal shaped a bit like pita bread. It has no symmetry and no mouth. It has several thousand cells organized in two layers. It briefly humps up when its body glides over food (Figure 23.5). Gland cells in the lower layer secrete digestive enzymes onto the food, and individual cells absorb breakdown products. Reproduction may be asexual (by budding or fission) or sexual, by means not yet understood. In sum, structurally and functionally, *Trichoplax* is as simple as animals get.

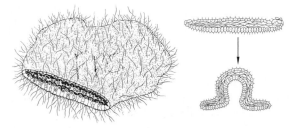

Figure 23.5 Cutaway views of *Trichoplax adhaerens*, an animal with a two-layer body measuring about three millimeters across.

Possibly the question of origins requires more than one answer. It may be that different lineages descended from more than one group of protistan-like ancestors.

Multicelled animals arose from protistan-like ancestors that may have resembled ciliates, colonial flagellates, or both.

SPONGES—SUCCESS IN SIMPLICITY

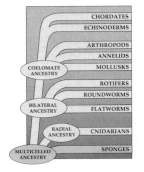

CHORDATES
ECHINODERMS
ARTHROPODS
ANNELIDS
COELOMATE ANCESTRY
MOLLUSKS
ROTIFERS
ROUNDWORMS
BILATERAL ANCESTRY
FLATWORMS
RADIAL ANCESTRY
CNIDARIANS
MULTICELLED ANCESTRY
SPONGES

Sponges (Porifera) are animals with no symmetry, no tissues, and no organs. Even so, they are one of nature's success stories. Sponges have been abundant ever since the precambrian, particularly in coral reefs and coastal waters. Of 8,000 or so known species, only about 100 live in fresh water. Sponges are home to tiny worms, shrimps, and many other animals. Some species are big enough for you to sit in; others are as small as a fingernail. Their shapes are flattened, sprawling, lobed, compact, tubelike, and vaselike (Figures 23.6 and 23.7).

The sponge body has no symmetry. Flattened cells line its outer surface and inner cavities. But the linings aren't much more specialized than those cell layers of *Trichoplax*, and they also differ from the tissues of other animals. Amoeboid cells live in a gelatin-like substance between the two linings (Figure 23.7*b*).

Typically, tough protein fibers and sharp, glasslike spicules of calcium carbonate or silica stiffen the body. These skeletal elements may be a reason why sponges as a group have endured so long. Cleveland Hickman put it this way: Most potential predators discover that sampling any sponge is about as pleasant as eating a mouthful of glass splinters embedded in fibrous gelatin. Besides, chemically speaking, many sponges stink.

Water flows into a sponge's body through many microscopic pores and chambers, then out through one or more large openings. It does so when thousands or millions of phagocytic **collar cells** beat their flagella. The cells are part of the inner lining (Figure 23.7*c*). The "collars" are food-trapping structures. Bacteria and other edibles get trapped in them, then are engulfed. Some of the food is transferred to the amoebalike cells for further breakdown, storage, and distribution.

Figure 23.6 A sprawling, red-orange sponge, one of many types that encrust underwater ledges in temperate seas.

Figure 23.7 (**a**, **b**) Body plan of a simple sponge. The outer lining consists of flattened cells. It also has some contractile cells arranged around the large opening at the top. Their slow contraction directs water flow through the body. In the gelatin-like matrix between the inner and outer linings, amoeba-like cells secrete the materials for spicules and fibers. Other amoeboid cells digest and transport food. They don't lose the capacity to divide, and their cellular descendants can differentiate into any other type of sponge cell. They also have roles in asexual processes, such as gemmule formation.

(**c**) Many flagellated, phagocytic cells line the body's inner canals and chambers. Each has a collar of food-trapping structures called microvilli. Fine filaments connect microvilli to each other to form a "sieve" that strains food particles from the water. Cells at the base of the collar engulf the trapped food.

(**d**) A marine sponge, Venus's flower basket (*Euplectella*). Its silica spicules are fused in a rigid network. A thin layer of cells stretches over them. At the base of the body is an anchoring tuft of spicules. (**e**) Basket sponge releasing a cloud of sperm into the water.

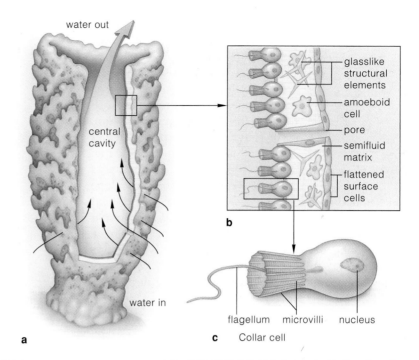

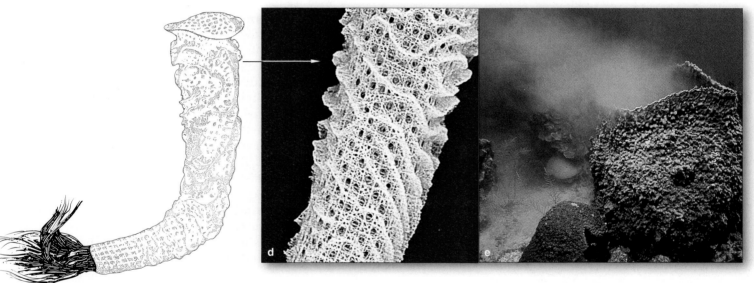

When most sponges reproduce sexually, they simply release sperm into the water (Figure 23.7e). But they retain the fertilized eggs and the embryos as they start developing. Each new, young sponge proceeds through a microscopic, swimming larval stage. A **larva** (plural, larvae) is a sexually immature stage that grows and develops into an adult, the sexually mature form of the species. As you will see, the life cycle of many animal species includes larval stages.

Some kinds of sponges also reproduce asexually by fragmentation; small fragments break away from the parent and grow into new sponges. Most freshwater species also reproduce asexually by way of gemmules. These are clusters of sponge cells, some of which form a hard covering around others. The clusters inside are protected from extreme cold or drying out. Later, when favorable conditions return, the gemmules germinate and establish a new colony of sponges.

Sponges have no symmetry, tissues, or organs; they are at the cellular level of construction. Yet they have successfully endured through time, possibly because most predators find their spicule-rich and often stinky bodies unappetizing.

CNIDARIANS—TISSUES EMERGE

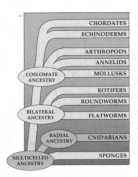

Cnidarians (Cnidaria) are radial, tentacled animals, most of which live in the seas. Jellyfishes belong to the group called scyphozoans. The anthozoans (including sea anemones) and hydrozoans (such as *Hydra*) also are in the phylum. Of roughly 11,000 species, fewer than 50 are found in fresh water. All cnidarians are at the tissue level of organization. They alone make **nematocysts**, capsules with tubular threads that can be discharged. Many threads entangle prey or trap it in a sticky substance oozing from their tip. Some have prey-piercing barbs; others have toxin-delivering tips that sometimes inflict painful stings (Figure 23.8). Hence the name Cnidaria, after the Greek word for nettle.

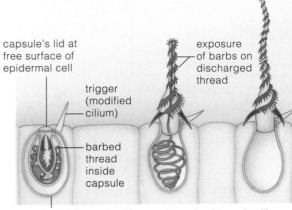

nematocyst (capsule at free surface of epidermal cell)

Figure 23.8 One type of nematocyst before and after a prey organism (not shown) touched its trigger. The contact made the capsule more "leaky" to water. Water diffused inward. Turgor pressure built up in the capsule and forced the thread to turn inside out. The thread tip swiftly pierced the prey's body.

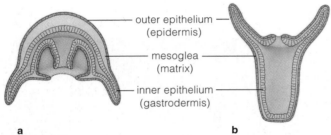

outer epithelium (epidermis)

mesoglea (matrix)

inner epithelium (gastrodermis)

a

b

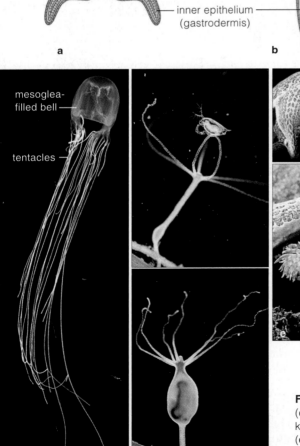

mesoglea-filled bell

tentacles

c

d

Cnidarian Body Plans and Life Cycles

The **medusa** (plural, medusae) and **polyp** are the most common body forms of cnidarians. Both have a saclike gut, as in Figure 23.9. Medusae float and look like bells or saucers. The mouth, centered under the bell, often has tentacles and other extensions used in prey capture and feeding. The tubelike polyps have a tentacle-ringed mouth at one end; the other end attaches to a substrate.

Trichoplax draped over and digesting food is like a gut on the run. But the cnidarian gut is a permanent food-processing chamber. Its sheetlike inner lining, the gastrodermis, has glandular cells that secrete digestive enzymes. Epidermis lines the body surfaces (Figure 23.9). Each lining is an **epithelium** (plural, epithelia), a tissue with one free surface bathed in some body fluid or exposed to the outside. All animals more complex than sponges have this tissue. In cnidarians, it incorporates a "nerve net" of interacting **nerve cells**. These receive signals from receptors, then signal **contractile cells** to carry out responses. Such cells can shorten and go back to their original length. The nerve net, a simple nervous system, only controls movement and shape changes.

Between the outer and inner linings is a layer of gelatinous secreted material called mesoglea ("middle jelly"). The volume of mesoglea in jellyfishes imparts buoyancy and serves as a firm yet deformable skeleton

Figure 23.9 The common cnidarian body plans: (**a**) medusa and (**b**) polyp, midsection. (**c**) Sea wasp (*Chironix*), a jellyfish with tentacles up to fifteen meters long. Its toxin can kill a human within minutes. (**d**) Hydrozoan polyp (*Hydra*) capturing and digesting its prey. (**e**) Sea anemone using its hydrostatic skeleton to escape from a sea star. It closes its mouth, and contractile cells in epithelial tissues generate force against water inside the gut. The body changes shape and thrashes about, generally away from the predator.

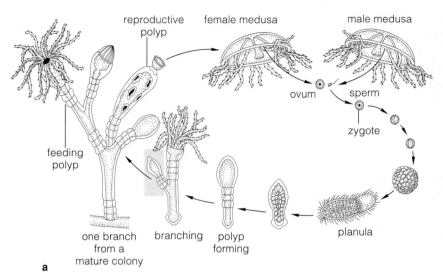

reproductive polyp

female medusa

male medusa

feeding polyp

one branch from a mature colony

branching

polyp forming

ovum

sperm

zygote

planula

a

gas-filled medusa

tentacle of one polyp

b

c

three polyps (reproductive, feeding, and prey-capturing) attached one after the other

interconnected skeletons of polyps of a colonial coral

Figure 23.10 (**a**) Life cycle of a hydrozoan (*Obelia*) that includes medusa and polyp stages. Established colonies have thousands of feeding polyps. (**b**) Portuguese man-of-war (*Physalia*). (**c**) A colonial coral.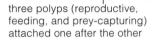

against which contractile cells act. Visualize many such cells in a jellyfish bell. Their contraction narrows the bell, forcing water to jet out from beneath it. The bell returns to its resting position, cells contract again, and the jellyfish moves. What about polyps, most of which have little mesoglea? Contractile cells act against water in their gut cavity. Any fluid-filled cavity or cell mass against which contractile cells can act is a **hydrostatic skeleton**. The cavity's volume or mass doesn't change; it is simply shunted about to change the body's shape (Figure 23.9*e*). As you will read in Chapter 33, nearly all animals have some type of skeletal system against which the force of contraction is applied.

Medusae, polyps, or both grow and develop during the life cycle of different cnidarians. Consider *Obelia* (Figure 23.10*a*). Its medusa is a sexual stage. Embedded in the medusa's epithelium are **gonads**, the primary (gamete-producing) reproductive organs of animals. *Obelia* gonads simply rupture to release gametes. The zygotes develop into **planulas**, a type of swimming or creeping larva, usually with ciliated epidermal cells. In time, a mouth opens at one end of the planula, forming a polyp or medusa that starts the cycle again.

Examples of Cnidarian Diversity

Cnidarians that form colonies are fine representatives of this phylum's diversity. You may know about one of the colonial hydrozoans, the Portuguese man-of-war (*Physalia*). Its nematocyst toxin is dangerous to bathers, fishermen, and fish prey. *Physalia* lives mainly in warm waters, but currents also move it to the cooler Atlantic coastal waters of North America and Europe. A blue,

gas-filled float develops from the planula's body. It keeps the colony near the water's surface, where winds move it (Figure 23.10*b*). Beneath the float, groups of polyps and medusae interact as "teams" in specialized tasks, such as feeding, reproduction, and defense.

Another example: Colonial anthozoans such as reef-forming corals interconnect by their external skeletons (Figure 23.10*c*). In large part, reefs are the accumulated and compressed piles of these skeletons. Reef-forming corals do best in warm, clear seas, where they feed on nutrients stirred up by tides. Dinoflagellates, a type of photosynthetic protistan, are mutualists with the corals. They supply their hosts with oxygen, recycle mineral wastes, and adjust the pH of seawater in a way that helps calcium deposits build up their host's skeleton. In return, corals give the photosynthesizers a relatively safe habitat exposed to the sun's rays, dissolved carbon dioxide, and minerals. Their interaction cycles scarce nutrients quickly, directly, and efficiently.

If you like to snorkel, you won't come across coral reefs in shallow, still water, which can get too hot for corals to survive. Also, corals can go into osmotic shock after heavy rains freshen seawater. That is one reason why you won't come across reefs where a river drains into the sea or near rainfall-drenched land.

Cnidarians are radial animals equipped with tentacles, a saclike gut, epithelia, a simple nerve net, and a hydrostatic skeleton. They alone produce nematocysts.

Cnidarians are at the tissue level of construction; compared with sponges, they have layers of cells interacting in more coordinated fashion to carry out specific tasks.

ACOELOMATE ANIMALS—AND THE SIMPLEST ORGAN SYSTEMS

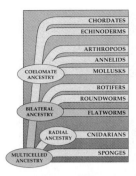

Moving beyond the cnidarians in our survey, we find animals that range from flatworms to humans. Most are bilateral, and all contain organs. Each **organ** is a structural unit of two or more tissues that are arrayed in a specific pattern and that perform a common task. Most organs interact with others. By definition, an **organ system** is composed of two or more organs interacting chemically, physically, or both in ways that contribute to the survival of the whole organism.

Flatworms are the simplest animals at the organ-system level of construction. As you will see, a few of the parasitic types have the capacity to make our lives miserable. Parasites, recall, reside in or on living hosts and feed on their tissues. Most do not kill their hosts, at least not until after they have reproduced. The *definitive* host harbors the mature stage of a parasite's life cycle. One or more *intermediate* hosts harbor immature stages.

Characteristics of Flatworms

Among the 15,000 or so known species of **flatworms** (phylum Platyhelminthes) are turbellarians, flukes, and tapeworms. Most of these bilateral, cephalized animals have simple organ systems inside a flattened body, as in Figure 23.11. The digestive system is only a pharynx and a saclike gut, often with branchings. A **pharynx** is just a muscular tube (which flatworms use for feeding). Species differ in their reproductive systems. Most often, individuals are **hermaphrodites**, with female *and* male gonads. Each has a penis (a sperm-delivery structure), so two flatworms can exchange sperm and reproduce sexually. Secretions from certain flatworm glands form a protective capsule around the fertilized eggs.

Classes of Flatworms

TURBELLARIANS Most turbellarians (class Turbellaria) live in the seas; only planarians and a few others live in fresh water. Some eat tiny animals or suck juices from dead or wounded ones. A planarian is able to reproduce asexually by *transverse* fission. First it divides in half at its midsection, then each half regenerates missing parts. Like you, a planarian adjusts the composition and the volume of its body fluids. Its water-regulating system has one or more branched tubes called protonephridia (singular, protonephridium). These tubes extend from pores at the body surface to bulb-shaped flame cells in tissues. When excess water diffuses into the flame cells, a tuft of cilia "flickering" in the bulb drives the water through the tubes to the outside (Figure 23.11b).

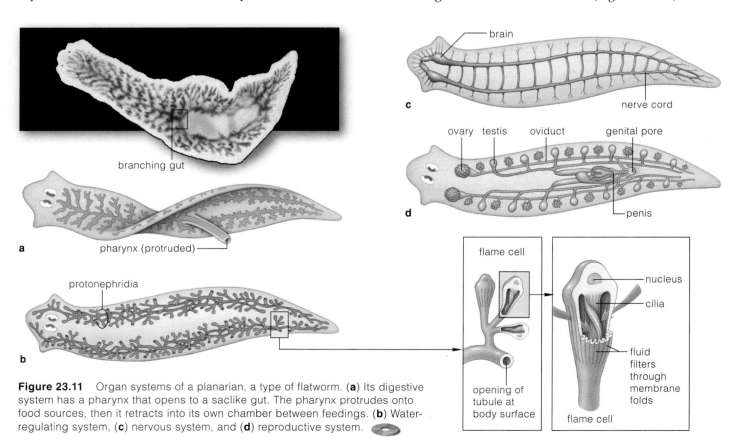

Figure 23.11 Organ systems of a planarian, a type of flatworm. (**a**) Its digestive system has a pharynx that opens to a saclike gut. The pharynx protrudes onto food sources, then it retracts into its own chamber between feedings. (**b**) Water-regulating system, (**c**) nervous system, and (**d**) reproductive system.

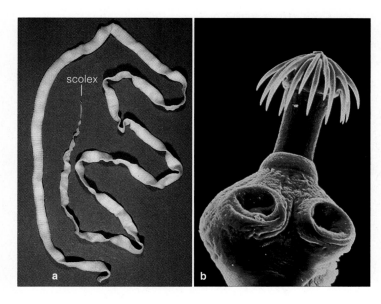

Figure 23.12 (**a**) Scolex of a tapeworm that uses a shorebird as its primary host. (**b**) Sheep tapeworm.

FLUKES The flukes (class Trematoda) are parasitic worms. Their life cycle has sexual and many asexual phases, and one to four hosts. Almost always, a snail or a clam is the initial host for larval or juvenile stages, and some vertebrate is the definitive host. Section 23.7 provides a detailed look at a blood fluke (*Schistosoma*). Each year, schistosomes parasitize 200 million of us.

TAPEWORMS Tapeworms (class Cestoda) are parasites of vertebrate intestines. Ancestral tapeworms probably had a gut, then lost it when they evolved in intestines, which are habitats rich in predigested food. Existing descendants attach to the intestinal wall with a scolex, a structure with suckers, hooks, or both (Figure 23.12).

Proglottids bud behind a scolex; each is a new unit of the tapeworm body. The units are hermaphroditic; they mate and transfer sperm to one another. The older proglottids (the ones farthest from the scolex) store the fertilized eggs. They break off from younger ones, then leave the body in feces. Later on, the eggs may meet an intermediate host. Section 23.7 shows how proglottids form in the life cycle of a representative tapeworm.

In some respects, the simplest turbellarians, larval flukes, and larval tapeworms resemble the planulas of cnidarians. The resemblance inspires speculation that bilateral animals evolved from planula-like ancestors. They may have done so by increased cephalization and the development of tissues derived from mesoderm.

Flatworms are among the simplest bilateral, cephalized, acoelomate animals with organ systems.

ROUNDWORMS

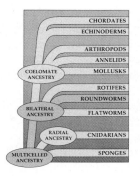

Roundworms (Nematoda) thrive nearly everywhere. They may be the most abundant animals alive. Shallow saltwater or freshwater sediments typically hold a million roundworms per square meter. In a handful of rich topsoil, maybe thousands of the scavenging types make fast work of rotting plant parts and dead earthworms. We have identified 20,000 species, but there may be a hundred times more.

A roundworm is bilateral, yet its body is cylindrical. It's usually tapered at both ends, and a protective cuticle covers it. Animal **cuticles** are tough and often flexible body coverings. A roundworm is the simplest animal equipped with a complete digestive system. Between the gut and body wall is a false coelom filled mainly with reproductive organs. Cells in all tissues absorb nutrients from the coelomic fluid and give up wastes to it.

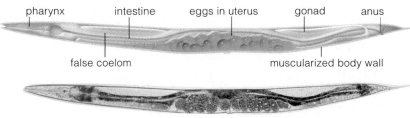

pharynx   intestine   eggs in uterus   gonad   anus

false coelom   muscularized body wall

Figure 23.13 Body plan and micrograph of the roundworm *Caenorhabditis elegans*. More complex animals have the same basic tissues and body plan (cephalized, bilateral, complete digestive system). Some hermaphroditic individuals self-fertilize and make clones of themselves, which is very useful for genetic studies. So are the worm's short generation time, transparency, small size (adults are about a millimeter long), and rather small genome (19,900 genes; only 3,000 are vital for normal structure and function). Researchers have deciphered each cell's lineage with electron microscopy and serial sections at different times of development, from egg to adult.

Parasitic species often do great harm to humans, cats, dogs, cows, sheep, soybeans, potatoes, and other hosts. But most roundworms are free-living and either harmless or beneficial, as when their populations cycle nutrients for communities. One species, *Caenorhabditis elegans*, is used in studies of inheritance, development, and aging (Figure 23.13). Its genome was the first of any multicelled organism to be fully sequenced.

Roundworms are cylindrical, bilateral, cephalized animals with a false coelom and a complete digestive system. Most cycle nutrients in communities; many are parasites.

A Rogue's Gallery of Worms

A number of parasitic flatworms and roundworms call the human body home. Consider the Southeast Asian blood fluke *Schistosoma japonicum*, which causes *schistosomiasis*. It requires a human definitive host, standing water for its swimming larvae, and an aquatic snail as an intermediate host. Figure 23.14*a* shows its life cycle. In infected humans, white blood cells that defend the body attack the masses of fluke eggs. Masses of grainy debris form in tissues. In time, the liver, spleen, bladder, and kidneys deteriorate.

Some tapeworms, too, parasitize humans. Different species use pigs, freshwater fish, or cattle as intermediate hosts. Humans get infected by eating raw, improperly pickled, or insufficiently cooked pork, fish, or beef that contains tapeworm larvae (Figure 23.15).

One parasitic roundworm (guinea worm) causes thin, serpentlike ridges in human skin. For several thousand years, healers have extracted the "serpents" from the body, painfully, by slowly winding them out around a stick. The roundworms called pinworms and hookworms cause other problems. *Enterobius vermicularis*, a pinworm of temperate regions, parasitizes humans. It lives in the large intestine, but at night the centimeter-long females migrate to the anal region of the host and lay eggs. Their presence causes itching, and scratchings made in response transfer some eggs to hands, then to other objects. Newly laid eggs contain embryonic pinworms. In less than a few hours, they develop into juveniles and are ready to hatch if another human inadvertently ingests them.

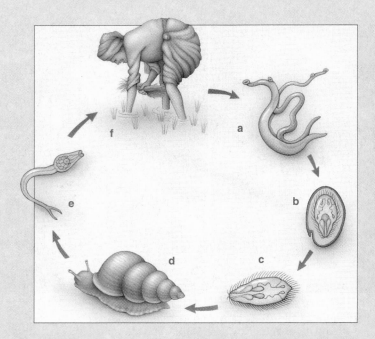

Figure 23.14 Life cycle of *Schistosoma japonicum*. (**a**) This dangerous blood fluke grows, matures, and mates in human hosts. (**b,c**) Fertilized eggs leave the body in feces and then hatch into ciliated, swimming larvae. (**d**) The larvae burrow into an aquatic snail and multiply asexually. (**e,f**) Fork-tailed larvae develop next. These leave the snail and swim about until they bore into human skin. They migrate to thin-walled intestinal veins, then the cycle begins anew.

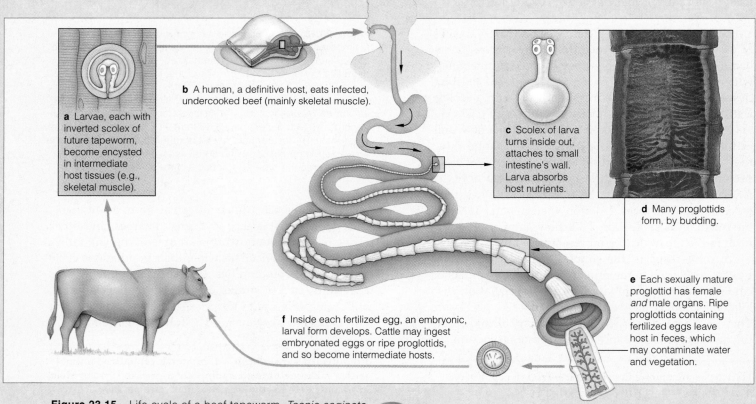

a Larvae, each with inverted scolex of future tapeworm, become encysted in intermediate host tissues (e.g., skeletal muscle).

b A human, a definitive host, eats infected, undercooked beef (mainly skeletal muscle).

c Scolex of larva turns inside out, attaches to small intestine's wall. Larva absorbs host nutrients.

d Many proglottids form, by budding.

e Each sexually mature proglottid has female *and* male organs. Ripe proglottids containing fertilized eggs leave host in feces, which may contaminate water and vegetation.

f Inside each fertilized egg, an embryonic, larval form develops. Cattle may ingest embryonated eggs or ripe proglottids, and so become intermediate hosts.

Figure 23.15 Life cycle of a beef tapeworm, *Taenia saginata*.

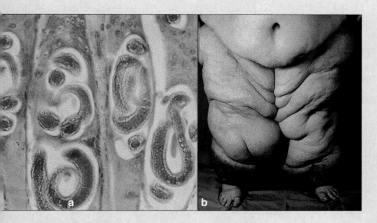

Figure 23.16 (**a**) Juveniles of the roundworm *Trichinella spiralis* in the muscle tissue of a host animal. (**b**) Severe edema in the legs of a woman who was infected by the roundworm *Wuchereria bancrofti*.

Adult hookworms live inside a host's small intestine. Using toothlike devices or sharp ridges around their mouth, they cut into the intestinal wall, feed on blood and other tissues, and so compete with their host for nutrients. Adult females, about a centimeter long, can release a thousand eggs daily. After leaving the body in feces, the eggs hatch into juveniles. When a juvenile hookworm contacts bare human skin, it cuts its way inside. It travels the bloodstream to the lungs, works its way into the air spaces, and moves up the windpipe. When the host swallows, the parasite is transported to the stomach and then to the small intestine. And there it may mature and live for several years.

The roundworm *Trichinella spiralis* causes painful and sometimes fatal symptoms. Adults live in the lining of the small intestine, where females release juveniles (Figure 23.16*a*). These work their way into blood vessels, then into muscles. There they become encysted; they secrete a covering around themselves and enter a resting stage. Humans become infected mainly by eating undercooked meat from pigs or certain game animals. It isn't easy to detect the encysted juveniles when inspecting fresh meat.

Figure 23.16*b* shows the results of prolonged, repeated infections by *Wuchereria bancrofti*, another roundworm. Adult worms become lodged in lymph nodes, organs that filter lymph (excess tissue fluid) that normally flows into the bloodstream. The worms obstruct lymph flow. When an obstruction causes fluid to back up and accumulate in tissues, the legs and other body regions may enlarge grossly. We call this condition *elephantiasis*.

A mosquito is *Wuchereria*'s intermediate host. Females of this parasite produce active young that travel about at night in the bloodstream. When a mosquito sucks blood from an infected person, the juveniles enter the insect's tissues. In time they move close to the insect's sucking device and enter a new host when it draws blood again.

TWO MAJOR DIVERGENCES

Late in the precambrian, bilateral animals not much more complex than modern flatworms evolved. Soon afterward, they gave rise to two lineages of coelomate animals (Figure 23.2). We call these two great lineages **protostomes** and **deuterostomes**. Mollusks, annelids, and arthropods are protostomes. The echinoderms and chordates are deuterostomes.

Different mutations accumulated in these separate lineages, and some influenced how an embryo develops from a fertilized egg. For instance, mitotic cell divisions cut an egg's cytoplasm along prescribed planes to form a tiny ball of cells, the early embryo (Section 39.3). In protostomes, the pattern is *spiral* cleavage: the earliest cuts occur at oblique angles relative to the main body axis. In deuterostomes, however, the pattern is *radial* cleavage: the earliest cuts are parallel with and also perpendicular to the original body axis:

Early protostome embryo. Its four cells are undergoing cleavages *oblique to* the original body axis:

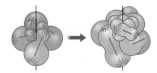

Early deuterostome embryo. Its four cells are undergoing cleavages *parallel with* and *perpendicular to* the original body axis:

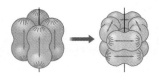

As other examples, the first opening that forms in an early protostome embryo becomes a mouth; an anus forms elsewhere. The first opening in a deuterostome embryo becomes the anus and the second becomes the mouth. Also, a protostome's coelom forms from spaces that open in mesoderm, but a deuterostome's coelom forms from outpouchings of the gut wall:

How a coelom forms in a protostome embryo: pouch will form mesoderm around coelom — developing gut —

How a coelom forms in a deuterostome embryo: solid mass of mesoderm — developing gut —   coelom

Such modifications to the embryonic stages of the two kinds of animals led to major differences in body plans.

Soon after coelomate animals evolved in Cambrian times, two great lineages—the protostomes and deuterostomes—evolved through mutations that affected how their embryos develop, and this led to major differences in body plans.

A SAMPLING OF MOLLUSKS

Molluscan Diversity

Few have trouble recognizing a snail. Yet few of us know much about its 110,000 relatives in one of the largest phyla (Mollusca). A **mollusk** is a bilateral animal with a small coelom and a fleshy soft body. Most have a shell of calcium carbonate and proteins secreted from the cells of a tissue that drapes like a skirt over the body mass. This tissue, the **mantle**, is unique to mollusks. Their organs of gas exchange are a type of gill that has thin-walled leaflets. Most mollusks have a fleshy foot. Many have a radula, a tonguelike, toothed organ that rasps algae and other tidbits from substrates and draws it into the mouth. Mollusks with a well-developed head have eyes and tentacles, but not all have a head.

Beyond these generalizations we find tremendous diversity, ranging from tiny snails in treetops to giant predators of the seas. Here we sample four classes: the gastropods, chitons, bivalves, and cephalopods.

Gastropods ("belly foots") make up the largest class. The 90,000 species of snails and slugs got this name because their foot spreads out when they crawl. Many species have spirally coiled or conical shells. Coiling compacts the organs into a mass that can be balanced above the body, rather like a backpack. Other species have a reduced shell or none at all (Figure 23.17b,e).

Chitons are slow-moving or sedentary grazers with a dorsal shell divided into eight plates (Figure 23.17c). The bivalves—animals having a "two-valved shell"—include clams, scallops, oysters, and mussels (Figure 23.17d). Some bivalves are only a millimeter across. A few giant clams are over a meter across and weigh 225 kilograms (close to 500 pounds). Humans have eaten one type of bivalve or another since prehistoric times.

Cephalopods are highly active predators of the seas. This class includes the swiftest invertebrates (squids, Figure 23.17e), the largest known invertebrate (*Loligo*, the giant squid), and the smartest (octopuses). Show an octopus an object with a distinct shape and give it a mild electric shock, and it thereafter avoids that object. In terms of memory and the capacity to learn, octopuses and some squids are the most complex invertebrates.

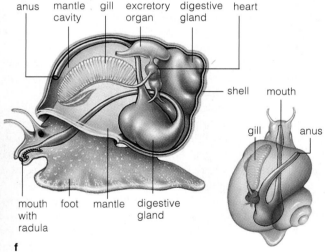

Figure 23.17 Gastropods, which creep, swim, or float as they graze on, prey upon, or parasitize other species. (**a**) Land snail. (**b**) Two sea slugs. (**c**) Chiton. With its broad foot, it creeps on and clings to rocks. (**d**) Scallop, one of the bivalves. Light-sensitive "eyes" (dark dots) fringe the two halves of its shell. (**e**) A diver and a large squid (*Dosidiscus*) inspecting each other. Being highly active, cephalopods demand lots of oxygen. They are the only mollusks with a closed circulatory system. Their main heart pumps blood to two gills, each with a booster (accessory) heart at its base that speeds blood flow. This arrangement enhances the uptake of oxygen (for muscle cells especially) and the removal of carbon dioxide. (**f**) Body plan of an aquatic snail.

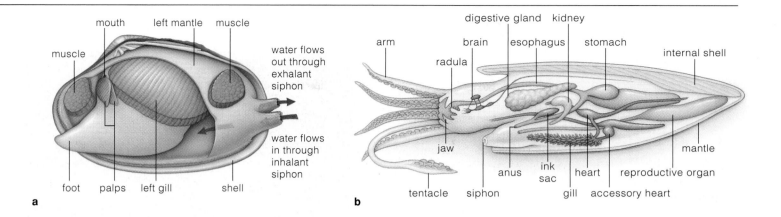

a

mouth left mantle muscle

muscle

water flows
out through
exhalant
siphon

water flows
in through
inhalant
siphon

foot palps left gill shell

b

digestive gland kidney

arm

brain esophagus stomach

internal shell

radula

jaw

anus ink sac heart reproductive organ

tentacle siphon gill accessory heart

mantle

Evolutionary Experiments With Body Plans

Maybe it was their soft, fleshy bodies—so forgiving of chance evolutionary changes in morphology—that gave the ancestors of mollusks the potential to diversify in so many ways and radiate into so many habitats. Let's explore this idea by looking at a few species.

TWISTING AND DETWISTING OF SOFT BODIES Notice, in Figure 23.17*f*, how evolution put an odd twist in the soft snail body. The anus dumps wastes near the mouth! As a gastropod embryo develops, a cavity between the mantle and shell twists 180° counterclockwise. So does nearly all of the visceral mass (the gut, heart, gills, and other internal organs). This process of *torsion* operates only in gastropods. It must have created an unpleasant sanitation problem in ancestral torsioned species.

We find evolutionary compensations for this state of affairs. Most gastropods now have enough cilia in this region to create currents that sweep wastes away. Sea slugs have even undergone detorsion; the soft body of a larva twists, then it untwists before adulthood.

HIDING OUT, ONE WAY OR ANOTHER If you were small, edible, and soft of body, an external shell would be a distinct advantage, as it is for chitons and clams. When a chiton is disturbed by predators or the surf or exposed by a receding tide, it hunkers under its shell. Foot muscles pull the body mass down, and the mantle's edge around the shell's rim presses like a suction cup against a rock. Pull it away, and the chiton rolls up into a ball until it can unroll and reattach itself elsewhere.

Besides hiding in a shell, protection also can be had by hiding in sand or some other substrate. A bivalve's head is not much to speak of. Its foot is usually large and specialized for burrowing. Bivalves that burrow in sand or mud have a pair of siphons: extensions of the mantle edges, fused into tubes (Figure 23.18*a*). Water is drawn into the mantle cavity through one siphon and leaves through the other, carrying wastes. One wonders what preyed on the ancestors of geoducks of the Pacific Northwest, which have siphons more than a meter long.

Figure 23.18 (**a**) Body plan of a clam, with half of its shell removed. The gills of nearly all bivalves collect food and serve in respiration. As water moves through the mantle cavity, mucus on the gills traps food. Cilia move the mucus and food to palps, appendages that sort things out before suitable bits are driven to the mouth. (**b**) Generalized body plan of a cuttlefish, one of the cephalopods. Its tentacles specialize in prey capture.

ON THE CEPHALOPOD NEED FOR SPEED Five hundred million years ago, cephalopods had buoyant, chambered shells and were the supreme predators of the Ordivician seas (Section 19.5). Yet, of a lineage that once had more than 7,000 species, all but one existing species has no shell or a reduced one. What happened to the shell? This evolutionary trend coincided with an adaptive radiation of bony fishes—which preyed on cephalopods or were strong competitors for the same prey.

During the long race for speed and wits, cephalopods lost their thick external shell. They became streamlined and highly active. Of all mollusks, they now have the largest brain relative to body size and show the most complex behavior. Nerves connect the brain to muscles that respond quickly to food or to danger. Circulation of blood and respiration are very efficient. And except for the chambered nautilus, cephalopods can discharge dark fluid from an ink sac, maybe to confuse predators.

Long ago, *jet propulsion* became the name of the game. Cephalopods force a jet of water out from their mantle cavity and a funnel-shaped siphon. As mantle muscles relax, water is drawn into the cavity. As they contract, a water jet is squeezed out. When the mantle's free edge closes down on the head at the same time, a jet shoots out through the siphon. The brain controls the siphon's activity and, with it, the direction of pursuit or escape.

Mollusks are bilateral, soft-bodied, coelomate animals that differ tremendously in body details, size, and life-styles.

Lively stories emerge when evolutionary theory is used to interpret the fossil record and the range of existing species diversity, as we have done for the mollusks.

ANNELIDS—SEGMENTS GALORE

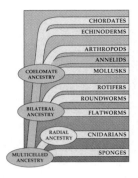

Maybe you've seen earthworms after a downpour. They wriggle out of their flooded burrows to avoid drowning. Earthworms are among the 15,000 or so species of bilateral, segmented animals we call **annelids** (Annelida, meaning ringed forms). Their relatives are the leeches and the less familiar but far more diverse polychaetes (Figures 23.19 and 23.20). Their body actually has repeating segments, not "rings" as the phylum name implies. The segmentation is highly pronounced. Except in leeches, nearly all segments have pairs or clusters of chitin-reinforced bristles on each side of the body. These are also called setae or chaetae, which are just formal names for bristles. When pushed into soil, the bristles provide traction for crawling and burrowing. Earthworms, which are oligochaetes, have a few bristles on their segments; the marine polychaete worms typically have many (*oligo–*, few; *poly–*, many).

Advantages of Segmentation

The segmented body apparently had great evolutionary potential for some annelid lineages. Most segments of an earthworm are still similar, but leeches have suckers at both ends, and polychaetes have an elaborate head and fleshy-lobed appendages called parapods ("closely resembling feet"). Existing species offer us glimpses of the kinds of modifications in body plans that permitted increases in size, more complex internal organs, and segments adapted for specialized tasks.

Annelid Adaptations—A Case Study

Let us use earthworms as our representative annelid. Partitions divide their body into a series of coelomic chambers, each with repeats of muscles, blood vessels,

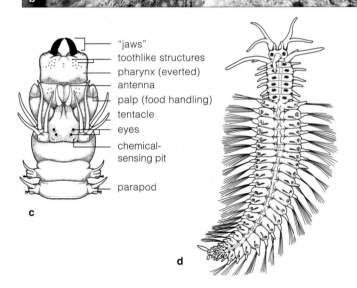

"jaws"
toothlike structures
pharynx (everted)
antenna
palp (food handling)
tentacle
eyes
chemical-sensing pit

parapod

c

d

before feeding

after feeding

Figure 23.19 Leeches. Most leeches have sharp jaws and a blood-sucking device. This one is shown before and after gorging on human blood. For at least 2,000 years, *Hirudo medicinalis*, a freshwater leech, has been employed as a blood-letting tool to "cure" problems ranging from nosebleeds to obesity.

Some surgeons still use leeches, but more selectively, as when they draw off pooled blood after a severed ear, lip, or fingertip is reattached. A patient's body can't do this on its own until the severed blood circulation routes are reestablished.

Figure 23.20 (**a**) A familiar annelid—one of several species of earthworms. (**b**–**d**) Polychaetes, which give us a sense of the dizzying variety of modifications that have evolved in this group, starting with a segmented, coelomic body plan.

The polychaete shown in (**b**) is one of the tube-dwellers. Featherlike structures at its head end are coated with mucus. After the mucus has trapped bacteria and other bits of food, coordinated beating of cilia sweeps them to the mouth. Most polychaetes live in marine habitats. They actually are one of the most common types of animals along coasts. Many are predators or scavengers; others dine on algae.

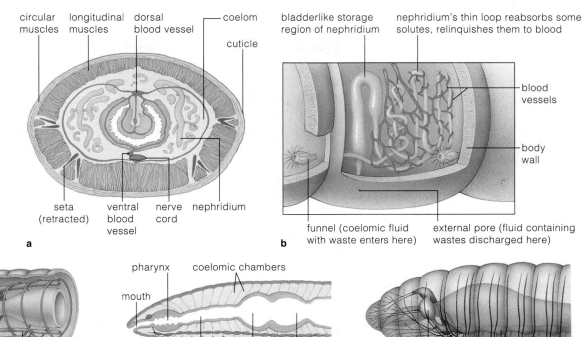

Figure 23.21 Earthworm body plan. (**a**) Midbody, transverse section. (**b**) A nephridium, one of many functional units that help maintain the volume and the composition of body fluids. (**c**) Portion of the closed circulatory system. The system is linked in its functioning with nephridia. (**d**) Part of the digestive system, near the worm's head end. (**e**) Part of the nervous system.

circular muscles · longitudinal muscles · dorsal blood vessel · coelom · cuticle · seta (retracted) · ventral blood vessel · nerve cord · nephridium

bladderlike storage region of nephridium · nephridium's thin loop reabsorbs some solutes, relinquishes them to blood · blood vessels · body wall · funnel (coelomic fluid with waste enters here) · external pore (fluid containing wastes discharged here)

a

b

c · hearts · blood vessels

d · pharynx · mouth · coelomic chambers · esophagus · crop (storage) · gizzard (mashing)

e · brain · nerve cord

branching nerves, and other organs. The gut extends through all chambers, from mouth to anus. As in all annelids, an earthworm's body surface is enclosed in a flexible cuticle of secreted material (Figure 23.21). The permeable cuticle is good for gas exchange but not for conserving water. That is one reason annelids cannot venture from aquatic habitats or moist habitats on land.

Earthworms are scavengers. They ingest moistened soil rich in detritus (decomposing organic bits, such as leaf litter). One worm can eat the equivalent of its own weight each day. When many worms burrow and feed, they collectively aerate the soil and lift nutrients to the surface, to the benefit of many plants.

Like other annelids, earthworms have a hydrostatic skeleton. Their muscles work against fluid-cushioned coelomic chambers. In each segment's wall is a layer of circular muscles (Figure 23.21a). Longitudinal muscles bridge several segments. As they contract, the circular muscles relax. The segments shorten and fatten, and their bristles anchor them by plunging into soil. The pattern reverses. Segments lengthen and their bristles retract and thereby free them for movement. The whole worm moves forward by muscle contractions in one segment after another, down the length of its body.

An earthworm burrows by closing off the coelomic fluid in segments near its head end, then contracting circular muscles to compress the fluid and to extend the longitudinal muscles. Segments near the back have

their bristles plunged into the wall of the burrow; these anchor the worm as its head is being pushed forward.

Figure 23.21*b* shows part of a system of **nephridia** (singular, nephridium). These units control the volume and composition of body fluids. In most annelids, each nephridium starts out as a funnel that collects excess fluid from one coelomic chamber. The funnel connects with a tubular part of the nephridium, which delivers fluid to a pore at the surface in the body wall of the next coelomic chamber.

The worm's head end has a rudimentary **brain**. This cluster of nerve cell bodies integrates sensory input and commands for muscle responses for the whole body. Paired **nerve cords**, each a bundle of long extensions of nerve cell bodies, lead out from the brain. These are pathways for fast communication. In each segment, the paired cords broaden into a ganglion (plural, ganglia), a cluster of nerve cell bodies that controls local activity.

Finally, as is the case for most annelids, earthworms have a closed circulatory system, with blood confined in hearts and muscularized blood vessels. Contractions keep blood circulating in one direction. Smaller blood vessels service the gut, nerve cord, and body wall.

Annelids are bilateral, coelomate, segmented worms that have complex organ systems. Some show the great degree of specialization possible with a segmented body plan.

ARTHROPODS—THE MOST SUCCESSFUL ORGANISMS ON EARTH

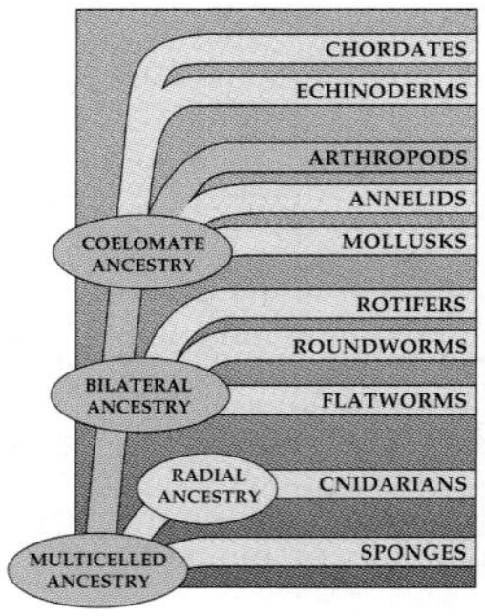

Arthropod Diversity

Evolutionarily speaking, "success" means having the most offspring, the greatest number of species, and the most habitats; fending off threats and competition efficiently; and having the capacity to exploit the greatest amounts and kinds of food. These are the features that come to mind when we attempt to characterize **arthropods** (Arthropoda). We have already identified more than a million species—mostly insects —and researchers are discovering new ones weekly.

Of four major lineages, trilobites (Section 19.5) are extinct. The other three are chelicerates (spiders and their relatives), crustaceans (such as crabs and shrimps), and uniramians (centipedes, millipedes, and insects).

Adaptations of Insects and Other Arthropods

Six important adaptations contributed to the success of arthropods in general and to the insects in particular: a hardened exoskeleton, jointed appendages, fused and modified segments, specialized respiratory structures, efficient nervous system and sensory organs, and often a division of labor in the life cycle.

HARDENED EXOSKELETONS Arthropods have a cuticle of chitin, proteins, and surface waxes that are often impregnated with calcium carbonate deposits. It acts as a rigid, protective external skeleton—an **exoskeleton**. Such cuticles might have evolved as defenses against predation. They took on added functions when some arthropods first invaded the land. They support a body deprived of water's buoyancy, and their waxy surface limits evaporative water loss. Hard cuticles restrict size increases, but arthropods grow in spurts, by **molting**. At certain stages of the life cycle, they secrete a soft new cuticle under the old one, which they shed (Figure 23.22). The body mass increases by the uptake of air or water, and by repeated, rapid cell division before the new cuticle can harden.

JOINTED APPENDAGES If an arthropod had a uniformly hardened cuticle, it wouldn't move much. But the arthropod cuticle is thinner at *joints,* where two body parts

abut. Muscles associated with joints make the thinner cuticle bend in certain directions and move body parts. The jointed exoskeleton was a key innovation; it led to appendages as diverse as wings, antennae, and legs (*arthropod* means "jointed foot").

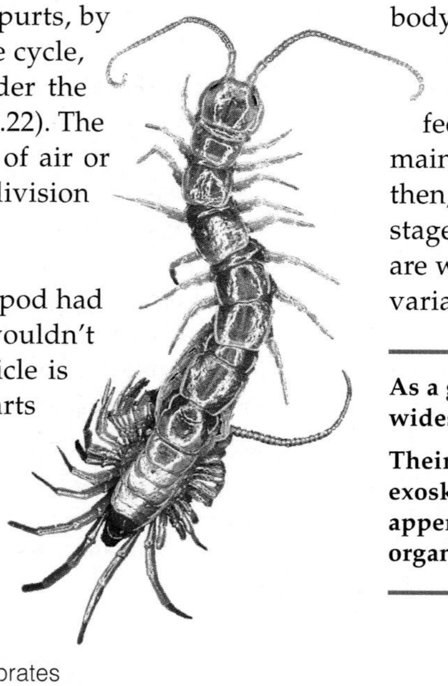

FUSED AND MODIFIED SEGMENTS The first arthropods were segmented, like the annelid stock that presumably gave rise to them. In most existing descendants, serial repeats of the body wall and organs are masked; many fused-together segments are modified to perform more specialized functions. For example, in the ancestors of insects, different segments fused to form three regions —head, thorax, and abdomen—which morphologically diverged from one another in remarkable ways.

RESPIRATORY STRUCTURES Many aquatic arthropods depend on gills for gas exchange. Air-conducting tubes evolved among insects and other land dwellers. Insect tracheas begin as pores on the body surface and branch into tubes that deliver oxygen directly to tissues. They support flight and other energy-consuming activities.

SPECIALIZED SENSORY STRUCTURES Intricate eyes and other sensory organs contributed to arthropod success. With their wide angle of vision, many species receive visual information coming from many directions.

DIVISION OF LABOR Moths, butterflies, beetles, flies, and many other species divide the job of surviving and reproducing among different stages of development. For many species, the new individual is a *juvenile*—a miniaturized version of the adult that simply changes in size and proportion until reaching sexual maturity. Other species show **metamorphosis**, meaning the body form changes from embryo to adult. Under hormonal commands, their size increases, tissues reorganize, and body parts are remodeled. Sections 23.13, 23.15, and 32.8 have examples of this transitional time. Immature stages such as caterpillars typically specialize in feeding and growing in size. Adult stages specialize mainly in dispersal and reproduction. For such species, then, the life cycle turns on a *division of labor*: different stages of development become specialized in ways that are well adapted to environmental conditions—seasonal variation in food resources, water supplies, and so on.

As a group, the arthropods are exceptionally abundant and widespread, and they have enormously different life-styles.

Their success arises largely from their hardened, jointed exoskeletons; fused, modified body segments; specialized appendages; specialized respiratory, nervous, and sensory organs; and often a division of labor in the life cycle.

Figure 23.22 Molting, demonstrated by a red-orange centipede wriggling out of its old exoskeleton.

A LOOK AT SPIDERS AND THEIR KIN

The chelicerates originated in shallow seas early in the Paleozoic. The only surviving marine species are a few mites, sea spiders, and horseshoe crabs (Figure 23.23). The familiar chelicerates on land—spiders, scorpions, ticks, and chigger mites—are all classified as arachnids. And we might say this about the whole group: Never have so many been loved by so few. True, scorpions and spiders are formidable predators that sting, bite, or subdue prey with venom. But rarely do they endanger us with their stings or bites, however painful. And in one respect spiders especially are beneficial: they prey on huge numbers of our greatest competitors—insects.

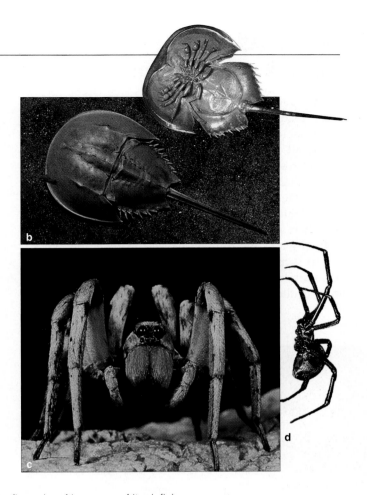

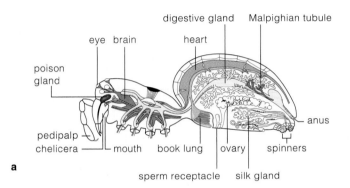

Figure 23.23 (**a**) A spider's internal organization. (**b**) Horseshoe crab. Its *five* pairs of legs, one of its defining features, are hidden beneath a hard, shieldlike cover. (**c**) Wolf spider. Like most spiders, it helps keep insect populations in check. Its bite is harmless to humans. (**d**) Female black widow. It prefers dark, undisturbed places, as in a rarely used closet or behind a clothes dryer. Its bite can be painful and sometimes dangerous.

Bites of ticks that parasitize deer, mice, and other vertebrates cause maddening itches and often serious diseases. For example, some ticks transmit bacteria that cause Rocky Mountain spotted fever and Lyme disease (Section 20.2). Most mites are free-living scavengers.

Arachnid segments are fused into a forebody and hindbody. The forebody's jointed appendages include four pairs of legs, a pair of pedipalps with sensory roles, and a pair of chelicerae that wound or put venom into prey. Hindbody appendages spin out silk threads for webs and egg cases. Most webs are netlike. One spider spins a thread with a ball of sticky material at the end, then uses a leg to swing it at insects flying past! The body has an *open* circulatory system. A heart pumps blood into tissues, then receives blood through small openings in its wall. Slowly circulating blood travels through moist folds of book lungs. These respiratory organs resemble book pages; they greatly increase the surface area for gas exchange with air (Figure 23.23*a*).

Scorpions prefer pincers to venom in fights with other arthropods, tiny lizards, and mice. They hunt at night. They wait out the day in burrows or under bark, stones, and logs. They can stay there for a year without starving. They live in Canada, the southern Andes, the

Figure 23.24 From Arizona, *Centruroides sculpturatus*, one of the most dangerous scorpions known.

southern Alps, even moist forests. But most live in hot, dry places. The ones that can hurt or kill people aren't the fiercest looking, and there aren't many of them. An Arizona scorpion is the most dangerous species in the United States (Figure 23.24). When in scorpion country, it is definitely not a good idea to walk about barefoot at night or to put on your shoes without first turning them upside-down and giving them a vigorous shaking.

Spiders, scorpions, and their relatives have a variety of appendages specialized for predatory or parasitic life-styles.

A LOOK AT THE CRUSTACEANS

Shrimps, lobsters, crabs, barnacles, pillbugs, and other crustaceans got their name because they have a hard yet flexible "crust" (an external skeleton)—but so do nearly all arthropods. Only some of the 35,000 species live in fresh water or on land. The vast majority live in marine habitats, where they are so abundant they have been dubbed the insects of the seas. Lobsters and crabs are "giants" of this subphylum; most crustaceans are less than a few centimeters long. All have major roles in food webs, and humans harvest many edible types.

The simplest crustaceans may resemble their annelid ancestors in having pairs of similar appendages along most of their body. Among the more complex lineages,

unspecialized appendages evolved into many diverse structures of the sort shown in Figure 23.25. The strong claws of lobsters and crabs, for example, are used to collect food, intimidate other animals, and sometimes dig burrows. Feathery appendages of barnacles comb microscopic bits of food from the water.

Many crustaceans have sixteen to twenty segments; some have more than sixty. In crabs, lobsters, and some other crustaceans, the dorsal cuticle extends back from the head as a shieldlike cover over some or all of the segments. The head has two pairs of antennae, one pair of jawlike appendages (mandibles), and two pairs of appendages for handling food. Crayfish, crabs, lobsters,

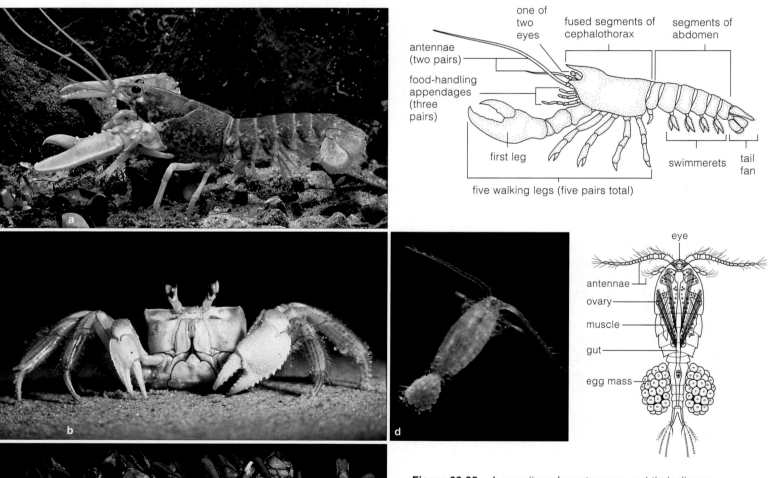

Figure 23.25 A sampling of crustaceans and their diverse life-styles. (**a**) Photograph and body plan of a lobster. For most of their lives, lobsters are secretive. (**b**) Crab. Guess why crabs skitter actively and openly across sand and rocks at night but not in the day. (**c**) Stalked barnacles. Adults cement themselves to one spot. You might mistake them for mollusks, but as soon as they open their hinged shell to filter-feed, you can observe their jointed appendages—the hallmark of arthropods. (**d**) Photograph and body plan of a copepod. Different copepods are free-living filter feeders, predators, and parasites. This female has a pair of long antennae and is carrying her eggs with her.

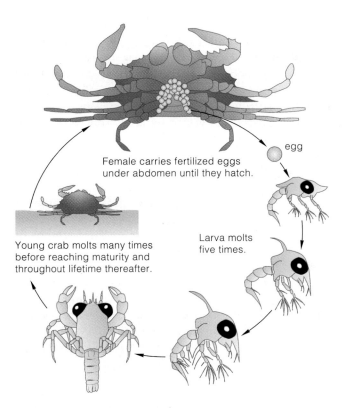

Figure 23.26 Life cycle of a crab. The larval and juvenile stages molt repeatedly and grow in size.

Female carries fertilized eggs under abdomen until they hatch.

egg

Larva molts five times.

Young crab molts many times before reaching maturity and throughout lifetime thereafter.

shrimps, and their various relatives are equipped with five pairs of walking legs.

Of all the arthropods, only barnacles have a calcified "shell." This modified external skeleton protects them from predators, drying winds, battering surf, and strong currents. Adult barnacles cement themselves to rocks, wharf pilings, and similar surfaces (Figure 23.25c). A few kinds attach themselves only to the skin of whales.

Figure 23.25d shows a copepod. Copepods are less than two millimeters long and are the most numerous animals in aquatic habitats, maybe even in the world. About 1,500 kinds parasitize various invertebrates and fishes. The majority—8,000 species—are consumers of phytoplankton, the "pastures" of aquatic habitats. Some prey on larval or small adult invertebrates, fish eggs, and fish larvae, which they grab with a pair of food-handling appendages. In turn, the copepods are food for different invertebrates, fishes, and baleen whales.

Like other arthropods, crustaceans repeatedly molt and shed the exoskeleton during their life cycle. Figure 23.26 shows a crab's larval stages and increases in size.

Crustaceans differ greatly in their number and kind of appendages. As is the case for arthropods in general, they repeatedly replace their external skeleton by molting.

HOW MANY LEGS?

The **millipedes** and **centipedes** have a long, segmented body with many legs. Of course, millipedes don't have "a thousand," as their name implies. Most have about 100 legs, although one impressive individual grew 752. Centipedes have between 15 and 177 pairs of legs, not a nicely rounded number of "one hundred."

As a millipede develops, its pairs of segments fuse, and each segment in its cylindrical body ends up with two pairs of legs (Figure 23.27a). Millipedes scavenge decaying plant material in soil and forest litter.

Figure 23.27 (a) Millipede. (b) A Southeast Asian centipede.

Centipedes have a flattened body, and all but two segments have a pair of walking legs. All species are fast-moving, aggressive predators outfitted with fangs and venom glands. They prey on insects, earthworms, and snails. The one in Figure 23.27b can subdue small lizards, toads, and frogs. A house centipede (*Scutigera*) often hides in buildings and hunts cockroaches, flies, and other pests. Although helpful in this respect, its vaguely terrifying body keeps it from being welcomed.

Mild-mannered, scavenging millipedes and aggressive, predatory centipedes do not lend themselves to leg counts as they walk by.

A LOOK AT INSECT DIVERSITY

As a group, insects share the adaptations listed earlier in Section 23.11. Here we expand the list. Insects have a head, a thorax, and an abdomen. The head has paired sensory antennae and mouthparts for biting, chewing, sucking, or puncturing (Figure 23.28). Three pairs of legs and usually two pairs of wings project from the thorax. Insects are the *only* winged invertebrates. Their abdomen has egg-laying devices and other reproductive structures. They have a foregut, midgut (where digestion occurs), and hindgut (where water is reabsorbed). They get rid of wastes through **Malpighian tubules**, small tubes that connect with the midgut. When proteins are digested, the nitrogen-containing wastes diffuse from blood into the tubules and are converted into harmless crystals of uric acid. These crystals are eliminated with feces. This system lets land-dwelling insects dispose of potentially toxic wastes without losing precious water.

We've already catalogued more than 800,000 species of insects. If we use sheer numbers and distribution as the yardstick, the most successful ones are small in size and have a staggering reproductive capacity. You may find some species growing and reproducing in great numbers on a single plant that would be an appetizer for another animal. By one estimate, if all offspring of one female fly survived and reproduced for six more generations, she would have over 5 trillion descendants!

Also, the most successful insects are winged. They move among food sources that are too widely scattered to be exploited by other kinds of animals. The capacity for flight contributed greatly to their success on land.

Finally, insect life cycles commonly proceed through stages that allow exploitation of different resources at different times. As an insect embryo develops, organs required for feeding and other vital activities form and become functional. Before an insect becomes an adult (the sexually mature form), it goes through immature, post-embryonic stages. **Nymphs** and **pupae**, as well as larvae, are such stages. Like human infants, nymphs of some insects have the form of miniature adults. Unlike children, the forms molt (Figure 23.29a). Other insects develop through post-embryonic stages of reactivated growth, tissue reorganization, and remodeling of body parts. Metamorphosis is the name for this resumption of growth and transformation of post-embryonic stages into the adult. That transformation is more drastic in some species than in others (Figure 23.29).

All the factors that contribute to insect success also make them our most aggressive competitors. Insects

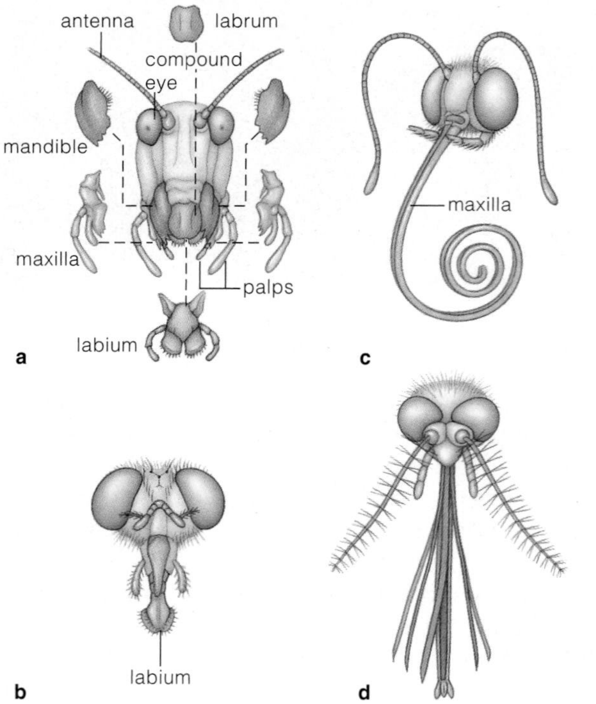

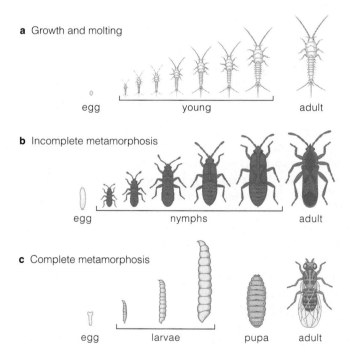

Figure 23.28 Examples of insect appendages. Head parts typical of (**a**) grasshoppers, a chewing insect; (**b**) flies, which sponge up nutrients; (**c**) butterflies, which siphon up nectar; and (**d**) mosquitoes, with piercing and sucking appendages.

Figure 23.29 Examples of post-embryonic development. (**a**) Young silverfish are adults in miniature, changing little except in size and proportion as they mature into adults. (**b**) True bugs show *incomplete* metamorphosis, which involves gradual, partial change from the first immature form until the last molt. (**c**) Fruit flies show *complete* metamorphosis. Tissues of immature forms are destroyed and replaced before the adult emerges.

Figure 23.31 (**a**) Damaged corn plants. (**b**) Western corn rootworm (*Diabrotica virgifera*). This insect makes the rounds in cornfields and is a representative of our major competitors for food. Its larvae (**c**) so damage corn roots that the plants keel over. Each year we lose about a billion dollar's worth of crops to rootworms alone. The astronomical losses compel farmers to spread about 30 million pounds of pesticides a year on cornfields, or about half the total for all row crops in the United States. However, many rootworms now show pesticide tolerance. Inventive researchers are targeting adults, which are fond of a bitter plant juice called curcurbitacin. This juice evolved as a natural pesticide in cucumbers, melons, squashes, and some other plants. Rootworms apparently appropriated the chemical as a defense against bird predators. (Birds learn to avoid rootworms after a taste trial that ends in vomiting.) Robert Schroder and others concocted a bait for the pests. In one test, they laced watermelon juice with red dye 28, a photoactive chemical. Adult rootworms gorged themselves, turned deep red, and after five minutes in sunlight, dropped dead. The sun's rays triggered oxidation reactions that destroyed the insect tissues. Beneficial insects avoid the dyed juice. In this ongoing contest, chalk one up for researchers.

Figure 23.30 Representative insects. (**a**) Stinkbugs (order Hemiptera), newly hatched. (**b**) Ladybird beetles (order Coleoptera) swarming. These commercially raised beetles will be released as biological controls of aphids and other pests. (**c**) Honeybees (order Hymenoptera) communicating through a remarkable dance, described in Section 41.4. (**d**) Scarab beetle (order Coleoptera). With more than 300,000 species, this is the largest order of the animal kingdom. (**e**) Dragonfly (order Odonata). It swiftly captures and eats insects in midflight. (**f**) Flea (order Siphonaptera), with strong legs for jumping onto and off animal hosts. (**g**) Duck louse (order Mallophaga). It eats bits of feathers and skin. (**h**) European earwig (order Dermaptera), a common household pest. (**i**) Mediterranean fruit fly (order Diptera). Its larvae destroy citrus fruit and other crops. (**j**) Luna moth (order Lepidoptera) of North America. Microscopic scales cover the wings and body of most butterflies and moths, including this one.

destroy crops, stored food, wool, paper, and timber. As some stealthily draw blood from us and from our pets, they commonly transmit pathogens. Yet many pollinate flowering plants, including highly valued crop plants. And many "good" insects attack or parasitize the ones we would rather do without. Figures 23.30 and 23.31 show representatives of major orders of insects.

As a group, insects show immense variation on the basic arthropod body plan. Many have wings, and their life cycles have stages that allow exploitation of different and often widely scattered food sources. Many species produce great numbers of small individuals that pass through immature, post-embryonic stages before the adult form emerges.

THE PUZZLING ECHINODERMS

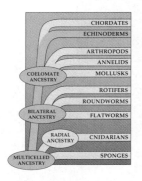

We arrive, finally, at the second lineage of coelomate animals, the deuterostomes. The **echinoderms** (Echinodermata) are invertebrate members of this lineage. Species in this phylum include the feather stars, sea urchins, sea cucumbers, and brittle stars (Figure 23.32), as well as sea lilies (crinoids), sand dollars, and sea biscuits. Sea lilies look a bit like stalked plants and flourished in Silurian times (Figure 19.5). About 13,000 echinoderm species are known from the fossil record, but most of these are extinct. Nearly all of the 6,000 or so existing species live in marine habitats.

An echinoderm body wall bears a number of spines, spicules, or plates made rigid with calcium carbonate. (Echinodermata means spiny-skinned.) The structures serve defensive functions, as you might suspect if you have ever stepped barefoot on a sea urchin. Its spines trigger painful swelling when they break off under the skin. Most echinoderms have a well-developed internal skeleton, which is composed of calcium carbonate and other substances secreted from specialized cells. Oddly, the adults are radial with some bilateral features. Most species even have bilateral larvae. Did the echinoderms evolve from bilateral ancestors, in which some radial features appeared at a later time? Maybe.

Adult echinoderms have no brain. However, their decentralized nervous system allows them to respond to information about food, predators, and so forth that is coming from different directions. For instance, any arm of a sea star that touches the shell of a tasty scallop

tube feet   spine

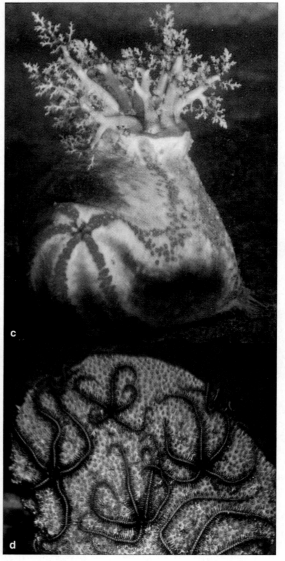

Figure 23.32 Representative echinoderms. (**a**) Feather star, with finely branched food-gathering appendages. (**b**) Sea urchin, which moves about on spines and tube feet. (**c**) Sea cucumber, with rows of tube feet along its body. (**d**) Brittle stars. Their arms (rays) make rapid, snakelike movements.

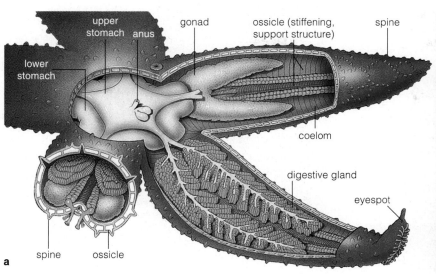

upper stomach
anus
gonad
ossicle (stiffening, support structure)
spine
lower stomach
coelom
digestive gland
eyespot
spine
ossicle

a

part of the water-vascular system
sieve plate
ring canal
ampulla

b

c

d

tube feet on the ventral surface of a sea star arm

Figure 23.33 (**a**) Key aspects of the radial body plan of a sea star. Its water-vascular system, combined with great numbers of tube feet, is the basis of locomotion. (**b–d**) Five-armed sea star, with closer views of tube feet.

can become the leader, directing the rest of the body to move in a direction suitable for capturing its prey.

Figures 23.32 and 23.33 show examples of tube feet. These fluid-filled, muscular structures have suckerlike adhesive disks. Sea stars use their tube feet for walking, burrowing, clinging to rocks, or gripping a clam or snail about to become a meal. Tube feet are components of a **water-vascular system** unique to echinoderms. In sea stars, that system includes a main canal in each arm. Short side canals extend from them and deliver water to the tube feet. Each tube foot contains a fluid-filled, muscular structure shaped rather like the rubber bulb on a medicine dropper. As the bulb contracts, it forces fluid into the foot and causes it to lengthen.

Tube feet change shape constantly as muscle action redistributes fluid through the water-vascular system. Hundreds of tube feet may move at a time. After being released, each one swings forward, reattaches to the substrate, then swings backward and is released before swinging forward again. Their motions are splendidly coordinated, so sea stars glide rather than lurch along.

On their ventral surface, sea stars have a formidable feeding apparatus, such as the one shown here:

Some eager sea stars simply swallow their prey whole. Others push part of their stomach outside the mouth and around their prey, then start digesting their meal even before swallowing it. Sea stars get rid of coarse, undigested residues through the mouth. They do have a small anus, but this is of no help in getting rid of an empty clam shell or snail shell.

With their curious traits, echinoderms are a suitable point of departure for this chapter. Even though broad trends have been identified in animal evolution, keep in mind that there are confounding exceptions to the perceived macroevolutionary patterns.

Echinoderms are coelomate animals with spines, spicules, or plates in the body wall. From the evolutionary perspective, they are a puzzling mix of bilateral and radial features.

SUMMARY   *Gold* indicates text section

1. Animals are multicelled, aerobic heterotrophs that ingest or parasitize other organisms. Most kinds have diploid body cells organized into tissues, organs, and organ systems. Animals reproduce sexually and often asexually. They undergo embryonic development. Most are motile during at least part of the life cycle. *23.1*

2. Animals range from structurally simple placozoans and sponges to vertebrates. *23.1*

   a. By comparing major animal phyla and integrating the information with the fossil record, biologists have identified major evolutionary trends among them.

   b. Revealing aspects of body plans are the type of symmetry, gut, and cavity (if any) between the gut and body wall; whether there is a head end; and whether the body is divided into segments (Figure 23.34).

3. *Trichoplax*, the only known placozoan, is the simplest animal. It is composed of little more than two layers of cells with a fluid matrix in between. *23.2*

4. A sponge body has no symmetry. It is at the cellular level of construction. Although it consists of several kinds of cells, these are not organized like the epithelia and other tissues seen in complex animals. *23.3*

5. Cnidarians include the jellyfishes, sea anemones, and hydras. They have radial symmetry, and they are at the tissue level of construction. Cnidarians alone produce nematocysts, capsules with dischargeable threads that are used mainly in prey capture. *23.4*

6. Most animals more complex than cnidarians have bilateral symmetry, tissues, organs, and organ systems. Their gut is saclike, as in flatworms, or complete, with anus and mouth. Those more complex than flatworms have a cavity between the gut and body wall (a coelom, with its peritoneum, or a false coelom). *23.5, 23.6*

7. Two major lineages diverged shortly after flatworms evolved. One (protostomes) gave rise to the mollusks, annelids, and arthropods. The other (deuterostomes) gave rise to echinoderms and chordates. *23.8*

8. All mollusks have a fleshy, soft body and a mantle. Most have either a shell or a remnant of one. Mollusks vary greatly in size, body details, and life-styles. *23.9*

9. Annelids (earthworms, polychaetes, leeches) have complex organs in repeats of coelomic chambers. Their segmented body works as a hydrostatic skeleton. *23.10*

10. Collectively, arthropods are the most successful of all groups in diversity, numbers, distribution, defenses, and capacity to exploit food resources. *23.11–23.15*

    a. All arthropods, including arachnids, crustaceans, and insects, have hardened, jointed exoskeletons. They also have modified segments, specialized appendages, specialized respiratory, nervous, and sensory organs, and (in insects only) wings.

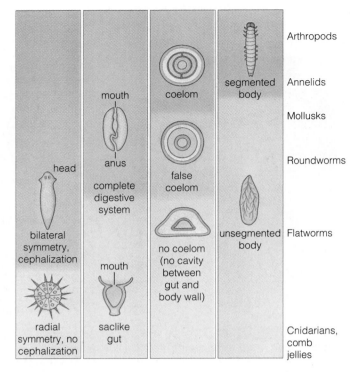

Figure 23.34 Summary of key trends in the evolution of animals, as identified by comparing body plans of major phyla. Not all of these features appeared in every group.

    b. Arthropods develop by growth in size and molting or develop through a series of immature stages, such as larvae and nymphs. Many metamorphose; the tissues of immature forms undergo major reorganization, and body parts are remodeled before the adult emerges.

11. Echinoderms have spines, spicules, or plates in their body wall. Evolutionarily, the phylum is puzzling. The larvae of most species form bilateral features, but they go on to develop into basically radial adults. *23.16*

Review Questions

1. List the six main features that characterize animals. *23.1*

2. When attempting to discern evolutionary relationships among major groups of animals, which aspects of their body plans provide the most useful clues? *23.1*

3. What is a coelom? Why was it important in the evolution of certain animal lineages? *23.1*

4. Name some animals with a saclike gut. Evolutionarily, what advantages does a complete gut afford? *23.1, 23.4–23.6*

5. Choose a species of insect that lives in your neighborhood and describe some of the observable adaptations that underlie its success. *23.15*

Self-Quiz   ANSWERS IN APPENDIX III

1. Which is *not* a general characteristic of the animal kingdom?
   a. multicellularity; most have tissues, many form organs
   b. exclusive reliance on sexual reproduction
   c. motility at some stage of the life cycle
   d. embryonic development during the life cycle

Figure 23.35 From a tropical reef, one of the soft, branching corals (*Telesto*).

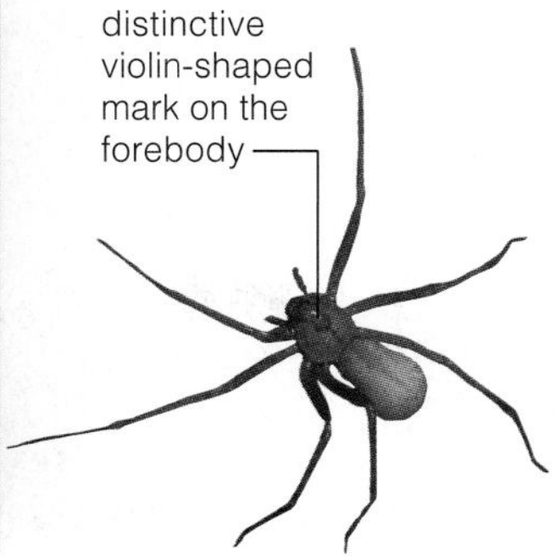

distinctive violin-shaped mark on the forebody

Figure 23.36 Brown recluse; avoid it.

2. Between the gut and body wall of most animals is a _____.
   a. pharynx   c. coelom
   b. pseudocoelom   d. archenteron

3. Jellyfishes, sea anemones, and their relatives have _____ symmetry, and their cells form _____ .
   a. radial; mesoderm   c. radial; tissues
   b. bilateral; tissues   d. bilateral; mesoderm

4. Most animals that are more complex than cnidarians have _____ symmetry, and _____ forms in their embryos.
   a. radial; mesoderm   c. bilateral; mesoderm
   b. bilateral; endoderm d. radial; endoderm

5. Which phylum contains members that are notorious for causing serious diseases in humans?
   a. cnidarians   c. segmented worms
   b. flatworms   d. chordates

6. _____ have a coelom and pronounced segmentation, and a dizzying variety of modifications to the segments.
   a. arthropods   c. sponges   e. sea stars
   b. annelids   d. snails and clams f. vertebrates

7. _____ are the most evolutionarily successful animals.
   a. arthropods   c. sponges   e. sea stars
   b. annelids   d. snails and clams f. vertebrates

8. Match the terms with the appropriate groups.
   ____ sponges   a. spiny-skinned
   ____ cnidarians   b. vertebrates and kin
   ____ flatworms   c. flukes and tapeworms
   ____ roundworms   d. no tissue organization
   ____ rotifers   e. no males for some
   ____ mollusks   f. nematocysts, radial symmetry
   ____ annelids   g. hookworms, elephantiasis
   ____ arthropods   h. jointed exoskeleton
   ____ echinoderms   i. "belly-foots" and kin
   ____ chordates   j. segmented worms

Critical Thinking

1. A carnivorous sponge was found in an underwater cave in the Mediterranean Sea. Unlike other sponges, it has no pores or canals. Hooklike spicules on surface branchings trap shrimp and other animals. The branchings envelop prey, which cells digest. Which body parts had to evolve for such a feeding strategy?

2. Tapeworms are hermaphroditic. What selective advantages might this feature offer, and in what kinds of environments?

3. People who eat raw oysters or clams harvested from sewage-polluted waters develop mild to severe gastrointestinal ailments. Think about the feeding modes of these mollusks and develop a hypothesis to explain why mollusk eaters can get sick.

4. You are diving in the calm, warm waters behind a large tropical reef. You see something that looks like a plant with many tentacles (Figure 23.35). This is a branching, soft coral with many polyps. You won't see it growing on reef surfaces exposed to the open sea. Formulate two hypotheses that might explain its distribution.

5. Bites of a brown recluse spider (Figure 23.26) can be severe to fatal. Search the web and list its habitats, behavior, and some possible effects of its bite on humans.

6. The small animal in Figure 23.37a is packed with organs, and its head end has a crown of cilia. Females of most species can reproduce without help from males. The marine worm in Figure 23.37b has a segmented body. Most segments are similar to one another, and on their sides are bristles that are used to dig burrows in sediments. Referring to Table 23.1, to which groups do you think these two animals belong?

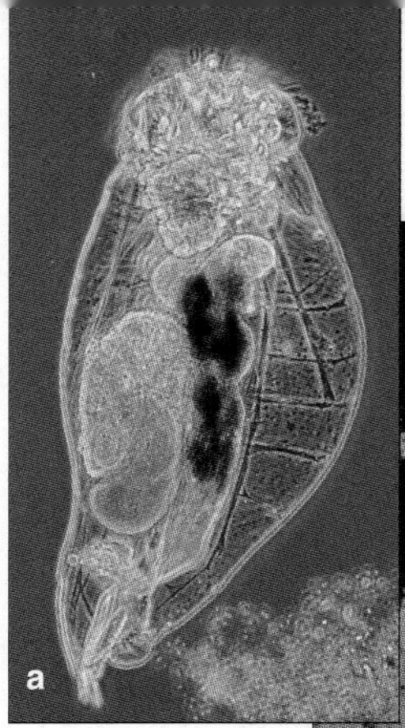

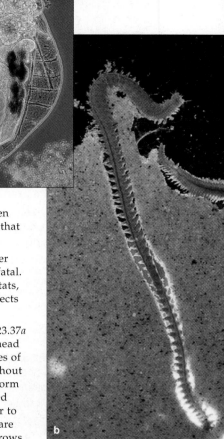

Figure 23.37 Go ahead, name the mystery animals.

Selected Key Terms

animal 23.1
annelid 23.10
arthropod 23.11
bilateral
 symmetry 23.1
brain 23.10
centipede 23.14
cephalization 23.1
cnidarian 23.4
coelom 23.1
collar cell 23.3
contractile cell 23.4
cuticle 23.6
deuterostome 23.8
echinoderm 23.16
ectoderm 23.1
endoderm 23.1
epithelium 23.4
exoskeleton 23.11

flatworm 23.5
gonad 23.4
gut 23.1
hermaphrodite 23.5
hydrostatic
 skeleton 23.4
invertebrate 23.1
larva 23.3
Malpighian
 tubule 23.15
mantle 23.9
medusa 23.4
mesoderm 23.1
metamorphosis 23.11
millipede 23.14
mollusk 23.9
molting 23.11
nematocyst 23.4
nephridium 23.10

nerve cell 23.4
nerve cord 23.10
nymph 23.15
organ 23.5
organ system 23.5
pharynx 23.5
placozoan 23.2
planula 23.4
polyp 23.4
proglottid 23.5
protostome 23.8
pupa 23.15
radial symmetry 23.1
roundworm 23.6
sponge 23.3
vertebrate 23.1
water-vascular
 system 23.16

Readings

Pearse, V., et al. 1987. *Living Invertebrates*. Palo Alto, California: Blackwell.

Pechenik, J. 1995. *Biology of Invertebrates*. Third edition. Dubuque: W. C. Brown.

Raloff, J. July 10, 1999. "The Bitter End: Enticing Agricultural Pests to Their Last Repast." *Science News* 156:24–26.

On-Line readings at Student Guide for InfoTrac:
www.brookscole.com/biology

24

ANIMALS: THE VERTEBRATES

Making Do (Rather Well) With What You've Got

In 1798, a few naturalists were skeptically probing a specimen delivered to the British Museum in London, looking for signs that a prankster had slyly stitched the bill of an oversized duck onto the pelt of a small furry mammal. They didn't know it, but they were examining the remains of a platypus, a web-footed mammal about half the size of a housecat. Like other mammals, the duck-billed platypus (*Ornithorhynchus anatinus*) has hair, and the females have mammary glands. Like birds and reptiles, it has a cloaca, an enlarged duct through which gametes, feces, and excretions from kidneys pass. It lays shelled eggs, as birds and most reptiles do. Its young hatch pink and unfinished, as embryonic stages too helpless to fend for themselves (Figure 24.1*a*). Its fleshy bill does appear ducklike, and its broad, flat, furry tail looks like the one on a beaver (Figure 24.1*b*).

With its unusual traits, the platypus invites you to challenge preconceived notions of what constitutes "an animal" and what made it that way. The platypus belongs to a lineage that began after the supercontinent Pangea broke apart. For 150 million years, its ancestors were isolated on a huge, drifting fragment that became Australia. Geographic separation was a springboard for genetic divergence. Unique mutations and selection pressures resulted in the platypus body. Some might ridicule that body plan, but it's scarcely a failure. It has endured far longer than the one we humans inherited.

The platypus is well adapted to remote streams and lagoons in Australia and Tasmania. At night it slips into the water and hunts for tiny invertebrates. Dense fur keeps its body warm in cool water all night, just as it does all day when the platypus rests in a cool burrow. A broad, thick tail serves as a rudder for underwater maneuvers and as a storehouse for energy-rich reserves of fat. A broad, flattened bill is shaped for scooping up aquatic snails, shrimps, worms, mussels, and insect larvae. Horny pads lining the jaws grind up and make short work of shelled and hard-cuticled meals. Strong

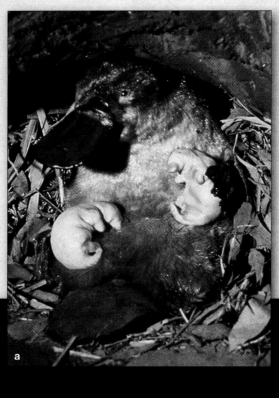

a

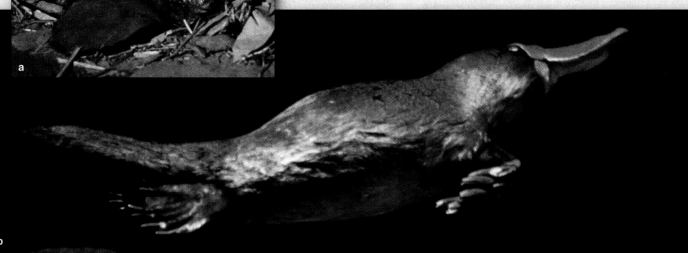

b

Figure 24.1   One of evolution's success stories—the platypus, (**a**) raising its incompletely formed offspring in a burrow and (**b**) underwater, eyes and ears shut, yet homing in on prey.

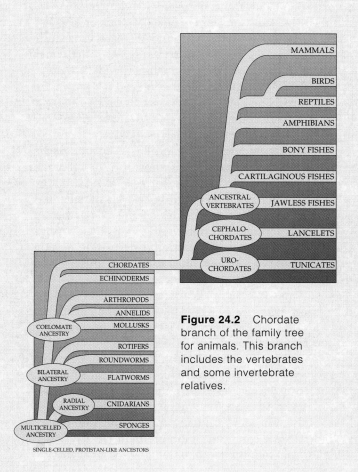

MAMMALS

BIRDS

REPTILES

AMPHIBIANS

BONY FISHES

CARTILAGINOUS FISHES

ANCESTRAL VERTEBRATES — JAWLESS FISHES

CEPHALO-CHORDATES — LANCELETS

URO-CHORDATES — TUNICATES

CHORDATES

ECHINODERMS

ARTHROPODS

ANNELIDS

COELOMATE ANCESTRY — MOLLUSKS

ROTIFERS

ROUNDWORMS

BILATERAL ANCESTRY — FLATWORMS

RADIAL ANCESTRY — CNIDARIANS

MULTICELLED ANCESTRY — SPONGES

SINGLE-CELLED, PROTISTAN-LIKE ANCESTORS

Figure 24.2 Chordate branch of the family tree for animals. This branch includes the vertebrates and some invertebrate relatives.

claws projecting from its hind feet are well adapted for digging out underground chambers.

Like a submarine, a platypus closes the hatches, so to speak, when it dives. Its nostrils and a fleshy groove around the eyes and ears snap shut. No need for eyes or ears to zero in on food; many of the 800,000 sensory receptors embedded in its bill detect tiny oscillations in water pressure as prey swim past. Other receptors detect changes in electric fields, even those generated by the weak flicks of a shrimp tail.

The platypus is one of 4,500 kinds of mammals that now occupy the vertebrate branch of the animal family tree (Figure 24.2). Its vertebrate relatives include fishes, amphibians, reptiles, and birds, which you will read about in this chapter's survey of the chordates. The chapter concludes with a a brief evolutionary history of the human species, starting with its mammalian and primate stocks. As you consider each group, keep this key concept in mind: *Each animal is a mosaic of traits, many conserved from remote ancestors and others unique to its branch on the animal family tree.*

Key Concepts

1. The chordate branch of the family tree for animals includes invertebrates and vertebrate species. All are bilateral animals. Five features develop in their embryos and set them apart from other animals: a supporting rod for the body (notochord), a dorsal nerve cord, a pharynx, gill slits in the pharynx wall, and a tail that extends past the anus. Some or all of these features persist in adults.

2. Existing invertebrate chordates include the tunicates and the lancelets.

3. Of the eight traditional classes of vertebrates, seven have living representatives. These are jawless fishes, cartilaginous fishes, bony fishes, amphibians, reptiles, birds, and mammals. The other class, the placoderms, became extinct early in vertebrate history.

4. We can identify four major trends in the evolution of certain vertebrate lineages. Body support and movement came to depend on a backbone, not the notochord. Jaws evolved, sparking the evolution of the nerve cord into a spinal cord and brain. As some lineages moved onto land, lungs became more important than gills for gas exchange, and increasingly efficient blood circulation enhanced their function. Among those same lineages, fleshy fins with skeletal supports evolved into the diverse limbs of amphibians, reptiles, birds, and mammals.

5. The primate branch of the mammalian lineage includes prosimians, tarsioids, and anthropoids (monkeys, apes, and humans). Apes and humans are hominoids. Humans and some extinct species with a mosaic of apelike and humanlike traits are further classified as hominids.

6. A long-term cooling trend that started in the Miocene led to seasonal changes in habitats and food sources. As food sources became scarcer, early hominoids spread out through Africa and, later, entered Europe and southern Asia. One lineage gave rise to the hominids.

7. Unlike earlier hominids, *Homo erectus* and *H. sapiens* displayed exceptional behavioral flexibility and creative experimentation with their environment, as when they started using fire. These characteristics allowed them to survive the challenges of dispersing into novel and often harsh environments around the world.

THE CHORDATE HERITAGE

Characteristics of Chordates

The preceding chapter left off with echinoderms, one of the most ancient lineages of the deuterostome branch of the animal family tree. Dominating this branch are more recently evolved species, the **chordates** (phylum Chordata). Of these bilateral, coelomate animals, about 2,100 are, like the echinoderms, invertebrates. The vast majority—about 48,000 species—are vertebrates.

Vertebrates are chordates with a backbone, of either cartilage or bone, and a brain enclosed in a protective cartilaginous or bony chamber. "Invertebrate chordates" share certain features with these dominant members of the phylum, but a backbone isn't one of them.

Four features are evident in chordate embryos, and in many species these features persist into adulthood. *First*, a **notochord**, a long rod of stiffened tissue (not cartilage or bone), helps support the body. *Second*, the nervous system of chordate embryos develops from a tubular, dorsal **nerve cord**, which runs parallel to the notochord and gut. The cord's anterior end increases in mass and undergoes modifications to form the brain. *Third*, the embryos have distinctive slits in the wall of their **pharynx**, which is a muscularized tube. A pharynx functions in feeding, respiration, or both. *Fourth*, a tail forms in embryos and extends past the anus.

Chordate Classification

Biologists group nearly all of the chordates into three subphyla. These are named Urochordata (tunicates and their kin), Cephalochordata (lancelets), and Vertebrata (vertebrates). There are eight classes of vertebrates:

Agnatha	*Jawless fishes*
Placodermi	*Jawed, armored fishes (extinct)*
Chondrichthyes	*Cartilaginous fishes*
Osteichthyes	*Bony fishes*
Amphibia	*Amphibians*
Reptilia	*Reptiles*
Aves	*Birds*
Mammalia	*Mammals*

Appendix I has an expanded classification scheme for the vertebrates. Unit VI provides details of their body plans and functions. Here we become acquainted with major trends in their evolution. We find clues to those trends among the invertebrate chordates.

The embryos of chordates alone have this combination of features: a notochord, a tubular dorsal nerve cord, a pharynx with slits in its wall, and a tail extending past the anus.

INVERTEBRATE CHORDATES

Tunicates

The 2,000 species of existing urochordates are baglike animals one to several centimeters long. Most often we call them the **tunicates**, after the gelatinous or leathery "tunic" that adults secrete around themselves. Adults of sea squirts, the most common species, spurt water out through a siphon when something irritates them.

You'll find tunicates from the intertidal zone all the way to surprising depths in the seas. Some live solitary lives; others are colonial. The adults of most species are attached to hard substrates, such as rocks.

Sea squirts undergo metamorphosis. Drastic tissue remodeling and reorganization transform the bilateral, swimming larvae into the adult form (Figure 24.3). The larva's firm, flexible notochord is a series of cells that acts like a torsion bar. As muscles on one side of the tail or the other contract, the notochord bends, then springs back as muscles relax. The strong, side-to-side motion propels the larva forward. Most fishes use muscles and their backbone for the same kind of motion.

Like some other animals, tunicates are **filter feeders**: they direct a current of water through some part of the body and filter food from it. Water flows in through a siphon and on through **gill slits**: openings in the thin pharynx wall. It flows out through a different siphon. The tunicate pharynx also acts as a respiratory organ. Dissolved oxygen is less concentrated in blood vessels adjoining the pharynx than in the surrounding water. It follows the gradient, diffusing from water into blood. Carbon dioxide's concentration gradient is such that it diffuses from blood into water leaving the pharynx.

Lancelets

About twenty-five species of fish-shaped, translucent animals, the cephalochordates, live in nearshore marine sediments around the world. Most are shorter than your little finger. Much of the time these animals burrow in sand or sediments, almost up to the mouth. Their body tapers sharply at both ends; hence the common name, **lancelets**. The lancelet body plan sketched in Figure 24.4*a* clearly shows the four chordate features. Notice the segmented pattern of muscles on both sides of the lancelet's notochord.

Like tunicate larvae, lancelets use their notochord and muscles for swimming motions. Unlike tunicates, lancelets have a closed circulatory system, although you won't find any red blood cells. They don't have a complex brain. But the anterior end of the dorsal nerve cord is expanded, and pairs of nerves extend into each muscle segment. Their mode of respiration is simple yet effective for their small body. Dissolved oxygen and carbon dioxide simply diffuse across their thin skin.

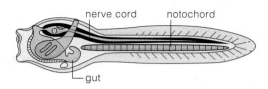

b *Right:* Generalized body plan of the tadpole-like larva, midsagittal section.

nerve cord   notochord

gut

c *Right:* A new larva briefly swims about. Metamorphosis starts when its head attaches to a substrate. The tail, notochord, and most of the nervous system are resorbed (recycled and used to form new tissues). Slits multiply in the pharynx wall. Organs rotate until the openings through which water flows into and out of the pharynx become directed away from the substrate.

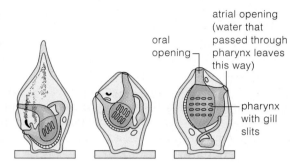

atrial opening (water that passed through pharynx leaves this way)

oral opening

pharynx with gill slits

Figure 24.3 Example of a tunicate. (**a**) Adult form of a sea squirt. (**b**) Diagram of a sea squirt larva. (**c**) Metamorphosis of the larva into the adult.

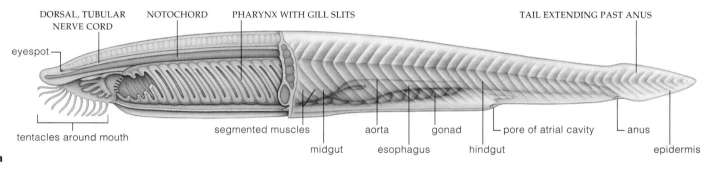

DORSAL, TUBULAR NERVE CORD   NOTOCHORD   PHARYNX WITH GILL SLITS   TAIL EXTENDING PAST ANUS

eyespot

tentacles around mouth

segmented muscles   midgut   aorta   esophagus   gonad   hindgut   pore of atrial cavity   anus   epidermis

a

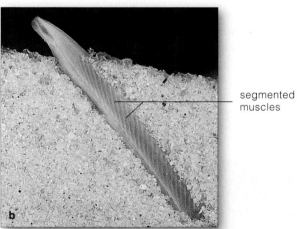

segmented muscles

b

Figure 24.4 (**a**) Cutaway view and (**b**) photograph of a lancelet, which has burrowed into sediments.

cannot deliver sufficient food to a filter-feeding animal. But a lancelet pharynx has a food-trapping surface area that is large, relative to the overall body length, and it helps secure food (Figure 24.4). Also, as many as 200 ciliated, food-trapping gill slits perforate the pharynx wall in lancelets. As you will read in sections to follow, the pharynx turned out to be an organ with interesting evolutionary possibilities for the vertebrates.

Like tunicates, lancelets are filter-feeding animals. The coordinated beating of cilia that line their mouth cavity produces a current that draws water through it. Water then moves through the pharynx, where food gets trapped in mucus. The trapped food is delivered to the rest of the gut. By itself, the beating of tiny cilia

Tunicate larvae and the lancelets use their notochord and muscles for fishlike swimming movements. Their pharynx, a muscularized tube, has finely divided openings across its thin wall.

Tunicates and lancelets are filter feeders. They use their pharynx to filter microscopic food from a current of water, which they draw through the body. Tunicate larvae also use their pharynx as a respiratory organ, for gas exchange.

EVOLUTIONARY TRENDS AMONG THE VERTEBRATES

Puzzling Origins, Portentous Trends

Did vertebrates arise from tadpole-shaped chordates? Hemichordates, obscure invertebrates of the seas, offer clues to whether this happened (*hemi*– means half, as in "halfway to chordates"). Their body plan is midway between echinoderms and chordates. They don't have a notochord, but they do have a pharynx with gill slits and a dorsal, tubular nerve cord. Their larvae resemble echinoderm *and* tunicate larvae. Did mutations in the forerunners of vertebrates alter the rates at which sex organs developed? If the sex organs matured earlier in tadpole-shaped larvae, were metamorphosis and the previous adult form dispensed with? Maybe. We know that adults of a few existing tunicate species resemble

larvae but have sex organs. Also, some amphibians are like larvae in certain respects but are sexually mature.

Assuming tadpole-shaped chordates did evolve, how did they give rise to vertebrates? The key trends started in fishes (Figures 24.5 and 24.6). One involved a shift from the notochord to reliance on a column of separate, hardened segments: **vertebrae** (singular, vertebra). This was the start of an endoskeleton—an internal skeleton—that muscles could work against and initiate motion. *A vertebral column evolved in fast-moving predators, some of which were ancestral to all other vertebrates.*

In a related trend, part of the nerve cord expanded into a brain. It did so after **jaws** evolved from the first of a series of structural supports for the gill slits. Jaws were a key innovation, an evolutionary foot in the door for novel feeding possibilities. Prey and jawed predators now began a coevolutionary race for defenses and new ways to overcome them. As part of this, selective agents favored sensory organs that could detect, say, snapping jaws, and brains that could figure out fast getaways. *A trend toward complex sensory organs and nervous systems did start among fishes and continued in land vertebrates.*

Another trend began when paired fins evolved. **Fins** are appendages that help propel, stabilize, and guide the body through water (compare Figure 24.9). In some lineages, fleshy ventral fins had skeletal supports that turned out to be forerunners of limbs. *Paired, fleshy fins were a starting point for all the legs, arms, and wings that evolved among amphibians, reptiles, birds, and mammals.*

Another trend started with a change in respiration. In lancelets, recall, oxygen and carbon dioxide simply diffuse across the skin. By contrast, in most vertebrate lineages, gills evolved. **Gills** are respiratory organs with a moist, thin, and greatly folded surface. They are richly endowed with blood vessels, and they represent a large surface area for gas exchange.

For instance, five to seven pairs of a shark's gill slits are continuous with gills extending from the pharynx to the body's surface. When a shark opens its mouth

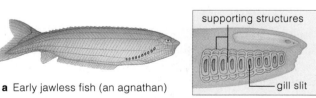

Figure 24.5 Evolutionary history of the fishes.

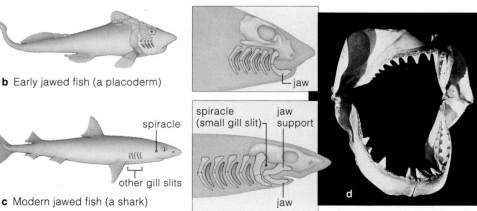

a Early jawless fish (an agnathan)

b Early jawed fish (a placoderm)

c Modern jawed fish (a shark)

Figure 24.6 Comparison of gill-supporting structures in jawless fishes and jawed fishes. In the placoderms and other early jawed vertebrates, cartilage supported the mouth's rim. In modern jawed fishes, water is drawn through a gill slit between the jaws and a nearby supporting element. In (**a**), notice the gill supports just under the skin. In (**b**) and (**c**), the gill supports are internal to the gill surface. (**d**) Jaws of a modern shark.

and closes external gill openings, the pharynx expands. Oxygen diffuses *from* water in the mouth into the gills as carbon dioxide diffuses *into* the water. Then muscles constrict the pharynx and force the oxygen-depleted, carbon dioxide–enriched water out through gill slits.

As some fishes became larger and more active, gills became more efficient. But gills can't work out of water; their thin surfaces stick together without a flow of water to keep them moist. In some fishes ancestral to the land vertebrates, pouches developed from the gut wall. The pouches evolved into **lungs**—internally moistened sacs for gas exchange. In a related trend, modifications to the heart enhanced the pumping of oxygen and carbon dioxide through the body. *Ancestors of land vertebrates relied less on gills and more on lungs. And more efficient circulatory systems accompanied the evolution of lungs.*

The First Vertebrates

Figure 24.5 has a time frame for vertebrate evolution. Free-swimming species originated in Cambrian times. They gave rise to two kinds of fishes—those without and those with jaws. Among them were ancestors of all vertebrate lineages that followed. The earliest jawless fishes (Agnatha) included **ostracoderms**. Figure 24.6a shows one of those bottom-dwelling filter feeders. Their skeleton probably was a notochord plus a protective case for an enlarged brain. The body's armorlike plates consisted of bony tissue and dentin, a hardened tissue that still persists in vertebrate teeth. Maybe the armor was useful against pincers of giant sea scorpions, but it wasn't good against jaws. When jawed fishes started their first adaptive radiation, ostracoderms vanished.

Placoderms were among the earliest fishes with jaws and paired fins (Figure 24.6b). Those bottom-dwelling scavengers and predators had a notochord reinforced with bony elements. The first gill-supporting structures were enlarged and had bony projections something like teeth, and they functioned as jaws. Before placoderms, feeding strategies had been limited to filtering, sucking, or rasping away at edible material. When placoderms started biting and tearing off large chunks of prey, they set off an evolutionary race of offensive and defensive adaptations that has continued to the present.

Placoderms diversified, but they all became extinct during the Carboniferous. New kinds of predators—the cartilaginous and bony fishes—replaced them in the seas. We will consider the living descendants of those new predators after a look at existing jawless fishes.

The vertebral column, jaws, paired fins, and lungs were pivotal developments in the evolution of lineages that gave rise to fishes, amphibians, reptiles, birds, and mammals.

EXISTING JAWLESS FISHES

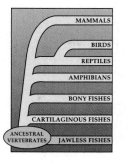

Hagfishes and **lampreys** are the jaw-free descendants of the early jawless fishes. All seventy-five existing species have a cylindrical body and a skeleton of cartilage (Figure 24.7). None has the paired fins of jawed fishes (Figure 24.6). Most are less than 1 meter long.

Hagfishes live in groups on sediments of continental shelves. They scavenge weak or dead organisms, often eating them from the inside out. Sensory tentacles surround their fleshy mouth, and their tongue rasps soft tissues from their food of choice. Within seconds after being attacked, hagfishes can secrete up to a gallon of slimy, smelly mucus from a series of glands along their body.

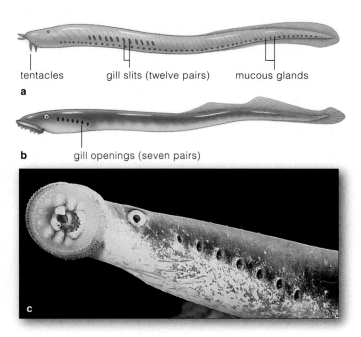

tentacles   gill slits (twelve pairs)   mucous glands

a

b   gill openings (seven pairs)

c

Figure 24.7 Body plan of (**a**) a hagfish and (**b**) a lamprey. (**c**) A lamprey's oral disk, pressed against aquarium glass.

Lampreys are specialized predators that are almost parasitic. Their suckerlike oral disk has horny, toothlike parts that rasp flesh from prey. Some types latch onto salmon, trout, and other commercially valuable fishes, then suck out juices and tissues. Just before the turn of the century, lampreys began invading the Great Lakes of North America. Their introduction resulted in the collapse of populations of lake trout and other large, valued fishes, as it has done in other aquatic habitats.

Hagfishes and lampreys get along without jaws; they latch onto prey with their efficient, specialized mouthparts.

EXISTING JAWED FISHES

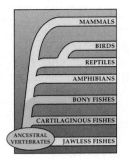

Unless you study underwater life, you may not know that fishes are the world's dominant vertebrates. Their numbers exceed those of all other vertebrate groups combined. And they show far more diversity; we already know of at least 21,000 kinds of bony fishes alone.

The form and behavior of a fish tell us about the challenges it faces in water. For instance, being 800 times denser than air, water resists fast motion through it. As an adaptation to this constraint, many marine fishes are streamlined for pursuit or escape. Sharks are like this (Figure 24.8*a*). Their long, trim body reduces friction, and tail muscles are organized for propulsive force and forward motion. By contrast, the rays and some other bottom-dwelling fishes have a flattened body (Figure 24.8*b*). Their body plan is good for hiding, not for high-speed runs.

A trout suspended in shallow water shows another adaptation to water's density. Like many fishes, it can maintain neutral buoyancy with a **swim bladder**. This adjustable flotation device exchanges gases with blood inside the body. When a trout gulps air at the water's surface, it is adjusting the volume of its swim bladder.

Cartilaginous Fishes

Cartilaginous fishes (Chondrichthyes) include about 850 species of skates, sharks, and chimaeras (Figure 24.8). These marine predators have prominent fins, a skeleton of cartilage, and five to seven gill slits on both sides. Most have a few scales or many rows of them. **Scales** are small, bony plates that typically protect the body surface without weighing it down.

The mainly bottom-dwelling skates and rays have flattened teeth for crushing hard-shelled prey. Manta rays, the largest species, can be six meters across. The venom gland in a stingray's tail might deter predators. In the tail or fins of other rays are electric organs that stun prey with as much as 200 volts of electricity.

At fifteen meters head to tail, some sharks are among the largest living vertebrates. Shark jaws bearing sharp teeth grab and rip off chunks of prey. Sharks continually shed and replace their teeth, which are modified scales. Attacks on humans have given the entire group a bad reputation. Yet sharks have been hunting invertebrates, fishes, and marine mammals for many millions of years. Surfboards with human legs dangling over the sides are a relatively new temptation for some of them.

The thirty or so species of chimaeras feed mostly on mollusks. With their bulky body and slender tail, they look a bit like a rat; hence the common name, ratfishes. They have a venom gland in front of the dorsal fin.

Figure 24.8 Cartilaginous fishes: (**a**) shark, (**b**) blue-spotted reef ray, and (**c**) chimaera, sometimes called a ratfish.

Bony Fishes

Bony fishes (Osteichthyes) are the most numerous and diverse vertebrates. They make up all but 4 percent of existing fish species. Their ancestors arose in Silurian times and gave rise to three lineages: ray-finned fishes, lobe-finned fishes, and lungfishes. Descendants of the early forms radiated into nearly all aquatic habitats.

Body plans differ greatly. Marine predators typically have a torpedo shape, a flexible body, and strong tail fins used in swift pursuit. Many reef dwellers are small finned and box shaped; they can move easily in narrow spaces. Those with an elongated, flexible body, such as the moray eel, wriggle through mud and into crevices that conceal them. The cryptic body shape of sea horses and many bottom-dwelling species conceals them from prey, predators, or both. Figure 24.9 shows examples.

Ray-finned fishes have projecting rays, derived from part of the skin, that support pairs of fins. Most have highly maneuverable fins and light, flexible scales that don't hamper complicated swimming motions. In their ancestors, lungs evolved as sac-shaped outpouchings of the wall of the esophagus, a tube to the stomach. The sacs supplemented the gills for gas exchange. In certain species, those sacs evolved into swim bladders, which help some fishes hold their position at different depths.

Figure 24.9 Some common features of the teleosts, which are the most diverse bony fishes: (**a**) Fins of a soldierfish. (**b**) Internal organs of a perch.

A few variations on the basic body plan: (**c**) Sea horse, which uses its tail to attach itself to substrates. (**d**) Long-nose gar. (**e**) A coelacanth (*Latimeria*). This "living fossil" shares traits with the early lobe-finned fishes. (**f**) Sketch of a moray eel. You can see a photograph of this reef dweller later, in Section 25.3.

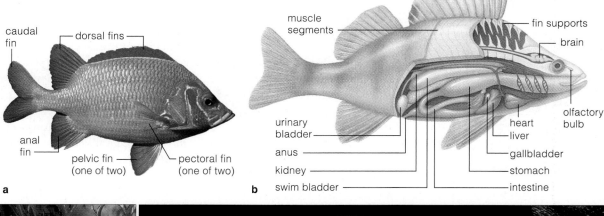

One group of the ray-finned fishes bears resemblance to their early ancestors. It includes the sturgeons and paddlefishes of the Mississippi River basin. Teleosts, the most abundant fishes, include salmon, tuna, rockfish, catfish, perch, minnows, moray eels, flying fish, sculpins, blennies, scorpionfish, and pikes. All are thin scaled or scaleless.

By contrast, only one species of lobe-finned fishes and three genera of lungfishes made it to the present. As their name suggests, a **lobe-finned fish** is unique in having paired fins that incorporate fleshy extensions from the body. As you will see next, neither it nor the lungfishes have changed much from their ancient stock.

Take a look at the coelacanth in Figure 24.9*e*. This lobe-finned fish is a "living fossil," a relic of an early time when some vertebrates first moved onto land. It has supporting, skeletal elements inside thickly lobed ventral fins. Remember the earlier description of living conditions in the Devonian? The sea level rose and fell repeatedly, and vast swamps along the coasts drained

many times. Ancestors of lobe-finned fishes evolved during prolonged intervals of drought. They probably used their lobed fins to pull themselves from dried-up ponds to ones that still had water and were habitable. What else made their pond-to-pond lurchings possible? They gulped air, and they had lungs.

Existing lungfishes provide more clues to how the ancestral forms might have made it through stressful times. They live in stagnant water but surface to gulp air. In dry seasons, when streams dwindle to mud, they encase themselves in a slathering of mud and slime that keeps them from drying out until the next rainy season.

Of all existing vertebrates, the bony fishes are the most diverse and abundant. Ray-finned fishes, lobe-finned fishes, and lungfishes are the three major lineages.

Lobe-finned fishes have fleshy ventral fins reinforced with internal skeletal elements, and they have lungs. These traits evolved in swamps that drained repeatedly during major shifts in climate during the Devonian.

AMPHIBIANS

Origin of Amphibians

As you just read in the preceding section, the Devonian lobe-finned fishes lurched over land only as a way of reaching more hospitable ponds. Even so, the very act of traveling out of water favored the evolution of supporting elements in ventral fins. It also favored the evolution of more efficient lungs. Among the evolving adventurers were the ancestors of amphibians.

We generally define an **amphibian** (Amphibia) as a vertebrate with a body plan and reproductive mode somewhere between fishes and reptiles. Most kinds of amphibians have a largely bony endoskeleton. They also have four legs, or four-legged ancestors.

Early amphibians were spending time on land by the close of the Devonian (Figure 24.10a). For them, life in those drier habitats was dangerous—and promising. Temperatures shifted more on land than in the water, air didn't support the body as well as water did, and water was not always plentiful. However, air is richer in oxygen. Amphibian lungs continued to be modified in ways that enhanced oxygen uptake. Also, circulatory systems became better at rapidly distributing oxygen to living cells throughout the body. Both modifications increased the energy base for more active life-styles.

New sensory information also challenged the early amphibians. Swamp forests supported vast numbers of edible insects and other invertebrate prey. Animals with good vision, hearing, and balance—the senses that are most advantageous on land—were favored. And brain regions concerned with those senses expanded.

All of the frogs, toads, salamanders, and caecilians alive today are descended from those first amphibians. None has escaped the water entirely. Even when they have gills or lungs, amphibians still use their thin skin as a respiratory surface—that is, for gas exchange. But respiratory surfaces must be kept moist, and skin dries easily in air. Some species still spend their whole life in water; others lay eggs in water or have aquatic larvae. Species adapted to land lay eggs in moist places, but embryos of some develop inside the moist adult body.

Frogs and Toads

Frogs and toads are the most familiar amphibians, and with close to 4,000 species, they also have been the most successful (Figure 24.11a,b). They use long hindlimbs and powerful muscles to catapult into the air or barrel through water. Most frogs capture prey by flipping a sticky-tipped tongue out of their mouth. An adult eats just about any animal it can catch and stuff into its oral cavity. As is the case for other amphibians, frog skin contains mucous glands, poison glands, and antibiotics that afford protection against many pathogens in their aquatic habitats. The body surface of highly poisonous frogs often has bright warning coloration. In Unit VI, you will have opportunities to take a look at how frogs are put together and how they function.

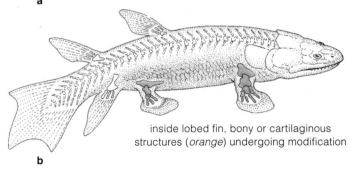

inside lobed fin, bony or cartilaginous structures (*orange*) undergoing modification

b

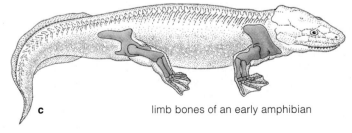

c   limb bones of an early amphibian

Figure 24.10 (**a**) *Ichthyostega*, one of the early Devonian amphibians. Fossils of this species have been recovered in Greenland. The skull, deep tail, and fins were decidedly fishlike. Unlike fishes, this species had four limbs adapted for moving across land, and a short neck intervened between its head and the rest of the body. Its vertebral column and rib cage were adapted to support the body's weight out of water. (**b**,**c**) Proposed evolution of skeletal elements inside the lobed fins of certain fishes into the limb bones of early amphibians.

Figure 24.11 Amphibians. (**a**) A frog, splendidly jumping. (**b**) An American toad. (**c**) Terrestrial stage in the life cycle of a red-spotted salamander. (**d**) Salamander versus fish mode of locomotion. (**e**) A caecilian. How do you suppose this long, thin amphibian burrows through soil? (Reflect on Section 23.10.)

a fish swimming

a salamander walking

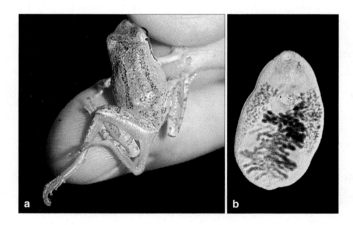

Figure 24.12 (**a**) Deformed frog with extra legs, an outcome of a parasitic infection. Stanley Sessions of Hartwick College and Stanford graduate Pieter Johnson identified a parasitic infection that affects frog limbs. (**b**) Trematodes (*Ribeiroia*) are the culprits. They burrow into tadpole limb buds and physically or chemically alter cells. Infected tadpoles grow extra legs or none at all. With heavy exposure to *Ribeiroia*, the number of tadpoles that successfully complete metamorphosis declines. Trematode cysts have been found in weirdly legged frogs and salamanders in California, Oregon, Arizona, and New York.

Frog population sizes are plummeting. Increases in parasitic attacks, predation, ultraviolet radiation, habitat losses, and chemical pollution are contributing factors (Figure 24.12). Even mud-caked hiking boots are known to import parasites into remote frog habitats.

Salamanders

About 380 species of salamanders and their kin, the newts, live in north temperate zones and tropical parts of Central and South America. Most are less than fifteen centimeters long. The forelimbs and hindlimbs are about the same size, and most project at right angles from the body. As they walk, all salamanders bend from side to side, like fishes and early amphibians (Figure 24.11*d*). Probably the first four-legged vertebrates also walked this way. Larval and adult salamanders are carnivores. Adults of some species retain some larval features. For example, an adult Mexican axolotl retains the larva's external gills. Also, its teeth and bones stop developing at an early stage. In addition, as in some other species, axolotl larvae are sexually precocious; they breed.

Caecilians

As some amphibians evolved, they lost their limbs and their vision, but not their prey-capturing jaws. They gave rise to worm-shaped caecilians (Figure 24.11*e*). Most of the 160 or so species burrow through moist soil, using senses of touch and smell to pursue insects and earthworms. The few aquatic types use electrical cues.

In body form and behavior, amphibians show resemblances to fishes and certain reptiles. Regardless of how far they venture onto land, most have not fully escaped dependency on aquatic or moist habitats to complete the life cycle.

THE RISE OF REPTILES

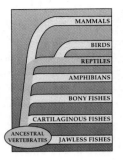

Divergences from the amphibian lineage in the late Carboniferous gave rise to **reptiles** (Reptilia). Of all vertebrates, reptiles were first to escape dependency on standing water. They did so mainly by four adaptations. *First*, they have tough, dry, scaly skin that restricts water loss from the body (Figure 24.13). *Second*, fertilization is internal, as it is in only some amphibians. A copulatory organ delivers sperm inside a female's body; sperm don't require free water to reach eggs. *Third*, reptilian kidneys are good at conserving water. *Fourth*, during the life cycle of most species, **amniote eggs** form, with internal membranes that conserve water and protect or provide metabolic support to the embryo (Figure 24.14 and Section 24.8). Most reptile eggs have a leathery or calcified shell. The embryos develop to an advanced stage before being hatched or born in dry habitats. Chapter 39 has more on the structure and development of amniote eggs.

The name Reptilia is from the Latin *repto* (to creep). Maybe the first reptiles did creep about, but a capacity to move fast and with agility evolved among some of their descendants, which race, lumber, or slither about. Compared to amphibians, even the early reptiles chased prey with far greater cunning and speed. With their well-muscled jawbones and formidable teeth, reptiles seized and applied sustained, crushing force on prey. Their prey included insects and other vertebrates. Also, their limbs generally were better at supporting the body's trunk on land. For many species, the nervous system became more complex. A reptile's brain is small compared to the rest of the body mass, but it governs forms of behavior unknown among amphibians. The cerebral cortex, where information from sensory organs is integrated and stored, first evolved among reptiles.

Reptiles called crocodilians were the first animals with a muscular, four-chambered heart fully separated

Figure 24.14
(**a**) Reconstruction of the nest of a duck-billed, plant-eating dinosaur (*Maiasaura*), discovered in Montana. This species lived 80 million years ago. It showed social behavior, as by traveling in herds and caring for the young. (**b**) Eastern hognose snakes, shown emerging from leathery shelled amniote eggs.

into two halves. Blood enters the first chamber of each half, and the second pumps it out. The separation lets oxygen-rich blood travel from lungs to the rest of the body, and oxygen-poor blood from the body to lungs, in two separated circuits (Section 34.1). Gas exchange across skin, so vital for amphibians, was abandoned by reptiles, nearly all of which have well-developed lungs.

Adaptive radiations of early reptiles led to fabulous diversity. The group called dinosaurs evolved during the Triassic. For the next 125 million years they were the dominant land vertebrates. Fossilized nests of one kind, *Maiasaura*, hold eggs and juveniles a few months old, at most (Figure 24.14a). Thus, at least some dinosaurs showed parental behavior; they took care of the young through a period of dependency. The last dinosaurs, recall, became extinct at the K–T boundary (Section 19.7). Crocodilians, turtles, tuataras, snakes, and lizards are reptilian groups that made it to the present.

Crocodiles and alligators, the closest relatives of the birds and dinosaurs, live in or near water. They have

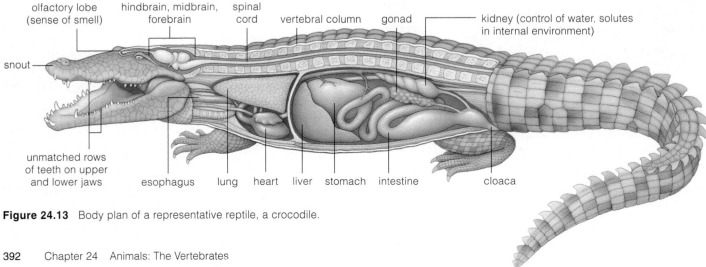

olfactory lobe (sense of smell) · hindbrain, midbrain, forebrain · spinal cord · vertebral column · gonad · kidney (control of water, solutes in internal environment)

snout

unmatched rows of teeth on upper and lower jaws · esophagus · lung · heart · liver · stomach · intestine · cloaca

Figure 24.13 Body plan of a representative reptile, a crocodile.

Figure 24.15 (**a**) Galápagos tortoise. Like most turtles, it pulls its head and limbs into its shell when threatened. (**b**) Frilled lizard, flaring a large ruff of neck skin as defensive behavior. (**c**) A rattlesnake in mid-strike. Snakes usually aren't aggressive toward people. Even so, rattlesnakes and other poisonous types bite 8,000 or so Americans annually and kill about 12 of them. King cobras and other snakes bite 200,000 or so people in India and kill about 9,000. (**d**) Tuatara (*Sphenodon*). The two existing species live on small, windswept islands near New Zealand.

powerful jaws, a lengthy snout, and sharp teeth (Figure 24.13). Large crocodiles yank mammals and birds into water, rip them apart by violently spinning their body, then gulp down the chunks. Like the other reptiles and birds, crocodilians adjust their body temperature with behavioral and physiological mechanisms. Like birds, they show complex social behaviors, as when parents guard nests and assist hatchlings into water. Such traits suggest that they share a common ancestor with birds.

The 250 existing species of turtles live inside a shell attached to its skeleton (Figure 24.15*a*). Only sea turtles and other highly mobile types have a reduced shell. The body plan works well; it's been around ever since the Triassic. Turtles use tough, horny plates and strong jaws to chew food. Often they have a fierce disposition. All lay eggs on land, then leave them. Predators eat most of the unattended eggs, so few new turtles ever hatch. Sea turtles have been hunted to the brink of extinction.

About 95 percent of modern-day reptiles, the lizards and snakes, are distantly related to dinosaurs. Most are small. But the Komodo monitor lizard is large enough to prey on young water buffalo, and the longest snake would stretch across ten yards of a football field.

Most lizards, including iguanas and geckos, live in deserts and tropical forests. Most grab insect prey with small, peglike teeth; chameleons capture them with flicks of a tongue that is longer than the body (Section 29.4). Some lizards startle predators and intimidate rivals by flaring a throat fan (Figure 24.15*b*). Many give up their tail when grabbed. The detached tail wriggles and may distract a predator enough to permit a getaway.

In the late Cretaceous, elongated, limbless snakes evolved from short-legged, long-bodied lizards. Some have bony remnants of ancestral hindlimbs (Section 16.1). Snakes are carnivores with flexible skull bones and notably moveable jaws (Figures 24.15*c* and 37.4). Pythons and boas coil around prey, suffocate it into submission, and gulp it down. Fanged types, including coral snakes, rattlesnakes, cottonmouths, copperheads, and king cobras, subdue prey with deadly venom.

Even the most feared snakes are vulnerable during their life cycle; birds and other predators relish snake eggs. Female snakes will store sperm and lay several clutches of fertilized eggs at intervals after they mate, which improves the odds that some eggs will hatch.

Tuataras are like amphibians in some brain features and in how they walk. Their body plan hasn't changed much for 140 million years (Figure 24.15*d*). Like lizards, they have a third "eye," with retina, lens, and nerves to the brain. Being covered with skin, it can only register changes in daylength and light intensity. Does it have roles in reproduction? Maybe (compare Section 32.8). Tuataras, like turtles, can live longer than sixty years. They engage in sex only after they are twenty years old.

Reptiles, with their tough, scaly skin, reliance on internal fertilization, water-conserving kidneys, and amniote eggs, were the first vertebrates to escape dependency on free-standing water in their habitats. The move onto land also involved major modifications in the nervous, circulatory, and respiratory systems.

BIRDS

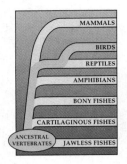

| |
| MAMMALS |
| BIRDS |
| REPTILES |
| AMPHIBIANS |
| BONY FISHES |
| CARTILAGINOUS FISHES |
| ANCESTRAL VERTEBRATES   JAWLESS FISHES |

Four groups—insects, pterosaurs (now extinct), birds, and bats— are masters of flight. And of those groups, only **birds** grow feathers. **Feathers** are lightweight structures derived from skin that are used for flight and insulation. There are about 28 orders of birds and close to 9,000 named species. They vary enormously in size, proportions, coloration, and capacity for flight. One of the smallest known birds barely tips the scales at 2.25 grams (0.08 ounce). The ostrich, the largest living bird, weighs as much as 150 kilograms (330 pounds). It can't fly but is an impressive sprinter (Figure 16.1). Many birds, such as warblers, parrots, and other perching types, differ markedly in feather coloration and territorial behavior. Bird song and other social behaviors are topics of later chapters. Figures 24.16 through 24.18 highlight just a few basic features of this diverse group of vertebrates.

The first birds evolved during an adaptive radiation of reptiles in the Mesozoic. They apparently diverged from a lineage of small reptiles that ran around on two legs. Feathers evolved as highly modifed reptilian scales.

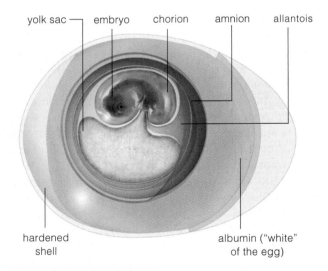

yolk sac — embryo   chorion   amnion   allantois

hardened shell

albumin ("white" of the egg)

Figure 24.16 Generalized sketch of an amniote egg. All birds lay these eggs. The amnion, chorion, allantois, and yolk sac are its four extraembryonic membranes with roles in development. Chapter 39 has an example of how an amniote egg develops. *Left:* Speckled, hard-shelled eggs of a magpie, unhatched in the nest.

a

b

c

Figure 24.17 (**a**) Flight. Of all living vertebrates, only birds and bats fly by flapping their wings. (**b**) Feathers, the key defining characteristic of birds. The flamboyant plumage of this male pheasant is an outcome of sexual selection. This bird is a native of the Himalaya Mountains of India, and it is an endangered species. As is the case for many other kinds of birds, its bright, jewel-colored feathers end up adorning humans—in this case, on the caps of native tribespeople.

(**c**) Many birds, including these Canada geese, display migratory behavior. *Animal migration* is a recurring pattern of movement between two or more regions in response to environmental rhythms. For instance, seasonal change in daylength is a cue that acts on internal timing mechanisms, or biological clocks. The cues trigger physiological and behavioral changes in individuals. Such changes induce migratory birds to make round trips between distant regions that differ in climate. Canada geese spend the summer nesting in marshes and lakes in the northern United States and Canada. Their wintering grounds are in New Mexico and other parts of the southern United States.

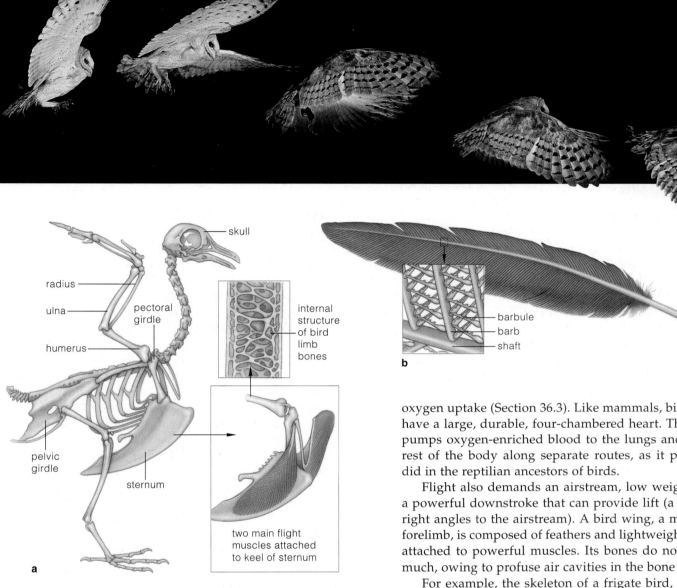

Figure 24.18 (a) Body plan of a typical bird. Flight muscles attach to the large, keeled breastbone (sternum). A bird wing is a complicated system of lightweight bones and feathers. (b) Feathers gain strength from a hollow central shaft and from tiny barbules that are interlocked in a latticelike array. See Section 18.4 for a look at the evolution of bird forelimbs.

In Figure 24.18a labels:

skull

radius

ulna

pectoral girdle

humerus

internal structure of bird limb bones

pelvic girdle

sternum

two main flight muscles attached to keel of sternum

a

In Figure 24.18b labels:

barbule

barb

shaft

b

Archaeopteryx was in or near that lineage. As you read in Section 18.8, it had reptilian *and* avian traits, including feathers. Birds still share some traits with their closest relatives, the dinosaurs and crocodilians. For example, they still have scales on their legs and some of the same internal structures. They, too, lay amniote eggs. Most show parenting behavior, such as nest guarding.

Bird flight involves more than feathers. It involves bones with a lightweight, honeycombed structure and efficient modes of respiration and circulation. High rates of metabolism sustain flight, and those rates depend on a strong flow of oxygen through the body. Bird lungs connect to elastic ventilating sacs that greatly enhance

oxygen uptake (Section 36.3). Like mammals, birds also have a large, durable, four-chambered heart. The heart pumps oxygen-enriched blood to the lungs and to the rest of the body along separate routes, as it probably did in the reptilian ancestors of birds.

Flight also demands an airstream, low weight, and a powerful downstroke that can provide lift (a force at right angles to the airstream). A bird wing, a modified forelimb, is composed of feathers and lightweight bones attached to powerful muscles. Its bones do not weigh much, owing to profuse air cavities in the bone tissue.

For example, the skeleton of a frigate bird, with its seven-foot wingspan, weighs a mere four ounces. That is less than the feathers weigh! A bird's flight muscles attach to an enlarged breastbone (sternum) and to the upper limb bones adjacent to it (Figure 24.18*a*). Muscle contraction creates the powerful downstroke required for flight. Wings, with their long flight feathers, serve as airfoils. Usually, a bird spreads out long feathers on a downstroke and thus increases the size of the surface pushing against air. On the upstroke, a bird folds its feathers somewhat, so that each wing presents the least possible resistance to air.

Like mammals, birds rely on metabolic heat to help maintain a warm body temperature even when outside temperatures decline. Feathers help conserve that heat, and the elastic air sacs help dissipate excess heat.

Of all animals, birds alone have feathers, which they use in flight, in heat conservation, and in socially significant communication displays.

THE RISE OF MAMMALS

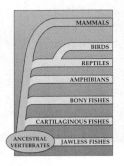

Mammals (Mammalia) are the only organisms with hair and mammary glands. Females feed their young with milk, a nutritious fluid that mammary glands secrete into ducts that open onto the body's surface (Figure 24.19a). Mammals also are distinctive in that most care for the young for extended periods, with adults serving as models for their behavior. Young mammals are born with a capacity to learn and to repeat behaviors with survival value. It is among mammals that we see the most stunning shows of **behavioral flexibility**, a capacity to expand on basic activities with novel forms of behavior and learning. This trait is more notable in some species than others. It coevolved with the expansion of brain regions, and it became greatest among the mammals called primates.

Unlike reptiles, which generally swallow prey whole, most mammals secure, cut, and in some cases chew food before swallowing it. They also differ in **dentition**, the type, number, and size of teeth. (Figure 24.19d). Their incisors (flat chisels or cones) nip or cut food and are large in grazing mammals. Canines have piercing points and are longest in meat eaters. Premolars and molars (cheek teeth) have a bumpy (cusped) platform used to crush, grind, and shear food. If a mammal has large, flat-surfaced cheek teeth, then its early ancestors relied on fibrous plants as the mainstay of their diet.

More than 200 million years ago, a divergence from small reptiles (synapsids) gave rise to therapsids, the ancestors of mammals. By the Jurassic, plant-eating and meat-eating mammals—therians—had evolved. Most were mouse-sized. They had hair and modifications in jaws, teeth, and body form. For instance, their four limbs were positioned upright under their trunk. This skeletal arrangement made it easier to walk erect, but a trunk higher from the ground was not as stable. At that time

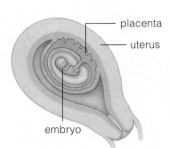

Figure 24.20 Location of the placenta in a pregnant human female.

the cerebellum, a brain region dealing with the body's balance and spatial positioning, started expanding.

Therians and dinosaurs coexisted in the Cretaceous. Then the last dinosaurs vanished at the K–T boundary (Section 19.7). With their extinction, new adaptive zones opened up for mammals. *Critical Thinking* question 4 at the chapter's end invites you to take a look at what happened. For now, it is enough to recognize that three major lineages became geographically isolated from one another on distant land masses. Even so, many of their members evolved in similar ways in similar habitats, and in time they came to resemble one another in their form and function. These species are a prime example of **convergent evolution** (Section 18.4 and Table 24.1).

We call the three lineages the egg-laying mammals (monotremes), the pouched mammals (marsupials), and the placental mammals (eutherians). Females of all three lineages lay amniote eggs. Placental mammals do have the competitive edge. They have higher metabolic rates, more precise ways to control body temperature, and a **placenta**. This spongy, blood-filled tissue of maternal and fetal membranes forms in a uterus, a chamber in which embryos develop in relative freedom from harsh external conditions (Figure 24.20). The placenta delivers nutrients and oxygen to the embryo and disposes of its metabolic wastes. Embryos of placental mammals also grow faster in a uterus than marsupials do in a pouch. Many species are fully formed—and less vulnerable to predators—at birth.

Figure 24.19 Three distinctly mammalian traits. (**a**) A human baby busily demonstrating a key defining feature: it derives nourishment from mammary glands. (**b**) Juvenile raccoons, with a thick, furry coat. Unlike reptiles, such as speckled caimans (**c**), mammals (**d**) have an upper row of four kinds of teeth that match up with four kinds on the lower jaw.

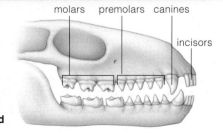

Figure 24.21
(**a**) Spiny anteater (*Tachyglossus*), an egg-laying mammal. Its sharp spines are modified hairs. (**b**) Australian pouched mammals: A female koala and her albino offspring. They eat eucalyptus leaves only, but farmland is replacing eucalyptus forests in Australia. Placental mammals: (**c**) A herd of camels, adapted to crossing hot, sandy deserts. (**d**) Manatee, which lives underwater and feeds on submerged vegetation. (**e**) Arctic fox. Its thick fur coat camouflages it from prey. In summer, its light-brown fur blends with golden-brown grasses. In winter, the fur turns white and blends with snow-covered ground.

<table>
<tr><td colspan="3">**Table 24.1**   *Examples of Morphological Convergence Between Mammalian Groups*</td></tr>
</table>

Life-Style	Homeland	Mammalian Family
Aquatic invertebrate eater	North America Central America Australia	Water shrew (Soricidae) Water mouse (Cricedidae) Platypus (Ornithorhynchidae)
Land-dwelling carnivore	North America Australia	Wolf (Canidae) Tasmanian wolf (Thylacinedae)
Land-dwelling anteater	South America Africa Australia	Giant anteater (Myrmecophagidae) Aardvark (Orycteropodidae) Spiny anteater (Tachyglossidae)
Ground-dwelling leaf, tuber eater	North America South America Eurasia	Pocket gopher (Geomyidae) Tuco-tuco (Ctenomyidae) Mole rat (Spalacidae)
Tree-dwelling leaf eater	South America Africa Madagascar Australia	Howler monkey (Cebidae) Colobus monkey (Cercopithecidae) Woolly lemur (Indriidae) Koala (Phascolarctidae)
Tree-dwelling nut, seed eater	Southeast Asia Africa Australia	Flying squirrel (Sciuridae) Flying squirrel (Anomaluridae) Flying squirrel (Phalangeridae)

The only living egg-laying mammals are the duck-billed platypus, described in the chapter introduction, and two species of spiny anteaters (Figure 24.21*a*). They incubate one egg. The hatchling suckles and finishes its development in a skin pouch that forms *temporarily* (by muscle contractions) on the mother's ventral surface. Of 260 existing species of marsupials, most are native to Australia and nearby islands; only a few live in the Americas (Figure 24.21*b*). The tiny, blind, and hairless newborns suckle and finish developing in a *permanent* pouch on the mother's ventral surface.

All other existing mammals are placental. Figure 24.21 shows a few representative species, and Appendix I lists the major groups. The body form and function, behavior, and ecology of the mammals will occupy our attention later in the book. Chapter 25, for example, offers insight into why so many existing mammalian species are threatened by a current mass extinction that we humans are bringing about.

Mammals alone have hair and mammary glands. They have distinctive dentition, a highly developed nervous system, and a notable capacity for behavioral flexibility.

Much of mammalian history has been a matter of luck, of species with particular traits being in the right or wrong places on a changing geologic stage at particular times.

24.10

FROM EARLY PRIMATES TO HOMINIDS

Let's turn now to an evolutionary story about primates. **Primates** include prosimians, tarsioids, and anthropoids (Appendix I). The first prosimians were tree dwellers. They dominated northern forests for millions of years before being displaced by monkeys and apes. The only living tarsioids are somewhere between prosimians and anthropoids in traits (Figure 24.22). Monkeys, apes, and humans are all anthropoids. In biochemistry and body form, the apes are closer to humans than to monkeys. Apes, humans, and extinct species of their lineages are hominoids. **Hominids** are all humanlike and human species of a lineage that started with a divergence from the last shared ancestor of apes and humans.

Most primates live in tropical or subtropical forests, woodlands, or savannas (grasslands with scattered trees). The vast majority are tree dwellers. Yet no one feature sets them apart from other mammals, and each lineage has its defining traits. Five trends define one that led to humans. They were set in motion as primates adapted to life up in the trees. *First,* there was less reliance on the sense of smell and more on daytime vision. *Second,* skeletal modifications promoted **bipedalism**—upright walking—which freed the hands for novel tasks. *Third,* bone and muscle changes led to refinements in hand movements. *Fourth,* teeth became less specialized. *Fifth,* changes in the brain became interlocked with changes in behavior and the evolution of culture. **Culture** is the sum of behavior patterns of a social group, passed on to generations through learning and symbolic behavior. In brief, *"uniquely" human traits emerged by modification of traits that had evolved earlier, in ancestral forms.*

Primates evolved from mammals about 60 million years ago, in tropical Paleocene forests. Like the small rodents and tree shrews they resembled (Figure 24.22*e*), they had huge appetites and foraged at night for eggs, insects, seeds, and buds under the trees. They had a long snout and a good sense of smell, suitable for snuffling food and predators. They clawed their way up stems and branches, although not with much speed or grace. In the Eocene (from 54 to 38 million years ago), some primates had moved into the trees. They had a shorter snout, enhanced daytime vision, a larger brain, and far better ways to grasp objects. *How did these traits evolve?*

Consider the trees. They offered abundant food and safety from ground-dwelling predators. They also were habitats of uncompromising selection. Picture an Eocene morning with dappled, sunlit boughs swaying in the breeze, colorful fruit, and predatory birds. A long, odor-sensitive snout wouldn't have been of much use in the trees, where air currents disperse odors. But a brain able to evaluate motion, depth, shape, and color—and one that worked fast as its owner ran and leaped about—would have been a plus. (The brain simultaneously had to estimate body weight, distance, wind speed, and the

Figure 24.22 (**a**) Tarsiers, which are vertical climbers and leapers. (**b**) A gibbon can swing arm over arm in the trees. Gibbons, orangutans, chimpanzees, and gorillas are the only existing apes. (**c**) Spider monkey, a four-legged climber, leaper, and runner. (**d**) Bonobos, our closest primate relatives, walk upright. (**e**) Tree shrew. Although not in the order Primates, it resembles early ancestors of primates.

suitability of a destination. Adjustments had to be quick.) Also, skeletal changes that increased fitness would have been favored. For instance, forward-directed eye sockets would have contributed to enhanced depth perception.

By 36 million years ago, tree-dwelling anthropoids had evolved. They were on or close to the lineage that led to monkeys and apes. One had forward-directed eyes, a snoutless, flattened face, and an upper jaw with shovel-shaped front teeth. It must have used its hands to grab food. Some early anthropoids lived in swamps infested with predatory reptiles. Was that one reason why it became imperative to think fast, grip strongly,

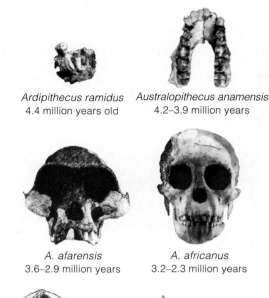

Ardipithecus ramidus
4.4 million years old

Australopithecus anamensis
4.2–3.9 million years

A. afarensis
3.6–2.9 million years

A. africanus
3.2–2.3 million years

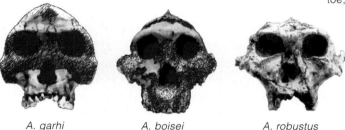

A. garhi
2.5 million years

A. boisei
2.3–1.4 million years

A. robustus
1.9–1.5 million years

Figure 24.23 Representative fossils of hominids that lived in Africa between 4.4 and 1.4 million years ago.

Figure 24.24 (**a**) Remains of "Lucy" (*A. afarensis*), dated at 3.2 million years ago. (**b**) In Tanzania, Mary Leakey found these footprints, made 3.7 million years ago. (**c,d**) The arch, big toe, and heel marks of the footprints are signs of bipedalism. Unlike apes, early hominids didn't have a splayed big toe, such as the one obligingly demonstrated by the chimpanzee in (**c**).

and avoid adventures on the ground? Maybe. Slip-ups were always possible; primates still fall out of trees.

Between 25 and 5 million years ago, in the Miocene, apelike forms—*the first hominoids*—evolved and spread into Africa, Asia, and Europe. Continents were on the move, and shifts in ocean circulation patterns triggered a long-term cooling trend. African climates grew cooler, drier, and more seasonal. Tropical forests, with their edible soft fruits, leaves, and abundant insects, started giving way to woodlands and later to open grasslands. Food was drier and harder (and harder to find). Now the hominoids that had evolved in the forests had two options: Move into new adaptive zones or die out. Not all made the transition; most became extinct. But two lineages branched from a common ancestor between 6 million and 4 million years ago. One gave rise to the great apes, and the other gave rise to the first *hominids*. Fossil fragments of early hominids, some older than 5 million years, have been found in eastern and southern Africa (Figure 24.23 and 24.24). One species, *Ardipithecus ramidus*, lived 4.4 million years ago. It was more apelike than humanlike. Its small molars, large canines, and thin tooth enamel suggest that its diet consisted of easily chewed fruits and vegetables. Paleontologist Tim White puts it in the minor leagues of brain development.

Diverse hominids evolved over the next 2 million years. We still don't know how they are related, so we informally call them **australopiths**, or "southern apes." At this writing, the oldest known australopith species is *Australopithecus anamensis*. Like *A. afarensis* and *A. africanus*, it was gracile (slightly built). The ones called *A. boisei* and *A. robustus* were robust (muscular and heavily built). Like apes, australopiths had a large face, protruding jaws, and a small skull (and brain) size. Yet they differed in key respects from hominoids. Example: Their molars had thicker enamel and could grind harder foods. Also, australopiths could walk upright. We know this from fossilized hip bones and limb bones and from footprints. Around 3.7 million years ago, *A. afarensis* walked on newly fallen volcanic ash, which a light rain turned into quick-drying cement (Figure 24.24*b*).

Bipedal hominids had started to evolve late in the Miocene. In the trees, their hands had become adapted to grip objects strongly and with precision. When their descendants left the trees for life on the ground, rather than becoming specialized in running fast on all fours, they used manipulative skills to advantage. They kept their hands free to hold offspring and probably to carry precious food during their foraging expeditions.

Primates evolved from small, rodentlike mammals that moved into the trees 60 million years ago. In the Miocene, the first hominoids (apelike forms) evolved and radiated through Africa, Europe, and southern Asia. Divergences led to the australopiths—the first hominids. They were humanlike in a crucial respect: They walked upright.

These key adaptations evolved along the pathway from arboreal primates to modern humans: complex, forward-directed vision; bipedalism; refined hand movements; generalized dentition; and an interlocked elaboration of brain regions, behavior, and culture.

EMERGENCE OF EARLY HUMANS

What might the fossilized fragments of early hominids tell us about our own origins? The record is still too sketchy for us to know how the diverse australopiths were related to one another, let alone which ones may have been ancestral to humans. Besides, which traits do characterize **humans**—members of the genus *Homo*?

Well, what about brains? The modern human brain is the basis of fine analytical and verbal skills, complex social behavior, and technological innovation. It clearly sets us apart from apes, which have a far smaller skull volume and brain size. Yet this feature alone can't tell us when certain hominids made the leap to being human. Why? Their brain size fell within the ape range. And even though they made simple tools, so do chimps and some parrots. We have no clues to their social behavior.

We are left to speculate on physical traits, indicated by many fossils—a skeleton adapted for bipedalism; manual dexterity; larger brain volume; a smaller face; and smaller, more thickly enameled teeth. These traits, which apparently originated during the late Miocene, are signs of what may have been the earliest humans—*Homo habilis,* a name that means "handy man."

Between 2.4 and 1.6 million years ago, early forms of *Homo* lived in dry woodlands adjoining the savannas of eastern and southern Africa. Their teeth tell us they could eat hard-shelled nuts, dry seeds, soft fruits, leaves, and insects—all seasonal foods. Most likely, *H. habilis* had to think ahead, to plan when to gather and store foods that would help it survive the cold dry season.

H. habilis shared its habitat with saber-tooth cats and other predators. The cats' teeth could impale prey and shear off flesh but couldn't crush open marrow bones. Carcasses with meat shreds clinging to marrow bones afforded nutrients in nutrient-stingy places. *H. habilis* was not a full-time carnivore. But it may have enriched its diet opportunistically, by scavenging carcasses.

Fossil hunters have found many stone tools dating to the time of *H. habilis.* They cannot say for sure that *H. habilis* was the only species that made them. Perhaps australopiths as well as *H. habilis* used sticks and other perishable materials for tools before then, as modern apes do, but we have no way of knowing.

Figure 24.25 (**a**) Artist's view of *Homo habilis* males alert to australopiths in an East African woodland. Two of the 37,000+ stone tools from Olduvai Gorge: (**b**) chopper and (**c**) cleaver.

Maybe ancestors of modern humans started down a toolmaking road by picking up rocks to crack marrow bones. Maybe they scraped flesh from bones with sharp flakes split naturally from rocks. However it happened, at some point they started *shaping* stone implements. Paleontologist Mary Leakey was the first to discover evidence of stone toolmaking in sedimentary layers at Africa's Olduvai Gorge. In the deepest layers are the oldest tools—crudely chipped pebbles (Figure 24.25). Early humans might have used them to smash marrow bones, dig for roots, and poke insects from bark. More recent layers have more complex tools.

At such sites we find fossils of an early form of *Homo* that was twice as brainy as australopiths and obviously ate well. There apparently was no selection pressure for more creativity in securing food resources; stone tools did not change much for the next 500,000 years.

Ancestors of modern humans stayed put in Africa until about 2 million years ago. Then, a divergence led to *Homo erectus*—a species related to modern humans (Figures 24.26 and 24.27). Its name means upright man. Although its forerunners also were upright walkers, *H. erectus* populations did the name justice. They trekked

Figure 24.26 Representative fossils of early humans (*Homo*), Reconstructed portions are the dark areas on which the fossil fragments are mounted.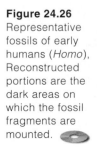

Homo rudolfensis
2.4–1.8 million years

H. habilis
1.9–1.6 million years

H. erectus
2 million–53,000? years

H. neanderthalensis
200,000–30,000 years

H. sapiens
100,000–?

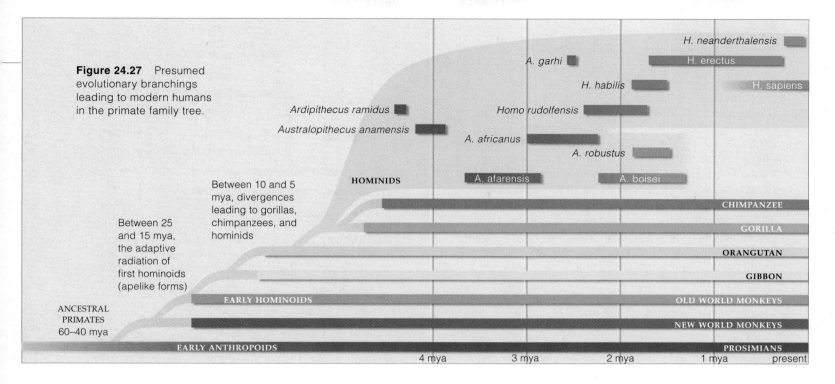

Figure 24.27 Presumed evolutionary branchings leading to modern humans in the primate family tree.

Ardipithecus ramidus

Australopithecus anamensis

A. garhi

H. neanderthalensis

H. erectus

H. habilis

H. sapiens

Homo rudolfensis

A. africanus

A. robustus

Between 10 and 5 mya, divergences leading to gorillas, chimpanzees, and hominids

HOMINIDS

A. afarensis

A. boisei

Between 25 and 15 mya, the adaptive radiation of first hominoids (apelike forms)

CHIMPANZEE

GORILLA

ORANGUTAN

GIBBON

EARLY HOMINOIDS

OLD WORLD MONKEYS

ANCESTRAL PRIMATES 60–40 mya

NEW WORLD MONKEYS

EARLY ANTHROPOIDS

PROSIMIANS

4 mya   3 mya   2 mya   1 mya   present

out of Africa, turned left into Europe, and right into Asia. Some even walked to China; fossils from Southeast Asia and the former Soviet republic of Georgia are 1.8 million and 1.6 million years old. *H. erectus* survived as ice sheets advanced—more than once—into northern Europe, southern Asia, and North America.

Whatever pressures triggered the far-flung travels, this was a time of physical changes, as in skull size and leg length. It also was a time of cultural lift-off for the human lineage. *H. erectus* had a larger brain and was a more creative toolmaker. Its social organization and communication skills must have been well developed. How else can we explain its successful dispersal? From southern Africa to England, different populations used the same kinds of hand axes and other tools to pound, scrape, shred, chop, cut, and whittle material. *H. erectus* withstood environmental challenges by building fires and using furs for clothing. Clear evidence of fire use dates from an ice age in the early Pleistocene.

As Middle Eastern fossils show, *Homo sapiens* evolved by 100,000 years ago. Early *H. sapiens* had smaller teeth and jaws than *H. erectus* (Figure 24.26). Many had a novel feature: a chin. Facial bones were smaller, the skull higher and rounder, the brain larger. *H. erectus* might have developed complex language. But their origin and geographic dispersal are hotly debated (Section 24.12).

One group of early humans, the Neandertals, lived in Europe and in the Near East from 200,000 to 30,000 years ago. Massively built and large brained, some were the first to adapt to the coldest regions (Figure 24.28). Their disappearance coincided with the appearance of anatomically modern humans in the same areas about 40,000 to 30,000 years ago. We have no evidence that Neandertals warred or interbred with the later arrivals. We still do not know what happened to them. We do know that Neandertal DNA has unique sequences, so

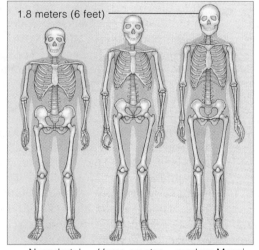

Figure 24.28 Human body build correlated with climate. The body of early and modern humans that became adapted to cold climates are heat-conserving (for example, stocky, with short legs) compared to the body of humans adapted to hot climates.

1.8 meters (6 feet)

Neandertal   *Homo erectus*   modern Masai

they might not have contributed to the gene pools of modern European populations.

From 40,000 years ago to today, human evolution has been almost entirely cultural, not biological. And so we leave our story with these conclusions: Humans spread rapidly through the world by devising *cultural* means to deal with a broader range of environments. Compared to their predecessors, they developed rich, varied cultures. Although hunters and gatherers persist in parts of the world, others moved from "stone-age" technology to the age of "high tech," attesting to the great plasticity and depth of human adaptations.

Cultural evolution has outpaced the biological evolution of the only remaining human species, *H. sapiens*. Today, humans everywhere rely on cultural innovation to adapt rapidly to a broad range of environmental challenges.

Out of Africa—Once, Twice, Or . . .

If researchers are interpreting the fossil record of human evolution correctly, then it would seem Africa was the cradle for us all. At this writing, at least, no one has any fossils of humans older than 2 million years *except* in Africa. *H. erectus* coexisted for a time with earlier humans (*H. habilis*) before dispersing from Africa to the cooler grasslands, forests, and mountains of Europe and Asia. This form apparently left Africa in waves between about 2 million and 500,000 years ago. Judging from *H. erectus* fossils from Java, isolated populations may have endured until some time between 53,000 and 37,000 years ago.

Where, on the larger geologic stage, do we place the origin of *H. sapiens*? *Here we find a good example of how the same body of evidence can be interpreted in different ways.* The interpretations are called the **multiregional model** and **African emergence model** for modern human origins. Both attempt to explain the world distribution of fossils of particular ages. Both are based on measurements of genetic distances among existing human populations.

For example, biochemical and immunological studies both indicate that the greatest genetic distance separates *H. sapiens* populations native to Africa from all other human populations. The next greatest distance separates Southeast Asian and Australian populations from others (Figure 24.29). Figure 24.30 summarizes another line of evidence; it correlates *H. sapiens* fossils to specific times.

MULTIREGIONAL MODEL   By this model, *H. erectus* populations spread through much of the world by about 1 million years ago. The geographically separated groups faced different selection pressures, so their traits evolved in regionally distinctive ways. Then the subpopulations ("races") of *H. sapiens* evolved from them in different places. They differed phenotypically but did not evolve into separate species because gene flow continued among them, even to the present. Thus, for example, while the armies of Alexander the Great were sweeping eastward, they also contributed "blue-eye genes" from Greeks to the allele pool of generally brown-eyed subpopulations in Africa, the Near East, and Asia.

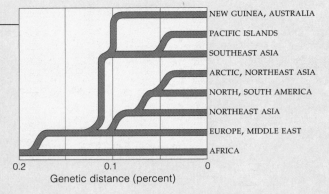

Figure 24.29   One proposed family tree for populations of modern humans (*Homo sapiens*) that are native to different regions. Branch points show presumed genetic divergences. The tree is based on nucleic acid hybridization studies of many genes (including those for mitochondrial DNA and the ABO blood group) and immunological comparisons.

AFRICAN EMERGENCE MODEL   This model does not dispute fossil evidence that populations of *H. erectus* evolved in different ways in different regions. However, it holds that *H. sapiens*—modern humans—originated in sub-Saharan Africa somewhere between 200,000 and 100,000 years ago. Only later did *H. sapiens* populations move out of Africa, into regions along the routes shown in the Figure 24.30 map. In each region that populations of *H. sapiens* settled, they replaced the archaic *H. erectus* populations that had preceded them. Only then did regional phenotypic differences become superimposed on the original *H. sapiens* body plan.

In support of this model, the oldest known *H. sapiens* fossils are from Africa. Also, in Zaire, a finely wrought barbed-bone harpoon and other exquisite tools suggest that African populations were as skilled at making tools as *Homo* populations known earlier from Europe. Also, in 1998, researchers at the University of Texas and in China announced findings from the Chinese Human Genome Diversity Project. Detailed analysis of gene patterns from forty-three ethnic groups in Asia suggest that modern humans moved from Central Asia, along the coast of India, then on into Southeast Asia and southern China. Populations later moved north and northwest into China, then into Siberia, and then on down into the Americas.

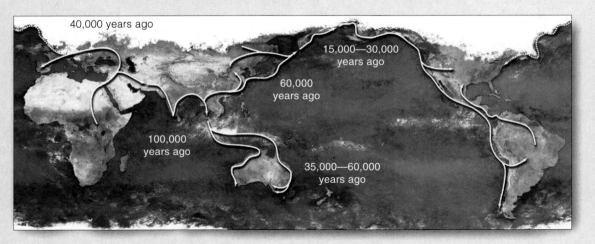

Figure 24.30
Estimated times when populations of early *H. sapiens* were colonizing different regions of the world, based on radiometric dating of fossils. The presumed dispersal routes (*white* lines) seem to support the African emergence model.

SUMMARY

Gold indicates text section

1. Generally, chordate embryos have a notochord, a dorsal hollow nerve cord, a pharynx with gill slits (or hints of these), and a tail extending past the anus. Some or all of these traits persist in adults. Chordates with a backbone are vertebrates. Tunicates (e.g., sea squirts) and lancelets are invertebrate chordates. *24.1, 24.2*

2. The eight vertebrate classes are jawless fishes, jawed armored fishes (now extinct), cartilaginous fishes, bony fishes, amphibians, reptiles, birds, and mammals. *24.1*

3. The earliest vertebrates, the jawless fishes that arose in the Cambrian, included ostracoderms. Their modern descendants are lampreys and hagfishes. Jawed fishes also arose in Cambrian times. A once-dominant lineage, the placoderms, became extinct. Other lineages gave rise to the cartilaginous fishes and bony fishes. *24.3–24.5*

4. Four evolutionary trends are associated with certain vertebrate lineages: *24.3, 24.5*

   a. A vertebral column supplanted the notochord as a structural element against which muscles can act. This development led to fast-moving predatory animals.

   b. Jaws evolved from gill-supporting elements. This led to increased predator–prey competition; it favored more efficient nervous systems and sensory organs.

   c. In one lineage of bony fishes, paired fins evolved into fleshy lobes having internal structural elements. The lobes were the forerunners of paired limbs.

   d. In some fish lineages, respiration by gills came to be supplemented by paired lungs, which proved to be adaptive during the invasion of the land. In a related development, the circulatory system became far more efficient at distributing oxygen through the body.

5. The amphibian body plan and reproductive mode are between those of fishes and reptiles. Amphibians were the first vertebrates to invade the land, but they never fully escaped the water. Their skin dries out, and aquatic stages persist in most life cycles. *24.6*

6. Unlike amphibians, reptiles escaped dependence on free-standing water. Key adaptations that permitted them to do so are a toughened, scaly skin that restricts water loss; a reliance upon internal fertilization; more efficient, water-conserving kidneys; and amniote eggs, often leathery or shelled, that protect and metabolically support the embryos. *24.7*

7. Like reptiles, the birds and mammals (both of which descended from reptiles) have efficient circulatory and respiratory systems. They also have a highly developed nervous system and sensory organs. *24.8, 24.9*

   a. Of all organisms, birds alone have feathers, used in flight, heat conservation, and social displays.

   b. Of all organisms, only mammals have mammary glands that produce milk. They have hair or thick skin that functions in insulation. They also have distinctive dentition and a highly developed cerebral cortex, the brain's most complex region. Adults of most species nurture their offspring through an extended period of dependency and learning.

   c. The first mammals evolved from small, hairless reptiles in the Triassic. After the extinction of the last dinosaurs and geologic isolation following the breakup of Pangea, three lineages radiated into new adaptive zones. All three have persisted to the present. They are egg-laying mammals (monotremes), pouched mammals (marsupials), and placental mammals (eutherians).

   d. Placental mammals have higher metabolic rates, more precise control of body temperature, and a more efficient way to nourish and protect embryos. In many places, they outcompeted monotremes and marsupials, which retain many archaic traits. In similar habitats in different parts of the world, some members of the three lineages also show morphological convergence.

8. Primates include the prosimians (lemurs and related forms), tarsioids, and anthropoids (monkeys, apes, and humans). Apes and humans alone are hominoids. Only australopiths and anatomically modern humans (*Homo sapiens*) and others of their lineage, including *H. habilis* to *H. erectus*, are currently classified as hominids. *24.10*

   a. The first primates were tiny, rodentlike mammals that evolved by 60 million years ago in tropical forests.

   b. Hominoids (apelike forms) evolved in the Miocene between 25 to 5 million years ago in Africa. Some gave rise to hominids by at least 5 million years ago.

   c. The origin and early evolution of hominids might coincide with selection pressures that emerged during a long-term change in climate, brought about by shifts in land masses and ocean circulation patterns. Tropical forest gave way to dense woodlands, then grasslands. The hominids were bipedal (upright walkers), they had modified teeth that handled diverse foods. They also had a more complex brain that gave them a capacity to plan for securing scarcer and seasonally available food.

9. Modern humans emerged by modification of traits in ancestral primate forms. Their tree-dwelling ancestors relied less on the sense of smell and more on enhanced daytime vision. Manipulative skills increased when the hands began to be freed from load-bearing functions. Starting with Miocene apes, there were shifts from four-legged climbing to bipedalism, from specialized to omnivorous eating habits, and to remarkable increases in brain complexity and behavior. *24.10*

10. *H. habilis*, the earliest known species of the genus *Homo*, evolved by 2.5 million years ago. It was one of the first stone toolmakers. *Homo erectus*, ancestral to modern humans, evolved by 2 million years ago. *H. erectus* populations radiated out of Africa, into Asia and Europe. The oldest known fossils of early modern humans (*H. sapiens*) date from about 100,000 years ago. About 40,000 years ago, cultural evolution outstripped biological evolution of the human form. *24.10, 24.11*

1. Which features distinguish chordates from other groups of animals? *24.1*

2. List four major trends that apparently developed during the evolution of at least some vertebrate lineages. *24.3*

3. Describe some of the defining characteristics of jawless fishes, cartilaginous fishes, and bony fishes. Which group is the most abundant? *24.3–24.5*

4. Which traits distinguish reptiles from amphibians? *24.6, 24.7*

5. Birds, crocodilians, and dinosaurs share what traits? *24.7, 24.8*

6. List some characteristics that distinguish each of the three mammalian lineages from the reptiles. *24.7, 24.9*

7. What are the key differences between a "hominoid" and a "hominid"? *24.10*

8. Briefly describe some of the conserved physical traits that connect anatomically modern humans with their mammalian ancestors, then their primate ancestors. *24.10*

Self-Quiz   ANSWERS IN APPENDIX III

1. Only _____ have a notochord, a tubular dorsal nerve cord, a pharynx with slits in the wall, and a tail extending past the anus.
   a. echinoderms   d. both b and c
   b. tunicates and lancelets   e. all of the above
   c. vertebrates

2. Gill slits function in _____ .
   a. respiration   d. water regulation
   b. circulation   e. both a and c
   c. food trapping

3. A shift from a reliance on _____ to reliance on _____ was pivotal in the evolution of all vertebrates.
   a. the notochord; a backbone   c. gills; lungs
   b. filter feeding; jaws   d. all of the above

4. The first vertebrates included _____ .
   a. bony fishes   c. lobe-finned fishes
   b. jawless fishes   d. both a and b

5. Of all existing vertebrates, _____ are the most diverse.
   a. cartilaginous fishes   d. reptiles
   b. bony fishes   e. birds
   c. amphibians   f. mammals

6. Generally, the only amphibian groups that entirely escaped dependency on free-standing water are _____ .
   a. salamanders   c. caecilians
   b. toads   d. all require it

7. Reptiles moved fully onto land owing to _____ .
   a. tough skin   d. amniote eggs
   b. internal fertilization   e. none of the above
   c. good kidneys   f. all of the above

8. A four-chambered heart evolved first in _____ .
   a. bony fishes   d. mammals
   b. amphibians   e. crocodilians
   c. birds   f. both c and d

9. Amniote eggs form during the life cycle of most _____ .
   a. reptiles   c. mammals
   b. birds   d. all of the above

10. Birds use feathers in _____ .
    a. flight   c. social functions
    b. heat conservation   d. all of the above

11. Various mammals _____ .
    a. hatch
    b. finish embryonic development in a pouch
    c. finish embryonic development in a uterus
    d. both b and c
    e. all of the above

12. Early humans _____ .
    a. became adapted to a wide range of environments
    b. became adapted to a narrow range of environments
    c. had flexible bones that cracked easily
    d. were limber enough to swing through the trees

13. Match the organisms with the appropriate features.
    ____ jawless fishes   a. milk glands, behavioral
    ____ cartilaginous   flexibility, thick skin or hair
         fishes   b. respiration by skin and lungs
    ____ bony fishes   c. include coelacanths
    ____ amphibians   d. include hagfishes
    ____ reptiles   e. include sharks and rays
    ____ birds   f. complex social behavior, feathers
    ____ mammals   g. first with amniote eggs

Critical Thinking

1. List factors that could contribute to the collapse of native fish populations in a lake after the accidental introduction of lampreys to it.

2. Think about the flight muscles of birds and their demands for oxygen and ATP energy. What type of organelle would you expect to be profuse in such muscles? Explain your reasoning.

3. Fossil hunters Kathie and Gary unearthed the complete fossilized remains of an early mammal. How can they determine whether the mammal was a herbivore, carnivore, or omnivore?

4. Select one example of morphological convergence listed in Table 24.1. Then use the family names to search the literature and the web for information on their similarities in body form, behavior, and habitats. Write up what you find, and include appropriate illustrations.

5. As you read in this chapter, three mammalian lineages made it to the present: the monotremes (egg-laying mammals), marsupials (pouched mammals), and eutherians (placental mammals). Again, monotremes and marsupials retain many archaic traits. Placental mammals have higher metabolic rates, more precise ways of regulating body temperature, and an efficient way to nourish embryos inside the mother's body.

   Now take a look at Figure 24.31*a* on the facing page. The fossil record indicates the ancestors of the monotremes and marsupials had entered the southern part of the supercontinent Pangea by the late Jurassic. After the breakup of Pangea, the species on the land mass that would become Australia were already isolated from ancestors of placental mammals, which were evolving elsewhere (Figure 24.19*b*).

   In the future South America, monotremes were replaced by marsupials and early placental mammals that had evolved independently on other continents. A land bridge rejoined South and North America in the Pliocene. Highly evolved placental mammals radiated southward and rapidly drove many of the previously isolated mammals to extinction. Only opossums and a few other species successfully invaded land to the north.

   Today in Australia, many marsupial species compete with recently introduced placental mammals for resources but are not winning. You can read about one example of this in Section 42.8, which describes the introduction of rabbits into Australia. Write a brief paragraph on the reasons why placental mammals such as rabbits that did not evolve in Australian habitats are showing greater fitness than native mammals.

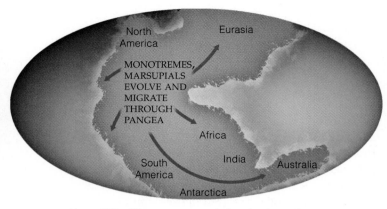

a About 150 million years ago, during the Jurassic

North America
Eurasia
MONOTREMES, MARSUPIALS EVOLVE AND MIGRATE THROUGH PANGEA
Africa
South America
India
Australia
Antarctica

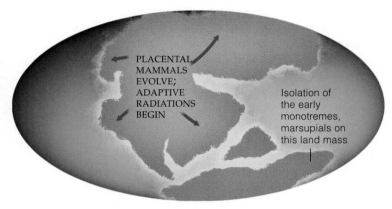

b Between 100 and 85 million years ago, during the Cretaceous

PLACENTAL MAMMALS EVOLVE; ADAPTIVE RADIATIONS BEGIN
Isolation of the early monotremes, marsupials on this land mass

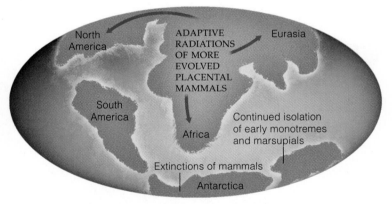

c About 20 million years ago, during the Miocene

North America
ADAPTIVE RADIATIONS OF MORE EVOLVED PLACENTAL MAMMALS
Eurasia
South America
Africa
Continued isolation of early monotremes and marsupials
Extinctions of mammals
Antarctica

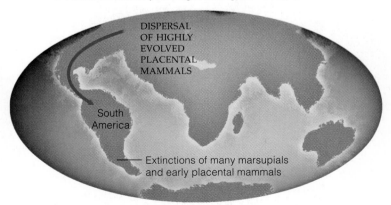

d About 5 million years ago, during the Pliocene

DISPERSAL OF HIGHLY EVOLVED PLACENTAL MAMMALS
South America
Extinctions of many marsupials and early placental mammals

Figure 24.31 Adaptive radiations of mammals.

Figure 24.32
Model for a long-term shift in the global climate in the Miocene on into the Pliocene.

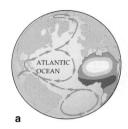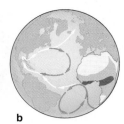

ATLANTIC OCEAN

a   **b**

6. Before the Isthmus of Panama formed (Figure 24.32*a*), the salinity levels of ocean surface currents were much the same around the world. Circulation patterns kept Arctic waters warm. After the isthmus formed (Figure 24.32*b*), North Atlantic surface currents became saltier and heavier, so they sank before reaching the Arctic. With colder water in polar regions, the Arctic ice cap formed and ushered in a long-term trend toward a cooler, drier climate in Africa. Referring to Section 24.10, speculate on how the change may have been a factor in hominid evolution.

7. The first mammals spread their toes apart to support the body as they walked or ran on four legs. Primates still spread fingers or toes. Many also cup their fingers, as when monkeys lift food to the mouth. In ancient tree-dwelling primates, modifications to hand bones allowed them to wrap their fingers around objects (prehensile movements) and touch a thumb to the tip of a finger (opposable movements). Later, hands were freed of load-bearing functions, and the power grip and precision grip evolved:

power grip   precision grip

Write a paragraph on the roles these skeletal modifications may have played in the cultural evolution of early humans.

Selected Key Terms

Vertebrates

amniote egg 24.7
amphibian 24.6
behavioral flexibility 24.9
bird 24.8
bony fish 24.5
cartilaginous fish 24.5
chordate 24.1
convergent evolution 24.9
dentition 24.9
feather 24.8
filter feeder 24.2
fin 24.3
gill 24.3

gill slit 24.2
hagfish 24.4
jaw 24.3
lamprey 24.4
lancelet 24.2
lobe-finned fish 24.5
lung 24.3
mammal 24.9
nerve cord 24.1
notochord 24.1
ostracoderm 24.3
pharynx 24.1
placenta 24.9
placoderm 24.3
reptile 24.7
scale (fish) 24.5

swim bladder 24.5
tunicate 24.2
vertebra 24.3
vertebrate 24.1

Human Evolution

African emergence model 24.12
australopith 24.10
bipedalism 24.10
culture 24.10
hominid 24.10
human 24.11
multiregional model 24.12
Primates (order) 24.12

Readings

Gould, S. J. (general editor). 1993. *The Book of Life*. New York: Norton. Splendid easy-to-read essays, gorgeous illustrations.

Romer, A. S., and T. S. Parsons. 1986. *The Vertebrate Body*. Sixth edition. Philadelphia: Saunders. Still one of the best.

Tattersall, I. 1998. *Becoming Human: Human Evolution and Human Uniqueness*. Pennsylvania: Harvest Books.

On-Line readings at Student Guide for InfoTrac:
www.brookscole.com/biology

BIODIVERSITY IN PERSPECTIVE

The Human Touch

In 1722, on Easter morning, a European explorer landed on a small volcanic island and found a few hundred skittish, hungry Polynesians living in caves. The island had no trees whatsoever, only dry, withered grasses and scorched, shrubby plants. There were about 200 massive stone statues near the coast and 700 unfinished, abandoned ones in inland quarries (Figure 25.1). Some weighed fifty tons. Without trees for wood and fibrous plants to make ropes, how were they erected? There were no wheeled carts or strong animals. How did the statues get from the quarries to the coast?

Two years later James Cook visited the island. When islanders paddled out to his ships, he noted that they didn't have the knowledge or materials to keep their canoes from leaking; they spent half the time bailing water. He saw only four canoes on the whole island. Nearly all of the statues had been tipped over, often onto spikes that shattered the faces on impact.

Later, researchers solved the mystery of the statues. Easter Island, as it came to be called, has a mere 165 square kilometers (64 square miles) of land. Voyagers from the Marquesas discovered this eastern outpost of Polynesia around A.D. 350, possibly after being blown off course by storms. The place was a paradise. Its fertile volcanic soil supported dense palm forests, hauhau trees, toromino shrubs, and lush grasses. The new arrivals built large, buoyant, ocean-going canoes of long, straight palms strengthened with rope made of fibers from hauhau trees. They used toromino wood as fuel to cook fishes and dolphins; they cleared forests to plant taro, bananas, sugarcane, and sweet potatoes.

By 1400, between 10,000 and 15,000 descendants were living on the island. Society had become highly structured, given the need to cultivate, harvest, and distribute food for so many. Crop yields had declined; precious soil nutrients had been carried off by each harvest and by ongoing, neglected erosion. Edible species vanished from the waters around the island, so fishermen had to build larger canoes to sail out ever farther, into the open and more dangerous sea.

Figure 25.1 On Easter Island, a few of the massive stone tributes to the gods. Apparently islanders erected the statues as a plea for divine intervention after their once-large population wiped out the biodiversity of their tropical paradise. The plea didn't have any effect whatsoever on reversing the losses on land and in the surrounding sea. The human population didn't recover, either.

Survival was at stake. Those in power appealed to the gods. They directed the community to carve divine images of unprecedented size and power, and to move the newly carved statues over miles of rough terrain to the coast. Jo Anne Van Tilburg recently demonstrated that they probably lashed the statues to canoe-shaped rigs, then rolled them along a horizontal "ladder" of greased logs. She got the idea after observing how existing islanders haul canoes from water onto land.

By now, islanders had all of the arable land under cultivation. They had eaten all the native birds, and no new birds came to nest. They were raising and eating rats, the descendants of hitchhikers on the first canoes. They coveted palm seeds, now a scarce delicacy.

Wars broke out over dwindling food and space. By about 1550 no one was venturing offshore to harpoon fishes or dolphins. They couldn't build canoes because the once-rich forests were gone. They had cut down all of the palms. Hauhau trees were extinct; islanders had used every last one as firewood for cooking.

The islanders turned to the only remaining source of animal protein. They started to hunt and eat one another.

Central authority crumbled, and gangs replaced it. As gang wars raged, those on the rampage burned the remaining grasses to destroy hideouts. The rapidly dwindling population retreated to caves and launched raids against perceived enemies. Winners ate the losers and tipped over the statues. Even if the survivors had wanted to leave the island, they no longer had a way to do so. What possibly could they have been thinking when they chopped down the last palm?

Easter Island initially sustained a society that rose to greatness amid abundant resources, then abruptly fell apart. In life's greater evolutionary story, its loss of biodiversity might not even warrant a footnote. After all, compare the loss to the many millions of species that appeared and became extinct in the distant past!

Yet there is a lesson here. The near-total destruction that 15,000 Polynesians wrought on biodiversity was confined to a small bit of land in the vastness of the Pacific Ocean. What are the requirements and whims of *6 billion* people doing to global biodiversity today? Is the loss of species on Easter Island an isolated case? Or should we take what happened there as a warning of a worldwide extinction crisis of our own making, one that is even now under way? Let's take a look.

Key Concepts

1. The current range of global biodiversity is largely an outcome of an overall pattern of abrupt extinction crises and slow recoveries. It takes tens of millions of years to reach the level of biodiversity that prevailed before such mass extinctions.

2. The loss of individual species as well as extinctions of major groups may have complicated causes, the understanding of which requires scientific analysis.

3. Species found only in a restricted geographic region and extremely vulnerable to extinction are classified as endangered species. Over the past four decades, rates of extinction have been rising through a combination of habitat loss and fragmentation, species introductions, overharvesting, and illegal trading in wildlife.

4. The rapidly growing human population is expected to reach 9 billion by 2050. Its demands for food, materials, and living space are threatening biodiversity around the world and are the basis of a new extinction crisis.

5. Conservation biology takes a three-pronged approach to preserving biodiversity. It includes the systematic survey of the full range of biodiversity, analysis of its evolutionary and ecological origins, and identification of methods to maintain and use biodiversity for the benefit of the human population.

6. Systematic surveys focus initially on identifying hot spots, which are limited areas where many species are in danger of extinction owing to human activities. At the next research level, broader or multiple hot spots are researched. Data are combined into a global picture of ecoregions where biodiversity is most threatened.

ON MASS EXTINCTIONS AND SLOW RECOVERIES

Based on many lines of evidence accumulated over the past few centuries, an estimated 99 percent of all species that have ever lived are extinct. Even so, the full range of biodiversity is greater now than it has ever been at any time in the past.

Reflect on the evolutionary stories of the preceding chapters in this unit. For at least the first 2 billion years of life's history, single-celled prokaryotes dominated the evolutionary stage. They did so until the Cambrian period, when atmospheric oxygen started to approach current levels. At that time, some of the single-celled eukaryotes began their genetic divergences and intricate species interactions, which led to the origin of diverse protistans, plants, fungi, and animals.

You saw how mass extinctions reduced biodiversity on land and in the seas. Each global episode spurred evolutionary change and radiations into newly vacated adaptive zones. However, recovery to the same level of biodiversity was exceedingly slow, requiring 20 to 100 million years. Figure 25.2a is a review of the pattern of five major extinctions and slow recoveries.

That pattern is only a composite of what happened to the major taxa. Lineages, remember, differ in their time of origin, the extent to which the member species diverged, and how long they persisted. If you consider ongoing survival and reproduction to be the measures of success, then each became a loser or a winner when environmental conditions changed in drastic or novel ways. To appreciate this point, reflect on Figure 25.2b, which shows the evolutionary history of representative lineages. Many such histories were combined to give us the overall pattern of Figure 25.2a.

Era	Period	MASS EXTINCTION UNDER WAY
CENOZOIC	QUATERNARY	With high population growth rates and cultural practices (e.g., agriculture, deforestation), humans become major agents of extinction.
CENOZOIC	—1 mya— TERTIARY	
	—65—	*MASS EXTINCTION*
MESOZOIC	CRETACEOUS —138— JURASSIC —205— TRIASSIC	Slow recovery after Permian extinction, then adaptive radiations of some marine groups and plants and animals on land. Asteroid impact at K–T boundary, 85% of all species disappear from land and seas.
	—240—	*MASS EXTINCTION*
PALEOZOIC	PERMIAN —290— CARBONIFEROUS	Pangea forms; Land area exceeds ocean surface area for first time. Asteroid impact? Major glaciation, colossal lava outpourings, 90%–95% of all species lost.
	—360—	*MASS EXTINCTION*
PALEOZOIC	DEVONIAN —410— SILURIAN	More than 70% of marine groups lost. Reef builders, trilobites, jawless fishes, and placoderms severely affected. Meteorite impact, sea level decline, global cooling?
	—435—	*MASS EXTINCTION*
PALEOZOIC	ORDOVICIAN —505— CAMBRIAN	Second most devastating extinction in seas; nearly 100 families of marine invertebrates lost
	—570—	*MASS EXTINCTION*
	(precambrian)	Massive glaciation. 79% of all species lost, including most marine microorganisms

a

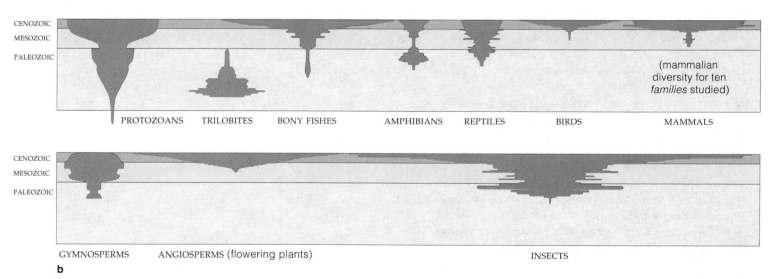

b

Figure 25.2 (**a**) Review of the five greatest mass extinctions and subsequent slow recoveries in the past. (More mass extinctions occurred than are shown here.) Compare Figure 18.5, in Section 18.2. (**b**) Within the framework of this generalized graph, patterns of extinction and recoveries differed significantly for different major taxa, as these selected examples indicate.

a

Figure 25.3 Two extinct and one threatened species. (**a**) Charles Knight's magnificent painting of *Tylosaurus*, one of the mosasaurs. This powerful marine lizard flourished in shallow, nearshore waters of the Cretaceous seaways. Indonesia's Komodo dragon, which grows 10 feet (3 meters) long, is a living relative. We think the Komodo dragon is a giant, yet imagine meeting up with a mosasaur. Some were 40 feet (12 meters) long.

(**b**) The dodo (*Raphus cucullatus*), a large, flightless bird that evolved on Mauritius, then vanished more than 300 years ago. Certain trees (*Calvaria major*) also evolved on that island, as did tortoises. Did the tree coevolve with dodos, tortoises, or both? Either may have fed on the tree's large fruits. For example, the large dodo gizzard could have partially digested the thick-walled coat of seeds inside fruits, just enough to help the seeds germinate after leaving the gut. Either way, after the dodo became extinct, the seeds stopped germinating. Only thirteen trees were still standing by the mid-1970s, and by some estimates they are more than 300 years old. Each year they still produce seeds, but these apparently cannot break out of the seed coats without help. Today, botanists employ turkeys or gem polishers as substitute grinders.

b

What causes mass extinctions? The answer may not be obvious. For example, considerable evidence suggests that many lineages never made it past the K–T boundary (Section 19.7). Many believe "the asteroid did it." It did deliver the coup de grace for some lineages, including the dinosaurs and mosasaurs (Figure 25.3*a*). But other factors were at work. Biodiversity had been declining for 10 million years. Pterosaurs were gone before the K–T event. The event had little effect on insects. Was tectonic change a factor? When a land mass destined to become Australia was being rafted away from Antarctica, deep, cold currents from the south were able to move into the warmer, equatorial seas—where many groups were hit. Seawater composition, sea levels, and the climate itself shifted. The consequences affected life on land as well as in the seas. The K–T asteroid impact did indeed end some lineages abruptly. But for others, it may simply have been the final blow in a long streak of bad luck.

Now extend this thought about hidden causes of extinctions to individual species. Long ago, Dutch sailors clubbed to death every last dodo, a flightless bird that lived only on Mauritius. After that, a tree species also native to the island simply stopped reproducing. It was not until the 1970s that a hypothesis emerged: If the tree depended intimately on a coevolved species that became extinct, then the tree would be vulnerable to extinction, also. Was its partner the dodo? Maybe (Figure 25.3*b*).

Biodiversity is greater now than it has ever been in the past.

The current range of global biodiversity is an outcome of an overall pattern of mass extinctions and slow recoveries in the history of life. Within that pattern, lineages differ in which member species persisted and which became extinct.

The loss of individual species, as well as mass extinctions, may have obvious or complicated causes.

THE NEWLY ENDANGERED SPECIES

No biodiversity-shattering asteroids have hit the Earth for 65 million years. *Yet the sixth major extinction event is under way*. Throughout the world, human activities are swiftly driving many species to extinction. Think of the mammalian lineage, which outlasted the dinosaurs. About 2 million years ago, early humans started to hunt mammals in earnest. About 11,000 years ago, they started to encroach on mammalian habitats in a big way; with agriculture, domesticated species became favored at the expense of wild stock. Think of the once-stupendous herds of bison, which nearly vanished in the 1800s. As adventurers moved westward, they shot too many bison for sport, sometimes from the platforms of trains.

Only 4,500 or so species of mammals made it to the present. Of those, 300 (including most whales, wild cats, otters, and primates other than humans) have the bad luck to be enrolled in the endangered species club. An **endangered species** is an endemic species extremely vulnerable to extinction. *Endemic* means it originated in only one geographic region and lives nowhere else.

We are just one species among millions. Yet because of our rapid population growth, we are threatening others by way of habitat losses, species introductions, overharvesting, and illegal wildlife trading. For the past forty years, combinations of these factors have raised extinction rates. We give a few examples in Figures 25.4 through 25.6. In time, nature probably will again heal itself. But recoveries after mass extinctions are known to take millions of years. We humans probably won't be around to congratulate ourselves on not having caused irrevocable harm, if and when healing is complete.

Figure 25.5 Confiscated products made from endangered species. All but 10 percent of the illegal wildlife trade, which threatens more than 600 species of animals and plants, goes undetected. Examples from North America and their black market prices: live large saguaro cactus ($5,000–15,000), bighorn sheep head ($10,000–60,000), polar bear ($6,000), grizzly ($5,000), bald eagle ($2,500), and peregrine falcon ($10,000). Other examples: live chimpanzee ($50,000), live mountain gorilla ($150,000), live imperial Amazon macaw ($30,000), Bengal tiger hide ($100,000), snow leopard hide ($14,000), and rhinoceros horn ($28,600 per kilogram).

Habitat Loss and Fragmentation

Edward O. Wilson defines **habitat loss** as the physical reduction in suitable places to live, as well as the loss of suitable habitat as an outcome of chemical pollution. Habitat loss is one of the major threats to more than 90 percent of endemic species facing extinction.

Biodiversity is greatest in the tropics. There, habitats on land are assaulted daily by deforestation, a topic you will read about in Sections 45.4 and 45.5. Also under attack are grasslands, wetlands, and aquatic habitats

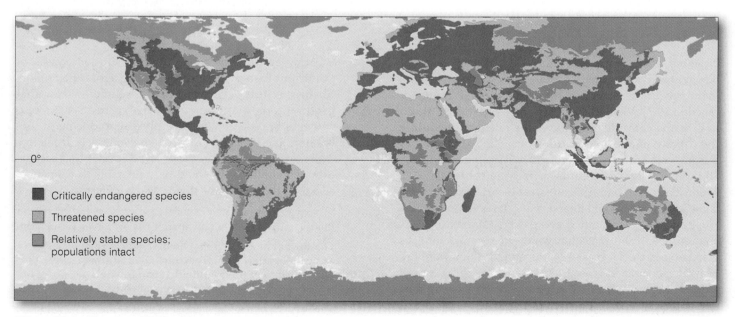

- Critically endangered species
- Threatened species
- Relatively stable species; populations intact

Figure 25.4 Threats to biodiversity: major regions of the world in which species are critically endangered, threatened, or relatively stable, as projected from 1998 through 2018.

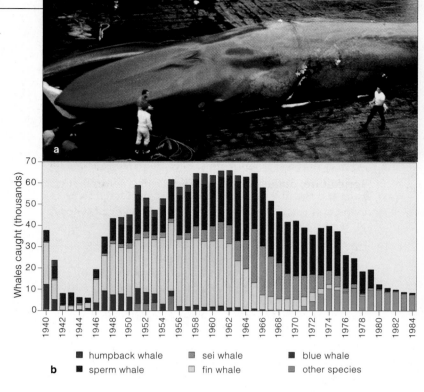

Figure 25.6 (**a**) Blue whale being butchered in British Columbia. (**b**) Graph of whale harvesting that reflects the declining numbers of great whales, now at the brink of extinction.

(Section 25.3). Other examples: 98 percent of tallgrass prairies, 50 percent of wetlands, and 85–95 percent of old-growth forests of the United States alone are gone.

Habitats also may become chopped up into isolated patches. This **habitat fragmentation** has three effects on biodiversity. First, it increases a habitat's boundaries (edges), so species are more vulnerable to predators, temperature changes, winds, fires, and disease. Second, the patches might not be large enough to support the population sizes required for successful breeding. (Each 90 percent reduction in a habitat's area generally leads to a 50 percent reduction in the number of its species.) Third, patches might not have enough food and other resources to sustain a population of the species.

ISLAND BIOGEOGRAPHY AND HABITAT ISLANDS Section 42.9 describes factors shaping biodiversity on islands, which are but a fraction of the Earth's surface. Even so, about half of the plant and animal species that have become extinct since 1600 were native to islands, mainly because they had nowhere else to go. R. MacArthur and Wilson's model of island biogeography is one way to estimate the number of current and future extinctions in regions surrounded by logging, mining, urbanization, and other destructive events. National parks, tropical forests, reserves, and lakes are examples. Think of them as **habitat islands** in seas of unsuitable habitat. By this model, destruction of 50 percent of an island habitat or a habitat island will drive about 10 percent of endemic species to extinction. A 90 percent loss will drive about 50 percent of endemic species to extinction.

INDICATOR SPECIES Conservation biologists often study **indicator species**, which offer warning of changes in habitat and impending widespread loss of biodiversity. Birds are one example. Different types live in all major land regions and climate zones, they respond quickly to changes in habitats, and they are relatively easy to track.

Think of the 700 migratory species that make annual trips to and from summer breeding grounds in North America. Many live most of the year in tropical forests. Biologists surveyed 64 species between 1978 and 1987. Population sizes of 44 insect-eating, migratory songbird species fell. Contributing factors were the deforestation of winter habitats and fragmenting of summer habitats. Intrusion of farms, freeways, and suburbs also chopped forests into patches. This made it easier for skunks, opossums, squirrels, raccoons, and blue jays to feast on the eggs and juveniles of migratory songbirds. About 69 percent of the 9,600 known species of birds are now confronted with habitat loss.

Whales caught (thousands)

b ■ humpback whale ■ sei whale ■ blue whale
■ sperm whale □ fin whale ■ other species

Other Contributing Factors

Whether by accident or deliberate importation, a species sometimes moves from its home range and successfully insinuates itself into a different habitat. Each habitat has only so many resources, and its occupants compete for their share. Introduced species have adaptations that might make them highly competitive, so they often displace one or more endemic species. You will read about these *exotic* species in Sections 42.6 and 42.7. For now, it is enough to know that species introductions are a major factor in almost 70 percent of the cases where endemic species are being driven to extinction.

In a sad commentary on human nature, the more rare a wild animal becomes, the more its value soars in the black market. Figure 25.5 gives examples.

Overharvesting also is reducing biodiversity. Whales are an example (Figure 25.6). Humans have killed them for food, lubricating oil, fuel, cosmetics, fertilizer, even corset stays. Substitutes exist for all whale products. Yet fishermen of several nations still ignore a moratorium on slaughtering large whales, and they routinely cross boundaries of sanctuaries set aside for their recovery.

An endangered species is any endemic species extremely vulnerable to extinction.

For the past forty years, human activities have raised rates of extinction through habitat losses, species introductions, overharvesting, and illegal wildlife trading.

Habitat loss is the physical reduction in suitable living spaces and habitat closure by chemical pollution.

Case Study: The Once and Future Reefs

Coral reefs are wave-resistant formations that develop mainly in clear, warm waters between latitudes 25° north and south (Figures 25.7 and 25.8). All show spectacular biodiversity. All suffer losses from hurricanes, warming of ocean currents, and other natural calamities. Usually they recover within a few decades. Reefs also suffer from human assaults and are less likely to recover from them.

Each coral reef began with the accumulated remains of countless organisms. Hard parts of corals became the structural foundation. Coralline algae contributed to it; deposits of calcium and magnesium carbonate hardened cell walls of the red algae, including *Corallina*. Secretions from other organisms helped cement things together.

The massive, pocketed spine of reefs is home to corals and a staggering variety of red algae and other organisms. You might observe 50 coral species on some reefs; about 500 live on the Great Barrier Reef. So, for example, do 3,000 species of fishes, 1,000 species of mollusks, and 40 species of sea snakes. Figure 25.8 hints at the wealth of warning colors, spines, tentacles, and stealthy behavior—all clues to dangers and fierce competition for resources among species that are packed together in limited space.

Dinoflagellates often live as symbionts in the tissues of reef-building corals (Section 23.4). In return for protection,

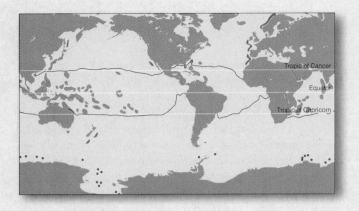

Figure 25.8   (**a**) Distribution map for coral reefs (color-coded *orange*), coral banks (*green*), and solitary corals (*red*). Nearly all reef-building corals are confined to regions with warm sea surface temperatures, here enclosed in dark lines. (**b**) *Facing page:* A small sampling of biodiversity on coral reefs.

they provide a coral polyp with oxygen and recycle its mineral wastes. When stressed, the polyps expel their protistan symbionts. When stressed for more than a few months, they die, and only bleached hard parts remain.

Abnormal, widespread bleaching in the Caribbean and tropical Pacific began in the 1980s. So did increases in sea surface temperature, which may be a major stress factor. Is the damage one outcome of human activities that are contributing to global warming (Section 43.10)? If so, as marine biologists Lucy Bunkley-Williams and Ernest Williams suggest, the future looks grim for reefs.

Human activities are destroying reefs in more direct ways than this. For example, pollutants such as raw sewage are being discharged into the nearshore waters around populated islands. Massive oil spills, as occurred during the Persian Gulf war, have calamitous effects. So do dredging operations and mining for coral rock.

Where commercial fishermen from Japan, Indonesia, and Kenya move in, reef life is vanishing. No simple nets for these fellows. They drop dynamite in the water, so fish hiding in the coral are blasted out and float dead to the surface. They also squirt sodium cyanide into the water to stun fish, which float to the surface. Some survivors end up as tropical fish in pet stores. The endangered species are transported to Asian restaurants, where they are dispatched and served up as exorbitantly expensive status symbols. On small, native-owned islands, fishing rights are traded for paltry sums. Then fishermen destroy the reefs, which can no longer sustain the small human populations that once depended on them for survival.

Reef biodiversity is in great danger around the world, in places extending from Australia and Southeast Asia to such distant places as the Hawaiian Islands, Galápagos Islands, the Gulf of Panama, Florida, and Kenya. Reef formations off Florida's Key Largo have already been diminished by one-third, mostly since 1970.

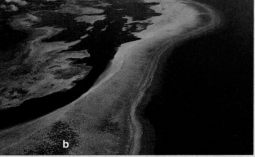

Figure 25.7   Three types of coral reef formations. *Fringing* reefs form near land in regions of light rainfall and runoff, as on the downwind side of the most recently formed volcanic islands. Many reefs in the Hawaiian Islands and Moorea (**a**) are like this. *Barrier* reefs form parallel with the shore of continents and volcanic islands, as in Bora Bora (**b**). Behind them are calm lagoons a few meters to sixty meters deep. Ring-shaped *atolls* are coral reefs and coral debris. They partially or completely enclose a shallow lagoon that is typically linked by a channel to the open ocean (**c**). Biodiversity isn't great in shallow water, which can get too hot for corals to survive (Section 23.4).

LIONFISH

CORAL REEF IN THE RED SEA

CORAL REEF IN THE SOUTH PACIFIC

CRAB

MORAY EEL

CHAMBERED NAUTILUS

SEA ANEMONE

DAISY CORAL

PILLAR CORAL

CROWN-OF-THORNS SEA STAR

Rachel's Warning

In 1951 Rachel Carson (Figure 25.9), an oceanographer and marine biologist, wrote a book about human impact on the ocean. *The Sea Around Us* was translated into thirty-two languages and sold more than 2 million copies. Seven years later, officials happened to spray mosquito-controlling DDT near the home and private bird sanctuary of one of her friends, who became agitated over the subsequent agonizing deaths of several birds. That friend begged Carson to find someone who might study the effects of pesticides on birds and other wildlife.

Carson discovered that independent, critical research on the subject was almost nonexistent. She conducted an extensive survey of the literature and then methodically developed information about the harmful effects of the widespread use of pesticides. In 1962, she published her findings. *Silent Spring*, the book's title, was an allusion to the silencing of "robins, catbirds, doves, jays, wrens, and scores of other bird voices" following pesticide exposure.

Many scientists, politicians, and policymakers read *Silent Spring*, and the public embraced it. Manufacturers of chemicals viewed Carson as a threat to their booming pesticide sales, and they swiftly mounted a campaign to discredit her. Critics and industry scientists claimed the book was full of inaccuracies, made selective and biased use of research findings, and failed to balance the picture with an account of the benefits of pesticides. Some even claimed that, as a woman, Carson simply was incapable of understanding such a highly technical subject. Others charged she was a hysterical woman and a radical nature lover trying to scare the public in order to sell books.

During those intense attacks, Carson knew she had terminal cancer. Yet she strongly defended her research and successfully countered her critics. She died in 1964—eighteen months after publication of *Silent Spring*—without knowing that her efforts would be the start of what is now known as the environmental movement in the United States. The new field of conservation biology is one outgrowth of that movement.

Figure 25.9 Rachel Carson, who helped awaken the public's interest in human impact on nature. She died without knowing she helped start the environmental movement.

CONSERVATION BIOLOGY

All countries can count their wealth in three forms: material, cultural, and biological. Until recently, many undervalued their biological wealth—*their biodiversity*, which is a source of food, medicine, and other goods. No one knows how much countries have already lost.

For instance, in the 1970s, a Mexican college student poking around in Jalisco discovered *Zea diploperennis*, a wild species of maize. Unlike domesticated corn, it resists disease and lives for more than one growing season. Gene transfers from the wild species to *Z. mays* have the potential to enormously boost production of corn for hungry people in Mexico and elsewhere. That species was growing in a mountain habitat no larger than ten hectares (twenty-five acres). The student saved it in the nick of time. A week later, humans started to clear the habitat with machetes and controlled burns. They would have pushed *Z. diploperennis* into oblivion.

There is a monumental problem. In the next fifty years, human population size may reach 9 billion, with most growth proceeding in the developing countries. Like all organisms, each of those 9 billion people must have raw materials, energy, and living space. How many species will they crowd out? The richest nations have the smallest biotas; the poorest nations, with the fastest population growth, are stewards of the largest. Resource consumption knows few boundaries, however. A huge share of natural resources from around the world helps meet the demands of the richest nations (Section 43.9).

Awareness of the impending extinction crisis gave rise to **conservation biology**. We define this field of pure and applied research as (1) a systematic survey of the full range of biological diversity, (2) an attempt to decipher the evolutionary and ecological origins of diversity, and (3) an attempt to identify methods that might maintain and use biodiversity for the good of the human population. Its goal is to conserve and use, in sustainable fashion, as much biodiversity as is feasible.

The Role of Systematics

Realistically, we simply don't have enough time, money, or people to complete a global survey. Possibly 99 percent of existing species have not even been discovered, and we don't know where most of them live. Thus initial work has focused on identifying **hot spots**. These are habitats with many species found nowhere else and in greatest danger of extinction because of human activities.

At the first survey level, researchers target a limited area, such as an isolated valley. A complete survey is out of the question, so they inventory birds, mammals, fishes, butterflies, and other indicator species for the habitat. At the next level, broader areas that are major or multiple hot spots are systematically explored. The widely separated forests of Mexico are an example of a

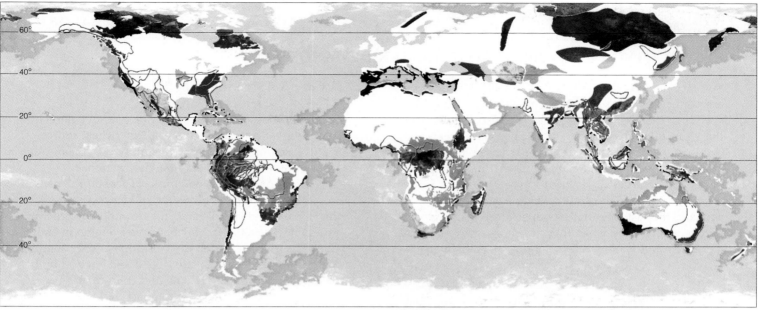

Figure 25.10 One map of the most vulnerable areas of land and seas, compiled by the World Wildlife Fund.

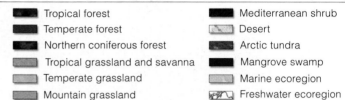

■ Tropical forest	■ Mediterranean shrub
■ Temperate forest	▨ Desert
■ Northern coniferous forest	■ Arctic tundra
▨ Tropical grassland and savanna	■ Mangrove swamp
▨ Temperate grassland	▨ Marine ecoregion
▨ Mountain grassland	▨ Freshwater ecoregion

major hot spot. Research stations are set up at different latitudes and elevations across the region.

At the highest survey level, hot spot inventories are combined with existing information on biodiversity of ecoregions. An **ecoregion** is a broad land or ocean region defined by climate, geography, and producer species. For example, Figure 25.10 is a work in progress by the World Wildlife Fund. Currently this organization has identified 233 regions crucial to global biodiversity (136 land, 36 fresh water, and 61 marine). Of these, it selected 25 for immediate conservation efforts. Such efforts on behalf of biodiversity will be refined over the coming decades.

Bioeconomic Analysis

The systematic expansion of species inventories serves as a baseline for economic analysis—that is, for assigning future value to ecoregions. This concept is something new in human history. For example, when the ancestors of Polynesians journeyed eastward from Southeast Asia, they literally ate their way across the Pacific, through New Zealand, Tonga, and Easter Island, all the way to the Hawaiian Islands. Flightless birds, eagles, and other endemic species that had no evolutionary experience with humans didn't stand a chance. Another wave of extinction accompanied the expansion of Paleo-Indians throughout the Americas. What counted in the past, as counts today, is food on the table for self and family. It counted for the Mexican truck driver who said, after shooting one of the two remaining imperial woodpeckers on Earth, "It was a great piece of meat."

Probably, short-sighted thinking can be countered only by showing economically pressed individuals that sustaining biodiversity has more economic value than destroying it. Should governments—and individuals—protect a habitat, withdraw resources in a sustainable

way, or obliterate it? Answers will require analysis of what a threatened species may offer. For example, some may be sources of new medicines and other chemical products. Developing countries do not have advanced laboratories for chemical analysis. Large pharmaceutical companies do. One of the largest has been paying Costa Rica's National Institute of Biodiversity to collect and identify promising species and send in chemical samples extracted from them. If natural chemicals end up being marketed, Costa Rica will get a share of the royalties, which are earmarked for conservation programs.

Sustainable Development

Ultimately, biodiversity will be best protected when its species can be used over the long term for the good of local economies. Identifying and implementing such uses is easier said than done. The technical problems and social barriers are huge. However, there are a few success stories, as the next section describes.

Conservation biology entails a systematic survey of the full range of biodiversity, analysis of its evolutionary and ecological origins, and identification of methods to maintain and use biodiversity for the benefit of humans.

Researchers identify local hot spots (limited areas with many endemic species in greatest danger of extinction by human activities), then broader or multiple hot spots, which they combine into a global picture of biodiversity.

To counter economic demands of the human population, the future economic value of biodiversity must be determined.

RECONCILING BIODIVERSITY WITH HUMAN DEMANDS

The fact remains that, when people locked in poverty are confronted with the choice of saving endangered species or themselves, the species will lose. Deprived of the education and resources that people in developed countries take for granted, they destroy tropical forests. They hunt animals and dig up plants in "protected" parks and reserves. They try to raise crops and herds on marginal land. They will take the last tree, whale, or primate, the last imperial woodpecker. We can expect people to stop only when they are shown ways to earn a living from the biodiversity in threatened habitats.

A Case for Strip Logging

Section 45.4 describes the staggering pace of destruction of once-great tropical rain forests, the richest sources of biodiversity on land. For now, simply think about one concept of how to minimize the destruction.

Besides providing wood for local economies, trees of tropical rain forests yield diverse, exotic woods prized by developed countries. What if they could be logged in a profitable, sustainable way that preserved biodiversity? Gary Hartshorn was the first to propose **strip logging** in portions of forests that are sloped and have a number of streams (Figure 25.11). The idea is to clear a narrow corridor that parallels contours of the land, and use the upper part of it as a road to haul away logs. After a few years, saplings start to grow in the corridor. Another corridor is cleared above the road. Precious nutrients that runoff leaches from the exposed soil trickle into the first corridor. There they are taken up by saplings, which benefit from all the nutrient input by growing more rapidly. Later, a third corridor is cut above the second one —and so on in a profitable cycle of logging, which the habitat sustains over time.

Ranching and the Riparian Zones

Reflect on the Figure 25.10 map, and you sense at once that the biodiversity crisis is not confined to Third World countries. Also take a look at Figure 25.12, which shows a riparian zone

uncut forest

cut 1 year ago

dirt road

cut 3–5 years ago

cut 6–10 years ago

uncut forest

stream in watershed

Figure 25.11 Strip logging, a practice that might sustain biodiversity even as it enhances the regeneration of economically valued forest trees.

Figure 25.12 Riparian zone along the San Pedro River in Arizona, shown before and after restoration.

in the western United States. Each **riparian zone** is a relatively narrow corridor of vegetation along a stream or river. Its plants afford a major line of defense against flood damage by sponging up water during spring runoffs and summer storms. Shade from the canopy of taller shrubs and trees helps conserve water during droughts. Riparian zones also offer food, shelter, and shade for wildlife, particularly in arid and semiarid regions. For example, in the western United States, 67 to 75 percent of the endemic species must spend all or part of their life cycle in riparian zones. They include more than 136 species of songbirds, some of which nest only in riparian vegetation.

Cattle raised in the American West provide beef for much of the human population. Compared with wild ungulates, cattle drink a lot more water, so they tend to congregate at rivers and streams. There they trample and feed until grasses and herbaceous shrubs are gone. It takes only a few head of cattle to destroy riparian vegetation. All but 10 percent of the riparian vegetation of Arizona and New Mexico has already disappeared, mainly into the stomachs of grazing cattle.

Restricted access and development of watering sites for cattle away from streams can help conserve riparian zones. So can the rotation of livestock and provision of supplemental feed at different grazing areas. Cattle ranchers resist implementing such measures. Putting in more fencing, for example, is costly.

Jack Turnell is one rancher who knows almost as much about riparian biodiversity as he knows about cattle. He runs cattle on a 32,000-hectare (80,000-acre) ranch south of Cody, Wyoming, and on 16,000 hectares of Forest Service land for which he has grazing rights. Ten years after he took over the ranch, he decided to raise cattle in an unconventional way by systematically rotating them away from riparian areas. Turnell also crossed Hereford and Angus cattle with a breed from France that does not consume as much water. He made most of his ranching decisions in consultation with specialists in range and wildlife management.

Willows and other plants are sustaining diversity in the restored zones. And Turnell's ranching approach is profitable; grasses maintained in the riparian zones help put more meat on his cattle. His may be only a small step toward sustainable ranching, but it appears to be a step in the right direction.

Throughout much of the world, the unprecedented rate of human population growth is the elephant in the room that no one wants to talk about.

Protecting biodiversity depends on finding ways for people to make a living from it without destroying it. Can such a balance be struck among so many billions of people?

SUMMARY

Gold indicates text section

1. Global biodiversity is greater now than ever, but the current rate of species losses is high enough to suggest that an extinction crisis is under way. *25.1*

2. The history of life indicates that, after global mass extinctions, 20 million to 100 million years pass before biodiversity recovers to the preceding level. *25.1*

   a. Different taxa have different evolutionary histories. Some were eliminated by major extinction events. Others passed through them relatively unscathed.

   b. Mass extinctions, and extinction of single species, may have obvious or hidden, complicated causes.

3. Human population size, expected to reach 9 billion within fifty years, is causing the sixth major extinction crisis. Human population growth is most rapid in areas with the richest, most vulnerable biodiversity. *25.2*

4. Habitat losses, habitat fragmentation, introduction of species into novel habitats, overharvesting, and illegal wildlife trading threaten endemic species. An endemic species originated in and is limited to one geographic region. An endangered species is defined as an endemic species extremely vulnerable to extinction. *25.2*

5. Habitat loss refers to physical reduction or chemical pollution of suitable places for species to live. Habitat fragmentation is carving a habitat into isolated patches. It puts species at risk, as by chopping up populations to sizes that cannot promote successful breeding. *25.2*

6. Island biogeography models help predict the number of current and future extinctions. A habitat island is a habitat in a sea of possibly destructive human activities, such as logging. Generally, destruction of 50 percent of an island habitat (or habitat island) will drive one-tenth of its species to extinction. Destruction of 90 percent will drive one-half to extinction. *25.2*

7. Indicator species are birds and other easily tracked species that can provide warning of changes in habitats and impending widespread loss of biodiversity. *25.2*

8. Humans are directly destroying coral reefs and other regions, as when commercial fishermen dynamite reefs to harvest fish. They may be destroying them indirectly, as by contributing to global warming (with concurrent rises in sea surface temperatures and sea levels). *25.3*

9. Conservation biology is a field of pure and applied research. Its goal is to conserve and use biodiversity in sustainable ways. The field encompasses: *25.5*

   a. A systematic survey at three ever more inclusive levels of the full range of biodiversity.

   b. Analysis of biodiversity's origins in evolutionary and ecological terms.

   c. Identification of methods that might maintain and use biodiversity for the good of the human population, which may otherwise destroy it.

10. Today the systematic survey of conservation biology is proceeding at three levels: *25.5*

　a. Local hot spots (for example, an isolated valley) are identified and indicator species inventoried. The hot spots are habitats with many species in greatest danger of extinction because of human activities.

　b. Major hot spots (or multiple ones) are inventoried. Research stations are set up across these broader areas to gather data by latitude and elevation.

　c. Data gathered at the first two levels are combined with data on ecoregions; these are the most vulnerable, broad regions of land and seas throughout the world.

　d. Data gathered for such maps will be refined into an increasingly detailed picture of global biodiversity.

11. Protecting biodiversity depends on finding ways for people to make a living from it without destroying it. To counter growing economic demands, biodiversity's future economic value must be determined. Methods by which local economies can exploit that biodiversity in sustainable ways must be developed. *25.6*

12. When people are given a choice between saving a species or themselves, that species will lose. Millions of people are making that choice as an outcome of their demands for food, shelter, and material goods. *25.6*

Review Questions

1.　How many major mass extinctions have occurred, including the one that is now under way? *25.1*

2.　List four human activities that are major contributors to the current extinction crisis. Which threaten coral reefs? *25.2, 25.3*

3.　Define endangered, endemic, and indicator species. *25.2*

4.　Distinguish between island habitat and habitat island. *25.2*

5.　State the goal of conservation biology and briefly describe its three-pronged approach to achieving that goal. *25.5*

6.　Define hot spot. Why are conservation biologists focusing on hot spots rather than quickly completing a global survey? *25.5*

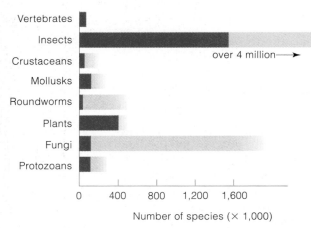

Figure 25.13 Current species diversity for a few major taxa. *Red* indicates the number of named species; *gold* indicates the estimated number of species not yet discovered.

Self-Quiz　ANSWERS IN APPENDIX III

1.　Following mass extinctions in the past, recovery to the same level of biodiversity has taken many _____ of years.
　a. hundreds　　b. millions　　c. billions

2.　Species may be driven to extinction by _____ .
　a. obvious factors　　　　　　c. human activities
　b. complicated, hidden factors　d. all of the above

3.　The goal(s) of conservation biology is (are) to _____ .
　a. conduct a three-level, systematic survey of all biodiversity
　b. analyze biodiversity's evolutionary and ecological origins
　c. identify ways to maintain and use biodiversity for people
　d. all of the above

4.　Strip logging _____ .
　a. can sustain forests　　c. destroys wild habitats
　b. is profitable　　　　　d. both b and c

Critical Thinking

1.　Figure 25.13 compares the current estimated number of species for some major taxa. Given what you learned about taxa from chapters in this unit, which are most vulnerable in the current extinction crisis? Which are likely to pass through it with much of their biodiversity intact? List some reasons why (for example, compare the distribution of member species).

2.　Visit or study a riparian zone near where you live. Imagine visiting it five years from now, then ten years after that. Given its location, what kinds of changes do you predict for it?

3.　Mentally transport yourself to a tropical rain forest of South America. Imagine you don't have a job. There are no jobs in sight, not even in overcrowded cities some distance away. You have no money or contacts to get you anywhere else. Yet you are the sole supporter of your large family. Today a stranger approaches you. He tells you he will pay good money if you can capture alive a certain brilliantly feathered parrot in the forest. You know the parrot is rare; it is seldom seen. However, you have an idea of where it lives. What will you do?

4.　In his review of this chapter, ecologist Robin Tyser offered these comments: Students might find the current biodiversity crisis too overwhelming to contemplate. But people are making a difference. For example, they helped reestablish viable bald eagle populations in the continental United States and wolves in northern Wisconsin. Daniel Janzen is working to recreate a dry forest ecosystem in Costa Rica. There is much good in the world. Do a literature search or computer research and report on one or two success stories you find inspiring.

Selected Key Terms

conservation
　biology *25.5*
coral reef *25.3*
ecoregion *25.5*

endangered species *25.2*
habitat fragmentation *25.2*
habitat island *25.2*
habitat loss *25.2*

hot spot *25.5*
indicator species *25.2*
riparian zone *25.6*
strip logging *25.6*

Readings

Cox, G. 1999. *Alien Species in North America and Hawaii: Impacts on Natural Ecosystems*. Washington, D.C.: Island Press.

Wilson, Edward O. 1992. *The Diversity of Life*. Cambridge, Massachusetts: Belknap Press. This slender book proves that Wilson is one of the most eloquent defenders of life's diversity.

On-Line readings at Student Guide for InfoTrac:
www.brookscole.com/biology

V Plant Structure and Function

A flowering plant (Prunus) busily doing what it does best: producing flowers for the fine art of reproduction.

26

PLANT TISSUES

Plants Versus the Volcano

On a clear spring day in 1980, in a richly forested region of the Cascade Range of southwestern Washington, Mount Saint Helens erupted and 500 million metric tons of ashes were blown skyward. Within minutes, shock waves from the blast flattened or incinerated hundreds of thousands of mature trees growing near the northern flanks of the mountain.

Rivers of hot volcanic ash surged down the slopes at rates exceeding forty-four meters per second. They turned into nightmarish rivers of cementlike mud when intense heat from the blast melted and released more than 75 billion liters of water that had been locked up in the mountain's snowfields and glacial ice.

In one mind-numbing moment, about 40,500 hectares—100,000 acres—of magnificent forests dominated by hemlock and Douglas fir were transformed into barren sweeps of land (Figure 26.1a,b). In the aftermath of the violent eruption, we gained stunning insight into what the world must have looked like long ago, before the first plants started to colonize habitats on land.

Yet it did not take long for existing plants to move back into habitats that their ancestors had claimed. In less than a year's time, the seeds of a variety of flowering plants, including fireweed and blackberry, sprouted near the grayed trunks of fallen trees around Mount Saint Helens.

Even before ten years had passed, young willows and alders took hold near riverbanks, and low shrubs cloaked the land (Figure 26.1c). Their presence afforded pockets of shade, which favored the germination of seeds of the slower growing but ultimately dominant species, the hemlocks and Douglas fir (Figure 26.1d). In less than a century, the forest will be as it once was.

Figure 26.1 (**a,b**) Grim reminder of what the world would be like without plants: the aftermath of the violent eruption of Mount Saint Helens in 1980. Nothing remained of the vast forest that surrounded this Cascade volcano. (**c**) In less than a decade, however, seed-bearing vascular plants were making a comeback. (**d**) Twelve years after the volcanic eruption, seedlings of a dominant species, Douglas fir, were starting to reclaim the land.

With this example, we open a unit dedicated to the seed-bearing vascular plants, with emphasis on the flowering types. In terms of distribution and diversity, they are the most successful plants on Earth.

This first chapter provides an overview of plant tissues and body plans. Chapter 27 explains how the seed-bearing plants absorb water and mineral ions, conserve water, and distribute organic substances among roots, stems, and leaves. Chapter 28 takes a closer look at their patterns of growth, development, and reproduction. As you will see, their structure and physiology (that is, the functioning of the plant body) help them survive sometimes hostile conditions on land—even momentary takeovers by volcanoes.

d

Key Concepts

1. Angiosperms (flowering plants) and, to a lesser extent, gymnosperms are groups that now dominate the plant kingdom. These seed-bearing vascular plants all have complex aboveground shoot systems, which consist of stems, leaves, and some other structures. They also have complex root systems that typically grow downward and outward through soil.

2. We find three major categories of tissue systems in all seed-bearing vascular plants. A ground tissue system makes up the bulk of the plant body. A vascular tissue system distributes water, dissolved minerals, and the products of photosynthesis through roots and shoots. A dermal tissue system covers and protects plant surfaces exposed to the surroundings.

3. The simple plant tissues—parenchyma, collenchyma, and sclerenchyma—are each composed of no more than one type of cell.

4. Complex plant tissues incorporate two or more types of cells. Xylem and phloem, which are vascular tissues, are like this. So are the dermal tissues called epidermis and periderm.

5. Plants lengthen and thicken only by way of mitotic cell divisions at meristems. At these localized regions of the plant body, undifferentiated cells give rise to all specialized cell lineages that form mature tissues.

6. Each growing season, shoots and roots lengthen. The lengthening, called primary growth, originates only at apical meristems, in the tips of shoots and roots.

7. Each growing season, the shoots and roots of many plants also thicken. Typically, lateral meristems inside shoots and roots give rise to an increase in diameter, which is called secondary growth. Wood is one outcome of secondary growth.

OVERVIEW OF THE PLANT BODY

Earlier, in Chapter 22, we surveyed representatives of the 295,000 known species of plants. Even that sprint through diversity revealed why no one species can be used as a typical example of plant body plans. When we hear the word "plant," however, we usually think of well-known species of seed-bearing vascular plants—gymnosperms (including pine trees) and angiosperms (flower-producing plants, such as roses, corn, cactuses, and elms). With 260,000 species, angiosperms dominate the plant kingdom, and they will be our main focus.

Flowering plant life cycles extend from germination to seed formation, then death. *Annuals* complete the life cycle in a single growing season, and they generally are "nonwoody," or herbaceous, plants. Marigolds and alfalfa are like this. *Biennials* such as carrots live for two consecutive growing seasons. Their roots, stems, and leaves form the first season; flowers form, seeds form, and the plant dies the next season. *Perennials* continue vegetative growth and seed formation year after year. Roots and stems thicken in many of them.

Shoots and Roots

Many flowering plants have a body plan similar to that shown in Figure 26.2. Its aboveground parts, or **shoots**, are stems, leaves, flowers (reproductive shoots), and other structures. Its stems offer structural support for upright growth, and some parts of the stems conduct water and solutes. Upright growth gives photosynthetic cells in young stems and leaves favorable exposure to sunlight. **Roots** are specialized structures that typically grow downward and outward through soil. A system of roots absorbs water and dissolved minerals, and it commonly anchors a plant's aboveground parts. A root also stores food, then releases it as required for its own cells and for distribution to aboveground parts.

Three Plant Tissue Systems

A flowering plant's stems, branches, leaves, and roots are alike in that each has three major tissue systems (Figure 26.2). The **ground tissue system** makes up the plant's bulk. The **vascular tissue system** has two kinds of tissues that distribute water and solutes. The **dermal tissue system** covers and protects the plant's surfaces.

Some tissues in each system are simple, in that they have only one type of cell. Parenchyma, collenchyma, and sclerenchyma are in this category. Other tissues are complex, with organized arrays of two or more types of cells. Xylem, phloem, and epidermis are like this.

The next sections describe the tissue organization of shoots and roots. You may find it easier to follow them by studying Figure 26.3. It shows the different ways in which botanists cut tissue specimens from plants.

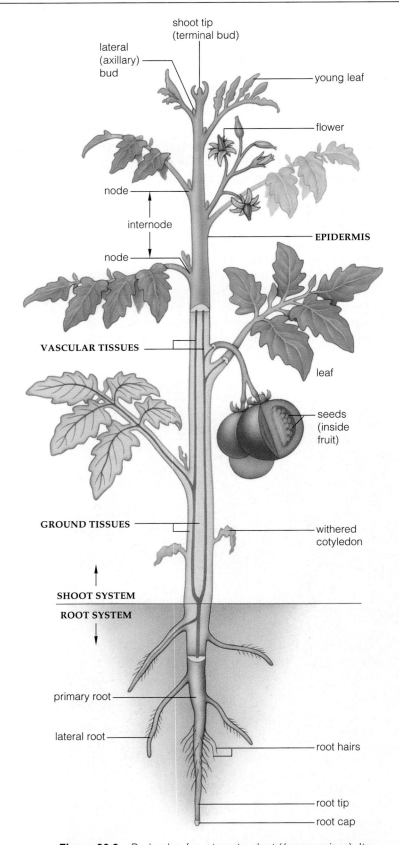

Figure 26.2 Body plan for a tomato plant (*Lycopersicon*). Its vascular tissues (*purple*) conduct water, dissolved minerals, and organic substances. They thread through ground tissues, which make up most of the plant. A dermal tissue (epidermis in this case) covers surfaces of the root and shoot systems.

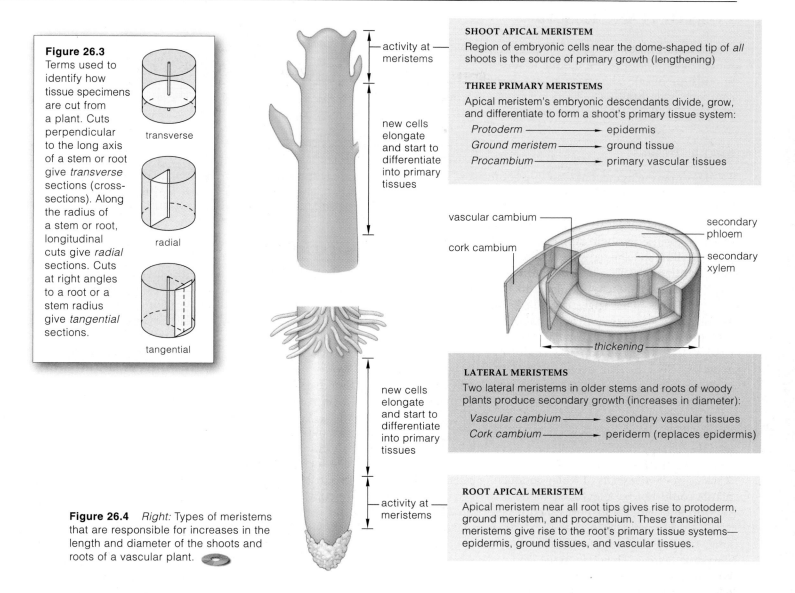

Figure 26.3 Terms used to identify how tissue specimens are cut from a plant. Cuts perpendicular to the long axis of a stem or root give *transverse* sections (cross-sections). Along the radius of a stem or root, longitudinal cuts give *radial* sections. Cuts at right angles to a root or a stem radius give *tangential* sections.

transverse

radial

tangential

activity at meristems

new cells elongate and start to differentiate into primary tissues

new cells elongate and start to differentiate into primary tissues

activity at meristems

Figure 26.4 *Right:* Types of meristems that are responsible for increases in the length and diameter of the shoots and roots of a vascular plant.

SHOOT APICAL MERISTEM

Region of embryonic cells near the dome-shaped tip of *all* shoots is the source of primary growth (lengthening)

THREE PRIMARY MERISTEMS

Apical meristem's embryonic descendants divide, grow, and differentiate to form a shoot's primary tissue system:

Protoderm ——————→ epidermis
Ground meristem ——————→ ground tissue
Procambium ——————→ primary vascular tissues

vascular cambium

cork cambium

secondary phloem

secondary xylem

thickening

LATERAL MERISTEMS

Two lateral meristems in older stems and roots of woody plants produce secondary growth (increases in diameter):

Vascular cambium ——————→ secondary vascular tissues
Cork cambium ——————→ periderm (replaces epidermis)

ROOT APICAL MERISTEM

Apical meristem near all root tips gives rise to protoderm, ground meristem, and procambium. These transitional meristems give rise to the root's primary tissue systems—epidermis, ground tissues, and vascular tissues.

Meristems—Where Tissues Originate

During a growing season, vascular plants do not grow everywhere at the same time. Most growth proceeds at **meristems**, which are localized regions of dividing cells (Figure 26.4). In other regions, cellular descendants of meristems are maturing or have already matured.

Apical meristem, in the tip of all shoots and roots, is where plant parts start to lengthen. Some populations formed here become protoderm, ground meristem, and procambium. These are immature forms of the primary tissues—epidermis, ground tissue, and vascular tissues respectively. Taken as a whole, the *lengthening* of stems and roots represents the plant's primary growth.

Also during a growing season, the older stems and roots of many plants thicken. Increases in girth start with lateral meristems, each a cylindrical array of cells that forms in stems and roots. One lateral meristem, the **vascular cambium**, produces secondary vascular

tissues. The other, **cork cambium**, produces a sturdier covering (periderm) that replaces epidermis. Taken as a whole, the *thickening* of stems and roots represents secondary growth.

Vascular plants have stems that support upright growth and conduct substances, leaves that function in photosynthesis, shoots specialized for reproduction, and other structures. They also have roots that absorb water and solutes. Roots often anchor aboveground parts and store food.

A ground tissue system makes up most of the plant body. A vascular tissue system distributes water, dissolved ions, and photosynthetic products through it. A dermal tissue system covers and protects plant surfaces.

Shoots and roots lengthen (put on primary growth) when their apical and primary meristems are active. In many plants, older stems and roots also thicken (add secondary growth) when lateral meristems called vascular cambium and cork cambium are active.

TYPES OF PLANT TISSUES

We turn now to an overview of the organization and functions of plant tissues. Simple tissues are composed of one type of cell. The vascular and dermal tissues are complex, with different cell types. Figures 26.5 through 26.9 show examples from these tissue categories.

Simple Tissues

Tissues of **parenchyma** make up most of the soft, moist, primary growth of roots, stems, leaves, flowers, and fruits. Most parenchyma cells are pliable, thin-walled, and many-sided. Mature cells are alive and retain the capacity to divide; their divisions often heal wounds. Mesophyll, a specialized parenchyma inside leaves, is photosynthetic. Air spaces between its cells enhance gas exchange. Parenchyma also functions in storage, secretion, and other specialized tasks. Even vascular tissue systems contain parenchyma cells.

Collenchyma provides flexible support for primary tissues. Its living cells often occur in patches or as a cylinder below the lengthening stem surface. They form many leaf stalk ribs. Most of its cells are long. Pectin, a polysaccharide, glues cellulose fibrils together in their unevenly thickened walls and makes the tissue pliable.

Most cells in **sclerenchyma** have lignin-impregnated thick walls. Lignin stiffens walls and gives them gravity-resisting compressive strength. It resists fungal attacks. It waterproofs the walls of water-conducting cells. Land plants could not have evolved without its mechanical support and water transport functions (Section 22.1).

Sclerenchyma cells are fibers or sclereids. *Fibers* are long, tapered cells in vascular tissue systems of some stems and leaves. They flex, twist, and resist stretching. We use certain fibers to make cloth, rope, paper, and other valued commodities (Figure 26.7*a,b*). The *sclereids*

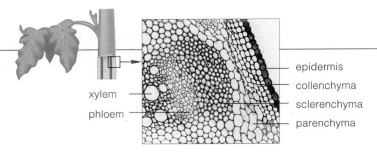

Figure 26.5 Locations of simple tissues and complex tissues in one kind of plant stem, transverse section.

are stubbier cells. Think about a hard seed coat, coconut shell, and peach pit or a pear's gritty texture; sclereids are the source of such features (Figure 26.7*c*).

Complex Tissues

VASCULAR TISSUES Two vascular tissues—xylem and phloem—distribute substances through plants. Fibers and parenchyma often sheath their conducting cells.

Xylem conducts water and dissolved mineral ions. It also helps mechanically support plants. Figure 26.8*a,b* shows examples of its conducting cells. The cells, *vessel members* and *tracheids*, are dead at maturity, and their lignified walls interconnect. Collectively, the cell walls form water-conducting pipelines and strengthen plant parts. Water flows into and out of the adjoining cells through numerous pits in the cell walls.

Phloem conducts sugars and other solutes. Its main conducting cells, called *sieve-tube members*, are alive at maturity (Figure 26.8*c*). Adjoining cells interconnect at openings in their side walls and at open or perforated end walls. Sugars synthesized inside leaves are loaded into sieve-tube members. *Companion cells* (specialized, living parenchyma cells) commonly assist the loading process. Sugars moving through phloem's pipelines are unloaded in regions where cells are growing or storing food. The chapter to follow describes this process.

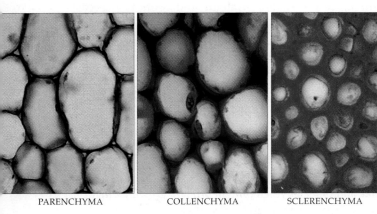

PARENCHYMA   COLLENCHYMA   SCLERENCHYMA

Figure 26.6 Examples of simple tissue from a sunflower plant (*Helianthus*) stem, transverse section. Parenchyma makes up the bulk of the plant body. Collenchyma and sclerenchyma strengthen and support plant parts.

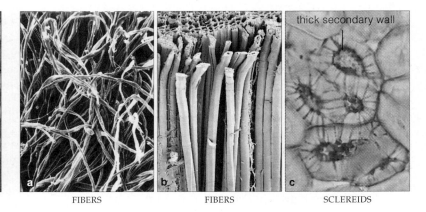

FIBERS   FIBERS   SCLEREIDS

Figure 26.7 Examples of sclerenchyma. (**a**) From cotton plant seeds, loosely twisted fibers that can interlock and form threads when spun. (**b**) Strong fibers from flax stems. Compare Figure 4.22*c,d*. (**c**) From a pear, one type of sclereid: stone cells, each with a thick, lignified wall.

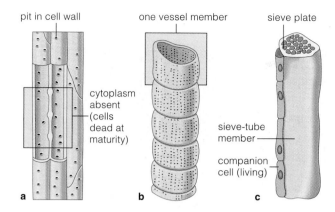

Figure 26.8 From xylem, portions of (**a**) tracheids and (**b**) a vessel. Pipelines made of such cells conduct water and dissolved ions. (**c**) One type of phloem cell. Long tubes of many cells conduct sugars and other organic compounds.

cuticle   epidermal cell

Figure 26.9 Light micrograph of a section through the upper surface of a kaffir lily leaf. The plant cuticle is made of secretions from epidermal cells. Inside the leaf are many photosynthetic parenchyma cells.

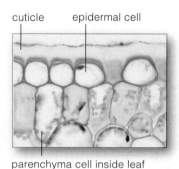

parenchyma cell inside leaf

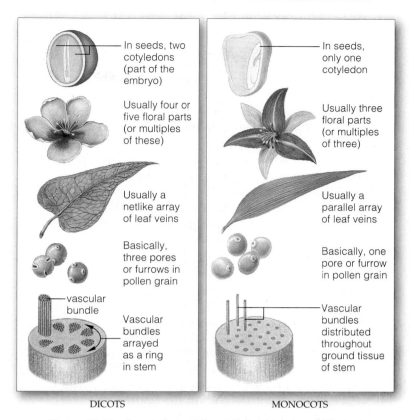

Figure 26.10 Comparison of the defining features of dicots and monocots. Both classes of flowering plants consist of the same simple and complex tissues, but their body parts show some differences in structural organization.

DERMAL TISSUES A dermal tissue system, **epidermis**, covers surfaces of primary plant parts. In most plants, it is mainly a single layer of unspecialized cells. Waxes and the fatty substance cutin coat the outermost cell walls. The surface coating is a **cuticle**. A plant cuticle functions in restricting water loss and often in resisting attacks by certain microorganisms (Figure 26.9).

Stem and leaf epidermis contains many specialized cells. For instance, pairs of guard cells change shape in response to changing conditions. As they do, a gap—or **stoma** (plural, stomata)—closes or opens between them. The next chapter looks at how stomata work as control points for the movement of water vapor, oxygen, and carbon dioxide across the epidermis. **Periderm** replaces epidermis in stems and roots with secondary growth. Dead cork cells in this protective covering have a wall heavily impregnated with suberin, a fatty substance.

Dicots and Monocots—Same Tissues, Different Features

The **dicots** and **monocots**, recall, are the two classes of flowering plants (Section 22.8). Most trees and shrubs other than conifers—such as maples, elms, roses, cacti, peas, beans, lettuces, cotton, and carrots—are dicots.

Palm trees, lilies, orchids, ryegrass, bamboos, wheat, corn, sugarcane, and pineapples are familiar monocots.

Dicots and monocots are similar in structure and function, but they differ in some distinctive ways. For instance, dicot seeds have two cotyledons and monocot seeds have only one. Cotyledons are leaflike structures commonly known as seed leaves. They form in seeds as part of a plant embryo, and they store or absorb food for it. After the seed germinates, cotyledons wither and new leaves grow and start to make food. Figure 26.10 shows other differences between dicots and monocots.

Most of the plant body (ground tissue system) consists of parenchyma, collenchyma, and sclerenchyma. Each of these simple tissues is composed of only one type of cell.

Xylem and phloem are vascular tissues. In xylem, pipelines made of tracheids and vessel members conduct water and dissolved ions. In phloem, sieve-tube members interact with companion cells to distribute organic compounds.

Of two dermal tissues, epidermis covers the surfaces of the primary plant body. Periderm replaces epidermis on plant parts with extensive secondary growth.

Monocots and dicots consist of the same tissues, but each has some of the tissues organized in distinctive ways.

PRIMARY STRUCTURE OF SHOOTS

How Stems and Leaves Form

Next time you or a friend eats a bundle of bean sprouts or alfalfa sprouts, pull one aside to look at its structure. That seedling started forming while it was still inside a seed coat. It already has a primary root and shoot. In the primary shoot's tip, apical meristem and its descendant tissues are laying out orderly frameworks for the stem primary structure (Figure 26.11). Below the shoot tip, tissues become specialized as cells divide at different rates in prescribed directions, and differentiate in size, shape, and function. These tissues make up distinctive stem regions, leaves, and lateral (axillary) buds from which lateral shoots develop. Lateral shoots give rise to side branches and reproductive structures.

Briefly, as a typical shoot lengthens, bulges of tissue develop along the sides of apical meristem. Each bulge is one immature leaf (Figures 26.11 and 26.12). While growth continues, the stem lengthens between tier after tier of new leaves. Each part of the stem where one or more leaves are attached is a node. As shown in Figure 26.2, the stem region between two successive nodes is an internode. Buds develop in the upper angle where leaves attach to the stem. Each **bud** is an undeveloped shoot of mostly meristematic tissue, often protected by bud scales (modified leaves). As Chapter 28 describes, buds give rise to new stems, leaves, and flowers.

Internal Structure of Stems

While the primary plant body of a monocot or dicot is forming, the ground, vascular, and dermal tissues of its stems become organized in distinctive ways. Most often, primary xylem and phloem develop inside the same sheath of cells, as **vascular bundles**. The bundles are multistranded cords threading lengthwise through the ground tissue system of primary and lateral shoots.

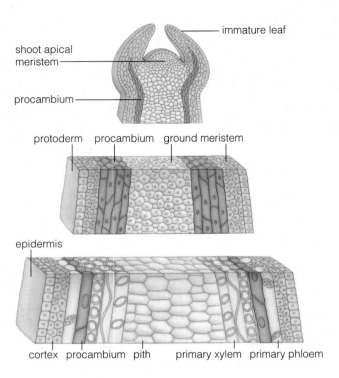

Figure 26.11 Successive stages in primary growth that started with activity at the shoot apical meristem of a typical dicot, then continued at the primary meristems derived from it. Notice the progressive differentiation of most of the tissue regions.

They commonly develop in two genetically dictated patterns. In most dicot stems, long bundles form a ring that divides the ground tissue into a cortex and pith (Figure 26.13a). A stem's **cortex** is the region between the vascular bundles and the epidermis. Its **pith** is the center of the stem, inside the ring of vascular bundles. Ground tissue of the plant's roots becomes similarly divided, into root cortex and pith. A different pattern is common inside the stems of most monocots and some

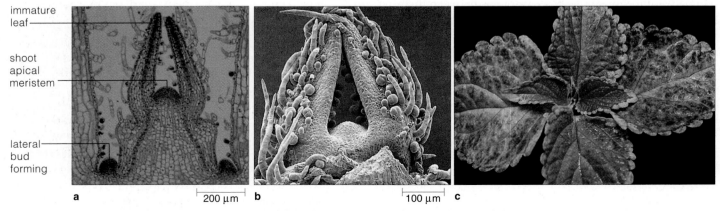

a   200 µm   b   100 µm   c

Figure 26.12 (a) Light micrograph of a *Coleus* shoot tip, cut longitudinally through its center. (b) Scanning electron micrograph of its surface. (c) New *Coleus* leaves.

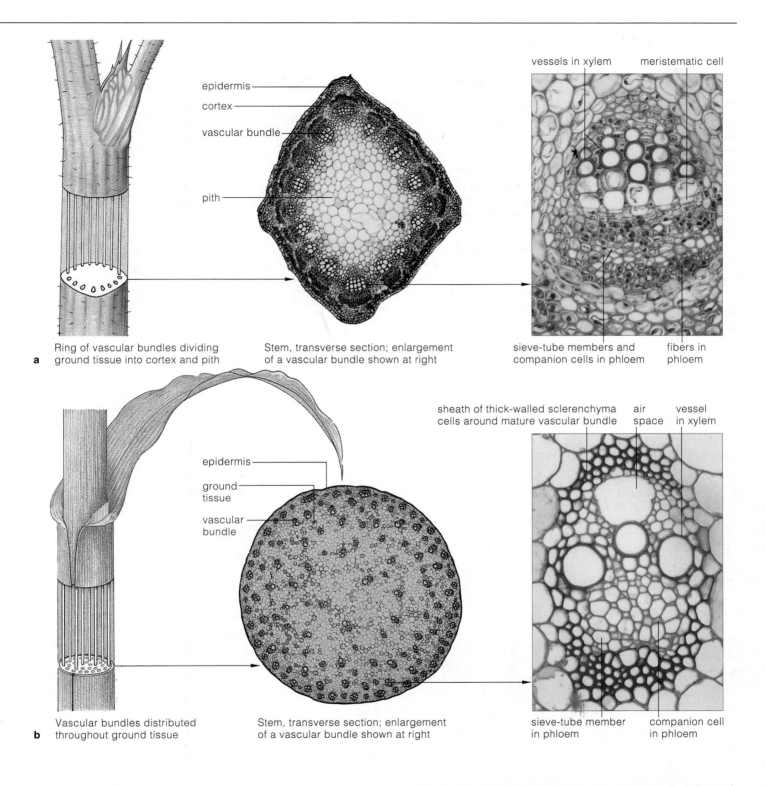

epidermis
cortex
vascular bundle
pith

vessels in xylem   meristematic cell

sieve-tube members and   fibers in
companion cells in phloem   phloem

a Ring of vascular bundles dividing ground tissue into cortex and pith

Stem, transverse section; enlargement of a vascular bundle shown at right

sheath of thick-walled sclerenchyma   air   vessel
cells around mature vascular bundle   space   in xylem

epidermis
ground tissue
vascular bundle

sieve-tube member   companion cell
in phloem   in phloem

b Vascular bundles distributed throughout ground tissue

Stem, transverse section; enlargement of a vascular bundle shown at right

Figure 26.13 Internal organization of cells and tissues inside the stems from a dicot and a monocot. (**a**) Part of a stem from alfalfa (*Medicago*), a dicot. In many species of dicots and conifers, the vascular bundles develop in a more or less ringlike array in the ground tissue system, as shown here. The portion of the ground tissue between the ring and the surface of the stem is the cortex. The portion enclosed within the ring is the pith. (**b**) Part of a stem from corn (*Zea mays*), a monocot. In most monocots and some nonwoody dicots, vascular bundles are scattered through the ground tissue, as shown.

dicots. The long vascular bundles are scattered through the ground tissue (Figure 26.13*b*). How substances are conducted through such vascular systems is a topic of the next chapter.

The shoot apical meristem gives rise to the primary plant body, which develops a distinctive internal structure (as in the pattern in which its vascular bundles are arranged).

A CLOSER LOOK AT LEAVES

Similarities and Differences Among Leaves

Every **leaf** that forms is a metabolic factory equipped with many photosynthetic cells. Yet leaves vary greatly in size, shape, surface details, and internal structure. A duckweed leaf is no more than 1 millimeter (0.04 inch) across; the leaves of some water lilies are 2 meters (6.5 feet) across. Different leaves look like blades, spikes, cups, needles, feathers, tubes, and other structures. They differ greatly in color, odor, and edibility; many produce toxins. The leaves of birches and other species of *deciduous* plants wither and then drop from stems as winter approaches. The leaves of camellias and other *evergreen* plants also drop, but not all at the same time.

A typical leaf has a flat blade, as in Figures 26.14*a* and 26.15. It has a stalk (petiole) that attaches it to the stem. *Simple* leaves are undivided, but many are lobed. *Compound* leaves have blades divided into leaflets, all oriented in the same plane. Leaves of most monocots, such as ryegrass and corn, are flat surfaced like a knife blade. The blade's base encircles and sheaths the stem.

Leaves of most species are thin, with a high surface-to-volume ratio. Their flat surface grows and orients itself perpendicular to light. A leaf often projects from a stem in patterns that minimize shading of other leaves. For example, a clover leaf stalk is attached to the stem at right angles to its neighbors (Figure 26.15*c*).

Such leaf adaptations intercept as much of a plant's energy source —sunlight—as possible. They also help oxygen diffuse out and carbon dioxide diffuse in. And when a leaf

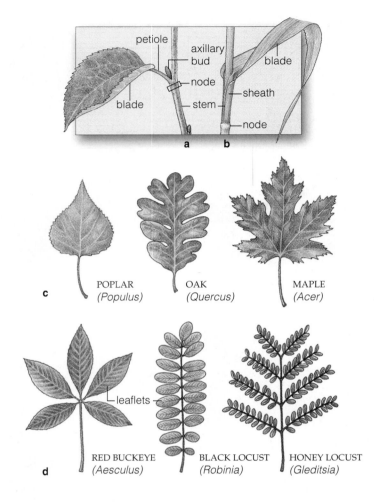

Figure 26.14 Common leaf forms of (**a**) dicots and (**b**) monocots. Examples of (**c**) simple leaves and (**d**) compound leaves.

is thick, you can assume it belongs to a succulent or some other plant of arid habitats and serves in water storage as well as photosynthesis. And still another example: Leaves of many desert plants orient themselves parallel with the sun's rays to reduce heat absorption to tolerable levels.

Leaf Fine Structure

In its fine structure, too, each leaf is adapted to intercept energy from the sun's rays and promote gas exchange. In addition, many have distinctive surface specializations.

Figure 26.15 (**a**) Decaying dicot leaf, with its netlike veins. (**b**) Parallel veins of a monocot leaf (*Agapanthus*). (**c**) Leaf orientation of a four-leaf clover (*Trifolium*).

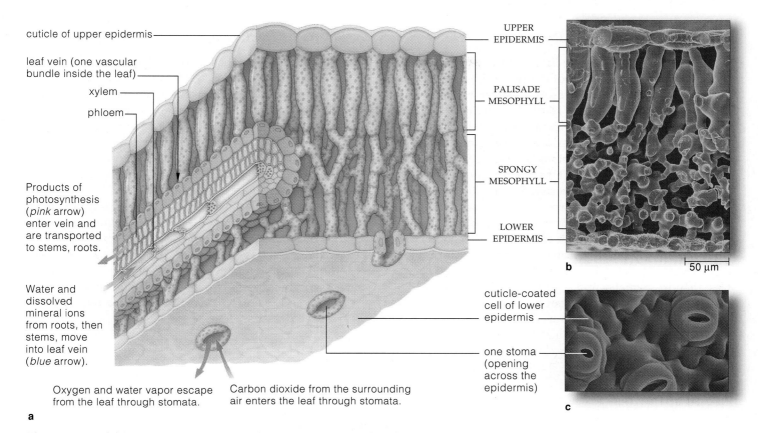

cuticle of upper epidermis

leaf vein (one vascular bundle inside the leaf)

xylem

phloem

Products of photosynthesis (*pink* arrow) enter vein and are transported to stems, roots.

Water and dissolved mineral ions from roots, then stems, move into leaf vein (*blue* arrow).

Oxygen and water vapor escape from the leaf through stomata.

Carbon dioxide from the surrounding air enters the leaf through stomata.

UPPER EPIDERMIS

PALISADE MESOPHYLL

SPONGY MESOPHYLL

LOWER EPIDERMIS

50 µm

b

cuticle-coated cell of lower epidermis

one stoma (opening across the epidermis)

c

a

Figure 26.16 (**a**) Diagram of leaf structure for many kinds of flowering plants. (**b**) Scanning electron micrograph of the tissue organization of a leaf from the kidney bean plant (*Phaseolus*). Notice the compact organization of the epidermal cells. (**c**) Stomata, each a tiny opening across the epidermis, appear when paired guard cells are in their plumped configuration. See also Figure 6.15.

LEAF EPIDERMIS Epidermis covers every leaf surface exposed to the air. It may be smooth, sticky, or slimy, with "hairs," scales, spikes, hooks, glands, and other surface specialties. The Chapter 27 introduction has two examples. A cuticle covers the sheetlike, compact array of epidermal cells; it restricts the loss of precious water (Figures 26.9 and 26.16). Most leaves have many more stomata on their lower surface. In arid or cold habitats, the stomata are often located at depressions in the leaf surface, together with thickly coated epidermal hairs. Both leaf adaptations help conserve water.

MESOPHYLL—PHOTOSYNTHETIC GROUND TISSUE As you read earlier, **mesophyll**, a type of parenchyma with photosynthetic cells and air spaces, extends through a leaf's interior (Figure 26.16). Most cell walls are exposed to air. Plasmodesma connect the cytoplasm of adjoining cells and allow rapid exchange of substances (Section 4.11). Carbon dioxide from the outside reaches cells by diffusing through stomata and the air spaces; oxygen diffuses in the opposite direction.

Leaves oriented perpendicular to the sun have two mesophyll layers. Attached to the upper epidermis is

palisade mesophyll—columnar parenchymal cells with more chloroplasts and more photosynthetic potential, compared to cells of the *spongy* mesophyll layer below them (Figure 26.16). Monocot leaves grow vertically and intercept light from all directions; as Figure 6.15 shows, their mesophyll is not organized as two layers.

VEINS—THE LEAF'S VASCULAR BUNDLES Leaf **veins** are vascular bundles, usually strengthened with fibers. Their continuous strands of xylem rapidly move water and dissolved nutrients to all mesophyll cells, and the continuous strands of phloem carry the photosynthetic products—especially sugars—away from them. In most dicots, the veins branch lacily into a number of minor veins embedded inside mesophyll. In most monocots, veins are more or less similar in length, and they run parallel with the leaf's long axis (Figure 26.15).

A leaf's structure is adapted for sunlight interception, gas exchange, and distribution of water, dissolved nutrients, and photosynthetic products. Leaves of each species have a characteristic size, shape, and often surface specializations.

PRIMARY STRUCTURE OF ROOTS

Taproot and Fibrous Root Systems

When a seed germinates, the first part to poke through the seed coat is a **primary root** (Figure 26.17). In nearly all dicot seedlings, it increases in diameter as it grows downward. Later on, **lateral roots** form in the primary root's tissues, perpendicular to its axis, and then erupt through epidermis. Youngest lateral roots are closest to the root tip. A **taproot system** is a primary root and its lateral branchings. Dandelions, carrots, and oak trees are examples of plans with this system (Figure 26.17c).

By comparison, the primary root of most monocots, such as grasses, is short-lived. In its place, adventitious roots arise from the stem, and then lateral roots branch from these. (*Adventitious* means the structures form at an unusual location.) The lateral roots are all more or less alike in diameter and length. Together, roots that form this way are a **fibrous root system** (Figure 26.17d).

Internal Structure of Roots

In Figure 26.17a, notice the meristems inside the tip of one root. Many cellular descendants of these meristems divide, enlarge, elongate, and become cells of primary tissue systems. Notice also the root cap, a dome-shaped mass of cells at the tip. Apical meristem produces the cap, which in turn protects the meristem.

Protoderm gives rise to root epidermis, the plant's absorptive interface with the soil. Some epidermal cells send out extensions called **root hairs**. Collectively, root hairs enormously increase the surface area available for taking up water and dissolved nutrients. Only first-time or foolish gardeners would yank a plant from the ground when transplanting it. Such yanking would tear off too much of the highly fragile absorptive surface.

Apical meristem also gives rise to the ground tissue system and to a **vascular cylinder**. A vascular cylinder

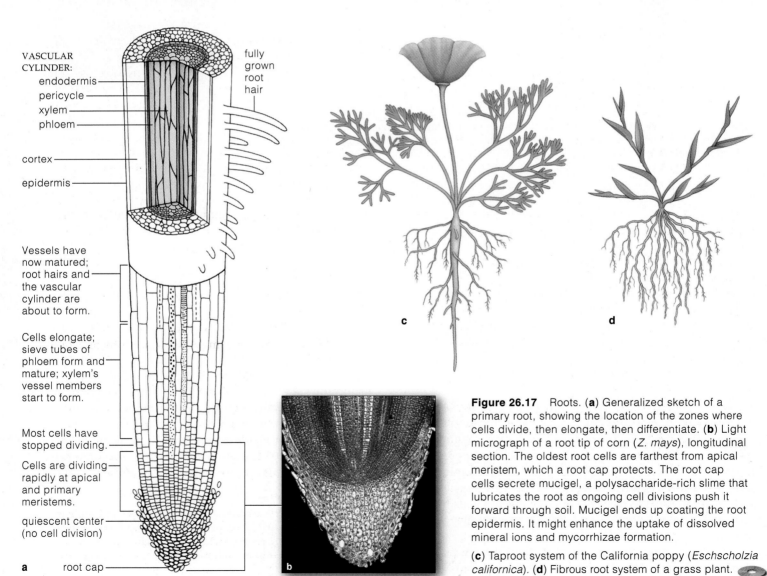

VASCULAR CYLINDER:
endodermis
pericycle
xylem
phloem

fully grown root hair

cortex

epidermis

Vessels have now matured; root hairs and the vascular cylinder are about to form.

Cells elongate; sieve tubes of phloem form and mature; xylem's vessel members start to form.

Most cells have stopped dividing.

Cells are dividing rapidly at apical and primary meristems.

quiescent center (no cell division)

a root cap

c

d

Figure 26.17 Roots. (**a**) Generalized sketch of a primary root, showing the location of the zones where cells divide, then elongate, then differentiate. (**b**) Light micrograph of a root tip of corn (*Z. mays*), longitudinal section. The oldest root cells are farthest from apical meristem, which a root cap protects. The root cap cells secrete mucigel, a polysaccharide-rich slime that lubricates the root as ongoing cell divisions push it forward through soil. Mucigel ends up coating the root epidermis. It might enhance the uptake of dissolved mineral ions and mycorrhizae formation.

(**c**) Taproot system of the California poppy (*Eschscholzia californica*). (**d**) Fibrous root system of a grass plant.

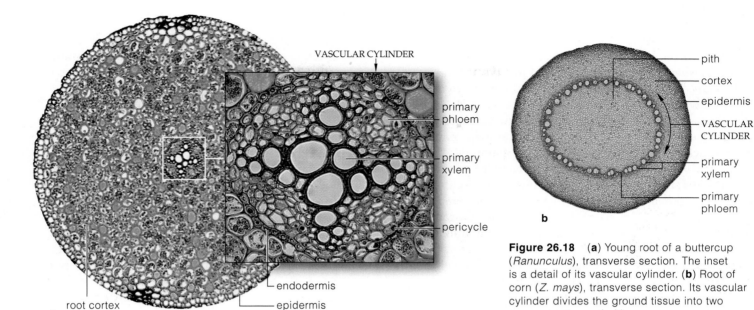

VASCULAR CYLINDER

primary phloem

primary xylem

pericycle

endodermis

epidermis

root cortex

a

pith

cortex

epidermis

VASCULAR CYLINDER

primary xylem

primary phloem

b

Figure 26.18 (**a**) Young root of a buttercup (*Ranunculus*), transverse section. The inset is a detail of its vascular cylinder. (**b**) Root of corn (*Z. mays*), transverse section. Its vascular cylinder divides the ground tissue into two zones—cortex and pith.

consists of primary xylem and phloem and one or more layers of cells called the pericycle. Figure 26.18*a* shows a vascular cylinder at the center of a dicot root's cortex. Figure 26.18*b* shows how the vascular cylinder divides the ground tissue system of one type of monocot into cortex and pith regions. With either pattern, there are plenty of air spaces between cells of the ground tissue system, so oxygen easily diffuses through them. Like

other cells in the plant, all living root cells depend on oxygen for aerobic respiration.

When water enters a root, it moves from cell to cell until it reaches the endodermis, a layer of cells around the vascular cylinder. Where endodermal cells abut, their walls are waterproofed, so incoming water is forced to pass through their cytoplasm. As described in Chapter 27, this arrangement controls the movement of water and dissolved substances into the vascular cylinder.

The pericycle is located just inside the endodermis. Its cells divide repeatedly and form lateral roots, which erupt through the cortex and epidermis (Figure 26.19).

Regarding the Sidewalk-Buckling, Record-Breaking Root Systems

Unless tree roots start to buckle a sidewalk or choke off a sewer line, most of us do not pay much attention to flowering plant root systems. Roots mine the soil for water and minerals, and most reach a depth of 2 to 5 meters. In hot deserts, where free water is scarce, one hardy mesquite shrub sent its roots down 53.4 meters (175 feet) near a stream bed. Some cacti have shallow roots radiating outward for 15 meters. Someone once measured the roots of a young rye plant that had been growing for four months in 6 liters of soil water. If the surface area of that root system were laid out as one sheet, it would occupy more than 600 square meters!

Roots provide a plant with a tremendous surface area for absorbing water and solutes. Taproot systems consist of a primary root and lateral branchings. Fibrous root systems consist of adventitious roots that replace the primary root.

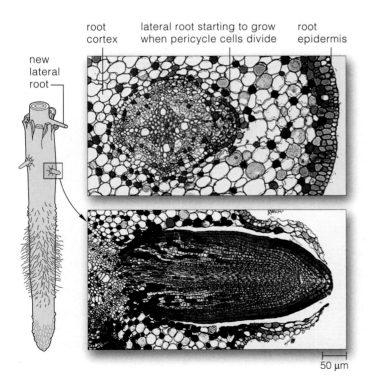

root cortex

lateral root starting to grow when pericycle cells divide

root epidermis

new lateral root

50 μm

Figure 26.19 Lateral root formation in a primary root from a willow tree (*Salix*), transverse section.

ACCUMULATED SECONDARY GROWTH—THE WOODY PLANTS

Annual plants, recall, complete their life cycle in one growing season. Most are *nonwoody,* or herbaceous. Like all gymnosperms, some monocots and many dicots are perennials. They put on secondary growth over two or more growing seasons; they are *woody* plants. Early in life, their stems and roots are like those of nonwoody plants. The differences emerge after lateral meristems become active and start giving rise to large amounts of secondary vascular tissues, especially secondary xylem.

Massive stems and roots originate at the vascular cambium. Look at Figure 26.20*a.* Each spring, primary growth resumes at this stem's buds; secondary growth builds up *inside* it. When the stem vascular cambium is formed, it's like a cylinder one cell or a few cells thick. Some cells give rise to secondary xylem and phloem that extend longitudinally through the stem. Other cells of vascular cambium give rise to rays of parenchyma cells. The rays deliver water "sideways" through the stem, in a radial pattern a bit like a sliced pie. Without these vascular tissues, substances could not travel up, down, and sideways in enlarging woody stems.

Figure 26.20*b* shows a simple way to think about the pattern of secondary growth at vascular cambium. Secondary xylem forms on this meristematic tissue's *inner* face. Secondary phloem forms on its *outer* face. As the inner core of xylem slowly thickens, it displaces meristematic cells toward the stem's surface. The cells also maintain a ring of vascular cambium by dividing sideways, in a widening circle.

Secondary xylem (wood) makes up 90 percent or so of a tree. Secondary phloem, restricted to a thin zone outside the vascular cambium, has thin-walled, living parenchyma cells and sieve tubes, often between thick-walled fibers. Tubes close to the vascular cambium are functional; the rest are dead but protect the living cells.

We have been focusing on how stems thicken. But bear in mind, secondary xylem and phloem form at vascular cambium in the plant's roots, also. As seasons pass and a tree ages, its expanding inner core of xylem exerts pressure on stem and root surfaces. Epidermis splits off; periderm replaces it. This new cover is made of cork cambium and its products—parenchyma and cells that die and form cork. **Cork** has densely packed rows of cell walls thickened with suberin. It insulates, waterproofs, and protects woody stems and roots.

All tissues external to the vascular cambium make up **bark** (Figure 26.21*a*). Like all living plant cells, the ones in woody stems and roots use oxygen for aerobic respiration and give off carbon dioxide wastes. Both gases cross bark's suberized, corky surface at lenticels, where cell walls are not as packed. Those dark spots on a wine bottle cork are all that's left of lenticels.

Wood's appearance and function change as a stem or root ages. The core becomes **heartwood**. This is a dry

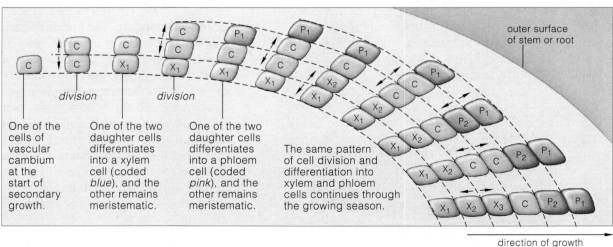

Figure 26.20 (a) Twig of a walnut tree in winter, after its leaves dropped. Each spring, *primary* growth resumes at its terminal and lateral buds. *Secondary* growth resumes at the vascular cambium (*1* and *2*). (b) Pattern of activity at the twig's vascular cambium. Reading left to right, ongoing cell divisions enlarge the inner core of secondary xylem and displace vascular cambium toward the stem or root surface.

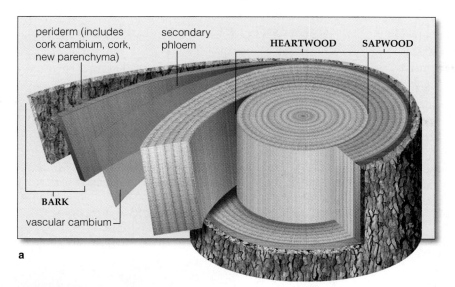

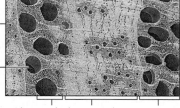

 vessel in xylem

early wood   late wood   early wood

Figure 26.22 A scanning electron micrograph of early and late wood in a block cut from red oak (*Quercus rubra*).

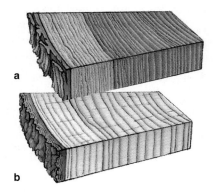

a

b

Figure 26.23 (a) Hickory, one of the hardwood dicots, is durable, strong, and yet resilient. It is used to make handles of hammers and other high-impact tools. (b) Pine, a softwood, is lightweight yet resists warping. Pine grows faster than hardwoods and is commercially "farmed" as a source of relatively cheap lumber.

periderm (includes cork cambium, cork, new parenchyma)

secondary phloem

HEARTWOOD   SAPWOOD

BARK

vascular cambium

a

Figure 26.21 (a) Structure of a stem with extensive secondary growth. When you look at this sketch, think of *Luna*, a redwood that became a symbol for anti-logging activism. An anti-activist girdled Luna by buzz-sawing most of the trunk and cutting through much of the vertical phloem. Without phloem, food from leaves can't reach roots. If the roots die, so, in time, will the tree. (b) Radial cut through a woody stem having three annual rings. The first year, the stem put on primary and some secondary growth; the next two years, it added secondary growth.

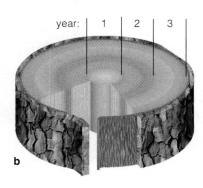

year: 1   2   3

b

tissue that no longer transports water and solutes. It helps the tree defy gravity and is a dumping ground for some metabolic wastes, such as resins, tannins, gums, and oils. Eventually, wastes clog and fill in the oldest xylem pipelines. They often darken heartwood, make it more aromatic, and strengthen it. By contrast, **sapwood** is secondary growth in between the vascular cambium and heartwood (Figure 26.21*a*). Unlike heartwood, it is wet, usually pale, and not as strong. Maple trees are an example. Each spring, from early March through early April, New Englanders insert metal tubes into sugar maple sapwood. Sap, the sugar-rich fluid in secondary phloem, drips through the tubes, into buckets.

Vascular cambium becomes inactive for part of the year in regions having cool winters or prolonged dry spells. *Early* wood, formed at the start of the growing season, has large-diameter, thin-walled xylem cells. But *late* wood, formed in dry summer, has xylem cells with smaller diameters and thicker walls (Figure 26.22). In a transverse section from a trunk, alternating bands of early and late wood reflect light differently. We call the differences **growth rings** or, informally, "tree rings."

Oak, hickory, and other dicot trees that evolved in temperate and tropical regions are all **hardwoods**; they have vessels, tracheids, and fibers in their xylem. Pines,

redwoods, and other conifers are **softwood** trees; their xylem contains tracheids and rays of parenchyma, but no vessels or fibers. Lacking fibers, the trees are weaker and less dense than the hardwoods (Figure 26.23).

What are the selective advantages of having woody stems and roots? Remember, like all other organisms, plants compete for resources. Plants with taller stems or wider canopies that resist the pull of gravity can intercept more of the energy streaming in from the sun. With a greater energy supply for photosynthesis, they have the metabolic means to develop larger root and shoot systems. And being more competitive in securing resources ultimately contributes to reproductive success.

In woody plants, secondary vascular tissues form at a ring of vascular cambium inside older stems and roots. Wood is an accumulation of secondary xylem especially. It may be classified by its location, functions, and type of plant.

With their sturdier tissues, woody plants defy gravity and grow taller and broader. Where competition for sunlight is intense, the ones that intercept the most sunlight win. Other factors being equal, having more energy to drive photosynthesis provides advantages in terms of metabolic capabilities, growth, and reproductive success.

SUMMARY   *Gold* indicates text section

1. Seed-bearing vascular plants include gymnosperms and angiosperms (the flowering plants). Their shoots (stems, leaves, and other structures) and roots consist of dermal, ground, and vascular tissue systems. *26.1*

2. All plant growth originates at meristems, localized regions of cells that retain the capacity to divide. *26.1*

   a. Primary growth (lengthening of stems and roots) originates at apical meristems in root and shoot tips.

   b. In many plants, secondary growth (increases in diameter) originates inside stems and roots, at lateral meristems called vascular cambium and cork cambium.

3. Parenchyma, sclerenchyma, and collenchyma are the simple tissues; each has only one cell type (Table 26.1).

   a. Parenchyma cells, alive and metabolically active at maturity, make up the bulk of ground tissue systems. They function in a variety of tasks. The ones making up mesophyll, for example, are photosynthetic. *26.2*

   b. Collenchyma supports growing plant parts. With its thick, lignified cell walls, sclerenchyma functions in mechanical support and water transport. *26.2*

4. Complex tissues include vascular tissues (xylem and phloem) and dermal tissues (epidermis and periderm). Each has two or more cell types (Table 26.1). *26.2*

   a. Vascular tissues distribute water and dissolved substances throughout a plant. Vascular bundles (each having xylem and phloem clustered together inside a cellular sheath) thread through the ground tissue.

   b. Water-conducting cells of xylem are not alive at maturity. Their lignified, pitted walls interconnect and serve as pipelines for water and dissolved minerals.

   c. Phloem's conducting cells are alive at maturity. The cytoplasm of adjoining cells interconnects across perforated end and side walls. In leaves, sugars and other photosynthetic products are loaded into the cells; often companion cells assist in this. They are unloaded where cells are growing or storing food.

   d. Epidermis covers and protects the outer surfaces of primary plant parts. Periderm replaces epidermis on plants showing extensive secondary growth. *26.2, 26.6*

5. Stems support upright growth and conduct water and solutes through their vascular bundles. Monocot stems often have vascular bundles distributed through ground tissue. Most dicot stems have a ring of bundles dividing the ground tissue into cortex and pith. *26.3*

6. Leaves have veins and mesophyll (photosynthetic parenchyma) between the upper and lower epidermis. Air spaces around the photosynthetic cells enhance gas exchange. Water vapor and gases cross the epidermis through numerous tiny openings called stomata. *26.4*

7. Roots absorb water and dissolved mineral ions for distribution to aboveground parts. Most anchor plants and store food. Some help support the shoot. *26.5*

Table 26.1 *Summary of Flowering Plant Tissues and Their Components*

SIMPLE TISSUES	
Parenchyma	Parenchyma cells
Collenchyma	Collenchyma cells
Sclerenchyma	Fibers or sclereids

COMPLEX TISSUES	
Xylem	Conducting cells (tracheids, vessel members); parenchyma cells; sclerenchyma cells
Phloem	Conducting cells (sieve-tube members); parenchyma cells; sclerenchyma cells
Epidermis	Undifferentiated cells; also guard cells and other specialized cells
Periderm	Cork; cork cambium; new parenchyma

Figure 26.24 Flower of (**a**) St. John's wort (*Hypericum*) and (**b**) an iris (*Iris*).

8. Wood (secondary xylem) is classified by location and function (as in heartwood or sapwood) and plant type (as in hardwood of many dicots, softwood of conifers). Bark consists of secondary phloem and periderm. *26.6*

Review Questions

1. Choose a flowering plant and list some functions of its roots and shoots. *26.1*

2. Name and define the basic functions of a flowering plant's three main tissue systems. *26.1*

3. Describe the differences between:
   a. apical, transitional, and lateral meristems *26.1*
   b. parenchyma and sclerenchyma *26.2*
   c. xylem and phloem *26.2*
   d. epidermis and periderm *26.2, 26.6*

4. Is the plant with the yellow flower in Figure 26.24 a dicot or a monocot? What about the plant with the purple flower? *26.2*

5. Which of the two stem sections below is typical of most dicots? Which is typical of most monocots? Label the main tissue regions of both sections. *26.2, 26.3*

6. Label the components of the three-year-old tree section to the right. Then make a rough count of the number of growth rings in the photograph of a stem section below it. *26.6*

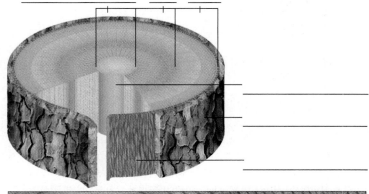

Self-Quiz ANSWERS IN APPENDIX III

1. Roots and shoots lengthen through activity at _____ .
   a. apical meristems   c. vascular cambium
   b. lateral meristems   d. cork cambium

2. Older roots and stems thicken through activity at _____ .
   a. apical meristems   c. vascular cambium
   b. cork cambium   d. both b and c

3. Soft, moist plant parts consist mostly of _____ cells.
   a. parenchyma   c. collenchyma
   b. sclerenchyma   d. epidermal

4. Xylem and phloem are _____ tissues.
   a. ground   b. vascular   c. dermal   d. both b and c

5. _____ conducts water and ions; _____ conducts food.
   a. Phloem; xylem   c. Xylem; phloem
   b. Cambium; phloem   d. Xylem; cambium

6. Buds give rise to _____ .
   a. leaves   c. stems
   b. flowers   d. all of the above

7. Mesophyll consists of _____ .
   a. waxes and cutin   c. photosynthetic cells
   b. lignified cell walls   d. cork but not bark

8. In early wood, cells have _____ diameters, _____ walls.
   a. small; thick   c. large; thick
   b. small; thin   d. large; thin

9. Match the plant parts with the most suitable description.
   ____ apical meristem   a. masses of xylem
   ____ lateral meristem   b. source of primary growth
   ____ xylem, phloem   c. corky surface covering
   ____ periderm   d. source of secondary growth
   ____ vascular cylinder   e. distribution of water, food
   ____ wood   f. central column in roots

Critical Thinking

1. Think about the kinds of conditions that might prevail in a hot desert in New Mexico or Arizona; on the floor of a shady, moist forest in Georgia or Oregon; and in the arctic tundra in Alaska. Also consider the kinds of animals living in each type of environment. Then "design" a type of flowering plant that would be suitably adapted to each place.

2. Fitzgerald, known for his highly underdeveloped sense of nature, sneaks into a forest preserve late at night and then maliciously girdles an old-growth redwood. He settles down and watches the tree, waiting for it to die. But the tree will not die that night or any time soon. Explain why.

3. Sylvia lives in Santa Barbara, where droughts are common and a long-term abundance of water is not. She replaced most of her garden with drought-tolerant plants and cut back the size of the lawn. The lawn does not get a light sprinkling every day. Sylvia waters it only twice a week in the evening, after the sun goes down. Then the lawn gets a good soak to a depth of several inches. Why is her strategy good for lawn grasses?

4. Oscar and Lucinda meet on a trip through a tropical rain forest and fall in love. In the exuberance of the moment, he carves their initials into the bark of a small tree. They never do get together, though. Ten years later, the still-heartbroken Oscar

searches for the tree. Given what you know about primary and secondary growth, will he find the carved initials higher relative to ground level? If he goes berserk and cuts down the tree, what kind of growth rings will he see?

Selected Key Terms

bark *26.6*
bud *26.3*
collenchyma *26.2*
cork *26.6*
cork cambium *26.1*
cortex *26.3*
cuticle (plant) *26.2*
dermal tissue system *26.1*
dicot *26.2*
epidermis *26.2*
fibrous root system *26.5*
ground tissue system *26.1*
growth ring *26.6*
hardwood *26.6*
heartwood *26.6*
lateral root *26.5*
leaf *26.4*
meristem *26.1*
mesophyll *26.4*
monocot *26.2*
parenchyma *26.2*
periderm *26.2*
phloem *26.2*
pith *26.3*
primary root *26.5*
root *26.1*
root hair *26.5*
sapwood *26.6*
sclerenchyma *26.2*
shoot *26.1*
softwood *26.6*
stoma (stomata) *26.2*
taproot system *26.5*
vascular bundle *26.3*
vascular cambium *26.1*
vascular cylinder *26.5*
vascular tissue system *26.1*
vein (leaf) *26.4*
xylem *26.2*

Readings

Esau, K. 1977. *Anatomy of Seed Plants.* Second edition. New York: Wiley. Well, sometimes the oldies are still goodies.

Rost, T., M. Barbour, C. R. Stocking, and T. Murphy. 1998. *Plant Biology.* Belmont, California: Wadsworth.

Simpson, B., and M. Conner-Ogorzaly. 1995. *Economic Botany.* Second edition. New York: McGraw-Hill. Fascinating book on uses and abuses of plants.

On-Line readings at Student Guide for InfoTrac:
www.brookscole.com/biology

PLANT NUTRITION AND TRANSPORT

Flies for Dinner

How often do we think that plants actually do anything impressive? Being mobile, intelligent, and emotional, we humans tend to be fascinated more with ourselves than with immobile, expressionless plants. Yet plants don't just stand around soaking up sunlight. Consider the Venus flytrap (*Dionaea muscipula*), a flowering plant native to the bogs of North and South Carolina. Its two-lobed, spine-fringed leaves open and close like a steel trap (Figure 27.1*a–d*). Like all other plants, it cannot grow properly without nitrogen and other nutrients, which happen to be scarce in the soil of bogs. However, plenty of insects fly in from places around the bogs.

Sticky sugars ooze from epidermal glands onto the surface of the flytrap's leaf. The sugars entice insects to land. As they do, they brush against hairlike structures that project from the leaf surface. These are triggers for the trap. When an insect touches two hairs at the same time or the same hair twice in rapid succession, the two lobes of the leaf snap shut. Now digestive juices pour out from cells of the leaf. They pool around the insect,

dissolve it, and so release nutrients from it. In this way the Venus flytrap makes its own nutrient-rich water, which it proceeds to absorb!

The Venus flytrap is only one of several species of **carnivorous plants**. We call them this even though it takes a leap of the imagination to put their mode of nutrient acquisition—a form of extracellular digestion and absorption—in the same category as the chompings of lions, dogs, and other meat eaters. Besides, not all carnivorous plants actively spring traps. Some types have fluid-filled traps into which prey slip, slide, or fall and then simply drown (Figure 27.2).

All carnivorous plants evolved in habitats where nitrogen and other nutrients are hard to come by. For instance, you may also come across plants with bizarre nutrient-acquiring habits in shallow freshwater lakes and streams, which contain only dilute concentrations of dissolved minerals.

Given the variety and numbers of insects and other animals that attack plants, you can just imagine how endearing the carnivorous plants are to botanists. With their plucky modes of nutrition, these plants also are a fine way to start thinking about **plant physiology**—the study of adaptations by which plants function in their environment. As you already know, nearly all plants are photoautotrophs that use energy from sunlight to drive the synthesis of organic compounds from water, carbon dioxide, and some minerals. Like people, they do not have unlimited supplies of the resources necessary to nourish themselves. Of every 1 million molecules of air, only 350 are carbon dioxide. Unlike the soggy habitats

VENUS FLYTRAP, OPEN FOR DINNER

Figure 27.1 Do plants take nutrition seriously? You bet. (**a**) A Venus flytrap (*Dionaea muscipula*). This carnivorous plant makes up for scarce nutrients in its habitat by absorbing them from animals that land on its leaves. (**b**) A fly stuck in sugary goo on a lobed leaf. (**c**) It brushes against hairlike triggers projecting from the leaf; the base of one is shown here. (**d**) The leaf, once activated, snaps shut.

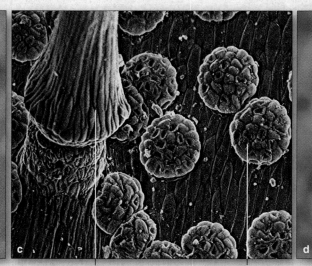

base of epidermal hair   epidermal gland

Figure 27.2 Cobra lily (*Darlingtonia californica*). Its leaves form a "pitcher" that is partly filled with digestive juices. Insects lured in by irresistible odors often cannot find the way back out; light shining through the pitcher's patterned dome confuses them. They just wander around and down, adhering to downward-pointing leaf hairs—which are slickened with wax above the potent vat.

of Venus flytraps, most soils are frequently dry. And nowhere except in overfertilized gardens does soil water hold lavish amounts of dissolved minerals. As you continue with these chapters on vascular plants, keep this point in mind: *Many aspects of plant structure and function are responses to low concentrations of vital environmental resources.*

Key Concepts

1. Many aspects of the structure and function of plants are adaptive responses to low concentrations of water, minerals, and other environmental resources.

2. A plant's root system takes up water from soil and also mines the soil for nutrients. For many land plants, mycorrhizae and bacterial symbionts assist in nutrient uptake. In a given habitat, soil properties influence the availability of water and nutrients for plants.

3. A plant's cuticle and its many stomata function in the conservation of water, a scarce resource in most habitats on land. Stomata are passageways across the epidermis of leaves and, to a lesser extent, stems. When open, stomata permit gas exchange. When closed, they help control water loss.

4. Stomata open during the day, when photosynthesis proceeds. Carbon dioxide diffuses into leaves, oxygen diffuses out—and water loss is rapid. Most plant species conserve water by closing stomata at night.

5. In flowering plants and other vascular plants, the flow of water and solutes through xylem and phloem functionally connects all living cells of roots, stems, and leaves. Xylem serves in the uptake and distribution of water and dissolved mineral ions. Phloem serves in the distribution of photosynthetically produced sugars and other organic compounds through the plant body.

6. Water absorbed from soil moves on up through xylem and into leaves. By the process of transpiration, dry air around leaves promotes evaporation through stomata. The force of evaporation is enough to pull continuous columns of water molecules, which are hydrogen-bonded to one another, from roots to aboveground parts.

7. By the energy-requiring process of translocation, sucrose and other organic compounds are distributed throughout the plant. Organic compounds produced by photosynthetic cells in leaves are loaded into conducting cells of phloem. They are unloaded at actively growing regions or storage regions of the plant body.

SOIL AND ITS NUTRIENTS

Properties of Soil

At the surface of most habitats on land is a thin cloak of soil. **Soil** consists of particles of minerals mixed with variable amounts of decomposing organic material, or **humus**. Its minerals come from the weathering of hard rocks. Dead organisms and organic litter (fallen leaves, feces, and so on) make up humus. Water and oxygen occupy spaces between the particles and organic bits.

In different regions, even in different parts of the same habitat, soils differ in the proportions of mineral particles and in the extent to which these have become compacted together. Generally, mineral particles come in three sizes, known as sand, silt, and clay. If you have ever let beach sand dribble between your fingers, you already have an idea that sand particles are large (0.05 to 2 millimeters across). Rub some silt from pond mud between your fingers. You won't be able to distinguish among individual particles; they are only 0.002 to 0.05 millimeters across. Clay particles are the finest of all.

How suitable is a given soil for plant growth? Is it gummy when wet because it does not have enough air spaces? Does it form hard clods when dry? The answer depends partly on the relative proportions of sand, silt, and clay. The more clay, the finer the soil's texture.

Each clay particle consists of thin, stacked layers of aluminosilicates with negatively charged ions at their surfaces. Clay attracts and binds (adsorbs) positively charged mineral ions dissolved in water that trickles through soil as well as the water molecules themselves.

Ions and water cling reversibly to clay. This chemical behavior is vital for all plants. With its high adsorption capacity, clay latches on to many nutrients for plants even as water percolates on past and drains away.

Yet too much clay is bad for plants. The packing of clay particles is so tight, there aren't enough air spaces to supply root cells with all the oxygen they need for aerobic respiration. Packing also retards penetration of water into the soil. Runoff is pronounced in heavy clay soils, so water and the nutrients dissolved in it are less available to help sustain plant growth. Plants do best in **loams**—soils having roughly the same proportions of sand, silt, and clay.

The amount of humus in a given soil also affects plant growth. Generally, humus has an abundance of negatively charged organic acids. It weakly binds with and retains dissolved mineral ions of opposite charge. Humus also has a high capacity to absorb and swell with water, then shrink when water is slowly released. Its alternating swelling and shrinking aerates soil. And as decomposers gradually work it over, humus releases nutrients that plants can take up.

In general, soils that are 10 to 20 percent humus are most favorable for plant growth. The worst have less than 10 percent humus or more than 90 percent humus, which is characteristic of swamps and bogs.

Soils can be classified by *profile* properties. The word refers to the layered characteristics of soils. It happens that the soils in different places are in different stages of development. Figure 27.3 has an example. **Topsoil**, the uppermost layering of soil, is called the A horizon. This is the most essential layer for plant growth, and its depth is variable from one habitat to the next.

Which Nutrients Are Essential for Plant Growth?

We have mentioned nutrients in passing. But what does the word mean? **Nutrients** are elements that are essential for a given organism because they directly or indirectly have roles in metabolism—hence in growth and survival—that no other element is able to fulfill. For plants, the essential elements are oxygen and hydrogen (from water) and carbon (from carbon dioxide fixed during photosynthesis). Besides carbon, hydrogen, and oxygen, plants rely on the uptake of at least thirteen other elements (Table 27.1).

— O HORIZON
Fallen leaves and other organic material littering the surface of mineral soil

— A HORIZON
Topsoil, which contains some percentage of decomposed organic material and which is variably deep (only a few centimeters deep in deserts, but elsewhere extending as far as thirty centimeters below the soil surface)

— B HORIZON
Compared with the A horizon, larger soil particles, not much organic material, but greater accumulation of minerals; extends thirty to sixty centimeters below soil surface

— C HORIZON
No organic material, but partially weathered fragments and grains of rock from which soil forms; extends to underlying bedrock

— BEDROCK

Figure 27.3 Some soil horizons that developed in one habitat in Africa.

Table 27.1 Essential Elements and Plant Function

MACRONUTRIENT	Some Functions	Some Deficiency Symptoms
Carbon, Hydrogen, Oxygen	Raw materials for second stage of photosynthesis	None; all are abundantly available (water, carbon dioxide are sources)
Nitrogen	Protein, nucleic acid, coenzyme, chlorophyll component	Stunted growth; light-green older leaves; older leaves yellow and die (these are symptoms that define a condition called chlorosis)
Potassium	Activation of enzymes; contributes to water–solute balances that influence osmosis*	Reduced growth; curled, mottled, or spotted older leaves; burned leaf edges; weakened plant
Calcium	Component in control of many cell functions; cementing of cell walls	Terminal buds wither, die; deformed leaves; poor root growth
Magnesium	Chlorophyll component; activation of enzymes	Chlorosis; drooped leaves
Phosphorus	Phospholipid, nucleic acid, ATP component	Purplish veins; stunted growth; fewer seeds, fruits
Sulfur	Component of most proteins, two vitamins	Light-green or yellowed leaves; reduced growth

MICRONUTRIENT	Some Functions	Some Deficiency Symptoms
Chlorine	Role in root and shoot growth and photolysis	Wilting; chlorosis; some leaves die
Iron	Roles in chlorophyll synthesis and in electron transport	Chlorosis; yellow and green striping in leaves of grass species
Boron	Roles in germination, flowering, fruiting, cell division, nitrogen metabolism	Terminal buds, lateral branches die; leaves thicken, curl, become brittle
Manganese	Chlorophyll synthesis; coenzyme action	Dark veins, but leaves whiten and fall off
Zinc	Role in forming auxin, chloroplasts, starch; enzyme component	Chlorosis; mottled or bronzed leaves; root abnormalities
Copper	Component of several enzymes	Chlorosis; dead spots in leaves; stunted growth
Molybdenum	Part of enzyme used in nitrogen metabolism	Pale green, rolled or cupped leaves

* All mineral elements contribute to water–solute balances, but potassium is notable because there is so much of it.

These usually are dissolved in soil water in ionic forms that reversibly bind with clay. Ions of calcium (Ca^{++}) and potassium (K^+) are examples. Plants easily exchange hydrogen ions for these weakly bound elements in clay.

Nine essential elements are *macro*nutrients. Normally they are required in amounts above 0.5 percent of the plant's dry weight (weighed after all the water has been

Figure 27.4 Erosion in an agricultural field. Water's erosive force forms gullies that channel runoff from the land. As gullies grow deeper and wider, erosion becomes more and more rapid. When topsoil is depleted, productivity declines and fertilizers usually must be trucked in to replace lost nutrients.

removed from the plant). The other elements listed are *micro*nutrients; they make up traces (usually a few parts per million) of the dry weight. Even the trace amounts are essential for normal growth.

Leaching and Erosion

Leaching refers to the removal of some of the nutrients in soil as water percolates through it. Leaching is most pronounced in sandy soils, which are not as good as clay at binding nutrients. It is not the same as **erosion**, which is the movement of land under the force of wind, running water, and ice (Figure 27.4). For example, each year, erosion from farmlands in the Mississippi River watershed puts about 25 billion metric tons of topsoil into the Gulf of Mexico. Whether by leaching or erosion, the loss of nutrients from soil is bad for plants and for all organisms that depend on plants for survival.

The mineral component of soil includes particles ranging from large-grained sand to silt and fine-grained clay. These particles, clay especially, reversibly bind water molecules and dissolved mineral ions and thereby make them more accessible for uptake by plant roots.

Soil also contains humus, which is a reservoir of organic material, rich in organic acids, in different stages of decay.

Most plants grow best in soils having equal proportions of sand, silt, and clay, as well as 10 to 20 percent humus.

Nutrients are essential elements. No other element can substitute for their direct or indirect roles in the metabolic activities that sustain growth and keep organisms alive.

HOW DO ROOTS ABSORB WATER AND MINERAL IONS?

In terms of energy outlays, mining the soil for mineral ions and water molecules clinging to clay particles is an expensive task. Plants spend considerable energy on building extensive root systems. Wherever the soil's texture and composition change, new roots must form to replace old ones and snake out in different regions. It isn't that roots "explore" soil for resources. Rather, the patches of soil where concentrations of water and mineral ions are greater stimulate outward growth.

Absorption Routes

Think back on the preceding chapter's discussion of a typical root's structure (Section 26.5). Water molecules in soil are only weakly bound to clay particles, so they readily cross the root epidermis and continue into the **vascular cylinder**, a column of vascular tissue. There, a cylindrical layer of cells, an **endodermis**, wraps around the column. A band of waxy deposits—the **Casparian strip**—is embedded in abutting endodermal cell walls (Figure 27.5). Water molecules can't penetrate it. They infiltrate only the unwaxed wall regions, pass through cells, then cross unwaxed wall regions on the opposite

side. This is the only way water and solutes move into the vascular cylinder. Like all cells, endodermal cells have many transport proteins embedded in the plasma membrane. The proteins let some solutes but not others cross it (Section 5.7). *The transport proteins of endodermal cells are control points where plants adjust the quantity and types of solutes absorbed from soil water.*

Roots of many plants also have an **exodermis**, a cell layer just beneath their surface (Figure 27.5a). Walls of exodermal cells commonly have a Casparian strip that functions like the one next to a root vascular cylinder.

Specialized Absorptive Structures

ROOT HAIRS   Vascular plants require great amounts of water. Roots of a mature corn plant absorb more than three liters of water daily. They could not do so without **root hairs**. Recall, from the preceding chapter, that root hairs are slender extensions of specialized epidermal cells. They greatly increase the surface area available for absorption (Section 26.5 and Figure 27.6). When a plant is putting on primary growth, its system of roots may develop millions or billions of root hairs.

ROOT NODULES   Certain bacteria and fungi help many plants absorb dissolved nutrients and receive something in return. As you read in Chapter 22, any two-way flow of benefits between species is a symbiotic interaction we call **mutualism**. As an example, think about how nitrogen deficiency limits plant growth. There is an abundance of gaseous nitrogen ($N\equiv N$) in the air. But plants do not have the

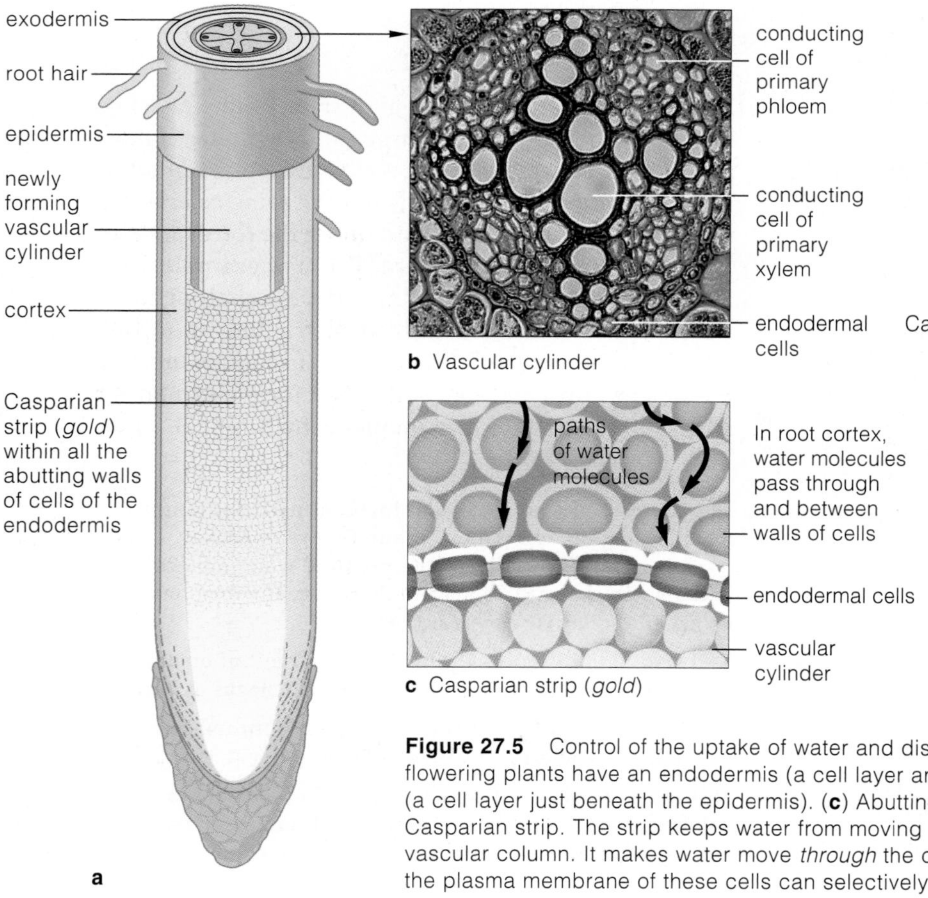

exodermis
root hair
epidermis
newly forming vascular cylinder
cortex
Casparian strip (*gold*) within all the abutting walls of cells of the endodermis

conducting cell of primary phloem
conducting cell of primary xylem
endodermal cells

b Vascular cylinder

paths of water molecules

In root cortex, water molecules pass through and between walls of cells
endodermal cells
vascular cylinder

c Casparian strip (*gold*)

Casparian strip

d Two endodermal cells. For clarity, the cytoplasm and most of the cell wall (*blue*) are not shown. Water and solutes can only move into the vascular cylinder by passing through the cells and unwaxed parts of their walls, not in between the cells.

a

Figure 27.5   Control of the uptake of water and dissolved nutrients inside roots. (**a,b**) Roots of most flowering plants have an endodermis (a cell layer around the vascular cylinder) and an exodermis (a cell layer just beneath the epidermis). (**c**) Abutting walls of cells of both layers contain a waxy Casparian strip. The strip keeps water from moving indiscriminately *around* the cells and into the vascular column. It makes water move *through* the cells. (**d**) That way, transport proteins that span the plasma membrane of these cells can selectively control water and nutrient uptake.

Figure 27.6 Scanning electron micrograph of root hairs (extremely fine extensions of specialized root epidermal cells) of a young root.

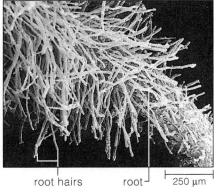

root hairs   root |— 250 μm

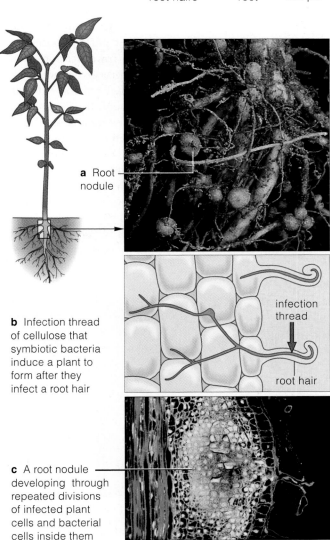

a Root nodule

b Infection thread of cellulose that symbiotic bacteria induce a plant to form after they infect a root hair

infection thread

root hair

c A root nodule developing through repeated divisions of infected plant cells and bacterial cells inside them

Figure 27.7 (**a**) Nutrient uptake at root nodules of legumes that are mutualists with nitrogen-fixing bacteria (*Rhizobium* and *Bradyrhizobium*). When infected by the bacteria, root hair cells form a thread of cellulose deposits. Bacteria use the thread as a highway to invade plant cells in the root cortex. (**b,c**) Infected plant cells and the bacterial cells inside them divide repeatedly. Together they form a swollen mass that becomes a root nodule. The bacteria start fixing nitrogen when membranes of plant cells surround them. The plant takes up some of the nitrogen; the bacteria take up some photosynthetic compounds.

Figure 27.8 Demonstration of the effect of root nodules on plant growth. At *left*, rows of soybean plants growing in nitrogen-poor soil. At *right*, growing in the same soil, plants in these rows were inoculated with *Rhizobium* cells and developed root nodules.

metabolic means to engage in **nitrogen fixation**. With this process, certain enzymes break all three covalent bonds in gaseous nitrogen, then attach nitrogen atoms to organic compounds. To get high crop yields, farmers either apply nitrogen-rich fertilizers or encourage the growth of certain nitrogen-fixing bacteria that naturally inhabit soil. Such bacteria can convert gaseous nitrogen to forms that they—and plants—can use. In this respect, string beans, peas, alfalfa, clover, and other legumes have an advantage. Nitrogen-fixing bacteria reside as symbionts in their roots, in localized swellings called **root nodules** (Figures 27.7 and 27.8). The bacterial cells withdraw some photosynthetically produced organic compounds that were distributed from leaves to the roots. But in return they provide the plants with some of the nitrogen that they secured.

MYCORRHIZAE   Also think back on the **mycorrhizae** (singular, mycorrhiza). As described in Chapter 22, a mycorrhiza is a symbiotic interaction between a young root and a fungus. Hence the name, meaning "fungus-root." Fungal filaments—hyphae—form a velvety cover around roots or penetrate the root cells. Collectively, hyphae have a large surface area that absorbs mineral ions from a larger volume of soil than the roots can do. The fungus can absorb sugars and nitrogen-containing compounds from root cells. The root cells obtain some scarce minerals that the fungus is better able to absorb.

Gymnosperms and flowering plant roots control the uptake of water and dissolved nutrients at the vascular cylinder's endodermis and at a similar layer near the root surface.

Root hairs, root nodules, and mycorrhizae greatly enhance the uptake of water and dissolved nutrients.

HOW IS WATER TRANSPORTED THROUGH PLANTS?

Transpiration Defined

By now, you have a sense of how the distribution of water and dissolved mineral ions to all living cells is central to plant growth and functioning. Let's turn now to a model for how water actually moves from a plant's roots to its stems, then into leaves.

Plants, recall, use only a fraction of the water they absorb for growth and metabolism. Most of that water is lost mainly through the numerous stomata in leaves. Evaporation of water molecules from leaves, stems, and other plant parts is a process called **transpiration**.

Cohesion–Tension Theory of Water Transport

This brings up an interesting question. Assuming that plants lose most of the absorbed water from their leaves, how does water actually get *to* the leaves? What gets it to the uppermost leaves of plants, including redwoods and other trees that may be more than 100 meters tall?

In a plant's vascular tissues, water moves through a complex tissue called **xylem**. Recall, from Section 26.2, that the water-conducting cells of xylem are **tracheids** and **vessel members**. Figure 27.9 provides examples. These cells are dead at maturity, and only their lignin-impregnated walls are left. This means the conducting cells in xylem can't be actively pulling water "uphill."

Figure 27.10 *Facing page:* Diagram illustrating the cohesion–tension theory of water transport in vascular plants.

Some time ago, the botanist Henry Dixon came up with a useful way to explain water transport in plants. By his **cohesion–tension theory**, the water inside xylem is pulled upward by air's drying power, which creates continuous negative pressures—that is, tensions. These tensions extend downward from leaves to roots. Figure 27.10 illustrates Dixon's theory. Think about these three points as you review the illustration:

First, air's drying power causes transpiration: the evaporation of water from all plant parts exposed to air—but most notably at stomata. Transpiration puts the water confined in xylem's waterproofed conducting tubes into a state of tension. And that tension extends from veins inside leaves, down through the stems, and on into young roots where water is being absorbed.

Second, the unbroken, fluid columns of water show *cohesion*; they resist rupturing while they are pulled up under *tension*. (Here you may wish to reflect on Section 2.5.) The collective strength of all the hydrogen bonds between water molecules in the narrow, tubular xylem cells imparts this cohesion.

Third, for as long as molecules of water continue to escape from a plant, the continuous tension inside the

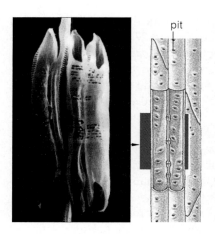 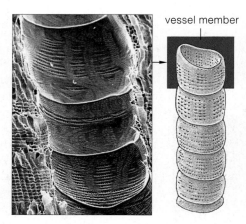 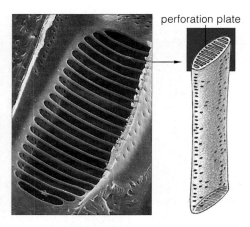

a Tracheids have unperforated, tapered end walls. In intact xylem, small pits in the walls of adjoining tracheids match up. The pits are too small to permit rapid flow, but they do confine air bubbles to individual tracheids in the system. Bubbles can obstruct water transport.

b Close-up of three of the adjoining vessel members that make up a vessel. The thick, finely perforated walls of these dead cells connect one after another to form vessels, another type of water-conducting tube in xylem. The walls of all these tubes are impregnated with lignin, which strengthens them and also makes them waterproof.

c Perforation plate at the end wall of one type of vessel member. The perforated ends permit water and air bubbles to flow unimpeded through the conducting tube. This may be a reason why natural selection favored conserving tracheids and vessel members in the same plants.

Figure 27.9 Scanning electron micrographs and sketches of a few types of tracheids and vessel members from xylem. These water-conducting tubes are made of the walls of cells that are dead at maturity. The cell walls still remain interconnected, and they form the tubes. Tracheids probably evolved before vessel members. Both occur in nearly all vascular plants.

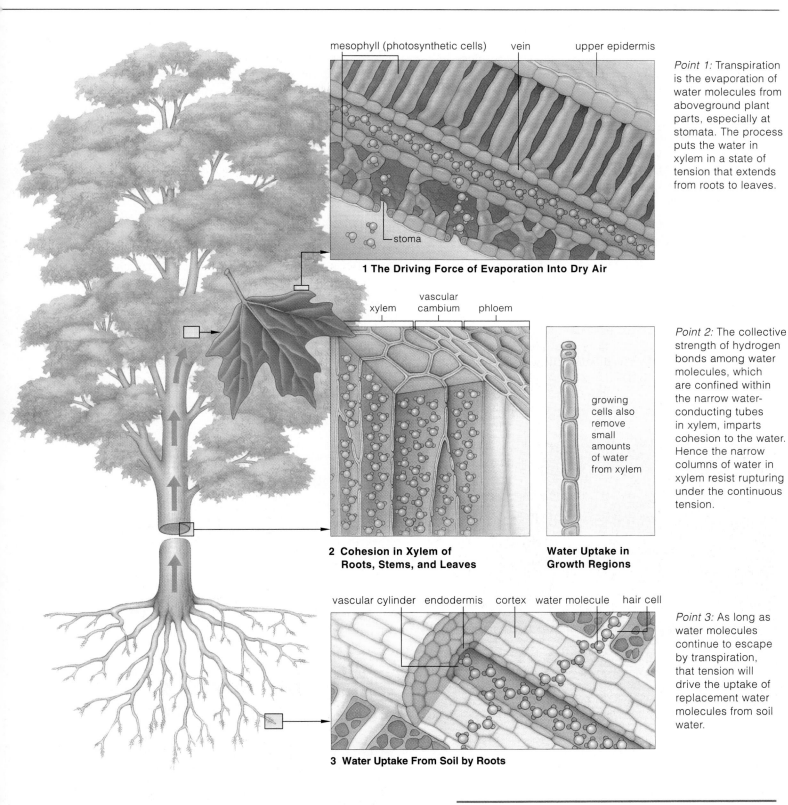

mesophyll (photosynthetic cells)　　vein　　upper epidermis

stoma

1 The Driving Force of Evaporation Into Dry Air

Point 1: Transpiration is the evaporation of water molecules from aboveground plant parts, especially at stomata. The process puts the water in xylem in a state of tension that extends from roots to leaves.

vascular
xylem　　cambium　　phloem

growing cells also remove small amounts of water from xylem

2 Cohesion in Xylem of Roots, Stems, and Leaves

Water Uptake in Growth Regions

Point 2: The collective strength of hydrogen bonds among water molecules, which are confined within the narrow water-conducting tubes in xylem, imparts cohesion to the water. Hence the narrow columns of water in xylem resist rupturing under the continuous tension.

vascular cylinder　endodermis　　cortex　water molecule　　hair cell

3 Water Uptake From Soil by Roots

Point 3: As long as water molecules continue to escape by transpiration, that tension will drive the uptake of replacement water molecules from soil water.

xylem permits more molecules to be pulled upward from the roots, and therefore to replace them.

Hydrogen bonds are strong enough to hold water molecules together inside the xylem. However, they are not strong enough to prevent the water molecules from breaking away from one another during transpiration and then escaping from leaves, through stomata.

Transpiration is the process of evaporation from plant parts.

By the cohesion–tension theory of water transport, this process is the key source of tensions in water in xylem. The tensions extend from leaves to roots, and they allow columns of water molecules that are hydrogen-bonded to one another to be pulled upward through the plant body.

HOW DO STEMS AND LEAVES CONSERVE WATER?

At least 90 percent of the water that enters a leaf goes right on through it and evaporates into the surrounding air. Cells use only 2 percent of the water that stays in the leaf for photosynthesis, membrane functions, and other events. That tiny amount is vital. Plants wilt and water-dependent events are severely disrupted when water loss exceeds water uptake at roots for extended periods (Figure 27.11). Yet land plants are not entirely at the mercy of changes in the availability of soil water. They have a cuticle, and they have stomata.

The Water-Conserving Cuticle

Even mildly water-stressed plants would rapidly wilt and die without their **cuticle** (Figure 27.12). Epidermal cells secrete this translucent, water-impermeable layer, which coats cell wall regions exposed to the air. At the cuticle surface are deposits of waxes, which are water-insoluble lipids having long fatty-acid tails. The cuticle itself consists of waxes embedded in **cutin**, an insoluble lipid polymer. Beneath all the waxes, cellulose fibers weave through the cutin. A layer of polysaccharides (pectins) often helps bind the cuticle to the cell walls.

A cuticle does not bar the passage of light rays into photosynthetic parts of the plant. It does restrict water loss. It also restricts the *inward* diffusion of the carbon dioxide necessary for photosynthesis and the *outward* diffusion of oxygen by-products. Recall, from Section 6.6, that a buildup of oxygen in the air spaces inside a leaf has bad effects on the rate of photosynthesis.

Controlled Water Loss at Stomata

How do carbon dioxide and oxygen get past the cuticle-covered, water-conserving epidermis? They cross it at many **stomata** (singular, stoma). At all these openings, carbon dioxide enters and water evaporates from the plant. A pair of specialized cells, the **guard cells**, define each opening (Figure 27.13). As each swells with water, the force of internal fluid pressure distorts its wall. An osmotically induced, internal fluid pressure that builds up against cell walls is called **turgor pressure**. It makes

Figure 27.11 Osmosis and the wilting of a plant with soft green leaves. When such a plant is growing well, you can bet soil water is dilute (hypotonic) compared to fluids in the plant's living cells.

In response to solute concentration gradients, water osmotically moves into a plant cell. Internal fluid pressure builds up against the cell's wall (Section 5.9). Water also is squeezed out when this turgor pressure becomes great enough to counter the attractive force of cytoplasmic fluid. (Cytoplasm usually has more solutes compared to soil water.) Both forces have the potential to cause water to move directionally. The sum of the two opposing forces is the "water potential."

When soft plant parts are erect, as much water is moving into cells as is moving out. The constant pressure keeps cell walls plump. When the soil dries or gets too salty, water's concentration gradient reverses and the cells lose water. Osmotically induced shrinkage of cytoplasm in all of the young cells of soft plant parts results in wilting; the parts droop with the loss of turgor pressure.

A simple experiment demonstrates the wilting effect. Put 10 grams of table salt (NaCl) in 60 milliliters of water. Pour the salty solution into the soil around a tomato plant. The plant starts to collapse after 5 minutes. In less than 30 minutes, you see severe wilting.

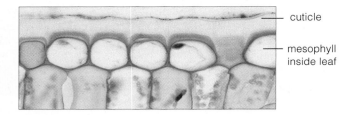

cuticle

mesophyll inside leaf

Figure 27.12 Waxy cuticle on upper leaf epidermis. Water, carbon dioxide, and oxygen cross a cuticle mainly at stomata. In general, plants in deserts and other seasonally dry habitats have thick cuticles, which provide greater protection against water loss. Aquatic plants have thin cuticles or none at all.

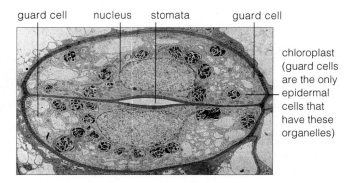

guard cell   nucleus   stomata   guard cell

chloroplast (guard cells are the only epidermal cells that have these organelles)

Figure 27.13 Guard cells in stem epidermis from a prickly pear (beavertail cactus). Cacti have stomata on stems and tiny leaves, if any. Their spines, modified nonphotosynthetic shoots, deter browsing animals while reducing the surface area from which water can evaporate.

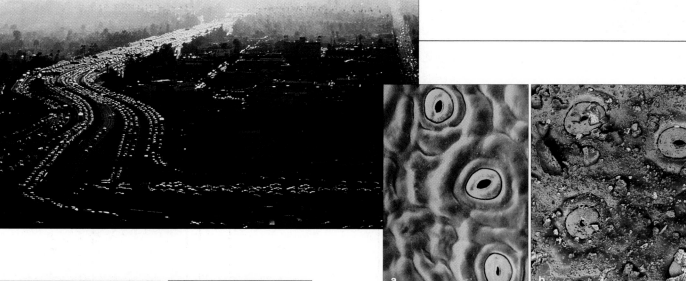

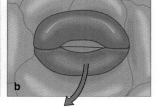

Figure 27.14 Stomatal action. Whether a stoma is open or closed during a given interval depends on changes in the shape of the two guard cells that define this small opening across the cuticle-covered epidermis. (**a**) Stomatal opening. Water enters collapsed guard cells, which swell under turgor pressure and move apart and so form a stoma. (**b**) Stomatal closing. Water leaves swollen guard cells, which collapse against each other and close the stoma.

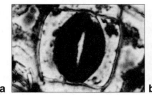 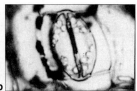

Figure 27.15 Experimental evidence that potassium ions accumulate in guard cells that are expanding with incoming water. Strips from leaf epidermis of a dayflower (*Commelina communis*) were immersed in solutions with dark-staining substances that bind preferentially with potassium ions. (**a**) In leaf samples with opened stomata, most of the ions became concentrated in guard cells. (**b**) In leaf samples with closed stomata, very little potassium entered guard cells; most was present in other epidermal cells.

guard cells bend in such a way that a gap (one stoma) forms between them. When two guard cells lose water and their internal fluid pressure drops, they collapse against each other and close the gap (Figure 27.14).

In most plants, stomata remain open during the day, when photosynthesis takes place. They do lose water—but they gain carbon dioxide for the reactions. Stomata stay closed at night. Then, plants conserve water, and precious carbon dioxide accumulates inside the leaf as the cells busily engage in aerobic respiration.

Figure 27.16 (**a**) Stomata at holly leaf surface. (**b**) What they look like when a holly plant is growing in industrialized regions. Gritty airborne pollutants clog its stomata and prevent much of the sun's rays from reaching photosynthetic cells in the leaf.

Guard cells are the only epidermal cells that contain chloroplasts. Whether a stoma opens or closes depends on how much water and carbon dioxide its guard cells hold. Photosynthesis starts after the sun comes up. As the morning progresses, carbon dioxide levels drop in cells—including guard cells. The decrease helps trigger active transport of potassium ions into the guard cells (Figure 27.15). So do blue wavelengths of light, which penetrate the atmosphere better as the sun arcs higher in the sky. Water follows potassium ions into the cells, by osmosis. That inward movement of water provides the fluid pressure to open the stoma.

Carbon dioxide levels in the cells rise when the sun sets and photosynthetic activity ceases. Potassium, then water, move out of guard cells. And so stomata close.

CAM plants, including most cacti, conserve water differently. They open stomata at night, when they fix carbon dioxide by the metabolic C4 pathway described in Section 6.6. The next day, when their stomata close, CAM plants use carbon dioxide in photosynthesis.

As this section makes clear, plant survival depends on stomata. Think about it when you are out and about on smog-shrouded days (Figure 27.16).

Water-dependent events in plants are severely disrupted when water loss exceeds uptake at roots for extended periods. Wilting is one observable outcome.

Transpiration and gas exchange occur mainly at stomata. These numerous small openings span the waxy cuticle, which covers all plant epidermal surfaces exposed to air.

Plants open and close stomata at different times to control water loss, carbon dioxide uptake, and oxygen disposal, all of which affect rates of photosynthesis and plant growth.

HOW ARE ORGANIC COMPOUNDS DISTRIBUTED THROUGH PLANTS?

Whereas xylem distributes water and minerals through the plant, the vascular tissue called **phloem** distributes organic products of photosynthesis. Like xylem, phloem consists of many conducting tubes, fibers, and strands of parenchyma cells, all of which extend through the plant body. Unlike xylem, phloem contains conducting tubes of a type called **sieve tubes**. These tubes consist of *living* cells, positioned side by side and end to end in vascular tissues. Figure 27.17 is an example of the cells, which are called sieve-tube members. Dissolved sugars and other organic compounds flow swiftly through the large pores in their abutting end walls.

Adjoining the long sieve tubes are **companion cells**. Like the sieve-tube members, these are differentiated parenchyma cells. As you will see, companion cells help load organic compounds into sieve tubes.

To understand what goes on inside phloem, think of what happens to sucrose and other organic products of photosynthesis. Leaf cells use some of the products for their own activities. The rest travels to roots, stems, buds, flowers, and fruits. In most cells, carbohydrates get stored as starch, in plastids. The proteins and fats, which the cells synthesize from carbohydrates and the amino acids distributed to them, often get stored in seeds. Avocados and some other fruits also accumulate fats.

Starch molecules are too large for transport across the plasma membrane of cells that produce them. Also, they are too insoluble for transport to other parts of the plant. Overall, proteins are too large and fats are too insoluble for transport from the storage sites. *However, the cells convert the storage forms of organic compounds to solutes of smaller size that are far more easily transported through phloem.*

For instance, the cells degrade starch to glucose monomers. When one of these monomers combines with fructose, the result is sucrose —an easily transportable sugar.

Simple experiments with aphids revealed that sucrose is the main carbohydrate that is transported inside phloem. These insects were anesthetized by exposure to high

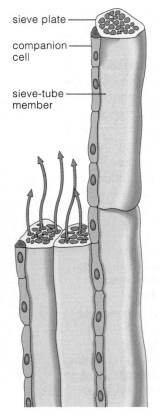

sieve plate

companion cell

sieve-tube member

Figure 27.17 Examples of sieve-tube members, living cells that interconnect to form a conducting tube in phloem. Living companion cells expend ATP energy to load organic compounds into adjacent sieve-tube members.

Figure 27.18 A honeydew droplet exuding from the end of an aphid gut. The tubular mouthpart of the insect penetrated a conducting tube in phloem. The sugary fluid was under high pressure in phloem and was forced out through the gut's terminal opening.

levels of carbon dioxide as they were feeding on juices in conducting tubes of phloem. The body of the aphids was detached from their mouthparts—which were still embedded in the sieve tubes. For most of the plants studied, sucrose was the most abundant carbohydrate in the fluid that was being forced out of the tubes.

Translocation

Translocation is the technical name for the transport of sucrose and all other organic compounds through the phloem of a vascular plant. High pressure drives this process. Often the pressure is five times higher than in automobile tires. Aphids demonstrate the magnitude of it when they force their mouthparts into sieve tubes and feed upon the dissolved sugars. The high pressure can force fluid through an aphid gut and out the other end, as "honeydew" (Figure 27.18). Park a car under some trees being attacked by aphids, and it might get spattered by sticky honeydew droplets, thanks to the high fluid pressure in phloem.

Pressure Flow Theory

Phloem translocates the photosynthetic products along decreasing pressure and solute concentration gradients. The *source* of the flow is any region of the plant where organic compounds are being loaded into sieve tubes. Common sources are mesophylls—the photosynthetic tissues in leaves. The flow ends at a *sink*, which is any plant region where the products are being unloaded, used, or stored. Example: While flowers and fruits are growing during development, they are sink regions.

Why do organic compounds flow from a source to a sink? According to the **pressure flow theory**, internal pressure builds up at the source end of the sieve tube system and *pushes* the solute-rich solution on toward any sink, where solutes are being removed.

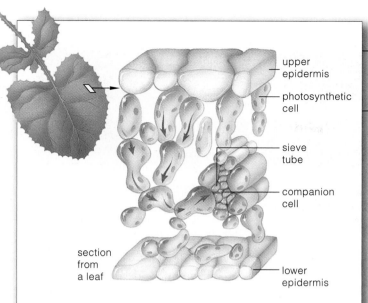

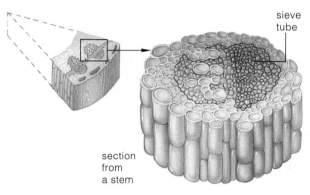

a *Loading at a source.* Photosynthetic cells in leaves are a common source of organic compounds that must be distributed through a plant. Small, soluble forms of these compounds move from the cells into phloem (a leaf vein).

b *Translocation along a distribution path.* Fluid pressure is greatest inside sieve tubes at the source. It pushes the solute-rich fluid to a sink, which is any region where cells are growing or storing food. There, the pressure is lower because cells are withdrawing solutes from the tubes.

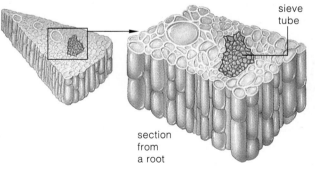

c *Unloading at the sink.* Solutes are unloaded from sieve tubes into cells at the sink; water follows. Translocation continues as long as solute concentration gradients and a pressure gradient exist between the source and the sink.

Figure 27.19 Translocation of organic compounds in sow thistle (*Sonchus*). Review Figure 6.3 to get an idea of how translocation relates to photosynthesis in vascular plants.

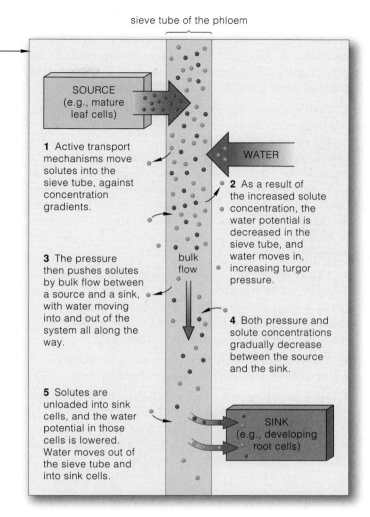

sieve tube of the phloem

SOURCE (e.g., mature leaf cells)

WATER

1 Active transport mechanisms move solutes into the sieve tube, against concentration gradients.

2 As a result of the increased solute concentration, the water potential is decreased in the sieve tube, and water moves in, increasing turgor pressure.

3 The pressure then pushes solutes by bulk flow between a source and a sink, with water moving into and out of the system all along the way.

bulk flow

4 Both pressure and solute concentrations gradually decrease between the source and the sink.

5 Solutes are unloaded into sink cells, and the water potential in those cells is lowered. Water moves out of the sieve tube and into sink cells.

SINK (e.g., developing root cells)

Experimental evidence for the pressure flow theory was obtained from studies of sow thistle (*Sonchus*). Use Figure 27.19 to track what happens after sucrose moves from photosynthetic cells into small veins in the leaves of this plant. Companion cells in veins spend energy to load sucrose into adjacent sieve-tube members. As the sucrose concentration increases in the tubes, water also moves in, by osmosis. More internal fluid pressure is exerted against the sieve-tube walls. When the pressure increases, it pushes the sucrose-laden fluid out of the leaf, into the stem, and on to the sink.

Plants store organic compounds in the form of starch, fats, and proteins. They convert the storage forms to sucrose and other small units that are soluble and easily translocated.

Translocation is the distribution of organic compounds to different plant regions. It depends on concentration and pressure gradients in the sieve-tube system of phloem.

Gradients exist as long as companion cells load compounds into the sieve tubes at sources, such as mature leaves, and as long as compounds are removed at sinks, such as roots.

SUMMARY

1. Vascular plants depend on the distribution of water, dissolved mineral ions, and organic compounds to its cells. Figure 27.20 summarizes how that distribution sustains plant growth and survival. **27.1–27.5**

2. Root systems efficiently take up water and nutrients, which commonly are scarce in soil. *CI, 27.1*

   a. Collectively, a vascular plant's root hairs (slender extensions of specialized root epidermal cells) greatly increase the surface area available for absorption. *27.2*

   b. Bacterial and fungal symbionts help many plants take up mineral ions. They benefit from the interaction by using some products of photosynthesis. Examples of these mutualistic interactions are root nodules and mycorrhizae. *27.2*

3. Plants distribute water and dissolved mineral ions through water-conducting tubes of xylem, a vascular tissue. Cells called tracheids and vessel members form the tubes but are dead at maturity. Their waterproofed walls interconnect as narrow pipelines. *27.3*

4. Plants lose water by transpiration, or evaporation of water from leaves and other parts exposed to air. *27.3*

5. Here are the points of the cohesion–tension theory of water transport in plants: *27.3*

   a. Inside xylem's water-conducting cells, continuous negative pressures (tensions) extend from leaves to the roots. Transpiration causes the tension.

   b. When water molecules escape from the leaves, replacements are pulled into the leaf under tension.

   c. The collective strength of numerous hydrogen bonds between water molecules imparts cohesion that allows water molecules to be pulled up as continuous fluid columns. The "cohesion" part of the theory refers to the capacity of the hydrogen bonds between water molecules to resist rupturing when put under tension.

6. Most plants have a waxy, water-impermeable cuticle covering their aboveground parts. They lose water and take up carbon dioxide at stomata. A pair of guard cells defines each of these microscopically small openings across the epidermis of leaves and stems. Guard cells are specialized parenchyma cells. *27.4*

7. Stomata open and close at different times. Control of their action balances water conservation with carbon dioxide uptake and the release of oxygen. *27.4*

   a. In many kinds of plants, stomata open during the day. Such plants lose water but take in carbon dioxide for photosynthesis. Stomata close at night; thus, water and the carbon dioxide released by aerobically respiring cells are conserved for the next day.

   b. CAM plants open stomata and fix carbon dioxide by the C4 pathway at night. During the day, they close their stomata and use the carbon fixed the night before for photosynthesis.

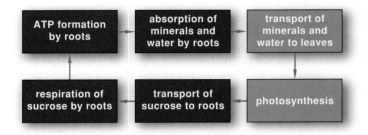

Figure 27.20 Summary of the interdependent processes that sustain the growth of vascular plants. All living cells in plants require oxygen, carbon, hydrogen, and at least thirteen mineral ions. They all produce ATP to drive metabolic activities.

8. Plants distribute organic compounds through sieve tubes of phloem, a vascular tissue. Translocation is the distribution of organic compounds in phloem. *27.5*

9. According to the pressure flow theory, translocation is driven by differences in solute concentrations and pressure between source and sink regions. A source is any site where organic compounds are being loaded into sieve tubes (e.g., mature leaves). A sink is any site where compounds are being unloaded from sieve tubes. Roots are examples. Solute concentration gradients and pressure gradients exist for as long as companion cells expend energy to load solutes into the sieve tubes and for as long as solutes are removed at the sink. *27.5*

Review Questions

1. Define soil, then distinguish between: *27.1*
   a. humus and loam
   b. leaching and erosion
   c. macronutrient and micronutrient (for plants)

2. Define nutrient. What are signs that a plant is deficient in one of the essential nutrients listed in Table 27.1? *27.1*

3. What is the function of the Casparian strip in roots? *27.2*

4. Using Dixon's model, explain how water moves from soil upward through tall plants. *27.3*

5. Describe the structure and function of a plant cuticle. *27.4*

6. Which type of ion influences stomatal action? *27.4*

7. Explain translocation according to the pressure flow theory described in this chapter. *27.5*

Self-Quiz   ANSWERS IN APPENDIX III

1. Carbon, hydrogen, oxygen, nitrogen, and potassium are examples of _____ for plants.
   a. macronutrients   d. essential elements
   b. micronutrients   e. both a and d
   c. trace elements

2. A _____ strip in endodermal cell walls forces water and solutes to move through root cells, not around them.
   a. cutin   b. lignin   c. Casparian   d. cellulose

3. The nutrition of some plants depends on a root–fungus association known as a _____ .
   a. root nodule   c. root hair
   b. mycorrhiza   d. root hypha

4. The nutrition of some plants depends on a root–bacterium association known as a _____ .
   a. root nodule   c. root hair
   b. mycorrhiza   d. root hypha

5. Water can be pulled up through a plant by the cumulative strength of _____ between water molecules.

6. Water evaporation from plant parts is called _____ .
   a. translocation   c. transpiration
   b. expiration   d. tension

7. Water transport from roots to leaves is explained by _____ .
   a. the pressure flow theory
   b. differences in source and sink solute concentrations
   c. the pumping force of xylem vessels
   d. the cohesion–tension theory

8. In daytime, most plants lose _____ and take up _____ .
   a. water; carbon dioxide   c. oxygen; water
   b. water; oxygen   d. carbon dioxide; water

9. At night, most plants conserve _____ , and _____ accumulates.
   a. carbon dioxide; oxygen   c. oxygen; water
   b. water; oxygen   d. water; carbon dioxide

10. In phloem, organic compounds flow through _____ .
    a. collenchyma cells   c. vessels
    b. sieve tubes   d. tracheids

11. Match the concepts of plant nutrition and transport.
    _____ stomata   a. evaporation from plant parts
    _____ nutrient   b. response to scarce soil nutrients
    _____ sink   c. balancing water loss with
    _____ root system   carbon dioxide requirements
    _____ hydrogen   d. cohesion in water transport
           bonds   e. sugars unloaded from sieve tubes
    _____ transpiration   f. organic compounds distributed
    _____ translocation   through the plant body
                            g. element with roles in metabolism
                               that no other element can fulfill
                               for an organism

Critical Thinking

1. Home gardeners, like farmers, must ensure that their plants have access to nitrogen from either nitrogen-fixing bacteria or fertilizer. Insufficient nitrogen stunts plant growth; leaves turn yellow and die. Which major classes of biological molecules incorporate nitrogen? How would a low nitrogen level in plants affect biosynthesis and cause symptoms of nitrogen deficiency?

2. When moving a plant from one place to another, it helps to include some native soil around the roots. Explain why, given what you know about mycorrhizae and root hairs.

3. In the sketch at *right*, label the actual stoma. Now think about how Henry discovered a way to keep all of a plant's stomata open at all times. He also figured out how to keep those of another plant closed all the time. Both plants died. Explain why.

4. Allen is studying the rate of transpiration from tomato plant leaves. He notices that several environmental factors, including wind and relative humidity, affect the rate. Explain why.

5. You have just returned home from a three-day vacation. Your plants tell you, by their severe wilting, that you forgot to water them before you departed. Being aware of the cohesion–tension theory of water transport, explain what happened to them.

6. Not having a green thumb, your friend Stephanie decides to grow zucchini plants, which are very forgiving of poor soils

Figure 27.21   Middle fork of the Salmon River, Idaho.

and amateur gardeners. She plants too many seeds and ends up with far too many plants. Among them you happen to notice a stunted plant and decide to find out what happened to it. After many experiments, you decide that its leaves are producing a mutated, malfunctioning form of an enzyme that is necessary for the formation of sucrose. Knowing what you do about the pressure flow theory, explain why plant growth was hindered.

7. At the middle fork of the Salmon River in Idaho, clear water from a wilderness area (visible at *right* in Figure 27.21) converges on brown-colored water (visible at *left* and *center*). The brown water is enriched with silt from a wilderness area that was disturbed by cattle ranching and some other human activities. Knowing what you do about the nature of topsoil, what is probably happening to plant growth in the disturbed habitat? Knowing how fishes acquire oxygen for aerobic respiration, how might the silt be affecting their survival?

Selected Key Terms

CAM plant 27.4	leaching 27.1	root nodule 27.2
carnivorous plant *CI*	loam 27.1	sieve tube 27.5
Casparian strip 27.2	mutualism 27.2	soil 27.1
cohesion–tension	mycorrhiza 27.2	stoma (stomata) 27.4
theory 27.3	nitrogen	topsoil 27.1
companion cell 27.5	fixation 27.2	tracheid 27.3
cuticle 27.4	nutrient 27.1	translocation 27.5
cutin 27.4	phloem 27.5	transpiration 27.3
endodermis 27.2	plant	turgor pressure 27.4
erosion 27.1	physiology *CI*	vascular cylinder 27.2
exodermis 27.2	pressure flow	vessel member 27.3
guard cell 27.4	theory 27.5	xylem 27.3
humus 27.1	root hair 27.2	

Readings

Galston, A., P. Davies, and R. Satter. 1980. *The Life of a Green Plant.* Englewood Cliffs, New Jersey: Prentice-Hall.

Hausenbuiller, R. 1985. *Soil Science: Principles and Practices.* Third edition. Dubuque, Iowa: W. C. Brown.

Hopkins. W. G. 1995. *Introduction to Plant Physiology.* New York: Wiley.

Salisbury, F., and C. Ross. 1992. *Plant Physiology.* Fourth edition. Belmont, California: Wadsworth.

On-Line readings at Student Guide for InfoTrac:
www.brookscole.com/biology   449

PLANT REPRODUCTION AND DEVELOPMENT

A Coevolutionary Tale

We find flowering plants almost everywhere, from icy tundra to deserts to oceanic islands. What accounts for their distribution and diversity? Consider the **flower**, a specialized reproductive shoot. About 435 million years ago, when plants first invaded the land, insects that ate decaying plant parts and spores probably weren't far behind. The plant smorgasbord seems to have favored natural selection of winged insects with an amazing array of sucking, piercing, and chewing mouthparts.

By 390 million years ago, in humid coastal forests, seed-bearing plants were making pollen grains. These tiny, sperm-bearing packages can travel to eggs, which develop in ovules in female plant parts. Pollen is rich in nutrients. At first, air currents may have dispersed it to the ovules. Then hungry insects made the connection between "plant parts with pollen" and "food." Plants lost some pollen to insects but gained a reproductive advantage. How? Unlike air currents, pollen-dusted insects clambering over plants could accurately deliver pollen to ovules. The tastier the pollen, the more home deliveries, and the more seeds. *The greater the number of seeds, the greater the chances of reproductive success.*

What we are describing is a case of **coevolution**. The word refers to two or more species jointly evolving as an outcome of close ecological interactions. A heritable change in one species affects selection pressure operating between them, so the other species evolves, too.

In our coevolutionary tale, new or modified plant structures that were more enticing to pollen-delivering insects were favored. Insects quicker to recognize and locate specific plants also enjoyed a competitive edge. Hence the evolution of plants with distinctive flowers, fragrances, and sugar-rich nectar—and of pollinators.

A **pollinator** is any agent that transfers pollen from male to female reproductive parts of flowers of the same plant species. Besides insects, pollinating agents include air currents, water currents, bats, birds, and other animals (Figures 28.1 and 28.2).

You can correlate many floral features with specific mutualists. For instance, a flower's reproductive parts are positioned so that pollinating agents will brush past them. Many parts are positioned above nectar-filled floral tubes the same length as the feeding device of a preferred pollinator. Red and yellow flowers attract birds, which have great daytime vision but a poor sense of smell. As you might suspect, plants that birds visit don't divert metabolic resources to making fragrances. Red flowers don't attract beetles. Neither do flowers

Figure 28.1 Example of adaptations uniting a flowering plant with its pollinator. The giant saguaro of Arizona's Sonoran Desert has large, showy white flowers at the tips of its spiny arms. Insects and birds visit the flowers by day, and bats visit by night. The plant offers these animals nectar, and the animals transport pollen grains that stick to their body from one cactus plant to another.

How we see it   How bees see it

Figure 28.2 (**a**) Bahama woodstar sipping nectar from a hibiscus blossom. Like other hummingbirds, it forages for nectar in midflight. Its long, narrow bill coevolved with long, narrow floral tubes. (**b,c**) Shine ultraviolet light on a gold-petaled marsh marigold to reveal its bee-attracting pattern.

with nectar cups large and deep enough for beetles to drown in. Like flies, beetles pollinate flowers that smell like rotten meat, moist dung, or decaying litter on the forest floor—where beetles first evolved.

Daisies and other fragrant flowers with distinctive patterns, shapes, and red or orange components attract butterflies, which forage by day. Nectar-sipping bats and most moths forage by night. They pollinate intensely sweet-smelling flowers with white or pale petals that are more visible than colored petals in the dark. Long, thin mouthparts of moths and butterflies reach nectar in narrow floral tubes or floral spurs. The Madagascar hawkmoth uncoils a mouthpart the same length as a narrow floral spur of an orchid, *Angraecum sesquipedale*. It is 22 centimeters (more than 8–1/2 inches) long!

Flowers with sweet odors and yellow, blue, or purple parts attract bees. Their ultraviolet-light-absorbing pigments form patterns that say, "Nectar here!" Unlike us, bees see ultraviolet light and find nectar guides alluring (Figure 28.2*b,c*). Unlike beetles, they also have long mouthparts that extend into floral tubes.

In short, far from being merely decorative, flowers contribute to reproductive success. This chapter focuses on the modes of reproduction and development of the plants that bear them.

Key Concepts

1. The premier reproductive mode of flowering plant life cycles is sexual reproduction. It involves the formation of spores and gametes, both of which develop inside the specialized reproductive shoots called flowers.

2. Microspores form in a flower's male reproductive parts and develop into pollen grains, the sperm-bearing male gametophytes. Animals, air currents, and other pollinating agents transfer pollen grains to the female reproductive parts of flowers.

3. Megaspores develop inside ovules, structures that form on the inner ovary wall of the flower's female reproductive parts. Within each ovule, a mature female gametophyte forms from a megaspore. One of its cells is the egg.

4. After sperm fertilize the eggs, ovules mature into seeds. Each seed consists of an embryo sporophyte and tissues that function in its nutrition and protection.

5. While seeds are developing, tissues of the ovary and sometimes neighboring tissues mature into fruits, which function in seed dispersal. Air currents, water currents, and animals function as dispersal agents.

6. Many species of flowering plants also can reproduce asexually by various mechanisms of vegetative growth, parthenogenesis, and tissue culture propagation.

7. Plant hormones help bring about predictable patterns of development, including the extent and direction of growth and cell differentiations in particular plant parts.

8. Plant hormones also help adjust patterns of growth and development in response to environmental rhythms, including seasonal changes in daylength and temperature. Additionally, they help adjust the patterns in response to environmental circumstances in which an individual plant finds itself, such as the amount of sunlight or shade, moisture, and so on.

28.1

REPRODUCTIVE STRUCTURES OF FLOWERING PLANTS

Think Sporophyte and Gametophyte

Like us, flowering plants engage in sex. They produce, nourish, and protect sperm and eggs in reproductive systems. Like human females, they nourish and protect developing embryos. They put out floral invitations to third parties—pollinators that get sperm to eggs. Long before we thought of it, they were using perfumes and colors to improve the odds for sexual success.

When you hear the word "plant," you may think of something like a cherry tree (Figure 28.3a). The tree is a typical **sporophyte**, a vegetative body that grows, by mitotic cell divisions, from a fertilized egg. During the life cycle, this sporophyte bears flowers—floral shoots specialized for reproduction. In flowers, haploid spores gives rise to haploid bodies called **gametophytes**. Sperm develop inside male gametophytes, and eggs develop inside female gametophytes. Fusion of a haploid sperm with a haploid egg at fertilization results in a cell with two sets of genetic instructions.

Components of Flowers

Flowers form at floral shoots of the primary plant body. As they form, they differentiate into sepals and petals (nonfertile parts) and stamens and carpels (fertile parts). Figures 28.3b and 28.4 show how these components are arranged. All connect to a receptacle, the modified base of a floral shoot. Peel open a rosebud. Sepals, the outer whorl of leaflike parts, surround leaflike petals, which surround fertile parts. This is a common pattern. Sepals are a flower's calyx, and petals are its corolla.

Like leaves, sepals and petals have ground tissues, vascular tissues, and epidermis. Why are many flowers sweet smelling? Some epidermal cells of petals produce fragrant oils. What gives the petals their shimmer and color? Cells of the ground tissue contain pigments, such as carotenoids (yellow to red-orange) and anthocyanins (red to blue), plus small, light-refracting crystals. What does a flower's fragrance or its coloration, patterning, or arrangement of petals do? It attracts pollinators.

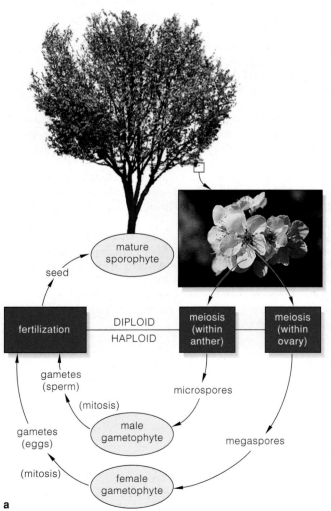

a

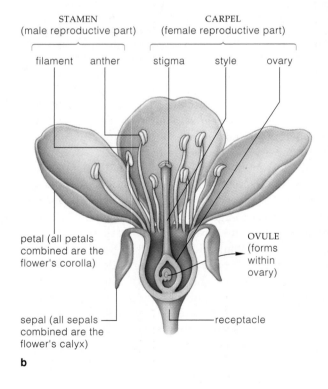

b

Figure 28.3 (a) Overview of flowering plant life cycles, using a cherry tree (*Prunus*) as the example. (b) Structure of one flower, a cherry blossom. Like the flowers of many plants, this blossom has a single carpel (a female reproductive part) and stamens (the male reproductive parts). Flowers of other plants have two or more carpels, often fused into a single structure. Whether single or fused, carpels have an ovary and a stigma. Often the ovary extends upward as a slender column, the style.

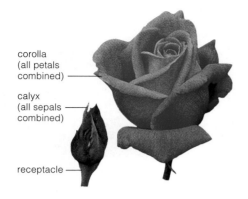

Figure 28.4 Location of floral parts in a prized cultivated plant, the rose (*Rosa*).

corolla (all petals combined)

calyx (all sepals combined)

receptacle

Where Pollen and Eggs Develop

Nearly all **stamens**, a flower's male reproductive parts, consist of an anther and a one-veined stalk (filament). An anther is divided into pollen sacs, the chambers in which walled, haploid spores form and give rise to the male gametophytes called **pollen grains** (Figure 28.5).

A flower's female reproductive parts are located at its center. You may have heard someone call these parts pistils, but their more recent name is **carpels**.

Many flowers have one carpel. Others have more, often fused as a compound structure. The lower part of single or fused carpels—the **ovary**—is where the eggs develop, fertilization takes place, and seeds mature. The more formal name for flowering plants, angiosperm, refers to the carpel (from the Greek *angeion*, meaning vessel; and *sperma*, seed). The upper portion of a carpel is the stigma. This sticky or hairy surface tissue captures pollen grains and favors their germination. Commonly the stigma is elevated on a style, a slender extension of the upper ovary wall.

Not every flower has stamens and carpels. Some species have *perfect* flowers, with both female and male parts. Others have *imperfect* flowers; they have female or male parts. In certain species such as oaks, the same individual has male and female flowers. In willows and other species, they are on separate, individual plants.

Sexual reproduction is the dominant reproductive mode of flowering plant life cycles. Such cycles alternate between the production of sporophytes (spore-producing bodies) and gametophytes (gamete-producing bodies).

Flowering plant sporophytes grow by mitotic cell divisions from a fertilized egg. A sporophyte is all of the plant body except for its male and female gametophytes.

Gametophytes arise from haploid spores, which form in stamens and carpels—the male and female reproductive parts of the flower.

Each male gametophyte gives rise to a sperm-producing pollen grain. Each female gametophyte develops inside a carpel's ovary; one of its cells is an egg. Also inside the ovary, fertilization takes place and seeds mature.

Pollen Sets Me Sneezing

In the year 1835 Sidney Smith wrote, "I am suffering from my old complaint, the hay-fever . . . that sets me sneezing; and if I begin sneezing at twelve, I don't leave off till two o'clock, and am heard distinctly in Taunton [six miles distant] when the wind sets that way."

Sidney suffered what is now called *allergic rhinitis*. This is the name for hypersensitivity to a normally harmless substance. Such hypersensitivity is common.

The pollen of ragweed, shown in Figure 28.5, brought on Sidney's dreaded hay fever, but it wasn't out to get him. Every spring and summer, flowering plants release pollen grains. In many millions of people, white blood cells respond by mounting an immune response against some proteins that project from the surface of pollen grain walls. The pollen grains from different plants have differences in their wall proteins. The immune system of a person who is hypersensitive might chemically respond to some of these but not others. The misdirected response results in a profusely runny nose, reddened and itchy eyelids, congestion, and bouts of sneezing.

Hay fever is a genetic abnormality; it runs in families. Some individuals simply are genetically predisposed to overreact to some kinds of pollen. Infections, emotional stress, and changes in temperature may also trigger overreactions to pollen.

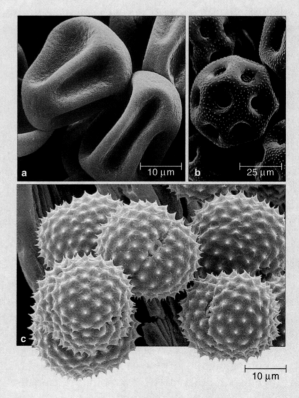

a   10 µm   b   25 µm

c   10 µm

Figure 28.5 Pollen grains from (**a**) grass, (**b**) a rose, and (**c**) ragweed plants. Pollen grains of most families of plants differ in size, wall sculpturing, and number of wall pores.

A NEW GENERATION BEGINS

From Microspores to Pollen Grains

We turn now to pollen grain formation. While anthers are growing, four masses of spore-producing cells form by mitotic cell divisions. Walls develop around them. Each anther now has four chambers, called pollen sacs (Figure 28.6*a*). Haploid **microspores** form when cells in the sacs undergo meiosis and cytoplasmic division. They go on to develop an elaborately sculpted wall. The walled microspores divide either once or twice by way of mitosis to form pollen grains, which enter a period of arrested growth. In time, these will be released from the anther. Components of their wall will help protect them from decomposers in their surroundings.

As soon as many types of pollen grains form, they produce sperm nuclei, which are the male gametes of flowering plants. Other types don't do this until after they reach a carpel and start growing toward its ovule. In short, each pollen grain is a mature *or* an immature male gametophyte, depending on the plant species.

From Megaspores to Eggs

Meanwhile, one or more masses of cells are forming on the inner wall of the flower's ovary. Each is the start of an **ovule**, a structure that houses a female gametophyte and that may become a seed. As each cell mass grows, a tissue forms within it, and one or two protective layers called integuments form around it. Inside the mass, a cell divides by meiosis, and four haploid spores form. Spores formed in flowering plant ovaries are generally larger than microspores and are called **megaspores**.

Commonly, all megaspores but one disintegrate. That one undergoes mitosis three times without cytoplasmic division. At first, it has eight nuclei (Figure 28.6*b*). Each nucleus migrates to a prescribed location, and then the cytoplasm divides. The result, a seven-celled embryo sac, is a female gametophyte. One cell has two nuclei. This endosperm mother cell will help form the **endosperm**, a nutritive tissue for the embryo. Another cell is the egg.

From Pollination to Fertilization

Flowering plants release pollen in spring. You are very aware of this reproductive event if you are one of the millions of people who experience hay fever, an allergic reaction to wall proteins of pollen grains (Section 28.2).

Specifically, **pollination** is a transfer of pollen grains to a receptive stigma. Air or water currents, birds, and other agents of pollination carry out the transfer, as you read in the chapter introduction.

Once a pollen grain lands on a receptive stigma, it germinates. For this event, germination means a pollen grain resumes growth and then develops into a tubular

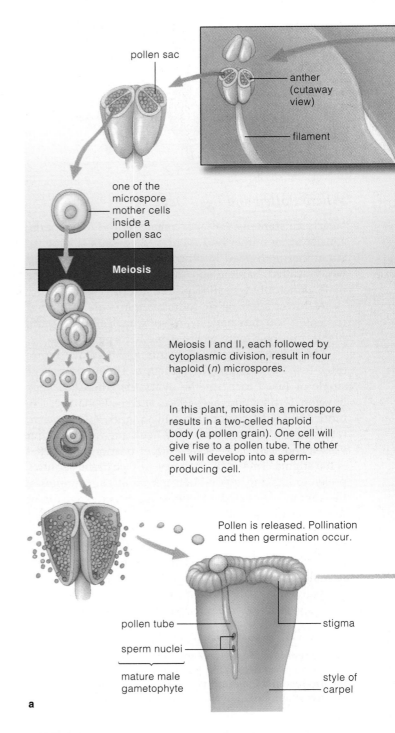

pollen sac

anther (cutaway view)

filament

one of the microspore mother cells inside a pollen sac

Meiosis

Meiosis I and II, each followed by cytoplasmic division, result in four haploid (*n*) microspores.

In this plant, mitosis in a microspore results in a two-celled haploid body (a pollen grain). One cell will give rise to a pollen tube. The other cell will develop into a sperm-producing cell.

Pollen is released. Pollination and then germination occur.

pollen tube — stigma

sperm nuclei

mature male gametophyte

style of carpel

a

Figure 28.6 Life cycle of cherry (*Prunus*), one of the flowering plants classified as a dicot. (**a**) How pollen grains develop and germinate. (**b**) Events inside the ovule in this ovary.

structure. This "pollen tube" grows down through the tissues of the ovary and carries the sperm nuclei with it (Figure 28.6*b*). Chemical and molecular cues guide the growth of the pollen tube through the ovary's tissues,

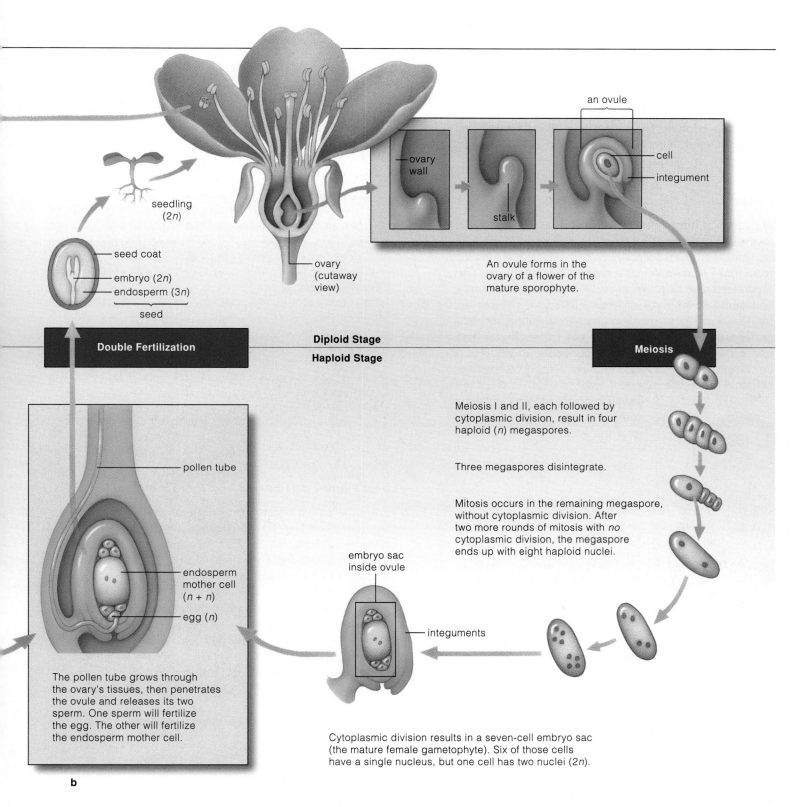

seedling
(2n)

seed coat
embryo (2n)
endosperm (3n)
seed

ovary
(cutaway
view)

an ovule

ovary
wall

stalk

cell
integument

An ovule forms in the
ovary of a flower of the
mature sporophyte.

Double Fertilization

Diploid Stage

Haploid Stage

Meiosis

pollen tube

endosperm
mother cell
(n + n)

egg (n)

embryo sac
inside ovule

integuments

Meiosis I and II, each followed by
cytoplasmic division, result in four
haploid (n) megaspores.

Three megaspores disintegrate.

Mitosis occurs in the remaining megaspore,
without cytoplasmic division. After
two more rounds of mitosis with *no*
cytoplasmic division, the megaspore
ends up with eight haploid nuclei.

The pollen tube grows through
the ovary's tissues, then penetrates
the ovule and releases its two
sperm. One sperm will fertilize
the egg. The other will fertilize
the endosperm mother cell.

Cytoplasmic division results in a seven-cell embryo sac
(the mature female gametophyte). Six of those cells
have a single nucleus, but one cell has two nuclei (2n).

b

toward the egg chamber and sexual destiny. When the
pollen tube reaches an ovule, it penetrates the embryo
sac, and its tip ruptures to release two sperm.

"Fertilization" generally means fusion of a sperm
nucleus with an egg nucleus. But **double fertilization**
occurs in flowering plants. In species having a diploid
chromosome number, one sperm nucleus fuses with an
egg nucleus to form a diploid (2n) zygote. Meanwhile,
the other sperm nucleus fuses with both nuclei of the

endosperm mother cell. The resulting cell has a triploid
(3n) nucleus and will give rise to the nutritive tissue.

**In flowering plants, sperm cells form within pollen grains,
the male gametophytes. Ovules form inside ovaries, and
female gametophytes (with egg cells) develop within them.**

**After pollination and double fertilization, an embryo and
nutritive tissue form in the ovule, which becomes a seed.**

FROM ZYGOTES TO SEEDS AND FRUITS

After fertilization, the newly formed zygote embarks on a course of mitotic cell divisions that lead to a mature embryo sporophyte. It develops as part of an ovule and is accompanied by formation of a **fruit**: a mature ovary, with or without other, neighboring tissues.

Formation of the Embryo Sporophyte

Let's look at how a shepherd's purse (*Capsella*) embryo forms. By the time it reaches the stage in Figure 28.7e, two **cotyledons**, or seed leaves, have started to develop from lobes of meristematic tissue. Cotyledons are part of all flowering plant embryos. Dicot embryos have two, and monocot embryos have one. The *Capsella* embryo, like those of most dicots, absorbs nutrients from the endosperm and stores them in its cotyledons. By contrast, in corn, wheat, and most other monocots, endosperm is not tapped until a seed germinates. Digestive enzymes get stockpiled in a monocot embryo's thin cotyledons. When activated, they will help transfer stored endosperm to the growing seedling.

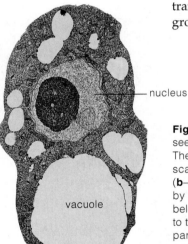

nucleus

vacuole

a

Figure 28.7 Development of a dicot seed—shepherd's purse (*Capsella*). The micrographs are not to the same scale. (**a**) The single-celled zygote. (**b–d**) The early embryo is identified by the *yellow* boxes. The row of cells below each box transfers nutrients to the embryo sporophyte from the parent plant. The embryo in (**e**) is well developed, and the one in (**f**) is mature. (**g**) *Capsella* fruit.

Table 28.1 *Categories and Examples of Fruits*

SIMPLE FRUITS From one ovary of one flower.

Dry fruit

Dehiscent. Fruit wall splits along definite seams to release seeds. Legume (e.g., pea, bean), poppy, larkspur, mustard

Indehiscent. Fruit wall does not split on seams to release seeds. Acorn, grains (e.g., corn), sunflower, carrot, maple

Fleshy fruit

Berry. Compound ovary, many seeds. Tomato, grape, banana
  Pepo. Ovary wall has hard rind. Cucumber, watermelon
  Hesperidium. Ovary wall has leathery rind. Orange, lemon

Drupe. One or two seeds. Thin skin, part of flesh around a usually hard seed. Peach, cherry, apricot, almond, olive

AGGREGATE FRUITS

Many ovaries of one flower, all attached to the same receptacle. Many seeds. Raspberry, blackberry (not really "berries")

MULTIPLE FRUITS

Combined from ovaries of many flowers. Pineapple, fig

ACCESSORY FRUITS

Most tissues of the flesh are not derived from the ovary; e.g., mainly from the receptacle. Pome (apple, pear), strawberry

Seeds and Fruit Formation and Dispersal

From the time a zygote forms until an embryo matures, a parent plant transfers nutrients to ovule tissues. Food reserves accumulate in endosperm or cotyledons. The ovule separates from the ovary's wall; its integuments thicken and harden into a seed coat. The embryo, food reserves, and coat are a self-contained package, a **seed**, which is defined as a mature ovule.

As seeds form, other floral parts start to form fruits. Depending on the species, they are simple, aggregate, or multiple fruits (Table 28.1). When mature, simple dry fruits such as acorns are intact; others, such as pea pods,

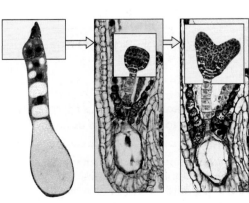

b Upper part of the zygote gives rise to embryo

c Globular stage of embryo

d Heart-shaped stage of embryo

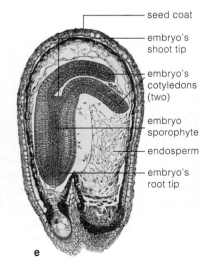

seed coat

embryo's shoot tip

embryo's cotyledons (two)

embryo sporophyte

endosperm

embryo's root tip

e

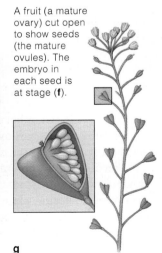

mature embryo within ovule

f

A fruit (a mature ovary) cut open to show seeds (the mature ovules). The embryo in each seed is at stage (**f**).

g

seed

fleshy fruit, from ovary tissue

a

b

one of many individual fruits

receptacle

ovary

c

remnants of sepals, petals,

expanding receptacle

d

ovary tissue
seed
enlarged receptacle

e

wing

seed (in carpel)

Figure 28.8 Fruits. (**a**) Peach, a drupe. (**b**) Pineapple (*Ananas*), a multiple fruit. Native Americans cultivated pineapples. To them, it was symbol of hospitality. They used the juice as the base for an alcoholic beverage and to poison arrow tips. Christopher Columbus thought the fruit looked like pine cones, hence the name. (**c**) Strawberry (*Fragaria*), a fragrant accessory fruit. (**d**) Fruit formation on an apple (*Malus*) tree. When petals drop, eggs are fertilized. (**e**) Winged seeds of maple (*Acer*).

unzip and so release seeds. Figure 28.8 shows a simple fleshy fruit (a peach) and a multiple fruit (a pineapple, with ovaries of many separate flowers grown together into a fleshy mass). A strawberry, an aggregate fruit, has many seeds on a fleshy, enlarged receptacle. An apple, an accessory fruit, is mainly an enlarged receptacle and calyx.

Fruits have a common function—*seed dispersal*. To gain insight into their structure, think about how they coevolved with dispersing agents—air or water currents and animals. Wind-dispersed maple fruits have fleshy winglike projections (Figure 28.8*e*). They may spin away from parent plants that outcompete seedlings for light, water, and nutrients. As other examples, air currents lift the plumelike "parachute" of dandelion fruits and also disperse orchid seeds as fine as dust particles.

Other fruits taxi to new locations on or in animals, including many insects, birds, and mammals. Some can adhere to feathers, feet, or fur with hooks, spines, hairs, and sticky surfaces. The fruits of cocklebur, bur clover, and bedstraw are like this. Embryo sporophytes inside the seed coats of fleshy fruits withstand the attacks by potent digestive enzymes inside an animal gut. Besides digesting the fruit's flesh, the enzymes digest some of the seed coat. The digested portions make it easier for

embryo sporophytes to break through their hard coat after seeds are expelled from the animal body and start to germinate. We return to this topic in Section 28.6.

Some water-dispersed fruits have heavy wax coats, and others, including sedges, have sacs of air that help them float. With their thick, water-repelling skin, fruits of coconut palms are adapted to travel across the open ocean. If they don't beach soon enough, saltwater may penetrate the skin and kill their embryos, but coconuts can travel hundreds of kilometers before that happens.

Humans are grand dispersing agents. Past explorers carried seeds all over the world. Cacao, oranges, corn, and many other plants were domesticated and currently reproduce in their new habitats, in far greater numbers than they did on their own. And in evolutionary terms, reproductive success is what life is all about.

A mature ovule, which encases an embryo sporophyte and food reserves inside a protective coat, is a seed.

A mature ovary, with or without additional floral parts that have become incorporated into it, is a fruit.

Seeds and fruits are structurally adapted for dispersal by air currents, water currents, and many kinds of animals.

ASEXUAL REPRODUCTION OF FLOWERING PLANTS

Asexual Reproduction in Nature

The features of sexual reproduction that we considered in earlier sections dominate flowering plant life cycles. Bear in mind, many species also reproduce asexually by modes of **vegetative growth**, as listed in Table 28.2. In essence, new roots and shoots grow right out from extensions or fragments of parent plants. Such asexual reproduction proceeds by way of mitosis, so that a new generation of offspring is genetically identical to the parent; it is a clone.

One "forest" of quaking aspen (*Populus tremuloides*) provides us with an impressive example of vegetative reproduction. The leaves of this flowering plant species tremble even in the slightest breeze, hence the name. Figure 28.9 is a panoramic view of the shoot systems of a single individual. The root system of the parent plant gave rise to adventitious shoots, which became separate shoot systems. Barring rare mutations, individuals at the north end of this vast clone are genetically identical to individuals at the south end. Water travels from roots near a lake all the way to shoot systems in much drier soil. Dissolved ions travel in the opposite direction.

As long as environmental conditions favor growth and regeneration, such clones are about as close as one can get to being immortal. No one knows how old the aspen clones are. The oldest known clone is a ring of creosote bushes (*Larrea divaricata*) growing in the Mojave Desert. It has been around for the past 11,700 years.

The reproductive possibilities are amazing. Watch strawberry plants send out horizontal, aboveground stems (runners), then watch the new roots and shoots develop at every other node. The oranges you eat may be descended from a single tree in southern California that reproduced by parthenogenesis. With this type of process, an embryo develops from an unfertilized egg.

Parthenogenesis can be stimulated when pollen has contacted a stigma even though a pollen tube has not grown through the style. Maybe certain hormones from the stigma or pollen grains diffuse to the unfertilized egg and trigger formation of an embryo. That embryo is $2n$ as a result of fusion of the products of mitotic cell division in the egg. A $2n$ cell outside the gametophyte also may be stimulated to develop into an embryo.

Induced Propagation

Most houseplants, woody ornamentals, and orchard trees are clones. People typically propagate them from cuttings or fragments of shoot systems. For example, with suitable encouragement, a severed African violet leaf may form a callus from which adventitious roots develop. A callus is one type of meristem. Remember, a meristem is a localized region of plant cells that retain the potential for mitotic cell division.

Or consider a twig or bud from one plant, grafted onto a different variety of some closely related species. Vintners in France, for instance, graft prized grapevines onto disease-resistant root stock from America.

Frederick Steward and his coworkers pioneered in **tissue culture propagation**. They cultured small bits of phloem from the differentiated roots of carrot plants (*Daucus carota*) in rotating flasks. They used a liquid growth medium that contained sucrose, minerals, and vitamins. The liquid also contained coconut milk, which Steward knew was rich in then-unidentified growth-inducing substances. As the flasks rotated, individual cells that were torn away from the tissue bits divided and formed multicelled clumps, which sometimes gave rise to new roots (Figure 28.10a). Steward's experiments were among the first to demonstrate that some cells of

Mechanism	Examples	Characteristics
Table 28.2 *Asexual Reproductive Modes of Flowering Plants*		
VEGETATIVE REPRODUCTION ON MODIFIED STEMS		
1. Runner	Strawberry	New plants arise at nodes along aboveground, horizontal stems.
2. Rhizome	Bermuda grass	New plants arise at nodes of underground horizontal stems.
3. Corm	Gladiolus	New plants arise from axillary buds on short, thick, vertical underground stems.
4. Tuber	Potato	New shoots arise from axillary buds on tubers, which are the enlarged tips of slender underground rhizomes.
5. Bulb	Onion, lily	New bulbs arise from axillary buds on short underground stems.
PARTHENOGENESIS		
	Orange, rose	An embryo develops without nuclear or cellular fusion (for example, from an unfertilized haploid egg or by developing adventitiously, from tissue surrounding the embryo sac).
VEGETATIVE PROPAGATION		
	Jade plant, African violet	A new plant develops from tissue or structure (a leaf, for instance) that drops from the parent plant or is separated from it.
TISSUE CULTURE PROPAGATION		
	Orchid, lily, wheat, rice, corn, tulip	A new plant is induced to arise from a parent plant cell that has not become irreversibly differentiated.

Figure 28.9 A mere portion of Pando the Trembling Giant, so named by Michael Grant and his coworkers at the University of Colorado, who studied its genetic makeup. This "forest" in Utah is actually a single asexually reproducing male organism, of a type called quaking aspen (*Populus tremuloides*). Its root system extends through about 106 hectares (about 262 acres) of soil and functionally supports its 47,000 genetically identical shoots; that is, the trees. By one estimate, this clone weighs more than 5,915,000 metric tons.

Figure 28.10 (**a**) Cultured cells from a carrot plant. At the time this photograph was taken, roots and shoots of embryonic plants were already forming. (**b**) Young orchid plants, developed from cultured meristem and young leaf primordia. Orchids are one of the most highly prized cultivated plants. Before the meristems were cloned, they were difficult to hybridize. From the time seeds form, it can take seven years or more until a new plant bears flowers. In natural habitats, the seeds normally will not germinate unless they interact with a specific fungus.

specialized tissues still house the genetic instructions required to produce an entire individual.

Researchers now use shoot tips and other parts of individual plants for tissue culture propagations. The techniques prove useful when an advantageous mutant arises. Such a mutant may show resistance to a disease that is crippling to wild-type plants of the same species. Tissue culture propagation can result in hundreds, even thousands of identical plants from merely one mutant

specimen. The techniques are already being employed in efforts to improve major food crops, including corn, wheat, rice, and soybeans. They also are being used to increase production of hybrid orchids, lilies, and other prized ornamental plants (Figure 28.10*b*).

Besides reproducing sexually, flowering plants engage in asexual reproduction, as by vegetative growth.

PATTERNS OF EARLY GROWTH AND DEVELOPMENT—AN OVERVIEW

Seed Germination

Let's first consider the overall patterns of growth and development for dicots and monocots, starting with the events that take place inside a seed. Figure 28.11 shows the embryo sporophyte inside a grain of corn. (A grain, remember, is a seed-containing dry fruit.) The growth of the embryo idles before or after the seed is dispersed from the parent plant. Later, if all goes well, the seed germinates. **Germination** is the process by which some immature stage in the life cycle of a species resumes growth after a period of arrested development.

Germination depends on environmental factors, such as soil temperature, moisture, and oxygen level, and the number of daylight hours. The factors vary with the seasons. For instance, mature seeds don't hold enough water for cell expansion or metabolism. In many land habitats, ample water is available on a seasonal basis, so seed germination coincides with the spring rains. By the process of imbibition, water molecules move into a seed, being attracted mainly to hydrophilic groups of proteins stored in endosperm or cotyledons. As more water moves in, the seed swells and its coat ruptures.

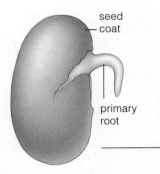

Figure 28.12 (a,b) Pattern of growth and development of a dicot, the common bean plant (*Phaseolus vulgaris*). When this seed germinates, the embryo in it resumes growth. A hypocotyl (a hook-shaped part of the shoot below the cotyledons) forces a channel through soil so the food-storing cotyledons are pulled up without being torn apart. At the surface, sunlight exposure makes the hook straighten. For several days, photosynthetic cells make food in the seedling's cotyledons. These wither and fall off. Foliage leaves take over photosynthesis. Flowers form in buds at nodes. (c) Cotyledons breaking through a seed coat.

seed coat

primary root

a Bean seedling at the close of germination.

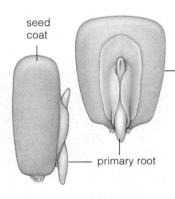

Figure 28.13 Pattern of growth and development of a monocot, using corn (*Zea mays*) as the example. (a,b) After germination of a corn grain, the coleoptile protects new leaves as the seedling grows through the soil. In corn seedlings, adventitious roots develop at the coleoptile's base. (c) Coleoptile and primary root of a corn seedling. (d) Coleoptile and first foliage leaf of two seedlings poking above the soil surface.

seed coat

primary root

a Corn grain at the close of germination.

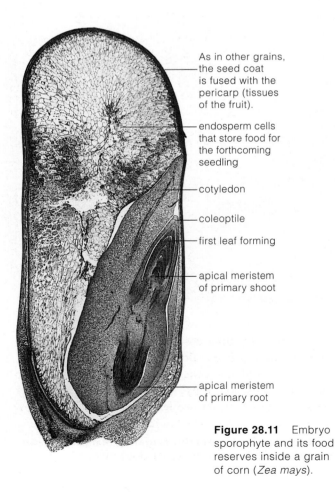

As in other grains, the seed coat is fused with the pericarp (tissues of the fruit).

endosperm cells that store food for the forthcoming seedling

cotyledon

coleoptile

first leaf forming

apical meristem of primary shoot

apical meristem of primary root

Figure 28.11 Embryo sporophyte and its food reserves inside a grain of corn (*Zea mays*).

Once the seed coat splits, more oxygen reaches the embryo, and aerobic respiration moves into high gear. The embryo's meristematic cells now divide rapidly. The root meristem is usually activated first. Its descendants divide, elongate, and give rise to the first, primary root of a seedling sporophyte. Germination is over when the seedling's primary root breaks through the seed coat.

Genetic Programs, Environmental Cues

Figures 28.12 and 28.13 show patterns of germination, growth, and development for dicots and monocots. The patterns have a heritable basis, being dictated by genes. All cells in a plant arise from the same cell (a zygote), so they generally inherit the same genes. But unequal divisions between daughter cells and their positions in

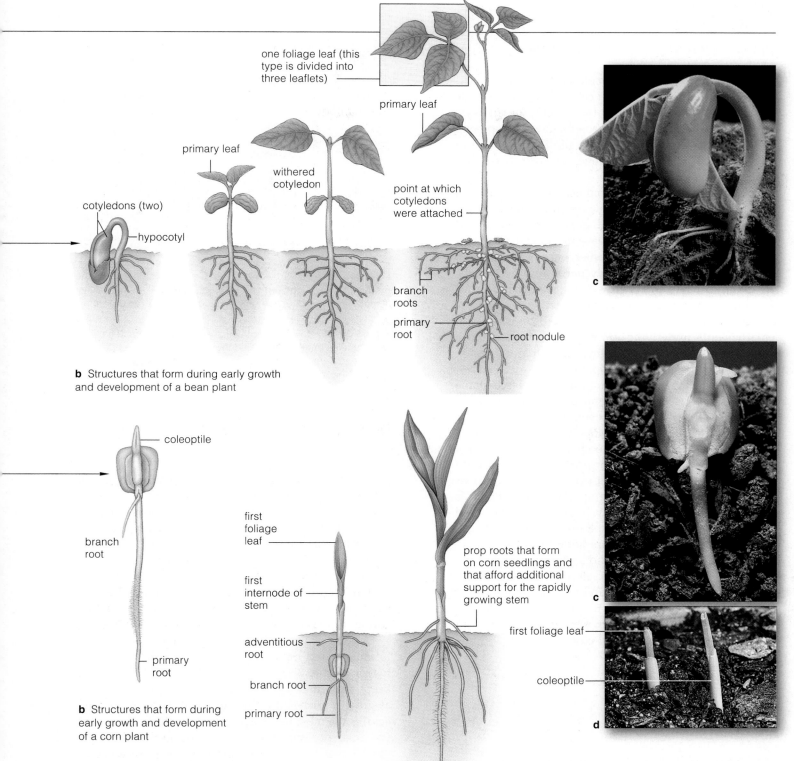

one foliage leaf (this type is divided into three leaflets)

primary leaf

primary leaf

withered cotyledon

point at which cotyledons were attached

cotyledons (two)

hypocotyl

branch roots

primary root

root nodule

c

b Structures that form during early growth and development of a bean plant

coleoptile

branch root

primary root

first foliage leaf

first internode of stem

prop roots that form on corn seedlings and that afford additional support for the rapidly growing stem

adventitious root

branch root

primary root

c

first foliage leaf

coleoptile

d

b Structures that form during early growth and development of a corn plant

the developing plant lead to differences in metabolic equipment and output. Cell activities start to vary with the onset of selective gene expression (Chapter 14). For example, genes that govern the synthesis of growth-stimulating hormones are activated in some cells but not others. As you will see shortly, such events seal the developmental fate of different cell lineages.

Bear in mind, plants commonly adjust prescribed growth patterns in response to unusual environmental pressures. Suppose a seed germinates in a vacant lot and a heavy paper bag blows on top of it. The primary shoot will quickly bend and grow out from under the bag, in the direction of sunlight impinging on it. Interactions among enzymes, hormones, and other gene products in its cells result in this growth response.

Plant growth involves cell divisions and cell enlargements. Plant development requires cell differentiation, as brought about by selective gene expression.

Interactions among genes, hormones, and the environment govern how each individual plant grows and develops.

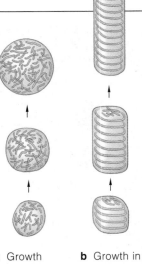

| **a** Growth proceeds in all directions | **b** Growth in longitudinal direction only | **c** Cellulose rings thicken the secondary wall of this tracheid, a conducting tube of xylem |

Figure 28.14 How different cell shapes arise. In each cell, microtubules in the cytoplasm inherited from the parent cell are already oriented in prescribed patterns. The patterns govern how cellulose microfibrils will be oriented in the cell wall. With random orientation (**a**), a primary wall is elastic all over, so the cell can expand in all directions and assume a spherical shape. (**b,c**) When cellulose strands form in a transverse orientation, the cell can only lengthen.

28.7

EFFECTS OF PLANT HORMONES

Hormones are signaling molecules secreted by specific cells that travel to other, target cells and influence gene activity. Any cell having receptors for a hormone is its target. Hormones orchestrate growth and development. They respond to each other's signals. They respond to cues afforded by seasonal change, as when days grow longer and warmer in spring.

Before getting into hormonal effects on growth and development, be sure you know the difference between these two processes. In general, **growth** of a multicelled organism means its cells increase in number, size, and volume. **Development** is the emergence of specialized, morphologically different body parts. In other words, we must measure plant growth in *quantitative* terms, and plant development in *qualitative* terms.

As you know from Section 26.1, cell divisions and enlargements underlying growth occur at meristems. About half the resulting daughter cells do not increase in size but rather retain the capacity to divide. The rest enlarge, often by twenty times. Meristem descendants divide in prescribed planes and expand in particular directions (Figure 28.14). The eventual outcome will be plant parts having specific shapes and functions.

When young cells of an embryo sporophyte or a seedling take up water, turgor pressure (internal fluid pressure) increases against their primary wall and drives enlargement. Imagine blowing up a balloon. While it is soft, the balloon inflates easily. Similarly, a soft-walled cell under turgor pressure expands quickly. That wall thickens as polysaccharides are added to it, and more cytoplasm forms between it and a central vacuole.

Hormones help control the selective gene expression that underlies cell differentiation and the emergence of basic patterns of development. At least five different classes of hormones are known to have predictable and important effects on flowering plants, starting with the germination of the new seed. As with animals, genetic instructions in a plant's DNA govern the synthesis of enzymes and other proteins necessary for metabolism, hence for all cell activities. *And hormonal action affects how and when each type of enzyme functions.*

The main plant hormones are gibberellins, auxins, cytokinins, abscisic acid, and ethylene (Table 28.3).

Gibberellins influence stem lengthening. They help buds and seeds break dormancy to resume growth in spring. In at least some cases they influence flowering.

Auxins play roles in stem lengthening and responses to gravity and light. They also make coleoptiles grow longer (Figure 28.15). A coleoptile is a thin sheath that keeps the primary shoot of some plants from shredding as growth pushes it up through soil. IAA (indoleacetic acid) is the most pervasive auxin in nature. Orchardists use auxins to thin flowers in spring so that their trees yield fewer but larger fruits. They also use auxins to prevent premature fruit drop, which cuts labor costs; all fruit in the orchard can be picked at the same time.

Some synthetic auxins serve as *herbicides*. At proper concentrations, these compounds kill some plants but not others (Section 3.7). 2,4-D controls broadleaf weeds that can outcompete cereal plants for nutrients. It does not seem to harm humans when properly handled. But mixing it with equal parts of a related compound, 2,4,5-T, yields *Agent Orange*. This herbicide was used to clear forested war zones during the Vietnam conflict. Later, laboratory experiments linked trace amounts of dioxin, a

Table 28.3 *Plant Hormones and Their Effects*

GIBBERELLINS. Promote stem elongation; help end dormancy of seeds and buds; contribute to flowering. Notable amounts in apical meristems of buds, roots, and leaves and in embryos.

AUXINS. Promote cell elongation in coleoptiles and stems; roles in phototropism and gravitropism. Notable amounts in bud and leaf apical meristems and in embryos in seeds.

CYTOKININS. Promote cell division and leaf expansion; retard leaf aging. Synthesized in roots and diffuses elsewhere.

ABSCISIC ACID. Promotes stomatal closure; promotes bud, seed dormancy. Notable amounts in leaves, stems, unripened fruit.

ETHYLENE. Promotes fruit ripening and abscission of leaves, flowers, fruits. Notable amounts in fruit, stems, leaves, roots, seeds.

Figure 28.15
Example of auxin's effects: The taller cutting at *left* is from a gardenia plant four weeks after an auxin had been applied to its base. Another cutting, used as a control (*right*), was untreated.

contaminant of 2,4,5-T, to miscarriages, birth defects, leukemia, and liver disorders. Besides this, dioxin is a carcinogen. 2,4,5-T is now banned in the United States.

Cytokinins stimulate cell division (the name refers to cytoplasmic division). They are abundant in root and shoot meristems and in maturing fruits. Cytokinins are used for basic research and to prolong the shelf life of cut flowers and other horticultural goods.

Abscisic acid (ABA) helps plants adapt to seasonal changes, as by inducing bud dormancy and inhibiting cell growth and premature seed germination. ABA also contributes to stomatal closure when a plant is water stressed. Growers typically apply ABA to nursery stock before shipping it, because plants are not as vulnerable to injury when they are dormant.

Ethylene induces fruit ripening, leaf drop, and other aging responses. Ancient Chinese knew to burn incense to speed up fruit ripening. By the early 1900s, growers were ripening citrus fruits by storing them in sheds with kerosene stoves. Food distributors now use ethylene to ripen tomatoes and other green fruit after they ship it to grocery stores. Fruit picked green doesn't bruise or deteriorate as fast. Ethylene exposure brightens citrus rinds before the fruit is displayed in the market.

Besides the hormones just described, root and leaf cells make their own hormones. One or more unknown hormones might induce flowering. Another, associated with shoot tips, blocks lateral bud growth. This form of growth inhibition is called apical dominance. Gardeners pinch off shoot tips to stop the hormone from diffusing through stems and inhibiting growth. Lateral buds are free to branch out, and gardeners get bushier plants.

Growth is a quantitative increase in the number, size, and volume of cells. Development, the emergence of qualitative differences in body parts, is basically a result of selective gene expression, hormone action, and cell differentiation.

The main categories of plant hormones are gibberellins, auxins, cytokinins, abscisic acid, and ethylene.

Foolish Seedlings!

A few years before the American stock market feverishly grew and then collapsed in 1929, a researcher in Japan came across a substance that caused runaway growth and collapse of rice plants. Ewiti Kurosawa was studying what the Japanese call *bakane*, or the "foolish seedling" effect on rice plants. Stems of rice seedlings infected with *Gibberella fujikuroi*, a fungus, grew twice as long as the stems of uninfected plants. The elongated stems were weak and spindly. When they collapsed, the plants died. Kurosawa discovered he could trigger the disease by applying extracts of the fungus to plants. Years later, other researchers purified the disease-causing substance from fungal extracts. It was named gibberellin.

Researchers have isolated more than eighty different forms of gibberellin from flowering plant seeds and from fungi. Possibly the same hormone also occurs in other plant groups. The changes that gibberellins trigger can make stems lengthen. For example, expose a radish plant to a suitable concentration of a gibberellin and it may grow as large as a beach ball. Do the same to a cabbage plant and it may well astound you (Figure 28.16).

Apply gibberellins and you can get longer, crispier celery stalks. You can stop the skin of navel oranges from maturing too quickly in an orchard. Walk past plump seedless grapes stacked inside the produce bins of a grocery store and observe how the fleshy fruits of the grape plant (*Vitis*) grow in packed-together bunches along the stems. Gibberellin applications made the stems lengthen between internodes. With more physical space between them, individual grapes grew far larger. Also, air circulated more freely between them, which made it harder for disease-causing, grape-juice-loving fungi to take hold and do damage.

Figure 28.16 Effects of gibberellin on the growth of cabbage plants.

two cabbages used as controls (untreated)

cabbages treated with gibberellins

ADJUSTMENTS IN THE RATE AND DIRECTION OF GROWTH

What Are Tropisms?

Generally, the young roots of land plants grow down through soil, and the shoots grow upright through air. Both also can adjust the direction of growth in response to environmental stimuli, as when a new shoot turns toward sunlight. When any root or shoot turns toward or away from an environmental stimulus, we call this a plant tropism (after the Greek *trope*, for turning). As the following examples illustrate, these responses are an outcome of hormone-mediated shifts in the rates at which different cells in the plant grow and elongate.

GRAVITROPISM   The first root to emerge from a seed coat always curves downward, and the coleoptiles and new stems curve upward. Such growth responses to the Earth's gravitational force are forms of **gravitropism**.

Auxin, together with a growth-inhibiting hormone in roots, may trigger such responses. Turn a young root on its side and remove its root cap, and it will *not* curve down. Put the cap back on, and the root curves down. Elongating root cells won't stop growing if you remove the root cap; if anything, cells will grow faster. Suppose a growth inhibitor in root cap cells gets *redistributed* in a root turned on its side. If gravity somehow causes the inhibitor to leave the cap and accumulate in cells on the root's lower side, the cells won't elongate as much as cells on the upper side. The root will curve down.

The gravity-sensing mechanisms of plants are based on **statoliths**, which generally are clusters of particles in a number of different cells. In plants, they are clusters of unbound starch grains in modified plastids. Figure 28.17 gives one example. These plastids, made denser by the statoliths, collect near the bottom of root cells in response to gravity. They settle downward until they rest in the lowest region of cytoplasm. Redistribution of the statoliths may trigger a redistribution of auxin in the cells, and thereby initiate a gravitropic response.

Similarly, Figure 28.18 shows how to track what can happen after you turn a potted seedling on its side in a dark room. The stem curves up, even in the absence of light. Why? In that horizontally positioned stem, cell elongations markedly slow down on the side facing up but elongations on the lower side rapidly increase. The different rates of elongation on opposing sides of the stem are enough to make it bend upward. It appears that something makes the cells on the bottom of a stem turned on its side *more* sensitive to auxin or some other hormone, and those on top *less* so.

PHOTOTROPISM   When stems or leaves adjust the rate and direction of growth in response to light, they are showing a form of **phototropism**. Phototropism clearly is adaptive; the light-dependent reactions of photosynthesis stop in the dark. Plants that can reorient themselves to maximize light interception have an advantage.

Charles Darwin pondered on phototropism after it dawned on him that a coleoptile was growing toward light that was striking one side of its tip. But it wasn't until the 1920s that Fritz Went, a graduate student in Holland, linked phototropism with a growth-promoting substance. He was the person who named the substance auxin (after the Greek *auxein*, meaning "to increase"). Went demonstrated that auxin moves from the tip of a coleoptile into cells less exposed to light and makes them elongate faster than cells on the illuminated side.

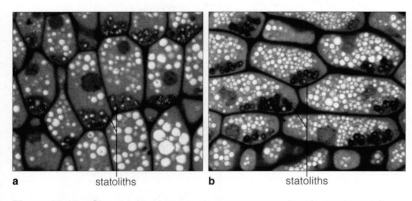

a   statoliths   b   statoliths

Figure 28.17   Observational test to demonstrate gravitropic responses by young roots. (**a**) Examine the root cap on a corn root. Normally, plastids of root cap cells that contain statoliths settle downward. This light micrograph shows their normal orientation. (**b**) Turn the root sideways. Five to ten minutes later, the plastids will settle to the "bottom" of the cells. A gravity-sensing mechanism based on statoliths may be sensitive to auxin redistribution in a root tip. A difference in auxin concentrations may cause cells on the "top" of a root turned sideways to elongate faster than cells on the bottom. Different elongation rates will make the root tip curve downward.

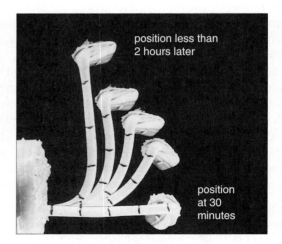

position less than 2 hours later

position at 30 minutes

Figure 28.18   Observational test to demonstrate gravitropic responses by young shoots. Force a sunflower seedling to grow in the dark for five days. Then turn it on its side and mark the shoot at 0.5-centimeter intervals. A gravity-sensing mechanism will cause the stem to turn upright.

bean   pea   oat

a

Figure 28.19 (**a**) Phototropism in seedlings. The plant physiologist Frank Salisbury grew these seedlings in darkness, then allowed light to strike their right side for a few hours before photographing them. (**b,c**) Hormone-mediated differences in the rates of cell elongation induce coleoptiles and stems to bend toward light.

b Rays of sunlight strike one side of a coleoptile.

c The coleoptile bends after auxin diffuses down from its tip to cells on its shaded side.

Figure 28.20 Passion flower (*Passiflora*) tendril busily twisting thigmotropically.

Figure 28.21 Effect of mechanical stress on tomato plants. (**a**) This plant, the control, grew in a greenhouse. (**b**) Each day for twenty-eight days, this plant was mechanically shaken for thirty seconds. (**c**) This one had two shakings each day.

The difference in their rates of growth brings about the bending toward light (Figure 28.19).

You can observe a phototropic response by putting seedlings of sun-loving plants in a dark room next to a window through which rays of sunlight are streaming in. Bean sprouts will do. They will start curving in the direction of the most light (Figure 28.19*c*).

As we now know, blue wavelengths of light induce plants to make their strongest phototropic response. Therefore, flavoprotein, a yellow pigment molecule that absorbs blue wavelengths, may be a central component of the phototropic bending mechanism.

THIGMOTROPISM Plants also shift their direction of growth when they contact solid objects. This response is called **thigmotropism** (after the Greek *thigma*, which means "touch"). Auxin and ethylene may have roles in the response. Vines, which are stems too slender or soft to grow upright without support, make such a contact response. So do tendrils, which are modified leaves and stems that wrap around objects and thus help support the plant (Figure 28.20). You can observe what happens when they grow against a stem of another plant. Within minutes, cells on the contact side stop elongating and the vine or tendril starts curling around the stem, often more than once. Afterward, the cells on both sides will resume growth at the same rate.

Responses to Mechanical Stress

Mechanical stress, as inflicted by prevailing winds and grazing animals, can inhibit stem elongation and plant growth. You can see such effects on trees growing near the snowline of windswept mountains; they are stubby compared to trees of the same species that are growing at lower elevations. Similarly, plants grown outdoors commonly have shorter stems than plants grown inside a greenhouse. You can observe this response to stress by shaking a plant daily for a brief period. Doing so will inhibit growth of the whole plant (Figure 28.21).

Plants adjust the direction and rate of their growth in response to environmental stimuli.

BIOLOGICAL CLOCKS AND THEIR EFFECTS

Like other organisms, flowering plants have internal mechanisms that preset the time for recurring changes in biochemical events. The internal timing mechanisms—**biological clocks**—trigger shifts in daily activities. They also help induce seasonal adjustments in basic patterns of growth, development, and reproduction.

The alarm button for some plant biological clocks is **phytochrome**. This blue-green pigment absorbs red and far-red wavelengths of light, with different results. At sunrise, when red wavelengths dominate the sky, phytochrome converts to Pfr, its active molecular form (Figure 28.22). Activation may induce cells to take up free calcium ions (Ca^{++}) or induce certain organelles to release them. When the ions combine with calcium-binding proteins in cells, it starts rhythmic movements of leaves and some other responses to light. At sunset, at night, or in the shade, phytochrome reverts to its inactive molecular form (Pr). At such times, wavelengths are predominantly far-red.

Rhythmic Leaf Movements

Each day, some plants position their leaves horizontally, then fold them closer to the stem at night (Figure 28.23). Keep such a plant in full sun or darkness for a few days and it will continue to move its leaves into and out of the "sleep" position!

The rhythmic leaf movements are one type of **circadian rhythm**, a biological activity that is repeated in cycles, each lasting for close to twenty-four hours. *Circadian* means "about a day." Some experiments by Ruth Satter, Richard Crain, and their University of Connecticut colleagues showed that phytochrome has a role in leaf movements. Satter, a pioneer in the study of timing mechanisms, was one of the first to correlate sleep movements of plants with "hands of a biological clock."

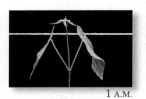
1 A.M.

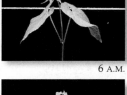
6 A.M.

NOON

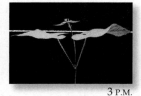
3 P.M.

10 P.M.

MIDNIGHT

Figure 28.23 Rhythmic movements of bean plant leaves. The investigator kept a bean plant in full darkness for twenty-four hours. Its leaves continued to move independently of sunrise (6 A.M.) and sunset (6 P.M.). Leaves folded close to stems may reduce heat loss from leaves exposed to cold night air.

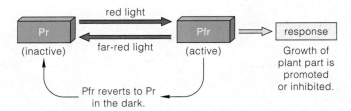

Figure 28.22 Interconversion of the phytochrome molecule from active form (Pfr) to inactive form (Pr). This blue-green pigment is part of a switching mechanism that promotes or inhibits the growth of a variety of plant parts.

Flowering—A Case of Photoperiodism

As another example of internal timekeeping, let's look at one way in which a plant resets biological clocks as seasons change. At a certain time of year, a plant starts diverting more energy to forming flowers. Figure 28.24 indicates how this activity and others are predictable responses to rhythmic environmental cues, such as more hours of light and warmer temperatures in summer.

Photoperiodism is a biological response to change in the length of daylight relative to darkness in a cycle of twenty-four hours. Phytochrome's active form, Pfr, may be a switching mechanism for the process. It may trigger synthesis of specific enzymes in different plant cells; different enzymes have roles in seed germination, stem branching and lengthening, leaf expansion, and formation of flowers, fruits, and seeds.

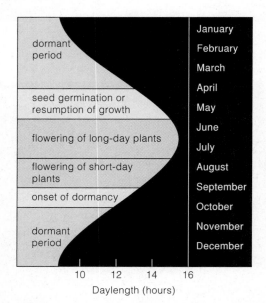

Figure 28.24 Plant growth and development as correlated with the number of hours of light available each day. The number changes with the passing of the seasons. The data shown reflect photoperiodic responses of plants that grow in temperate regions of North America. In such regions, rainfall and temperature shift with the seasons.

Figure 28.25 (a–c) Results of experiments demonstrating that short-day plants flower by measuring the length of night, not daylength. Each horizontal bar signifies twenty-four hours. Daylight hours are coded *yellow*; night hours are *dark blue*. (a) Plants grown in a room where simulated nights were shorter than required for flowering didn't flower; they flowered when nights were longer (b). Plants exposed to a pulse of white light in the middle of the night didn't flower (c). Phytochrome is the receptor in short-day plants that responds to light. (d) A short pulse of red light at night will inhibit flowering. (e) Flowering did occur when 10 minutes of far-red light followed a pulse of red light.

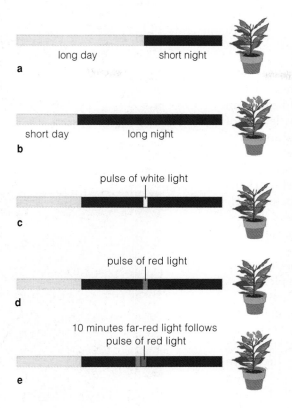

a long day short night

b short day long night

c pulse of white light

d pulse of red light

e 10 minutes far-red light follows pulse of red light

flowers

Figure 28.26 Experiment involving flowering responses of (a) spinach, a long-day plant, and (b) chrysanthemum, a short-day plant. The plant at the left grew under short-day conditions. The plant at the right grew under long-day conditions.

Plants do not all flower in response to the same environmental cues. Poinsettias, chrysanthemums, and cockleburs are examples. They are *short-day* plants that flower in late summer or early fall, when daylength is shorter than a critical value (Figures 28.24 and 28.25). Spinach, irises, and other *long-day* plants form flowers in spring, when daylength exceeds some critical value (Figure 28.26). *Day-neutral* plants simply flower when they are mature enough to do so.

Spinach plants won't flower and form seeds unless they are exposed to ten hours of darkness (or fourteen hours of daylight) for two weeks. That's why you would not want to start, say, a spinach seed farm in the tropics (where this set of cues is not available). By contrast, chrysanthemum growers stall the flowering process by exposing plants to a flash of light at night to break up one long night into two "short" nights. That's why you can buy chrysanthemums in spring even though they naturally bloom in autumn. Cockleburs normally flower after only one night that's longer than eight and one-half hours. If artificial light interrupts their dark period for even a minute, they won't flower. And why didn't poinsettias planted by California's interstate highways put out flowers? Headlights from vehicles zipping past during the night blocked the flowering response.

Besides phytochrome, other kinds of hormones that are not yet identified probably influence flowering and other growth responses. Some of the hormones may be produced in leaves and transported to new buds. For example, if you trim all but one leaf from a cocklebur, and then cover the remaining leaf with black paper for eight and one-half hours, the plant will flower. If you cut off that one leaf right after the dark period is over, you will not see any cocklebur flowers.

Like other organisms, flowering plants have biological clocks, which are internal time-keeping mechanisms.

Phytochrome, a blue-green pigment, is part of a switching mechanism for phototrophic responses to light of red and far-red wavelengths. Its active form, Pfr, might interact with other hormones to control which kinds of enzymes are being produced in particular cells.

Different enzymes are required to complete flowering and other growth responses, which are influenced by daylength and other environmental cues.

LIFE CYCLES END, AND TURN AGAIN

Senescence

While leaves and fruits are growing, cells inside them produce auxin (IAA), which moves into stems. There it interacts with cytokinins and gibberellins to maintain growth. When autumn approaches and the length of daylight decreases, plants start to withdraw nutrients from leaves, stems, and roots, then distribute them to flowers, fruits, and seeds. Deciduous plants, which shed leaves when a growing season ends, channel nutrients to storage sites in twigs, stems, and roots before their leaves die and drop. The dropping of flowers, leaves, fruits, and other plant parts is known as **abscission**. An abscission zone is made of thin-walled parenchyma cells at the base of a petiole or some other part about to drop from the plant. Figure 28.27 shows one example.

Senescence is the sum total of processes that lead to death of a plant or some of its parts. The recurring cue is a decrease in daylength. But other factors, such as drought, wounds, and nutrient deficiencies, also bring it about. The cue causes a decline in IAA production in leaves and fruits. A different signal that the plant itself may produce stimulates abscission zone cells to make ethylene. Cells enlarge, deposit suberin in their walls, and produce enzymes that digest cellulose and pectin in middle lamellae. A middle lamella is the cementing layer between plant cell walls (Section 4.11). While cells continue to enlarge and their walls become digested, they separate from one another. Eventually the leaf or other plant part above the abscission zone drops away.

Interrupt the diversion of nutrients into flowers, seeds, or fruits, and you can prevent senescence inside a plant's leaves, stems, and roots. For example, if you remove each new flower or seed pod from a plant, its leaves and stems will remain vigorous and green much longer. Gardeners routinely remove flower buds from many kinds of plants to maintain vegetative growth.

Entering and Breaking Dormancy

As autumn approaches and days grow shorter, many perennials and biennials start to shut down growth. They do so even when temperatures are still mild, the sky is bright, and water is plentiful. When a plant stops growing under conditions that seem (to us) suitable for growth, it has entered a state of **dormancy**, in which its metabolic activities idle. Ordinarily, the plant's buds will not resume growth until there is a convergence of precise environmental cues in early spring.

Short days and long, cold nights are strong cues for dormancy. (So is dry, nitrogen-deficient soil.) Test this for yourself by interrupting the long dark period of, say, Douglas firs with a short period of red light. The plants will respond as if nights are shorter and days are

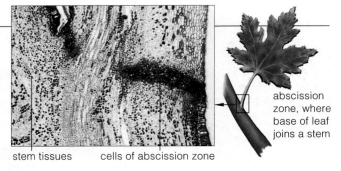

stem tissues   cells of abscission zone

Figure 28.27 Abscission zone in maple (*Acer*). The section is through the base of a leaf petiole.

Three Douglas fir plants exposed to different day-night cycles during three year-long experiments

Figure 28.28 Effect of the relative length of day and night on Douglas firs. The plant at left was exposed to 12-hour light/12-hour dark cycles for a year. Its buds became dormant; daylength was too short. The plant at right was exposed to 20-hour light/4-hour dark cycles; growth continued. The middle plant was exposed to 12-hour light/11-hour dark cycles with 1 hour of light in the middle of the dark period. The light interruption prevented bud dormancy; it caused Pfr formation at a sensitive time in the normal day–night cycle.

longer. They will continue to grow taller (Figure 28.28). In this experiment, conversion of Pr to Pfr by red light during the dark period prevents dormancy. In nature, buds might enter dormancy because less Pfr can form when daylength shortens in late summer.

A dormancy-breaking process is at work between fall and spring. The temperature may become milder, and rains and nutrients become available again. With the return of favorable conditions, the cycling of life begins anew. Seeds germinate; buds resume growth and give rise to new leaves, then to flowers. Depending on the species of plant, breaking dormancy probably involves gibberellins and abscisic acid, and it requires exposure to low winter temperatures at specific times.

Multiple cues from the environment influence hormonal secretions that stimulate or inhibit processes of growth and development during the life cycle of plants. The cues include changes in daylength, temperature, moisture, and nutrient availability.

SUMMARY   *Gold* indicates text section

1. Sexual reproduction is the main reproductive mode of all flowering plant life cycles. A diploid sporophyte (spore-producing plant body) alternates with haploid gametophytes (gamete-producing bodies). *28.1*

   a. The sporophyte is a multicelled vegetative body with roots, stems, leaves, and, at some point, flowers. Air and water currents, as well as animals, coevolved with flowers as pollinating agents. Gametophytes form in male and female floral parts. *28.1*

   b. Many flowering plants also reproduce asexually. They do so naturally, as with runners, rhizomes, and bulbs; and artificially, as by cuttings and grafting. *28.5*

2. Flowers typically have sepals, petals, one or more stamens (male reproductive parts), and carpels (female reproductive parts), most or all attached to a receptacle, the modified end of a floral shoot. *28.1–28.2*

   a. Anthers of stamens contain pollen sacs in which cells divide by meiosis. A wall develops around each resulting haploid cell (a microspore), which gives rise to a sperm-bearing pollen grain (male gametophyte).

   b. A carpel (or two or more fused carpels) has an ovary where eggs develop, fertilization takes place, and seeds mature. A stigma (sticky or hairy surface tissue) above the ovary captures pollen grains and promotes their germination. Ovules form on the ovary's inner wall. Each ovule is a female gametophyte with an egg cell, an endosperm mother cell, a surrounding tissue, and one or two protective layers called integuments.

3. At double fertilization, one sperm nucleus fuses with an egg nucleus and so forms a diploid zygote. The other sperm nucleus and both nuclei of another cell in the female gametophyte fuse to form a cell that will give rise to endosperm, a nutritive tissue. *28.3*

4. After fertilization, the endosperm forms, the ovule expands, the embryo sporophyte develops, and the integuments harden and often thicken. A fully matured ovule is a seed (its integuments have become the seed coat). While seeds form, the ovaries develop into fruits, which will help protect and disperse the seeds. *28.3*

5. After dispersal, seeds germinate—the embryo inside absorbs water, resumes growth, and breaks through the seed coat. The seedling increases in size; its tissues and organs develop. Fruits and seeds form. Older leaves drop. Plant hormones govern such patterns of growth and development. They adjust patterns in response to unpredictable and recurring environmental cues, such as seasonal changes in daylength. *28.4–28.11*

6. We know of five classes of plant hormones. Auxins and gibberellins promote stem elongation. Cytokinins promote cell division and leaf expansion, and retard leaf aging. Abscisic acid promotes bud and seed dormancy, and limits water loss by triggering stomatal closure. Ethylene promotes fruit ripening and abscission. *28.7*

7. Plant parts make tropic responses to light, gravity, and other environmental conditions. Hormones induce a difference in the rate and direction of growth on two sides of the part, which causes it to turn or move. *28.9*

8. A biological clock is a type of internal mechanism that measures time. It has a biochemical basis. *28.10*

   a. In photoperiodism, plants respond to a change in the relative length of daylight and darkness, as occurs seasonally. The blue-green pigment phytochrome is the switching mechanism of the clock that promotes and inhibits germination, stem elongation, leaf expansion, stem branching, and flower, fruit, and seed formation.

   b. Long-day plants flower during spring or summer, when there are more hours of daylight than darkness. Short-day plants flower when daylength is less. Day-neutral plants flower regardless of daylength.

9. Senescence is the sum of processes leading to the death of a plant or plant structure. Dormancy is a state in which a biennial or perennial plant stops growing even if conditions seem suitable for continued growth. A decrease in Pfr levels may trigger dormancy. *28.11*

Review Questions

1. Label the floral parts. Explain floral function by relating the parts to events in the life cycle of flowering plants. *28.1*

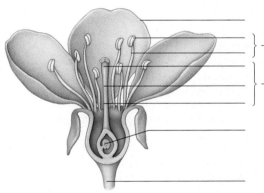

2. Distinguish between:
   a. Sporophyte, gametophyte *28.1*
   b. Megaspore, microspore *28.3*
   c. Pollination, fertilization *28.3*
   d. Pollen grain, pollen tube *28.1, 28.3*
   e. Ovule, female gametophyte *28.3*
   f. Seed, fruit *28.4*
   g. Senescence, dormancy *28.11*

3. Describe the steps by which a seven-cell, eight-nucleate embryo sac (a type of female gametophyte) forms. *28.3*

4. Define and describe one mode of vegetative growth. *28.5*

5. Name the three key factors that interact to dictate patterns of plant growth and development. *28.6*

6. List five types of plant hormones and briefly describe the known functions of each. *28.7*

7. Define plant tropism and give a specific example. *28.9*

8. What is phytochrome, and what is its role in flowering or some other process? *28.10*

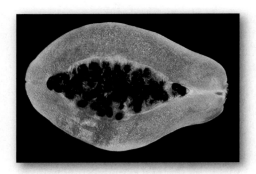

Figure 28.29
Seeds inside the fruit of a papaya (*Carica papaya*).

Self-Quiz ANSWERS IN APPENDIX III

1. The flowers of many species coevolved with insects, birds, and other agents that function as _____ .

2. The _____ , which bears flowers, roots, stems, and leaves, dominates the life cycle of flowering plants.
   a. sporophyte   b. gametophyte

3. A _____ is a vessel, the lower portion of which is an ovary in which eggs develop, fertilization occurs, and seeds mature.
   a. pollen sac   b. carpel   c. receptacle   d. sepal

4. After meiosis within pollen sacs, haploid _____ form.
   a. megaspores   c. stamens
   b. microspores   d. sporophytes

5. After meiosis in ovules, _____ megaspores form.
   a. two   b. four   c. six   d. eight

6. Cotyledons develop as part of all flowering plant _____ .
   a. seeds   b. fruits   c. embryos   d. ovaries

7. Seeds are mature _____ ; fruits are mature _____ .
   a. ovaries; ovules   c. ovules; ovaries
   b. ovules; stamens   d. stamens; ovaries

8. Which of the following statements is false?
   a. Auxins and gibberellins promote stem elongation.
   b. Cytokinins promote cell division but retard leaf aging.
   c. Abscisic acid promotes water loss and dormancy.
   d. Ethylene promotes fruit ripening and abscission.

9. Plant hormones _____ .
   a. interact with one another
   b. are influenced by environmental cues
   c. are active in plant embryos within seeds
   d. are active in adult plants
   e. all of the above

10. Light of _____ is the strongest stimulus for phototropism.
    a. red wavelengths   c. green wavelengths
    b. far-red wavelengths   d. blue wavelengths

11. The flowering process is a _____ response.
    a. phototropic   c. photoperiodic
    b. gravitropic   d. thigmotropic

12. Match the terms with the most suitable description.
    ____ double   a. formation of zygote and first
         fertilization   cell of endosperm
    ____ ovule   b. outcome of two species
    ____ mature female   interacting in close ecological
         gametophyte   fashion over geologic time
    ____ asexual   c. contains a female gametophyte
         reproduction   and has potential to be a seed
    ____ coevolution   d. an embryo sac, commonly with
                             seven cells (one has two nuclei)
                          e. mitotic cell division at bud or
                             node produces a new plant

Critical Thinking

1. Observe flowers growing in the area where you live. Given the coevolution of flowering plants and their pollinators, name the kinds of pollinators your floral neighbors depend on.

2. Wanting to impress her many friends with her sophisticated botanical knowledge, Dixie Bee is preparing a plate of tropical fruits for a party and cuts open a papaya (*Carica papaya*) for the first time. Inside are great numbers of slime-covered seeds, surrounded by soft flesh and soft skin (Figure 28.29). Knowing that her friends will ask what sort of fruit this is, she panics, runs to her biology book, and opens it to Section 28.4. What does she find out? What will she tell them when they ask what kinds of agents disperse the seeds of such a fruit from the parent plant?

3. Before cherries, apples, peaches, and many other fruits ripen and the seeds inside them mature, their flesh is bitter or sour. Only later does it become tasty to animals that assist in seed dispersal. Develop a hypothesis of how this feature improves the odds for the plant's reproductive success.

4. Given what you know about the primary growth of plants (Section 26.1), propose an explanation of why plant hormones need not travel very far from the cells that secrete them.

5. Plant growth depends on photosynthesis, which depends on sunlight energy. How, then, can seedlings germinated in a dark room grow taller than seedlings that germinated in the sun?

6. *Solar tracking* refers to the observation that many plants are able to maintain the flat blades of their leaves at right angles to the sun throughout the day. Sunflowers are an example. This tropic response maximizes light harvest by leaves. Name one type of molecule that might be involved in the response.

7. Belgian scientists isolated a mutant of wall cress (*Arabidopsis thaliana*) that produces excess amounts of auxin. Predict what some of this mutant plant's phenotypic traits might be.

8. Cattle receive somatotropin, an animal hormone that makes them grow bigger (added weight means more profits). There is growing concern that such hormones may have unforeseen effects on beef-eating humans. Would you think plant hormones applied to crop plants can affect humans also? Why or why not?

Selected Key Terms

abscisic acid *28.7*	endosperm *28.3*	phototropism *28.9*
abscission *28.11*	ethylene *28.7*	phytochrome *28.10*
auxin *28.7*	flower *CI*	pollen grain *28.1*
biological	fruit *28.4*	pollination *28.3*
clock *28.10*	gametophyte *28.1*	pollinator *CI*
carpel *28.1*	germination *28.6*	seed *28.4*
circadian	gibberellin *28.7*	senescence *28.11*
rhythm *28.10*	gravitropism *28.9*	sporophyte *28.1*
coevolution *CI*	growth *28.7*	stamen *28.1*
cotyledon *28.4*	hormone *28.7*	statolith *28.9*
cytokinin *28.7*	megaspore *28.3*	thigmotropism *28.9*
development *28.7*	microspore *28.3*	tissue culture
dormancy *28.11*	ovary *28.1*	propagation *28.5*
double	ovule *28.3*	vegetative
fertilization *28.3*	photoperiodism *28.10*	growth *28.5*

Readings

Grant, M. October 1993. "The Trembling Giant." *Discover* 14(10): 82–89.

Rost, T., M. Barbour, C. R. Stocking, and T. Murphy. 1998. *Plant Biology*. Belmont, California: Wadsworth.

On-Line readings at Student Guide for InfoTrac: www.brookscole.com/biology

VI Animal Structure and Function

How many and what kinds of body parts does it take to function as a lizard in a tropical forest? Make a list of what comes to mind as you start reading Unit VI, then see how resplendent the list can become at the unit's end.

TISSUES, ORGAN SYSTEMS, AND HOMEOSTASIS

Meerkats, Humans, It's All the Same

After a cold night in Africa's Kalahari Desert, animals small enough to fit inside a coat pocket emerge stiffly from their burrows. These "meerkats" are a type of mongoose. They stand on their hind legs and face east, exposing their chilled bodies to the warm rays of the morning sun (Figure 29.1). Meerkats don't know it, but sunning behavior helps their enzymes. If their body's internal temperature were to fall below some tolerable range, the action of countless enzyme molecules in their cells would falter and metabolism would suffer.

Once meerkats warm up, they fan out from their burrows and search for food. Into the meerkat gut go insects and the occasional lizard. These are pummeled, dissolved, and digested into glucose and other nutritious bits small enough to move across the gut wall, into the blood, and on to the body's cells. Aerobic machinery in the cells cracks apart molecules of glucose and other organic compounds and releases energy. A respiratory system supplies the machinery with oxygen and takes away carbon dioxide leftovers.

All of this activity changes the composition and volume of the **internal environment**—blood and the interstitial fluid (tissue fluid) that bathes all living cells of any complex animal. Drastic changes in these fluids would kill the cells, but a urinary system works to keep

that from happening. Governing this system and the others are the nervous and endocrine systems. They interact as a central command post to mobilize the body as a whole for everything from simple housekeeping tasks to heart-thumping flights from predators.

And so meerkats start us thinking about this unit's topics: how the animal body is structurally put together (its *anatomy*) and the adaptations by which it functions in the environment (its *physiology*). This chapter is an overview of the animal tissues and organ systems we will be considering. It also introduces the key concept of **homeostasis**. With respect to animals, the word means stable operating conditions in the internal environment, as brought about by the coordinated activities of cells, tissues, organs, and organ systems.

Amazingly, all complex animals consist of only four basic types of tissues. These are epithelial, connective, muscle, and nervous tissues. A **tissue** is a group of cells and intercellular substances, all interacting in one or more tasks. As an example, muscle tissue takes part in contraction. An **organ** consists of different tissues that are organized in specific proportions and patterns. Thus every vertebrate heart has predictable proportions and arrangements of epithelial, connective, muscle, and nervous tissues. An **organ system** consists of two or

more organs that are interacting physically, chemically, or both in a common task, as when interconnected arteries and other vessels transport blood through the body under the driving force of a beating heart.

Cells, tissues, organs, and organ systems split up the work, so to speak, in ways that contribute to survival of the body as a whole. This is sometimes referred to as a **division of labor**. By the end of this unit, you may have an abiding appreciation of the sheer magnitude of the division of labor among the body's separate parts and the extent to which their activities are integrated.

As you will see, whether you look at a flatworm or salmon, a meerkat or human being, the animal body is structurally and physiologically adapted to perform four overriding tasks:

1. *Maintain conditions in the internal environment within ranges that are most favorable for cell activities.*

2. *Acquire nutrients and other raw materials, distribute them through the body, and dispose of wastes.*

3. *Afford protection against injury or attack from viruses, bacteria, and other agents of disease.*

4. *Reproduce, then often help nourish and protect the new individuals during their early growth and development.*

Figure 29.1 In the Kalahari Desert, gray meerkats (*Suricata suricatta*) face the sun's warming rays, as they do every morning. This simple behavior helps them maintain internal body temperature. How animals function in their environment is the focus of this unit.

Key Concepts

1. The cells of most animals interact at three levels of organization—in tissues, many of which are combined into organs, which are components of organ systems.

2. Most animals are constructed of four types of tissues: epithelial, connective, muscle, and nervous tissues.

3. Each animal cell engages in basic metabolic activities that ensure its own survival. At the same time, animal cells of a given tissue perform one or more activities that contribute to the survival of the animal as a whole.

4. The internal environment consists of all fluids that are not inside the body's cells—that is, blood and interstitial fluid.

5. The combined contributions of cells, tissues, organs, and organ systems help maintain stability in the internal environment, which is required for the survival of each individual cell. This concept helps us understand the functions of any organ or organ system.

6. Homeostasis is the formal name for stable operating conditions in the internal environment.

EPITHELIAL TISSUE

General Characteristics

We commonly refer to an epithelial tissue as **epithelium** (plural, epithelia). This tissue has a free surface, which faces a body fluid or the outside environment. *Simple epithelium*, with only a single layer of cells, functions as a lining for body cavities, ducts, and tubes. *Stratified epithelium*, which has two or more cell layers, typically functions in protection, as it does in your skin. Figure 29.2 shows examples of this type of animal tissue.

All cells in epithelium are positioned close together, with little intervening material. They make, absorb, and secrete a variety of substances. As is the case for cells in nearly all animal tissues, specialized junctions serve as structural and functional links between them.

Cell-to-Cell Contacts

In epithelium and other tissues, we see three classes of cell junctions. **Tight junctions** prevent substances from leaking across the tissue. **Adhering junctions** are like spot welds; they cement neighboring cells together. **Gap junctions** are open channels between the cytoplasm of abutting cells. They facilitate the rapid transfer of ions and small molecules from one cell to its neighbors.

Consider tight junctions. The cells in some epithelia have parallel rows of proteins that form tight seals and fuse each cell with its neighbors. The junctions prevent most substances from leaking across the tissue's free surface. Substances reach tissues below only by passing

free surface
of epithelium

simple
squamous
epithelium

basement
membrane

a　connective
tissue

Figure 29.2 (**a**) Some basic characteristics of epithelium. Epithelium has a free surface exposed to some body fluid or to the outside environment. Between the opposite surface and the underlying connective tissue is a basement membrane. All epithelial cells rest on this type of membrane. The sketch below the athlete shows this arrangement for one type of epithelium that is a single layer of cells. The light micrograph below her arm is a section of the upper portion of stratified epithelium. Stratified epithelium consists of more than one layer of cells, which are flattened out near the free surface.

(**b**) Light micrographs and sketches of three simple epithelia, showing three common shapes of cells for this tissue.

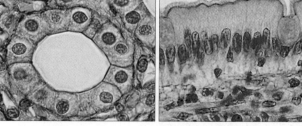

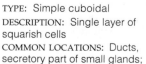

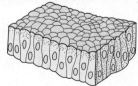

TYPE: Simple squamous
DESCRIPTION: Friction-reducing slick, single layer of flattened cells
COMMON LOCATIONS: Lining of blood and lymph vessels, heart; air sacs of lungs; peritoneum
FUNCTION: Diffusion; filtration;
b　secretion of lubricants

TYPE: Simple cuboidal
DESCRIPTION: Single layer of squarish cells
COMMON LOCATIONS: Ducts, secretory part of small glands; retina; kidney tubules; ovaries, testes; bronchioles
FUNCTIONS: Secretion, absorption

TYPE: Simple columnar
DESCRIPTION: Single layer of tall cells; free surface may have cilia, mucus-secreting glandular cells, microvilli
COMMON LOCATIONS: Glands, ducts; gut; parts of uterus; small bronchi
FUNCTION: Secretion; absorption; ciliated types move substances

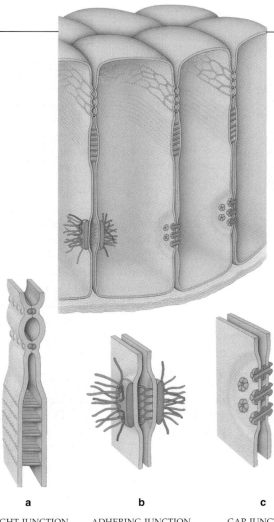

a	b	c
TIGHT JUNCTION	ADHERING JUNCTION	GAP JUNCTION
Strands (rows of proteins) running parallel with the free surface of the tissue; they block leaking between adjoining cells.	Adjoining cells adhere at a mass of proteins (a plaque) anchored beneath their plasma membrane by many intermediate filaments of the cytoskeleton.	Cylindrical arrays of proteins span the plasma membrane of adjoining cells. They pair up as open channels for signals between cells.

Figure 29.3 Examples of cell junctions.

through the cytoplasm of epithelial cells (Figure 29.3). Built-in controls make the plasma membrane of these cells selectively permeable. They allow some substances but not others to move across, through the interior of transport proteins (Sections 4.1, 5.6, and 5.7).

Example: Gastric fluid inside the stomach is highly acidic. If it were to leak across the stomach's epithelial lining, it would digest the proteins of your own body instead of those brought in with meals. Actually, that happens in people who have peptic ulcers (Section 37.4). Many tight junctions in such linings act as a leakproof barrier between all cells near their free surface.

Adhering junctions (spot welds or collars) act as a unit to hold cells together in all animal tissues. They are profuse in skin's surface layer and other tissues that are subjected to ongoing abrasion (Figure 29.3).

Figure 29.4 Section through the glandular epithelium of a frog. This frog, of the genus *Dendrobates*, makes one of the most lethal glandular secretions known. (Natives of one tribe in Colombia use its exocrine gland secretion to poison tips of darts, which they shoot through blowguns.) Pigment-rich cells branching through this epithelium impart color to the skin. The striking coloration of all poisonous frogs evolved as a clear warning signal to predators. In essence, it tells them "Don't even think about it."

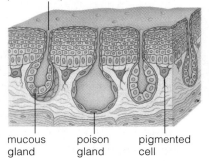

pore that opens at skin surface

mucous gland   poison gland   pigmented cell

Gap junctions let small molecules and ions diffuse swiftly among cells. They enhance chemical or electrical communication between cells when activities must be rapidly coordinated. They are abundant in the stomach, heart, skeletal muscles, and other highly active organs.

Glandular Epithelium and Glands

Epithelia of structurally simple invertebrates contain **gland cells**, which secrete (release) products, unrelated to their own metabolism, that are to be used elsewhere. In more complex animals, such cells occur in glandular epithelium and in glands, which are secretory organs derived from epithelia. Note that *secretion* isn't the same as *excretion*, the concentration and removal of metabolic wastes or excess substances of no use to the body.

Exocrine glands secrete mucus, saliva, earwax, oil, milk, digestive enzymes, and other cell products. Most exocrine products are released onto the free surface of epithelium through ducts or tubes, as in Figure 29.4.

By contrast, **endocrine glands** have no ducts. Their products are hormones, which they secrete directly into fluid bathing the gland. Molecules of animal hormones typically enter the bloodstream, which distributes them to target cells elsewhere in the body (Chapter 32).

Epithelia are sheetlike tissues lining the body's surface and its cavities, ducts, and tubes. Epithelia have one free surface facing a body fluid or the outside environment. Junctions structurally and functionally connect the adjoining cells.

Glands are secretory organs derived from epithelium.

CONNECTIVE TISSUE

Of all tissues in complex animals, connective tissues are the most abundant and widely distributed. They range from soft connective tissues to the specialized types called cartilage, bone tissue, adipose tissue, and blood (Table 29.1). In all connective tissues except blood, cells secrete fibers of structural proteins: collagen or elastin. (This is the collagen that plastic surgeons use to plump wrinkled skin and lips.) These cells also secrete modified polysaccharides, which accumulate between cells and fibers as the connective tissue's "ground substance."

Soft Connective Tissues

Generally, these tissues have the same components but different proportions of them. **Loose connective tissue** contains fibers and fibroblasts (cells that produce and secrete the fibers), all loosely arranged in a semifluid ground substance (Figure 29.5*a*). The tissue often acts as a framework for epithelium. White blood cells patrol it and mount early counterattacks against pathogens—as when bacteria invade skin at an abrasion or breach the digestive, respiratory, or urinary tract lining and enter the internal environment.

    Dense, irregular connective tissue has fibroblasts and many fibers (mostly collagen-containing ones) that are oriented every which way. This tissue is present in skin and forms protective capsules around organs that

Table 29.1	*Categories of Connective Tissues*	
SOFT	**SPECIALIZED**
Loose connective tissue	Cartilage
Dense, irregular connective tissue	Bone tissue
Dense, regular connective tissue (ligaments, tendons)	Adipose tissue
	Blood

do not stretch much (Figure 29.5*b*). In **dense, regular connective tissue**, the fibroblasts occur in rows between many parallel bundles of fibers. Tendons, which attach skeletal muscle to bone, have this tissue. Its bundles of collagen fibers help tendons resist being torn (Figure 29.5*c*). Dense, regular connective tissue also is present in elastic ligaments, which attach one bone to another. Its elastic fibers facilitate movement around joints.

Specialized Connective Tissues

Like rubber, **cartilage** is a pliable yet solid intercellular material that resists compression. Its cells make and secrete cartilage, then get imprisoned in cavities inside their secretions (Figure 29.5*d*). In vertebrate embryos, bones develop on structural models made of cartilage deposits. They replace most of the models. In adults, cartilage still maintains the shape of the nose, outer ear,

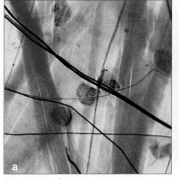 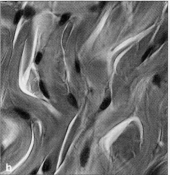 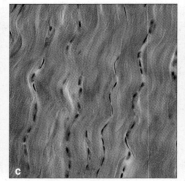 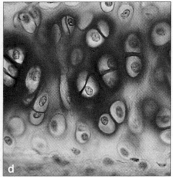

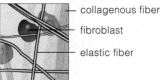
— collagenous fiber
— fibroblast
— elastic fiber

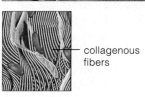
— collagenous fibers

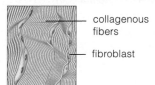
— collagenous fibers
— fibroblast

— ground substance with very fine collagen fibers
— cartilage cell (chondrocyte)

TYPE: Loose connective tissue
DESCRIPTION: Fibroblasts, other cells, plus fibers loosely arranged in semifluid ground substance
COMMON LOCATIONS: Under the skin and most epithelia
FUNCTION: Elasticity, diffusion

TYPE: Dense, irregular connective tissue
DESCRIPTION: Collagenous fibers, fibroblasts, less ground substance
COMMON LOCATIONS: In skin and capsules around some organs
FUNCTION: Support

TYPE: Dense, regular connective tissue
DESCRIPTION: Collagen fibers in parallel bundles, long rows of fibroblasts, little ground substance
COMMON LOCATIONS: Tendons, ligaments
FUNCTION: Strength, elasticity

TYPE: Cartilage
DESCRIPTION: Cells embedded in pliable, solid ground substance
COMMON LOCATIONS: Ends of long bones, nose, parts of airways, skeleton of vertebrate embryos
FUNCTION: Support, flexibility, low-friction surface for joint movement

Figure 29.5 Examples of soft connective tissues and specialized connective tissues.

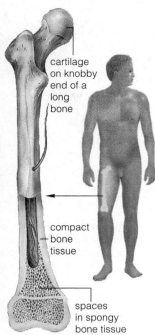

cartilage on knobby end of a long bone

compact bone tissue

spaces in spongy bone tissue

Figure 29.6 Cartilage and bone tissue. Spongy bone tissue has tiny, needlelike hard parts with spaces in between. Compact bone tissue is more dense. Bone is a load-bearing tissue that resists compression. Over time, it served as a basis for increases in size for many land vertebrates, such as giraffes. It gives large animals selective advantages. Among other things, they can ignore most predators with impunity, roam farther for food and water, and gain or lose heat more slowly than small animals. They can do so because of their lower surface-to-volume ratio and greater heat production.

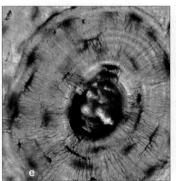

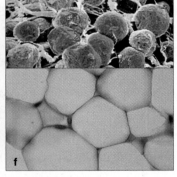

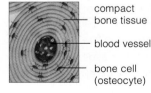

compact bone tissue

blood vessel

bone cell (osteocyte)

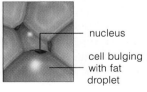

nucleus

cell bulging with fat droplet

TYPE: Bone tissue

DESCRIPTION: Collagen fibers, ground substance hardened with calcium

COMMON LOCATIONS: Bones of vertebrate skeleton

FUNCTION: Movement, support, protection

TYPE: Adipose tissue

DESCRIPTION: Large, tightly packed fat cells occupying most of ground tissue

COMMON LOCATIONS: Under skin, around heart, kidneys

FUNCTION: Energy reserves, insulation, padding

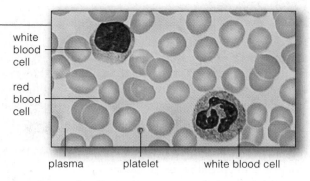

white blood cell

red blood cell

plasma   platelet   white blood cell

Figure 29.7 Some components of human blood. This tissue's straw-colored, liquid matrix (plasma) is primarily water in which many nutrients, diverse proteins, oxygen, carbon dioxide, ions, and other substances are dissolved.

and some other body parts. Cartilage also protects and cushions the joints between bones of the limbs, vertebral column, and elsewhere.

Bone tissue is mineral hardened; its collagen fibers and ground substance are strengthened with calcium salts (Figures 29.5e and 29.6). This is the main tissue of bones. In vertebrate skeletons, different bones support and protect softer tissues and organs. Limb bones, such as the long bones of your two legs, have weight-bearing functions. They interact with skeletal muscles attached to them to bring about movements. Parts of some bones also are production sites for blood cells.

Excess carbohydrates and proteins that cells do not use at once are converted to fats and other stored forms of energy. A few fat droplets collect in the cytoplasm of a variety of cells. However, the cells in **adipose tissue**, located mainly beneath skin, specialize in fat storage. Fat droplets nearly fill their cytoplasm (Figure 29.5f). Fine blood vessels threading through adipose tissue are like highways for swiftly moving fats to and from cells.

Because **blood** is derived primarily from connective tissue, many biologists classify it as a connective tissue. Blood serves transport functions. Circulating within its *plasma*, a fluid medium, are great numbers of red blood cells, white blood cells, and platelets (Figure 29.7). Red blood cells efficiently deliver oxygen to metabolically active tissues and then carry carbon dioxide and other wastes away from them. Plasma is largely water, but it has a great number of different kinds of proteins, ions, and other substances dissolved in it. We will be taking a closer look at this complex tissue in Section 34.2.

Diverse types of connective tissues bind together, support, strengthen, protect, and insulate other tissues in the body.

Soft connective tissues consist of protein fibers as well as a variety of cells arranged in a ground substance.

Cartilage, bone, blood, and adipose tissue are specialized connective tissues. Cartilage and bone are both structural materials. Blood is a fluid connective tissue with transport functions. Adipose tissue is a reservoir of stored energy.

MUSCLE TISSUE

In all muscle tissues, cells *contract*–that is, shorten–in response to stimulation from the outside, then lengthen and so return to their uncontracted state. These tissues have many cells arranged in parallel arrays. Layers of muscles and muscular organs contract, then relax, in a coordinated fashion. The three tissue types are skeletal, smooth, and cardiac muscle tissues.

Skeletal muscle tissue is the main tissue of muscles that are attached to bones. It functions in maintaining

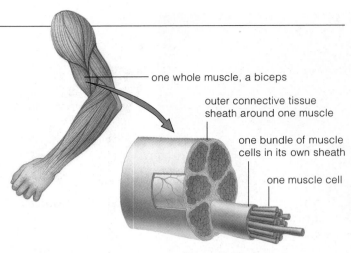

— one whole muscle, a biceps

outer connective tissue sheath around one muscle

one bundle of muscle cells in its own sheath

one muscle cell

Figure 29.9 Location and general arrangement of the muscle cells in a typical skeletal muscle. Its many muscle cells are bundled together in parallel to direct the force of contraction against a bone.

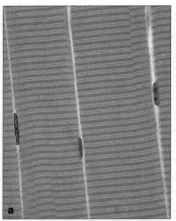

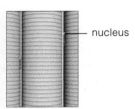

— nucleus

TYPE: Skeletal muscle
DESCRIPTION: Bundles of cylindrical, long, striated contractile cells; many mitochondria; often reflex-activated but can be consciously controlled
LOCATIONS: Partner of skeletal bones, against which it exerts great force
FUNCTION: Locomotion, posture; head, limb movements

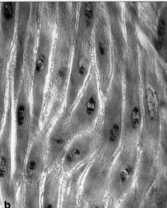

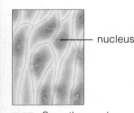

— nucleus

TYPE: Smooth muscle
DESCRIPTION: Contractile cells tapered at both ends; not striated
LOCATIONS: Wall of arteries, sphincters, stomach, intestines, urinary bladder, many other soft internal organs
FUNCTION: Controlled constriction; motility (as in gut); arterial blood flow

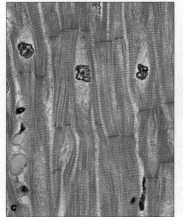

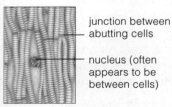

junction between abutting cells

nucleus (often appears to be between cells)

TYPE: Cardiac muscle
DESCRIPTION: Unevenly striated, fused-together cylindrical cells that contract as a unit owing to signals at gap junctions between them
LOCATIONS: Heart wall
FUNCTION: Pump blood forcefully through circulatory system

Figure 29.8 Characteristics and examples of skeletal muscle tissue, smooth muscle tissue, and cardiac muscle tissue.

posture and in moving the body and its assorted parts. The many transverse stripes (or striations) of its long, cylindrical skeletal muscle cells mark many repeating units of contractile proteins—actin and myosin (Figure 29.8*a* and 29.9). These rapidly contracting cells demand a great deal of energy. Each has many mitochondria and enzymes of glycolysis. You will read more about this tissue's structure and functioning in Chapter 33.

Smooth muscle tissue is present in the wall of the stomach, bladder, lungs, and other soft internal organs. Its contractile cells have many mitochondria. Unlike skeletal muscle, contractile proteins in its tapered cells are not arranged in a series of repeating units, so this tissue is not striated (Figure 29.8*b*). Its contractions are slower than in skeletal muscle but they can be sustained longer. They function in gut motility, urinary bladder emptying, sphincter closure, blood flow in arteries, and other internal organ activities. Smooth muscle action is said to be "involuntary" because we usually can't make it contract just by thinking about it. We can do this with skeletal muscle; its action is said to be "voluntary."

Cardiac muscle tissue is a contractile tissue present only in the heart (Figure 29.8*c*). Its muscle cells contract as a unit owing to signals at gap junctions where their plasma membranes fuse together. When one cell gets a signal to contract, others are stimulated to contract, too. The tissue's cells hold more mitochondria, so it is not as evenly striated as skeletal muscle tissue. Chapter 34 has a close look at this tissue's structure and function.

Muscle tissue, which can contract (shorten) in response to stimulation, helps move the body and specific body parts.

Skeletal muscle is the only muscle tissue attached to bones. Smooth muscle tissue occurs in many soft internal organs. Cardiac muscle alone makes up the contractile walls of the heart. Connective tissue sheathes all three types of tissues.

NERVOUS TISSUE

Of all tissues, **nervous tissue** exerts the greatest control over the body's responsiveness to changing conditions. In your own body, **neuroglia** makes up more than one-half the volume of nervous tissue. That word refers to a variety of cells that protect, structurally support, and metabolically support neurons, which make up the rest of your nervous tissue. **Neurons** are excitable cells, the communication units of most nervous systems.

When a neuron is suitably stimulated, an electrical disturbance swiftly travels along its plasma membrane. Arrival of the disturbance at the neuron's endings, or output zone, triggers events that may cause stimulation or inhibition of adjacent neurons and other cells.

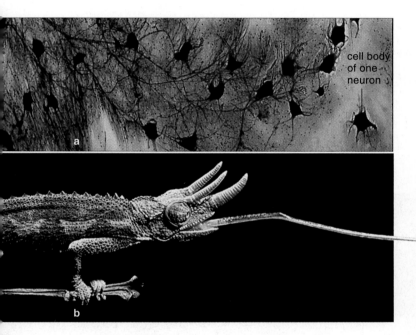

cell body of one neuron

Figure 29.10 (**a**) Motor neurons, which relay signals from the brain or spinal cord to muscles and glands. Diverse neurons interact to detect and process information about internal and external conditions, and to initiate responses. (**b**) For instance, without neurons, this chameleon could not detect an edible insect, calculate its distance, and command a long, sticky, prey-capturing tongue to uncoil with stunning speed.

For example, more than a hundred billion neurons are organized as communication lines throughout your body. Some detect specific changes in environmental conditions. Others coordinate immediate and long-term responses to change. The type shown in Figure 29.10*a* delivers signals from the brain to muscles and glands. How these cells function is a topic of later chapters.

Neurons are the basic units of communication in nervous tissue. Different kinds detect specific stimuli, integrate information, and issue or relay commands for response.

Frontiers in Tissue Research

As you know by now, a tissue is more than the sum of its cells. When each new animal grows and develops, its cells interact and become organized in prescribed ways to give rise to the body's diverse tissues. Also, cells of each new tissue synthesize specific gene products that are vital for normal body functioning.

For decades, medical researchers attempted to grow artificial tissues in quantity in the laboratory. *Lab-grown epidermis* could be used to speed the healing of extensive burns, ulcers, cancerous lesions, and blistering disorders. And now, lab-grown epidermis is a reality. One company makes it by mixing cells from foreskins (discarded earlier from circumcised male infants) and proteins (from cattle tendons). The mixture is added to rows of small, shallow dishes holding a culture medium that is enriched with nutrients and growth factors. The cells multiply to form round, paper-thin grafts. A bit of foreskin no larger than a postage stamp contains enough undifferentiated cells to make 200,000 grafts, each with a five-day shelf life. Surgeons apply such grafts to a wound and bind it with a gauze dressing. Over the next few weeks, the graft's cells interact biochemically and structurally with a patient's cells to regenerate damaged or missing tissue. Use of such lab-grown skin is less costly and less risky than surgery to obtain a patch of the patient's own skin.

On the horizon are *designer organs*: capsules holding preselected groups of living cells that synthesize specific hormones, enzymes, growth factors, and other substances. The idea is to surgically snip a bit of epithelium from a patient and use it to encapsulate the preselected cells. The capsule would be derived from the patient's own epithelial cells, so it wouldn't be chemically recognized as "foreign" and attacked by the immune system. As you will see in Chapter 35, that is what happens when the body rejects organ or tissue implants. Such rejection can have severe medical consequences.

Biotechnologists are close to understanding how to synthesize molecular cues that will help designer organs stick to suitable sites in the body. Once attached, they might become integral parts of normal body functioning.

Ultimately, the goal of this research is to put together packages of cells able to make life-saving substances that are absent in patients who suffer from severe genetic disorders or chronic diseases. For example, imagine the potential for people with type 1 *diabetes mellitus*. This metabolic disorder results in an elevated concentration of glucose in the blood. Affected people produce little if any insulin, the hormone that signals cells to take up glucose from the blood. Unless they get insulin injections on a regular basis, they die. However, if a customized, insulin-secreting organ could be successfully installed inside the body of such individuals, daily injections of insulin may become a thing of the past.

ORGAN SYSTEMS

Overview of Major Organ Systems

Figure 29.11 is an overview of eleven organ systems of a typical vertebrate, an adult human. Figure 29.12 lists some terms used when describing positions of various organs. It also shows the major body cavities in which a number of important organs are located.

Each organ system contributes to the survival of all living cells in the animal body. You might think this is stretching things a bit. For example, how could muscles and bones be helping each microscopically small cell stay alive? Yet interactions between the skeletal system and muscular system let us move, say, toward sources of nutrients and water. Parts of these organ systems even help keep blood circulating to cells, as when leg muscle contractions help move blood in veins back to the heart. The circulatory system rapidly transports oxygen and other substances dissolved in blood to cells, and moves metabolic products and wastes away from them. The respiratory system swiftly delivers oxygen from the air to the circulatory system and takes up carbon dioxide wastes from it, skeletal muscles assist the respiratory system—and so it goes, throughout the entire body.

Figure 29.12 (**a**) Major cavities in the human body. (**b,c**) Directional terms and planes of symmetry for the vertebrate body. Notice how the *midsagittal* plane divides the body into right and left halves. Most vertebrates, such as fishes and rabbits, move with the main body axis parallel with the Earth's surface. For them, *dorsal* pertains to their back or upper surface, and *ventral* pertains to the opposite, lower surface.

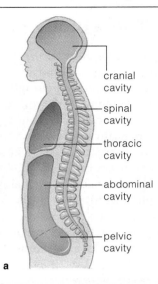

cranial cavity
spinal cavity
thoracic cavity
abdominal cavity
pelvic cavity

a

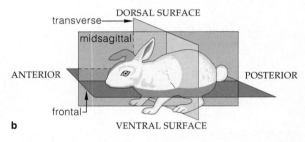

transverse
DORSAL SURFACE
midsagittal
ANTERIOR
POSTERIOR
frontal
VENTRAL SURFACE
b

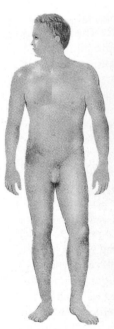

INTEGUMENTARY SYSTEM	MUSCULAR SYSTEM	SKELETAL SYSTEM	NERVOUS SYSTEM	ENDOCRINE SYSTEM	CIRCULATORY SYSTEM
Protects body from injury, dehydration, and some pathogens; controls its temperature; excretes some wastes; receives some external stimuli.	Moves body and its internal parts; maintains posture; generates heat by increases in metabolic activity.	Supports and protects body parts; provides muscle attachment sites; produces red blood cells; stores calcium, phosphorus.	Detects external and internal stimuli; controls and coordinates the responses to stimuli; integrates all organ system activities.	Hormonally controls body functioning; works with nervous system to integrate short-term and long-term activities.	Rapidly transports many materials to and from cells; helps stabilize internal pH and temperature.

Figure 29.11 Overview of human organ systems and their functions.

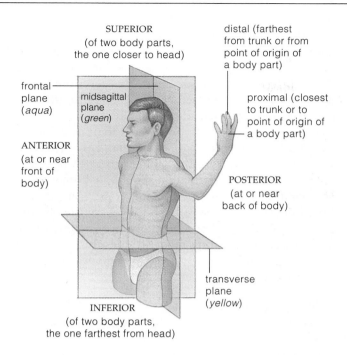

SUPERIOR
(of two body parts,
the one closer to head)

distal (farthest
from trunk or from
point of origin of
a body part)

frontal
plane
(*aqua*)

midsagittal
plane
(*green*)

proximal (closest
to trunk or to
point of origin of
a body part)

ANTERIOR
(at or near
front of
body)

POSTERIOR
(at or near
back of body)

transverse
plane
(*yellow*)

INFERIOR
(of two body parts,
the one farthest from head)

c Unlike quadrupedal animals, humans walk upright, with the main body axis perpendicular to the ground. *Anterior* refers to the front of an upright walker; it corresponds to ventral, as in (**b**). *Posterior* refers to the back; it corresponds to dorsal.

Tissue and Organ Formation

Where do tissues of organ systems come from? To get a sense of how they originate, start with sperm and eggs. (These form from *germ* cells, or immature reproductive cells. All other cells in the body are *somatic*, after a Greek word for body.) A zygote forms after a sperm fertilizes an egg, then mitotic cell divisions form an early embryo. In vertebrates, cells become arranged as three primary tissues—ectoderm, mesoderm, and endoderm. These three are embryonic forerunners of tissues in the adult. **Ectoderm** gives rise to skin's outer layer and to tissues of the nervous system. **Mesoderm** gives rise to tissues of muscle, bone, and most of the circulatory, reproductive, and urinary systems. **Endoderm** gives rise to the lining of the digestive tract and to organs derived from it.

In general, all vertebrates have the same kinds of organ systems. Each organ system serves specialized functions, such as gas exchange, blood circulation, and locomotion.

Vertebrate tissues, organs, and organ systems arise from three primary tissues that form in the early embryo. The primary tissues are ectoderm, mesoderm, and endoderm.

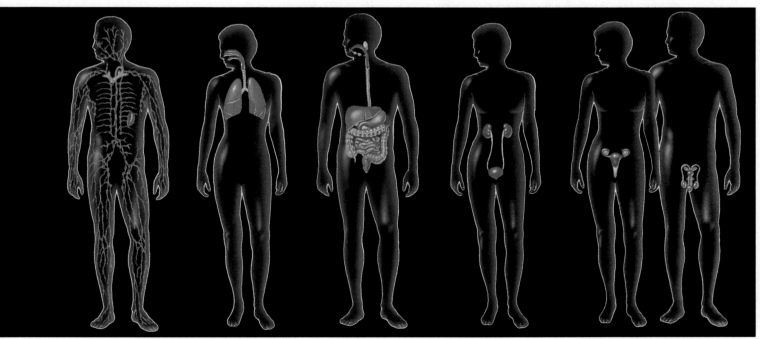

LYMPHATIC SYSTEM

Collects and returns some tissue fluid to the bloodstream; defends the body against infection and tissue damage.

RESPIRATORY SYSTEM

Rapidly delivers oxygen to the tissue fluid that bathes all living cells; removes carbon dioxide wastes of cells; helps regulate pH.

DIGESTIVE SYSTEM

Ingests food and water; mechanically, chemically breaks down food, and absorbs small molecules into internal environment; eliminates food residues.

URINARY SYSTEM

Maintains the volume and composition of internal environment; excretes excess fluid and blood-borne wastes.

REPRODUCTIVE SYSTEM

Female: Produces eggs; after fertilization, affords a protected, nutritive environment for the development of new individual. *Male:* Produces and transfers sperm to the female. Hormones of both systems also influence other organ systems.

HOMEOSTASIS AND SYSTEMS CONTROL

Concerning the Internal Environment

To stay alive, your cells must remain bathed in a fluid that offers nutrients and carries away metabolic wastes. In this they are no different from an amoeba or some other free-living, single-celled organism. The difference is, trillions of cells coexist in your body. They all must draw nutrients from and dump wastes into the same fifteen liters of fluid, which is less than sixteen quarts.

The fluid *not* inside cells is **extracellular fluid**. Much of it is *interstitial*, meaning it occupies spaces between cells and tissues. The remainder is *plasma*, which is the fluid portion of blood. The interstitial fluid exchanges substances with the cells it bathes and with blood.

In functional terms, changes in extracellular fluid cause changes in cells. That is why drastic changes in the composition and volume of extracellular fluid have drastic effects on cell activities. The type and number of ions are especially crucial, for they must be maintained at concentrations that are compatible with metabolism. Otherwise, the animal itself cannot survive.

It makes no difference whether an animal is simple or complex. *The component parts of every animal work together to maintain the stable fluid environment that all of its living cells require.* This concept is absolutely central to understanding the structure and function of animals, and its key points may be summarized as follows. First, each cell of the animal body engages in basic metabolic activities that ensure its own survival. Second, the cells of a given tissue perform one or more activities that contribute to survival of the whole organism. Third, the combined contributions of individual cells, tissues, organs, and organ systems that are engaged in a division of labor help maintain a stable internal environment—extracellular fluid—required for individual cell survival.

Mechanisms of Homeostasis

Homeostasis, recall, means stable operating conditions in the internal environment. Three components interact to maintain this state: sensory receptors, integrators, and effectors. **Sensory receptors** are cells or parts of cells that detect forms of energy, such as light and pressure. Any specific form of energy that a receptor detects is a **stimulus**. When someone kisses you, for example, the pressure on your lips changes. Receptors in the skin of your lips translate the stimulus into signals that are sent to the brain. The brain is an **integrator**, a central command post that pulls together different bits of information to select a response. Integrators send out signals to muscles or glands (or both). Muscles and glands are the body's **effectors**—they carry out suitable responses. In this case,

the response might involve flushing with pleasure and kissing the person back. Of course, you can't engage in a kiss indefinitely. Doing so would stop you from eating and carrying out other activities necessary to maintain operating conditions inside your body.

So how does your brain reverse the physiological changes induced by the kiss? Receptors only provide it with information about how things *are* operating. The brain also receives information about how things *should be* operating—that is, information from "set points." When physical or chemical conditions deviate sharply from a set point, the brain functions to bring them back to an effective operating range. It does this by way of signals that cause specific muscles and specific glands to increase or decrease their activity.

NEGATIVE FEEDBACK Feedback mechanisms are among the controls that operate to keep physical and chemical aspects of the body within tolerable ranges. One type is a **negative feedback mechanism**: An activity gets under way and changes some condition—until the condition is altered enough to trigger a response that reverses the change. Figure 29.13 is one model of such control.

Think of a furnace with a thermostat. A thermostat can sense the temperature of the surrounding air and "compare" it against a preset point on a thermometer built into the furnace's control system. Suppose the temperature falls below the preset point. The thermostat sends signals to a switching mechanism that turns on the furnace. When the air becomes heated enough to match the prescribed level, the thermostat again signals the switching mechanism, which shuts off the furnace.

Similarly, feedback mechanisms help keep the body temperature of meerkats, humans, huskies, and many other animals near 37°C (98.6°F) even during hot or cold weather. Visualize a young husky running around on a hot summer day. Soon its body gets hot, and receptors trigger events that slow down the whole dog *and* its cells. The husky searches for shade and rests under a tree. Moisture from its respiratory system evaporates from its tongue and carries away some body heat with

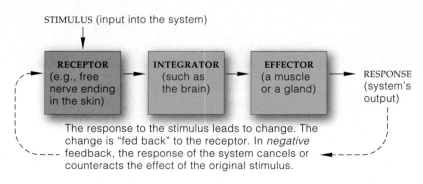

Figure 29.13 Components required for negative feedback at the organ level.

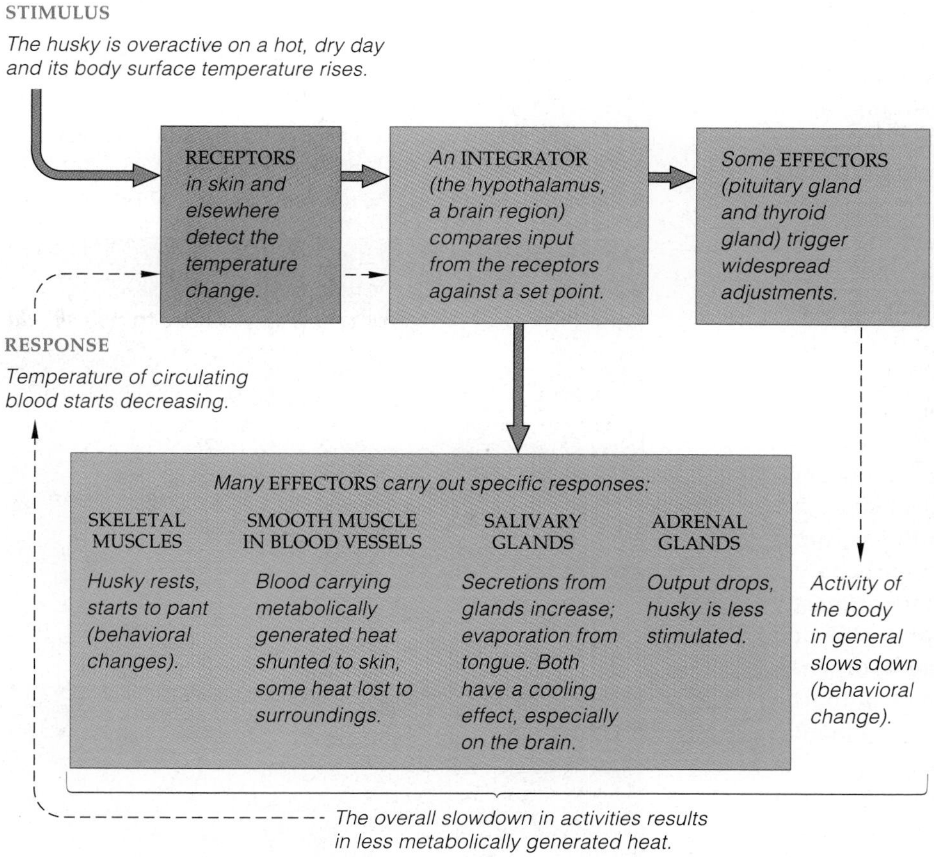

STIMULUS

The husky is overactive on a hot, dry day and its body surface temperature rises.

RECEPTORS in skin and elsewhere detect the temperature change.	An INTEGRATOR (the hypothalamus, a brain region) compares input from the receptors against a set point.	Some EFFECTORS (pituitary gland and thyroid gland) trigger widespread adjustments.

RESPONSE

Temperature of circulating blood starts decreasing.

Many EFFECTORS *carry out specific responses:*

SKELETAL MUSCLES	SMOOTH MUSCLE IN BLOOD VESSELS	SALIVARY GLANDS	ADRENAL GLANDS	Activity of the body in general slows down (behavioral change).
Husky rests, starts to pant (behavioral changes).	Blood carrying metabolically generated heat shunted to skin, some heat lost to surroundings.	Secretions from glands increase; evaporation from tongue. Both have a cooling effect, especially on the brain.	Output drops, husky is less stimulated.	

The overall slowdown in activities results in less metabolically generated heat.

Figure 29.14 Homeostatic controls over the internal temperature of a husky's body. *Blue* arrows indicate the main control pathways. The dashed line shows how a feedback loop is completed.

it, as Figure 29.14 indicates. These control mechanisms and others counter overheating by curbing the activities that naturally generate metabolic heat and by giving up the body's excess heat to the surrounding air.

POSITIVE FEEDBACK In some cases, **positive feedback mechanisms** operate. These controls set in motion a chain of events that *intensify* a change from an original condition—and after a limited time, the intensification reverses the change. Such positive feedback is associated with instability in a system. For example, during sexual intercourse, chemical signals from the nervous system of a human female might induce her to make intense physiological responses to stimulation from her partner. Those responses may stimulate changes in her partner that stimulate the female even more, and so on until she reaches an explosive, climax level of excitation. Normal conditions now return, and homeostasis prevails.

As another example, at childbirth, the fetus exerts pressure on the wall of its mother's uterus. Pressure stimulates the production and secretion of oxytocin, a hormone. Oxytocin causes wall muscles to contract and exert pressure on the fetus, which exerts more pressure on the wall, and so on until the fetus is expelled.

What we have been describing is a general pattern of detecting, evaluating, and responding to a continual flow of information about the animal's internal and external environments. During all of this activity, organ systems operate together in astoundingly coordinated fashion. Throughout this unit of the book, you will be asking the following questions about their operation:

1. *Which physical or chemical aspects of the internal environment are organ systems working to maintain as conditions change?*

2. *By what means are organ systems kept informed of the various changes?*

3. *By what means do they process incoming information?*

4. *What mechanisms are set in motion in response?*

As you will read in later chapters, the operation of all organ systems is under neural and endocrine control.

Each living cell of an animal body executes basic metabolic activities that ensure its own survival. Concurrently, the cells of any given tissue perform one or more activities that contribute to the survival of the whole animal.

The combined contributions of cells, tissues, organs, and organ systems maintain the stable internal environment (extracellular fluid) required for individual cell survival.

Homeostatic control mechanisms help maintain physical and chemical aspects of the body's internal environment within ranges that are most favorable for cell activities.

SUMMARY
Gold indicates text section

1. A tissue is an aggregation of cells and intercellular substances that perform a common task. An organ is a structural unit of different tissues combined in definite proportions and patterns that allow them to perform a common task. An organ system has two or more organs interacting chemically, physically, or both in ways that contribute to the survival of the body as a whole. *CI*

2. Epithelial tissues cover external body surfaces and line internal cavities and tubes. All have a free surface exposed to a body fluid or to the environment. *29.1*

3. Different connective tissues bind together, support, strengthen, protect, and insulate other tissues. Most have fibers of structural proteins (especially collagen), fibroblasts, and other cells in a ground substance. *29.2*

   a. Loose connective tissue, with a semifluid ground substance, is present under skin and most epithelia.

   b. Dense, irregular connective tissue contains mostly collagen fibers and fibroblasts. It is present in skin and forms protective capsules around a number of organs.

   c. Dense, regular connective tissue, such as that in tendons, contains parallel bundles of collagen fibers. It protects and structurally supports organs.

   d. Cartilage, with its solid yet pliable intercellular material, has structural and cushioning roles. Bone, the weight-bearing tissue of a vertebrate skeleton, interacts with skeletal muscle to bring about movement.

   e. Blood, a specialized connective tissue, consists of plasma, cellular components, and dissolved substances. Adipose tissue, another specialized connective tissue, is a reservoir of energy; it consists mainly of fat cells.

4. Muscle tissues contract (shorten), then return to the resting position. They help move the body or parts of it. The three types of muscle tissue are skeletal muscle, smooth muscle, and cardiac muscle tissue. *29.3*

5. Nervous tissue intercepts and integrates information about internal and external conditions, and governs the body's responses to change. Its neurons are basic units of communication in nervous systems. *29.4*

6. Tissues, organs, and organ systems work together to maintain a stable internal environment (extracellular fluid) needed for individual cell survival. At homeostasis, conditions in the internal environment are balanced at levels most favorable for cell activities. *29.6, 29.7*

7. Feedback controls help maintain internal operating conditions for the body's cells. With negative feedback, for example, a change in a particular condition triggers a response that results in reversal of the change. *29.7*

8. Homeostasis depends on receptors, integrators, and effectors. Stimuli are specific forms of energy detected by receptors. Integrating centers such as a brain receive and process signals from receptors and direct effectors (muscles and glands) to carry out responses. *29.7*

Review Questions

1. Describe the characteristics of epithelial tissue in general. Then describe the various types of epithelial tissues in terms of specific characteristics and functions. *29.1*

2. List the major types of connective tissues; add the names and characteristics of their specific types. *29.2*

3. Identify and describe the following tissues: *29.1–29.3*

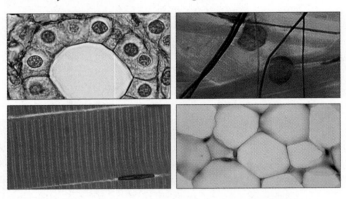

4. Identify this category of tissue and its characteristics: *29.4*

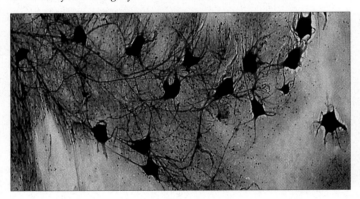

5. What type of cell serves as the basic unit of communication in nervous systems? *29.4*

6. Define animal tissue, organ, and organ system. List and define the functions of the eleven major organ systems of the human body. *CI, 29.6*

7. Define extracellular fluid, interstitial fluid, and plasma. *29.7*

8. Define homeostasis. *CI, 29.7*

9. Briefly describe two major categories of the homeostatic mechanisms operating in the human body. *29.7*

Self-Quiz   ANSWERS IN APPENDIX III

1. _____ tissues have closely linked cells and one free surface.
   a. Epithelial   c. Nervous
   b. Connective   d. Muscle

2. In most _____ , cells secrete fibers of collagen and elastin.
   a. epithelial tissue   c. muscle tissue
   b. connective tissue   d. nervous tissue

3. _____ has a semifluid ground substance and occurs under most epithelia.
   a. Dense, irregular   c. Dense, regular
       connective tissue   connective tissue
   b. Loose connective tissue   d. Cartilage

4. _____ , a specialized connective tissue, is mostly plasma with cellular components and various dissolved substances.
   a. Irregular connective tissue   c. Cartilage
   b. Blood   d. Bone

5. After you eat too many carbohydrates and proteins, your body converts the excess to storage fats, which accumulate in _____ .
   a. connective tissue proper   c. adipose tissue
   b. dense connective tissue   d. both b and c

6. In your own body, _____ can shorten (contract).
   a. epithelial tissue   c. muscle tissue
   b. connective tissue   d. nervous tissue

7. Components of _____ detect and coordinate information about changes and control responses to those changes.
   a. epithelial tissue   c. muscle tissue
   b. connective tissue   d. nervous tissue

8. Cells of complex animals _____ .
   a. survive by their own metabolic activities
   b. contribute to the survival of the whole animal
   c. help maintain extracellular fluid
   d. all of the above

9. At _____ , physical and chemical conditions in the internal environment are within tolerable ranges.
   a. the time of positive feedback
   b. the time of negative feedback
   c. homeostasis
   d. metastasis

10. With negative feedback mechanisms, _____ .
    a. a stimulus brings about a response that tends to return internal operating conditions to the original state
    b. a stimulus suppresses internal operating conditions to levels below a set point for the body
    c. a stimulus raises internal operating conditions to levels above a set point for the body
    d. fewer solutes are fed back to the affected cells

11. Of the components that exert feedback control of organ activity, _____ detect specific changes in the environment, an _____ pulls together bits of information and selects a suitable response, and then _____ carry out the response.

12. Match the terms with the suitable description.
    _____ exocrine gland   a. strong, pliable; like rubber
    _____ endocrine gland   b. secretion through duct
    _____ cartilage   c. stable internal environment
    _____ homeostasis   d. integrating center
    _____ muscles and glands   e. most common homeostatic
    _____ positive feedback   control mechanism
    _____ negative feedback   f. effectors
    _____ brain   g. chain of events intensifies
                                  original condition in body
                               h. ductless secretion

Critical Thinking

1. *Anhidrotic ectodermal dysplasia* is a disorder associated with a recessive allele on the mammalian X chromosome (Section 14.3). As one symptom of this genetic disorder, affected males and females have no sweat glands in tissues where the recessive allele is expressed. What type of tissue are we talking about?

2. Adipose tissue and blood are often said to be "atypical" connective tissues. Compared with other connective tissues, which of their features are *not* typical?

3. *Porphyria*, a genetic disorder, shows up in about 1 in 25,000 individuals. Affected people lack certain enzymes that are part of a metabolic pathway leading to the formation of heme, which is the iron-containing group of hemoglobin. An accumulation of porphyrins (intermediates of the pathway) results in visually jarring symptoms, especially after exposure to sunlight. Lesions and scars form on the skin. Hair grows thickly on the face and hands. Gums retreat from the teeth, and this makes the canines take on a fanglike appearance.

Symptoms worsen upon exposure to a variety of substances, including garlic and alcoholic beverages. Affected individuals avoid sunlight and aggravating substances. They also may get injections of heme from normal red blood cells.

If you are familiar with vampire stories, which began before the Middle Ages, speculate on how they may have evolved among superstitious folk who lacked medical knowledge of porphyria.

4. After graduating from high school, Jeff and Ryan set out on a trip along the desert roads of California and Arizona. One hot, dry morning in Joshua Tree National Monument, they saw an unusual rock formation that didn't appear to be too far from the road. They left the car and started hiking toward it.

Their destination turned out to be farther away than they thought, and the sun's rays were more relentless than they had anticipated. They reached the shade of the rocks in the early afternoon. Their canteen was nearly empty, however, and the physiological meaning of "thirst" made itself known in a scary way. They knew they had to locate and drink water (or some other fluid), which is what the brain usually tells us to do when dehydration sets in.

From what you read in this chapter, would you suspect that Ryan and Jeff's *thirst behavior* is part of a positive or negative feedback control mechanism?

Selected Key Terms

adhering junction *29.1*	homeostasis *CI*
adipose tissue *29.2*	integrator *29.7*
blood *29.2*	internal environment *CI*
bone tissue *29.2*	loose connective tissue *29.2*
cardiac muscle tissue *29.3*	mesoderm *29.6*
cartilage *29.2*	negative feedback
dense, irregular	mechanism *29.7*
connective tissue *29.2*	nervous tissue *29.4*
dense, regular	neuroglia *29.4*
connective tissue *29.2*	neuron *29.4*
division of labor *CI*	organ *CI*
ectoderm *29.6*	organ system *CI*
effector *29.7*	positive feedback
endocrine gland *29.2*	mechanism *29.7*
endoderm *29.6*	sensory receptor *29.7*
epithelium *29.1*	skeletal muscle tissue *29.3*
exocrine gland *29.2*	smooth muscle tissue *29.3*
extracellular fluid *29.7*	stimulus *29.7*
gap junction *29.1*	tight junction *29.1*
gland cell *29.2*	tissue *CI*

Readings

Bloom, W., and D. W. Fawcett. 1995. *A Textbook of Histology.* Twelfth edition. Philadelphia: Saunders.

Leeson, C. R., T. Leeson, and A. Paparo. 1988. *Textbook of Histology.* Philadelphia: Saunders.

Telford, I., and C. Bridgman. 1995. *Introduction to Functional Histology.* Second edition. New York: HarperCollins.

Wright, C. November 1999. "Ready-to-Wear Flesh." *Discover.*

On-Line readings at Student Guide for InfoTrac:
www.brookscole.com/biology

INTEGRATION AND CONTROL: NERVOUS SYSTEMS

Why Crack the System?

Suppose your biology instructor asks you to volunteer for an experiment. You'll get a microchip implanted in your brain. It will make you feel really good. But it may mess up your health, lop ten years off your life, and destroy part of your brain. Your behavior will change for the worse. You may have trouble finishing school, keeping a job, and having a normal family life.

The longer the chip is implanted, the less you will want to give it up. You won't get paid. You will pay the experimenter—first at bargain rates, then a little more each week. The chip is illegal. Get caught using it, and you and the experimenter will go to jail.

Sometimes Jim Kalat, a professor at North Carolina State University, proposes this experiment, which of course is hypothetical. Hardly any students volunteer. Then he substitutes *drug* for microchip and *dealer* for experimenter, and a surprising number of students come forward! Like 30 million other Americans, the "volunteers" seem ready to engage in self-destructive uses of drugs that alter emotional and behavioral states.

The destructiveness shows up in unexpected places. For instance, each year 300,000 or so newborns already are addicted to crack, thanks to their addicted mothers. *Crack*, a cheap form of cocaine, relentlessly stimulates

brain regions that govern the sense of pleasure. Crack dampens normal urges to eat and sleep, and it raises blood pressure. Elation and sexual desire intensify. In time, brain cells that produce the stimulatory chemicals can't keep up with the abnormal demand. The chemical vacuum makes crack users frantic and then profoundly depressed. Only crack makes them feel good again.

Addicted babies quiver with "the shakes" and respond to the world with chronic irritation. They are abnormally small. As they developed in their mother, their tissues did not get enough oxygen and nutrients. Why? As one side effect, crack causes blood vessels to constrict—and maternal blood vessels are the only supply lines that the developing individual has.

Crack babies can't respond to rocking and to other normally soothing forms of stimulation. A year or more may pass before they even recognize their own mother. Untreated, they are likely to grow up as emotionally unstable children, prone to aggressive outbursts and stony silences. This, then, is their legacy. Their mother's drug habit crippled their nervous system.

Think about it. The **nervous system** evolved as a way to sense and respond, with exquisite precision, to changing conditions inside and outside the body.

Figure 30.1
(**a**) Owners of an evolutionary treasure—a complex brain, the foundation for our memory and reasoning, and our future. (**b**) Overview of the communication lines of vertebrate nervous systems.

Awareness of sounds and sights, of odors, of hunger and passion, fear and rage—all begin with the flow of information along communication lines of the nervous system. Do the lines remain silent until they receive outside signals, much as telephone lines wait to carry calls from all over the country? Absolutely not. Even before you were born, excitable cells called neurons became organized into gridworks in your newly forming tissues and started chattering among themselves. All through your life, in moments of danger or reflection, supreme excitement or sleep, their chattering has never stopped and will not stop until the time you die.

Three classes of neurons interact in all vertebrate nervous systems. **Sensory neurons** respond to specific stimuli and relay information about them to the spinal cord and brain. A **stimulus** is a specific form of energy, such as light. **Interneurons** are in the spinal cord and brain. They receive and process sensory input, then influence the activity of other neurons. **Motor neurons** relay information away from the brain and spinal cord to the body's effectors—muscles or glands—which carry out the specified responses (Figure 30.1b). More than half the volume of the nervous system consists of cells, collectively called neuroglia, that metabolically assist, protect, and structurally support the neurons.

Our initial focus in this chapter will be on the structure and function of neurons. Later on, we will consider how neurons interact in nervous systems.

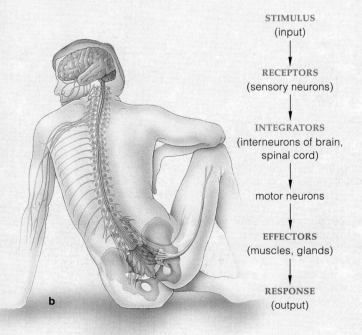

STIMULUS
(input)

↓

RECEPTORS
(sensory neurons)

↓

INTEGRATORS
(interneurons of brain,
spinal cord)

↓

motor neurons

↓

EFFECTORS
(muscles, glands)

↓

RESPONSE
(output)

b

Key Concepts

1. Neurons are the basic units of communication in nearly all nervous systems. Collectively, many neurons interact to detect and integrate information about external and internal conditions, then select or control muscles and glands in ways that produce suitable responses.

2. Neurons are a type of excitable cell, which means a suitable stimulus can disturb the distribution of electric charge across their plasma membrane. Such disturbances are the basis of messages that travel from the input zone of a neuron to its output zone near a neighboring cell.

3. Information flow through nervous systems depends on the moment-by-moment integration of excitatory and inhibitory signals that act upon each neuron in a given pathway.

4. The simplest nervous systems are the nerve nets of radial animals, such as hydras and sea anemones. The nervous systems of most animals show pronounced cephalization and bilateral symmetry.

5. Vertebrate nervous systems are functionally divided into central and peripheral regions. The brain and spinal cord make up the central nervous system. Paired nerves that thread through the remainder of the body are the main components of the peripheral nervous system.

6. The somatic nerves of the peripheral nervous system deal with skeletal muscles. The autonomic nerves deal with the heart, lungs, and other soft internal organs that are generally enclosed in body cavities.

7. The vertebrate brain has three functional divisions called the hindbrain, midbrain, and forebrain. Its most ancient parts deal with reflex control of breathing, blood circulation, and other basic functions that are essential for staying alive.

NEURONS—THE COMMUNICATION SPECIALISTS

Functional Zones of a Neuron

Look at a neuron's nucleated cell body and you see it has cytoplasmic extensions that differ in number and length. Typically, the cell body and slender extensions called **dendrites** are *input* zones for incoming signals. At a nearby patch of plasma membrane called the *trigger* zone, the incoming signals are added up. From there, signals may flow to an **axon**, a slender but often longer extension that's a *conducting* zone. An axon propagates very rapidly from a trigger zone down to its branched endings. Axon endings are the neuron's *output* zones, where information is sent to other cells (Figure 30.2).

A Neuron at Rest, Then Moved to Action

In the body's communication lines, each neuron is only a single unit. It can only accept messages about stimuli at its input zones and swiftly convey signals about them to its output zones. In essence, each "signal" is a self-propagating disturbance to the distribution of electric charge across that neuron's plasma membrane.

When a neuron is not being bothered with incoming messages, it is maintaining a difference in electric charge across its plasma membrane. Its cytoplasmic fluid next to the membrane is negatively charged, compared with interstitial fluid just outside. We measure such charges in millivolts. Think of the amount of energy inherent in the steady voltage difference across a neural membrane as the **resting membrane potential**. For many neurons, the amount is about −70 millivolts.

When a weak signal reaches a patch of membrane in the neuron's trigger zone, the voltage difference at the patch changes slightly, if at all. But a strong signal may trigger an **action potential**. We define this as a fleeting reversal in the steady voltage difference across the neural membrane. The cytoplasmic fluid inside the membrane becomes positive, relative to the fluid outside. As you will see, that reversal triggers a sequence of equivalent reversals, away from the neuron's trigger zone.

Gradients Required for Action Potentials

How does the neuron maintain readiness for an action potential? It does so by working with and against ion gradients across its plasma membrane.

First, every cell membrane has a lipid bilayer that blocks the passage of potassium ions (K^+), sodium ions (Na^+), and other charged substances. The bilayer allows a neuron to build up differences in ion concentrations across its plasma membrane. Second, such ions do flow from one side of the membrane to the other—but only through the interior of transport proteins that span the bilayer (Figure 30.3). And that flow is controlled.

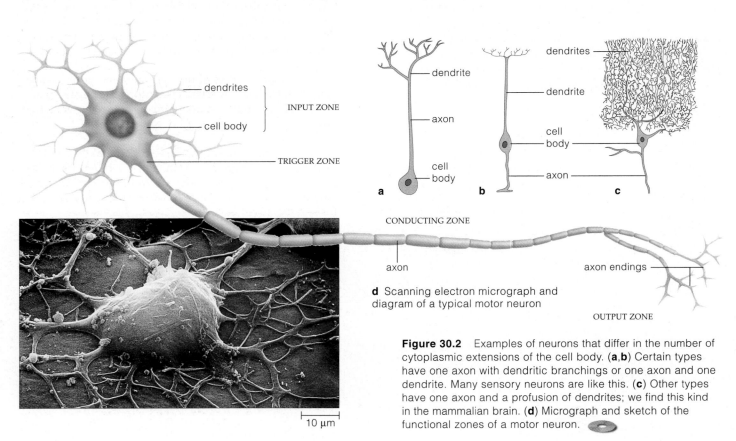

d Scanning electron micrograph and diagram of a typical motor neuron

Figure 30.2 Examples of neurons that differ in the number of cytoplasmic extensions of the cell body. (**a**,**b**) Certain types have one axon with dendritic branchings or one axon and one dendrite. Many sensory neurons are like this. (**c**) Other types have one axon and a profusion of dendrites; we find this kind in the mammalian brain. (**d**) Micrograph and sketch of the functional zones of a motor neuron.

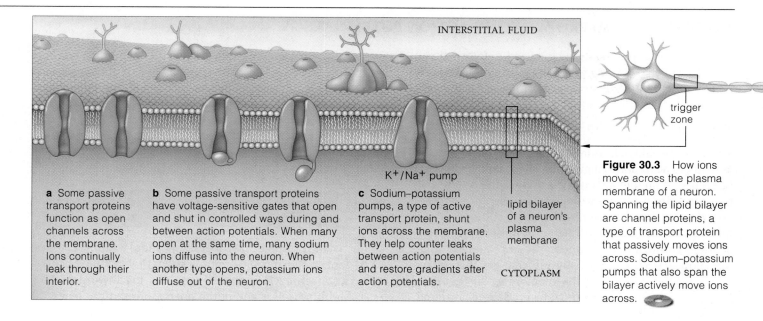

a Some passive transport proteins function as open channels across the membrane. Ions continually leak through their interior.

b Some passive transport proteins have voltage-sensitive gates that open and shut in controlled ways during and between action potentials. When many open at the same time, many sodium ions diffuse into the neuron. When another type opens, potassium ions diffuse out of the neuron.

c Sodium–potassium pumps, a type of active transport protein, shunt ions across the membrane. They help counter leaks between action potentials and restore gradients after action potentials.

K$^+$/Na$^+$ pump

lipid bilayer of a neuron's plasma membrane

INTERSTITIAL FLUID

CYTOPLASM

trigger zone

Figure 30.3 How ions move across the plasma membrane of a neuron. Spanning the lipid bilayer are channel proteins, a type of transport protein that passively moves ions across. Sodium–potassium pumps that also span the bilayer actively move ions across.

Suppose a motor neuron has 15 sodium ions inside the membrane for every 150 outside. Also suppose that it has 150 potassium ions inside for every 5 outside. You can depict each ion's concentration gradient in this way (from the large to the small letters):

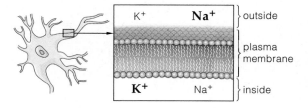

Gradients such as these determine the net direction in which sodium and potassium ions move. These ions diffuse through the interior of channel proteins, a type of transport protein in the membrane. Some channels never shut, so ions can leak through them all the time. But other transport proteins have molecular gates that open only after the neuron is adequately stimulated.

Suppose a motor neuron is at rest. Gated proteins in its membrane are closed, so sodium ions can't rush in. Some potassium leaks out through a few open channels and makes the cytoplasm a bit more negative, so some potassium is attracted back in. When the inward pull of electric charge balances the outward force of diffusion, potassium ions show no net movement. Concentration and electric gradients across the membrane now make the neuron ready to respond to stimulation.

Sodium–potassium pumps maintain the gradients and restore them after their reversal during an action potential. The pumps are transport proteins spanning the membrane (Section 5.7). When they get an energy boost from ATP, they actively transport potassium into the neuron and sodium out at the same time. Even in a resting neuron, a tiny fraction of the outward-leaking

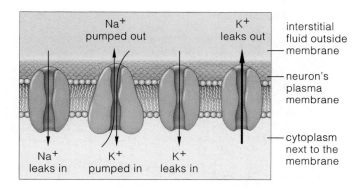

Figure 30.4 Pumping and leaking processes that dictate the distribution of sodium and potassium ions across the plasma membrane of a neuron at rest. The arrow widths indicate the magnitude of the movements. Notice how the total inward and outward movements for each kind of ion are balanced.

potassium isn't attracted back in. And a tiny fraction of the sodium is leaking in through a few open channels (Figure 30.4). If the leaks went unattended, the crucial gradients would gradually disappear.

With this bit of information on the gradients across the neural membrane, we are ready to look at how an action potential arises at a trigger zone and propagates itself, undiminished, to an output zone.

An undisturbed neuron maintains ion concentration and electric gradients across its plasma membrane. We measure this as a steady voltage difference, and energy inherent in it is the resting membrane potential. An abrupt, short-lived reversal in that voltage difference is an action potential.

Sodium–potassium pumps restore and maintain the resting membrane potential in between action potentials.

TRIGGERING AND PROPAGATING ACTION POTENTIALS

Action potential propagation isn't hard to follow if you already know something about the gradients across the neural membrane. And so we now build on Section 30.1.

Approaching Threshold

When you weakly stimulate a neuron at its input zone, you disturb the ion balance across its membrane, but not much. Imagine putting a bit of pressure on the skin of a snoozing cat by gently tapping a toe on it. Tissues beneath the skin surface have receptor endings—input zones of sensory neurons. Patches of plasma membrane at these endings deform under pressure and let some ions flow across. The flow slightly changes the voltage difference across the membrane. In this case, pressure has produced a graded, local signal.

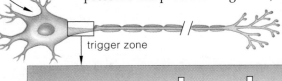

influx, the cytoplasmic side of the membrane becomes less negative. This causes more gates to open and more sodium to enter. The ever increasing, inward flow of sodium is a case of **positive feedback**, whereby an event intensifies as a result of its own occurrence:

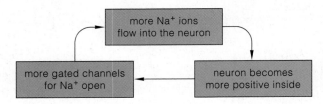

At threshold, the opening of more sodium gates no longer depends on stimulus strength. Now the positive-feedback cycle is under way, and the inward-rushing sodium itself is enough to open more sodium gates.

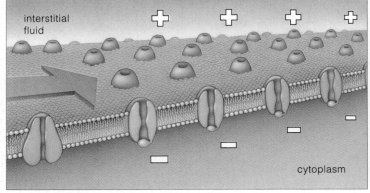
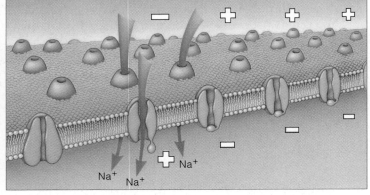

a Membrane at rest (inside negative with respect to the outside). An electrical disturbance (*red* arrow) spreads from an input zone to an adjacent trigger region of the membrane, which has a great number of gated sodium channels.

b A strong disturbance initiates an action potential. Sodium gates open. The sodium inflow decreases the negativity inside the neuron. The change causes more gates to open, and so on until threshold is reached and the voltage difference across the membrane reverses.

Figure 30.5 Propagation of an action potential along the axon of a motor neuron.

Graded means that signals arising at an input zone vary in magnitude. They are small to large, depending on the stimulus intensity or duration. *Local* means these signals do not spread far from the site of stimulation. Why? It takes certain kinds of ion channels to propagate a signal, and input zones simply don't have them.

When a stimulus is intense or long lasting, graded signals spread from the input zone into an adjoining trigger zone. This patch of membrane is richly endowed with voltage-sensitive gated channels for sodium ions. *And this is where a certain amount of change in the voltage difference across the plasma membrane triggers an action potential.* The amount is the neuron's threshold level.

The stimulus causes positively charged sodium ions to flow into the neuron, as in Figure 30.5. With the ion

An All-or-Nothing Spike

Figure 30.6 shows a recording of the voltage difference across the plasma membrane before, during, and after an action potential. Notice how the membrane potential spikes once threshold is reached. Every action potential in a neuron spikes to the same level above threshold as an *all-or-nothing* event. Once a positive-feedback cycle starts, nothing will stop the full spiking. If threshold is not reached, the disturbance to the plasma membrane will subside as soon as the stimulation stops.

Each spike lasts only for a millisecond or so. Why? At the patch where the charge reversed, gated sodium channels closed and shut off the sodium inflow. Also, about halfway into that reversal, potassium channels

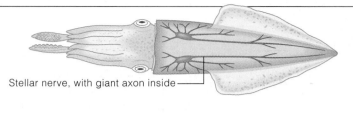

Stellar nerve, with giant axon inside

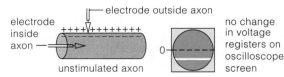

electrode inside axon — electrode outside axon

no change in voltage registers on oscilloscope screen

a unstimulated axon

Figure 30.6 Action potentials. (**a**) When researchers started to study neural function, a squid (*Loligo*) provided them with evidence of action potential spiking. This squid's "giant" axons are large enough to slip electrodes inside. (**b**) The researchers put electrodes inside and outside the axon, then stimulated it. The electrodes detected the voltage change, which showed up as deflections in a beam of light across the screen of an oscilloscope connected to the electrodes. (**c**) A typical waveform (*yellow* line) for an action potential on an oscilloscope screen.

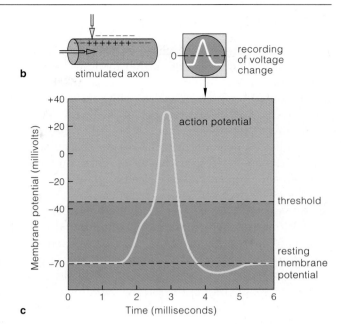

b stimulated axon

recording of voltage change

c

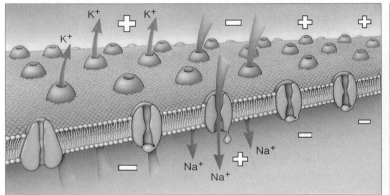

c With the reversal, sodium gates shut and potassium gates open (*pink* arrows). Potassium follows its gradient out of the neuron. Voltage is restored. The disturbance triggers an action potential at the adjacent site, and so on, away from the point of stimulation.

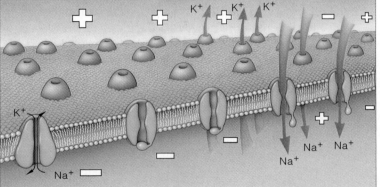

d Following each action potential, the inside of the plasma membrane becomes negative once again. However, the sodium and potassium concentration gradients are not yet fully restored. Active transport at sodium–potassium pumps restores them.

opened, so potassium ions flowed out. This restored the voltage difference at the patch but not the original ion gradients. Sodium–potassium pumps actively transport sodium back outside and potassium inside. Once this is done, most potassium gates close and sodium gates are in their original position—ready to be opened with the arrival of a suitable signal at the membrane patch.

The Direction of Propagation

During an action potential, the inward rush of sodium ions affects the charge distribution across the adjacent membrane patch, where an equivalent number of gated channels open. Gated channels open in the *next* patch, and the next, and so on. This positive feedback event is self-propagating and does not diminish in magnitude. You might be wondering: Do action potentials spread

back to the trigger zone? No. For a brief period after the inward rush of sodium, a membrane patch remains insensitive to stimulation. The transport proteins having the sodium gates are inactivated, so ions cannot move through them. This is one reason why action potentials do not spread back to the site where they were initiated but rather propagate themselves away from it.

Ions cross the neural membrane through transport proteins that serve as gated or open channels. At a suitably disturbed trigger zone, sodium gates open in an ever increasing way, and the inward-rushing sodium causes an action potential. Sodium–potassium pumps restore the original ion gradients.

Sodium gates across the membrane are briefly inactivated after an action potential, which is one reason why an action potential is self-propagating away from a trigger zone.

CHEMICAL SYNAPSES

When action potentials reach a neuron's output zone, they usually do not proceed farther than this. But their arrival may induce the neuron to release one or more **neurotransmitters**, which are signaling molecules that diffuse across chemical synapses. A **chemical synapse** is a narrow cleft between a neuron's output zone and an input zone of a neighboring cell (Figures 30.7 and 30.8). Some clefts are between two neurons. Others intervene between a neuron and a muscle cell or gland cell.

One of the two cells stores neurotransmitter inside certain vesicles. This *pre*synaptic cell has gated channels for calcium ions, which are more concentrated on the outside of the resting membrane. The gates open when an action potential arrives. As ions follow the gradient, into the cell, the flow induces synaptic vesicles to fuse with and become part of the plasma membrane, and so they dump neurotransmitter into the synaptic cleft.

Neurotransmitter molecules diffuse across the cleft. The *post*synaptic cell's membrane has protein receptors that bind to specific neurotransmitters. Binding changes the receptor shape, so a passage opens up through the protein's interior. Ions now cross the plasma membrane by diffusing through the passageway (Figure 30.8c).

How does the postsynaptic cell respond? It depends on the type and concentration of neurotransmitter, the types of receptors, and the number and responsiveness of gated channels in that cell's membrane. Such factors

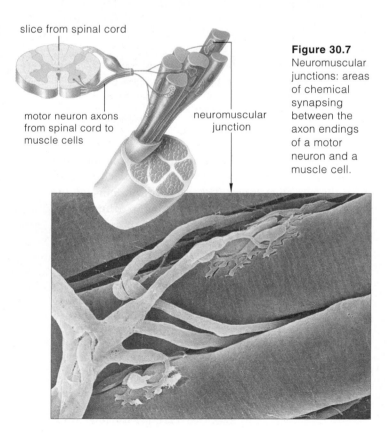

slice from spinal cord

motor neuron axons from spinal cord to muscle cells

neuromuscular junction

Figure 30.7
Neuromuscular junctions: areas of chemical synapsing between the axon endings of a motor neuron and a muscle cell.

Figure 30.8 *Facing page:* Example of a chemical synapse. (**a**,**b**) A thin cleft is all that separates a presynaptic cell from a postsynaptic cell. The arrival of an action potential at an axon ending triggers the release of molecules of neurotransmitter from the presynaptic cell. (**c**) The molecules diffuse across the cleft. They bind to receptors that are part of gated proteins of the postsynaptic cell membrane. The gates open. Ions flow in and trigger a graded potential at the membrane site.

influence whether a given neurotransmitter will have an *excitatory* effect and help drive the postsynaptic cell's membrane toward the threshold of an action potential. They influence whether it will have an *inhibitory* effect and drive the membrane away from threshold.

Consider the neurotransmitter **acetylcholine** (ACh). It has both excitatory and inhibitory effects on the brain, spinal cord, glands, and muscles. For example, it acts at chemical synapses between a motor neuron and muscle cell, as in Figure 30.7. ACh is released from the motor neuron, diffuses across the cleft, and binds to receptors on the muscle cell membrane. In this kind of cell it has excitatory effects; it can trigger action potentials, which in turn initiate muscle contraction (Section 33.7).

A Smorgasbord of Signals

Acetylcholine is only one of a veritable smorgasbord of signals that neurons deliver to target cells. For example, serotonin's targets in brain regions help control sensory perception, sleeping, body temperature, and emotions. Norepinephrine's targets influence emotions, dreaming, and waking up. Dopamine, too, works in brain regions that deal with emotions. GABA (gamma aminobutyric acid) is the most common inhibitory signal working in the brain. Valium and other antianxiety drugs work by enhancing GABA's effects.

Neuromodulators are signaling molecules that reduce or magnify a neurotransmitter's effects on neighboring or distant neurons. Examples: Substance P induces pain perception, and endorphins (natural painkillers) inhibit the release of substance P from sensory nerves. Also, neuromodulators may influence memory and learning, sexual activity, body temperature, and emotions.

Synaptic Integration

Anywhere from 1,000 to 10,000 lines of communication synapse on a typical neuron in your brain, which holds at least 100 billion neurons. As long as you are alive, those neurons hum with messages about doing what it takes to be a human. At any moment, a great number of excitatory and inhibitory signals may wash over input zones of a postsynaptic cell. Some drive its membrane

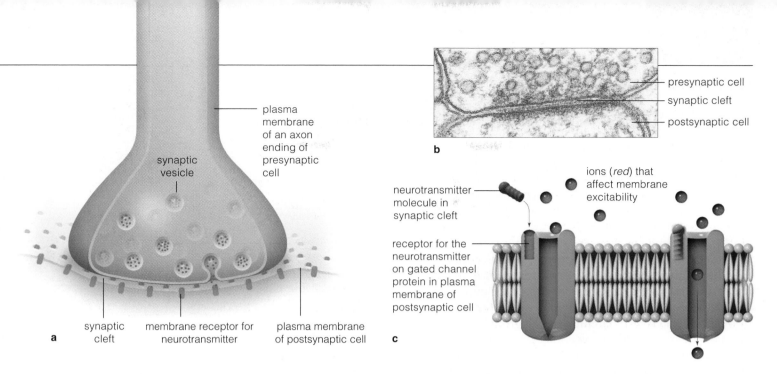

a

synaptic vesicle

plasma membrane of an axon ending of presynaptic cell

synaptic cleft

membrane receptor for neurotransmitter

plasma membrane of postsynaptic cell

b

presynaptic cell

synaptic cleft

postsynaptic cell

c

neurotransmitter molecule in synaptic cleft

receptor for the neurotransmitter on gated channel protein in plasma membrane of postsynaptic cell

ions (*red*) that affect membrane excitability

closer to threshold; others maintain the resting level or drive it away from threshold. Said another way, signals compete for control of the neuron's membrane.

All synaptic signals are graded potentials. The type we call an EPSP (for excitatory postsynaptic potential) has a *depolarizing* effect. This simply means it drives a membrane closer to threshold. An IPSP (for inhibitory postsynaptic potential) may have a *hyperpolarizing* effect and drive a membrane away from threshold, or it may help maintain the membrane at its resting level.

With **synaptic integration**, competing signals that reach an input zone of a neuron at the same time are summed. By this process, two or more signals arriving at a neuron may be dampened, suppressed, reinforced, or sent onward to other cells in the body. Figure 30.9 is a composite of an EPSP recording, an IPSP recording,

and a recording of their summation. Integration occurs when neurotransmitter molecules from more than one presynaptic cell reach a neuron's input zone at the same time. Integration also occurs after neurotransmitter is released swiftly and repeatedly from a presynaptic cell responding to a rapid series of action potentials.

How Is Neurotransmitter Removed From the Synaptic Cleft?

Information flow through the nervous system depends on the prompt, controlled removal of neurotransmitter molecules from synaptic clefts. Some of the molecules simply diffuse away from the cleft. Enzymes inside the cleft cleave others, as when acetylcholinesterase breaks apart ACh. Also, transport proteins in the membrane of presynaptic cells actively pump molecules back inside or into neighboring neuroglial cells.

What happens if neurotransmitter accumulates in the cleft? As one example, cocaine blocks the uptake of dopamine. Molecules of this neurotransmitter linger in synaptic clefts and keep on stimulating target cells. At first the abnormal stimulation invites euphoria (intense pleasure). Its later effects are disastrous (Section 30.11).

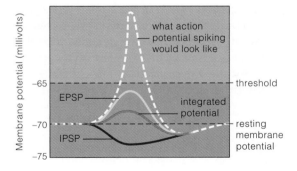

Figure 30.9 Example of synaptic integration. The *yellow* line shows how an EPSP of a certain magnitude would register on an oscilloscope screen if it were acting alone. The *purple* line shows the effect of an IPSP if *it* were acting alone. Suppose both signals arrive at a postsynaptic cell membrane at the same time. The *red* line shows their effect. In this case, when the two signals are integrated, threshold is not reached. So an action potential (the *white* dashed line) cannot be initiated in the target cell.

Neurotransmitters are signaling molecules that bridge a synaptic cleft. The cleft is a thin gap between two neurons or between a neuron and a muscle cell or gland cell.

Neurotransmitters have excitatory or inhibitory effects on different kinds of target cells. Synaptic integration is the moment-by-moment combining of excitatory signals and inhibitory signals acting on the same postsynaptic cell.

The summation process allows messages traveling through the nervous system to be reinforced or downplayed, sent onward or suppressed. The process is vital for normal body functioning.

PATHS OF INFORMATION FLOW

Blocks and Cables of Neurons

By synaptic integration, messages arriving at a neuron may be reinforced and sent on to its neighbors. What determines the direction in which a message travels? It depends on how neurons are organized in the body.

For example, your brain's staggering numbers of neurons engage in something like block parties. Regional blocks of hundreds or thousands of them get excitatory and inhibitory signals. They integrate signals entering their block, then send out new ones in response. In the regions of *divergent* circuits, the processes of neurons in one block fan out to form connections with other blocks. In regions of *convergent* circuits, signals from many are relayed to just a few. In regions of *reverberating* circuits, neurons synapse back on themselves and repeat signals like gossip that just won't go away. Such circuits make your eye muscles twitch rhythmically as you sleep.

In the cablelike **nerves**, long axons of many sensory neurons, motor neurons, or both permit long-distance communication between the brain and spinal cord and the rest of the body. Connective tissue bundles most of the axons in parallel array (Figure 30.10). Each long axon has a myelin sheath, which enhances the rate of action potential propagation. The sheath is a series of neuroglial cells (Schwann cells) wrapped like jelly rolls around the axon. It hampers ion movements across the

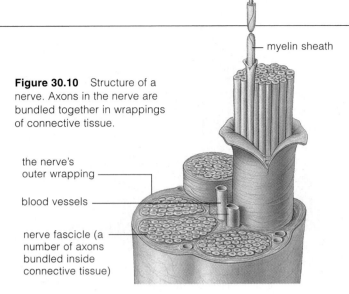

Figure 30.10 Structure of a nerve. Axons in the nerve are bundled together in wrappings of connective tissue.

axon

myelin sheath

the nerve's outer wrapping

blood vessels

nerve fascicle (a number of axons bundled inside connective tissue)

membrane. But between each cell is an exposed gap or node with voltage-sensitive, gated sodium channels in the neural membrane (Figure 30.11). Ion disturbances tend to flow until reaching a node, where they initiate a new action potential. Action potentials are propagated at 120 meters per second in large sheathed axons.

Multiple sclerosis is an autoimmune disorder. White blood cells wrongly attack the myelin of sheathed axons in the spinal cord. Chronic inflammation damages the sheaths, and then the axons they enclose. Some people have a genetic predisposition to the disorder, but viral infection may trigger it. The disruption in information flow leads to muscle weakening, pain, fatigue, uncontrolled movements, and other symptoms. The disorder affects more than a million people.

Reflex Arcs

Figure 30.12 is one example of the direction of information flow through a nervous system. It shows the roles of sensory and motor neurons of certain nerves in the stretch reflex. **Reflexes** are automatic movements made in response to stimuli. Sensory neurons synapse directly upon motor neurons in the simplest reflex arcs.

By the *stretch reflex*, a muscle contracts after gravity or some other load stretches it. Suppose

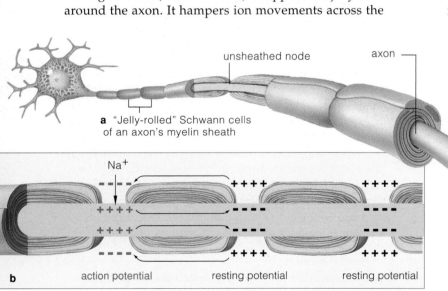

unsheathed node

axon

a "Jelly-rolled" Schwann cells of an axon's myelin sheath

Na⁺

++++ ++++

++++ ---- ----

++++ ---- ----

---- ++++ ++++

b action potential   resting potential   resting potential

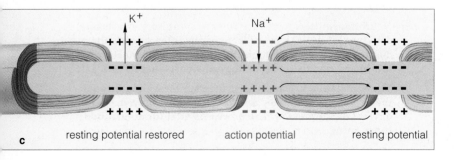

K⁺   Na⁺

++++ ---- ++++

---- ++++ ----

---- ++++ ----

++++ ---- ++++

c resting potential restored   action potential   resting potential

Figure 30.11 Action potential propagation in a sheathed neuron. (**a**) A myelin sheath (a series of cells wrapped like jelly rolls around an axon) blocks ion flow across the neural membrane. Ions cross only at unsheathed nodes between the jelly rolls. Each node contains dense arrays of gated sodium channels. (**b**) A disturbance caused by an action potential spreads along the axon. When it reaches a node, sodium gates open in the positive feedback cycle that triggers a new action potential. (**c**) The disturbance spreads swiftly to the next node, where it triggers a new action potential, and so on down the line.

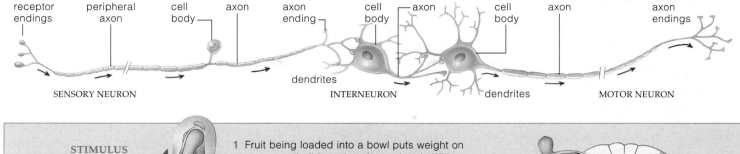

a   SENSORY NEURON   INTERNEURON   MOTOR NEURON

receptor endings   peripheral axon   cell body   axon   axon ending   cell body   axon   cell body   axon   axon endings   dendrites   dendrites

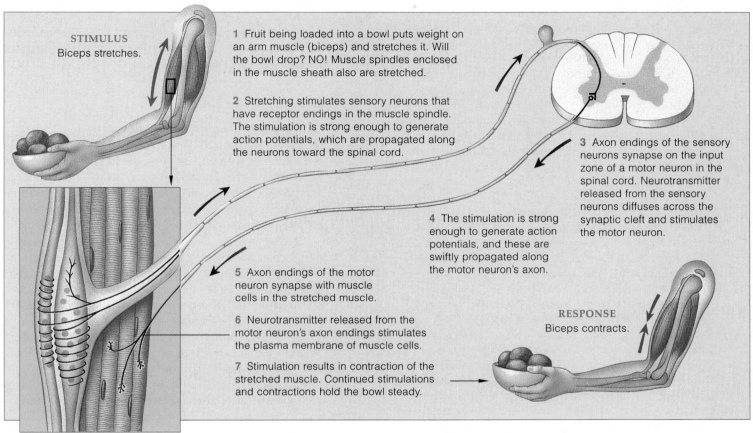

STIMULUS
Biceps stretches.

1 Fruit being loaded into a bowl puts weight on an arm muscle (biceps) and stretches it. Will the bowl drop? NO! Muscle spindles enclosed in the muscle sheath also are stretched.

2 Stretching stimulates sensory neurons that have receptor endings in the muscle spindle. The stimulation is strong enough to generate action potentials, which are propagated along the neurons toward the spinal cord.

3 Axon endings of the sensory neurons synapse on the input zone of a motor neuron in the spinal cord. Neurotransmitter released from the sensory neurons diffuses across the synaptic cleft and stimulates the motor neuron.

4 The stimulation is strong enough to generate action potentials, and these are swiftly propagated along the motor neuron's axon.

5 Axon endings of the motor neuron synapse with muscle cells in the stretched muscle.

6 Neurotransmitter released from the motor neuron's axon endings stimulates the plasma membrane of muscle cells.

7 Stimulation results in contraction of the stretched muscle. Continued stimulations and contractions hold the bowl steady.

RESPONSE
Biceps contracts.

muscle spindle   muscle cell

b

Figure 30.12 (**a**) General direction of information flow in nervous systems. Sensory neurons relay information *into* the spinal cord and brain, where they synapse with interneurons. Interneurons *within* the spinal cord and brain integrate signals. Many synapse with motor neurons, which carry signals *away* from the spinal cord and brain.

(**b**) Organization of nerves in a reflex arc that deals with muscle stretching. In a skeletal muscle, stretch-sensitive receptors of a sensory neuron are located in muscle spindles. The stretching generates action potentials, which reach axon endings in the spinal cord. These synapse with a motor neuron that carries signals to contract, from the spinal cord back to the stretched muscle.

you hold out a bowl and keep it stationary as someone puts some peaches into it. The peaches add weight to the bowl. Your hand starts to drop, which causes the biceps, a muscle in your arm, to stretch. Stretching activates the receptor endings of sensory organs called muscle spindles. The endings, enclosed in a sheath running parallel with the muscle, are input zones of sensory neurons. Inside the spinal cord, axons of the neurons synapse with motor neurons—the axons of which extend back to the stretched muscle (Figure 30.12). Action potentials that arrive at the axon endings trigger the release of ACh, which initiates contraction in the manner described in Chapter 33. As long as the receptor activity continues, motor neurons are excited even more, and the stimulation allows them to hold your hand's position against the force of gravity.

In nearly all reflex pathways, sensory neurons also interact with interneurons, which activate or suppress all motor neurons required for a coordinated response.

Vertebrates have interneurons organized in information-processing blocks. Cablelike nerves that have long axons of sensory neurons, motor neurons, or both connect the brain and spinal cord with the rest of the body.

Reflex arcs, in which sensory neurons synapse directly on motor neurons, are the simplest paths of information flow.

Skewed Information Flow

What happens if something disrupts information flow in nervous systems? Ask *Clostridium botulinum*, an anaerobic, endospore-forming bacterium that normally lives in soil. It can cause the disease *botulism*. Its endospores may be present in improperly preserved, canned, or stored food, and these resting structures can germinate in anyone who eats the food. One metabolic product of the bacterium is toxic. It enters neurons that synapse with muscle cells and blocks the release of acetylcholine (ACh). Without ACh, muscles cannot contract. They become more and more flaccid and paralyzed. Unless a poisoned individual receives antitoxins within ten days, breathing becomes impossible and the heart stops beating.

A related bacterium, *C. tetani*, lives in the gut of horses, cattle, and other grazing animals, even many people. Its endospores survive in soil, especially manure-rich ones. They persist for years if sunlight and oxygen do not reach them. They resist disinfectants, heat, and boiling water. When they enter the body through a deep puncture or a deep cut, they germinate in anaerobic, dead tissues. The bacteria do not spread from dead tissues. They produce a toxin that the blood or nerves deliver to the spinal cord and brain. In the spinal cord, the toxin acts on interneurons that help control motor neurons. It blocks the release of inhibitory neurotransmitters (GABA and glycine) and frees the motor neurons from normal inhibitory control. Symptoms of the disease *tetanus* are about to begin.

After four to ten days, the overstimulated muscles stiffen and go into spasm. They cannot be released from contraction, and prolonged, spastic paralysis follows. Fists and jaws may remain clenched (the disease is also called lockjaw). The back may arch severely and permanently. Once respiratory and cardiac muscles become paralyzed, death nearly always follows.

Vaccines were not available for soldiers of early wars, when dead cavalry horses and manure littered battlefields (Figure 30.13). Today, vaccines have all but eradicated tetanus in the United States.

Figure 30.13 Painting of a young victim of a contaminated battle wound, as he lay dying in a military hospital.

30.6

INVERTEBRATE NERVOUS SYSTEMS

We turn now from the messages sent through nervous systems to the systems themselves. At this point in the book, you know all animals except sponges have some type of nervous system in which nerve cells, such as neurons, are oriented relative to one another in signal-conducting and information-processing highways. At a minimum, the cells making up the communication lines receive information about changing conditions outside and inside the body, then they elicit suitable responses from muscle and gland cells.

Regarding the Nerve Net

To appreciate the diversity of nervous systems, start by recalling that animals first evolved in the seas. It is in the seas that we still find animals having the simplest nervous systems. They are the cnidarians, including sea anemones. These invertebrates have *radial* symmetry. Their body parts are arranged around a central axis, a bit like the spokes of a bike wheel (Section 23.1).

Radial animals rely on a **nerve net**, a loose mesh of nerve cells intimately associated with epithelial tissue (Figure 30.14). The nerve cells interact with sensory cells and contractile cells along reflex pathways in the same epithelial tissue. In reflex pathways, remember, sensory stimulation triggers simple, stereotyped movements.

In all cnidarians, one pathway dealing with feeding behavior extends from sensory receptors in the tentacles, along nerve cells, to contractile cells organized around the mouth (Figure 30.14*a*). In jellyfishes, other reflexes are the basis of movements necessary to swim (slowly) and to keep the soft body right-side up.

The nerve net itself extends through the animal's body, but information flow through it is not focused. It simply commands the body wall to slowly contract and expand or tentacles to move through water. Its owners will never dazzle you with bursts of speed or precision acrobatics. Either they are weak swimmers or they are sedentary types that spend most of their life attached to substrates. Yet in their watery world, the nerve net is adaptive. It is equally responsive to signals about food or danger that might come from any direction.

On the Importance of Having a Head

Flatworms are the simplest animals having a bilateral nervous system (Figure 30.14). *Bilateral* symmetry means having equivalent body parts on the left and right sides of the body's midsagittal plane. (Imagine sliding down a staircase banister that turns into a razor and you may never forget the location of the midsagittal plane.) Both sides have the same array of muscles that function in moving the body forward. Both have the same array of nerves to control the muscles, and so on.

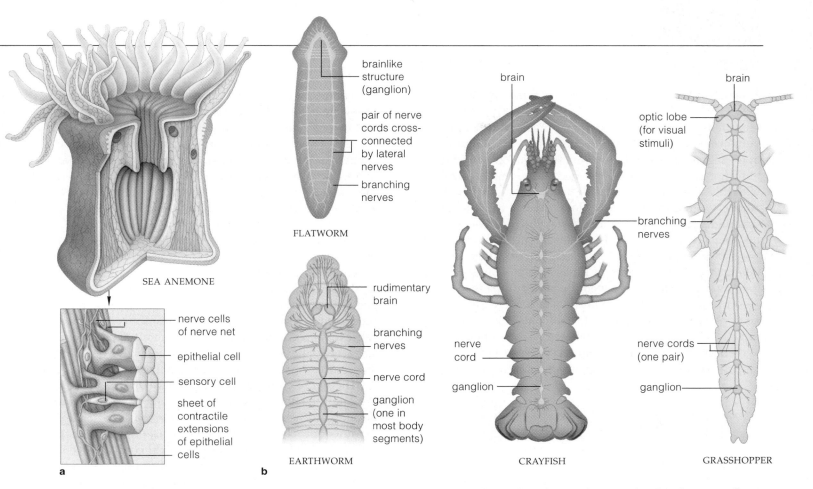

FLATWORM
- brainlike structure (ganglion)
- pair of nerve cords cross-connected by lateral nerves
- branching nerves

SEA ANEMONE
- nerve cells of nerve net
- epithelial cell
- sensory cell
- sheet of contractile extensions of epithelial cells

EARTHWORM
- rudimentary brain
- branching nerves
- nerve cord
- ganglion (one in most body segments)

CRAYFISH
- brain
- branching nerves
- nerve cord
- ganglion

GRASSHOPPER
- brain
- optic lobe (for visual stimuli)
- nerve cords (one pair)
- ganglion

a b

Figure 30.14 (**a**) Nerve net of a sea anemone, a cnidarian. Its nerve cells interact with sensory and contractile cells. Both cell types are embedded in epithelium between the outer epidermis and a jellylike midlayer (mesoglea) of the body wall. (**b**) Bilateral nervous systems of a few invertebrates. The sketches are not to scale relative to one another.

The flatworm's ladderlike nervous system has two cordlike nerves. The two run longitudinally through the body and have many side branches. In their head end, some flatworms have two ganglia. A ganglion (plural, ganglia) is a cluster of nerve cell bodies that is a local integrating center. Those in a flatworm head coordinate signals from paired sensory organs (two eyespots, for example), and provide a bit of control over the nerves.

Did bilateral nervous systems evolve from nerve nets? Maybe. In nearly all animals more complex than flatworms, we find local nerve nets—plexuses—such as the one in your intestinal wall. More intriguing, a self-feeding larval stage called a planula develops in some cnidarian life cycles. Like the flatworms, a planula has a flattened body and uses cilia to swim or crawl about:

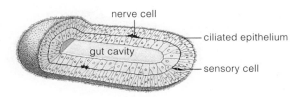

- nerve cell
- ciliated epithelium
- gut cavity
- sensory cell

Visualize a planula crawling on the Cambrian seafloor. A mutation blocks its metamorphosis into the adult yet doesn't stop reproductive organs from maturing. This does happen in some larvae (Sections 18.5 and 24.3). The planula kept on crawling but now had the capacity to reproduce. Its offspring inherited the mutant gene for forward mobility. They crawled longer and entered new parts of their habitat—food-rich, food-poor, and maybe dangerous. The mutation had survival value, because it led to more rapid, more effective responses to diverse stimuli. So a concentration of sensory cells in the body's leading end—not the trailing end—was favored.

Regardless of how it happened, cephalization (the formation of a head) and bilateral symmetry developed in most invertebarate lineages (Chapter 23). Patterns of cephalization and bilateral symmetry also are evident in the paired sensory structures, brain centers, nerves, and skeletal muscles of all vertebrates, including you.

All animals except sponges have a nervous system, in which nerve cells are oriented relative to one another in signal-conducting and information-processing highways.

Radial animals have a nerve net, a diffuse mesh of nerve cells that take part in simple reflex pathways involving sensory cells and contractile cells of an epithelial tissue.

Cephalized, bilateral animals have a nervous system with a brain or ganglia at the head end, as well as paired nerves and paired sensory structures.

VERTEBRATE NERVOUS SYSTEMS—AN OVERVIEW

Evolution of the Spinal Cord and Brain

Hundreds of millions of years ago, the earliest fishlike vertebrates were evolving (Section 24.3). A column of bony segments was taking over the functions of their notochord, a long rod of stiffened tissue that worked with their segmented muscles to bring about movement. Above the notochord, the hollow, tubular nerve cord was undergoing modification. It was the forerunner of the spinal cord and brain. These developments were the basis of different life-styles. Not long afterward, genetic divergences from lineages of filter-feeding, scavenging fishes gave rise to swift, jawed predators of the seas.

At first, simple reflex pathways prevailed. Sensory neurons synapsed directly with motor neurons, which directly signaled muscles to contract. No other neurons altered the flow of signals from reception of a stimulus to the response. In the changing world of fast-moving vertebrates, however, the types of predators and prey that were better equipped to sense the presence of food and danger had the competitive edge.

The senses of smell, hearing, and balance became keener among the vertebrates that invaded land. Bones and muscles evolved in ways that allowed specialized movements. Also, the brain became variably thickened with nervous tissue that could integrate rich sensory information and issue orders for complex responses.

The oldest regions of the vertebrate brain still deal with reflex coordination of respiration and other vital functions. But a great many interneurons now synapse on the sensory and motor neurons of ancient pathways and with one another in the newer brain regions. In the most complex vertebrates, interneurons receive, store, retrieve, and compare information about experiences. They weigh possible responses. And they give our own species the capacity to reason, remember, and learn.

The nerve cord persists in all vertebrate embryos. We call it the **neural tube**. While an embryo grows and develops, its neural tube expands. It becomes regionally modified into the brain and spinal cord (Figure 30.15*a*). A vertebral column encloses the spinal cord. Adjacent tissues give rise to nerves that thread through all body regions and connect with the spinal cord and brain.

The System's Functional Divisions

Figure 30.16 gives you some idea of the expansions of nervous tissue in the human nervous system. It also shows the major paired nerves of this bilateral system. What it cannot show are the 100 billion interneurons in the brain alone. To be sure, humans do have the most intricately wired nervous system in the animal world. Even so, you still find similar patterns of organization and information flow among other vertebrates.

FOREBRAIN. Receives, integrates sensory information from nose, eyes, and ears; in land-dwelling vertebrates, contains the highest integrating centers

MIDBRAIN. Coordinates reflex responses to sight, sounds

HINDBRAIN. Reflex control of respiration, blood circulation, other basic tasks; in complex vertebrates, coordination of sensory input, motor dexterity, and possibly mental dexterity

(start of spinal cord)

a Diagram of how the anterior end of the dorsal, hollow nerve cord of vertebrates expanded into functionally distinct regions, which also increased in complexity in certain lineages

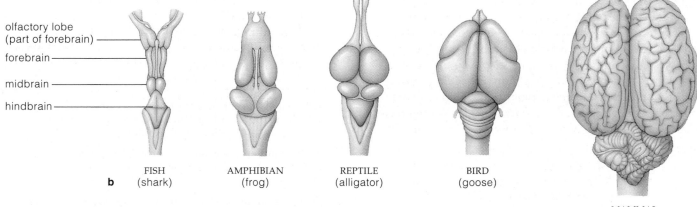

olfactory lobe (part of forebrain)
forebrain
midbrain
hindbrain

b

| FISH (shark) | AMPHIBIAN (frog) | REPTILE (alligator) | BIRD (goose) | MAMMAL (horse) |

Figure 30.15 Evolutionary trend toward an expanded, more complex brain. The trend became apparent after morphological comparisons were made of the brains of some vertebrates. These dorsal views are not to the same scale.

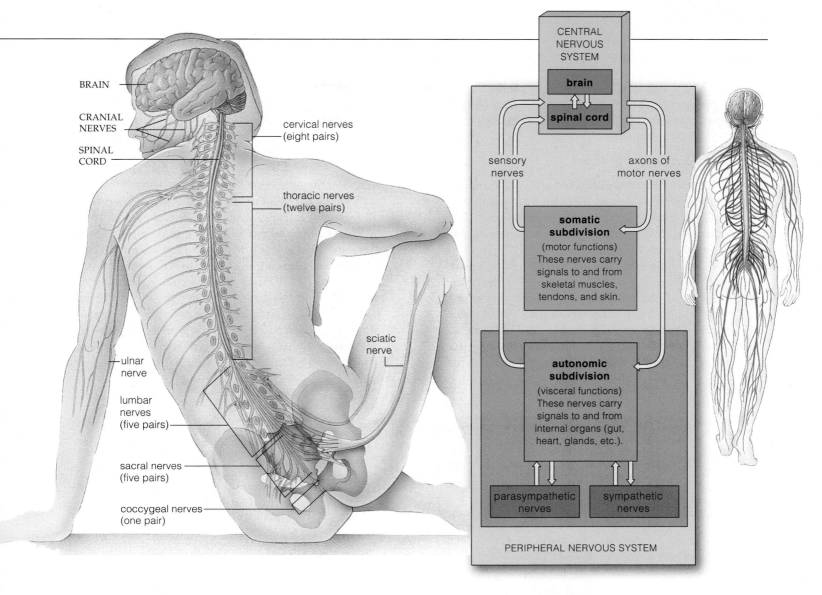

Figure 30.16 Sketch of the brain, spinal cord, and some major peripheral nerves of the human nervous system. The system also incorporates twelve pairs of cranial nerves that originate from the brain. Other vertebrates have a similar nervous system.

Figure 30.17 Functional divisions of the human nervous system. The central nervous system is color-coded *blue*, somatic nerves *green*, and autonomic nerves *red*. Sometimes nerves that carry sensory input to the central nervous system are called *afferent* (a word meaning "to bring to"). Nerves that carry motor output away from the central nervous system to muscles and glands are *efferent* ("to carry outward").

Investigators typically approach the complexity of the vertebrate nervous system by functionally dividing it into central and peripheral regions (Figure 30.17). All of the interneurons are confined to the **central nervous system**, which consists of the spinal cord and brain. The **peripheral nervous system** consists mainly of nerves that extend through the rest of the body. Nerves carry signals into and out of the central nervous system.

Inside the brain and spinal cord, the communication lines are called tracts, not nerves. The tracts making up *white* matter have axons with glistening white myelin sheaths; they function in rapid signal transmission. By contrast, *gray* matter consists of unmyelinated axons, dendrites, and cell bodies of neurons, plus neuroglial cells. Neuroglia, remember, protects or structurally and functionally supports neurons. It makes up more than half the volume of vertebrate nervous systems.

The coevolution of nervous, sensory, and motor systems made more complex life-styles possible among vertebrates.

The vertebrate nervous system has become so intricately wired that the structure and functions of its central and peripheral regions are described separately.

The central nervous system consists of the brain and spinal cord. The peripheral nervous system consists of nerves that thread through the rest of the body and carry signals into and out of the central region.

Myelinated axons make up the white matter of tracts inside the spinal cord and the brain. Neuroglia, unmyelinated axons, dendrites, and cell bodies of neurons make up the gray matter of those tracts.

THE MAJOR EXPRESSWAYS

Let's now take a look at the peripheral nervous system and the spinal cord. The two interconnect as the main expressways for information flow through the body.

Peripheral Nervous System

SOMATIC AND AUTONOMIC SUBDIVISIONS In humans, the peripheral nervous system includes thirty-one pairs of *spinal* nerves, which connect with the spinal cord. The system also includes twelve pairs of *cranial* nerves that connect directly with the brain.

Cranial and spinal nerves are further classified by function. Those carrying signals about moving the head, trunk, and limbs are **somatic nerves**. Their sensory axons deliver information from receptors in the skin, skeletal muscles, and tendons into the central nervous system. Their motor axons deliver the commands issued by the brain and spinal cord to the body's skeletal muscles.

By contrast, spinal and cranial nerves dealing with smooth muscle, cardiac (heart) muscle, and glands are **autonomic nerves**. They deal with the viscera—internal organs and structures. Autonomic nerves carry signals to and from these organs and structures.

PARASYMPATHETIC AND SYMPATHETIC NERVES Figure 30.18 shows the two categories of autonomic nerves. We call them parasympathetic and sympathetic. Normally they work antagonistically, with signals from one kind opposing signals from the other. However, both carry excitatory and inhibitory signals to the internal organs. Often their signals arrive at the same time at muscle or gland cells and compete for control over them. In such cases, synaptic integration at the cellular level leads to minor adjustments in the organ's level of activity.

Parasympathetic nerves dominate when the body is not receiving much outside stimulation. They tend to

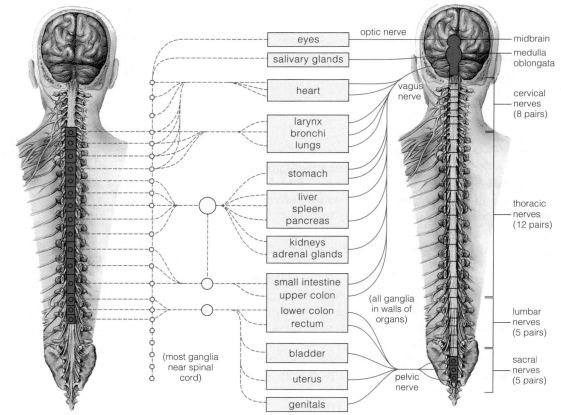

SYMPATHETIC OUTFLOW FROM THE SPINAL CORD

Examples of Responses
Heart rate increases
Pupils of eyes dilate (widen, let in more light)
Glandular secretions in airways to lungs decrease
Salivary gland secretions thicken
Stomach and intestinal movements slow down
Sphincters (rings of muscle) contract

PARASYMPATHETIC OUTFLOW FROM THE SPINAL CORD AND BRAIN

Examples of Responses
Heart rate decreases
Pupils of eyes constrict (keep more light out)
Glandular secretions in airways to lungs increase
Salivary gland secretions become dilute
Stomach and intestinal movements increase
Sphincters (rings of muscle) relax

Figure 30.18 Autonomic nervous system. This diagram shows the major sympathetic nerves and parasympathetic nerves leading from the central nervous system to some major organs. Remember, there are *pairs* of both kinds of nerves, servicing the right and left halves of the body. The ganglia are clusters of the cell bodies of neurons that have their axons bundled together in nerves.

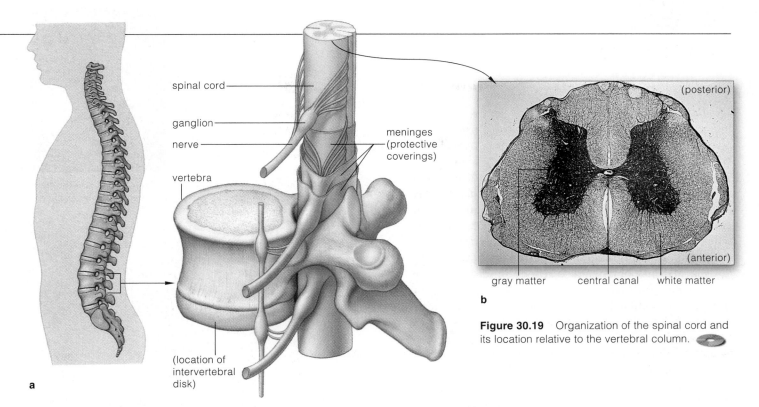

spinal cord

ganglion

nerve

vertebra

meninges (protective coverings)

(location of intervertebral disk)

a

(posterior)

(anterior)

gray matter   central canal   white matter

b

Figure 30.19 Organization of the spinal cord and its location relative to the vertebral column.

slow down the body overall and divert energy to basic "housekeeping" tasks, such as digestion.

Sympathetic nerves dominate in times of sharpened awareness, excitement, or danger. They tend to shelve housekeeping tasks. At the same time, they prepare the animal to fight or escape when threatened or to frolic, as in play or sexual behavior. Right now, sympathetic nerves are commanding your heart to beat a bit faster, and parasympathetic nerves are commanding it to beat a bit slower. Integration of the opposing signals adjusts the heart rate. If something scares or excites you, the parasympathetic input to the heart drops. Sympathetic signals trigger the release of epinephrine, which makes your heart beat faster. They also incite you to breathe faster and start to sweat. In this state of intense arousal, you are primed either to fight (or play) hard, or to get away fast. Hence the name *fight–flight response*.

Suppose that the stimulus for a fight–flight response ends. Sympathetic activity may decrease abruptly, and parasympathetic activity may rise suddenly. You might observe this "rebound effect" after someone has been instantly mobilized to rush onto a road to save a child from an oncoming car. The person might well faint as soon as the child has been swept out of danger.

The Spinal Cord

By definition, the **spinal cord** is a vital expressway for signals between the peripheral nervous system and the brain. Also, sensory and motor neurons make direct reflex connections in the spinal cord, as they do for the stretch reflex. The spinal cord threads through a canal made of bones of the vertebral column (Figure 30.19a).

The bones, and ligaments attached to them, protect the cord. So do meninges, three tubelike coverings around the spinal cord and brain. The coverings are tough but vulnerable. *Meningitis*, an often-fatal disease, is brought on by certain viral and bacterial infections. Symptoms include severe headaches, fever, a stiff neck, and nausea.

Signals swiftly travel up and down the spinal cord in bundles of myelinated axons. These axons, which make up the cord's white matter, surround the gray matter (Figure 30.19b). Gray matter, again, consists of dendrites and cell bodies of neurons, and neuroglial cells. It plays a key role in controlling reflexes for limb movements, as when you dance or wave your arms about, and for organ activity, such as bladder emptying.

Experiments tell us about such pathways. Examples: Between a frog's spinal cord and brain are circuits that command bent legs to straighten. Sever the circuits at the base of the brain, and the legs become paralyzed— but only for about a minute. Extensor reflex pathways in the spinal cord recover quickly and have the frog hopping about in no time. The recovery time is minimal after similar damage in humans and other primates— the vertebrates with the greatest cephalization.

***

Nerves of the peripheral nervous system connect the brain and spinal cord with the rest of the body.

The somatic nerves of the peripheral nervous system deal with skeletal muscle movements. Its autonomic nerves deal with smooth muscle, cardiac muscle, and glands.

The spinal cord is a vital expressway for signals between the brain and peripheral nerves. Some of its interneurons also exert direct control over certain reflex pathways.

THE VERTEBRATE BRAIN

Evolutionary High Points

The **brain**—the body's master control center—receives, integrates, stores, and retrieves sensory information. It also issues commands to adjust activities in response to stimuli. Skull bones, fluid, and membranes protect it. The brain has three functional divisions, and the **brain stem**—the most ancient nervous tissue with basic reflex centers—persists in all of them. The gray matter of that ancient tissue expanded and became far more complex when the senses of smell, hearing, and vision started to become more important among early vertebrates.

The medulla oblongata, cerebellum, and pons are parts of the **hindbrain** (Figure 30.20). Reflex centers for respiration, circulation, and other basic tasks are in the medulla oblongata, which coordinates motor responses with complex reflexes, such as coughing, and influences sleeping and arousal. The cerebellum deals with motor skills; it integrates sensory input from the eyes, ears, and muscle spindles with motor signals from the forebrain. Its expansion in humans might relate to language and some other forms of mental dexterity. Many bands of axons extend from both sides of the cerebellum to the pons (meaning bridge). The pons is a traffic center for information flow between the cerebellum and forebrain.

The **midbrain** coordinates reflex responses to sights and sounds. In fishes and amphibians, its roof of gray matter—the tectum—is a center for coordinating nearly all sensory input and initiating motor responses. (If you surgically remove a frog's highest brain center, its activities won't change much if its tectum is intact.) In most vertebrates (not mammals), the midbrain has a pair of optic lobes: centers for input from the eyes. In mammals, the eyes formed functional connections with the forebrain, and the tectum became a reflex center that only relays sensory signals to the forebrain.

What about the **forebrain**? For much of vertebrate history, chemical odors in aquatic habitats were the key clues to survival. An olfactory lobe dealing with odors from prey, predators, and mates was a major forebrain structure. So were paired outgrowths of the brain stem where olfactory input and responses to it could become integrated. Especially in lineages that invaded land, the outgrowths expanded and became the cerebrum. A thin layer of gray matter evolved into the **cerebral cortex**.

Another forebrain region, the thalamus, evolved as a coordinating center for sensory input and as a relay station to the cerebrum. Below this, the hypothalamus evolved into the premier center for homeostatic control of the internal environment. The hypothalamus became central to behaviors related to internal organs, such as thirst, hunger, and sex, and to emotional expression.

The brain isn't tidily subdivided into three regions. An ancient mesh of interneurons extends from the upper

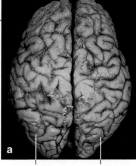

Figure 30.20 (**a**) Human brain. This dorsal view shows how a longitudinal fissure separates the two cerebral hemispheres. (**b**) Sagittal view of the right cerebral hemisphere. Not visible is the reticular formation; it extends between the upper spinal cord and the cerebrum.

left cerebral hemisphere   right cerebral hemisphere

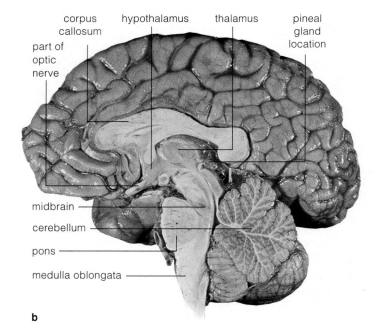

corpus callosum   hypothalamus   thalamus   pineal gland location

part of optic nerve

midbrain

cerebellum

pons

medulla oblongata

b

spinal cord through the brain stem, to the cerebrum's higher integrative centers. This network is the reticular formation. It persists as a low-level pathway to motor centers of the medulla oblongata and spinal cord. It also influences activities of the nervous system as a whole.

Functional Divisions of the Human Brain

An average-sized person's brain weighs 1,300 grams (3 pounds). Over half of it is neuroglia. Your "humanness" arises in the 100 billion interneurons of the cerebrum. A fissure divides the cerebrum into two hemispheres, each having frontal, occipital, temporal, and parietal lobes (Figure 30.21). The left hemisphere deals mostly with analytical skills, speech, and mathematics. It typically dominates the right hemisphere, which deals more with visual–spatial relationships and music. Each responds to sensory input mainly from the opposite side of the body. For example, pressure signals from the right arm travel to the left hemisphere. A band of nerve tracts (corpus callosum) coordinates activities of the two hemispheres.

What we comprehend, communicate, remember, and consciously act upon arises in the cerebral cortex. Here

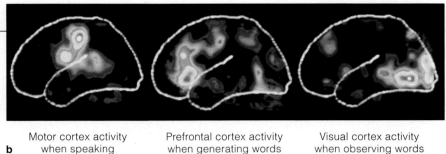

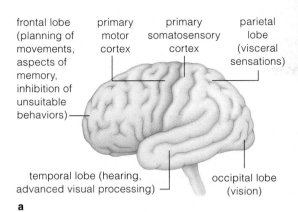

frontal lobe (planning of movements, aspects of memory, inhibition of unsuitable behaviors)

primary motor cortex

primary somatosensory cortex

parietal lobe (visceral sensations)

temporal lobe (hearing, advanced visual processing)

occipital lobe (vision)

a

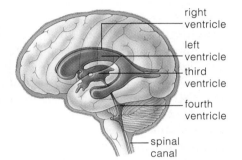

b Motor cortex activity when speaking

Prefrontal cortex activity when generating words

Visual cortex activity when observing words

Figure 30.21 (**a**) Primary receiving and integrating centers of the human cerebral cortex. Primary cortical areas receive signals from receptors on the body's periphery. Association areas coordinate and process sensory input from different receptors. (**b**) Three PET scans identifying which brain regions were active when an individual performed three tasks: speaking, generating words, and observing words.

we find *motor* areas for the control of voluntary motor activities, *sensory* areas for perceiving what sensations mean, and *association* areas for integrating information that precedes conscious action. For example, the frontal lobe's premotor cortex is a domain of learned patterns of motor skills. Dribble a basketball, use a keyboard— all repetitive motions arise through the coordination of simultaneous and sequential activities of many muscle groups. The parietal lobe's somatosensory cortex, a map of the body, handles input from the skin and joints. The prefrontal cortex—the most complex of the association areas—is the basis of complex learning, intellect, and personality. Without it, we'd be incapable of abstract thought, judgment, planning, and concern for others.

Encircling the upper brain stem is a **limbic system**. It includes the hypothalamus—the clearinghouse for emotions and visceral activity. Its other brain structures mediate emotional stability, interpretation of social cues, the will to act, and conversion of stimuli to long-term memory. This system is distantly related to olfactory lobes and still deals with the sense of smell. (That's one reason why you may feel pleasure when you recall the cologne of a special person.) The limbic system's ties with other brain centers help correlate organ activities with self-gratifying behavior, including eating and sex. For example, it can make you feel your stomach or heart is "on fire" with indigestion or passion. Signals based on reasoning in the cerebral cortex often override or dampen rage, hatred, and other "gut reactions." Hence we call the limbic system our emotional–visceral brain.

Protection at the Blood–Brain Barrier

The hollow neural tube that forms in embryos persists in adults, as a continuous system of fluid-filled cavities and canals. Inside the system is cerebrospinal fluid, a clear extracellular fluid that cushions the brain and the spinal cord against jarring movements (Figure 30.22).

A mechanism called the **blood–brain barrier** protects the brain and spinal cord by exerting some control over which solutes enter the cerebrospinal fluid. No other

Figure 30.22 Cerebrospinal fluid (*blue*). This is an extracellular fluid that fills the four interconnected cavities in the brain and in the spinal cord's central canal.

right ventricle

left ventricle

third ventricle

fourth ventricle

spinal canal

portion of extracellular fluid has solute concentrations kept within such narrow limits. Even normal changes in fluid composition that accompany, say, eating and exercising are opposed here. Why? Some blood-borne hormones and amino acids can change the functioning of neurons. Also, shifting levels of certain ions (such as K^+) can skew the threshold for action potentials.

The barrier works at the plasma membrane of cells making up the wall of blood capillaries that service the brain. In most brain regions, tight junctions fuse all the abutting cell walls—so water-soluble substances must move *through* these cells to reach the brain. Membrane transport proteins let glucose and other vital nutrients as well as some ions move into and out of cells. They bar some toxins and metabolic wastes, such as urea. The barrier does not keep out fat-soluble substances, such as oxygen, carbon dioxide, anesthetics, alcohol, caffeine, and nicotine. It is nonexistent around the brain stem's vomiting center and the hypothalamus (guess why).

The vertebrate brain develops from a hollow neural tube, which persists in adults as a system of cavities and canals filled with cerebrospinal fluid. The fluid cushions nervous tissue from sudden, jarring movements. The nervous tissue is subdivided into the hindbrain, forebrain, and midbrain.

The brain stem is the most ancient nervous tissue in all three regions. It affords reflex control over basic functions necessary for survival. The highest integrative centers are in the forebrain, especially in the cerebral cortex.

MEMORY

Memory refers to the brain's ability to store and recall information about past sensory experience. Without it, learning and adaptive modifications of behavior would be impossible. Information is stored in stages. In *short-term* storage, neural excitation lasts a few seconds to a few hours. This stage is limited to a few bits of sensory information—numbers, words of a sentence, and so on. In *long-term* storage, seemingly unlimited amounts of information get tucked away more or less permanently.

Only part of the sensory input reaching the cerebral cortex is selected for the short-term memory bins where information is processed for relevance. If irrelevant, it's forgotten. Otherwise it is consolidated in long-term bins. Emotional states influence the outcome. So does having time to repeat or rehearse the input.

Our brain processes facts and skills separately. *Facts*, soon forgotten or stored in long-term bins, include faces, names, dates, words, smells, and other bits of explicit information, plus the circumstance in which they were learned. That is why you might associate, say, the smell of sun-warmed watermelon with a long-ago great picnic at the beach. By contrast, *skills* are gained by practicing specific motor activities. A skill such as slam-dunking a basketball or playing a violin concerto is best recalled by performing it, not by rehashing the circumstances under which the skill was initially learned.

Memory starts at the sensory cortex, then flows in two circuits. Facts flow to limbic system structures that mediate emotions, learning, and spatial relations, then to memory banks in the prefrontal cortex. A feedback loop between this system and the cortex reinforces the input until it gets consolidated in long-term storage.

All motor skills require muscle conditioning. As you might suspect, the skill circuit includes the cerebellum, which coordinates motor activity. It includes structures deep in the brain that promote motor responses.

Amnesia is a loss of memory. Its severity depends on whether the hippocampus, amygdala, or both have been damaged, as by a severe head blow. But the loss doesn't affect the ability to learn new skills. With *Parkinson's disease*, basal ganglia are destroyed and learning ability is lost, yet skill memory is retained. *Alzheimer's disease* usually starts late in life and involves structural changes in the cerebral cortex and hippocampus. Often, affected people can recall long-known facts, such as their Social Security number. But they have trouble remembering what has just happened to them. In time they become confused, depressed, and incoherent.

Memory, the storage and retrieval of sensory information, involves neural excitation along the circuits between the cerebral cortex and parts of the limbic system. Sensory input is processed through short-term and long-term storage.

Drugging the Brain

Broadly speaking, a drug is a substance introduced into the body to provoke a specific physiological response. Some drugs help a person cope with illness or emotional stress. Others artificially fan pleasure associated with sex and other self-gratifying behaviors.

Many drugs are habit-forming. That is, even if the body can function well without them, a person continues to use such drugs for the real or imagined relief they afford. Often the body develops tolerance of such drugs; it takes larger or more frequent doses to produce the same effect. Habituation and tolerance are signs of **drug addiction**, a chemical dependence on a drug (Table 30.1). *In cases of addiction, a drug has assumed an "essential" biochemical role in the body.* Abruptly deprive addicts of the drug, and they go through physical pain and mental anguish. The entire body goes through an episode of biochemical upheaval.

STIMULANTS   Stimulants make you more alert, then they depress you. *Caffeine* in coffee, tea, chocolate, and many soft drinks is a stimulant. Low doses act at the cerebral cortex to increase alertness. Higher doses act at the medulla oblongata to make you a bit clumsy and mentally incoherent. Another stimulant, the *nicotine* in tobacco, mimics ACh. By directly stimulating a variety of sensory receptors, its effects on the nervous system are widespread. Nicotine addiction also imposes staggering health, social, and financial burdens (Section 36.7).

An estimated 1.5 to 3.7 million Americans are *cocaine* abusers. This stimulant gives a rush of pleasure by blocking reabsorption of norepinephrine, dopamine, and other neurotransmitters (Figure 30.23). These molecules collect in synaptic clefts and incessantly stimulate postsynaptic cells. Heart rate, blood pressure, and sexual appetite rise. In time the molecules diffuse away. But neurons can't synthesize replacements fast enough to counter the loss. The sense of pleasure evaporates as the hypersensitized postsynaptic cells demand stimulation. After extended, heavy cocaine use, "pleasure" is impossible.

Table 30.1 *Warning Signs of Drug Addiction**
1. Tolerance—it takes increasing amounts of the drug to produce the same effect.
2. Habituation—it takes continued drug use over time to maintain the self-perception of functioning normally.
3. Inability to stop or curtail use of the drug, even if there is persistent desire to do so.
4. Concealment—not wanting others to know of the drug use.
5. Extreme or dangerous behavior to get and use a drug, as by stealing, asking more than one doctor for prescriptions, or jeopardizing employment by drug use at work.
6. Deterioration of professional and personal relationships.
7. Anger and defensive behavior when someone suggests there may be a problem.
8. Preference of drug use over previously customary activities.

* Three or more of these signs may be cause for concern.

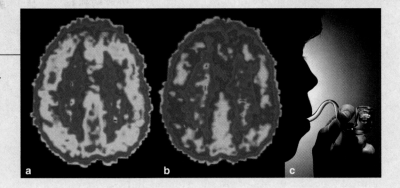

Figure 30.23 (**a**) Normal brain activity revealed by a PET scan. (**b**) PET scan of a comparable section that reveals cocaine's effect. *Red* indicates greatest activity; and *yellow*, *green*, and *blue* indicate increasingly inhibited activity. (**c**) In less than eight seconds, smoking crack puts cocaine in the brain.

Granular cocaine is inhaled (snorted). Abusers burn crack cocaine and inhale the smoke. As suggested at the start of this chapter, crack is extremely addictive. Its highs are higher and its crashes more devastating. The social and economic tolls are extreme. In 1999, about 170,000 people ended up in emergency rooms for cocaine-related disorders. There is no antidote for overdoses, which induce seizures, respiratory failure, stroke, or heart failure.

Amphetamines induce the massive release of dopamine and norepinephrine. Addicts often smoke, snort, inject, or gulp a form called *crank*. Its effects range from euphoria and sexual arousal to a pounding heart, agitation, dry mouth, tremors, and paranoia. Crank kills the appetite (people have become addicts after using it to lose weight). In time, dopamine and norepinephrine synthesis dwindles as the brain depends more on artificial stimulation. The long-term outcomes? Malnutrition, psychosis, depression, memory loss, and damage to the brain, heart, lungs, and liver. Crank is made cheaply in home kitchens (dubbed Beavis and Butthead labs) from such caustic ingredients as drain cleaners. Abuse of this "poor man's cocaine" is epidemic, especially in the American West and Midwest.

Another synthetic amphetamine is MDMA (or *Ecstasy*, Adam, or XTC). Users say it enhances sex, trust, and tranquility. They don't say as much about the confusion, depression, severe anxiety, blurred vision, nausea, and hypertension. They may not care that its long-term use permanently damages the brain and liver. By the way, destroy dopamine-secreting neurons and you get the tremors, lack of coordination, and often paralysis of the sort seen in patients with Parkinson's disease.

Ecstasy is one of many *club drugs* that young adults use at raves, dance clubs, and bars. All of these drugs are dangerous; especially when mixed with alcohol, they can kill you. As if that weren't bad enough, sexual predators are sneaking colorless, tasteless, and odorless club drugs into the drinks of individuals they wish to assault.

DEPRESSANTS, HYPNOTICS These addictive drugs act on nerves; some act at the reticular formation and thalamus. Responses vary. You might sink into emotional relief, drowsiness, sleep, coma, or death, depending on the dose and your physiological and emotional states. Low doses have the most effect on inhibitory synapses. Initially, users feel excited or euphoric. Increased doses suppress excitatory synapses and lead to depression. Depressants and hypnotics amplify each other, as when alcohol plus barbiturates heightens depression.

Alcohol, or ethyl alcohol, acts directly on the plasma membrane to alter cell function. Like nicotine and cocaine, it's lipid soluble and swiftly crosses the blood–brain barrier. Some people mistakenly think it is a harmless stimulant because of the initial "high" it produces. But alcohol, one of the most potent psychoactive drugs, is a factor in many deaths. Even low intake leads to diminished judgment, disorientation, and uncoordinated movements. Long-term addiction can lead to *cirrhosis*. Connective tissue replaces damaged liver cells. The liver's self-regenerating capacity declines; chronic liver failure and death follow.

ANALGESICS When severe stress leads to physical or emotional pain, the brain produces natural *analgesics*, or pain relievers. Two kinds, endorphins and enkephalins, influence many parts of the nervous system. *Codeine*, *heroin*, and other narcotic analgesics sedate the body and relieve pain. They are among the most addictive drugs known. Deprivation of heroin following massive doses is followed by hyperactivity, anxiety, fever, chills, severe vomiting, cramping, and diarrhea.

PSYCHEDELICS, HALLUCINOGENS These drugs skew sensory perception by interfering with the action of acetylcholine, norepinephrine, or serotonin. *LSD* (lysergic acid diethylamide) alters serotonin's roles in inducing sleep, controlling the body's core temperature, and mediating sensory perception. Even in small doses, LSD warps perceptions, as when some users "perceived" they could fly and "flew" off buildings.

Marijuana, another hallucinogen, is made from crushed leaves, flowers, and stems of the plant *Cannabis*. In low doses it is like a depressant. It slows down but does not impair motor activity; it relaxes the body and elicits mild euphoria. It also causes disorientation, anxiety bordering on panic, delusions, and hallucinations. Like alcohol, it interferes with the performance of complex tasks, such as driving a car. In one study, pilots showed a marked deterioration in instrument-flying ability for more than two hours after smoking marijuana. In time, marijuana impairs the immune system and mental functions.

Each of us possesses a body of great complexity. Its architecture, its functioning are legacies of millions of years of evolution. Its nervous system is unparalleled in the living world. One of its most astonishing products is language, the encoding of shared experiences of groups of individuals in time and space. Through the evolution of our nervous system, the sense of history was born, and the sense of destiny. Through this system we can ask how we have come to be what we are and where we are headed from here. Perhaps the sorriest consequence of drug abuse is its implicit denial of this legacy—the denial of self when we cease to ask, and cease to care.

SUMMARY   *Gold* indicates text section

1. Nervous systems sense and interpret specific aspects of the environment and issue commands for responses to them. Most communication lines consist of sensory neurons, interneurons, and motor neurons. *CI, 30.2*

   a. Sensory neurons are receptors that detect specific stimuli (specific forms of energy, such as light).

   b. Interneurons are integrators in the brain and spinal cord. They receive and interpret signals from sensory receptors, then issue commands for suitable responses.

   c. Motor neurons carry commands away from the brain and spinal cord to the body's effectors. Effectors are muscle and gland cells that carry out responses.

Table 30.2 *Summary of the Central Nervous System**

FOREBRAIN	Cerebrum	Localizes and processes sensory inputs; initiates and controls skeletal muscle activity. Governs memory, emotions, and abstract thought in the most complex vertebrates
	Olfactory lobe	Relays sensory input from the nose to olfactory centers of cerebrum
	Thalamus	Has relay stations for conducting sensory signals to and from cerebral cortex; has role in memory
	Hypothalamus	With the pituitary gland, a homeostatic control center that adjusts the volume, composition, and temperature of internal environment. Governs behaviors affecting organ functions (e.g., thirst and hunger) and expression of emotion
	Limbic system	A complex of brain structures that governs emotions; has roles in memory
	Pituitary gland (Chapter 32)	With hypothalamus, provides endocrine control of metabolism, growth, and development
	Pineal gland (Chapter 32)	Helps control some circadian rhythms; also has role in mammalian reproductive physiology
MIDBRAIN	Tectum	In fishes and amphibians, its centers coordinate sensory input (as from optic lobes) and motor responses. In mammals, its centers (mainly reflex centers) swiftly relay sensory input to forebrain
HINDBRAIN	Pons	Is a "bridge" of tracts between the cerebrum and cerebellum; its other tracts connect spinal cord with forebrain. Works with the medulla oblongata to control the rate and depth of respiration
	Cerebellum	Coordinates motor activity for moving limbs and maintaining posture, and for spatial orientation
	Medulla oblongata	Its tracts relay signals between the spinal cord and pons; its reflex centers help control heart rate, adjustments in blood vessel diameter, respiratory rate, vomiting, coughing, other vital functions
SPINAL CORD		Makes reflex connections for limb movements. Many of its tracts carry signals between the brain and the peripheral nervous system

* The reticular formation extends from the spinal cord to the cerebral cortex.

2. A neuron's dendrites and cell body are input zones. If arriving signals spread to a trigger zone (e.g., the start of an axon), they may trigger an action potential that can be propagated to an output zone (axon endings). *30.1*

   a. Transport proteins pepper the plasma membrane from the trigger zone to axon endings of a neuron. They serve as gated or open channels for the passage of ions.

   b. The controlled flow of ions across the membrane is the basis of an action potential: an abrupt, short-lived reversal in the voltage difference across the membrane in response to adequate stimulation. *30.1, 30.2*

   c. Sodium–potassium pumps counter small ion leaks and help maintain ion gradients across the membranes. They also restore the gradients after an action potential.

3. At output zones, action potentials trigger the release of neurotransmitters, which diffuse across the chemical synapse between a neuron and another neuron, a muscle cell, or a gland cell. They may excite the postsynaptic cell membrane (drive it closer to threshold) or inhibit it (drive it away from threshold). *30.3*

4. Integration is the moment-by-moment summation of excitatory and inhibitory signals reaching synapses on a neuron. It is a way to play down, suppress, reinforce, or send on information through a nervous system. *30.3*

5. The simplest nervous systems are nerve nets, such as those of cnidarians and other radial animals. Their meshwork of nerve cells forms reflex connections with contractile and sensory cells of the epithelium. Most animals have a bilateral, cephalized nervous system with a brain or ganglia at their anterior end. These animals have cordlike nerves: axons of sensory neurons, motor neurons, or both bundled inside a sheath. *30.6*

6. The vertebrate central nervous system is composed of a spinal cord and brain (Table 30.2). The peripheral nervous system is mostly nerves that connect all body regions with the spinal cord and brain. *30.7–30.9*

   a. Skeletal muscles are serviced by somatic nerves. Soft internal organs (viscera) are serviced by autonomic nerves (parasympathetic and sympathetic) that work in opposition continuously to adjust organ activities.

   b. Parasympathetic nerve outflow causes energy to be diverted to basic housekeeping tasks if stimulation is low. Sympathetic nerve outflow stimulates activities during times of heightened awareness or danger.

Review Questions

1. Describe sensory neuron, interneuron, and motor neuron in terms of their structure and functions. Label this motor neuron's functional zones: *CI, 30.1, 30.4*

2. Define resting membrane potential, graded potential, and action potential. *30.1, 30.2*

3. A neuron at rest is controlling the ion distribution across its plasma membrane. Identify two kinds of ions. Do they leak across the membrane, are they pumped across, or both? *30.1*

4. With respect to action potentials, explain threshold level, all-or-nothing spikes, and self-propagation. *30.2*

5. Define chemical synapse and neurotransmitter. Choose an example of a neurotransmitter and state where it acts. *30.3*

6. Define synaptic integration. *30.3*

7. What is a myelin sheath? Do all neurons have one? *30.4*

8. Define reflex, then give an example of a reflex arc. *30.4*

9. Contrast the nervous system of a radial animal with that of a bilateral animal. *30.6*

10. Distinguish between the following:
    a. brain and brain stem *30.9*
    b. central and peripheral nervous system *30.7*
    c. cranial and spinal nerves *30.8*
    d. somatic and autonomic nerves *30.8*
    e. parasympathetic and sympathetic nerves *30.8*

11. Explain the limbic system in terms of some of its component parts and their functions. *30.9*

12. Define cerebrospinal fluid and the blood–brain barrier. *30.9*

Self-Quiz   ANSWERS IN APPENDIX III

1. Action potentials occur when _____ .
   a. a neuron receives adequate stimulation
   b. sodium gates open in an ever accelerating way
   c. sodium–potassium pumps kick into action
   d. both a and b

2. The resting membrane potential is maintained by _____ .
   a. ion leaks   c. neurotransmitters
   b. ion pumps   d. both a and b

3. Neurotransmitters diffuse across a _____ .
   a. chemical synapse   c. myelin sheath
   b. membrane pump   d. both a and b

4. A nerve may consist of bundled-together axons of _____ .
   a. sensory neurons   c. sensory and motor neurons
   b. motor neurons   d. all of the above

5. Is this statement true or false: White matter and gray matter are components of the spinal cord alone.

6. Match the terms with their most suitable description.
   ____ synaptic   a. at input zone of excitable cell
        integration   b. summation of all signals arriving
   ____ muscle spindle   at any neuron at the same time
   ____ graded, local   c. at trigger zone
        potential   d. stretch-sensitive receptor
   ____ action potential

7. Match the component with its main functions.
   ____ spinal cord   a. motor, sensory, association areas
   ____ medulla   for most complex integration
        oblongata   b. homeostatic control of internal
   ____ hypothalamus   environment, internal organs
   ____ limbic system   c. our "emotional brain"
   ____ cerebral cortex   d. reflex control of respiration, blood
                              circulation, other basic activities
                           e. expressway between brain and
                              peripheral nervous system; also,
                              direct reflex connections

Critical Thinking

1. With *multiple sclerosis*, recall, myelin sheaths around axons in the spinal cord slowly degenerate. Affected individuals progressively lose the ability to control movements of their skeletal muscles. Reflect on the function of myelin sheaths in the nervous system, then formulate a hypothesis about why their degeneration leads to a loss of muscle control.

2. A mud-covered nail punctured Evita's foot. To protect Evita from tetanus, her doctor administered an antitoxin. This one had enough protein molecules that could specifically bind with and neutralize the *C. tetanus* toxin molecules. Will Evita now have long-lasting protection against future attacks by this bacterium?

3. In human newborns and premature babies, the blood–brain barrier is not fully developed. Explain why this might be reason enough to pay careful attention to their diet.

4. When Jennifer was six years old, a man lost control of his car and hit a tree in front of her house. She ran over to him and screamed when she saw blood from a head wound dripping on a bouquet of red roses. Thirty-five years later, someone gave her a bottle of *Tea Rose* perfume. When she sniffed the perfume, she became frightened and extremely anxious. A few minutes later she also had a vivid recollection of the accident. Explain this incident in terms of what you learned about memory.

5. Eric typically drinks one cup of coffee nearly every hour, all day long. By midafternoon he has trouble concentrating on his studies, and he feels tired and more than a little clumsy. Drinking another cup of coffee doesn't make him more alert. Explain how caffeine in coffee might produce such symptoms.

6. *Epilepsy* is a neurological disorder characterized by brief, recurring episodes of sensory and motor malfunctioning. The episodes, or seizures, may arise when reverberating circuits in the brain are wrongly activated. Muscles contract involuntarily and lights, sounds, and odors are often sensed even when receptors in the eyes, ears, and nose aren't being stimulated.

   The drug valproic acid can eliminate or lessen the severity of the seizures by stimulating the body's synthesis of GABA. Why would stepping up GABA production help?

Selected Key Terms

acetylcholine (ACh) *30.3*	motor neuron *CI*
action potential *30.1*	nerve *30.4*
autonomic nerve *30.8*	nerve net *30.6*
axon *30.1*	nervous system *CI*
blood–brain barrier *30.9*	neural tube *30.7*
brain *30.9*	neurotransmitter *30.3*
brain stem *30.9*	parasympathetic nerve *30.8*
central nervous system *30.7*	peripheral nervous system *30.7*
cerebral cortex *30.9*	positive feedback *30.2*
chemical synapse *30.3*	reflex *30.4*
dendrite *30.1*	resting membrane potential *30.1*
drug addiction *30.11*	sensory neuron *CI*
forebrain *30.9*	sodium–potassium pump *30.1*
hindbrain *30.9*	somatic nerve *30.8*
interneuron *CI*	spinal cord *30.8*
limbic system *30.9*	stimulus *CI*
memory *30.10*	sympathetic nerve *30.8*
midbrain *30.9*	synaptic integration *30.3*

Readings

Romer, A., and T. Parsons. 1986. *The Vertebrate Body.* Sixth edition. Philadelphia: Saunders. It's been around a while but has fine insights into the evolution of vertebrate nervous systems.

31

SENSORY RECEPTION

Different Strokes for Different Folks

You might be reluctant to pet a python or scratch a bat behind its ears. But you have to give them credit for being vertebrates with special traits (Figure 31.1). In rows of pits above and below a python's mouth are thermoreceptors that detect infrared energy—in this case, the body heat of small, night-foraging mammals, the prey of choice. When stimulated, the receptors send messages to the brain, which processes them and issues commands to muscle cells. In no time at all, a strike is aimed and executed with stunning accuracy.

A motionless, edible frog wouldn't stimulate the receptors, so the snake would slither past it. Frog skin isn't warm, and it blends with colors of a frog's habitat. A python doesn't have receptors that can detect frogs or a neural program for responding to them.

Or consider the bats. Nearly all of these mammalian species sleep during the day and spread their webbed wings at dusk. Different kinds take to the air in search of nectar, fruit, frogs, or insects. Many sensory receptors in their eyes, nose, ears, mouth, and skin are not all that different from yours. But other receptors provide bats with a sense of hearing that you cannot even begin to match. Even tiny-eyed, nearly blind species navigate and capture flying insects with precision in the dark!

Bats are masters of **echolocation**. They emit calls, as the bat in Figure 31.1*b* is doing. When sound waves of the calls bounce off insects, trees, and other objects,

acoustical receptors inside the bat's ears detect the echoes and send signals about them to the bat brain.

As an echolocating bat flies, it emits a steady stream of about ten clicking sounds per second. You can't hear the clicks. They are intense "ultrasounds," above the range of sound waves that receptors in human ears are able to detect. When a bat hears a pattern of distant echoes from, say, an airborne mosquito or moth, it increases the rate of ultrasonic clicks to as many as 200 per second, which is faster than a machine gun fires bullets. In the few milliseconds of silence between the clicks, sensory receptors detect the echoes and deliver messages about them to the bat brain. The brain rapidly constructs a "map" of the sounds, which the bat follows during its maneuvers through the night world.

With this chapter, we turn to **sensory systems**, the means by which animals receive signals from inside and outside the body, then decode them in ways that give rise to awareness of sounds, sights, odors, and other sensations. Information about different stimuli also is integrated to give rise to compound sensations. For instance, "wetness" isn't a single stimulus, for our perception of it arises from simultaneous inputs that concern pressure, touch, and temperature. Two points to keep in mind: **Perception** is an understanding of what a stimulus means. It isn't the same as **sensation**, which is simply conscious awareness of a stimulus.

Figure 31.1 Examples of sensory receptors. (**a**) Thermoreceptors in pits above and below a python's mouth detect body heat, or infrared energy, of nearby prey. (**b**) Some bats listen to echoes of their own high-frequency sounds. Their brain constructs a sound map from the echoes that bounce back from prey and other objects in the surroundings. Such maps help this night-flying bat capture insects in midair without the help of eyes.

The system components are sensory neurons, nerve pathways, and specific brain regions. The receptors are the front doors of the nervous system. We define each type in terms of a **stimulus**, which is a specific form of energy that a sensory receptor is able to detect:

Mechanoreceptors detect forms of mechanical energy (changes in pressure, position, or acceleration).

Thermoreceptors detect infrared energy (heat).

Pain receptors (nociceptors) detect tissue damage.

Chemoreceptors detect chemical energy of specific substances dissolved in the fluid surrounding them.

Osmoreceptors detect changes in water volume (solute concentration) in the surrounding fluid.

Photoreceptors detect visible and ultraviolet light.

Depending upon the kinds and numbers of their sensory receptors, animals sample the environment in different ways and differ in their awareness of it. Unlike bees, you have no receptors for ultraviolet light and do not see many flowers the way they do. Unlike many bats, you and bees have no receptors for ultrasound. Unlike pythons, you, bees, and bats have no receptors for detecting warm-blooded prey in the dark. In this chapter, you will encounter processes and molecular structures responsible for these fascinating differences.

Key Concepts

1. Sensory systems are portions of the nervous system. Each consists of specific types of sensory receptors, nerve pathways from receptors to the brain, and brain regions that receive and process the sensory information.

2. A stimulus is a form of energy that activates a specific type of sensory receptor, which is either a sensory neuron or a specialized cell adjacent to it. Photoreceptors detect light energy, thermoreceptors detect infrared energy, and so on.

3. We define a sensation as conscious awareness of change in some aspect of the external or internal environment. It begins when sensory receptors detect a specific stimulus. The stimulus energy becomes converted to a graded, local signal that may contribute to initiating an action potential.

4. Information concerning the stimulus becomes encoded in the number and frequency of action potentials sent to the brain along particular nerve pathways. Specific brain regions translate the information into a sensation.

5. The somatic sensations include touch, pressure, pain, temperature, and muscle sense.

6. Taste, smell, hearing, and vision are special senses.

OVERVIEW OF SENSORY PATHWAYS

Recall, from Chapter 30, that sensory axons carry signals from receptors to the brain. They do so by converting stimulus energy into action potentials, the signals that travel on excitable cell membranes. When a stimulus disturbs the plasma membrane of a receptor's ending, certain ions flow across a local patch of the membrane. The disturbance may trigger a graded potential, a type of signal that doesn't spread far from the site of stimulation and that can vary in magnitude. Intense or rapidly repeated stimulation may give rise to action potentials, which can propagate themselves from a receptor to axon endings of a sensory neuron (Figure 31.2). There, the release of a neurotransmitter may trigger action potentials in a neighboring cell that is part of a pathway leading to the brain.

Action potentials moving along a sensory neuron aren't like a wailing ambulance siren. *They do not vary in amplitude.* How, then, does the brain assess the nature of the stimulus? The answer lies in *which* nerve pathways are carrying action potentials, the *frequency* of action potentials on each axon that is part of the pathway, and the *number* of axons that the stimulus recruited.

First, the wiring in an animal's brain is genetically programmed to interpret action potentials in certain ways. That is why you "see stars" when an eye gets poked, even in a dark room. Photoreceptors in that eye were mechanically disturbed enough to send messages that traveled along one of two optic nerves to the brain. Your brain always interprets any signals arriving from an optic nerve as "light."

Second, when a stimulus is strong, the receptors fire action potentials more often than they do with a weak stimulus. The same receptor can detect sounds of a throaty whisper and a wild screech. The brain senses the difference through frequency variations in signals that the receptor has sent to it.

Third, a stronger stimulus recruits more sensory receptors than a weaker stimulus does. Tap a spot of skin on an arm and you activate only a few receptors. Press hard and you activate more receptors in

Message will be sent on from spinal cord to interneurons in the brain.

Message travels from stimulated sensory neuron to interneurons in spinal cord.

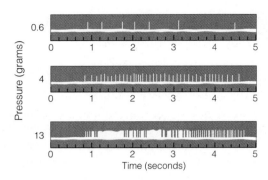

Figure 31.3 *Above:* Recordings of action potentials from one pressure receptor with endings in a human hand. They correspond to variations in stimulus strength. A thin rod was pressed against skin with the amount of pressure indicated to the left of each diagram. Vertical bars above each thick horizontal line record individual action potentials. Increases in their frequency correspond to increases in the strength of the stimulus.

a larger area. The increase in the disturbance leads to action potentials in many sensory axons. And the brain interprets that combined commotion as an increase in stimulus intensity. Figure 31.3 gives an example.

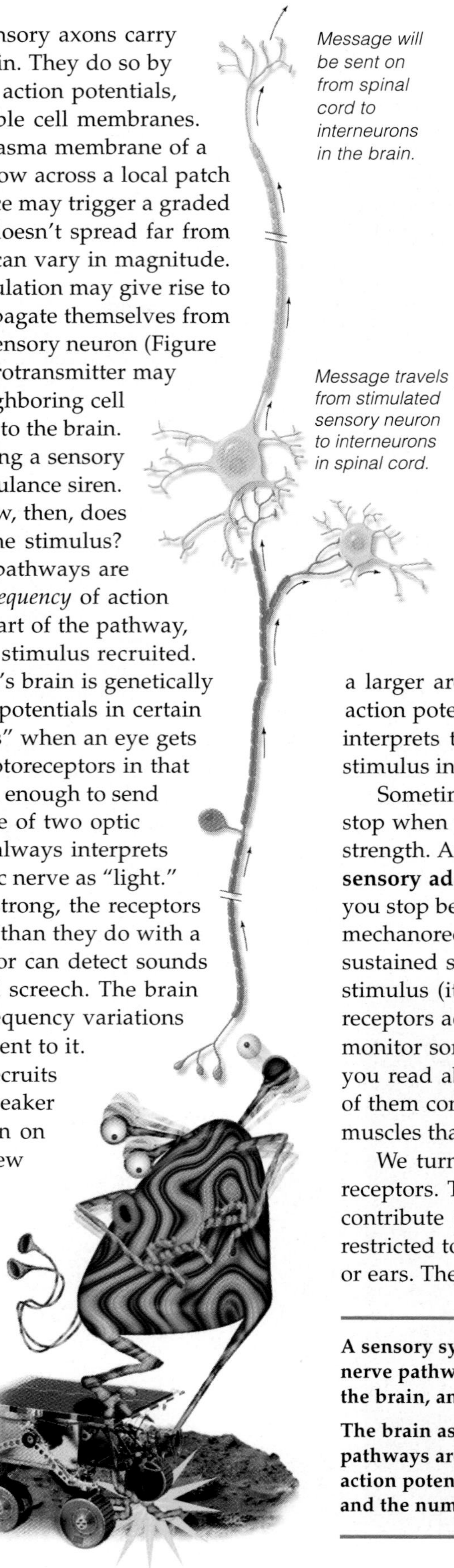

Sometimes action potentials become less frequent or stop when the stimulus is being maintained at constant strength. Any decrease in the response to a stimulus is a **sensory adaptation**. Example: After you put on clothes, you stop being aware of its pressure on your skin. Some mechanoreceptors in your skin adapted rapidly to the sustained stimulation; they only signal a change in the stimulus (its onset and its removal). By contrast, other receptors adapt slowly or not at all; they help the brain monitor some stimuli all the time. The stretch receptors you read about in Section 30.4 are like this. A number of them continually signal the brain about the length of muscles that help maintain balance and posture.

We turn next to some specific examples of sensory receptors. Types at more than one location in the body contribute to the **somatic sensations**. Other types are restricted to particular locations, such as inside the eyes or ears. They contribute to the **special senses**.

Figure 31.2 NASA Rover running over a Martian's foot and stimulating the receptor endings of a sensory neuron. This is a case of a sensory nerve pathway leading away from a sensory receptor to a brain. The sensory neuron is coded *red*, and interneurons are coded *yellow*.

A sensory system has sensory receptors for specific stimuli, nerve pathways that conduct information from receptors to the brain, and brain regions that receive the information.

The brain assesses a given stimulus based on which nerve pathways are carrying action potentials, the frequency of action potentials traveling along each axon of that pathway, and the number of axons recruited into action.

SOMATIC SENSATIONS

Somatic sensations begin with receptors in the body's surface tissues, skeletal muscles, and walls of internal organs. The receptors are most highly developed in birds and mammals; amphibians have a few, and apparently fishes have none. Receptor inputs travel to the spinal cord and on into the **somatosensory cortex**, part of the gray matter of the cerebral hemispheres (Section 30.9). The interneurons in this brain region are organized like maps corresponding to particular parts of the body's surface. Map regions with the largest areas correspond to body parts that have the greatest sensory acuity and that require the most intricate control. Such body parts include the fingers, thumbs, and lips (Figure 31.4).

You and other mammals discern sensations of touch, pressure, cold, warmth, and pain. **Pain** is perception of injury to some body region. Sensations of *somatic* pain start with signals from pain receptors in skin, skeletal muscles, joints, and tendons. The sensations of *visceral* pain, associated with the internal organs, are related to excessive chemical stimulation, muscle spasms, muscle fatigue, inadequate blood flow to organs, distension of digestive tract regions, and other abnormal conditions.

Free nerve endings are the simplest receptors. These thinly myelinated or unmyelinated (naked) branched endings of sensory neurons are distributed in skin and internal tissues. Different types are mechanoreceptors, thermoreceptors, and pain receptors. All adapt slowly to stimulation. One subpopulation gives rise to a sense of prickling pain, as when you jab a finger with a pin. Another contributes to itching or warming sensations that are provoked by chemicals, including histamine. Two thermoreceptive types have peak sensitivities that are higher and lower than normal body temperature, respectively. One mechanoreceptive type coils around hair follicles and detects hair movements (Figure 31.5).

Capsules encase and shield some receptors near the body surface from all stimuli except pressure changes. One type, Meissner's corpuscle, adapts very slowly to low-frequency vibration. It is abundant in fingertips, lips, eyelids, nipples, and genitals. Another, the bulb of Krause, is a thermoreceptor that kicks in at 20°C (68°F) or lower. Below 10°C, it contributes to painful freezing sensations. Slowly adapting, encapsulated types called Ruffini endings respond to steady touching, pressure, and temperatures above 45°C (113°F).

The Pacinian corpuscle, also encapsulated, is widely distributed deep in the skin, where it responds to fine textures. It also resides near freely movable joints and in some internal organs. Onionlike layers of membrane alternating with fluid-filled spaces enclose this sensory ending (Figure 31.5). They help it detect rapid pressure changes associated with touch and vibrations.

Sensing limb movements and the body's position in space requires mechanoreceptors in skeletal muscle,

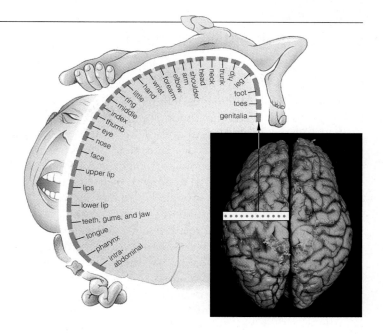

Figure 31.4 Differences in the representation of different body parts in the primary somatosensory cortex. This region is a strip of cerebral cortex, a little wider than 2.5 centimeters (about an inch), from the top of the head to above the ear.

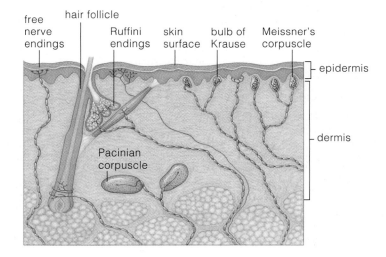

Figure 31.5 A sampling of receptors in human skin.

joints, tendons, ligaments, and skin. Examples include the stretch receptors of muscle spindles. As described in Section 30.4, these sensory organs run parallel with skeletal muscle cells. The degree of their responses to stimulation depends on how much and how fast the muscle is being stretched.

Diverse sensory receptors near the body surface and in internal organs detect touch, pressure, temperature, pain, motion, and positional changes of body parts. Their input travels through the spinal cord to the somatosensory cortex and other brain regions where somatic sensations arise.

SENSES OF HEARING AND BALANCE

Evolution of the Vertebrate Ear

Many arthropods and most vertebrates have a sense of **hearing**—sound perception. Sounds are vibrations, or wavelike forms of mechanical energy. You might wish to review Figure 31.6, which describes their properties.

Your sense of hearing evolved from structures that at one time had nothing at all to do with hearing. Here is the story, in brief: Most animals assess and respond to displacements from a natural, *equilibrium* position in which the body is balanced relative to gravity, velocity, acceleration, and other forces that act on its positions and movements. For instance, animals right themselves after tilting or turning upside-down. Most vertebrates have a pair of **inner ears** on both sides of their brain (Figure 31.7). Inner ears initially evolved as organs of equilibrium. Each has fluid-filled sacs and canals used to detect rotational, linear, and accelerated motions of the head. In all amphibians, birds, and mammals, the brain now integrates sensory input from the inner ears, eyes, skin, and joints to arrive at sensations of balance.

After vertebrates invaded the land, hearing became far more crucial than it had been in aquatic habitats,

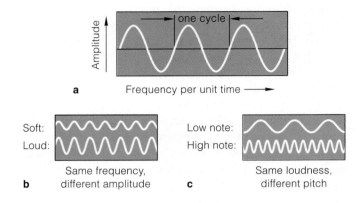

a

b Same frequency, different amplitude

c Same loudness, different pitch

Figure 31.6 Properties of sound—waves of compressed air. (**a**) Sounds differ in pressure, or *amplitude*, which we show as wave forms. We can measure amplitude in decibels. Every ten decibels is a tenfold increase in intensity above the faintest sound humans hear. (**b**) A sound's intensity, or loudness, depends on amplitude. (**c**) A sound's *frequency* is the number of wave cycles per second. Each cycle extends from the start of one wave peak to the start of the next. The more cycles per second, the higher the frequency and perceived *pitch*. Unlike a tuning fork's pure tone, most sounds are combinations of waves of different frequencies. Combinations of overtones give them their timbre (quality), a property that helps you recognize, say, voices of people you know on a phone.

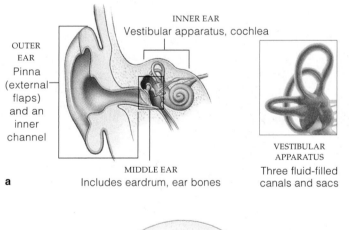

a

OUTER EAR
Pinna (external flaps) and an inner channel

INNER EAR
Vestibular apparatus, cochlea

MIDDLE EAR
Includes eardrum, ear bones

VESTIBULAR APPARATUS
Three fluid-filled canals and sacs

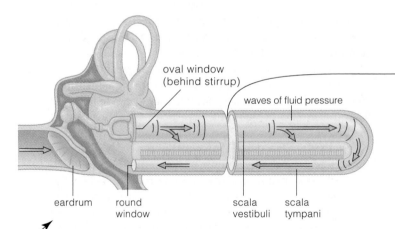

oval window (behind stirrup)

waves of fluid pressure

eardrum   round window   scala vestibuli   scala tympani

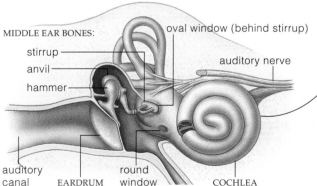

MIDDLE EAR BONES:
stirrup
anvil
hammer

oval window (behind stirrup)

auditory nerve

b auditory canal   EARDRUM   round window   COCHLEA

Figure 31.7 Sensory reception at the human ear. This sensory organ collects, amplifies, and sorts out sound waves (acoustical stimuli). The external flaps of the outer ear collect sound waves, which move through an auditory canal to the eardrum. The eardrum is also called the tympanic membrane.

c Sound reception in the inner ear. An eardrum vibrates with incoming sound waves. Middle ear bones pick up the vibrations. They amplify the stimulus by transmitting the force of pressure waves to a smaller surface, the oval window. The "window" is an elastic membrane over the entrance to the cochlea, part of the inner ear. Here, we "uncoiled" the cochlea for clarity.

The oval window bows in and out from the sound waves, thus producing waves of fluid pressure in two ducts (scala vestibuli and scala tympani). The waves reach another membrane, the round window, which bulges under the pressure and forces fluid to move back and forth in the inner ear.

Pressure waves are sorted out in the cochlear duct, the third duct inside the inner ear. One of its membranes (the basilar membrane) is narrow and stiff, but then it broadens and becomes more flexible deeper in the coil. At different points along its length, the basilar membrane vibrates more strongly to sounds of different frequencies.

Figure 31.8 Results of an experiment on the effect of intense sound on the inner ear. (**a**) From a guinea pig ear, three rows of hair cells projecting into the tectorial membrane in the organ of Corti. (**b**) Hair cells in that organ after exposure for twenty-four hours to noise levels comparable to extremely loud music.

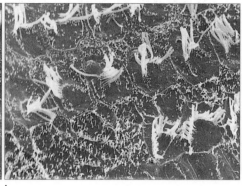

a   b

To give you a sense of what "loud" is, a ticking watch measures 10 decibels (100 times louder than the threshold of hearing for humans). A normal conversation is about 60 decibels (about a million times louder), a food blender operating at high speed is about 90 decibels (a billion times louder), and a hard rock or heavy metal concert, about 120 decibels (a trillion times louder).

where sounds are muffled. The parts of the internal ear dealing with equilibrium didn't change much. But other parts evolved into structures that were able to receive and process even faint sounds traveling through air.

A **middle ear** evolved that amplified and sent air waves to the inner ear. In reptiles, a depression on each side of the head became an eardrum, a thin membrane that could vibrate swiftly in response to air waves. In existing crocodiles, birds, and mammals is an air-filled cavity and small bones that transmit vibrations to the inner ear. Those bones had structurally supported gill pouches and evolved as part of the jaw joint in ancient

fishes (Section 24.3). In other words, bones that once had roles in gas exchange and feeding in fishes became specialized for hearing in reptiles, birds, and mammals.

Among mammals, paired **external ears** that collect sound waves became well developed. Each external ear typically has a pinna, or sound-collecting, skin-covered and folded flaps of cartilage projecting from the side of the head. It also has a channel to the middle ear.

Figure 31.7 shows the external, middle, and inner ear of humans. Like other mammals (and birds), it has a cochlea, a part of the inner ear that receives and sorts out sound waves. Its acoustical receptors—**hair cells**—bend in response to pressure waves. This starts a flow of information along an auditory nerve that leads from the receptors to brain centers. Prolonged exposure to intense sounds permanently damages hair cells. These cells are not adapted to extremely loud music, roars of jet planes, and other recent developments (Figure 31.8).

A Closer Look at Inner Ear Functions

In each of your inner ears is a **vestibular apparatus** (Figure 31.7). When your head rotates, fluid in a canal corresponding to the direction of rotation is displaced the opposite way. The resulting fluid pressure activates an array of hair cells that bend in response to pressure. Deforming their plasma membrane may trigger action potentials, and so messages may reach adjacent sensory neurons. A different part of the vestibular apparatus is activated when you start and stop moving in a straight line. In *motion sickness*, a monotonous linear, angular, or vertical motion has overstimulated its hair cells.

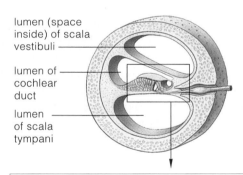

lumen (space inside) of scala vestibuli

lumen of cochlear duct

lumen of scala tympani

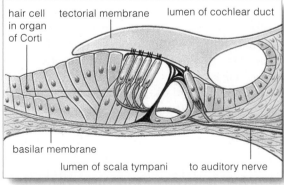

hair cell in organ of Corti

tectorial membrane

lumen of cochlear duct

basilar membrane

lumen of scala tympani   to auditory nerve

ORGAN OF CORTI
(organ of hearing that rests on top of basilar membrane)

d At the organ of Corti, 16,000 hair cells project into a jellylike flap (tectorial membrane) over the basilar membrane. When the flap vibrates suitably, hair cells bend, action potentials arise, and messages travel along an auditory nerve to the brain.

Organs of equilibrium help keep the body balanced in relation to gravity, velocity, acceleration, and other forces that influence its position and movement. Humans have such organs in the vestibular apparatus of each inner ear.

Hearing among land vertebrates involves structures that collect, amplify, and sort out sound waves traveling through air. In the inner ear, sound waves produce fluid pressure variations that hair cells transduce into action potentials.

SENSE OF VISION

All organisms are sensitive to light. Sunflowers track the sun as it arcs across the sky; single-celled amoebas abruptly stop moving when you shine a light on them. However, **vision** requires eyes and image formation in brain centers that receive and interpret signals about shapes, brightness, positions, and movements of visual stimuli. **Eyes** are sensory organs with a dense array of photoreceptors. Each photoreceptor contains pigments that can absorb photon energy, which can be converted into excitation energy in sensory neurons.

A Sampling of Invertebrate Eyes

Some eyeless invertebrates respond to light by way of photoreceptors embedded in their integument. Others have dense arrays of photoreceptors in eyelike patches or depressions in the integument. Vision evolved first among fast-moving predators that had to discriminate quickly among prey and other objects in their rapidly changing visual field. A **visual field** is the part of the outside world that an animal sees. Different groups of photoreceptors sample light intensity of different parts of the field. Most invertebrates don't have good vision; they are too small to have enough photoreceptors. For instance, a planarian's pigment cup only has about 200. Each square millimeter of an octopus eye has 70,000.

Image formation benefits from a **lens**, a transparent body that bends all light rays from a given point in the visual field so they converge onto photoreceptors. The round abalone lens bends light rays but not to the same focal point, so images are blurred (Figure 31.9b). Things improve with a **cornea**. This transparent cover directs light rays onto a lens, as in octopus eyes (Figure 31.9d).

Spiders have *simple* eyes, in which one lens services all of the photoreceptors. Crustaceans and insects have *compound* eyes. These eyes contain photosensitive units,

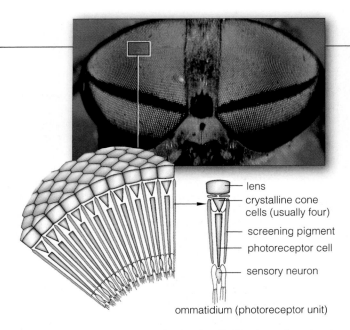

Figure 31.10 Compound eyes of a deerfly. The lens of each photosensitive unit directs light onto a crystalline cone, which focuses light onto photoreceptor cells below it.

each with a lens and a bundle of photoreceptors (Figure 31.10). Some have thousands of units called ommatidia (singular, ommatidium). Each unit samples a small part of the visual field and contributes a tiny bit of sensory information that helps the brain form a visual mosaic.

Of all invertebrates, octopuses and squids have the most complex eyes. Like vertebrates, they have **camera eyes**, so named because the eyeball is structured like a camera. Inside it is a darkened chamber (Figure 31.9d). Light enters at a pupil, an opening in a contractile tissue ring called an iris (equivalent to a camera diaphragm). Behind the pupil, a lens focuses light on a **retina**, a tissue packed with photoreceptors (in the camera, onto light-sensitive film). Similarities in cephalopod and vertebrate eyes might be a result of convergent evolution.

Vertebrate Eyes

In most vertebrates, the eyeball has a three-layer wall. The *outer* layer has a sclera of dense, white fibers that protect most of the eyeball, and a cornea of transparent collagen fibers that cover the front of the eyeball.

The *middle* layer has a choroid, ciliary body, iris, and pupil. The choroid, a dark-pigmented and vascularized tissue, absorbs light the photoreceptors miss and stops it from scattering in the eye. Its iris is a pigmented ring suspended behind the cornea. The pupil at the center is an adjustable entrance for light. In bright light, circular muscles contract and shrink the diameter of the pupil. In dim light, radial muscles contract and enlarge it.

The *inner* layer, including the retina, is located at the back of the human eyeball (Figure 31.11). In some birds, it is located at the top of the eyeball (Figure 31.12).

The eye's interior incorporates a lens constructed of layers of transparent proteins. Bathing the lens is a clear

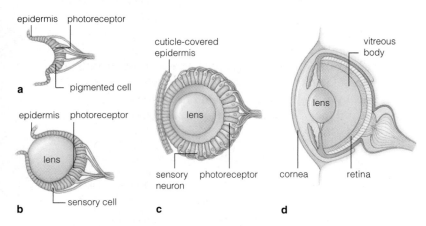

Figure 31.9 Organization of invertebrate eyes, longitudinal section. There are far more photoreceptors than are shown. (**a**) Limpet ocellus, a shallow depression in epidermis that has light-sensitive receptors. (**b**) Abalone eye, with its spherical, transparent lens. (**c**) Snail eye. (**d**) Octopus eye.

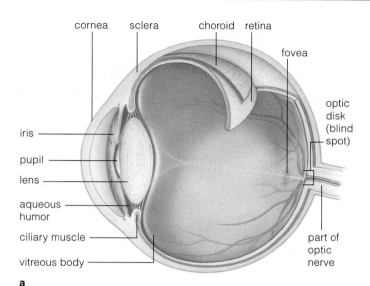

cornea   sclera   choroid   retina

fovea

optic disk (blind spot)

iris

pupil

lens

aqueous humor

ciliary muscle

vitreous body

part of optic nerve

a

Figure 31.11 (**a**) Structure of the human eye.

(**b**) Pattern of retinal stimulation in the human eye. Light rays converge at the back of the eye and stimulate the retina in a distinct way. The cornea has a curved surface, so light rays coming from a given point in the visual field hit it at different angles and their trajectories change. (Rays of light bend at the boundaries between substances of different densities, and the bending sends them in new directions.) Because of their newly angled trajectories, light rays that converge at the back of the eyeball stimulate the retina in a pattern that is upside-down and reversed left to right, relative to the original source of the rays.

Focusing mechanisms use a ciliary muscle that encircles and attaches to a lens. (**c**) When it relaxes, the lens flattens; the focal point moves farther back and brings distant objects into focus. (**d**) When the muscle contracts, the lens bulges; the focal point moves closer and brings close objects into focus.

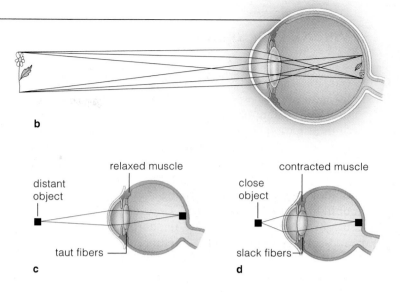

b

relaxed muscle

distant object

taut fibers

c

contracted muscle

close object

slack fibers

d

Figure 31.12 Here's looking at you! In owls and some other birds of prey, photoreceptors are more concentrated on top of the inner eyeball, not at back. These birds look down more than up when they fly and scan the ground for a meal. When on the ground, they cannot easily see objects above them unless they turn their head almost upside-down.

fluid, the aqueous humor. Behind the lens is a chamber filled with a jellylike substance, the vitreous body.

Light rays from sources at varying distances strike the cornea at different angles. But adjustments in lens position or shape usually focuses all of them onto the retina. This **visual accommodation** keeps the light rays from distant objects from being focused in front of the retina. Visual accommodation also prevents light rays from close objects from being focused behind it.

Fish and reptile eye muscles move the lens forward or back, like a camera's focusing device. Extending the distance between the lens and retina moves the focal point forward; shrinking the distance moves it back. The lens shape gets adjusted in birds and mammals. It flattens and moves the focal point back when a muscle encircling it relaxes. It bulges and moves the focal point forward when the muscle contracts (Figure 31.11*c,d*).

When an eyeball is not shaped quite right, the lens is positioned too close to or too far from the retina, so accommodation cannot bend the incoming rays of light to the same focal point. This happens with *astigmatism*, a condition in which the cornea has an uneven curvature.

Nearsightednesss (myopia) is common when the eyeball's horizontal axis is longer than the vertical. It is common when the ciliary muscle that adjusts the lens contracts too strongly and thus focuses light from distant objects in front of the retina, not onto it. *Farsightedness* is the opposite of myopia. The lens is "lazy" or the eyeball's vertical axis is longer than its horizontal axis. Either way, light from close objects gets focused behind the retina. Eyeglasses typically correct the visual disorders.

Vision requires eyes (sensory organs containing a dense array of photoreceptors) and a capacity for image formation in the brain, based on patterns of visual stimulation.

For most vertebrate eyes, the eyeball has three layers, and it encloses a lens, an aqueous humor, and a vitreous body.

A protective sclera and a light-focusing cornea make up the outer layer. The middle layer has a vascularized, pigmented choroid and other parts that admit and control incoming light. Photoreception occurs at the retina of the inner layer.

Adjustments in the positioning or shape of the lens focus incoming visual stimuli onto the retina.

VISUAL PERCEPTION

A sensory pathway from the retina to the brain detects, transmits, and also combines raw visual information. It leads to awareness of light and shadows, of colors, of near and distant objects in the outside world. The flow of information starts when light strikes dense arrays of **rod cells** and **cone cells**, two classes of photoreceptors (Figure 31.13). Rod cells detect very dim light. At night or in dark places, they contribute to coarse perception of movement by reporting changes in the intensity of light across the visual field. Cone cells detect bright light; they contribute to sharp vision and color perception in the daytime.

Part of a rod cell's membrane is folded into several hundred disks. Each disk is loaded with 10^8 molecules of rhodopsin, a visual pigment. The stacked membrane increases the density of pigments, hence the odds of intercepting packets of light energy—photons—of specific wavelengths. Action potentials that arise from absorption of even a few photons can lead to conscious awareness of objects in dimly lit surroundings.

Rhodopsin effectively absorbs photons of blue-to-green wavelengths (Section 6.2). Absorption changes the pigment's shape, which triggers a cascade of reactions that affect ion channels and pumps across the rod cell's plasma membrane. The voltage difference across the membrane shifts as gated sodium channels close, and a graded potential results. This graded potential reduces the ongoing release of a neurotransmitter that inhibits neighboring sensory neurons. Released from inhibition, the neighbors fire off signals about the visual stimulus that will travel to the brain.

The sense of color and of daytime vision starts with photon absorption by red, green, and blue cone cells, each having a distinct visual pigment. Here, too, photon absorption slows the release of a neurotransmitter that otherwise inhibits nearby sensory neurons. At the back of the eyeball is the fovea, a funnel-shaped depression near the retina's center. This retinal area has the highest density of photoreceptors and offers the greatest visual acuity. Its cone cells are the ones that discriminate most precisely between adjacent points in space.

Distinct layers of neurons are located above the rods and cones. Bipolar sensory neurons, then ganglion cells, accept visual information from these layers. Input from 125 million or so rods and cones converges on a mere 1 million ganglion cells. Then the axons of ganglion cells converge to form two optic nerves to the brain (Figure 31.13). But synaptic integration and visual processing start even before the information is sent to the brain.

Integration starts with the retina's organization. Its surface is divided into receptive fields: restricted areas that affect the activity of individual sensory neurons.

Figure 31.13 (**a**) Rods (*red*) and cones (*yellow*). (**b**) The sensory pathway from retinas to the brain.

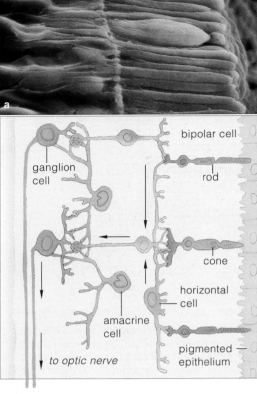

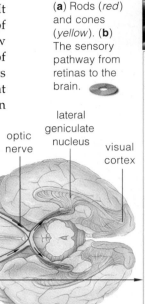

For instance, each ganglion cell's field is a tiny circle. Some cells respond best to a spot of light with a dark ring in the field's center. Others respond to a fast change in light intensity, to a spot of one color, or to motion. Some respond to diagonal objects and ignore diffuse, uniform illumination. Lateral interactions among other players—horizontal cells and amacrine cells—dampen or enhance signals before the ganglion cells get them.

Remember, an individual's visual field is the part of the outside world that it actually sees. The right side of both retinas intercepts light from the *left* half of the visual field; the left side intercepts light from the *right* half. An optic nerve leading out from each eye delivers signals about a stimulus from the *left* visual field to the right cerebral hemisphere, and delivers signals from the *right* visual field to the left hemisphere (Figure 31.13).

Axons of the optic nerves end in a layered brain region, the lateral geniculate nucleus. Each layer has a map corresponding to the receptive fields, and it deals with one aspect of a stimulus—form, movement, depth, color, texture, and so on. After initial processing, signals rapidly and simultaneously reach different portions of the visual cortex. There, final integration organizes the electrical activity and produces the sensation of sight.

Eyes collect and integrate information on distance, shape, brightness, position, and movement of visual stimuli. Their sensory pathways start at receptive fields in the retina and proceed along optic nerves to the brain, where signals are processed and integrated in the visual cortex.

SENSES OF TASTE AND SMELL

We leave this chapter with two final examples of the special senses: taste and smell. Both are *chemical* senses. Their sensory pathways start at chemoreceptors, which are activated when they bind a chemical substance that is dissolved in the fluid bathing them. These receptors wear out and new ones replace them on an ongoing basis. In both pathways, sensory input travels from the receptors through the thalamus and on to the cerebral cortex, where perceptions of the stimulus take shape and undergo fine-tuning. The input also travels to the limbic system, which integrates it with emotional states and stored memories (Sections 30.9 and 30.10).

Different animals taste with their mouth, antennae, legs, tentacles, or fins, depending on where their **taste receptors** are. In your mouth, throat, and especially the upper surface of your tongue, these chemoreceptors are located in about 10,000 sensory organs called taste buds (Figure 31.14). In a taste bud, fluids in the mouth move through a pore to receptor cells. Thousands of perceived tastes are some combination of five primary sensations: sweet (as elicited by glucose and other simple sugars), sour (acids), salty (NaCl and other salts), bitter (plant toxins, including alkaloids), and *umami* (glutamate, the brothy or savory taste typical of aged cheese and meat).

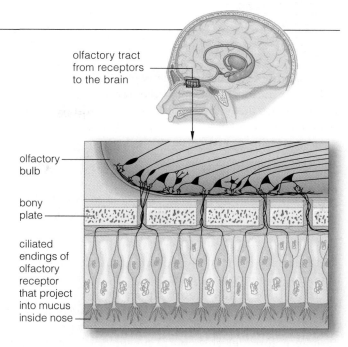

Figure 31.15 Sensory pathway from the sensory endings of olfactory receptors in the nose to the cerebral cortex and limbic system. Receptor axons pass through holes in a bony plate between the nasal lining and the brain. In the earliest vertebrates, an olfactory bulb and olfactory lobe dominated the forebrain. In certain lineages, the olfactory bulb became less important when the cerebrum expanded greatly in size and functions, as Section 30.9 indicates.

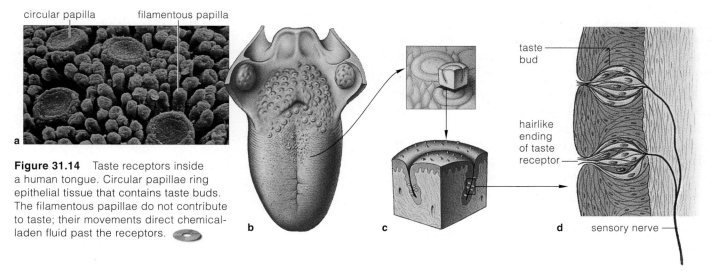

Figure 31.14 Taste receptors inside a human tongue. Circular papillae ring epithelial tissue that contains taste buds. The filamentous papillae do not contribute to taste; their movements direct chemical-laden fluid past the receptors.

Olfactory receptors detect water-soluble or volatile (easily vaporized) substances. A human nose has about 5 million; a bloodhound nose has more than 200 million. The receptor axons lead into one of two olfactory bulbs. In these small brain structures, the axons synapse with groups of cells that sort out the components of a given scent and relay the information along an olfactory tract for further processing in the cerebrum (Figure 31.15).

Pheromones, signaling molecules secreted from the exocrine glands of one individual, influence the social behavior of other individuals of its species, including potential sex partners (Chapter 41). Bombykol is one of the sex pheromones. Olfactory receptors of a male silk moth hit by only one bombykol molecule per second fire action potentials. They help a male find a female in the dark, even if she is more than a kilometer upwind. Human pheromones may influence sexual preferences, menstrual cycles, and menopause. Receptors for them, and the olfactory pathway itself, are not yet identified.

The senses of taste and smell both start at chemoreceptors. Both involve sensory pathways that lead to processing regions in the cerebral cortex and in the limbic system.

SUMMARY

Gold indicates text section

1. A stimulus is a specific form of energy that the body detects by means of sensory receptors. A sensation is an awareness that stimulation has occurred. Perception is understanding what the sensation means. *CI*

2. Sensory receptors are endings of sensory neurons or specialized cells next to them. They respond to specific forms of energy, such as mechanical pressure and light. Animals can only respond to aspects of the internal or external environment when they have receptors that are sensitive to the energy of specific stimuli. *CI*

    a. Mechanoreceptors, including free nerve endings and hair cells, detect mechanical energy related to touch, pressure, and motion and changes in position.

    b. Thermoreceptors detect radiant energy (heat).

    c. Pain receptors (nociceptors) detect tissue damage.

    d. Chemoreceptors, such as olfactory receptors and taste receptors, detect chemical substances dissolved in the fluids bathing them.

    e. Osmoreceptors detect changes in water volume (hence solute concentrations) in the surrounding fluid.

    f. Photoreceptors detect light energy. The rods and cones of the retina in the human eye are examples.

3. A sensory system has receptors for specific stimuli and nerve pathways from those receptors to receiving and processing centers in the brain. The brain assesses a particular stimulus based on which nerve pathway is delivering signals, the frequency of signals traveling on individual axons of that pathway, and the number of axons that were recruited into action. *31.1*

4. Somatic sensations include touch, pressure, pain, temperature, and muscle sense. The receptors associated with these sensations are not localized in a single organ or tissue. Special senses include taste, smell, hearing, balance, and vision. Receptors associated with these senses typically reside in sensory organs, such as eyes, or some other particular body region. *31.2*

5. Organs of equilibrium, as in the vertebrate inner ear, detect gravity, velocity, acceleration, and other forces that affect body positions and movements. The sense of hearing (sound perception) in vertebrates requires components of the outer, middle, and inner ear that collect, amplify, and sort out sound waves. *31.3*

6. Vision requires eyes: sensory organs having a dense array of photoreceptors, as in a retina (Table 31.1). It requires a capacity for image formation in the brain, based on patterns of visual stimulation. The sense of vision and discrimination among objects first evolved among fast-moving, predatory animals. *31.4, 31.5*

7. The senses of taste and smell both involve sensory pathways from chemoreceptors to processing regions in the cerebral cortex and limbic system. *31.6*

Table 31.1 *Summary of Vertebrate Eye Components*

THREE LAYERS FORMING WALL OF EYEBALL		
Outer layer	*Sclera.* Protects eyeball	
	Cornea. Focuses light	
Middle layer	*Choroid.* Blood vessels nutritionally support wall's cells; pigments prevent light scattering	
	Ciliary body. Its muscles control lens shape; its fine fibers hold lens in upright position	
	Iris. Adjustments here control incoming light	
	Pupil. Serves as entrance for light	
Inner layer	*Retina.* Absorbs and transduces light energy	
	Fovea. Increases visual acuity	
	Start of optic nerve. Carries signals to brain	
INTERIOR OF EYEBALL		
Lens	Focuses light onto photoreceptors	
Aqueous humor	Transmits light, maintains pressure	
Vitreous body	Transmits light, supports lens and eyeball	

Review Questions

1. Define stimulus, sensation, and perception. *CI*

2. Name six categories of sensory receptors and the type of stimulus energy that each kind detects. *CI*

3. What are the basic components of a sensory system? How does the brain assess the nature of a given stimulus? *31.1*

4. How do somatic sensations differ from special senses? *31.2*

5. What is pain? Describe one type of pain receptor. *31.2*

6. Which evolved first, a sense of balance or a sense of hearing? Do both involve the outer, middle, and inner ear? *31.3*

7. Label the component parts of the human ear: *31.3*

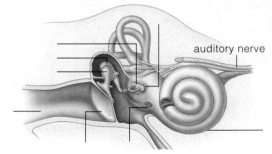

auditory nerve

8. How does vision differ from light sensitivity? What sensory organs and structures does vision require? *31.4, 31.5*

9. Label the component parts of the human eye: *31.4*

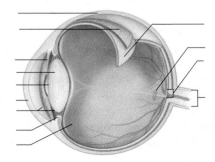

10. What are the five primary taste sensations? *31.6*

Figure 31.16 Flamingos at a large lake in Tanzania, East Africa.

what Tim sees

what Francesca sees

Self-Quiz ANSWERS IN APPENDIX III

1. A _____ is a specific form of energy that is detected by a sensory receptor.

2. Conscious awareness of a stimulus is called a _____ .

3. _____ is understanding what particular sensations mean.

4. Each sensory system consists of _____ .
   a. nerve pathways from specific receptors to the brain
   b. sensory receptors
   c. brain regions that deal with sensory information
   d. all of the above

5. _____ is (are) a decrease in the response made to an ongoing stimulus.
   a. Perception   c. Visual accommodation
   b. Sensory adaptation   d. both b and c

6. _____ detect mechanical energy associated with changes in pressure, position, or acceleration.
   a. Chemoreceptors   c. Photoreceptors
   b. Mechanoreceptors   d. Thermoreceptors

7. Detecting light energy is the function of _____ .
   a. chemoreceptors   c. photoreceptors
   b. mechanoreceptors   d. thermoreceptors

8. Vision requires _____ .
   a. a tissue with dense arrays of photoreceptors
   b. eyes
   c. image-forming centers in the brain
   d. all of the above

9. The outer layer of the human eyeball includes the _____ .
   a. lens and choroid   c. retina
   b. sclera and cornea   d. both a and c

10. The inner layer of the human eyeball includes the _____ .
    a. lens and choroid   c. retina
    b. sclera and cornea   d. both a and c

11. Match each term with the most suitable description.
    ____ somatic senses   a. evolved first as organ of
    ____ umami   equilibrium
    ____ special senses   b. taste, smell, hearing, balance,
    ____ variations in   and vision
         intensity of   c. one of five primary taste
         stimulus   sensations
    ____ inner ear   d. encoded in the frequency and
                               number of action potentials
                            e. touch, pressure, temperature,
                               pain, and muscle sense

Critical Thinking

1. Wayne, on standby for the last flight from San Francisco to New York, was assigned the last available seat on the plane. It was in the last row, where vibrations and noise from the engines are most pronounced. When Wayne got off the plane in New York, he was speaking very loudly and having trouble hearing what other people were saying. The next day, things were back to normal. Speculate on what happened to his sense of hearing during the flight.

2. Juanita made an appointment with her doctor because she was experiencing recurring episodes of dizziness. Her doctor immediately asked whether "dizziness" meant she had a sense of lightheadedness, as if she were going to faint, or whether it meant she had sensations of *vertigo*—that is, a feeling that she herself or objects near her were spinning around. Why do you suppose her doctor considered this clarification important early in his evaluation of her condition?

3. To Tim, a graduate student studying the social behavior of flamingos in Tanzania, the birds closest to him along the shore of a lake are in sharp focus, but the ones flying some distance away look fuzzy (Figure 31.16*a*). Francesca, another student, says the birds closest to her look fuzzy but those in the distance are in focus (Figure 31.16*b*).

   Is Tim nearsighted or farsighted, and is the focal point of light rays from the flamingos falling in front of or behind his retinas? What about Francesca?

Selected Key Terms

camera eye *31.4*
chemoreceptor *CI*
cone cell *31.5*
cornea *31.4*
echolocation *CI*
external ear *31.3*
eye *31.4*
hair cell *31.3*
hearing *31.3*
inner ear *31.3*
lens *31.4*
mechanoreceptor *CI*
middle ear *31.3*
olfactory
   receptor *31.6*
osmoreceptor *CI*
pain *31.2*
pain receptor *CI*
perception *CI*
pheromone *31.6*
photoreceptor *CI*
retina *31.4*
rod cell *31.5*
sensation *CI*
sensory
   adaptation *31.1*
sensory system *CI*
somatic sensation *31.1*
somatosensory
   cortex *31.2*
special sense *31.1*
stimulus *CI*
taste receptor *31.6*
thermoreceptor *CI*
vestibular
   apparatus *31.3*
vision *31.4*
visual
   accommodation *31.4*
visual field *31.4*

Readings

Delcomyn, F. 1998. *Foundations of Neurobiology*. New York: Freeman.

Romer, A., and T. Parsons. 1986. *The Vertebrate Body*. Sixth edition. New York: Saunders.

Sherwood, L. 1997. *Human Physiology*. Third edition. Belmont, California: Wadsworth.

Zeki, S. September 1992. "The Visual Image in Mind and Brain." *Scientific American* 267(3): 68–76.

On-Line readings at Student Guide for InfoTrac: www.brookscole.com/biology

32

ENDOCRINE CONTROL

Hormone Jamboree

In the 1960s, at her camp in a forest by the shore of Lake Tanganyika in Tanzania, the primatologist Jane Goodall let it be known that bananas were available. One of the first chimpanzees attracted to the delicious food was a female—Flo, as she came to be called (Figure 32.1). Flo brought along her offspring, an infant female and a juvenile male, and exhibited commendable parental behavior toward them.

Three years passed, and Goodall observed that Flo's earlier preoccupation with motherhood gave way to a preoccupation with sex. She also observed that male chimpanzees followed Flo to the camp by the lake and stayed for more than the bananas.

Sex, Goodall discovered, is *the* premier binding force in the social life of chimpanzees. These primates do not mate for life as eagles do, or wolves. Before the rainy season begins, mature females that are entering their

fertile cycle become sexually active. Changing hormone levels drive behavioral and physical changes. Sexual organs, for instance, become swollen and vivid pink. The swellings are strong visual signals to males. They are the flags of sexual jamborees, of great gatherings of highly stimulated chimps in which any males present may have a turn at copulating with the same female.

The gathering of many flag-waving females draws together chimpanzees, which forage alone or in small family groups for most of the year. It reestablishes bonds that unite a rather fluid community.

Infants and juveniles now play with one another and with the adults. Their aggressive and submissive jostlings help map out future dominance hierarchies. Flo happened to be a high-ranking female in the social hierarchy. Through her sexual attractiveness and direct solicitations, she built strong alliances with many males.

Figure 32.1 (**a**) Primatologist Jane Goodall in Gombe National Park, near the shores of Lake Tanganyika, scouting for chimpanzees. (**b**) Flo and two of her offspring, all subjects of long-term field observations. Flo helped Goodall clarify the central role of sex—and of the hormones that orchestrate it—in the social life of these primate relatives of humans.

b

Through her high status and aggressive behavior, she helped her male offspring win confrontations with other young male chimps.

The hormone-induced swelling during the fertile period of female chimps lasts somewhere between ten and sixteen days. Yet females are fertile for only one to five days. Sex hormones induce swelling even after a female gets pregnant. We can hypothesize that sexual selection has favored prolonged swelling. Males groom a sexually attractive female more often, protect her, give her more food, and let her tag along to new foraging sites. The more that males accept a female, the higher she rises in the social hierarchy—and, evolutionarily speaking, the more her individual offspring benefit.

Through their effects, hormones help orchestrate the growth, development, and reproductive cycles of nearly all animals, from the invertebrate worms to chimpanzees and humans. They influence minute-by-minute and day-to-day metabolic functions. Through their interplays with one another and with the nervous system, hormones profoundly influence the physical appearance, well-being, and behavior of individuals. Behavior, too, affects whether individuals will survive, either on their own or as part of social groups.

This chapter focuses primarily on hormones—on their sources, targets, and interactions as well as the mechanisms involved in their secretion. If the details start to seem remote, remember that this is the stuff of life. Hormones underwrote Flo's appearance, behavior, and rise through the chimpanzee social hierarchy. Just imagine what they've been doing for you.

Key Concepts

1. For nearly all animals, hormones and other signaling molecules have central roles in integrating the activities of individual cells in ways that benefit the whole body.

2. Only the cells with molecular receptors for a specific hormone are its targets. Hormones operate by serving as signals for change in the activities of their targets.

3. Many types of hormones influence gene transcription and protein synthesis in target cells. Other types call for alterations in existing molecules and structures in cells.

4. Some hormones work by binding with and altering membrane characteristics, such as the permeability of a plasma membrane to a particular solute. Others interact with a cell's DNA, either directly or after first binding with a receptor in the cytoplasm.

5. Some hormones help the body adjust to short-term shifts in diet and levels of activity. Others help induce long-term adjustments in cell activities that bring about the body's growth, development, and reproduction.

6. Among vertebrates, the hypothalamus and pituitary gland interact in ways that coordinate the activities of a number of endocrine glands. Together, they exert control over many aspects of the body's functioning.

7. Hormones, neural signals, shifts in local chemical conditions, and environmental cues function as triggers for the secretion of particular hormones.

THE ENDOCRINE SYSTEM

Hormones and Other Signaling Molecules

Throughout their lives, cells must respond to changing conditions by taking up and releasing various chemical substances. In vertebrates, the responses of millions to many billions of cells must become integrated in ways that help keep the whole body alive and functioning.

Signaling molecules help integrate activities within and between cells, and between tissues. We call them hormones, neurotransmitters, local signaling molecules, and pheromones. Each acts on target cells. A "target" is any cell that has receptors for a signaling molecule and that may change its activities in response to it. Targets may or may not be adjacent to cells that send signals.

By definition, **animal hormones** are the secretory products of endocrine glands, endocrine cells, and some neurons that the bloodstream delivers to nonadjacent target cells. They are this chapter's focus.

Neurotransmitters, released from axon endings of neurons, act swiftly on target cells by diffusing across a tiny synaptic cleft between them. You read about them in Section 30.3. **Local signaling molecules** released by many types of cells alter conditions in local tissues. For instance, some prostaglandins restrict or enhance blood flow to certain tissues. They do this by causing smooth muscle cells in blood vessel walls to contract or relax.

Pheromones, nearly odorless secretions of certain exocrine glands, diffuse through water or air to targets outside the animal body. These hormone-like secretions act on cells of other individuals of the same species and help integrate social behavior. For example, female silk moths secrete bombykol as a sex attractant, and soldier termites secrete alarm signals when ants attack their colony (Chapter 41). In vertebrates, the vomeronasal organ is a common pheromone detector. Even humans have one, but we don't know much about its function. Human pheromones or other chemical signals do affect our sexual behavior (Section 31.6). Do they also work at the subconscious level to trigger those inexplicable impressions, such as instant good or bad "feelings" about somebody you have just met? Maybe.

Discovery of Hormones and Their Sources

The word "hormone" dates to the early 1900s. W. Bayliss and E. Starling were trying to find out what triggers the secretion of pancreatic juices when food is traveling through the canine gut. As they knew, acids are mixed with food in the stomach, and the pancreas secretes an alkaline solution after the acidic mixture is propelled forward into the small intestine. Was the nervous system or something else stimulating the pancreatic response?

To find an answer, Bayliss and Starling blocked the nerves—but not blood vessels—leading to a laboratory

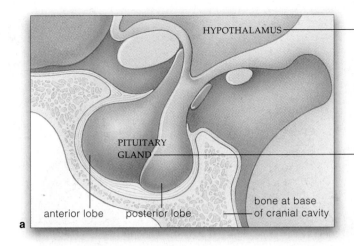

Figure 32.2 (**a**) A major neural–endocrine control center. The pituitary gland interacts closely with the hypothalamus, a part of the brain that also secretes some hormones. (**b**) *Facing page:* Overview of the key components of the human endocrine system and of the primary effects of their main hormonal secretions. The system also includes endocrine cells in many organs, such as the liver, kidneys, heart, small intestine, and skin.

animal's upper small intestine. Later, when acidic food entered the small intestine, the pancreas still secreted the alkaline solution. Even more telling, extracts of cells from the intestinal lining—a glandular epithelium— also induced the response. Glandular cells had to be the source of the pancreas-stimulating substance.

The substance came to be called secretin. Proof of its existence and mode of action confirmed a centuries-old idea: *The bloodstream picks up internal secretions that can influence the activities of organs inside the body.* Starling coined the word "hormone" for such internal glandular secretions (after *hormon*, meaning to set in motion).

Later on, researchers identified many other kinds of hormones and their sources. Figure 32.2 is a simplified picture of the locations of the following major sources of hormones for the human body. Bear in mind, most vertebrates have the same hormone sources:

> Pituitary gland
> Adrenal glands (*two*)
> Pancreatic islets (*numerous cell clusters*)
> Thyroid gland
> Parathyroid glands (*in humans, four*)
> Pineal gland
> Thymus gland
> Gonads (*two*)
> Endocrine cells of the hypothalamus, stomach, small intestine, liver, kidneys, heart, placenta, skin, adipose tissue, and other organs

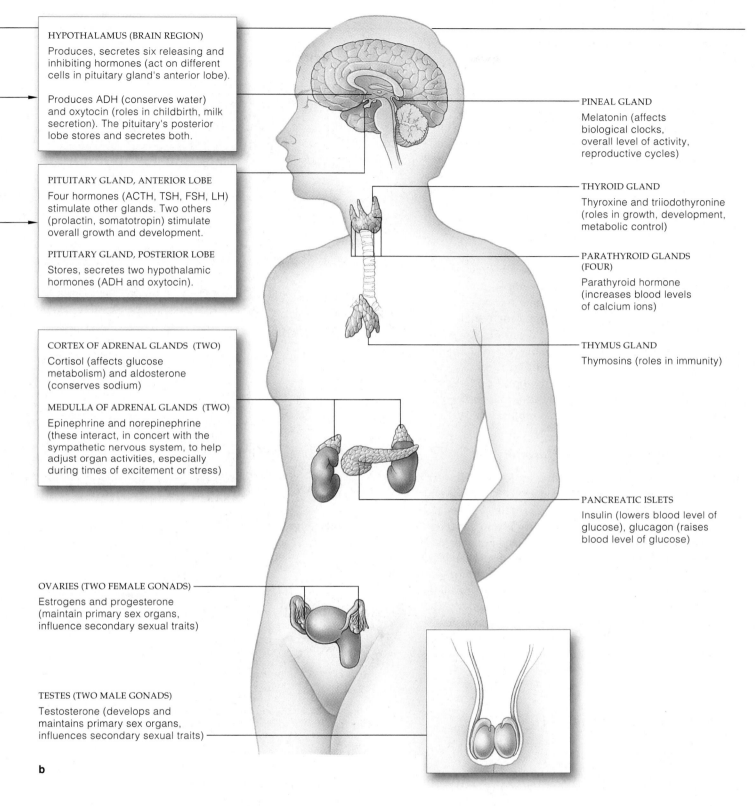

HYPOTHALAMUS (BRAIN REGION)

Produces, secretes six releasing and inhibiting hormones (act on different cells in pituitary gland's anterior lobe).

Produces ADH (conserves water) and oxytocin (roles in childbirth, milk secretion). The pituitary's posterior lobe stores and secretes both.

PITUITARY GLAND, ANTERIOR LOBE

Four hormones (ACTH, TSH, FSH, LH) stimulate other glands. Two others (prolactin, somatotropin) stimulate overall growth and development.

PITUITARY GLAND, POSTERIOR LOBE

Stores, secretes two hypothalamic hormones (ADH and oxytocin).

CORTEX OF ADRENAL GLANDS (TWO)

Cortisol (affects glucose metabolism) and aldosterone (conserves sodium)

MEDULLA OF ADRENAL GLANDS (TWO)

Epinephrine and norepinephrine (these interact, in concert with the sympathetic nervous system, to help adjust organ activities, especially during times of excitement or stress)

OVARIES (TWO FEMALE GONADS)

Estrogens and progesterone (maintain primary sex organs, influence secondary sexual traits)

TESTES (TWO MALE GONADS)

Testosterone (develops and maintains primary sex organs, influences secondary sexual traits)

PINEAL GLAND

Melatonin (affects biological clocks, overall level of activity, reproductive cycles)

THYROID GLAND

Thyroxine and triiodothyronine (roles in growth, development, metabolic control)

PARATHYROID GLANDS (FOUR)

Parathyroid hormone (increases blood levels of calcium ions)

THYMUS GLAND

Thymosins (roles in immunity)

PANCREATIC ISLETS

Insulin (lowers blood level of glucose), glucagon (raises blood level of glucose)

b

Collectively, the body's sources of hormones came to be called the **endocrine system**. The name implies there is a separate control system for the body, apart from the nervous system. (*Endon* means "within" and *krinein* is taken to mean "secrete.") Later, biochemical research and electron microscopy studies revealed that endocrine sources and the nervous system function in intricately connected ways, as you will see shortly.

Integration of cell activities depends on hormones and other signaling molecules. Each type of signaling molecule acts on target cells, which are any cells that have receptors for it and that may alter their activities in response to it.

Components of the endocrine system and certain neurons produce and secrete hormones, which the bloodstream takes up and distributes to nonadjacent target cells.

SIGNALING MECHANISMS

The Nature of Hormonal Action

Hormones and other signaling molecules interact with protein receptors of target cells. Their interactions have diverse effects on physiological processes. Some kinds of hormones induce a target cell to increase its uptake of glucose, calcium, or some other substance. Others stimulate or inhibit a target in ways that alter its rates of protein synthesis, modify the structure of proteins or other elements in the cytoplasm, or change cell shape.

To a large extent, responses to hormones depend on two factors. *First*, different hormones act on different mechanisms in target cells. *Second*, not all types of cells can respond to a given signal. For example, many have receptors for cortisol, so this hormone has widespread effects through the body. By contrast, only certain cells of small tubes in the kidneys have receptors for ADH, a hormone that initiates a highly directed response to a decrease in the volume of extracellular fluid.

With these points in mind, let's look briefly at the effects of two main categories of signaling molecules, the steroid and peptide hormones (Table 32.1).

Characteristics of Steroid Hormones

Steroid hormones are lipid-soluble molecules, derived from cholesterol, that are secreted by cells in adrenal glands and in primary reproductive organs, or gonads. Testosterone, one of the sex hormones, is an example.

Think of how testosterone affects secondary sexual traits associated with "maleness." These traits develop normally only if target cells have working receptors for testosterone. In *testicular feminization syndrome*, a genetic disorder, these receptors are defective. Genetically, the affected person is male (XY). He has functional testes that are secreting testosterone. But the target cells just can't respond to the hormone, so the secondary sexual traits that do develop are like those of females.

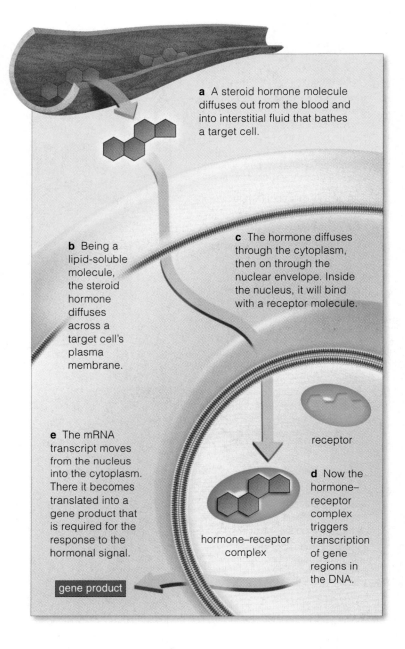

a A steroid hormone molecule diffuses out from the blood and into interstitial fluid that bathes a target cell.

b Being a lipid-soluble molecule, the steroid hormone diffuses across a target cell's plasma membrane.

c The hormone diffuses through the cytoplasm, then on through the nuclear envelope. Inside the nucleus, it will bind with a receptor molecule.

e The mRNA transcript moves from the nucleus into the cytoplasm. There it becomes translated into a gene product that is required for the response to the hormonal signal.

receptor

d Now the hormone–receptor complex triggers transcription of gene regions in the DNA.

hormone–receptor complex

gene product

Figure 32.3 Example of a mechanism by which a steroid hormone initiates changes in a target cell's activities.

Table 32.1 *Two Main Categories of Hormones*	
Type	Examples
Steroid and steroid-like hormones	Estrogens (feminizing effects), progestins (related to pregnancy), androgens (such as testosterone; masculinizing effects), cortisol, aldosterone. Thyroid hormones and vitamin D act like steroid hormones.
Peptide hormones	
Peptides	Glucagon, ADH, oxytocin, TRH
Proteins	Insulin, somatotropin, prolactin
Glycoproteins	FSH, LH, TSH

How does a steroid hormone operate? Being lipid soluble, it can diffuse directly across the lipid bilayer of a cell's plasma membrane (Figure 32.3). Once inside the cytoplasm, the hormone molecule commonly moves into the nucleus, where it binds to a receptor. In other cases, it binds with a receptor in the cytoplasm, then the hormone–receptor complex moves into the nucleus. The complex interacts with a specific region of DNA. Different complexes inhibit or stimulate transcription of gene regions into mRNA. Translation of such mRNA

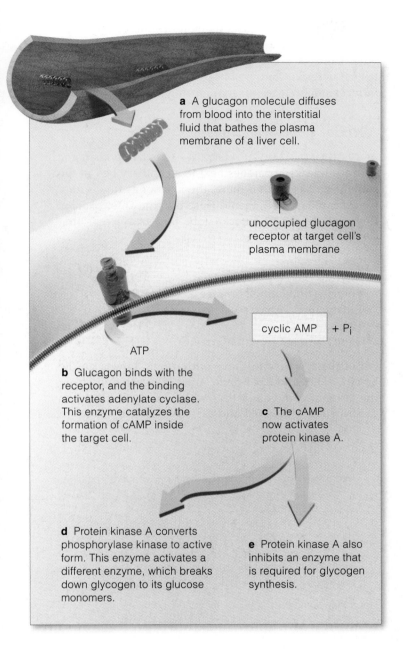

a A glucagon molecule diffuses from blood into the interstitial fluid that bathes the plasma membrane of a liver cell.

unoccupied glucagon receptor at target cell's plasma membrane

ATP

cyclic AMP + P$_i$

b Glucagon binds with the receptor, and the binding activates adenylate cyclase. This enzyme catalyzes the formation of cAMP inside the target cell.

c The cAMP now activates protein kinase A.

d Protein kinase A converts phosphorylase kinase to active form. This enzyme activates a different enzyme, which breaks down glycogen to its glucose monomers.

e Protein kinase A also inhibits an enzyme that is required for glycogen synthesis.

Figure 32.4 Example of a mechanism by which a peptide hormone initiates changes in a target cell's activities. When the hormone glucagon binds at a receptor, it initiates reactions in the cell. In this case cyclic AMP, which is one type of second messenger, relays the signal into the cell interior.

transcripts results in enzymes and other proteins that can carry out a response to the hormonal signal.

Steroid hormones may exert effects in still another way. Some types have receptors on plasma membranes. They modify some function in a target cell by altering the properties of the membrane itself.

Part of a group of genes that specifies receptors for steroid hormones also specifies thyroid hormone and vitamin D receptors. Thyroid hormones and vitamin D are not steroid hormones, but they behave like them.

Characteristics of Peptide Hormones

Traditionally, **peptide hormones** have been defined as water-soluble signaling molecules with 3 to 180 amino acids. Many peptides, polypeptides, and glycoproteins are in this category. When peptide hormones bind to receptors at the plasma membrane, specific membrane-bound enzyme systems become activated. These systems initiate reactions that lead to a cellular response.

Think of a liver cell. Glucagon, a peptide hormone, binds with a receptor that spans the plasma membrane and extends into the cytoplasm. Binding action triggers formation of a **second messenger**—a small molecule in the cytoplasm that relays signals from hormone–receptor complexes at the plasma membrane into the cell. cAMP (cyclic adenosine monophosphate) is this messenger.

An enzyme system at the plasma membrane starts the response to glucagon binding. Adenylate cyclase, an enzyme, is activated. It causes a cascade of reactions by converting ATP to many cyclic AMP molecules (Figure 32.4). These products serve as signals to convert many molecules of a protein kinase, another type of enzyme, to its active form. The kinases activate other enzymes, and so on until a final reaction converts the glycogen stored inside the cell to glucose. In short order, a huge number of molecules execute the hormonal response.

Or think of a muscle cell. It has receptors that bind insulin, a protein hormone. As one of its functions, an insulin–receptor complex induces glucose transporters to move through the cytoplasm and insert themselves into the plasma membrane. The transporters, a type of protein, help the muscle cell take up glucose faster. The signal also activates enzymes that help store glucose.

Bear in mind, there are other hormone categories, including the catecholamines such as epinephrine. Like glucagon, epinephrine binds with membrane receptors and triggers the release of a second messenger, cyclic AMP, that assists in the cellular response.

Hormones interact with receptors at the plasma membrane or inside the cytoplasm of target cells. Hormone binding directly influences protein synthesis in target cells.

Steroid hormones interact with a cell's DNA after entering the nucleus directly or after binding with a receptor in the cytoplasm. Some act by altering membrane properties.

Peptide hormones do not exert their effects by entering a cell. When they bind to a membrane receptor, the binding itself is a signal for enzyme-mediated, intracellular events. Often a second messenger inside the cytoplasm relays the signal into the cell interior.

THE HYPOTHALAMUS AND PITUITARY GLAND

From Section 30.9, you already know the **hypothalamus** is part of the forebrain's centers for homeostatic control of the internal environment, the viscera, and emotional states. Secretory neurons extend down into a slender stalk at its base, then into the lobed, pea-size **pituitary gland**. The hypothalamus and the pituitary interact as a major neural–endocrine control center.

The *posterior* lobe of the pituitary gland secretes two hormones that the hypothalamus produces. Its *anterior* lobe produces and secretes its own hormones. Most of these help control the release of hormones from other endocrine glands (Table 32.2). The pituitary of many vertebrates—*not* humans—has an intermediate lobe as well. In many cases, the third lobe secretes a hormone that governs reversible changes in skin or fur color.

Posterior Lobe Secretions

Figure 32.5*a* shows how hypothalamic neurons extend into the posterior lobe and end next to a capillary bed. The neurons produce antidiuretic hormone (ADH) and oxytocin, then store them in their axon endings. They secrete these hormones into interstitial fluid. Hormone molecules diffuse into capillaries, and the bloodstream distributes them through the body. ADH acts inside the

kidneys. Its action adjusts the volume and composition of extracellular fluid in ways that help conserve water when necessary. Oxytocin has roles in reproduction. It triggers uterine contractions during childbirth and also induces milk release when offspring are being nursed.

Anterior Lobe Secretions

ANTERIOR PITUITARY HORMONES   Other hypothalamic neurons release hormones next to a capillary bed in the stalk at its base. From there, the hormones flow into a *second* capillary bed, then diffuse into the anterior lobe. (Figure 32.6). Their anterior lobe targets are cells that secrete these hormones:

Corticotropin	ACTH
Thyrotropin	TSH
Follicle-stimulating hormone	FSH
Luteinizing hormone	LH
Prolactin	PRL
Somatotropin (growth hormone)	STH (or GH)

All of these hormones have widespread effects on the body. ACTH and TSH orchestrate the secretions from the adrenal glands and thyroid gland, respectively, as described shortly. FSH and LH have essential roles in

Table 32.2 *Hormones Released From the Mammalian Pituitary Gland*

Pituitary Lobe	Secretions	Designation	Main Targets	Primary Actions
POSTERIOR Nervous tissue (extension of hypothalamus)	Antidiuretic hormone	ADH	Kidneys	Induces water conservation as required during control of extracellular fluid volume and solute concentrations
	Oxytocin	OCT	Mammary glands	Induces milk movement into secretory ducts
			Uterus	Induces uterine contractions during childbirth
ANTERIOR Glandular tissue, mostly	Corticotropin	ACTH	Adrenal cortex	Stimulates release of cortisol, an adrenal steroid hormone
	Thyrotropin	TSH	Thyroid gland	Stimulates release of thyroid hormones
	Follicle-stimulating hormone	FSH	Ovaries, testes	In females, stimulates estrogen secretion, egg maturation; in males, helps stimulate sperm formation
	Luteinizing hormone	LH	Ovaries, testes	In females, stimulates progesterone secretion, ovulation, corpus luteum formation; in males, stimulates testosterone secretion, sperm release
	Prolactin	PRL	Mammary glands	Stimulates and sustains milk production
	Somatotropin (or growth hormone)	STH (GH)	Most cells	Promotes growth in young; induces protein synthesis, cell division; roles in glucose, protein metabolism in adults
INTERMEDIATE* Glandular tissue, mostly	Melanocyte-stimulating hormone	MSH	Pigmented cells in skin and other integuments	Induces color changes in response to external stimuli; affects some behaviors

* Present in most vertebrates (not humans). MSH is associated with the anterior lobe in humans.

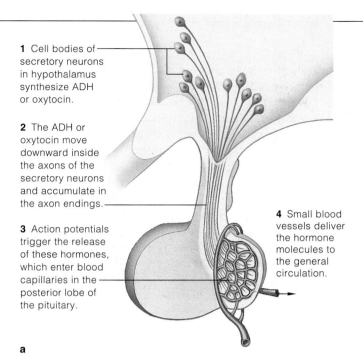

1 Cell bodies of secretory neurons in hypothalamus synthesize ADH or oxytocin.

2 The ADH or oxytocin move downward inside the axons of the secretory neurons and accumulate in the axon endings.

3 Action potentials trigger the release of these hormones, which enter blood capillaries in the posterior lobe of the pituitary.

4 Small blood vessels deliver the hormone molecules to the general circulation.

a

1 Cell bodies of different secretory neurons in the hypothalamus secrete releasing and inhibiting hormones.

2 The hormones are picked up by a capillary bed at the base of the hypothalamus.

3 Bloodstream delivers hormones to a second capillary bed in anterior lobe of pituitary.

5 Hormones secreted from anterior lobe cells enter small blood vessels that lead to the general circulation.

4 Molecules of the releasing or inhibiting hormone diffuse out of the capillaries and act on endocrine cells in the anterior lobe.

a

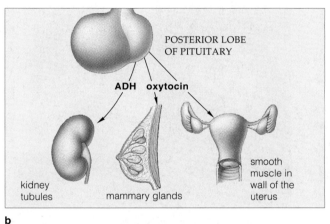

POSTERIOR LOBE OF PITUITARY

ADH   oxytocin

kidney tubules   mammary glands   smooth muscle in wall of the uterus

b

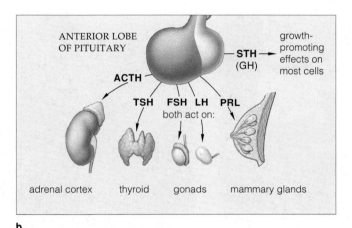

ANTERIOR LOBE OF PITUITARY

STH (GH) — growth-promoting effects on most cells

ACTH

TSH   FSH   LH   PRL
both act on:

adrenal cortex   thyroid   gonads   mammary glands

b

Figure 32.5 (**a**) Functional links between the hypothalamus and the posterior lobe of the pituitary gland. (**b**) Main targets of the posterior lobe secretions.

Figure 32.6 (**a**) Functional links between the hypothalamus and the anterior lobe of the pituitary gland. (**b**) Main targets of the anterior lobe secretions.

gamete formation and hormonal secretions required in sexual reproduction of animals, a topic of Chapter 39. Somatotropin affects metabolism in many tissues and triggers liver secretions that affect growth of bone and soft tissues (Table 32.2 and Figure 32.6*b*). Prolactin acts on different cell types. It's best known for stimulating and sustaining milk production in mammary glands, after other hormones prime the tissues. It also affects hormone production in the ovaries of some species.

ABOUT THE HYPOTHALAMIC TRIGGERS   Most of the hypothalamic hormones acting in the anterior lobe of the pituitary are **releasers**. This means they stimulate secretion of hormones from target cells. For example, GnRH (gonadotropin-releasing hormone) brings about secretion of FSH and LH, which are gonadotropins. As

another example, TRH (thyrotropin-releasing hormone) stimulates secretion of thyrotropin. Other hypothalamic hormones acting in the anterior lobe are **inhibitors** of their targets. Somatostatin is one inhibitor. It is a signal to slow down somatotropin and thyrotropin secretion.

The hypothalamus and pituitary gland produce eight kinds of hormones and interact to control their secretion.

The posterior lobe of the pituitary stores and secretes two hypothalamic hormones, ADH and oxytocin, both of which target specific cell types.

The anterior lobe of the pituitary produces and secretes six hormones, ACTH, TSH, FSH, LH, PRL, and STH. These trigger release of other hormones from other glands, with a variety of effects throughout the body.

EXAMPLES OF ABNORMAL PITUITARY OUTPUT

The body does not churn out enormous quantities of hormone molecules. Two researchers, Roger Guilleman and Andrew Schally, realized this when they isolated the first known releasing hormone. In their four-year attempt to secure TRH, they dissected 500 metric tons of brains and 7 metric tons of hypothalamic tissue from sheep and ended up with only a single milligram of it.

Yet normal body functions depends on tiny but significant amounts of hormone secretions, which commonly are released in short bursts. The frequency of secretion must be controlled to block the underproduction or overproduction of a hormone. When something interferes with controls over hormonal output, the form and function of the body is altered, sometimes in abnormal ways.

For instance, *gigantism* results from overproduction of somatotropin. Affected adults are proportionally like average-size people but larger (Figure 32.7a,b). *Pituitary dwarfism* develops as an outcome of underproduction of somatotropin. Proportionally, affected adults are like an average person but much smaller (Figure 32.7b).

Sometimes somatotropin output becomes excessive in adults, when long bones have stopped lengthening. *Acromegaly* follows. In time, cartilage, bone, and other connective tissues in the jaws, feet, and hands thicken abnormally. So do epithelia of the skin, nose, eyelids, lips, and tongue. Figure 32.7c shows an affected female.

Another example: ADH secretion may dwindle or stop if the pituitary's posterior lobe is damaged, as by a

Figure 32.7 Examples of the outcome of abnormalities in the secretion of a hormone. (**a**) Manute Bol, an NBA center, is 7 feet 6–3/4 inches tall, an outcome of excessive secretion of somatotropin (STH) in childhood. (**b**) More examples of the effect of STH on body growth. The male at the center of this photograph is affected by gigantism, which resulted from excessive STH production in childhood. The male at the right is affected by pituitary dwarfism, an outcome of an underproduction of STH in childhood. The male at the left is of average height.

(**c**) Acromegaly, which resulted from excessive production of STH during adulthood. Before this female reached maturity, she was symptom-free.

blow to the head. This is one cause of *diabetes insipidus*. Symptoms include excessive excretion of dilute urine and life-threatening dehydration. Patients may respond to hormone replacement therapy involving injections or nasal spray applications of synthetic ADH.

Generally, endocrine glands release very small amounts of hormones at frequencies governed by control mechanisms. When controls fail, oversecretion or undersecretion follows and may cause alterations in the body's form or functions.

SOURCES AND EFFECTS OF OTHER HORMONES

Table 32.3 lists hormones from endocrine sources other than the pituitary. The rest of this chapter gives a few examples of their effects and of the controls over their output. As you read, keep the following points in mind.

First, hormones often interact with one another. One or more may oppose, add to, or prime target cells for another hormone's effects. *Second*, negative feedback mechanisms often control secretion. When a hormone's concentration increases or decreases in a body region, the change triggers events that respectively dampen or stimulate its further secretion. *Third*, a target cell may react differently at different times. Responses depend on the hormone's concentration and on the functional state of the cell's receptors. *Fourth*, environmental cues may be important mediators of hormonal secretion.

Hormone secretion and its effects depend on hormone interactions, feedback mechanisms, variations in the state of target cells, and sometimes environmental cues.

Table 32.3 *Hormone Sources Other Than the Mammalian Hypothalamus and Pituitary Gland*

Source	Secretion(s)	Main Targets	Primary Actions
ADRENAL CORTEX	Glucocorticoids (including cortisol)	Most cells	Promote breakdown of glycogen, fats, and proteins as energy sources; thus help raise blood level of glucose
	Mineralocorticoids (including aldosterone)	Kidney	Promote sodium reabsorption (sodium conservation); help control the body's salt–water balance
ADRENAL MEDULLA	Epinephrine (adrenaline)	Liver, muscle, adipose tissue	Raises blood level of sugar, fatty acids; increases heart rate and force of contraction
	Norepinephrine	Smooth muscle of blood vessels	Promotes constriction or dilation of certain blood vessels; thus helps control the flow of blood volume to different body regions
THYROID	Triiodothyronine, thyroxine	Most cells	Regulate metabolism; have roles in growth, development
	Calcitonin	Bone	Lowers calcium level in blood
PARATHYROIDS	Parathyroid hormone	Bone, kidney	Elevates calcium level in blood
GONADS			
Testes (in males)	Androgens (including testosterone)	General	Required in sperm formation, development of genitals, maintenance of sexual traits, growth, and development
Ovaries (in females)	Estrogens	General	Required for egg maturation and release; preparation of uterine lining for pregnancy and its maintenance in pregnancy; genital development; maintenance of sexual traits; growth, development
	Progesterone	Uterus, breasts	Prepares, maintains uterine lining for pregnancy; stimulates development of breast tissues
PANCREATIC ISLETS	Insulin	Liver, muscle, adipose tissue	Promotes cell uptake of glucose, thus lowers glucose level in blood
	Glucagon	Liver	Promotes glycogen breakdown, raises glucose level in blood
	Somatostatin	Insulin-secreting cells	Inhibits digestion of nutrients, hence their absorption from gut
THYMUS	Thymopoietin	T lymphocytes	Poorly understood regulatory effect on T lymphocytes
PINEAL	Melatonin	Gonads (indirectly)	Influences daily biorhythms, seasonal sexual activity
STOMACH, SMALL INTESTINE	Gastrin, secretin, etc.	Stomach, pancreas, gallbladder	Stimulate activities of stomach, pancreas, liver, gallbladder required for food digestion, absorption
LIVER	Somatomedins	Most cells	Stimulate cell growth and development
KIDNEYS	Erythropoietin	Bone marrow	Stimulates red blood cell production
	Angiotensin*	Adrenal cortex, arterioles	Helps control secretion of aldosterone (hence sodium reabsorption, and blood pressure)
	1,25-hydroxyvitamin D_6* (calcitriol)	Bone, gut	Enhances calcium reabsorption from bone and calcium absorption from gut
HEART	Atrial natriuretic hormone	Kidney, blood vessels	Increases sodium excretion; lowers blood pressure

* Kidneys produce *enzymes* that modify precursors of this substance, which enters the general circulation as an activated hormone.

FEEDBACK CONTROL OF HORMONAL SECRETIONS

What controls hormonal secretion? A look at a few of the endocrine glands listed in Table 32.3 will give you a sense of some of the control mechanisms. Briefly, the hypothalamus, pituitary, or both signal a gland to step up or slow down its secretory activity, so a hormone's concentration changes in the blood or elsewhere. With that change, a feedback mechanism trips into action.

By **negative feedback**, a rise in the concentration of a secreted hormone triggers events that *inhibit* further secretion. By **positive feedback**, a rising concentration of a secreted hormone triggers events that *stimulate* its further secretion.

Negative Feedback From the Adrenal Cortex

Humans have a pair of adrenal glands, one above each kidney (Figure 32.2b). Some cells of the **adrenal cortex**, an adrenal gland's outer layer, secrete glucocorticoids and other hormones. Glucocorticoids raise the blood level of glucose. One type, cortisol, acts when the body is so stressed that the glucose level declines below a set point that acts as an alarm. It starts a stress response, which a negative feedback mechanism can later cut off.

Figure 32.8 shows what goes on. The hypothalamus secretes the releasing hormone CRH when a too-low level of glucose in blood sets off the alarm. CRH makes the anterior pituitary secrete corticotropin (ACTH). In turn, ACTH causes cells of the adrenal cortex to secrete

cortisol. In response to that cortisol, liver cells break down their stored glycogen, adipose cells break down fat, and skeletal muscle cells break down proteins. The breakdown products—glucose, fatty acids, and amino acids—enter blood. Except for brain cells, cells take up all three, not just glucose, as energy sources (Section 7.6). With more glucose and alternative energy sources available, the glucose level rises above the set point. If the threat to homeostasis ends, the hypothalamus and pituitary will issue signals to inhibit cortisol secretion.

Chronic stress, injury, or illness causes the nervous system to keep this response going. Why? Prolonged inflammation damages tissues, and cortisol suppresses the inflammation. Hence the use of cortisol-like drugs for asthma and other chronic inflammatory disorders.

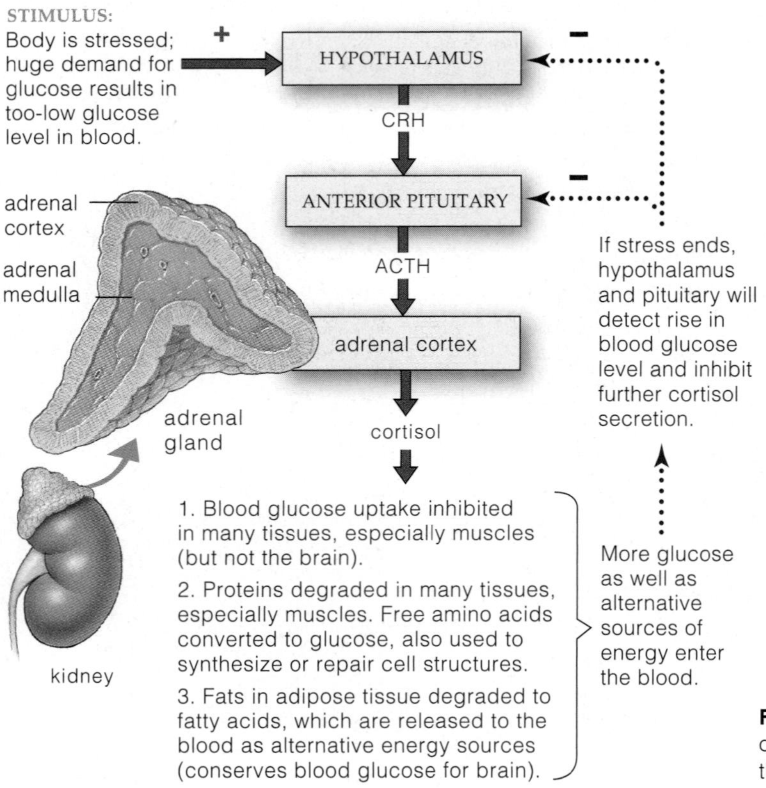

Local Feedback in the Adrenal Medulla

Some neurons reside in the **adrenal medulla**, the adrenal gland's inner region. Their secretions, epinephrine and norepinephrine, are neurotransmitters in some contexts and hormones in others. Sympathetic nerves deliver hypothalamic signals to the adrenal medulla. Suppose they call for norepinephrine's release. Soon, hormone molecules collect in the synaptic cleft between target cells and axon endings from the nerve. A local negative feedback mechanism now operates at receptors on the axon endings. Excess amounts of norepinephrine bind to the receptors and shut down its further release.

When you are excited or stressed, epinephrine and norepinephrine help adjust blood circulation, and fat and carbohydrate metabolism. They boost heart rate, make arteriole diameters widen or narrow in different regions, and dilate airways to the lungs. These highly controlled events direct more of the total blood volume to heart and muscle cells that are to increase oxygen flow to cells demanding a lot of energy. These are the features of the *fight–flight response* (Section 30.8).

Cases of Skewed Feedback From the Thyroid

The human **thyroid gland** is located at the base of the neck in front of the trachea, or windpipe (Figures 32.2b and 32.9a–c). Thyroxine and triiodothyronine, its main hormones, have widespread effects. In their absence, many tissues cannot develop normally. Also, metabolic rates of warm-blooded animals depend on them. The importance of feedback control of their secretion comes into sharp focus in cases of abnormal thyroid output.

STIMULUS:
Body is stressed; huge demand for glucose results in too-low glucose level in blood.

+ → HYPOTHALAMUS → − (to hypothalamus)

CRH

ANTERIOR PITUITARY → −

ACTH

adrenal cortex

cortisol

adrenal cortex
adrenal medulla
adrenal gland
kidney

If stress ends, hypothalamus and pituitary will detect rise in blood glucose level and inhibit further cortisol secretion.

1. Blood glucose uptake inhibited in many tissues, especially muscles (but not the brain).

2. Proteins degraded in many tissues, especially muscles. Free amino acids converted to glucose, also used to synthesize or repair cell structures.

3. Fats in adipose tissue degraded to fatty acids, which are released to the blood as alternative energy sources (conserves blood glucose for brain).

More glucose as well as alternative sources of energy enter the blood.

Figure 32.8 Structure of the human adrenal gland. One gland rests on top of each kidney. The diagram shows a negative feedback loop that governs cortisol secretion from the adrenal cortex.

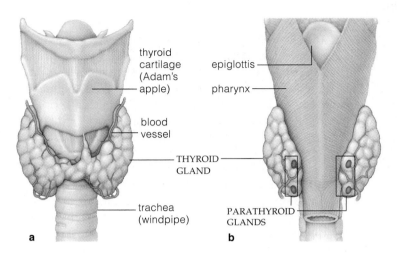

Figure 32.9 (**a**) Human thyroid gland, anterior view. (**b**) Posterior view, showing the location of four parathyroid glands. (**c**) Position of the thyroid in the body. (**d**) A mild case of goiter, displayed by Maria de Medici in the year 1625. A rounded neck was considered to be a sign of great beauty during the late Renaissance. It showed up regularly in parts of the world where iodine supplies were not sufficient for normal thyroid function.

Example: Thyroid hormones cannot be synthesized without iodide, a form of iodine. Iodine-deficient diets cause one or both lobes of the thyroid gland to enlarge (Figure 32.9d). The enlargement, a *simple goiter*, occurs after low blood levels of thyroid hormones cause the anterior pituitary to secrete THS (thyroid-stimulating hormone). The thyroid attempts to make the hormones but cannot do so, which leads to continued secretion of TSH, and so on in a sustained, abnormal feedback loop. *Hypothyroidism* is the clinical name for low blood levels of thyroid hormones. Affected adults may be sluggish, overweight, intolerant of cold, dry-skinned, confused, and depressed. Simple goiter is no longer common in places where people use iodized salt.

Grave's disorder and other forms of *toxic goiter* result from excessive levels of thyroid hormones in blood, a condition we call *hyperthyroidism*. Symptoms include heat intolerance, irritability, anxiety, tremors, fatigue, difficulty sleeping, protruding eyes, and an irregular, pounding heartbeat. Some of the cases are autoimmune disorders; antibodies wrongly stimulate thyroid cells (Section 35.9). Other cases arise after the thyroid has become inflamed or develops nodules or tumors.

Feedback Control of the Gonads

Gonads are *primary* reproductive organs, which make and secrete gametes and certain sex hormones. Testes (singular, testis) in males and ovaries in females are examples. Testes secrete testosterone; ovaries secrete estrogens and progesterone. These hormones influence secondary sexual traits (as they did for those chimps described earlier), and feedback controls govern their secretion. Figure 32.10 is a preview of the feedback loops from ovaries to the hypothalamus and pituitary during the menstrual cycle, a key topic of Chapter 39.

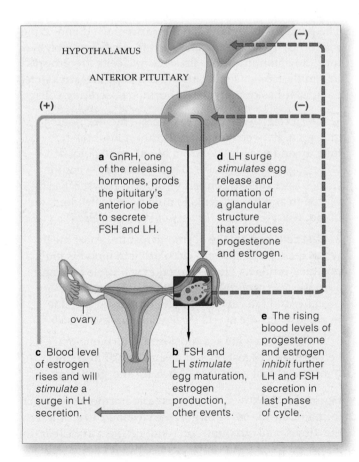

Figure 32.10 Feedback loops to the hypothalamus and the pituitary gland from the ovaries during the menstrual cycle, a recurring reproductive event. Positive feedback triggers egg release from an ovary. Negative feedback after its release prevents release of another egg until the cycle is over.

Feedback mechanisms control secretions from endocrine glands. In many cases, feedback loops to the hypothalamus, pituitary, or both govern the secretory activity.

Negative feedback slows down the further secretion of a hormone. Positive feedback enhances its further secretion.

RESPONSES TO LOCAL CHEMICAL CHANGES

Hormones and neural signals are not the main controls over some endocrine glands or cells. Chemical changes in the local neighborhood stimulate or inhibit hormone secretions, too, as the following examples illustrate.

Secretions From Parathyroid Glands

Humans have four **parathyroid glands** on the posterior surface of the thyroid gland (Figure 32.9b). The glands secrete parathyroid hormone (PTH), the key regulator of calcium levels in blood. Calcium ions have roles in muscle contraction, enzyme action, blood clotting, and other activities. PTH is secreted when the calcium level is low. Its secretion slows when the calcium level rises. PTH acts on cells present in the skeleton and kidneys.

PTH induces living bone cells to secrete enzymes that digest bone tissue, and this releases calcium and other minerals into interstitial fluid. Calcium enters blood, then it enters nephrons in the kidneys, where PTH enhances its reabsorption. PTH also induces some kidney cells to secrete enzymes that act on blood-borne precursors of an active form of vitamin D_3, a hormone (Section 33.2). The activated form stimulates intestinal cells to absorb much more calcium from the lumen of the gut. In children who have vitamin D deficiency, not enough calcium and phosphorus are absorbed, so rapidly growing bones develop improperly. The resulting bone abnormality, *rickets*, is characterized by bowed legs, a malformed pelvis, and in many cases a malformed skull and rib cage (Figure 32.11).

Figure 32.11 A child who is affected by rickets.

Effects of Local Signaling Molecules

Many cells detect changes in the surrounding chemical environment and alter their activity, often in ways that counteract or amplify those changes. The cells secrete various local signaling molecules, the effects of which are confined to the immediate vicinity of change. Target cells take up most signaling molecules so rapidly that few enter the general circulation.

At least sixteen types of prostaglandins, each having a twenty-carbon fatty acid structure with a five-carbon ring, occur in many tissues. Production and secretion of these local signaling molecules rise as local chemical conditions change, with potent results. Some make rings of smooth muscle in arteriole walls constrict or dilate; this helps redirect blood flow through the body. Others have roles in inflammation, control of stomach acid and intestinal motility, and childbirth. They also function in overall endocrine control. They might have therapeutic use for asthma, ulcers, hypertension, and heart attacks.

Those growth factors you read about in Section 14.4 are local signaling molecules. They affect transcription of genes required in development. Example: Rita Levi-Montalcini discovered one nerve growth factor (NFG) that helps neurons survive and guides their direction of growth in an embryo. We know that in the presence of NFG, immature neurons can survive indefinitely in tissue culture. Without it, they die within a few days.

Secretions From Pancreatic Islets

The pancreas is a gland having exocrine and endocrine functions. Its *exocrine* cells secrete digestive enzymes into the small intestine. Its *endocrine* cells are grouped in 2 million or so small clusters called **pancreatic islets**. Each islet has three types of hormone-secreting cells:

1. *Alpha* cells in the pancreas secrete the hormone glucagon. Between meals, cells throughout the body take up and use glucose from blood, so the blood level of glucose decreases. Then, glucagon secretion causes liver cells to convert its storage polysaccharide glycogen and amino acids to glucose, which enters the blood. *Glucagon raises the level of glucose in blood.*

2. *Beta* cells secrete the hormone insulin. After meals, when the level of glucose circulating in blood is high, insulin stimulates glucose uptake by muscle cells and adipose cells especially. It promotes the synthesis of proteins and fats, and it inhibits protein conversion to glucose. *Insulin lowers the level of glucose in blood.*

3. *Delta* cells secrete somatostatin, a hormone that helps control digestion and absorption of nutrients. It also can block secretion of insulin and glucagon.

Figure 32.12 shows how pancreatic hormones interact to maintain the level of glucose in blood even though the times and amounts of food intake vary. Bear in mind, insulin is the only hormone that causes cells to take up and store glucose in forms that can be swiftly tapped when required. Its crucial role in carbohydrate, protein, and fat metabolism becomes clear when we consider people who cannot produce enough insulin or whose target cells cannot respond to it.

Figure 32.12 Some of the homeostatic controls over glucose metabolism.

Following a meal, glucose enters the blood faster than cells can take it up. The blood glucose level rises, and the chemical change stimulates pancreatic beta cells to secrete insulin. The main targets, liver and muscle cells, use glucose at once and store excess amounts as glycogen. In this way, insulin *lowers* the glucose level in blood.

Between meals, the blood glucose level decreases, and the chemical change stimulates pancreatic alpha cells to secrete glucagon. Target cells with receptors for this hormone convert glycogen back to glucose, which enters the blood. In this way, glucagon *raises* the glucose level in blood.

The nervous system helps control glucose metabolism. For example, as you read earlier, the hypothalamus orders the adrenal medulla to secrete glucocorticoids when the body is being stressed. As one effect, glycogen is broken down to glucose in the liver, and glycogen synthesis slows, in liver and muscle tissue especially.

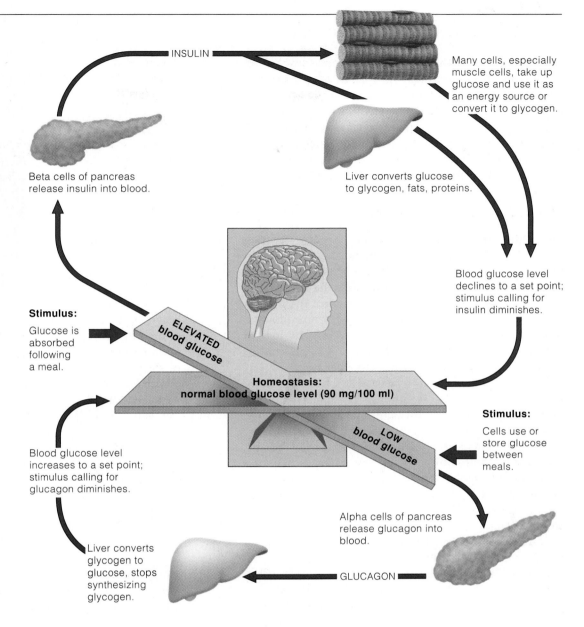

INSULIN

Many cells, especially muscle cells, take up glucose and use it as an energy source or convert it to glycogen.

Beta cells of pancreas release insulin into blood.

Liver converts glucose to glycogen, fats, proteins.

Blood glucose level declines to a set point; stimulus calling for insulin diminishes.

Stimulus:
Glucose is absorbed following a meal.

ELEVATED blood glucose

Homeostasis:
normal blood glucose level (90 mg/100 ml)

LOW blood glucose

Stimulus:
Cells use or store glucose between meals.

Blood glucose level increases to a set point; stimulus calling for glucagon diminishes.

Liver converts glycogen to glucose, stops synthesizing glycogen.

Alpha cells of pancreas release glucagon into blood.

GLUCAGON

For example, insulin deficiency may lead to *diabetes mellitus*. In this disorder, excess glucose accumulates in blood, then in urine. Urination becomes excessive, so the body's water–solute balance is disrupted. Affected people become dehydrated and thirsty, abnormally so. Without a steady supply of glucose, their body cells start depleting their own fats and proteins as sources of energy. Weight loss is one outcome. Another is that ketones accumulate in the blood and urine. Ketones are normal acidic products of fat breakdown. When they accumulate, they contribute to water losses and alter the body's acid–base balance. Such imbalances disrupt brain function. In extreme cases, death follows.

In *type 1 diabetes*, the body mounts an autoimmune response against its insulin-secreting beta cells. White blood cells mistakenly identify the beta cells as foreign and kill them. A combination of genetic susceptibility

and environmental factors gives rise to this disorder, which is less common than other forms of diabetes but more dangerous in the short term. Usually, symptoms first appear in childhood and adolescence (the disorder is also called juvenile-onset diabetes). Type 1 diabetic patients survive with insulin injections.

In *type 2 diabetes*, insulin levels are close to or above normal, but target cells cannot respond to it. Symptoms usually develop in middle age, for beta cells of affected people produce less insulin over time. Affected people lead normal lives by controlling diet and weight. Often prescription drugs enhance insulin action or secretion.

Secretions from some endocrine glands and endocrine cells are direct homeostatic responses to chemical change in the local neighborhood.

HORMONES AND THE ENVIRONMENT

This last section of the chapter invites you to reflect on a key point. An individual's growth, development, and reproduction begin with genes and hormones, and so does behavior. *But certain environmental factors commonly influence gene expression and hormonal secretion, and they do so in predictable ways.* Later chapters invite analysis of specific environmental effects on animals. For now, it is enough to consider the following examples.

Daylength and the Pineal Gland

Inside the brain is the **pineal gland**, a photosensitive organ (Figure 32.2). It secretes the hormone melatonin in the absence of light, so the blood level of melatonin varies from day to night and through the seasons. In many species, these variations help control the growth and development of gonads, the primary reproductive organs. How? Their melatonin is a piece of a **biological clock**, an internal timing mechanism. It helps control reproductive cycles and reproductive behavior.

Think of a hamster in winter, when there are more hours of darkness than in summer. The melatonin level in its blood is high, and this suppresses sexual activity. In summer, when daylength is longest, the blood level is low and hamster sex peaks. Similarly, in the fall and winter, melatonin indirectly suppresses the growth of a male white-throated sparrow's gonads (Figure 32.13a). When daylight increases in spring, stepped-up gonadal activity leads to production of hormones that influence singing behavior, as described in Section 41.1. With his distinctive song, the male sparrow defines his territory.

Does melatonin influence human behavior as well? Perhaps. Clinical observations and studies suggest that decreased melatonin levels may trigger **puberty**, the age at which reproductive organs and structures start to mature. For example, some patients who did not have a functional pineal gland entered puberty prematurely.

Figure 32.13 (**a**) Male white-throated sparrow, belting out a song that began, indirectly, with an environmentally induced decrease in melatonin secretion from the pineal gland. (**b**) Annie blanketing her winter blues.

Melatonin also acts on neurons that make the body temperature decline and make you drowsy after sunset, when light wanes. At sunrise, less melatonin is secreted and body temperature rises. In response, you wake up and become active. A biological clock governs the cycle of sleep and arousal. This one ticks in synchrony with daylength. We have indirect evidence of its action from night workers who cannot get to sleep in the morning. The same thing happens to travelers from the United States to Paris who go through four days of "jet lag." Two or three hours past midnight they sit up in bed, but two hours past noon they are ready for sleep.

Seasonal affective disorder (SAD) hits some people in winter. They become exceedingly depressed, binge on carbohydrates, and have an almost overwhelming need to sleep (Figure 32.13b). Such "winter blues" might develop when a biological clock is out of sync with the seasonally shorter daylengths. Clinically administered melatonin worsens the seasonal symptoms. Exposure to intense light, which shuts down pineal activity, often leads to dramatic improvement.

Thyroid Function and Frog Habitats

Since 1994, in natural habitats around the world, the number of deformed frogs has been skyrocketing. The reasons are not fully understood. In 1999, researchers decided to expose *Xenopus laevis* embryos to Minnesota and Vermont lakewater. Half the samples came from lakes where deformity rates were low. The other half came from "hot spots" with as many as twenty kinds of dissolved pesticides—and with high deformity rates.

When the frog embryos developed into tadpoles, the ones that had been raised in the hot-spot water had bent spines, malformed eyes, and a malformed mouth. Some did not metamorphose into adults (Figure 32.14). The other frog embryos of the sampling were normal.

Thyroid hormones orchestrate much of vertebrate development. For instance, a surge in thyroid hormone output stimulates a tadpole to metamorphose into an

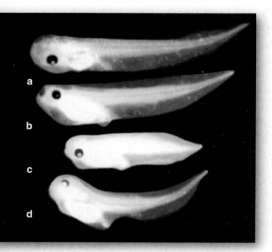

Figure 32.14 *Xenopus laevis* tadpoles showing possible environmental effects on the thyroid, hence on development. (**a**) Tadpole raised in water taken from a lake where there were few deformed frogs. (**b–d**) Three tadpoles raised in water taken from three "hot spot" lakes that had increasingly higher concentrations of dissolved chemical compounds, both natural and synthetic.

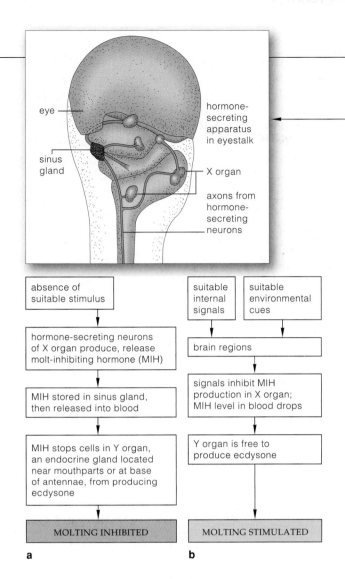

Figure 32.15 (**a,b**) Steps in the hormonal control of molting in crustaceans, including crabs (**c**). The steps differ a bit in insects, which do not use a molt-inhibiting hormone. Rather, stimulation of the insect brain makes certain neurons secrete ecdysiotropin. This hormone induces other neurons to make and secrete still another hormone, which targets ecdysone-producing cells in prothoracic glands. (**d**) This insect, one of the cicadas, is emerging from its old cuticle.

adult. Are conditions in hot-spot habitats interfering with thyroid functions? Maybe. Embryos that developed in hot-spot water displayed few symptoms, or none at all, when they were given extra thyroid hormones.

Other studies correlated some frog deformities with ultraviolet radiation and with infections by a parasitic protistan (Section 24.6). "Chemical soup" habitats that disrupt thyroid function are now added to the list.

Comparative Look at a Few Invertebrates

Although this chapter's focus has been on vertebrates, do not lose sight of the fact that all organisms produce signaling molecules of one sort or another. Let's look at hormonal control of **molting**, a periodic discarding and replacement of a hardened cuticle that otherwise would limit increases in body mass. Molting occurs in the life cycle of all insects, crustaceans, and other invertebrates with thick cuticles (Sections 23.11, 23.13, and 23.15).

Although details vary from group to group, molting is largely under the control of **ecdysone**. This steroid hormone is derived from cholesterol and is chemically related to many crucial vertebrate hormones. In insects and crustaceans, molting glands synthesize and store

ecdysone and then release it for distribution through the body at molting time. Hormone-secreting neurons in the brain seem to control its release. Apparently they respond to a combination of environmental cues, such as light and temperature, as well as internal signals.

Figure 32.15 provides examples of the control steps, which differ in crustaceans and insects. During premolt and molting periods, coordinated interactions among ecdysone and other hormones bring about structural and physiological changes. The interactions make the old cuticle separate from the epidermis and muscles. They induce changes that dissolve inner layers of the cuticle and recycle the remnants. The interactions also trigger changes in metabolism and in the composition and volume of the internal environment. They promote cell divisions, secretions, and pigment formation, all of which go into making the new cuticle. Simultaneously, hormonal interactions control heart rate, muscle action, color changes, and other physiological processes.

Environmental cues, such as changes in light intensity from day to night and seasonal changes in daylength, influence certain hormonal secretions.

SUMMARY   *Gold* indicates text section

1. The cells of complex animals continually exchange substances with the body's internal environment. Their withdrawals and secretions are integrated in ways that ensure cell survival through the whole body. *32.1*

2. Integration of cell activities requires the stimulatory or inhibitory effects of signaling molecules. *32.1, 32.2*

a. Signaling molecules are chemical secretions from a cell that adjust the behavior of other, target cells.

b. Any cell with molecular receptors for a signaling molecule is its target. Target cells may or may not be adjacent to the cell that sends the signal.

c. There are different kinds of signaling molecules. Hormones as well as neurotransmitters, local signaling molecules, and pheromones are the main kinds.

3. In target cells, hormones influence gene activation, protein synthesis, and alterations in existing enzymes, membranes, and other cellular components. Hormones exert physiological effects by interacting with specific protein receptors at the plasma membrane or inside the cytoplasm of target cells. *32.2*

a. Steroid hormones enter the target cell's nucleus directly, or after binding with an intracellular receptor, or by binding with plasma membrane receptors.

b. Being water soluble, the protein hormones cannot enter target cells. They bind to membrane receptors at the cell surface. Their effect is exerted with the help of transport proteins as well as second messengers in the cytoplasm, some of which trigger the actual response.

4. The posterior lobe of the pituitary stores and secretes two hypothalamic hormones, ADH and oxytocin. ADH targets kidney cells, thus influencing extracellular fluid volume. Oxytocin acts on cells in mammary glands and the uterus to influence reproductive events. *32.3*

5. Hypothalamic hormones known as releasing and inhibiting hormones control secretions from a variety of cells of the anterior lobe of the pituitary gland. *32.3*

6. The anterior lobe makes and secretes six hormones: ACTH, TSH, FSH, LH, PRL, and STH. These trigger secretion from the adrenal cortex, the thyroid gland, gonads, and mammary glands. They elicit a variety of responses throughout the body. *32.3*

7. The vertebrate body has other sources of hormones (e.g., adrenal medulla; parathyroid, thymus, and pineal glands; the pancreatic islets; and endocrine cells in the stomach, small intestine, liver, and heart). *32.5*

8. Interactions among hormones, feedback mechanisms, the number and kind of target cell receptors, variations in the state of target cells, and often environmental cues influence secretion of a hormone and its effects. *32.6*

9. The secretion of local signaling molecules, such as prostaglandins, is a direct response to a change in the localized chemical environment. *32.7*

a. In general, secretion of hormones such as insulin and parathyroid hormone can change rapidly when the extracellular concentration of some substance must be homeostatically controlled.

b. Hormones such as somatotropin have prolonged, slow, and often irreversible effects, as on development.

10. Environmental cues, such as the change in sunlight intensity from day to night and the seasonal changes in daylength, influence some hormonal secretions. *32.8*

Review Questions

1. Name the endocrine glands typical of most vertebrates and state where each is located in the human body. *32.1*

2. Distinguish among hormones, neurotransmitters, local signaling molecules, and pheromones. *32.1*

3. A hormone molecule binds with a receptor on a plasma membrane. It does not enter the cell. Binding activates a second messenger in the cell, which triggers an amplified response to the hormonal signal. State whether the molecule is a steroid hormone or a peptide hormone. *32.2*

4. Which secretions of the posterior lobe of the pituitary gland have the targets indicated? (Fill in the blanks.) *32.3*

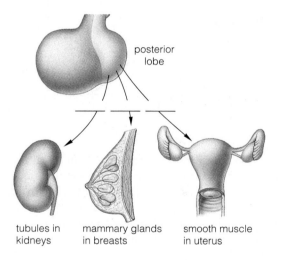

tubules in kidneys   mammary glands in breasts   smooth muscle in uterus

5. Which secretions of the anterior lobe of the pituitary gland have the targets indicated? (Fill in the blanks.) *32.3*

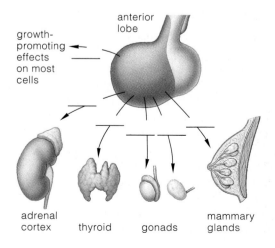

adrenal cortex   thyroid   gonads   mammary glands

Self-Quiz   ANSWERS IN APPENDIX III

1. _____ are molecules released from a signaling cell that have effects on target cells.
   - a. Hormones
   - b. Neurotransmitters
   - c. Pheromones
   - d. Local signaling molecules
   - e. both a and b
   - f. a through d

2. Hormones are products of _____ .
   - a. endocrine glands
   - b. some neurons
   - c. exocrine cells
   - d. a and b
   - e. a and c
   - f. a, b, and c

3. Second messengers include _____ .
   - a. steroid hormones
   - b. protein hormones
   - c. cyclic AMP
   - d. both a and b

4. ADH and oxytocin are hypothalamic hormones secreted from the _____ lobe of the pituitary gland.
   - a. anterior
   - b. posterior
   - c. intermediate
   - d. secondary

5. GnRH is a _____ secreted by hypothalamic neurons.
   - a. releasing hormone
   - b. inhibiting hormone
   - c. corticotropin
   - d. somatotropin

6. Which do *not* stimulate hormone secretions?
   - a. neural signals
   - b. local chemical changes
   - c. hormonal signals
   - d. environmental cues
   - e. All of the above can stimulate secretion.

7. _____ lowers blood sugar levels; _____ raises it.
   - a. Glucagon; insulin
   - b. Insulin; glucagon
   - c. Gastrin; insulin
   - d. Gastrin; glucagon

8. The pituitary detects a rising hormone concentration in blood and inhibits the gland secreting the hormone. This is a _____ feedback loop.
   - a. positive
   - b. negative
   - c. long-term
   - d. b and c

9. Match the hormone source with the closest description.
   - _____ adrenal medulla
   - _____ thyroid gland
   - _____ parathyroids
   - _____ pancreatic islets
   - _____ pineal gland
   - _____ prostaglandin
   - a. affected by daylength
   - b. potent local effects
   - c. raise blood calcium level
   - d. epinephrine source
   - e. insulin, glucagon
   - f. hormones require iodide

Critical Thinking

1. The zebra offspring being nursed in Figure 32.16 is too young to nourish itself by eating grasses. Its source of nutrients is its mother's milk. Outline how secretions from the hypothalamus and the pituitary gland influence milk production and secretion.

2. In winter, with its fewer daylight hours than in the summer, Maxine became exceedingly depressed, craved carbohydrate-rich foods, and stopped exercising regularly. She put on a great deal of weight. Her doctor diagnosed her condition as *seasonal affective disorder* (SAD), the winter blues. Maxine was advised to purchase a cluster of intense, broad-spectrum lights and sit near them for at least an hour each day. The treatment quickly lifted the cloud of depression. Use your knowledge of the secretory activity of the pineal gland to explain why Maxine's symptoms appeared and why the prescribed therapy worked.

3. Marianne is affected by *type 1 insulin-dependent diabetes*. One day, after injecting herself with too much insulin, she starts to shake and feels confused. Her doctor recommends a glucagon injection. What caused her symptoms? How would an injection of glucagon help?

Figure 32.16   Female zebra nursing her offspring.

4. In one application of recombinant DNA technology, growth hormone—somatotropin—is now commercially available for treating pituitary dwarfism. Although it's illegal to do so, some athletes use somatotropin instead of anabolic steroids. (Refer to Section 33.9, *Critical Thinking* question 4.) Why? Somatotropin can't be detected by drug test procedures currently employed in sports medicine. Explain how athletes might believe this hormone can improve their performance.

5. *Osteoporosis* is a condition in which loss of calcium results in thin, brittle bones. Combined with other treatments, vitamin D_3 injections are sometimes recommended. Explain why.

Selected Key Terms

adrenal cortex 32.6	neurotransmitter 32.1
adrenal medulla 32.6	pancreatic islet 32.7
biological clock 32.8	parathyroid gland 32.7
ecdysone 32.8	peptide hormone 32.2
endocrine system 32.1	pheromone 32.1
gonad 32.6	pineal gland 32.8
hormone, animal 32.1	pituitary gland 32.3
hypothalamus 32.3	positive feedback 32.6
inhibitor (hypothalamic) 32.3	puberty 32.8
local signaling	releaser (hypothalamic) 32.3
molecule 32.1	second messenger 32.2
molting 32.8	steroid hormone 32.2
negative feedback 32.6	thyroid gland 32.6

Readings

Goodall, J. 1986. *The Chimpanzees of Gombe*. Cambridge, Massachusetts: Belknap Press of Harvard University Press.

Goodman, H. 1994. *Basic Medical Endocrinology*. Second edition. New York: Raven Press.

Hadley, M. 1995. *Endocrinology*. Fourth edition. Englewood Cliffs, New Jersey: Prentice-Hall.

Raloff, J. 2 October 1999. "Thyroid Linked to Some Frog Defects." *Science News* 156:212.

Sherwood, L. 1997. *Human Physiology*. Third edition. Belmont, California: Wadsworth.

On-Line readings at Student Guide for InfoTrac: www.brookscole.com/biology

PROTECTION, SUPPORT, AND MOVEMENT

Of Men, Women, and Polar Huskies

In 1989 Will Steger and his dog-sled team walked on ice for seven months, enduring temperatures of −113°F and a blizzard that lasted for more than seven weeks. They crossed Antarctica, all 6,023 kilometers (3,741 miles) of it. In 1995 this legendary polar explorer set out with four men, two women, and thirty-three sled dogs to cross 3,220 kilometers of the Arctic Ocean in one season. Ice blankets this northernmost ocean in winter and thins treacherously during the spring thaw. The sled dogs stayed with the team for two-thirds of the journey. Steger flew them out only when the team had to cross too much melting ice and switched to using canoes.

To Steger's mind, polar huskies were the heroes of the crossings, the members of the team that worked hardest and pulled all the weight (Figure 33.1). The traits of this mixed breed were modified through years of artificial selection among Canadian and Greenland huskies (bred for size and strength), Siberian huskies (bred for intelligence), and Alaskan racing dogs (bred for spirit and endurance).

A husky's leg bones are sturdy yet lightweight. Its forelegs move freely, thanks to a deep but not too broad rib cage. You won't see the massive muscles of its hind legs in sprinting greyhounds or cheetahs. They are the muscles of a load-pulling, long-distance runner. The husky has tough, calloused foot pads—cushions against sharp ice and frozen rock. Like many other mammals, it has a fur coat with underhair: a dense, soft, insulative layer that traps heat. Its coarser, longer, slightly oily guard hairs protect the underhair from wear and tear. On winter nights, a husky gets comfortable and covers its nose with its furry tail, oblivious of drifting snow.

Steger and his teammates, Victor Boyarsky, Julie Hanson, Martin Hignell, Paul Pregont, and Takako Takano, did not even approach the polar husky's stamina and built-in protection against the elements. Long before the polar crossings, the team followed a regimen of diet and exercise to put their arm and leg muscles in peak condition. Human legs are adapted for long-distance walking, not load-pulling motion.

Figure 33.1 In Ely, Minnesota, Will Steger and his polar huskies warming up for an Arctic crossing.

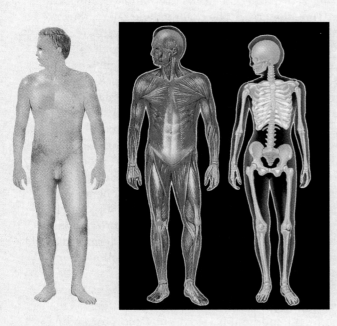

Figure 33.2 From left to right, overview of the human integumentary, muscle, and skeletal systems.

And our skin can't withstand bitter cold. Lacking the fur coat of mammals that evolved in polar climates, team members used clothes that could insulate and protect them from cold without restricting movement. From this perspective, it was human ingenuity that let humans keep company with polar huskies, which are supremely adapted for the challenges of life on ice.

With this chapter, we turn to three systems that give the body of most animals its superficial features, shape, and capacity to move. We call them the integumentary system, muscle system, and skeletal system. Figure 33.2 shows examples from a representative vertebrate.

Animals ranging from worms to humans have an outer covering, or **integument** (after the Latin *integer*, meaning to cover). Most are tough, pliable barriers against many environmental threats. Roundworms and insects, crabs, and other arthropods have a chitin-hardened cuticle (Chapter 23). Vertebrate **skin** is an integument that includes various structures derived from epidermal cells of its outer tissue layers.

Also, regardless of the species, movement of the body or parts of it requires contractile cells and some medium or structure against which contractile force is applied. As you will see, hydrostatic skeletons, exoskeletons, and endoskeletons receive the applied force among different animal groups.

Key Concepts

1. Nearly all animals have an integument, some type of skeleton, and muscles. An integument is the outer covering of the animal body. Vertebrate skin is a prime example.

2. Skin protects the body from abrasion, ultraviolet radiation, bacterial attack, and other environmental assaults. It also contributes to overall body functioning, as when it helps control moisture loss.

3. Three categories of skeletal systems are common in the animal kingdom. We call them hydrostatic skeletons, exoskeletons, and endoskeletons. Each has body fluids or structural elements, such as bones, against which a contractile force can be applied.

4. Bones are collagen-rich, mineralized organs. They function in movement, protection and support of soft organs, and mineral storage. Blood cells form only in certain bones. Ligaments or cartilage bridges the joints between bones, and tendons attach them to skeletal muscles.

5. Many responses to changes in external and internal conditions involve muscles that move the animal body or parts of it. In response to suitable stimulation, the cells of muscle tissue contract, or shorten.

6. Smooth muscle and cardiac muscle are responsible for the motions of internal organs. Skeletal muscle helps move the body's limbs and other structural elements and maintain their spatial positions.

7. In each muscle cell, many threadlike structures called myofibrils are divided into sarcomeres. The sarcomere is the basic unit of contraction. It has parallel arrays of actin and myosin filaments. ATP-driven interactions between the actin and myosin shorten the sarcomeres of a muscle and collectively account for its contraction.

VERTEBRATE SKIN

No garment comes close to having the qualities of skin. Skin can hold its shape after repeated stretchings and washings, block harmful rays from the sun, kill many bacteria on contact, hold in moisture, fix small cuts or burns, *and* last as long as its owner. Skin makes vitamin D, required for calcium metabolism. It helps adjust body temperature when the nervous system rapidly directs the flow of blood—which transports metabolic heat—to and from blood vessels in skin, the body's largest organ. And signals from sensory receptor endings in skin help the brain assess what's going on in the outside world.

Vertebrate skin consists of a dermis and epidermis. Like puff pastry, **epidermis** has sheetlike layers; it is a stratified epithelium. Many junctions structurally and functionally join its cells (Section 29.1). Ongoing, rapid

mitotic cell divisions push epidermal cells from deeper layers to the sheet's free surface. Because of wear and tear at the surface, along with pressure exerted by the growing cell mass, older cells are flattened and dead by the time they reach the outer layers. There, they are abraded off or flake away on an ongoing basis.

Keratinocytes, the most abundant epidermal cells, make keratin, a water-insoluble protein. Its **melanocytes** produce and donate melanin, a brown-black pigment, to the keratinocytes. Melanin blocks harmful ultraviolet radiation. Humans generally have the same number of melanocytes, but skin color varies owing to differences in the distribution and metabolic activity of these cells. For example, melanocytes in *albinos* cannot produce all of the enzymes necessary for melanin production. Pale skin contains little melanin, so the pigment hemoglobin inside red blood cells is not masked. Skin appears pink because hemoglobin's red color shows through thin-walled blood vessels and through the epidermis itself, all of which are transparent. Carotene, a yellow-orange pigment, also contributes to skin color.

The **dermis** is mainly a dense connective tissue with stretch-resisting elastin fibers and supportive collagen fibers. Blood and lymph vessels, and receptor endings of sensory nerves, thread through it. Below the dermis, a hypodermis anchors skin to underlying structures, yet lets it move a bit (Figure 33.3). Fats stored here help insulate the body and cushion some of its parts.

Human skin contains sweat glands, oil glands, and hair follicles (husklike cavities). Fluid secreted by sweat glands is 99 percent water, with dissolved salts, traces of ammonia, vitamin C, and other substances. Except on foot soles and the palms of hands, sebaceous glands (oil glands) lubricate and soften skin and hair. Their secretions kill many bacteria. *Acne*, an inflammation of skin, develops after bacteria infect oil gland ducts.

Beyond the typical array of epithelia and glands, structures derived from skin vary within and between vertebrates. For instance, birds have feathers, bills, and claws. Porcupines have quills. Plant-eating mammals have hooves and horns, you have nails, and so on.

Each **hair** is a flexible structure rooted in skin, with a shaft above the skin's surface (Section 3.5). Near its base, its mostly keratinized cells divide, are pushed upward, then flatten and die. Flattened cells of the shaft's outer layer overlap like roof shingles. If mechanically abused, these frizz out as "split ends." An average human scalp has about 100,000 hairs, but genes, nutrition, hormones, and stress influence hair growth and density.

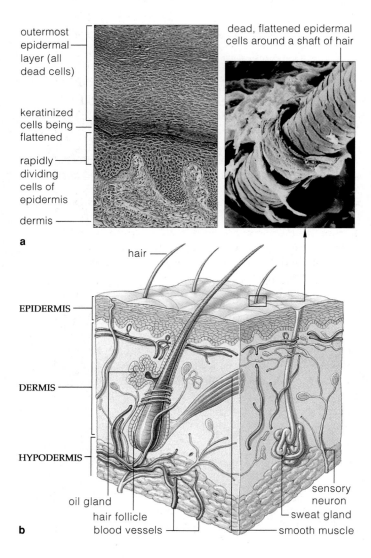

outermost epidermal layer (all dead cells)

keratinized cells being flattened

rapidly dividing cells of epidermis

dermis

dead, flattened epidermal cells around a shaft of hair

a

hair

EPIDERMIS

DERMIS

HYPODERMIS

oil gland
hair follicle
blood vessels

sensory neuron
sweat gland
smooth muscle

b

Figure 33.3 (**a**) Micrograph of the thickened skin structure typical of human palms and soles. (**b**) Sketch of the skin structure typical of body parts with hair, oil glands, and sweat glands.

Like most animals, vertebrates have an integument, a protective covering that typically is tough yet pliable. Their integument, skin, consists of epidermis and dermis.

Sunlight and Skin

THE VITAMIN CONNECTION Sit outside in the sun and you give some of your skin's epidermal cells the chance to make cholecalciferol, a form of vitamin D. **Vitamin D** is a generic name for steroid-like compounds that help the body absorb calcium from food. Expose one type of skin cell to sunlight and it produces vitamin D from a precursor molecule related to cholesterol. The cells then secrete vitamin D, which the bloodstream takes up and transports to absorptive cells of the intestinal lining. This is a hormone-like action. It means that skin acts like an endocrine gland when exposed to sunlight.

As you know, the sun's rays also contain ultraviolet (UV) wavelengths, which are harmful. Dark skin is able to block UV radiation while still making vitamin D. It also blocks the breakdown of a B vitamin, **folate**. Among other things, folate happens to be essential for proper embryonic development, hence for reproduction.

Humans first evolved beneath the intense sun of the African savanna. Skin alone would have supplied our early ancestors with enough vitamin D. What happened when some populations moved out of the tropics, to higher latitudes? What happened when they moved into caves and bundled up in clothing? Dietary sources of vitamin D became more important. In addition, selection pressure for embryo-protecting dark skin probably eased up. Nina Jablonksi and George Chaplin compared satellite data on annual UV exposure levels with skin reflectance (the higher the reflectance, the lighter the skin). In more than fifty countries throughout the world, they matched up increasingly light skin with decreasing UV levels.

SUNTANS AND SHOE-LEATHER SKIN When exposed to UV radiation, the melanocytes in skin are stimulated to make melanin. Production accelerates slowly, then peaks about ten days after the initial exposure. Before then, unprotected skin can get mildly to severely burned.

Ongoing exposure produces the "tan" that so many light-skinned people covet (Figure 33.4). Dark-skinned people have better initial protection from UV radiation. Yet even in naturally dark skin, prolonged UV exposure causes elastin fibers in connective tissue of the dermis to clump together. Skin loses its resiliency as a result. In time, it starts looking like old shoe leather.

Skin will age anyway. As any person grows older, epidermal cells divide less often. Skin gets thinner and more vulnerable to injury; glandular secretions that once kept it soft and moistened dwindle. Collagen and elastin fibers in the dermis break down and become sparser, so skin loses its elasticity and wrinkles deepen. However, you can accelerate the process and look older faster in several ways, as by continually exposing your skin to dry winds or tobacco smoke—and by excessive tanning.

SUNLIGHT AND THE FRONT LINE OF DEFENSE Besides having cells that make melanin and keratin, skin also has cells that help protect the body against pathogens and against cancer. The two defending cell types of the epidermis are Langerhans and Granstein cells.

Langerhans cells are phagocytes that develop in bone marrow, then take up stations in skin. After engulfing virus particles or bacteria, they display molecular alarm signals at their plasma membrane. The signals mobilize the body's immune system. UV radiation damages these cells. This may be why sunburns often trigger *cold sores*, small, painful blisters that announce the recurrence of a *Herpes simplex* infection. Nearly everyone harbors the *H. simplex* virus. It remains hidden in the face, inside a ganglion. (A ganglion, recall, is a cluster of neuron cell bodies.) Sunburns and other stress factors can activate the virus. When that happens, virus particles move down the neurons to the axonal endings in skin. There they infect epithelial cells and cause eruptions.

When UV radiation damages Langerhans cells, it weakens one of your first lines of defense. It may activate proto-oncogenes and initiate the cancerous transformation of skin cells. (You read about this in the Chapter 14 introduction and Section 14.4.) Skin cancers grow rapidly and may spread to adjacent lymph nodes unless they are surgically removed.

In some as-yet undetermined way, **Granstein cells** interact with the white blood cells that can put the brakes on immune responses in the skin. By issuing suppressor signals, they help keep responses from spiraling out of control. Although their roles are not completely understood, researchers found that these cells are less vulnerable than Langerhans cells to the damaging effects of ultraviolet radiation.

Figure 33.4 Demonstration of how shoe-leather skin forms.

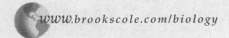

SKELETAL SYSTEMS

Animals move by activating, contracting, and relaxing muscle cells. But muscle cells alone can't move the body or parts of it. *Muscles require the presence of some medium or structural element against which the force of contraction is applied.* A skeletal system fulfills this requirement.

Three types of animal skeletons are common. With a **hydrostatic skeleton**, muscles work against an internal body fluid and redistribute it within a confined space. The confined fluid resists compression, like a waterbed. With an **exoskeleton**, rigid, *external* body parts such as a cuticle or a shell receive the applied force of muscle contraction. With an **endoskeleton**, *internal* body parts receive the applied force of muscle contraction.

We see hydrostatic skeletons among the soft-bodied invertebrates, such as a sea anemone with a soft body and saclike gut (Figure 33.5). When longitudinal muscles in its body wall are contracted (shortened) and radial ones are relaxed (lengthened), its body is squat. When longitudinal muscles relax and radial muscles contract (thus forcing fluid out of the gut), the body lengthens upright. An earthworm, too, has a hydrostatic skeleton. It uses the coelomic chambers of its segmented body (Section 23.10). Contracting and relaxing the fluid-filled segments in sequence moves the body forward. The body thrashes forward and back by alternating contractions of muscles on one side and then the other.

Recall how an arthropod has a hinged exoskeleton? Its hard parts are moved like levers by sets of muscles attached to them. Small contractions bring about large movements. This is especially true of a winged insect's cuticle. It extends over certain body segments and gaps between segments. The cuticle is pliable at the gaps. It

Figure 33.5 A sea anemone's hydrostatic skeleton. Contractile cells in the body wall run longitudinal to the main body axis and radially around the gut. *Left:* Radial cells are relaxed; longitudinal ones are contracted. This resting position is assumed at low tide, when currents don't deliver food. *Right:* The radial cells are contracted; longitudinal ones are relaxed. The body extends upward, to its feeding position.

resting position   feeding position

functions like a hinge when certain muscles raise either the wing or body parts to which the wings are attached (Sections 18.4 and 23.11).

Most vertebrates have a bony endoskeleton (Figures 33.6 and 33.7). In humans, it has 206 bones. Its pectoral girdle (at the shoulders), pelvic girdle (at the hips), and paired arms, hands, legs, and feet are the *appendicular* portion. Its pectoral girdles include slender collarbones and flat shoulder blades. Fall on an outstretched arm and you might dislocate a shoulder blade or fracture a collarbone, which is the most frequently broken bone.

The *axial* portion of the human skeleton has skull bones, twelve pairs of ribs, a breastbone, and twenty-six vertebrae. **Vertebrae** (singular, vertebra) are bony

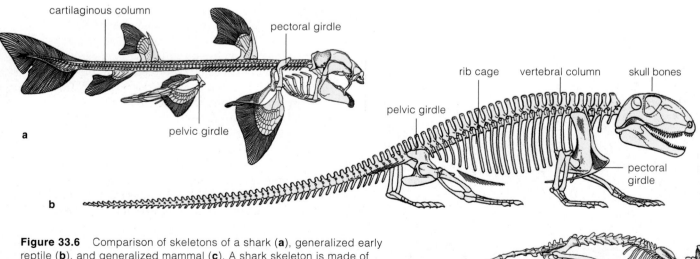

cartilaginous column

pectoral girdle

pelvic girdle

a

b

rib cage   vertebral column   skull bones

pelvic girdle

pectoral girdle

c

Figure 33.6 Comparison of skeletons of a shark (**a**), generalized early reptile (**b**), and generalized mammal (**c**). A shark skeleton is made of calcium-hardened cartilage. As in bony fishes, its pelvic and pectoral girdles serve as a stable base for fin motion. In four-legged vertebrates, these girdles transfer body weight to limbs. They have more surface area for attaching limb muscles and are tied closer to the body. Thus they offer better support and assist in the thrust of locomotion.

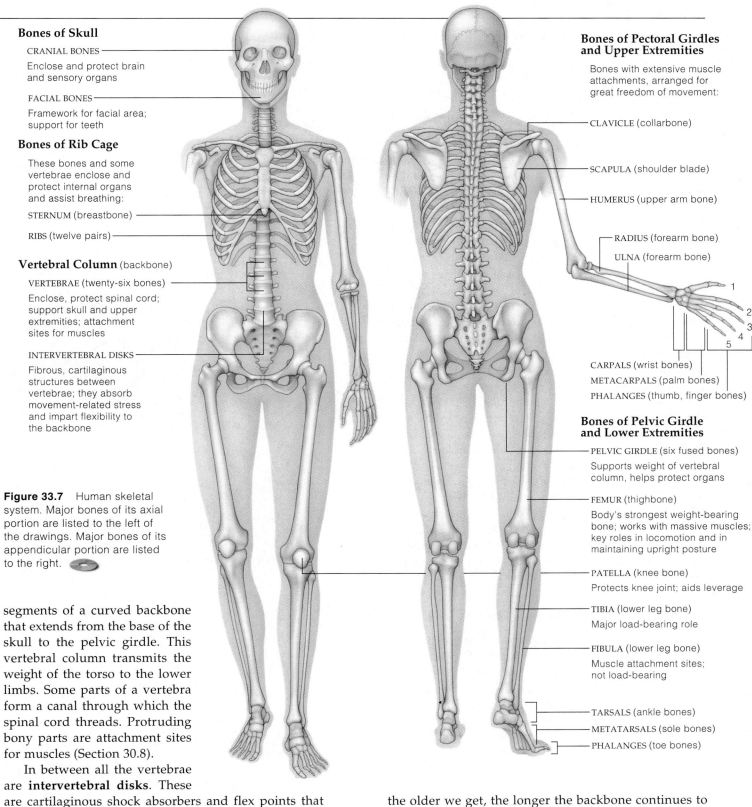

Bones of Skull

CRANIAL BONES

Enclose and protect brain and sensory organs

FACIAL BONES

Framework for facial area; support for teeth

Bones of Rib Cage

These bones and some vertebrae enclose and protect internal organs and assist breathing:

STERNUM (breastbone)

RIBS (twelve pairs)

Vertebral Column (backbone)

VERTEBRAE (twenty-six bones)

Enclose, protect spinal cord; support skull and upper extremities; attachment sites for muscles

INTERVERTEBRAL DISKS

Fibrous, cartilaginous structures between vertebrae; they absorb movement-related stress and impart flexibility to the backbone

Figure 33.7 Human skeletal system. Major bones of its axial portion are listed to the left of the drawings. Major bones of its appendicular portion are listed to the right.

Bones of Pectoral Girdles and Upper Extremities

Bones with extensive muscle attachments, arranged for great freedom of movement:

CLAVICLE (collarbone)

SCAPULA (shoulder blade)

HUMERUS (upper arm bone)

RADIUS (forearm bone)

ULNA (forearm bone)

CARPALS (wrist bones)

METACARPALS (palm bones)

PHALANGES (thumb, finger bones)

Bones of Pelvic Girdle and Lower Extremities

PELVIC GIRDLE (six fused bones)

Supports weight of vertebral column, helps protect organs

FEMUR (thighbone)

Body's strongest weight-bearing bone; works with massive muscles; key roles in locomotion and in maintaining upright posture

PATELLA (knee bone)

Protects knee joint; aids leverage

TIBIA (lower leg bone)

Major load-bearing role

FIBULA (lower leg bone)

Muscle attachment sites; not load-bearing

TARSALS (ankle bones)

METATARSALS (sole bones)

PHALANGES (toe bones)

segments of a curved backbone that extends from the base of the skull to the pelvic girdle. This vertebral column transmits the weight of the torso to the lower limbs. Some parts of a vertebra form a canal through which the spinal cord threads. Protruding bony parts are attachment sites for muscles (Section 30.8).

In between all the vertebrae are **intervertebral disks**. These are cartilaginous shock absorbers and flex points that permit some movement. Severe or rapid shock often forces a disk to slip out of place or rupture. *Herniated disks* are a painful, less-than-advantageous outcome of bipedalism. Remember, our early hominid ancestors walked about on four legs; later ones were walking upright before 4 million years ago. The change put a pronounced S-shaped curve in the backbone. So today,

the older we get, the longer the backbone continues to fight gravity in a compromised way, and the more back pain we may suffer.

Hydrostatic skeletons, exoskeletons, and endoskeletons all have structural elements or body fluids against which the force of contraction can be applied.

A CLOSER LOOK AT BONES AND JOINTS

Bone Structure and Function

By definition, **bones** are complex organs that function in movement, protection, support, mineral storage, and formation of blood cells (Table 33.1). Bones that support and anchor skeletal muscles help maintain or change the positions of body parts. Some form hard compartments that enclose and protect the brain, the lungs, and other internal organs. Bones are reservoirs for calcium and phosphorus ions. Continual deposits and withdrawals of ions from bone help maintain blood levels of calcium and phosphorus, thus supporting metabolic activities. Some bones (not all) are sites of blood cell formation.

Human bones range in size from middle ear bones smaller than lentils to clublike femurs, or thighbones (Figure 33.8). They have long, flat, short or cubelike, or irregular shapes. All consist of connective tissues (bone tissue especially) and epithelia. Bone tissue consists of **osteocytes**—mature bone cells—and collagen fibers in a calcium-hardened ground substance.

The shaft and both ends of thighbones have *compact* bone tissue, which resists mechanical shock. This tissue develops as Haversian systems: many thin, cylindrical, dense layers around interconnecting canals for blood vessels and nerves that service living bone cells. Also in the bone ends and shaft is *spongy* bone tissue, which imparts strength but doesn't weigh much. Abundant spaces make it look spongy, but its flattened parts are firm. **Red marrow**, a major site of blood cell formation, fills spaces in some bones, such as the breastbone. The cavities in most mature bones hold **yellow marrow**, a mostly fatty region that converts to red marrow and makes more red blood cells in times of severe blood loss.

Bone Formation and Remodeling

Bone tissue first forms in an embryo, where many bones are constructed upon cartilage models. **Osteoblasts** are the bone-forming cells. Using the models, they form and secrete organic substances, which become mineralized. Cartilage in the shaft breaks down, which opens up a marrow cavity. Once osteoblasts have been imprisoned by their own secretions, we call them osteocytes.

The total bone mass in healthy young adults doesn't change much, but minerals and osteocytes are removed and replaced on an ongoing basis. In **bone remodeling**, minerals are normally deposited and removed at the same time. The process adjusts bone strength and helps maintain blood levels of calcium and phosphorus. Two cell types interact in this process. Osteoblasts deposit bone, and **osteoclasts** secrete enzymes that digest bone's organic matrix. Thanks to these bone-degrading cells, liberated ions enter interstitial fluid, from which they can be reabsorbed by the bloodstream.

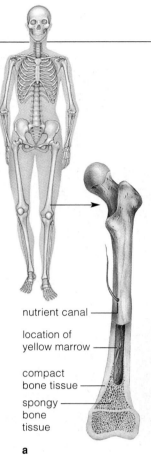

Figure 33.8 (**a**) Human thighbone structure, and (**b**) appearance of its spongy and compact bone tissue. Thin, dense layers of compact bone tissue form cylindrical, interconnected arrays around the canals for blood vessels and nerves. Each array is a Haversian system. The blood vessel inside services osteocytes, living bone cells in small tissue spaces connected by tunnels.

nutrient canal

location of yellow marrow

compact bone tissue

spongy bone tissue

a

space occupied by living bone cell

blood vessel

HAVERSIAN SYSTEM

55 μm

spongy bone tissue

compact bone tissue

blood vessel

outer layer of dense connective tissue

b

Table 33.1 *Functions of Bone*

1. *Movement.* Bones interact with skeletal muscle to change or maintain the position of the body and its parts.

2. *Support.* Bones support and anchor muscles.

3. *Protection.* Many bones form hard compartments that enclose and protect soft internal organs.

4. *Mineral storage.* Bones are a reservoir for calcium and phosphorus, the deposits and withdrawals of which help maintain ion concentrations in body fluids.

5. *Blood cell formation.* Only certain bones contain regions where blood cells are produced.

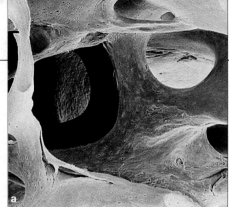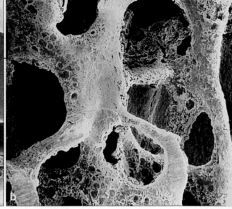

Figure 33.9 Bone affected by osteoporosis. (**a**) Section through normal bone tissue; mineral deposits continually replace withdrawals. (**b**) After the onset of osteoporosis, mineral replacements lag behind withdrawals. The tissue erodes; bones become hollow and brittle.

Bones and the Blood Level of Calcium

Calcium is central to neural function, contraction, and other vital activities, and its level in blood is one of the most tightly controlled aspects of metabolism. Bones and teeth store all but 1 percent or so of a body's calcium. Negative feedback loops to glands from the blood govern the rates of calcium release and uptake. The thyroid gland releases calcitonin when the calcium level rises. This hormone suppresses osteoclast action and slows calcium's release into the blood. Parathyroid glands release parathyroid hormone (PTH) when the calcium level falls. PTH promotes calcium release from bones and the kidneys, enhances osteoclast action, and helps activate vitamin D. Vitamin D stimulates calcium absorption from the small intestine (Section 33.2).

When a bone is stressed, as when it is subjected to compression, mineral deposition by osteoblasts exceeds withdrawals by osteoclasts. That's why bones of people who exercise rigorously are denser and stronger than bones of people who don't. In injured bones, mineral withdrawals predominate. Also, as a person ages, the backbone and other bones decline in mass. We call this condition *osteoporosis* (Figure 33.9). Decreased physical activity, declining activity of the bone-forming cells, loss of calcium, excessive protein intake, and sex hormone deficiencies contribute to the disorder.

Where Bones Meet—Skeletal Joints

Joints, areas of contact or near-contact between bones, have a distinct bridge of connective tissue. Very short connecting fibers bridge *fibrous* joints. Cartilage straps bridge *cartilaginous* joints. **Ligaments**, or long straps of dense connective tissue, bridge *synovial* joints.

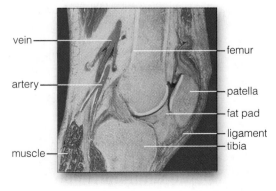

vein — femur
artery — patella
— fat pad
— ligament
muscle — tibia

Figure 33.10 Human knee joint, longitudinal section.

Fibrous joints hold teeth in their sockets. They join the flat skull bones of a fetus. At childbirth, the loose connections allow bones to slide over each other a bit and prevent skull fractures. A newborn's skull still has fibrous joints and membranous areas called soft spots, or fontanels. By childhood, the fibrous tissue hardens and the skull bones are fused together as a single unit.

Cartilaginous joints bridge the vertebrae, ribs, and breastbone, and allow slight movements. Synovial joints move freely; ligaments stabilize them. The knee joint is an example (Figures 33.7 and 33.10). Cartilage cushions the abutting bones and helps absorb shocks. A flexible capsule of dense connective tissue surrounds the area of contact. Cells of a membrane that lines the interior of the capsule secrete a fluid that lubricates the joint.

Like other joints, knee joints are vulnerable to stress. They let you swing, bend, and turn the long bones below them. As you run, they absorb the force of your weight. Abruptly stretch or twist a knee joint too far and you *strain* it. Tear its ligaments or tendons and you *sprain* it. Move the wrong way and you may dislocate attached bones. In collision sports such as football, a blow to the knee often severs ligaments. These must be reattached within ten days. Why? Phagocytes in a lubricating fluid inside the joint usually clean up after everyday wear and tear. When presented with torn ligaments, these phagocytes indiscriminately turn the tissue to mush.

Joint inflammation and degenerative disorders are collectively called "arthritis." In *osteoarthritis*, cartilage at knees and other freely movable joints wears off as a person ages. Most often, the joints in the fingers, knees, hips, and vertebral column are affected. In *rheumatoid arthritis*, synovial membranes in joints become inflamed and thicken, cartilage degenerates, and bone deposits accumulate. These are the outcomes of an autoimmune response, whereby the body is abnormally triggered into attacking itself (Chapter 35). A bacterial or viral infection may trigger this degenerative disorder, which has a genetic basis. Rheumatoid arthritis can begin at any age, but symptoms usually emerge before age fifty.

Bones are collagen-rich, mineralized organs that function in movement, protection, support, storage of calcium and other minerals, and blood cell formation.

SKELETAL–MUSCULAR SYSTEMS

How Do Muscles and Bones Interact?

Skeletal muscles are the functional partners of bones. Each skeletal muscle contains bundles of hundreds to many thousands of muscle cells, which look like long, striped fibers. In muscle tissue, remember, muscle cells contract (shorten) in response to adequate stimulation. They lengthen in response to gravity and other loads. When you run, dance, breathe, squint, or scribble notes, many muscle cells are working to move your body or change the positions of some of its parts.

Connective tissue bundles muscle cells together and extends past them to form **tendons**. Each tendon, a cord or strap of dense connective tissue, attaches a muscle to bone (Figure 33.11). Most of these attachment sites are like a gearshift in a car. *They form a lever system in which a rigid rod is attached to a fixed point and can move about it.* Muscles connect to bones (rigid rods) near a joint (fixed point). When they contract, they transmit force to bones and so make them move. Tendons often rub against bones, as at the knees and wrists. They slide inside fluid-filled sheaths that help reduce friction.

Skeletal muscles interact with one another, also. Some work in pairs or groups to promote a movement. Others work in opposition; the action of one opposes

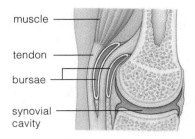

Figure 33.11 Tendon. A bursa is a flattened sac filled with synovial fluid. Bursae are located between tendons and bones (or some other structure). They help reduce friction during movements.

muscle
tendon
bursae
synovial cavity

or reverses the action of another. Figure 33.12 shows how opposing muscle groups work to move frog legs.

Also look at Figure 33.13. Extend your right arm forward, then place your left hand over the biceps in the upper right arm and slowly bend the elbow. Feel the biceps contract? Even when a biceps contracts only a bit, it causes a large motion of the bone connected to it. This is the case for most leverlike arrangements.

Bear in mind, only *skeletal* muscle is the functional partner of bone. As mentioned earlier, smooth muscle

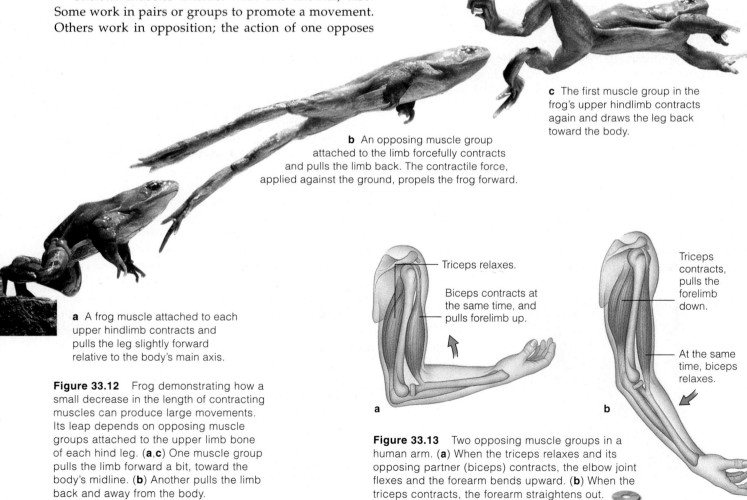

c The first muscle group in the frog's upper hindlimb contracts again and draws the leg back toward the body.

b An opposing muscle group attached to the limb forcefully contracts and pulls the limb back. The contractile force, applied against the ground, propels the frog forward.

a A frog muscle attached to each upper hindlimb contracts and pulls the leg slightly forward relative to the body's main axis.

Figure 33.12 Frog demonstrating how a small decrease in the length of contracting muscles can produce large movements. Its leap depends on opposing muscle groups attached to the upper limb bone of each hind leg. (**a,c**) One muscle group pulls the limb forward a bit, toward the body's midline. (**b**) Another pulls the limb back and away from the body.

Triceps relaxes.

Biceps contracts at the same time, and pulls forelimb up.

Triceps contracts, pulls the forelimb down.

At the same time, biceps relaxes.

Figure 33.13 Two opposing muscle groups in a human arm. (**a**) When the triceps relaxes and its opposing partner (biceps) contracts, the elbow joint flexes and the forearm bends upward. (**b**) When the triceps contracts, the forearm straightens out.

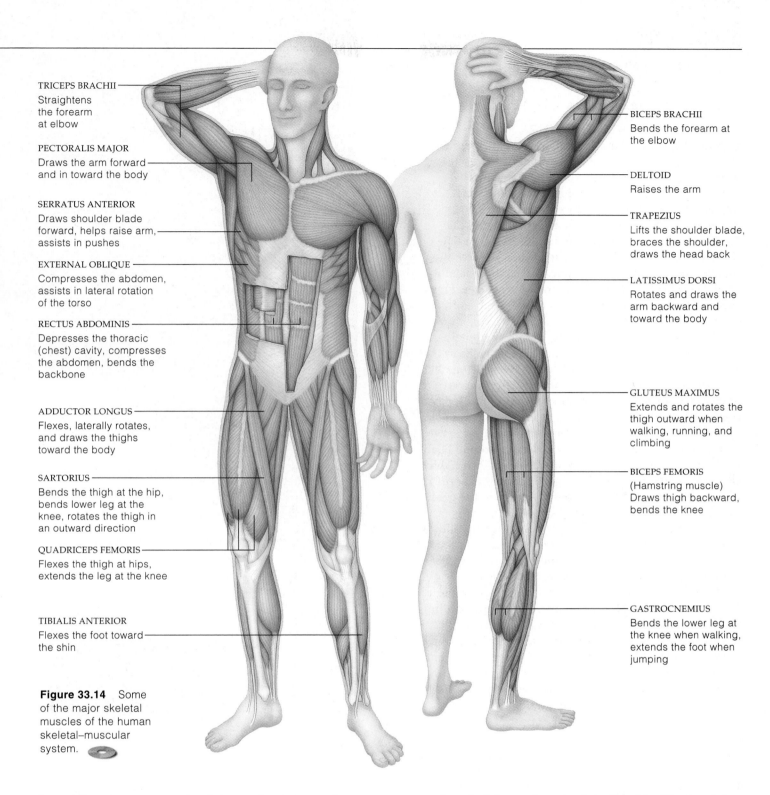

TRICEPS BRACHII
Straightens the forearm at elbow

PECTORALIS MAJOR
Draws the arm forward and in toward the body

SERRATUS ANTERIOR
Draws shoulder blade forward, helps raise arm, assists in pushes

EXTERNAL OBLIQUE
Compresses the abdomen, assists in lateral rotation of the torso

RECTUS ABDOMINIS
Depresses the thoracic (chest) cavity, compresses the abdomen, bends the backbone

ADDUCTOR LONGUS
Flexes, laterally rotates, and draws the thighs toward the body

SARTORIUS
Bends the thigh at the hip, bends lower leg at the knee, rotates the thigh in an outward direction

QUADRICEPS FEMORIS
Flexes the thigh at hips, extends the leg at the knee

TIBIALIS ANTERIOR
Flexes the foot toward the shin

BICEPS BRACHII
Bends the forearm at the elbow

DELTOID
Raises the arm

TRAPEZIUS
Lifts the shoulder blade, braces the shoulder, draws the head back

LATISSIMUS DORSI
Rotates and draws the arm backward and toward the body

GLUTEUS MAXIMUS
Extends and rotates the thigh outward when walking, running, and climbing

BICEPS FEMORIS
(Hamstring muscle) Draws thigh backward, bends the knee

GASTROCNEMIUS
Bends the lower leg at the knee when walking, extends the foot when jumping

Figure 33.14 Some of the major skeletal muscles of the human skeletal–muscular system.

is mainly a component of soft internal organs, such as the stomach (Section 29.3). Cardiac muscle is present only in the heart wall. We will look at the structure and functioning of smooth muscle as well as cardiac muscle in later chapters in this unit.

Human Skeletal–Muscular System

The human body has more than 600 skeletal muscles, some superficial, others deep in the body wall. Some, such as facial muscles, attach to the skin. The trunk has muscles of the thorax, backbone, abdominal wall, and pelvic cavity. Other groups of muscles attach to upper and lower limb bones. Figure 33.14 shows a few of the main skeletal muscles and lists their functions. We turn next to the mechanisms underlying their contraction.

Skeletal muscles transmit contractile force to bones and make them move. Tendons strap skeletal muscles to bones.

33.6

A CLOSER LOOK AT MUSCLES

Skeletal Muscle Structure and Function

Consider the ballerina in Figure 33.15. Her bones move in some direction when skeletal muscles attached to them shorten. When a muscle shortens, its muscle cells are shortening. When a muscle cell shortens, many units of contraction inside it are shortening. The basic units of contraction are **sarcomeres**.

Muscle cells run parallel with a skeletal muscle. Each cell has parallel arrays of myofibrils, threadlike structures divided by cross-bands. When stained, the bands have an alternating light–dark pattern. Bands of all myofibrils in the cell are in close register, which gives a muscle cell a cross-striated appearance. That is why skeletal (and cardiac) muscles have cross-stripes.

The dark bands, or Z bands, define sarcomeres. A sarcomere has many thin and thick filaments. Each *thin* filament is like two strands of pearls twisted together. It is attached at one end to a Z band. The "pearls" are globular molecules of **actin** and other proteins located next to them. Each *thick* filament consists of **myosin**, a motor protein having contractile properties (Section 4.8). A myosin molecule has a tail and a double head. Many of the molecules are bundled together in a thick filament, with their heads projecting outward. All of the precisely angled heads point away from the center of the sarcomere (Figures 33.15 and 33.16).

Thus muscle bundles, muscle cells, and myofibrils and their filaments run in the same direction. What's the point of the parallel orientation? *It focuses the force of contraction on a bone in a particular direction.*

How do sarcomeres shorten and contract a muscle? The answer lies with coordinated sliding and pulling motions. Two sets of actin filaments, each anchored at a different Z band, extend into each sarcomere. They overlap a parallel, stationary set of myosin filaments that is not attached to the Z bands. During contraction, both actin sets slide over the fixed myosin neighbors, toward the center of the sarcomere (Figure 33.16).

By a **sliding-filament model** of contraction, myosin and actin interact by **cross-bridge formation**, whereby activated myosin heads briefly attach to binding sites on an adjacent actin filament. Driven by ATP energy, they tilt in a short stroke toward the sarcomere's center. The heads pull the actin filament along with them. Another energy input makes the heads let go, attach to another binding site, tilt in another stroke, and so on. A single contraction of each sarcomere involves many myosin

Figure 33.15 From the Dance Theatre of Harlem, an example of exquisite control of skeletal muscle movements. (**a–e**) This sequence shows skeletal muscle components from a biceps down to molecules with contractile properties.

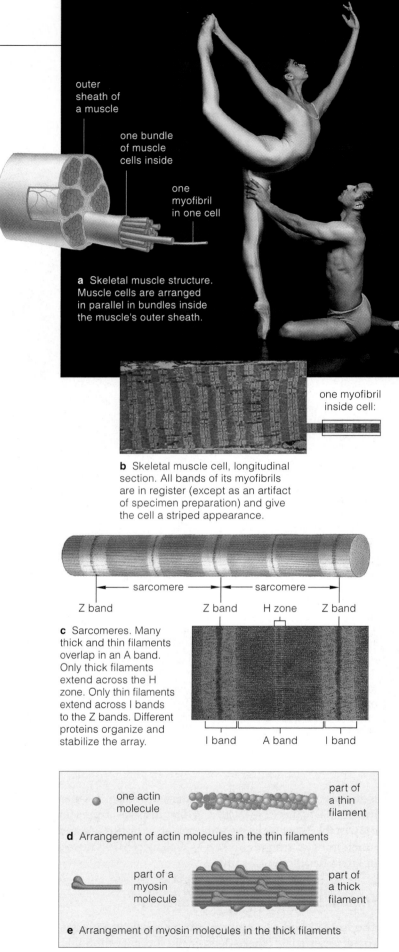

a Skeletal muscle structure. Muscle cells are arranged in parallel in bundles inside the muscle's outer sheath.

b Skeletal muscle cell, longitudinal section. All bands of its myofibrils are in register (except as an artifact of specimen preparation) and give the cell a striped appearance.

c Sarcomeres. Many thick and thin filaments overlap in an A band. Only thick filaments extend across the H zone. Only thin filaments extend across I bands to the Z bands. Different proteins organize and stabilize the array.

d Arrangement of actin molecules in the thin filaments

e Arrangement of myosin molecules in the thick filaments

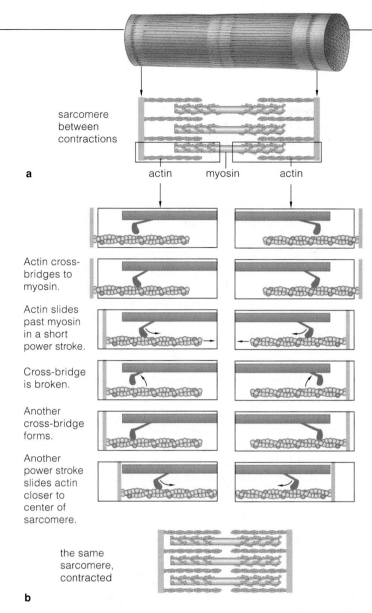

a

sarcomere between contractions

actin   myosin   actin

Actin cross-bridges to myosin.

Actin slides past myosin in a short power stroke.

Cross-bridge is broken.

Another cross-bridge forms.

Another power stroke slides actin closer to center of sarcomere.

the same sarcomere, contracted

b

Figure 33.16 (**a**) Arrangement of actin filaments and myosin filaments that interact to shorten the width of each sarcomere. (**b**) Diagram of the sliding-filament model of contraction as it proceeds in each sarcomere of a muscle cell. For simplicity, we show the action of only one myosin head. Many others form cross-bridges at the same time.

heads performing a series of short power strokes down the length of their neighbors, the actin filaments.

Sources of Energy for Contraction

All cells need ATP. But only in muscle cells does the demand skyrocket in so short a time. If a muscle cell at rest is called on to contract, phosphate donations from ATP must occur twenty to a hundred times faster. But a cell has only a small supply of ATP when contractile activity starts. At such times, it forms ATP by a quick reaction. An enzyme simply transfers phosphate from the organic compound **creatine phosphate** to ADP. A

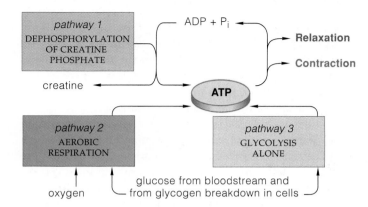

Figure 33.17 Three metabolic routes by which ATP forms in muscle cells in response to the demands of physical exercise.

cell has about five times as much creatine phosphate as ATP, so this reaction is good for a few contractions. And that is enough to buy time for a relatively slower ATP-forming pathway to kick in (Figure 33.17).

During prolonged, moderate exercise, the oxygen-requiring reactions of aerobic respiration typically can supply most of the ATP for contraction. Suppose that a muscle cell taps its store of glycogen for glucose, the starting substrate, during the first five to ten minutes. For the next half hour or so of sustained activity, the cell depends on glucose and fatty acid deliveries from the blood. For contractile activity longer than this, the fatty acids are the main fuel source (Section 7.6).

What if you exercise so intensely that you exceed the capacity of your respiratory and circulatory systems to deliver oxygen for aerobic reactions? At such times, glycolysis alone will contribute more of the total ATP being produced. Remember, a glucose molecule is not fully degraded by the anaerobic reactions of glycolysis, so the net ATP yield is small. Still, muscle cells use this metabolic route for as long as glycogen stores continue to provide glucose, before fatigue sets in.

After intense exercise, deep, rapid breathing helps repay the body's **oxygen debt**, incurred when ATP use by muscles exceeded the aerobic pathway's deliveries.

A skeletal muscle shortens through combined decreases in the length of its numerous sarcomeres. Sarcomeres are the basic units of skeletal and cardiac muscle contraction.

The parallel orientation of a skeletal muscle's component parts directs the force of contraction toward a bone that must be pulled in some direction.

By energy-driven interactions between myosin and actin filaments, the many sarcomeres of a muscle cell shorten and collectively account for its contraction.

During exercise, ATP availability in muscle cells affects whether contraction will proceed, and for how long.

CONTROL OF MUSCLE CONTRACTION

When skeletal muscles contract, they move the body and its parts at certain times, in certain ways. They do so in response to commands from the nervous system. The commands stimulate or inhibit the release of ACh, a neurotransmitter, from motor neurons. Muscle cells are a key target of this neurotransmitter (Section 30.3).

Like all cells, each muscle cell shows a difference in electric charge across its plasma membrane. That is, the cytoplasm just beneath the membrane is a bit more negative than fluid outside it. But only in muscle cells, neurons, and other *excitable* cells does this difference in charge reverse suddenly, though briefly, in response to sufficient stimulation.

That abrupt reversal in charge, an **action potential**, occurs as charged ions flow across the membrane in an ever accelerating way. The excitation spreads outward and travels along the membrane without diminishing.

a Commands from nervous system travel along spinal cord, then a motor neuron.

section from spinal cord

motor neuron

section from a skeletal muscle

b Endings of motor neuron terminate next to a muscle cell. Signals from endings excite the cell's plasma membrane.

part of a muscle cell

c

Figure 33.18 Pathway for signals from the nervous system that stimulate contraction of skeletal muscle. The plasma membrane of muscle cells connects with their myofibrils and with inward-threading T tubules. These membranous tubes are close to the sarcoplasmic reticulum, a calcium-storing system used in the control of contraction. Action potentials arise and spread to T tubules, where they cause calcium to be released. Calcium diffuses into myofibrils, where it binds to sites in thin filaments that allow cross-bridge formation and contraction.

When action potentials arise in a muscle cell, they quickly spread away from the stimulation point along small, tubelike extensions of the plasma membrane. The tubes connect with a system of membrane-bound chambers threading lacily around the myofibrils inside a muscle cell. That system, the **sarcoplasmic reticulum**, repeatedly takes up, stores, and releases many calcium ions in controlled ways (Figure 33.18).

The arrival of action potentials causes an outward flow of calcium ions from the sarcoplasmic reticulum. The released ions diffuse into the myofibrils and reach actin filaments. Before this happened, the muscle was at rest; it wasn't contracting. The myosin binding sites on the actin filaments were blocked—so cross-bridges couldn't form. But now, with calcium's arrival, cross-bridge attachments can form and contraction can occur. Afterward, calcium ions are actively transported back into the membrane storage system, so muscles relax.

What blocks cross-bridge binding sites in a resting muscle? Tropomyosin and troponin are two accessory proteins located in or near surface grooves of the actin filaments (Figure 33.15*d*). When the calcium level is *low*, the proteins are tightly joined. Tropomyosin is forced out of the groove, and in its new position it blocks the cross-bridge binding site. When the calcium level *rises*, calcium binds to the troponin and alters its shape. With

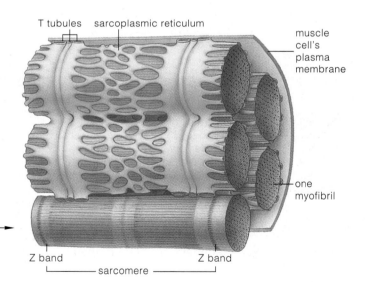

T tubules   sarcoplasmic reticulum

muscle cell's plasma membrane

one myofibril

Z band   Z band
└──────── sarcomere ────────┘

this alteration, troponin has a different molecular grip on tropomyosin—which becomes free to move into the groove and expose the binding site.

Commands from the nervous system trigger action potentials in muscle cells by way of ACh. These action potentials are signals for cross-bridge formation—hence for contraction.

PROPERTIES OF WHOLE MUSCLES

Muscle Tension and Muscle Fatigue

Whether a muscle actually shortens during cross-bridge formation depends on the external forces acting on it. Collectively, the cross-bridges exert **muscle tension**. By definition, this is a mechanical force that a contracting muscle exerts on an object, such as a bone. Opposing it is a load, either the weight of an object or gravity's pull on the muscle. Only when muscle tension exceeds the load does a stimulated muscle shorten.

Isotonically contracting muscles shorten and move a load (Figure 33.19). An *isometrically* contracting muscle develops tension but doesn't shorten. With *lengthening* contraction, a muscle lengthens when an external load is greater than its tension for the period of contraction. This happens to leg muscles as you walk down stairs.

A muscle's tension relates to the formation of cross-bridges in its cells and to the number of cells recruited into action. Consider a **motor unit**: a motor neuron and all muscle cells that form junctions with its endings. By stimulating that motor unit with an electrical impulse, we can induce an action potential and make a recording of an isometric contraction. It takes a few milliseconds for tension to increase, then it peaks and declines. This response is a **muscle twitch** (Figure 33.20*a*). Its duration depends on the load and on cell type. For example, fast-acting muscle cells rely on glycolysis—not efficient, but fast—and use up ATP faster than slow-acting cells do.

If another stimulus is applied before the response is finished, the muscle twitches again. **Tetanus** is a large contraction that results from the repeated stimulation of a motor unit, so twitches mechanically run together. (In *tetanus*, a disease by the same name, toxins disrupt muscle relaxation, as described in Section 30.5.) Figure 33.20*c* shows a recording of tetanic contraction.

Continuous, high-frequency stimulation that keeps a muscle in a state of tetanic contraction leads to *muscle fatigue*, or a decline in tension. After a few minutes of rest, a fatigued muscle will contract again in response to stimulation. The extent of recovery depends largely on how long and how often it was stimulated before. Muscles associated with brief, intense exercise (such as weight lifting) fatigue fast but recover fast. The muscles associated with prolonged, moderate exercise fatigue slowly but take longer to recover, often up to twenty-four hours. The molecular mechanisms causing muscle fatigue are unknown, but glycogen depletion is a factor.

Effects of Exercise and Aging

A muscle's properties depend on how often, how long, and how intensely it is put to use. With regular **exercise** (that is, increased levels of contractile activity), muscle cells do not increase in number. Rather, they increase in

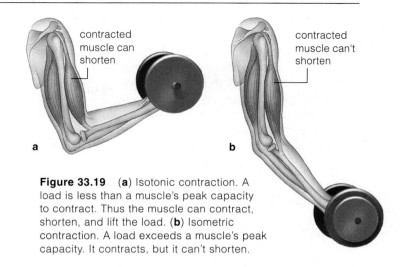

Figure 33.19 (**a**) Isotonic contraction. A load is less than a muscle's peak capacity to contract. Thus the muscle can contract, shorten, and lift the load. (**b**) Isometric contraction. A load exceeds a muscle's peak capacity. It contracts, but it can't shorten.

contracted muscle can shorten

contracted muscle can't shorten

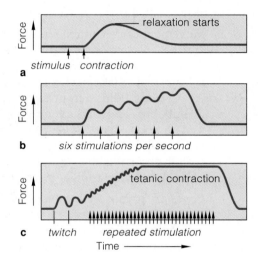

Figure 33.20
Three recordings of twitches in muscles artificially stimulated in three different ways.
(**a**) A single twitch.
(**b**) Summation of twitches following six stimulations per second. (**c**) Tetanic contraction following twenty stimulations per second.

Force — relaxation starts — stimulus contraction — a

Force — six stimulations per second — b

Force — tetanic contraction — twitch — repeated stimulation — c — Time

size and metabolic activity, and become more resistant to fatigue. Think of *aerobic exercise*—not intense, but long in duration. Aerobic exercise increases the number of mitochondria in both fast and slow muscle cells, and increases the blood capillaries that service them. Such physiological changes improve endurance. By contrast, *strength training* (intense, short-duration exercise such as weight lifting) makes fast-acting muscle cells form more myofibrils and enzymes for glycolysis. Strong, bulging muscles may develop, although these muscles do not have much endurance. They fatigue rapidly.

Muscle tension declines in adult humans after age thirty or forty. Older people might exercise just as long and intensely as younger ones, but their muscles cannot adapt (change) in response to the same extent. Even so, some adaptation can be beneficial. Aerobic exercise can improve blood circulation. And, as it turns out, modest strength training slows the loss of muscle tissue that is an inevitable part of the aging process.

Properties of muscles vary with age and levels of activity.

1. Most animals have an integumentary system, which covers the body's surface. Examples are insect cuticles and vertebrate skin. Skin protects against abrasion, UV radiation, dehydration, and many pathogens. It helps control body temperature; blood flow to skin dissipates heat. Its sensory receptors detect external stimuli. Skin exposed to sunlight has an endocrine function; it helps produce vitamin D, a hormone-like substance required for absorption of calcium from food. *CI, 33.1, 33.2*

2. Skin consists of two regions: an outer epidermis and underlying dermis, where rapid divisions replace cells continually being shed or abraded away. Keratinocytes (keratin producers) are the most abundant cells. Others are melanocytes (melanin producers), and Langerhans cells and Granstein cells (both types defend the body against pathogens and cancer cells). *33.1, 33.2*

3. Movement of the animal body or parts of it requires contractile cells and some medium or structure against which contractile force can be applied. *33.3*

   a. With hydrostatic skeletons, such as that of a sea anemone, body fluids accept the contractile force and, as a result, are redistributed within a confined space.

   b. With exoskeletons, such as that of an insect, rigid external body parts accept the force of contraction.

   c. With endoskeletons, rigid internal structures such as bones receive the applied force of contraction.

4. Bones are organs with osteocytes (living bone cells) in a mineralized, collagen-rich ground substance. They have roles in moving, protecting, and supporting body parts; storing minerals; and, in bones that contain red and yellow marrow, forming blood cells. *33.4*

5. Tendons attach skeletal muscles to bones. Skeletal muscles and bones interact as a system of levers, with rigid rods (bones) moving at fixed points (joints). Many muscles work together or in opposition to bring about movement or positional changes in body parts. *33.5*

6. A skeletal joint, a gap between bones, is bridged by connective tissues. Short fibers bridge fibrous joints. Cartilage bridges cartilaginous joints. Ligaments bridge synovial joints. *33.4*

7. Smooth, cardiac, and skeletal muscle cells contract (shorten) in response to adequate stimulation. Many soft internal organs incorporate smooth muscle. Only the heart incorporates cardiac muscle. Skeletal muscle is the functional partner of bones. *33.5*

8. Inside each skeletal muscle cell are many myofibrils arranged parallel with the long axis. These threadlike structures contain filaments of actin and myosin, also organized in parallel. Every myofibril is transversely divided into sarcomeres, the basic units of contraction. The parallel orientation of all parts directs the contractile force at a bone to be pulled in some direction. *33.6*

9. In response to stimulation from the nervous system, skeletal muscles shorten by decreases in the length of all of its sarcomeres. Key points for the sliding-filament model of muscle contraction follow: *33.6, 33.7*

   a. Action potentials cause the release of calcium ions from a membrane system (sarcoplasmic reticulum) that threads around the cell's myofibrils. Calcium diffuses into sarcomeres, and binds to and pulls aside accessory proteins on actin filaments to expose binding sites for myosin. Cross-bridges can now form.

   b. Each cross-bridge is a brief attachment between a myosin head and an actin binding site. Cross-bridges form during repeated, ATP-driven power strokes. The repeated strokes make actin filaments slide past myosin filaments, and collectively they shorten the sarcomere.

10. Muscle cells obtain the ATP for contraction in three ways: *Dephosphorylation of creatine phosphate*, a direct, fast pathway good for a few seconds of contraction. *Aerobic respiration*, a pathway that predominates during prolonged, moderate exercise. *Glycolysis*, which takes over if intense exercise exceeds the body's capacity to deliver oxygen to muscle cells. *33.7*

11. Muscle tension is a mechanical force caused by cross-bridge formation. The load (gravity or weight) is an opposing force. Stimulated muscles shorten if tension exceeds the load and lengthen if tension is less than the load. Exercise and aging affect muscle properties. *33.8*

Review Questions

1. List the functions of skin. Describe the distinctive features of its regions. Is the hypodermis part of skin? *33.1, 33.2*

2. Identify four cell types in vertebrate skin and briefly describe their functions. *33.1, 33.2*

3. Distinguish between:
   a. hydrostatic skeleton, exoskeleton, and endoskeleton *33.3*
   b. vertebra and intervertebral disk *33.3*
   c. ligament and tendon *33.4, 33.5*

4. What are the functions of bones? What is a joint? *33.4*

5. Which hormones help control calcium ion concentrations in blood? Name their general effects on bone tissue turnover. *33.4*

6. In what respect are tendons like a car's gearshift? *33.5*

7. Review Figure 33.15. On your own, sketch and label the fine structure of a muscle, down to one of its individual myofibrils. Identify the basic unit of contraction in the myofibrils. *33.6*

8. What role does calcium play in control of contraction? What role does ATP play, and by what routes does it form? *33.6, 33.7*

Self-Quiz   ANSWERS IN APPENDIX III

1. Which is *not* a function of skin?
   a. resist abrasion   c. initiate movement
   b. restrict dehydration   d. help control temperature

2. _____ are shock pads and flex points.
   a. Vertebrae   c. Marrow cavities
   b. Femurs   d. Intervertebral disks

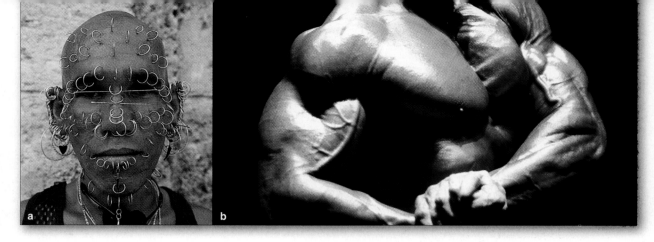

Figure 33.21 (**a**) A demonstration of the curious invasions of the integument that have no immediately obvious adaptive value. (**b**) Extremely pumped-up muscles of a human male.

3. Blood cells form in _____ .
   a. red marrow   c. certain bones only
   b. all bones   d. a and c

4. The _____ is the basic unit of contraction.
   a. myofibril   c. muscle fiber
   b. sarcomere   d. myosin filament

5. Muscle contraction requires _____ .
   a. calcium ions   c. action potential arrival
   b. ATP   d. all of the above

6. ATP for muscle contraction can be formed by _____ .
   a. aerobic respiration   c. creatine phosphate breakdown
   b. glycolysis   d. all of the above

7. Match the M words with their defining feature.
   ____ muscle   a. actin's partner
   ____ muscle twitch   b. all in the hands
   ____ muscle tension c. blood cell production
   ____ melanin   d. decline in tension
   ____ myosin   e. brownish-black pigment
   ____ marrow   f. motor unit response
   ____ metacarpals   g. force exerted by cross-bridges
   ____ myofibrils   h. bundles of muscle cells in a
   ____ muscle fatigue   connective tissue sheath
                        i. threadlike parts in muscle cell

Critical Thinking

1. The nose, lips, tongue, navel, and private parts are common targets for *body piercing*, cutting holes into the body so jewelry and other objects can be threaded through them (Figure 33.21*a*). *Tattooing*, or using permanent dyes to create patterns in skin, is another fad. Besides being painful, both skin invasions invite bacterial infections, chronic viral hepatitis, AIDS, and other nasty diseases if piercers and tattooers reuse unsterilized needles, dye, razors, gloves, swabs, or trays. Months may pass before problems arise, so the cause-and-effect connection isn't always obvious. If you dismiss the risks and think unregulated tissue invasions are okay, how could you make sure the equipment used is sterile?

2. For young women, the recommended daily allowance (RDA) of calcium is 800 milligrams/day. During pregnancy, the RDA is 1,200 milligrams/day. Why the increase? What might happen to a pregnant woman's bones without the larger amount?

3. Compared to most people, Joe and other long-distance runners have a greater number of muscle fibers with more mitochondria. Sprinters have muscle fibers that contain more of the enzymes necessary for glycolysis but fewer mitochondria. Think about how these two forms of exercise differ. Then explain why the muscle fibers differ between the two kinds of runners.

4. Extreme sexual dimorphism in body size involves differences in muscle mass (compare Section 16.9). This may be a measure of how much male mammals invest in fighting capacity. Maybe the cost of securing and using resources to grow and maintain a massive body is offset by reproductive rewards. Now think of the social rewards for millions of athletes who use anabolic steroids to increase muscle mass and strength (Figure 33.21*b*). *Anabolic steroids*, synthetic hormones, mimic testosterone. This sex hormone governs secondary sexual traits and also increases aggressive behavior, which is often associated with maleness.

By stimulating protein synthesis, anabolic steroids induce rapid gains in muscle mass and strength during weight-training and exercise programs. However, at high concentrations, they trigger a sharp decline in testosterone production, resulting in acne, baldness, shrinking testes, infertility, and maybe early heart disease in men. Even occasional use may damage kidneys or set the stage for cancers of the liver, testes, and prostate gland. Among women, anabolic steroids deepen the voice and produce pronounced facial hair. Breasts may shrink and the menstrual cycle may become irregular.

Not every steroid user develops severe side effects. More common are mental difficulties, called '*roid rage* or *body-builder's psychosis*. Users become irritable and increasingly aggressive. Some men have become uncontrollably aggressive, manic, and delusional. One steroid user accelerated his car to high speed and deliberately drove it into a tree.

Using natural selection theory as the basis for argument, would you say the advantages of anabolic steroid use outweigh the disadvantages? This is not a trick question.

Selected Key Terms

actin *33.6*
action potential *33.7*
bone *33.4*
bone remodeling *33.4*
creatine
  phosphate *33.6*
cross-bridge
  formation *33.6*
dermis *33.1*
endoskeleton *33.3*
epidermis *33.1*
exercise *33.8*
exoskeleton *33.3*
folate *33.2*
Granstein cell *33.2*
hair *33.1*

hydrostatic
  skeleton *33.3*
integument *CI*
intervertebral
  disk *33.3*
joint *33.4*
keratinocyte *33.1*
Langerhans cell *33.2*
ligament *33.4*
melanocyte *33.1*
motor unit *33.8*
muscle tension *33.8*
muscle twitch *33.8*
myosin *33.6*
osteoblast *33.4*
osteoclast *33.4*

oxygen debt *33.6*
red marrow *33.4*
sarcomere *33.6*
sarcoplasmic
  reticulum *33.7*
skeletal muscle *33.5*
skin *CI*
sliding-filament
  model *33.6*
tendon *33.5*
tetanus *33.8*
vertebra
  (vertebrae) *33.3*
vitamin D *33.2*
yellow marrow *33.4*

Readings

Sherwood, L. 1997. *Human Physiology*. Third edition. Belmont, California: Wadsworth.

Travis, J. 15 January 2000. "Boning Up." *Science News* 157: 41–43.

On-Line readings at Student Guide for InfoTrac:
www.brookscole.com/biology

34

CIRCULATION

Heartworks

For Dr. Augustus Waller, Jimmie the bulldog was no ordinary pooch. Connected to wires and soaked to his ankles in buckets of salty water, Jimmie was on a four-footed exploration of the heart (Figure 34.1a). Press your fingers to your chest a few inches left of center, between the fifth and sixth ribs, and feel the repetitive thumpings of your heart. The same rhythms intrigued Waller and other nineteenth-century physiologists. They wondered: Is there a pattern of electrical currents that corresponds to each heartbeat? Could they find out by devising a painless way to record currents at the body surface?

That's where Jimmie and the buckets of salty water came in. Saltwater happens to be an efficient conductor of electricity. In Waller's experiment, it picked up faint signals from Jimmie's beating heart through the skin of his legs and conducted them to a crude monitoring device. With that device, Waller made one of the first recordings of heart activity (Figure 34.1c). Today we call such a recording an ECG. That's the abbreviation for electrocardiogram.

A graph of your heart's normal electrical activity would look very much the same. The pattern emerged a few weeks after you started growing, by mitotic cell divisions, from a fertilized egg in your mother. Early on, some embryonic cells differentiated into cardiac muscle cells and started to contract on their own. One small patch of the cells took the lead. It's been functioning as your heart's pacemaker ever since.

If all goes well, that patch of cardiac muscle cells will continue to contract as it should until the day you die. It is a natural pacemaker; it sets the baseline rate at which blood is pumped out of the heart, through blood vessels, then back to the heart. The rate, about seventy beats every minute, is moderate. But commands from the nervous and endocrine systems continually adjust it. When you run, skeletal muscle cells demand more blood-borne oxygen and glucose than they do when

JIMMIE   DR. WALLER

a   BUCKET   BUCKET   b

Figure 34.1 A bit of history in the making. (**a**) Jimmie the bulldog taking part in a painless experiment. (**b**) The physiologist Augustus Waller and his beloved pet bulldog sharing a moment in Waller's study after the experiment, which yielded one of the world's first electrocardiograms (**c**).

c

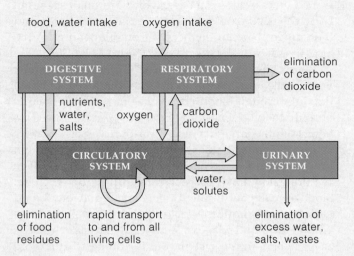

Figure 34.2 Diagram of the functional connections between the circulatory, respiratory, and digestive systems, which interact in transporting substances to and from all living cells in the animal body. Their integrated activities help maintain favorable operating conditions in the internal environment.

you sleep. At such times of intense exercise, your heart starts pounding more than twice as fast, and this helps deliver sufficient blood to muscle tissues.

We've come a long way from Waller and Jimmie in monitoring the heart. Sensors now detect faint signals characteristic of an impending heart attack. Internists use computers to analyze a patient's beating heart, and ultrasound probes to get images of it on video screens. Cardiologists routinely substitute battery-powered pacemakers for faltering natural ones.

With this chapter, we turn to the **circulatory system**, the means by which substances move rapidly to and from the interstitial fluid that bathes all living cells in nearly all animals. The system continually accepts the oxygen, nutrients, and other substances that an animal secures, through its respiratory and digestive systems, from the environment (Figure 34.2). Simultaneously it picks up carbon dioxide and other wastes from cells and gets them to the respiratory and urinary systems for disposal. Its operation is central to maintaining the operating conditions of the internal environment within a tolerable range. And that is the state we call homeostasis.

CIRCULATORY SYSTEMS—AN OVERVIEW

General Characteristics

Imagine that an earthquake has closed off the highways around your neighborhood. Grocery trucks can't enter and waste-disposal trucks can't leave, so food supplies dwindle and garbage piles up. Cells would face similar predicaments if your body's highways were disrupted. The highways are part of a circulatory system, which swiftly moves substances to and from cells. The system helps maintain a favorable neighborhood, so to speak, and this is vital work. Your differentiated cells are just too specialized to fend for themselves. They interact to maintain the volume, composition, and temperature of **interstitial fluid**, a tissue fluid bathing them. **Blood**, a circulating connective tissue, interacts with interstitial fluid. Its continuous deliveries and pickups help keep conditions tolerable for enzymes and other molecules that carry out cell activities. Interstitial fluid and blood are the body's "internal environment."

Blood flows inside blood vessels, or tubes that differ in wall thickness and diameter. A muscular pump, the **heart**, generates pressure that keeps blood flowing. As is the case for many animals, you have a *closed* circulatory

system; it confines blood inside continuously connected walls of a heart and blood vessels. This isn't the case in *open* circulatory systems of arthropods, such as insects, and most mollusks. In an open system, blood flows out from the vessels into sinuses, which are small spaces in body tissues. Blood mingles with the tissue fluids and then moves back to the heart through openings in the blood vessels or the heart wall (Figure 34.3*a,b*).

Think about the overall "design" of a closed system. Because the heart pumps incessantly, the *volume* of blood flowing through blood vessels has to equal the heart's output at any given time. The flow's *velocity* (speed) is highest in the large-diameter transport vessels. It slows in specific body regions, where there must be enough time for blood to exchange substances with cells. The required slowdown proceeds at **capillary beds**. There, blood spreads out through many small-diameter blood vessels called **capillaries**. During any interval, the same volume of blood is moving forward through the beds as elsewhere in the system. But it's doing so at a more leisurely pace because of the great total cross-sectional area of the blood capillaries. The simple analogy given in Figure 34.3*e* may help you grasp this concept.

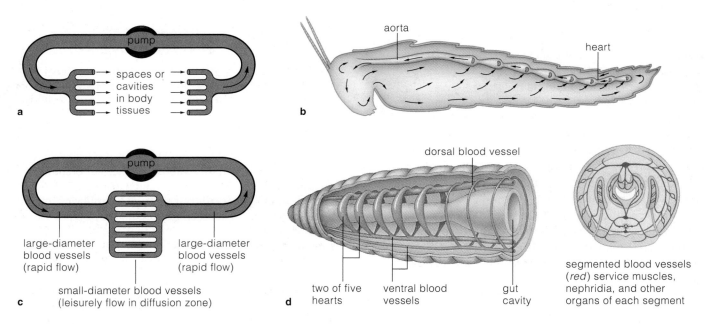

Figure 34.3 Open and closed circulatory systems. (**a,b**) In a grasshopper's open system, a heart (not like yours) pumps blood through a vessel, the aorta. Blood moves into tissue spaces and mingles with fluid bathing cells. It reenters the heart at openings in the heart wall. (**c,d**) An earthworm's closed system confines blood in several pairs of muscular hearts near the head end and inside blood vessels.

(**e**) Relationship between flow velocity and total cross-sectional area in any closed circulatory system. Visualize two fast rivers flowing into and out from a lake. The flow *rate* is the same in all three places; an identical volume of water moves from point *1* to point *3* during the same interval. Flow *velocity* decreases within the lake. Why? The volume spreads out through a larger cross-sectional area and moves forward a shorter distance during the same length of time.

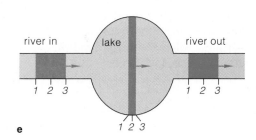

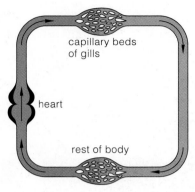 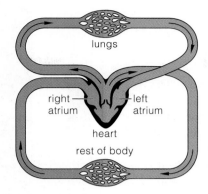 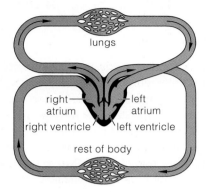

a In fishes, a two-chambered heart (atrium, ventricle) pumps blood in one circuit. Blood picks up oxygen in gills, delivers it to rest of body. Oxygen-poor blood flows back to heart.

b In amphibians, a heart pumps blood through two partially separate circuits. Blood flows to lungs, picks up oxygen, returns to heart. It mixes with oxygen-poor blood still in heart, flows to rest of body, returns to heart.

c In birds and mammals, the heart is fully partitioned into two halves. Blood circulates in two circuits: from the heart's right half to lungs and back, then from the heart's left half to oxygen-requiring tissues and back.

Figure 34.4 Comparison of the closed circulatory system of four vertebrate groups.

Evolution of Vertebrate Circulatory Systems

Humans and other existing vertebrates have a closed circulatory system, although the pump and plumbing for fishes, amphibians, birds, and mammals differ in their details. The differences evolved over hundreds of millions of years and arose during the move of some vertebrate lineages onto land. Recall from Section 27.3 that fishes, the first vertebrates, had gills. Gills, like all respiratory structures, have a thin, moist surface that oxygen and carbon dioxide can diffuse across. Later, in the ancestors of all land vertebrates, lungs evolved that supplemented gas exchange. Being *internally moistened* sacs, lungs had advantages for the move onto dry land. Also advantageous were concurrent modifications in circulatory systems, which pick up oxygen from lungs and deliver carbon dioxide wastes to them.

Consider this: In fishes, blood flows in *one* circuit (Figure 34.4*a*). Pressure generated by a two-chambered heart forces it from gill capillary beds into the largest artery, through capillary beds of organs, and back to the heart. Extensive gill capillaries offer so much resistance to flow that pressure drops considerably before blood enters the main artery. Blood delivery is enough for the activity level of most fishes. It would not be good enough for the more active life-styles of most land vertebrates.

As amphibians were evolving, their heart became partitioned into right and left halves—only partly so, but enough to pump blood through *two* partially separated circuits (Figure 34.4*b*). The flow separation continued in reptiles called crocodilians. It became complete in the two side-by-side pumps of birds and mammals (Figure 34.4*c*). Their heart's *right* half pumps oxygen-poor blood to lungs, where blood picks up oxygen and gives up carbon dioxide. The freshly oxygenated blood flows to the heart's left half. This route is the **pulmonary circuit**.

In the **systemic circuit**, the heart's *left* half pumps the freshly oxygenated blood to every tissue and organ where oxygen is used and carbon dioxide forms. Then the oxygen-poor blood flows to the heart's right half.

The double circuit is a rapid and efficient mode of blood delivery. It supports the high levels of activity typical of vertebrates whose ancestors evolved on land.

Links With the Lymphatic System

The heart's pumping puts pressure on blood flowing through the circulatory system. Partly because of this pressure, small amounts of water and a few proteins dissolved in blood are forced out of the capillaries and become part of interstitial fluid. Drainage vessels pick up excess interstitial fluid and reclaimable solutes and return them to the circulatory system. These vessels are part of the **lymphatic system**. Later, you will see how other parts of the system help cleanse fluid inside it.

Closed circulatory systems confine blood in one or more hearts and a network of blood vessels. In open systems, blood also intermingles with tissue fluids.

The closed system of vertebrates transports substances to and from interstitial fluid and is functionally connected with the lymphatic system.

Blood flows rapidly in large-diameter vessels between the vertebrate heart and capillary beds. At these beds, the flow velocity drops and exchanges are made between blood and interstitial fluid, and between interstitial fluid and cells.

In fishes, blood flows in one circuit from and back to the heart. In birds and mammals, blood flows in two circuits, through a heart partitioned as two side-by-side pumps. The double circuit supports the high levels of activity typical of vertebrates that evolved on land.

CHARACTERISTICS OF BLOOD

Functions of Blood

Blood is a connective tissue with multiple functions. It transports oxygen, nutrients, and other solutes to cells. It carries away metabolic wastes and secretions such as hormones. It helps stabilize internal pH and serves as a highway for phagocytic cells that fight infections and scavenge debris in tissues. In birds and mammals, blood helps equalize body temperature by moving excess heat from regions of high metabolic activity (such as skeletal muscles) to the skin, where heat can be dissipated.

2 μm

8 μm
average
diameter

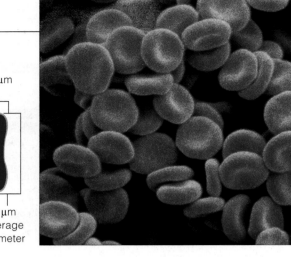

Figure 34.6 Size and shape of red blood cells.

Blood Volume and Composition

The volume of blood depends on body size and on the concentrations of water and solutes. Blood volume for average-size adult humans is about 6 to 8 percent of the total body weight. That amounts to about four or five quarts. As is the case for all vertebrates, human blood is a sticky fluid, thicker than water and slower flowing. Its components are the **plasma**, **red blood cells**, **white blood cells**, and **platelets**. Normally, plasma accounts for 50 to 60 percent of the total blood volume.

PLASMA Prevent a blood sample in a test tube from clotting, and it separates into a red, cellular portion and the plasma, a straw-colored liquid, which floats on the cellular portion (Figure 34.5). Plasma is mostly water, and it serves as a transport medium for blood cells and platelets. Plasma also functions as a solvent for ions and molecules, including hundreds of different plasma proteins. Some of the plasma proteins transport lipids and fat-soluble vitamins through the body. Others have

roles in blood clotting or in defense against pathogens. Collectively, the concentrations of the plasma proteins influence the fluid volume of blood, for it affects the movement of water between the blood and interstitial fluid. Glucose and other simple sugars as well as lipids, amino acids, vitamins, and hormones are dissolved in plasma. So are oxygen, carbon dioxide, and nitrogen.

RED BLOOD CELLS Erythrocytes, or red blood cells, are biconcave disks (Figure 34.6). They transport the oxygen used in aerobic respiration and carry away some carbon dioxide wastes. When oxygen first diffuses into blood, it binds to hemoglobin. It's this iron-containing pigment that makes red blood cells look red. Section 3.5 shows its structure. Oxygenated blood is bright red. Poorly oxygenated blood is darker red, but it appears blue in blood vessel walls near the body surface.

Stem cells in bone marrow give rise to red blood cells (Figure 34.7). Generally speaking, **stem cells** stay

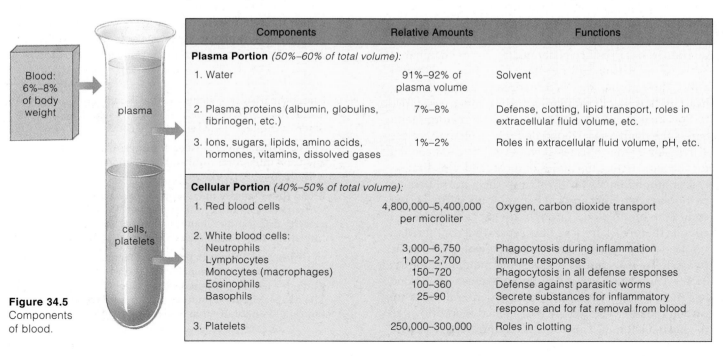

Figure 34.5
Components
of blood.

Blood:
6%–8%
of body
weight

plasma

cells,
platelets

Components	Relative Amounts	Functions
Plasma Portion (50%–60% of total volume):		
1. Water	91%–92% of plasma volume	Solvent
2. Plasma proteins (albumin, globulins, fibrinogen, etc.)	7%–8%	Defense, clotting, lipid transport, roles in extracellular fluid volume, etc.
3. Ions, sugars, lipids, amino acids, hormones, vitamins, dissolved gases	1%–2%	Roles in extracellular fluid volume, pH, etc.
Cellular Portion (40%–50% of total volume):		
1. Red blood cells	4,800,000–5,400,000 per microliter	Oxygen, carbon dioxide transport
2. White blood cells:		
Neutrophils	3,000–6,750	Phagocytosis during inflammation
Lymphocytes	1,000–2,700	Immune responses
Monocytes (macrophages)	150–720	Phagocytosis in all defense responses
Eosinophils	100–360	Defense against parasitic worms
Basophils	25–90	Secrete substances for inflammatory response and for fat removal from blood
3. Platelets	250,000–300,000	Roles in clotting

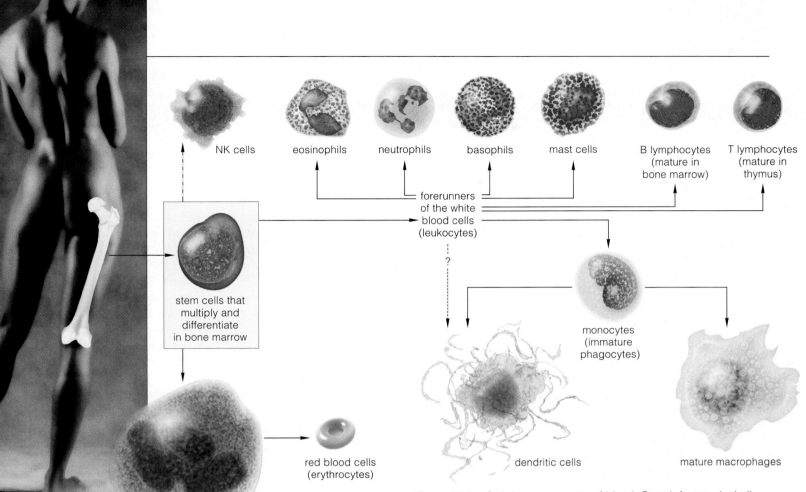

NK cells   eosinophils   neutrophils   basophils   mast cells   B lymphocytes (mature in bone marrow)   T lymphocytes (mature in thymus)

forerunners of the white blood cells (leukocytes)

stem cells that multiply and differentiate in bone marrow

monocytes (immature phagocytes)

red blood cells (erythrocytes)

dendritic cells

mature macrophages

megakaryocytes

platelets

Figure 34.7 Cellular components of blood. Growth factors, including erythropoietin and interleukin 3, stimulate growth and differentiation in the cell lineages. The next chapter looks at white blood cell functions. Section 34.10 describes the role of platelets in blood clotting.

unspecialized and retain their capacity for mitotic cell division. Their daughter cells divide, also, but only a portion go on to differentiate into specialized types.

Mature red blood cells no longer have their nucleus, nor do they require it. They have enough hemoglobin, enzymes, and other proteins to function for about 120 days. The body's own phagocytes engulf the oldest red blood cells and those already dead. Normally, ongoing replacements keep the cell count fairly stable.

A **cell count** is a measure of the number of cells of a given type in a microliter of blood. For example, the average number of red blood cells per cubic centimeter is 5.4 million in males and 4.8 million in females.

WHITE BLOOD CELLS   Leukocytes, or white blood cells, arise from stem cells in bone marrow. They function in daily housekeeping and defense. Some types patrol tissues. They target or engulf damaged or dead cells and anything chemically recognized as foreign to the body. Many others are massed together in lymph nodes and the spleen, which are organs of the lymphatic system. There they divide to produce armies of cells that battle specific viruses, bacteria, and other threats to health. Figure 34.7 shows the main categories of white blood

cells, which differ in size, nuclear shape, and staining traits. Their numbers change, depending on whether an individual is active, healthy, or under siege. You'll read about them in the next chapter. For now, a few examples will give you a sense of what they do. Neutrophils are fast-acting phagocytes. Like basophils, they have roles in inflammatory responses to invasions and tissue damage. All immune responses against specific threats use two classes of lymphocytes: B and T cells. Macrophages and dendritic cells can call up immune responses. NK cells destroy virus-infected and cancerous body cells.

PLATELETS   Some stem cells in bone marrow give rise to megakaryocytes. These "giant" cells shed cytoplasmic fragments wrapped in a bit of plasma membrane. The membrane-bound fragments are what we call platelets. Each platelet lasts only five to nine days, but hundreds of thousands are always circulating in blood. They can release substances that initiate blood clotting.

Vertebrate blood has major roles in transporting substances, defense, clotting, and maintaining the volume, composition, and temperature of the internal environment.

Blood Disorders

The body continually replaces blood cells, for good reason. Besides aging and dying off regularly, blood cells typically encounter a variety of pathogens that use them as places to complete their life cycle. Besides this, sometimes blood cells malfunction as a result of gene mutations.

RED BLOOD CELL DISORDERS   Consider the **anemias**. These disorders result from too few red blood cells or deformed ones. Oxygen levels in blood cannot be kept high enough to support normal metabolism. Shortness of breath, fatigue, and chills follow. *Hemorrhagic* anemias result from sudden blood loss, as from a severe wound; *chronic* anemias result from ongoing but slight blood loss, as from an undiagnosed bleeding ulcer or hemorrhoids.

Some bacteria and protozoans replicate in blood cells. They escape by lysis and kill the cells, thus causing some *hemolytic* anemias. Insufficient iron in the diet results in *iron deficiency* anemia; red blood cells can't make enough normal hemoglobin without iron. B_{12} *deficiency* anemia is a potential hazard for strict vegetarians and alcoholics. Red blood cells form but cannot divide without vitamin B_{12}, which you can get from meats, poultry, and fish.

As described elsewhere in the book, a mutation that gives rise to an abnormal form of hemoglobin can result in *sickle-cell* anemia. Another mutation partially or fully blocks synthesis of the globin chains of hemoglobin; it causes *thalassemias*. Too few red blood cells form, and those that do are thin and fragile.

Polycythemias—symptoms of far too many red blood cells—make blood flow sluggish. Some bone marrow cancers lead to this condition. So can *blood doping*. Some athletes about to compete in strenuous events withdraw and store their own red blood cells, then reinject them a few days prior to competition. The withdrawal triggers red blood cell formation as the body attempts to replace "lost" cells, so when withdrawn cells are put back in the body, the cell count bumps up. The idea is to increase oxygen-carrying capacity and endurance. Blood doping leads to temporary high blood pressure and also lowers blood's viscosity. Some believe blood doping works, but others call it unethical. It is banned from the Olympics.

WHITE BLOOD CELL DISORDERS   Possibly you've heard of *infectious mononucleosis*. An Epstein–Barr virus causes this highly contagious disease, which results from the formation of too many monocytes and lymphocytes. After a few weeks of fatigue, aches, low-grade fever, and a chronic sore throat, patients usually recover.

Recovery is chancy for *leukemias*. As you read in Chapter 11, these cancers suppress or impair the formation of white blood cells in bone marrow. Cancer cells can be killed by radiation therapy and chemotherapy, but side effects are usually severe. Remissions may last for months or years. A "remission" is a symptom-free period of a chronic illness. New gene therapies are leading to remissions for some leukemias, so far with only mild side effects.

BLOOD TRANSFUSION AND TYPING

Concerning Agglutination

Whenever blood volume or blood cell counts decline, automatic countermeasures kick in. If the volume were to decrease by more than 30 percent, circulatory shock would follow and could cause death.

Blood from donors can be transfused into patients affected by a blood disorder or severe blood loss. Such **blood transfusions** can't be hit-or-miss. Why not? *Red blood cells of a potential donor and recipient may not have the same kind of recognition proteins at their surface.* Some of the proteins are "self" markers. They identify cells as belonging to one's own body. If the donor's cells have the "wrong" marker, a recipient's immune system will recognize them as foreign, with serious consequences.

Figure 34.8 shows what happens when blood from incompatible donors and recipients intermingle. In a defense response called **agglutination**, proteins called antibodies that are circulating in the plasma act against

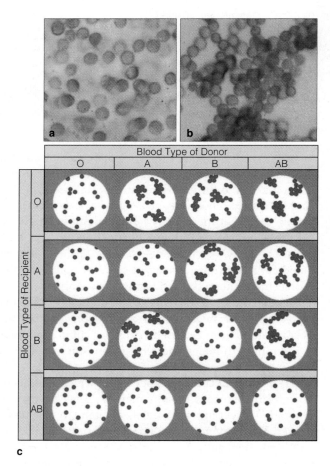

Figure 34.8   Light micrographs showing (**a**) an absence of agglutination in a mixture of two different yet compatible blood types and (**b**) agglutination in a mixture of incompatible types. (**c**) Responses in blood types A, B, AB, and O when mixed with samples of the same and different types.

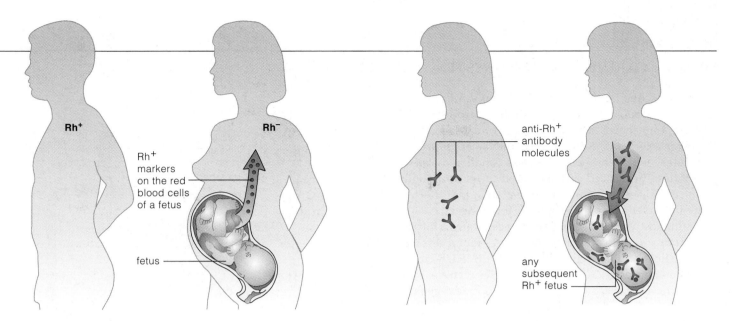

a A forthcoming child of an Rh$^-$ woman and Rh$^+$ man inherits the gene for the Rh$^+$ marker. During childbirth, some of its cells bearing the marker may leak into the maternal bloodstream.

b The foreign marker stimulates antibody formation. If the same woman becomes pregnant again and if her second fetus (or any other) inherits the gene for the marker, her circulating anti-Rh$^+$ antibodies will act against it.

Figure 34.9 Antibody production in response to Rh$^+$ markers on red blood cells of a fetus.

the foreign cells and cause them to clump together. As the next chapter describes, antibodies bind to specific foreign markers and target any cell or particle bearing them for destruction by the immune system. Also, the large number of foreign cells transfused into a patient end up in numerous clumps that can clog small blood vessels and damage tissues. Without treatment, death may follow. The same thing can happen during certain pregnancies when antibodies diffuse from a mother's circulatory system into that of her unborn child.

Based on an understanding of cell surface markers and antibodies, scientists have devised ways to analyze the forms of self markers on a person's red blood cells. Blood typing is based on these analyses.

ABO Blood Typing

Molecular variations in one kind of self marker on red blood cells are analyzed with **ABO blood typing**. The genetic basis of such variation is described in Section 10.4. People with one form of the marker are said to have type A blood; those with another form have type B blood. Those with both forms of the marker on their red blood cells have type AB blood. Others do not have either form of the marker; they have type O blood.

If you are type A, your antibodies ignore A markers but will act against B markers. If you are type B, your antibodies will ignore B markers but will act against A markers. If you are type AB, your antibodies ignore both forms of the marker, so you can tolerate donations of type A, B, AB, or O blood. If you are type O, you have antibodies against both forms of the marker, so your options are limited to type O donations.

Rh Blood Typing

Rh blood typing is based on the presence or absence of the Rh marker (first identified in blood of *Rh*esus monkeys). If you are type Rh$^+$, your blood cells bear this marker. If you are type Rh$^-$, they do not. Usually people don't have antibodies against Rh markers. But an Rh$^-$ recipient of transfused Rh$^+$ blood will make antibodies against them, and these stay in the blood.

If an Rh$^-$ woman becomes impregnated by an Rh$^+$ man, there is a chance that the fetus will be Rh$^+$. At childbirth, some fetal red blood cells typically leak into the woman's bloodstream. At that time her body will produce antibodies against Rh (Figure 34.9). And if she becomes pregnant again, Rh antibodies will enter the bloodstream of the new fetus. If its blood is type Rh$^+$, her antibodies will cause fetal red blood cells to swell, rupture, and release hemoglobin.

Erythroblastosis fetalis is a severe outcome of mixing Rh$^+$ and Rh$^-$ types. Too many cells are killed, and the fetus dies. If the condition is diagnosed before birth, a newborn can survive when its blood is slowly replaced with transfusions that are free of Rh antibodies. Today, a known Rh$^-$ woman can be treated right after her first pregnancy with an anti-Rh gamma globulin (RhoGam) that will protect her next fetus. The drug will inactivate any Rh$^+$ fetal blood cells that are circulating through her bloodstream before she can become sensitized and start producing potentially dangerous antibodies.

To avoid symptoms of blood incompatibilities, red blood cells should be typed before transfusions or pregnancies.

HUMAN CARDIOVASCULAR SYSTEM

"Cardiovascular" comes from the Greek *kardia* (heart) and Latin *vasculum* (vessel). In a human cardiovascular system, a muscular heart pumps the blood into large-diameter **arteries**. The blood flows into small, muscular **arterioles**, which branch into even smaller diameter capillaries (introduced earlier). Blood flows continually from capillaries into small **venules**, and from there to large-diameter **veins** that return it to the heart.

As in most vertebrates, a partition separates a human heart into a double pump, which drives blood through two cardiovascular circuits (Figure 34.10). Each circuit has its own arteries, arterioles, capillaries, venules, and veins. The *pulmonary* circuit, a short loop, oxygenates blood. It leads from the heart's right half to capillary beds in both lungs, then returns to the heart's left half. The *systemic* circuit, a longer loop, starts at the heart's left half. Its main artery, the **aorta**, accepts oxygenated blood, which flows through arterioles, capillary beds in all regions, then veins that deliver the oxygen-poor blood to the heart's right half. Figure 34.11 identifies key blood vessels of both circuits and their functions.

For most of the systemic branchings, a given volume of blood flows through one capillary bed. The branch to the intestines is one of the exceptions. First blood picks up glucose and other absorbed substances from one bed, then moves through another capillary bed in the liver —a very important organ in nutrition. The second bed gives the liver time to process absorbed substances.

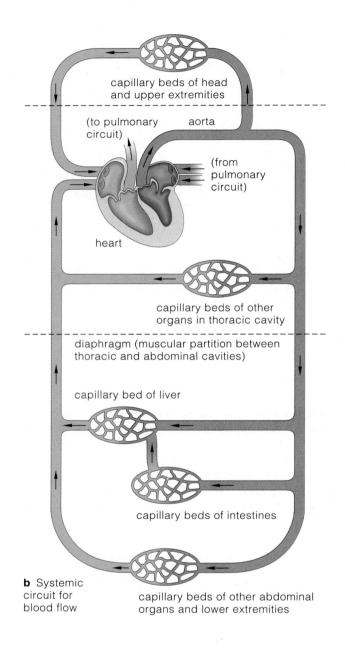

b Systemic circuit for blood flow

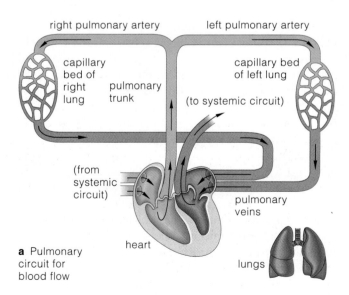

a Pulmonary circuit for blood flow

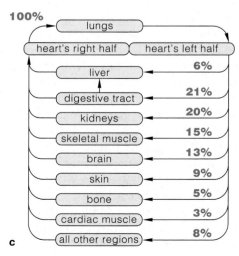

Figure 34.10 (**a**,**b**) Pulmonary and systemic circuits for blood flow through the human cardiovascular system. Blood vessels carrying oxygenated blood are color-coded *red*. Those carrying oxygen-poor blood are color-coded *blue*. (**c**) Distribution of the heart's output in a person at rest. The lungs receive all blood pumped out of the heart's right half. Organs serviced by the systemic circuit receive the portions indicated from the heart's left half. Control mechanisms adjust the flow distribution when required.

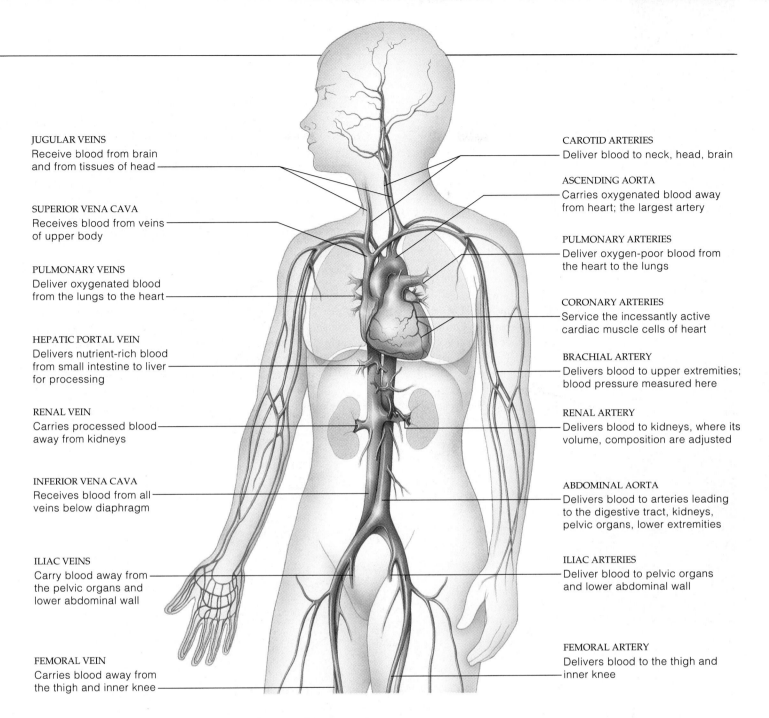

JUGULAR VEINS
Receive blood from brain and from tissues of head

SUPERIOR VENA CAVA
Receives blood from veins of upper body

PULMONARY VEINS
Deliver oxygenated blood from the lungs to the heart

HEPATIC PORTAL VEIN
Delivers nutrient-rich blood from small intestine to liver for processing

RENAL VEIN
Carries processed blood away from kidneys

INFERIOR VENA CAVA
Receives blood from all veins below diaphragm

ILIAC VEINS
Carry blood away from the pelvic organs and lower abdominal wall

FEMORAL VEIN
Carries blood away from the thigh and inner knee

CAROTID ARTERIES
Deliver blood to neck, head, brain

ASCENDING AORTA
Carries oxygenated blood away from heart; the largest artery

PULMONARY ARTERIES
Deliver oxygen-poor blood from the heart to the lungs

CORONARY ARTERIES
Service the incessantly active cardiac muscle cells of heart

BRACHIAL ARTERY
Delivers blood to upper extremities; blood pressure measured here

RENAL ARTERY
Delivers blood to kidneys, where its volume, composition are adjusted

ABDOMINAL AORTA
Delivers blood to arteries leading to the digestive tract, kidneys, pelvic organs, lower extremities

ILIAC ARTERIES
Deliver blood to pelvic organs and lower abdominal wall

FEMORAL ARTERY
Delivers blood to the thigh and inner knee

Figure 34.11 Location and functions of major blood vessels of the human cardiovascular system.

Figure 34.10*c* shows you how the two blood circuits distribute the heart's output to different organs. The percentages listed are for a person at rest. As you will see, change in levels of physical activity, plus shifting internal and external conditions, lead to adjustments in the flow distribution by the systemic route. Example: Blood flow to and from a skeletal muscle varies greatly, depending on what it's being called on to do at a given time. The same is true of skin. When the body gets too cold, blood flow—hence metabolic heat—gets diverted away from the skin's extensive capillary beds. When the body is too hot, blood flow to these beds increases and heat radiates away from the skin's surface. Only the brain is exempt; it cannot tolerate flow variations.

The human cardiovascular system consists of two separate circuits, pulmonary and systemic, for blood flow.

The pulmonary circuit is a short loop from the heart's right half, through both lungs, to the heart's left half. Oxygen-poor blood flowing into the circuit rapidly picks up oxygen and gives up carbon dioxide at capillary beds in the lungs.

The longer systemic circuit starts at the heart's left half and aorta. It delivers oxygen and accepts carbon dioxide at capillary beds of all metabolically active regions. Then its veins deliver oxygen-poor blood to the heart's right half.

34.6

THE HEART IS A LONELY PUMPER

Heart Structure

A human heart beats 2.5 billion times during a seventy-year life span, so you know it must be a durable pump. Its structure speaks of its durability. The pericardium, a double sac of tough connective tissue, protects and anchors the heart to nearby structures (*peri*, around). A fluid between its layers lubricates the heart during its perpetual twisting motions. The inner layer is actually part of the heart wall (Figure 34.12). The myocardium, the bulk of the wall, has cardiac muscle cells tethered to elastin and collagen fibers. The fibers are so densely crisscrossed, they serve as a "skeleton" against which contractile force can be applied. Cardiac muscle cells need so much oxygen, they have a separate, *coronary* circulation; coronary arteries branch off the aorta and lead to the heart's own capillary bed. The heart wall's glistening, inner layer is composed of connective tissue and endothelium. This sheet of single cells, a type of epithelium, occurs only in the heart and blood vessels.

Each half of the heart has two chambers: an atrium (plural, atria) and a ventricle. Blood flows into the atria, then the ventricles, then out through the great arteries: the aorta and pulmonary trunk. Between each atrium and ventricle is an AV valve (short for atrioventricular). Between each ventricle and the artery leading out from it is a semilunar valve. Both types are *one-way* valves. They are alternately forced open and then shut to help keep the blood moving in the forward direction only.

Cardiac Cycle

Each time the heart beats, its four chambers go through phases of systole (contraction) and diastole (relaxation). The sequence of contraction and relaxation is a **cardiac cycle**. First, relaxed atria fill with blood. Increasing fluid pressure forces both AV valves open. Blood flows into the ventricles, which fill when the atria contract (Figure 34.13). As the filled ventricles contract, the rising fluid pressure forces the AV valves shut. It rises so sharply

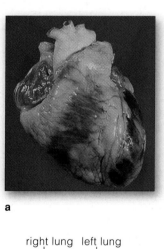

a

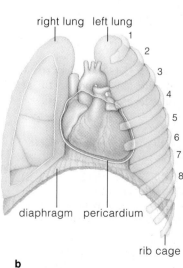

right lung left lung

1
2
3
4
5
6
7
8

diaphragm   pericardium

rib cage

b

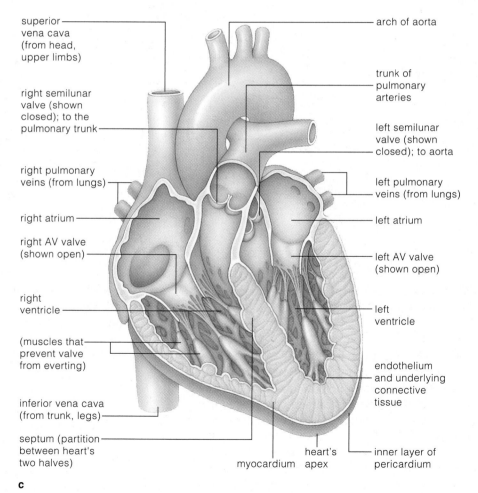

superior vena cava (from head, upper limbs)

right semilunar valve (shown closed); to the pulmonary trunk

right pulmonary veins (from lungs)

right atrium

right AV valve (shown open)

right ventricle

(muscles that prevent valve from everting)

inferior vena cava (from trunk, legs)

septum (partition between heart's two halves)

arch of aorta

trunk of pulmonary arteries

left semilunar valve (shown closed); to aorta

left pulmonary veins (from lungs)

left atrium

left AV valve (shown open)

left ventricle

endothelium and underlying connective tissue

heart's apex

inner layer of pericardium

myocardium

c

Figure 34.12 (**a**) Photograph of the human heart and (**b**) its location in the thoracic cavity. (**c**) Cutaway view of the heart, showing its wall and internal organization.

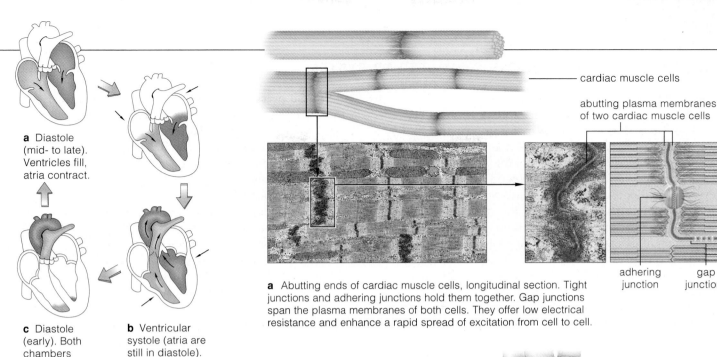

a Diastole (mid- to late). Ventricles fill, atria contract.

c Diastole (early). Both chambers relaxed.

b Ventricular systole (atria are still in diastole). Ventricles eject.

Figure 34.13 Blood flow during part of a cardiac cycle. Blood and heart movements generate a "lub-dup" sound at the chest wall. At each "lub," AV valves are closing as the ventricles contract. At each "dup," the semilunar valves are closing as ventricles relax.

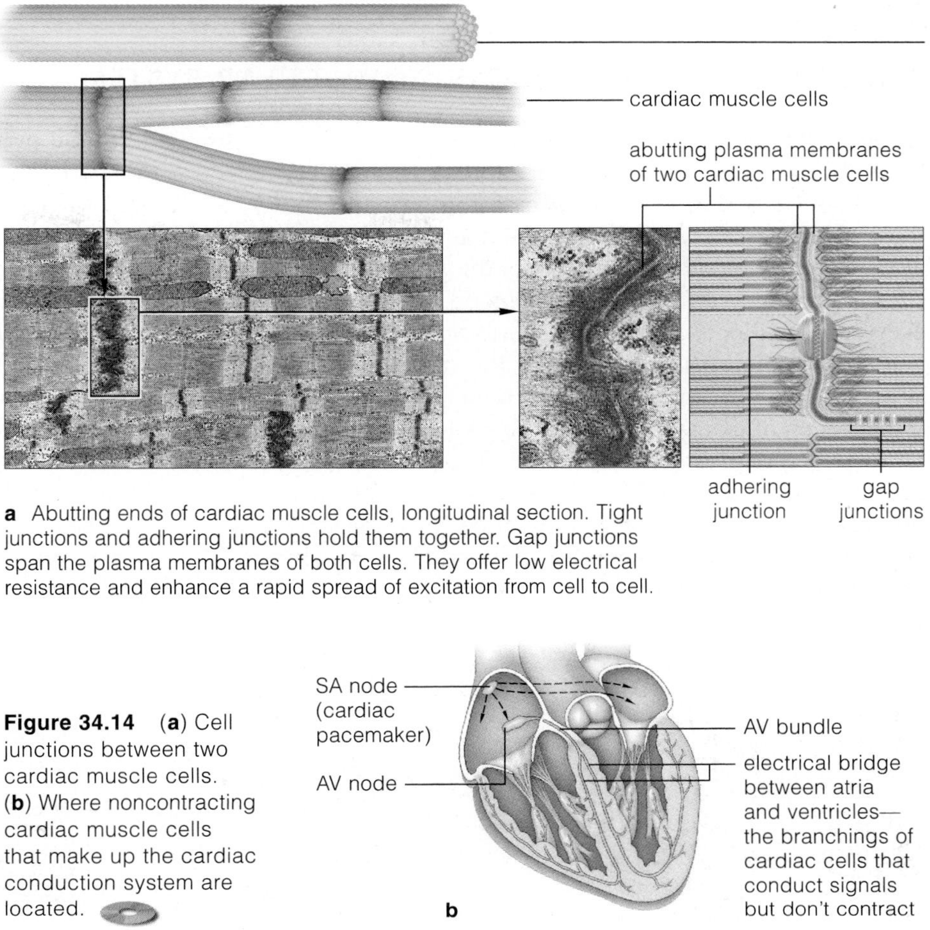

cardiac muscle cells

abutting plasma membranes of two cardiac muscle cells

adhering junction

gap junctions

a Abutting ends of cardiac muscle cells, longitudinal section. Tight junctions and adhering junctions hold them together. Gap junctions span the plasma membranes of both cells. They offer low electrical resistance and enhance a rapid spread of excitation from cell to cell.

Figure 34.14 (**a**) Cell junctions between two cardiac muscle cells. (**b**) Where noncontracting cardiac muscle cells that make up the cardiac conduction system are located.

SA node (cardiac pacemaker)

AV node

AV bundle

electrical bridge between atria and ventricles—the branchings of cardiac cells that conduct signals but don't contract

b

above the pressure in the great arteries that it forces the semilunar valves open—and blood leaves the heart. The ventricles relax, semilunar valves close, and the already filling atria are ready to repeat the cycle. And so, atrial contraction simply helps fill the ventricles. *Contraction of the ventricles is the driving force for blood circulation.*

Cardiac Muscle Contraction

Like skeletal muscle, cardiac muscle is striated (striped). In response to action potentials, the sarcomeres of its cells contract in the manner predicted by the sliding-filament model (Section 33.6). This tissue, too, requires ATP for contraction, which a great many mitochondria in the myocardium provide. But cardiac muscle cells are structurally unique. They are branching, short, and connected at their end regions. Gap junctions span both plasma membranes of adjoining cells. They let action potentials spread swiftly through cardiac tissue in waves of excitation that wash over the heart (Figure 34.14).

In addition, about 1 percent of the cardiac muscle cells *do not* contract. They function instead as a **cardiac conduction system**. About seventy times each minute, these specialized cells initiate and propagate waves of excitation that travel in a rhythmic, orderly sequence from atria to ventricles. Their synchronized excitation directs the heart's efficient pumping. Each wave starts

at the SA node, a cluster of cell bodies in the wall of the right atrium (Figure 34.14). It passes through the wall to another cluster of cell bodies, the AV node. This is the only electrical bridge between the atria and ventricles, which connective tissue insulates everywhere else. After the AV node, conducting cells are arranged as a bundle in the partition between the heart's two halves. These cells branch, and their branches deliver the excitatory wave up the ventricle walls. The ventricles contract in response with a twisting movement, upward from the heart's apex, that ejects blood into the great arteries.

The SA node fires action potentials faster than the rest of the system and serves as the **cardiac pacemaker**. Its spontaneous, rhythmic signals are the basis for the normal rate of heartbeat. The nervous system can only *adjust* the rate and strength of contractions dictated by the pacemaker. Even if all nerves leading to a heart are severed, the heart will keep on beating!

The heart's construction reflects its role as a durable pump. Under the spontaneous, rhythmic signals from the cardiac pacemaker, its branching, abutting cardiac muscle cells contract in synchrony, almost as if they were a single unit.

Although the heart has four chambers (each half has one atrium and one ventricle), contraction of the ventricles is the driving force for blood circulation away from the heart.

BLOOD PRESSURE IN THE CARDIOVASCULAR SYSTEM

As you have seen, blood flows to and from the human heart through arteries, arterioles, capillaries, venules, and then veins. Figure 34.15 shows the structure of these blood vessels. Arteries are the main transporters of oxygenated blood through the body. Arterioles are sites of control over the volume of blood flow through every organ. Blood capillaries and, to a lesser extent, venules are the diffusion zones of the cardiovascular system. Veins function as blood volume reservoirs and transporters of oxygen-poor blood back to the heart.

Two factors greatly influence the flow rate through each type of blood vessel. First, the flow rate is directly proportional to the pressure gradient between the start and end of the vessel. Second, the flow rate is inversely proportional to the vessel's resistance to flow.

Blood pressure is fluid pressure imparted to blood by heart contractions. In the pulmonary and systemic circuits, the *difference* in pressure between two points determines the flow rate. The beating heart establishes high pressure at the start of a circuit. Flowing blood rubs against the vessel walls, and the friction impedes flow. The major factor determining resistance to flow is blood vessel radius. Simply put, resistance increases as the tubes narrow down. Even a twofold decrease in radius increases the resistance sixteenfold.

Mainly because of frictional losses after the heart pumps out blood, pressure differs along the circuits. It is highest in contracting ventricles and still high at the start of arteries. It continues to drop until the relaxed atria, where blood pressure is lowest (Figure 34.16).

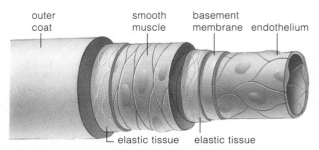

a ARTERY

outer coat / smooth muscle / basement membrane / endothelium / elastic tissue / elastic tissue

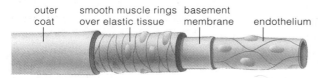

b ARTERIOLE

outer coat / smooth muscle rings over elastic tissue / basement membrane / endothelium

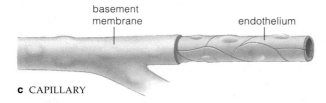

c CAPILLARY

basement membrane / endothelium

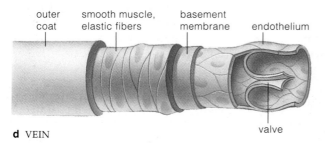

d VEIN

outer coat / smooth muscle, elastic fibers / basement membrane / endothelium / valve

Figure 34.15 Structure of blood vessels. The basement membrane around each vessel's endothelium is a noncellular layer, rich with proteins and polysaccharides. Small venules resemble capillaries in structure and function.

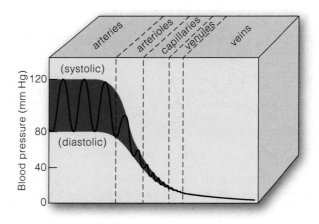

Figure 34.16 Plot of the decline in fluid pressure for a volume of blood moving through the systemic circuit.

Arterial Blood Pressure

With their large diameter and low resistance to flow, arteries are fast transporters of oxygenated blood. They also are pressure reservoirs that smooth out pulsations in pressure caused by each cardiac cycle. Their thick, muscular, elastic wall bulges from the large volume of blood that ventricular contraction forces into them. The wall recoils, and this forces the blood forward through the circuit while the heart is relaxing.

Suppose you want to measure your blood pressure daily. Typically, you'd take a reading from the brachial artery of an upper arm while resting (Figure 34.17). For one cardiac cycle, *systolic* pressure is the peak pressure that contracting ventricles exert against the artery wall. *Diastolic* pressure is the lowest arterial blood pressure of a cardiac cycle, reached when ventricles are relaxing. You can estimate the mean arterial pressure as *diastolic pressure + 1/3 pulse pressure*. (The difference between the highest and lowest readings is the pulse pressure.)

Figure 34.17 Measuring blood pressure.

A hollow cuff attached to a pressure gauge is wrapped around the upper arm. The cuff is inflated with air to a pressure above the highest pressure of the cardiac cycle—at systole, when ventricles contract. Above this pressure, you cannot hear sounds through a stethoscope positioned below the cuff and above the artery, for no blood is flowing through the vessel. Air in the cuff is slowly released, so some blood flows into the artery. The turbulent flow causes soft tapping sounds. When the tapping starts, the gauge's value is the systolic pressure, or about 120 mm mercury (Hg) in young adults at rest. This value means the measured pressure would force mercury to move upward 120 millimeters in a narrow glass column.

More air is released from the cuff. Just after the sounds grow dull and muffled, blood is flowing continuously, so the turbulence and tapping end. The silence corresponds to diastolic pressure at the end of a cardiac cycle, just before the heart pumps out blood. The reading is usually about 80 mm Hg. In this case, pulse pressure (the difference between the highest and lowest pressure readings) is 120 – 80, or 40 mm Hg. Assuming you're an adult in good health, this average resting value stays fairly constant over a few weeks or months.

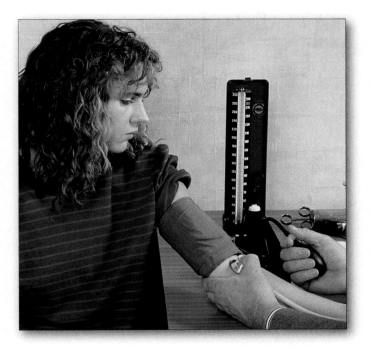

How Can Arterioles Resist Flow?

Track a volume of blood through the systemic circuit and you notice the greatest pressure drop at arterioles (Figure 34.16), for they offer the greatest resistance to flow. The drop allows time for controls to divert more or less of the total volume of flow to different organs. Control proceeds at rings of smooth muscle in arteriole walls. Some control signals cause smooth muscle cells to relax, which results in **vasodilation**. The word means an enlargement, or dilation, of blood vessel diameter. Other signals cause contraction of the smooth muscle cells, which brings about a decrease in the blood vessel diameter, or **vasoconstriction**.

The nervous and endocrine systems can control the arteriole diameter. For example, sympathetic nerves end on smooth muscle cells of many arterioles. Increased activity in these nerves initiates action potentials that stimulate contraction and widespread vasoconstriction. Epinephrine and angiotensin are two of the signaling molecules that trigger changes in arteriole diameter.

Arteriole diameter also is adjusted when changes in metabolic activity shift the localized concentrations of substances in a tissue. Such local chemical changes are "selfish" in that they invite or divert the blood flow to meet the tissue's own metabolic needs. For example, when you run, the oxygen level falls in skeletal muscle tissue, and levels of carbon dioxide, potassium and hydrogen ions, and other substances rise. The changed conditions make arterioles in the vicinity dilate. Now more blood can flow through active muscles to deliver more raw materials and carry away cell products and metabolic wastes. When the muscles relax and demand fewer blood deliveries, the oxygen level increases and makes the arterioles constrict.

Controlling Mean Arterial Blood Pressure

Mean arterial pressure depends on cardiac output and on total resistance through the vascular system. Cardiac output is adjusted by controls over the rate and strength of heartbeats. The total resistance is adjusted mainly by coordinating vasoconstriction at arterioles. Of course, organs and tissues vary in their demands for blood's pickups and deliveries, so the balance between cardiac output and total resistance is juggled all the time.

A **baroreceptor reflex** is the key short-term control over arterial pressure. It starts as baroreceptors detect changes in arterial mean pressure and pulse pressure. The most vital of these mechanoreceptors are in carotid arteries (which deliver blood to the brain) and in the aortic arch (which delivers it to the rest of the body). Baroreceptors continually generate signals that reach a control center in the part of the hindbrain called the medulla oblongata. In response, the center adjusts its commands, which flow along sympathetic nerves and parasympathetic nerves to the heart and blood vessels. As an example, when the mean arterial pressure rises, the center commands the heart to beat more slowly and contract less forcefully.

Long-term controls over blood pressure are exerted at kidneys, which adjust the volume and composition of the blood. And that is a topic of another chapter.

Mean arterial pressure is an outcome of controls over the heart's output and over the total resistance to blood flow through the vascular system, as exerted mainly at arterioles.

FROM CAPILLARY BEDS BACK TO THE HEART

Capillary Function

THE NATURE OF THE EXCHANGES Capillary beds, again, serve as diffusion zones for exchanges between blood and interstitial fluid. Living cells in any body tissue die quickly when deprived of these exchanges. Brain cells die within four minutes.

Between 10 billion and 40 billion capillaries service the human body. Collectively, these components of the circulatory system offer a tremendous surface area for gas exchanges. They extend into nearly all tissues. At least one threads as close as 0.001 centimeter to each living cell. Proximity to cells is vital, because shorter distances mean faster diffusion rates. Think about this: It would take years for oxygen to diffuse on its own from your lungs all the way down your legs. By then your toes, and the rest of you, would long be dead.

Red blood cells (8 micrometers wide) squeeze single file through capillaries, which are 3 to 7 micrometers wide and deform only a bit. The squeeze puts oxygen-transporting red blood cells, and substances dissolved in plasma, in direct contact with the exchange area—the capillary wall—or only a short distance away from it.

And remember, capillaries collectively present a greater cross-sectional area than the arterioles leading into them, so the flow velocity cannot be as great in the capillary beds. The slowdown means there is plenty of time for interstitial fluid to exchange substances with the 5 percent or so of the total volume of blood moving forward through these beds during a specified interval. Here you may wish to reflect on Figure 34.3e.

Each capillary is a single sheet of endothelial cells in the form of a tube (Figure 34.15c). The cells abut one another, but the "fit" differs in different body regions. In most cases, narrow, water-filled clefts occur between endothelial cells. In capillaries that service the brain, tight junctions join the cells and clefts are nonexistent; substances cannot "leak" between cells but must move through them. Thus the junctions are a functional part of the blood–brain barrier described in Section 30.9.

ARTERIOLE END OF CAPILLARY BED	
Outward-Directed Pressure:	
Hydrostatic pressure of blood in capillary:	35 mm Hg
Osmosis due to plasma proteins:	28 mm Hg
Inward-Directed Pressure:	
Hydrostatic pressure of interstitial fluid:	0
Osmosis due to interstitial proteins:	3 mm Hg
Net Ultrafiltration Pressure:	
(35 – 0) – (28 – 3) = 10 mm Hg	
ULTRAFILTRATION FAVORED	

VENULE END OF CAPILLARY BED	
Outward-Directed Pressure:	
Hydrostatic pressure of blood in capillary:	15 mm Hg
Osmosis due to plasma proteins:	28 mm Hg
Inward-Directed Pressure:	
Hydrostatic pressure of interstitial fluid:	0
Osmosis due to interstitial proteins:	3 mm Hg
Net Reabsorption Pressure:	
(15 – 0) – (28 – 3) = – 10 mm Hg	
REABSORPTION FAVORED	

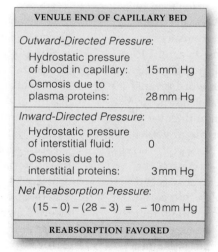

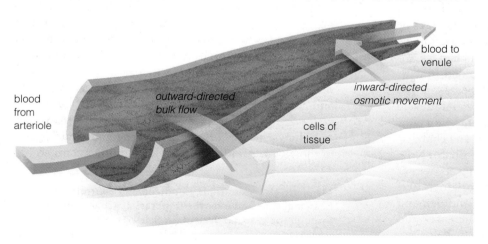

blood from arteriole

outward-directed bulk flow

inward-directed osmotic movement

blood to venule

cells of tissue

Figure 34.18 Bulk flow in an idealized capillary bed. The fluid movement plays no significant role in diffusion of solutes. However, it is important in maintaining the distribution of extracellular fluid between blood and interstitial fluid.

The fluid movements across a capillary wall result from the two opposing effects of ultrafiltration and reabsorption.

At a capillary's arteriole end, the difference between blood pressure and interstitial fluid pressure forces some plasma, but very few plasma proteins, to leave the capillary. Plasma moves by bulk flow through clefts between endothelial cells of the capillary wall. Ultrafiltration is the bulk flow of fluid *out* of the capillary.

Reabsorption is an osmotic movement of some interstitial fluid *into* the capillary as a result of a difference in the water concentration between interstitial fluid and plasma. Plasma, with its dissolved protein components, has a greater solute concentration and therefore a lower water concentration.

Reabsorption near the end of a capillary bed tends to balance ultrafiltration at the beginning. Normally there is only a small *net* filtration of fluid, which the lymphatic system returns to the blood.

Oxygen, carbon dioxide, and most other small, lipid-soluble substances cross the capillary wall by diffusing through the lipid portion of an endothelial cell's plasma membrane and through its cytoplasm. Certain proteins enter and leave the cells by endocytosis or exocytosis. Small, water-soluble substances such as ions enter and leave at the clefts between endothelial cells. So do white blood cells, as you will see in the chapter to follow.

MECHANISMS OF EXCHANGE   Diffusion is not the only way substances are exchanged across a capillary wall. When fluid pressures acting on the wall are imbalanced, there may be bulk flow one way or the other across it. Bulk flow, recall, is a movement of water and solutes in the same direction in response to fluid pressure.

As Figure 34.18 shows, water and solutes in blood and interstitial fluid affect the direction of flow. At the start of the capillary bed, the outward-directed force of blood pressure is greater than the inward-directed osmosis associated with the plasma proteins. A small amount of protein-free plasma is pushed out in bulk at the clefts in the capillary wall. This process is called **ultrafiltration**. Farther on, the balance shifts. Whereas blood pressure has been steadily declining, osmosis has remained the same. When the inward-directed osmosis exceeds the outward force of that pressure, tissue fluid moves on through the clefts between cells and into the capillary. This process is called **reabsorption**.

Normally, the outcome is a very small *net* outward movement of fluid from a capillary bed. The lymphatic system returns fluid to the blood. Such bulk flow helps maintain the fluid balance between interstitial fluid and blood. This is important, because blood pressure is maintained only when there is adequate blood volume. When blood volume plummets during hemorrhaging and some other abnormal events, interstitial fluid may be reabsorbed to help counter the loss.

High blood pressure can cause excess ultrafiltration. When too much fluid collects in interstitial spaces, *edema* follows. Some edema occurs during physical exercise, when arterioles dilate in many tissues. It also results from obstructions in veins or from heart failure. Edema is extreme during *elephantiasis*. This, disease, recall, is brought on by roundworm infection and a subsequent obstruction of lymphatic vessels (Section 23.7).

Venous Pressure

What happens to the flow velocity after it continues on past the vast cross-sectional area of the capillary beds? Remember, the capillaries merge into venules, or "little veins." These in turn merge into large-diameter veins. Once again the total cross-sectional area is reduced, and flow velocity increases as blood returns to the heart.

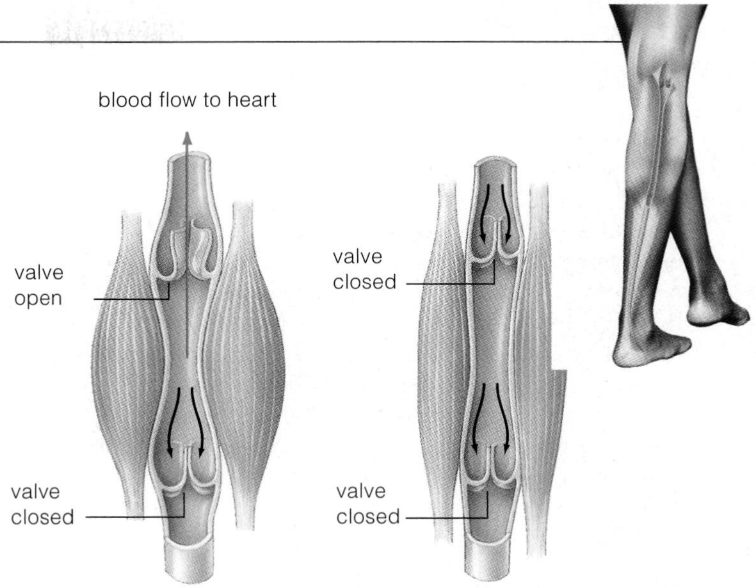

blood flow to heart

valve open

valve closed

valve closed

valve closed

a Contracted skeletal muscle next to vein helps blood flow forward.

b Skeletal muscles relaxed; venous valves shut. There is no backflow.

Figure 34.19   Vein function. Valves in medium-sized veins prevent the backflow of blood. Bulging of adjacent contracting skeletal muscles helps increase fluid pressure inside a vein.

In functional terms, venules are a bit like capillaries. Some solutes diffuse across their wall, which isn't much thicker than the thin capillary wall. Some control over capillary pressure also is exerted at these vessels.

Veins are large-diameter, low-resistance transport tubes to the heart (Figure 34.15*d*). Their valves prevent backflow. When gravity beckons, venous flow reverses direction and pushes the valves shut. The vein wall can bulge greatly under pressure, more so than an arterial wall. Thus veins are reservoirs for variable volumes of blood. Taken together, the veins of an adult human can hold up to 50 to 60 percent of the total blood volume.

The vein wall contains some smooth muscle. When blood must circulate faster, as during physical exercise, the smooth muscle contracts. The wall stiffens and the vein bulges less, so that venous pressure increases and drives more blood back to the heart. Also, when limbs move, skeletal muscles bulge against the veins in their vicinity. They help raise the venous pressure and drive blood back to the heart (Figure 34.19). Rapid breathing also contributes to increased venous pressure. Inhaled air pushes down on internal organs and thereby alters the pressure gradient between the heart and veins.

Capillary beds are diffusion zones for exchanges between blood and interstitial fluid. Here, too, bulk flow contributes to the fluid balance between blood and interstitial fluid.

Venules overlap somewhat with capillaries in function.

Veins are highly distensible blood volume reservoirs, and they help adjust flow volume back to the heart.

Cardiovascular Disorders

Each year, cardiovascular disorders disrupt the lives of 40 million people and kill a million of them. All affect blood circulation; they are the leading cause of death in the United States. The two most common are *hypertension*, or sustained high blood pressure, and *atherosclerosis*, a progressive thickening of the arterial wall that narrows the lumen. Both damage or destroy heart muscle, this being a *heart attack*. Both cause *strokes*, or brain damage.

Warning signs of a heart attack are pain or a sensation of squeezing behind the breastbone, numbness or pain in the left arm, sweating, and nausea. Women, more often than men, also experience neck and back pain, tiredness, shortness of breath, anxiety, a sense of indigestion, a fast heartbeat, and low blood pressure.

RISK FACTORS   Researchers correlate these factors with an increased risk of developing cardiovascular disorders:

1. Smoking (Section 36.7)
2. Genetic predisposition to heart failure
3. High level of cholesterol in the blood
4. High blood pressure
5. Obesity (Section 37.10)
6. Lack of regular exercise
7. Diabetes mellitus (Section 32.7)
8. Age (the older you get, the greater the risk)
9. Gender (until age fifty, males are at much greater risk than females)

Exercising regularly, eating properly, and not smoking help lower the risk. With too little exercise and too much food, adipose tissue masses increase in the body, more capillaries develop to service them, and the heart has to work harder to pump blood through the increasingly divided vascular circuits. In people who smoke, nicotine in tobacco makes the adrenal glands secrete epinephrine. This hormone causes blood vessel diameters to constrict, thus boosting heart rate and blood pressure. Also, carbon monoxide in cigarette smoke competes with oxygen for binding sites on hemoglobin. As a result, the heart has to pump harder to deliver enough oxygen to cells. Smoking also predisposes people to atherosclerosis even if their blood level of cholesterol is normal.

The next paragraphs describe some of the damage inflicted by cardiovascular disorders.

HYPERTENSION   Gradual increases in flow resistance through the small arteries leads to hypertension. In time, blood pressure stays above 140/90 even when a person is resting. Heredity may be a factor; the disorder tends to run in families. Diet is also a factor for some people; high salt intake raises their blood pressure and increases the heart's workload. In time the heart may enlarge and fail to pump blood effectively.

High blood pressure may contribute to a "hardening" of arterial walls, which hampers the delivery of oxygen to the brain, heart, and other vital organs.

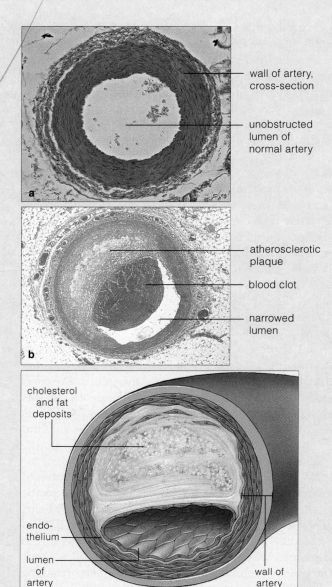

Figure 34.20   Sections from (**a**) a normal artery and (**b**) one with a narrowed lumen. (**c**) Atherosclerotic plaque.

Hypertension is called a "silent killer" because affected individuals may show no outward symptoms. Even when they know their blood pressure is high, some people tend to resist helpful medication, changes in diet, and regular exercise. Of 23 million hypertensive Americans, most don't seek treatment. About 180,000 die each year.

ATHEROSCLEROSIS   With *arteriosclerosis*, arteries thicken and lose their elasticity. With atherosclerosis, the condition worsens as cholesterol and other lipids build up in the arterial wall, thus narrowing the lumen (Figure 34.20).

Recall, from the Chapter 15 introduction, that the liver produces enough cholesterol to satisfy the body's needs. A cholesterol-rich diet typically increases the cholesterol concentration in blood. When cholesterol circulates in

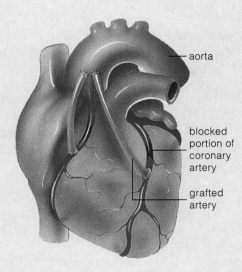

Figure 34.21 Two coronary bypasses (color-coded *green*), which extend from the aorta and past two clogged parts of the coronary arteries.

- aorta
- blocked portion of coronary artery
- grafted artery

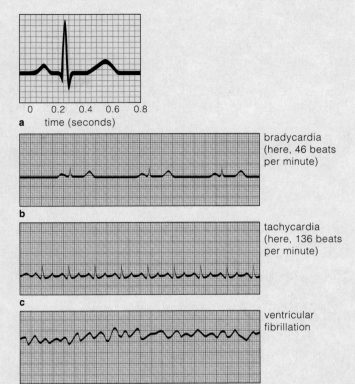

a time (seconds)

bradycardia (here, 46 beats per minute)

b

tachycardia (here, 136 beats per minute)

c

ventricular fibrillation

d

Figure 34.22 (**a**) ECG of a single, normal human heartbeat. (**b–d**) Three examples of recordings of arrhythmias.

blood, it is bound to proteins as *low-density* lipoproteins, or **LDLs**. Cells throughout the body have LDL receptors, and they take up LDL—with its cholesterol cargo—for use in metabolism. They do not take up all of it. Some of the excess cholesterol becomes attached to proteins as *high-density* lipoproteins, or **HDLs**. The blood transports HDLs back to the liver, where they are metabolized.

For a variety of reasons, some people just can't sweep enough LDL from their blood. As the blood level of LDL rises, so does the risk of atherosclerosis. LDLs infiltrate the wall of arteries. In the wall, abnormal smooth muscle cells multiply, and connective tissue components increase in mass. Cholesterol accumulates in the endothelial cells and in clefts between them. Calcium deposits actually form microscopic slivers of bone on top of the lipids! Then a fibrous net forms over the whole mass. This *atherosclerotic plaque* sticks out into the arterial lumen (Figure 34.20c).

The bony slivers shred the endothelium. Platelets gather at the damaged site and initiate clot formation. Conditions worsen as fatty globules in the plaque become oxidized. The globules change shape, and many end up resembling molecules at the surface of common bacteria. One of the bacteria is a type that triggers the formation of bonelike calcium deposits in the lungs. And so the call goes out to bacteria-fighting monocytes. An inflammatory response gets under way, and certain chemicals released during the response activate the genes for bone formation. Normally, this is a good thing; it helps wall off invaders and keeps an infection from spreading. It is bad news for arteries.

Growing plaques and clots narrow or block arteries. Blood flow to tissues that an artery services dwindles or stops. A clot that stays in place is a *thrombus*. When it gets dislodged and travels the bloodstream, it is an *embolus*. Coronary arteries, which have very narrow diameters, are especially vulnerable to clogging. When they narrow to one-quarter of their former diameter, the outcome ranges from *angina pectoris*, or mild chest pains, to a heart attack.

Physicians can use *stress electrocardiograms* to diagnose atherosclerosis in coronary arteries. These are recordings of electrical activity of the cardiac cycle made as a person

exercises on a treadmill. They also diagnose the condition by *angiography*. By this procedure, an opaque dye injected into blood reveals blockages of blood flow in x-ray film.

Surgery may alleviate severe blockage. With *coronary bypass surgery*, a section of an artery from the chest is stitched to the aorta and to the coronary artery below the narrowed or blocked region, as shown in Figure 34.21. With *laser angioplasty*, laser beams directed at the plaques vaporize them. With *balloon angioplasty*, a small balloon inflated within a blocked artery breaks up the plaques.

ARRHYTHMIAS ECGs detect irregular heart rhythms, or *arrhythmias* (Figure 34.22). Arrhythmias aren't always abnormalities. For example, endurance athletes may show *bradycardia*, a below-average resting cardiac rate. As an adaptation to ongoing strenuous exercise, their nervous system has adjusted the cardiac pacemaker's rate of contraction downward. Exercise or stress often causes 100+ heartbeats a minute, or *tachycardia*.

Atrial fibrillation, an abnormal, irregular heartbeat, affects over 10 percent of the elderly and young people with heart diseases. Coronary occlusions or some other disorder may cause irregular rhythms that rapidly lead to a dangerous condition called *ventricular fibrillation*. Cardiac muscle contracts haphazardly in parts of the ventricles, so blood pumping falters. Within seconds, a person loses consciousness, which may be a sign of impending death. A strong electric shock to the chest sometimes restores normal cardiac function.

HEMOSTASIS

Small blood vessels are vulnerable to ruptures, cuts, and similar injuries. **Hemostasis** is a process that stops blood loss from them and constructs a framework for repairs. It involves blood vessel spasm, platelet plug formation, and blood coagulation (Figure 34.23).

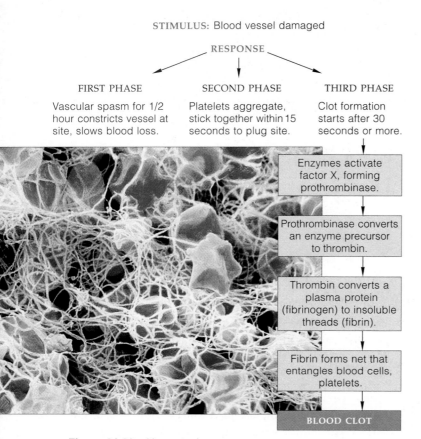

STIMULUS: Blood vessel damaged

RESPONSE

FIRST PHASE	SECOND PHASE	THIRD PHASE
Vascular spasm for 1/2 hour constricts vessel at site, slows blood loss.	Platelets aggregate, stick together within 15 seconds to plug site.	Clot formation starts after 30 seconds or more.

Enzymes activate factor X, forming prothrombinase.

Prothrombinase converts an enzyme precursor to thrombin.

Thrombin converts a plasma protein (fibrinogen) to insoluble threads (fibrin).

Fibrin forms net that entangles blood cells, platelets.

BLOOD CLOT

Figure 34.23 Hemostasis.

In the vascular phase of the process, smooth muscle located in the damaged wall contracts in an automatic response called a spasm. Vasoconstriction may stanch the blood flow as a temporary fix. In the platelet phase, platelets clump together to temporarily fill the breach. They release substances that help prolong the spasm and attract more platelets. In the coagulation phase, plasma proteins convert blood to a gel and form a clot. First, rod-shaped fibrinogens react to collagen fibers exposed by the damage. They stick together as long, insoluble threads that adhere to the exposed collagen, forming a net that traps blood cells and platelets. The entire mass is a clot. Finally, the clot retracts and forms a compact mass, thus sealing the breach in the wall.

The body routinely repairs damage to small blood vessels and prevents blood loss. The repair process involves blood vessel spasm, platelet plug formation, and coagulation.

LYMPHATIC SYSTEM

We conclude this chapter with a brief section on the manner in which the lymphatic system supplements blood circulation. Think of this section as a bridge to the next chapter, on immunity. Why? The lymphatic system also helps defend the body against injury and attack. This important system is composed of drainage vessels, lymphoid organs, and lymphoid tissues. When tissue fluid moves into the vessels, we call this lymph.

Lymph Vascular System

The portion of the lymphatic system called the **lymph vascular system** consists of many tubes that collect and deliver water and solutes from the interstitial fluid to ducts of the circulatory system. Its main components are lymph capillaries and lymph vessels (Figure 34.24).

The lymph vascular system serves three functions. First, its vessels are drainage channels for water and plasma proteins that have leaked out from the blood at capillary beds and must be delivered back to the blood circulation. Second, the system also takes up fats that the body has absorbed from the small intestine and delivers them to the general circulation, in the manner described in Section 37.5. Third, it delivers pathogens, foreign cells and material, and cellular debris from the

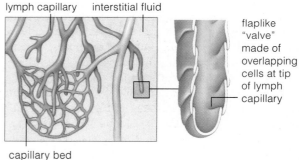

lymph capillary   interstitial fluid

flaplike "valve" made of overlapping cells at tip of lymph capillary

a   capillary bed

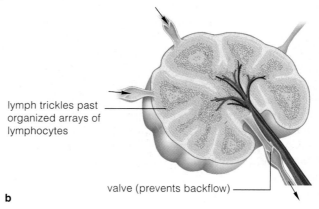

lymph trickles past organized arrays of lymphocytes

valve (prevents backflow)

b

Figure 34.24 (**a**) Diagram of lymph capillaries at the start of a drainage network, the lymph vascular system. (**b**) Cutaway view of a lymph node. Its inner compartments are packed with organized arrays of infection-fighting white blood cells.

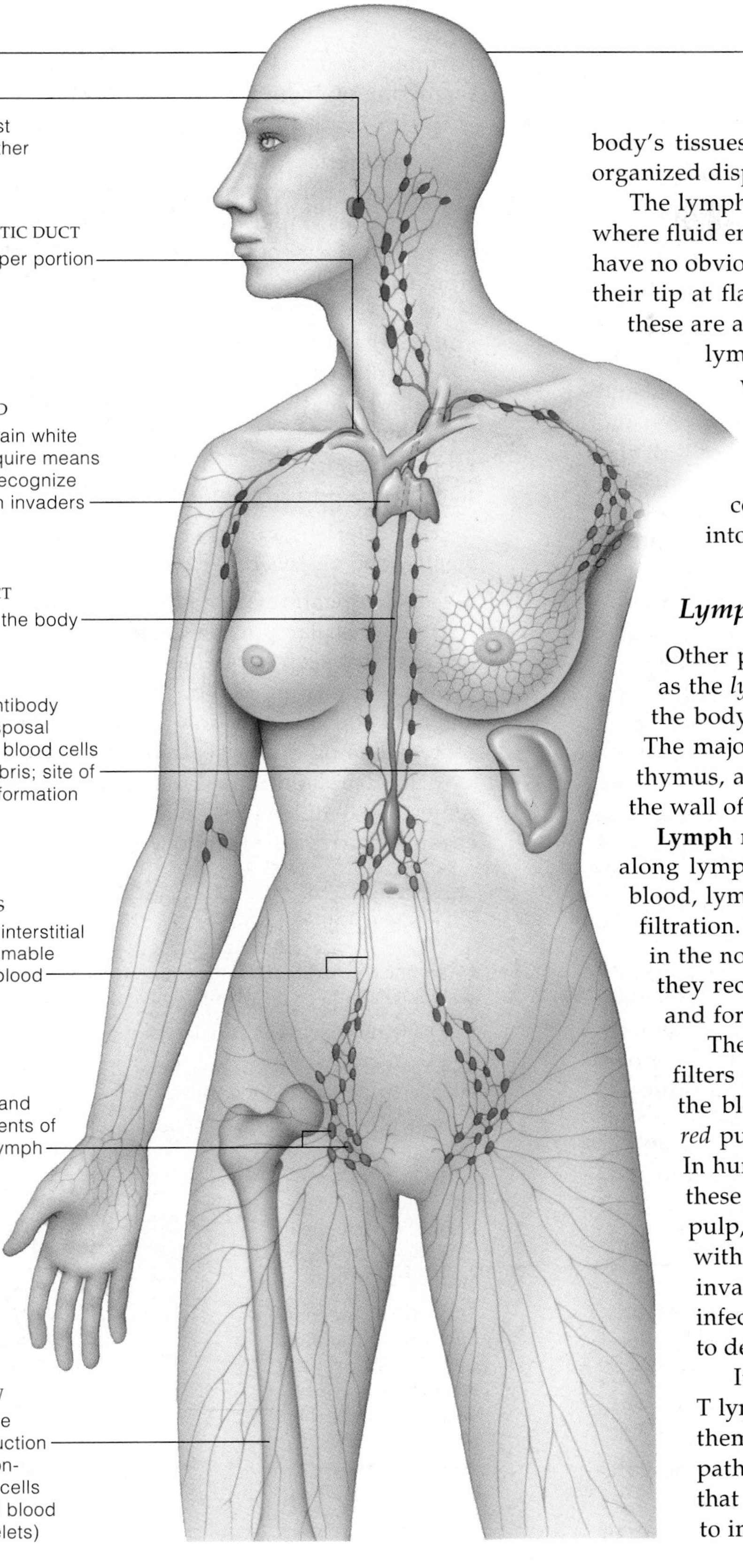

TONSILS
Defense against bacteria and other foreign agents

RIGHT LYMPHATIC DUCT
Drains right upper portion of the body

THYMUS GLAND
Site where certain white blood cells acquire means to chemically recognize specific foreign invaders

THORACIC DUCT
Drains most of the body

SPLEEN
Major site of antibody production; disposal site for old red blood cells and foreign debris; site of red blood cell formation in the embryo

SOME OF THE LYMPH VESSELS
Return excess interstitial fluid and reclaimable solutes to the blood

SOME OF THE LYMPH NODES
Filter bacteria and many other agents of disease from lymph

BONE MARROW
Marrow in some bones is production site for infection-fighting blood cells (as well as red blood cells and platelets)

Figure 34.25 Components of the human lymphatic system and their functions.

The small *green* ovals represent some of the major lymph nodes. Not shown are the patches of lymphoid tissue in the small intestine and in the appendix, which also are part of the lymphatic system.

body's tissues to the lymph vascular system's highly organized disposal centers, the lymph nodes.

The lymph vascular system starts at capillary beds, where fluid enters the lymph capillaries. The capillaries have no obvious entrance; water and solutes move into their tip at flaplike "valves." As Figure 34.24*a* shows, these are areas where endothelial cells overlap. The lymph capillaries merge into lymph vessels, which have a larger diameter. The lymph vessels contain some smooth muscle in their wall as well as valves in their lumen that can prevent backflow. They converge into collecting ducts that drain into veins in the lower neck (Figure 34.25).

Lymphoid Organs and Tissues

Other portions of the lymphatic system, known as the *lymphoid* organs and tissues, are central to the body's defense responses to injury or attack. The major ones are the lymph nodes, spleen, and thymus, as well as tonsils and patches of tissue in the wall of the small intestine and appendix.

Lymph nodes are strategically located at intervals along lymph vessels (Figure 34.25). Before entering blood, lymph trickles through at least one node for filtration. Masses of lymphocytes take up stations in the nodes after forming in bone marrow. When they recognize an invader, they multiply rapidly and form large armies to destroy it.

The **spleen** is the largest lymphoid organ. It filters pathogens and used-up blood cells from the blood itself. One of its compartments, the *red* pulp, is a huge reservoir of red blood cells. In human embryos, the red pulp also produces these cells. The other compartment, the *white* pulp, has masses of lymphocytes associated with blood vessels. If and when a particular invader reaches the spleen during a serious infection, the lymphocytes become mobilized to destroy it, just as they are in lymph nodes.

It is in the **thymus gland** that immature T lymphocytes differentiate in ways that allow them to recognize and respond to particular pathogens. The thymus produces hormones that influence these vital actions. It is central to immunity, the focus of the next chapter.

The lymph vascular portion of the lymphatic system returns water and solutes from tissue fluid to blood, and delivers fats and foreign material to lymph nodes. The system's lymph nodes and other lymphoid organs help defend the body against tissue damage and infectious diseases.

SUMMARY

Gold indicates text section

1. The closed circulatory system of humans and other vertebrates consists of a heart (muscular pump), many blood vessels (arteries, arterioles, capillaries, venules, and veins), and blood. It functions in the rapid internal transport of substances to and from cells. In birds and mammals, blood flows in two separate circuits. One is pulmonary and the other is systemic. *CI, 34.1*

2. Blood helps maintain favorable conditions for cells. This fluid connective tissue consists of plasma, red and white blood cells, and platelets. It delivers oxygen and other substances to interstitial fluid around cells. It also picks up cell products and wastes from that fluid. *34.2*

   a. Plasma, the liquid portion of blood, is a transport medium for blood cells and platelets. Plasma is a solvent for plasma proteins, simple sugars, lipids, amino acids, mineral ions, vitamins, hormones, and several gases.

   b. Red blood cells transport oxygen from the lungs to all body regions. They are packed with hemoglobin, an iron-containing pigment that binds reversibly with oxygen. In addition, red blood cells also transport some carbon dioxide from interstitial fluid to the lungs.

   c. Phagocytic white blood cells cleanse tissues by engulfing dead cells, cellular debris, and anything else recognized as not being part of the body. White blood cells called lymphocytes form great armies that destroy specific bacteria, viruses, and other disease agents.

3. Blood flows in these two circuits in humans: *34.5*

   a. A pulmonary circuit loops between the heart and lungs. Oxygen-poor blood from systemic veins enters the heart's *right* atrium, is pumped through pulmonary arteries to the two lungs, picks up oxygen, then flows through pulmonary veins to the heart's left atrium.

   b. A systemic circuit loops between the heart and all tissues. Oxygenated blood in the *left* atrium flows into the left ventricle, is pumped into the aorta, then is distributed to capillary beds. There, the blood gives up oxygen and picks up carbon dioxide. Systemic veins return the blood to the heart's right atrium.

4. The human heart, a double pump, beats incessantly. Each half has two chambers: atrium and ventricle. Blood flows into the atria, ventricles, then the great arteries. Ventricular contraction drives blood circulation. One-way valves enforce the forward-directed flow. *34.6*

5. The cardiac conduction system is the basis for the heart's rhythmic, spontaneous contractions. *34.6*

   a. One percent of the cardiac muscle cells are not contractile. They are specialized to initiate and distribute action potentials, independently of the nervous system. The nervous system only adjusts the rate and strength of the basic contractions; it does not initiate them.

   b. The system's SA node fires action potentials the fastest and is the cardiac pacemaker. Waves of excitation starting here wash over the atria, then down the heart's

partition, then up the ventricles. The ventricles contract in a wringing motion that ejects blood from the heart.

6. Blood pressure is highest in contracting ventricles. It drops in arteries, then in arterioles, capillaries, venules, and veins. It is lowest in relaxed atria. *34.7, 34.8*

   a. Arteries, rapid-transport vessels and a reservoir for pressure, smooth out pressure changes that result from heartbeats and thereby smooth out blood flow.

   b. Arterioles are major sites of control over the flow volume through different body regions.

   c. Capillary beds are diffusion zones where blood and interstitial fluid exchange substances.

   d. Venules overlap capillaries in function. Veins are rapid-transport vessels and a blood volume reservoir that is tapped to adjust flow volume back to the heart.

7. Hemostasis is a process of stopping blood loss from small blood vessels that have been injured. *34.10*

8. The lymphatic system has these functions: *34.11*

   a. Its vascular portion takes up water and plasma proteins that seep out of blood capillaries, then returns them to circulating blood. It transports absorbed fats and delivers pathogens and foreign material to disposal centers. Lymph capillaries and vessels are components.

   b. Its lymphoid organs and tissues have production centers for lymphocytes. Some are battlegrounds where organized arrays of lymphocytes fight disease agents.

Review Questions

1. Define the functions of the circulatory system and lymphatic system. Distinguish between blood and interstitial fluid. *34.1*

2. Distinguish between systemic and pulmonary circuits. *34.1*

3. Describe the cellular components of blood. Then describe the plasma portion of blood. *34.2*

4. First distinguish between the functions of the human heart's atria and ventricles. Label the components of the heart diagram at *right*. *34.6*

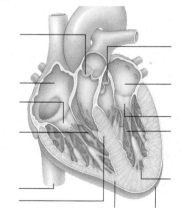

5. State the functions of arteries, arterioles, capillaries, veins, and venules, plus lymph vessels. Identify the four types of blood vessels shown *below*. *34.1, 34.5, 34.7, 34.8*

1. Cells directly exchange substances with _____ .
   a. blood vessels   c. interstitial fluid
   b. lymph vessels   d. both a and b

2. Which are *not* components of blood?
   a. plasma
   b. blood cells and platelets
   c. gases and other dissolved substances
   d. All of the above are components of blood.

3. The _____ produces red blood cells, which transport _____ and some _____ .
   a. liver; oxygen; mineral ions
   b. liver; oxygen; carbon dioxide
   c. bone marrow; oxygen; hormones
   d. bone marrow; oxygen; carbon dioxide

4. The _____ produces white blood cells, which function in _____ and _____ .
   a. liver; oxygen transport; defense
   b. lymph nodes; oxygen transport; pH stabilization
   c. bone marrow; housekeeping; defense
   d. bone marrow; pH stabilization; defense

5. In the pulmonary circuit, the heart's _____ half pumps blood to lungs, then _____ blood flows to the heart.
   a. right; oxygen-poor   c. right; oxygen-rich
   b. left; oxygen-poor   d. left; oxygen-rich

6. In the systemic circuit, the heart's _____ half pumps _____ blood to all body regions.
   a. right; oxygen-poor   c. right; oxygen-rich
   b. left; oxygen-poor   d. left; oxygen-rich

7. Blood pressure is high in _____ and lowest in _____ .
   a. arteries; veins   c. arteries; ventricles
   b. arteries; relaxed atria d. arterioles; veins

8. _____ contraction drives blood circulation; and blood pressure is highest in contracting _____ .
   a. Atrial; ventricles   c. Ventricular; arteries
   b. Atrial; atria   d. Ventricular; ventricles

9. Which is *not* a function of the lymphatic system?
   a. delivers disease agents to disposal centers
   b. produces lymphocytes
   c. delivers oxygen to cells
   d. returns water and plasma proteins to blood

10. Match the type of blood vessel with its major function.
    _____ arteries   a. diffusion
    _____ arterioles   b. control of blood volume distribution
    _____ capillaries   c. transport, blood volume reservoirs
    _____ venules   d. overlap capillary function
    _____ veins   e. transport, pressure reservoirs

11. Match the components with their most suitable description.
    _____ capillary beds   a. two atria, two ventricles
    _____ lymph vascular system   b. bathes body's living cells
    _____ human heart chambers   c. driving force for blood
    _____ blood   d. zones of diffusion
    _____ heart contractions   e. starts at capillary beds
    _____ interstitial fluid   f. fluid connective tissue

Critical Thinking

1. Shirelle, who is using a light microscope to examine a human tissue specimen, sees red blood cells moving single file through thin-walled tubes. She makes a photomicrograph (Figure 34.26). What type of blood vessel has she captured on film?

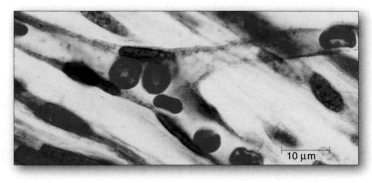

10 μm

Figure 34.26   Light micrograph of a branching blood vessel.

2. In individuals who have weak or leaky valves in their veins, fluid pressure associated with the backflow of blood causes venous walls below the valves to bulge outward. In time, the walls become stretched and flabbily distorted, a condition called *varicose veins*. Some people are genetically predisposed to develop the condition, but the cumulative mechanical stress associated with prolonged standing, pregnancy, and aging can contribute to it. With chronic varicosity, the legs themselves become swollen. Speculate on why this swelling might happen. Also speculate on why veins close to the leg surface are more susceptible to varicosity than those deeper in the leg tissues.

3. Infection by *Streptococcus pyogenes*, a hemolytic bacterium, may trigger an inflammation that ultimately damages valves in the heart. The disease symptoms of *rheumatic fever* follow. Explain how this disease must affect the heart's functioning and what kinds of symptoms would arise as a consequence.

Selected Key Terms

ABO blood typing *34.4*	LDL *34.9*
agglutination *34.4*	lymph node *34.11*
anemia *34.3*	lymph vascular
aorta *34.5*	system *34.11*
arteriole *34.5*	lymphatic system *34.1*
artery *34.5*	plasma *34.2*
baroreceptor reflex *34.7*	platelet *34.2*
blood *34.1*	pulmonary circuit *34.1*
blood pressure *34.7*	reabsorption *34.8*
blood transfusion *34.4*	red blood cell *34.2*
capillary (blood) *34.1*	Rh blood typing *34.4*
capillary bed *34.1*	spleen *34.11*
cardiac conduction system *34.6*	stem cell *34.2*
cardiac cycle *34.6*	systemic circuit *34.1*
cardiac pacemaker *34.6*	thymus gland *34.11*
cell count *34.2*	ultrafiltration *34.8*
circulatory system *CI*	vasoconstriction *34.7*
HDL *34.9*	vasodilation *34.7*
heart *34.1*	vein *34.5*
hemostasis *34.10*	venule *34.5*
interstitial fluid *34.1*	white blood cell *34.2*

Readings

Baraga, M. 3 May 1996. "Finding New Drugs to Treat Stroke." *Science*, 274.

Little, R., and W. Little. 1989. *Physiology of the Heart and Circulation*. Fourth edition. Chicago: Year Book Medical.

Zamir, M. September–October 1996. "Secrets of the Heart." *The Sciences*. Effects of exercise on heart function.

IMMUNITY

Russian Roulette, Immunological Style

Until about a century ago, smallpox epidemics swept repeatedly through the world's cities. Some outbreaks were so severe that only half of the stricken survived. Survivors had permanent scars on the face, neck, arms, and shoulders. Yet seldom did they contract the disease again; they were said to be "immune" to smallpox.

No one knew what caused smallpox, but the idea of acquiring immunity was dreadfully appealing. In twelfth-century China, people in good health gambled with deliberate infections. They sought out individuals with mild cases of smallpox, who displayed only mild scarring. They removed bits of crusts from the scars, ground them up, and inhaled the powder.

By the seventeenth century Mary Montagu, wife of the English ambassador to Turkey, was championing inoculation. She went so far as to poke bits of smallpox scabs into her children's skin. Others soaked threads in fluid from smallpox sores, then poked them into skin. Some people survived the chancy practices and they

acquired immunity to smallpox. Yet many developed raging infections. As if the odds weren't dangerous enough, the crude inoculation practices also made the body vulnerable to several other infectious diseases.

While this immunological version of Russian roulette was going on, Edward Jenner was growing up in the countryside of England. At the time, it was common knowledge that people who contracted cowpox never got smallpox. (Cowpox is a mild disease that can be transmitted from cattle to humans.) No one thought

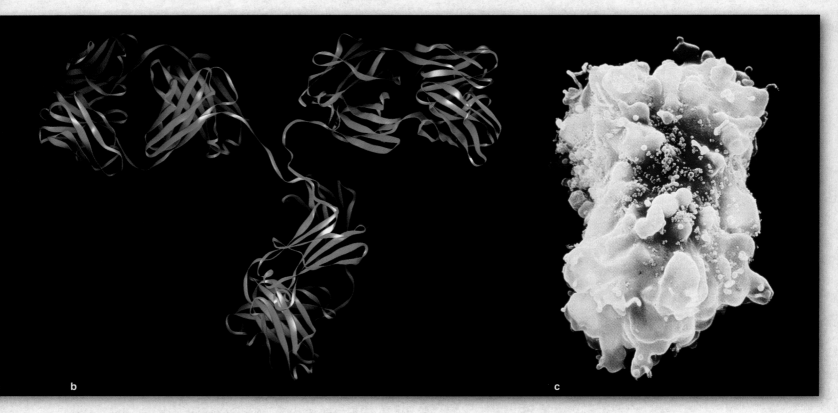

Figure 35.1 (a) Statue honoring Edward Jenner's work to develop a vaccine against smallpox, one of the most dreaded diseases in human history. (b) Computer model for an antibody molecule, a type of weapon that Jenner's procedure mobilized. (c) White blood cell being attacked by HIV (*blue particles*), the virus that causes AIDS. Immunologists are working to develop effective weapons against this modern-day scourge.

much about this until 1796, when Jenner, by now a physician, injected material from a cowpox sore into the arm of a healthy boy. Six weeks later, after the reaction subsided, Jenner injected some material from *smallpox* sores into the boy (Figure 35.1*a*). He had hypothesized that the earlier injection might provoke immunity to smallpox. Fortunately he was right; the boy remained free of smallpox. Jenner had developed an effective immunization procedure against a specific pathogen.

The French mocked Jenner's procedure by calling it **vaccination**. The word literally means "encowment." Much later Louis Pasteur, an influential French chemist, developed similar immunization procedures for other diseases. In acknowledgment of Jenner's pioneering work, Pasteur also called his procedures vaccinations, and only then did the word become respectable.

By Pasteur's time, improved light microscopes had revealed bacteria, fungal spores, and other previously invisible forms of life. As Pasteur himself discovered, microorganisms abound in ordinary air. Did some cause diseases? Probably. Could they drift onto food or drink and spoil it? He demonstrated that they could and did.

Pasteur found a way to kill most of the suspected pathogens (disease agents) in food or beverages. As he and others knew, boiling killed the suspects. He also knew that boiling changes the properties of wine—or beer or milk, for that matter. He devised a way to heat beverages at a temperature low enough not to ruin them but high enough to kill most of the pathogens suspended in them. We still rely on his antimicrobial method, which was named pasteurization in his honor.

In the late 1870s Robert Koch, a German physician, linked a specific pathogen to a specific disease: anthrax. In one experiment, Koch injected blood from animals with anthrax into healthy animals. Later, the blood of the recipient animals teemed with cells of the bacterium *Bacillus anthracis*—and the animals developed anthrax. More convincing, injections of *B. anthracis* cells cultured outside of the animal body also caused the disease!

Thus, by the beginning of the twentieth century, the promise of understanding the basis of infectious disease and immunity loomed large—and the battles against those diseases were about to begin in earnest. Since that time, spectacular advances in microscopy, biochemistry, and molecular biology have increased our understanding of the body's defenses. Today we have greater insight into responses to tissue damage in general and immune responses to specific pathogens or tumor cells. The responses are the focus of this chapter.

Key Concepts

1. The vertebrate body has physical, chemical, and cellular defenses against pathogenic microorganisms, malignant tumor cells, and other agents capable of destroying tissues and the individual itself.

2. In the early stages of tissue invasion and damage, white blood cells and certain proteins in the plasma portion of blood escape from capillaries. They execute a rapid counterattack in response to a general alarm, not to the presence of a particular pathogen, so we call this a nonspecific inflammatory response. Phagocytic white blood cells ingest invading agents and clean up tissue debris. Plasma proteins promote phagocytosis, and some also destroy invaders directly.

3. If the invasion persists, certain white blood cells make immune responses. The cells can chemically recognize distinct configurations on molecules that are abnormal or foreign to the body, such as those on bacterial cells and viruses. If the foreign or abnormal molecule triggers an immune response, it is called an antigen.

4. In one type of immune response, activated B cells produce and secrete enormous quantities of antibodies. Antibodies are molecules that bind to a specific antigen, thus tagging it for destruction.

5. In another type of immune response, cytotoxic T cells directly destroy body cells already infected by a variety of intracellular pathogens as well as tumor cells.

THREE LINES OF DEFENSE

You continually cross paths with astoundingly diverse **pathogens**—the viruses, bacteria, fungi, protozoans, parasitic worms, and other agents that cause diseases. Having coevolved with most of them, you and other vertebrates have three lines of defense, so you need not lose sleep over this. Most pathogens cannot breach the body surface. If they do, they face chemical weapons and white blood cells, or leukocytes, that attack anything perceived as foreign. Other white blood cells zero in on specific targets. Table 35.1 lists the three lines of defense.

Surface Barriers to Invasion

Most often, pathogens can't get past the linings of body surfaces, including skin. Think of skin as a habitat of low moisture, low pH, and thick layers of dead cells. Normally harmless bacteria tolerate these conditions. Few pathogens can compete with dense populations of established types unless conditions change. Repeatedly enclose your toes in warm, damp shoes, for instance, and you may be inviting certain fungi to penetrate the sodden, weakened tissues and cause *athlete's foot.*

Similarly helpful bacteria live on the mucous lining of the gut or vagina. Think of *Lactobacillus* populations in the vagina. Lactate, their acidic fermentation product, helps maintain a low pH that most bacteria and fungi can't tolerate. Think of the branching, tubular airways into the lungs. When air rushes in, it flings airborne bacterial cells into the sticky, mucus-coated tube walls. That mucus contains protective substances, especially a type of **lysozyme**. Enzymes in this category digest cell walls of bacteria and thus invite their death. As a final touch, broomlike cilia lining the airways sweep away trapped and enzymatically destroyed pathogens.

There are more defenses. Tears, saliva, and gastric fluid contain lysozyme and other protective substances. Tears give eyes a sterile wash. Urine's flushing action and low pH help remove pathogens from the urinary tract. Diarrhea flushes pathogens from the gut, which is why stopping diarrhea can prolong infection. (Even so, severe cases lead to dehydration, especially in children. Oral hydration therapy uses a glucose–salt solution to reverse such dangerous losses of water and ions.)

Nonspecific and Specific Responses

All animals react to tissue damage. Even small aquatic invertebrates have phagocytic cells and antimicrobial substances, including lysozymes. But the more complex the animal, the more complex are the systems to defend it. Think back on the evolution of the circulatory and lymphatic systems of vertebrates that invaded the land (Section 34.1). As the circulation of body fluids became more efficient, so did the means to defend the body. Phagocytic cells and plasma proteins could be swiftly transported to tissues under attack and could intercept pathogens trickling along the vascular highways.

Vertebrates came to be equipped with sets of plasma proteins. Some proteins promote rapid clot formation after tissue damage; others destroy invading pathogens or target them for phagocytosis. In addition, exquisitely focused responses to specific dangers evolved.

In short, *internal defenses against a tremendous variety of pathogens are in place even before damage occurs.* Ready and waiting are white blood cells and plasma proteins. They make *nonspecific* responses to damaged tissue in general, not to a particular pathogen. Other white blood cells may recognize unique molecular configurations of a *specific* pathogen. The resulting "immune" response will run its course whether or not tissues are damaged.

Table 35.1	*The Vertebrate Body's Three Lines of Defense Against Pathogens*

BARRIERS AT BODY SURFACES (*nonspecific* targets)

Intact skin; mucous membranes at other body surfaces

Infection-fighting chemicals (e.g., lysozyme in tears, saliva)

Normally harmless bacterial inhabitants of skin and other body surfaces that can outcompete pathogenic visitors

Flushing effect of tears, saliva, urination, and diarrhea

NONSPECIFIC RESPONSES (*nonspecific* targets)

Inflammation:

1. Fast-acting white blood cells (neutrophils, eosinophils, and basophils)
2. Macrophages (also take part in immune responses)
3. Complement proteins, blood-clotting proteins, and other infection-fighting substances

Organs with pathogen-killing functions, such as lymph node

Some cytotoxic cells (e.g., NK cells) with a range of targets

IMMUNE RESPONSES (*specific* targets only)

T cells and B cells; dendritic cells, macrophages alert them

Communication signals (e.g., interleukins and interferons) and chemical weapons (e.g., antibodies, perforins)

Intact skin, mucous membranes, antimicrobial secretions, and other barriers at the body surface constitute the first line of defense against invasion and tissue damage.

Inflammation and other internal, nonspecific responses to invasion are the second line of defense.

Immune responses against specific invaders, as executed by armies of white blood cells and their chemical weapons, are the third line of defense.

COMPLEMENT PROTEINS

A set of plasma proteins has roles in nonspecific *and* specific defenses. Collectively, they are the **complement system**. About twenty kinds circulate in the blood, in inactive form. If even a few molecules of one kind are activated, they trigger cascading reactions that activate many molecules of another complement protein. Each molecule activates many molecules of another kind of protein at the next reaction step, and so on. Deploying huge numbers of molecules has the following effects.

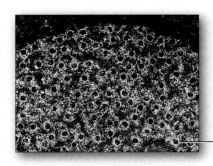

Figure 35.2 Micrograph of the surface of a pathogenic cell. Membrane attack complexes made the pores.

pore

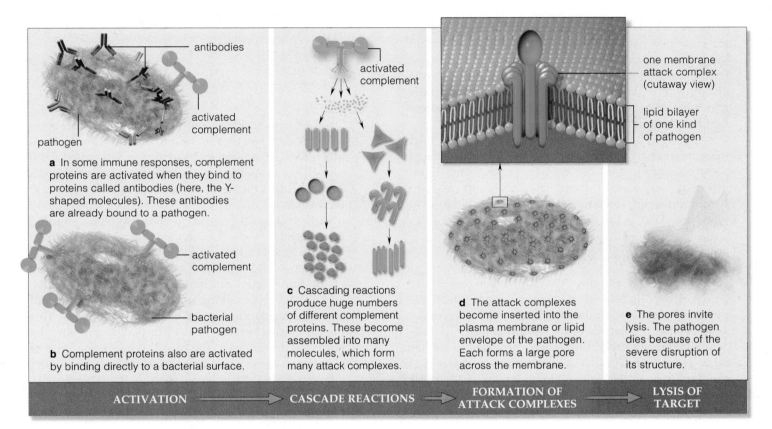

antibodies

activated complement

pathogen

a In some immune responses, complement proteins are activated when they bind to proteins called antibodies (here, the Y-shaped molecules). These antibodies are already bound to a pathogen.

activated complement

bacterial pathogen

b Complement proteins also are activated by binding directly to a bacterial surface.

activated complement

c Cascading reactions produce huge numbers of different complement proteins. These become assembled into many molecules, which form many attack complexes.

one membrane attack complex (cutaway view)

lipid bilayer of one kind of pathogen

d The attack complexes become inserted into the plasma membrane or lipid envelope of the pathogen. Each forms a large pore across the membrane.

e The pores invite lysis. The pathogen dies because of the severe disruption of its structure.

ACTIVATION ⟶ CASCADE REACTIONS ⟶ FORMATION OF ATTACK COMPLEXES ⟶ LYSIS OF TARGET

Figure 35.3 Formation of membrane attack complexes. One reaction pathway starts when complement proteins bind to bacterial surfaces. Another operates in immune responses to specific invaders. Both pathways result in membrane pore complexes that induce lysis in pathogens.

Some complement proteins form attack complexes: molecular structures with an inner channel (Figures 35.2 and 35.3). The complexes are inserted into the plasma membrane of many pathogens and initiate lysis. They also are inserted into the wall of certain bacterial cells. **Lysis**, remember, is the gross structural disruption of a plasma membrane or cell wall; it results in cell death.

Certain activated proteins promote inflammation, a nonspecific defense response described next. By cascades of synthesis reactions, they form concentration gradients in irritated or damaged tissues. Phagocytic white blood cells follow the gradients to invasion sites. In addition, complement proteins bind to the surface of many types of invaders. Together, they form a complement "coat" on the invaders that promotes phagocytosis.

In such ways, the complement proteins target many bacteria, parasitic protistans, and enveloped viruses.

The complement system, a set of about twenty kinds of plasma proteins circulating in blood, takes part in cascades of reactions that defend the body against many bacteria, some parasitic protistans, and enveloped viruses.

Complement proteins take part in nonspecific defenses. They also take part in immune responses, as when some are activated by antibody molecules bound to pathogens.

INFLAMMATION

The Roles of Phagocytes and Their Kin

Certain types of white blood cells take part in an initial response to a damaged tissue. White blood cells, recall, arise from stem cells in bone marrow. Many of the cells circulate in blood and lymph. A great many take up stations in lymph nodes as well as in the spleen, liver, kidneys, lungs, and brain. Here you may wish to scan Sections 34.2 and 34.11 once more; they introduce the components of blood and the lymphatic system.

Like SWAT teams, three kinds of white blood cells react swiftly to danger in general but are not adapted for sustained battles. **Neutrophils**, the most abundant kind, phagocytize bacteria. They ingest, kill, and digest bacterial cells into molecular bits. **Eosinophils** secrete enzymes that put holes in parasitic worms. **Basophils** and **mast cells** make and secrete chemical substances that help keep inflammation going after it starts.

Although slower to act, the white blood cells called **macrophages** are "big eaters." Figure 35.4 shows one of these phagocytic cells. Macrophages engulf and digest just about any foreign agent. They also help clean up damaged tissues. Immature macrophages circulating in blood are classified as monocytes.

The Inflammatory Response

An inflammatory response develops when something damages or kills cells of any tissue region. Infections, punctures, burns, and other assaults are the triggers. By a mechanism known as **acute inflammation**, the fast-acting phagocytes and complement proteins, as well as other plasma proteins, escape from the bloodstream at capillary beds in the vulnerable tissue. Localized signs that they have entered interstitial fluid and that acute inflammation is proceeding include redness, warmth, swelling, and pain, as listed in Table 35.2.

Table 35.2	*Localized Signs of Inflammation and Their Causes*
Redness	Arteriolar vasodilation increases blood flow to the affected tissue.
Warmth	Arteriolar vasodilation delivers more blood, carrying more metabolic heat, to the tissue.
Swelling	Chemical signals increase capillary permeability; plasma proteins escape and disrupt fluid balance across capillary walls; localized edema results.
Pain	Nociceptors (pain receptors) are stimulated by increased fluid pressure, local chemical signals.

Mast cells reside in connective tissues. In response to tissue damage, they make and release **histamine** and other local signaling molecules into interstitial fluid. These secretions trigger vasodilation of arterioles that snake through the tissue. Vasodilation, remember, is an increase in the diameter of a blood vessel when smooth muscle in its wall is made to relax. So blood engorges the arterioles—which reddens the tissue and warms it up with blood-borne metabolic heat.

The released histamine also increases permeability of the thin-walled capillaries in the tissue. It induces endothelial cells making up the capillary wall to pull apart farther at the narrow clefts between them. Thus the capillaries become "leaky" to plasma proteins that normally do not leave the blood. When some proteins leak out, osmotic pressure increases in the surrounding interstitial fluid. Also, blood pressure is higher because of the increased flow of blood to the tissue. As a result, ultrafiltration rises and reabsorption declines across the capillary wall. Localized edema is the outcome of the

Figure 35.4
(**a**) White blood cell squeezing out of a blood capillary at a cleft between endothelial cells. (**b**) Macrophage, caught in the act of engulfing a yeast cell.

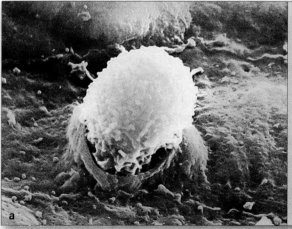
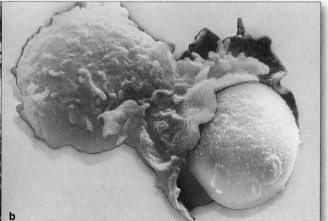

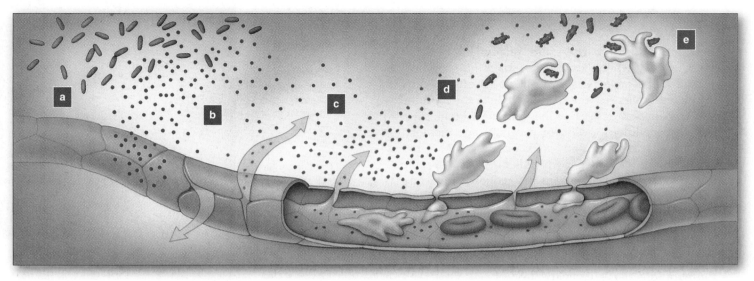

a Bacteria invade a tissue and directly kill cells or release metabolic products that damage tissue.

b Mast cells in tissue release histamine, which then triggers arteriole vasodilation (hence redness and warmth) as well as increased capillary permeability.

c Fluid and plasma proteins leak out of capillaries; localized edema (tissue swelling) and pain result.

d Complement proteins attack bacteria. Clotting factors wall off inflamed area.

e Neutrophils, macrophages, and other phagocytes engulf invaders and debris. Macrophage secretions attract even more phagocytes, directly kill invaders, and call for fever and for T and B cell proliferation.

Figure 35.5 Acute inflammation in response to a bacterial invasion. The response involves delivering phagocytes and plasma proteins to the tissue. Collectively, these components of blood inactivate, destroy, or wall off the invaders, remove chemicals and cellular debris, and prepare the tissue for repair. These are their functions in all inflammatory responses.

shift in the fluid balance across the wall. (Here you may wish to review Section 34.8.) The tissue swells with fluid, and certain free nerve endings threading through it give rise to sensations of pain. Voluntary movements aggravate the pain, so an affected person tends to avoid movements and, in doing so, promotes tissue repair.

Within hours of the first physiological responses to the damage, neutrophils are squeezing across capillary walls. They swiftly go to work. Monocytes arrive later, differentiate into macrophages, and engage in sustained action (Figure 35.5). While macrophages are engulfing the invaders and debris, they secrete a number of local signaling molecules that act as chemical mediators.

Chemical mediators called *chemotaxins* attract more phagocytes. *Interleukins* stimulate the formation of B cell and T cell armies, as described shortly. *Lactoferrin* directly kills bacteria. *Endogenous pyrogen* triggers the release of certain prostaglandins, which have the effect of raising the set point on the hypothalamic thermostat that controls the body's core temperature. A **fever** is a core temperature that reaches the higher set point.

A fever of about 39°C (100°F) is not a bad thing. It increases body temperature to a level that is "too hot" for the functioning of most pathogens. It also promotes an increase in the host's defense activities. *Interleukin-1* induces drowsiness, which reduces the body's demands for energy. And so more energy may be diverted to the

tasks of defense and tissue repair. Macrophages take part in the cleanup and repair operations.

Besides complement proteins, the plasma proteins leaking into the tissue include clotting factors, which are activated by chemicals that the phagocytes secrete. Fibrin threads form, stick to exposed collagen fibers in the damaged tissue, and trap blood cells and platelets to form a clot (Section 34.10). By walling off inflamed areas, clots can prevent or delay the spread of invaders and toxic chemicals into surrounding tissues. After the inflammation subsides, anticlotting factors, which also escaped from the capillaries, dissolve the clots.

An inflammatory response develops in a local tissue when cells are damaged or killed, as by infection. It proceeds during both nonspecific and specific defenses of tissues.

Mast cells in damaged or invaded tissues secrete histamine, which causes arterioles to vasodilate and increases capillary permeability to fluid and to plasma proteins. The tissue warms and reddens because of the localized vasodilation. Edema results from the fluid imbalance across the capillary wall. The tissue swelling causes pain.

The response involves phagocytes such as macrophages, which engulf invaders and debris and secrete chemical mediators. It requires plasma proteins, such as complement proteins that target invaders for destruction, as well as clotting factors that wall off the inflamed tissue.

OVERVIEW OF THE IMMUNE SYSTEM

Defining Features

Sometimes physical barriers and inflammation are not enough to overwhelm an invader, so an infection may become well established. Then, the body calls upon its third line of defense—the B and T cells of the immune system. Four features define the **immune system**: self/nonself recognition, specificity, diversity, and memory. *Self/nonself recognition* and *specificity* mean that B and T cells ignore the body's own cells but attack specific foreign agents. *Diversity* means that, collectively, B and T cell receptors have the potential to respond to a billion specific threats. *Memory* means a portion of the B and T cells formed during a primary response to a foreign agent is set aside for a future battle with that agent.

Self/nonself recognition starts at unique molecular configurations that give every kind of cell, virus, and substance a unique identity. An individual's own cells bear self-marker configurations. That individual's B and T cells chemically recognize self markers and normally don't attack them. If they come across nonself markers on specific foreign agents or abnormal configurations on altered body cells, they divide again and again, and form huge populations. Any molecular configuration that incites formation of lymphocyte armies and is their target is an **antigen**. The most important antigens are proteins at the surface of pathogens and tumor cells.

All cells resulting from the divisions are sensitized to the same antigen. But some subpopulations become *effectors*: fully differentiated types that kill the enemy. Others become *memory* cells. Instead of joining the first response against a specific antigen, they enter a resting phase. But they remember that antigen. If it shows up again, they will join a larger, faster response to it.

In short, immune responses involve three events: recognition of antigen, repeated cell divisions that form huge populations of lymphocytes, and differentiation into subpopulations of effector and memory cells that have receptors for one kind of antigen.

The Key Defenders

Interactions among antigen-presenting cells, activated T helper cells, B cells, and T cytotoxic cells are central to all immune responses. Here, you can start thinking about how some of them process and display antigen, and how others recognize and attack its owners.

MHC markers are named after genes that code for them (the *Major Histocompatibility Complex*). They are among protein molecules projecting above the plasma membrane of body cells. When paired up with antigen, a cell's MHC markers represent a call to arms.

Any cell that can (1) process and display antigen in association with MHC markers and (2) activate T cells

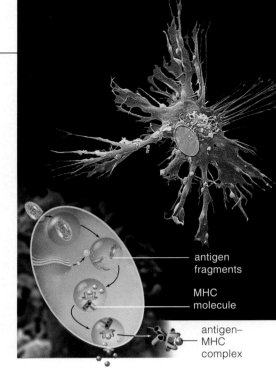

Figure 35.6 A dendritic cell —one of the major antigen-presenting cells. This class of cells patrol the blood, internal organs, and skin. (The Langerhans cells described in Section 33.2 are dendritic.) They ingest, process, and then display antigen with MHC molecules.

antigen fragments

MHC molecule

antigen– MHC complex

with a chemical signal is an **antigen-presenting cell**. Macrophages, B cells, and dendritic cells (Figure 35.6) are professionals in this enterprise. They ingest antigen, then the endocytic vesicle that forms around it fuses with a lysosome. The lysosome's potent enzymes cleave antigen molecules into fragments. Most of the fragments are exocytosed, but some associate with MHC markers and form **antigen–MHC complexes**. These move to the plasma membrane, where they are displayed.

Helper T cells can bind to antigen–MHC complexes. Binding stimulates them to produce and secrete signals that promote immune responses. The signals call for mitotic cell divisions of any B and T cell sensitive to the antigen, and for differentiation into subpopulations of effector and memory cells (Figure 35.7).

Effector **cytotoxic T cells** and **natural killer** (NK) **cells** carry out *cell-mediated* responses against infected body cells and tumor cells. All recognize antigen–MHC complexes, provided the MHC marker has a coreceptor called CD8 bound to it. They release chemical weapons that form pores in the plasma membrane of the targets.

Only antigen-sensitized **B cells** carry out *antibody-mediated* responses. They alone make and secrete many antigen-binding receptor molecules called **antibodies**. When macrophages, NK cells, or neutrophils contact a foreign agent with antibody bound to it, they destroy it.

The Main Targets

Any cell with antigen–MHC complexes at its surface turns on cytotoxic cells with receptors for that antigen. This includes body cells infected with an intracellular pathogen; they, too, display bits of antigen with MHC markers at their surface. Tumor cells and cells of organ transplants also are targets. Besides this, antibodies can directly inactivate antigen-bearing cells and toxins in

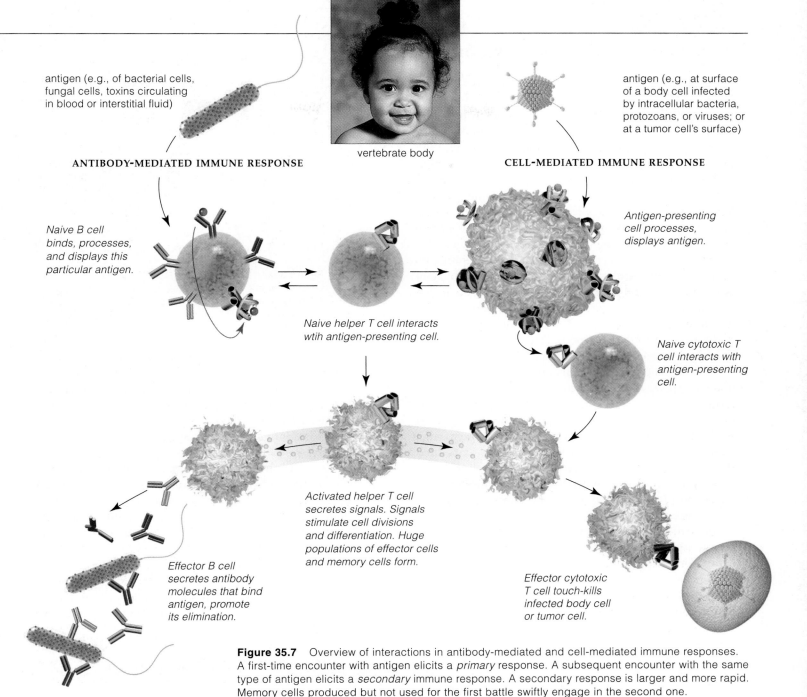

antigen (e.g., of bacterial cells, fungal cells, toxins circulating in blood or interstitial fluid)

vertebrate body

antigen (e.g., at surface of a body cell infected by intracellular bacteria, protozoans, or viruses; or at a tumor cell's surface)

ANTIBODY-MEDIATED IMMUNE RESPONSE

CELL-MEDIATED IMMUNE RESPONSE

Naive B cell binds, processes, and displays this particular antigen.

Antigen-presenting cell processes, displays antigen.

Naive helper T cell interacts wtih antigen-presenting cell.

Naive cytotoxic T cell interacts with antigen-presenting cell.

Activated helper T cell secretes signals. Signals stimulate cell divisions and differentiation. Huge populations of effector cells and memory cells form.

Effector B cell secretes antibody molecules that bind antigen, promote its elimination.

Effector cytotoxic T cell touch-kills infected body cell or tumor cell.

Figure 35.7 Overview of interactions in antibody-mediated and cell-mediated immune responses. A first-time encounter with antigen elicits a *primary* response. A subsequent encounter with the same type of antigen elicits a *secondary* immune response. A secondary response is larger and more rapid. Memory cells produced but not used for the first battle swiftly engage in the second one.

the blood or interstitial fluid. Antibodies also promote inflammation and activate complement proteins.

Control of Immune Responses

Antigen provokes an immune response—and removal of antigen stops it. For example, by the time the tide of battle turns, effector cells and their secretions already have killed most of the antigen-bearing agents in the body. With less antigen around to stimulate cells, the response declines, then stops. In addition, **suppressor T cells** release signals that help shut down both cell-mediated and antibody-mediated responses.

The immune system is defined by self/nonself recognition, specificity, diversity, and memory. Collectively, T and B cells bear receptors for millions of specific antigens: foreign molecular configurations. Recognition of antigen, not self markers on the body's own cells, triggers immune responses.

Huge T and B cell armies form by repeated mitotic cell divisions after antigen recognition. These differentiate into subpopulations of effector and memory cells, all of which are sensitized to that one kind of antigen.

Antigen-sensitized cytotoxic cells execute cell-mediated responses. Antigen-sensitized B cells execute antibody-mediated responses.

HOW LYMPHOCYTES FORM AND DO BATTLE

B and T Cell Formation

Your body is continually exposed to a mind-boggling array of antigens. However, antigen receptors of all of its B and T cell populations combined show staggering diversity—enough to recognize a billion or so different antigens. How does such diversity arise? *In each new T and B cell, gene sequences with instructions for building the antigen-binding parts of receptor molecules get shuffled, extensively and randomly, before they are expressed.* Figure 35.8 shows what happens before the antigen receptors called antibodies are synthesized. Antigen receptors are mostly proteins—polypeptide chains. While each chain is forming, its antigen-binding region folds up into a unique array of bumps and grooves having a certain distribution of charge. The only antigen in the world that will bind to that region will have complementary grooves, bumps, and distribution of charge.

As you read earlier, B cells arise from stem cells in bone marrow, and each acquires its unique antigen receptors before leaving it. As a B cell matures, it starts making many copies of a unique, typically Y-shaped antibody molecule. Each molecule moves to the plasma membrane, where the tail of the Y becomes embedded in the lipid bilayer and its two arms project above it. Soon, the cell bristles with antigen receptors—bound antibodies. It is ready to defend the body as a "naive" B cell, meaning it has not yet met up with the antigen it is genetically programmed to detect. Its membrane-bound antibodies will not recognize any antigen–MHC complexes. They will recognize antigen alone.

T cells also arise from stem cells in bone marrow. Then they migrate to the thymus gland, and there they mature and acquire unique antigen-binding receptors

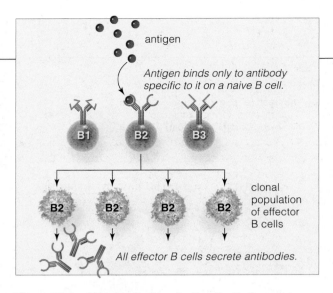

Figure 35.9 Clonal selection of a B cell with the one antibody able to combine with a specific antigen. Only antigen-selected B cells and T cells give rise to a clonal population of immunologically identical cells.

called **TCRs** (short for *T-Cell Receptors*). Bristling with antigen receptors, naive T cells leave the thymus and enter the general circulation. Their TCRs will ignore free antigen; they will ignore unadorned MHC markers at the surface of the body's own cells. Their TCRs will recognize antigen–MHC complexes only.

And so each B cell or T cell bears only one type of antigen receptor. By the theory of *clonal selection*, of all cells in the body, antigen "chooses" (binds to) only the B or T cell that is displaying a receptor specific for it (Figure 35.9). Once activated, that cell will be the start of the repeated mitotic cell divisions that form a huge population of genetically identical cells—a clone—all with the same antigen receptor.

The clonal selection theory also explains how an individual acquires immunological memory of the first encounter with antigen. The term refers to the body's

Figure 35.8 Generation of antigen receptor diversity, using the antibody molecule as an example. Antibodies are proteins. Protein-building instructions are encoded in genes. In chromosomes with genes for antibodies, long DNA stretches contain different versions of the segments coding for variable regions of an antibody molecule. (Compare Figure 35.13a.)

Different V and J segments code for the variable region. A recombination event occurs in this region when each B cell is maturing. Any V segment may be joined to any of the J segments. Afterward, DNA intervening between them is excised. The new sequence is attached to a C (*C*onstant) segment. This completes a rearranged antibody gene. That gene will be present in every descendant of the B cell.

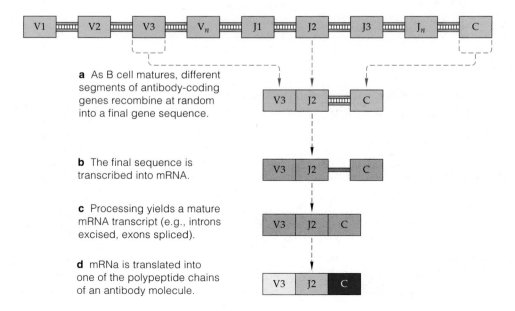

a As B cell matures, different segments of antibody-coding genes recombine at random into a final gene sequence.

b The final sequence is transcribed into mRNA.

c Processing yields a mature mRNA transcript (e.g., introns excised, exons spliced).

d mRNa is translated into one of the polypeptide chains of an antibody molecule.

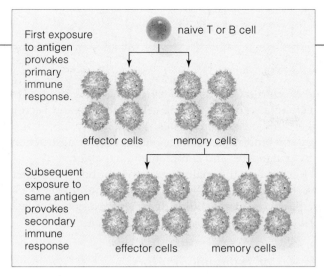

Figure 35.10 Immunological memory. Not all B and T cells are activated during a primary immune response to an antigen. A large number continue to circulate as memory cells, which become activated during a secondary immune response.

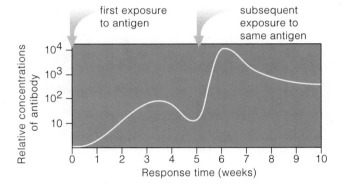

Figure 35.11 Example of the differences in magnitude and duration between a primary and a secondary immune response to the same antigen. The primary response peaked twenty-four days after it started. The secondary response peaked after seven days (the span shown between weeks five and six). Antibody concentration during the secondary response was 100 times greater, 10^4 compared to 10^2.

capacity to make a secondary immune response to any subsequent encounter with the same type of antigen that provoked the primary response (Figure 35.10).

Memory cells that form during a primary immune response don't engage in that battle. They circulate for years or for decades. Compared to a primary response, antigen is intercepted sooner by battalions that already formed and are patrolling tissues. Thus populations of effector cells form faster, so the infection is terminated before the individual gets sick. Far greater numbers of memory B and C cells form in a secondary response. Figure 35.11 gives an example. In evolutionary terms, advance preparations against further encounters with the same pathogen may bestow a survival advantage on the individual.

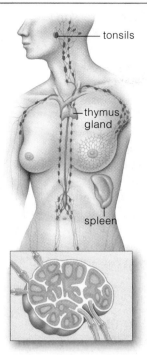

Figure 35.12 Lymph nodes, which contain organized arrays of antigen-presenting cells and lymphocytes. Compare Section 34.11).

Lymphocyte Battlegrounds

Where is antigen actually captured and dealt with? In the skin, mucous membranes, and internal organs, the macrophages and dendritic cells that form an early line of defense pick up antigen by means of phagocytosis or endocytosis. Then they migrate into regional lymph nodes (Section 34.11), where macrophages and dendritic cells present antigen to naive B and T cells.

Antigen in interstitial fluid enters the lymph vascular system. Because lymph drains into the blood, there is always a chance that antigen will be circulated throughout the body. However, lymph nodes trap most of it. Also, antigen that does manage to enter the blood filters through the spleen (Figure 35.12). Each day, more circulating lymphocytes move through the spleen than through all lymph nodes combined.

In lymph nodes, antigen first percolates through the region occupied by B cells, macrophages, and dendritic cells. There it is captured, processed, and presented to helper T cells, thereby activating them. Other dendritic cells that bear receptors for antibody and complement may activate B cells in the nodes. Deeper in the node, T cells start multiplying and maturing into effector and memory cells. So do B cells, which mature and start to secrete antibodies. Quantities of antibodies leave the node in lymph. So do enormous numbers of newly formed lymphocytes. At the same time, the antigen has been stimulating great numbers of lymphocytes to slip out of neighboring venules and enter the lymph node. Hence those swollen nodes you've probably felt from time to time under your jawline.

By recombination of segments drawn at random from the receptor-encoding regions of DNA, each B cell or T cell receives a gene sequence for just one of a billion possible kinds of antigen receptors.

Immunological specificity means the clonal descendants of an antigen-"chosen" cell bind only to the selecting antigen.

Immunological memory means B and T cells set aside during a primary response can make a faster and greater secondary response to another encounter with the same antigen.

Macrophages, dendritic cells, B cells, and T cells mount immune responses in lymphoid organs. They are organized for efficient interactions in the lymph nodes and spleen.

ANTIBODY-MEDIATED RESPONSE

B cells without helper T cells would be like a symphony orchestra without the conductor. (The same goes for cytotoxic T cells.) Activated helper T cells unleash both the antibody-mediated and cell-mediated responses to antigen. Their activation must be tightly controlled to put the immune fighters on suitable courses of action. Limiting how they recognize antigen is a key control. Helper T cells recognize antigen *only* when a fragment of it is associated with an MHC molecule at the surface of a B cell, macrophage, or dendritic cell. Also, after helper T cells interact with the antigen–MHC complex, they must be stimulated to divide by a chemical signal that the antigen-presenting cell releases. The resulting clone of helper T cells drives the immune responses.

The Roles of Antibodies

Let's now follow a naive B cell bristling with bound antibodies. All of its antibody molecules are alike; they are specific for only one antigen. Figure 35.13 shows a typical antibody's antigen-binding sites. When antigen binds with some of the molecules, it cross-links them together. The cross-linking activates receptor-mediated endocytosis, and so the antigen is moved into the cell (Section 5.8). There, antigen–MHC complexes form and move to the B cell surface, where they are displayed.

Suppose a responsive helper T cell meets up with a B cell's antigen–MHC complex. The two cells exchange a costimulatory signal, then disengage. When the B cell again encounters unprocessed antigen, its antibodies bind to it. The binding, in conjunction with interleukins secreted from nearby activated helper T cells, drives the B cell into mitosis. Its clonal descendants differentiate

into B effector and B memory cells (Figure 35.14). The effectors (also called plasma cells) produce and secrete huge numbers of antibodies. When freely circulating antibody binds to it, antigen gets tagged for destruction, as by phagocytes and complement activation.

Remember, the main targets of antibody-mediated responses are extracellular pathogens and toxins freely circulating in extracellular fluid. *Antibodies cannot bind to pathogens or toxins that are hidden inside a host cell.*

Classes of Immunoglobulins

B effector cells produce five classes of antibodies that are known as **immunoglobulins**, or **Igs**. These protein products of gene shufflings form when B cells mature and when they are engaged in immune responses. We abbreviate them as IgM, IgD, IgG, IgA, and IgE. Each has antigen-binding sites *and* other sites having special roles. When secreted, they assume these forms:

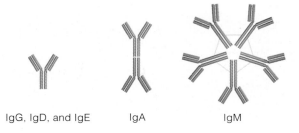

IgG, IgD, and IgE   IgA   IgM

IgM is the first one to be secreted during a primary response and the first one produced by newborns. IgM molecules aggregate into a structure having ten antigen binding sites. This makes it more efficient at binding

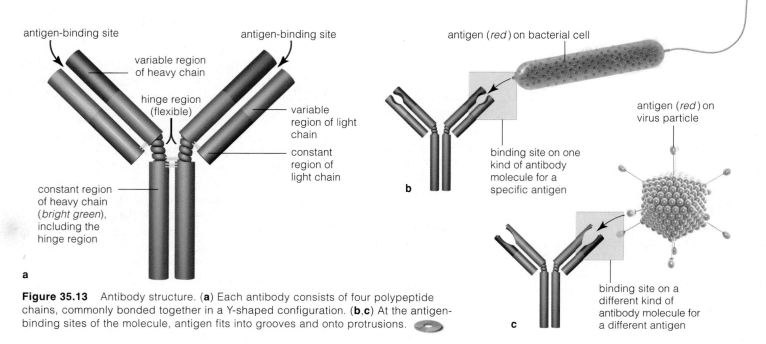

Figure 35.13 Antibody structure. (**a**) Each antibody consists of four polypeptide chains, commonly bonded together in a Y-shaped configuration. (**b,c**) At the antigen-binding sites of the molecule, antigen fits into grooves and onto protrusions.

antigen-binding site

antigen-binding site

variable region of heavy chain

hinge region (flexible)

variable region of light chain

constant region of light chain

constant region of heavy chain (*bright green*), including the hinge region

a

antigen (*red*) on bacterial cell

binding site on one kind of antibody molecule for a specific antigen

b

antigen (*red*) on virus particle

binding site on a different kind of antibody molecule for a different antigen

c

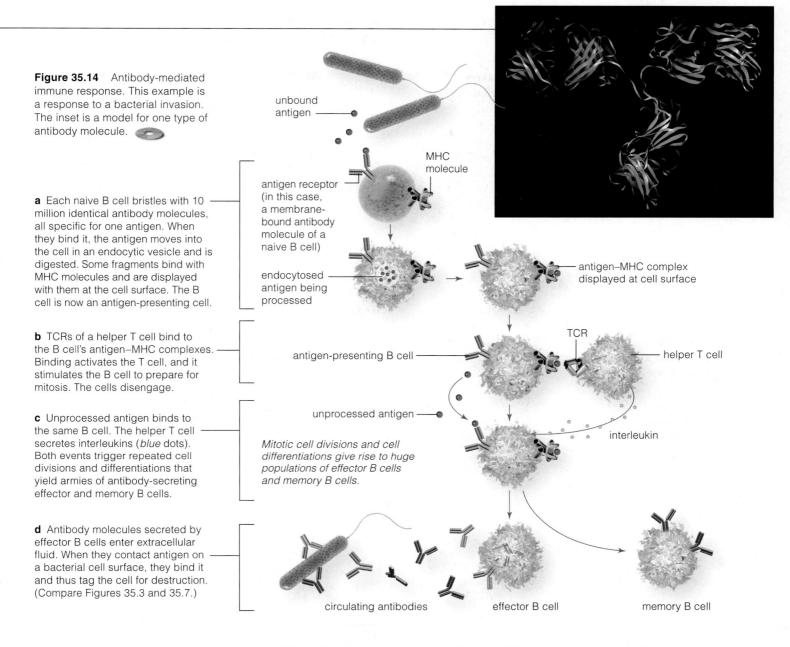

Figure 35.14 Antibody-mediated immune response. This example is a response to a bacterial invasion. The inset is a model for one type of antibody molecule.

unbound antigen

MHC molecule

antigen receptor (in this case, a membrane-bound antibody molecule of a naive B cell)

endocytosed antigen being processed

antigen–MHC complex displayed at cell surface

a Each naive B cell bristles with 10 million identical antibody molecules, all specific for one antigen. When they bind it, the antigen moves into the cell in an endocytic vesicle and is digested. Some fragments bind with MHC molecules and are displayed with them at the cell surface. The B cell is now an antigen-presenting cell.

TCR

antigen-presenting B cell

helper T cell

b TCRs of a helper T cell bind to the B cell's antigen–MHC complexes. Binding activates the T cell, and it stimulates the B cell to prepare for mitosis. The cells disengage.

unprocessed antigen

interleukin

Mitotic cell divisions and cell divisions and cell differentiations give rise to huge populations of effector B cells and memory B cells.

c Unprocessed antigen binds to the same B cell. The helper T cell secretes interleukins (*blue* dots). Both events trigger repeated cell divisions and differentiations that yield armies of antibody-secreting effector and memory B cells.

d Antibody molecules secreted by effector B cells enter extracellular fluid. When they contact antigen on a bacterial cell surface, they bind it and thus tag the cell for destruction. (Compare Figures 35.3 and 35.7.)

circulating antibodies

effector B cell

memory B cell

clumped targets, such as agglutinating red blood cells (Section 34.4) and dense aggregations of virus particles.

Along with IgM, *IgD* is the most common antibody bound at the plasma membrane of naive B cells. It may take part in the activation of helper T cells.

IgG makes up 80 percent or so of the blood-borne immunoglobulins. It's the most efficent one at turning on complement proteins, and it neutralizes many toxins. This long-lasting antibody easily crosses the placenta. It helps protect the developing fetus with the mother's acquired immunities. IgG secreted into early milk is also absorbed into a suckling newborn's bloodstream.

IgE triggers inflammation after attacks by parasitic worms and other pathogens. The tails of IgE antibodies lock on to basophils and mast cells, with their antigen receptors facing outward. The antigen binding induces basophils and mast cells to release histamine and thus promote inflammation. As you will see shortly, IgE also

is involved in allergic reactions, including asthma, hay fever, and hives.

IgA is the main immunoglobulin in exocrine gland secretions, including tears, saliva, and breast milk. It also is in mucus that coats surfaces of the respiratory, digestive, and reproductive tracts—the areas that are vulnerable to most infectious agents. Like IgM, it can form larger structures that can bind larger antigens. Bacteria and viruses can't bind to the cells of mucous membranes when secreted IgA is bound to them. IgA is known to be effective in fighting the agents that cause salmonella, cholera, gonorrhea, influenza, and polio.

Antibodies secreted by B effector cells bind to antigens of extracellular pathogens or toxins and tag them for disposal, as by phagocytes and complement. Five antibody classes (immunoglobulins) offer defenses against diverse threats.

CELL-MEDIATED RESPONSE

If every antigen stayed out in the open in the internal environment, maybe the antibody-mediated response would be sufficient to dispose of them. But a number of pathogens evade antibodies. They hide in body cells, drain the life out of them, and often reproduce in them. They are exposed only briefly after they slip out of one cell and before they infect others. You've already read about some of the intracellular viruses, bacteria, fungi, protozoans, and sporozoans. You know about genetically altered body cells, such as tumor cells, that pose other threats. All are targets of cell-mediated responses.

Helper T cells and cytotoxic T cells, along with NK cells, macrophages, neutrophils, eosinophils, and some other types, are the executioners. Local concentrations of interleukins, interferons, and other signals get their attention and stimulate them to divide, differentiate, and attack. Interferons also call for increases in making MHC molecules, thus enhancing the capacity of some cells to display antigen. If antibodies bind with exposed antigen, they, too, will call in some of the defenders.

Cytotoxic T cells are so sensitive to antigen–MHC complexes and to altered configurations on body cells,

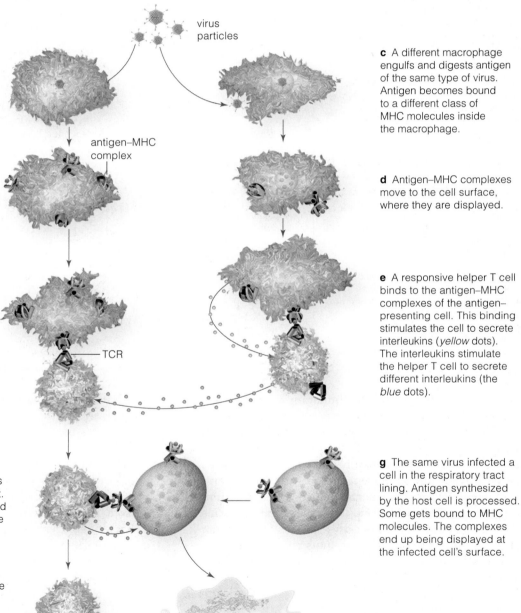

a A virus infects a macrophage. The host cell's pirated metabolic machinery makes viral proteins, which are antigenic. Enzymes of the macrophage cleave the antigens into fragments. Some fragments bind to a certain class of MHC molecules.

b Some antigen fragments are bound to class I MHC molecules. (These class I proteins occur on all nucleated cells and present antigen that originates inside infected cells.) Antigen–MHC complexes move to the cell surface and are displayed.

f The TCRs of a responsive cytotoxic T cell bind to the antigen–MHC complexes of the macrophage. Interleukins being secreted from the helper T cell stimulate the cytotoxic T cell to initiate cell divisions and differentiations. Large populations of effector and memory T cells form.

h An effector cytotoxic T cell contacts the infected cell and then touch-kills it. The effector cell releases perforins and toxic substances (*green* dots) onto the cell, which programs it for death.

i The effector disengages from the doomed cell and reconnoiters for more targets. Meanwhile, perforins make holes in the infected cell's plasma membrane. Toxins move into the cell, disrupt its organelles, and make the DNA disassemble. The cell dies.

virus particles

antigen–MHC complex

TCR

c A different macrophage engulfs and digests antigen of the same type of virus. Antigen becomes bound to a different class of MHC molecules inside the macrophage.

d Antigen–MHC complexes move to the cell surface, where they are displayed.

e A responsive helper T cell binds to the antigen–MHC complexes of the antigen-presenting cell. This binding stimulates the cell to secrete interleukins (*yellow* dots). The interleukins stimulate the helper T cell to secrete different interleukins (the *blue* dots).

g The same virus infected a cell in the respiratory tract lining. Antigen synthesized by the host cell is processed. Some gets bound to MHC molecules. The complexes end up being displayed at the infected cell's surface.

Figure 35.15 Diagram of a T cell–mediated immune response.

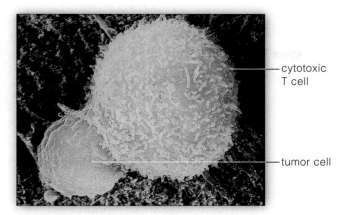

Figure 35.16 Cytotoxic T cell caught in the act of touch-killing a tumor cell.

they don't need further signals to start dividing. Also, they display and secrete molecules used in touch-killing infected and altered body cells. Example: Perforins are protein molecules that form doughnut-shaped pores in a target's plasma membrane. The pores look similar to those shown in Figure 35.2. T cytotoxic cells persuade target cells to commit suicide by **apoptosis**, a process described in Section 14.4. The cell's cytoplasm dribbles out, its organelles are disrupted, and its DNA becomes fragmented. After making its lethal hit, the cytotoxic T cell disengages from the doomed cell, and it resumes reconnoitering (Figures 35.15 and 35.16).

Cytotoxic T cells also contribute to the rejection of tissue and organ transplants. Parts of MHC markers on donor cells are different enough from the recipient's to be recognized as antigens. Other parts are enough alike to call for the costimulatory signal. MHC typing and matching donors to recipients minimize the risk.

What about NK cells? Antigen–MHC complexes do not activate them, so other receptors must be involved. Some of their receptors block attacks on cells with MHC markers, but apparently the inhibition is lifted for cells with altered or deficient numbers of MHC molecules. However they become activated, NK cells are mobilized when infections by some viruses, intracellular bacteria, and tumor cells lead to rapidly rising concentrations of interferons and interleukins. Being the first immune cells called into battle, they buy time while others are being activated, dividing, and differentiating.

Intracellular pathogens and altered body cells are targets of cell-mediated immune responses.

Helper T cells and cytotoxic T cells are antigen-specific defenders. NK cells, macrophages, neutrophils, eosinophils, and other cells make nonspecific responses. All cytotoxic types touch-kill infected cells and tumor cells.

Cancer and Immunotherapy

Carcinomas, sarcomas, leukemia—these chilling words refer to malignant tumors in skin, bone, and other tissues. Such tumors arise when viral attack, irradiation, or chemicals alter genes and cells turn cancerous (Section 14.4). The transformed cells divide repeatedly. Unless they can be destroyed or surgically removed, they kill the individual.

Transformed cells often bear abnormal proteins and protein fragments bound to MHC. Immune responses to them may make a tumor regress but may not be enough to kill it. Also, some tumors "hide" by releasing copies of the abnormal proteins, which saturate antigen receptors. Hidden tumors that reach a critical mass may overwhelm the immune system. The idea behind *immunotherapy* is to enlist white blood cells to destroy this threat and others.

MONOCLONAL ANTIBODIES   Two decades ago, Cesar Milstein and Georges Kohler made limited amounts of antibodies that home in on tumor-specific antigens. They injected an antigen into mice, which made antibodies against it. Then they fused antibody-producing B cells from the mice with cells extracted from the B cell tumors. Clonal descendants of such hybrid cells made *monoclonal antibodies* (identical copies of antibodies). However, mice cannot produce useful amounts. Genetic engineers have now designed cows that can secrete antibodies into milk, but it may be difficult to isolate these molecules from bacteria or other pathogens that may be in the milk.

Today, genetically engineered plants can manufacture monoclonal antibodies. Cornfields might mass-produce them. Besides being cost-effective, *plantibodies* are safer than antibodies from cattle. (Few plant pathogens infect people.) The first plantibody used on human volunteers prevented infection by a bacterial agent of tooth decay.

MULTIPLYING THE TUMOR KILLERS   Lymphocytes often infiltrate tumors. Researchers have removed them from a tumor and exposed them to lymphokine, an interleukin. Large populations of tumor-infiltrating lymphocytes with enhanced killing abilities have formed. These *LAK* cells (short for *Lymphokine-Activated Killers*) appear to be somewhat effective when injected back into a patient.

THERAPEUTIC VACCINES   On the horizon are *therapeutic vaccines*—wake-up calls against specific tumor cells. One approach has been to genetically engineer antigen so it becomes obvious to the killer lymphocytes. Example: Remember those Langerhans cells (Section 33.2)? Like other dendritic cells, these phagocytes prowl on many pseudopods. Once activated, they swiftly migrate to the lymph nodes and sound the immune alarm. Dirk Schadendorf harvested dendritic cells from melanoma patients. He cultured them with ground-up tumor cells, then injected them at intervals into his patients' lymph nodes or skin. A year later, two patients had no trace of melanoma. In three others, tumors shrank more than half their size. Such vaccines may soon be available.

DEFENSES ENHANCED, MISDIRECTED, OR COMPROMISED

Immunization

Immunization refers to various processes that promote immunity against diseases. With *active* immunization, an antigen-containing preparation called a **vaccine** is taken orally or is injected into the body, as in Figure 35.17. An initial injection triggers a primary immune response. A subsequent injection—a booster—elicits a secondary response; more effector and memory cells form rapidly and provide long-lasting protection.

Many vaccines are manufactured from weakened or killed pathogens or use inactivated natural toxins. Still others are made of harmless viruses that have genes from three or more pathogens inserted into their DNA or RNA. After the vaccination, body cells use the novel genes to make antigens, and immunity is established.

Passive immunization protects individuals infected by agents of hepatitis B, diphtheria, tetanus, measles, and some other diseases. At-risk patients receive injections of a purified antibody. The best source is someone who has already made a lot of the required antibody. The effects don't last; a patient hasn't made antibodies or memory cells. But injections help counter the initial attack.

A vaccine may fail or have harmful effects. In a few reported cases, vaccines have caused immunological problems; a few have also resulted in neurological disorders. It's a good idea to assess the risks and benefits before agreeing to any immunization procedure.

Allergies

In many people, normally harmless substances provoke inflammation, excess mucus secretion, and sometimes immune responses. Such substances are **allergens**, and hypersensitivity to them is called an **allergy**. Common allergens are pollen, many drugs and foods, dust mites, fungal spores, insect venom, perfumes, and cosmetics.

Some people are genetically predisposed to develop allergies. Infections, emotional stress, or changes in air temperature also trigger reactions that otherwise might not occur. Upon exposure to an antigen, IgE is secreted and binds to mast cells. When it then binds antigen, the mast cells secrete prostaglandins, histamine, and other substances that fan inflammation. Copious amounts of mucus are secreted, and airways constrict. Stuffed-up sinuses, labored breathing, sneezing, and a drippy nose are symptoms of allergic responses in *asthma* and *hay fever* (Figure 35.18). Each year in the United States, more than 14.5 million people suffer asthma attacks.

In a few cases, the inflammatory reactions wash through the body in a life-threatening event called *anaphylactic shock*. Example: someone allergic to bee or wasp venom may die within minutes of a sting. Airways constrict massively and fluid escapes rapidly from grossly permeable capillaries. Blood pressure plummets; the circulatory system may collapse.

Antihistamines (anti-inflammatory drugs) often relieve symptoms. With desensitization programs, skin tests may identify offending allergens. Larger doses of them are given slowly. Each time, the body makes more circulating IgG, and B memory cells form. Thus IgG, not IgE, binds with the allergen.

Figure 35.17 From the Centers for Disease Control and Prevention, guidelines for immunizing children as of December 2000. Doctors routinely immunize infants and children. Low-cost or free vaccinations are available at many community clinics and health departments. You can look up more information on website http://aap.org/family/parents/immunize/htm.

RECOMMENDED VACCINES	RECOMMENDED AGES
Hepatitis B	Birth–2 months
Hepatitis B booster	1–4 months
Hepatitis B booster	6–18 months
Hepatitis B assessment	11–12 years
DTP (Diphtheria; Tetanus; and Pertussis, or whooping cough)	2, 4, and 6 months
DTP booster	15–18 months
DTP booster	4–6 years
DT	11–16 years
HiB (*Haemophilus influenzae*)	2, 4, and 6 months
HiB booster	12–15 months
Polio	2 and 4 months
Polio booster	6–18 months
Polio booster	4–6 years
MMR (Measles, Mumps, Rubella)	12–15 months
MMR booster	4–6 years
MMR assessment	11–12 years
Varicella	12–18 months
Varicella assessment	11–12 years
Hepatitis A (in selected areas)	1–12 years

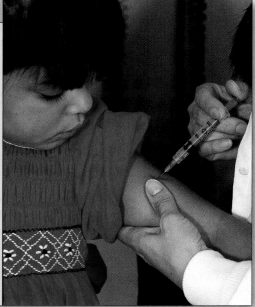

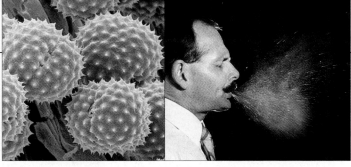

RAGWEED POLLEN . . .   . . . AND SOMETHING IT CAN PROVOKE

Figure 35.18 Example of the effects of allergens in sensitive people. Allergy sufferers who moved to deserts to escape pollen brought it with them. Half of those in Tucson, Arizona, are now sensitized to pollen from olive and mulberry trees, planted far and wide in cities and the suburbs.

Autoimmune Disorders

Sometimes self recognition fails and the immune system mounts an inappropriate attack against the body's own components. We call this **autoimmunity**. Antibodies or T cells wrongly activated can severely damage tissues.

In some cases, brakes on immune responses fail, and B and T cells become hyperactive. *Rheumatoid arthritis*, for instance, is a chronic inflammation of the joints and, often, of cardiovascular and respiratory tissues. T cells attack joints and cause an inflammatory response. Often auto-antibodies, such as IgM–IgG complexes, collect in the joints. They fan complement cascades that invite chronic inflammation. In time, joints fill with synovial membrane cells and become immobilized.

In other cases, antibodies wrongly bind to hormone receptors and stimulate cell division. *Grave's disorder* is like this. The body makes too many thyroid hormone molecules, which control metabolic rates and growth in many tissues. The antibodies don't respond to normal feedback controls over thyroid hormone production. Symptoms include elevated rates of metabolism, heart fibrillations, and excessive sweating (Section 32.6). The most common cause of neurological disorders, *multiple sclerosis*, is another case. It arises after autoreactive T cells trigger inflammation of myelin sheaths and also slip into cerebrospinal fluid, thereby disrupting nerves (Section 30.4). Symptoms range from mild numbness to paralysis and blindness. More than eight genes that influence immune function may heighten susceptibility to multiple sclerosis, but the activation of lymphocytes against viral infections may trigger its development.

Immune responses are stronger in women than men. And autoimmunity is far more frequent in women. We know that the estrogen receptor is involved in controls over gene expression throughout the body. Is it wrongly occupied in autoimmune responses? Maybe.

Deficient Immune Responses

Loss of immune function can have lethal consequences. *Primary* immune deficiencies—present at birth—result from altered genes or abnormal developmental steps.

Figure 35.19 A case of severe combined immunodeficiency (SCID). Ashanthi DeSilva was born without an immune system. She has a mutated gene for ADA (adenosine deaminase). Without this enzyme, her cells couldn't break down adenosine, so a reaction product that's toxic to lymphocytes accumulated. Infections can't be controlled. High fever, severe ear and lung infections, diarrhea, and an inability to gain weight result.

Ashanthi's parents consented to the first federally approved human gene therapy. Genetic engineers spliced the ADA gene into the genetic material of a harmless virus. The modified virus was used to deliver copies of the "good" gene into Ashanthi's bone marrow cells. Some cells incorporated the gene in their DNA and started to synthesize the missing enzyme. Ashanthi is now in her teens. Like other ADA-deficient patients who have undergone this gene therapy, she is doing well.

Researchers also take bone marrow stem cells from blood in the umbilical cord of affected newborns. (The cord connects the fetus and placenta during pregnancy and is discarded after birth.) They expose the cells to viruses that deliver copies of the ADA gene into them and to factors that stimulate mitotic cell division and growth. The cells are reinserted into newborns.

SCID (severe combined immunodeficiency) is like this. Figure 35.19 describes one type, called ADA deficiency. *Secondary* immune deficiencies are losses of immune function following exposure to outside agents, such as viruses. Severe immune deficiencies make individuals more vulnerable to infections by opportunistic agents, which are otherwise harmless to those in good health.

AIDS (acquired immunodeficiency syndrome) is the most common secondary immune deficiency. It results from viral attack and almost always results in death. The next section describes how the viral agent, HIV, replicates inside lymphocytes and destroys the body's capacity to fight infections. We also will return to this topic in Section 39.19.

Immunization programs are designed to boost immunity to specific diseases.

Some heritable disorders, developmental defects, or attacks by viruses and other outside agents result in misdirected, compromised, or nonexistent immunity.

AIDS—The Immune System Compromised

Figure 35.20 Lesions that are a sign of Kaposi's sarcoma.

CHARACTERISTICS OF AIDS AIDS is a constellation of disorders that follow infection by HIV (short for *Human Immunodeficiency Virus*). The virus cripples the immune system—which makes the body extremely susceptible to usually harmless infections and rare forms of cancer. Worldwide, more than 34 million are infected (Table 35.3). At this writing 18 million have died, 3.8 million of them children. The death rate has declined in countries where people can afford improved, expensive treatments. It is appalling in sub-Saharan Africa. That is where all but 5 percent of the 13.2 million children orphaned by this epidemic reside. Because of the stigma associated with AIDS, the children are vulnerable to malnutrition, abuse, disease, and sexual exploitation. In seven of the region's countries, at least one in five is infected.

There is no way to rid the body of known forms of the virus (HIV-I and HIV-II). *There is no cure for those already infected.* At first an infected person appears to be in good health, suffering "just a bout of the flu." Then symptoms that foreshadow AIDS emerge. Fever, many enlarged lymph nodes, fatigue, chronic weight loss, and drenching night sweats are typical. Opportunistic infections strike. They include yeast infections of the mouth, esophagus, vagina, and elsewhere, and a form of pneumonia caused by *Pneumocystis carinii*. Bruises often develop on the legs and feet especially (Figure 35.20). The bruises are signs of Kaposi's sarcoma, a cancer of blood vessel endothelium. The infections or cancers end in death.

HOW HIV REPLICATES HIV infects antigen-presenting macrophages and helper T cells, which are also called CD4 lymphocytes. HIV is a retrovirus. Each particle has an outer lipid envelope, a bit of plasma membrane acquired when it escaped from an infected cell. Various proteins spike from the envelope, span it, or line its inner surface. Inside the envelope, viral coat proteins enclose two RNA strands and several copies of reverse transcriptase, a retroviral enzyme (Sections 15.1 and 20.5).

Once it is inside an effector or memory T cell, the viral enzyme uses the RNA as a template to make DNA, which gets inserted into a host chromosome (Figure 35.21a). In some cells, inserted viral genes remain silent but are activated in a later round of infection. Whenever it occurs, transcription yields copies of viral RNA. Some transcripts are translated into viral proteins. Others get enclosed in viral coat proteins as new particles are put together; they will be the viral hereditary material. The particles are released by budding from the plasma membrane

Table 35.3 *AIDS Cases**	
Sub-Saharan Africa:	24,500,000
South/Southeast Asia:	6,000,000
Latin America:	1,300,000
North America:	920,000
East Asia and Pacific:	530,000
Western Europe:	520,000
Caribbean Islands:	360,000
East Europe/Central Asia:	360,000
Middle East/North Africa:	220,000
Australia/New Zealand:	12,000

* Global estimates as of December 1999.

(Figure 35.21b). These start a new round of infection in other cells. With every round, more macrophages, antigen-presenting cells, and helper T cells are impaired or killed.

A TITANIC STRUGGLE BEGINS Infection marks the onset of a titanic battle between HIV and the immune system. In response to HIV antigenic proteins, B cells synthesize antibodies. (These are used in diagnostic tests to identify HIV infection.) Armies of helper T and cytotoxic T cells form. But during some infection phases, the virus infects about 2 billion helper T cells and makes 100 million to 1 billion virus particles each day. Every two days, about half of the particles are destroyed and half of the helper T cells lost in battle are replaced. Huge reservoirs of HIV and masses of infected T cells accumulate in the lymph nodes. Gradually, the number of virus particles in the general circulation rises and the battle tilts. The body produces fewer and fewer replacement helper T cells. It may take a decade or more, but with the erosion of the helper T cell count, the body inevitably loses its capacity to mount effective immune responses.

Some viruses, including the measles virus, produce far more virus particles in a given day, but the immune system usually wins out. Other viruses, including herpes viruses, can lurk in the body for a lifetime, but the immune system keeps them in check. With HIV, the system almost always loses; infections and tumors kill the individual.

HOW HIV IS TRANSMITTED Like any other virus that infects humans, HIV requires a medium that allows it to leave one host, survive in the outside environment, then enter another host. It's transmitted when some body fluid of an infected person enters tissues of another person. In the United States, transmission initially occurred most often between homosexual males, as by anal intercourse, then among drug abusers who share blood-contaminated syringes and needles. It also spread in the heterosexual population, increasingly by vaginal intercourse.

Infected mothers can transmit HIV to offspring during pregnancy, birth, and breast-feeding. Also, blood supplies contaminated before 1985 accounted for some AIDS cases, although careful screening procedures have been in place since then. Tissue transplants resulted in a few infections. Health care providers in some developing countries have inadvertently spread HIV by contaminated transfusions and reuse of unsterile needles and syringes.

The molecular structure of HIV does not remain stable outside the human body. This is why the virus must be directly transmitted from one host to another. It has been

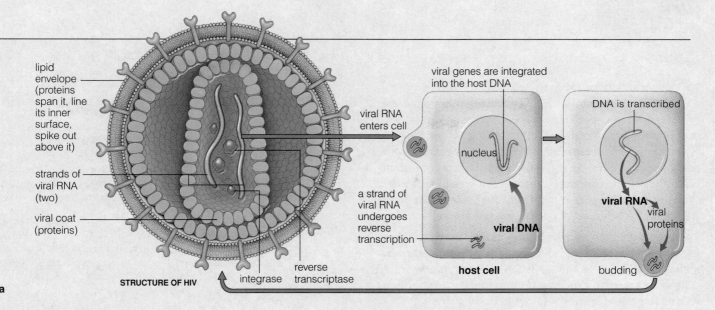

lipid envelope (proteins span it, line its inner surface, spike out above it)

strands of viral RNA (two)

viral coat (proteins)

STRUCTURE OF HIV

integrase

reverse transcriptase

viral RNA enters cell

a strand of viral RNA undergoes reverse transcription

viral genes are integrated into the host DNA

nucleus

viral DNA

host cell

DNA is transcribed

viral RNA

viral proteins

budding

Figure 35.21 (**a**) Replication cycle of HIV, a retrovirus. (**b**) Electron micrographs of an HIV particle budding from a host cell.

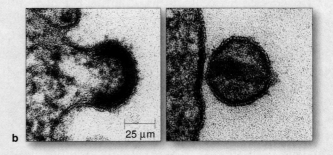

b

25 µm

isolated from human blood, semen, vaginal secretions, saliva, tears, breast milk, amniotic fluid, urine, and cerebrospinal fluid. Probably it is present in other fluids, secretions, and excretions. Until recently, only infected blood, semen, vaginal secretions, and breast milk were thought to contain HIV particles in concentrations high enough for successful transmission. Transmission by oral sex is now known to have resulted in some cases of AIDS. HIV is not effectively transmitted by way of food, air, water, casual contact, or insect bites.

DRUG TREATMENTS The drugs currently available can't cure infected people because they can't attack the HIV genes already incorporated in someone's DNA.

Given the replication mechanisms and the staggering number of replications in an infected person, the HIV genome acquires new mutations at a high rate. Natural selection inevitably operates in patients undergoing drug therapies. It favors drug resistance and thereby fans the evolution of drug-resistant HIV populations. Also, long-term use of the currently available drugs has serious side effects.

AZT (azidothymidine) and ddI (dideoxyinosine) are among those that can block reverse transcription. Over the long term, however, it becomes incorporated in the DNA of the body's cells and kills them. Ritonavir and Indinavir are among the protease inhibitors, which block the cleavages that convert new viral proteins into the building blocks for new virus particles. Potential side effects include nausea and vomiting, headache, diarrhea, blurred vision, and disruption of fat metabolism.

The expense of current treatments—as much as 15,000 dollars a year—effectively puts them out of reach where the AIDS epidemic is raging at its fullest, in developing countries. At present, the best option appears to be the development of a safe, effective, low-cost vaccine.

REGARDING VACCINES Developing vaccines against HIV is a formidable challenge for several reasons. Here are just a few of them.

First, the most effective vaccines mimic a natural infection from which a person can recover. There are no recovered AIDS patients. A few rare individuals have displayed a long-term capacity to fight the infection, but whatever is going on is simply slowing the assault and does not lend itself to vaccine development.

Second, most vaccines prevent disease, not infection. HIV already integrated into an infected person's DNA cannot be eradicated easily. The best an HIV vaccine may do is help prolong life by lowering the viral load.

Third, unusually high mutation rates rapidly give rise to variations in HIV antigens. Mutated forms that evade the immune system are favored. This makes it hard for researchers to select effective antigens for vaccines.

Fourth, there is a scarcity of HIV strains with poor cell-killing abilities. Such strains are central to efforts to make antigen that can stimulate formation of cytotoxic T cell armies, the best protection against HIV.

Fifth, most vaccines are carefully tested for safety through clinical trials with laboratory animals before testing them on human volunteers. There are no suitable laboratory animals that mirror HIV infections in humans.

In short, *until researchers develop effective vaccines and treatments, checking the spread of HIV depends absolutely on persuading people to avoid or modify social behaviors that put them at risk.* We return to this topic in Section 39.19.

35.11

1. Vertebrates fend off many pathogens with physical and chemical barriers at body surfaces. They also are protected by nonspecific and specific defense responses of the white blood cells listed in Table 35.4. *35.1*

　a. Nonspecific responses to irritation or damage of a tissue include inflammation and participation of organs with phagocytic functions, such as the spleen and liver.

　b. Immune responses are mounted against specific pathogens, foreign cells, or abnormal body cells.

2. Skin and mucous membranes lining body surfaces are physical barriers to infection. The chemical barriers include glandular secretions (such as lysozyme in tears, saliva, and gastric fluid) and some metabolic products of bacteria that normally reside on body surfaces. *35.1*

3. Diverse plasma proteins make up the complement system. When activated, they take part in nonspecific reactions and in immune responses to many bacteria, some parasitic protistans, and enveloped viruses. *35.2*

4. An inflammatory response develops in body tissues that have become damaged, as by infection. *35.3*

　a. Each inflammatory response starts with arteriolar vasodilation that increases the blood flow to the tissue, which reddens and becomes warmer as a result. Blood capillary permeability increases and the resulting local edema causes swelling and pain.

　b. Pathogens and dead or damaged body cells release the substances that trigger increased permeability of capillaries. White blood cells leave the blood and enter the tissue. There, they release a number of chemical mediators and engulf invaders. Also, plasma proteins enter the tissue. Complement proteins bind pathogens and induce their lysis, and they also attract phagocytes. Blood-clotting proteins wall off the damaged tissue.

5. An immune response has these features: *35.5, 35.6*

　a. Self/nonself recognition. B and T cells ignore the body's cells but mount a counterattack against antigen. Antigen is a molecular configuration that lymphocytes perceive as foreign (nonself).

　b. Specificity. One antigen triggers the response.

　c. Diversity. Collectively, unique receptors on each B and T cell may detect millions of kinds of antigen.

　d. Memory. A later encounter with the same antigen triggers a secondary response (greater and more rapid).

6. Antigen-presenting cells process and bind fragments of antigen to their own MHC markers. T cell receptors can bind to displayed antigen–MHC complexes. Binding is a start signal for immune responses. *35.4–35.7*

7. After antigen recognition, repeated cell divisions form clones of B and T cells. These differentiate into subpopulations of effector and memory cells. Chemical mediators such as interleukins secreted by white blood cells drive the responses. *35.4–35.7*

Table 35.4 *Summary of White Blood Cells and Their Roles in Defense*

Cell Type	Main Characteristics
MACROPHAGE	Phagocyte; acts in nonspecific and specific responses; presents antigen to T cells, and cleans up and helps repair tissue damage
NEUTROPHIL	Fast-acting phagocyte; takes part in inflammation, not in sustained responses, and is most effective against bacteria
EOSINOPHIL	Secretes enzymes that attack certain parasitic worms
BASOPHIL AND MAST CELL	Secrete histamine, other substances that act on small blood vessels to produce inflammation; also contribute to allergies
DENDRITIC CELLS	Process and directly present antigen to helper T cells
LYMPHOCYTES:	*All take part in most immune responses; following antigen recognition, form clonal populations of effector and memory cells.*
B cell	Effectors secrete five types of antibodies (IgM, IgG, IgA, and IgD known to protect a host in specialized ways; also IgE)
Helper T cell	Effectors secrete interleukins that stimulate rapid divisions and differentiation of both B cells and T cells
Cytotoxic T cell	Effectors kill infected cells, tumor cells, and foreign cells by a touch-kill mechanism
NATURAL KILLER (NK) CELL	Cytotoxic cell of undetermined affiliation; kills virus-infected cells and tumor cells by a touch-kill mechanism

8. B cells, which form and mature in bone marrow, make antibody-mediated responses. They alone make and secrete antibodies (typically Y-shaped proteins with binding sites for a specific antigen). Antibody binding neutralizes toxins, tags pathogens for destruction, or prevents pathogens from binding to body cells. *35.6*

9. Helper T cells and cytotoxic cells make cell-mediated responses. T cells arise in bone marrow but mature in the thymus, where they acquire T-cell receptors that recognize and bind antigen–MHC complexes on antigen-presenting cells. Activated cytotoxic cells directly kill virus-infected cells, tumor cells, and cells making up tissue or organ transplants. *35.7*

10. By active immunization, vaccines provoke immune responses involving production of effector and memory cells. By passive immunization, injections of purified antibodies help counter an initial infection. *35.9*

11. Allergic reactions are misguided immune responses to harmless substances. In autoimmune responses, B and T cells mount inappropriate attacks on the body's cells. Immune deficiency is a nonexistent or weakened capacity to mount a response. *35.9, 35.10*

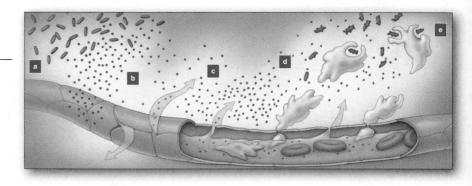

Review Questions

1. While jogging barefoot in the surf, your toes accidentally land on a jellyfish. Soon they are swollen, red, and warm to the touch. Using the diagram at right as a guide, describe events that result in these signs of inflammation. *35.3*

2. Distinguish between:
   a. neutrophil and macrophage *35.3*
   b. cytotoxic T cell and natural killer cell *35.4, 35.6*
   c. effector cell and memory cell *35.4*
   d. antigen and antibody *35.4*

3. Describe the events by which a macrophage turns into an antigen-presenting cell. *35.4*

4. Why is a vaccine to control AIDS so elusive? *35.10*

Self-Quiz (ANSWERS IN APPENDIX III)

1. _____ can deter many pathogens at body surfaces.
   a. Intact skin and mucous   c. Resident bacteria
      membranes   d. Urine flow
   b. Tears, saliva, gastric fluid e. all of the above

2. Complement proteins functions in defense. They _____ .
   a. neutralize toxins   d. form pore complexes that
   b. enhance resident   cause lysis of pathogens
      bacteria   e. both a and b
   c. promote inflammation   f. both c and d

3. Macrophages are derived from _____ .
   a. basophils   c. neutrophils
   b. monocytes   d. eosinophils

4. _____ are certain molecules that lymphocytes recognize as foreign and that elicit an immune response.
   a. Interleukins   d. Antigens
   b. Antibodies   e. Histamines
   c. Immunoglobulins

5. The most important antigens are _____ .
   a. nucleotides   c. steroids
   b. triglycerides   d. proteins

6. Immunological specificity is based on _____ .
   a. antigen receptor diversity c. mast cell proliferation
   b. recombinations of   d. both a and b
      receptor genes   e. all of the above

7. Antibody-mediated responses work best against _____ .
   a. intracellular pathogens   d. both b and c
   b. extracellular pathogens   e. all of the above
   c. extracellular toxins

8. Immunoglobulins _____ increase antimicrobial activity in mucus-coated surfaces of some organ systems.
   a. IgA   b. IgE   c. IgG   d. IgM   e. IgD

9. _____ would be a target of an effector cytotoxic T cell.
   a. Extracellular virus particles in blood
   b. A virus-infected body cell or tumor cell
   c. Parasitic flukes in the liver
   d. Bacterial cells in pus
   e. Pollen grains in nasal mucus

10. Match the immunity concepts.
    ___ inflammation   a. neutrophil
    ___ antibody secretion   b. effector B cell
    ___ fast-acting phagocyte c. nonspecific response
    ___ immunological   d. improper immune response
        memory   to body's own components
    ___ allergy   e. basis of secondary response
    ___ autoimmunity   f. hypersensitivity to allergen

Critical Thinking

1. As described in the chapter introduction, Edward Jenner lucked out. He performed a potentially harmful experiment on a boy who managed to survive it. What would happen if a would-be Jenner tried to do the same thing today?

2. Researchers have been attempting to develop a way to get the immune system to accept foreign tissue as "self." Speculate on some of the clinical applications of such a development.

3. Before each flu season, you get an influenza vaccination. This year you come down with "the flu" anyway. What do you suppose happened? (There are at least three explanations.)

4. Infection by a deadly strain of *Ebola* virus starts with high fever and flu-like aches, nausea, vomiting, and diarrhea. Blood vessels are destroyed. Blood seeps into tissues and out through the body's orifices. The liver and kidneys rapidly turn to mush. There is no vaccine. Suppose a patient is given a blood serum transfusion from another who survived the disease. Explain why a transfusion might increase chances of survival.

5. Ellen developed *chicken pox* when she was in kindergarten. Later in life, when her children developed chicken pox, she remained healthy even though she was exposed to countless virus particles daily. Explain why.

6. Write a short essay on how immune responses contribute to stability in the internal environment. (Compare Section 29.7.)

Selected Key Terms

allergen *35.9*
allergy *35.9*
antibody *35.4*
antigen *35.4*
antigen–MHC complex *35.4*
antigen-presenting cell *35.4*
apoptosis *35.7*
autoimmunity *35.9*
B lymphocyte (B cell) *35.4*
basophil *35.3*
complement system *35.2*
cytotoxic T cell *35.4*
eosinophil *35.3*
fever *35.3*
helper T cell (CD4 lymphocyte) *35.4*
histamine *35.3*
immune system *35.4*
immunization *35.9*
immunoglobulin (Ig) *35.6*
inflammation, acute *35.3*
lysis *35.2*
lysozyme *35.1*
macrophage *35.3*
mast cell *35.3*
MHC marker *35.4*
natural killer (NK) cell *35.4*
neutrophil *35.3*
pathogen *35.1*
suppressor T cell *35.4*
T lymphocyte (T cell) *35.4*
TCR *35.5*
vaccination *CI*
vaccine *35.9*

Readings

Goldsby, R., T. Kindt, and B. Osborne. 2000. *Kuby Immunology*. Fourth edition. New York: Freeman.

Pines, M. (editor). 1998. *Arousing the Fury of the Immune System*. Chevy Chase, Maryland: Howard Hughes Medical Institute.

Stine, G. 2001. *AIDS Update 2001*. Upper Saddle River, New Jersey: Prentice-Hall.

On-Line readings at Student Guide for InfoTrac:
www.brookscole.com/biology

RESPIRATION

Conquering Chomolungma

To experienced climbers, Chomolungma may be the ultimate challenge (Figure 36.1). The summit of this Himalayan mountain, also known as Everest, is 9,700 meters (29,128 feet) above sea level. It is the highest place on Earth. Iced-over vertical rock, driving winds, blinding blizzards, and heart-stopping avalanches await the mountain's challengers. So does the threat that oxygen-poor air poses to the brain.

Most of us live at low elevations. Of the air we breathe, one molecule in five is oxygen. When we travel 2,400 meters (about 8,000 feet) or more above sea level, there is less air pressure pushing down from above. Gas molecules are more spread out, and the breathing game changes. We face *hypoxia*, or cellular oxygen deficiency.

Sensing the oxygen deficiency, the brain responds by making us hyperventilate; it makes us breathe faster and deeper than usual. Above 3,300 meters (10,000 feet), hyperventilating severely disrupts ion balances in cerebrospinal fluid. It may lead to heart palpitations, shortness of breath, headaches, nausea, and vomiting. These are signs that the body's cells are, in a manner of speaking, screaming for oxygen.

Living for months at high elevations helps climbers adapt physiologically to the thinner air. For example, mechanisms raise the red blood cell count, hence the body's oxygen-carrying capacity. Professional climbers know that, above 7,000 meters, oxygen scarcity and low air pressure combine to make blood capillaries leaky.

Figure 36.1 A climber inching toward the summit of Chomolungma, where oxygen is brutally scarce.

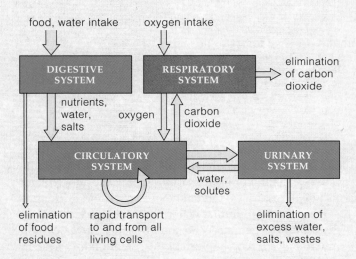

food, water intake   oxygen intake

DIGESTIVE SYSTEM   RESPIRATORY SYSTEM → elimination of carbon dioxide

nutrients, water, salts   oxygen   carbon dioxide

CIRCULATORY SYSTEM   URINARY SYSTEM

water, solutes

elimination of food residues   rapid transport to and from all living cells   elimination of excess water, salts, wastes

Figure 36.2 Interactions between the respiratory system and other organ systems, the foundation for homeostasis in complex animals.

More plasma escapes through clefts between endothelial cells of the capillary wall. Brain and lung tissues swell with excess fluid. Climbers will become comatose and die if this severe form of edema is not reversed. When prepared rescuers reach stricken climbers in time, they often zip them up in airtight bags. They pump bottled oxygen into the bag and remove carbon dioxide from it until the "air" inside is more like the air at 2,400 meters.

Few of us will ever find ourselves near the peak of Chomolungma, pushing our reliance on oxygen to the limits. Here in the lowlands, disease, smoking, and other environmental assaults push it in more ordinary ways, although the risks can be just as great.

The point is this: *Each animal has a body plan adapted to the oxygen levels of a particular habitat.* That body plan allows for **respiration**—movement of oxygen into the internal environment and movement of carbon dioxide out of it. Why is this crucial? Animals, remember, have high energy demands. Their cells require a great deal of oxygen, especially for aerobic respiration. And as this metabolic pathway releases energy from organic compounds, it produces carbon dioxide wastes.

This chapter samples a few **respiratory systems**, which function in the exchange of gases between the body and the environment. Together with other organ systems, they also contribute to homeostasis—that is, to maintaining internal operating conditions for all of the body's living cells (Figure 36.2).

Key Concepts

1. Of all organisms, multicelled animals require the most energy to drive their metabolic activities. At the cellular level, the energy comes mainly from aerobic respiration, an ATP-producing metabolic pathway that uses oxygen and produces carbon dioxide wastes.

2. By a physiological process called respiration, animals move oxygen into their internal environment and give up carbon dioxide to the external environment.

3. Oxygen diffuses into the animal body as a result of a pressure gradient. The pressure of this gas is higher in the outside air than it is in metabolically active tissues, where cells rapidly use oxygen. Carbon dioxide follows its own gradient in the opposite direction. Its pressure is higher in tissues, where it is a by-product of metabolism, than it is in the air.

4. In most respiratory systems, oxygen and carbon dioxide diffuse across a respiratory surface, such as the thin, moist respiratory membrane inside human lungs. Blood flowing through the body's circulatory system picks up oxygen and gives up carbon dioxide at this respiratory membrane.

5. Gas exchange is most efficient when the rate of air flow matches the rate of blood flow. The nervous system brings the rates into balance by controlling the rhythmic pattern and magnitude of breathing.

THE NATURE OF RESPIRATION

The Basis of Gas Exchange

A concentration gradient, recall, is a difference in the number of molecules (or ions) of a substance between two regions. Like other substances, oxygen and carbon dioxide tend to diffuse down a concentration gradient; each shows a net outward movement from the region where molecules of it are colliding more frequently (Section 5.6). The process of respiration is based on the tendency of these two gases to follow their respective concentration gradients—or, as we say for gases, to diffuse down **pressure gradients**—that form between the animal and its surroundings.

The gases don't exert the same pressure. Pump air into a flat tire near a beach in San Diego or Miami or anywhere else at sea level. You will fill it with about 78 percent nitrogen, 21 percent oxygen, 0.04 percent carbon dioxide, and 0.96 percent of some other gases. By mercury barometer measures, atmospheric pressure at sea level is about 760 mm Hg (Figure 36.3). Oxygen exerts just part of that total pressure on the tire wall. Its "partial" pressure is greater than carbon dioxide's. And so oxygen's **partial pressure**—its contribution to the total atmospheric pressure—is 760 × 21/100, or about 160 mm Hg. When similarly measured, carbon dioxide's partial pressure is about 0.3 mm Hg.

Gases enter and leave an animal body by crossing a **respiratory surface**, a thin layer of epithelium or some other body tissue. The surface must be kept moist at all times; gaseous molecules can't diffuse across it unless they are dissolved in fluid. What dictates the *number* of gas molecules that can move across a respiratory surface during any specified interval? According to **Fick's law**, the greater the surface area and the larger the partial pressure gradient, the faster the diffusion rate will be.

Which Factors Influence Gas Exchange?

SURFACE-TO-VOLUME RATIO   All animal body plans promote favorable rates of inward diffusion of oxygen and outward diffusion of carbon dioxide. For example, animals without respiratory organs are tiny, tubelike, or flattened, and gases diffuse directly across the body surface. These body plans meet the constraint imposed by the surface-to-volume ratio, as Section 4.1 describes. To get a sense of the ratio's effects, imagine a flatworm growing in all directions like an inflating balloon. The worm's surface area does not increase at the same rate as its volume. Once its girth exceeds even a millimeter, the diffusion distance between the body's surface and internal cells will be so great that the worm will die.

VENTILATION   Large-bodied, active animals have huge demands for gas exchange, more than diffusion alone

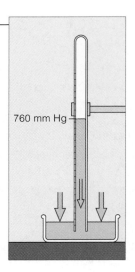

Figure 36.3   Atmospheric pressure as measured with a device called a mercury barometer. Part of the device is a glass tube in which the height of a column of mercury (Hg) increases or decreases, depending on air pressure outside the device. At sea level, the mercury rises to about 760 millimeters (29.91 inches) from the base of the tube. At this level, the pressure that the column of mercury exerts inside the tube is equal to the atmospheric pressure on the outside.

760 mm Hg

can satisfy. Diverse adaptations make the exchange rate more efficient. For example, a flap of tissue above many fish gills (respiratory organs) moves back and forth, thus stirring the surrounding water. The stirring brings more dissolved oxygen closer to the gills and moves carbon dioxide away. Among vertebrates, a circulatory system rapidly transports oxygen to cells and carbon dioxide to gills or lungs for disposal. As another example, you breathe in and out to ventilate your lungs.

TRANSPORT PIGMENTS   Rates of gas exchange get a boost with transport pigments—mainly **hemoglobin**—that help maintain the steep pressure gradients across the respiratory surface. For example, at the respiratory surface in a human lung, the oxygen concentration is high, and each hemoglobin molecule in blood weakly binds up to four oxygen molecules (Section 3.5.) The circulatory system swiftly transports hemoglobin away from the respiratory surface. In oxygen-poor tissues, oxygen follows its gradient and diffuses away from the hemoglobin, thereby helping maintain the gradient that entices oxygen into blood. **Myoglobin**, a similar but smaller pigment, has a higher oxygen-binding capacity. It's notably abundant in heart muscle cells and skeletal muscle cells that contract slowly and resist fatiguing.

As the chapter introduction states, respiration is a process by which animals move oxygen into the internal environment, especially for use in aerobic respiration, then rids the body of this pathway's carbon dioxide wastes.

Oxygen and carbon dioxide enter and leave the internal environment by diffusing across a moist respiratory surface. Like other atmospheric gases, they tend to move down their respective pressure gradients. Each type of gas exerts only part of the total pressure across a respiratory surface.

Gas exchange depends on steep partial pressure gradients between the outside and inside of the animal body. The greater the area of the respiratory surface and the larger the partial pressure gradient, the faster diffusion will proceed.

INVERTEBRATE RESPIRATION

Flatworms, earthworms, and many other invertebrates are not massive, and their life-styles do not depend on high metabolic rates (Figure 36.4a). Demands for gas exchange are simply met by **integumentary exchange**, in which gases diffuse directly across the body's surface covering (integument). This mode of respiration only works when the surface is moist, and invertebrates that rely exclusively on it are restricted to aquatic or damp habitats. (By contrast, amphibians and some other large animals use integumentary exchange also, but only as a supplement to other modes of respiration.)

Many invertebrates of aquatic habitats have moist, thin-walled respiratory organs called **gills**. Extensively folded gill walls have an increased respiratory surface area that enhances the rates of exchange between blood or some other body fluid and the surroundings. Figure 36.4b shows the much-folded gill of a sea hare (*Aplysia*). By supplementing integumentary exchange, the gill helps provide enough oxygen for this big mollusk. Some sea hares are 40 centimeters (nearly 16 inches) long.

Spiders and other invertebrates of dry habitats have a small, thick, or hardened integument that is not well

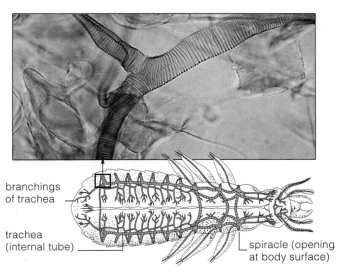

Figure 36.5 General plan of insect tracheal systems. Chitin rings reinforce many of the branching tubes of the system.

endowed with blood vessels. Their integument holds in precious water, but it is not a good respiratory surface. Such animals have an *internal* respiratory surface. For instance, most spiders have internal respiratory organs with thin, folded walls that resemble book pages; hence the name, book lungs (Section 23.12). Like most insects, millipedes, and centipedes, some other spiders have a system of internal tubes for **tracheal respiration**.

Figure 36.5 shows the tracheal system of an insect. Small openings perforate the insect's integument. Each opening, a spiracle, is the start of a tube that branches within the body. Each of the last branchings dead-ends at its fluid-filled tip, where gases diffuse directly into tissues. The tube tips are especially profuse in muscle and other tissues with high oxygen demands.

We find hemoglobin or other respiratory pigments in many invertebrates, although they are rare in insects. Respiratory pigments increase the capacity of body fluids to transport oxygen, as they do in vertebrates. In species having a well-developed head, oxygenated blood tends to circulate first through the head end and then through the rest of the body.

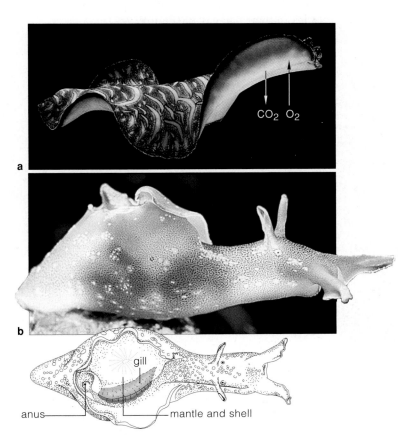

Figure 36.4 Invertebrates of aquatic habitats. (**a**) A flatworm, small enough to get along well without an oxygen-transporting circulatory system. Dissolved oxygen in such habitats reaches individual cells simply by diffusing across the body surface. (**b**) Gill of a sea hare (*Aplysia*), one of the gastropods.

Flatworms and some other invertebrates of aquatic or moist habitats do not have a massive body. By their simple mode of respiration, called integumentary exchange, oxygen and carbon dioxide diffuse directly across the body surface.

Most marine invertebrates and many freshwater types have gills of one sort or another. These respiratory organs have moist, thin, and often highly folded walls.

Most insects, millipedes, centipedes, and some spiders use tracheal respiration. Gases flow through open-ended tubes that start at the body surface and end directly in tissues.

VERTEBRATE RESPIRATION

Gills of Fishes and Amphibians

Many vertebrates use gills as respiratory organs. A few kinds of fish larvae and a few amphibians have *external* gills projecting into water. Adult fishes have a pair of *internal* gills: rows of slits or pockets at the back of the mouth that extend to the body's surface (Figure 36.6*a*). Whatever its form, each gill has a wall of moist, thin, vascularized epithelium.

In fishes, water flows into the mouth and pharynx, then over arrays of filaments in the gills (Figure 36.6*b*). Blood vessels thread through the respiratory surfaces in each filament. First water flows past a vessel leading to the rest of the body. The blood inside it has less oxygen than the water does, so oxygen diffuses into the blood. The same volume of water flows over a vessel leading into the gills. Although it already gave up some oxygen, it still holds more than the blood in this vessel does. So more oxygen diffuses into the filament. Any movement of two fluids in opposing directions is a **countercurrent flow**. With such a mechanism, a fish can extract about 80 to 90 percent of the oxygen dissolved in the water flowing past. That is more than the fish would get from a one-way flow mechanism at far less energy cost.

Lungs

Some fishes and all amphibians, birds, and mammals have a pair of **lungs**, internal respiratory surfaces in the shape of a cavity or sac. More than 450 million years ago, lungs started to develop in some fish lineages as outpouchings of the gut wall. They evolved rapidly by way of natural selection, possibly because they offered survival advantages. In oxygen-poor habitats, a pair of lungs afforded a larger surface area for gas exchange. Also, when some vertebrates entered dry habitats on land, lungs were better than gills (Section 24.6). Why? Gill filaments stick together and can't function unless water flows through them and keeps them moist.

Lungfishes of oxygen-poor habitats still have gills. They also use tiny lungs as a backup for gas exchange.

Amphibians never completed the transition to land. Salamanders especially use their skin as a respiratory surface for integumentary exchange. Frogs and toads rely more on small lungs for oxygen uptake, but most of their carbon dioxide wastes diffuse outward across the skin. Frogs also are heavy-duty breathers; they *force* air into their lungs, which they empty by contracting muscles in their body wall (Figure 36.7).

Water flows in through open mouth

FISH GILL

Water flows over gills, then out.

a

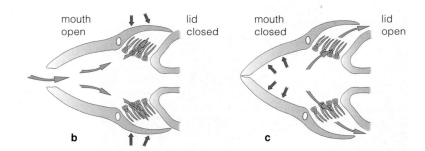

mouth open / lid closed

mouth closed / lid open

b   **c**

Figure 36.6 Example of fish gills. (**a**) One of a pair of gills. A bony lid, removed for this sketch, protects each gill.

(**b**) Gill ventilation. Opening the mouth and closing the lid pulls water across the gill.
(**c**) Closing the mouth and opening the lid forces water to the outside.

(**d**) Gas exchange at a gill. Filaments in a gill have vascularized respiratory surfaces. The filament organization directs incoming water past the gas exchange surfaces. (**e**) A blood vessel delivers oxygen-poor blood from the body into each filament. Another delivers oxygenated blood into the body. Blood flows from one into the other and runs counter to the direction of water flow over respiratory surfaces. This countercurrent flow favors the movement of oxygen (down its partial pressure gradient) from water into the blood.

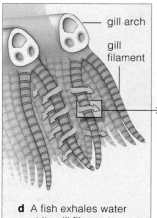

gill arch

gill filament

d A fish exhales water past its gill filaments, which are organized to direct the water past a gas exchange surface.

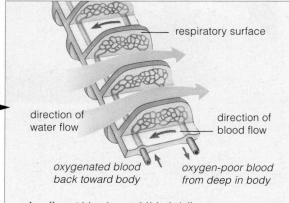

respiratory surface

direction of water flow

direction of blood flow

oxygenated blood back toward body

oxygen-poor blood from deep in body

e An efferent blood vessel (*blue*) delivers oxygen-poor blood to the filament. An afferent vessel (*red*) transports oxygenated blood back to the body. Blood in capillaries connecting the two vessels flows counter to the direction of water flowing over the gas exchange surface.

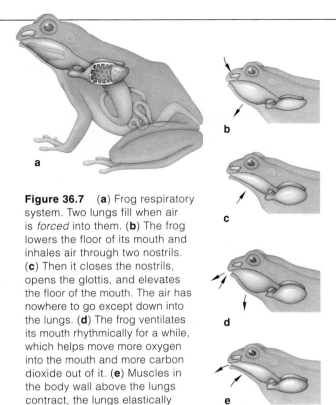

Figure 36.7 (**a**) Frog respiratory system. Two lungs fill when air is *forced* into them. (**b**) The frog lowers the floor of its mouth and inhales air through two nostrils. (**c**) Then it closes the nostrils, opens the glottis, and elevates the floor of the mouth. The air has nowhere to go except down into the lungs. (**d**) The frog ventilates its mouth rhythmically for a while, which helps move more oxygen into the mouth and more carbon dioxide out of it. (**e**) Muscles in the body wall above the lungs contract, the lungs elastically recoil, and air is forced out.

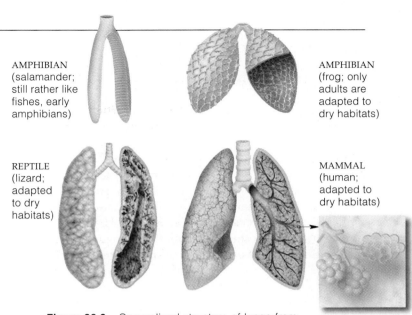

Figure 36.8 Generalized structure of lungs from four vertebrates, suggestive of an evolutionary trend from sacs for simple gas exchange to a larger, more complex respiratory surface area.

AMPHIBIAN (salamander; still rather like fishes, early amphibians)

AMPHIBIAN (frog; only adults are adapted to dry habitats)

REPTILE (lizard; adapted to dry habitats)

MAMMAL (human; adapted to dry habitats)

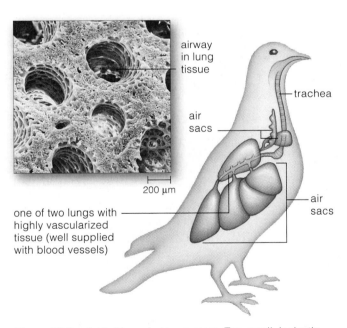

Figure 36.9 A bird's respiratory system. Two small, inelastic lungs have a number of attached air sacs. When the bird inhales, it draws air into the sacs through tubes, open at both ends, that thread through vascularized lung tissue. This tissue is the respiratory surface, where gases are exchanged.

When the bird exhales, it forces air out of the sacs, through the tubes, and out of the trachea. Thus, air isn't just drawn into bird lungs. It is continuously drawn through them and across respiratory surfaces. This unique ventilating system supports the high metabolic rates that birds require for flight and other energy-intensive activities.

Frogs also use the lungs to make sounds. So do all mammals except whales. Sounds start near the entrance to a larynx. Part of a mucous membrane folds into this airway to the lungs. The folds are **vocal cords**; the gap between them is a **glottis**. Frogs force air back and forth through the glottis, and between the lungs and a pair of pouches on the floor of the mouth. Air flow makes the vocal cords vibrate. Like other vertebrates, frogs make different sounds by controlling the vibrations.

Paired lungs are the dominant respiratory organs in reptiles, birds, and mammals (Figure 36.8). Breathing moves air by bulk flow into and out of the lungs, where capillaries thread lacily around the respiratory surface. Concentration gradients for oxygen and carbon dioxide are so steep that both gases diffuse rapidly across the surface. Oxygen enters the capillaries and is circulated quickly through the body. In regions with low oxygen levels, it diffuses into interstitial fluid, then into cells. Carbon dioxide moves quickly in the other direction and is expelled from the lungs.

This mode of gas exchange is embellished a bit only in birds. As described in Figure 36.9, birds are unique in that air not only flows into and out from their lungs; it also flows *through* their lungs. With that exception in mind, we turn next to the human respiratory system, for its operating principles apply to most vertebrates.

A countercurrent flow mechanism in fish gills compensates for low oxygen levels in aquatic habitats. Internal air sacs—lungs—are more efficient in dry habitats on land.

Amphibians use integumentary exchange and force air into and out of small lungs. Ventilation of paired lungs is the major mode of respiration in reptiles, birds, and mammals.

HUMAN RESPIRATORY SYSTEM

What Are the System's Functions?

Getting oxygen from air and expelling carbon dioxide are the key functions of the human respiratory system. Breathing, a form of ventilation, alternately moves air into and out of its paired lungs. Each lung has about 300 million outpouchings. These tiny air sacs are called **alveoli** (singular, alveolus). Control mechanisms adjust breathing rates so that the inflow and outflow of air at alveoli match metabolic demands for gas exchange.

The system has other functions. Breathing is used in vocalizations, such as speech. It enhances the return of venous blood to the heart and helps dispose of excess heat and water. Breathing controls adjust the body's acid–base balance. Carbon dioxide, recall, helps form carbonic acid (H_2CO_3), which works with bicarbonate (HCO_3^-) as a buffer system. HCO_3^- accepts or releases hydrogen ions (H^+), depending on blood's pH. Deep, rapid breathing expels more carbon dioxide, so less carbonic acid forms and the body loses acid. Shallow, slow breathing has the opposite effect. Carbon dioxide builds up, more H_2CO_3 forms, so the body gains acid.

The respiratory system also has built-in mechanisms for dealing with airborne foreign agents and substances

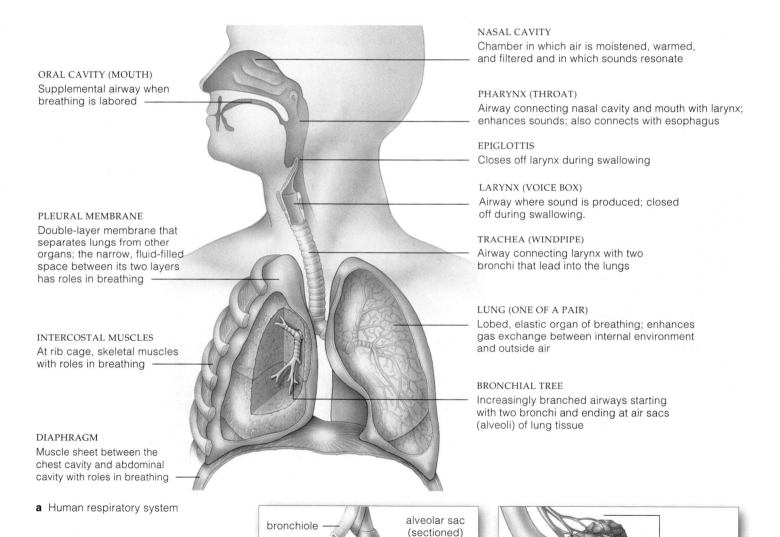

ORAL CAVITY (MOUTH)
Supplemental airway when breathing is labored

PLEURAL MEMBRANE
Double-layer membrane that separates lungs from other organs; the narrow, fluid-filled space between its two layers has roles in breathing

INTERCOSTAL MUSCLES
At rib cage, skeletal muscles with roles in breathing

DIAPHRAGM
Muscle sheet between the chest cavity and abdominal cavity with roles in breathing

a Human respiratory system

NASAL CAVITY
Chamber in which air is moistened, warmed, and filtered and in which sounds resonate

PHARYNX (THROAT)
Airway connecting nasal cavity and mouth with larynx; enhances sounds; also connects with esophagus

EPIGLOTTIS
Closes off larynx during swallowing

LARYNX (VOICE BOX)
Airway where sound is produced; closed off during swallowing.

TRACHEA (WINDPIPE)
Airway connecting larynx with two bronchi that lead into the lungs

LUNG (ONE OF A PAIR)
Lobed, elastic organ of breathing; enhances gas exchange between internal environment and outside air

BRONCHIAL TREE
Increasingly branched airways starting with two bronchi and ending at air sacs (alveoli) of lung tissue

bronchiole

alveolar sac (sectioned)

alveolar duct

alveoli

alveolar sac

pulmonary capillary

Figure 36.10 (**a**) Components of the human respiratory system and their functions. Muscles, including the diaphragm, and parts of the axial skeleton have secondary roles in respiration. (**b**,**c**) Location of alveoli relative to bronchioles and to the pulmonary (lung) capillaries.

b

c

inhaled with air. Finally, it removes and inactivates or otherwise modifies numerous blood-borne substances before they can circulate through the rest of the body.

The respiratory system's role in respiration ends at alveoli. There, the circulatory system takes over the task of moving gases. Oxygen and carbon dioxide diffuse across both the alveoli and the pulmonary capillaries next to them. (The Latin *pulmo* means lung.)

From Airways Into the Lungs

It will take at least 300 million breaths to get you to age seventy-five. You might find yourself going without food for a few hours or days. But stop breathing even for five minutes and normal brain function is over.

Take a deep breath. Now look at Figure 36.10 to get an idea of where the air will travel in your respiratory system. Unless you are out of breath and panting, air has entered two nasal cavities, not your mouth. There, secretions of mucus warm and moisten it. In the cavities, ciliated epithelium and hairs filter dust and particles from air, and olfactory receptors function in the sense of smell (Section 31.6). Now air is poised at the **pharynx**, or throat. The pharynx is the entrance to the **larynx**, an airway having two paired folds of mucous membrane. The lower pair are vocal cords (Figure 36.11). When you breathe, air is forced in and out through the glottis, the gap between these cords. As in frogs (Section 36.3), the air flow makes the vocal cords vibrate in ways that can be controlled to make different sounds.

Inside the folds are thick bands of elastic ligaments attached to cartilage. When the larynx muscles contract and relax, these ligaments tighten or slacken to change how much the folds are stretched. The nervous system coordinates the narrowing and widening of the glottis. For example, when it calls for increased tension in the larynx muscles, the gap between the vocal cords shrinks and you make high-pitched sounds and squeaks. The lips, teeth, tongue, and the soft roof above the tongue are enlisted to modify different sounds into patterns of vocalization, such as speech and song.

When vocal cords are inflamed as an outcome of an infection or irritation, swelling of their mucous lining interferes with their capacity to vibrate. If the swelling causes hoarseness, this condition is called *laryngitis*.

The **epiglottis** is a tissue flap at the entrance to the larynx. When it points upward, air can move into the **trachea**, or windpipe. The epiglottis points downward and shuts off the trachea's entrance when you swallow. Food and fluids enter a different tube (the esophagus) that connects the pharynx with the stomach.

The trachea branches into two airways, which lead into tissues of the lungs. Each airway is a **bronchus** (plural, bronchi). Its epithelial lining has a profusion of

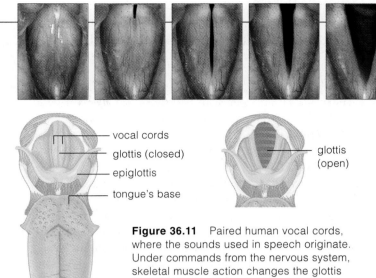

vocal cords
glottis (closed)
epiglottis
tongue's base
glottis (open)

Figure 36.11 Paired human vocal cords, where the sounds used in speech originate. Under commands from the nervous system, skeletal muscle action changes the glottis (the gap between them). The photograph series gives you an idea of how the glottis changes during quiet breathing.

cilia and mucus-secreting cells. The lining is a barrier to infection. Bacteria and airborne particles stick to the mucus, and cilia sweep the debris-laden mucus toward the mouth. The mucus is expelled or swallowed.

Human lungs are elastic, cone-shaped organs of gas exchange. They are located inside the rib cage to the left and right of the heart and above the diaphragm. The **diaphragm** is a muscular partition between the thoracic and abdominal cavities. The saclike pleural membrane lines the outer surface of the lungs and inner surface of the thoracic cavity's wall. The thoracic wall and lungs press its thin, opposing surfaces together.

A thin film of lubricating fluid cuts friction between the pleural membrane surfaces. In *pleurisy*, a respiratory ailment, the membrane becomes inflamed and swollen. Its two surfaces rub each other, so breathing is painful.

Inside each lung, air moves through finer and finer branchings of a "bronchial tree" (Figure 36.10*a*). These airways are **bronchioles**. Their finest branchings, the *respiratory* bronchioles, end in the cup-shaped alveoli. Alveoli usually are clustered as larger pouches called alveolar sacs, as in Figure 36.10*b,c*. Collectively, alveolar sacs offer a tremendous surface area for gas exchange with blood. If all the alveolar sacs were stretched out in one layer, they would cover a racquetball court floor!

Oxygen uptake and carbon dioxide removal are the major functions of the human respiratory system. Inside its pair of lungs, the circulatory system takes over the remaining tasks of respiration.

The respiratory system also has roles in moving venous blood to the heart, in vocalization, in adjusting the body's acid–base balance, in defense against harmful airborne agents or substances, in removing or modifying a number of blood-borne substances, and in the sense of smell.

CYCLIC REVERSALS IN AIR PRESSURE GRADIENTS

The Respiratory Cycle

There is a cyclic pattern to breathing, which ventilates the lungs. Each **respiratory cycle** consists of two actions: *inhalation* (a single breath of air drawn into the airways) and *exhalation* (a single breath out).

Look at Figure 36.12. *Inhalation always is an active, energy-requiring action.* When you are breathing quietly, inhalation is brought about by the contraction of the diaphragm and, to a lesser extent, external intercostal muscles. The outcome is an increase in the thoracic cavity volume. Breathe hard, and the volume increases more because neck muscles contract, thereby elevating the sternum and first two ribs attached to them.

During each respiratory cycle, the thoracic cavity's volume increases, then decreases. So pressure gradients between the air inside and outside the respiratory tract change. *Atmospheric* pressure, 760 mm Hg at sea level, is exerted by the combined weight of all the atmospheric gases on all airways. Before inhalation, *intrapulmonary* pressure (the pressure inside all alveoli) is also 760 mm Hg (Figure 36.12d).

A different gradient helps keep the lungs close to the thoracic cavity wall during the cycle even through exhalation, when lungs have a smaller volume than the thoracic cavity (Figure 36.12e,f). As the cavity expands, the lungs expand also, as a result of a pressure gradient across the lung wall.

In a person who is resting, the *intrapleural* pressure (inside the pleural sac) averages 756 mm Hg, which is less than atmospheric pressure. Intrapleural pressure is exerted outside the lungs (inside the thoracic cavity). As intrapleural pressure pushes inward on a lung wall, the intrapulmonary pressure is pushing outward. The difference between them—4 mm Hg—is large enough to make the lungs stretch and fill the thoracic cavity.

The cohesiveness of water molecules present in the intrapleural fluid (fluid in the pleural sac) also helps keep lungs close to the thoracic wall. By analogy, wet two small panes of glass and press them together. The panes easily slide back and forth but strongly resist being pulled apart. Similarly, intrapleural fluid "glues" lungs to the wall. As a result, when the thoracic cavity expands by inhalation, the lungs must expand, also.

Figure 36.12b shows what happens as you start to inhale. The diaphragm flattens and moves downward, and the rib cage is lifted upward and outward. As the thoracic cavity is expanding, the lungs expand along with it. At this time, the air pressure in all alveolar sacs combined is lower than the atmospheric pressure. Fresh air follows this pressure gradient and flows down into the airways, then into the alveoli.

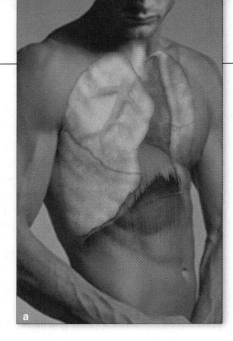

Figure 36.12 (**a**) The location of lungs relative to the diaphragm. (**b,c**) Changes in the thoracic cavity's size during one respiratory cycle. The X-ray images show how the maximum inhalation possible changes the thoracic cavity volume. (**d–f**) Changes in the intrapulmonary pressure and lung volume in a respiratory cycle.

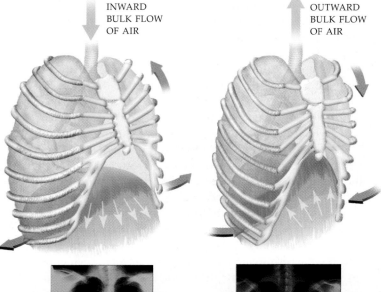

INWARD BULK FLOW OF AIR

OUTWARD BULK FLOW OF AIR

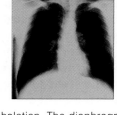

b Inhalation. The diaphragm contracts, moves down. External intercostal muscles contract and lift rib cage upward and outward. The lung volume expands.

c Exhalation. Diaphragm, external intercostal muscles return to resting positions. Rib cage moves down. Lungs recoil passively.

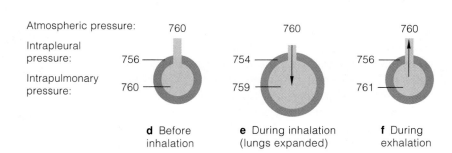

Atmospheric pressure:	760	760	760
Intrapleural pressure:	756	754	756
Intrapulmonary pressure:	760	759	761

d Before inhalation

e During inhalation (lungs expanded)

f During exhalation

The second part of the respiratory cycle, exhalation, is passive when you are breathing quietly. The muscles that brought about inhalation relax and the lungs passively recoil, with no further energy outlays. The decrease in lung volume compresses air in the alveolar sacs. At this time, pressure inside the sacs is greater than atmospheric pressure. Air follows the gradient, flowing out from the lungs, as in Figure 36.12*c*.

Exhalation becomes active and energy-requiring only when you vigorously exercise and must expel more air. When exhalation is active, abdominal wall muscles contract. Abdominal pressure rises and exerts an upward-directed force on the diaphragm, which is thereby pushed upward. Also, the internal intercostal muscles contract, and so pull the thoracic wall inward and downward. Their action flattens the chest wall. In such ways, the dimensions of the thoracic cavity decrease. The lung volume decreases as well when elastic tissue of the lungs passively recoils.

Breathing at High Altitudes

From the chapter introduction, you already know that breathing becomes an agonizing challenge for climbers of Mount Everest and other cloud-piercing peaks. Even so, like llamas and other species that evolved at high elevations, millions of people live quite comfortably 4,800 meters (16,000 feet) or more above sea level.

Compared with people living at lower elevations, the permanent residents of high mountains have lungs with far more alveoli and blood vessels, which formed as they were growing up. Also, their heart developed larger ventricles, so it pumps larger volumes of blood. And more mitochondria formed in their muscle tissue.

Llamas have an advantage, too. Compared to human hemoglobin, llama hemoglobin has greater affinity for oxygen. It picks up oxygen more efficiently at the lower pressures characteristic of high altitudes (Figure 36.13).

Does this mean a healthy person who grew up by the sea can't pick up roots and move to the mountains? No. By mechanisms of **acclimatization**, people make long-lasting physiological and behavioral adaptations to a new habitat that differs from the one left behind. At high altitudes, gradual changes in the pattern and magnitude of breathing and heart output can supplant the initial and acute compensations someone makes to cellular oxygen deficiency (hypoxia).

Within a few days, the reduction in oxygen delivery stimulates kidney cells to secrete more **erythropoietin**, a hormone that induces stem cells in bone marrow to

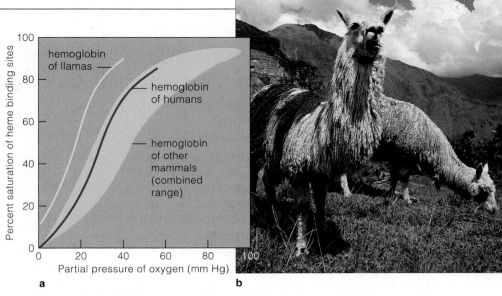

Figure 36.13 (**a**) Comparison of hemoglobin's binding and releasing capacity for humans, llamas, and other animals. (**b**) Llamas in the Peruvian Andes.

divide repeatedly and give rise to red blood cells. Each second in adults, 2 million to 3 million red blood cells form as replacements for ones that continually die off. Stepped-up erythropoietin secretion can increase that pace by six times during episodes of extreme stress. Increased numbers of circulating red blood cells raise the blood's oxygen-delivery capacity. When the oxygen concentration in the blood increases sufficiently, the kidneys slow down erythropoietin secretion.

Erythropoietin is the main hormonal mediator of red blood cell production. But men have a larger muscle mass and greater demands for oxygen. They depend as well on a sex hormone, testosterone, which stimulates increases in the basic rate of red blood cell production. That's why a blood sample from males normally has a greater percentage of red blood cells than an equivalent sample from females. But the compensatory increase in red blood cells comes at a cost. Having many more cells in the bloodstream increases resistance to blood flow, because the blood is more viscous, or "thicker." The heart must work harder to pump blood.

Breathing, which ventilates the lungs, has a cyclic pattern. The respiratory cycle consists of inhalation (one breath of air in) and exhalation (one breath of air out).

Inhalation is always an active, energy-requiring process involving contractions mainly of the diaphragm and the external intercostal muscles.

During quiet breathing, exhalation is a passive process. Muscles relax, the thoracic cavity volume decreases, and the lungs recoil elastically. Forceful exhalation is an active process that requires abdominal muscle contraction.

Breathing reverses pressure gradients between the lungs and the air outside the body.

GAS EXCHANGE AND TRANSPORT

Exchanges at the Respiratory Membrane

An alveolus is a cupped sheet of epithelial cells with a basement membrane. A pulmonary capillary consists of endothelial cells, and its basement membrane fuses with that of the alveolus. Together, the alveolar epithelium, capillary endothelium, and their basement membranes form a respiratory membrane (Figure 36.14). Gases can diffuse across this thin membrane very quickly.

Oxygen and carbon dioxide passively diffuse across the membrane in response to partial pressure gradients. An inward-directed gradient for oxygen is maintained because inhalations replenish oxygen and cells take it up an ongoing basis. An outward-directed gradient for carbon dioxide is maintained because cells continually produce carbon dioxide and exhalation removes it.

Oxygen Transport

Oxygen, like carbon dioxide, does not dissolve well in blood, so it can't be efficiently transported on its own. In vertebrates, as in other large animals, demands for oxygen are met with the help of hemoglobin molecules. These respiratory pigments are packed inside red blood cells. A hemoglobin molecule, recall, is a compact array of four polypeptide chains and four **heme groups**, each incorporating an iron atom that binds reversibly with oxygen. *Of all the oxygen inhaled into the human body, 98.5 percent of it is bound to heme groups of hemoglobin.*

Normally, inhaled air that reaches alveoli has plenty of oxygen. The opposite is true of blood in pulmonary capillaries. Thus, in the lungs, oxygen tends to diffuse into the plasma portion of blood. Then it diffuses into red blood cells where it binds with hemoglobin. In this way, it forms **oxyhemoglobin**, or HbO_2.

The amount of HbO_2 that forms in a given interval depends on oxygen's partial pressure. The higher the pressure, the greater will be the oxygen concentration. When plenty of oxygen molecules are available, they tend to randomly collide with the heme binding sites at a faster rate. The encounters continue until all four of the binding sites in hemoglobin are saturated.

HbO_2 molecules bind oxygen weakly and give it up where oxygen's partial pressure is lower than in the lungs. They give it up faster in tissues where blood is warmer, the pH is lower, and carbon dioxide's partial pressure is high. Such conditions prevail in contracting muscle and other metabolically active tissues.

Carbon Dioxide Transport

Carbon dioxide diffuses into blood capillaries in any tissue where its partial pressure is higher than it is in the blood flowing past. From there, three mechanisms transport it to the lungs. About 10 percent of the gas stays dissolved in blood. Another 30 percent binds with hemoglobin to form **carbamino hemoglobin** ($HbCO_2$). Most of it—60 percent—is transported in the form of bicarbonate (HCO_3^-). How do these HCO_3^- molecules form? First, carbonic acid forms when carbon dioxide combines with water in blood, but it quickly separates into bicarbonate and hydrogen ions (H^+):

$$CO_2 + H_2O \rightleftharpoons \underset{\text{CARBONIC ACID}}{H_2CO_3} \rightleftharpoons \underset{\text{BICARBONATE}}{HCO_3^-} + H^+$$

In blood plasma, that reaction converts only 1 of every 1,000 carbon dioxide molecules—which doesn't amount to much. It is a different story in red blood cells, which have the enzyme **carbonic anhydrase**. Enzyme action enhances the reaction rate 250 times! Most of the carbon dioxide not bound to hemoglobin becomes converted to carbonic acid this way. Conversion makes the blood level of carbon dioxide decline rapidly. Thus it helps maintain a gradient that promotes diffusion of carbon dioxide from interstitial fluid into the bloodstream.

What about the HCO_3^- that forms in the reactions? It tends to move out of red blood cells and into blood

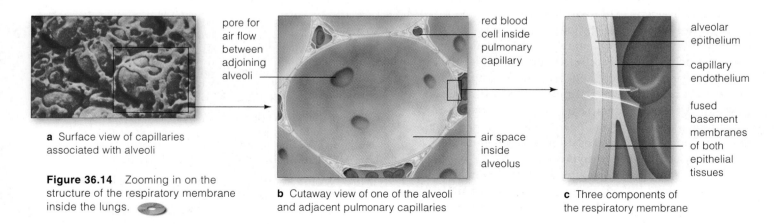

a Surface view of capillaries associated with alveoli

Figure 36.14 Zooming in on the structure of the respiratory membrane inside the lungs.

pore for air flow between adjoining alveoli

b Cutaway view of one of the alveoli and adjacent pulmonary capillaries

red blood cell inside pulmonary capillary

air space inside alveolus

alveolar epithelium

capillary endothelium

fused basement membranes of both epithelial tissues

c Three components of the respiratory membrane

plasma. What about the H^+ ions? Hemoglobin acts as a buffer for them and keeps blood from getting too acidic. A *buffer* is any molecule that combines with or releases H^+ in response to a shift in cellular pH (Section 2.6).

The reactions are reversed in alveoli, where carbon dioxide's partial pressure is lower than it is in the lung capillaries. Carbon dioxide and water form here. The gas diffuses into the alveolar sacs, then it is exhaled.

At this point, reflect on Figure 36.15. It summarizes the partial pressure gradients for oxygen and carbon dioxide through the human respiratory system.

Matching Air Flow With Blood Flow

Gas exchange is most efficient when the rate of air flow matches the rate of blood flow. The nervous system acts to balance them by controlling the *rhythm* of breathing and its *magnitude*—that is, its rate and depth.

A respiratory center inside the medulla oblongata controls rhythmic patterns of breathing. A group of its neurons sets the pace for neurons that signal muscles involved in inhalation. The signals trigger contraction. In their absence, the muscles relax and air is exhaled. When exhalation must be active and strong, different neurons in the respiratory center issue signals to the muscles involved in exhalation.

Higher in the brain stem, in the pons, other centers smooth out the breathing rhythm set by the respiratory pacemakers. An apneustic center prolongs inhalation; a pneumotaxic center curtails it.

Controlling the magnitude of breathing depends on the partial pressures of oxygen and carbon dioxide as well as the H^+ concentration in arterial blood. Carbon dioxide has the greatest effect on ongoing adjustments. Chemoreceptors in the brain are sensitive to a H^+ rise in the cerebrospinal fluid bathing them. (Remember, H^+ is also a by-product of the reactions that kick in when blood has too much carbon dioxide.) The brain responds by ordering the diaphragm and other muscles to alter activities and so adjust the rate and depth of breathing.

Aortic bodies (in the aortic arch) and carotid bodies (at a branching of the carotid artery) notify the brain when oxygen's partial pressure plummets below 60 mm Hg in arterial blood. Such life-threatening decreases occur at very high altitudes and during severe lung diseases.

In some situations, a person can "forget" to breathe. Breathing that is briefly interrupted and then resumes spontaneously is known as *apnea*. It may stop for one or two seconds or minutes—in a few cases as often as 500 times a night. With *sudden infant death syndrome* (SIDS), a sleeping infant cannot awaken from an apneic episode. SIDS might have a heritable basis. It has been linked to mutations that cause an irregular heartbeat. Also, the risk is three times as high among infants of

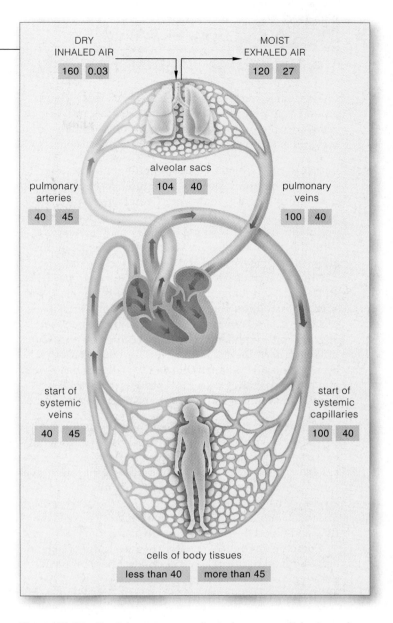

Figure 36.15 Partial pressure gradients for oxygen (*blue* boxes) and carbon dioxide (*pink* boxes) in the respiratory tract.

women who smoked cigarettes or had been exposed to secondhand smoke while pregnant. Infants who sleep on their back or sides are less vulnerable to SIDS than those who sleep on their stomach.

Driven by its partial pressure gradient, oxygen diffuses from alveolar air spaces, through interstitial fluid, and into lung capillaries. Carbon dioxide, driven by its partial pressure gradient, diffuses in the opposite direction.

Hemoglobin in red blood cells enormously enhances the oxygen-carrying capacity of blood. Most carbon dioxide is transported in blood in the form of bicarbonate, nearly all of which forms by an enzyme action in red blood cells.

Respiratory centers in the brain stem control the rhythmic pattern of breathing and the magnitude of breathing, in ways that maintain appropriate levels of carbon dioxide, oxygen, and hydrogen ions in arterial blood.

When the Lungs Break Down

In large cities, in certain workplaces, even in the cloud around a cigarette smoker, airborne particles and certain gases are present in abnormally high concentrations. And they put extra workloads on the respiratory system.

BRONCHITIS Ciliated, mucous epithelium lines the inner wall of your bronchioles (Figure 36.16). It is one of the built-in defenses that protect you from respiratory infections. Toxins in cigarette smoke and other airborne pollutants irritate the lining and may lead to *bronchitis*. With this respiratory ailment, epithelial cells along the airways become irritated and secrete excessive mucus. The mucus accumulates, and so do bacteria and particles stuck in it. Coughing brings up some of the gunk. If the source of irritation persists, so does the coughing.

Initial attacks of bronchitis are treatable. When the aggravation continues, bronchioles become chronically inflamed. Bacteria, chemical agents, or both attack the bronchiole walls. Ciliated cells in the walls are destroyed, and mucus-secreting cells multiply. Fibrous scar tissue forms and in time may narrow or obstruct the airways.

EMPHYSEMA Thick mucus clogs the airways during persistent bronchitis. Inside the lungs, tissue-destroying bacterial enzymes attack the thin, stretchable walls of alveoli. The walls break down, and inelastic fibrous tissue forms around them. Now gas exchange proceeds at fewer alveoli, which become enlarged. In time, the lungs remain distended and inelastic, so the fine balance between air flow and blood flow is permanently compromised.

Compare Figure 36.17*a* with *b* to get a sense of why it becomes hard to run, walk, and exhale. These problems are symptoms of *emphysema*, a respiratory ailment that affects over a million people in the United States alone.

A few people are genetically predisposed to developing emphysema. They do not have a workable gene for the enzyme antitrypsin, which can inhibit bacterial attack on alveoli. Poor diet and persistent or recurring colds and

Figure 36.16 False-color scanning electron micrograph of ciliated and mucus-secreting cells that line a bronchus.

free surface of a mucus-secreting cell

free surface of a cluster of ciliated cells

other respiratory infections also invite emphysema later in life. However, *smoking is the major cause of the disease*. Emphysema may develop slowly, over twenty or thirty years. When severe damage is not detected in time, the lung tissues cannot be repaired.

EFFECTS OF SMOKING Worldwide, 3 million people die each year as a result of complications arising from tobacco smoking. The World Health Organization, the Harvard School of Health, and the World Bank estimate that deaths from smoking may surpass even the AIDS epidemic by the year 2020. Every thirteen seconds in the United States, 1 of every 50 million smokers dies from emphysema, chronic bronchitis, or heart disease, and one nonsmoker dies of ailments brought on by *secondhand smoke*—prolonged exposure to tobacco smoke in the surrounding air. Children who breathe secondhand smoke at home and elsewhere are far more vulnerable to allergies and lung problems.

Smoking is the major cause of lung cancer. Yet every day, 3,000 to 5,000 Americans light a cigarette for the first time. Even children spend a billion dollars a year on cigarettes. Each year, direct medical costs of treating smoke-induced respiratory disorders drain 22 billion dollars from the economy.

How does cigarette smoke damage lungs? Noxious particles in the smoke from just one cigarette immobilize cilia in the bronchioles for several hours. The particles also trigger mucus secretions, which eventually clog the airways. They also can kill infection-fighting

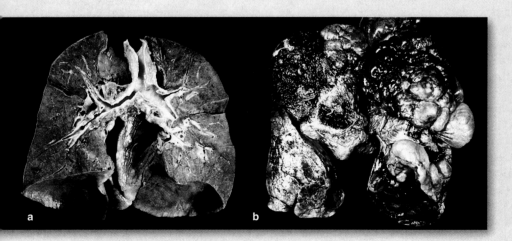

a b

Figure 36.17 (**a**) Normal appearance of tissues of human lungs. (**b**) Lungs from someone affected by emphysema.

RISKS ASSOCIATED WITH SMOKING	REDUCTION IN RISKS BY QUITTING
SHORTENED LIFE EXPECTANCY: Nonsmokers live 8.3 years longer on average than those who smoke two packs daily from the midtwenties on.	Cumulative risk reduction; after 10 to 15 years, life expectancy of ex-smokers approaches that of nonsmokers.
CHRONIC BRONCHITIS, EMPHYSEMA: Smokers have 4–25 times more risk of dying from these diseases than do nonsmokers.	Greater chance of improving lung function and slowing down rate of deterioration.
LUNG CANCER: Cigarette smoking is the major cause of lung cancer.	After 10 to 15 years, risk approaches that of nonsmokers.
CANCER OF MOUTH: 3–10 times greater risk among smokers.	After 10 to 15 years, risk is reduced to that of nonsmokers.
CANCER OF LARYNX: 2.9–17.7 times more frequent among smokers.	After 10 years, risk is reduced to that of nonsmokers.
CANCER OF ESOPHAGUS: 2–9 times greater risk of dying from this.	Risk proportional to amount smoked; quitting should reduce it.
CANCER OF PANCREAS: 2–5 times greater risk of dying from this.	Risk proportional to amount smoked; quitting should reduce it.
CANCER OF BLADDER: 7–10 times greater risk for smokers.	Risk decreases gradually over 7 years to that of nonsmokers.
CORONARY HEART DISEASE: Cigarette smoking is a major contributing factor.	Risk drops sharply after a year; after 10 years, risk reduced to that of nonsmokers.
EFFECTS ON OFFSPRING: Women who smoke during pregnancy have more stillbirths, and weight of liveborns averages less (hence, babies are more vulnerable to disease, death).	When smoking stops before fourth month of pregnancy, risk of stillbirth and lower birthweight eliminated.
IMPAIRED IMMUNE SYSTEM FUNCTION: Increase in allergic responses, destruction of defensive cells (macrophages) in respiratory tract.	Avoidable by not smoking.
BONE HEALING: Evidence suggests that surgically cut or broken bones require up to 30 percent longer to heal in smokers, possibly because smoking depletes the body of vitamin C and reduces the amount of oxygen reaching body tissues. Reduced vitamin C and reduced oxygen interfere with production of collagen fibers, a key component of bone. Research in this area is continuing.	Avoidable by not smoking.

Figure 36.18 From the American Cancer Society, a list of the risks incurred by smoking and the benefits of quitting. The photograph shows swirls of cigarette smoke poised at the entrance to the two bronchi that lead into the lungs.

macrophages in the respiratory tract. What starts out as "smoker's cough" can end in bronchitis and emphysema.

Or consider how cigarette smoke contributes to lung cancer. Inside the body, certain compounds in coal tar and in cigarette smoke become converted to highly reactive intermediates. These are the carcinogens; they provoke uncontrolled cell divisions in lung tissues. On average, 90 of every 100 smokers who develop lung cancer will die from it. If you now smoke, or are thinking about starting or quitting, you may wish to give serious thought to the information in Figure 36.18.

EFFECTS OF MARIJUANA SMOKE In 1994, 17 million in the United States smoked *pot*—marijuana (*Cannabis*)—at least once to induce light-headed euphoria. The number is rising, especially in junior high and high schools. About 1.5 million are chronic users who are enamored of the euphoria. However, the sense of well-being does not last long; it fades to apathy, depression, and fatigue. Users keep smoking to avoid the negative effects.

Besides causing psychological dependency, long-term use may result in chronic throat irritations, persistent coughing, bronchitis—and emphysema.

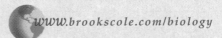

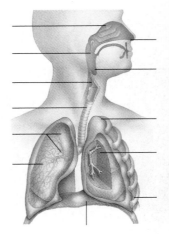

SUMMARY

1. Aerobic respiration is the main metabolic pathway that yields enough energy for active life-styles. It uses oxygen and has carbon dioxide by-products. Respiration is a process by which the animal body takes in oxygen and disposes of carbon dioxide wastes. *CI, 36.1*

2. Air is a mixture of oxygen, carbon dioxide, and other gases, each exerting a partial pressure. Each gas tends to move from areas of higher to lower partial pressure. Respiratory systems make use of this tendency. *36.1*

3. Gases diffuse across a respiratory surface in animals. In human lungs, a thin respiratory membrane consists of alveolar epithelium, lung capillary epithelium, and the fused basement membranes of both. *36.1, 36.6*

4. Animals have diverse modes of respiration. *36.2–36.3*
   a. Invertebrates with a small body mass depend only on integumentary exchange; oxygen and carbon dioxide simply diffuse across the body surface. This mode also persists in some large animals, including amphibians.
   b. Many marine and some freshwater invertebrates have gills: respiratory organs with moist, thin, and often highly folded walls. Most insects and some spiders use tracheal respiration. Gases flow through open-ended tubes leading from the body surface directly to tissues. Most spiders have book lungs, with leaflike folds.
   c. Fishes have greater energy demands than small invertebrates. At their paired gills, a countercurrent flow mechanism compensates for the low oxygen levels of aquatic habitats. Paired lungs are the dominant means of respiration in reptiles, birds, and mammals.

5. The airways of the human respiratory system are the nasal cavities, pharynx, larynx, trachea, bronchi, and bronchioles. Millions of alveoli at the end of terminal bronchioles are the main sites of gas exchange. *36.4*

6. Breathing ventilates the lungs. Each respiratory cycle consists of inhalation (one breath in) and exhalation (one breath out). During inhalation, the chest cavity expands, lung pressure decreases below atmospheric pressure, and air flows into the lungs. The events are reversed during normal exhalation. *36.5*

7. Driven by its partial pressure gradient, oxygen in lungs diffuses from alveolar air spaces into pulmonary capillaries, then diffuses into red blood cells and binds weakly with hemoglobin. Hemoglobin gives up oxygen at capillary beds in metabolically active tissues. Oxygen diffuses across the interstitial fluid, then into cells. *36.6*

8. Driven by its partial pressure gradient, carbon dioxide in tissues diffuses from cells, across interstitial fluid, and into the blood. Most of it reacts with water to form bicarbonate. The reactions are reversed in the lungs; carbon dioxide diffuses from the pulmonary capillaries into air spaces of alveoli, then is exhaled. *36.6*

Review Questions

1. Distinguish between respiration and aerobic respiration. *CI*

2. Define respiratory surface. Why must the oxygen and carbon dioxide partial pressure gradients across it be steep? *36.1*

3. What is the name of the main respiratory pigment? *36.1*

4. A few of your friends who have not taken a biology course ask you what insect lungs look like. How do you answer them (assuming your instructor is listening)? *36.2*

5. Briefly describe how countercurrent flow through a fish gill is so efficient at taking up dissolved oxygen from water. *36.3*

6. Does the respiratory system of fishes, amphibians, reptiles, birds, or mammals have air sacs that ventilate the lungs? *36.3*

7. Distinguish between:
   a. tracheal respiration and trachea (windpipe) *36.2, 36.4*
   b. pharynx and larynx *36.4*
   c. bronchiole and bronchus *36.4*
   d. pleural sac and alveolar sac *36.4, 36.5*

8. Define the functions of the human respiratory system. In the diagram above, label its components and the major bones and muscles with which it interacts during breathing. Also explain why humans can't survive for long underwater. *36.4*

Self-Quiz   ANSWERS IN APPENDIX III

1. Insects and most spiders depend on _____ .
   a. tracheal respiration   c. integumentary exchange
   b. lungs   d. both a and c

2. _____ have an abundance of respiratory pigments.
   a. Invertebrates   c. All photoautotrophs
   b. Vertebrates   d. both a and b

3. In birds, air flows _____ .
   a. into and out of lungs   c. into air sacs
   b. through lungs   d. all of the above

4. Each human lung encloses a _____ .
   a. diaphragm   c. pleural sac
   b. bronchial tree   d. both b and c

5. In human lungs, gas exchange occurs at the _____ .
   a. two bronchi   c. alveolar sacs
   b. pleural sacs   d. both b and c

6. When you breathe quietly, inhalation is _____ and exhalation is _____ .
   a. passive; passive   c. passive; active
   b. active; active   d. active; passive

7. Hypoxia triggers increased _____ secretion.
   a. carbonic anydrase   c. erythropoietin
   b. carbon monoxide   d. myoglobin

8. After oxygen diffuses into pulmonary capillaries, it diffuses into _____ and binds with _____ .
   a. interstitial fluid; red blood cells
   b. interstitial fluid; carbon dioxide
   c. red blood cells; hemoglobin
   d. red blood cells; carbon dioxide

9. Oxyhemoglobin (HbO_2) gives up oxygen faster where the _____ than it is in the lungs.
   a. pH is lower   c. O_2 partial pressure is higher
   b. blood is cooler   d. CO_2 partial pressure is lower

10. Sixty percent of the carbon dioxide in human blood is in the form of _____ and thirty percent in the form of _____ .
    a. carbon dioxide; carbonic acid
    b. carbonic anhydrase; bicarbonate
    c. bicarbonate; carbamino hemoglobin
    d. carbamino hemoglobin; bicarbonate

11. Match the descriptions with their components in humans.
    _____ trachea   a. airway leading into a lung
    _____ pharynx   b. gap between vocal cords
    _____ alveolus   c. fine branch of bronchial tree
    _____ hemoglobin   d. windpipe
    _____ bronchus   e. respiratory pigment
    _____ bronchiole   f. site of gas exchange
    _____ glottis   g. throat

Critical Thinking

1. Some cigarette manufacturers are urging people to "smoke responsibly." What social and biological issues are involved here? For example, what about risks to a nonsmoking spouse or to children of a smoker? To nonsmoking patrons in restaurants? To the unborn child of a pregnant smoker? In your opinion, what behavior would constitute "responsible smoking"?

2. People sometimes poison themselves with carbon monoxide by building a charcoal fire in an enclosed area. Assuming help arrives in time, what would be the best treatment: (1) placing the victim outdoors in fresh air or (2) rapidly administering pure oxygen? Explain how you arrived at your answer.

3. When you swallow food, contracting muscles force the epiglottis down, to its closed position. This prevents food from going down the trachea and blocking gas flow. Yet each year, several thousand people choke to death after food enters the trachea and blocks air flow for as little as four or five minutes. The *Heimlich maneuver*—an emergency procedure only—often dislodges food stuck in the trachea (Figure 36.19).

When correctly performed, the maneuver forcibly elevates the diaphragm, causing a sharp decrease in the thoracic cavity volume and an abrupt rise in alveolar pressure. Air forced up into the trachea by the increased pressure may be enough to dislodge the obstruction. Once the obstacle is dislodged, a doctor must see the person at once because an inexperienced rescuer can inadvertently cause internal injuries or crack a rib.

Reflect now on our current social climate in which lawsuits abound. Would you risk performing the Heimlich maneuver to save a relative's life? A stranger's life? Why or why not?

4. Weddell seals can dive 600 meters and stay submerged for a half hour or more. Sperm whales can dive 2,250 meters and stay there well over an hour. If they cannot inhale oxygen and exhale carbon dioxide while under water, how do these and other air-breathing mammals routinely stay submerged for prolonged periods? Research and describe their adaptations for

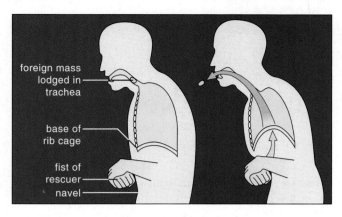

Figure 36.19 Heimlich maneuver to dislodge food stuck in the trachea. Stand behind the victim, make a fist with one hand, then position the fist, thumb-side in, against the victim's abdomen. The fist must be slightly above the navel and well below the rib cage. Now press the fist into the abdomen with a sudden upward thrust. Repeat the thrust several times if needed. The maneuver can be performed on someone who is standing, sitting, or lying down.

deep-sea diving. (*Hints:* These include increased reliance on myoglobin, selective distribution of oxygen to certain organs, and oxygen-collecting aspects of the respiratory system.)

5. Water pressure increases greatly with depth. Professional divers do not descend into very deep water without tanks of compressed air (air under pressure). Why the caution? Because of the increased pressure at depths, more gaseous nitrogen (N_2) than usual has become dissolved in body tissues. The outcome is *nitrogen narcosis*, or "raptures of the deep."

At depths of 45 meters (150 feet), divers become euphoric and drowsy, as if they were tipsy with alcohol. Some have been known to offer their airtank mouthpiece to fishes. At lower depths, divers become clumsy and weak. Below 105 meters, they can slip into a coma. Gaseous nitrogen, a lipid-soluble gas, readily dissolves in the lipid bilayer of cell membranes. Given what you have learned about human body cells in preceding chapters, which type of cell is most likely to be a key player in nitrogen narcosis when excess nitrogen disrupts its membrane properties? (*Hint*: Reflect on the nature of the symptoms.)

Selected Key Terms

acclimatization 36.5
alveolus (alveoli) 36.4
bronchiole 36.4
bronchus 36.4
carbamino hemoglobin 36.6
carbonic anhydrase 36.6
countercurrent flow 36.3
diaphragm 36.4
epiglottis 36.4
erythropoietin 36.5
Fick's law 36.1
gill (fish) 36.2
glottis 36.3
heme group 36.6
hemoglobin 36.1

integumentary exchange 36.2
larynx 36.4
lung 36.3
myoglobin 36.1
oxyhemoglobin 36.6
partial pressure 36.1
pharynx 36.4
pressure gradient 36.1
respiration CI
respiratory cycle 36.5
respiratory surface 36.1
respiratory system CI
trachea 36.4
tracheal respiration 36.2
vocal cord 36.3

Readings

Gorman, J. 31 March 2001. "Breathing on the Edge." *Science News*, 159: 202–204.

Sherwood, L. 1997. *Human Physiology*. Second edition. Belmont, California: Wadsworth.

On-Line readings at Student Guide for InfoTrac:
www.brookscole.com/biology

DIGESTION AND HUMAN NUTRITION

Lose It—And It Finds Its Way Back

America's fixation on Beautiful People who have nary an ounce of extra fat on their utterly perfect selves is bad enough. After all, we might rationalize our own extra ounces with the truism that beauty is only skin deep. But we can't rationalize away the Fat Connection to atherosclerosis, heart attacks, strokes, colon cancer, and other dangerous ailments.

The proportion of body fat relative to total tissue mass is now supposed to be 18 to 24 percent for women less than thirty years old. For men, it is supposed to be no more than 12 to 18 percent. An estimated 34 million Americans don't even come close to the standards.

No matter how much we diet, the lost weight seems to find its way back. This physiological dilemma is an outcome of our evolutionary heritage. Like most other mammals, we have an abundance of fat-storing cells in adipose tissue. Collectively, the cells are an adaptation for survival, an energy warehouse that opens up when food is scarce. Once the fat-storing cells have formed, they are in the body to stay. Variations in food intake only affect how empty or full each cell gets.

Dieting opens the fat warehouse. The brain interprets this as STARVATION! and calls for a metabolic slowdown. In response, the body uses energy far more efficiently even for basic tasks, such as breathing and digesting food. It now takes less food to do the same things!

As you've probably heard, dieting is useless without long-term commitment to physical exercise. Why? Your skeletal muscles adapt to STARVATION! signals and burn less energy than before. Jog for four hours or play tennis for eight hours straight and you might lose a pound of fat. Meanwhile, your appetite surges. If and when you do stop dieting, your "starved" fat cells quickly refill. Unless you eat less and exercise moderately all through your life, you can't keep off extra weight (Figure 37.1).

What about gaining weight? A weight gain triggers increases in metabolism. You use 15 to 20 percent more energy than before until you lose the excess.

As if this were not enough, emotions affect weight gains and losses. As an extreme case, *anorexia nervosa* is a potentially fatal eating disorder based on a flawed assessment of body weight. Most anorexics dread being fat and hungry. Many starve themselves, overexercise, and have fears about growing up and of being sexually mature. They often have irrationally high expectations about their personal performance.

Another extreme is called *bulimia*, an out-of-control "oxlike appetite." A bulimic on an hour-long eating

Figure 37.1 A few organisms positively engaged in countering imbalances between caloric intake and energy output.

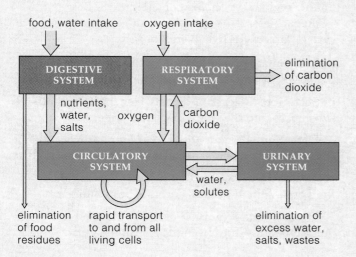

food, water intake   oxygen intake

```
                                              elimination
  DIGESTIVE          RESPIRATORY          ⇒   of carbon
  SYSTEM             SYSTEM                    dioxide

    nutrients,
    water,        oxygen    carbon
    salts                   dioxide

  CIRCULATORY                       URINARY
  SYSTEM                            SYSTEM
                      water,
                      solutes

elimination   rapid transport                elimination of
of food       to and from all                excess water,
residues      living cells                   salts, wastes
```

Figure 37.2 Functional links between the digestive, respiratory, circulatory, and urinary systems. These organ systems and others work together to supply cells with raw materials and eliminate wastes.

binge may take in 50,000 kilocalories' worth of food, then vomit or use laxatives to get rid of it. Some follow a binge–purge routine as an "easy" way to lose weight. Others may not even like to eat but purge themselves to relieve anger and frustration. The binge–purge routines range from once a month to a few times a day. Purgings damage the gut. Chronic vomiting brings up gastric fluid that erodes teeth to stubs. In severe cases, the stomach can rupture and the heart and kidneys fail.

Are severe eating disorders rare? No. In the United States, an estimated 7 million females and 1 million males are anorexic or bulimic. Most are in their teens and early twenties, but the number also is increasing among preadolescents. Each year, complications kill 5 to 6 percent of the weight-obsessed individuals.

With these sobering thoughts, we start our tour of **nutrition**. The word encompasses all those processes by which an animal ingests and digests food, then absorbs the released nutrients for later conversion to the body's own carbohydrates, lipids, proteins, and nucleic acids.

The nutritional processes proceed at the **digestive system**. This body cavity or tube mechanically and chemically reduces food to particles, then to molecules that are small enough to be absorbed into the internal environment. It also eliminates unabsorbed residues. Other organ systems, especially those shown in Figure 37.2, contribute to the nutritional processes.

Key Concepts

1. In complex animals, interactions among digestive, circulatory, respiratory, and urinary systems supply the body's living cells with raw materials, dispose of wastes, and maintain the volume and composition of extracellular fluid.

2. Most digestive systems include specialized regions for food transport, processing, and storage. Different regions mechanically break apart and chemically break down food, absorb breakdown products, and eliminate the unabsorbed residues.

3. "Nutrition" refers to all processes by which an animal ingests food, digests food, and then absorbs the released nutrients, which become converted to the body's own carbohydrates, lipids, proteins, and nucleic acids.

4. A subcategory of nutritional processes is the intake of foods that are good sources of vitamins, minerals, and a number of amino acids and fatty acids that the body itself cannot produce.

5. To maintain suitable body weight and overall health, energy intake must balance energy output by way of metabolic activity, regular physical exertion, and so on. Complex carbohydrates are the main source of dietary glucose. Typically, glucose is the body's main source of instant energy.

THE NATURE OF DIGESTIVE SYSTEMS

Incomplete and Complete Systems

Recall, from Chapter 23, that we do not see many organ systems until we get to the flatworms. Flatworms are among the invertebrates with an **incomplete digestive system**, a saclike, branching gut cavity having a single opening at the start of a tubular pharynx (Figure 37.3a). Food enters the sac, where it is partly digested, then is circulated to cells even as residues are sent out.

The saclike gut of flatworms, cnidarians, and a few other small invertebrates has a single opening for food intake *and* for waste disposal. The two-way traffic does not favor regional specialization. Other animals have a **complete digestive system**, a tube with an opening at one end for taking in food and an opening at the other end for eliminating unabsorbed residues (for example, mouth and anus). In between, the tube is subdivided into specialized regions, which function in the one-way transport, processing, and storage of raw materials.

Consider a frog's complete digestive system (Figure 37.3b). Between its mouth and the anus are a pharynx,

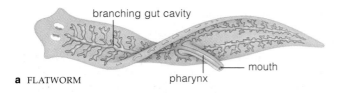

a FLATWORM

branching gut cavity

mouth

pharynx

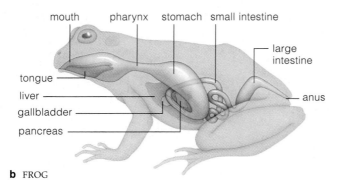

b FROG

mouth pharynx stomach small intestine

large intestine

tongue

liver

gallbladder

pancreas

anus

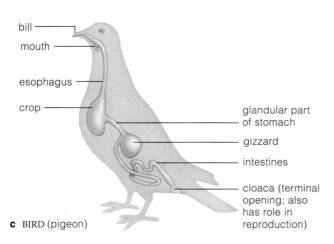

c BIRD (pigeon)

bill

mouth

esophagus

crop

glandular part of stomach

gizzard

intestines

cloaca (terminal opening; also has role in reproduction)

stomach, small intestine, and large intestine. A liver, gallbladder, and pancreas—all organs with accessory roles in digestion—connect with the small intestine. Birds, too, have a complete digestive system with a few unique regional specializations (Figure 37.3c).

Regardless of its complexity, a complete digestive system carries out five overall tasks:

1. **Mechanical processing and motility**: Movements that break up, mix, and propel food material.

2. **Secretion**: Release of digestive enzymes and other substances into the space inside the tube.

3. **Digestion**: Breakdown of food into particles, then into nutrient molecules small enough to be absorbed.

4. **Absorption**: Passage of digested nutrients and fluid across the tube wall and into body fluids.

5. **Elimination**: Expulsion of the undigested and unabsorbed residues from the end of the gut.

Correlations With Feeding Behavior

We can correlate any digestive system's specializations with feeding behavior. Consider the pigeon in Figure 37.3c. Its *food-gathering* region is a bill adapted to peck seeds off the ground. Seeds enter a mouth, a long tube (esophagus), and then a *food-processing* region, which is compactly centered in the body mass. The compactness helps balance the bird during flight. As in other seed eaters, part of the esophagus bulges out as a *crop*. This stretchable storage organ allows the bird to gulp a lot of food fast and make quick getaways from predators. Like all birds, a pigeon eats during the day and fills its crop before the sun sets. The first few hours after dark, it releases food to lessen the time of overnight fasting.

The first part of the pigeon stomach has a glandular lining, which secretes enzymes and other substances that function in digestion. The second part, the gizzard, is a muscular organ that grinds up food much as teeth and jaws do. A pigeon's intestines are proportionally shorter than those in ducks or ostriches. Such birds eat plant parts rich in tough, fibrous cellulose—which requires a much longer processing time than seeds do.

And what about the digestive system of pronghorn antelope (*Antilocapra americana*)? Fall through winter, on the mountain ridges from central Canada down into northern Mexico, pronghorn antelope browse on wild

Figure 37.3 (**a**) Incomplete digestive system of a flatworm, which has a two-way traffic of food and undigested material through one opening. (**b,c**) Examples of a complete digestive system, a tube with specialized regions and an opening at each end.

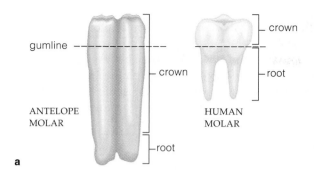

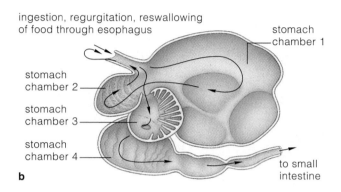

a

ingestion, regurgitation, reswallowing
of food through esophagus

stomach
chamber 1

stomach
chamber 2

stomach
chamber 3

stomach
chamber 4

to small
intestine

b

Figure 37.4 Some regional specializations of the complete digestive system of pronghorn antelope (*Antilocapra americana*).

(**a**) Comparison of antelope and human molars. (**b**) Multiple-chambered antelope stomach. The first chamber is a large pouch. The second is smaller, with a honeycombed inner surface (what butchers call "tripe"). In both chambers, food is mixed with fluid, kneaded, and exposed to fermentation activities of bacterial and protozoan symbionts. Some of the symbionts degrade cellulose; others synthesize organic compounds, fatty acids, and vitamins. The host uses a portion of these substances. Kneaded food is regurgitated into the mouth, rechewed, then swallowed again. It enters the third chamber, where it is pummeled once more before entering the last stomach chamber.

Figure 37.5 Exceptionally Big Mac of the snake world.

sage. Come spring, they move to open grasslands and deserts, there to browse on new growth (Figure 37.4).

Now think of your own cheek teeth, or molars, each having a flattened crown that is a grinding platform. The crown of an antelope's molars dwarfs yours (Figure 37.4*a*). Why the difference? You probably don't brush your mouth against dirt as you eat. An antelope does. Abrasive bits of soil enter its mouth along with tough plant material, so the crown wears out quickly. Natural selection has favored more crown to wear down.

Antelopes are **ruminants**, a type of hoofed mammal having multiple stomach chambers in which cellulose is slowly broken down. The breakdown gets under way inside the first two of four stomach sacs (Figure 37.4*b*). There, bacterial symbionts release digestive enzymes to access the nutrients in cellulose. As they do, their host regurgitates and rechews the contents of the first two sacs, then swallows again. (That's what "chewing cud" means.) Pummeling the plants more than once exposes more surface area to enzymes, which gives them more time to release nutrients that the antelope (as well as the bacteria) can use. The ruminant stomach system can steadily accept food during extended feeding times and slowly liberate nutrients when the animal rests.

Predatory and scavenging mammals have different specializations. They gorge on food when they can get it and might not eat again for some time (Figure 37.5). Part of their digestive system stores food, which they digest and absorb at a more leisurely pace. As you will

see, other organs with accessory roles in digestion help ensure that there will be an adequate distribution of nutrients between meals.

An incomplete digestive system is a saclike body cavity. Both food and residues enter and leave through the same opening. A complete digestive system is a tube that has two openings and regional specializations in between.

Complete digestive systems carry out the following tasks in controlled ways: movements that break up, mix, and propel food material; secretion of digestive enzymes and other substances; breakdown of food; absorption of nutrients; and expulsion of undigested and unabsorbed residues.

VISUAL OVERVIEW OF THE HUMAN DIGESTIVE SYSTEM

Humans have a complete digestive system, a tube with many specialized organs between two openings (Figure 37.6). If the tube of an adult were fully stretched out, it would extend 6.5 to 9 meters (21 to 30 feet). From start to finish, mucus-coated epithelium lines every surface facing the lumen. (A *lumen* is the space within a tube.) The thick, moist mucus helps protect the tube wall and promotes diffusion across its inner lining.

Ingested substances advance in one direction, from the mouth through the pharynx and esophagus, then the gastrointestinal tract, or gut. The human **gut** starts at the stomach and extends through the small intestine, large intestine (colon) and rectum, to the anus. Salivary glands and a gallbladder, liver, and pancreas function as accessory organs; they secrete a variety of substances into different regions of the tube.

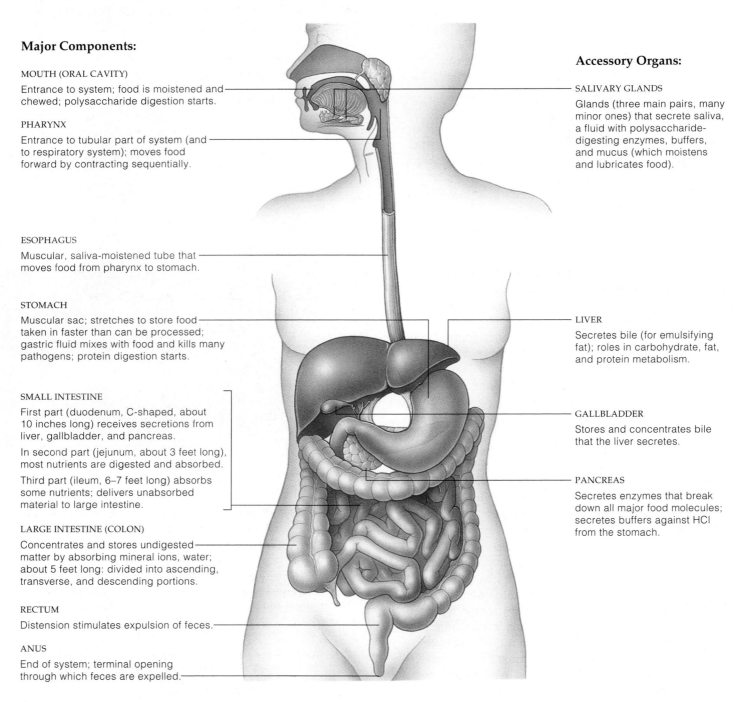

Major Components:

MOUTH (ORAL CAVITY)

Entrance to system; food is moistened and chewed; polysaccharide digestion starts.

PHARYNX

Entrance to tubular part of system (and to respiratory system); moves food forward by contracting sequentially.

ESOPHAGUS

Muscular, saliva-moistened tube that moves food from pharynx to stomach.

STOMACH

Muscular sac; stretches to store food taken in faster than can be processed; gastric fluid mixes with food and kills many pathogens; protein digestion starts.

SMALL INTESTINE

First part (duodenum, C-shaped, about 10 inches long) receives secretions from liver, gallbladder, and pancreas.

In second part (jejunum, about 3 feet long), most nutrients are digested and absorbed.

Third part (ileum, 6–7 feet long) absorbs some nutrients; delivers unabsorbed material to large intestine.

LARGE INTESTINE (COLON)

Concentrates and stores undigested matter by absorbing mineral ions, water; about 5 feet long: divided into ascending, transverse, and descending portions.

RECTUM

Distension stimulates expulsion of feces.

ANUS

End of system; terminal opening through which feces are expelled.

Accessory Organs:

SALIVARY GLANDS

Glands (three main pairs, many minor ones) that secrete saliva, a fluid with polysaccharide-digesting enzymes, buffers, and mucus (which moistens and lubricates food).

LIVER

Secretes bile (for emulsifying fat); roles in carbohydrate, fat, and protein metabolism.

GALLBLADDER

Stores and concentrates bile that the liver secretes.

PANCREAS

Secretes enzymes that break down all major food molecules; secretes buffers against HCl from the stomach.

Figure 37.6 Overview of the component parts of the human digestive system and their specialized functions. Organs with accessory roles in digestion are also listed.

INTO THE MOUTH, DOWN THE TUBE

Food is chewed and polysaccharide breakdown begins in the oral cavity, or mouth. Thirty-two teeth typically project into an adult's mouth (Figure 37.7). Each **tooth** has an enamel coat (hardened calcium deposits), dentin (a thick, bonelike layer), and an inner pulp with nerves and tiny blood vessels. It is an engineering marvel, able to withstand years of chemical and mechanical stress. Chisel-shaped incisors shear off chunks of food. Cone-shaped canines tear food. Premolars and molars, with broad crowns and rounded cusps, grind and crush it.

Also in the mouth is a **tongue**, an organ consisting of membrane-covered skeletal muscles that function in positioning food in the mouth, swallowing, and speech. On that tongue's surface are many circular structures with taste buds embedded in their tissues (Figure 37.8). A taste bud contains sensory receptors that respond to chemical differences in dissolved substances. The brain uses this information to give rise to our sense of taste.

Chewing mixes food with **saliva**. This fluid contains an enzyme (salivary amylase), a buffer (bicarbonate, or HCO_3^-), mucins, and water. Salivary glands beneath and in back of the tongue produce and secrete saliva through ducts to the free surface of the mouth's lining. Salivary amylase breaks down starch. The HCO_3^- helps maintain the mouth's pH when you eat acidic foods. Modified proteins called mucins help form the mucus that binds food into a softened, lubricated ball (bolus).

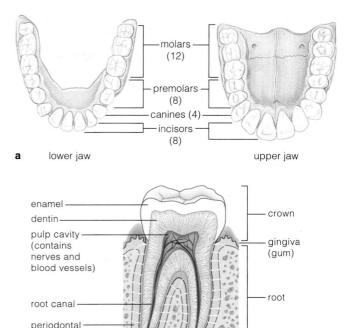

Figure 37.7 (**a**) Number and arrangement of human teeth in the upper and lower jaw. (**b**) A molar's main regions: the crown and the root. Enamel capping the crown is made of calcium deposits; it is the hardest substance in the body.

Normally harmless bacteria live on and between teeth. Daily flossing, gentle brushing, and avoiding too many sweets help keep bacterial populations in check. Without such preventive measures, conditions favor infection that may lead to *caries* (tooth decay), *gingivitis* (inflamed gums), or both. Infections also can spread to the periodontal membrane that anchors teeth to the jawbone. In *periodontal disease*, infection slowly destroys the bone tissue around a tooth.

When you swallow, contractions of tongue muscles force boluses into the **pharynx**, the tubular entrance to the esophagus *and* trachea, an airway to the lungs. To block breathing as food leaves the pharynx, a flaplike valve (the epiglottis) and the vocal cords close off the trachea. That's why you normally don't choke on food (Sections 36.4 and 36.8, *Critical Thinking* question 3).

The tubular **esophagus** connects the pharynx with the stomach. Contractions of its muscular wall propel food past a sphincter, into the stomach. A **sphincter** is a ring of smooth muscles, the contractions of which close off a passageway or an opening to the body surface.

By the action of the mouth's teeth and tongue, food gets chewed, mixed with saliva, and bound into soft, lubricated balls that will be propelled through tubes to the stomach. Enzymes in saliva start the digestion of polysaccharides.

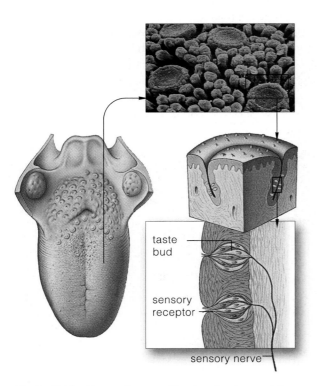

Figure 37.8 Human tongue. Circular structures at its surface house many taste buds. Filamentous structures next to them help keep food moving in the mouth.

DIGESTION IN THE STOMACH AND SMALL INTESTINE

We arrive now at the premier food-processing organs, the stomach and small intestine. Both have layers of smooth muscles, the contractions of which break apart, mix, and directionally move food. The lumen of each organ receives digestive enzymes and other secretions that help break down nutrients into fragments and then molecules small enough to be absorbed. Table 37.1 lists the names and sources of the major digestive enzymes. Carbohydrate breakdown *starts* in the mouth. Protein breakdown *starts* in the stomach. Digestion of nearly all carbohydrates, lipids, proteins, and nucleic acids in food is *completed* in the small intestine.

The Stomach

The **stomach**, a muscular, stretchable sac (Figure 37.9*a*), has three important functions. First, it mixes and stores ingested food. Second, its secretions help dissolve and degrade the food, particularly proteins. Third, it helps control passage of food into the small intestine.

The stomach wall surface exposed to the lumen is lined with glandular epithelium. Each day, the lining's glandular cells secrete about two liters of hydrochloric acid (HCl), mucus, pepsinogens, and other substances that make up **gastric fluid**, or stomach fluid. Its potent acidity, in combination with strong contractions of the stomach wall, converts food into a thick, liquid mixture called **chyme**. Chyme's acidity kills many pathogens that enter the stomach with food. It also can cause *heartburn* when gastric fluid backs up into the esophagus.

Protein digestion starts as the high acidity alters the structure of proteins and exposes their peptide bonds. It also converts pepsinogens to pepsins: active enzymes

that cleave the bonds. Protein fragments accumulate in the stomach's lumen. Meanwhile, glandular cells of the stomach lining secrete gastrin. This hormone stimulates cells of the lining that secrete HCl and pepsinogen.

Peptic ulcers form when the digestive enzymes and gastric fluid erode through the lining of the stomach or small intestine. Erosion follows when the body doesn't secrete enough mucus and buffers or secretes too much pepsin. Heredity, chronic stress, smoking, and excessive intake of alcohol or aspirin make things worse. At least 80 percent of these ulcers form after *Helicobacter pylori* infections (Section 20.1). Antibiotic therapy cures them.

The stomach empties by waves of contraction and relaxation, of a type called peristalsis. These waves mix chyme and gather force as they approach a sphincter at the stomach's base (Figure 37.9*b*). The arrival of a strong contraction closes the sphincter, which squeezes most of the chyme back. All of the chyme does move into the small intestine, but only in small amounts at a time.

The Small Intestine

The small intestine has three regions: the duodenum, jejunum, and ileum (Figure 37.6). On average, each day 9 liters of fluid enters the first region from the stomach and through the ducts of three organs: the **pancreas**, **liver**, and **gallbladder**. At least 95 percent of the fluid is absorbed across the intestine's epithelial lining.

Enzymes secreted by cells of the intestinal lining and pancreas digest food to monosaccharides (such as glucose), monoglycerides (each having a glycerol head and fatty acid tail), free fatty acids, free amino acids, nucleotides, and nucleotide bases. Example: Like the

Table 37.1 *Major Digestive Enzymes and Their Breakdown Products*

Enzyme	Source	Where Active	Substrate	Main Breakdown Products
CARBOHYDRATE DIGESTION				
Salivary amylase	Salivary glands	Mouth, stomach	Polysaccharides	Disaccharides
Pancreatic amylase	Pancreas	Small intestine	Polysaccharides	Disaccharides
Disaccharidases	Intestinal lining	Small intestine	Disaccharides	MONOSACCHARIDES* (such as glucose)
PROTEIN DIGESTION				
Pepsins	Stomach lining	Stomach	Proteins	Protein fragments
Trypsin and chymotrypsin	Pancreas	Small intestine	Proteins	Protein fragments
Carboxypeptidase	Pancreas	Small intestine	Protein fragments	AMINO ACIDS*
Aminopeptidase	Intestinal lining	Small intestine	Protein fragments	AMINO ACIDS*
FAT DIGESTION				
Lipase	Pancreas	Small intestine	Triglycerides	FREE FATTY ACIDS, MONOGLYCERIDES*
NUCLEIC ACID DIGESTION				
Pancreatic nucleases	Pancreas	Small intestine	DNA, RNA	NUCLEOTIDES*
Intestinal nucleases	Intestinal lining	Small intestine	Nucleotides	NUCLEOTIDE BASES, MONOSACCHARIDES*

* Breakdown products small enough to be absorbed into the internal environment.

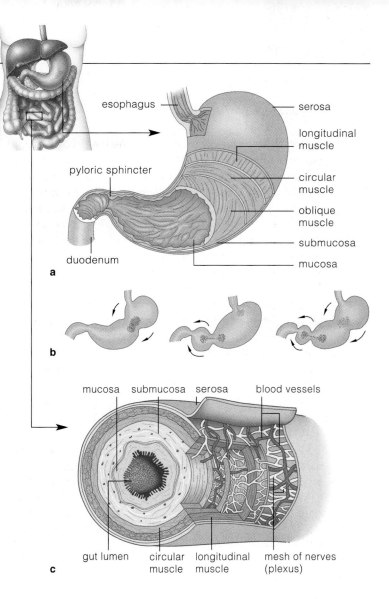

Figure 37.9 (**a**) Stomach structure. (**b**) Peristaltic wave down the stomach. (**c**) Small intestine structure. The gut wall usually has an innermost mucosa: epithelium and an underlying connective tissue layer. Next is the submucosa, a connective tissue layer with blood and lymph vessels and a mesh of nerves that locally control digestion. Next are smooth muscle layers that differ in orientation, hence in the direction of contraction. The gut wall's outermost layer of connective tissue is the serosa.

pepsin secreted from cells in the stomach lining, two pancreatic enzymes (trypsin and chymotrypsin) digest proteins to peptides. Another cleaves peptides to free amino acids. The pancreas also secretes bicarbonate, a buffer that helps neutralize HCl from the stomach.

The Role of Bile in Fat Digestion

Besides enzymes, fat digestion requires **bile**. This fluid, which the liver secretes continually, contains bile salts, bile pigments, cholesterol, and lecithin, a phospholipid. One of the pigments is iron-rich bilirubin from degraded red blood cells. After the stomach empties, a sphincter closes the main bile duct from the liver. Bile backs up into the gallbladder, which stores and concentrates it.

By a process of **emulsification**, bile salts accelerate fat digestion. Most fats in food are triglycerides. Being insoluble in water, the triglycerides tend to cluster into large fat globules in chyme. But the intestinal wall has muscle layers (Figure 37.9c). As these move and agitate chyme, the fat globules break apart into small droplets that get coated with bile salts. Bile salts carry negative charges, so these coated droplets repel each other and stay separated. The suspension of fat droplets, formed by mechanical and chemical action, is the "emulsion."

Compared with fat globules, the emulsion droplets offer the fat-digesting enzymes a much greater surface area. And so the enzymes can break down triglycerides more rapidly to fatty acids and monoglycerides.

Controls Over Digestion

Homeostatic controls counter changes in the internal environment. By contrast, controls over digestion act *before* food is absorbed into the internal environment. The nervous system, endocrine system, and meshes of nerves (plexuses) in the gut wall exert the control.

When food enters the stomach, it distends the wall and stimulates mechanoreceptors. Signals travel along short reflex pathways to smooth muscles and glands (longer reflex pathways also signal the brain). As one response, gut wall muscles contract and glandular cells secrete enzyme-rich fluid into the lumen or hormones into blood. The response depends partly on the chyme volume and composition. A large meal activates more mechanoreceptors in the stomach wall, so contractions

become more forceful and the stomach empties faster. High acidity or a high fat content in the small intestine triggers hormonal secretions that lead to a slowdown in stomach emptying, so chyme is not moved forward faster than it can be processed. Fear, depression, and other emotional upsets also trigger slowdowns.

What are the gastrointestinal hormones? If chyme contains amino acids and peptides, cells of the stomach lining will secrete the gastrin that induces secretion of acid into the stomach. Secretin induces the pancreas to secrete bicarbonate. CCK (cholecystokinin) calls for the secretion of pancreatic enzymes and contraction of the gallbladder. When the small intestine holds glucose and fat, GIP (glucose insulinotropic peptide) causes cells to take up more glucose by stimulating insulin secretion.

Carbohydrate breakdown starts in the mouth, and protein breakdown starts in the stomach. In the small intestine, most large organic compounds are digested to molecules small enough to be absorbed into the internal environment.

Signals from the nervous system and a nerve plexus in the gut wall, along with endocrine secretions, control digestion.

ABSORPTION IN THE SMALL INTESTINE

Structure Speaks Volumes About Function

Most nutrient absorption occurs in the small intestine. (The stomach absorbs few substances, alcohol being one of them.) Figure 37.10 offers glimpses into the structure of the small intestine's wall. Focus first on the profuse folds of the mucosa that project into the lumen. Look closer and you see large numbers of tinier projections from each large fold. Look closer still and you see that epithelial cells at the surface of these tiny projections have a brushlike crown of great numbers of even tinier projections, all exposed to the intestinal lumen.

What is the significance of so much convolution? It has a highly favorable surface-to-volume ratio (Section 4.1). Collectively, all the projections from the intestinal mucosa ENORMOUSLY increase the surface area available to interact with chyme and absorb nutrients. Without that immense surface area, absorption would proceed hundreds of times more slowly, and that would not be enough to sustain human life.

Figure 37.10*c,d* shows **villi** (singular, villus), which are absorptive structures on mucosal folds. Each villus is about one millimeter long. Millions project from the mucosa. Their sheer density gives the mucosa a velvety appearance. Inside each villus is an arteriole, a venule, and a lymph vessel that function in moving substances to and from the general circulation (Figure 37.10*e*).

Most cells in the epithelial lining of a villus have **microvilli** (singular, microvillus), which are ultrafine, threadlike projections from their free surface. Each of these cells has 1,700 or so microvilli. Hence the name, "brush border" cell. Glandular cells in the lining make and secrete digestive enzymes (Table 37.1), Also, some phagocytic cells patrol and help protect the lining.

Figure 37.10 (**a**,**b**) Structure of the mammalian small intestine. Notice the circular folds of the intestinal mucosa, which are permanent. (**c**) Projecting from the free surface of each fold are many absorptive structures called villi. (**d**) Fine structure of one villus. Monosaccharides and most amino acids that cross the intestinal lining enter blood capillaries in the villus. Fats enter the lymph vessels. (**e**) Epithelial cell types at the free surface of a villus. The absorptive cell has a thick crown of microvilli facing the intestinal lumen.

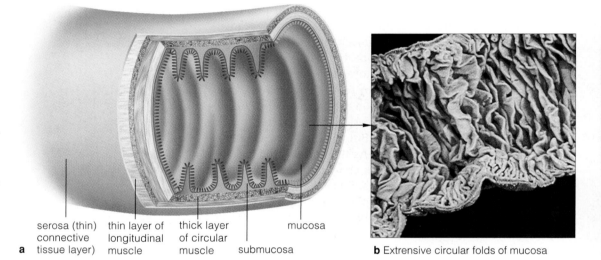

a serosa (thin connective tissue layer) thin layer of longitudinal muscle thick layer of circular muscle submucosa mucosa

b Extensive circular folds of mucosa

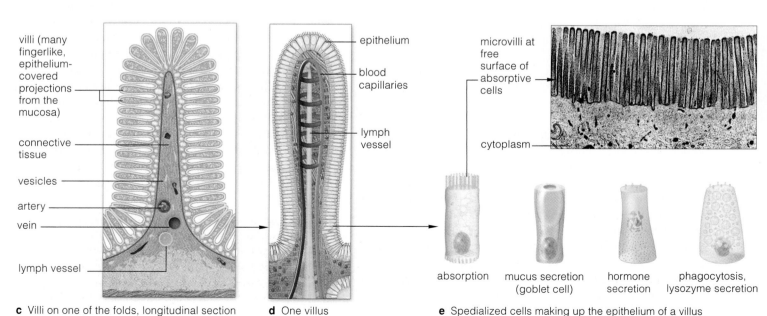

villi (many fingerlike, epithelium-covered projections from the mucosa)

connective tissue

vesicles

artery

vein

lymph vessel

epithelium

blood capillaries

lymph vessel

microvilli at free surface of absorptive cells

cytoplasm

absorption   mucus secretion (goblet cell)   hormone secretion   phagocytosis, lysozyme secretion

c Villi on one of the folds, longitudinal section   **d** One villus   **e** Spedialized cells making up the epithelium of a villus

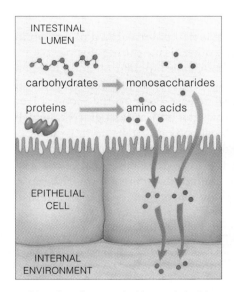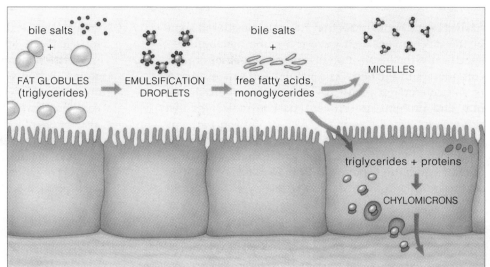

a Digestion of carbohydrates to monosaccharides, and proteins to amino acids, is completed with enzymes secreted by the pancreas and by cells of the epithelial lining.

b Monosaccharides and amino acids are actively transported across the plasma membrane of the epithelial cells, then out of the same cells and into the internal environment.

c Emulsification: The constant movement of the intestinal wall breaks up fat globules into small emulsification droplets. Bile salts prevent the fat globules from re-forming. Pancreatic enzymes digest the droplets to fatty acids and monoglycerides.

d Micelles form as bile salts combine with digestion products and phospholipids. Products can readily slip into and out of the micelles.

e Concentration of monoglycerides and fatty acids in micelles enhances gradients that lead to diffusion of both substances across the lipid bilayer of the plasma membrane of cells making up the lining.

f Products of fat digestion reassemble into triglycerides in cells of the intestinal lining. These become coated with proteins, then are expelled (by exocytosis from the cells) into the internal environment.

Figure 37.11 Digestion and absorption in the small intestine.

What Are the Absorption Mechanisms?

Absorption, recall, is the passage of nutrients, water, salts, and vitamins into the internal environment. The intestinal wall's large absorptive surface facilitates the process, but so does the action of its smooth muscle. In **segmentation**, rings of circular muscle contract and relax repeatedly. This creates an oscillating (back and forth) movement that constantly mixes and forces the lumen contents against the wall's absorptive surface:

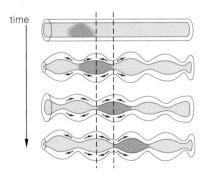

By the time the meal is halfway through the small intestine, most is broken apart and digested. Molecules of water cross the intestinal lining by osmosis. Mineral ions are selectively absorbed. Transport proteins in the plasma membrane of brush border cells actively shunt some breakdown products, including monosaccharides and amino acids, across the lining. By contrast, being lipid-soluble, fatty acids and monoglycerides diffuse right across the plasma membrane's lipid bilayer. And bile salts help them do this (Figure 37.11).

By a process called **micelle formation**, the bile salts combine with fat digestion products and phospholipids to form tiny droplets (micelles). The product molecules in the micelles continuously exchange places with those dissolved in chyme. However, the micelles concentrate them next to the intestinal lining, and when they are concentrated enough, gradients promote their diffusion out of the micelles and into epithelial cells. Fatty acids and monoglycerides recombine inside the cells to form triglycerides. Then triglycerides combine with proteins into particles (chylomicrons) that leave the cells by way of exocytosis and enter the internal environment.

Once absorbed, the glucose and amino acids enter blood vessels directly. Triglycerides enter lymph vessels that eventually drain into blood vessels.

With its richly folded intestinal mucosa, millions of villi, and hundreds of millions of microvilli, the small intestine has a vast surface area for absorbing nutrients.

Substances pass through the brush border cells that line the free surface of each villus by active transport, osmosis, and diffusion across the lipid bilayer of plasma membranes.

DISPOSITION OF ABSORBED ORGANIC COMPOUNDS

Earlier in the book, in Section 7.6, you considered some of the mechanisms that govern organic metabolism—specifically, the disposition of glucose and other organic compounds in the body as a whole. You saw examples of the conversion pathways by which carbohydrates, fats, and proteins are broken apart to molecules that serve as intermediates in the ATP-producing pathway of aerobic respiration. Here, Figure 37.12 rounds out the picture by showing all the major routes by which organic compounds obtained from food can be shuffled and reshuffled in the body as a whole.

energy source. There is no net breakdown of protein in muscle tissue or any other tissue during this period.

Between meals, body cells tap their fat and glycogen stores. Adipose tissue cells dismantle fats to glycerol and fatty acids, and release these to the blood. Liver cells break apart glycogen and release glucose, which also enters blood. Most body cells take up fatty acids as well as glucose and use them for ATP production.

Bear in mind, the liver does more than store and convert organic compounds to different forms. Example: it helps maintain their concentrations in blood. It also

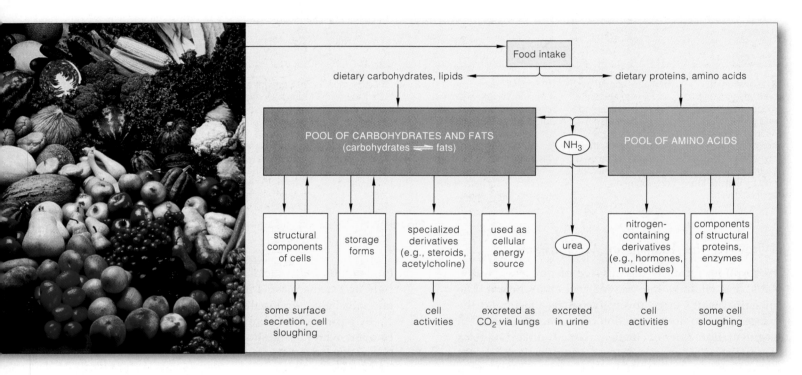

Figure 37.12 Summary of the major pathways of organic metabolism. Cells continually synthesize and tear down carbohydrates, fats, and proteins. Urea forms mainly in the liver. See also Sections 8.6 and 32.7.

All living cells in the body continually break apart most of their carbohydrates, lipids, and proteins, then use the breakdown products as energy sources or as building blocks. The nervous and endocrine systems interact to integrate this massive molecular turnover, but a treatment of how they do this would be beyond the scope of this book.

For now, simply reflect on a few key points about organic metabolism. When you eat, your body adds to its pools of materials. Excess amounts of carbohydrates and other organic compounds absorbed from the gut are transformed mostly into fats, which become stored in adipose tissue. Some are converted to glycogen in the liver and muscle tissue. *While* compounds are being absorbed and stored, most cells use glucose as the main

inactivates most hormone molecules, which are sent to the kidneys for excretion, in urine. The liver removes worn-out blood cells from the circulation and shuts down many potentially toxic compounds. For example, ammonia (NH_3), produced by amino acid breakdown, can be toxic in high concentrations. The liver converts NH_3 to urea, which is a less toxic waste product. Urea is excreted from the kidneys, as a component of urine.

Right after a meal, cells take up the glucose being absorbed from the gut and use it as a quick energy source. Excess amounts of glucose and other organic compounds become converted mainly to fats, which get stored in adipose tissue. Some of the excess is converted to glycogen, which gets stored mainly in the liver and in muscle tissue.

Between meals, cells tap fat reservoirs as their main energy source. Fats are converted to glycerol and fatty acids, both of which enter ATP-producing pathways.

THE LARGE INTESTINE

What happens to the material *not* absorbed in the small intestine? It moves into the large intestine, or **colon**. The colon concentrates and stores feces: a mixture of water, undigested and unabsorbed matter, and bacteria. The colon starts at a cup-shaped pouch (cecum). It ascends on the right side of the abdominal cavity and continues across to the other side. Then it descends and connects with a short tube, the rectum (Figures 37.6 and 37.13).

Colon Functioning

Material becomes concentrated as water moves across the colon's lining. Cells of the lining actively transport sodium ions out of the lumen. Ion concentrations fall, which increases the water concentration, so water also moves out of the lumen, by osmosis. These same cells secrete mucus that lubricates feces and helps keep them from mechanically damaging the colon wall. They also secrete bicarbonate, which buffers acidic fermentation products of ingested bacteria that managed to colonize the colon. The colonizers generally are harmless unless they breach the wall and enter the abdominal cavity.

Short, longitudinal bands of smooth muscle in the colon wall are gathered at their ends, like a series of full skirts nipped in at elastic waistbands. As they contract and relax, they move the lumen's contents back and forth against the colon wall's absorptive surface. Nerve plexuses largely control the motion, which is similar to segmentation in the small intestine but much slower.

Whereas the quick transit time through the small intestine does not favor the rapid population growth of ingested bacteria, the colon's slower motion favors it. The volume of cellulose fibers and other undigested material that cannot be decreased by absorption in the colon is called **bulk**. It adds to the volume of material and influences the transit time through the colon.

Following each new meal, secretion of gastrin and commands from the nervous system (along autonomic nerves) stimulate large portions of the ascending and transverse colon to contract together. Within seconds, the lumen's contents move as much as three-fourths of the colon's length, thus making way for incoming food.

The contents are stored in the last part of the colon until they distend the rectal wall enough to trigger a reflex action that leads to their expulsion. The nervous system controls defecation by stimulating or inhibiting the contraction of a muscle sphincter at the anus, the terminal opening of the gut.

Colon Malfunctioning

The normal frequency of defecation ranges from three times a day to once a week. Aging, emotional stress, a low-bulk diet, injury, or disease can result in delayed

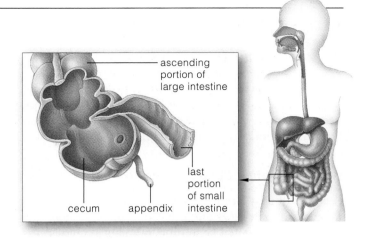

Figure 37.13 Cecum and appendix of the large intestine (colon).

defecation, called *constipation*. The longer the delay, the more water gets absorbed, so feces harden and dry out. The abdominal discomfort is accompanied by loss of appetite, headaches, and often nausea and depression.

Hard feces may become lodged inside the **appendix**, a narrow projection from the cecum (Figure 37.13). The appendix has no known digestive functions, but, like the ileum, it holds a concentration of bacteria-fighting white blood cells. Feces that obstruct the normal blood flow and mucus secretion to this projection can cause *appendicitis*, or an inflamed appendix. Unless removed, an inflamed appendix can rupture. Then bacteria, even otherwise harmless ones, would be free to enter the abdominal cavity and cause life-threatening infections.

The colon also is vulnerable to cancer, which occurs most often among the world's wealthiest and best-fed populations. Too many of these people skip meals, eat too much and too fast when they do sit down at the table, and generally give their gut erratic workouts. Their diet tends to be rich in sugar, cholesterol, and salt, and low in bulk. Too little bulk extends the transit time of feces through the colon. The longer that irritating, potentially carcinogenic material is in contact with the colon wall, the more damage it may cause. Symptoms of *colon cancer* include a change in bowel functioning, rectal bleeding, and blood in feces. Some people appear to be genetically predisposed to develop colon cancer, but a diet too low in fiber might also be a factor.

Colon cancer is almost nonexistent in rural India and Africa, where people can't afford to eat much more than fiber-rich whole grains. When these same people move to cities in the affluent nations and change their eating habits, the incidence of colon cancer increases.

The large intestine, or colon, functions in the absorption of water and mineral ions from the gut lumen. It also functions in the compaction of undigested residues into feces, for expulsion at the terminal opening of the gut.

HUMAN NUTRITIONAL REQUIREMENTS

You grow and maintain yourself by eating foods that are appropriate sources of energy and raw materials. Nutritionists measure food energy in **kilocalories**. A kilocalorie is 1,000 calories of heat energy, the amount needed to raise the temperature of 1 kilogram of water by 1°C. Raw materials in food will now be described.

New, Improved Food Pyramids

A few million years ago, our hominid ancestors dined mostly on fresh fruits and other fibrous plant material. From that nutritional beginning, humans in many parts of the world came to prefer low-fiber, high-fat foods. They still do even when they study **food pyramids**, or charts of well-balanced diets, from elementary school onward. Nutritionists revise the charts on an ongoing basis. Figure 37.14 is one version of daily portions for an average-sized adult male, expressed as percentages of energy intake. It suggests 55–60 percent complex carbohydrates, 15–20 percent proteins (less for females), and 20–25 percent fats. Alternative pyramids abound. Example: The *Mediterranean diet* pyramid has a large

base of grain products, too. But fruits and vegetables are the next largest groups, then legumes and nuts, then *olive oil* as a fat group, followed by cheese and yogurt. This pyramid limits the weekly intake of fish, poultry, eggs, simple sugars, and—at its tiny top—red meat. Olive oil is as much as 40 percent of the total energy intake. But it's monounsaturated and less likely than saturated fats to raise cholesterol levels. It also is a fine antioxidant, good at eliminating free radicals.

Carbohydrates

Starch and, to a lesser extent, glycogen should be the main carbohydrates in the human diet. These complex carbohydrates are easily degraded to glucose, the body's main energy source. Starch is abundant in fleshy fruits, cereal grains, and legumes, including beans and peas.

Foods rich in complex carbohydrates have another advantage: they typically are high in fiber. Section 37.7 touched upon the consequences of a low-fiber diet. As epidemiologist Denis Burkitt put it, when people pass small volumes of feces, you have large hospitals.

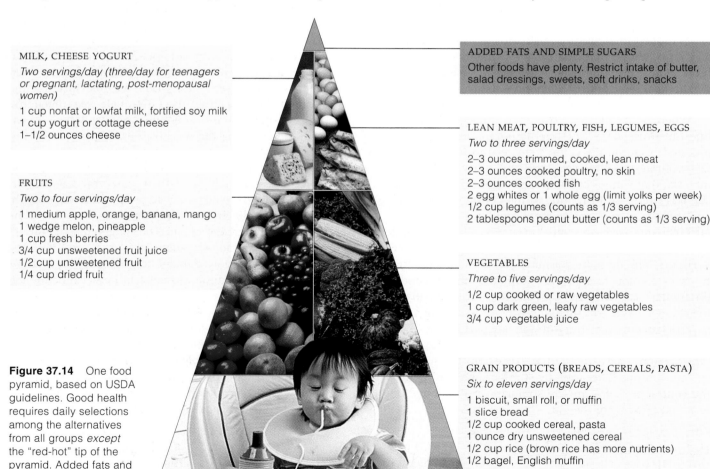

MILK, CHEESE YOGURT

Two servings/day (three/day for teenagers or pregnant, lactating, post-menopausal women)

1 cup nonfat or lowfat milk, fortified soy milk
1 cup yogurt or cottage cheese
1–1/2 ounces cheese

FRUITS

Two to four servings/day

1 medium apple, orange, banana, mango
1 wedge melon, pineapple
1 cup fresh berries
3/4 cup unsweetened fruit juice
1/2 cup unsweetened fruit
1/4 cup dried fruit

ADDED FATS AND SIMPLE SUGARS
Other foods have plenty. Restrict intake of butter, salad dressings, sweets, soft drinks, snacks

LEAN MEAT, POULTRY, FISH, LEGUMES, EGGS

Two to three servings/day

2–3 ounces trimmed, cooked, lean meat
2–3 ounces cooked poultry, no skin
2–3 ounces cooked fish
2 egg whites or 1 whole egg (limit yolks per week)
1/2 cup legumes (counts as 1/3 serving)
2 tablespoons peanut butter (counts as 1/3 serving)

VEGETABLES

Three to five servings/day

1/2 cup cooked or raw vegetables
1 cup dark green, leafy raw vegetables
3/4 cup vegetable juice

GRAIN PRODUCTS (BREADS, CEREALS, PASTA)

Six to eleven servings/day

1 biscuit, small roll, or muffin
1 slice bread
1/2 cup cooked cereal, pasta
1 ounce dry unsweetened cereal
1/2 cup rice (brown rice has more nutrients)
1/2 bagel, English muffin

Figure 37.14 One food pyramid, based on USDA guidelines. Good health requires daily selections among the alternatives from all groups *except* the "red-hot" tip of the pyramid. Added fats and simple sugars pile on calories but have few vitamins and minerals.

No Limiting Amino Acid	Low Lysine	Low Methionine, Cysteine	Low Tryptophan
Legumes: soybeans (e.g., tofu, soy milk) Grains: wheat germ Milk Cheeses (not cream) Yogurt Eggs (Meats)	Legumes: peanuts Grains: barley, rice, buckwheat, oats, corn, rye, wheat Nuts: almonds, cashews, coconut, walnuts, hazelnuts, pecans Seeds: pumpkin, sunflower	Legumes: beans (dried), black-eyed peas, garbanzos, lentils, lima beans, mung beans, peanuts Nuts: hazelnuts Fresh vegetables: asparagus, green peas, broccoli, mushrooms, parsley, potatoes, soybeans, Swiss chard	Legumes: beans (dried), garbanzos, lima beans, mung beans, peanuts Grains: cornmeal Nuts: almonds, English walnuts Fresh vegetables: corn, green peas, mushrooms, Swiss chard

Figure 37.15 Vegetarian diets. All eight essential amino acids, a small portion of the total protein intake, must be available at the same time, in certain amounts, if cells are to build their own proteins. Milk and eggs are among the *complete* proteins; they have high amounts of all eight in proportions that humans require.

Nearly all plant proteins are *incomplete*. Vegetarians must be careful to avoid protein deficiency. For instance, they might combine different foods from the three columns of incomplete proteins shown. Strict vegetarians (who also avoid dairy products and eggs) must take vitamins B_{12} and B_2 (riboflavin) supplements. Animal protein is a luxury in most traditional societies. Their cuisines have good combinations of plant proteins, including rice/beans, chili/cornbread, tofu/rice, and lentils/wheat bread.

Unlike complex carbohydrates, the simple sugars don't offer fiber. Unlike whole foods, they don't offer vitamins and minerals. Yet each week, we in the United States typically consume up to two pounds of refined sugar (sucrose). Is this a far-fetched statistic? No. Look critically at the ingredients listed on the containers for dry cereals, frozen foods, soft drinks, and many other prepared foods. Sugars often are "hidden" in labels as corn syrup, corn sweeteners, dextrose, and so on.

Lipids

Your body can't function without fats and other lipids. For example, cell membranes require the phospholipid lecithin; and fats serve as energy reserves, cushion many internal organs, and act as insulation beneath the skin. Dietary fats also help you store fat-soluble vitamins. But the body can synthesize most of its own fats, including cholesterol, from proteins and carbohydrates. The ones it can't make are **essential fatty acids**, but whole foods offer plenty of these. Linoleic acid is an example. One teaspoon each day of olive oil, corn oil, or some other polyunsaturated fat in food provides enough of it.

On average, butter and other fats now make up 40 percent of the American diet. It should be 30 percent or less, according to most of the medical community.

Like other animal fats, butter is a saturated fat that tends to raise the level of cholesterol in the blood. As described in earlier chapters, cholesterol is used in the synthesis of bile acids and steroid hormones, and it is a required component of animal cell membranes. But too much cholesterol often interferes seriously with blood circulation (Section 34.9). Certain potato chips and other snack foods are made with edible but nondigestible oils such as *sucrose polyester*. They are aimed at people who crave fats enough to put up with the digestive upsets.

Proteins

Amino acid components of dietary proteins are needed for the body's own protein-building programs. Of the twenty common types, eight are **essential amino acids**.

That is, our body cells cannot synthesize them, so we must get enough of them from food. The eight amino acids are either methionine or cysteine (its metabolic equivalent), isoleucine, leucine, lysine, phenylalanine (or tyrosine), threonine, tryptophan, and valine.

Most animal proteins are *complete*: their amino acid ratios match human nutritional needs. Plant proteins are *incomplete*; they lack one or more essential amino acids. To get enough amino acids, vegetarians must eat certain combinations of different plants (Figure 37.15).

A measure called the net protein utilization (NPU) is used to compare proteins from different sources. NPU values range from 100 (all essential amino acids occur in ideal proportions) to 0 (one or more amino acids are absent; when eaten alone, the protein is not complete).

Protein-energy malnutrition (PEM), the world's most prevalent form of malnutrition, affects more than 500 million children today. Enzymes and other proteins are crucial for the body's structure and functions. PEM is especially worrisome for fetuses and infants. The brain grows and develops most rapidly early in life. Unless a mother-to-be takes in enough protein, especially just before and after birth, her child can expect irreversible mental retardation.

Ironically, people in developed countries often get too much protein, thus inviting acidosis, dehydration, diarrhea, fever, and excess ammonia and urea in blood. Diets high in purified protein may also promote calcium loss from bone. Fluid intake and kidney functions set the upper limit for safe protein intake. To calculate the most reasonable intake, use the guidelines in Section 37.10 to find the midpoint for your acceptable weight range. Assuming you are in good health, convert the pounds to kilograms (divide them by 2.2), and multiply the value by 0.8 (or by 0.9 for males up to 18 years old). This will give the energy intake in grams per day.

Individuals grow and maintain themselves in good health by eating certain amounts of complex carbohydrates, lipids, and proteins that provide them with adequate raw materials and energy, as measured in kilocalories.

VITAMINS AND MINERALS

Vitamins are *organic* substances that are essential for growth and survival, for no other substances can play their metabolic roles. Most plants are able to synthesize vitamins on their own. Most kinds of animals lost the ability to do so and must get vitamins from food.

At the minimum, human cells require the thirteen vitamins listed in Table 37.2. Each vitamin has specific metabolic functions. Many reactions use several types, and the absence of one affects the functions of others.

Minerals are *inorganic* substances that are essential for growth and survival; no other substances can play their metabolic roles. For instance, all of your cells need iron for their electron transport chains. Red blood cells don't function at all without iron in hemoglobin. And neurons stop functioning in the absence of sodium and potassium (Table 37.3).

People who are in good health get all the vitamins and minerals they need from a balanced diet of whole

Table 37.2 *Vitamins: Sources, Functions, and Effects of Deficiencies or Excesses**

Vitamin	Common Sources	Main Functions	Effects of Chronic Deficiency	Effects of Extreme Excess
FAT-SOLUBLE VITAMINS				
A	Its precursor comes from beta-carotene in yellow fruits, yellow or green leafy vegetables; also in fortified milk, egg yolk, fish liver	Used in synthesis of visual pigments, bone, teeth; maintains epithelia	Dry, scaly skin; lowered resistance to infections; night blindness; permanent blindness	Malformed fetuses; hair loss; changes in skin; liver and bone damage; bone pain
D	D_3 formed in skin and in fish liver oils, egg yolk, fortified milk; converted to active form elsewhere	Promotes bone growth and mineralization; enhances calcium absorption	Bone deformities (rickets) in children; bone softening in adults	Retarded growth; kidney damage; calcium deposits in soft tissues
E	Whole grains, dark green vegetables, vegetable oils	Counters effects of free radicals; helps maintain cell membranes; blocks breakdown of vitamins A and C in gut	Lysis of red blood cells; nerve damage	Muscle weakness, fatigue, headaches, nausea
K	Enterobacteria form most of it; also in green leafy vegetables, cabbage	Blood clotting; ATP formation via electron transport	Abnormal blood clotting; severe bleeding (hemorrhaging)	Anemia; liver damage and jaundice
WATER-SOLUBLE VITAMINS				
B_1 (thiamin)	Whole grains, green leafy vegetables, legumes, lean meats, eggs	Connective tissue formation; folate utilization; coenzyme action	Water retention in tissues; tingling sensations; heart changes; poor coordination	None reported from food; possible shock reaction from repeated injections
B_2 (riboflavin)	Whole grains, poultry, fish, egg white, milk	Coenzyme action	Skin lesions	None reported
Niacin	Green leafy vegetables, potatoes, peanuts, poultry, fish, pork, beef	Coenzyme action	Contributes to pellagra (damage to skin, gut, nervous system, etc.)	Skin flushing; possible liver damage
B_6	Spinach, tomatoes, potatoes, meats	Coenzyme in amino acid metabolism	Skin, muscle, and nerve damage; anemia	Impaired coordination; numbness in feet
Pantothenic acid	In many foods (meats, yeast, egg yolk especially)	Coenzyme in glucose metabolism, fatty acid and steroid synthesis	Fatigue, tingling in hands, headaches, nausea	None reported; may cause diarrhea occasionally
Folate (folic acid)	Dark green vegetables, whole grains, yeast, lean meats; enterobacteria produce some folate	Coenzyme in nucleic acid and amino acid metabolism	A type of anemia; inflamed tongue; diarrhea; impaired growth; mental disorders	Masks vitamin B_{12} deficiency
B_{12}	Poultry, fish, red meat, dairy foods (not butter)	Coenzyme in nucleic acid metabolism	A type of anemia; impaired nerve function	None reported
Biotin	Legumes, egg yolk; colon bacteria produce some	Coenzyme in fat, glycogen formation and in amino acid metabolism	Scaly skin (dermatitis), sore tongue, depression, anemia	None reported
C (ascorbic acid)	Fruits and vegetables, especially citrus, berries, cantaloupe, cabbage, broccoli, green pepper	Collagen synthesis; possibly inhibits effects of free radicals; structural role in bone, cartilage, and teeth; role in carbohydrate metabolism	Scurvy, poor wound healing, impaired immunity	Diarrhea, other digestive upsets; may alter results of some diagnostic tests

* Guidelines for appropriate daily intakes are being worked out by the Food and Drug Administration.

Table 37.3 *Major Minerals: Sources, Functions, and Effects of Deficiencies or Excesses***

Mineral	Common Sources	Main Functions	Signs of Severe Long-Term Deficiency	Signs of Extreme Excess
Calcium	Dairy products, dark green vegetables, dried legumes	Bone, tooth formation; blood clotting; neural and muscle action	Stunted growth; possibly diminished bone mass (osteoporosis)	Impaired absorption of other minerals; kidney stones in susceptible people
Chloride	Table salt (usually too much in diet)	HCl formation in stomach; contributes to body's acid–base balance; neural action	Muscle cramps; impaired growth; poor appetite	Contributes to high blood pressure in susceptible people
Copper	Nuts, legumes, seafood, drinking water	Used in synthesis of melanin, hemoglobin, and some transport chain components	Anemia, changes in bone and blood vessels	Nausea, liver damage
Fluorine	Fluoridated water, tea, seafood	Bone, tooth maintenance	Tooth decay	Digestive upsets; mottled teeth and deformed skeleton in chronic cases
Iodine	Marine fish, shellfish, iodized salt, dairy products	Thyroid hormone formation	Enlarged thyroid (goiter), with metabolic disorders	Goiter
Iron	Whole grains, green leafy vegetables, legumes, nuts, eggs, lean meat, molasses, dried fruit, shellfish	Formation of hemoglobin and cytochrome (transport chain component)	Iron-deficiency anemia, impaired immune function	Liver damage, shock, heart failure
Magnesium	Whole grains, legumes, nuts, dairy products	Coenzyme role in ATP–ADP cycle; roles in muscle, nerve function	Weak, sore muscles; impaired neural function	Impaired neural function
Phosphorus	Whole grains, poultry, red meat	Component of bone, teeth, nucleic acids, ATP, phospholipids	Muscular weakness; loss of minerals from bone	Impaired absorption of minerals into bone
Potassium	Diet provides ample amounts	Muscle and neural function; roles in protein synthesis and body's acid–base balance	Muscular weakness	Muscular weakness, paralysis, heart failure
Sodium	Table salt; diet provides ample to excessive amounts	Key role in body's salt–water balance; roles in muscle and neural function	Muscle cramps	High blood pressure in susceptible people
Sulfur	Proteins in diet	Component of body proteins	None reported	None likely
Zinc	Whole grains, legumes, nuts, meats, seafood	Component of digestive enzymes; roles in normal growth, wound healing, sperm formation, and taste and smell	Impaired growth, scaly skin, impaired immune function	Nausea, vomiting, diarrhea; impaired immune function and anemia

* The guidelines for appropriate daily intakes are being worked out by the Food and Drug Administration.

foods. Generally, vitamin and mineral supplements are necessary only for strict vegetarians, the elderly, and people suffering a chronic illness or taking medication that interferes with utilization of specific nutrients. As examples, vitamin K supplements aid calcium retention and lessen osteoporosis in elderly women. Vitamin C, vitamin E, and the beta-carotene precursor for vitamin A may lessen some aging effects and improve immune functioning by inactivating free radicals. A free radical, remember, is an atom or group of atoms that is highly reactive because it has an unpaired electron.

However, metabolism varies in its details from one person to the next, so no one should take massive doses of any vitamin or mineral supplement except under medical supervision. Also, excessive amounts of many

vitamins and minerals can harm everyone (Tables 37.2 and 37.3). Large doses of vitamins A and D are good examples. As is the case for all fat-soluble vitamins, excess amounts of these vitamins accumulate in tissues and interfere with normal metabolic activity. Similarly, sodium is present in plant and animal tissues and is a component of table salt. Sodium has roles in the body's salt–water balance, muscle activity, and nerve function. However, prolonged, excessive intake of sodium may contribute to high blood pressure in some people.

Severe shortages or self-prescribed, massive excesses of vitamins and minerals can disturb the delicate balances in body function that promote health.

Weighty Questions, Tantalizing Answers

Have you ever asked yourself *why* you want to weigh a certain amount? Are you merely fearful of "being fat"— that is, obese? By definition, **obesity** is an excess of fat in adipose tissues, most often caused by imbalances between caloric intake and energy output. One standard of what constitutes obesity is simply cultural, and it varies from one culture to the next. For example, one female student despaired over her plumpness until her graduate studies program in Africa, where many males found her to be one of the most desirable females on the planet. However, a lot depends on how healthy you want to be and how long you want to live. Thinner people generally live longer.

In the United States, obesity is now the second leading cause of death; 300,000 or so people die annually from preventable, weight-related conditions. Each year, the weight-related cases of type 2 diabetes, heart disease, hypertension, breast cancer, colon cancer, gout, gallstones, and osteoarthritis add up to a 100-billion-dollar drain on the economy. The future doesn't look any rosier: children today are 42 percent fatter than they were in 1980.

IN PURSUIT OF THE "IDEAL" WEIGHT Figure 37.16 shows charts for estimating the "ideal" weight for adults. Another indicator of obesity-related health risk is the *body mass index* (BMI), as determined by this formula:

$$\text{BMI} = \frac{\text{weight (pounds)} \times 700}{\text{height (inches)}^2}$$

Shape Up America, a national initiative for promoting health, will quickly do the calculation for you. Their web site is http://www.shapeup.org. If your BMI value is 27 or higher, the health risk rises dramatically. Other factors that affect the risk are smoking habits, family history of heart disorders, use of sex hormones after menopause, and fat distribution. Fat stored above the belt, as with "beer bellies," definitely is not good.

Dieting alone can't lower a BMI value. Eat less, and the body slows its metabolic rate to conserve energy. So how do you function normally over the long term while maintaining acceptable weight? *You must balance caloric intake with energy output.* For most of us, this means eating controlled portions of low-calorie, nutritious foods *and* exercising regularly.

To figure out how many kilocalories you should take in daily to maintain a desired weight, multiply that weight (in pounds) by 10 if you are not active physically, by 15 if moderately active, and by 20 if highly active. From the value you get this way, subtract the following amount:

Age:		Subtract:	
25–34			0
35–44			100
45–54			200
55–64			300
Over 65			400

For instance, if you want to weigh 120 pounds and are very active, 120 × 20 = 2,400 kilocalories. If you're thirty-five years old, then (2,400 − 100), or 2,300 kilocalories. Such calculations give a rough estimate of caloric intake. Other factors, including height, must also be considered. An active person 5 feet, 2 inches tall doesn't need as much energy as an active 6-footer who weighs just the same.

Figure 37.16 How to estimate the "ideal" weight for an adult. The values shown are consistent with a long-term Harvard study into the link between excessive weight and increased risk of cardiovascular disorders. Depending on certain factors (such as having a small, medium, or large skeletal frame), the "ideal" might vary by plus or minus 10 percent.

WEIGHT GUIDELINES FOR WOMEN

Starting with an ideal weight of 100 pounds for a woman who is 5 feet tall, add five additional pounds for each additional inch of height. Examples:

Height (feet)	Weight (pounds)
5' 2"	110
5' 3"	115
5' 4"	120
5' 5"	125
5' 6"	130
5' 7"	135
5' 8"	140
5' 9"	145
5' 10"	150
5' 11"	155
6'	160

WEIGHT GUIDELINES FOR MEN

Starting with an ideal weight of 106 pounds for a man who is 5 feet tall, add six additional pounds for each additional inch of height. Examples:

Height (feet)	Weight (pounds)
5' 2"	118
5' 3"	124
5' 4"	130
5' 5"	136
5' 6"	142
5' 7"	148
5' 8"	154
5' 9"	160
5' 10"	166
5' 11"	172
6'	178

MY GENES MADE ME DO IT   On the average, one out of three Americans has far more trouble keeping off excess weight than others do, and genes have a lot to do with it. Researchers suspected as much for a long time. Studies of identical twins offered clues. (*Identical twins* are born with identical genes.) As an outcome of family problems, some of the twins had been separated at birth and were raised apart, in different households. Yet, at adulthood, the separated twins had similar body weights! People, it seemed, were born with a "set point" for body fat. Could it be that whatever set point someone inherits is the one he or she will be stuck with for life?

Many experiments confirmed this suspicion. The first started in 1950 with a seriously obese mouse (Figure 37.17). In 1995, molecular geneticists identified one of its genes that affects the set point for body fat. They named it the ***ob* gene** (guess why). Which clues led to its discovery? The nucleotide sequence in the gene region of obese mice differs from that in normal mice. The gene is active in adipose tissue, nowhere else. Its biochemical message translates into a protein released from cells and picked up by the bloodstream; the gene specifies a hormone.

Jeffrey Friedman discovered the hormone in 1994 and called it **leptin**. A nearly identical hormone was isolated in humans. Leptin is just one of a number of factors that mediate the brain's commands to suppress or whip up appetite, but it's a crucial one. By assessing the blood levels of incoming hormonal signals, an appetite control center in the hypothalamus "decides" whether the body has taken in enough food to provide enough fat for the day. If so, hypothalamic commands go out to increase metabolic rates—and to stop eating.

If the *ob* gene is mutated, then its expression alone may be enough to disrupt the control center so a mouse's appetite skyrockets and metabolic furnaces burn less. In one experiment that supported this hypothesis, obese mice were injected with leptin. The mice quickly shed their excess weight.

Will genetic researchers go on to develop therapies for obesity in humans? Maybe. As you read in Chapter 32, however, hormonal control is tricky business. It now appears that, besides its role in controlling body weight, leptin is a selective inhibitor of bone formation, as mediated by the central nervous system. Thus the danger is that anti-obesity therapies will bring on osteoporosis.

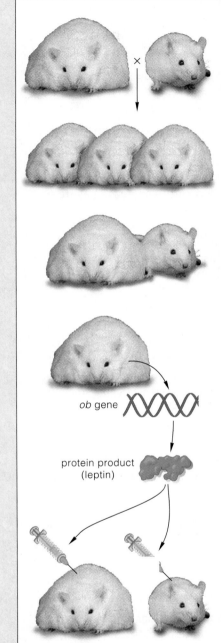

ob gene

protein product (leptin)

a *1950.* Researchers at the Jackson Laboratories in Maine notice that one of their laboratory mice is extremely obese, with an uncontrollable appetite. Through crossbreeding of this apparent mutant individual with a normal mouse, they produce a strain of obese mice.

b *Late 1960s.* Douglas Coleman of the Jackson Laboratories surgically joins the bloodstreams of an obese mouse and a normal one. The obese mouse now loses weight. Coleman hypothesizes that a factor circulating in blood may be influencing its appetite, but he is not able to isolate it.

c *1994.* Late in the year, Jeffrey Friedman of Rockefeller University discovers a mutated form of what is now called the *ob* gene in obese mice. Through DNA cloning and gene sequencing, he defines the protein that the mutated gene encodes. The protein, now called leptin, is a hormone that influences the brain's commands to suppress appetite and increase metabolic rates.

d *1995.* Three different research teams develop and use genetically engineered bacteria to produce leptin, which, when injected in obese and normal mice, triggers significant weight loss, apparently without harmful side effects.

Figure 37.17 Chronology of research developments that revealed the identity of a key factor in the genetic basis of body weight.

SUMMARY   *Gold* indicates text section

1. For animals, *nutrition* refers to processes by which the body takes in, digests, absorbs, and uses food. *CI*

2. Most animals have a complete digestive system—a tube that has two openings (a mouth and an anus), and regional specializations between them. Mucus-coated epithelium lines and protects every exposed surface of the tube and facilitates diffusion across its wall. *37.1*

3. These activities proceed in digestive systems: *37.1*
   a. Mechanical processing and motility: Movements that break up, mix, and propel food material.
   b. Secretion: Release of digestive enzymes and other substances from the pancreas and liver, as well as from glandular epithelium, into the gut lumen.
   c. Digestion: Breakdown of food into particles, then into nutrient molecules small enough to be absorbed.
   d. Absorption: The diffusion or transport of digested organic compounds, fluid, and ions from the gut lumen into the internal environment.
   e. Elimination: The expulsion of undigested as well as unabsorbed residues at the end of the system.

4. The human digestive system has a mouth, pharynx, esophagus, stomach, small intestine, large intestine (colon), rectum, and anus. The salivary glands and the liver, gallbladder, and pancreas have accessory roles in the system's functions (Table 37.4). *37.2–37.5, 37.7*

5. Starch digestion starts in the mouth, and protein digestion starts in the stomach. Digestion is completed and most nutrients are absorbed in the small intestine. The pancreas secretes the main digestive enzymes. Bile from the liver assists in fat digestion. *37.3*

6. The nervous and endocrine systems, and meshes of neurons (plexuses) in the gut wall, interact to govern the digestive system. Many controls operate in response to the volume and composition of food in the gut. They cause changes in muscle activity and in secretion rates for hormones and enzymes. *37.4*

7. In absorption, cells of the intestinal lining actively transport glucose and most amino acids out of the gut lumen. Fatty acids and monoglycerides diffuse across the lipid bilayer of these cells. They are recombined in the cytoplasm as triglycerides, which then are released by exocytosis, into interstitial fluid. *37.5–37.6*

8. Nutritionists advise a daily food intake in certain proportions. Example: One diet for healthy adult males of average body weight recommends 55–60 percent complex carbohydrates, 15–20 percent proteins (less for females), 20–25 percent fats. A well-balanced diet of whole foods normally supplies all necessary vitamins and minerals. *37.8–37.9*

9. To maintain health and a given body weight, caloric (energy) intake must balance energy output. *37.10*

Table 37.4	*Summary of the Digestive System*
MOUTH (oral cavity)	Start of digestive system, where food is chewed, moistened; polysaccharide digestion begins
PHARYNX	Entrance to tubular parts of digestive and respiratory systems
ESOPHAGUS	Muscular tube, moistened by saliva, that moves food from pharynx to stomach
STOMACH	Sac where food mixes with gastric fluid and protein digestion begins; stretches to store food taken in faster than can be processed; gastric fluid destroys many microbes
SMALL INTESTINE	The first part (duodenum) receives secretions from the liver, gallbladder, and pancreas
	Most nutrients are digested, absorbed in second part (jejunum)
	Some nutrients absorbed in last part (ileum), which delivers unabsorbed material to colon
COLON (large intestine)	Concentrates and stores undigested matter (by absorbing mineral ions and water)
RECTUM	Distension triggers expulsion of feces
ANUS	Terminal opening of digestive system

Accessory Organs:

SALIVARY GLANDS	Glands (three main pairs, many minor ones) that secrete saliva, a fluid with polysaccharide-digesting enzymes, buffers, and mucus (which moistens and lubricates ingested food)
PANCREAS	Secretes enzymes that digest all major food molecules and buffers against HCl from stomach
LIVER	Secretes bile (used in fat emulsification); roles in carbohydrate, fat, and protein metabolism
GALLBLADDER	Stores and concentrates bile from the liver

Review Questions

1. Define the five key tasks carried out by a complete digestive system. Then correlate some organs of such a system with the feeding behavior of a particular kind of animal. *37.1*

2. Using the sketch to the right, list the organs and accessory organs shown and the main functions of each. *Table 37.4*

3. With respect to digestive systems, define segmentation. Does segmentation occur in the stomach? Does it occur in the small intestine, colon, or both? *37.5, 37.7*

4. Using the *black* lines shown in Figure 37.18 as a guide, name the types of breakdown products small enough to be absorbed across the small intestine's lining, into the internal environment. *37.5*

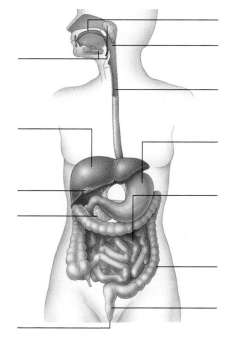

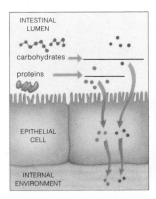

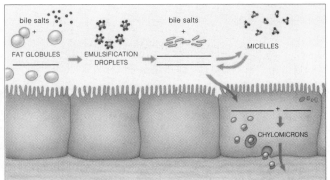

Figure 37.18 Fill in the blanks for substances that cross the lining of the small intestine.

Self-Quiz   ANSWERS IN APPENDIX III

1. The _____ maintains the internal environment, supplies cells with raw materials, and disposes of metabolic wastes.
   a. digestive system   d. urinary system
   b. circulatory system   e. interaction of all of the
   c. respiratory system   systems listed

2. Most digestive systems have regions for _____ food.
   a. transporting   c. storing
   b. processing   d. all of the above

3. _____ secretions do not assist in digestion and absorption.
   a. Salivary gland   c. Liver
   b. Thymus gland   d. Pancreas

4. Digestion is completed and most nutrients are absorbed in the _____ .
   a. mouth   c. small intestine
   b. stomach   d. colon

5. Bile has roles in _____ digestion and absorption.
   a. carbohydrate   c. protein
   b. fat   d. amino acid

6. Monosaccharides and most amino acids are absorbed _____ .
   a. by active transport   c. at lymph vessels
   b. by diffusion   d. as fat droplets

7. Most of our caloric intake should come from _____ .
   a. complex carbohydrates   c. proteins
   b. simple carbohydrates   d. lipids

8. On its own, the human body cannot produce all of the _____ it requires.
   a. vitamins and minerals   d. a through c
   b. fatty acids   e. a and c
   c. amino acids

9. Maintaining good health and normal body weight requires that _____ intake be balanced by _____ output.

10. Match the organ with its key digestive function(s).
   _____ gallbladder   a. secrete bile; many metabolic roles
   _____ stomach   b. digest, absorb most nutrients
   _____ colon   c. store, mix, dissolve food; start
   _____ pancreas   protein breakdown
   _____ salivary   d. store, concentrate bile
        gland   e. concentrate undigested matter
   _____ small   f. secrete substances that moisten food,
        intestine   start polysaccharide breakdown
   _____ liver   g. secrete digestive enzymes and
                            bicarbonate

Critical Thinking

1. A glassful of whole milk contains lactose, proteins, butterfat (mostly triglycerides), vitamins, and minerals. Explain what will happen to each component in your digestive tract.

2. As people age, the number of their body cells steadily decreases and energy needs decline. If you were planning an older person's diet, which foods would you emphasize, and why? Which ones would you deemphasize?

3. Using Section 37.10 as a reference, determine your ideal weight and then design a well-balanced program of diet and exercise that will help you achieve or maintain that weight.

4. Holiday meals often are larger than everyday ones and have a high fat content. After stuffing themselves at an early dinner on Thanksgiving Day, Richard and other members of his family feel uncomfortably full for the rest of the afternoon. Based on what you have learned about controls over digestion, propose a biochemical explanation for their discomfort.

5. Explain some of the ways in which your digestive system helps protect you against many pathogenic bacteria that may have contaminated the kinds of food you eat. (You may wish to refer to Section 35.1, also.)

6. Angelina is chronically anorexic. Clinical tests show that she has very low blood pressure and a below-average cardiac rate (bradycardia). The heart itself has become smaller. How did anorexia bring about these life-threatening conditions?

Selected Key Terms

appendix *37.7*
bile *37.4*
bulk *37.7*
chyme *37.4*
colon *37.7*
complete digestive system *37.1*
digestive system *CI*
emulsification *37.4*
esophagus *37.3*
essential amino acid *37.8*
essential fatty acid *37.8*
food pyramid *37.8*
gallbladder *37.4*
gastric fluid *37.4*
gut *37.2*
incomplete digestive system *37.1*
kilocalorie *37.8*
leptin *37.10*

liver *37.4*
micelle formation *37.5*
microvillus (microvilli) *37.5*
mineral *37.9*
nutrition *CI*
ob gene *37.10*
obesity *37.10*
pancreas *37.4*
pharynx *37.3*
ruminant *37.1*
saliva *37.3*
segmentation *37.5*
sphincter *37.3*
stomach *37.4*
tongue *37.3*
tooth *37.3*
villus (villi) *37.5*
vitamin *37.9*

Readings

Blaser, M. J. February 1996. "The Bacteria Behind Ulcers." *Scientific American*, 274.

Kassirer, J. 17 September 1998. *New England Journal of Medicine.* Editorial introduces reports on unproven and harmful uses of dietary supplements, some of which were contaminated.

Sherwood, L. 1997. *Human Physiology.* Third edition. Monterey, California: Brooks-Cole.

On-Line readings at Student Guide for InfoTrac:
www.brookscole.com/biology

38

THE INTERNAL ENVIRONMENT

Tale of the Desert Rat

Look closely at a fish or some other marine animal, and you'll find that its body cells are exquisitely adapted to life in a salty fluid. And yet, about 375 million years ago, some lineages of animals that evolved in the seas moved onto dry land. They were able to do so partly because they brought salty fluid along with them, as an *internal* environment for their cells. Even so, it was not a simple transition. The pioneers on land and their descendants faced intense sunlight, dry winds, more pronounced swings in temperature, water of dubious salt content, and sometimes no water at all.

How did the pioneers conserve or replace the water and salts lost as a result of everyday activities? How did they manage to stay comfortably warm when their surroundings became too cold or too hot? They must have done these things. Otherwise the composition, volume, and temperature of their internal environment would have spun out of control. In short, the question is this: *How did the land-dwelling descendants of marine animals maintain internal operating conditions and so prevent cellular anarchy?*

Observe any of their existing descendants and you get a sense of what some answers might be. Suppose you focus on a tiny mammal, a kangaroo rat living in an isolated desert of New Mexico (Figure 38.1). After a brief rainy season, the sun bakes the desert sand for months. The only obvious water is imported, sloshing about in the canteens of the occasional researcher and tourist. Yet with nary a sip of free water, a kangaroo rat routinely counters threats to its internal environment.

It waits out the heat of the day inside a burrow, then forages in the cool of night for dry seeds and maybe a succulent. It is not sluggish about this. It hops rapidly and far, searching for seeds and fleeing from coyotes and snakes. All that hopping requires ATP energy and water. Seeds, chockful of energy-rich carbohydrates, provide both. Metabolic reactions that release energy from carbohydrates and other organic compounds also yield water. Each day, such "metabolic water" makes up a whopping 90 percent of a kangaroo rat's total water intake. By comparison, metabolic water is only about 12 percent of the total water intake for your body.

	KANGAROO RAT	HUMAN
WATER GAIN (milliliters)		
by ingesting solids	6.0	850
by ingesting liquids	0.0	1,400
by metabolism	54.0	350
	60.0	2,600
WATER LOSS (milliliters)		
in urine	13.5	1,500
in feces	2.6	200
by evaporation	43.9	900
	60.0	2,600

Figure 38.1 Kangaroo rat, master of water conservation in a New Mexico desert. The chart will give you an idea of how kangaroo rats and humans gain and lose water. They do so in different ways. Even so, in both cases, *water losses balance out the gains* —just as they do in all animals. How this balancing happens, and why it absolutely must happen, is this chapter's initial focus.

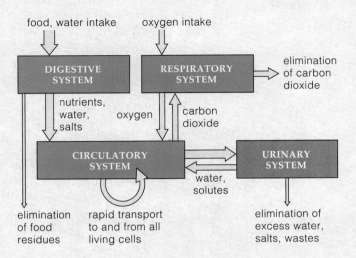

food, water intake   oxygen intake

DIGESTIVE SYSTEM → RESPIRATORY SYSTEM → elimination of carbon dioxide

nutrients, water, salts   oxygen   carbon dioxide

CIRCULATORY SYSTEM ⇄ URINARY SYSTEM

water, solutes

elimination of food residues   rapid transport to and from all living cells   elimination of excess water, salts, wastes

Figure 38.2 Links between the urinary system and other organ systems that contribute to homeostasis, or stability in favorable operating conditions in the animal body.

A kangaroo rat resting in a cool burrow conserves and recycles water. As it inhales cool air into its warm lungs, water vapor condenses onto epithelium in its nose. Some of that water diffuses back into the body. Also, after a busy night of foraging, it empties its cheek pouches of seeds. As seeds leave its mouth, they soak up water dripping from its nose. When the kangaroo rat eats the dripped-upon seeds, it reclaims water.

A kangaroo rat can't lose water by perspiring; it has no sweat glands. It loses water when it urinates, but its two specialized kidneys don't let it piddle away much. Kidneys filter the blood's water and solutes, including dissolved salts. They continually adjust *how much water* and *which solutes* return to the blood or depart as urine.

Overall, kangaroo rats and all other animals take in enough water and solutes to replace daily losses (Figure 38.1). How they do so will be our initial focus. Later, we will consider how mammals withstand hot, cold, and often unpredictable temperature changes on land.

As a starting point, think back on the fluids inside most animals. **Interstitial fluid** fills the spaces between living cells and other components of tissues. Another fluid, **blood**, moves substances to and from all tissue regions by way of a circulatory system. Together, the interstitial fluid and blood are **extracellular fluid**. In many animals, a well-developed urinary system helps keep the composition and volume of extracellular fluid within tolerable ranges. As you'll see, the organ systems shown in Figure 38.2 interact with the urinary system in the performance of this homeostatic task.

Key Concepts

1. Animals are continually gaining and losing water and solutes, or dissolved substances. They continually produce metabolic wastes. Even with all of the inputs and outputs, the overall composition and volume of the extracellular fluid in the body remain relatively constant.

2. In humans as in other vertebrates, a urinary system is crucial to balancing the intake and output of water and solutes. This system filters water and solutes from blood on an ongoing basis. It reclaims some amount of both and eliminates the rest. Different amounts are reclaimed at different times, depending on what's required to maintain extracellular fluid.

3. Kidneys are blood-filtering organs, and the urinary system of vertebrates has a pair of them. Packed inside each kidney are a great number of nephrons.

4. At its receiving end, each nephron cups around a tuft of blood capillaries and accepts water and solutes from it. The nephron returns most of the filtrate to the blood by giving it up to capillaries that intertwine around the nephron's tubular parts.

5. Water and solutes not returned to the blood leave the body as a fluid called urine. During any interval, control mechanisms make adjustments that influence whether urine is concentrated or dilute. Two hormones, ADH and aldosterone, have key roles in the adjustments.

6. The internal body temperature of animals depends on a balancing of heat produced through metabolism, heat absorbed from the environment, and heat lost to the environment.

7. The internal body temperature is maintained within a favorable range through controls over metabolic activity and adaptations in body form and behavior.

URINARY SYSTEM OF MAMMALS

The Challenge—Shifts in Extracellular Fluid

Different solid foods and fluids intermittently enter a mammal's gut. Afterward, variable amounts of absorbed water, nutrients, and other substances move into the blood, then into interstitial fluid and on into cells. Such events could easily shift the volume and composition of extracellular fluid beyond tolerable limits. But the body makes compensatory adjustments that balance out the gains and losses. Within a given time frame, it takes in as much water and solutes as it gives up.

WATER GAINS AND LOSSES To start, think about how humans and other mammals *gain* water, mainly by two processes:

Absorption from the gut
Metabolism

Considerable water is absorbed from solids and liquids inside the gut. As you know, water also forms as a normal by-product of many metabolic reactions. In land mammals, how much water enters the gut in the first place depends on a thirst mechanism. When the body loses too much water, land mammals seek out streams, waterholes, and so on. We will be looking at the thirst mechanism later in the chapter.

Normally, the mammalian body *loses* water mainly by way of four physiological processes—especially the first process listed here:

Urinary excretion
Evaporation from lungs and the skin
Sweating, by mammals that sweat
Elimination, in feces

Urinary excretion affords the most control over water loss. The process eliminates excess water and solutes as urine. This fluid forms in a urinary system such as the one in Figure 38.3. Water evaporates from respiratory surfaces, too. Some species lose it in sweat. A mammal in good health tends to lose very little water from the gut; most is absorbed, not eliminated in feces.

SOLUTE GAINS AND LOSSES The mammalian body *gains* solutes mainly by the following four processes:

Absorption from the gut
Secretion from cells
Respiration
Metabolism

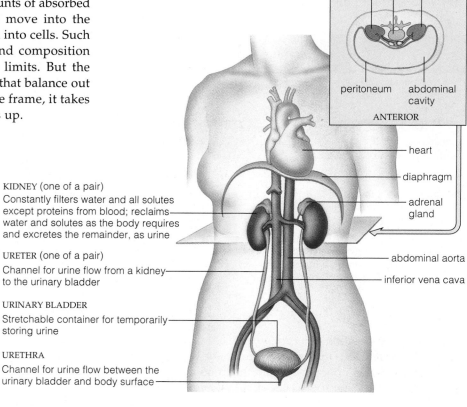

KIDNEY (one of a pair)
Constantly filters water and all solutes except proteins from blood; reclaims water and solutes as the body requires and excretes the remainder, as urine

URETER (one of a pair)
Channel for urine flow from a kidney to the urinary bladder

URINARY BLADDER
Stretchable container for temporarily storing urine

URETHRA
Channel for urine flow between the urinary bladder and body surface

Figure 38.3 Organs of the human urinary system and their functions. The two kidneys, two ureters, and urinary bladder are located *outside* the peritoneum, the membranous lining of the abdominal cavity. Compare Section 23.1.

For example, you absorb food, mineral ions, drugs, and food additives from the gut. Cell secretions and wastes, such as carbon dioxide, enter interstitial fluid and then blood. The respiratory system puts oxygen into blood; aerobically respiring cells put carbon dioxide into it.

Mammals typically *lose* solutes by three processes:

Urinary excretion
Respiration
Sweating, by some species

The urine of mammals includes various wastes formed by the breakdown of organic compounds. For example, ammonia forms as amino groups are split from amino acids. Then **urea**, a major waste, forms in the liver when two ammonia molecules join with carbon dioxide. Also in urine are uric acid (from nucleic acid breakdown), hemoglobin breakdown products (which give the urine much of its color), drugs, and food additives. Besides these solute losses, some mammals also lose mineral ions in sweat. And all mammals lose carbon dioxide, the most abundant waste, by way of respiration.

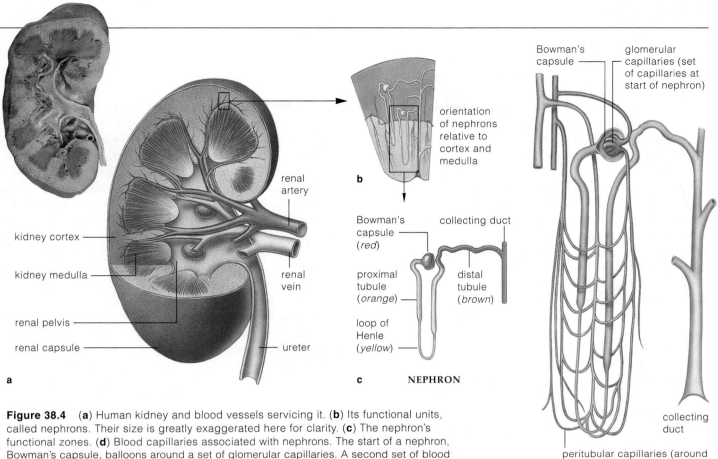

renal artery

kidney cortex

kidney medulla

renal pelvis

renal capsule

renal vein

ureter

a

b

orientation of nephrons relative to cortex and medulla

Bowman's capsule (*red*)

proximal tubule (*orange*)

loop of Henle (*yellow*)

collecting duct

distal tubule (*brown*)

c   **NEPHRON**

Bowman's capsule

glomerular capillaries (set of capillaries at start of nephron)

collecting duct

peritubular capillaries (around tubular parts of nephron)

d

Figure 38.4 (**a**) Human kidney and blood vessels servicing it. (**b**) Its functional units, called nephrons. Their size is greatly exaggerated here for clarity. (**c**) The nephron's functional zones. (**d**) Blood capillaries associated with nephrons. The start of a nephron, Bowman's capsule, balloons around a set of glomerular capillaries. A second set of blood vessels, the peritubular capillaries, threads around tubular parts of the nephron.

Components of the Urinary System

Mammals routinely counter shifts in the composition and volume of extracellular fluid, mainly with a **urinary system**. This consists of two kidneys, two ureters, one urinary bladder, and one urethra. **Kidneys** are a pair of bean-shaped organs, about as large as a fist in an adult human (Figures 38.3 and 38.4). Each has an outer capsule of connective tissue. Blood capillaries penetrate its two inner regions, the cortex and medulla.

Kidneys filter water, mineral ions, organic wastes, and other substances from blood. Then they adjust the filtrate's composition and return all but about 1 percent to the blood. The small portion of unreclaimed water and solutes is urine. By definition, **urine** is a fluid that rids the body of water and solutes that are in excess of the amounts required to maintain extracellular fluid.

Urine flows from each kidney into a **ureter**, a tubular channel between it and the **urinary bladder**, a muscular sac. Urine is briefly stored here before flowing into the **urethra**, a muscular tube that opens at the body surface.

Flow from the bladder (urination) is a reflex action. When the bladder is full, the sphincter around its neck opens, smooth muscle in its balloonlike wall contracts, and so urine is forced out through the urethra. Skeletal muscle surrounds the urethra. Its contraction, which is under voluntary control, prevents urination.

Nephrons—Functional Units of Kidneys

Each human kidney has more than a million **nephrons**, slender tubules packed inside lobes that extend from the kidney's cortex down through the medulla. *It is at the nephrons that water and solutes are filtered from blood and adjustments are made in the amounts to be reclaimed.*

A nephron starts as a **Bowman's capsule**, where its wall cups around a set of *glomerular* capillaries. The cup and capillaries interact as one blood-filtering unit, the **renal corpuscle** (Figure 38.4c). Next comes a **proximal tubule** (tubular region closest to the capsule), a hairpin-shaped **loop of Henle**, and a **distal tubule** (most distant from the capsule). A **collecting duct**, the nephron's last region, is part of a duct system leading to the kidney's central cavity (renal pelvis), then into a ureter.

Blood does not give up all of its water and solutes. The unfiltered part flows into capillaries threading lacily around the nephron's tubular parts. In these *peritubular* capillaries, the blood reclaims water and solutes. Then it flows into veins and back to the general circulation.

A urinary system counters unwanted shifts in the volume and composition of extracellular fluid. In its paired kidneys, water and solutes are filtered from blood. The body reclaims most of this, but the excess leaves the kidneys as urine.

URINE FORMATION

Urine forms in nephrons by three processes: filtration, tubular reabsorption, and tubular secretion. All three processes depend on properties of cells of the nephron wall. Cells in different wall regions differ in membrane transport mechanisms and in their permeability.

Blood pressure generated by the heart's contractions drives **filtration**—the forcing of water and all solutes except proteins out from glomerular capillaries into the fluid-filled space of Bowman's capsule. The protein-free filtrate moves into the proximal tubule (Figure 38.5).

Next, with **tubular reabsorption**, most of the water and solutes that moved down into the proximal tubule are returned to the peritubular capillaries (Figure 38.6). As you will see shortly, hormonal controls adjust the amounts conserved or excreted during a given interval.

Blood still inside peritubular capillaries holds excess ions (mainly H^+ and K^+), metabolites such as urea, and neurotransmitters, histamine, and maybe drugs or toxins. Such substances diffuse into the surrounding interstitial fluid. By **tubular secretion**, they now enter the nephron through proteins that function as channels and transporters across tubular portions of the nephron wall. Then they are excreted.

Factors Influencing Filtration

Each minute, about 1.5 liters (1–1/2 quarts) of blood flow through an adult's kidneys, and 120 milliliters of water and small solutes are filtered into nephrons. That's 180 liters of filtrate per day! A high filtration rate is possible mainly because the glomerular capillaries are 10 to 100 times more permeable to water and small solutes than other capillaries are. Also, blood pressure is high inside them. Why? Arterioles delivering blood to the nephron have a wider lumen and less flow resistance than those taking blood away. As a result, blood dams up inside the renal corpuscle, adding to the pressure in its capillaries.

Filtration rates also depend on flow volume to the kidneys. Neural, endocrine, and local controls maintain the flow even when blood pressure changes. Example: Dance until dawn and your nervous system commands an above-normal volume of blood to be diverted away from your kidneys and sent instead toward the energy-demanding cells in skeletal muscles and the heart. Such redistribution of blood volume involves coordinating the vasoconstriction and vasodilation of arterioles throughout the body.

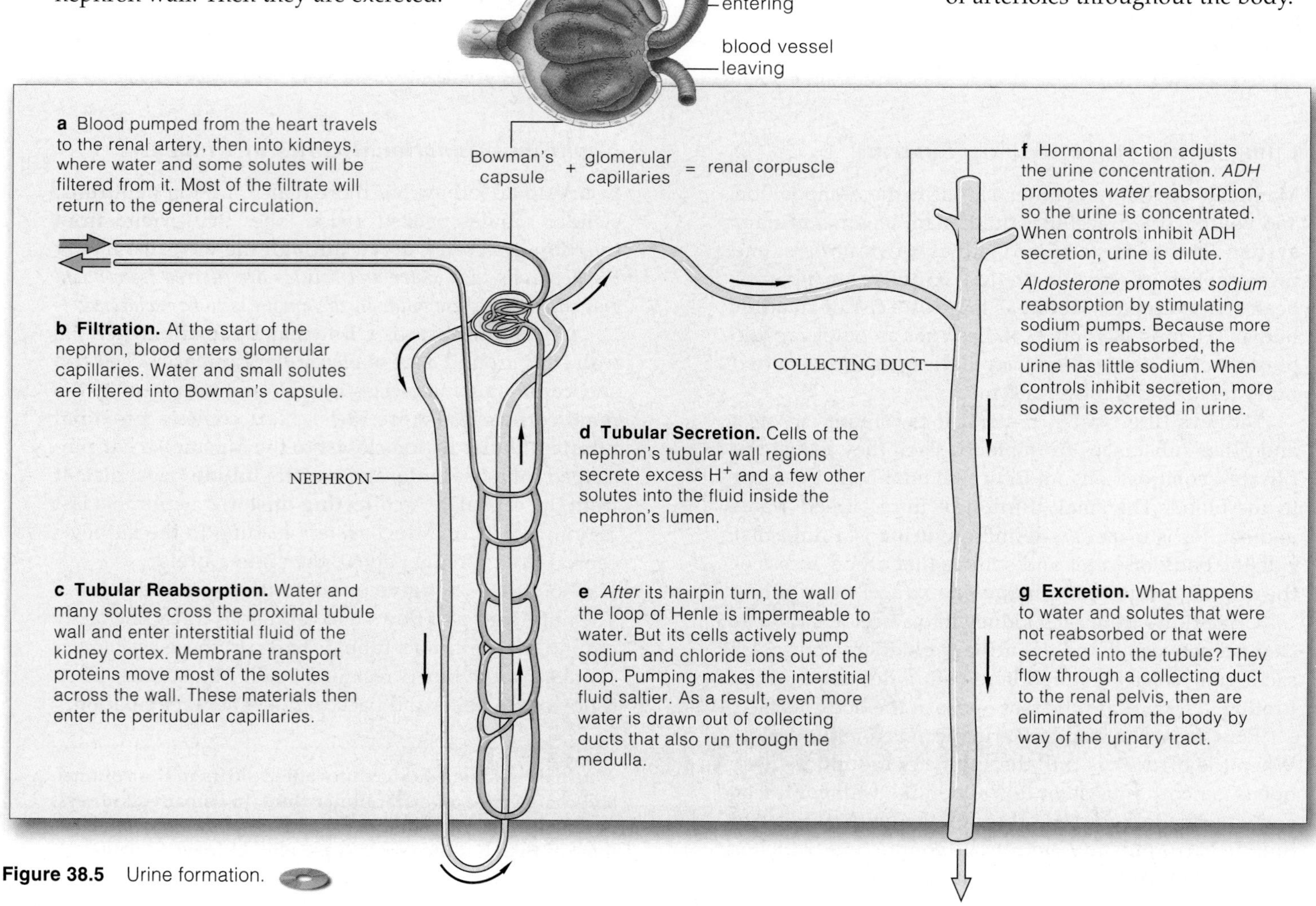

blood vessel entering

blood vessel leaving

a Blood pumped from the heart travels to the renal artery, then into kidneys, where water and some solutes will be filtered from it. Most of the filtrate will return to the general circulation.

Bowman's capsule + glomerular capillaries = renal corpuscle

f Hormonal action adjusts the urine concentration. *ADH* promotes *water* reabsorption, so the urine is concentrated. When controls inhibit ADH secretion, urine is dilute.

Aldosterone promotes *sodium* reabsorption by stimulating sodium pumps. Because more sodium is reabsorbed, the urine has little sodium. When controls inhibit secretion, more sodium is excreted in urine.

b **Filtration.** At the start of the nephron, blood enters glomerular capillaries. Water and small solutes are filtered into Bowman's capsule.

COLLECTING DUCT

NEPHRON

d **Tubular Secretion.** Cells of the nephron's tubular wall regions secrete excess H^+ and a few other solutes into the fluid inside the nephron's lumen.

c **Tubular Reabsorption.** Water and many solutes cross the proximal tubule wall and enter interstitial fluid of the kidney cortex. Membrane transport proteins move most of the solutes across the wall. These materials then enter the peritubular capillaries.

e *After* its hairpin turn, the wall of the loop of Henle is impermeable to water. But its cells actively pump sodium and chloride ions out of the loop. Pumping makes the interstitial fluid saltier. As a result, even more water is drawn out of collecting ducts that also run through the medulla.

g **Excretion.** What happens to water and solutes that were not reabsorbed or that were secreted into the tubule? They flow through a collecting duct to the renal pelvis, then are eliminated from the body by way of the urinary tract.

Figure 38.5 Urine formation.

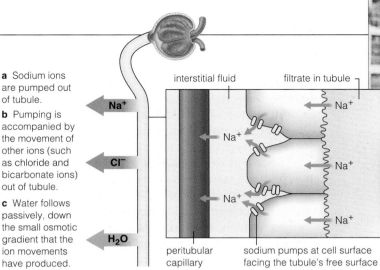

a Sodium ions are pumped out of tubule.

b Pumping is accompanied by the movement of other ions (such as chloride and bicarbonate ions) out of tubule.

c Water follows passively, down the small osmotic gradient that the ion movements have produced.

Na⁺

Cl⁻

H₂O

interstitial fluid   filtrate in tubule

Na⁺

Na⁺   Na⁺

Na⁺

Na⁺   Na⁺

peritubular   sodium pumps at cell surface
capillary   facing the tubule's free surface

Figure 38.6 Membrane pumps in the nephron wall. Sodium and water are reabsorbed here in response to, say, gulping too much water or ingesting too much salt.

Also, the walls of arterioles leading to nephrons are sensitive to pressure. When blood pressure falls, the arterioles vasoconstrict, so less of the total blood volume is sent to the kidneys. When blood pressure rises, these arterioles vasodilate, so more blood flows in.

Reabsorption of Water and Sodium

REABSORPTION MECHANISM Kidneys precisely adjust how much water and sodium ions the body excretes or saves. Drink too much or too little water, wolf down salty potato chips, or lose too much sodium in sweat, and responses are quick. As filtrate enters the proximal tubules, cells actively transport some sodium out of it. Other ions follow sodium into interstitial fluid. Water leaves the filtrate, too, by osmosis. The proximal tubule wall is so highly permeable that about two-thirds of the filtrate's water is reabsorbed here (Figure 38.5*c,d*).

Interstitial fluid is saltiest around the hairpin turn of the loop of Henle. Water moves out of the filtrate by osmosis *before* the turn. The fluid left behind gets saltier until it matches interstitial fluid. The loop wall *after* the turn is impermeable to water, but sodium gets pumped out by active transport mechanisms. Interstitial fluid gets even saltier—so it attracts even more water out of the filtrate that is just entering the loop (Figure 38.5*e*).

Fluid arriving at a distal tubule is dilute. The stage is set for adjustments, which can lead to urine that is highly dilute, concentrated, or anywhere in between.

HORMONE-INDUCED ADJUSTMENTS When the volume of extracellular fluid drops, the hypothalamus calls for the release of antidiuretic hormone, or **ADH**, from the pituitary's posterior lobe (Section 32.3). ADH binds to receptors of distal tubules and collecting ducts, making their wall more permeable to water. More water gets reabsorbed so urine is more concentrated (Figure 38.5*f*). But less ADH is released when there is too much water. The wall of the tubes becomes less permeable, so less water is reabsorbed and the urine remains dilute.

Aldosterone promotes reabsorption of sodium. The extracellular fluid volume falls when too much sodium is lost. Sensory receptors in the heart and blood vessels detect the drop and stimulate renin-secreting gland cells in the wall of arterioles at glomeruli. The enzyme renin splits a plasma protein. A fragment is converted to the hormone **angiotensin II**. Among its many targets are aldosterone-secreting cells of the adrenal cortex, part of a gland on each kidney (Figure 38.3). The cells secrete more aldosterone, which stimulates cells of the distal tubules and collecting ducts to reabsorb sodium faster and excrete less sodium. Conversely, when there is too much sodium in the body, aldosterone secretion slows. Less sodium is reabsorbed, so more is excreted.

THIRST BEHAVIOR A **thirst center** in the hypothalamus induces water-seeking behavior. Osmoreceptors nearby warn of decreases in blood volume and rising blood solute levels. The signals activate the thirst center *and* ADH-secreting cells in the brain. While water intake is being encouraged, urinary output is reduced.

Angiotensin II also stimulates thirst by direct effects on brain cells controlling ADH secretion, so it helps to adjust the sodium balance *and* the water balance. Thirst also is initiated when free nerve endings detect dryness in the mouth, an early sign of dehydration.

Filtration, reabsorption, and secretion in the kidneys help maintain the composition and volume of extracellular fluid.

Blood filtration rates depend on high hydrostatic pressure (generated by heart contractions) and on neural, endocrine, and local control over how much blood flows into kidneys.

During tubular reabsorption, hormonal controls adjust the amounts of water and solutes in urine. ADH promotes water conservation and concentrated urine. Aldosterone promotes sodium conservation and dilute urine.

Certain excess solutes diffuse out of peritubular capillaries and into interstitial fluid. Then, by tubular secretion, they move into the nephron for subsequent excretion.

When Kidneys Break Down

By this point in the chapter, you probably have figured out that good health depends on nephron functioning. Whether by illness or accident, when the nephrons of both kidneys become damaged and no longer perform their regulatory and excretory functions, we call this **renal failure**. Chronic renal failure is irreversible.

Renal failure may occur after infectious agents reach the kidneys through the bloodstream or the urethra. It may occur after someone ingests lead, arsenic, pesticides, or other toxins. Ongoing high doses of aspirin and some other drugs also can bring it about. Abnormal retention of metabolic wastes, such as the by-products of protein breakdown, results in *uremic toxicity*. Atherosclerosis, heart failure, hemorrhage, and shock diminish blood flow and disrupt filtration pressure in the kidneys.

In *glomerulonephritis*, antibody–antigen complexes get trapped in glomeruli. Unless phagocytes remove them, the complexes keep activating complement and other agents that cause widespread inflammation and tissue damage. Fortunately, this type of disease is rare.

Kidney stones form when uric acid, calcium salts, and other wastes settle out of urine and collect in the renal pelvis. These hard deposits are usually passed in urine, but they can become lodged in the ureter or the urethra. When they disrupt urine flow, they must be medically or surgically removed to prevent renal failure.

About 13 million people in the United States alone suffer from renal failure. A *kidney dialysis machine* often restores the proper solute balances. Like the kidneys, it helps maintain extracellular fluid by selectively adjusting the solutes in blood. "Dialysis" refers to an exchange of substances across an artificial membrane interposed between two solutions that differ in composition.

In *hemodialysis*, the machine is connected to an artery or a vein. Then the patient's blood is pumped on through tubes that have been submerged in a warm saline bath. The bath's mix of salts, glucose, and other substances sets up the correct concentration gradients with blood. Then the blood flows back into the body. For kidney dialysis to have optimum effect, it must be performed three times a week. Each time, the procedure takes about four hours, because the patient's blood must circulate repeatedly to improve solute concentrations in her or his body.

In *peritoneal dialysis*, fluid of suitable composition is introduced into the patient's abdominal cavity, left in there for a specific length of time, then drained out. In this case, the cavity's lining itself, the peritoneum, acts as the membrane for dialysis.

Bear in mind, kidney dialysis is used as a temporary measure in reversible kidney disorders. In chronic cases, the procedure must be used for the rest of the patient's life or until a transplant operation provides her or him with a functional kidney. With treatment and controlled diets, many patients resume fairly normal activity.

THE BODY'S ACID–BASE BALANCE

In addition to maintaining the volume and composition of extracellular fluid, kidneys help keep it from getting too acidic or too basic (alkaline). The overall **acid–base balance** of that fluid is an outcome of controls over its concentrations of H^+ and other dissolved ions. *Metabolic acidosis* hints at the importance of the balancing acts. This condition results when the kidneys cannot secrete enough H^+ to keep pace with all the H^+ that is forming during metabolism. It is life-threatening.

Buffer systems, respiration, and urinary excretion all work in concert to provide control over the acid–base balance. A buffer system, remember, consists of weak acids or bases that can reversibly bind and release ions, thus helping to minimize shifts in pH (Section 2.6).

Normally, the extracellular pH of the human body should be maintained between 7.37 and 7.38. As you know, acids lower the pH and bases raise it. A variety of acidic and basic substances enter blood after being absorbed from the gut and as by-products of normal metabolism. Cell activities typically produce an excess of acids. These dissociate into H^+ and other fragments, and pH declines. The effect is minimized when excess hydrogen ions react with buffer molecules. An example is the *bicarbonate–carbon dioxide* buffer system:

$$H^+ + HCO_3^- \rightleftharpoons H_2CO_3 \rightleftharpoons CO_2 + H_2O$$
$$\text{BICARBONATE} \qquad \text{CARBONIC ACID}$$

In this case, the buffer system neutralizes excess H^+, and the carbon dioxide that forms during the reactions is exhaled from the lungs. Like other buffer systems in the body, however, this one has temporary effects only; it does not *eliminate* excess H^+. Only the urinary system can do so and thereby restore the buffers.

The same reactions proceed in reverse in cells of the nephron's tubular walls. HCO_3^- formed by the reverse reactions moves into interstitial fluid, then peritubular capillaries. Afterward, it enters the general circulation and buffers excess acid. The H^+ formed in the cells is secreted into the nephron and may join with HCO_3^-. The CO_2 that formed may be returned to blood, then exhaled. H^+ also may combine with phosphate ions or ammonia (NH_3), then leave the body in urine.

The kidneys work in concert with buffering systems that neutralize acids and with the respiratory system to help keep the extracellular fluid from becoming too acidic or too basic (alkaline).

A bicarbonate–carbon dioxide buffer system temporarily neutralizes excess hydrogen ions. The urinary system alone eliminates the excess ions and thus restores these buffers.

The bicarbonate–carbon dioxide buffer system is one of the key mechanisms that help maintain the acid–base balance.

ON FISH, FROGS, AND KANGAROO RATS

Now that you have a general sense of how your body maintains water and solute levels, consider what goes on in some other vertebrates, including that kangaroo rat hopping about at the start of the chapter.

Bony fishes and amphibians of freshwater habitats gain water and lose solutes (Figure 38.7a). Water moves into their internal environment by osmosis; they don't gain water by drinking it. Water diffuses across thin gill membranes in fishes and across the skin in adult amphibians. Excess water leaves as dilute urine, formed inside a pair of kidneys. In both groups of vertebrates, solute losses are balanced when more solutes come in with food and when gill or skin cells pump in sodium.

Body fluids of herring, snapper, and other marine fishes are about three times less salty than seawater. These fishes lose water by osmosis, then replace it by drinking more. They excrete ingested solutes against concentration gradients (Figure 38.7b). Fish kidneys do not have loops of Henle, so urine cannot ever become saltier than body fluids. Cells in the fish gills actively pump out most of the excess solutes in blood.

Figure 38.7c describes the water–solute balancing act in a salmon. This type of fish spends part of its life cycle in fresh water and another part in seawater.

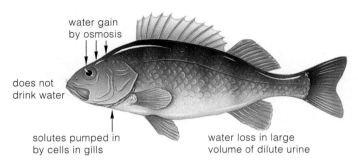

a Freshwater bony fish (body fluids far saltier than surroundings)

water gain by osmosis

does not drink water

solutes pumped in by cells in gills

water loss in large volume of dilute urine

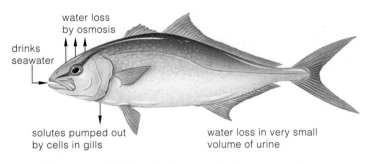

b Marine bony fish (body fluids less salty than surroundings)

water loss by osmosis

drinks seawater

solutes pumped out by cells in gills

water loss in very small volume of urine

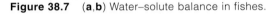

Figure 38.7 (**a**,**b**) Water–solute balance in fishes.

(**c**) Water–solute balance by salmon. This type of fish lives in saltwater *and* fresh water. Salmon hatch in streams and later move downstream to the open sea, where they feed and mature. They return to home streams to spawn.

For most salmon, salt tolerance is an outcome of changing hormone levels that appear to be triggered by increasing daylength in spring. Prolactin, a pituitary hormone, plays a key role in sodium retention in freshwater habitats. We know this because a freshwater fish that has its pituitary gland removed will die from sodium loss—but that fish will live if prolactin is administered to it.

A salmon's salt tolerance depends on cortisol. Secretion of this steroid hormone from the adrenal cortex correlates with rises in sodium excretion, sodium–potassium pumping by cells in the salmon's gills, and absorption of ions and water from the gut. In young salmon, cortisol secretion increases prior to the seaward movement—and so does salt tolerance.

salmon avoiding grizzly while maintaining water–solute balance

c

And what about that kangaroo rat! Proportionally, the loops of Henle of its nephrons are amazingly long, compared to yours. This means a great deal of sodium is pumped out from the nephron. Therefore, the solute concentration in the interstitial fluid around the loops becomes very high. The osmotic gradient between fluid in the loops of Henle and urine is *so* steep that nearly all the water that does reach the equally long collecting ducts gets reabsorbed. The kangaroo rats gives up only a tiny volume of urine. And that urine is three to five times more concentrated than the most concentrated urine of humans.

The urinary systems of vertebrates differ in their details, such as the length of the nephron's loop of Henle. They are adapted to balance the body's gains in water and solutes with its losses of water and solutes in particular habitats.

MAINTAINING BODY TEMPERATURE

We turn now to another major aspect of the internal environment, its temperature. Many physiological and behavioral responses to change help maintain a **core temperature** within the tolerance range for the body's enzymes. *Core* means the body's internal temperature, as opposed to temperatures of tissues near its surface.

Figure 38.8 In hot desert sand at dusk, the J-shaped track of a sidewinder. As near as you can tell, how might this rattlesnake be gaining and losing heat?

Heat Gains and Heat Losses

Each metabolic reaction generates heat. If heat were to accumulate internally, the core temperature would rise. But a warm body loses heat to cool surroundings. Its core temperature stabilizes when the rate of heat loss balances the rate of metabolic heat production. The heat content of any complex animal depends on a balancing between heat gains and losses, as summarized here:

$$\text{CHANGE IN BODY HEAT} = \text{HEAT PRODUCED} + \text{HEAT GAINED} - \text{HEAT LOST}$$

Heat is gained and lost through exchanges at skin and other surfaces. Four processes—radiation, conduction, convection, and evaporation—drive the exchanges.

With **radiation**, the animal gains heat after being exposed to radiant energy—typically, sunlight—or to a surface warmer than its body surface temperature.

With **conduction**, an animal and an object in direct contact with it exchange heat in response to a thermal gradient between them. It loses heat when the object is cooler than its body, and it gains heat when that object is warmer than its body, as in Figure 38.8.

With **convection**, movement of air or water next to the body aids conductive heat loss. For example, as air heated by conduction moves away from a body, cooler air moves in to replace it. In turn, that air gains heat from the body and moves away, and so on. Heated air moves this way because it is less dense than cool air.

With **evaporation**, heat is lost as a liquid converts to a gas. Energy for the conversion comes from the heat content of the liquid (Section 2.5). Evaporation from a body surface has a cooling effect because the escaping water molecules carry away some energy with them.

Evaporation rates depend on humidity and on the rate of air movement. If air next to an animal is already saturated with water (when the local relative humidity is 100 percent), water will not evaporate. When air next to the body is hot and dry, evaporation may be the only means of countering metabolic heat production and the heat gains from radiation and convection.

Animals adjust heat losses and gains by changing their behavior and physiology, although some do this better than others. Like most animals, reptiles have low metabolic rates and poor body insulation; small ones especially absorb and gain heat quickly. They maintain the core temperature mostly by gaining environmental heat. Hence reptiles are among the **ectotherms**, which means "heat from outside."

When outside temperatures change, ectotherms use *behavioral* temperature regulation. Thus a lizard basks on warm rocks and gains heat by conduction. It keeps reorienting its body to expose the most surface area to the sun's infrared radiation. It loses heat after sunset. Before its metabolic rate slows, it hides under rocks or in crevices to conserve heat and avoid predators.

Most birds and mammals are **endotherms,** meaning "heat from within." They all have high metabolic rates; a foraging mouse uses up to thirty times more energy than a foraging lizard of the same weight. But that rate lets them stay active under a wide temperature range. Endotherms balance metabolism, heat gains and losses, and behavioral control of temperature. Their body form also is adapted to conserve or dissipate metabolic heat. Examples: Some mammals in cold habitats that have close relatives in warmer habitats have a more massive body. A large body mass, fur, fat layers, feathers, and clothing all help reduce heat loss (Figure 38.9).

Other birds and mammals are **heterotherms**. They are mainly endotherms but reduce metabolic rates by allowing the core temperature to drop at certain times. Activities idle, but the energy cost of maintaining body temperature is lowered. Example: Very high metabolic rates sustain tiny-bodied hummingbirds as they search for food during the day. Those rates plummet at night, when they can't feed; then, hummingbirds save energy by becoming almost as cool as their surroundings.

Warm, humid regions favor ectotherms, which need not spend much energy to maintain core temperatures. They can devote more energy to reproducing and other tasks. In tropical regions, reptiles exceed mammals in numbers and diversity. Endotherms have the edge in cool or cold regions. Thus arctic foxes, snowshoe hares, and some other endotherms do well in polar habitats, where you would never find a lizard.

Responses to Stressful Temperatures

The hypothalamus has control centers for maintaining the core temperature of mammals (Figure 29.14). They constantly get input from peripheral thermoreceptors in skin and from central thermoreceptors. Suppose the

Figure 38.9 How might the intrepid tourist relaxing on a deck chair of a ship off the coast of Antarctica be gaining and losing heat?

core temperature deviates from a set point. The centers integrate responses made by skeletal muscles, arterioles in skin and, in certain species, sweat glands. Negative feedback loops to the hypothalamus stop the responses when a suitable temperature returns.

RESPONSES TO COLD STRESS In mammals, cold stress brings on **peripheral vasoconstriction.** The diameter of arterioles constricts, so the blood's convective delivery of heat to body surfaces lessens. When your fingers are chilled, all but 1 percent of blood that would otherwise flow to your skin is diverted to other body regions.

Cold stress makes hairs and feathers "stand up." This **pilomotor response** creates a layer of still air next to skin and reduces convective and radiative heat loss. Behavioral changes also can minimize exposed surface areas and reduce heat loss, as when a cat curls up and when you hold both arms tightly against your body.

A **shivering response** to prolonged cold isn't useful for long. Rhythmic tremors begin as muscles contract ten to twenty times a second. Heat production rises by several times, but at a high energy cost. Prolonged or severe cold also leads to a hormonally induced rise in metabolic rates. This **nonshivering heat production** is greatest in *brown* adipose tissue. Hibernating animals, cold-acclimated animals, and human infants have this connective tissue. So do the ama of Japan and Korea who dive for shellfish six hours a day in frigid water. Adults who are not similarly adapted have little of it.

In *hypothermia*, core temperature is below normal. In humans, a drop of a few degrees alters brain function; more cooling leads to coma and death. Many mammals can recover from hypothermia. Frozen human cells die unless tissues thaw under medical supervision. Tissue destruction by localized freezing is called *frostbite.*

RESPONSES TO HEAT STRESS If a mammal gets too hot, the hypothalamus calls for **peripheral vasodilation**. It signals blood vessels in the skin to dilate. More blood flows to the skin, which can dissipate excess heat.

Evaporative heat loss is another response to heat stress. It occurs at moist respiratory surfaces and across skin. Animals that sweat lose some water this way. For instance, humans and some other mammals have sweat glands that release water and solutes through pores at the skin's surface. An average-sized adult human has 2–1/2 million or more sweat glands. For every liter of sweat produced, 600 kilocalories of heat energy are lost through evaporative heat loss.

Bear in mind, sweat that is dripping from skin does not dissipate heat. Outside temperatures must be high enough to cause water from sweat to *evaporate*. On hot, humid days, evaporation rates cannot match the rate of sweat secretion; air's high water content slows it down. During strenuous exercise, sweating may help balance heat production in skeletal muscle. Extreme sweating, as in a marathon race, forces the body to lose a vital salt—sodium chloride—as well as water. Such losses disrupt the composition and the volume of extracellular fluid. Runners collapse and faint with large losses.

What about mammals that sweat little or not at all? Some make behavioral responses, such as licking fur, or panting. "Panting" is shallow, rapid breathing that increases evaporative water loss from the respiratory tract. The body cools when water evaporates from the nasal cavity, mouth, and tongue (Section 29.7).

Sometimes peripheral blood flow and evaporative heat loss cannot counter heat stress, and *hyperthermia* results. In this condition, the core temperature rises above normal. For humans and other endotherms, even a few degrees above normal can be dangerous.

A *fever*, recall, is part of an inflammatory response to damaged tissues (Section 35.3). The hypothalamus resets the body's thermostat, which dictates what the core temperature is supposed to be. Mechanisms that raise metabolic heat production and cut heat loss are working; however, their operation maintains a higher temperature. A person feels chilled at a fever's onset. When the fever "breaks," peripheral vasodilation and sweating increase; the body is attempting to restore the normal core temperature and the person feels warm.

By bringing down a fever, anti-inflammatory drugs such as aspirin may actually prolong the healing time. Only when fevers approach dangerous levels should drugs be prescribed under medical supervision.

The internal, core temperature of an animal's body is being maintained when heat gains and heat losses are in balance.

Metabolic reactions generate heat inside the body. Radiation, conduction, and convection can move heat down thermal gradients that exist between the body and its surroundings. Evaporative heat loss carries heat away from the body.

Besides being morphologically adapted to their habitats, animals can make behavioral and physiological adjustments to environmental temperatures.

SUMMARY   *Gold* indicates text section

Control of Extracellular Fluid

1. The extracellular fluid in the animal body consists of certain types and amounts of substances dissolved in water. It is distributed as interstitial fluid (in tissue spaces) and blood. Its volume and composition are maintained when an animal's daily intake and output of water as well as solutes are in balance. The following processes maintain the balance in mammals: *CI, 38.1*

   a. Water is gained by absorption from the gut and by metabolism. It is lost by urinary excretion, evaporation from lungs and skin, sweating, and elimination of feces.

   b. Solutes are gained by absorption from the gut and by secretion, respiration, and metabolism. Solutes are lost by excretion, respiration, and sweating.

   c. Losses of water and solutes are controlled mainly by adjusting the volume and composition of urine.

2. The vertebrate urinary system has a pair of kidneys, a pair of ureters, a urinary bladder, and a urethra. *38.1*

3. Kidneys have many nephrons that filter blood and form urine. A nephron interacts with two sets of blood capillaries: glomerular and peritubular. *38.1, 38.2*

   a. The start of a nephron is cup-shaped (Bowman's capsule). It continues as three tubular regions (proximal tubule, loop of Henle, and distal tubule, which empties into a collecting duct).

   b. Together, Bowman's capsule and the set of highly permeable glomerular capillaries within it are a blood-filtering unit: a renal corpuscle. Blood pressure forces water and small solutes from capillaries, into the fluid-filled cup. Most of the filtrate gets reabsorbed at the nephron's tubular regions and is returned to the blood. A portion is excreted as urine.

4. Urine forms in the nephron by three processes: *38.2*

   a. Filtration of blood at the glomerulus, which puts water and small solutes into the nephron.

   b. Reabsorption. Water and solutes to be conserved leave the nephron's tubular parts and enter capillaries that thread around them. A small volume of water and solutes remains in the nephron.

   c. Secretion. A few substances can leave peritubular capillaries and enter the nephron, for disposal in urine.

5. Two hormones that act on cells of the wall of distal tubules and collecting ducts adjust urine concentration. ADH conserves water by enhancing reabsorption across the wall. In its absence, more water is excreted (urine is dilute). Aldosterone enhances sodium reabsorption. In its absence, sodium is excreted. Angiotensin II stimulates aldosterone secretion (sodium conservation) and thirst (water conservation by calling for more ADH). *38.2*

6. The urinary system acts in concert with buffers and with the respiratory system to maintain the acid–base balance of extracellular fluid. *38.4*

Control of Body Temperature

1. Maintaining an animal's core (internal) temperature depends on balancing metabolically produced heat and heat absorbed from and lost to the environment. *38.6*

2. Animals exchange heat with their environment by four processes: *38.6*

   a. Radiation. Emission from the body of infrared and other wavelengths. Radiant energy can be absorbed at the body surface, then converted to heat energy.

   b. Conduction. Direct transfer of heat energy from one object to another object in contact with it.

   c. Convection. Heat transfer by air or water currents; involves conduction and mass transfer of heat-bearing currents away from or toward the animal body.

   d. Evaporation. Conversion of liquid to a gas, driven by energy inherent in the heat content of the liquid. Some animals lose heat by evaporative water loss.

3. Core temperatures depend on metabolic rates and on anatomy, behavior, and physiology. *38.6*

   a. For ectotherms, core temperature depends more on heat exchange with the environment than on heat generated by metabolism.

   b. For endotherms, core temperature depends more on high metabolic rates and precise controls over heat produced and heat lost.

   c. For heterotherms, the core temperature fluctuates some of the time, and controls over heat balance come into play at other times. *38.6*

Review Questions

1. State the function of the urinary system in terms of gains and losses for the internal environment. State the components of the mammalian urinary system and their functions. *38.1*

2. Label the component parts of this kidney and nephron. *38.1*

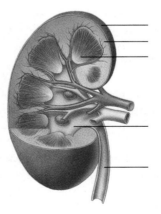 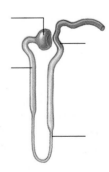

3. Define filtration, tubular reabsorption, and secretion. How does urine formation help maintain an internal environment? *38.2*

4. Which hormone or hormones promote (a) water conservation, (b) sodium conservation, and (c) thirst behavior? *38.2*

5. Name and define the physical processes by which animals gain and lose heat. What are the main physiological responses to cold stress and to heat stress in mammals? *38.6*

1. In mammals, water intake depends on _____ .
   a. absorption from gut   c. a thirst mechanism
   b. metabolism   d. all of the above

2. In mammals, water is lost by way of the _____ .
   a. skin   d. urinary system
   b. respiratory system   e. c and d
   c. digestive system   f. a through d

3. Water and small solutes enter nephrons during _____ .
   a. filtration   c. tubular secretion
   b. tubular reabsorption   d. both a and c

4. Kidneys return water and small solutes to blood by _____ .
   a. filtration   c. tubular secretion
   b. tubular reabsorption   d. both a and b

5. Certain excess solutes move out of peritubular capillaries that thread around the tubular parts of the nephron. These solutes are moved into the nephron during _____ .
   a. filtration   c. tubular secretion
   b. tubular reabsorption   d. both a and c

6. A nephron's reabsorption mechanism depends on _____ .
   a. osmosis across nephron wall
   b. active transport of sodium across nephron wall
   c. a steep solute concentration gradient
   d. all of the above

7. _____ promotes water conservation.
   a. ADH   c. Low extracellular fluid volume
   b. Aldosterone   d. both a and c

8. _____ enhances sodium reabsorption.
   a. ADH   c. Low extracellular fluid volume
   b. Aldosterone   d. both b and c

9. Match the term with the most suitable description.
   ____ renal corpuscle   a. surrounded by saltiest fluid
   ____ distal tubule   b. extra-long loops of Henle
   ____ loop of Henle   c. involves buffer systems
   ____ acid–base balance   d. blood-filtering unit
   ____ kangaroo rat   e. ADH, aldosterone act here

10. Match the term with the most suitable description.
   ____ ectotherm   a. heat transfer by air or water currents
   ____ endotherm   b. body temperature fluctuates some of
   ____ evaporation   the time, is controlled at other times
   ____ heterotherm   c. emission of radiant energy
   ____ radiation   d. direct heat transfer between an object
   ____ conduction   and another object in contact with it
   ____ convection   e. metabolism dictates core temperature
                        f. environment dictates core temperature
                        g. conversion of liquid to gas

Critical Thinking

1. Fatty tissue holds kidneys in place. Rarely, extremely rapid weight loss may cause the tissue to shrink and kidneys to slip from their normal position. If slippage puts a kink in one or both ureters and blocks urine flow, what may happen to the kidneys?

2. Drink one quart of water in one hour. What changes might you expect in your kidney function and in urine composition?

3. In 1912, the ocean liner *Titanic* left Europe on her maiden voyage to America—and a chunk of the leading edge of a glacier in Greenland broke away and floated out to sea. Late at night, off the Newfoundland coast, the iceberg and the *Titanic* made an ill-fated rendezvous (Figure 38.10). The *Titanic* was said to be unsinkable. Survival drills were neglected. There weren't enough

Figure 38.10   Sinking of the *Titanic*, based on eyewitness accounts.

lifeboats to hold even half the 2,200 passengers. The *Titanic* sank in about 2–1/2 hours. Rescue ships were on the scene in less than two hours, but 1,517 bodies were recovered from a calm sea. All the dead had on life jackets; none had drowned. They probably died from _____ . If so, how did their blood flow, metabolism, and skeletal muscle action change prior to death?

4. When iguanas have an infection, they rest for a prolonged period in the sun. Propose a hypothesis to explain why.

5. Out on a first date, Jon takes Geraldine's hand in a darkened theater. "*Aha!*" he thinks. "*Cold hands, warm heart!*" What does this tell us about the regulation of core temperature, let alone Jon?

Selected Key Terms

Salt–Water Balance   renal corpuscle *38.1*   core temperature
acid–base balance *38.4*   renal failure *38.3*   ectotherm
ADH *38.2*   thirst center *38.2*   endotherm
aldosterone *38.2*   tubular   evaporation
angiotensin II *38.2*   reabsorption *38.2*   evaporative
blood *CI*   tubular secretion *38.2*   heat loss
Bowman's capsule *38.1*   urea *38.1*   heterotherm
collecting duct *38.1*   ureter *38.1*   nonshivering
distal tubule *38.1*   urethra *38.1*   heat production
extracellular fluid *CI*   urinary bladder *38.1*   peripheral
filtration *38.2*   urinary excretion *38.1*   vasoconstriction
interstitial fluid *CI*   urinary system *38.1*   peripheral
kidney *38.1*   urine *38.1*   vasodilation
loop of Henle *38.1*   pilomotor response
nephron *38.1*   *Body Temperature* *38.6*   radiation
proximal tubule *38.1*   conduction   shivering response
                             convection

Readings

Sherwood, L. 1997. *Human Physiology*. Second edition. Monterey, California: Brooks-Cole.

Smith, H. 1961. *From Fish to Philosopher*. New York: Doubleday.

On-Line readings at Student Guide for InfoTrac:
www.brookscole.com/biology   643

ANIMAL REPRODUCTION AND DEVELOPMENT

From Frog to Frog and Other Mysteries

With a quavering, low-pitched call that only a female of its kind could find seductive, a male frog proclaims the onset of warm spring rains, of ponds, of sex in the night. By August the summer sun will have parched the earth, and his pond dominion will be gone. But tonight is the hour of the frog!

Through the dark, a hormone-primed female moves toward the vocal male. They meet; they dally in the behaviorally prescribed ways of their species. He clamps his forelegs above her swollen abdomen and gives her a prolonged squeeze (Figure 39.1a). Out into the water streams a ribbon of hundreds of eggs, which the male blankets with a milky cloud of sperm. Soon afterward, fertilized eggs—**zygotes**—are suspended in the water.

For the leopard frog, *Rana pipiens*, a drama begins to unfold that has been reenacted each spring, with only minor variations, for many millions of years. Within a few hours after fertilization, each zygote divides into two cells, the two divide into four, then the four into eight. In less than twenty hours after fertilization, the mitotic cell divisions have produced a ball of cells no

larger than the zygote. It is an early embryonic stage, of a type known as a blastocyst.

The cells continue to divide, but now they start to interact by way of their surface structures and chemical secretions. At prescribed times, many change shape and migrate to prescribed positions. They all inherited the same genetic instructions from the zygote—yet now they start to differ in appearance and function!

Through their associations, the cells form layers of embryonic tissues and then embryonic organs. A pair of tissue regions at the embryo's surface interact with the tissues beneath them. Together they give rise to a pair of eyes. Within the embryo a heart is forming, and soon it starts an incessant, rhythmic beating. In less than a week, events have transformed the frog embryo into a swimming, algae-eating larva called a tadpole.

Several months pass. Legs form; the tail shortens and disappears. The mouth develops jaws that snap shut on insects and worms. Eventually the transformations lead to an adult frog. With luck the frog will avoid predators, disease, and other threats in the months ahead. In time

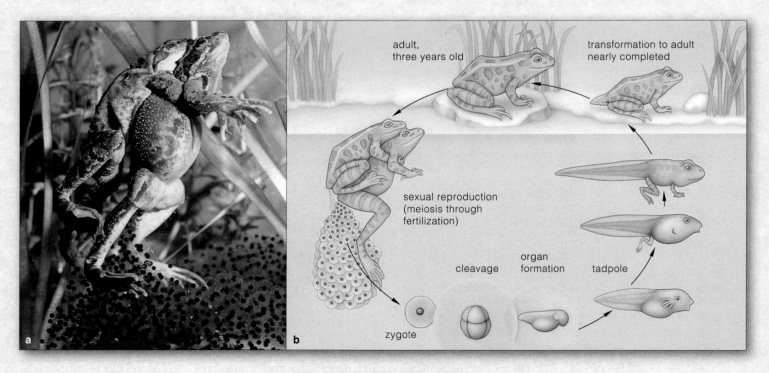

Figure 39.1 Reproduction and development of *Rana pipiens*, the leopard frog. (**a,b**) We zoom in on the life cycle as a male clasps a female in a reproductive behavior called amplexus. The female releases eggs into the water. The male releases sperm over the eggs. A zygote forms when an egg nucleus and a sperm nucleus fuse at fertilization. (**c**) Frog embryos suspended in the water. (**d**) A tadpole. (**e**) A transitional form between a tadpole and the young adult frog (**f**).

it may even call out quaveringly across a moonlit pond, and the life cycle will turn again.

Many years ago you, too, started a developmental journey when a zygote carved itself up. Three weeks into the journey, your embryonic body had the stamp of "vertebrate" on it. A mere five weeks after that, it was a recognizable human in the making!

With this chapter we turn to one of life's greatest dramas—the development of offspring in the image of sexually reproducing parents. The guiding question is this: *How does a single-celled zygote of a frog, human, or any other complex animal become transformed into all the specialized cells and structures of the adult form?* Some answers will become apparent as we move through a survey of basic principles, then through a case study of human reproduction and development.

DEVELOPING EMBRYO

Key Concepts

1. Sexual reproduction dominates the life cycles of nearly all animals. The separation into sexes requires specialized reproductive structures, hormonal control mechanisms, and forms of behavior. Having two distinct sexes affords a great selective advantage—variation in traits among the offspring. This advantage offsets the biological cost of the separation.

2. The life cycles of humans and many other animals proceed through six stages of embryonic development— gamete formation, fertilization, cleavage, gastrulation, organ formation, and growth and tissue specialization. Each stage builds on tissues and structures that formed during the stage preceding it.

3. In a developing embryo, the fate of each type of cell depends partly on cleavage, which distributes different regions of the fertilized egg's cytoplasm to different daughter cells. It also depends on interactions among cells of the embryo. These activities are the foundation for cell differentiation and morphogenesis.

4. With cell differentiation, each cell type selectively uses certain genes and synthesizes proteins not found in other types, and thereby becomes unique in structure and function. With morphogenesis, tissues and organs change in size, shape, and proportion. They also become organized relative to one another in prescribed patterns.

5. All complex, multicelled animals that show extensive cell differentiation undergo aging. Their cells gradually break down in structure and function, which leads to the decline of tissues, organs, and eventually the body.

6. Human males continually produce sperm from puberty onward. The sex hormones testosterone, LH, and FSH, are central to controlling male reproductive functions.

7. From puberty onward, human females are fertile on a cyclic basis. Each month during their reproductive years, an egg is released from one of a pair of ovaries, and the lining of the uterus is primed for a possible pregnancy. The hormones estrogen, progesterone, FSH, and LH are central to controlling the cyclic activity.

THE BEGINNING: REPRODUCTIVE MODES

Sexual Versus Asexual Reproduction

In earlier chapters, you learned about the cellular basis of **sexual reproduction**. Briefly, by this reproductive mode, meiosis and gamete formation typically proceed in two prospective parents. At fertilization, a gamete from one parent fuses with a gamete from the other to form the zygote, the first cell of the new individual. You also read about **asexual reproduction**, in which a single parent organism produces offspring by various means (but not by gamete formation). We now turn to a few structural, behavioral, and ecological aspects of the two reproductive modes.

Picture a scuba diver accidentally kicking a sponge. A tissue fragment breaks away from the sponge body, then grows and develops by mitotic cell divisions and cell differentiation into a new sponge. Or picture one of the flatworms undergoing transverse fission while it glides along through the water. First its body constricts below the midsection. The part behind the constriction grips a substrate and starts a tug-of-war with the part in front, and it splits off a few hours later. Both parts go their separate ways, regenerate what's missing, and thus become a whole worm. Only some species do this.

In such cases of *asexual* reproduction, all offspring are genetically the same as their individual parent, or nearly so. Phenotypically they are much the same, too. We can speculate that phenotypic uniformity is useful when each individual's gene-encoded traits are highly adapted to a limited and more or less consistent set of environmental conditions. Most variations introduced into the finely tuned gene package would not do much good, and often they could do harm.

Most animals live where opportunities, resources, and danger are highly variable. For the most part, such animals reproduce sexually, meaning female and male parents bestow different mixes of alleles on offspring. As you already learned from Section 9.1, the resulting variation in traits improves the odds that some of the offspring, at least, should survive and reproduce even if conditions change in the environment.

Costs and Benefits of Sexual Reproduction

Among animals, separation into sexes is not without cost. Some cells that can serve as gametes must be set aside and nurtured. Housing and delivering gametes close to or into a prospective mate require specialized reproductive structures. Mating often requires special forms of behavior, such as courtship, that can promote fertilization. Mating also requires built-in controls that can synchronize the timing of gamete formation, sexual readiness, even parental behavior in two individuals.

Figure 39.2 *Facing page:* Examples of where the embryos of invertebrates and vertebrates develop, how they are nourished, and how (if at all) parents protect them.

Snails (**a**) and spiders (**b**) are *oviparous*. They release eggs from which the young later hatch (*ovi–*, egg; *parous*, produce). Snails are hermaphroditic parents that leave eggs unprotected. Spider eggs develop in a silk egg sac that the female anchors to a substrate or carts around with her. The females often die soon after making the sac. But some guard the sac, then cart spiderlings about for a few days and even feed them.

(**c**) Birds also are oviparous. Their fertilized eggs have large yolk reserves, and they, too, develop and hatch outside the mother's body. Unlike snails, one or both parent birds expend considerable energy feeding and caring for the young.

Some fishes, all lizards, and many snakes are *ovoviviparous*. Their fertilized eggs develop inside the mother, then offspring are born live (*vivi–*, alive). Yolk reserves, not the mother's own tissues, sustain the eggs. As one example, the liveborn copperhead offspring in (**d**) are still in their egg sacs.

Most mammals are *viviparous*, meaning their young are born live. (**e**, **f**) In kangaroos and some other species, embryos are born in unfinished form. The young of this marsupial complete development inside a pouch on the mother's ventral surface. The juvenile stages (joeys) continue to draw nourishment from mammary glands in the mother's pouch. (**g**) By contrast, the human female retains the fertilized egg in her body. Maternal tissues nourish the developing individual until the time of its birth, in the manner described later in this chapter.

Just look at the question of *reproductive timing*. How do mature sperm in one individual become available exactly when the eggs mature in a different individual? Timing depends upon energy outlays for constructing, maintaining, and operating neural as well as hormonal control mechanisms in each parent. Also, parents must produce mature gametes in response to the same cues, such as a seasonal change in daylength, that mark the onset of the most appropriate time of reproduction for their species. Example: Male and female moose become sexually active only during late summer and early fall. The coordinated timing means that their offspring will be born the following spring—a time when the weather improves and food is plentiful for many months.

Locating and then actually recognizing a potential mate of the same species is another challenge. Different species invest energy when they synthesize signaling molecules—pheromones. They construct visual mating signals such as feathers of certain colors and patterns, and complex sensory receptors to detect the specific signals being sent. The males often expend astonishing amounts of energy to execute tricky courtship routines, as you will read in Chapter 41.

Ensuring the survival of offspring is costly (Figure 39.2). Many invertebrates, bony fishes, and frogs simply

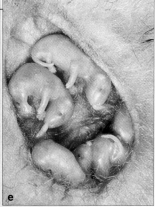

liveborn snake inside egg sac

release eggs and motile sperm into the surrounding water, as in Figure 39.1a. If each adult were to produce only *one* sperm or *one* egg each season, the chances for fertilization would not be good. These species invest energy in making numerous gametes, often thousands of them. As another example, nearly all land animals rely on internal fertilization, the union of sperm and egg *within* the body of a female. They invest metabolic energy to construct elaborate reproductive organs, such as a penis (by which certain males deposit sperm in a female) and a uterus (a chamber inside the females of certain species where the embryo grows and develops).

Finally, animals set aside energy in forms that can *nourish the developing individual* until it has developed enough to feed itself. For instance, nearly all animal eggs contain **yolk**. This substance, rich in proteins and lipids, nourishes embryonic stages. The eggs of some species have much more yolk than others. Sea urchins make tiny eggs with little yolk, release large numbers of them, and thus limit the biochemical investment in each one. Within twenty-four hours, each fertilized egg has developed into a freely moving, self-feeding larva. Sea stars and other predators eat most of the eggs. So for sea urchins, bestowing as little metabolic energy as possible on each one of many individual gametes pays off, in terms of reproductive success.

By contrast, mother birds lay truly yolky eggs. Yolk nourishes the bird embryo through an extended period of development inside an eggshell that forms after the egg is fertilized. Your mother put tremendous demands on her body to protect and nourish you through nine

months of development inside her, starting from the time you were an egg with almost no yolk. After you implanted yourself into her uterus, physical exchanges with her tissues supported you through that extended pregnancy (Figure 39.2g).

As these few examples suggest, animals show great diversity in reproduction and development. However, as you will see in the sections to follow, some patterns are widespread throughout the animal kingdom, and they can serve as a framework for our reading.

Separation into male and female sexes requires special reproductive cells and structures, neural and hormonal control mechanisms, and forms of behavior. A selective advantage—variation in traits among offspring—offsets biological costs associated with the separation into sexes.

STAGES OF DEVELOPMENT—AN OVERVIEW

Embryos are a class of transitional forms between the fertilized egg and an adult. Although they all start out as a single cell, embryos of different species often look different as they grow and develop. For example, you do not look like a frog now, and you did not look like one when you and the frog were early embryos, either. However, despite such differences in appearance, it is possible to identify certain patterns in how embryos of nearly all animal species develop.

Figure 39.3 is an overview of the stages of animal development. During **gamete formation**, the first stage, eggs or sperm develop inside the reproductive organs of a parent body. **Fertilization**, the second stage, starts when sperm's plasma membrane fuses with the plasma membrane of an egg. Fertilization is over when the egg nucleus and sperm nucleus fuse and form a zygote.

The third stage, **cleavage**, is a program of mitotic cell divisions that divide the volume of egg cytoplasm into a number of **blastomeres**—smaller cells, each with its own nucleus. There is no growth during this stage. Cleavage only increases the number of cells; it does not change the original volume of the egg cytoplasm.

As cleavage draws to a close, the pace of mitotic cell division slackens. The embryo enters **gastrulation**. This fourth stage of animal development is a time of major cellular reorganization. The new cells become arranged into a gastrula with two or three primary tissues, often called germ layers. Cellular descendants of the primary tissues give rise to all tissues and organs of the adult:

1. **Ectoderm.** This is the *outermost* primary tissue layer, the one that forms first in the embryos of every animal. Ectoderm is the embryonic forerunner of the cell lineages that give rise to tissues of the nervous system and to the integument's outer layer.

2. **Endoderm.** Endoderm is the *innermost* primary tissue layer. It is the embryonic forerunner of the gut's inner lining and organs derived from the gut.

3. **Mesoderm.** This *intermediate* primary tissue layer gives rise to muscle, most of the skeleton; circulatory, reproductive, and excretory organs; and connective tissues of the gut and integument. It evolved hundreds of millions of years ago, and it was a pivotal step in the evolution of nearly all large, complex animals.

After the primary tissue layers form in embryos, they give rise to subpopulations of cells. This marks the onset of **organ formation**. The subpopulations become unique in structure and functions, and their descendants give rise to different kinds of tissues and organs.

During the last stage of animal development, **growth and tissue specialization**, organs increase in size, and they gradually assume specialized functions. This stage continues into adulthood.

Now take a look at Figure 39.4. Its photographs and sketches show several stages in the embryonic development of a typical animal, a frog. Take a moment to study this

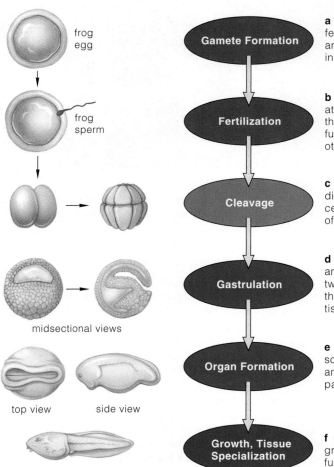

frog egg

frog sperm

midsectional views

top view   side view

Gamete Formation

a Eggs form and mature in female reproductive organs, and sperm form and mature in male reproductive organs.

Fertilization

b A sperm and an egg fuse at their plasma membrane, then the nucleus of one fuses with the nucleus of the other to form the zygote.

Cleavage

c By a series of mitotic cell divisions, different daughter cells receive different regions of the egg cytoplasm.

Gastrulation

d Cell divisions, migrations, and rearrangements produce two or three primary tissues, the forerunners of specialized tissues and organs.

Organ Formation

e Subpopulations of cells are sculpted into specialized organs and tissues in prescribed spatial patterns at prescribed times.

Growth, Tissue Specialization

f Organs increase in size and gradually assume specialized functions.

Figure 39.3 Overview of the stages of animal development. We use a few forms that appear during the frog life cycle as examples.

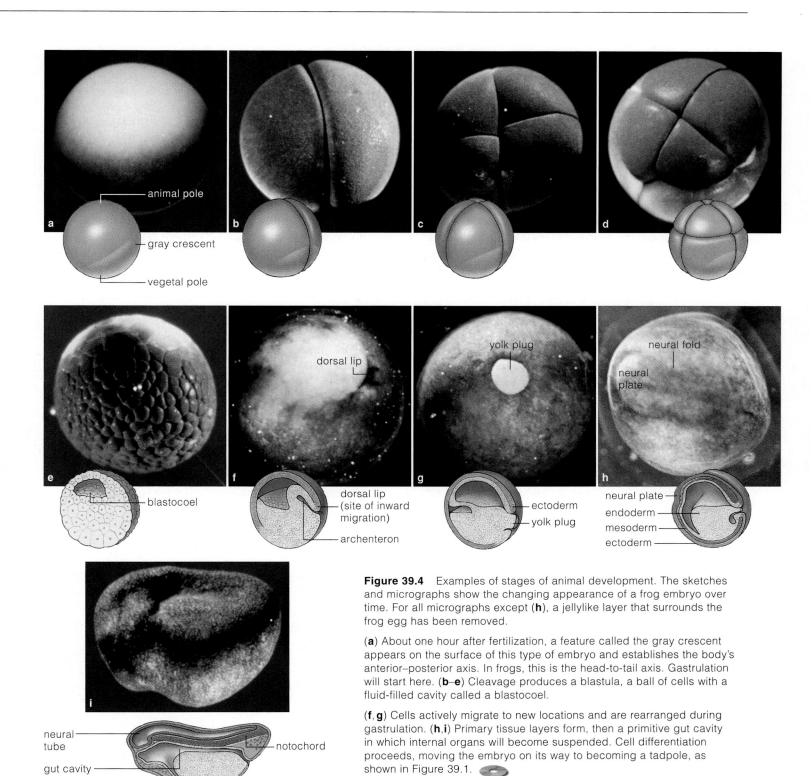

Figure 39.4 Examples of stages of animal development. The sketches and micrographs show the changing appearance of a frog embryo over time. For all micrographs except (**h**), a jellylike layer that surrounds the frog egg has been removed.

(**a**) About one hour after fertilization, a feature called the gray crescent appears on the surface of this type of embryo and establishes the body's anterior–posterior axis. In frogs, this is the head-to-tail axis. Gastrulation will start here. (**b**–**e**) Cleavage produces a blastula, a ball of cells with a fluid-filled cavity called a blastocoel.

(**f**, **g**) Cells actively migrate to new locations and are rearranged during gastrulation. (**h**, **i**) Primary tissue layers form, then a primitive gut cavity in which internal organs will become suspended. Cell differentiation proceeds, moving the embryo on its way to becoming a tadpole, as shown in Figure 39.1.

illustration, for it reinforces an important concept. The structures that form during one stage of development serve as the foundation for the stage that comes after it. Successful development depends on whether all of the structures develop according to normal patterns, in a prescribed sequence.

Animal development proceeds from gamete formation and fertilization through cleavage, gastrulation, then organ formation, and finally growth and tissue specialization.

Development cannot proceed properly unless each stage is successfully completed before the next begins.

EARLY MARCHING ORDERS

Information in the Egg Cytoplasm

Why don't you have an arm attached to your nose or toes growing from your navel? The patterning of body parts for all complex animals, including yourself, starts with messages built into the cytoplasm of immature eggs—or **oocytes**—even before sperm enter the picture. A **sperm**, remember, consists only of paternal DNA and a bit of equipment, including a tail, that helps it reach and penetrate an egg. Compared to a sperm, an oocyte is much larger and more complex (Section 9.5).

While an oocyte is maturing, its volume increases. Enzymes, mRNA transcripts, and other factors become stockpiled in different parts of the cytoplasm. Typically they will be activated, after fertilization, for use in the early rounds of DNA replication and cell division. Also present are tubulin molecules and factors that will help control the angle and timing of tubulin assembly into microtubules for mitotic spindles. How the cytoplasm gets cut depends on such factors and on the yolk in the cytoplasm. The amount of yolk and its distribution will influence how large the resulting blastomeres will be.

Such regionally localized aspects of the oocyte are "maternal messages." We find evidence of their effects as early as fertilization. As an example, a frog egg has pigment granules concentrated near one pole and yolk near the other. The egg's cortex (the plasma membrane and the cytoplasm just under it) undergoes structural reorganization when a sperm fertilizes the egg. Part of the cortex shifts toward the point of sperm entry, and it exposes a crescent-shaped area of yolky cytoplasm. This area of intermediate pigmentation, a gray crescent, forms near the frog egg's midsection. It establishes the body's anterior–posterior axis. Section 29.6 depicts the directional planes for this body axis and others.

In itself, a gray crescent is not evidence of regional differences in maternal messages. We get such evidence from experiments and by observing embryos in which localized cytoplasmic differences are obvious enough to be tracked during development (Figure 39.5). Thus, if you were to track the development of fertilized frog eggs, you would observe that the gray crescent is the site where gastrulation normally begins.

Cleavage—The Start of Multicellularity

Once an egg is fertilized, the zygote enters cleavage. Beneath its plasma membrane, its midsection has a ring of microfilaments made of the contractile protein actin. As the microfilaments slide past one another, the ring tightens and pinches in the cytoplasm. The cell surface above the shrinking ring is drawn inward, as a cleavage furrow (Section 9.5). It is this force of contraction that splits the cytoplasm into two blastomeres.

Simply by virtue of where they form, blastomeres end up with different maternal messages. This outcome

Figure 39.5 Examples of experiments that illustrate how the cytoplasm of a fertilized egg has localized differences that help determine the fate of cells in a developing embryo. The cortex of frog eggs contains granules of dark pigment concentrated near one pole. At fertilization, a portion of the granule-containing cortex shifts toward the point of sperm entry. The shift exposes lighter colored, yolky cytoplasm in a crescent-shaped gray area:

pigmented cortex
yolk-rich cytoplasm
sperm penetrating frog egg
gray crescent

The first cleavage normally puts part of the gray crescent in both of the resulting blastomeres.

(**a**) For one experiment, the first two blastomeres that formed were physically separated from each other. Each blastomere still gave rise to a whole tadpole.

(**b**) For another experiment, a fertilized egg was manipulated so the cut through the first cleavage plane missed the gray crescent. Of two resulting blastomeres, only one received the gray crescent. It alone developed into a normal tadpole. The other, deprived of maternal messages present in the gray crescent, gave rise to a ball of undifferentiated cells.

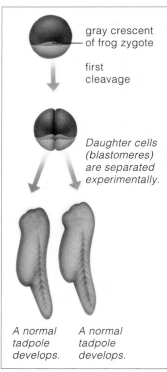

gray crescent of frog zygote

first cleavage

Daughter cells (blastomeres) are separated experimentally.

A normal tadpole develops.

A normal tadpole develops.

a EXPERIMENT 1

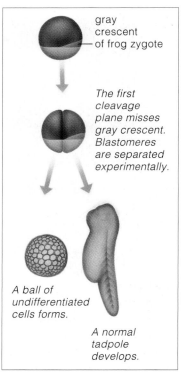

gray crescent of frog zygote

The first cleavage plane misses gray crescent. Blastomeres are separated experimentally.

A ball of undifferentiated cells forms.

A normal tadpole develops.

b EXPERIMENT 2

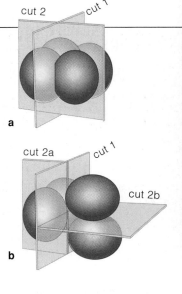

Figure 39.6 Comparison of the early cleavage planes for (**a**) sea urchins and (**b**) mammals. The early cuts of a fertilized egg are radial in sea urchins and rotational in mammals. (**c**) To get a sense of the impact of cleavage, consider Sabra and Nina, identical twins who started out life from the same zygote. The first two blastomeres formed at cleavage, the inner cell mass, or another early embryonic stage split. The split was the start of two genetically identical, look-alike individuals.

of cleavage, called **cytoplasmic localization**, helps seal the developmental fate of each cell's descendants. Its cytoplasm alone may have molecules of a protein that can activate, say, a gene coding for a certain hormone. And its descendants alone will make that hormone.

Cleavage Patterns

The simplest cleavage pattern has full cuts perpendicular to the mitotic spindle. The cuts parcel out a copy of the nucleus to each blastomere, so all the blastomeres are genetically equivalent. The cuts made during cleavage only divide the volume of cytoplasm into increasingly smaller cells; the blastomeres don't grow in size.

In most animals, a zygote's genes are silent during early cleavage; proteins and mRNAs stockpiled in the cytoplasm control how fast the cuts proceed and how the blastomeres become arranged. In mammals, certain genes must be activated first. Cleavage cannot even be completed without the protein products of those genes.

We can correlate cleavage patterns of major animal groups with the amount and distribution of yolk, which usually restricts cleavage. Cuts may go right through a nearly yolkless egg. Early cuts go only partway through an egg with yolk stockpiled at one end. That egg has polarity. Its *vegetal* pole is the yolk-rich end; its *animal* pole is closest to the nucleus. Frog eggs are like this.

Other heritable factors affect the cuts. Frog and sea urchin eggs undergo *radial* cleavage; cleavage furrows run horizontal and vertical to the animal–vegetal axis (Figure 39.6*a*). Successive cuts produce a **blastula**, an embryonic stage in which the blastomeres surround a fluid-filled cavity (a blastocoel). But a sea urchin egg is nearly yolkless, so the cuts produce horizontal rows of blastomeres. A frog egg's yolk impedes cuts near the vegetal pole (Figure 39.4*e*). More, smaller blastomeres form near its animal pole, and the blastocoel is offset. Between the time that 16 to 64 blastomeres form, any amphibian blastula looks like a mulberry. It's called a morula (after the Latin word for mulberry).

What about eggs of reptiles, birds, and most fishes? They undergo *incomplete* cleavage. The large volume of yolk restricts the early cuts to a small, caplike region near the animal pole. The result is two flattened layers of cells with a narrow cavity in between.

What about a mammal's egg? It undergoes *rotational* cleavage, with blastomeres dividing slowly at different times. The first cut is vertical through both poles. In the next cleavage, one blastomere is cut vertically and one horizontally (Figure 39.6*b*). The third cleavage yields a loose arrangement with space between eight cells. The cells undergo compaction; they abruptly huddle into a compact ball. Tight junctions stabilize the ones outside and seal off those inside. Gap junctions form between the inner cells and facilitate chemical communication among them. Descendants of the outer cells form a thin surface layer. Their secretions form a fluid-filled cavity in the ball, and the inner cells mass together against one side of it. We call this type of blastula a **blastocyst**. In all mammals, including humans, the inner cell mass gives rise to the embryo proper (Section 39.12).

Once in a while, the first two blastomeres, the inner cell mass, or even a later stage splits. Such a split may result in *identical twins*, which have the same genetic makeup and share a common placenta (Figure 39.6*c*). By contrast, *fraternal twins* arise from two oocytes that matured and then became fertilized during the same menstrual cycle. Each is serviced by its own placenta.

The egg cytoplasm contains maternal instructions in the form of regionally distributed enzymes and other proteins, mRNAs, cytoskeletal elements, yolk, and other factors.

Cleavage divides a zygote into blastomeres, each with a localized part of maternal messages inherent in the egg cytoplasm. We call this outcome cytoplasmic localization.

Differences in the amount and distribution of yolk and other heritable factors give rise to different patterns of cleavage. Those patterns contribute to differences in body plans among different animal groups.

HOW DO SPECIALIZED TISSUES AND ORGANS FORM?

Nearly all animals have a gut, with tissues and organs that function in digestion and absorption of nutrients. They have surface parts that protect internal parts and detect what is going on outside. In between, most have organs, such as those dealing with structural support, movement, and blood circulation. This three-layer body plan emerges as cleavage ends and gastrulation begins. The embryo's size increases little, if any. But cells start migrating to new, predictable locations to form primary tissue layers—ectoderm, endoderm, and mesoderm.

Figure 39.7 shows steps in the formation of a sea urchin gastrula. Figure 39.8 shows the emergence of a bird embryo's anterior–posterior axis. In all vertebrates, this axis defines where a **neural tube**, the forerunner of a brain and spinal cord, will form. All such organs now start to form by cell differentiation and morphogenesis.

Cell Differentiation

All cells of an embryo have the same number and kinds of genes (they all descend from the same zygote). They all activate the genes that specify proteins necessary for cell survival, such as histones and glucose-metabolizing enzymes. From gastrulation onward, certain groups of genes are used in some cells but not in others. When a cell selectively activates genes and synthesizes proteins not found in other cell types, we call this process **cell differentiation**. You read about the molecular basis of selective gene expression in Chapter 14. Basically, the process yields the proteins required for distinctive cell structures, products, and functions.

For instance, when your eye lenses developed, only some cells could activate genes for crystallins, a family of proteins that are assembled into transparent fibers for each lens. Long crystallin fibers formed in the cells,

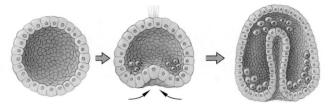

a Blastula   **b** Cell migrations in early gastrula

Figure 39.7 (**a**,**b**) A few early stages of development of a sea urchin (*Lytechinus*), cross-section. After cleavage, the resulting blastula becomes transformed into a gastrula. (**c**) Scanning electron micrograph of cells at the gastrula surface cells. Some cells have begun their inward migration.

c endoderm cells migrating inward

forcing them to lengthen and flatten. The differentiated cells provided each lens with unique optical properties. And crystallin-producing cells are only 1 of 150 or so differentiated cell types now present in your body.

As many experiments show, nearly all cells become differentiated with no loss of genetic information. For instance, John Gurdon stripped unfertilized eggs of an African clawed frog (*Xenopus laevis*) of their nucleus. He ruptured the plasma membrane of intestinal cells from tadpoles of the same species. He left the nucleus and much of the cytoplasm of these fully differentiated cells intact and inserted them into the enucleated eggs. Some eggs gave rise to a complete frog! The intestinal cells still had the same number and kinds of genes as

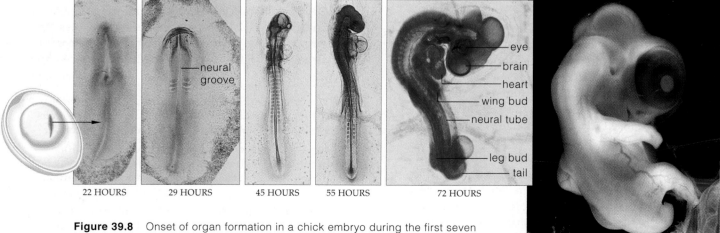

22 HOURS   29 HOURS   45 HOURS   55 HOURS   72 HOURS

neural groove

eye
brain
heart
wing bud
neural tube
leg bud
tail

168 HOURS (SEVEN DAYS OLD)

Figure 39.8 Onset of organ formation in a chick embryo during the first seven days of development. The heart begins to beat between thirty and thirty-six hours. You may have observed such embryos at the yolk surface of raw, fertilized eggs.

Figure 39.9 Examples of morphogenesis. (**a**) Cell migration. A nerve cell (*orange*) "climbs" through a developing brain to its final position, using a glial cell (*yellow*) as a highway. (**b**) How a neural tube forms. As gastrulation ends, ectoderm is a uniform sheet of cells. In some cells, microtubules lengthen. The elongating cells form a neural groove. In other cells, rings of microfilaments at one end constrict, so the cells become wedge shaped. Their part of the sheet folds over to form the tube.

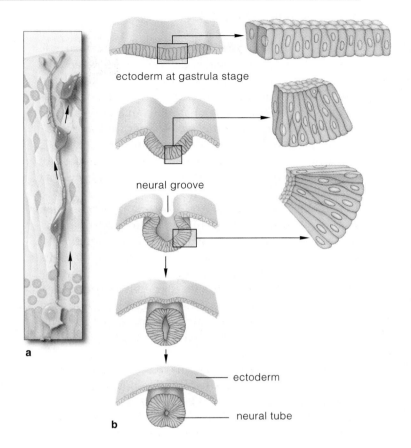

ectoderm at gastrula stage

neural groove

a

ectoderm

neural tube

b

the zygote; its nucleus still had all the genes required to make all cell types that make up a frog.

Embryonic cells of humans also retain the capacity to give rise to a whole individual. This is what happens after spontaneous separation of human blastomeres at the two-cell stage. The split does not result in two half-embryos, but rather in identical twins (Figure 39.6).

Morphogenesis

Morphogenesis refers to a program of orderly changes in an embryo's size, shape, and proportions, the result being specialized tissues and early organs. As part of the program, cells divide, grow, migrate, and change in size. Tissues expand and fold, and the cells in some of them die in controlled ways at prescribed locations.

Think of active cell migration. *Cells send out and use pseudopods that move them along prescribed routes.* When they reach their destination, they establish contact with cells already there. For instance, forerunners of neurons interconnect this way as a nervous system is forming.

How do cells "know" where to move? They respond to adhesive cues, as when migrating nerve cells stick to adhesion proteins on glial cells but not blood vessels (Figure 39.9*a*). They respond to chemical gradients, too. Their migrations may be coordinated by the synthesis, release, deposition, and removal of specific chemicals in the extracellular matrix. And adhesive cues tell cells when to stop. Cells will migrate to regions of strongest adhesion, but once there, further migration is impeded. Section 20.7 describes the chemical-induced migration of cells of the slime mold *Dictyostelium discoideum*.

Also, *whole sheets of cells expand and fold inward and outward.* Microtubules lengthen and microfilament rings constrict inside the cells. The assembly and disassembly of these components of the cytoskeleton are aspects of a controlled program of changes in cell shape.

Through such controlled, localized events, the size, shape, and proportions of body parts emerge. Example: Figure 39.9*b* shows what happens after three primary tissues form in embryos of amphibians, reptiles, birds, and mammals. At the embryo midline, ectodermal cells elongate and form a neural plate—the first sign that a region of ectoderm is on its way to becoming nervous tissue. Microtubules lengthen in some cells; other cells become wedge shaped as a microfilament ring contracts

and constricts them at one end. The changes in shape cause tissue flaps to fold over and meet at the midline, thus forming the neural tube (compare Section 30.7).

As a last example, *programmed cell death helps sculpt body parts.* Cells that function for a limited time only in embryos execute themselves. By this form of cell death, called apoptosis, molecular signals from certain cells activate weapons of self-destruction stockpiled inside other, target cells. Section 14.6 describes what happens to targets at the molecular level. For now, think about a human embryo's hand when it is still a paddle (Section 8.4). Cartilage models of the digits form inside it. Cells in thin tissue zones between digits have receptors for proteins, each a product of selective gene expression. When the proteins bind to their receptors, cells commit mass suicide, and digits separate. In rare cases, a gene mutation blocks apoptosis and the digits stay webbed.

In cell differentiation, a cell selectively uses certain genes and makes certain proteins, not found in other cell types, for distinctive cell structures, products, and functions.

Morphogenesis is a program of orderly changes in body size, shape, and proportions. It results in specialized tissues and organs at prescribed locations, at prescribed times.

Morphogenesis involves cell division, active cell migration, tissue growth and foldings, changes in cell size and shape, and programmed cell death by way of apoptosis.

PATTERN FORMATION

In developing embryos, differences between cells arise in spatially organized ways. Cells divide, differentiate, and live or die; they migrate and stick to like cells in tissues. Their cellular descendants fill in body details in ordered patterns. Cells of amphibian embryos make a head at one end and a tail at the other, and so on.

A cell locked in at a given place in the embryo has chemical memory; it selectively reads some genes but not others. It makes and secretes molecular signals that change the activities of its neighbors, and it responds to neighborhood signals. And its memory is passed on to all of its descendants, which go on to form tissues in places where we expect them to be. Now is the time of **embryonic induction**—the developmental fates of the embryonic cell lineages change when exposed to gene products from adjacent tissues. Through these changes, specialized tissues and organs emerge from clumps of embryonic cells in predictably ordered, spatial patterns. We call their emergence **pattern formation**.

We have experimental evidence of cell memory. For instance, researchers excised the dorsal lip of a normal axolotl embryo, then grafted it into a novel location in a different axolotl embryo (Figure 39.10). Gastrulation proceeded at the recipient's own dorsal lip *and at the graft*. Siamese twins—a double embryo having two sets of body parts—developed. Such studies showed that the dorsal lip is central to organizing this amphibian's main axis; how else would the two-tailed, two-headed tadpole in Figure 39.10 form? In fact, the dorsal lip was the first embryonic signaling center to be discovered.

a Dorsal lip excised from donor embryo is grafted to an abnormal site in another embryo.

b Graft induces another site of invagination.

c Gastrula develops into a "double" tadpole. Most of its tissues originated from the host embryo.

Figure 39.10 Experimental evidence that the dorsal lip of an embryonic axolotl (a type of amphibian) controls gastrulation. When transplanted to a different site in another axolotl embryo, it organized the formation of a second set of body parts.

Embryonic Signals in Pattern Formation

Cytoplasmic localization during cleavage ensures that different cells will send and receive different chemical signals. While the embryo forms, cells that started out as neighbors use the signals. So do cells that meet up by gastrulation and other morphogenetic movements.

For example, cells of the dorsal lip are the source of a morphogen. **Morphogens** are degradable molecules that diffuse as long-range signals from an embryonic signaling center. The concentration gradient helps cells assess their position and affects how they differentiate. The signal is strongest at the start of the gradient and weakens with distance. Cells in different body regions affected by the gradients receive different information, so they selectively express different genes as a result.

Other signals are short range, involving cell-to-cell contacts. Example: Changes in the expression of genes for cadherins, integrins, and other glycoproteins at the cell surface alter how cells associate with one another when the gastrula, then organs, form. When a cell starts to make a cytoskeletal protein, it may move or lengthen in some direction. If adhesion proteins are affected, a cell may cohere with neighbors, break free and migrate, or become segregated from cells of an adjoining tissue.

Also, cells respond differently to signals depending on when they receive them. Graft a bit of animal pole epithelium from an early gastrula over a rudimentary eye of a later embryo, and it gives rise to something like neural tube tissue. Leave it alone in vitro for a few hours before the graft, and it differentiates into a lens. Leave it even longer in vitro and it will not respond at all to the inductive signals from the eye tissue.

A Theory of Pattern Formation

Do the same classes of genes govern spatial patterning in all animals? It now appears so. After studies of many diverse animals, researchers have put together a theory of pattern formation. Here are its key points:

1. The formation of tissues and organs in ordered, spatial patterns starts with cytoplasmic localization. It unfolds through short-range cell-to-cell contacts, and through long-range signals—chemical gradients— that weaken with distance and thus provide different information to cells in different locations.

2. Morphogens and other inducer molecules diffuse through embryonic tissues, sequentially activating classes of **master genes**, such as the homeotic genes found in all major animal groups.

3. Master gene products and control elements interact to map out the overall body plan. Products appear as blocks of genes are activated and suppressed in cells along the embryo's anterior–posterior axis and dorsal–ventral axis. Other gene products interact to fill in the details of specific body parts.

Figure 39.11 (**a**) Fate map for the surface of a *Drosophila* zygote, showing the origin of each kind of differentiated cell in the adult. An unfertilized egg shows polarity. The embryo develops a head end and a tail end, with a series of body segments in between. Dashed lines mark regions that become different body segments. Genes specify whether each segment will grow legs, or legs and wings, and so on.

(**b**) Pattern formation in *Drosophila*. Products of *maternal effect* genes for regulatory proteins become localized in the egg cytoplasm and influence gene expression in the zygote. Products of those genes activate or inhibit *gap* genes, which map out broad body regions. One gap gene product creates a chemical gradient that affects others. Differences in concentrations of gap gene products activate *pair-rule* genes, the products of which accumulate in bands. These products activate *segment polarity* genes, which divide the embryo into segment-sized units. Interactions among products of gap, pair-rule, and segment polarity genes control expression of *homeotic* genes that govern how each segment will develop.

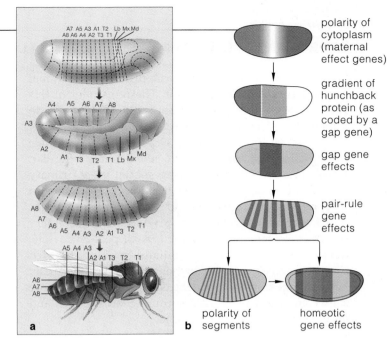

When organs first form, gene products and control elements activate and inhibit blocks of genes—and they do so in similar ways among all major animal groups. The inductions they direct map out the overall body plan, including its major axes, as in Figure 39.11. When master genes fail to do so, disaster follows, as when a heart forms at the wrong location. Once the basic body plan is locked in, the many inductions remaining have only localized effects. For instance, if a lens fails to form, then only the eye will be affected.

Evolutionary Constraints on Development

How long have the master genes been operating? The basic body plan for sponges, flatworms, mollusks, flies, vertebrates, and all other major animal groups hasn't changed much for about 500 million years. *These many millions of species are variations on a few dozen plans!*

Why is this so? Apparently there are very few new master genes, so variations in body plans may be more of an outcome of how control genes control each other. For example, the *eyeless* gene regulates eye formation in fruit flies. Humans have a nearly identical gene which, in mutated form, results in eyeless babies. The *distalless* gene in fruit flies causes legs to grow from leg buds. The same gene determines the fate of cells that give rise to crab legs, beetle legs, butterfly wings, sea star arms, fish fins, and mouse feet. As you know, legs, fins, and wings are different structures with different functions. For example, beetle legs have their skeletal structures on the outside, your legs have them on the inside. Yet the structures all start out as buds from the main body axis. Similarly, insect, squid, and vertebrate eyes differ greatly in their structure, and yet the genes that govern their formation are nearly identical.

And so, rather than starting from scratch with new genes, diverse animals are using identical or similar master genes to control development. By analogy, they have old software programs in more recent computers.

For a long time, we have known about the *physical* constraints (including the surface-to-volume ratio) and *architectural* constraints (as imposed by body axes). The kinds of events described in this chapter indicate there also are *phyletic* constraints on change. These are the constraints imposed on each lineage by interactions of the master organizer genes, which operate when organs form and control induction of the basic body plan.

That might be why we have so many species and so few body plans. Once master genes had evolved and were engaged in intricate interactions, it would have been difficult to change the body's basic characteristics without destroying the embryo. Mutations have indeed added marvelous variations to animal lineages. But the basic body plans have prevailed through time; maybe it simply proved unworkable to start all over again.

Pattern formation is the orderly, sequential sculpting of embryonic cells into specialized tissues and organs. Cells differentiate partly through cytoplasmic localization and, later, according to chemical memory of their position in the developing embryo.

The pattern starts with different chemical signals inside blastomeres having different maternal messages. Cell-to-cell inductive interactions fill in the details of the pattern.

Inductive interactions among classes of master genes map the basic body plan and specify where and how body parts develop. Products of their selective expression are like long-range and short-range beacons that help cells assess their position in the embryo. Cells respond differently to the signals at different times of development.

Physical, architectural, and phyletic constraints limit the evolution of animal body plans. Of millions of species, we see only a few basic plans. The master genes that guide development are similar or identical in all major groups.

REPRODUCTIVE SYSTEM OF HUMAN MALES

So far, you've looked at the basic principles of animal reproduction and development. The remainder of this chapter shows how you can apply these principles to humans, starting with the male reproductive system.

As Figure 39.12 and Table 39.1 show, an adult male has a pair of gonads, of a type called **testes** (singular, testis). In addition to producing sperm, these primary reproductive organs make sex hormones, which govern male reproductive function and secondary sexual traits.

Mature gametes start to be produced and secondary sexual traits emerge during the time of post-embryonic development called **puberty**. Enlarging testes are often the first sign that boys have entered puberty, usually between ages twelve and sixteen. Additional signs are a growth spurt, more hair sprouting on the face and elsewhere, and a deepening voice.

Where Sperm Form

Recall, from Section 11.3, that testes form on the wall of an XY embryo's abdominal cavity. Before birth, they descend into the scrotum, an outpouching of skin that is suspended beneath the pelvic region. At the time of birth, testes are fully formed miniatures of the adult organs. They start to produce sperm at puberty.

Figure 39.12a shows the scrotum's position in an adult male. If sperm cells are to develop properly, the temperature within the scrotum must be kept a few degrees cooler than the core temperature of the rest of the body. A control mechanism works by stimulating and inhibiting the contraction of smooth muscles located in the scrotum's wall. It helps ensure that the internal temperature of the scrotum stays close to 95°F. When air just outside the body is too cold, contractions draw the pouch up, closer to the body mass, which is warmer. When it is warmer outside, the muscles relax and lower the pouch.

Packed within each testis are a large number of small, highly coiled tubes, called **seminiferous tubules**. Sperm formation begins in these tubules, in the manner described in Section 39.7.

Where Semen Forms

Mammalian sperm travel from each testis through a series of ducts, which lead to the urethra. They are not quite mature when they enter the epididymis, the first duct, which is lengthy and coiled. The secretions from glandular cells located in the duct wall trigger events that put the finishing touches on the maturing sperm cells. When fully matured, sperm are stored in the last stretch of each epididymis. There they will be stored until they are ejaculated from the body.

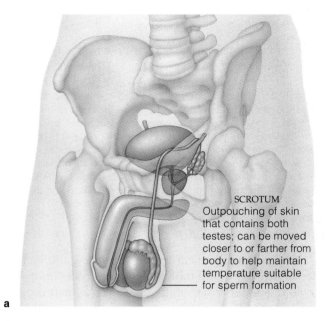

SCROTUM
Outpouching of skin that contains both testes; can be moved closer to or farther from body to help maintain temperature suitable for sperm formation

a

Figure 39.12 *Above and facing page:* (**a**) Position of the human male reproductive system relative to the pelvic girdle and urinary bladder. The midsagittal section in (**b**) shows the components of the system and lists their functions.

Table 39.1	*Organs and Accessory Glands of the Male Reproductive Tract*
REPRODUCTIVE ORGANS	
Testis (2)	Production of sperm, sex hormones
Epididymis (2)	Sperm maturation site and sperm storage
Vas deferens (2)	Rapid transport of sperm
Ejaculatory duct (2)	Conduction of sperm to penis
Penis	Organ of sexual intercourse
ACCESSORY GLANDS	
Seminal vesicle (2)	Secretion of large part of semen
Prostate gland	Secretion of part of semen
Bulbourethral gland (2)	Production of lubricating mucus

When the male is sexually aroused, muscles in the walls of reproductive organs contract and propel the mature sperm into and through a pair of thick-walled tubes, the vasa deferentia (singular, vas deferens). The contractions propel the sperm even further, through a pair of ejaculatory ducts, then on through the urethra. This last tube in the series extends through the penis, the male sex organ, and opens at its tip. The urethra is a duct that also functions in urinary excretion.

As sperm travel to the urethra, they are mixed with glandular secretions. The result is semen, a thick fluid expelled from the penis during sexual activity. While

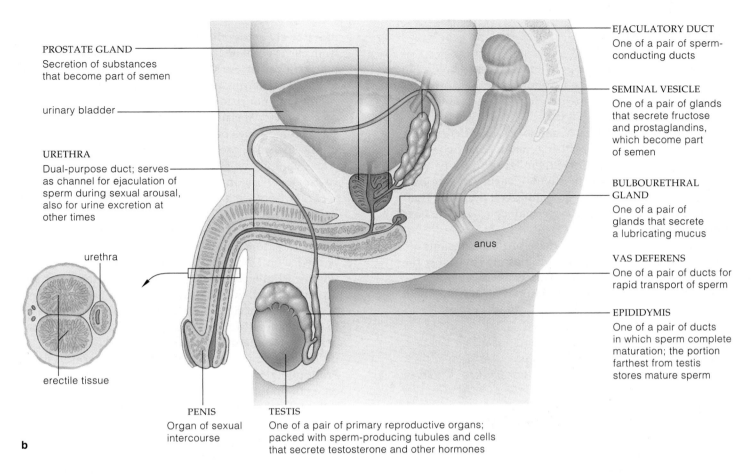

PROSTATE GLAND
Secretion of substances
that become part of semen

urinary bladder

URETHRA
Dual-purpose duct; serves
as channel for ejaculation of
sperm during sexual arousal,
also for urine excretion at
other times

urethra

erectile tissue

EJACULATORY DUCT
One of a pair of sperm-
conducting ducts

SEMINAL VESICLE
One of a pair of glands
that secrete fructose
and prostaglandins,
which become part
of semen

BULBOURETHRAL
GLAND
One of a pair of
glands that secrete
a lubricating mucus

VAS DEFERENS
One of a pair of ducts for
rapid transport of sperm

EPIDIDYMIS
One of a pair of ducts
in which sperm complete
maturation; the portion
farthest from testis
stores mature sperm

anus

PENIS
Organ of sexual
intercourse

TESTIS
One of a pair of primary reproductive organs;
packed with sperm-producing tubules and cells
that secrete testosterone and other hormones

b

semen forms, paired seminal vesicles secrete fructose. Sperm use this sugar as an energy source. The vesicles also secrete prostaglandins that can cause muscles to contract. Possibly, these signaling molecules exert their effects during sexual activity. At that time, they might induce contractions in the female's reproductive tract and thereby assist sperm movement through it.

Secretions from a prostate gland may help buffer the acidic conditions in the female reproductive tract. (The pH of vaginal fluid is about 3.5–4.0, but sperm are able to swim more efficiently at pH 6.) Two bulbourethral glands secrete some mucus-rich fluid into the urethra when a male is sexually aroused.

Cancers of the Prostate and Testis

Until recently, cancers of the male reproductive tract did not get much media coverage. Yet *prostate cancer* is the second leading cause of death among men, exceeded only by cancers of the respiratory tract. At this writing, an estimated 180,400 new cases may be diagnosed in 2001, and 31,900 men may die. The death rate for breast cancer, which is highly publicized, is not much higher. Even so, this is a dramatic decline from prostate cancer rates for 1989–1992, possibly as a result of widespread and earlier diagnostic screenings. Also, *testicular cancer*

is a frequent cause of death among young men. About 5,000 cases are diagnosed each year.

Both cancers are painless in the early stages. When they are not detected in time, they spread silently into lymph nodes of the abdomen, chest, neck, and then the lungs. Once the cancer metastasizes, prospects are not good. Testicular cancer, for example, now kills as many as half of those stricken.

Doctors can detect prostate cancer by blood tests for increases in prostate-specific antigen (PSA) and by physical examinations. Once a month from high school onward, men should examine each testis after a warm bath or shower, when scrotal muscles are relaxed. The testis should be rolled gently between the thumb and the forefinger to check for an enlargement, hardening, or lump. Such changes may or may not cause obvious discomfort. But they must be reported so a physician may order a full examination. Treatment of testicular cancer has one of the highest success rates, provided that the cancer is caught before it starts to spread.

Human males have a pair of testes—primary reproductive organs that produce sperm and sex hormones—as well as accessory glands and ducts. The hormones influence sperm formation and the development of secondary sexual traits.

MALE REPRODUCTIVE FUNCTION

Sperm Formation

Although smaller than a golfball, each testis holds 125 meters or so of seminiferous tubules. Up to 300 wedge-shaped lobes partition the interior, and each holds two or three coiled tubules (Figures 39.13 and 39.14).

Inside the wall of each tubule are undifferentiated cells called spermatogonia (singular, spermatogonium). Ongoing cell divisions force them away from the wall, toward the interior of the tubule. During their forced departure, the cells undergo mitotic cell division. Their daughter cells, the primary spermatocytes, are the ones that enter meiosis.

As Figure 39.14a shows, these nuclear divisions are accompanied by *incomplete* cytoplasmic divisions; the developing cells remain interconnected by cytoplasmic bridges. Ions and molecules required for development

move freely across the bridges, so successive divisions from each spermatogonium result in a clone of cells that mature together. Sertoli cells, the only other type of cell in the seminiferous tubules, provide the forerunners of sperm with nourishment and molecular signals.

Secondary spermatocytes form by way of meiosis I. The chromosomes of these haploid cells are still in the duplicated state; each consists of two sister chromatids. (Here you may wish to check the general overview of spermatogenesis in Section 9.5.) The sister chromatids separate from each other during meiosis II, and then daughter cells form. These are haploid spermatids that gradually develop into sperm, the male gametes.

Each mature sperm is a flagellated cell. Its head is packed with DNA and its tail has a microtubular core (Figure 39.14b). Extending over most of the head is a cap (acrosome). Enzymes in the cap will help the sperm penetrate extracellular material around the egg prior to fertilization. Mitochondria in a midpiece behind the head supply energy for the tail's whiplike motions.

It takes nine to ten weeks for each sperm to form. From puberty onward, a human male produces sperm on a continuous basis, so many millions of cells are in different stages of development on any given day.

Hormonal Controls

Coordinated secretions of LH, FSH, testosterone, and other hormones govern reproductive function in males. Take a look at Figure 39.13b and notice the Leydig cells between lobes in the testes. They secrete **testosterone**, a steroid hormone essential for the growth, form, and function of the male reproductive tract. Besides having the key role in sperm formation, testosterone promotes development of male secondary sexual traits. It also stimulates sexual behavior and aggressive behavior.

LH and **FSH**, recall, are secretions of the anterior lobe of the pituitary gland (Section 32.3). Initially, both hormones were named for their effects in females. (One

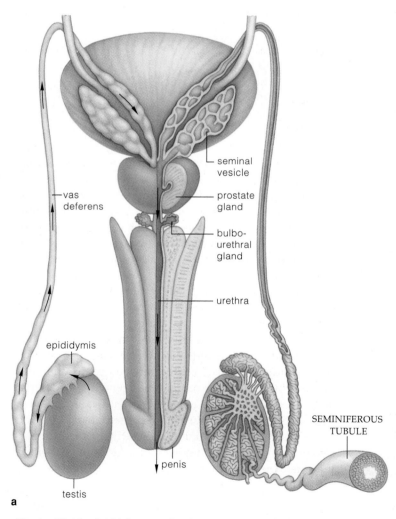

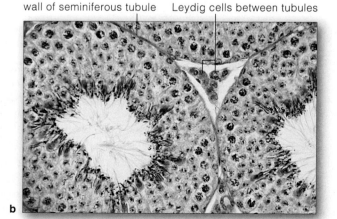

Figure 39.13 (a) Male reproductive tract, posterior view. The arrows show the route that sperm take before ejaculation from a sexually aroused male. (b) Light micrograph of cells inside three adjacent seminiferous tubules, cross-section. Leydig cells occupy tissue spaces between the tubules.

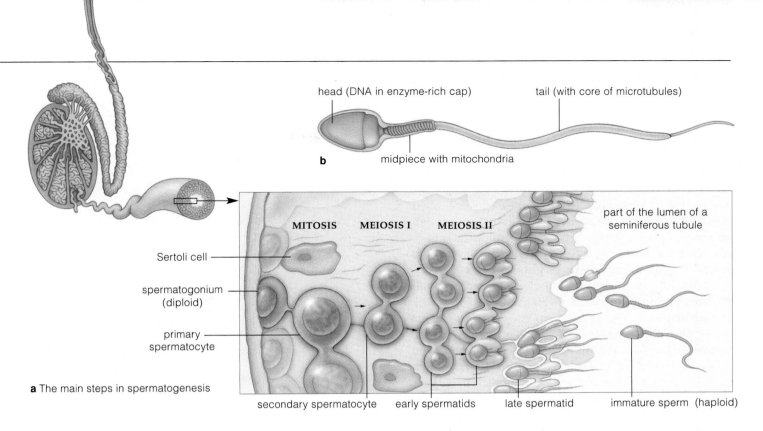

head (DNA in enzyme-rich cap)   tail (with core of microtubules)

b   midpiece with mitochondria

MITOSIS   MEIOSIS I   MEIOSIS II

part of the lumen of a
seminiferous tubule

Sertoli cell

spermatogonium
(diploid)

primary
spermatocyte

a The main steps in spermatogenesis

secondary spermatocyte   early spermatids   late spermatid   immature sperm (haploid)

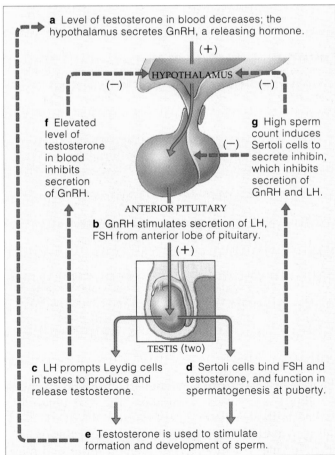

a Level of testosterone in blood decreases; the
hypothalamus secretes GnRH, a releasing hormone.

(+)

HYPOTHALAMUS

(−)   (−)

f Elevated
level of
testosterone
in blood
inhibits
secretion
of GnRH.

g High sperm
count induces
Sertoli cells to
secrete inhibin,
which inhibits
secretion of
GnRH and LH.

(−)

ANTERIOR PITUITARY

b GnRH stimulates secretion of LH,
FSH from anterior lobe of pituitary.

(+)

TESTIS (two)

c LH prompts Leydig cells
in testes to produce and
release testosterone.

d Sertoli cells bind FSH and
testosterone, and function in
spermatogenesis at puberty.

e Testosterone is used to stimulate
formation and development of sperm.

Figure 39.15   Negative feedback loops to the hypothalamus
and to the pituitary gland's anterior lobe from the testes. With
these loops, excess levels of testosterone in blood slow down
the mechanisms leading to its production. This helps maintain
the testosterone level in amounts required for sperm formation.

Figure 39.14   (**a**) Sperm formation. The process starts with a
spermatogonium, a type of diploid germ cell. Mitosis, meiosis,
and incomplete cytoplasmic divisions result in a clone of immature
haploid cells that differentiate and develop into mature sperm.
(**b**) Structure of a mature sperm from a human male.

abbreviation stands for *Luteinizing Hormone*, the other
for *Follicle-Stimulating Hormone*.) We now know their
molecular structure is identical in males and females.

The hypothalamus controls secretion of LH, FSH,
and testosterone, and thus controls sperm formation.
When blood levels of testosterone and other factors are
low, it secretes the releasing hormone GnRH (Figure
39.15). GnRH calls on the anterior lobe to step up the
release of LH and FSH, which have targets in testes.

LH prods Leydig cells to secrete testosterone, which
assists in stimulating the formation and development
of sperm. Sertoli cells have receptors for FSH, which is
required to jump-start spermatogenesis at puberty. We
don't know whether FSH is also necessary for normal
functioning of mature human testes.

Figure 39.15 also shows how feedback loops to the
hypothalamus can cut back on testosterone secretion
and sperm formation. An elevated testosterone level in
blood slows down the release of GnRH. Also, when the
sperm count is high, Sertoli cells release inhibin. This
protein hormone acts on the hypothalamus and pituitary
to cut back on the release of GnRH and FSH.

**Sperm formation depends on the hormones LH, FSH, and
testosterone. Negative feedback loops from the testes to the
hypothalamus and pituitary gland control their secretion.**

REPRODUCTIVE SYSTEM OF HUMAN FEMALES

The Reproductive Organs

We turn next to the reproductive system of the human female. Figure 39.16 shows its components, and Table 39.2 summarizes their functions. The female's *primary* reproductive organs, her pair of ovaries, produce eggs and secrete sex hormones. Her immature eggs are called oocytes. An oocyte released from an ovary moves into one of a pair of oviducts (also called Fallopian tubes). An oviduct is a channel into the **uterus**, a pear-shaped, hollow organ in which embryos grow and develop.

A thick layer of smooth muscle, the myometrium, makes up most of the uterine wall. As you will see, the **endometrium**, the inner lining of the wall, is vital for development. It consists of connective tissues, glands, and blood vessels. The narrowed-down portion of the uterus is the cervix. A vagina, a muscular tube, extends from the cervix to the surface of the body. The vagina receives sperm and functions as part of the birth canal.

At the body surface are external genitals (vulva), including organs for sexual stimulation. Outermost are a pair of fat-padded skin folds (the labia majora). They enclose a smaller pair of skin folds (the labia minora), which are highly vascularized but have no fatty tissue. The smaller folds partly enclose the clitoris, a sex organ highly sensitive to stimulation. At the body surface, the urethra's opening is positioned about midway between the clitoris and the vaginal opening.

Overview of the Menstrual Cycle

Females of most mammalian species follow an *estrous* cycle. They are fertile and in heat (sexually receptive to males) only at certain times of year. By contrast, female primates, including humans, follow a **menstrual cycle**. They become fertile intermittently, on a cyclic basis, and heat is not synchronized with fertile periods. Said another way, female primates of reproductive age can become pregnant only at certain times of year, but they may be receptive to sex at any time.

Briefly, an oocyte matures and escapes from an ovary during each menstrual cycle. Also during the cycle, the endometrium becomes primed to receive and nourish a forthcoming embryo *if* a sperm penetrates the oocyte and fertilization occurs. However, if that oocyte does not get fertilized, then about four to six tablespoons of blood-rich fluid from the uterus will start flowing out through the vaginal canal. Such a recurring blood flow is called menstruation. It means "there is no embryo at this time," and it marks the first day of a new cycle. The uterine nest is being sloughed off and is about to be constructed once again.

The events just sketched out proceed through three phases. The cycle starts with a *follicular* phase. This is

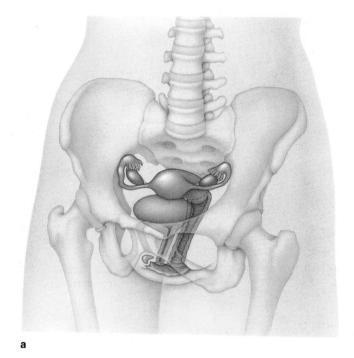

a

Figure 39.16 *Above and facing page:* (**a**) Position of the human female reproductive system relative to the pelvic girdle and urinary bladder. The midsagittal section in (**b**) shows the components of the system and lists their functions.

Table 39.2	*Female Reproductive Organs*
Ovaries	Oocyte production and maturation, sex hormone production
Oviducts	Ducts for conducting oocyte from ovary to uterus; fertilization normally occurs here
Uterus	Chamber in which new individual develops
Cervix	Secretion of mucus that enhances sperm movement into uterus and (after fertilization) reduces embryo's risk of bacterial infection
Vagina	Organ of sexual intercourse; birth canal

the time of menstruation, endometrial breakdown and rebuilding, and oocyte maturation. The next phase is limited to the release of an oocyte from the ovary. We call it **ovulation**. During the *luteal* phase of the cycle, an endocrine structure (corpus luteum) forms and the endometrium is primed for pregnancy (Table 39.3).

All three phases are governed by feedback loops to the hypothalamus and pituitary gland from the ovaries. FSH and LH promote cyclic changes in the ovaries. As you will see, FSH and LH also stimulate the ovaries to

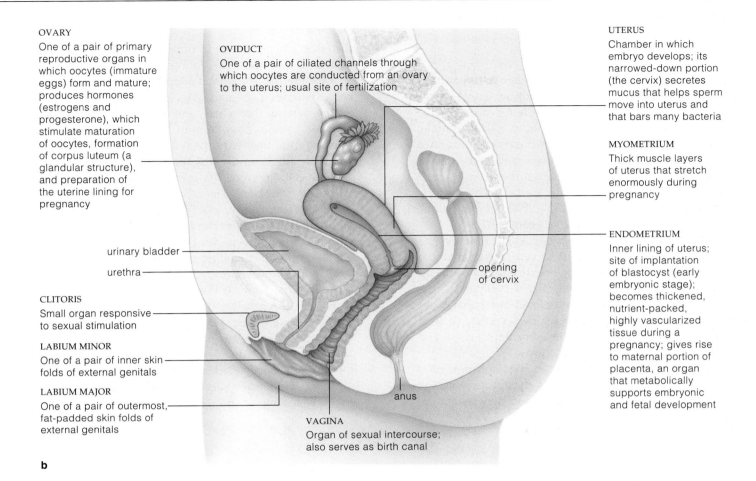

OVARY
One of a pair of primary reproductive organs in which oocytes (immature eggs) form and mature; produces hormones (estrogens and progesterone), which stimulate maturation of oocytes, formation of corpus luteum (a glandular structure), and preparation of the uterine lining for pregnancy

OVIDUCT
One of a pair of ciliated channels through which oocytes are conducted from an ovary to the uterus; usual site of fertilization

UTERUS
Chamber in which embryo develops; its narrowed-down portion (the cervix) secretes mucus that helps sperm move into uterus and that bars many bacteria

MYOMETRIUM
Thick muscle layers of uterus that stretch enormously during pregnancy

ENDOMETRIUM
Inner lining of uterus; site of implantation of blastocyst (early embryonic stage); becomes thickened, nutrient-packed, highly vascularized tissue during a pregnancy; gives rise to maternal portion of placenta, an organ that metabolically supports embryonic and fetal development

urinary bladder

urethra

opening of cervix

CLITORIS
Small organ responsive to sexual stimulation

LABIUM MINOR
One of a pair of inner skin folds of external genitals

LABIUM MAJOR
One of a pair of outermost, fat-padded skin folds of external genitals

anus

VAGINA
Organ of sexual intercourse; also serves as birth canal

b

Table 39.3	*Events of the Menstrual Cycle*	
Phase	Events	Days of Cycle*
Follicular phase	Menstruation; endometrium breaks down	1–5
	Follicle matures in ovary; endometrium rebuilds	6–13
Ovulation	Oocyte released from ovary	14
Luteal phase	Corpus luteum forms, secretes progesterone; the endometrium thickens and develops	15–28

* Assuming a 28-day cycle.

secrete sex hormones—**estrogens** and **progesterone**—to promote the cyclic changes in the endometrium.

A human female's menstrual cycles start between ages ten and sixteen. Each cycle lasts for about twenty-eight days, but this is only the average. It runs longer for some women and shorter for others. The menstrual cycles continue until a woman is in her late forties or early fifties, when her supply of eggs is dwindling and secretions of hormones slow down. This is the onset of *menopause*, the twilight of reproductive capacity.

You may have heard of *endometriosis*, a condition that arises after endometrial tissue abnormally spreads and grows outside the uterus. Endometrial scar tissue may form on ovaries or oviducts and lead to infertility. In the United States, 10 million women may be affected annually. Perhaps the condition arises when menstrual flow backs up through the oviducts and spills into the pelvic cavity. Or perhaps some embryonic cells became positioned in the wrong place before birth and were stimulated to grow when sex hormones became active at puberty. Whatever the case, estrogen still stimulates cells in the mislocated tissue. The resulting symptoms include pain during menstruation, sex, and urination.

Ovaries, the female primary reproductive organs, produce oocytes (immature eggs) and sex hormones. Endometrium lines the uterus, a chamber in which embryos develop.

Secretion of sex hormones (estrogens and progesterone) is coordinated on a cyclic basis through the reproductive years.

A menstrual cycle starts with menstruation, breakdown and rebuilding of the endometrium, and maturation of an oocyte. After an oocyte escapes from an ovary (ovulation), the cycle ends with the formation of a corpus luteum and an endometrium that is primed for pregnancy.

FEMALE REPRODUCTIVE FUNCTION

Cyclic Changes in the Ovary

Take a moment to review Section 10.5, the generalized picture of meiosis in an oocyte. A normal baby girl has about 2 million primary oocytes in her ovaries. By the time she is seven years old, only about 300,000 remain; her body resorbed the rest. Her primary oocytes have already started meiosis, but the nuclear division process was arrested in a genetically programmed way. Meiosis resumes in one oocyte at a time, starting with the first menstrual cycle. Only about 400 to 500 oocytes will be released during the reproductive years.

Figure 39.17 shows a primary oocyte near an ovary's surface. A cell layer (of granulosa cells) surrounds and nourishes it. We call a primary oocyte and the cell layer around it a follicle. As the menstrual cycle starts, the hypothalamus is secreting GnRH in amounts that cause the anterior pituitary to step up *its* secretion of FSH and LH. The blood level of these hormones rises, and the increase causes the follicle to grow (Figure 39.18).

The oocyte starts to increase in size, and more cell layers form around it. Deposits of glycoproteins collect between the layers and the oocyte, and they widen the distance between them. In time they make up the zona pellucida, a noncellular coating around the oocyte.

Outside the zona pellucida, FSH and LH stimulate cells to secrete estrogens. An estrogen-containing fluid accumulates in the follicle, and estrogen levels in blood start to increase. About eight to ten hours before being released from the ovary, the oocyte completes meiosis I. Then its cytoplasm divides, forming two cells.

One cell, the *secondary* oocyte, ends up with nearly all of the cytoplasm. The other cell is the first of three **polar bodies**. The meiotic parceling of chromosomes among four cells gives the secondary oocyte a haploid chromosome number—the precise number required for gametes and for sexual reproduction.

About halfway through each menstrual cycle, the pituitary gland detects the rise in estrogens in blood. It responds with a brief outpouring of LH. The surge in LH causes rapid vascular changes that make the follicle swell, and it makes enzymes digest the bulging follicle wall. The weakened wall ruptures. Fluid, together with the secondary oocyte, escapes (Figures 39.17 and 39.18).

Thus, the midcycle surge of LH triggers ovulation—the release of a secondary oocyte from the ovary.

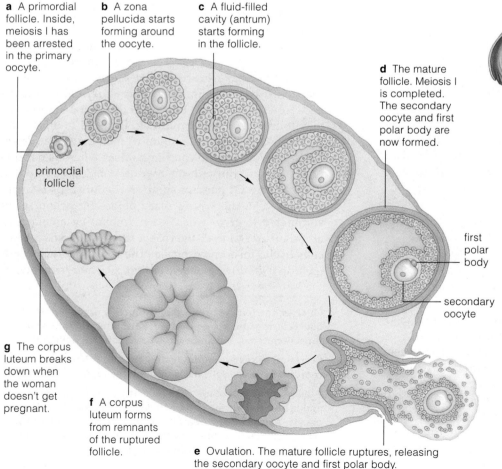

a A primordial follicle. Inside, meiosis I has been arrested in the primary oocyte.

b A zona pellucida starts forming around the oocyte.

c A fluid-filled cavity (antrum) starts forming in the follicle.

d The mature follicle. Meiosis I is completed. The secondary oocyte and first polar body are now formed.

primordial follicle

first polar body

secondary oocyte

g The corpus luteum breaks down when the woman doesn't get pregnant.

f A corpus luteum forms from remnants of the ruptured follicle.

e Ovulation. The mature follicle ruptures, releasing the secondary oocyte and first polar body.

oviduct

OVARY

uterus

vagina

Figure 39.17 Cyclic events in a human ovary, cross-section.

A follicle stays in the same place inside an ovary all through one menstrual cycle. It does not "move around" as in this diagram, which only shows the *sequence* in which events occur. In the cycle's first phase, a follicle grows and matures. At ovulation, the second phase, the mature follicle ruptures and releases a secondary oocyte. In the third phase, a corpus luteum forms from the follicle's remnants. If the woman doesn't get pregnant during the cycle, the corpus luteum self-destructs.

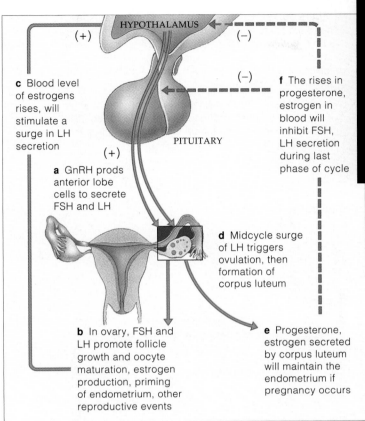

c Blood level of estrogens rises, will stimulate a surge in LH secretion

a GnRH prods anterior lobe cells to secrete FSH and LH

HYPOTHALAMUS

(+) (−)

(−)

PITUITARY

(+)

f The rises in progesterone, estrogen in blood will inhibit FSH, LH secretion during last phase of cycle

d Midcycle surge of LH triggers ovulation, then formation of corpus luteum

b In ovary, FSH and LH promote follicle growth and oocyte maturation, estrogen production, priming of endometrium, other reproductive events

e Progesterone, estrogen secreted by corpus luteum will maintain the endometrium if pregnancy occurs

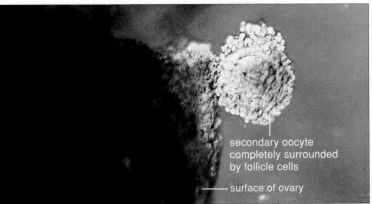

secondary oocyte completely surrounded by follicle cells

surface of ovary

Figure 39.18 Feedback control of hormonal secretion during a menstrual cycle. A positive feedback loop from an ovary to the hypothalamus causes a surge in LH secretion that triggers ovulation. The light micrograph shows a secondary oocyte being released from an ovary at this time. Afterward, negative feedback loops to the hypothalamus and pituitary inhibit FSH secretion. They block maturation of another follicle until the cycle is over.

Cyclic Changes in the Uterus

The estrogens released early in the menstrual cycle also help pave the way for a possible pregnancy. Estrogens stimulate growth of the endometrium and its glands. Just before the midcycle surge of LH, follicle cells start to secrete some progesterone as well as estrogens. Blood vessels grow quickly in the thickened endometrium. At ovulation, estrogens act on tissues around the cervical canal, which leads into the vagina. The cervix starts to secrete large quantities of a thin, clear mucus that is an ideal medium for sperm to swim through.

The midcycle surge of LH that triggers ovulation also triggers the formation of a **corpus luteum** in the ovary. This yellowish glandular structure forms by cell differentiation among granulosa cells left behind in the follicle. (Its name means yellow body.) Secretions from the corpus luteum influence the rest of the cycle.

The corpus luteum secretes progesterone and some estrogen. Progesterone prepares the reproductive tract for the arrival of a blastocyst, the type of blastula that forms from a fertilized mammalian egg (Section 39.3). For example, this hormone makes cervical mucus turn thick and sticky. The mucus may keep normal bacterial residents of the vagina out of the uterus. Progesterone also will maintain the endometrium during pregnancy.

A corpus luteum persists for about twelve days. All the while, the hypothalamus is calling for minimal FSH secretion, which stops other follicles from developing. If a blastocyst does not burrow into the endometrium, the corpus luteum will self-destruct in the last days of the cycle. It will secrete certain prostaglandins that can disrupt its own functioning.

After this, progesterone and estrogen levels in blood decline quickly, and so the endometrium starts to break down. Deprived of oxygen and nutrients, the lining's blood vessels constrict and tissues die. Blood escapes as weakened capillary walls start to rupture. Blood as well as the sloughed endometrial tissues make up the menstrual flow, which continues for three to six days. Then the cycle begins anew, with rising estrogen levels stimulating repair and growth of the endometrium.

By menopause, the supply of oocytes is dwindling, hormone secretions slow down, and in time menstrual cycles—and fertility—will be over. Eggs that are still to be released late in a woman's life are at some risk of alterations in chromosome number or structure when meiosis resumes. A newborn with Down syndrome, as described in Section 11.10, is one possible outcome.

During one menstrual cycle, FSH and LH stimulate growth of an ovarian follicle (a primary oocyte and the surrounding layer of cells). The first meiotic cell division results in one secondary oocyte and the first polar body.

A midcycle surge of LH triggers ovulation—the release of the secondary oocyte and the polar body from the ovary.

Early on, estrogens call for endometrial repair and growth. Then estrogens and progesterone prepare the endometrium and other parts of the reproductive tract for pregnancy.

VISUAL SUMMARY OF THE MENSTRUAL CYCLE

By now, you probably have come to the conclusion that the menstrual cycle is not a simple tune on a biological banjo. It's a full-blown hormonal symphony!

Before continuing with your reading, take a moment to review Figure 39.19. It correlates the cyclic events in the ovary and uterus with all the coordinated changes in hormone levels that bring about those events. The illustration may leave you with a better understanding of what goes on.

Figure 39.19 Changes in the ovary and uterus, as correlated with changing hormone levels during each turn of the menstrual cycle. *Green* arrows show which hormones dominate the cycle's first phase (when the follicle matures), and then the second phase (when the corpus luteum forms).

(**a–c**) FSH and LH secretions bring about the changes in ovarian structure and function. (**d,e**) Estrogen and progesterone secretions from the ovary stimulate the changes in the endometrium.

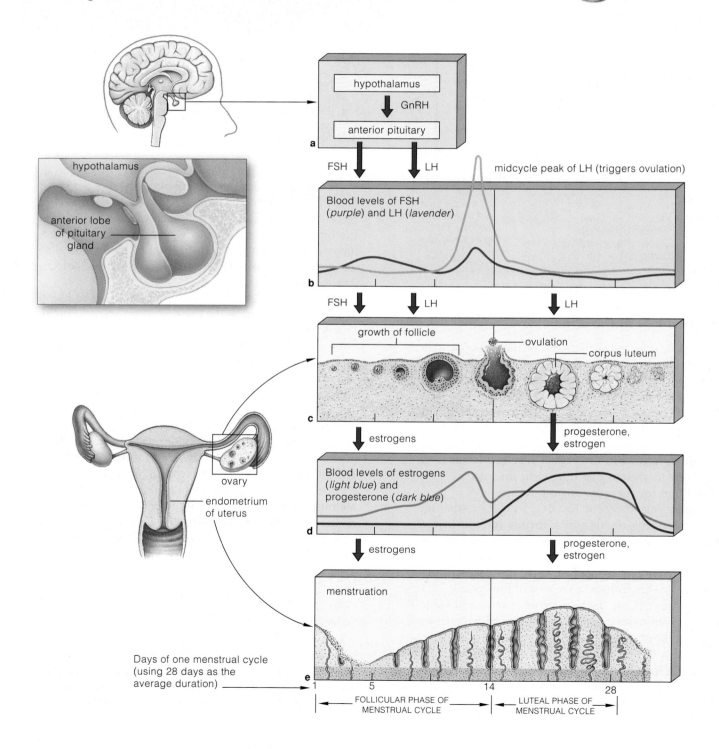

PREGNANCY HAPPENS

Sexual Intercourse

Suppose the secondary oocyte is on its way down an oviduct when a female and male are engaged in sexual intercourse, or coitus. The male sex act requires *erection*, whereby a normally limp penis stiffens and lengthens; and *ejaculation*, the forceful expulsion of semen into the urethra and out from the penis. As Figure 39.12 shows, the penis incorporates cylinders of spongy tissue. Many friction-activated sensory receptors are on its mushroom-shaped tip, the glans penis. In males *not* sexually aroused, large blood vessels that lead into the cylinders are vasoconstricted. In aroused males, they vasodilate and blood flows into the penis faster than it flows out, so blood collects in the spongy tissue. The resulting engorgement stiffens and lengthens the organ, this being helpful for penetration into the vaginal canal.

During coitus, pelvic thrusts stimulate the penis, the female's clitoris, and the vaginal wall. The mechanical stimulation induces involuntary contractions in the male reproductive tract. It forces sperm from the epididymes and the contents of seminal vesicles and the prostate gland into the urethra. The substances mix and form semen, which is ejaculated into the vagina. A sphincter closes the bladder's neck and prevents urination during ejaculation.

Emotional intensity, hard breathing, heart pounding, and the contraction of skeletal muscles in general accompany rhythmic throbbing of pelvic muscles. During *orgasm*, the end of the sex act, strong sensations of physical release, warmth, and relaxation dominate. Similar sensations typify female orgasm. It's a common misconception that a female cannot become pregnant if she doesn't reach orgasm. Don't believe it.

Fertilization

An ejaculation can put 150 million to 350 million sperm in the vagina. If they arrive a few days before or after ovulation or any time between, fertilization can occur. Less than thirty minutes after sperm arrive, contractions move them deep into the female's reproductive tract. Just a few hundred actually reach the upper part of the oviduct, where fertilization usually takes place.

Think back on the micrograph that opens Unit II, which depicts sperm around a secondary oocyte. When

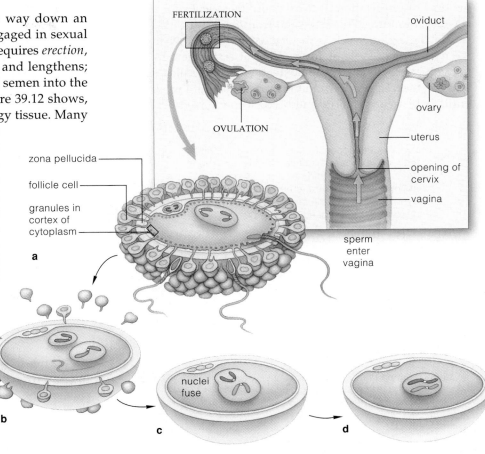

Figure 39.20 Fertilization. (**a**) Many sperm surround a secondary oocyte. Acrosomal enzymes clear a path through the zona pellucida. (**b**) When a sperm does penetrate the secondary oocyte, granules in the egg cortex release substances that make the zona pellucida impenetrable to other sperm. Penetration also stimulates meiosis II of the oocyte's nucleus. (**c**) The sperm's tail degenerates and its nucleus enlarges and fuses with the oocyte nucleus. (**d**) With fusion, fertilization is over. The zygote has formed.

sperm reach an oocyte, they release digestive enzymes that clear a path through the zona pellucida. Although many sperm might get this far, usually only one fuses with the oocyte (Figure 39.20). It degenerates inside the oocyte's cytoplasm until only its nucleus and centrioles remain. Penetration induces the secondary oocyte and the first polar body to finish meiosis II.

There are now three polar bodies and a mature egg, or **ovum** (plural, ova). The sperm nucleus and the egg nucleus fuse. Together, their chromosomes restore the diploid number for a brand new zygote.

The intense physiological events that accompany coitus have one function: to put sperm on a collision course with an egg. Fertilization is over with the fusion of a sperm nucleus and egg nucleus, which results in a diploid zygote.

39.12

FORMATION OF THE EARLY EMBRYO

Pregnancy lasts an average of thirty-eight weeks from the time of fertilization. It takes about two weeks for a blastocyst to form. The time span from the third to the end of the eighth week is the *embryonic* period, when the major organ systems form. When it ends, the new individual has distinctly human features and is called a **fetus**. In the *fetal* period from the start of the ninth week until birth, organs enlarge and become specialized.

d DAY 5. A blastocoel (a fluid-filled cavity) forms in the morula as a result of secretions from the surface cells. By the thirty-two-cell stage, cells of an inner cell mass are already differentiating. They will give rise to the embryo proper. This embryonic stage is called a blastocyst.

c DAY 4. By 96 hours, there is a ball of sixteen to thirty-two cells that is shaped like a mulberry. It is a morula (after *morum*, Latin for mulberry). Cells of the surface layer will function in implantation and will give rise to a membrane, the chorion.

b DAY 3. After the third cleavage, the cells suddenly huddle together into a compacted ball, which becomes stabilized by numerous tight junctions among the outer cells. Gap junctions form among the interior cells and enhance intercellular communication.

a DAYS 1–2. Cleavage begins within 24 hours after fertilization. The first cleavage furrow extends between the two polar bodies. Subsequent cuts are rotational, so the resulting cells are not symmetrically arranged (compare Section 39.3). Until the eight-cell stage forms, the cells are loosely arranged, with considerable space between them.

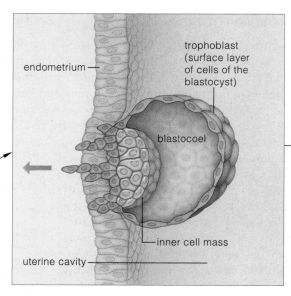

inner cell mass

e DAYS 6–7. Surface cells of the blastocyst attach to the endometrium and start to burrow into it. Implantation is under way.

actual size

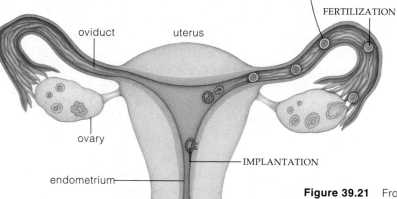

oviduct uterus

FERTILIZATION

ovary

IMPLANTATION

endometrium

We typically call the first three months of pregnancy the *first* trimester. A *second* trimester extends from the start of the fourth month to the end of the sixth. The *third* trimester extends from the seventh month until birth. Beginning with Figure 39.21, the next series of illustrations shows characteristic features of the new individual at progressive stages of development.

Cleavage and Implantation

Three to four days after fertilization, a zygote is already undergoing cleavage as it tumbles through the oviduct. Genes are being expressed; early divisions depend on their products. At the eight-cell stage, the cells huddle into a compact ball. By the fifth day, there is a surface layer of cells (the trophoblast), a cavity filled with their secretions (a blastocoel), and a cluster of interior cells (an inner cell mass). These are the defining features of a human blastocyst (Figure 39.21*d*).

Six or seven days after fertilization, **implantation** is under way. By this process, the blastocyst adheres to the uterine lining, some of its cells send out projections that invade the mother's tissues, and connections start to form that will metabolically support the developing embryo through the months ahead. While the invasion

Figure 39.21 From fertilization through implantation. A blastocyst forms at cleavage. Within the blastocyst, the inner cell mass gives rise to the disk-shaped early embryo. Three extraembryonic membranes (the amnion, chorion, and yolk sac) start forming. A fourth membrane (allantois) forms after the blastocyst is implanted.

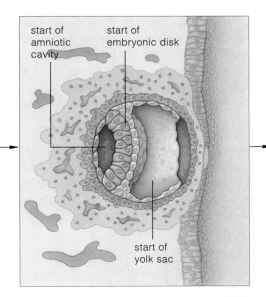

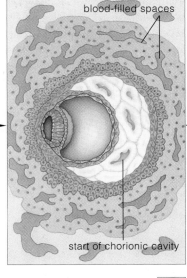

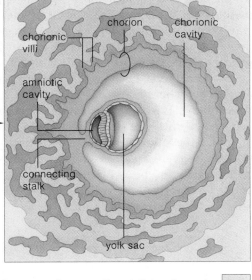

start of
amniotic
cavity

start of
embryonic disk

start of
yolk sac

blood-filled spaces

start of chorionic cavity

chorion

chorionic
cavity

chorionic
villi

amniotic
cavity

connecting
stalk

yolk sac

f DAYS 10–11. The yolk sac, embryonic disk, and amniotic cavity have started to form from parts of the blastocyst.

actual size

g DAY 12. Blood-filled spaces form in maternal tissue. The chorionic cavity starts to form.

actual size

h DAY 14. A connecting stalk has formed between the embryonic disk and chorion. Chorionic villi, which will be features of a placenta, start to form.

actual size

is proceeding, the inner cell mass develops into two cell layers of a flattened and somewhat circular shape. The two layers make up the embryonic disk—and in short order they will give rise to the embryo proper.

Extraembryonic Membranes

As implantation progresses, membranes start to form outside the embryo. First a fluid-filled *amniotic* cavity opens up between the embryonic disk and part of the blastocyst's surface (Figure 39.21*f*). Then cells migrate around the wall of the cavity and form the **amnion**, a membrane that will enclose the embryo. Fluid inside the cavity will function as a buoyant cradle where the embryo can grow, move freely, and be protected from abrupt temperature changes and mechanical impacts.

While the amnion forms, other cells migrate around the inner wall of the blastocyst's first cavity. They form a lining that becomes a **yolk sac**. This extraembryonic membrane speaks of the evolutionary heritage of land vertebrates (Section 24.7). For most animals that make shelled eggs, the sac holds nutritive yolk. In humans, part of it becomes a site of blood cell formation. Part will give rise to germ cells, the forerunners of gametes.

Before a blastocyst is fully implanted, spaces open in maternal tissues and fill with blood seeping in from ruptured capillaries. Inside the blastocyst, a different cavity opens up around the amnion and yolk sac. Now fingerlike projections start to form on the cavity lining, which is the **chorion**. This new membrane will become part of the placenta, a spongy, blood-engorged tissue.

After the blastocyst is implanted, an outpouching of the yolk sac will become the third extraembryonic membrane—the **allantois**. The allantois has different functions in different animal groups. Among reptiles, birds, and some mammals, it serves in respiration and in storing metabolic wastes. In humans, blood vessels for the placenta and the urinary bladder form from it.

One more point should be made here. Cells of the blastocyst secrete a hormone, HCG (*Human Chorionic Gonadotropin*), which stimulates the corpus luteum to keep on secreting progesterone and estrogen. In this way, the blastocyst itself prevents menstrual flow and works to avoid being sloughed off until the placenta takes over the function of secreting HCG, about eleven weeks later. By the start of the third week, HCG can be detected in samples of the mother's blood or urine. At-home *pregnancy tests* have a treated "dip-stick" that changes color when HCG is present in urine.

A human blastocyst is composed of a surface layer of cells around a fluid-filled cavity (blastocoel) and an inner cell mass, which will give rise to the embryo proper.

Six or seven days after fertilization, the blastocyst implants itself in the endometrium. Now projections from its surface invade maternal tissues, and connections start to form that in time will metabolically support the developing embryo.

Some parts of the blastocyst give rise to an amnion, yolk sac, chorion, and allantois. These extraembryonic membranes serve different functions. Together they are vital for the structural and functional development of the embryo.

EMERGENCE OF THE VERTEBRATE BODY PLAN

By the time a woman has missed her first menstrual period, cleavage is completed and gastrulation is under way. Gastrulation, recall, is a stage when extensive cell divisions, migrations, and rearrangements produce the primary tissue layers (Sections 39.2 and 39.4).

By now, the embryonic disk is surrounded by the amnion and chorion except where a stalk connects it to the chorion's inner wall. Earlier, the inner cell mass had behaved *as if* it were still developing on the large ball of yolk that evolved in reptilian ancestors of mammals. (Thus it becomes a *flattened* two-layer embryonic disk, as in reptiles and birds.) Cell divisions and migrations of one layer formed the yolk sac lining. The other layer now starts to generate the embryo proper.

For example, by the eighteenth day, the embryonic disk has two neural folds, which will merge to form a neural tube (Figure 39.22b). Early in the third week, a sausage-shaped outpouching forms on the yolk sac. It is the allantois (Greek *allas*, meaning sausage). In humans, recall, it only helps form the urinary bladder and blood vessels for the placenta. Some mesoderm folds into a tube that becomes the notochord. A notochord is only a structural framework; bony segments of the vertebral column soon form around it. Toward the end of the third week, some mesoderm gives rise to the **somites**. These paired segments are the embryonic source of most bones, of skeletal muscles of the head and trunk, and of the dermis overlying these regions.

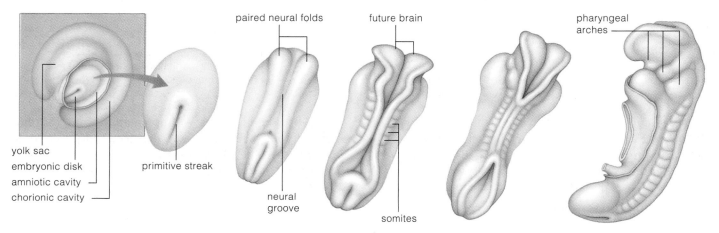

a DAY 15. A faint band appears around a depression along the axis of the embryonic disk. This is the primitive streak, and it marks the onset of gastrulation in vertebrate embryos.

b DAYS 18–23. Organs start to form through cell divisions, cell migrations, tissue folding, and other events of morphogenesis. Neural folds will merge to form the neural tube. Somites (bumps of mesoderm) appear near the embryo's dorsal surface. They will give rise to most of the skeleton's axial portion, skeletal muscles, and much of the dermis.

c DAYS 24–25. By now, some embryonic cells have given rise to pharyngeal arches. These will contribute to the formation of the face, neck, mouth, nasal cavities, larynx, and pharynx.

Figure 39.22 Hallmarks of the embryonic period of humans and other vertebrates. A primitive streak, then a notochord form. Neural folds, somites, and pharyngeal arches form later. (**a,b**) Dorsal views, of the embryo's back. (**c**) Side view. Compare the diagrams with the Figure 39.24 photographs.

Through cell divisions and migrations, the midline of that layer thickens faintly around a depression at its surface. This "primitive streak" lengthens and thickens further the next day. It marks the onset of gastrulation (Figure 39.22a). It also defines the anterior–posterior axis of the embryo and, in time, its bilateral symmetry. Endoderm and mesoderm form from cells migrating inward at this axis. Now pattern formation begins, as described in Section 39.5. Embryonic inductions and interactions among classes of master genes map out the basic body plan characteristic of all vertebrates. And specialized tissues and organs start forming in orderly sequences, in prescribed parts of the embryo.

Pharyngeal arches start to form that will contribute to the face, neck, mouth, nose, larynx, and pharynx. Tiny spaces open up in parts of the mesoderm. In time the spaces will interconnect as the coelomic cavity.

During the third week after fertilization—when a pregnant woman has missed her first menstrual period—the basic vertebrate body plan emerges in the new individual.

A primitive streak, neural tube, somites, and pharyngeal arches form during the embryonic period of all vertebrates. Formation of the primitive streak establishes the body's anterior–posterior axis and its bilateral symmetry.

WHY IS THE PLACENTA SO IMPORTANT?

Even before the beginning of the embryonic period, extraembryonic membranes have been collaborating with the uterus to sustain the embryo's rapid growth. By the third week, tiny fingerlike projections from the chorion have grown profusely into the maternal blood that has pooled in endometrial spaces. The projections —chorionic villi—enhance the exchange of substances between the mother and the new individual. They are functional components of the placenta.

The **placenta**, a blood-engorged organ, consists of endometrial tissue and extraembryonic membranes. At full term, a placenta will make up a fourth of the inner surface of the uterus (Figure 39.23).

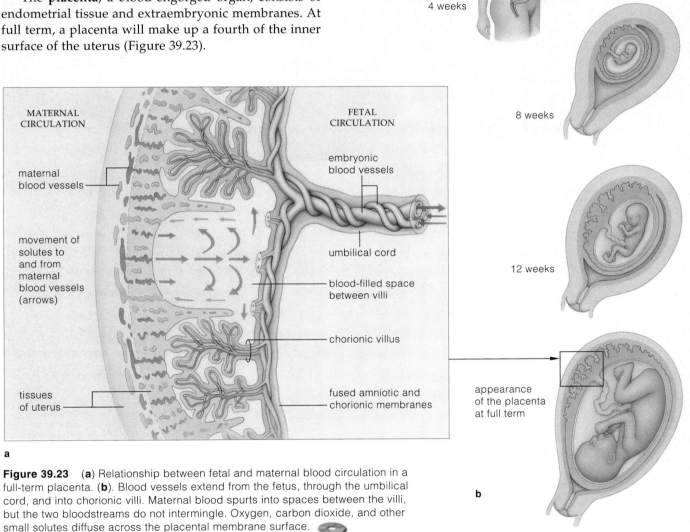

Figure 39.23 (**a**) Relationship between fetal and maternal blood circulation in a full-term placenta. (**b**). Blood vessels extend from the fetus, through the umbilical cord, and into chorionic villi. Maternal blood spurts into spaces between the villi, but the two bloodstreams do not intermingle. Oxygen, carbon dioxide, and other small solutes diffuse across the placental membrane surface.

A placenta is the body's way of sustaining the new individual while allowing its blood vessels to develop apart from the mother's. Oxygen and nutrients diffuse out of the maternal blood vessels, across the placenta's blood-filled spaces, then into embryonic blood vessels. (The vessels converge in the umbilical cord, the lifeline between the placenta and the new individual.) Carbon dioxide and other wastes diffuse in the other direction. The mother's lungs and kidneys quickly dispose of them.

After the third month, the placenta secretes progesterone and estrogens—and so maintains the uterine lining.

The placenta is a blood-engorged organ of endometrial and extraembryonic membranes. It permits exchanges between the mother and the new individual without intermingling their bloodstreams. Thus it sustains the individual and allows its blood vessels to develop apart from the mother's.

EMERGENCE OF DISTINCTLY HUMAN FEATURES

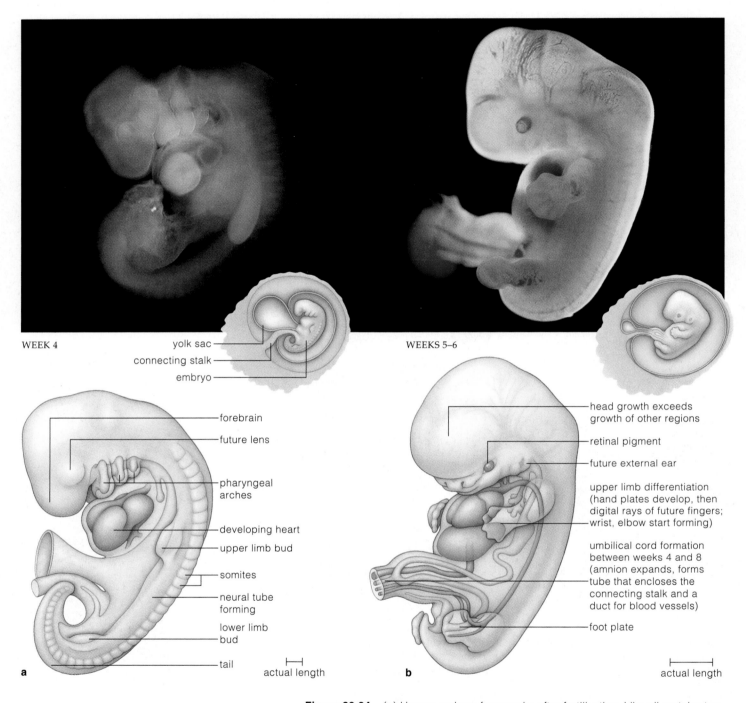

WEEK 4

yolk sac
connecting stalk
embryo

WEEKS 5–6

forebrain

future lens

pharyngeal
arches

developing heart

upper limb bud

somites

neural tube
forming

lower limb
bud

tail

a

actual length

head growth exceeds
growth of other regions

retinal pigment

future external ear

upper limb differentiation
(hand plates develop, then
digital rays of future fingers;
wrist, elbow start forming)

umbilical cord formation
between weeks 4 and 8
(amnion expands, forms
tube that encloses the
connecting stalk and a
duct for blood vessels)

foot plate

b

actual length

When the fourth week of the embryonic period ends, the embryo is 500 times its starting size. The placenta sustained the growth spurt, but now the pace slows as details of organs fill in. Limbs form; toes and fingers are sculpted from embryonic paddles. The umbilical cord also forms, and the circulatory system becomes intricate. Growth of the all-important head now surpasses that of any other body region (Figure 39.24). The embryonic period ends after the eighth week is completed. No longer is

Figure 39.24 (**a**) Human embryo four weeks after fertilization. Like all vertebrates, it has a tail and pharyngeal arches. (**b**) Embryo at five to six weeks. (**c**) Embryo at the boundary between the embryonic and fetal periods. It now has human features. It floats in amniotic fluid. The chorion covers the amniotic sac but has been pulled aside here. (**d**) Fetus at sixteen weeks. Movements begin as nerves make functional connections with forming muscles. Legs kick, arms wave, fingers grasp, the mouth puckers. These reflex actions will be vital skills in the world outside the uterus.

the embryo merely "a vertebrate." By now, its features clearly define it as a human fetus.

During the second trimester, the fetus is moving its facial muscles. It frowns; it squints. It busily practices

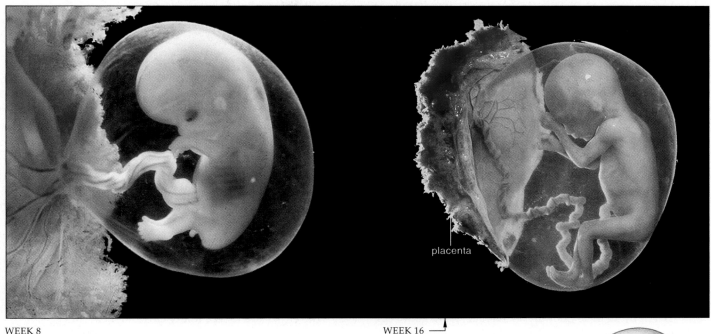

WEEK 8

final week of embryonic period; embryo looks distinctly human compared to other vertebrate embryos

upper and lower limbs well formed; fingers and then toes have separated

primordial tissues of all internal, external structures now developed

tail has become stubby

c |——— actual length ———|

placenta

WEEK 16
| Length: | 16 centimeters (6.4 inches) |
| Weight: | 200 grams (7 ounces) |

WEEK 29
| Length: | 27.5 centimeters (11 inches) |
| Weight: | 1,300 grams (46 ounces) |

WEEK 38 (full term)
| Length: | 50 centimeters (20 inches) |
| Weight: | 3,400 grams (7.5 pounds) |

During fetal period, length measurement extends from crown to heel (for embryos, it is the longest measurable dimension, as from crown to rump).

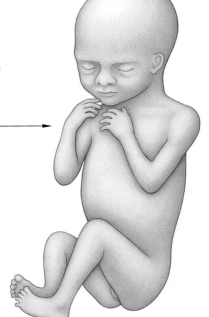

d

the sucking reflex, as shown in the next section. Now the mother can easily sense movements of the fetal arms and legs. When the fetus is five months old, she can hear its heart through a stethoscope positioned on her abdomen. Soft, fuzzy hair (the lanugo) covers the fetal body. The skin is wrinkled, reddish, and protected from abrasion by a thick, cheesy coating. In the sixth month, delicate eyelids and eyelashes form. During the seventh month, the eyes of the fetus open.

A fetus born prematurely (before 22 weeks) cannot survive. The situation also is grave for births before 28 weeks, mainly because the lungs have not developed sufficiently. The risks start to drop after this. But a fetus born before the optimal birthing time (about 38 weeks after the estimated time of fertilization) still has trouble breathing and maintaining a normal core temperature even with the best medical care. By 36 weeks, however, the survival rate is 95 percent.

In the fetal period, primary tissues that formed in the early embryo become sculpted into distinctly human features.

Mother as Provider, Protector, Potential Threat

A woman who decides to become pregnant is committing a large part of her body's resources and functions to the development of a new individual. From fertilization until birth, her future child is at the mercy of her diet, health habits, and life-style (Figure 39.25).

Nutritional considerations: A well balanced diet for a pregnant woman usually provides the embryo with all required carbohydrates, lipids, and proteins. (Chapter 37 outlines human nutritional requirements.) Her demands for vitamins and minerals rise as the placenta absorbs enough for the embryo from her blood. She must get a lot more folate (folic acid). Taking more B-complex vitamins (under a doctor's supervision) before conception and in early pregnancy reduces the risk that her embryo will develop severe neural tube defects. Smoking adversely affects nutrient utilization. In one study, pregnant women who smoked lowered the blood level of vitamin C for themselves *and* the developing fetuses, even when their intake of that vitamin matched that of a control group.

A pregnant woman must eat enough to increase her body weight by 20 to 25 pounds, on average. If she does not, her newborn may be significantly underweight. It will be at risk of more postdelivery complications and impaired mental functions later in life.

Most fetal organs are highly vulnerable to nutritional deficiencies. For example, the brain undergoes its greatest expansion in the weeks just before and after birth. Poor nutrition then will have repercussions on intelligence and other neural functions later in life.

Risk of infections: Antibodies circulating in a pregnant woman's blood continually cross the placenta and protect the new individual from all but the most serious bacterial infections. Some virus-induced diseases can be dangerous in the first six weeks after fertilization, a critical time of organ formation. Suppose a pregnant woman contracts *rubella* (German measles) in this critical period. There is a 50 percent chance that some organs of her embryo will not form properly. For example, if she becomes infected when embryonic ears are forming, her newborn may be deaf. If she becomes infected during or after the fourth month of pregnancy, the disease will have no notable effect. However, a woman can avoid this risk entirely, because vaccination before pregnancy can prevent rubella.

Prescription drug effects: Pregnant women should not take any drugs except under close medical supervision. Think of what happened when the tranquilizer *thalidomide* was routinely prescribed in Europe. Infants of women who used thalidomide in the first trimester were missing

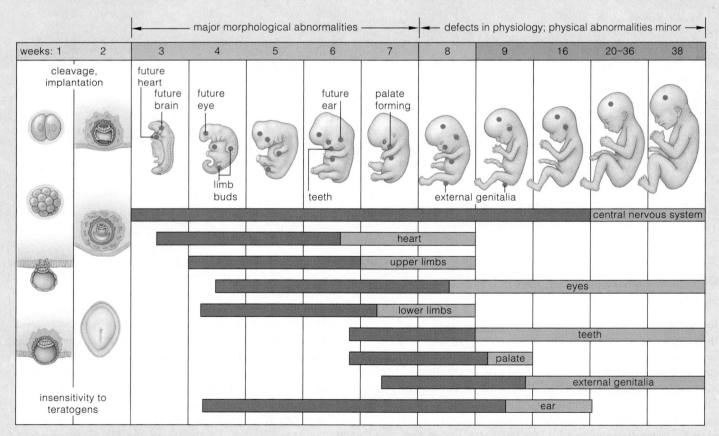

Figure 39.25 Teratogen sensitivity. *Teratogens* are drugs and other environmental factors that can induce deformities in an embryo or fetus. They usually have no effect before organs start to form. After that, they can adversely affect growth, tissue remodeling, and tissue resorption. For this graph, *dark blue* shows the highly sensitive period; *light blue* shows less severe sensitivity to teratogens.

Figure 39.26 An infant with fetal alcohol syndrome (FAS). Obvious symptoms are low and prominent ears, poorly formed cheekbones, and an abnormally wide, smooth upper lip. Growth problems and nervous system abnormalities are likely. This disorder now affects 1 in 750 newborns in the United States.

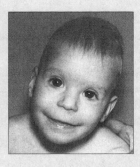

arms and legs or had grossly deformed ones. That drug was withdrawn from the market as soon as its connection with the deformities was apparent. But other tranquilizers, sedatives, and barbiturates are still being prescribed. They may cause similar, although less severe, damage. Even some *anti-acne drugs* increase the risk of facial and cranial deformities. Tetracycline, one of the most overprescribed antibiotics, yellows teeth. Streptomycin causes hearing problems and may adversely affect the nervous system.

Effects of alcohol: While a fetus grows, its physiology becomes increasingly like its mother's. Because it is so small and grows so rapidly, alcohol is more harmful to it—and alcohol passes freely across the placenta. Drinking alcohol while pregnant invites *fetal alcohol syndrome* (FAS). Symptoms include reduced brain and head size, mental retardation, facial deformities, poor growth and poor coordination, and often heart defects (Figure 39.26). One in every 750 newborns shows these abnormalities. The symptoms are irreversible; children affected by FAS never do "catch up," physically or mentally. Many researchers suspect that there is no "safe" drinking level; any alcohol may harm the fetus. More doctors are now urging total abstinence during pregnancy.

Effects of cocaine: A pregnant woman who uses cocaine, including crack, disrupts the nervous system of her future child as well as her own. Here you might wish to reflect again on the introduction to Chapter 30, which describes the kind of future that a crack addict's newborn faces.

Effects of smoking: Simply put, cigarette smoke impairs fetal growth and development. One long-term study at Toronto's Hospital for Sick Children showed that toxic elements in tobacco accumulate even inside the fetuses of pregnant nonsmokers who are exposed to *secondhand smoke* at home or work.

Daily smoking during pregnancy leads to underweight newborns even if the woman's weight, nutritional status, and all other relevant variables match those of pregnant nonsmokers. Smoking has other effects. In Great Britain, all infants born during the same week were tracked for seven years. Infants of smokers were smaller, had twice as many heart abnormalities, and died of more postdelivery complications. At age seven, they were nearly half a year behind children of nonsmokers in average "reading age."

The mechanisms by which smoking affects a fetus are not known. Its demonstrated effects are evidence that the placenta, marvelous structure that it is, cannot prevent all assaults on the fetus that the human mind can dream up.

GIVING BIRTH

Pregnancy ends thirty-eight weeks after fertilization, give or take a few weeks. We call the birth process labor (or delivery). The cervical canal dilates, and the fetus moves out from the uterus, then the vagina, and into the outside world. Strong uterine contractions are the driving force behind the expulsion.

Mild uterine contractions start in the last trimester. Relaxin, a hormone secreted by the corpus luteum and placenta, makes the cervical connective tissues soften and the bridges between pelvic bones looser. The fetus "drops" (shifts down), usually with its head touching the cervix (Figure 39.27). Rhythmic contractions herald the onset of labor. Contractions increase in frequency and intensity over the next two to eighteen hours.

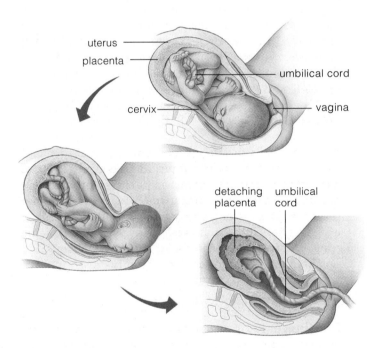

Figure 39.27 Expulsion of a human fetus and the afterbirth (the placenta, tissue fluid, and blood) during labor.

The amnion typically ruptures just before birth, so "water" (amniotic fluid) rushes out from the vagina. Contractions expel the fetus, usually less than an hour after the cervix has fully dilated. Strong contractions detach the placenta from the uterus and expel it as the afterbirth. They constrict blood vessels at the ruptured attachment site and stop hemorrhaging. The umbilical cord is cut and tied. A few days after shriveling up, the stump has become the newborn's navel.

On average, thirty-eight weeks after fertilization, strong, programmed uterine contractions drive the birth process.

FROM BIRTH ONWARD

Nourishing the Newborn

Once the lifeline to the mother is severed, a newborn enters an extended time of dependency and learning that is typical of all primates. Its early survival requires an ongoing supply of milk or its nutritional equivalent. Lactation (milk production) occurs in mammary glands in a mother's breasts (Figure 39.28). Before pregnancy, breasts consist mainly of adipose tissue and a system of undeveloped ducts. Their size depends on how much fat they contain, not on milk-producing ability. During pregnancy, estrogen and progesterone act to develop a glandular system of milk production.

For the first few days after birth, mammary glands produce a fluid rich in proteins and lactose. Prolactin, a hormone that calls for synthesis of the enzymes used in milk production, is secreted by pituitary's anterior lobe (Section 32.3). When a newborn suckles, the pituitary releases oxytocin. This hormone triggers contractions that force milk into breast tissue ducts and induce the uterus to shrink back to its pre-pregnancy size.

New mothers typically sink into *afterbaby blues*, or postpartum depression. CRH (corticotropin releasing hormone) produced by the placenta may contribute to it as well as to the timing of labor. Its level in blood rises by as much as three times in pregnancy. The rise may trigger secretion of cortisol, a stress hormone, to help the mother cope with extraordinary stresses that pregnancy and labor place on her. Once the placenta is expelled, her CRH levels crash to levels typical of some clinical depressions. Afterbaby blues end when CRH secretion from the hypothalamus returns to normal.

Regarding Breast Cancer

On average, 100,000 women develop *breast cancer* each year. Obesity, and high cholesterol and estrogen levels, contribute to the cancerous transformation. Prospects of being cured are excellent with early detection and treatment. Once a month, about a week after she has menstruated, a woman should examine her breasts. For recommended examination procedures, she can contact her physician or the American Cancer Society, which has listings in local telephone directories. It also posts information on its web site, http://www.cancer.org.

Postnatal Development

Like many species, humans start out as a miniaturized version of adults, then change in size and proportion until reaching sexual maturity. Figure 39.29 shows just a few of the proportional changes that take place as the life cycle unfolds during the prenatal ("before birth") stages and postnatal ("after birth") stages.

Postnatal growth is most rapid between thirteen and nineteen years. Secretions of sex hormones step up and bring about the emergence of secondary sexual traits as well as sexual maturity. Not until adulthood are the bones fully mature. Body tissues normally are maintained in top condition in early adulthood. As the years pass, tissues are harder to maintain and repair. They gradually deteriorate through aging processes.

Why Do Animals Age?

Every multicelled animal that undergoes extensive cell differentiation gradually deteriorates. Through processes collectively called **aging**, its cells gradually break down in structure and function, which leads to the decline of tissues, organs, and eventually the whole body.

THE PROGRAMMED LIFE SPAN HYPOTHESIS Does an internal, biological clock control aging? After all, each species has a maximum life span—20 years for dogs, 12 weeks for butterflies, and 35 days for fruit flies, and so on. No human has lived past 122 years. The consistency within species implies that genes have something to do with aging. Indeed, researchers have already doubled the life span of fruit flies by manipulating their genes.

Suppose the animal body is like a clock shop, with each type of cell, tissue, and organ ticking at its own genetically set pace. Many years ago, Paul Moorhead and Leonard Hayflick investigated such a possibility. They cultured normal human embryonic cells, which divided about fifty times, then died off. Hayflick also withdrew cultured cells that were partway through the series of divisions, then froze them for several years.

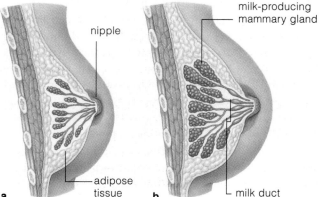

Figure 39.28 (**a**) Breast of a woman who is not pregnant. (**b**) Breast of a lactating woman.

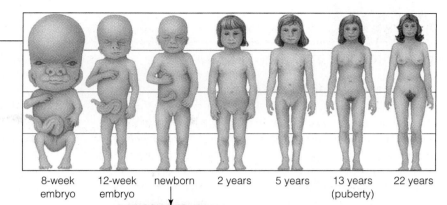

Figure 39.29 Observable, proportional changes in the human body during prenatal and postnatal growth. Changes in overall physical appearance are slow but noticeable until the teenage years. For example, the head becomes proportionally smaller, compared to the embryonic period. Legs become longer; the trunk gets shorter.

| 8-week embryo | 12-week embryo | newborn | 2 years | 5 years | 13 years (puberty) | 22 years |

After he thawed the cells and placed them in a culture medium, they proceeded to complete the in-vitro cycle of fifty doublings and died on schedule.

No cell in a human body divides more than eighty or ninety times. You may well wonder: If an internal clock ticks off their life span, then how can *cancer* cells go on dividing? The answer provides us with a clue to why *normal* cells can't beat the clock.

As you know, cells duplicate their chromosomes before they divide. Capping the chromosome ends are repetitive DNA sequences known as telomeres. A bit of each telomere is lost during each nuclear division, and when only a nub remains, cells stop dividing and die. The exceptions are cancer cells and reproductive cells. Both make telomerase. This enzyme causes telomeres to lengthen. After somatic cells were experimentally exposed to telomerase, they went on dividing well past their normal life span, with no apparent bad effects.

THE CUMULATIVE ASSAULTS HYPOTHESIS A different hypothesis holds that aging is an outcome of ongoing damage at the molecular and cellular levels. By this hypothesis, damage accumulates in DNA as a result of ongoing environmental assaults and declines in DNA's self-repair mechanisms.

For example, free radicals, those rogue molecular fragments described in Chapter 5, bombard DNA and all other biological molecules. This includes the DNA of mitochondria—the power plants of eukaryotic cells. Structural changes in DNA endanger the synthesis of functional enzymes and other proteins necessary for normal life processes. And free radicals are implicated in many age-related problems, such as atherosclerosis, cataracts, and Alzheimer's disease.

Problems in DNA replication and repair operations also have been implicated in aging. Researchers have correlated *Werner's syndrome*, an aging disorder, with a mutation in the gene that specifies a helicase. Such enzymes unwind nucleotide strands from each other. The mutated helicase probably doesn't compromise the replication of DNA; affected people don't die at once. (In their thirties, they start aging abnormally fast and die before reaching fifty.) But a nonmutated form may be vital for repairs. People with Werner's syndrome

accumulate mutations at rates that are above normal. Sooner or later, damage may interfere with cell division. Like the skin cells of very old folks, skin cells of Werner's patients will not divide many times at all.

It may be that both hypotheses have merit. Aging may be an outcome of many interconnected processes in which genes, hormones, environmental assaults, and a decline in DNA repair mechanisms come into play. Consider how living cells of all tissues depend upon exchanges of materials with extracellular fluid. Also consider how collagen is a structural element of many connective tissues. If something shuts down or mutates collagen-encoding genes, the missing or altered gene product might disrupt the flow of oxygen, nutrients, hormones, and so forth to and from cells through every connective tissue. Repercussions from such a mutation would ripple through the body.

Similarly, if mutations result in altered self markers on the body's cells, do T cells of the immune system perceive them as foreign and go on the attack? If such autoimmune responses were to increase over time, they certainly would promote the greater vulnerability to disease and stress associated with old age.

In evolutionary terms, reproductive success means surviving long enough to produce offspring. Humans can reach sexual maturity *and* help their own children reach adulthood in thirty years. We don't *need* to live longer. But we among all animals have the capacity to think about it, and we generally have decided that we like life better than the alternative. Eventually, though, even the most tenacious clingers to life must confront their own mortality; and perhaps in time we all might learn to accept the inevitable with wisdom and grace.

The human life cycle flows naturally from the time of birth, growth, and development, to production of the individual's own offspring, and on through aging to the time of death.

Aging may result from many factors, including an internal clock that ticks off the life spans of all cells, tissues, and organs, and accumulated errors and damaged areas in DNA.

CONTROL OF HUMAN FERTILITY

Some Ethical Considerations

The transformation of a zygote into an adult of intricate detail raises profound questions. *When does development begin?* As you have seen, major developmental events unfold even before fertilization. *When does life begin?* During her lifetime, a woman can produce as many as 500 eggs, all of which are alive. With one ejaculation, a man can release a quarter of a billion sperm, which are alive. Even before sperm and egg merge by chance and establish the genetic makeup of a new individual, they are as much alive as any other form of life. It is scarcely tenable, then, to say that life begins at fertilization. *Life began more than 3 billion years ago; and each gamete, each zygote, and each sexually mature individual is but a fleeting stage in the continuation of that beginning.*

This greater perspective on life cannot diminish the meaning of conception. It is no small thing to entrust a new individual with the gift of life, wrapped in the unique evolutionary threads of our species and handed down through an immense sweep of time.

Yet how can we reconcile the marvel of individual birth with growing awareness of the astounding birth rate for the human species? Almost 15,000 newborns are entering the world every hour. By the time you go to bed tonight, there may be 360,470 more newborns on Earth than there were last night at that hour. In three months there may be 32,892,930 more—almost as many people as there are now in the entire state of California. *Within three months.*

Worldwide, human population growth is rapidly outstripping resources. Each year, many millions face the horrors of starvation. Living as we do on one of the most productive continents on Earth, few of us know what it means to give birth to a child, to give it the gift of life, and have no food to keep it alive.

And how can we reconcile the marvel of birth with the confusion that surrounds unwanted pregnancies? Even highly developed countries do not have adequate educational programs concerning fertility. And many people aren't inclined to exercise control. In 1998 nearly 500,000 babies were born to teenage mothers in the United States. Many parents actually encourage early boy–girl relationships without thinking through the risks of premarital intercourse and unplanned pregnancies. Advice is often condensed to a terse "Don't do it. But if you do it, be careful!" And each year, there are between 1.2 and 1.3 million abortions among all age groups.

The motivation to engage in sex has been evolving for more than 500 million years. And a few centuries of moral and ecological arguments for its suppression have not stopped all that many unwanted pregnancies. Besides this, complex social factors have contributed to a population growth rate that is out of control.

How will we reconcile our biological past and the need for a stabilized cultural present? Whether and how human fertility is to be controlled is one of the most volatile issues of our time. We will return to this issue in the next chapter, in the context of principles that govern the growth and stability of all populations. Here, we briefly consider some control options.

Birth Control Options

The most effective method of birth control is complete *abstinence*, no sexual intercourse whatsoever. Data show that it's unrealistic to expect many people to practice it.

A less reliable variation of abstinence is the *rhythm method*. The idea is to avoid intercourse in a woman's fertile period, starting a few days before ovulation and ending a few days after. A woman identifies and tracks her fertile period. She also keeps records of the length of her menstrual cycles, takes her temperature every morning when she wakes up, or both. (The body's core temperature rises by one-half to one degree just before a fertile period.) But ovulation may not be regular, and miscalculations are frequent. Also, sperm deposited in the vagina a few days before ovulation may survive until ovulation. The rhythm method *is* inexpensive; it costs nothing after you buy a thermometer. It does not require fittings and periodic medical checkups. But its practitioners run a risk of pregnancy (Figure 39.30).

Withdrawal, or removing the penis from the vagina before ejaculation, dates back at least to biblical times. But withdrawal requires very strong willpower, and the practice may fail anyway. Fluid released from the penis just before ejaculation may contain some sperm.

Douching (rinsing the vagina with a chemical right after intercourse) is next to useless. Sperm move past the cervix, and out of reach of a douche, ninety seconds after ejaculation.

Controlling fertility by surgical intervention is less chancy. In *vasectomy*, a physician makes a tiny incision in a man's scrotum, then severs and ties off each vas deferens. The operation takes only twenty minutes and requires only a local anesthetic. After the operation, sperm can't leave the testes, so they can't be present in semen. So far, there is no firm evidence that vasectomy disrupts hormonal functions or adversely affects sexual activity. Vasectomies can be reversed. But half of those who submit to surgery later develop antibodies against sperm and may not be able to regain fertility.

Females may opt for a *tubal ligation*. By this surgical intervention, oviducts are cauterized or cut and tied off, usually in a hospital. When the operation is performed correctly, tubal ligation is the most effective means of birth control. A few women have recurring pain in the pelvic area. The operation sometimes can be reversed.

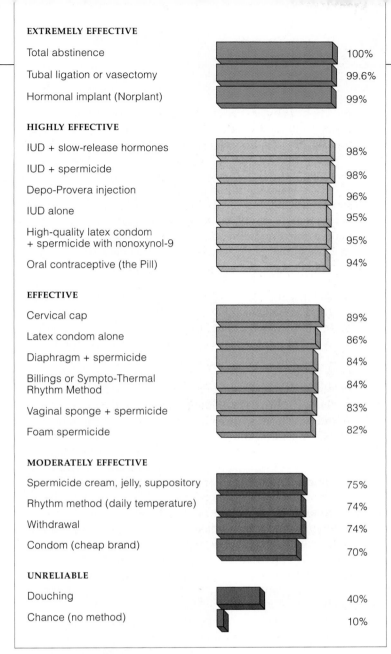

EXTREMELY EFFECTIVE	
Total abstinence	100%
Tubal ligation or vasectomy	99.6%
Hormonal implant (Norplant)	99%

HIGHLY EFFECTIVE	
IUD + slow-release hormones	98%
IUD + spermicide	98%
Depo-Provera injection	96%
IUD alone	95%
High-quality latex condom + spermicide with nonoxynol-9	95%
Oral contraceptive (the Pill)	94%

EFFECTIVE	
Cervical cap	89%
Latex condom alone	86%
Diaphragm + spermicide	84%
Billings or Sympto-Thermal Rhythm Method	84%
Vaginal sponge + spermicide	83%
Foam spermicide	82%

MODERATELY EFFECTIVE	
Spermicide cream, jelly, suppository	75%
Rhythm method (daily temperature)	74%
Withdrawal	74%
Condom (cheap brand)	70%

UNRELIABLE	
Douching	40%
Chance (no method)	10%

Figure 39.30 Comparison of the effectiveness of methods of contraception in the United States. The percentages also indicate the number of unplanned pregnancies per 100 couples who used only that method of birth control for a year. For example, "94% effectiveness" for oral contraceptives (the Pill) means that 6 of every 100 women will still become pregnant, on the average.

Less drastic methods of controlling fertility involve physical or chemical barriers that prevent sperm from entering the uterus and oviducts. *Spermicidal foam* and *spermicidal jelly* are toxic to sperm. They are transferred from an applicator into the vagina before intercourse. These products aren't always reliable unless used with another device, such as a diaphragm and a condom. *IUDs* are coils inserted into the uterus by a physician. They sometimes invite pelvic inflammatory disease.

A *diaphragm* is a flexible, dome-shaped device. It is inserted into the vagina and positioned over the cervix before intercourse. It is relatively effective when fitted initially by a doctor, used with foam or a jelly before each sexual contact, inserted correctly each time, and left in place only for a prescribed length of time.

Good brands of *condoms*—thin, tight-fitting sheaths worn over the penis during intercourse—are up to 95 percent effective if used with a spermicide. Only latex condoms offer protection against sexually transmitted diseases (Section 39.20). However, condoms often tear and leak, at which time they become absolutely useless.

The *birth control pill* is an oral contraceptive made of synthetic estrogens and progestins (progesterone-like hormones). Taken daily except for the last five days of a menstrual cycle, it suppresses oocyte maturation and ovulation. Often it corrects erratic menstrual cycles and reduces cramps, but in some women it causes nausea, weight gain, tissue swelling, and headaches. With more than 50 million women taking it, "the Pill" is the most-used fertility control method. It is 94 percent effective. Earlier formulations were linked to blood clots, high blood pressure, and maybe breast cancer. Lower doses of newer formulations might lessen the risk of breast and endometrial cancer. The possibility of a correlation between the Pill and breast cancer is still under study.

Progestin injections or implants block ovulation. A *Depo-Provera* injection works for three months and is 96 percent effective. *Norplant* (six rods implanted under skin) works for five years and is 99 percent effective. Both may cause sporadic, heavy bleeding, and doctors have some trouble surgically removing Norplant rods.

A pregnancy test won't register positive until after implantation. According to one view, a woman is not pregnant until that time, so *morning-after pills* intercept pregnancy. Actually, such pills work up to seventy-two hours following unprotected sex by interfering with the hormones that control events between ovulation and implantation. One is Preven, a kit with high doses of birth control pills that suppress ovulation and block corpus luteum function. RU-486, more commonly used as an abortion pill, also is used as a morning-after pill. It can block fertilization and progesterone-dependent implantation. It also may induce abortion when taken within seven weeks of implantation (progesterone also is vital to maintain pregnancy). It may trigger elevated blood pressure, blood clots, or breast cancer in some women. RU-486, now known as Mifeprex, is available in Europe. Although controversial, its use in the United States was granted FDA approval in 2000. By contrast, Preven had been used before as a contraceptive, and its repackaging as a morning-after pill has not been very controversial. It gained FDA approval in 1998.

Whether and how human fertility is to be controlled is a volatile issue. It centers on reconciling our biological past with seriously divergent views about our cultural present.

Sexually Transmitted Diseases

At some point in his or her life, at least one of every four individuals in the United States who engage in sexual intercourse will become infected by the pathogens that cause **sexually transmitted diseases**, or **STDs**. Tens of millions are already infected; STDs have reached epidemic proportions. Two-thirds are under age twenty-five; one-fourth are teenagers. Women and children are hardest hit. Worse yet, antibiotic-resistant strains of bacteria are on the rise, and some viral diseases simply cannot be cured.

Urban poverty, prostitution, intravenous drug abuse, and sex-for-drugs are fanning the epidemic. Yet STDs also are rampant in high schools and colleges, where students too often think, "It can't happen to me." They reject the idea that no sex—abstinence—is the only safe sex. In one poll of high school students, two-thirds of the respondents said they don't use condoms. More than 40 percent have two or more sex partners.

The economics of this health problem are staggering. In 1993, the Centers for Disease Control tallied the annual cost of treating the most prevalent STDs: *Herpes*, $759 million; gonorrhea, $1 billion; chlamydial infection, $2.4 billion; and pelvic inflammatory disease, $4.2 billion. This does not include the accelerating cost of treating patients with AIDS. In many developing countries, AIDS alone threatens to overwhelm health-care delivery systems and to unravel decades of economic progress.

The social consequences are sobering. Mothers bestow a chlamydial infection on 1 of every 20 newborns in the United States alone. They bestow type II *Herpes* virus on 1 in 10,000 newborns; one-half of the infected babies die, and one-fourth have severe neural defects. Every year, 1 million women develop pelvic inflammatory disease, and 100,000 to 150,000 become infertile.

AIDS Someone can become infected by HIV, the human immunodeficiency virus, and not even know it. However, the infection marks the start of a titanic battle that the immune system almost certainly will lose (Section 35.10). At first there may be no outward symptoms. Five to ten years later, a set of chronic disorders develops. They are evidence of *AIDS* (*A*cquired *I*mmune *D*eficiency *S*yndrome). When the immune system finally is weakened, the stage is set for opportunistic infections. Normally harmless resident bacteria are the first to take advantage of the lowered resistance. Dangerous pathogens also take their toll. In time, the immune-compromised person simply is overwhelmed.

Most commonly, HIV spreads through anal, vaginal, and oral intercourse and by IV drug users. Most infections occur by transfer of blood, semen, urine, or vaginal secretions between people. Cuts and abrasions on the penis, vagina, rectum, and oral membranes invite HIV into the internal environment of a new host.

Today there is no vaccine against HIV and no effective treatment for AIDS. If you get it, you die. *There is no cure*.

HIV was not identified until 1981, but it was present in some parts of Central Africa for at least several decades before that. In the 1970s and early 1980s, it spread to the United States and other developed countries, where most of those initially infected were male homosexuals. Today, many in the heterosexual population are infected or at risk.

Free or low-cost, confidential testing for HIV exposure is available at public health facilities and at many doctor's offices. People who are worried should know there may be a time lag between their first exposure and the first test to come out positive. It takes a few weeks to six months or more before detectable amounts of antibodies will form in response to infection. (The presence of antibodies in the blood indicates exposure to the virus.) Anyone who tests positive is capable of spreading the virus.

Public education programs attempt to stop the spread of HIV. Most health-care workers advocate safe sex, yet there is great confusion about what "safe" means. Many advocate the use of high-quality latex condoms *and* a spermicide that contains nonoxynol-9 to help prevent the transmission of the virus. Even then, there is a slight risk. As an unfortunate couple learned in 1997, no one should participate in open-mouthed kissing with someone who tests positive for HIV. Caressing is not risky *if* there are no lesions or cuts where body fluids that harbor the virus can enter the body. Pronounced lesions caused by other sexually transmitted diseases may increase susceptibility to HIV infection.

In sum, AIDS reached pandemic proportions mainly for three reasons. First, it took a while to discover that the virus can travel in semen, blood, and vaginal fluid and that *behavioral* controls can limit its spread. Second, it took time to develop ways to test symptom-free carriers, who infect others. Third, many still don't realize the medical, economic, and social consequences affect everyone.

GONORRHEA During intercourse, *Neisseria gonorrhoeae* (Figure 39.31) can enter the body at mucous membranes of the urethra, cervix, and anal canal. This bacterium causes *gonorrhea*. An infected female may notice a slight vaginal discharge or burning sensation while she is urinating. If the infection spreads to her oviducts, it may induce severe cramps, fever, vomiting, and scarring, which may cause sterility. Males have more obvious symptoms. Within a week of infection, the penis discharges yellow pus. Urinating is more frequent and may be painful.

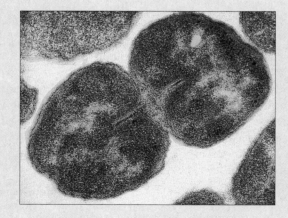

Figure 39.31 Bacterial agents of gonorrhea.

Figure 39.32
The bacterial agents of (**a**) syphilis and (**b**) chlamydia. (**c**) Genital warts on the external genitalia of a young woman.

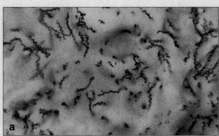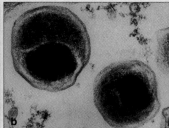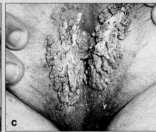

This STD is rampant even though prompt treatment quickly cures it. Why? Women experience no troubling symptoms during early stages. Also, infection does not confer immunity to the bacterium, perhaps because there are sixteen or more different strains of it. Contrary to common belief, someone can get gonorrhea over and over again. Also, oral contraceptives, used so widely, promote infection by altering vaginal pH. Populations of resident bacteria decline—and *N. gonorrhoeae* is free to move in.

SYPHILIS *Syphilis*, a dangerous STD, is caused by the spirochete *Treponema pallidum* (Figure 39.32a). Sex with an infected partner puts the motile, spiral bacterium on the surface of genitals or into the cervix, vagina, or oral cavity. It can enter the body through tiny epidermal cuts. One to eight weeks later, new treponemes are twisting about in a flattened, painless chancre (local ulcer). This first chancre is a symptom of the primary stage of syphilis. By then, treponemes are in the blood. Treponemes can cross the placenta of an infected, pregnant woman. The result will be a miscarriage, stillbirth, or syphilitic newborn.

Chancres usually heal, but treponemes multiply in the spinal cord, brain, eyes, mucous membranes, bones, and joints. More chancres and a skin rash develop during this infectious, secondary stage of syphilis. Immune responses succeed in about 25 percent of the cases, so the symptoms do subside. Another 25 percent remain infected but are symptom free. In the rest, minor to significant lesions and scars appear in the skin and liver, bones, and other internal organs. Few treponemes form during this tertiary stage. But the host's immune system is hypersensitive to them. Chronic immune reactions may severely damage the brain and spinal cord and lead to general paralysis.

Probably because the symptoms are so alarming, more people seek early treatment for syphilis than for gonorrhea. Later stages require prolonged treatment.

PELVIC INFLAMMATORY DISEASE *Pelvic inflammatory disease* (PID) is one of the most serious complications of gonorrhea, chlamydial infections, and other STDs. It also can arise when normal bacterial residents of the vagina ascend to the pelvic region. Most commonly, the uterus, oviducts, and ovaries are affected. There is bleeding and vaginal discharge. Pain in the lower abdomen may be as severe as an acute appendicitis attack. The oviducts may become scarred. Pelvic inflammatory disease invites abnormal pregnancies as well as sterility.

GENITAL HERPES About 25 million Americans have *genital herpes*, caused by the type II *Herpes simplex* virus. Infection requires direct contact with active *Herpes* viruses or sores that contain them. Mucous membranes of the mouth and the genitals are highly susceptible to invasion. The symptoms are often mild or absent. Among infected women, small, painful blisters may appear on the vulva, cervix, urethra, or anal tissues. Among men, blisters form on the penis and anal tissues. Within three weeks, the virus enters latency, and the sores crust over and heal.

The virus is reactivated sporadically. Each time, it causes painful sores at or near the original infection site. Sexual intercourse, menstruation, emotional stress, or other infections trigger *Herpes* flare-ups. Acyclovir, an antiviral drug, decreases healing time and often decreases the pain.

CHLAMYDIAL INFECTION *Chlamydia trachomatis* (Figure 39.32b), a parasitic bacterium, spends part of its life cycle in genital and urinary tract cells. It causes several diseases, including *NGU* (chlamydial nongonococcal urethritis). Chlamydial infections are now the most common reported STDs in the United States; 4 million are infected annually. Rates are highest for individuals between ages fifteen and nineteen. *C. trachomatis* and *N. gonorrhoeae* can enter new hosts at the same time by vaginal, oral, or anal intercourse. Prompt doses of penicillin cure the gonorrhea—but not NGU, which requires both tetracycline and sulfonamide.

NGU leads to inflammation of the cervix and, in both sexes, the urethra. Infected people may notice a burning sensation while urinating. When they are symptom-free, they do not seek treatment and complications develop. In males, the prostate becomes swollen and inflamed. In females, the infection may spread into the uterus and the oviducts to cause pelvic inflammatory disease.

Symptom-free infected individuals are unwittingly spreading destructive chlamydial infections through all ethnic groups, among poor and affluent alike.

GENITAL WARTS More than sixty types of the human papillomaviruses (HPV) are known. A few cause *genital warts*, or benign, bumplike growths (Figure 39.32c). HPV infection of the genitals and the anus is one of the most prevalent STDs in this country. Usually type 16 HPV does not cause obvious warts, but it may cause precancerous sores and cancers of the cervix, vagina, vulva, penis, and anus. In one Seattle study, 22 percent of female college students who were examined tested positive for the virus.

To Seek or End Pregnancy

Because of sterility or infertility, about 15 percent of all couples in the United States cannot conceive. For example, hormonal imbalances in the female body may prevent ovulation. Or the male's sperm count may be so low that fertilization would be next to impossible without medical intervention.

In vitro fertilization means conception outside the body (literally "in glass" petri dishes or test tubes). It may occur if the couple can produce normal sperm and oocytes. First the woman receives injections of a hormone that prepares her ovaries for ovulation. Afterward, the preovulatory oocyte is removed from her with a suction device. In the meantime, sperm from the male are placed in a solution that simulates fluid in the oviducts. A few hours after the sperm and suctioned oocyte make contact in a petri dish, fertilization may occur. Twelve hours later, the zygote is transferred to a solution that can sustain it through the initial cleavages. Two to four days later, the resulting ball of cells is transferred to the female's uterus.

Each attempt at in vitro fertilization costs about 8,000 dollars, and most attempts aren't successful. In 1994, each "test-tube" baby cost the nation's health-care system about 60,000 to 100,000 dollars, on average. The childless couple may believe no cost is too great. But in an era of increased population growth and shrinking medical coverage, is the cost too great for society to bear? Court battles are being waged over this issue.

At the other extreme is **abortion**, the dislodging and removal of the blastocyst, embryo, or fetus from the uterus. At one time in the United States, abortions were forbidden by law unless the pregnancy endangered the mother's life. Later the Supreme Court ruled that the government does not have the right to forbid abortions during the early stages of pregnancy (typically up to five months). Before this ruling, there were dangerous, traumatic, and often fatal attempts to abort embryos, either by pregnant women themselves or by quacks.

From a clinical standpoint, vacuum suctioning and other methods make abortion painless for the woman, rapid, and free of complications when performed during the first trimester. Abortions performed in the second and third trimesters are extremely controversial unless the mother's life is threatened. Even so, for both medical and humanitarian reasons, the majority of people in this country generally agree that the preferred route to birth control is not through abortion. Rather, it is through sexually responsible behavior that prevents unwanted pregnancy from happening in the first place.

This biology textbook cannot offer you the "right" answer to a moral question because of the reasons given in Section 1.7. It *can* provide you with a serious, detailed description of how a new human individual develops. Your choice of the "right" answer to the question of the morality of abortion will be just that—your choice—and one that can be based on objective insights into the nature of life.

SUMMARY
Gold indicates text section

Principles of Reproduction and Development

1. For animals, sexual reproduction is the dominant reproductive mode. It requires specialized reproductive structures, control mechanisms, and forms of behavior that assist fertilization and support the offspring. *39.1*

2. Embryonic development typically proceeds through six stages: *39.2–39.3*

   a. Gamete formation, when oocytes (immature eggs) and sperm form in reproductive organs. Molecular and structural components are stockpiled and are localized in different parts of the oocyte. *39.2*

   b. Fertilization, from the time a sperm penetrates an egg to the fusion of sperm and egg nuclei that results in a zygote (fertilized egg). *39.2*

   c. Cleavage, when mitotic cell divisions transform a zygote into smaller cells (blastomeres). Cleavage does not increase the original volume of egg cytoplasm; it only increases the number of cells. It regionally divides the yolk, mRNAs, proteins, cytoskeletal elements, and other "maternal messages" among new blastomeres, an outcome called cytoplasmic localization. *39.2*

   d. Gastrulation, when primary tissue layers (germ layers) form. Cell divisions, cell migrations, and other events lead to the formation of endoderm, ectoderm, and (in most species) mesoderm. All tissues of the adult body develop from primary tissue layers. *39.3*

   e. The onset of organ formation. Different organs start developing by a tightly orchestrated program of cell differentiation and morphogenesis. *39.2*

   f. Growth and tissue specialization, when organs enlarge overall and acquire their specialized chemical and physical properties. The maturation of tissues and organs continues into post-embryonic stages. *39.2*

3. Embryos cannot develop properly unless each stage of development is successfully completed before the next stage begins. *39.2*

4. In cell differentiation, a cell selectively uses certain genes and synthesizes proteins not found in other cell types. The outcome is subpopulations of specialized lineages of cells that differ from one another in their structure, biochemistry, and functioning. *39.4*

5. Morphogenesis starts at gastrulation. It changes the embryo's size, shape, and proportions. *39.4*

6. Pattern formation means the emergence of the basic body plan and body parts in specific regions of the embryo, in orderly sequence. Cytoplasmic localization helps seal the fate of cells. Cell determination depends also on interactions among classes of master genes that specify certain products. These products are spatially organized and create chemical gradients. The gradients influence differentiation, hence the identity of each cell lineage in the embryo. *39.5*

Human Reproduction and Development

1. Humans have a pair of primary reproductive organs (sperm-producing testes in males, and egg-producing ovaries in females), accessory ducts, and glands. Testes and ovaries also produce sex hormones that govern reproductive functions and secondary traits. *39.6–39.8*

    a. The hormones LH, FSH, and testosterone control sperm formation. They are part of feedback loops from the testes to the hypothalamus and anterior lobe of the pituitary gland. *39.7*

    b. The hormones estrogen, progesterone, FSH, and LH control the maturation and release of oocytes from the ovary, as well as cyclic changes in the endometrium (the inner lining of the uterus). Their secretion is part of feedback loops from ovaries to the hypothalamus and to the anterior lobe of the pituitary gland. *39.8–39.10*

2. A menstrual cycle is a recurring cycle of intermittent fertility during the reproductive years of female humans and other primates. *39.9, 39.10*

    a. Follicular phase: One of many follicles matures inside an ovary. Each follicle is an oocyte and the cell layer surrounding it. Meanwhile, the endometrium is prepared for a possible pregnancy. It breaks down each cycle if pregnancy does not occur.

    b. Ovulation: A midcycle surge of the blood level of LH triggers ovulation, the release of a secondary oocyte from the ovary.

    c. Luteal phase: After ovulation the corpus luteum, a glandular structure, develops from the remnants of the follicle. It secretes progesterone and some estrogen that prime the endometrium for fertilization. If fertilization occurs, the corpus luteum is maintained.

3. After fertilization, a blastocyst forms by cleavage and becomes implanted in the endometrium. Three primary tissue layers form that give rise to all organs. They are ectoderm, endoderm, and mesoderm. *39.11, 39.12*

4. Four extraembryonic membranes form as embryos of humans and other vertebrates develop: *39.12, 39.13*

    a. The amnion becomes a fluid-filled sac around the embryo, which it protects from drying out, mechanical shock, and abrupt temperature changes.

    b. The yolk sac stores nutritive yolk in most shelled eggs. In humans, part of the sac becomes a major site of blood formation, and some of its cells give rise to germ cells. (Germ cells later give rise to sperm and eggs.)

    c. The chorion is a protective membrane enclosing the embryo and the other extraembryonic membranes. It also becomes a major component of the placenta.

    d. In humans, the blood vessels for the placenta arise from the allantois, as does the urinary bladder.

5. The placenta, a blood-engorged organ, is composed of endometrium and extraembryonic membranes. The placenta permits embryonic blood vessels to develop independently of the mother's but also allows oxygen, nutrients, and wastes to diffuse between them. *39.14*

Review Questions

1. What is the main benefit of sexual reproduction? What are some of its biological costs? *39.1*

2. Briefly state the key events of gamete formation, fertilization, then cleavage, gastrulation, and organ formation. At which stage is the frog embryo in the photograph at right? *39.2*

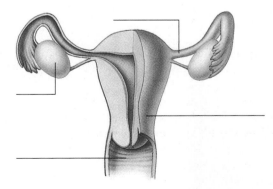

3. Does cleavage increase the volume of cytoplasm, the number of cells, or both, compared to the zygote? *39.2*

4. Give a brief example of cytoplasmic localization. *39.3*

5. Define cell differentiation and morphogenesis. *39.4*

6. Define pattern formation. Then explain the key points of the theory of pattern formation. *39.5*

7. Label the components of the human male reproductive system and state their functions: *39.6, 39.7*

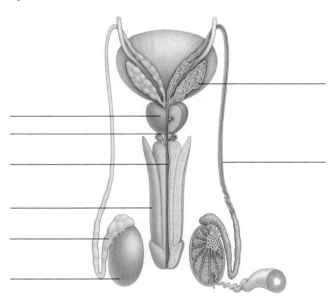

8. Label the components of the human female reproductive system and state their functions: *39.8, 39.9*

9. Does sperm formation require mitosis, meiosis, or both? *39.7*

10. Study Figure 39.15 on sperm formation. Then sketch the feedback loops to the hypothalamus and anterior pituitary from the testes. Include the names of releasing hormone, hormones, and cells in the testes involved in these loops. *39.7*

11. Study Figure 39.18. Then, sketch the feedback loops to the hypothalamus and the anterior pituitary from the ovaries. These loops govern the menstrual cycle. Name the releasing hormone, hormones, and ovarian structures involved in the loops. *39.9*

12. State the embryonic source of the amnion, yolk sac, chorion, and allantois. State the role each extraembryonic membrane plays in the structure or functioning of an embryo. *39.12*

Self-Quiz ANSWERS IN APPENDIX III

1. During cleavage, the new blastomeres are allocated different regions of the fertilized egg's cytoplasm. This is called _____ .
   a. cytoplasmic localization   c. cell differentiation
   b. embryonic induction   d. morphogenesis

2. Primary tissue layers first appear _____ .
   a. in the egg cortex   c. in the gastrula
   b. during cleavage   d. in primary organs

3. In organisms ranging from fruit flies to humans, the body's organization relative to its long axis depends on _____ .
   a. morphogens   c. embryonic induction
   b. homeotic genes   d. all of the above

4. The _____ , a fluid-filled sac, surrounds and protects the embryo from mechanical shocks and keeps it from drying out.
   a. yolk sac   b. allantois   c. amnion   d. chorion

5. At full term, a placenta _____ .
   a. is composed of extraembryonic membranes
   b. directly connects maternal and fetal blood vessels
   c. keeps maternal and fetal blood vessels separated

6. Match the development stage with its description.
   ____ cleavage
   ____ gamete formation
   ____ organ formation
   ____ growth, tissue specialization
   ____ gastrulation
   ____ fertilization

   a. egg and sperm mature in parents
   b. sperm nucleus, egg nucleus fuse
   c. formation of primary tissue layers
   d. cytoplasmic localization, not gene interactions, dominate in most animals (not in mammals)
   e. organs, tissues increase in size, acquire specialized properties
   f. starts when primary tissue layers split into subpopulations of cells

Critical Thinking

1. In normal *Drosophila* larvae, a cluster of cells is the embryonic source of antennae on the head. When a larva has a certain mutant gene, it develops legs instead of antennae on its head, as in Figure 39.33. Would you say that the mutated gene is first expressed during cleavage or organ formation?

2. Before an amphibian egg enters cleavage, you divide it so only one daughter cell gets the gray crescent. Separate the two cells and only the one with the gray crescent forms an embryo with a long axis, notochord, nerve cord, and back muscles. The other forms a shapeless mass of immature gut cells and blood cells. Does cytoplasmic localization

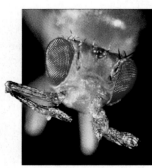

Figure 39.33 Mutant fruit fly with two legs (not two antennae) on its head.

or embryonic induction play a greater role in these outcomes? Explain your answer.

3. Despite obvious differences in early stages of development, chick and human embryos resemble one another once early organ formation is under way. Using the theory of pattern formation, speculate on the role of embryonic induction in bringing about the similarities between these vertebrate lineages.

4. Infection by the rubella virus apparently has an inhibitory effect on mitosis. Serious birth defects result when a woman is infected during the first trimester of pregnancy, but not later. Review the developmental events that unfold during pregnancy and explain why this might be so.

5. Suppose you are an obstetrician advising a woman who just learned she is pregnant. What instructions would you provide concerning her diet and behavior during pregnancy?

6. In the United States, teenage pregnancies and STD infections are rampant. Suppose the office of the Surgeon General asks you to join a task force that will recommend practices to reduce the incidence of both. What practices might be most successful? What kind of enthusiasm or resistance might they provoke among the teenagers in your community? Among adults? Explain why.

7. By the traditional view, separation of the sexes has selective advantages. By a highly provocative, alternative view, males and females have been waging a "gender war," at the level of genes, for billions of years. Matt Ridley's *Genome* (1999, New York: HarperCollins) may challenge your thinking on this topic.

Selected Key Terms

Principles
asexual reproduction *39.1*
blastocyst *39.3*
blastomere *39.2*
blastula *39.3*
cell differentiation *39.4*
cleavage *39.2*
cytoplasmic localization *39.3*
ectoderm *39.2*
embryo *39.2*
embryonic induction *39.5*
endoderm *39.2*
fertilization *39.2*
gamete formation *39.2*
gastrulation *39.2*
growth, tissue specialization *39.2*
master gene *39.5*
mesoderm *39.2*
morphogen *39.5*
morphogenesis *39.4*
neural tube *39.4*
oocyte *39.3*
organ formation *39.2*
pattern formation *39.5*
sexual reproduction *39.1*
sperm *39.3*
yolk *39.1*
zygote *CI*

Human Case Study
abortion *39.21*
aging *39.18*
allantois *39.12*
amnion *39.12*
chorion *39.12*
corpus luteum *39.9*
endometrium *39.8*
estrogen *39.8*
fetus *39.12*
FSH *39.7*
implantation *39.12*
in vitro fertilization *39.21*
LH *39.7*
menstrual cycle *39.8*
ovulation *39.8*
ovum *39.11*
placenta *39.14*
polar body *39.9*
progesterone *39.8*
puberty *39.6*
seminiferous tubule *39.6*
somite *39.13*
STD *39.20*
testis (testes) *39.6*
testosterone *39.7*
uterus *39.8*
yolk sac *39.12*

Readings

Nusslein-Volhard, C. August 1996. "Gradients That Organize Embryonic Development." *Scientific American*, 54–61.

On-Line readings at Student Guide for InfoTrac:
www.brookscole.com/biology

VII   Ecology and Behavior

Two organisms—a fox in the shadows cast by a snow-dusted spruce tree. What are the nature and consequences of their interactions with each other, with other kinds of organisms, and with their environment? By the end of this last unit, you may find worlds within worlds in such photographs.

40

POPULATION ECOLOGY

Tales of Nightmare Numbers

Across from Sausalito, California, the steep flanks of Angel Island rise from the waters of San Francisco Bay (Figure 40.1*a*). The island, set aside as a game reserve, escaped urban development but not the descendants of a few deer that well-meaning nature lovers shipped over in the early 1900s. With no natural predators to keep them in check, the few deer became many—too many for the limited food in the isolated habitat. Yet the island attracted a steady stream of picnickers from the mainland. They felt sorry for the malnourished deer and loaded picnic baskets with extra food for them.

The visitors imported so much food that scrawny deer kept on living and reproducing. In time, the herd nibbled away the native grasses, the roots of which had helped slow soil erosion on the steep hillsides. Hungry deer chewed off all the new leaves of seedlings; they killed small trees by stripping away bark and phloem. The herd was destroying the environment.

In desperation, game managers proposed using a few skilled hunters to thin the herd. They were denounced as cruel. They proposed importing a few coyotes to thin the herd naturally. Animal rights advocates said no.

As a compromise, about 200 of the 300+ deer were captured, loaded onto a boat, and shipped to suitable mainland habitats. Many were given collars with radio transmitters so that game managers could track them after the release. In less than sixty days, dogs, coyotes, bobcats, hunters, and speeding cars and trucks killed most of them. In the end, relocating each surviving deer cost taxpayers close to 3,000 dollars. The State of California refused to do it again. And no one else, anywhere, volunteered to pick up future tabs.

It's not difficult to define Angel Island's boundaries or track its inhabitants, so it's easy to draw a lesson from this tale: *A population's growth depends on environmental resources. And attempts to "beat nature" by altering the sometimes cruel outcome of limited resources only postpone the inevitable.* Does the lesson apply to other populations, in other places? Yes it does, as the next tale makes clear.

When 2000 drew to a close, there were over 6 billion people on Earth. About 2 billion already live in poverty. Each year 40 million more join the ranks of the starving. Next to China, India is now the most populous country (Figure 40.1*b*). Its population size exceeds 1 billion, and

Figure 40.1 (**a**) Angel Island, which turned out to be a good laboratory for studying population growth. (**b**) Bathers along a bank of the Ganges River in India— a small sampling of a human population that has exceeded 6 billion. In this chapter, we turn to principles that govern the growth and sustainability of all populations.

18 million are added annually. Forty percent of India's population live in rat-infested shantytowns. They don't have enough food or fresh water and are forced to wash clothes and dishes in open sewers. The land available to raise food shrinks by 365 acres a day, on average. Why? Irrigated soil becomes too salty when it drains poorly, and there isn't enough water to flush away the salts.

Can wealthier, less densely populated nations help? After all, they use most of the world's resources. Maybe they should learn to get by more efficiently, on less. For example, people might limit their meals to cereal grains and water; give up their private cars, living quarters, air conditioners, televisions, and dishwashers; stop taking vacations and stop laundering so much; close all of the malls, restaurants, and theaters at night; and so on.

Maybe wealthier nations also should donate more surplus food than they already donate to less fortunate ones. Then again, would huge donations help or would they encourage dependency and spur more increases in population size? And what if surpluses run out?

It is a monumental dilemma. At one extreme, global redistribution of resources would help the most people survive, but at the lowest comfort level. At the other extreme, foreign aid rationed only to nations that limit population growth might encourage fewer individuals to be born to help give each a better quality of life.

The United States foreign aid program is currently based on two premises: (1) that individuals of every nation have an irrevocable right to bear children, even if unrestricted reproduction ruins the environment that must sustain them; and (2) that because human life is precious above all else, the wealthiest nations have an absolute moral obligation to save lives everywhere.

Regardless of the positions that nations take on this issue, they ultimately must come to terms with this fact: *Certain principles govern the growth and sustainability of populations over time.* These principles are the bedrock of **ecology**—the systematic study of how organisms interact with one another and with their physical and chemical environment. Ecological interactions start within and between populations, and they extend on through communities, ecosystems, and the biosphere. They are the focus of this last unit of the book.

In this chapter we look first at relationships that influence the characteristics of populations. Later, we will apply the basic principles of population growth to the past, present, and future of the human species.

Key Concepts

1. Certain ecological principles govern the growth and sustainability of all populations, including our own.

2. In addition to genetic factors, a population's size, density, distribution, and the number of individuals in various age categories influence its patterns of growth.

3. When the birth rate exceeds the death rate, and when immigration and emigration are in balance, populations may show a pattern of exponential growth.

4. With exponential growth, a population expands by ever increasing increments during successive intervals, because the number of individuals in its reproductive age category (or about to enter it) becomes ever larger.

5. A shortage of any resource that individuals require is a limiting factor on population growth.

6. For any population, carrying capacity is the maximum number of individuals that can be sustained indefinitely by the resources of a given environment. That number may rise or fall with changes in resource availability.

7. A population may show a pattern of logistic growth. By this pattern, population density—that is, the number of individuals in a specified area at a given point in time—is initially low. Then the population size rapidly increases. It finally levels off as resource scarcity limits its further increase or causes a decline in numbers.

8. Some populations show fluctuations in numbers that cannot be explained by a single growth model.

9. All populations face limits to growth, because no environment can indefinitely sustain a successively increasing number of individuals. Competition, disease, predation, and other factors control population growth. The controls vary in their relative effects on populations of different species, and they vary over time.

CHARACTERISTICS OF POPULATIONS

By this point in the book you know that a population is a group of individuals of the same species occupying a given area. Its gene pool is the basis for characteristic ranges of morphological, physiological, and behavioral traits. When studying a population, ecologists consider the genetic make-up as well as reproductive modes and reproductive behavior. They consider the **demographics**, or vital statistics of a population, and that will be our focus here. The vital statistics include population size, density, distribution, and age structure.

Population size is the number of individuals that contribute to a population's gene pool. A population's **age structure** is the number of individuals in each of several to many age categories. For instance, all members may be grouped into *pre-reproductive*, *reproductive*, and *post-reproductive* ages. Individuals in the first category have the capacity to produce offspring when mature. Together with the actually and potentially reproducing individuals in the second category, they help make up the population's **reproductive base**.

Population density is the number of individuals in some specified area or volume of a habitat, such as the number of guppies per liter of water in a stream. (A *habitat*, remember, is the type of place where a species normally lives. We characterize a habitat by its physical features, chemical features, and the presence of other species.) **Population distribution** is the general pattern in which a population's individuals are dispersed in a specified area.

Crude density is a measured number of individuals in a specified area. Section 40.2 outlines how ecologists make their counts, which help them track changes in population density over time. The counts do not reveal how much of a habitat is used as living space. Even areas that appear to be uniform, such as a long, sandy beach, are more like fine tapestries of light, moisture, temperature, mineral composition, and other variables. Besides, the tapestry of environmental conditions often changes with the seasons. So a population might find one part of a habitat more suitable for occupancy than other parts, and it may do so some or all of the time.

Also, different species typically share the same area —and they compete for energy, nutrients, living space, and other resources. Most interact as predators, prey, or parasites. Later chapters describe these interactions. For now, simply note that species interactions influence population density and dispersion through a habitat.

Theoretically, populations show three distribution patterns. By these patterns, individuals are dispersed in clumps, nearly uniformly, or randomly (Figure 40.2). The individuals of most populations form aggregations at specific sites in a habitat. Why is clumping the most common pattern of dispersion? Three possible reasons come to mind.

Figure 40.2 Three generalized patterns of population distribution.

clumped

nearly uniform

random

First, each species is adapted to a limited set of ecological conditions, which usually are patchy through a habitat. For example, some parts of a habitat offer more or less shade, moisture, hiding spots, and hunting spots. Second, many animal species gather in social groups that afford advantages in terms of survival and reproduction. Such groups offer more opportunities for, say, mating and mutual defense against predators. Third, the adults of many species cannot disperse seeds, larvae, or other immature forms of the new generation over very large distances. For example, sponge larvae are able to swim, but not very far from their parent sponge. They simply settle down on substrates near the parents.

Where the individuals are more evenly spaced than they would be by chance alone, we call this a nearly uniform dispersal pattern. Evenly spaced fruit trees in orchards are often cited as examples. Few populations live this way in nature. You may see the pattern where habitat conditions are fairly uniform and where there is fierce competition for resources or territorial behavior.

Creosote bushes (*Larrea*) that live in arid habitats of the American Southwest may show this pattern. Large, mature plants deplete soil water all around them. Seed-eating ants and rodents forage near the plants, which offer cover from predators and the hot sun. Most seeds are devoured, which puts the seeds and seedlings at a competitive disadvantage. So seeds typically take hold only at patches where established, mature plants are weakened or have died. As one result, creosote bushes are not clumped together (Figure 40.3a). Yet neither are they dispersed at random; seeds that rodents and ants overlook and that await opportunity for life in the sun cannot travel far from the mature parent plants.

We observe random dispersion only when habitat conditions are nearly uniform, resource availability is fairly steady, and individuals of the population neither attract nor avoid one another. Example: Wolf spiders are solitary hunters on forest floors. Each generation might well be randomly spaced. However, a dispersion pattern such as this is extremely rare in nature.

Each population has its own gene pool and range of traits. It has a characteristic size, density, distribution pattern, and age structure. Environmental conditions and species interactions influence these characteristics.

Elusive Heads to Count

Does it seem like deer are just about everywhere, eating up gardens and smacking into more and more cars on the roads? The estimated number of deer in the United States has risen from fewer than 1 million at the turn of the century to more than 18 million today. Suppose you live in a northern California county and wonder how many deer are living there with you. How would you go about determining their population density?

To get some idea of what your study will entail, you start with a literature search. Ecologists have baselines of approximate population densities for many organisms. For example, in some habitats, baselines show that you can expect to find as many as 5 million diatoms per cubic meter, 500 trees per hectare, 250 field mice per hectare, and so on. (A hectare is 2.47 acres.) Your search turns up an estimate of 4 deer per square kilometer. This should tell you deer will be easier to count than, say, diatoms.

A full count is the most straightforward measurement of absolute population density. Census takers make such a count every ten years for the human population in the United States. Ecologists do the same for large organisms in small areas, such as songbirds in a forest, northern fur seals at their breeding grounds, sea stars in a tidepool, and creosote bushes in a patch of desert (Figure 40.3a). More often, they must content themselves with sampling a small part of a population and estimating total density.

You could get a map of your county and divide it into small plots, or quadrats. Quadrats are sampling areas of the same size and shape, such as rectangles, squares, and hexagons. Then you could count all the deer in several plots and extrapolate the average for the entire county. Ecologists conduct such counts for migrating herds and flocks. For example, the North American Breeding Bird Survey makes counts each year along more than 2,000 flyways for migratory waterfowl (Figure 40.3b).

Deer are among the animals that do not stay put in their habitat. So how can you be absolutely sure that the individuals you count in a given plot are not the same ones you counted earlier in a different plot?

For mobile animals, ecologists sample the population density with a **capture–recapture method**. The idea is to capture individuals and mark them in some way. Deer get bright collars, squirrels get tattoos, salmon get tags, migratory birds get leg rings, and so on. Marked animals are released at time 1. Animals are captured and checked for marks at time 2. In later samplings, the proportion of marked individuals should be representative of the proportion marked in the whole population:

$$\frac{\text{Marked individuals in sampling 2}}{\text{Total capture in sampling 2}} = \frac{\text{Marked individuals in sampling 1}}{\text{Total population size}}$$

Ideally, marked and unmarked individuals of the population are captured at random, no marked animal dies during the study interval, and none of the marked animals leaves the population or gets overlooked. In the real world, though, recapturing of marked individuals might *not* be random. Squirrels that were marked after being attracted to yummy bait in boxes might now be trap-happy or trap-shy, so they might overrepresent or underrepresent the population. Instead of mailing the tags of marked fish back to ecologists, some fishermen keep them as good-luck charms. Birds might lose their leg rings. Undocumented immigrants may not answer the door when the census taker comes knocking.

Your estimate also depends on the time of year when you do a sampling. Population distribution varies over time, as in migratory responses to environmental rhythms. Few places yield abundant resources all year long, and many animals move from one habitat to another with changes in the seasons (Figure 40.3b). Deer are like this, so you would probably use capture–recapture methods more than once a year, for several years.

Figure 40.3 (a) Example of nearly uniform spacing: creosote bushes near Death Valley, California. (b) Snow geese, one of many species of waterfowl that migrate through 3.4 million square kilometers of air space above Canada and the United States.

40.3

POPULATION SIZE AND EXPONENTIAL GROWTH

Gains and Losses in Population Size

Populations are dynamic units of nature. Depending on the species, they may add or lose individuals every half hour of every day, season, or year. Sometimes they glut portions of their habitat with individuals. Other times, individuals are scarce. Populations even drive themselves or are driven into extinction. We measure such changes in population size in terms of birth rates, death rates, and how many individuals are entering and leaving during a specified interval.

Population size increases as a result of (1) births and (2) **immigration**: the arrival of new residents from other populations of the species. Its size decreases because of (1) deaths and (2) **emigration**: individuals permanently moving out of the population. Population size for many species changes predictably owing to daily or seasonal migrations. However, **migration** is a recurring round trip between two regions, so we need not consider its transient effects in our initial study of population size.

From Ground Zero to Exponential Growth

For our purposes, assume that immigration is balancing emigration over time, so that we can ignore the effects of both on population size. Doing so allows us to define **zero population growth** as some interval during which the number of births is balanced out by the number of deaths. The population's size is stabilized during such an interval, with no overall increase or decrease.

Births, deaths, and other variables that might affect population size can be measured in terms of **per capita** rates, or rates per individual. (*Capita* means heads, as in head counts.) Visualize 2,000 mice living in a cornfield. Twenty or so days after their eggs are fertilized, the females produce a litter, then nurse the offspring for a month or so, and get pregnant again. Collectively, if all the females give birth to 1,000 mice per month, the birth rate would be 1,000/2,000 = 0.5 per mouse per month. If 200 of the 2,000 die during that interval, the death rate would be 200/2,000 = 0.1 per mouse per month.

If we assume the birth rate and death rate remain constant, we can combine both into a single variable—the **net reproduction per individual per unit time**, or *r* for short. For our mouse population in the cornfield, *r* is 0.5 − 0.1 = 0.4 per mouse per month. This example gives us a way to represent population growth:

population growth per unit time	=	net population growth rate per individual per unit time	×	number of individuals

or, more simply, $G = rN$.

As the next month begins, 2,800 mice are scurrying about. With the net increase of 800 fertile critters, the reproductive base is larger. Assume *r* doesn't change. The population size expands this month, too, for a net increase of 0.4 × 2,800 = 1,120 mice. The population is now 3,920. In the months ahead, it just so happens that

Figure 40.4 (**a**) Data showing net monthly increases in a population of field mice living in a cornfield. Start to finish, the numbers listed show a pattern typical of exponential growth. (**b**) Graph the data and you end up with a J-shaped growth curve.

		Net Monthly Increase:		New Population Size:
$G = r \times$	3,920 =	1,568 =		5,488
$r \times$	5,488 =	2,195 =		7,683
$r \times$	7,683 =	3,073 =		10,756
$r \times$	10,756 =	4,302 =		15,058
$r \times$	15,058 =	6,023 =		21,081
$r \times$	21,081 =	8,432 =		29,513
$r \times$	29,513 =	11,805 =		41,318
$r \times$	41,318 =	16,527 =		57,845
$r \times$	57,845 =	23,138 =		80,983
$r \times$	80,983 =	32,393 =		113,376
$r \times$	113,376 =	45,350 =		158,726
$r \times$	158,726 =	63,490 =		222,216
$r \times$	222,216 =	88,887 =		311,103
$r \times$	311,103 =	124,441 =		435,544
$r \times$	435,544 =	174,218 =		609,762
$r \times$	609,762 =	243,905 =		853,667
$r \times$	853,677 =	341,467 =		1,195,134

a

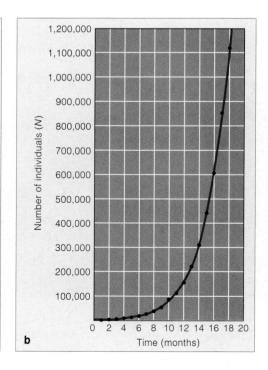

b

Figure 40.5 (**a**) Effect of deaths on the rate of increase in two bacterial populations. Growth curve *1* represents a population of bacterial cells that reproduced every half hour. Growth curve *2* represents a different population. This cell's population divided every half hour, but 25 percent died between cell divisions. Comparison of the two graph lines shows that deaths can slow the rate of increase but cannot in themselves prevent exponential growth.

(**b**) Do such growth curves seem far removed from your own experience? Think about the growth of populations of, say, the resident bacteria in your mouth after you present them with a smorgasbord of nutrients in candy or some other sugar-laden food.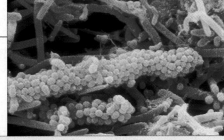

b Example of the types of bacterial cells that live in the human mouth

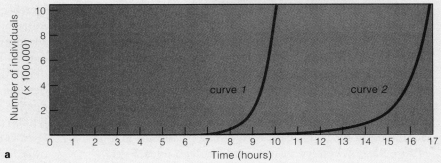

r remains constant, and a growth pattern emerges. As Figure 40.4*a* shows, *in less than two years from the time we started counting, the number of mice running around in the cornfield increased from 2,000 to more than a million!*

Plot the monthly increases against time, as in Figure 40.4*b,* and you end up with a graph line in the shape of a "J." When growth of any population over increments of time plots out as a J-shaped curve, you'll know that you are tracking **exponential growth**.

"Exponential" refers to a relationship in which one variable increases much faster than another in a specific mathematical way. In this case, a population's size is a variable that depends on how many individuals make up the reproductive base in successive increments of time. *The larger the reproductive base, the greater will be the expansion in population size during each specified interval.*

Now look at other aspects of exponential growth by supplying a single bacterium in a culture flask with all the nutrients required for growth. Thirty minutes later, the one cell divides in two. Thirty minutes pass, the two cells divide, and so on every thirty minutes. Assuming no cells die between divisions, the population size will double in each interval—from 1 to 2, then 4, 8, 16, 32, and so on. The length of time it takes for a population to double in size is its **doubling time**.

The larger the population gets, the more cells there are to divide. After 9–1/2 hours (nineteen doublings), it has more than 500,000 bacterial cells. After 10 hours (twenty doublings), it has more than 1 million. Plot the doublings in size against time, and you end up with the J-shaped curve typical of unrestricted, exponential growth. Figure 40.5, curve *1*, shows this.

To examine the effect of deaths on the growth rates, start over with one bacterium. Assume that 25 percent of the cells die in each thirty-minute interval. Because of deaths, it takes about seventeen hours (not ten) for population size to reach one million. *But deaths changed*

only the time scale for population growth. You still end up with a J-shaped curve, which is curve *2* in Figure 40.5.

What Is the Biotic Potential?

Finally, visualize a population occupying a place where conditions are ideal. Every one of its individuals has adequate shelter, food, and other vital resources. No predators, pathogens, or pollutants lurk anywhere in the habitat. That population might well display its **biotic potential**, which is the maximum rate of increase per individual under ideal conditions.

Each species has a characteristic maximum rate of increase. For many bacteria, it is 100 percent every half hour or so. For humans and other large mammals, it is between 2 and 5 percent per year. But the *actual* rate depends on the age at which each generation starts to reproduce, how often each individual reproduces, and how many offspring are produced. Now think about this. A human female is biologically capable of bearing twenty or more children, yet in each generation, many females do not reproduce at all. The human population has not been displaying its biotic potential. Even so, since the mid-eighteenth century, its growth has been exponential, for reasons that will soon be apparent.

During a specified interval, population size is generally an outcome of births, deaths, immigration, and emigration.

With exponential growth, population size expands by ever increasing increments during successive time intervals, because the reproductive base becomes ever larger.

Population size against time plots out as a characteristic J-shaped curve if the population is growing exponentially.

As long as the per capita birth rate remains even slightly above the per capita death rate, a population can grow exponentially.

LIMITS ON THE GROWTH OF POPULATIONS

What Are the Limiting Factors?

Most of the time, environmental circumstances prevent any population from fulfilling its biotic potential. That is why sea stars, the females of which could produce 2,500,000 eggs each year, do not fill up the oceans with sea stars. That also is why humans will never fill up the planet, even with our capacity for exponential growth.

In natural environments, complex interactions occur within and between populations of different species, so it is not easy to identify all the factors working to limit population growth. To get a sense of what some of the factors might be, start again with a lone bacterium in a culture flask, where you can control the variables. First you enrich the culture medium with glucose and other nutrients required for bacterial growth. Then you allow bacterial cells to reproduce for many generations. At first the growth pattern appears to be exponential. Then growth slows; and afterward, population size remains rather stable. After that stable period, population size declines rapidly and all the bacteria die.

What happened? When the population expanded by ever increasing amounts, cells tapped more and more nutrients. The dwindling nutrients signaled the cells to stop dividing (Section 8.2). And when the existing cells exhausted the nutrient supply, they starved to death.

Any essential resource that is in short supply is a **limiting factor** on population growth. Food, minerals of certain types, refuge from predators, living quarters, and an environment free of pollution are examples. The number of such factors can be huge, and their effects can vary. Even so, one factor alone is often enough to put the brakes on population growth at any given time.

Suppose you keep on freshening the supply of all required nutrients for that growing bacterial culture. After growing exponentially, the population still crashes. Like all other organisms, bacteria produce metabolic wastes. Wastes of that huge population were so great, they drastically altered living conditions in the culture. By its own metabolic activities, the population polluted the experimentally designed habitat and thus put a stop to any further exponential growth.

Carrying Capacity and Logistic Growth

Now visualize a small population in which individuals are dispersed through the habitat. As the population increases in size, more and more individuals must share nutrients, living quarters, and other resources. As the share available to each diminishes, fewer individuals may be born and more may die by starvation or nutrient deficiencies. Now the population's rate of growth will decline until births are balanced or outnumbered by deaths. Ultimately, the *sustainable* supply of resources will be the key factor determining population size. The term **carrying capacity** refers to the maximum number of individuals of a population (or species) that a given environment can sustain indefinitely.

The pattern of **logistic growth** is a fine example of how carrying capacity can affect a population. By this pattern, a low-density population starts growing slowly in size, then it grows rapidly, and finally its size levels off once the carrying capacity is reached. How can you represent the pattern? Starting with the exponential growth equation given earlier, in Section 40.3, add the term $(K - N)/K$, where K designates carrying capacity:

$$G = r_{max} N \frac{K - N}{K}$$

The term tells us how many individuals can be added to the population based on the proportion of resources still available. When population size is small, $(K - N)$ is close to 1. It approaches 0 when the size is close to the carrying capacity. Representing this another way,

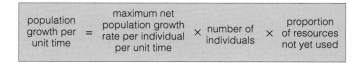

| population growth per unit time | = | maximum net population growth rate per individual per unit time | × | number of individuals | × | proportion of resources not yet used |

A plot of logistic growth gives an S-shaped curve (Figure 40.6). Such curves are just an approximation of what goes on in nature. For instance, a population that

Figure 40.6 Idealized S-shaped curve characteristic of logistic growth. After a rapid growth phase (time B to C), growth slows and the curve flattens out as the carrying capacity is reached (time C to D). S-shaped growth curves can show variations, as when changes in the environment lower the carrying capacity (time D to E). This happened to the human population of Ireland before 1900, when late blight, a disease caused by a water mold, destroyed the potato crops that were the mainstay of the diet (Section 20.7).

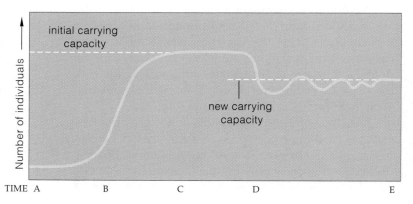

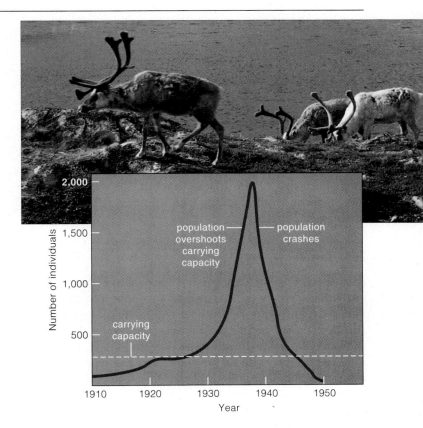

Figure 40.7 Carrying capacity and a reindeer herd. In 1910, four male and twenty-two female reindeer were introduced on St. Matthew Island in the Bering Sea. In less than thirty years, the size of the herd increased to two thousand. Individuals had to compete for dwindling vegetation, and overgrazing destroyed most of it. In 1950, herd size plummeted to eight. The growth curve reflects how the population size overshot the carrying capacity, then crashed.

grows too fast can overshoot the carrying capacity. The death rate skyrockets, and the birth rate plummets. These two responses drive the number of individuals down to the carrying capacity or lower (Figure 40.7).

Density-Dependent Controls

The logistic growth equation just described deals with **density-dependent control** of populations. That is, a small population may grow rapidly, but when it finally bumps up against its carrying capacity, so to speak, its size will stabilize or decline.

In addition to carrying capacity, other natural factors help control population density. When they come into play, they, too, put the number of individuals below a maximum sustainable level. For example, individuals are more likely to die under crowded conditions. Why? Predators, parasites, and pathogens can interact more intensely with their host population and bring about a decline in density. Once the population's size declines, the density-dependent interactions relax. And then the population size may increase once again.

Bubonic plague and *pneumonic plague*, two dangerous diseases, are examples of density-dependent controls. Both are caused by a bacterium, *Yersinia pestis*. A large reservoir of *Y. pestis* persists in rabbits, rats, and certain other small mammals. Fleas transmit it by biting new hosts. Bacterial cells reproduce so frequently in the flea gut that their numbers and metabolic activities disrupt digestion. Hunger sensations drive a flea to feed more often on more of its hosts, and so the disease spreads.

A devastating episode of bubonic plague occurred in the fourteenth century. The plague swept through European cities where humans were crowded together, sanitary conditions were poor, and rats were abundant. By the time the epidemic subsided, urban populations in Europe had declined by 25 million.

Both diseases are still threats. For example, in 1994 in India, they raced through rat-infested cities, where garbage and animal carcasses had piled up for months in the streets. Terrified residents fled by the thousands. Some carried the pathogen with them as far away as London. Global panic ensued before concerted efforts to burn the garbage, poison the rats and the fleas, and rapidly dispense antibiotics averted a pandemic.

Density-Independent Factors

Sometimes events result in more deaths or fewer births regardless of a population's density. For instance, each year, monarch butterflies migrate from Canada to spend the winter in Mexico's forested mountains. But logging opened up stands of the forest trees—which normally buffer temperature. In 1995 a sudden freeze, combined with deforestation, killed millions of butterflies. Both were **density-independent factors**, meaning the effects came into play independently of population density.

Similarly, heavy applications of pesticides in your backyard may also kill most insects, mice, cats, birds, and other animals. They will do this regardless of how uncrowded or dense the populations are.

Resources in short supply put limits on population growth. Together, all of the limiting factors acting on a population dictate how many individuals can be sustained.

Carrying capacity is the maximum number of individuals of a population that can be sustained indefinitely by the resources in a given environment. The number may rise or fall with changes in resource availability.

The size of a low-density population may increase slowly, go through a rapid growth phase, then level off once the carrying capacity for the population is reached. This is a logistic growth pattern.

Density-dependent controls and density-independent factors can bring about decreases in population size.

LIFE HISTORY PATTERNS

So far, we've looked at populations as if their members are identical all through a specified interval. For most species, however, individuals are at different stages of development and are interacting differently with other organisms and the environment. At different times in the life cycle, they may exploit different resources (as when caterpillars eat leaves and butterflies use nectar) and may be more or less vulnerable to danger. In short, each species has a **life history pattern**. Let's look at a few environmental variables that influence age-specific life history patterns.

Life Tables

Each species has a characteristic life span, but few of its individuals reach the maximum age possible. Death looms larger during some age intervals than others. In addition, the individuals of a species tend to reproduce and often emigrate during some expected interval.

Age-specific patterns first intrigued health and life insurance companies, then they appealed to ecologists. Investigators usually track a **cohort**, defined as a group of individuals from the time of birth until the last one dies. They also track the number of offspring born to individuals during each age interval. Life tables list the data gathered on a population's age-specific death schedule. Often they are converted to a more cheery "survivorship" schedule, or the number of individuals

that reach some specified age (x). Table 40.1 is a typical example of a life table.

Dividing a population into age classes and assigning birth rates and mortality risks to each one has practical applications. Unlike a crude census (head count), the data might be a basis for informed policy decisions in pest management, endangered species protection, social planning for human populations, and other areas. For instance, birth/death schedules for the northern spotted owl figured in federal court rulings to halt mechanized logging in old-growth forests, the owl's habitat.

Patterns of Survival and Reproduction

Evolutionarily, we measure an individual's reproductive success in terms of the number of surviving offspring. Yet the number differs among species, which differ in the size of offspring and in how much energy and time go into making gametes, getting mates, and parenting. Selection pressures, including species interactions and habitat conditions, have resulted in many tradeoffs.

For each species, a **survivorship curve** is a graph line that emerges when ecologists plot a cohort's age-specific survival in a habitat. Three types are common in nature.

Type I curves reflect high survivorship until fairly late in life, when deaths increase greatly. Such curves are typical of elephants and other large mammals that bear only one or a few large offspring at a time, then provide extended parental care (Figure 40.8a). A female elephant, for instance, gives birth to four or five calves and devotes several years to parenting each one. Type I curves also are typical of human populations that have access to good health care services. A sharp drop at the start of such curves usually reflects infant deaths in places where health care is poor. After that drop, the curve levels off from childhood to early adulthood.

Type II curves reflect a fairly constant death rate at all ages. They are typical of organisms just as likely to be killed or to die of disease at any age, such as lizards, small mammals, and some songbirds (Figure 40.8b).

Type III curves signify a death rate that is highest early in life. They are typical of species that produce many small offspring and show little, if any, parental behavior. Figure 40.8c shows how the curve plummets for sea stars. Sea stars produce mind-boggling numbers of tiny larvae, which must rapidly eat, grow, and finish developing on their own without support, protection, or guidance from the parents. Corals and other animals swiftly eat most of them. The plummeting survivorship curve is typical of many other marine invertebrates, most insects, and many fishes, plants, and fungi.

At one time, ecologists thought selection processes favored *either* an early, rapid production of many small offspring *or* the late production of just a few large ones.

Table 40.1 *Life Table for the United States Human Population in 1989**

Age Interval	Survivorship*	Mortality**	Life Expectancy***	Reported Live Births for Total Population
0–1	100,000	896	75.3	
1–5	99,104	192	75.0	
5–10	98,912	117	71.1	
10–15	98,795	132	66.2	11,486
15–20	98,663	429	61.3	506,503
20–25	98,234	551	56.6	1,077,598
25–30	97,683	606	51.9	1,263,098
30–35	97,077	737	47.2	842,395
35–40	96,340	936	42.5	293,878
40–45	95,404	1,220	37.9	44,401
45–50	94,184	1,766	33.4	1,599
50–55	92,418	2,727	28.9	
55–60	89,691	4,334	24.7	
60–65	85,357	6,211	20.8	
65–70	79,146	8,477	17.2	
70–75	70,669	11,470	13.9	
75–80	59,199	14,598	10.9	
80–85	44,601	17,448	8.3	
85 +	27,153	27,153	6.2	
				TOTAL: 4,040,958

* Number alive at start of age interval per 100,000 individuals.
** Number dying during the age interval.
*** Average lifetime remaining at start of age interval.

Figure 40.8 Three generalized survivorship curves. Type I populations show high survivorship until some age, and then high mortality. Type II populations show a fairly constant death rate. Type III populations show low survivorship early in life.

They now know that the two patterns are extremes at opposite ends of a range of possible life histories. Also, both life-history patterns—as well as intermediate ones—are sometimes evident in different populations of the same species, as the next section makes clear.

Tracking a cohort (a group of individuals from birth until the last one dies) reveals patterns of reproduction, death, and migration that typify the populations of a species.

Survivorship curves can reveal differences in age-specific survival among species. In some cases, such differences occur even between populations of the same species.

Natural Selection Among the Guppies

Several years ago, evolutionary biologists David Reznick and John Endler were netting guppies in the mountains of Trinidad, a Caribbean island. Populations of these live-bearing fishes inhabit different streams and even different parts of the same stream. They do not all face the same predators. One type of killifish shares some streams with guppies. It is not an especially large fish. It successfully preys on smaller, immature guppies but not on larger adults. A bigger fish, a pike-cichlid, shares other streams with guppies. It preys on larger, sexually mature guppies and tends to ignore small ones.

Using natural selection theory, Reznick and Endler hypothesized that predation influences guppy life history patterns. In streams where pike-cichlids reign supreme, individual guppies mature faster and their body size is smaller at maturity, compared with guppies in streams that killifish dominate. In addition, guppies targeted by pike-cichlids reproduce earlier in life, they produce far more offspring, and they do so more frequently.

Do other variables influence life history patterns in killifish streams versus pike-cichlid streams? To find out, the researchers shipped two groups of guppies from the two kinds of streams to a laboratory in the United States. There, guppies reproduced for two generations, free from predation. Aquarium conditions were identical for the experimental populations. In body size, generation time, and other features, offspring of both populations were like guppies in the wild, so the differences must have a genetic basis. After many generations, body size of the guppy lineage that was subjected to killifish predation became larger at maturity. The guppy lineage raised with pike-cichlids showed a trend toward earlier maturity.

For the first Trinidad field experiment, Reznick and Endler captured guppies that had evolved with pike-cichlids. They introduced them to a site upstream from a small waterfall, which had prevented the upstream dispersal of guppies and all predators except killifish. They designated that part of the stream the control site.

Eleven years later, other researchers visited the stream and discovered the guppies had evolved. They compared guppies from the experimental site and the control site. Just as Reznick and Endler had predicted, the body size, frequency of reproduction, and other aspects of guppy life history patterns correlated with preferences of the neighborhood predator. Later, a laboratory experiment involving two generations of guppies confirmed that the differences were genetic. The researchers concluded that the differences are the product of natural selection.

GUPPY FROM A
PIKE-CICHLID STREAM

GUPPY FROM A
KILLIFISH STREAM

HUMAN POPULATION GROWTH

The human population has now surpassed 6 billion. By midyear 2000, rates of increase for different countries ranged from more than 3 percent to below 1 percent, for an annual average of 1.26 percent. Visualize enough people to fill another New York City every month and you get the picture. Now think about this: The annual additions to that base population will result in a *larger* absolute increase each year into the foreseeable future.

Our staggering population growth continues even though as many as 2 billion are already malnourished or starving, without clean drinking water and adequate shelter. It continues when 1.5 billion are already going without the benefit of health care delivery and sewage treatment facilities. And it continues mainly in regions already overcrowded. Nearly all 6+ billion of us live on 10 percent of the land; 4 billion crowd together within 480 kilometers (about 300 miles) of the seas.

Suppose it were possible, by monumental efforts, to double the food supply to keep pace with growth. We would do little more than maintain marginal living conditions for most. The annual deaths from starvation could still be 20 million to 40 million. Even this would come at great cost, for we are severely modifying the environment that must sustain us. Salted-out cropland, desertification, deforestation, pollution—these are some of the consequences you will read about in Chapter 45, and they do not bode well for our future.

For a while, it will be like the Red Queen's garden in Lewis Carroll's *Through the Looking Glass*, where one is forced to run as fast as one can to remain in the same place. But what happens when our population doubles again? Can you brush the doubling aside as being too far in the future to warrant concern? *It is no further removed from you than the sons and daughters of the next generation.*

How We Began Sidestepping Controls

How did we get into this predicament? For most of its history, the human population grew slowly. But during the past two centuries, increases in growth rates became astounding. There are three possible reasons for this:

1. Humans steadily developed the capacity to expand into new habitats and new climate zones.

2. Humans increased the carrying capacity in their existing habitats.

3. Human populations sidestepped limiting factors.

The human population did all three of these things.

Reflect on the first point. The first humans evolved in woodlands and savannas. They were vegetarians, for the most part, but they also scavenged bits of meat. Small bands of hunter–gatherers started moving out of Africa about 2 million years ago. By 40,000 years ago,

different populations of their descendants had become established in much of the world (Section 27.15).

Most species could not have expanded into such a broad range of habitats. Humans, with their highly complex brain, could draw on learning and memory to figure out how to build fires, assemble shelters, make clothes and tools, and plan community hunts. Learned experiences did not die with individuals. They spread quickly from one band to another by language, which is the basis for extraordinary cultural communication. *And so the human population expanded into diverse, new environments in exceptionally short order, compared to the long-term geographic dispersal of other kinds of organisms.*

What about the second possibility? Starting about 11,000 years ago, many hunter–gatherer bands shifted to agriculture. Instead of following the migratory game herds, they settled in fertile valleys and other regions that favored seasonal harvesting of fruits and grains. By doing so, they developed a more dependable basis for life. A pivotal factor was the domestication of wild grasses, including species ancestral to modern wheats and rice. People harvested, stored, *and planted* seeds in one place. They domesticated animals and kept them close to home for food and for pulling plows. They dug ditches and diverted water to irrigate croplands.

Their agricultural practices increased productivity. And the larger, more dependable food supplies favored increases in population growth rates. Towns and cities developed. A social hierarchy emerged that provided a labor base for more intensive agriculture. Much later, food supplies increased again by the use of fertilizers, herbicides, and pesticides. Transportation improved, and so did food distribution. *Thus, even at its simplest, management of food supplies through agriculture increased the carrying capacity for the human population.*

What about the third possibility—the sidestepping of limiting factors? Think about what happened when medical practices and sanitary conditions improved. Until about 300 years ago, poor hygiene, malnutrition, and contagious disease kept death rates high enough to more or less balance birth rates. Contagious diseases, a type of density-dependent control, swept unchecked through overcrowded settlements and cities that were infested with fleas and rodents. Then came plumbing and methods of sewage treatment. Over time, vaccines, antitoxins, and antibiotics were developed as weapons against many of the pathogens. Death rates dropped sharply. Births began to exceed deaths—and extremely rapid population growth was under way.

In the mid-eighteenth century, people discovered how to harness energy stored in fossil fuels, starting with coal. Within a few decades, large industrialized societies began to form in western Europe and North America. Then efficient technologies developed after

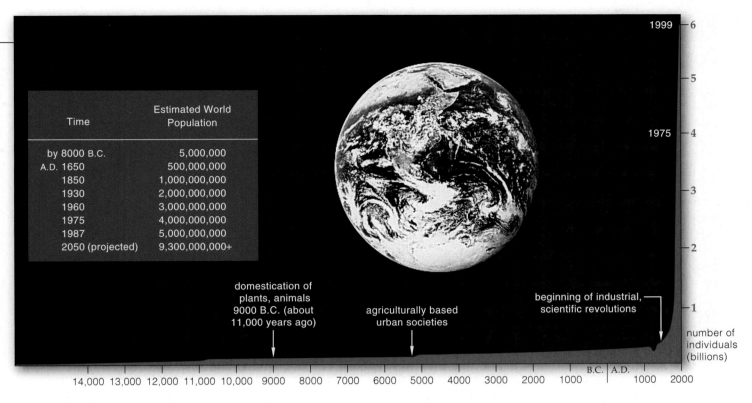

Time	Estimated World Population
by 8000 B.C.	5,000,000
A.D. 1650	500,000,000
1850	1,000,000,000
1930	2,000,000,000
1960	3,000,000,000
1975	4,000,000,000
1987	5,000,000,000
2050 (projected)	9,300,000,000+

domestication of plants, animals 9000 B.C. (about 11,000 years ago)

agriculturally based urban societies

beginning of industrial, scientific revolutions

number of individuals (billions)

14,000 13,000 12,000 11,000 10,000 9000 8000 7000 6000 5000 4000 3000 2000 1000 | B.C. A.D. | 1000 2000

World War I. Large factories mass-produced tractors, cars, and other affordable goods. Machines replaced many of the farmers who had produced the food, and fewer farmers could support a larger population.

And so, by controlling many disease agents and tapping into concentrated, existing forms of energy (fossil fuels), humans have managed to sidestep major factors that had previously limited their population growth.

Present and Future Growth

So where have the far-flung dispersals and spectacular advances in agriculture, industrialization, and health care taken us? Start with *Homo habilis* (Section 27.14). It took about 2.5 million years for human population size to get to 1 billion. As Figure 40.9 shows, it took only 80 years to reach 2 billion, another 30 to reach 3 billion, 15 more to get to 4 billion, and 12 more to get to 5 billion. And it took merely 11 more years to arrive at 6 billion!

From what we know of the principles governing population growth—and unless more breakthroughs in technology can increase the carrying capacity—we can expect a dramatic increase in death rates. *Although the stupendously accelerated growth of the human population continues, it cannot be sustained indefinitely.*

Besides having bad effects on resource supplies, the skyrocketing numbers invite horrid density-dependent controls. Example: Infection by *Vibrio cholerae* results in a disease called cholera. This bacterium enters hosts who drink water or eat food that is contaminated with raw sewage. It multiplies in the gut, where it produces a toxin that triggers severe diarrhea and massive fluid loss. Two to seven days afterward, infected, untreated people can die from extreme dehydration.

Figure 40.9 Growth curve (*red*) for the human population. Its vertical axis represents the world population, in billions. (The dip between the years 1347 and 1351 is when 60 million people died swiftly and horribly, first in Asia, then in Europe, during a bubonic plague.) Agricultural revolutions, industrialization, and health care improvements have sustained the accelerated growth pattern for the past two centuries. The *blue* box lists how long it took for the human population to increase from 5 million to 5 billion.

For centuries, *V. cholerae* has thrived and mutated into new strains in India's sewage-enriched rivers and estuaries (Figure 44.13). In the slums of Calcutta alone, millions are forced to bathe in polluted waterways and ponds. In 1992, cholera struck tens of thousands in that city. Since then, the largest cholera epidemic in 100 years has been sweeping through southern Asia. Like six previous epidemics, it began in India and spread through Africa, the Middle East, and Mediterranean countries. It reached Central America and the United States. This latest round of cholera resurgence might claim 5 million people. Existing vaccines do not work against the mutant bacterial strain causing it.

At this writing, a new era of human migration has begun. By some estimates, economic hardship and civil strife have put 50 million people on the move within and between many countries. Will their relocations be peaceable? Will they find sustainable supplies of food, clean water, and other basic resources?

Through expansion into new habitats, cultural intervention, and technological innovation, the human population has temporarily skirted environmental resistance to growth. Its accelerated growth cannot be sustained indefinitely.

CONTROL THROUGH FAMILY PLANNING

Figure 40.10 presents the annual rates of population increase for major regions in 1997. The current average rate of 1.4 percent adds another 77 million or so every year to the current total of 6.1+ billion. At this rate, world population may exceed 9 billion by 2050.

It is mind-numbing to think about all the natural resources required for that many people. We will have to increase food production and supplies of drinkable water, energy reserves, wood, and other materials to meet everyone's basic needs, something that we are not doing now. The gross manipulation of resources will probably intensify pollution and have harmful effects on water supplies, air quality, and food productivity.

Population growth, resource depletion, pollution, and the quality of life are all interconnected. Currently, most governments are attempting to lower birth rates, as through family planning programs. The programs help people decide how many children they will have, and when. Details vary from country to country, but all provide information on available methods of fertility control of the sort listed in Section 45.13. Thoughtfully developed and administered programs have the goal of bringing about a long-term decline in birth rates.

The **total fertility rate** (TFR) is the average number of children being born to women of a given population during their reproductive years. TFRs are estimated on the basis of current age-specific rates. The TFR for the world's population declined dramatically since 1950, when the average was 6.5. That number was far above the replacement level of 2.1, which is considered to be the level necessary to arrive at zero population growth. The average replacement level is slightly higher than two children per couple, because some female children die before reaching reproductive age.

At this writing the world population's TFR is 2.9. In many of the developed countries, TFRs are at or below the replacement level, but in developing countries it is about 3.2. Even if every couple on the planet decides that they will bear no more than two children, the world population will keep growing for another sixty years! Why? A huge number of existing children will soon reach reproductive age.

Figure 40.11 shows some age structure diagrams for populations growing at different rates. The central part of each diagram includes individuals of reproductive age. The lower part includes children who will move into the reproductive age category over the next fifteen years, with the average range for childbearing years being 15–44. Age structure diagrams for fast-growing populations, including Mexico, have a broad base. The human population of the United States has a relatively narrow base and is an example of slow growth. Figure 40.12 tracks its 78 million *baby-boomers*. This is a cohort that started to form in 1946, when American soldiers returned home after World War II and started families.

More than a third of the world population is now in the broad pre-reproductive base. This hints at the magnitude of what it will take to control world population growth.

The rate of birth slows when women bear children in their early thirties rather than in mid-teens or early twenties. Delayed reproduction slows the growth rate and lowers the average number of children in families.

China, for instance, supports the world's most far-reaching family planning. The government discourages premarital sex; it urges people to postpone marriage and limit families to one child. It makes available free abortions, contraceptives, and sterilization to married

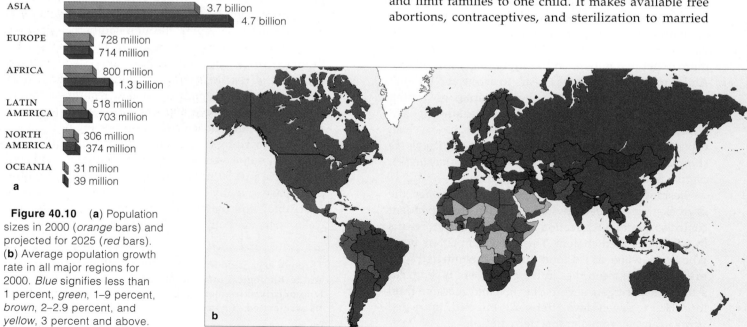

Figure 40.10 (a) Population sizes in 2000 (*orange* bars) and projected for 2025 (*red* bars). (b) Average population growth rate in all major regions for 2000. *Blue* signifies less than 1 percent, *green*, 1–9 percent, *brown*, 2–2.9 percent, and *yellow*, 3 percent and above.

Bar chart (a):

ASIA — 3.7 billion / 4.7 billion
EUROPE — 728 million / 714 million
AFRICA — 800 million / 1.3 billion
LATIN AMERICA — 518 million / 703 million
NORTH AMERICA — 306 million / 374 million
OCEANIA — 31 million / 39 million

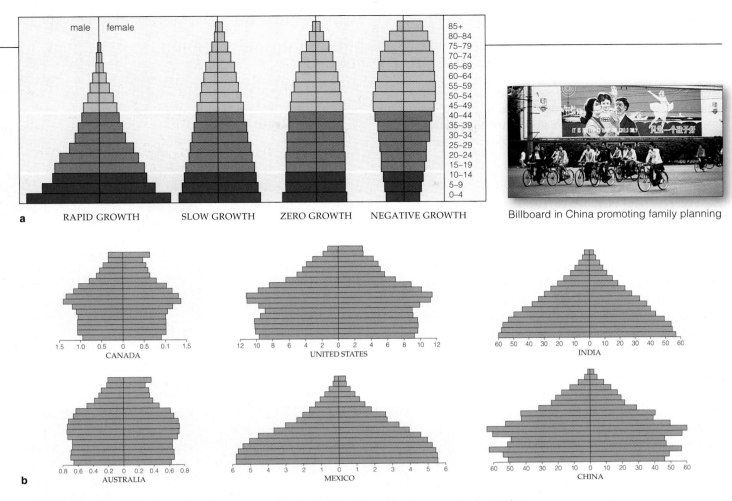

male | female

85+
80–84
75–79
70–74
65–69
60–64
55–59
50–54
45–49
40–44
35–39
30–34
25–29
20–24
15–19
10–14
5–9
0–4

a RAPID GROWTH   SLOW GROWTH   ZERO GROWTH   NEGATIVE GROWTH

Billboard in China promoting family planning

1.5 1.0 0.5 0 0.5 0.1 1.5
CANADA

12 10 8 6 4 2 0 2 4 6 8 10 12
UNITED STATES

60 50 40 30 20 10 0 10 20 30 40 50 60
INDIA

0.8 0.6 0.4 0.2 0 0.2 0.4 0.6 0.8
AUSTRALIA

6 5 4 3 2 1 0 1 2 3 4 5 6
MEXICO

60 50 40 30 20 10 0 10 20 30 40 50 60
CHINA

b

Figure 40.11 (**a**) General age structure diagrams for countries with rapid, slow, zero, and negative population growth rates. Pre-reproductive years are *green* bars; reproductive years, *purple*; and the post-reproductive years, *light blue*. A vertical axis divides each graph into males (*left*) and females (*right*). Bar widths correspond to the proportion of individuals in each age group. (**b**) Age structure diagrams for representative countries in 1997. Population sizes are measured in millions.

couples; paramedics and mobile units ensure access to these measures even in remote rural areas. Couples who pledge to have only one child receive more food,

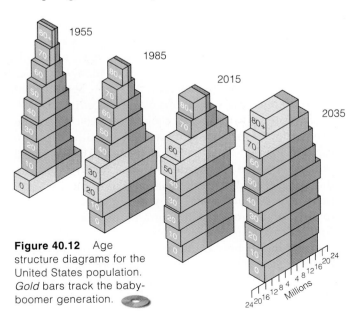

Figure 40.12 Age structure diagrams for the United States population. *Gold* bars track the baby-boomer generation.

free medical care, better housing, and salary bonuses. The child will get free tuition and preferred treatment when he or she is old enough to enter the job market. Breaking the pledge means losing benefits and paying more taxes. Do you find the measures inhumane? Then what realistic alternatives would you suggest? Between 1958 and 1962 alone, 30 million Chinese may have died of starvation because of widespread famines.

Since 1972, China's TFR has declined sharply, from 5.7 to 1.2. Even so, the population time bomb has not stopped ticking. Its current population is close to 1.27 billion—and 150 million of its young females are in the pre-reproductive age category. By 2050 the population in China is projected to surpass 1.36 billion.

Family planning programs on a global scale are designed to help stabilize the size of the human population.

Even if it reaches a level of zero population growth, the human population will continue to grow for sixty years, for its reproductive base already consists of a staggering number of individuals.

POPULATION GROWTH AND ECONOMIC DEVELOPMENT

More and more people are now aware that population growth rates and economic development interconnect. It seems economic security lifts pressure on individuals to produce a great many children to help them survive.

Demographic Transition Model

We can correlate changes in population growth with changes that often unfold in four stages of economic development. These four stages are at the heart of the **demographic transition model** (Figure 40.13).

By this model, living conditions are harshest during the *preindustrial* stage, before technology and medical advances become widespread. Birth and death rates are both high, so the population growth rate is low. Next, during the *transitional* stage, industrialization begins; food production and health care improve. Death rates drop, but birth rates stay fairly high. Thus the population grows rapidly. Its growth continues at high rates for an extended length of time. Annual growth rates tend to be 2.5 to 3 percent, on the average. When living conditions improve and birth rates begin to decline, the growth starts to level off.

During the *industrial* stage, when industrialization is in full swing, population growth slows dramatically. A slowdown starts primarily because people move from the country to cities; urban couples tend to control family size. Many couples get caught up in the accumulation of goods and often decide the time and cost of raising more than a few children conflict with that goal. As you will see, this is a sticky issue with respect to population size.

In the *postindustrial* stage, zero population growth is reached. The birth rate falls below the death rate, and population size slowly decreases.

The United States, Canada, most of western Europe, Australia, Japan, and the nations of the former Soviet Union are in the industrial stage. Their growth rate is slowly decreasing. In Germany, Bulgaria, Hungary, and some other countries, the death rates exceed the birth rates, and the populations are getting smaller.

Mexico and other less developed countries are in the transitional stage. They don't have enough skilled workers to complete the transition to a fully industrial economy. Fossil fuels and other resources that drive industrialization are being used up there as well as in the industrialized countries. Fuel costs might become

Figure 40.13
The demographic transition model for changes in the growth characteristics and size of populations, as correlated with changing economic development. The model helps explain previous changes in western Europe and other industrialized regions.

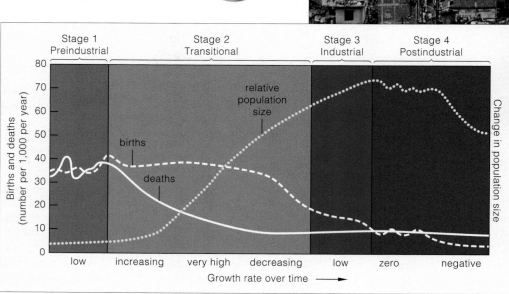

prohibitive for countries at the bottom of the economic ladder even before they enter the industrial stage.

If population growth continues to outpace economic growth, death rates will increase. Thus many countries might be stuck in the transitional stage. Some might return to the harsh conditions of the preceding stage.

Enormous disparities in economic development are driving forces for immigration and emigration. In 1996 close to 916,000 legal immigrants and 75,000 political refugees crossed the borders of the United States. So did 275,000 undocumented immigrants, mostly from Latin America and Asia. They brought the estimated total of undocumented immigrants in this developed country to 5 million.

Or think of how California's population increased by 12 *million* between 1970 and 1996. By July 1999, it exceeded 33 million. By 2025, it may be between 41.5 million to 51+ million. For some time, California has been a symbol of the good life, drawing people from all over the world. But its schools are overcrowded, funds for welfare and other social services are shrinking, and sewage treatment plants are near capacity. Shortages of water are frequent and often severe. Statewide electric power shortages and rolling blackouts started in 2001. The shortages, together with ongoing salinization of cropland and air pollution, are eroding the agricultural and technological bases of this state. At this writing, the economic repercussions of California's slowdown are affecting individuals throughout the nation.

Claiming that population growth affects economic health, many governments restrict immigration. Only the United States, Canada, Australia, and a few other countries are now accepting large annual increases. Elsewhere, appalling living conditions, extremely harsh governmental policies, and civil strife have combined to promote the exodus of more people than better-off nations can easily absorb.

A Question of Resource Consumption

This chapter opened with a brief look at the conditions endured by most people in India, with its whopping 16 percent of the human population. By comparison, the United States has merely 4.7 percent. Yet which country is the most "overpopulated"—not in terms of actual numbers, but rather in terms of resource consumption and instigation of environmental damage?

The highly industrialized United States produces 21 percent of the world's goods and services. However, its people consume fifty times more goods and services than the average person in India. Also, its people use 25 percent of the world's processed minerals, and as much of the available, nonrenewable sources of energy. They generate at least 25 percent of all pollution and trash. By contrast, India produces about 1 percent of all goods and services. It uses 3 percent of the available minerals and nonrenewable energy resources. It only generates about 3 percent of the pollution and trash.

Extrapolating from these numbers, G. Tyler Miller estimated that it would take 12.9 billion impoverished individuals living in India to have as much impact on the environment as 258 million people in the United States. Think about this estimate when you also reflect on the biodiversity crisis and conservation programs, as sketched out in Chapter 25.

Differences in population growth among countries correlate with levels of economic development, hence with economic security (or lack of it) of individuals. Growth rates are low in preindustrial, industrial, and post-industrial stages and are greatest during the transition to industrialization.

SOCIAL IMPACT OF NO GROWTH

For us, as for all species, the biological implications of extremely rapid growth are staggering. Yet so are the social implications of what will happen when (and if) the human population decreases to the point of zero population growth—and stays there.

For instance, in a growing population, most people are in younger age brackets. If living conditions ensure constant growth over time, the age distribution should guarantee availability of a future work force. This has social implications. Why? *It takes a large work force to support the older age brackets.* In the United States, older, nonproductive people expect the federal government to provide them with subsidized medical care, low-cost housing, and many other social programs. However, as a result of improved medicine and hygiene, people in these brackets are living far longer than the elderly did when the nation's social security program was set up. The current cash benefits exceed the contributions that they made to the program when *they* were younger.

If the population ever does reach and maintain zero growth, a larger proportion of individuals will end up in the older age brackets. Even slower growth will pose problems. What happens when baby-boomers retire? Will all those nonproductive members continue to get goods and services if productive ones must carry more and more of the economic burden? This isn't an abstract question. Put it to yourself. Just how much economic hardship are you willing to bear for the sake of your parents? For your grandparents? And how much will your children be willing or able to bear for you?

We have arrived at a major turning point, not only in our biological evolution but in our cultural evolution as well. The decisions awaiting us are among the most difficult we will ever have to make—yet it is clear that they must be made, and soon.

All species face limits to growth. We might think we are different from the rest, for our special ability to undergo rapid cultural evolution has enabled us to postpone the action of most of the factors that limit growth. But the crucial word is *postpone*. No amount of cultural intervention can hold back the ultimate check of limited resources and a damaged environment.

We have sidestepped a number of the smaller laws of nature. In doing so we have become more vulnerable to those laws which cannot be repealed. Today, there may be only two options available. Either we make a global effort to limit population growth in accordance with environmental carrying capacity, or we wait until the environment does it for us.

In the final analysis, no amount of cultural intervention can repeal the ultimate laws governing population growth as imposed by the carrying capacity of the environment.

SUMMARY   *Gold* indicates text section

1. A population is a group of individuals of the same species occupying a given area. It has a characteristic size, density, distribution, and age structure as well as characteristic ranges of heritable traits. *40.1*

2. The growth rate for a population during a specified interval may be determined by calculating the rates of birth, death, immigration, and emigration. To simplify calculations, we may put aside effects of immigration and emigration, and combine the birth and death rates into a variable *r* (net reproduction per individual per unit time). Then we may represent population growth (*G*) as $G = rN$, where *N* is the number of individuals during the interval specified. *40.3*

    a. In cases of exponential growth, the reproductive base of a population increases and its size expands by ever increasing increments during successive intervals. This trend plots out as a J-shaped growth curve.

    b. Any population will show exponential growth as long as its per capita birth rate stays even slightly above its per capita death rate.

    c. In logistic growth, a low-density population slowly increases in size, goes through a rapid growth phase, then levels off in size once carrying capacity is reached.

3. Carrying capacity is the name ecologists give to the maximum number of individuals in a population that can be sustained indefinitely by the resources available in their environment. *40.4*

4. The availability of sustainable resources as well as other factors that limit growth dictates population size during a specified interval. The limiting factors vary in their relative effects and vary over time, so population size also changes over time. *40.4*

5. Limiting factors such as competition for resources, disease, and predation are density-dependent. Density-independent factors, such as weather on the rampage, tend to raise the death rate or lower the birth rate more or less independently of population density. *40.4*

6. Patterns of reproduction, death, and migration vary over the life span of a species. Environmental variables help shape the life history (age-specific) patterns. *40.5*

7. The human population has now surpassed 6 billion. Its annual growth rate is below zero in a few developed countries and above 3 percent in some less developed countries. In 2000 the annual growth rate for the global human population was 1.26 percent. *40.7–40.8*

8. The human population's rapid growth in the past two centuries occurred through expansion into many diverse habitats and through agricultural, medical, and technological developments that raised the carrying capacity. Ultimately, we must confront the reality of the carrying capacity and limits to growth. *40.7–4.10*

Review Questions

1. Define population size, density, and distribution. Describe a typical population's age-structure categories. *40.1*

2. Define exponential growth. Be sure to state what goes on in the age category that is a foundation for its occurrence. *40.3*

3. Define carrying capacity. Describe its effect as evidenced by a logistic growth pattern. *40.4*

4. Give examples of the limiting factors that come into play when a population of mammals (for example, rabbits, deer, or humans) reaches very high density. *40.4, 40.5*

5. Define doubling time. In what year is the human population expected to surpass 9 billion? *40.3, 40.7*

6. How did earlier human populations expand steadily into new environments? How did they increase the carrying capacity in their habitats? Have we now avoided some limiting factors on population growth? Or is the avoidance an illusion? *40.7*

Self-Quiz   ANSWERS IN APPENDIX III

1. _____ is the study of how organisms interact with one another and with their physical and chemical environment.

2. A _____ is a group of individuals of the same species that occupy a certain area.

3. The rate at which a population grows or declines depends upon the rate of _____ .
    a. births   c. immigration   e. all of the above
    b. deaths   d. emigration

4. Populations grow exponentially when _____ .
    a. population size expands by ever increasing increments through successive time intervals
    b. size of low-density population increases slowly, then fast, then levels off once carrying capacity is reached
    c. Both a and b are characteristic of exponential growth.

5. For a given species, the maximum rate of increase per individual under ideal conditions is the _____ .
    a. biotic potential   c. environmental resistance
    b. carrying capacity   d. density control

6. Resource competition, disease, and predation are _____ controls on population growth rates.
    a. density-independent   c. age-specific
    b. population-sustaining   d. density-dependent

7. Which of the following factors does *not* affect sustainable population size?
    a. predation   c. resources   e. All of the above can
    b. competition   d. pollution   affect population size.

8. In 2000, the average annual growth rate for the human population at midyear was _____ percent.
    a. 0   b. 1.05   c. 1.26   d. 1.55   e. 2.7   f. 4.0

9. Match each term with its most suitable description.
    _____ carrying   a. maximum rate of increase per
         capacity   individual under ideal conditions
    _____ exponential   b. population growth plots out
         growth   as an S-shaped curve
    _____ biotic   c. the maximum number of
         potential   individuals sustainable by
    _____ limiting factor   an environment's resources
    _____ logistic growth   d. population growth plots out
                        as a J-shaped curve
                        e. short supply of any resource
                        essential to population growth

Critical Thinking

1. If house cats that have not been neutered or spayed live up to their biotic potential, two can be the start of many kittens—12 the first year, 72 the second year, 429 the third, 2,574 the fourth, 15,416 the fifth, 92,332 the sixth, 553,019 the seventh, 3,312,280 the eighth, and 19,838,741 kittens the ninth year. Is this a case of logistic growth? Exponential growth? Irresponsible cat owners?

2. A third of the world population is below age fifteen. Describe the effect of this age distribution on the future growth rate of the human population. If you conclude that it will have severe impact, what sorts of humane recommendations would you make to encourage individuals of this age group to limit family size? What are some social, economic, and environmental factors that might keep them from following the recommendations?

3. Write a short essay about a population having one of the age structures shown below. Describe what may happen to younger and older groups when individuals move into new categories.

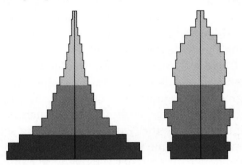

4. Figure 40.14 charts the legal immigration to the United States between 1820 and 1997. (The Immigration Reform and Control Act of 1986 accounted for the most recent dramatic increase; it granted legal status to undocumented immigrants who could prove they had lived in the country for years.) During the 1980s and 1990s, an economic downturn fanned resentment against newcomers. Many people now say legal immigration should be restricted to 300,000–450,000 annually and we should crack down on undocumented immigrants. Others argue that such a policy would diminish our reputation as a land of opportunity. They also say it would discriminate against legal immigrants during crackdowns on others of the same ethnic background. Do some research, then write an essay on the pros and cons of

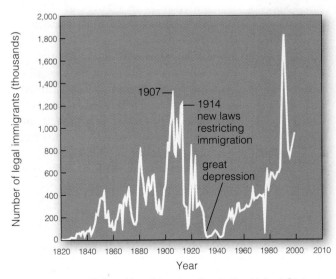

Figure 40.14 Chart of legal immigration to the United States between 1820 and 1997.

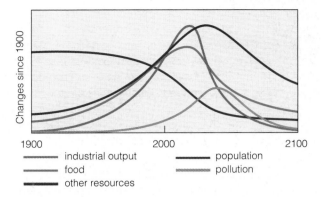

Figure 40.15 Computer-based projection of what may happen if human population size continues to skyrocket without dramatic policy changes and technological innovation. The assumptions used are that the population has already overshot the carrying capacity and current trends will continue unchanged.

both positions. (You may wish to start with web site http://www.census.gov.)

5. In his book *Environmental Science*, G. Tyler Miller points out that the projected increase in the human population to over 9 billion by the year 2050 raises serious questions. Will there be enough food, energy, water, and other resources to sustain that many people? Will governments be able to provide adequate education, housing, medical care, and other social services for all of them? Computer models suggest that the answers are no (Figure 40.15). Yet some people claim we can adapt politically and socially to a far more crowded world, assuming harvests improve through technological innovation, every inch of arable land is cultivated, and everybody eats only grain.

There are no easy answers to these questions. If you have not yet been doing so, start following the arguments in your local newspapers, in magazines, on television, and on the Web. This will allow you to become an informed participant in a global debate that surely will have impact on your future.

Selected Key Terms

age structure *40.1*
biotic potential *40.3*
capture–recapture method *40.2*
carrying capacity *40.4*
cohort *40.5*
demographic transition model *40.9*
demographics *40.1*
density-dependent control *40.4*
density-independent factor *40.4*
doubling time *40.3*
ecology *CI*
emigration *40.3*
exponential growth *40.3*
immigration *40.3*

life history pattern *40.5*
limiting factor *40.4*
logistic growth *40.4*
migration *40.3*
per capita *40.3*
population density *40.1*
population distribution *40.1*
population size *40.1*
r (net reproduction per individual per unit time) *40.3*
reproductive base *40.1*
survivorship curve *40.5*
total fertility rate *40.8*
zero population growth *40.3*

Readings

Bloom, D. and D. Canning. 18 February 2000. "The Health and Wealth of Nations." *Science* 287:1207–1209. Argues that poor health is more than a consequence of low income; that it is also one of its fundamental causes.

Cohen, J. E. 1995. *How Many People Can the Earth Support?* New York: Norton. No pat answer to title's question.

Miller, G. T. 2002. *Living in the Environment*. Twelfth edition. Belmont, California: Brooks/Cole.

41

SOCIAL INTERACTIONS

Deck the Nest With Sprigs of Green Stuff

In 1890, bird fanciers imported more than a hundred starlings (*Sturnus vulgaris*) from Europe and released them in New York City's Central Park. In less than a century, descendants of the introduced species had expanded their geographic range from coast to coast. They are so good at gleaning food from the agricultural fields of North America that they have outmultiplied native birds. They also have evicted great numbers of them from scarce nesting sites in the cavities of trees.

Once a male and female commandeer a tree cavity, they gather dry grass and twigs and use them to build or rebuild a nest (Figure 41.1). Theirs is no ordinary nest. They *decorate* the nest bowl with sprigs, freshly plucked.

Why do starlings do this? Do the sprigs of greenery camouflage the nest from predators? Not likely. The nests are already concealed in tree cavities. Do sprigs serve as insulative material that helps keep forthcoming eggs warm? Actually, the still-moist, green plant parts would promote heat loss, not heat conservation. Well, do the green sprigs combat parasites? Think of how mites parasitize birds and infest nest cavities. In short order, even a few tiny mites can produce thousands of descendants. In large numbers, mites can suck enough blood from a nestling to weaken it. They interfere with the nestling's growth, development, and survival.

Larry Clark and Russell Mason decided to test the third hypothesis. As they knew, starlings don't weave just any green plant material into the nests. Starlings choosily favor the leaves of certain plants, such as wild carrot (Figure 41.1). So the two biologists built a set of experimental nests, some with freshly cut wild carrot leaves and some without. They removed the natural nests that pairs of starlings had constructed and had already started using. Half of the nesting pairs got replacement nests decorated with sprigs of wild carrot. Replacement nests for the other pairs of starlings were sprigless.

Figure 41.2 shows the results. The number of mites in greenery-free nests was consistently greater than the number in nests decorated with greenery. At the end of one experiment, the sprig-free nests teemed with an average of 750,000 mites. Nests with sprigs contained a mere 8,000 mites.

Shoots of wild carrot happen to contain a highly aromatic steroid compound. The compound almost certainly repels herbivores and helps plants survive. By coincidence, the compound also prevents mites from maturing sexually—and so prevents mite population explosions in nests festooned with wild carrot.

Figure 41.1 A most excellent fumigator in nature—a European starling (*Sturnus vulgaris*). Starlings combat infestations of mites by decorating previously owned nests with fresh sprigs of wild carrot (*Daucus carota*), shown at left.

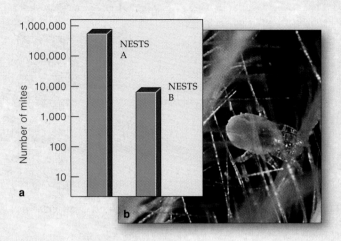

Figure 41.2 (**a**) Experimental results that point to the adaptive value of starling nest-decorating behavior. Nests designated *A* were kept free of fresh sprigs of wild carrot and other plants that contain aromatic compounds. The compounds prevent baby mites from developing into (**b**) adult mites. Other nests, designated *B*, had fresh sprigs tucked in every seven days. Head counts of mites (*Ornithonyssus sylviarum*) infesting the nests were made at the time the starling chicks left the nest, twenty-one days after this experiment was under way.

Far from being a trivial behavior, then, "decorating" a nest with aromatic greenery has adaptive value. It fumigates the nest and thus increases an individual's chance of producing healthy, surviving offspring.

And so starlings lead us into the fascinating world of research into behavior. As you will see, some of the studies focus on the adaptive value of behavioral traits to an individual's reproductive success. Others focus on the internal mechanisms that allow individuals to behave as they do.

As a starting point, recall that information encoded in genes directs the formation of tissues and organs that make up the animal body, including those of its nervous system. That system detects, processes, and integrates information about stimuli in the internal and external environments, then commands muscles and glands (also built according to genetic instructions) to make suitable responses. Other gene products called hormones contribute to the responses. Because genes specify all of the substances required for the structure and function of nervous and endocrine systems, *they are the heritable foundation for responses that animals make to stimuli.* We call these observable, coordinated responses to stimuli **animal behavior**.

Key Concepts

1. Animals make coordinated responses to stimuli. We call these responses animal behavior. The cooperative, interdependent relationships among individuals of a species are forms of social behavior.

2. Genes underlie the behavioral ability of individuals, as when instructions encoded in certain genes govern the development of the nervous and endocrine systems. These organ systems allow an animal to detect, process, and issue commands for behavioral responses to stimuli.

3. Most newly born or hatched animals have the neural wiring and motor systems necessary to perform some behaviors without having to learn them through actual experience. Sign stimuli, which are simple, well-defined environmental cues, trigger such instinctive behaviors.

4. For nearly all animal species, the nervous system has a capacity to process and retain information about specific experiences in the environment, then to use the information to vary or change behavioral responses. The outcomes are called learned behaviors.

5. Like other traits having a genetic basis, behavior has evolved by way of natural selection, which occurs when genetically different individuals in a population have different numbers of surviving offspring. Behavioral mechanisms that enhance the ability of individuals to pass on their genes to offspring have been favored.

6. Evolved modes of communication underlie social behavior. Communication signals, which are encoded in stimuli such as species-specific odors, body coloration and patterning, postures, and movements, hold clear meaning for both the sender and the receiver of signals.

7. Living in a social group has costs and benefits, as measured by an individual's ability to pass on its genes to offspring. Not every environment favors the evolution of such groups. In some cases, a solitary life-style allows an individual to leave more descendants than belonging to a large group would allow.

8. In altruism, an individual helps others in ways that require it to sacrifice its own reproductive success. The evolution of altruism requires special circumstances in which individuals can propagate their genes indirectly, by helping their relatives reproduce successfully.

BEHAVIOR'S HERITABLE BASIS

Genes and Behavior

An animal's nervous system, recall, is wired to detect, interpret, and issue commands for response to stimuli—that is, to specific aspects of the external and internal environments. One way or another, certain products of genes control each of the steps necessary to assemble and operate that system's receptors, nerve pathways, and brain. If we define animal behavior as coordinated responses to stimuli, then behavior starts with genes.

Stevan Arnold found evidence of the genetic basis of behavior by studying feeding preferences of coastal and inland populations of a snake species in California. Garter snakes near the coast prefer banana slugs (Figure 41.3a). Snakes living inland prefer tadpoles and small fishes. Offer them a banana slug and they ignore it.

In one set of experiments, Arnold offered captive newborn garter snakes a chunk of slug as the first meal. Coastal snake offspring usually ate the chunk and even flicked their tongue at cotton swabs drenched in essence of slug. (Snakes "smell" by tongue-flicking, which pulls chemical odors into the mouth.) However, inland snake offspring ignored the swabs and rarely ate slug meat. Here was an obvious difference between captive baby snakes that had no prior, direct experience with slugs. Arnold concluded the snakes were programmed before birth to accept or reject slugs; they certainly did not learn feeding preferences by taste trials. Maybe coastal and inland populations have allelic differences for the genes affecting how odor-detecting mechanisms form when a garter snake embryo is developing.

To test his hypothesis, Arnold crossbred coastal and inland snakes. If genes underlie the difference in food preferences, he figured that the hybrid offspring from

Figure 41.3 (**a**) Banana slug, food for (**b**) an adult garter snake of coastal California. (**c**) A newborn garter snake from a coastal population, tongue-flicking at a cotton swab drenched with tissue fluids from a banana slug.

snakes of both populations might make an *intermediate* response to slug chunks and odors. Results matched his prediction. Compared with a typical newborn inland snake, many baby snakes of mixed parentage tongue-flicked more often at the slug-scented cotton swabs—but less often than newborn coastal snakes did.

Hormones and Behavior

Gene products called hormones also guide behavior, including birdsong. As an example, each spring a male zebra finch (Figure 41.4) repeats a song with splendid consistency and clarity. In part, his singing arises from seasonal differences in how much melatonin is secreted from his pineal gland (Section 32.7). A high blood level of melatonin suppresses the growth and functions of a songbird's gonads. Suppression reverses when pineal gland photoreceptors absorb sunlight energy and issue signals that cause a slowdown in melatonin secretion.

Winter has fewer daylight hours than spring; there isn't enough light to slow melatonin secretion. Come

Figure 41.4 Hormones and the territorial song of zebra finches. (**a**) Sound spectrogram of the male's song. The spectrogram is a visual record of each note's pitch (that is, its frequency, measured in kilohertz). (**b**) Sound spectrogram of a female zebra finch that was experimentally converted to a singer. When she was a nestling, she received extra estrogen. At adulthood, she received an implant of testosterone—and started singing. Estrogen normally organizes the song system (certain parts of the brain) in developing songbird embryos. Testosterone activates the system in adult males.

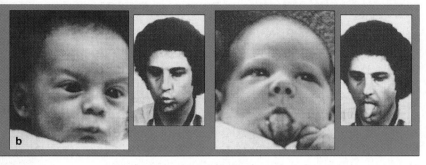

Figure 41.5 Examples of instinctive responses that human offspring make to sign stimuli. (**a**) A close-up face of an adult triggers smiling behavior in young infants. (**b**) Somewhat older infants instinctively imitate adult facial expressions.

spring, daylength increases. Hormonal suppression is lifted, so the gonads increase in size and secrete more estrogen and testosterone. These sex hormones are the start of gender-related differences in singing behavior. As most songbird embryos develop, estrogen controls the formation of their **song system**, which consists of several brain structures that will control the muscles of a vocal organ. Male and female songbirds differ in the size and structural components of this system.

Unlike humans, females of songbird species are XY and males are XX. Certain Y chromosome genes specify products that block estrogen production. Before a *male* bird hatches, its gonads secrete estrogen which, at high levels, stimulates development of a masculinized brain. The estrogen gets converted to testosterone. The gonads enlarge and secrete more testosterone as the breeding season starts. By acting on cells in the song system, this hormone induces metabolic changes that will prepare the male to sing when suitably stimulated.

Instinctive Behavior Defined

With their tongue-flicking and strikes at prey, newborn garter snakes can start us thinking about **instinctive behavior**. This term means the behavior is performed without having been learned through actual experience in the environment. Instead, the nervous system of a newly born or hatched animal is prewired to recognize **sign stimuli**—one or two simple, well-defined cues in the environment that can trigger a suitable response. The snake's response is a stereotyped motor program. When the snake recognizes certain sign stimuli, a **fixed action pattern** follows. It is a program of coordinated muscle activity that runs to completion independently of feedback from the environment.

Another example: When human infants are two or three weeks old, they tend to smile instinctively when an adult's face comes close to their own (Figure 41.5*a*). Infants make the same response to an overly simplified stimulus—a flat, face-sized mask with two dark spots

Figure 41.6 Example of a social parasite. European cuckoos lay eggs in nests of other species. In response to an environmental cue, this cuckoo hatchling is executing a behavior that is innate. It has inherited the knowledge of what to do without having to learn it. Even before its eyes open, it responds to the spherical shape of the host's eggs and shoves them out of the nest. The foster parents keep on feeding the usurper.

where eyes would be on a human face. A mask with one "eye" won't do the trick. As Figure 41.5*b* suggests, older infants continue to show instinctive behavior.

And remember the cuckoo, a social parasite? Adult females lay their eggs in nests of other species. If newly hatched cuckoos eliminate the natural-born offspring, they will receive the undivided attention of their foster parents (Section 42.4). They are blind at birth, but when they contact an egg or another round object, they carry out a fixed action pattern. They maneuver the egg onto their back, then push it from the nest (Figure 41.6).

Animals clearly have a genetically based capacity to respond automatically to certain environmental cues. But they also have the means to process information about specific experiences, then use the information to vary or change their responses. As you will see next, the outcome is what we call learned behavior.

Genes underlie animal behavior—coordinated responses to stimuli. Certain gene products are essential in constructing and operating the nervous system, which governs behavior. Other gene products called hormones also influence the mechanisms required for particular forms of behavior.

Animals start out life neurally wired to recognize important cues and to make an instinctively suitable response, one that has not been learned through actual experience.

LEARNED BEHAVIOR

Animals process and integrate information gained from experiences, then use it to vary or change responses to stimuli. We call this **learned behavior**. For example, a young toad's nervous system commands it to flip its sticky tongue instinctively at any dark object crossing its field of vision. In the toad world, such objects usually are edible insects. What if a dark object is a bumblebee that stings the toad tongue? Thereafter, that toad will avoid all "bumblebee-sized black-and-yellow-banded objects that sting." **Imprinting**, another example, is a time-dependent form of learning triggered by exposure to sign stimuli, most often at a sensitive period when

Figure 41.7 No one can tell these imprinted baby geese that Konrad Lorenz is not Mother Goose. (Refer to Table 41.1.)

Table 41.1 *A Few Categories of Learned Behavior*

IMPRINTING This time-dependent form of learning involves exposure to sign stimuli, most often early in development. For instance, in response to a moving object and probably to certain sounds, baby geese imprint on the mother and follow her during a short, sensitive period after hatching. They are neurally wired to learn crucial information—the identity of the individual that will protect them in the months ahead. Usually that individual is the mother or father. Imprinting occurs among many animals.

Konrad Lorenz, an early ethologist, must have presented the baby geese in Figure 41.7 with sign stimuli that made them form an attachment to him. As another outcome of imprinting, birds also direct sexual attention to members of the species they had been sexually imprinted upon when young.

CLASSICAL CONDITIONING Ivan Pavlov's early experiments with dogs are an example of classical conditioning. Just before eating, dogs will salivate. Pavlov's dogs were conditioned to salivate even in the absence of food. They did so in response to the sound of a bell or a flash of light that they initially associated with getting food. In this case, the dogs learned to associate an automatic, unconditioned response with a novel stimulus that does not normally trigger the response.

OPERANT CONDITIONING An animal learns to associate a voluntary activity with its consequences, as when a toad learns to avoid stinging insects after its first attempt to eat them.

HABITUATION An animal learns by experience *not* to respond to a situation if the response has neither positive nor negative consequences. Thus pigeons and other birds living in cities learn not to flee from people who pose no threat to them.

SPATIAL (LATENT) LEARNING By inspecting its environment, an animal acquires a mental map of a particular region, often by learning the position of local landmarks. For example, blue jays can store information about the position of dozens, if not hundreds, of places where they have stashed food.

INSIGHT LEARNING An animal abruptly solves some problem without trial-and-error attempts at the solution. Chimpanzees often exhibit insight learning in captivity when they suddenly solve a novel problem that their captors devise for them. Some chimps abruptly stacked and stood on several boxes *and* used a stick to reach bananas suspended out of their reach.

an animal is young. The imprinting of baby geese is a classic example (Table 41.1 and Figure 41.7).

Don't fall into the trap of thinking that instinct is governed only by genes and learned behavior, only by the environment. Behavior arises from gene expression *and* experiences. In different habitats, songbirds use variations, or dialects, of their species song. As Peter Marler found out, male songbirds acquire the full song ten to fifty days after hatching by listening to other birds sing it. The nervous system of, say, a male white-crown sparrow is prewired to recognize that song; its learning mechanism is primed to select and respond to acoustical cues in the environment. But its rendition of the species song is influenced by what it actually hears during the sensitive period.

The male birds learn parts of a song by picking up cues when other males sing. Marler raised white-crown nestlings in soundproof chambers so they could not hear adult males. The captives sang when mature, but not with the detailed structure of a typical adult song. Marler exposed other isolated captives to recordings of white-crown sparrows *and* song sparrows. At maturity, the captives sang just the white-crown song. They even mimicked the dialects. Marler's experiments support the hypothesis that birdsong requires *a genetically based capacity to learn* from acoustical cues.

This isn't the whole story. In another experiment, Marler let young, hand-reared white-crowns interact with a "social tutor" of a different species, as opposed to listening to taped songs. The males tended to learn the tutor's song. Their primed learning mechanism had responded to *social experience* as well as acoustical cues.

In instinctive behavior, animals make complex, stereotyped responses to specific, often simple environmental cues. In learned behavior, responses may vary or change as a result of individual experiences in the environment.

Whether instinctive or learned, behavior develops through interactions between genes and environmental experiences.

THE ADAPTIVE VALUE OF BEHAVIOR

If genes directly or indirectly govern forms of behavior, then forms of behavior must be subject to evolution by natural selection. **Natural selection**, recall, is a result of differences in reproductive success among individuals that differ in heritable traits. Alleles underlying the most adaptive versions of a trait tend to increase in frequency in a population, whereas other alleles do not. (Alleles are different molecular forms of the same gene.) In time, genetic changes that yield greater reproductive success for individuals spread through the population.

By using the theory of evolution by natural selection as a point of departure, we should be able to develop and test possible explanations of why a behavior has endured. We should be able to discern how some actions bestow reproductive benefits that offset reproductive costs (disadvantages) associated with them. If a behavior is adaptive, it must promote the *individual's* production of offspring. Keep this thought in mind while you read through the following terms, which you will encounter repeatedly in the remainder of this chapter:

1. **Reproductive success**: The number of surviving offspring that the individual produces.

2. **Adaptive behavior**: Any behavior that promotes propagation of an individual's genes; its frequency is maintained or increases in successive generations.

3. **Social behavior**: Cooperative, interdependent relationships among individuals of the species.

4. **Selfish behavior**: Within a population, any form of behavior that increases an individual's chance to produce or protect offspring of its own regardless of the consequences for the group as a whole.

5. **Altruistic behavior**: Within a population, a self-sacrificing behavior. The individual behaves in a way that helps others but lessens or precludes its own chance of producing offspring.

When biologists speak of selfish or altruistic behavior, they don't mean an individual is consciously aware of what it's doing or of the behavior's reproductive goal. A lion doesn't have to know that eating zebras is good for reproductive success. Its nervous system just calls for HUNTING BEHAVIOR! when the lion is hungry and sees a zebra. This behavior persists in the population because genes responsible for it are persisting also.

As a case in point, Norwegian lemmings disperse from a population when density skyrockets and food is scarce. Many die during the exodus. Are they helping the species by committing suicide to get rid of "excess" individuals? Or do some just happen to die as a result of starvation, predation, or accidental drowning while scurrying to new locations where they may reproduce? One of Gary Larson's instructive cartoons depicts a population of lemmings plunging over a cliff above water, presumably in the act of mass suicide. But one lemming has an inflated inner tube around its waist! If many lemmings were suicidally altruistic, then only the "selfish" ones would reproduce. Over time, genes underlying their altruism would disappear from the population. An alternative hypothesis is that individual lemmings disperse to less crowded places where each will have a better chance to survive and reproduce.

Another case: In northern forests, ravens scavenge carcasses of deer, elk, or moose, which are few and far between. When one of these large birds comes across a carcass—even in winter, when food is scarce—it often calls loudly and attracts a crowd of similarly hungry ravens. This calling behavior puzzled Bernd Heinrich. On the surface, it seemed to work against the caller's interests. Wouldn't a quiet raven eat more, increase its chance of surviving, and also leave more descendants compared to an "unselfish" vocalizer? If its behavior is an outcome of natural selection, then the biological cost in terms of lost calories and nutrients must be offset by some reproductive benefit for individual callers. What could be the benefit?

Maybe a lone bird picking at a carcass is vulnerable to predators that lie in wait for it. If that were so, then other ravens attracted by the calls could help keep an eye out for danger. But ravens are very big, agile birds that have very few known enemies. Heinrich stealthily watched ravens feeding alone and feeding in pairs at a carcass. He never saw a predator attack any of them.

Then he realized that territory may have something to do with it. A **territory** is an area that one or more individuals defend against competitors. After hauling a cow carcass into a Maine forest, Heinrich observed that single or paired ravens don't always vocally advertise food. Maybe the silent ravens were adults that had already staked out a large territory, which happened to include the spot where he put the carcass.

A pair of ravens would gain nothing by attracting others to their territory. But what if the Maine forest had been *subdivided* into territories and was defended by powerful adults? A wandering young bird would be lucky to eat at all in an aggressive pair's territory. But recruiting a gang of other, nonterritorial ravens might overwhelm the resident pair's defensive behavior.

As it turns out, only the wandering ravens advertise carcasses. Their calling behavior is basically selfish and adaptive. It gives them a shot at otherwise off-limits food and thereby promotes their reproductive success.

Behavioral biologists generally find it profitable to look for evidence of natural selection of the individual's traits rather than something that benefits the species as a whole.

41.4

COMMUNICATION SIGNALS

Competing for food, defending territory, alerting others to danger, advertising sexual readiness, forming bonds with a mate, caring for offspring—such *intraspecific* interactions depend on forms of communication among animals, as in the case of those vocalizing ravens.

The Nature of Communication Signals

Intraspecific interactions involve mixes of instinctive and learned behaviors by which individuals send and respond to information-laden cues, encoded in stimuli. These are **communication signals**, with unambiguous meaning for the species. They include specific odors, colors, patterning, sounds, postures, and movements.

Communication signals sent by one individual, the **signaler**, are meant to change the behavior of others of the species. Those responding are the **signal receivers**. A signal evolves or persists when it tends to help the reproductive success of the sender *and* receiver. When a signal proves disadvantageous to either party, then natural selection will favor individuals that don't send the signal or don't respond to it.

Remember **pheromones**? They are chemical signals between individuals of the same species. As you know, chemical odors from food and danger were the main stimuli that early animals had to deal with. As animals evolved, most started to rely on *signaling* pheromones to induce a receiver to respond swiftly. Some, including chemical alarm signals, call for aggressive or defensive behaviors. Others are sex attractants. Bombykol molecules released from female silk moths are like this (Section 32.1). The *priming* pheromones bring about generalized physiological responses. For instance, a volatile odor in the urine of certain male mice triggers and also enhances estrus in female mice.

Figure 41.8 A dog using a play bow to solicit a romp.

Acoustical signals also abound in nature, as when male songbirds sing to secure territory and attract a mate. At night, tungara frogs issue distinctive "whine-chuck" calls to attract females and warn rival males.

Some signals never vary. For example, zebra ears laid back flat against the head convey hostility, but ears pointing up convey its absence. Other signals convey the intensity of the signaler's message. A zebra with laid-back ears isn't too riled up when its mouth is just a bit open. When its mouth is gaping, watch out. That combination is a **composite signal**: a communication signal with information encoded in two cues (or more).

Signals often take on different meaning in different contexts. A lion emits a spine-tingling roar to keep in touch with its pride *or* to threaten rivals. Also, a signal can convey information about signals to follow. Dogs and wolves solicit play behavior with a play bow, as shown to the left in Figure 41.8. Because of the bow, subsequent behavioral patterns that a signal receiver would construe as aggressive, sexual, or exploratory in other contexts are construed as playful only.

Examples of Communication Displays

The play bow is a **communication display**—a pattern of behavior that is a social signal. Another common pattern, the **threat display**, unambiguously announces that a signaler is prepared to attack a signal receiver. When a rival for a receptive female confronts him, a dominant male baboon will role his eyes upward and "yawn" to expose his formidable canines (Figure 41.9). Often the rival backs down. The signaler benefits, for he retains access to the female without putting up a

Figure 41.9 *Above:* Exposed canines, part of a male baboon's threat display. *Below:* Part of a courtship display, with visual, tactile, and acoustical signals, that often precedes copulation. Here a male albatross spreads his wings, an information-laden cue for the female. See also Section 17.2.

Figure 41.10 Dances of the honeybees, as examples of tactile displays. (**a**) Honeybees that visit food sources close to their hive perform a *round* dance on the honeycomb. Worker bees that maintained contact with the forager through the dance go out and search for food near the hive.

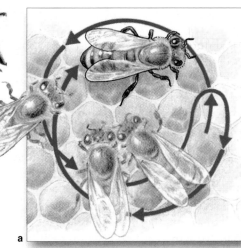

(**b**) Bees trained to visit feeding stations over 100 meters from the hive do a *waggle* dance. During the dance, a bee makes a straight run and waggles its abdomen. (**c**) As Karl von Frisch discovered, a straight run's orientation varies, depending on the direction in which food is located. When he put a dish of honey on a direct line between the hive and the sun, foragers that located it returned to the hive and oriented their straight runs right up the honeycomb. When he put the honey at right angles to a line between the hive and the sun, the foragers made their straight runs at 90 degrees to vertical. Thus, a honeybee "recruited" into foraging can orient its flight *with respect to the sun and the hive*. It will waste less time and energy during its food-gathering expedition.

Waggle dancers also vary the dance speed to convey even more information about the distance of a food source. When a site is 150 meters away, the dance is executed much faster, with more waggles per straight run, compared with a dance about a food source that is, say, 500 meters away.

When bee moves straight down comb, recruits fly to source directly away from the sun.

When bee moves to right of vertical, recruits fly at 90° angle to right of the sun.

When bee moves straight up comb, recruits fly straight toward the sun.

a

b   c

fight. The signal receiver benefits because he avoids a serious beating, infected wounds, and possibly death.

Such displays are *ritualized*, with intended changes in the function of common behavior patterns. Normal movements may be exaggerated yet simplified; postures may be frozen. Body parts such as feathers, manes, and claws may be conspicuously enlarged, patterned, and colored. Ritualization is often developed to an amazing degree in **courtship displays** between potential mates. For instance, food-related displays are common among many birds. A male bird may emit calls while bowing low, as if to peck the ground for food. When a female comes running, he may spread his wings, as if to focus attention on the ground in front of him (Figure 41.9). Here, one behavior (searching for food) has changed to attract a female, the intent being copulation. Another example: Firefly courtship displays use bioluminescent flashes (Section 6.9, *Critical Thinking* question 5). A male firefly uses a light-generating organ to emit a bright, flashing signal. A few seconds later, a receptive female of his species may answer him with a flash. Both may flash back and forth until they meet up.

Similarly, in **tactile displays**, a signaler touches the receiver in ritualized ways. After discovering a source of pollen or nectar, a foraging honeybee returns to its colony, a hive, and performs a complex dance. It moves about in a circle, jostling in the dark with a crowd of workers. Other bees may follow and maintain physical contact with the dancer. Honeybees that do so acquire information about the general location, distance, and direction of pollen or nectar (Figure 41.10).

Illegitimate Signalers and Receivers

Sometimes the wrong parties intercept communication signals. Termites will respond with defensive behavior if they catch a whiff of a scent from an invading ant. That scent is meant to identify the ant as a member of an ant colony, and it elicits cooperative behavior from other ants. The scent does have evolved functions, but announcing an invasion of a termite colony is not one of them. A termite also might detect the scent and kill the ant. In that case, it is an **illegitimate receiver** of a signal meant for individuals of a different species.

We even see **illegitimate signalers**. Certain assassin bugs hook dead and drained bodies of termite prey on their dorsal surface to acquire the odor of the victims. By deceptively signaling that they "belong" to a termite colony, they can more easily hunt termite prey. Another illegitimate signaler is the female of certain predatory fireflies. If one notices a male firefly's flash, she flashes back. If she can lure him into attack range, he becomes her meal—an evolutionary cost of having an otherwise useful response to a come-hither signal.

A communication signal between individuals of the same species is an action having a net beneficial effect on both the signaler and receiver. Natural selection tends to favor communication signals that promote reproductive success.

MATES, PARENTS, AND INDIVIDUAL REPRODUCTIVE SUCCESS

For reasons we need not explore here, most people find mating and parenting behaviors of animals fascinating. How useful is selection theory in helping us interpret such behavior? Let's take a look.

Sexual Selection Theory and Mating Behavior

Competition among members of one sex for access to mates is common. So is choosiness in selecting a mate. Recall, from Section 16.9, that such activities are forms of a microevolutionary process called **sexual selection**. Sexual selection favors traits that give the individual a competitive edge in attracting and holding on to mates.

But *whose* reproductive success is it—the male's or female's? Males, recall, produce large numbers of tiny sperm, and females produce far larger but fewer eggs. For the male, success generally depends on how many eggs he fertilizes. For the female, it depends more on how many eggs she produces or how many offspring she can care for. In most cases, *the key factor influencing a female's sexual preference is the quality of the mate—not a quantity of partners.* Hangingflies, sage grouse, and bison offer examples of how most females dictate the rules of male competition (Figures 41.11 through 41.13). They also show how males employ tactics that might help them fertilize as many eggs as possible.

Female hangingflies (*Harpobittacus apicalis*) choose males that offer superior material goods. And so you may observe some males capturing and killing a moth or some other insect. After they do so, they release a sex pheromone to attract females to their "nuptial gift,"

as in Figure 41.11*a*. A female will select the male that displays the larger calorie-rich offering. She lets him mate, but only *after* she has been eating a gift for five minutes or so. Then she accepts sperm and holds them in a storage sac, but only as long as the food holds out. Before twenty minutes are up, she can break off the mating at any point. If she does, she might well mate with another male and accept his sperm. Doing so will dilute the reproductive success of her first partner.

In western regions of North and South Dakota, we find sage grouse (*Centrocercus urophasianus*) dispersed among stands of sagebrush. They never roam far from its protective cover during their spring courtship and summer nesting. Males make no attempt to secure and hold on to a large territory. During the breeding season, they congregate in a type of communal display ground called a **lek**. Each male stakes out a few square meters as his territory. The females are attracted to the lek—not to feed or nest, but rather to watch the males. With tail feathers erect and large neck pouches puffed out, a male emits booming calls and stamps about rather like a wind-up toy on his display ground (Figure 41.11*b*). Females tend to select and mate with one male only. Then they go off to nest by themselves in sagebrush, unassisted by a sexual partner. Many often choose the same male, so most of the males never mate.

Another example: The females of a wolf pack court males. Why? Unlike most male mammals, which don't invest as much as females do in raising offspring, male wolves help pups by giving them food, defending the feeding territory, and driving off infanticidal intruders.

Figure 41.11 (**a**) Male hangingfly dangling a moth as a nuptial gift for a future mate. Females of some hangingfly species choose sexual partners on the basis of the size of prey offered to them. (**b**) Male sage grouse. Its ornamental feathers are visual signals that function in a courtship display—a dance performed at a lek. Males vigorously defend a small patch of this communal mating area. The females (the smaller brown birds) observe the prancing males before choosing the one they will mate with.

Figure 41.12 Dominant male of a wolf pack. Subordinate members of the pack surrounding him are engaged in a ritual display involving lowered heads and drooped tails.

Figure 41.13 Sexual competition between male bison, which are fighting for access to a cluster of females.

Unlike herbivores, which forage as individuals, wolves share captured prey with others of their social group. Therefore, a female's reproductive success increases if she monopolizes the benefits that males offer.

Most wolf packs have one dominant female and male that breed (Figure 41.12). The others are nonbreeding, altruistic brothers and sisters, aunts and uncles. They all hunt and bring back food to members that stay behind in the den and guard the young. The nonbreeding females may ovulate and court males, but they will fail to reproduce in the presence of a dominant female.

As a final example, females of some species cluster in defendable groups when they become sexually receptive. Where you come across such a group, most likely you'll see males competing for access to the clusters. Competition for ready-made harems favors extremely combative male lions, sheep, elk, elephant seals, and bison, just to name a few animals (Figures 1.7*g* and 41.13).

Costs and Benefits of Parenting

What happens after mating, when the offspring arrive? Until the offspring have developed enough to survive on their own, parents of some species care for them.

Example: Adult Caspian terns (Figure 41.14) will incubate the eggs, shelter the nestlings and feed them, then accompany and protect them after they start to fly. Such parental behavior comes at a reproductive cost. It drains time and energy that might otherwise be given to improving their own chances of living to reproduce another time. Yet for many species, parenting improves

Figure 41.14 Male and female Caspian terns protecting their chicks. Parental care has costs as well as benefits.

the likelihood that the current generation of offspring will survive. The benefit of devoting time and energy to immediate reproductive success outweighs the cost of reduced reproductive success at some later time.

Selection theory helps explain some aspects of mating behavior, for sexual selection favors behavioral traits that give the individual a competitive edge in reproductive success. Selection theory applies also to parental behavior that contributes to reproductive success.

COSTS AND BENEFITS OF LIVING IN SOCIAL GROUPS

Survey the animal kingdom and you observe a range of social groups. Individuals of some species spend most of their lives alone or in small family groups. Others live in groups of thousands of related individuals. Termite and honeybee societies are like this. Still others live in social units consisting mainly of nonrelatives. Human populations are like this. Evolutionary biologists often attempt to identify the costs and benefits of sociality in terms of individual reproductive success, as measured by contributions to next generation's gene pool.

Cooperative Predator Avoidance

Some groups act cooperatively against a predator and reduce the net risk to any one individual. A vulnerable flock or herd has more individuals that might scan for predators, join forces in a counterattack, or engage in defensive behavior (Figures 41.15 and 41.16).

Birgitta Sillén-Tullberg, a biologist, found evidence of such benefits. She was studying Australian sawfly caterpillars that live in clumps on tree branches. Figure 41.15 shows one clump. When something disturbs the caterpillars, they collectively rear up from the branch and writhe about, all the while regurgitating partially digested eucalyptus leaves. These leaves happen to be heavily impregnated with chemical compounds that are toxic to most kinds of predatory animals—most notably the songbirds that prey on the caterpillars.

Sillén-Tullberg hypothesized that individual sawfly caterpillars benefit from the coordinated act of repelling bird predators. She used her hypothesis to predict that birds are more likely to eat a solitary caterpillar. To test her prediction, she offered young, hand-reared Great Tits (*Parus major*) the chance to eat sawfly caterpillars. She offered the caterpillars one by one or in a group of twenty for a standard number of presentations. Ten birds that were offered one individual at a time ate an average of 5.6 caterpillars. But ten birds that were each offered a clump of caterpillars ate an average of 4.1. As predicted, individuals were somewhat safer in a group.

Figure 41.15 Example of social defensive behavior. Australian sawfly caterpillars form clumps in tree branches. They collectively regurgitate fluid (yellow blobs) into their mouth. The fluid is toxic to most animals and may repel predators. Once the danger passes, these caterpillars swallow the fluid.

The Selfish Herd

Simply by their physical position within a group, some individuals form a living shield against predation on the others. They all belong to a **selfish herd**, a simple society held together (not consciously) by reproductive self-interest. The selfish-herd hypothesis has been tested for male bluegill sunfishes, which build adjacent nests on lake bottoms. The females deposit eggs where males have used their fins to hollow out depressions in mud.

If a colony of bluegill males is a selfish herd, then we can predict competition for the "safe" sites—at the center of the colony. Compared to eggs at the periphery, eggs in nests at the center are less likely to be eaten by snails and largemouth bass. Competition does indeed exist. The largest, most powerful males tend to claim the central locations. Other, smaller males assemble around them and bear the brunt of predatory attacks. Even so, they are better off in the group than on their own, fending off a bass single-handedly, so to speak.

Dominance Hierarchies

In selfish herds, personal benefits of living with others simply seem to outweigh the costs. By contrast, some individuals may *help* others survive and reproduce at their own expense. Their helpful behavior might be a cost of belonging to the group. Or maybe they do not give up their chance to reproduce entirely.

We can find evidence of both possibilities in **dominance hierarchies**. In this type of social group, some members have subordinate status to others. In a baboon troop, individuals help one another, but reproductive opportunity is unequal. With a threat signal from a dominant member, subordinates relinquish safe sleeping

Figure 41.16 Cooperative defensive behavior among musk oxen (*Ovibos moschatus*). In the presence of wolf predators (and other threats), adults form a circle around their young and face outward. The "ring of horns" may deter the wolves.

Figure 41.17
Some benefits and costs of social behavior, illustrated.

(**a**) A baboon belonging to a troop is safer from leopards than a loner out in the open. This one is making its last stand with a threat display (it didn't work).

(**b**) Imagine how easily pathogens and parasites might spread through this colony of royal penguins living on Macquarie Island, between New Zealand and Antarctica.

sites, choice food, and sexually receptive females. Each knows its status in the troop. Dominance hierarchies of this sort also develop among chimpanzees (Chapter 32).

Subordinate adults probably stay in a group when long-term benefits offset their low status. Maybe it's impossible to survive alone; a solitary baboon out in the open surely quickens the pulse of the first leopard to spy it (Figure 41.17*a*). Challenging a strong member may invite injuries or death. Self-sacrificing behavior might give a subordinate a chance to reproduce if it lives long enough or if a predator or old age removes dominant peers. Some patient, subordinate wolves and baboons do indeed move up the social ladder that way.

Regarding the Costs

If social behavior is so advantageous, *then why are there so few social species among most groups of animals?* One answer is that, in some environments, costs outweigh benefits. Most obviously, when more individuals of a species live together in a habitat, competition for food increases. Royal penguins, herring gulls, cliff swallows, and prairie dogs are among the many animals that live in enormous colonies (Figure 41.17*b*). All individuals must compete for a share of the same ecological pie.

Crowded social groups invite contagious diseases and parasites that can be readily transmitted from host to host. In the past, plagues have spread like wildfire through densely crowded human populations. This is especially so for settlements and cities with recurring infestations of rats and fleas, with poor or nonexistent sewage treatment and medical care (Section 40.4).

Another cost is the risk of being killed or exploited by others in the group. When given the opportunity, breeding pairs of herring gulls quickly cannibalize the eggs and young chicks of neighbors. Lions compete for hunting territories even though they can survive for an extended time between kills. When three or more lions cooperate in a hunt, they are more likely to capture large prey. Yet individuals in prides of three or more actually eat *less* well than one or a pair of lions. Also, males intent on taking over a pride show infanticidal behavior; they kill cubs. So group living costs lionesses in terms of food intake and reproductive success. They still stick together. Maybe doing so helps them defend the territories against smaller, rival groups. Aggressive males almost always kill cubs of a lone lioness, but two or more lionesses occasionally can save some of them.

Individuals of a social group gain benefits, as through cooperative defenses or shielding effects against predators.

Individuals that belong to a social group pay costs in terms of increased competition for food, living quarters, mates, and other limited resources. They also are more vulnerable to contagious diseases and parasitic infections.

Individuals also may risk being exploited or having their offspring killed by others in the social group.

Why Sacrifice Yourself?

If altruistic individuals of a social group don't contribute their genes to the next generation, then how are genes for their altruistic behavior perpetuated over time? According to William Hamilton's theory of **indirect selection**, genes associated with caring for one's *relatives*, not one's direct descendants, may be favored in some situations. When a sexually reproducing, diploid parent cares for offspring, it isn't helping exact genetic copies of itself. Each gamete it produces, hence each offspring, inherits one-half of its genes. If other animals of the social group have the same ancestors, they share the same genes with parents. Two siblings (brothers, sisters) are as genetically similar as a parent is to one of its offspring. Nephews and nieces have inherited about one-fourth of their uncle's genes.

Maybe selective altruism is an extension of parenting. Suppose an uncle helps his niece survive long enough to reproduce. He's made an *indirect* genetic contribution to the next generation, as measured by genes he shares with his niece. Altruism costs him; he may lose his own chance to reproduce. But if the cost is less than the benefit, his actions will propagate his genes and favor the spread of his kind of altruism in the species. If an uncle saves two nieces, this is equivalent to saving his own daughter.

Figure 41.19
A peek at a few naked mole-rats.

Nonbreeding workers in insect societies indirectly promote their "self-sacrifice" genes through extreme altruistic behavior directed toward relatives. Honeybee, ant, and termite colonies are extended families (Figure 41.18). The family's sterile workers support their siblings, some of which are future kings and queens. When a guard bee drives her stinger into a raccoon, she dies, but siblings in the hive might perpetuate some of her genes.

Sterility and extreme self-sacrifice are rare in social groups of vertebrates. The known exceptions include the naked mole-rats (*Heterocephalus glaber*), which live in arid regions of eastern Africa in cooperatively excavated burrows. They always live in clans—small social units—of anywhere from 25 to 300 individuals (Figure 41.19).

One reproducing female dominates the clan. She mates with one to three males. Other, nonbreeding members live to protect and care for the "queen" and "king" (or kings) and their offspring. Nonreproducing "diggers" excavate extensive subterranean tunnels and chambers that serve as living rooms or waste-disposal centers. They locate and cut up large tubers growing underground and deliver the bits to the queen, her retinue of males, and her offspring.

Digger mole-rats also deliver food to other helpers that loaf about, shoulder to shoulder and belly to back, with the reproductive royals. The "loafers" actually spring to action when a snake or some other enemy threatens the clan. Collectively, and at great personal risk, they chase away or attack and kill the predator.

H. Kern Reeve and his colleagues wanted to find out if *H. glaber*'s helpful behavior is genetically advantageous. They used *DNA fingerprinting*, a method of establishing degrees of genetic relatedness among individuals (Section 15.3). Essentially, a set of identical twins has identical DNA fingerprints. Individuals with the same father and mother have similar DNA and similar DNA fingerprints. On average, genetically unrelated individuals should show far more differences in DNA fingerprints than siblings or some other relatives.

Reeve discovered that all individuals from the same clan are *very* close relatives. He also found that they are very different genetically from other clans. His findings suggest that each naked mole-rat clan is highly inbred as a result of generations of brother–sister, mother–son, and father–daughter matings. Inbreeding among individuals of a population results in extremely reduced genetic variability. Therefore, a self-sacrificing naked mole-rat is helping to perpetuate a high proportion of the alleles that it carries. As it turns out, the genotypes of the helpers and the helped may be as much as 90 percent identical!

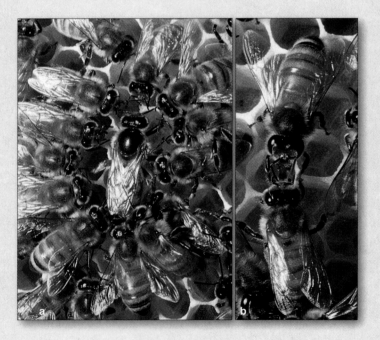

Figure 41.18   (a) A queen bee, the only fertile female in a honeybee colony. The sterile worker daughters feed the queen and relay her pheromones to influence others through the hive. At certain times of year stingless drones develop, then leave and attempt to mate with queens of other hives, thus perpetuating part of their family's genes. A hive's 30,000 to 50,000 workers feed larvae, clean and maintain the hive, construct honeycomb. They store honey or pollen in honeycomb, which holds eggs, larval stages, and pupae. Adult workers only live for about six weeks in spring and summer. (b) In other cooperative behaviors, bees transfer food to one another, and worker females guarding the hive entrance sacrifice themselves to repel intruders.

AN EVOLUTIONARY VIEW OF HUMAN SOCIAL BEHAVIOR

If we can analyze the evolutionary basis of the behavior of termites, naked mole-rats, and other animals, would it not be rewarding to analyze such a basis of human behavior, too? Many resist the idea. Apparently, they believe that attempts to identify the adaptive value of a human trait are attempts to define its moral or social advantage. However, there is a clear difference between trying to explain something in terms of its evolutionary history and attempting to justify it. "Adaptive" does not mean "morally right." To biologists, it only means valuable with respect to the transmission of the individual's genes.

Consider the case of altruistic behavior that we call **adoption**, or the acceptance of offspring of other individuals as one's own. If we wish to study adoptive behavior, we may use evolutionary theory to come up with a hypothesis about it.

This does not mean that we are attempting to judge whether adoption is a moral behavior or even a socially desirable behavior. "Morality" and "desirability" are issues about which biologists have no more to say than anybody else.

Figure 41.20 Two adult emperor penguins competing to adopt an orphan, which penguins accept as substitute offspring.

Start with a premise that all adaptations have costs and benefits. One cost is that some adaptive behaviors may be *redirected* under rare or unusual circumstances. In many species, adults that lost offspring do adopt a substitute. Some cardinals have fed goldfish. A whale tried to lift a log out of water as if it were a distressed infant that needed help to reach the water's surface. Emperor penguins fight to adopt orphans (Figure 41.20).

As John Alcock suggests, such examples constitute a test of a hypothesis about adoptions by us. Human adoptive behavior occurs "by mistake" when otherwise adoptive parental behavior gets focused on unrelated offspring. Suppose it arises through a frustrated desire to bear children (hypothesis). If so, then we can expect that couples who lost an only child or are sterile should be especially prone to adopt strangers (prediction).

Now consider this. For most of their evolutionary history, humans have lived in small groups and, later, small villages. They probably had few opportunities or showed little inclination to adopt strangers. They were likely to accept a child of a deceased relative. But how

would adoptive parents gain genetic representation in the next generation? According to theories of natural selection and indirect selection, individuals act in ways that promote genetic self-interest. So we might predict that parents with dependent, care-requiring children of their own are much less likely to adopt substitutes, compared with adults who have no children or who already raised children and are living by themselves. We could test the prediction by piecing together a large enough sampling of information, ideally from diverse existing human cultures, on a connection between childless conditions and adoption.

Indirect selection also favors adults who focus their parenting behavior on relatives and thereby perpetuate shared genes in an indirect way. We might even predict that such adults are far more likely to be related to an adopted child than we would expect by chance alone. Behaviorist Joan Silk tested the prediction in traditional societies. Her results showed that people will adopt related children far more often than nonrelated children. Particularly in large, industrialized societies that have welfare agencies and other means of adoption assistance, individuals become parents of nonrelatives. Such societies are evolutionary environments. Parenting mechanisms did evolve in the past. Perhaps their redirection toward nonrelatives says more about our evolutionary history than it does about the transmission of one's genes.

The bottom line is this: Evolutionary hypotheses about the adaptive value of behavior lend themselves to testing. Through testing, we can gain understanding about the evolution of human behavior.

It is possible to test evolutionary hypotheses regarding the adaptive value of different forms of human behavior.

There might be greater acceptance of such testing when more people come to understand that adaptive behavior and socially desirable behavior are separate issues.

In biology, "adaptive" means only that a specified trait of an individual has proved beneficial in the transmission of genes responsible for the trait.

---

1. Animal behavior (coordinated responses to stimuli) starts with genes that specify products required for the development and operation of nervous, endocrine, and skeletal–muscular systems. *CI, 41.1*

2. A behavior performed without having been learned by experience is instinctive, a prewired response to one or two simple, well-defined environmental cues (sign stimuli) that trigger a suitable response, such as a fixed action pattern. Experiences may lead to variations or changes in responses (learned behavior). Learning is a result of genetic and environmental inputs. *41.1, 41.2*

3. Behavior with a genetic basis is subject to evolution by natural selection. It evolved as a result of individual differences in reproductive success in past generations, when reproductive benefits of a behavior exceeded its reproductive costs (or disadvantages). *41.3*

4. Social groups require cooperative interdependency among individuals of a species. Sociality is promoted by communication signals, cues sent by one individual (a signaler) that can change the behavior of another of the same species (a signal receiver). *41.4*

5. Chemical, visual, acoustical, and tactile signals are components of communication displays. *41.4*

 a. Pheromones function in chemical communication. Signaling pheromones, such as sex attractants and alarm signals, can induce immediate change in the behavior of a receiver. Priming pheromones elicit a generalized physiological response by the receiver.

 b. Visual signals (observable actions or cues) are key components of courtship displays and threat displays.

 c. Acoustical signals (sounds with precise, species-specific information) include the mating calls of frogs.

 d. Tactile signals are specific forms of physical contact between a signaler and a receiver.

6. Members of the same species often create obstacles to one another's reproductive success, as by selective mate choice and by competing for access to mates. *41.5*

7. Social groups have costs, including competition for limited resources and greater exposure to contagious diseases and parasites. Benefits outweigh the costs in some cases, as when predation pressure is strong. *41.6*

8. In self-sacrificing (altruistic) behavior, individuals give up chances to reproduce while helping others of their group. In most social groups, individuals do not sacrifice reproductive chances to help others. *41.6*

 a. A dominance hierarchy is a social ranking in which some individuals give way to others of their group. Often, dominant individuals force subordinates to relinquish food or some other resource.

 b. When subordinates give in to dominant members of their group, they may receive compensatory benefits of group living, such as safety from predators. In some species, subordinates may reproduce if they live long enough and dominant ones slip in the hierarchy or die.

9. By one theory of indirect selection, genes associated with caring for relatives (not one's direct descendants but individuals that bear a fraction of the same genes) are favored in some cases. Indirect selection can lead to extreme altruism, as among workers of some species of social insects and naked mole-rats. The workers usually do not reproduce. Altruism helps reproducing relatives survive, so altruistic individuals pass on, "by proxy," genes underlying development of this behavior. *41.7*

10. Testing the adaptive value of human behavior starts with understanding that "adaptive" only means a trait is helpful in the transmission of an individual's genes; socially desirable behavior is a separate issue. *41.8*

Review Questions

1. Explain how genes and their products, including hormones, influence mechanisms required for forms of behavior. *41.1*

2. Define these terms: instinctive behavior, sign stimulus, fixed action pattern, and learned behavior. *41.1, 41.2*

3. Contrast altruistic behavior with selfish behavior. Why does either form of behavior persist in a population? *41.3*

4. Describe some characteristics of communication signals. Then give an example of a communication display. *41.4*

5. Describe some feeding behavior or mating behavior in terms of natural (individual) selection. *41.3, 41.5*

6. List some benefits and costs of sociality. *41.6, 41.7*

Self-Quiz   ANSWERS IN APPENDIX III

1. "Starlings minimize nest mites by festooning nests with wild carrot," said Clark and Mason. Their statement was _____ .
 a. an untested hypothesis   c. a test of a hypothesis
 b. a prediction   d. a proximate conclusion

2. Genes affect the behavior of individuals by _____ .
 a. influencing the development of nervous systems
 b. affecting the kinds of hormones in individuals
 c. governing the development of muscles and skeletons
 d. all of the above

3. Generally, living in a social group costs the individual, in terms of _____ .
 a. competition for food, other resources
 b. vulnerability to contagious diseases
 c. competition for mates
 d. a through c

4. A statement that overcrowding causes lemmings to disperse to areas that are more favorable for reproduction _____ .
 a. is consistent with Darwinian evolutionary theory
 b. is based on a theory of evolution by group selection
 c. is supported by the finding that most animals behave altruistically during their lives

5. The genetic similarity between an uncle and his nephew is _____ .
 a. the same as between a parent and its offspring
 b. greater than between two full siblings
 c. dependent on how many other nephews the uncle has
 d. less than that between a mother and her daughter

Figure 41.21
The puffed-up anterior segments of a caterpillar under siege.

Figure 41.23
Soldier termites guarding a break in a tunnel that leads from dead wood to a nest underground.

6. Match the terms with their most suitable description.

____ fixed action pattern	a. time-dependent form of learning requiring exposure to key stimulus
____ altruism	b. genes plus actual experience
____ basis of instinctive *and* learned behavior	c. stereotyped motor program that runs to completion independently of feedback from environment
____ imprinting	d. assisting another individual at one's own expense

Critical Thinking

1. A large caterpillar is crawling along a branch in a tropical forest. You poke it, and one of its responses is to let go of the branch and puff up anterior segments of its body (Figure 41.21). Propose an internal mechanism that may contribute to this defensive behavior. Also propose how the behavior may have adaptive value. How would you test your two hypotheses?

2. A cheetah scent-marks plants in its territory by releasing certain chemicals from exocrine glands. What evidence would you require to demonstrate that the cheetah's action is an evolved communication signal?

3. You observe mallard ducks in a small pond. Then you notice a rooster wading out to the ducks with amorous intent (Figure 41.22). What probably happened to the rooster early in life?

4. Develop an adaptive value hypothesis for this observation: Male lions kill offspring of females they acquire after chasing away previous pride holders that had mated with the females. How would you test your infanticide hypothesis?

5. What are the likely evolutionary costs and benefits to a male hangingfly of offering a nuptial gift to its potential mate? How might a female's choosiness depend on the costs and benefits?

6. In a eucalyptus forest in Queensland, Australia, you notice a narrow, brittle tube running along the trunk of a dead tree. You decide to chip a few fragments from it. Sunlight pours into the breached tube wall. Some small, nearly white insects bang their head against the wall, then scurry away. The head banging, an acoustical signal, makes vibrations that alert soldier termites. Soldiers run to the breach and make a defensive stand (Figure 41.23). Each has a swollen, eyeless, tapering head with a long,

Figure 41.22
Behaviorally confused rooster.

pointed "nose." Each disturbed soldier termite shoots thin jets of silvery goo out of its nose! The silvery strands release volatile compounds that attract more soldier termites that battle the danger, which more typically is an invasion by ants.

Insects in the ruptured tunnel belong to a complex termite colony. Whereas soldiers protect the colony, the pale ones are workers. They build tunnels that lead to places where they safely gather fibers of wood. They carry fibers to an underground nest where termites live and cultivate an edible fungus. Fungal hyphae grow into the wood fibers and absorb nutrients from them. The termites eat some portions of the fungus. They chew the wood into small particles that they can swallow. Unlike most termite species, they don't produce enzymes that can digest cellulose in wood. Symbiotic microorganisms in their gut do this, and the symbionts and termites both absorb the released nutrients.

Soldier termites and worker termites are sterile; neither can reproduce. They engage in self-sacrificing behavior that helps others survive. A single queen and one or more kings serve as "parents" for the entire society.

Identify some of the communication signals and the manner in which individuals in the colony respond to them. Speculate on the adaptive value of the social behaviors just described. How, for instance, could being sterile help an individual's chances for genetic success?

Selected Key Terms

adaptive behavior *41.3*	fixed action pattern *41.1*	selfish behavior *41.3*
adoption *41.8*	illegitimate receiver *41.4*	selfish herd *41.6*
altruistic behavior *41.3*	illegitimate signaler *41.4*	sexual selection *41.5*
animal behavior *CI*	imprinting *41.2*	sign stimulus *41.1*
communication display *41.4*	indirect selection *41.7*	signal receiver *41.4*
	instinctive behavior *41.1*	signaler *41.4*
communication signal *41.4*	learned behavior *41.2*	social behavior *41.3*
	lek *41.5*	song system *41.1*
composite signal *41.4*	natural selection *41.3*	tactile display *41.4*
courtship display *41.4*	pheromone *41.4*	territory *41.3*
dominance hierarchy *41.6*	reproductive success *41.3*	threat display *41.4*

Readings

Alcock, J. 1998. *Animal Behavior: An Evolutionary Approach*. Sixth edition. Sunderland, Massachusetts: Sinauer.

Frisch, K. von. 1961. *The Dancing Bees*. New York: Harcourt Brace Jovanovich. A classic on the behavior of honeybees.

Wilson, E. O. 1975. *Sociobiology: The New Synthesis*. Cambridge, Massachusetts: Harvard University Press.

On-Line readings at Student Guide for InfoTrac:
www.brookscole.com/biology

COMMUNITY INTERACTIONS

No Pigeon Is An Island

Flying through the rain forests of New Guinea is an extraordinary pigeon with cobalt blue feathers and lacy plumes on its head (Figure 42.1). It is about as big as a turkey, and it flaps so slowly and noisily that its flight sounds like an idling truck. Like eight species of smaller pigeons in the same forest, it perches on branches to eat fruit. How is it possible that *nine* species of large and small fruit-eating pigeons live in the space of the same forest? Wouldn't competition for food leave one species the winner? Actually, each species in that rain forest lives, grows, and reproduces in a characteristic way, as defined by its relationships with other organisms and with the surroundings.

Figure 42.1 A cobalt blue, turkey-sized Victoria crowned pigeon, one of nine species of pigeons living in the same tropical rain forest of New Guinea. In this habitat, each species has its own niche.

Big pigeons perch on the sturdiest branches when they eat. And they eat big fruit. Smaller pigeons, with their smaller bills, can't peck open big fruit. They eat small fruit hanging from slender branches that aren't sturdy enough to support the weight of a turkey-sized pigeon. Species of trees in the forest differ with respect to fruit size and the diameter of fruit-bearing branches. They attract pigeons with different traits. In such ways, the nine pigeon species *partition* the fruit supply.

And how do individual trees benefit from enticing pigeons to eat? Seeds in their fruits have tough coats, which resist the action of digestive enzymes in a pigeon gut. During the time it takes for ingested seeds to travel through the gut, pigeons fly about and dispense seed-containing droppings in more than one place. In this way, they tend to disperse seeds some distance from the parent plants. Later, when seedlings grow, the odds are better that at least some of them won't have to compete with their parents for sunlight, water, and mineral ions. The seeds that drop close to home can't compete in any significant way with the resource-gathering capacity of mature trees, which already have well-developed, extensive roots and leafy crowns.

Within the same forest, leaf-eating, fruit-munching, and bud-nipping insects interact with other organisms and their surroundings in certain ways. So do nectar-drinking, flower-pollinating bats, birds, and insects. And so do great numbers of beetles, worms, and other invertebrates that busily extract energy from remains and wastes of other organisms on the forest floor. By their activities, they cycle nutrients back to the trees.

Like humans, then, no pigeon is an island, isolated from the rest of the living world. The nine species of New Guinea pigeons eat fruit of different sizes. They disperse seeds from different sorts of trees. Dispersal influences where new trees will grow and where the decomposers will flourish. Ultimately, tree distribution and decomposition activities influence how the entire forest community is organized.

Directly or indirectly, interactions among coexisting populations organize the community to which they belong. With this chapter, we turn to community interactions that influence all populations over time and in the space of their environment.

Key Concepts

1. A habitat is the type of place where individuals of a species normally live. A community is an association of all the populations of species that occupy the same habitat.

2. Every species in the community has its own niche, defined as the sum of all activities and relationships in which its individuals engage as they secure and use the resources required for their survival and reproduction.

3. The structure of a community starts with adaptive traits that give individuals of each species the capacity to respond to the physical and chemical features of their habitat, and to levels and patterns of resource availability over time.

4. Interactions among species influence the structure of the community. They include mutually beneficial interactions, competition, predation, and parasitism.

5. Community structure also depends on the geographic location and size of the habitat, the rates at which the member species arrive and disappear, and the history of physical disturbances to the habitat.

6. The first species to occupy a particular type of habitat are replaced by others, which are replaced by still others, and so on in an expected sequence. This process, known as primary succession, produces a climax community. A climax community is a stable, self-perpetuating array of species in balance with one another and with the environment.

7. Different stages of succession often prevail within the same habitat owing to local differences in the soil and other habitat conditions, recurring disturbances such as seasonal fires, and chance events.

WHICH FACTORS SHAPE COMMUNITY STRUCTURE?

Think of a clownfish darting above a coral reef, a maple tree on a Vermont hillside, or a mole making a burrow in soft earth. The type of place where you will normally observe a clownfish, maple, or mole is its **habitat**. The habitat of an organism is characterized by physical and chemical features, such as its temperature and salinity, and by the array of other species living in it. Directly or indirectly, the populations of all species in a habitat associate with one another as a **community**.

Five factors shape the structure of a community. *First*, interactions among climate and topography help dictate the habitat's temperatures, rainfall, soil types, and other conditions. *Second*, the kinds and amounts of food and other resources that become available through the year influence which species can live there. *Third*, individuals of each species have adaptive traits that allow them to survive and exploit specific resources in the habitat. *Fourth*, species in the habitat interact by competition, predation, and mutually helpful actions. *Fifth*, community structure is influenced by the overall pattern of population size (and the history of changes in its size), by arrivals and disappearances of species, and by physical disturbances to the habitat.

Together, these factors help determine the number of species at several "feeding levels," starting with the producers and continuing through levels of consumers. They influence population sizes. They also help dictate the overall number of species. For example, high solar radiation, warm temperatures, and high humidity in tropical habitats favor growth of many kinds of plants, which support many kinds of animals. Conditions in arctic habitats do not favor great numbers of species.

Chapters to follow deal with energy flow through feeding levels and with geographic factors influencing community structure. Here we begin with interactions among species, using the niche concept as our guide.

The Niche

If the organisms of a community all share the same habitat—that is, if they all have the same "address"—in what respects do they differ? Each kind is distinct in terms of its "profession" in the community, meaning the sum of the activities and relationships in which it engages to secure and use the resources necessary for its survival and reproduction. This is its **niche**.

For each species, a *fundamental* niche is the one that might prevail in the absence of competition and other factors that could constrain its acquisition and use of resources. However, as you will see, such constraining factors do come into play in communities. They tend to bring about a more constrained, *realized* niche that shifts in large and small ways over time, as individuals of the species respond to a mosaic of changes.

Categories of Species Interactions

Dozens to hundreds of species interact in diverse ways even in simple communities. In spite of this diversity, we can identify six categories of interactions that have different effects on population growth (Table 42.1).

Table 42.1 *Categories of Two-Species Interactions**		
Type of Interaction	Direct Effect on Species 1	Direct Effect on Species 2
Neutral relationship	0	0
Commensalism	+	0
Mutualism	+	+
Interspecific competition	–	–
Predation	+	–
Parasitism	+	–

* 0 means no direct effect on population growth;
+ means positive effect; – means negative effect.

Each species has a neutral relationship with most species in its habitat. For example, Canadian lynx and grasses do not affect each other directly. Interactions with other species link them only indirectly. The lynx preys on and decreases the number of snowshoe hares that eat plants; plants fatten hares, the prey of lynx.

Commensalism directly helps one species but does not affect the other much, if at all. For instance, some birds use tree branches for roosting sites. The trees get nothing but are not harmed. With **mutualism**, benefits flow both ways between the interacting species. Don't think of this as cozy cooperation. Benefits flow from a two-way exploitation. With **interspecific competition**, disadvantages flow both ways between species. Finally, **predation** and **parasitism** are interactions that directly benefit one species (either the predator or the parasite) and directly hurt the other (the prey or host).

Commensalism, mutualism, and parasitism are cases of **symbiosis** ("living together"). For at least part of the life cycle, individuals of two or more species interact with neutral, positive, or negative effects on each other.

A habitat is the type of place where individuals of a species normally live. A community consists of all populations that live in the habitat.

Community structure arises from the habitat's physical and chemical features, resource availability over time, adaptive traits of its members, how the members interact, and the history of the habitat and its occupants. Species may have neutral, positive, or negative effects on one another.

A niche is the sum of all activities and relationships in which individuals of a species engage as they secure and use the resources necessary to survive and reproduce.

MUTUALISM

Mutualistic interactions in which positive benefits flow both ways abound in nature. When trees provide New Guinea pigeons with food and when the pigeons help disperse seeds from the trees to new germination sites, both function as mutualists. Many other kinds of plants and animals enter into such interactions. For example, most flowering plants, and the insects and birds, bats, and other animals that pollinate them, are mutualists. The introduction to Chapter 28 gives vivid examples.

of absorptive structures interact in two ways. Either the fungal hyphae penetrate root cells or they grow as a dense, velvety mat around them. The fungus is good at absorbing mineral ions from soil, and the plant comes to depend on taking some of them. In turn, the fungus withdraws some sugar molecules that the plant makes by photosynthesis. The fungus depends upon the plant for its reproductive success, too. When photosynthesis ceases, the fungus stops producing spores.

Figure 42.2 One mutualistic interaction on a rocky slope of Colorado's high desert.

(**a**) Different kinds of flowering plants of the genus *Yucca* are each pollinated exclusively by one species of yucca moth (**b**). This insect cannot complete its life cycle with any other plant.

The adult stage of the moth life cycle coincides with blossoming of yucca flowers. By using her specialized mouthparts, a female moth gathers sticky pollen and rolls it into a ball. Then she wings her way to another flower. She pierces the wall of the flower's ovary, where seeds form and develop, and lays her eggs inside. As she crawls out of the flower, she pushes a ball of pollen onto a pollen-receiving surface.

Pollen grains germinate and grow down through tissues of the ovary. They deliver sperm to the flower's eggs. The seeds develop after fertilization. Meanwhile, the moth eggs develop to the larval stage. (**c**) When larvae emerge, they eat a few seeds, then gnaw their way out of the ovary. The seeds that moth larvae do not eat give rise to new yucca plants.

Some forms of mutualism are *obligatory*. That is, the individuals of one species cannot grow and reproduce in the absence of intimate dependency with individuals of another species during the life cycle. This is the case for interactions between yucca plants and yucca moths. Each plant of this genus (*Yucca*) can be pollinated only by one species of the yucca moth genus (*Tegeticula*). In addition, the larval stages of the moth grow only in the yucca plant; they eat only yucca seeds (Figure 42.2).

We also find cases of obligatory mutualism between many fungi and plants. Mycorrhizae, remember, are intimate ecological interactions between fungal hyphae and young roots (Chapters 21 and 27). The two kinds

And what about the apparent endosymbiotic origin of eukaryotes? Long ago, phagocytic bacteria may have engulfed aerobic bacterial cells. These resisted digestion, tapped host nutrients, and reproduced independently. In time the hosts came to rely on ATP produced by the guests—which became mitochondria and chloroplasts. If those prokaryotic cells had not evolved as such close mutualists, you and all other eukaryotic species would not even be around today (Section 19.4).

In cases of mutualism, each of the participating species reaps benefits from the interaction.

COMPETITIVE INTERACTIONS

Where you find limited supplies of energy, nutrients, living space, and other natural resources, you observe organisms competing for a share of them. *Intraspecific* competition occurs between individuals of the same population or species. As you probably deduced from the preceding chapter, their interactions can be fierce. *Interspecific* competition occurs between populations of different species and usually isn't as intense. Why not? *Requirements of two species may be similar, but they never are as close as they are for individuals of the same species.*

Consider just two forms of competitive interactions. Sometimes individuals have equal access to a required but limited resource. If one is better than the other at exploiting it, competition can reduce the supply of that shared resource. (When you and a friend use straws to share a small milkshake, you might not get much at all if your friend uses a jumbo straw.) And sometimes individuals control access to a resource, letting others use only some of it or none at all, regardless of whether the resource is scarce or abundant. (For instance, even if you shared a ten-gallon milkshake, you still would get less if your friend pinched your straw.)

Competition abounds in nature. Chipmunks exclude other species of chipmunks from their habitats (Figure 42.3). A strangler fig tree wraps around other trees as a framework for its growth, and eventually it kills them.

From early spring until late summer, a male *broadtailed* hummingbird busily chases other males and females of its species from blossoms in his territory in the Rockies. In August, however, *rufous* hummingbirds of the Pacific Northwest migrate through the Rockies on their way to their wintering grounds in Mexico. Rufous males prove to be far more aggressive, hence stronger competitors for available food. As they follow the migratory route, they evict male broadtails from broadtail territories.

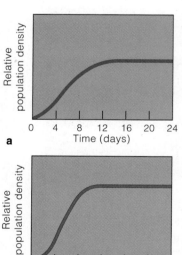

a

b

c

| ALPINE CHIPMUNK | LODGEPOLE CHIPMUNK | YELLOW PINE CHIPMUNK | LEAST CHIPMUNK |

Figure 42.3 Example of competition in nature. On the eastern slopes of the Sierra Nevada, different chipmunk species occupy different habitats. The alpine habitat is at the highest elevation. Below it are the lodgepole pine, piñon pine, and then sagebrush habitats. The least chipmunk lives at the base of the mountains in sagebrush. Its adaptations would allow it to move up into the piñon pine habitat, but the aggressively competitive behavior of yellow pine chipmunks in that habitat won't let it. Food preferences keep the yellow pine chipmunk out of the sagebrush habitat.

Competitive Exclusion

Any two species differ in their adaptations for getting food and avoiding enemies. In some way, one usually competes more effectively for scarce resources. The two are less likely to coexist in the same habitat if they use resources in very similar ways. G. Gause demonstrated this by growing two species of *Paramecium* separately, then together (Figure 42.4*a–c*). Both species used the same food (bacteria) and competed intensely for it. As results from experimental tests show, two species that use identical resources can't coexist indefinitely. Many other experiments have supported this concept, which ecologists now call **competitive exclusion**.

Gause also studied two other *Paramecium* species that didn't overlap as much in their requirements. When grown together, one species tended to feed on bacteria suspended in culture tube liquid. The other ate yeast cells at the bottom of the test tube. Population growth rates slowed for both of the species—but the overlap in resource use wasn't enough for one species to fully exclude the other. The two continued to coexist.

Field experiments also reveal effects of competition. Robert Paine used removal experiments to document effects of a **keystone species**, a dominant species that dictates community structure by affecting abundances

Figure 42.4 Results of competitive exclusion between two protistan species that compete for the same food. (**a**) *Paramecium caudatum* and (**b**) *P. aurelia* were grown in separate culture tubes and established stable populations. The S-shaped growth curves in these graphs indicate stability. (**c**) Then the populations were grown together. *P. aurelia* (*red* curve) drove the other species toward extinction (*blue* curve in **c**). This experiment and others suggest that two species cannot coexist indefinitely in the same habitat *when they require identical resources.* If their requirements do not overlap much, one might influence the population growth rate of the other, but they may still coexist.

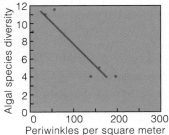

d Algal diversity in tidepools

(graph: Algal species diversity vs Periwinkles per square meter)

e Algal diversity on rocks that become exposed at high tide

(graph: Algal species diversity vs Periwinkles per square meter)

Figure 42.5 Effect of competition and predation on community structure. (**a**) Grazing periwinkles (*Littorina littorea*) affect the number of algal species in different ways in different marine habitats. (**b**) *Chondrus* and (**c**) *Enteromorpha*, two kinds of algae in their natural habitat. (**d**) In tidepools, periwinkles graze on the dominant alga (*Enteromorpha*); less competitive algae that might otherwise be overwhelmed have an advantage. (**e**) Algal diversity is lower on rocks exposed at low tide. There, periwinkles don't graze on *Chondrus* and other kinds of red algae, the dominant species.

of other species. All along North America's west coast, surf, tides, and storms repeatedly disturb the intertidal zones, and living space is scarce. Paine put a keystone species, the sea star *Pisaster ochraceus*, in control plots with its prey: mussels (*Mytilus*), barnacles, limpets, and chitons. He removed sea stars from experimental plots. Mussels took over those plots; they crowded out seven other invertebrate species. Mussels, the main prey of sea stars, are the strongest competitors in their absence. So sea star predation normally maintains the diversity of prey species because it blocks competitive exclusion by mussels. Remove all the sea stars, and the community shrinks from fifteen species to eight.

Similarly, Jane Lubchenco found that the periwinkle *Littorina littorea*, an alga-eating mollusk, promotes *or* limits diversity in different habitats. In tidepools, it eats dominant algal species, thereby helping many other, less competitive algae live. On rocks exposed at high tide only, *L. littorea* ignores tough, unpalatable species and favors the competitively weaker algal species it ignores in tidepools. And so *L. littorea* promotes algal diversity in tidepools—yet lowers it on rocks exposed at high tide (Figure 42.5).

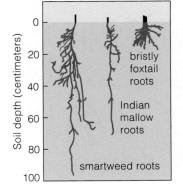

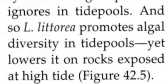

Figure 42.6 Partitioning of resources among three annual plants in a plowed, abandoned field. They differ in adaptations for securing soil water and mineral ions.

Resource Partitioning

Think back on the nine species of fruit-eating pigeons in the same New Guinea forest. They all use the same resource: fruit. Yet they overlap only slightly in their use of it, for each is adapted to eat fruits of a particular size. They are a fine example of **resource partitioning**—the *subdividing* of some category of similar resources that lets competing species coexist.

Another competitive situation arose among three species of annual plants in plowed, abandoned fields. Like other plants, all three require sunlight, water, and dissolved mineral ions. Yet each species is adapted to exploiting a different part of the habitat (Figure 42.6). Drought-tolerant foxtail grasses have a shallow, fibrous root system that quickly absorbs rainwater. They grow where moisture in soil varies from day to day. Mallow plants, with a taproot system, grow in deeper soil that is moist early in the growing season but drier later. Smartweed's taproot system branches in topsoil and soil below the roots of the other species. It grows where soil is continuously moist. In sum, species may coexist in the same habitat even when their niches do overlap.

In some competitive situations, all individuals have equal access to a resource that they all require, but some are better than others at exploiting it. In other competitive situations, some individuals control access to a shared resource.

The more that two species in the same habitat differ in their use of resources, the more likely they can coexist.

Two competing species also may coexist by sharing the same resource in different ways or at different times.

PREDATION AND PARASITISM

To keep things simple, let's use two broad definitions for interactions between consumers and their victims. **Predators** are animals that feed on living organisms—their **prey**—but do *not* take up residence on or in them. Prey may or may not die from the interaction. **Parasites** live in or on other living organisms—their **hosts**—and feed on specific host tissues for part of their life cycle. Hosts may or may not die from the interaction.

Many adaptations of predators (and parasites) and their victims arose by **coevolution**, the joint evolution of two or more species that exert selection pressure on each other as a result of close ecological interaction. Suppose a mutant prey organism has a new, heritable means of defense, which later spreads through the prey population. Some individual predators are better than others at countering the defense. They tend to eat more and have a better chance to survive and reproduce. In time the forms of traits *most* effective at overcoming the defense increase in frequency in the population. Now the better predators exert selection pressure that favors better prey defenses—and so on through time.

Dynamics of Predator–Prey Interactions

The outcome of predator–prey interactions in a specified interval depends partly on the carrying capacity of the prey population. **Carrying capacity**, remember, is the maximum number of individuals that the resources in a given environment can maintain indefinitely (Section 40.4). The reproductive rates of the predator and prey populations also influence the outcome. So do predator responses to increases in prey density.

When predation is keeping a prey population from exceeding the carrying capacity, both populations tend to coexist at fairly steady levels. More prey are around; predators reproduce fast and eat more prey. Population

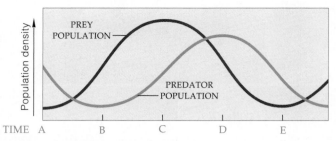

Figure 42.7 Idealized cycling of abundances of predator and prey. (For clarity, the diagram exaggerates predator density; predators are usually less common than prey during the cycle.) This pattern arises owing to time lags in predator responses to changes in abundance of prey. At time A, prey density is low; predators have a harder time getting food; their population is declining. In response to the decline, prey density rises. But predators don't start increasing until they start reproducing at time B. Both populations grow until the predator increase causes the prey population to decline (time C to E). Predators continue to increase and take out more prey. Having fewer prey leads to starving predators, so the predator population growth rate slows (starting at time D). At time E, a new cycle starts.

densities fluctuate when predators don't reproduce as fast as the prey, when they can eat only so many prey organisms in a given interval, and when the carrying capacity for the prey population is high.

Figure 42.7 shows an idealized pattern of the cyclic changes corresponding to time lags in the predator's response to changes in prey abundance. In nature, we sometimes observe such a correspondence—but other factors influence the pattern. Figure 42.8, for instance, shows a ten-year cycling in populations of the Canadian

Figure 42.8 Correspondence between abundances of Canadian lynx (*dashed* line) and snowshoe hares (*solid* line), based on counts of pelts sold by trappers to Hudson's Bay Company over ninety years. The photograph underscores the observation by Charles Krebs that predation causes mass paranoia among hares, which continually look over their shoulders during the declining phase of each cycle.

The graph is a good test of whether you will readily accept someone else's conclusions without questioning their basis in science. (Remember those sections in Chapter 1 on scientific methods?) What other factors may have impacted on the cycle? Did weather vary, with more severe winters imposing greater demand for hares (to keep lynx warmer) and higher death rates? Did the lynx compete with other predators, such as owls? Did predators turn to alternative prey during low points of the hare cycle? When fur prices rose in Europe, did trapping increase? When the supply of pelts outstripped the demand, did trapping decline?

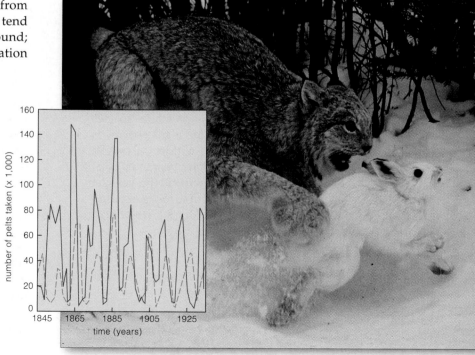

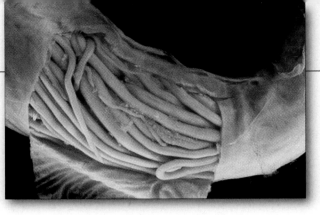

Figure 42.9
Adult roundworms (*Ascaris*), a type of endoparasite, packed inside a small intestine from a host pig.

lynx and showshoe hare. Charles Krebs led a long-term study in Alaska's Yukon to identify the basis of this cycling. For ten years, he tracked hare densities in 1-square-kilometer control plots and experimental plots that kept out mammalian predators (by electric fences), had more food, or had fertilizer to increase plant growth. The team placed radio collars on 1,000+ hares, lynx, squirrels, lynx, and coyotes. In predator-free plots, hare density doubled. In plots with extra food, it tripled. With fewer predators plus extra food, density increased elevenfold. But manipulations only delayed the cyclic declines; they didn't stop them. The fence could not keep out owls and other raptors. So predators got 83 percent of the collared hares, only 9 percent of which starved to death. Therefore, a simple predator–prey or plant–herbivore model can't explain the results. More variables are operating in a *three-level* interaction among plants, herbivores, and carnivores.

Dynamics of Parasite–Host Interactions

Parasites have pervasive influences on populations. By draining hosts of nutrients, they alter how much energy and how many nutrients the population is withdrawing from the habitat. Weakened hosts are more vulnerable to predators and less attractive to potential mates. Some infections result in sterility. Some may alter the ratio of host males to females. In such ways, parasitic infections lower the birth rate, raise the death rate, and influence intraspecific and interspecific competition.

The persistent drain of nutrients during a parasitic infection indirectly causes death if the weakened host just can't fight off secondary infections. Evolutionarily speaking, though, killing the host doesn't do much for the parasite's reproductive success. Infections of longer duration give the parasite more time to produce more offspring. So natural selection tends to favor parasite and host adaptations that promote a level of mutual tolerance and less-than-fatal effects. Death usually will result only when a parasite attacks a novel host, which has no coevolved defenses against it, or when there are too many individual parasites driving the infection.

*Ecto*parasites live on a host's surface; *endo*parasites live inside it. Some use one or more hosts throughout the life cycle. Others are free-living for part of it. Many use arthropods, including insects, to taxi to a new host. The **microparasites** among them include many bacteria, viruses, protozoans, and sporozoans (Chapter 20). Most reproduce rapidly. The larger, **macroparasites** include many flatworms, roundworms, and arthropods such as fleas and ticks (Chapter 23, and Figures 24.12 and 42.9).

Then there are parasitic plants. Nonphotosynthetic types withdraw nutrients and water from young roots of host plants. Species in the broomrape family do this to oak and beech tree roots. Other plants do so but still retain the capacity for photosynthesis. Mistletoe is like this; its extensions invade sapwood.

Not all parasites feed on tissues. **Social parasites** complete their life cycle by exploiting social behaviors of another species. For instance, a cowbird never builds a nest, incubates eggs, or cares for offspring. It removes an egg from another bird's nest and proceeds to lay a "replacement." Some birds hatch the alien egg and raise the hatchling. Usually, the young cowbird aggressively pushes the smaller, remaining rightful occupants out of the nest or demands and gets most of the food.

Many parasites are being commercially raised and selectively released as *biological controls*. They are touted as an alternative to chemical pesticides. But less than 20 percent of the existing selections qualify as effective defenses against pests. As outlined by C. Huffaker and C. Kennett, effective biological control agents display five attributes. They are well adapted to a host species and to their habitat. They are exceptionally efficient in searching for a host. Their rate of population growth is high relative to that of the host species. Their offspring are mobile enough to ensure adequate dispersal. And the lag time between their responses to changes in the numbers of the host population is minimal.

Releasing more than one kind of biological control agent in an area may trigger competition among them and eventually lessen their overall effectiveness. Also, a shotgun approach to biological control is risky. There is a possibility that the parasites will attack nontargeted species. In 1983, for example, F. Howarth reported that the populations of butterflies and moths native to the Hawaiian Islands are declining. Introductions of wasps that were supposed to serve as biological controls over something else are partly to blame. Later in the chapter, we will return to what can happen to native populations when a new species is introduced into a habitat.

Predators and their prey, as well as parasites and their hosts, interact in long-term, coevolutionary contests.

Predator and prey populations may coexist at fairly steady levels. Others may undergo recurring cycles of abundance and crashes, erratic cycles, or prey extinction.

Natural selection favors parasitic species that temper their demands in ways that ensure an adequate supply of hosts.

A Coevolutionary Arms Race

Populations of other species are part of any organism's environment, and the ones that interact as predators and prey exert continual selection pressure on each other. One must defend itself and the other must overcome the defenses. This is the basis of a coevolutionary arms race that has resulted in some truly amazing adaptations.

a

b

c

CAMOUFLAGE Consider prey species that **camouflage** themselves; they can hide in the open. Such organisms have adaptations in form, patterning, color, and behavior that help them blend with their surroundings and escape detection. Figure 42.10 shows classic examples, including a desert plant (*Lithops*) that resembles a small rock. Only during a brief rainy season does *Lithops* flower. That is when other plants grow profusely and distract herbivores from *Lithops*—and when free water instead of juicy plant tissues is available to quench an animal's thirst.

WARNING COLORATION Many prey species taste bad, are highly toxic, or inflict pain on attackers. Toxic types often have **warning coloration**, or conspicuous patterns and colors that predators learn to recognize as *AVOID ME!* signals. For example, maybe a young, inexperienced bird will spear a yellow-banded wasp or an orange-patterned monarch butterfly once. It will quickly learn to associate the distinctive colors and patterning with a painful sting or with the vomiting of foul-tasting butterfly toxins.

Truly dangerous or repugnant species make little or no attempt to conceal themselves. Skunks are like this. So are frogs of the genus *Dendrobates*; they are among the most vivid and most poisonous organisms (Section 29.1).

MIMICRY Many prey organisms bear close resemblance to dangerous, unpalatable, or hard-to-catch species. Any close resemblance in form, behavior, or both between one species that serves as a *model* for deception and another that is its *mimic* is called **mimicry**. Figure 42.11 shows how some tasty but weaponless mimics physically resemble species that predators have learned to ignore. In "speed" mimicry, sluggish prey species look like swift-moving species that predators have given up trying to catch.

MOMENT-OF-TRUTH DEFENSES When luck runs out, survival of prey organisms that are cornered or under attack may turn on a last-ditch trick. Suppose a leopard manages to run down a tasty baboon, as in Figure 41.17. By turning abruptly and displaying formidable canines, the baboon may startle and confuse this predator long enough for a getaway. Other cornered animals may spew irritating chemical repellents or toxins. Earwigs, skunks, and stink beetles produce disgusting odors. Some beetle species take aim and let loose with noxious sprays.

Similarly, many plants synthesize predator repellents. Tannins in the foliage and seeds of certain plants taste

Figure 42.10 A few prey organisms demonstrating the fine art of camouflage. (**a**) What bird??? When a predator approaches its nest, the least bittern stretches its neck (which is colored like the surrounding withered reeds), thrusts its beak upward, and sways gently like reeds in the wind. (**b**) An unappetizing bird dropping? No. This edible caterpillar's body coloration and the stiff body positions it assumes help camouflage it (hide it in the open) from predatory birds. (**c**) Find the plants (*Lithops*) hiding in the open from herbivores by their stonelike form, pattern, and coloring.

Figure 42.11 A few examples of mimicry among the insects. Many predators avoid prey that taste awful, secrete toxins, or inflict painful bites or stings. Such prey often display warning coloration (bright colors, bold markings, or both). Many don't even bother to hide. Many prey species unrelated to the dangerous or unpalatable ones have evolved striking behavioral and morphological resemblances to them. The inedible butterfly in (**a**) is a model for the edible mimic *Dismorphia* (**b**). The yellowjacket in (**c**) stings aggressively and probably is the model for nonstinging, edible wasps (**d**) and beetles (**e**) that have a similar appearance.

c The dangerous model . . .   **d** . . . one of its edible mimics . . .   **e** . . . and another edible mimic

bitter and make the plant tissues hard to digest. Make the mistake of nibbling on the visually luscious yellow petals of a buttercup (*Ranunculus*), and you'll inflict a chemical burn on the lining of your mouth.

PREDATOR RESPONSES TO PREY As part of the coevolutionary arms race, predators counter prey defenses with their own marvelous adaptations. Among their countermeasures are stealth, camouflage, and clever ways of avoiding repellents.

Consider the edible beetles that can direct sprays of noxious chemicals at their attackers. Grasshopper mice grab such beetles and plunge the "sprayer" end into the ground, then feast on the unprotected head (Figure 42.12*a,b*). Chameleons typically hold themselves motionless for extended intervals. Prey might not even notice them until the amazingly swift chameleon tongue zaps them (Section 29.4). And it is no accident that stealthy predators blend with backgrounds. Think of snow-white polar bears camouflaged against snow, golden tigers crouched in tall-stalked, golden grasses, and pastel predatory insects lurking in pastel flowers. And just hope you never step barefoot on a scorpionfish concealed on the seafloor.

Figure 42.12 Examples of adaptive responses of predators to prey defenses. (**a**) Some beetles spray noxious chemicals at attackers, which deters them some of the time. (**b**) At other times, grasshopper mice plunge the chemical-spraying tail end of their beetle prey into the ground and feast on the head end. (**c**) Find the scorpionfish, a venomous predator with camouflaging fleshy flaps, multiple colors, and profuse spines. (**d**) Where do the pink flowers end and the pink praying mantis begin?

FORCES CONTRIBUTING TO COMMUNITY STABILITY

A Successional Model

By now you might be asking, How does a community come into being? By the classical model for **ecological succession**, a community develops in sequence, from pioneer species to an end array of species that remain in equilibrium over some region. **Pioneer species** are opportunistic colonizers of vacant or vacated habitats. They enjoy high dispersal rates and rapid growth. In time, more competitive species replace the pioneers. They are themselves replaced until the array of species stabilizes under the prevailing habitat conditions. This persistent array of species is the **climax community**.

Primary succession is a process that begins when pioneer species colonize a barren habitat, such as a new volcanic island or land exposed when a glacier retreats (Figure 42.13). The pioneer species include lichens and plants that are small, have brief life cycles, and are adapted to exposed sites having intense sunlight, large temperature changes, and nutrient-deficient soil. In the early years, hardy flowering plants put out many small seeds, which are quickly dispersed.

Once established, the pioneers improve conditions for other species and often set the stage for their own replacement. Many new arrivals are mutualists with nitrogen-fixing bacteria, too, and outcompete the early species in nitrogen-poor habitats. Seeds of later species find shelter in the low-growing mats of pioneer species, which cannot shade out the seedlings. In time, wastes and remains accumulate, adding volume and nutrients to soil that help more species take hold. Successional species eventually crowd out the pioneers, whose spores and seeds travel as fugitives on the wind and water—destined, perhaps, for a new but temporary habitat.

In **secondary succession**, a disturbed area within a community recovers and moves again toward a climax state. The pattern is typical of abandoned fields, burned forests, and storm-battered intertidal zones. It occurs after a falling tree opens part of an established forest's canopy. Sunlight reaches seeds and seedlings that are already on the forest floor and spurs their growth.

Do colonizers facilitate their own replacement? By a different hypothesis, early colonizers compete against any species that could replace them, and the sequence of succession actually depends on who gets there first.

The Climax-Pattern Model

At one time, some ecologists thought the same general type of community always develops in a given region because of constraints imposed by climate. However, stable communities other than "the climax community" commonly persist in a region, as when tallgrass prairie extends from the west into Indiana's deciduous forests.

Figure 42.13 (**a**) Primary succession in Alaska's Glacier Bay area. The glacier in this photograph has been retreating since 1794. *Facing page:* (**b**) As a glacier retreats, meltwater leaches nitrogen and other minerals from the newly exposed soil. Less than ten years ago, ice buried this nutrient-poor soil. The first invaders are lichens, horsetails, and seeds of fireweed and mountain avens (*Dryas*). (**c**) *Dryas* is a pioneer species that relies on nitrogen-fixing activities of mutualistic microbes. It grows and spreads fast over the glacial till.

(**d**) Within twenty years, alder, cottonwood, and willow seedlings have taken hold in drainage channels. They, too, are symbionts with nitrogen-fixing microbes. (**e**) Within fifty years, alders have formed mature, dense thickets in which cottonwood, hemlock, and a few evergreen spruce can grow quickly. (**f**) After eighty years, cottonwood and spruce trees crowd out mature alders. (**g**) In areas deglaciated for more than a century, dense forests of Sitka spruce and western hemlock dominate.

By the **climax-pattern model**, a community is adapted to many environmental factors—topography, climate, soil, wind, species interactions, recurring disturbances, chance events, and so on—that vary in their influence over a region. As a result, even in the same region you may see one climax stage extending into another along gradients of influential environmental conditions.

Cyclic, Nondirectional Changes

Many small-scale changes occur over and over again in patches of a habitat. Such recurring changes contribute to the internal dynamics of the community as a whole. Observe a disturbed patch of some habitat, and you might conclude that great shifts in species composition are afoot. Observe the community on a larger scale, and you might discover that the overall composition includes all of the pioneer species that are colonizing the patch as well as the dominant climax species.

Think of a tropical forest. It slowly develops through phases of colonization by pioneer species, increases in

small fires are prevented. And so the underbrush slowly thickens with fire-susceptible species. But dense underbrush prevents the sequoia seeds from germinating —and it also fuels hotter fires. Hotter fires damage the giants. Park rangers now set controlled fires. By periodically removing underbrush, such controlled fires promote conditions that favor the community's cyclic replacements.

Restoration Ecology

Think back on the introduction to Chapter 26, which described the regrowth after the Mount Saint Helens eruption during 1980, and you already know that secondary succession can heal disturbances on a massive scale. Such *natural* restoration of communities often takes a long time. Today we also see *active* restoration—efforts to reestablish the biodiversity in key areas adversely affected or lost by agriculture, urban sprawl, and other human activities.

Chapter 25 included examples of this work. Here are a few more: Ecologists and volunteers are building artificial reefs along coasts, and repairing wetlands and grasslands. Species of natural prairie communities have been found in old cemeteries and other neglected patches of land in Illinois. Those species were transplanted by hand to a common restored site, which now extends over 180 hectares (445 acres). Volunteers maintain this small restored patch of prairie, as with controlled burns and hand weeding. Their goal is to find all of the 150–200 known species of the original community.

species diversity, and then reaches maturity. Whipping sporadically through the successional pattern, though, are heavy winds that cause treefalls. Where trees fall, gaps open in the forest canopy and more light reaches the forest floor. In that local patch, conditions favor the growth of previously suppressed small trees as well as germination of pioneers or shade-intolerant species.

Or think of the groves of sequoia trees in the Sierra Nevada of California. Some trees in this type of climax community are giants, more than 4,000 years old. Their persistence depends partly on recurring brush fires that sweep through parts of the forests. Sequoia seeds only germinate in the absence of far smaller, shade-tolerant plant species. Too much litter on the forest floor inhibits germination. Modest fires eliminate trees and shrubs that compete with young sequoias but do not damage the giant trees. Mature sequoias have thick bark that burns poorly and helps insulate their living phloem cells against modest heat damage.

At one time, fires were prevented in many sequoia groves in national and state parks—not just accidental fires from campsites and discarded cigarettes, but also natural, lightning-sparked fires. Litter builds up when

---

Community structure is an outcome of a balance of forces, such as predation and competition, that operate over time.

A climax community is a stable, self-perpetuating array of species in equilibrium with one another and their habitat. It develops as a sequence of species, starting with pioneers.

Similar climax stages can persist along gradients dictated by environmental factors and by species interactions. Small-scale, recurring changes help shape many communities.

Natural or deliberate ecological restoration often can repair a badly damaged climax community, provided that suitable species are available to reinstate the original biodiversity.

42.7

COMMUNITY INSTABILITY

The preceding sections might lead you to believe that all communities become stabilized in predictable ways. But this is not always the case. *Community stability is an outcome of forces that have come into uneasy balance.* Resources are sustained, as long as populations do not flirt dangerously with the carrying capacity. Predators and prey coexist, as long as neither wins. Competitors have no sense of fair play. Mutualists are stingy, as when plants make as little nectar as necessary to attract pollinators, and pollinators take as much nectar as they can for the least effort.

In the short term, disturbances can hurt the growth of some populations. You saw one example of this in Section 42.3. Long-term changes in climate or some other environmental variable often have destabilizing effects. If the instability is great enough, a community might change in ways that will persist even when the disturbance ends or is reversed. If some of its member species happen to be rare or don't compete well with others, they may become extinct.

For example, sometimes the residents of established communities move out from their home range and then successfully take up residence elsewhere. **Geographic dispersal** is the name for such directional movement, and it proceeds in three ways. First, over a number of generations, a population expands its home range by slowly invading outlying areas that prove hospitable. Second, some individuals are rapidly transported over great distances. This is called *jump* dispersal. It often takes an individual across a region where it could not survive on its own, as when an insect travels from the mainland to Maui in a ship's cargo. Third, a population moves away from its home range with imperceptible slowness, as brought about by continental drift.

The dispersal and colonization of vacant places can be amazingly rapid as well as successful. Consider one of Amy Schoener's experiments in the Bahamas. She set out small plastic sponges on the barren sandy floor of Bimini Lagoon. How fast did aquatic species take up residence in chambers of the artificial hotels? Schoener recorded occupancy by 220 species in thirty days.

Or think about the 4,500 or so exotic species (the non-natives) that successfully established themselves in the United States following jump dispersal. These are just the ones we happened to notice. Some, such as rice, soybeans, wheat, corn, and potatoes, have been put to good use as food resources. As the next section makes clear, however, most exotic species destabilize natural communities or adversely affect farmlands.

Long-term shifts in climate, the rapid introduction and successful establishment of a new species, and many other disturbances can permanently alter community structure.

42.8   FOCUS ON THE ENVIRONMENT

Exotic and Endangered Species

When you hear someone bubbling enthusiastically about an **exotic species,** you can safely bet the speaker isn't an ecologist. This is a name for a resident of an established community that was deliberately or accidentally moved from its home range and became established elsewhere. Unlike most imports, which can't take hold outside their home range, an exotic species permanently insinuates itself into a new community. Table 42.2 gives examples.

Sometimes the additions are harmless and even have beneficial effects. More often, they make native species **endangered species**, which by definition are extremely vulnerable to extinction (Section 25.2). Of all species on the rare or endangered lists or that recently became extinct, *close to 70 percent owe their precarious existence or demise to displacement by exotic species.*

THE PLANTS THAT ATE GEORGIA   A vine called kudzu (*Pueraria lobata*) was deliberately imported from Japan to the United States, where it faces no serious threats from herbivores, pathogens, or competitor plants (Figure 42.14). In temperate parts of Asia, it is a well-behaved legume with a well-developed root system. It *seemed* like a good idea to use it to control erosion on hills and highway embankments in the southeastern United States. With nothing to stop it, though, kudzu's shoots grew a third of a meter per day. Vines now blanket streambanks, trees, telephone poles, houses, and almost everything else in

Figure 42.14   Kudzu (*Pueraria lobata*) taking over part of Lyman, South Carolina. You can now find kudzu vines from East Texas to Florida and as far north as Pennsylvania.

Figure 42.15 Part of the fence against 200–300 million rabbits that are destroying Australia's vegetation.

their path. Attempts to dig up or burn kudzu are futile. Grazing goats and herbicides help, but goats eat other plants, too, and herbicides contaminate water supplies. Kudzu could reach the Great Lakes by the year 2040.

On the bright side, a Japanese firm is constructing a kudzu farm and processing plant in Alabama. Asians use a starch extract from kudzu in drinks, herbal medicines, and candy. The idea is to export the starch to Asia, where the demand currently exceeds the supply. Also, kudzu may eventually help reduce logging operations. At the Georgia Institute of Technology, researchers report that kudzu might be one alternative source of paper.

THE RABBITS THAT ATE AUSTRALIA During the 1800s, British settlers in Australia just couldn't bond with the koalas and kangaroos, so they started to import familiar animals from their homeland. In 1859, in what would be the start of a wholesale disaster, a northern Australia landowner imported and then released two dozen wild European rabbits (*Oryctolagus cuniculus*). Good food and good sport hunting, that was the idea. An ideal rabbit habitat with no natural predators—that was the reality.

Six years later, the landowner had killed 20,000 rabbits and was besieged by 20,000 more. The rabbits displaced livestock, even kangaroos. Now Australia has 200 to 300 million hippityhopping through the southern half of the country. They overgraze perennial grasses in good times and strip bark from shrubs and trees during droughts. You know where they've been; they transform grasslands and shrublands into eroded deserts (Figure 42.15). They have

been shot and poisoned. Their warrens have been plowed under, fumigated, and dynamited. Even when all-out assaults reduced their population size by 70 percent, the rapidly reproducing imports made a comeback in less than a year. Did the construction of a 2,000-mile-long fence protect western Australia? No. Rabbits made it to the other side before workers finished the fence.

In 1951, government researchers introduced myxoma virus by way of mildly infected South American rabbits, its normal hosts. This virus causes *myxomatosis*. The disease has mild effects on South American rabbits that coevolved with the virus but nearly always had lethal effects on *O. cuniculus*. Biting insects, mainly mosquitoes and fleas, quickly transmit the virus from host to host. Having no coevolved defenses against the novel virus, the European rabbits died in droves. But, as you might expect, natural selection has since favored rapid growth of populations of *O. cuniculus* resistant to the virus.

In 1991, on an uninhabited island in Spencer Gulf, Australian researchers released a population of rabbits that they had injected with a calicivirus. The rabbits died quickly and relatively painlessly from blood clots in their lungs, heart, and kidneys. In 1995, the test virus escaped from the island, possibly on insect vectors. It has been killing 80 to 95 percent of the adult rabbits in Australian regions. At this writing, researchers are now questioning whether the calicivirus should be used on a widespread scale, whether it can jump boundaries and infect animals other than rabbits (such as humans), and what the long-term consequences will be.

Table 42.2 *Detrimental Effects of Some Species Introduced Into the United States*			
Species Introduced	**Origin**	**Mode of Introduction**	**Outcome**
Water hyacinth	South America	Intentionally introduced (1884)	Clogged waterways; other plants shaded out
Dutch elm disease: *Ophiostoma ulmi* (fungus) Bark beetle (vector)	Europe	Accidental; on infected elm timber (1930) Accidental; on unbarked elm timber (1909)	Millions of mature elms destroyed
Chestnut blight fungus	Asia	Accidental; on nursery plants (1900)	Nearly all eastern American chestnuts killed
Argentine fire ant	Argentina	In coffee shipments from Brazil? (1891)	Crop damage; native ant communities; destroyed; ground-nesting birds killed
Japanese beetle	Japan	Accidental; on irises or azaleas (1911)	Over 250 plant species (e.g., citrus) defoliated
Sea lamprey	North Atlantic	Ship hulls, through canals (1860s, 1921)	Trout, whitefish destroyed in Great Lakes
European starling	Europe	Intentional release, New York City (1890)	Native songbirds displaced; crop damage; swine disease vector; huge flocks noisy, messy
House sparrow	England	Released intentionally (1853)	Crop damage; native songbirds displaced; transmission of some diseases

PATTERNS OF BIODIVERSITY

As you might well conclude from Chapter 25 and the preceding discussions, biodiversity differs greatly from one habitat to another. Also, as ecologists have found through many field investigations, biodiversity differs in significant ways between geographic regions.

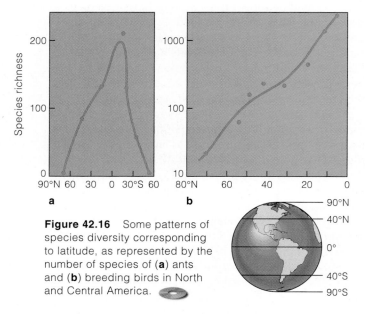

Figure 42.16 Some patterns of species diversity corresponding to latitude, as represented by the number of species of (**a**) ants and (**b**) breeding birds in North and Central America.

● MOSSES
● VASCULAR PLANTS

What Causes Mainland and Marine Patterns?

The most striking pattern of biodiversity corresponds to distance from the equator. For most groups of plants and animals, *the number of coexisting species on land and in the seas is greatest in the tropics, and it systematically declines from the equator to the poles*. Figure 42.16 shows two examples of this biodiversity pattern. What factors underlie it? Three are paramount.

First, as described in Chapter 44, tropical latitudes intercept more sunlight of consistently higher intensity, rainfall is heavier, and the growing season is longer. As one outcome, *resource availability tends to be greater and more reliable in the tropics than elsewhere*. All year long, different tree species in humid tropical forests put out new leaves, flowers, and fruit. Year in and year out, they support diverse herbivores, nectar foragers, and fruit eaters. Such specialization can't evolve to a comparable degree in temperate and arctic regions.

Second, *species diversity might be self-reinforcing*. The diversity of tree species in tropical forests is far greater than in comparable forests at higher latitudes. When more plant species compete and coexist, more herbivore species also evolve and coexist partly because no single herbivore can overcome all the chemical defenses of all the different plants. Typically, then, more predators and parasites evolve in response to a diversity of prey and hosts. The same applies to diversity on tropical reefs.

Third, *over geologic time, speciation rates in the tropics exceeded those of extinction rates*. Declines in biodiversity occurred, but mainly during mass extinctions. Global temperatures drop after such episodes. Species adapted to cool climates tend to survive, but those adapted to the temperatures of the tropics do not. Also, millions of species in tropical forests may disappear in the next decade, a topic addressed in Chapters 25 and 45.

What Causes Island Patterns?

Some islands are laboratories for biodiversity studies. For instance, in 1965, a volcanic eruption formed a new island southwest of Iceland. Within six months, bacteria, fungi, seeds, flies, and some seabirds were established on it. One vascular plant appeared after two years, and a moss two years after that (Figure 42.17). When soils improved, the number of plant species increased. All the colonists originated in Iceland. None originated on the island, which was named Surtsey. Yet the number

Figure 42.17 Aerial photograph of Surtsey, a volcanic island, at the time of its formation in the sea. Newly formed, isolated islands are natural laboratories for ecologists. The chart gives the number of colonizing species of mosses and vascular plants recorded on Surtsey between 1965 and 1973.

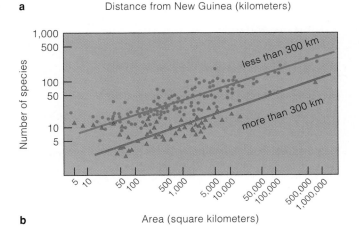

a Distance from New Guinea (kilometers)

Number of species (percent of sample studied)

b Area (square kilometers)

Number of species

less than 300 km

more than 300 km

c

Figure 42.18 Examples of island biodiversity, as correlated with (1) distance from a source of colonizing species and (2) the island's area.

(**a**) Distance effect. Biodiversity on islands of a specified size declines with increasing distance from the source of colonizing species. Each graph point denotes the number of species of land birds occupying lowland areas of South Pacific islands. The large island of New Guinea is the source of colonists for the smaller islands, each of which is at least 500 kilometers away. To correct for differences in island size (area), the number of species of birds on each of the distant islands is expressed as a percentage of the number of bird species on an island of equivalent size close to New Guinea.

(**b**) Area effect. Among islands similarly distant from their source of colonizing species, the larger ones support larger numbers of species. Graph points denote the number of bird species on subtropical and tropical islands. *Solid green* circles and the upper line are for islands less than 300 kilometers from the source of colonists. *Orange* triangles and the lower line denote islands more than 300 kilometers from the source areas. As you can see, this graph incorporates data on the distance effect as well as on the area effect.

(**c**) Wandering albatross, one travel agent for jump dispersals. Seabirds that island-hop long distances often have a few seeds stuck to their feathers. Seeds that successfully germinate in a new island community may give rise to a population of new immigrants.

of new species will not increase indefinitely. Why not? Studies of community patterns on islands around the world provide us with two models.

First, islands far from a source of potential colonists receive few colonizing species, and the few that arrive are adapted for long-distance dispersal (Figure 42.18). This is the **distance effect**. Second, larger islands tend to support more species than smaller islands at equivalent distances from source areas. This is the **area effect**. The larger islands tend to have more and varied habitats; most have more complex topography and extend higher above sea level. Thus they favor species diversity. Also, being bigger targets, they intercept more colonists.

Most importantly, extinctions suppress biodiversity on small islands. There, populations are small and far more vulnerable to storms, volcanic eruptions, disease, and random shifts in birth and death rates. As for any island, the number of species reflects a balance between

immigration rates for new species and extinction rates for established ones. Small islands that are distant from a source of colonists have low immigration rates and high extinction rates, so they support few species once the balance has been struck for their populations.

For most groups of organisms, the number of coexisting species is highest in the tropics and systematically declines from the equator to the poles.

A region's biodiversity depends on many factors, such as its climate, topographical variation, possibilities for dispersal, and evolutionary history—including extinctions and the time span available for speciation.

Disturbances to habitats tend to work against competitive exclusion and therefore to favor increases in biodiversity.

Island biodiversity is a balance between immigration rates for new species and extinction rates for established species.

SUMMARY *Gold* indicates text section

1. A habitat is the type of place where individuals of a given species normally live—that is, their "address." A community consists of all populations of all the species that occupy the habitat and that directly or indirectly associate with one another. *42.1*

2. Each species has its own niche, or "profession," in the community. This is defined as the sum of activities and relationships in which its members engage as they secure and use the resources they require to survive and reproduce. *42.1*

3. Mutualism, commensalism, competition, predation, and parasitism are species interactions that directly or indirectly link populations in a community. *42.1–41.4*

4. Two species requiring the same limited resource tend to compete, as by exploiting it as fast or efficiently as possible or interfering with the other's use of it. *42.3*

    a. According to the competitive exclusion concept, when two (or more) species require identical resources, they cannot coexist indefinitely.

    b. Species are more likely to coexist if they differ in their use of resources. They may also coexist by using a shared resource in different ways or at different times.

5. Some predator and prey populations may coexist at stable levels if predators keep prey from overshooting the carrying capacity. Others show recurring or erratic cycles of abundance and crashes. Delays in predator responses to changes in prey density, and changes in the prey's food supplies, influence the cycles. *42.4*

6. Predators and their prey coevolve. After a novel, heritable trait appears and gives prey an edge in their coevolutionary contest, selection pressure works on the predator populations, which may evolve in response. The same occurs when a novel, heritable trait arises in a predator population. *42.4, 42.5*

    a. Evolved defenses by prey include threat displays, chemical weapons, mimicry, and camouflaging.

    b. Predators overcome defenses of prey by adaptive behavior (including stealth), camouflaging, and so on.

7. Parasites and hosts coevolve; the resistant hosts and only moderately harmful parasites are favored. *42.4*

8. By the classical model for ecological succession, a community develops in predictable sequence, from its pioneer species to an end array of species that persists over an entire region. *42.6*

9. A stable, self-perpetuating array of species that are in equilibrium with one another and with a particular environment is called a climax community. *42.6*

10. Similar climax stages may persist within the same region yet show variation as a result of environmental gradients and species interactions. *42.6*

11. Recurring, small-scale changes are an aspect of the internal dynamics of communities. Other changes, such as long-term shifts in climate and species introductions, may permanently alter community structure. *42.7, 42.8*

    a. Community structure reflects an uneasy balance of forces, including predation (as by keystone species) and competition, that have been operating over time.

    b. Community structure can change permanently by the introduction of species that one way or another expanded their geographic range. Dispersals may occur when a population expands its home range gradually. Jump dispersals rapidly put individuals into distant habitats. Extremely slow dispersals also have occurred over geologic time as a result of plate tectonics.

12. The number of species in a community depends on the size of a region, the colonization rate, disturbances, and extinction rates. It depends also on the levels and patterns of resource availability. *42.9*

13. Biodiversity tends to be highest in the tropics and to decline systematically toward polar regions, except during times of mass extinction. *42.9*

Review Questions

1. Distinguish between the habitat and the niche of a species. Why is it difficult to define "the human habitat"? *42.1*

2. Describe competitive exclusion. How might two species that compete for the same resource coexist? *42.3*

3. Define the difference between a predator and a parasite. *42.4*

4. Define primary and secondary succession. *42.6*

5. What is a climax community? How does the climax-pattern model help explain its structure? *42.6*

Self-Quiz ANSWERS IN APPENDIX III

1. A habitat _____ .
    a. has distinguishing physical and chemical features
    b. is where individuals of a species normally live
    c. is occupied by various species
    d. both a and b
    e. a through c

2. A two-way flow of benefits in mutualistic interactions between species is an outcome of _____ .
    a. close cooperativeness   c. resource partitioning
    b. two-way exploitation   d. competitive coexistence

3. A niche _____ .
    a. is the sum of activities and relationships in a community by which individuals of a species secure and use resources
    b. is unvarying for a given species
    c. shifts in large and small ways
    d. both a and b
    e. both a and c

4. Two species in one habitat can coexist when they _____ .
    a. differ in their use of resources
    b. share the same resource in different ways
    c. use the same resource at different times
    d. all of the above

Figure 42.19 (a) Flesh fly and (b) weevil (*Zygops*).

5. A predator population and prey population _____ .
   a. always coexist at relatively stable levels
   b. may undergo cyclic or irregular changes in density
   c. cannot coexist indefinitely in the same habitat
   d. both b and c

6. Parasites _____ .
   a. tend to kill their hosts   c. feed on host tissues
   b. can kill novel hosts   d. both b and c

7. In _____ , a disturbed site in a community recovers and moves again toward the climax state.
   a. the area effect   c. primary succession
   b. the distance effect   d. secondary succession

8. The biodiversity of a given region is an outcome of _____ .
   a. climate and topography   d. both a and b
   b. possibilities for dispersal   e. a through c
   c. evolutionary history

9. Match the terms with the most suitable descriptions.
   _____ geographic   a. opportunistic colonizer of barren
         dispersal   or disturbed places
   _____ area effect   b. dominates community structure
   _____ pioneer   c. individuals leave home range, become
         species   successfully established elsewhere
   _____ climax   d. more biodiversity on large islands than
         community   small at same distance from source
   _____ keystone   e. stable, self-perpetuating array
         species   of species

Critical Thinking

1. Think of possible examples of competitive exclusion besides the ones used in this chapter, as by considering some of the animals and plants living in your own neighborhood.

2. Flesh flies (Figure 42.19*a*) have a gray and black body, red eyes, and red tail ends. They are fast fliers, and bird predators soon give up trying to catch them. Many sluggish insects, such as the weevil *Zygops rufitorquis*, resemble flesh flies (Figure 42.19*b*). This appears to be a case of _____ .

3. Frog populations all over the world declined dramatically during the 1990s. Individual frogs were being born with bizarre deformities, such as too few or too many legs (Sections 24.6 and 32.8). Should we view what is happening to frogs as a warning of drastically changing environments that will affect us, also? Are the same declines and deformities occurring in all aquatic habitats, or are there different problems in different regions? Use the library or a computer search engine to find information. (Try these key words: *frog deformities*.) Then write a brief list of possible answers to the preceding questions.

4. The water hyacinth (*Eichhornia crassipes*) is an aquatic plant native to South America. In the 1880s, someone fancied the plant's blue flowers and displayed it at an exposition in New Orleans. Other flower fanciers took home clippings and put them in ponds and streams. Unchecked by natural predators, the fast-growing hyacinths spread through nutrient-rich waters, displaced many native species, and choked rivers and canals

Figure 42.20 Water hyacinths choking a waterway in Florida.

(Figure 42.20). They have spread as far west as San Francisco. Do some research to see whether the United States now limits imports and, if so, how effective the restrictions have been.

5. Somewhere between predators and parasites are *parasitoids*, insect larvae that always kill what they eat. Peter Price studied the evolutionary effects of a parasitoid wasp that lays eggs on sawfly cocoons. Sawflies lay eggs in trees. Fertilized eggs give rise to fly larvae, which eat, grow, then drop to the forest floor. The larvae spin cocoons after burrowing into leaf litter. Some burrow deeper than others. The first adult sawflies emerge from cocoons that were the least deeply buried. Later in the season, more sawflies emerge from the more deeply buried ones.

These parasitoid wasps tend to lay eggs on the cocoons closest to the surface of leaf litter. They exert strong selection pressure on the sawfly population, for deep-burrowing sawfly individuals are more likely to escape detection. There are only so many flies near the surface, so the wasp able to locate cocoons deeper in the litter will be better at getting food for her larvae. Explain how the host species stays ahead in this coevolutionary contest.

Selected Key Terms

area effect (on island biodiversity) *42.9*
camouflage *42.5*
carrying capacity *42.4*
climax-pattern model *42.6*
cimax community *42.6*
coevolution *42.4*
commensalism *42.1*
community *42.1*
competitive exclusion *42.3*
distance effect (on island biodiversity) *42.9*
ecological succession *42.6*
endangered species *42.8*
exotic species *42.8*
geographic dispersal *42.7*
habitat *42.1*
host *42.4*
interspecific competition *42.1*

keystone species *42.3*
macroparasite *42.4*
microparasite *42.4*
mimicry *42.5*
mutualism *42.1*
niche *42.1*
parasite *42.4*
parasitism *42.1*
pioneer species *42.6*
predation *42.1*
predator *42.4*
prey *42.4*
primary succession *42.6*
resource partitioning *42.3*
secondary succession *42.6*
social parasite *42.4*
symbiosis *42.1*
warning coloration *42.5*

Readings

Krebs, C., et al. January 2000. "What Drives the 10-Year Cycle of Snowshoe Hares?" *Bioscience* 51(1): 25–35.

Smith, R. 1996. *Elements of Ecology.* Fifth edition. New York: HarperCollins.

On-Line readings at Student Guide for InfoTrac:
www.brookscole.com/biology

43

ECOSYSTEMS

Crêpes for Breakfast, Pancake Ice for Dessert

Think of Antarctica and you think of ice. Mile-thick slabs of the stuff hide all but a small fraction of a vast continent whipped by fierce winds and kept frozen by murderously low temperatures, on the order of –100°F. Yet, on patches of exposed rocky soil and on nearby islands, bacteria, lichens, and mosses endure. During the breeding season, a variety of penguins and seals form great noisy congregations, then reproduce and raise offspring. They cruise offshore or venture out in the open ocean in pursuit of krill, fishes, and squids.

The first explorers of Antarctica called it "The Last Place on Earth." For all of its harshness, however, the beauty of this ecosystem fired the imagination of those first travelers and others who followed them.

In 1961, thirty-eight nations signed a treaty to set aside Antarctica as a reserve for scientific research. This was the start of scientific outposts—and of the trashing of Antarctica. At first, researchers discarded a few oil drums and old tires. Then prefabricated villages went up. Onto the ice and into the water went garbage, used equipment, sewage, and chemical wastes.

Antarctica even became a destination of cruise ships (Figure 43.1a). Every summer thousands of tourists, fortified by three sumptuous meals a day plus snacks, are ferried from ship to land across channels glistening with pancake ice. They trample the sparse vegetation and bob around the penguins, cameras clicking.

We do not know what tourism's long-term impact will be, but it pales when compared to other assaults. Nations looking for new sources of food started licking their chops over Antarctica's krill. Word got out about potentially rich deposits of uranium, oil, and gold. By

Figure 43.1 (**a**) A boatload of tourists crossing a channel of "pancake ice" off the coast of Antarctica. (**b**) On the shore of an island near Antarctica, tourists get close to nature—maybe too close.

1988, treaty nations were poised to authorize digs and drillings. But oil spills or commercial harvesting of krill could easily destroy the fragile ecosystem. For instance, tiny, shrimplike krill are the only food source for Adélie penguins and a key food for baleen whales and other marine animals which are, in turn, food for still others.

In 1991, the treaty nations thought about all of this and imposed a fifty-year ban on mineral exploration. Research stations started to bury or incinerate wastes, treat raw sewage, and take other pollution-curbing steps. Tour operators promised to supervise tourists. Krill are not being harvested on a massive scale. Yet.

Because Antarctica seems remote from the rest of the world, it is easier to see how species interconnect there. We can ask whether harm to one species or one habitat will lead to collapse of the whole and be fairly sure of the answer. What about places not as sharply defined? Are they as vulnerable to disturbance or more resilient? We won't know for sure until researchers gain deeper insights into the evolutionary and ecological histories of their species, including the suspected great numbers of species we haven't even discovered yet.

In the meantime, biodiversity is rapidly declining in ecosystems on land and in the seas. Competition from exotic species is one threat; the human propensity to overexploit species and destroy habitats is another. *Conservation biology* is an attempt to counter the threat by working to lower current extinction rates and design management programs that will help sustain the most vulnerable ecosystems. Chapter 25 introduced the challenges. To gain more insight into the magnitude of the task, start with a few principles of how ecosystems work. They are the topic of this chapter.

b

Key Concepts

1. An ecosystem is an association of organisms and their physical environment, interconnected by an ongoing flow of energy and a cycling of materials through it.

2. An ecosystem is an open system, with inputs, internal transfers, and outputs of energy and nutrients.

3. Energy flows in only one direction through every ecosystem. Most commonly, energy flow begins when photosynthetic autotrophs harness sunlight energy and convert it to forms that they and other organisms of the ecosystem can use. Autotrophs are primary producer organisms for the ecosystem.

4. Energy-rich organic compounds that the primary producers synthesize become incorporated in their body tissues. They are stored forms of energy, and they serve as the foundation for the ecosystem's food webs. Such webs consist of a number of interconnected food chains.

5. Each chain in a food web extends in a straight-line sequence from producers through various consumers, decomposers, and detritivores, to the highest consumer level.

6. Over time, most of the energy that entered a food web is lost to the environment, mainly in the form of metabolic heat. Nutrients are cycled within food webs, although some amounts are lost to the environment.

7. The water, carbon, nitrogen, phosphorus, and other substances required for primary productivity move through biogeochemical cycles that are global in scale. Ions or molecules of the substances move slowly from environmental reservoirs, then rapidly among organisms of food webs, then back to their reservoirs.

8. Human activities are disrupting the natural cycles of materials and are thereby endangering ecosystems.

THE NATURE OF ECOSYSTEMS

Overview of the Participants

Diverse natural systems abound on the Earth's surface. In climate, landforms, soil, vegetation, animal life, and other features, deserts differ from hardwood forests, which differ from tundra and prairies. In biodiversity and physical properties, the open seas differ from reefs, which differ from lakes. *Yet despite the differences, such systems are alike in many aspects of structure and function.*

With few exceptions, these systems run on sunlight captured by autotrophs (self-feeders). Plants and other photosynthesizers are the most common autotrophs. They convert sunlight energy to chemical energy and use it to produce organic compounds from inorganic raw materials. Because they secure energy directly from their environment, they are **primary producers** for the entire system (Figure 43.2).

All other organisms in the system are heterotrophs, not self-feeders. They extract energy from compounds that primary producers put together. **Consumers** feed on tissues of other organisms. Those consumers called *herbivores* eat only plants. By converting energy stored in plant tissues into animal tissues, they become food for *carnivores* and *parasites*, which feed on living animal tissues. Nonliving products and remains of producers and consumers feed **decomposers**. These heterotrophic fungi and bacteria practice extracellular digestion and absorption: they secrete enzymes that digest organic compounds in the surroundings, and their cells absorb some breakdown products. In this respect, the decomposers are distinct from detritivores. **Detritivores** are heterotrophs that ingest decomposing bits of organic matter, such as leaf litter. Earthworms are examples.

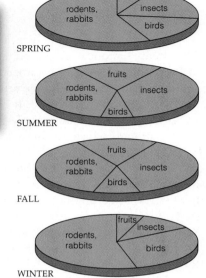

Figure 43.3 Red fox, an omnivore. Its diet shifts with seasonal changes in available food. In summer, it feeds mainly on fruits and insects. In spring and fall, it focuses mainly on rodents and rabbits.

Not all consumers fall into simple categories. Many are *omnivores*, which dine on assorted animals, plants, fungi, protistans, even bacteria. The red fox is like this (Figure 43.3). Also, a fox that comes across a dead bird may scavenge it. A *scavenger* is any animal that ingests nonliving plant and animal tissues all or some of the time. Vultures, termites, and many beetles do this.

What are the nutrients for the system? The primary producers take up small inorganic compounds—water and carbon dioxide—from the environment as sources for hydrogen, oxygen, and carbon. They also take up phosphorus, nitrogen, and other minerals. These are the raw materials for carbohydrates, lipids, proteins, and nucleic acids. Nutrients return to the environment as decomposers and detritivores work over the remains of organisms. And unless something removes nutrients from the system, as when minerals end up in a stream that flows away from a meadow, autotrophs typically take them up again.

What we have just described in broad outline is an ecosystem. An **ecosystem** is an array of organisms and their physical environment, interacting through a one-way flow of energy and a cycling of materials. It is an open system, unable to sustain itself. It runs on *energy inputs* (as from the sun) and often *nutrient inputs* (as from a creek delivering dissolved minerals into a lake). It has *energy outputs* and *nutrient outputs*. The energy cannot be recycled. Over time, most of the energy that autotrophs fixed is lost to the environment, mainly as metabolically generated heat. Nutrients are cycled, but some slip out. Most of this chapter explores ecosystem inputs, internal transfers, and outputs.

Figure 43.2
Simple model of ecosystems. Energy flows in one direction: into the ecosystem, through its living organisms, and then out from it. Its nutrients are cycled among autotrophs and heterotrophs. For this model, energy flow starts with autotrophs that can capture energy from the sun.

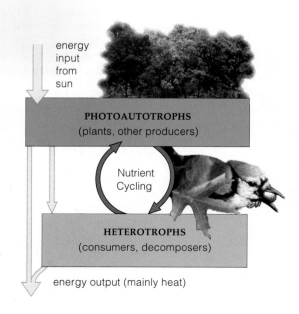

energy input from sun

PHOTOAUTOTROPHS
(plants, other producers)

Nutrient Cycling

HETEROTROPHS
(consumers, decomposers)

energy output (mainly heat)

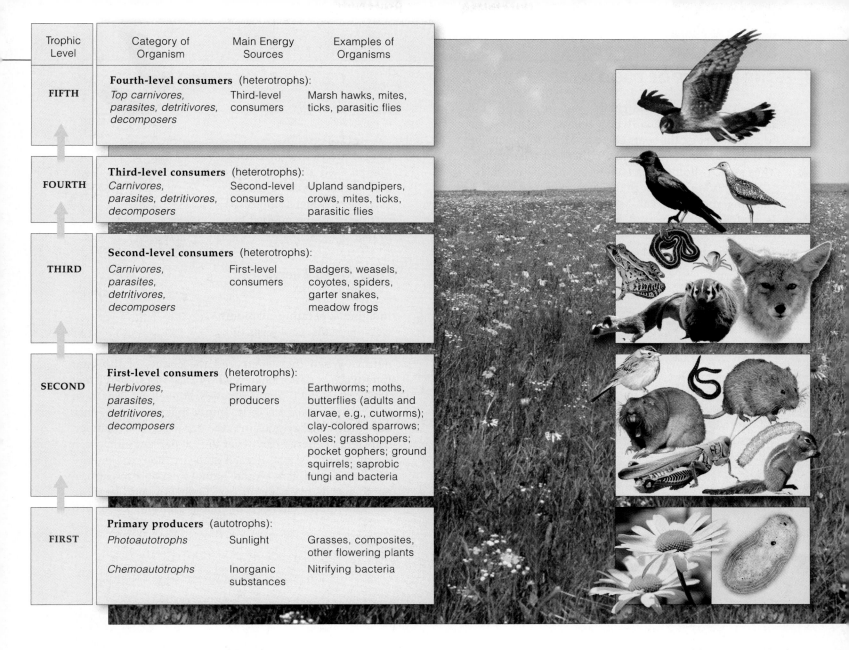

Trophic Level	Category of Organism	Main Energy Sources	Examples of Organisms
FIFTH	**Fourth-level consumers** (heterotrophs): *Top carnivores, parasites, detritivores, decomposers*	Third-level consumers	Marsh hawks, mites, ticks, parasitic flies
FOURTH	**Third-level consumers** (heterotrophs): *Carnivores, parasites, detritivores, decomposers*	Second-level consumers	Upland sandpipers, crows, mites, ticks, parasitic flies
THIRD	**Second-level consumers** (heterotrophs): *Carnivores, parasites, detritivores, decomposers*	First-level consumers	Badgers, weasels, coyotes, spiders, garter snakes, meadow frogs
SECOND	**First-level consumers** (heterotrophs): *Herbivores, parasites, detritivores, decomposers*	Primary producers	Earthworms; moths, butterflies (adults and larvae, e.g., cutworms); clay-colored sparrows; voles; grasshoppers; pocket gophers; ground squirrels; saprobic fungi and bacteria
FIRST	**Primary producers** (autotrophs): *Photoautotrophs*	Sunlight	Grasses, composites, other flowering plants
	Chemoautotrophs	Inorganic substances	Nitrifying bacteria

Structure of Ecosystems

We can classify all organisms of an ecosystem by their functional roles in a hierarchy of feeding relationships called **trophic levels** (*troph*, nourishment). "Who eats whom?" we may ask. When organism **B** eats organism **A**, energy gets transferred to **B** from **A**. All organisms at a given trophic level are the same number of transfer steps away from some energy input into an ecosystem.

Figure 43.4 shows organisms of a prairie ecosystem. Being closest to the energy input (sunlight), plants and other producers are at the first trophic level. They feed primary consumers, such as grasshoppers (herbivores) and earthworms (detritivores) at the next trophic level. Primary consumers feed carnivores and parasites at the third trophic level, and so on up the hierarchy.

At each trophic level, organisms interact with the same sets of predators, prey, or both. The omnivores feed at several levels, so they are partitioned among different levels or assigned one of their own.

Figure 43.4 Simplified picture of trophic levels in tallgrass prairie, a type of grassland in parts of the American Midwest.

A **food chain** is a straight-line sequence of steps by which energy stored in autotroph tissues passes on to higher trophic levels. We have a hard time finding such simple, isolated chains in nature. Why? Species usually belong to more than one food chain, especially species at a low trophic level. It is more accurate to visualize food chains *cross-connecting* with one another, as **food webs**. And that is the topic of the next section.

Producers, consumers, decomposers, detritivores, and their physical environment make up an ecosystem. A one-way energy flow and a cycling of materials interconnects them.

A food chain, a straight-line sequence of who eats whom in an ecosystem, starts with autotrophs. A food web is a network of cross-connected food chains.

43.2

THE NATURE OF FOOD WEBS

Make a sketch of a food chain—say, from a plant in a tallgrass prairie, to a cutworm that eats its juicy parts, to a baby snake that eats the cutworm, to a sandpiper that eats the snake, and a hawk that eats the sandpiper. Your sketch should look something like Figure 43.5.

Identifying a food chain is a simple way to start thinking about who eats whom in ecosystems, but this is just part of the picture. Most often, a complex array of species competes for food, particularly at the lower trophic levels. For instance, as the food web in Figure 43.6 shows, tallgrass

Figure 43.5 *Right:* Example of a simple food chain.

prairie producers (flowering plants) feed diverse insects and herbivorous mammals. Yet it is a hugely simplified picture of all the interactions in this ecosystem!

How Many Energy Transfers?

Imagine separating a spider web's threads. Ecologists did this for many food webs in nature. When they compared the chains of different food webs, a pattern emerged: *In most cases, energy that producers initially captured passes through no more than four or five trophic levels.* It simply makes no difference how much energy goes into the ecosystem. Remember, transfers of energy can never be 100 percent efficient; a bit of energy is lost at each step. In time, the energy that some organism would have to expend to capture another at an even higher trophic level would be more the amount of energy obtainable from it. Even rich

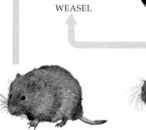

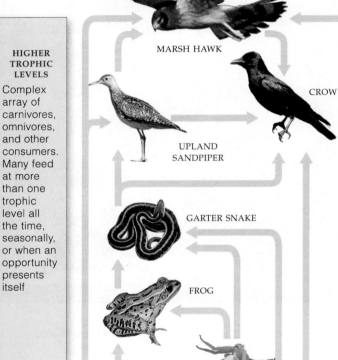

MARSH HAWK

UPLAND SANDPIPER

GARTER SNAKE

CUTWORM

PLANTS

CROW

GARTER SNAKE

FROG

SPIDER

WEASEL

BADGER

COYOTE

CLAY-COLORED SPARROW

EARTHWORMS, INSECTS (E.G., GRASSHOPPERS, CUTWORMS)

PRAIRIE VOLE

POCKET GOPHER

GROUND SQUIRREL

HIGHER TROPHIC LEVELS

Complex array of carnivores, omnivores, and other consumers. Many feed at more than one trophic level all the time, seasonally, or when an opportunity presents itself

SECOND TROPHIC LEVEL

Primary consumers (such as herbivores)

FIRST TROPHIC LEVEL

Primary producers

GRASSES, COMPOSITES

Figure 43.6 Small, highly simplified sampling of connecting interactions in a tallgrass prairie food web.

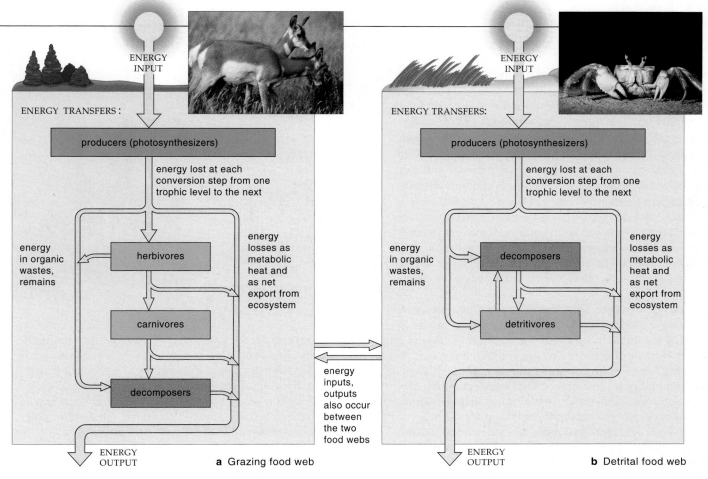

Figure 43.7 One-way flow of energy through (**a**) a grazing food web and (**b**) a detrital food web.

ecosystems that support complex food webs, including the one in Figure 43.6, don't have lengthy food chains.

Field studies and computer simulations of the food webs of marine, freshwater, and terrestrial ecosystems reveal more patterns. For instance, chains of food webs tend to be short where environmental conditions vary. Shifts in temperature, moisture, salinity, and pH are such variables. Chains are longer in stable environments, as in parts of the deep ocean. The most complex webs have the most herbivorous species, but their chains are shortest. You find such webs in grasslands. In contrast, the simplest food webs have more top carnivores.

Two Categories of Food Webs

In what direction does energy flow through ecosystems on land? Plants fix only a small part of the energy from the sun. They store half the energy in chemical bonds of new tissues but lose the rest as metabolic heat. Other organisms draw upon the energy that became stored in plant tissues, remains, and wastes. Then they, too, lose metabolic heat. Taken together, *all of these heat losses represent a one-way flow of energy out of the ecosystem.*

Energy from a primary source flows in one direction through two kinds of webs. In a **grazing food web**, it goes from photoautotrophs to herbivores, then through carnivores. By contrast, in a **detrital food web**, energy flows from photoautotrophs through detritivores and decomposers (Figure 43.7). Both webs cross-connect in nearly all ecosystems. For example, a connection forms when an opportunistic herring gull of a grazing food web gulps down a crab of a detrital food web.

The amount of energy moving through food webs differs from one ecosystem to the next and often varies with the seasons. In most cases, however, most of the net primary production moves through detrital food webs. You may doubt this. After all, when cattle graze heavily on pasture plants, about half of the net primary production enters a grazing food web. But cattle do not use all of the stored energy. A lot of undigested plant parts and feces become available for the decomposers and detritivores. Marshes are a similar case in point. In these aquatic ecosystems, most of the stored energy is not used until after parts of the marsh grasses die and become available for detrital food webs.

The loss of energy at each transfer in a food chain limits the number of trophic levels in ecosystems to four or five.

Tissues of living photosynthesizers are the basis of grazing food webs. Remains and wastes of photosynthesizers and consumers are the basis of detrital food webs.

Biological Magnification in Food Webs

ECOSYSTEM MODELING We turn now to a premise that opened this chapter—that disturbances to one part of an ecosystem often can have unexpected effects on other, seemingly unrelated parts. One approach to predicting unforeseen effects of disturbances is through **ecosystem modeling**. With this method, researchers identify crucial bits of data on different ecosystem components, and then use computer programs and models to combine the bits. Results help predict outcomes of the next disturbance.

A potential danger is that investigators may not have identified all of the key relationships in the ecosystem under study and incorporated them accurately into the computer model. The most crucial fact may be one that we do not yet know, as the following example makes clear.

DDT IN FOOD WEBS DDT, recall, is a synthetic organic pesticide (Section 3.7). This fairly stable hydrocarbon is nearly insoluble in water, so you might think that it would exert its effects only where applied. But winds carry DDT in vapor form; water moves fine particles of it. Being highly soluble in fats, DDT can accumulate in tissues. Hence DDT can show **biological magnification**. By this occurrence, a substance that degrades slowly or not at all gets more and more concentrated in tissues of organisms at higher trophic levels of a food web.

Most of the DDT that becomes concentrated in all the organisms that a consumer will eat during its lifetime will end up in the consumer's own tissues. DDT and modified forms of it disrupt metabolic activities and are toxic to many aquatic and terrestrial animals.

Figure 43.8 Peregrine falcon, a top carnivore of certain food webs. It almost became extinct as a result of biological magnification of DDT. A wildlife management program successfully increased its population sizes. Peregrine falcons were reintroduced into wild habitats. They also have adapted to cities. There, these raptors hunt pigeons, large populations of which are a messy nuisance.

Several decades ago, DDT started to infiltrate food webs and act on diverse organisms in ways that no one had predicted. Where it was sprayed to control Dutch elm disease, songbirds died. In small streams flowing in forests where it was sprayed to kill budworm larvae, fish died. In fields sprayed to control one kind of pest, new pests moved in. *DDT was indiscriminately killing the natural predators that keep pest populations in check!*

Then side effects of biological magnification showed up in habitats far removed from where DDT had been applied—*and much later in time*. Most vulnerable were brown pelicans, bald eagles, peregrine falcons, and other top carnivores of some food webs (Figures 43.8 and 43.9).

Why? A DDT breakdown product interferes with physiological processes. As one outcome, bird eggs developed brittle shells; many chick embryos did not even hatch. Some species faced extinction.

In the United States, DDT has been banned since the 1970s except where necessary to protect public health. Many species hit hardest have recovered somewhat. Some birds still lay thin-shelled eggs; they pick up DDT at their winter ranges in Latin America. Even as late as 1990, a fishery near Los Angeles was closed. DDT from industrial waste discharges that had stopped twenty years before was still contaminating that ecosystem.

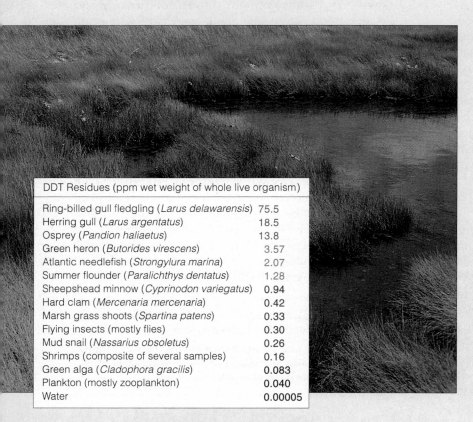

DDT Residues (ppm wet weight of whole live organism)	
Ring-billed gull fledgling (*Larus delawarensis*)	75.5
Herring gull (*Larus argentatus*)	18.5
Osprey (*Pandion haliaetus*)	13.8
Green heron (*Butorides virescens*)	3.57
Atlantic needlefish (*Strongylura marina*)	2.07
Summer flounder (*Paralichthys dentatus*)	1.28
Sheepshead minnow (*Cyprinodon variegatus*)	0.94
Hard clam (*Mercenaria mercenaria*)	0.42
Marsh grass shoots (*Spartina patens*)	0.33
Flying insects (mostly flies)	0.30
Mud snail (*Nassarius obsoletus*)	0.26
Shrimps (composite of several samples)	0.16
Green alga (*Cladophora gracilis*)	0.083
Plankton (mostly zooplankton)	0.040
Water	0.00005

Figure 43.9 Biological magnification in an estuary on the south shore of Long Island, New York, as reported in 1967 by George Woodwell, Charles Wurster, and Peter Isaacson. The researchers knew of broad correlations between the extent of DDT exposure and mortality. For instance, residues in birds known to have died from DDT poisoning were 30–295 ppm, and they were 1–26 ppm in several fish species. Some DDT concentrations measured during this study were below lethal thresholds but were still high enough to interfere with reproductive success.

STUDYING ENERGY FLOW THROUGH ECOSYSTEMS

What Is Primary Productivity?

To get an idea of how energy flow is studied, think of ecosystems on land, which typically have multicelled plants as primary producers. The rate at which primary producers secure and store some amount of energy in their own tissues during a specified interval is the **primary productivity** of the ecosystem. How much energy actually gets stored in their tissues depends on how many individual plants live there, and on the balance between photosynthesis (energy trapped) and aerobic respiration (energy used). The *gross* primary productivity is the total rate of photosynthesis for the whole ecosystem during the specified interval. The *net* amount is the rate of energy storage in plant tissues in excess of the rate of aerobic respiration by the plants.

Other factors impact on the amount of net primary production, its seasonal patterns, and its distribution through a habitat. They do so for marine ecosystems as well as on land (Figure 43.10). For example, how much energy ecosystems trap and store depends partly on the body size and form of primary producers, mineral availability, the temperature range, and the amount of sunlight and rainfall during each growing season. The harsher the conditions, the less new plant growth per season—and the lower the productivity.

What Are Ecological Pyramids?

Ecologists often will represent the trophic structure of an ecosystem in the form of an ecological pyramid. In such pyramids, the primary producers form a base for successive tiers of consumers above them.

A **biomass pyramid** depicts the weight of all of an ecosystem's organisms at each tier. Think about Silver Springs, Florida, a small aquatic ecosystem. A biomass pyramid for it, measured as grams per square meter during a specified interval, might look like this:

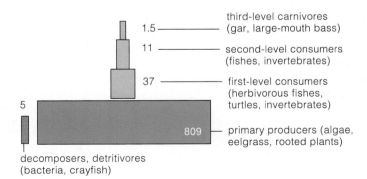

- 1.5 — third-level carnivores (gar, large-mouth bass)
- 11 — second-level consumers (fishes, invertebrates)
- 37 — first-level consumers (herbivorous fishes, turtles, invertebrates)
- 5
- 809 — primary producers (algae, eelgrass, rooted plants)

decomposers, detritivores (bacteria, crayfish)

Such biomass pyramids—mostly primary producers—are common. Top carnivores are few, which may be one reason why sighting them in the wild is so memorable.

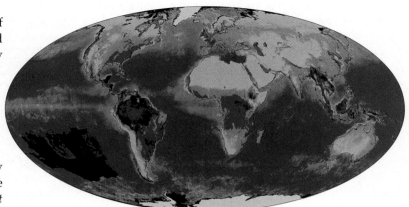

Figure 43.10 Three-year summary of satellite data on global primary productivity. The most highly productive regions, such as rain forests, are coded *dark green*. Deserts (low productivity) are *yellow*. Productivity near the ocean surface waters is *red* (highest) down through *orange, yellow, green,* and then *blue* (lowest). Compare the satellite maps in Section 6.8.

Some biomass pyramids are "upside-down." The smallest tier is on the bottom. A pond or the sea is like this. Its primary producers have less biomass than the consumers feeding on them. How is this possible? They consist of phytoplankton, the members of which grow and reproduce fast enough to support a much greater biomass of zooplankton. Organisms of zooplankton are bigger, grow slower, and consume less energy per unit weight. Phytoplankton fix more energy and produce more biomass than land plants (compare Section 6.8).

An **energy pyramid**, too, shows how usable energy diminishes as it flows through an ecosystem. Sunlight energy enters the pyramid base (first trophic level) and diminishes through successive levels to its tip (the top carnivores). Energy pyramids have a large energy base at the bottom and are always "right-side up." As you will see in the next section, they give a clearer picture of energy flows through successive trophic levels.

Energy flows into food webs of ecosystems from an outside source, usually the sun. Energy leaves ecosystems mainly by losses of metabolic heat, which each organism generates.

Gross primary productivity is an ecosystem's total rate of photosynthesis during a specified interval. The *net* amount is the rate at which primary producers store energy in tissues in excess of their rate of aerobic respiration. Heterotrophic consumption affects the rate of energy storage.

The loss of metabolic heat and the shunting of food energy into organic wastes mean that usable energy flowing through consumer trophic levels declines at each energy transfer.

Energy Flow at Silver Springs

Imagine you are with ecologists who are bent on gathering data to construct an energy pyramid for a small freshwater spring over the course of one year. You observe them as they measure the energy that each type of individual in the spring takes in, loses as metabolic heat, stores in its body tissues, and loses in waste products. You see that they multiply the energy per individual by population size, then they calculate energy inputs and outputs. In such ways, these ecologists are able to express the flow per unit of water (or land) per unit of time.

The energy pyramid shown in Figure 43.11 summarizes the data from a long-term study of a grazing food web in an aquatic ecosystem—Silver Springs, Florida. The larger diagram in Figure 43.12 shows some of the calculations that ecologists used when they constructed the pyramid.

Given the metabolic demands of organisms and the amount of energy lost in their organic wastes, only about 6 to 16 percent of the energy entering one trophic level becomes available for organisms at the next level.

Because the efficiency of the energy transfers is so low, most ecosystems can support no more than four or five consumer trophic levels.

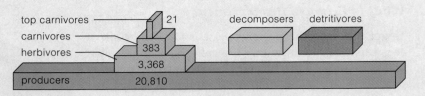

Figure 43.11 Pyramid of energy flow through Silver Springs, Florida, as measured in kilocalories/square meter/year.

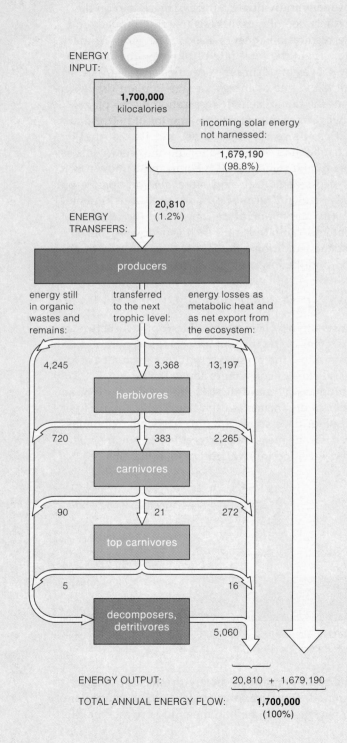

Figure 43.12 Breakdown of the annual energy flow through Silver Springs, Florida, as measured in kilocalories/square meter/year.

The primary producers in this small spring are mostly aquatic plants. The carnivores are insects and small fishes; top carnivores are larger fishes. The initial energy source, sunlight, is available all year long. The spring's detritivores and decomposers cycle organic compounds from the other trophic levels.

The producers trapped 1.2 percent of the incoming solar energy, and only a little more than a third of that amount became fixed in new plant biomass (4,245 + 3,368). The producers used more than 63 percent of the fixed energy for their own metabolism.

About 16 percent of the fixed energy was transferred to herbivores, and most of it was used for metabolism or transferred to detritivores and decomposers.

Of the energy that was transferred to herbivores, only 11.4 percent reached the next trophic level (carnivores). The carnivores used all but about 5.5 percent, which was transferred to top carnivores.

By the end of the specified time interval, all of the 5,060 kilocalories of energy that had been transferred through the system appeared as metabolically generated heat.

Bear in mind, this energy flow diagram is oversimplified, because no community is isolated from others. New individual organisms and substances continually drop from overhead leaves and branches into the springs. Also, organisms and substances are gradually lost by way of a stream that leaves the spring.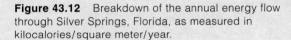

BIOGEOCHEMICAL CYCLES—AN OVERVIEW

Availability of nutrients as well as energy profoundly influences the structure of ecosystems. Photosynthetic producers need carbon, hydrogen, and oxygen, which they obtain from water and air. They require nitrogen, phosphorus, and other mineral ions. A scarcity of even one of these minerals has widespread adverse effects, for it lowers the ecosystem's primary productivity.

In a **biogeochemical cycle**, ions or molecules of a nutrient are transferred from the environment into organisms, then back to the environment, part of which functions as a vast reservoir for them. Transfer rates of nutrients into and from a reservoir are usually lower than the rates of exchange between and among organisms.

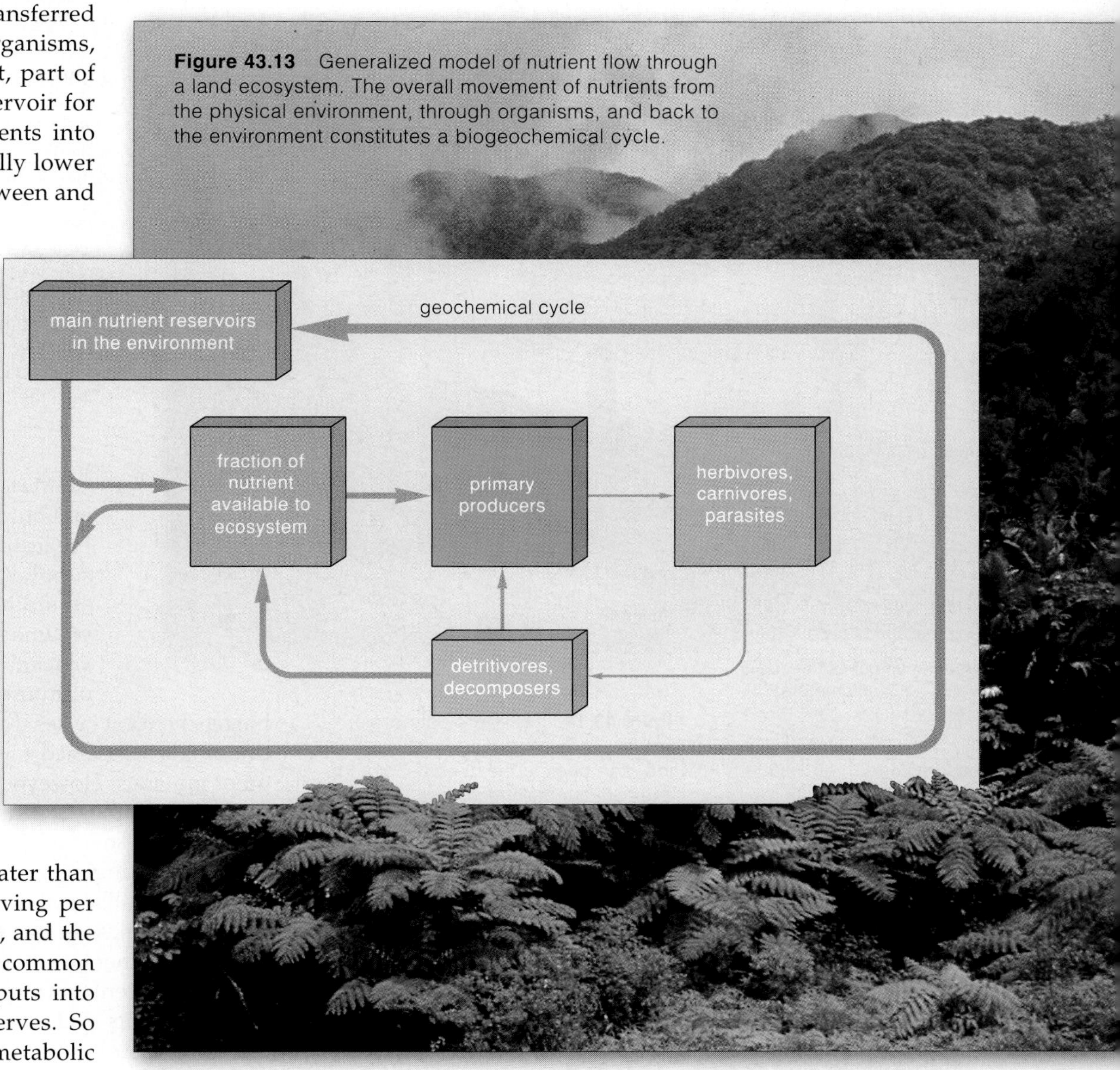

Figure 43.13 is a simple model of the relationship between most ecosystems and the geochemical part of the cycles. The model is based on four factors.

First, mineral elements that producer organisms use as nutrients typically are available in the forms of mineral ions, including ammonium (NH_4^+). *Second*, inputs from the physical environment, along with nutrient cycling activities of all of the decomposers and detritivores, maintain an ecosystem's reserves of nutrients. *Third*, the actual amount of a nutrient that is being cycled on through most major ecosystems is greater than the amount entering and leaving per year. *Fourth*, rainfall, snowfall, and the slow weathering of rocks are common sources of environmental inputs into the ecosystem's nutrient reserves. So are the combined effects of metabolic activities, such as nitrogen fixation.

Ecosystems on land also have typical outputs from the nutrient reserves. The loss of mineral ions from a river valley through ongoing erosion is an example.

There are three categories of biogeochemical cycles, based on the part of the environment that holds the largest portion of the specified ion or molecule. As you will see, in the *hydrologic* cycle, oxygen and hydrogen move in the form of water molecules. This movement is also known as the global water cycle. In *atmospheric*

cycles, a large percentage of the nutrient is in the form of an atmospheric gas. For instance, this is the case for gaseous forms of nitrogen and carbon (mainly carbon dioxide). *Sedimentary* cycles involve phosphorus and other solid nutrients that do not have gaseous forms. Solid nutrients move from land to the seafloor and return to dry land only by way of geological uplifting,

Figure 43.13 Generalized model of nutrient flow through a land ecosystem. The overall movement of nutrients from the physical environment, through organisms, and back to the environment constitutes a biogeochemical cycle.

which may take millions of years. The Earth's crust is the largest storehouse for sedimentary cycles.

Availability of nutrients profoundly influences the primary productivity on which ecosystems depend.

In a biogeochemical cycle, ions or molecules of a nutrient move slowly through the environment, then rapidly among organisms, and back to the environmental reservoir for them.

HYDROLOGIC CYCLE

Driven by ongoing inputs of solar energy, the Earth's waters move slowly from the ocean to the atmosphere, onto land, then back to the ocean—the main reservoir. Water evaporating into the lower atmosphere initially stays aloft as vapor, clouds, and ice crystals. It returns to Earth as precipitation, mainly rain and snow. Ocean currents and great prevailing wind patterns influence the global **hydrologic cycle**, as shown in Figure 43.14.

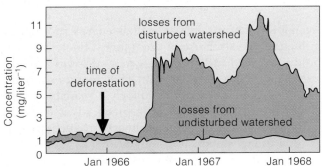

Figure 43.15 *Above:* Experimental results in disturbed forests of Hubbard Brook Valley watersheds. Researchers stripped away all plants but didn't disturb the soil. The arrow marks the time of deforestation. Herbicides applied to the soil for three years prevented regrowth. In each watershed, all water that drained away was measured as it flowed over V-notched concrete catchments. The concentrations of calcium and other minerals were compared against those in water flowing over a control catchment in an undisturbed area. Calcium losses were *six times* greater from the deforested areas.

a

Main Reservoirs	Volume (10^3 cubic kilometers)
Oceans	1,370,000
Polar ice, glaciers	29,000
Groundwater	4,000
Lakes, rivers	230
Soil moisture	67
Atmosphere (water vapor)	14

b

Figure 43.14 (**a**) Hydrologic cycle, from the ocean to the atmosphere, land, and back to the ocean. *Gold* boxes are the main reservoirs. Arrow labels identify processes involved in water movement between reservoirs, as measured in cubic kilometers per year. (**b**) Global water budget.

Measurements of watershed inputs and outputs have practical application. For instance, cities that draw on surface supplies of watersheds adjust usage according to seasonal change in water volume. Measurements also show how vegetation cover affects the movement of nutrients in the ecosystem phase of biogeochemical cycles. Example: You might think water draining a watershed rapidly leaches calcium ions and other minerals. However, in young, undisturbed forests in Hubbard Brook watersheds, each hectare lost only 8 kilograms or so of calcium. Rainfall and weathering of rock were bringing in calcium replacements. And tree roots "mined" the soil, so calcium was being stored in a growing biomass of tree tissues.

In experimental watersheds, deforestation caused a shift in nutrient outputs (Figure 43.15). Calcium and other nutrients cycle so slowly that deforestation may disrupt nutrient availability for entire ecosystems. This is the case for forests that cannot regenerate themselves over the short term. Coniferous forests are like this.

Water is vital for all organisms. It also is a transport medium, moving nutrients into and out of ecosystems. Its role in moving nutrients became clear in long-term watershed studies. A **watershed** is any region in which precipitation is funneled into a single stream or river. It may be any size of interest. For instance, the Mississippi River watershed extends across about one-third of the continental United States. Watersheds in the Hubbard Brook Valley of New Hampshire average 14.6 hectares (36 acres). Most water entering a watershed seeps into soil or becomes part of the surface runoff that enters streams. Plants take up water and dissolved minerals from the soil, then lose it by transpiration.

In the hydrologic cycle, water slowly moves on a global scale from the world ocean (the main reservoir), through the atmosphere, onto land, then back to the ocean.

In ecosystems on land, plants stabilize the soil and absorb dissolved minerals. By doing so, they minimize the loss of soil nutrients in runoff from land.

SEDIMENTARY CYCLES

A Look at the Phosphorus Cycle

We continue our survey of biogeochemical cycling with one of the sedimentary cycles. In the **phosphorus cycle**, phosphorus is passed through food webs as it moves from land, to ocean sediments, then back to land (Figure 43.16). The largest reservoir is the Earth's crust, just as it is for other minerals.

Most phosphorus in rock formations is in the form of phosphates. Weathering and erosion put phosphates into streams and rivers, which move them to sediments in the seas. Phosphorus collects mainly on the submerged continental shelves. Along with other minerals, it forms insoluble deposits. Millions of years go by. Where movements of crustal plates uplift part of the seafloor, phosphates are exposed on drained land surfaces. Over geologic time, weathering releases phosphates from the exposed rocks, so the cycle's geochemical phase begins again.

The cycle's ecosystem phase is brief, compared with the long-term geochemical phase. All organisms need phosphorus for synthesizing phospholipids, NADPH, ATP, nucleic acids, and other compounds. Plants take up dissolved, ionized forms of phosphorus so fast and efficiently that they often reduce its concentrations in soil to extremely low levels. Herbivorous animals get phosphates by dining on plants; carnivores obtain it by dining on the herbivores. Both types of animals excrete phosphates as a waste in urine and feces. Decomposers in soil release phosphates. Plants take up the mineral and conribute to its rapid cycling in the ecosystem.

Eutrophication

Of all minerals, phosphorus is the main limiting factor in natural ecosystems. Soils hold little of it; sediments tie up most of it in aquatic ecosystems. Phosphorus doesn't have a gaseous phase, so ecosystems lose little of their scarce supply to the atmosphere. It helps that phosphorus resists leaching. Producers are adapted to respond efficiently to low phosphorus concentrations.

Phosphorus gets concentrated in eroded sediments, agricultural runoff, and outflows from industries and sewage treatments. Dissolved phosphorus that enters streams, rivers, lakes, and estuaries can promote dense algal blooms. Like plants, photosynthetic algae cannot grow without phosphorus, nitrogen, and other minerals. Freshwater ecosystems usually have plenty of nitrogen-fixing bacteria, so nitrogen is not a limiting factor on growth of algae. Phosphorus is. Decomposition of the remains from algal blooms depletes water of oxygen, which kills off the fish and other organisms.

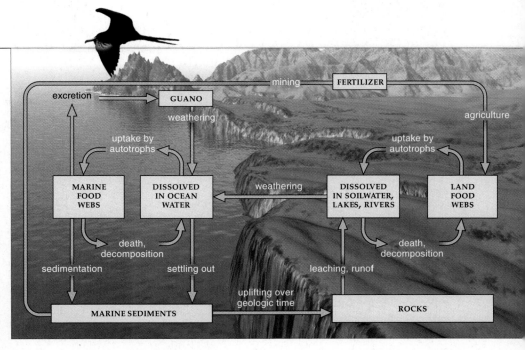

Figure 43.16 The phosphorus cycle. Most of the phosphorus moves in the form of phosphate ions (PO_4^{3-}).

Figure 43.17 Experiment demonstrating lake eutrophication in Ontario, Canada. Researchers stretched a plastic curtain across a channel between two basins of the same lake. They added phosphorus, carbon, and nitrogen to one basin (the basin in the *background*) and carbon and nitrogen to the other (the basin in the *foreground*). Within two months, the phosphorus-enriched basin showed signs of accelerated eutrophication; a dense algal bloom turned the water green.

Eutrophication is the name for nutrient enrichment of an ecosystem that is naturally low in nutrients. As you will be reading in Section 44.10, eutrophication is a natural process, but phosphorus inputs accelerate it. Field experiments, including the one shown in Figure 43.17, nicely demonstrate this outcome.

Sedimentary cycles, in combination with the hydrologic cycle, move phosphorus and most of the other mineral elements through terrestrial and aquatic ecosystems.

CARBON CYCLE

In an essential atmospheric cycle, carbon moves through the lower atmosphere and all food webs on its way to and from the ocean's water and sediments, and rocks. Its global movement is the **carbon cycle** (Figure 43.18). Sediments and rocks hold most of the carbon, followed by the ocean, then soil, the atmosphere, and biomass on land. It enters the atmosphere as cells engage in aerobic respiration, when fossil fuels burn, and when volcanoes erupt and so release the carbon from rocks that make up the Earth's crust. Most atmospheric carbon occurs in the form of carbon dioxide (CO_2). Carbon dissolved in the ocean is mainly in the forms of bicarbonate and carbonate.

When watching bubbles escape from a glass of carbonated soda left in the sun, you might wonder: Why doesn't all the CO_2 dissolved in the warm ocean surface waters escape to the atmosphere? Driven by winds and regional differences in water density, water makes a gigantic loop from the surface of the Pacific and Atlantic oceans to the seafloor of the Atlantic and Antarctic seafloors. There, the dissolved CO_2 moves into deep storage reservoirs before the seawater loops back up (Figure 43.19). This loop of ocean water is a key factor in carbon's distribution and global budget:

MAIN CARBON RESERVOIRS AND HOLDING STATIONS:

	p (10^{15} grams)
Sediments and rocks	
Ocean (dissolved forms)	39,700
Soil	1,500
Atmosphere	750
Biomass on land	715

ANNUAL FLUXES IN GLOBAL DISTRIBUTION OF CARBON:

From atmosphere to plants (carbon fixation)	120
From atmosphere to ocean	107
To atmosphere from ocean	105
To atmosphere from plants	60
To atmosphere from soil	60
To atmosphere from fossil fuel burning	5
To atmosphere from net destruction of plants	2
To ocean from runoff	0.4
Burial in ocean sediments	0.1

When photosynthetic autotrophs engage in carbon dioxide fixation, they lock up billions of metric tons of carbon atoms in organic compounds each year (Sections

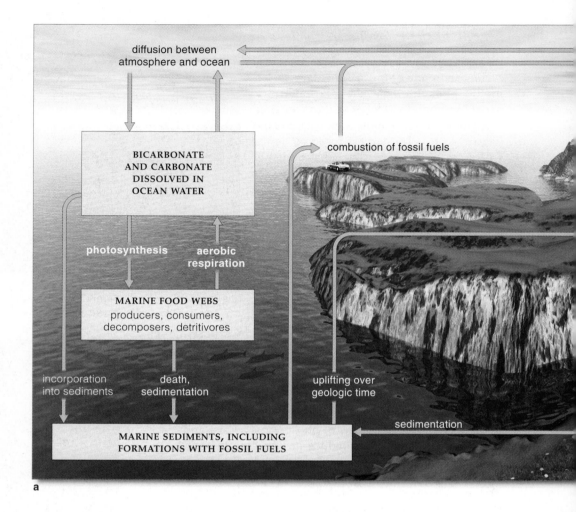

diffusion between atmosphere and ocean

BICARBONATE AND CARBONATE DISSOLVED IN OCEAN WATER

combustion of fossil fuels

photosynthesis   aerobic respiration

MARINE FOOD WEBS
producers, consumers, decomposers, detritivores

incorporation into sediments

death, sedimentation

uplifting over geologic time

sedimentation

MARINE SEDIMENTS, INCLUDING FORMATIONS WITH FOSSIL FUELS

a

Figure 43.18 (**a**) Global carbon cycle. The portion of the diagram on this page shows the movement of carbon through typical marine ecosystems. The *facing page* shows carbon's movement through ecosystems on land. The *gold* boxes indicate the main carbon reservoirs, the storage volumes of which are listed in the text.

Warm, less salty, shallow current

Cold, salty, deep current

Figure 43.19 Loop of ocean water that delivers carbon dioxide to its deep ocean reservoir. It sinks in the cold, salty North Atlantic and rises in the warmer Pacific.

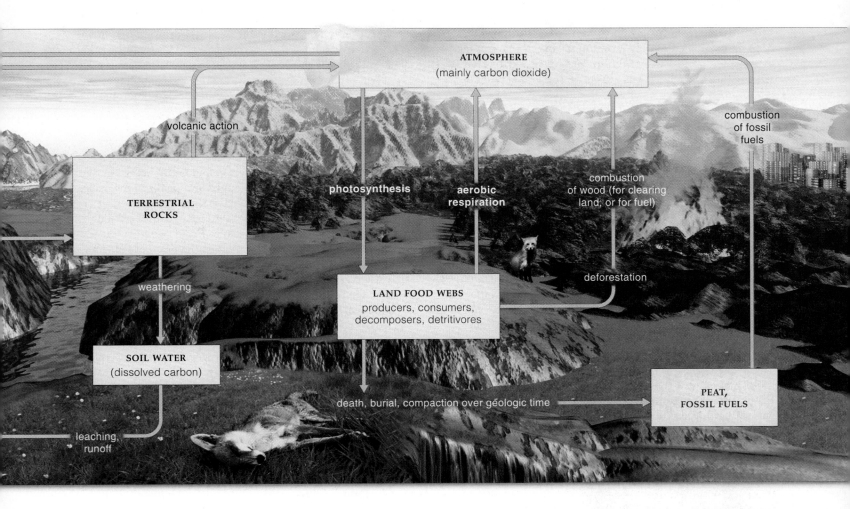

ATMOSPHERE
(mainly carbon dioxide)

volcanic action

combustion
of fossil
fuels

TERRESTRIAL
ROCKS

photosynthesis

aerobic
respiration

combustion
of wood (for clearing
land; or for fuel)

weathering

LAND FOOD WEBS
producers, consumers,
decomposers, detritivores

deforestation

SOIL WATER
(dissolved carbon)

death, burial, compaction over geologic time

PEAT,
FOSSIL FUELS

leaching,
runoff

6.8 and 43.4). Yet the average time that an ecosystem holds a given carbon atom varies greatly. For example, organic wastes and remains decompose so quickly in tropical rain forests that not much carbon accumulates at soil surfaces. In bogs, marshes, and other anaerobic sites, decomposers cannot degrade organic compounds to smaller bits, so carbon gradually accumulates in peat and other forms of compressed organic matter.

Also, in food webs of ancient aquatic ecosystems, carbon became incorporated in staggering numbers of shells and other hard body parts. Foramiferans and other shelled organisms died, sank through water, then were buried in sediments. The carbon remained buried for many millions of years in deep sediments until part of the seafloor was uplifted above the ocean surface by geologic forces of the sort described in Section 18.2. Remember those white cliffs of Dover (Section 20.8)?

As a final example, plants of the vast swamp forests of the Carboniferous also incorporated carbon atoms in their organic compounds. As you read in Section 22.4, those compounds were gradually converted into coal, petroleum, and gas reserves—which humans now tap as fossil fuels. Over the course of a few hundred years,

we have been extracting and burning fossil fuels that took millions of years to form. A major problem is that fossil fuels are nonrenewable resources (Chapter 45). Another problem is that they are the energy source of choice for a huge portion of the 6.1 billion people on Earth—mainly in the developed countries, which can afford them. At present, fossil fuel burning and other human activities are releasing more carbon into the atmosphere than can be cycled naturally to the ocean's storage reservoirs.

As you might deduce from the amounts listed on the preceding page, the ocean can remove only about 2 percent of the excess carbon entering the atmosphere. Most scientists now think the excess is amplifying the greenhouse effect, which means it may be contributing to global warming. The next section looks at this effect and some possible outcomes of modifications to it.

The ocean and atmosphere interact in the global cycling of carbon. Fossil fuel burning and other human activities appear to be contributing to imbalances in the global carbon budget.

From Greenhouse Gases to a Warmer Planet?

GREENHOUSE EFFECT Atmospheric concentrations of gaseous molecules play a profound role in shaping the average temperature near the surface of the Earth. That temperature has enormous effects on the global climate.

Countless molecules of carbon dioxide, water, ozone, methane, nitrous oxide, and chlorofluorocarbons are key players in interactions that dictate the global temperature. Collectively, the gases act somewhat like a pane of glass in a greenhouse—hence their name, "greenhouse gases." Wavelengths of visible light can pass around them and reach the Earth's surface. But greenhouse gases impede the escape of longer, infrared wavelengths—heat—from the Earth into space. How? Gaseous molecules can absorb these wavelengths and radiate much of the absorbed energy back toward the Earth, as shown in Figure 43.20.

The constant radiation of heat energy from greenhouse gases toward Earth proceeds lockstep with the constant bombardment and absorption of wavelengths from the sun, and so heat builds up in the lower atmosphere. The **greenhouse effect** is the name for this warming action.

GLOBAL WARMING DEFINED Without the action of greenhouse gases, the Earth's surface would be cold and lifeless. However, there can be too much of a good thing. Largely as an outcome of human activities, greenhouse gases are building to atmospheric levels that are higher than they were in the past. Figure 43.21 documents the increase. As many researchers suspect, greenhouse gases may be contributing to long-term higher temperatures at the Earth's surface, an effect called **global warming**. At present, the United States and some other developed countries are generating the most greenhouse gases.

EVIDENCE OF AN INTENSIFIED GREENHOUSE EFFECT In the 1950s, researchers on Hawaii's highest volcano started measuring the atmospheric concentrations of different greenhouse gases. That remote site is almost free of local airborne contamination and is representative of conditions for the Northern Hemisphere. The Chapter 3

introduction outlined what they found. Briefly, in the Northern Hemisphere, carbon dioxide levels follow the annual cycles of plant growth. They decline in summer when photosynthesis rates are highest. Then they rise in winter, when photosynthesis declines but aerobic respiration is still proceeding.

The troughs and peaks along the graph line in Figure 43.22a show the annual lows and highs. For the first time, scientists saw the integrated effects of carbon balances for land and water ecosystems of an entire hemisphere. Notice the midline of the troughs and peaks in the cycle. *And notice that the concentration is steadily increasing.*

Current data from many satellites, weather stations and balloons, research ships at sea, and supercomputer programs suggest that we are past the point of being able to reverse some impacts of climate change. In 2001, the United Nation's Intergovernmental Panel on Climate Change reported that global warming and climate change are real. They concluded that human activities, including fossil fuel burning, are contributing factors. In the past decade—the warmest on record—hurricanes, floods, tornadoes, and droughts were intense. Alaska saw rare thunderstorms; Hilo, Hawaii, saw homes washed away by 27 inches of rainfall in 24 hours.

CAUSES AND EFFECTS: SOME PREDICTIONS The lower atmosphere's temperature may spike higher by 2.5 to 10.4 degrees in this century. If that happens, then we might predict these outcomes:

Rising sea level. If the atmosphere warms by only 4°C (7°F), the sea level will rise by about 0.6 meter (2 feet). Why? The ocean surface temperature will increase, and water expands when heated. Also, glaciers and polar ice sheets will melt faster (*Critical Thinking* question 5 in Section 43.12). The volume of water released would flood low coastal regions. High tides and storm waves will exacerbate effects of a long-term rise in sea level. Low islands and coastal plains—which support many cities and agricultural lands—will be under water.

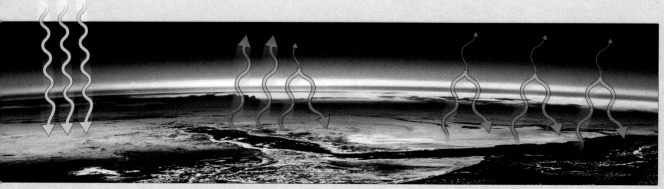

a The sun's rays penetrate the lower atmosphere, warm the Earth's surface.

b The surface radiates heat (infrared wavelengths) to the atmosphere. Some heat escapes into space. But greenhouse gases and water vapor absorb some infrared energy and radiate a portion of it back toward Earth.

c Increased concentrations of greenhouse gases trap more heat near Earth's surface. Sea surface temperature rises, more water evaporates into atmosphere. Earth's surface temperature rises.

Figure 43.20 The greenhouse effect, as executed mainly by carbon dioxide, water vapor, ozone, methane, nitrous oxide, and various chlorofluorocarbons present in the lower atmosphere.

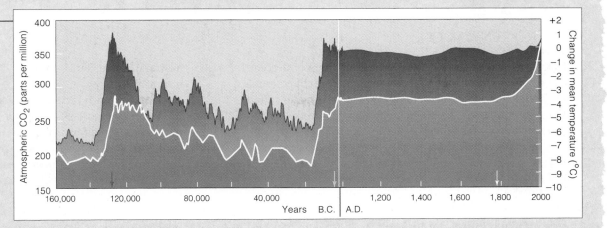

Figure 43.21 Correlation of changes in atmospheric CO₂ levels (*white* graph line) with ice ages and interglacials. The 1985 mean temperature was the baseline for tracking global changes in temperature (*red* graph line). Arrows show start of the last ice age (*red*), last interglacial (*yellow*), and the industrial revolution (*white*).

Increased rainfall and flooding. Global warming will change regional patterns of precipitation. Much of the Northern Hemisphere will get more rain. Snow will melt faster in its great mountain ranges, so spring runoff and summer droughts will be more severe. The already bad flooding and mudslides in the tropics and in California will worsen. Interiors of the great continents will become drier, and deserts will expand.

Heat waves and wildfires. Heat waves will intensify in the Northern Hemisphere, and deaths from heat stroke will probably climb in cities ranging from Shanghai to Chicago. Hotter, drier summers will also invite more wildfires in Alaska, California, Florida, Brazil, and Asia.

Crop failures. Droughts will intensify in the United States, Mexico, Brazil, Africa, Europe, and Asia. Major crop losses will increase global malnutrition and famine.

Contagious disease. Warmer, wetter conditions along the coasts will favor cholera, malaria, and other serious diseases. For example, the West Nile virus, introduced into the United States in 1999, is already expanding its range in the eastern United States.

Water wars. By 2015, an estimated 3 billion people will not have enough fresh water. We will be returning to this prospect in Section 45.7.

Some argue that the global climate is too complex and unpredictable for us to accept data gathered so far. But the majority now agree there is cause for concern. We already see actions being taken. Chapter 45 outlines some of them, such as developing energy alternatives to fossil fuels. Are the actions none too soon? Future climate changes might turn out to be twice as violent as scientists were predicting just five years ago.

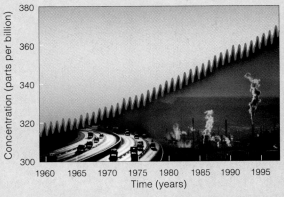

a Carbon dioxide (CO₂). Of all human activities, the burning of fossil fuels and deforestation contribute the most to increasing atmospheric levels.

c Methane (CH₄). Termite activity and anaerobic bacteria living in swamps, landfills, and the stomachs of cattle and other ruminants produce great amounts of methane as a by-product of metabolism.

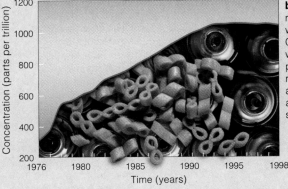

b CFCs. Until restrictions were in place, CFCs were widely used in plastic foams, refrigerators, air conditioners, and industrial solvents.

d Nitrous oxide (N₂O). Denitrifying bacteria produce N₂O in metabolism. Also, fertilizers and animal wastes release enormous amounts; this is especially the case for large-scale livestock feedlots.

Figure 43.22 Recent increases in atmospheric concentrations of four greenhouse gases.

NITROGEN CYCLE

Since the time of life's origin, the atmosphere and ocean have contained nitrogen. This component of all proteins and nucleic acids moves in an atmospheric cycle called the **nitrogen cycle**. Gaseous nitrogen (N_2) makes up 80 percent or so of the atmosphere, the largest reservoir. Successively smaller reservoirs are seafloor sediments, ocean water, soil, land biomass, atmospheric nitrous oxide (N_2O), and the smallest, marine biomass.

The atoms of N_2 are joined by triple covalent bonds ($N \equiv N$). Few organisms have the metabolic means to break them. Only certain bacteria, volcanic action, and lightning convert N_2 into forms that enter food webs.

Of all nutrients required for plant growth, nitrogen often is the scarcest. Nearly all of the nitrogen in soils was put there by nitrogen-fixing organisms. Ecosystems lose it when other bacteria "unfix" the fixed nitrogen. Land ecosystems lose more by leaching, although this is the basis of nitrogen inputs into aquatic ecosystems such as streams, lakes, and the seas (Figure 43.23).

Cycling Processes

Let's follow nitrogen atoms through the portion of the nitrogen cycle that proceeds in an ecosystem on land. They are objects of processes called nitrogen fixation,

assimilation and biosynthesis, decomposition, and then ammonification.

In **nitrogen fixation**, a few kinds of bacteria convert N_2 to ammonia (NH_3), which quickly dissolves in the cytoplasm, thus forming ammonium (NH_4^+). In aquatic ecosystems, *Anabaena*, *Nostoc*, and other cyanobacteria are nitrogen fixers. In many land ecosystems, *Rhizobium* and *Azotobacter* fix nitrogen. Collectively, these bacteria fix about 200 million metric tons of nitrogen each year!

Decomposition and **ammonification** are processes by which bacteria and fungi degrade nitrogenous wastes and remains of organisms. Decomposers use part of the released proteins and amino acids during metabolism. But most of the nitrogen remains in the decay products, in the form of ammonia or ammonium, which plants take up. Nitrifying bacteria also act on the ammonia or ammonium. In **nitrification**, they strip the compounds of electrons, and nitrite (NO_2^-) is the result. Other kinds of bacteria use the nitrite during their metabolism and produce nitrate (NO_3^-), which plants then take up.

Most plants growing in nitrogen-poor soil interact as mutualists with fungi. Their roots form mycorrhizae with fungal hyphae, the collective surface area of which absorbs a lot of mineral ions from soil. Clover, beans, peas, and other legumes are mutualists with nitrogen-

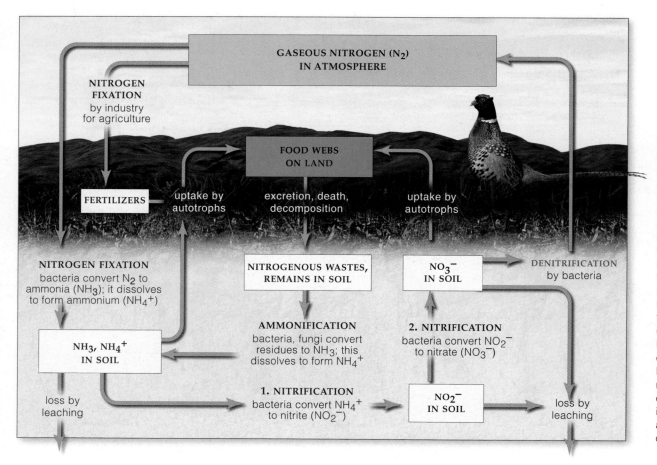

Figure 43.23 The nitrogen cycle in an ecosystem on land. Activities of nitrogen-fixing bacteria make nitrogen available to plants. Other bacterial species cycle nitrogen to plants when they convert organic wastes to ammonium and nitrates. The largest nitrogen reservoir by far is the atmosphere (mainly N_2), followed by ocean sediments, ocean water, and soil. Lesser amounts of nitrogen are in the form of nitrous oxide (N_2O) in the atmosphere and in the biomass of the seas.

fixing bacteria that form root nodules. Chapter 21 and Section 27.2 describe these interactions. How do plants use the nitrogen provided by bacterial activities? Their cells assimilate nitrogen and channel much of it into synthesizing amino acids, proteins, and nucleic acids. Plant tissues, remember, are the sole nitrogen source for animals, which directly or indirectly feed on plants.

Nitrogen Scarcity

Ammonium, nitrite, and nitrate that form during the nitrogen cycle are vulnerable to leaching and runoff. Leaching, recall, is the removal of some soil nutrients as water percolates through it (Section 27.1). Besides losses from leaching, some nitrogen is lost to the air by **denitrification**. By this process, certain bacteria convert nitrate or nitrite to N_2 and N_2O (nitrous oxide). Usually, most denitrifying bacteria rely on aerobic respiration. In waterlogged or poorly aerated soil, they use anaerobic pathways in which nitrate, nitrite, or N_2O (not oxygen) is the final electron acceptor. Section 7.5 describes this metabolic strategy. During the reactions, fixed nitrogen is converted to N_2, much of which escapes to the air.

Also, nitrogen fixation comes at high metabolic cost to plants that are mutualists with the nitrogen fixers. In exchange for nitrogen, plants give up sugars and other photosynthetic products that require large investments of ATP and NADPH. Such plants have a competitive edge in nitrogen-poor soil. In nitrogen-rich soil, species that don't pay the metabolic price often displace them.

Human Impact on the Nitrogen Cycle

Humans are having disproportionately bad effects on the nitrogen cycle, compared to other species. Where deforestation and grassland conversion for agriculture are rampant, losses of nitrogen from soil are huge. With each forest clearing and harvest, nitrogen stored in the plants is lost. Soil erosion and leaching take away more.

Many farmers counter nitrogen losses by rotating crops, as by alternating wheat with legumes. Rotation can help keep soil stable and productive. In developed countries, farmers also spread nitrogen-rich fertilizers. They even select new strains of crop plants that have a greater capacity to take up fertilizers from soil. With such practices, crop yields per hectare have doubled and even quadrupled over the past forty years.

But heavy fertilizer applications alter soil water and can harm plants. Remember, the surface charges of soil particles attract ions that weathering, irrigation, and fertilizers introduce to soils. By a pH-dependent process called **ion exchange**, ions dissociate from soil particles, and other ions dissolved in soil water replace them. Any change in the composition of soil water disrupts

Figure 43.24
Dead and dying spruce trees in a forest in Germany. They are cited as casualties of extremely high concentrations of nitrogen oxides and other forms of air pollution.

the number and kinds of ions in exchangeable form—which are the kinds that plants take up. Calcium and magnesium are the most abundant exchangeable ions. When nitrogen fertilizers enter soil, ammonium ions also assume importance. This is especially the case for ammonium sulfate $(NH_4)_2SO_4$, for all of its nitrogen is in the form of ammonium. Heavy applications of such a fertilizer greatly increase the acidity of soil water. Its many hydrogen ions compete for binding sites on soil particles. One outcome is that the levels of calcium and magnesium ions necessary for plant growth decline.

Humans introduce excess nitrogen into ecosystems in other ways, as when nitrogen-rich sewage typically enters rivers, lakes, and estuaries (Section 45.7). As another example, fossil fuel burning by power plants and vehicles having combustion engines releases NO_2 and other oxides of nitrogen into the atmosphere (Section 45.1). Winds carry the air pollutants to forests at high elevations and high northern latitudes, which have nitrogen-poor soils. Nitrogen is a limiting factor on tree growth, and the forest trees are adapted to storing and cycling it. With more nitrogen, they grow faster, well into the summer. But the late growth is vulnerable to autumn frost and dies during winter. In addition, their nitrogen-stimulated growth outstrips the availability of phosphorus and other nutrients, so the trees suffer the effects of nutrient deficiencies. Their leaves yellow and drop, so rates of photosynthesis decline. The decline has repercussions on the growth of all plant parts. The weakened trees are far more susceptible to disease and other environmental stresses (Figure 43.24).

The nitrogen cycle turns on processes of nitrogen fixation, assimilation and biosynthesis by all organisms, and on decomposition. Its natural cycling depends on the activity of nitrogen-fixing bacteria and on mycorrhizae.

Ecosystems lose nitrogen naturally by the activities of denitrifying bacteria, which convert nitrite or nitrate to forms that escape into the atmosphere.

Ecosystems often lose or gain too much nitrogen through human activities, in ways that compromise plant growth.

SUMMARY

Gold indicates text section

1. An ecosystem consists of an array of producers, consumers, detritivores, and decomposers along with their environment. It is an open system, with inputs, internal transfers, and outputs of energy and nutrients. There is a one-way flow of energy into and out from it and a cycling of materials among its organisms. *43.1*

2. Sunlight is the initial energy source for nearly all ecosystems. Photoautotrophs (the primary producers) convert that energy to other forms, such as chemical bond energy of ATP. They also assimilate nutrients that the ecosystem's heterotrophs require. *43.1, 43.2*

   a. Many heterotrophs are consumers. The herbivores feed upon algae and plants. Carnivores ingest animals. Parasites withdraw nutrients from the tissues of living hosts. Omnivores use a variety of food sources.

   b. Decomposers also are heterotrophs. Most (certain fungi and bacteria) obtain energy and nutrients from organic remains and wastes. Detritivores, such as crabs and earthworms, are heterotrophs that ingest bits of decomposing or dead organisms or organic products.

3. In ecosystems, feeding relationships are structured as a hierarchy of energy transfers (trophic levels). *43.2*

   a. Primary producers are at the first trophic level; consumers are at successively higher trophic levels.

   b. Herbivores are second-level consumers; carnivores and parasites are at higher levels. The detritivores and decomposers are assigned to all trophic levels above the first level. So are the parasites. Humans and other omnivores obtain energy from more than one source, so they cannot be assigned to a single trophic level.

4. A food chain is a straight-line sequence by which energy stored in autotroph tissues is passed on through higher trophic levels. Most cross-connect as food webs. In grazing food webs, energy is transferred from the primary producers to herbivores, then carnivores. In detrital food webs, energy is transferred from primary producers through detritivores and decomposers. *43.2*

5. The amount of useful energy that flows through successive levels of consumers declines at each energy transfer in a food web because of metabolic heat loss and shunting of food energy into organic wastes. This inefficiency typically limits the number of trophic levels in a food web to no more than four or five. *43.2*

   a. In environments that show variations in salinity, temperature, and other environmental conditions, food webs tend to have short chains. In stable environments, such as parts of the deep ocean, food chains are longer.

   b. The most complex food webs have short chains; the simplest have the most top carnivores.

6. Disturbance to one aspect of an ecosystem usually has unexpected effects on other, seemingly unrelated parts. To predict consequences of a disruption, as by computer modeling, researchers attempt to identify all the potential relationships and incorporate them into the model. *43.3*

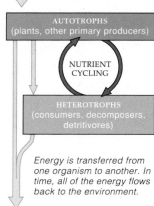

Producers trap, convert, and use or store some energy from the sun.

AUTOTROPHS (plants, other primary producers)

NUTRIENT CYCLING

HETEROTROPHS (consumers, decomposers, detritivores)

Energy is transferred from one organism to another. In time, all of the energy flows back to the environment.

ONE-WAY FLOW OF ENERGY

Figure 43.25 Summary of energy flow and nutrient cycling in ecosystems.

7. Primary productivity is the rate at which primary producers capture and store a given amount of energy in a given interval. Net primary productivity is the rate at which energy is stored in producers in excess of the rate of its release by way of aerobic respiration. *43.4, 43.5*

8. Primary productivity and ecosystem structure require cycling of nutrients as well as energy inputs (Figure 43.25). Water and minerals pass on through the environment, then organisms, and then back to the environment in biogeochemical cycles.

   a. Water moves through a hydrologic cycle. In land ecosystems, plants stabilize soil and minimize nutrient loss during the cycle, as by runoff. *43.7*

   b. In a sedimentary cycle, a nutrient (phosphorus, for example) is in solid form, and the main reservoir is the Earth's crust. *43.8*

   c. In atmospheric cycles, a nutrient prevails mainly in gaseous form. Examples: carbon in carbon dioxide, nitrogen in gaseous nitrogen. *43.9, 43.11*

9. Sedimentary cycles and the hydrologic cycle interact and move mineral nutrients on a global scale. *43.7*

   a. Certain human activities are depleting minerals from ecosystems, as when weathered soils of tropical forests are cleared for agriculture.

   b. Some human activities are accelerating the process of eutrophication. They are adding nutrients to aquatic ecosystems that are naturally low in these nutrients and thereby promote destructive algal blooms.

10. In the carbon cycle, carbon moves through the ocean (its main reservoir), the atmosphere (as carbon dioxide, mostly), and ecosystems. Fossil fuel burning, logging, and the conversion of natural ecosystems for crops and grazing affect the global carbon budget. *43.9, 43.10*

11. Like water, ozone, CFCs, methane, and nitrous oxide, carbon dioxide is a greenhouse gas. Fossil fuel burning and other human activities that release carbon to the atmosphere may be adding to global warming. *43.10*

12. For land ecosystems, nitrogen is a limiting factor in the total net primary productivity. Gaseous nitrogen is abundant in the atmosphere. Nitrogen-fixing bacteria convert N_2 to ammonia and nitrates, which producers take up. Mycorrhizae and root nodules, two symbiotic interactions, enhance the nitrogen uptake. *43.11*

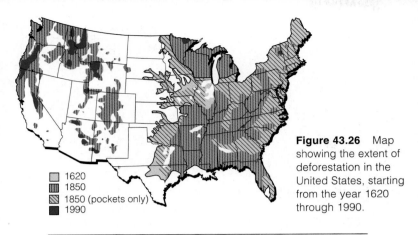

Figure 43.26 Map showing the extent of deforestation in the United States, starting from the year 1620 through 1990.

Legend:
- 1620
- 1850
- 1850 (pockets only)
- 1990

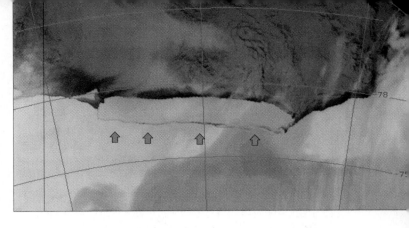

Figure 43.27 Satellite image of an iceberg roughly the same size as Connecticut. This one started to break away from Antarctica in 2000 and may be the longest ever recorded.

Review Questions

1. Define an ecosystem in terms of inputs and outputs. *43.1*

2. Define primary producer, consumer, decomposer, and detritivore. Give an example of each. *43.1*

3. Define and describe trophic levels. Which class of organisms is farthest from the energy input into an ecosystem? *43.1, 43.2*

4. Define food chain and food web. How do grazing food webs differ from detrital food webs? *43.1, 43.2*

5. Distinguish between biomass pyramid and energy pyramid. What did the pyramids for Silver Springs reveal? *43.4, 43.5*

6. Define three types of biogeochemical cycles and give an example of each. *43.6–43.11*

7. Define nitrogen fixation. What is the difference between nitrification and denitrification? *43.11*

Self-Quiz   ANSWERS IN APPENDIX III

1. Ecosystems have _____ .
   a. energy inputs and outputs
   b. one trophic level
   c. nutrient cycling but not outputs
   d. a and b

2. Trophic levels are _____ .
   a. structured feeding relationships
   b. who eats whom in an ecosystem
   c. a hierarchy of energy transfers
   d. all of the above

3. Primary productivity on land is affected by _____ .
   a. photosynthesis and respiration by plants
   b. how many plants are neither eaten nor decomposed
   c. rainfall and temperature
   d. all of the above

4. Match the ecosystem terms with the suitable description.
   ___ producers   a. herbivores, carnivores, omnivores
   ___ consumers   b. feed on partly decomposed matter
   ___ decomposers   c. degrade organic remains, wastes
   ___ detritivores   d. photoautotrophs

Critical Thinking

1. Imagine and describe an extreme situation in which you are a participant in a food chain rather than a food web.

2. Marguerite is growing a vegetable garden in Maine. What are the variables that can affect its net primary production?

3. List as many of the agricultural products and manufactured goods as you can identify that you depend upon. Are any of them implicated in the amplified greenhouse effect?

4. Biologists Reed Noss, J. Michael Scott, and Edward LaRoe issued a report on *endangered ecosystems* in the United States. They categorize thirty natural, once-vast domains as in danger of vanishing (Figure 43.26). Agricultural conversions, urban growth, and other human activities account for the losses. More biodiversity was lost at the ecosystem level than was generally recognized. Do some research on a habitat that is part of such an ecosystem. Get a sense of its decline in biodiversity, of the kinds of organisms that once flourished there. Would you take part in efforts to set aside land for restoration? Or do you consider such efforts too intrusive on individual rights of property ownership?

5. *Polar ice shelves* are vast, thickened sheets of ice floating on seawater. Measurements taken between 1978 and the present indicate that Antarctic ice shelves are retreating. A chunk of the Larsen Ice Shelf the size of Rhode Island broke away from Antarctica in 1995 and drifted off into the open ocean. Figure 43.27 shows another gargantuan chunk, 37 kilometers wide and 295 kilometers long, that started to split away from the Ross Ice Shelf in 2000. Also, in 1996, seawater 200 meters below the Arctic ice cap was 1°C warmer than just five years ago. Do you suppose that the retreating and fragmentation of the polar ice sheets are evidence of a long-term trend in global warming? If not, what other factors might cause such impressive events?

Selected Key Terms

ammonification *43.11*	food chain *43.1*
biogeochemical cycle *43.6*	food web *43.1*
biological magnification *43.3*	global warming *43.10*
biomass pyramid *43.4*	grazing food web *43.2*
carbon cycle *43.9*	greenhouse effect *43.10*
consumer *43.1*	hydrologic cycle *43.7*
decomposer *43.1*	ion exchange *43.11*
decomposition *43.11*	nitrification *43.11*
denitrification *43.11*	nitrogen cycle *43.11*
detrital food web *43.2*	nitrogen fixation *43.11*
detritivore *43.1*	phosphorus cycle *43.8*
ecosystem *43.1*	primary producer *43.1*
ecosystem modeling *43.3*	primary productivity *43.4*
energy pyramid *43.4*	trophic level *43.1*
eutrophication *43.8*	watershed *43.7*

Readings

Brady, N., and R. Weil. 1996. *The Nature and Properties of Soils.* Eleventh edition. New Jersey: Prentice-Hall.

Krebs, C. 1994. *Ecology.* Fourth edition. New York: HarperCollins.

On-Line readings at Student Guide for InfoTrac:
www.brookscole.com/biology

44

THE BIOSPHERE

Does a Cactus Grow in Brooklyn?

Suppose you live in the American Southwest but find yourself touring an African desert. There you notice a flowering plant with spines, tiny leaves, and fleshy, columnlike stems—just like some cactus plants back home (Figure 44.1). Or suppose you live in the coastal hills of California and decide to tour the Mediterranean coast, the southern tip of Africa, or even central Chile. There you come across many-branched, tough-leafed woody plants—very much like the many-branched, tough-leafed chaparral plants back home.

In both cases, vast geographic and evolutionary distances separate the plants. Why, then, are they alike? The question intrigues you, so you decide to compare their locations on a global map. You see that American and African desert plants live about the same distance from the equator. Chaparral plants and their distant look-alikes grow along the western and southern coasts of continents between latitudes 30° and 40°. As Charles Darwin and other naturalists did long ago, you have just stumbled onto one of many predictable patterns in nature.

What causes such patterns? With this question, we turn to **biogeography**—the study of the distribution of organisms, past and present, and of diverse processes that underlie their distribution patterns. Think back on those "accidents of history," including the colossal breakup of Pangea more than 100 million years ago. When chunks of that supercontinent began to slowly drift apart, most species had no choice but to go along for the ride. The travelers were dispersed to different isolated locations. And there they diverged genetically from their parent populations. Among the changing populations on the fragment that became Australia were ancestors of eucalyptus trees and platypuses, of wombats and kangaroos.

Species also owe their distribution to topography, climate, and species interactions. With diligence, you might grow a cactus under artificial lights in a heated room in Brooklyn or another New York City borough. Plant that cactus outside, and it won't last one winter.

As this last example reminds us, we humans tinker with species distributions. Not all of our tinkering is as harmless as growing a cactus in Brooklyn. Think of our predatory effects on the world's fisheries and the effects of our pesticide battles with insect competitors for food. Earlier chapters gave you a general picture of predation, competition, and other species interactions. This chapter considers the physical forces shaping the biosphere itself. Use it as a foundation for addressing the impact of the human species on the biosphere, the topic of the chapter to follow.

Start with the definition of the **biosphere**: the sum total of all places in which organisms live. Organisms occupy the Earth's waters—the hydrosphere—which includes the ocean, ice caps, and other forms of liquid and frozen water. They occupy soils and sediments of the lithosphere, the Earth's outer, rocky layer. And they occupy the lower **atmosphere**, which consists of gases and airborne particles that envelop the Earth.

Figure 44.1 Life in the biosphere—a case of morphological convergence. (**a**) *Echinocerus*, a member of the cactus family (Cactaceae), grows in hot deserts of the American Southwest. (**b**) In deserts of southwestern Africa, we find *Euphorbia*, of the spurge family (Euphorbiaceae). Although the lineages appear to be closely related, they are geographically and evolutionarily distant. Long ago, plants in both lineages put similar structures, including their leaves, to similar uses in similar habitats. Their descendants ended up resembling each other.

Air in the upper atmosphere, seventeen kilometers or so above the Earth's surface, is too thin to support life.

In this vast biosphere you find ecosystems ranging from continent-straddling forests to a rainwater pool in a small cup-shaped array of leaves. As you will see, climate profoundly influences all ecosystems except for a few at hydrothermal vents on the ocean floor.

Climate means average weather conditions, such as temperature, humidity, wind speed, cloud cover, and rainfall, over time. Many factors contribute to it. The main ones are variations in the amount of solar radiation streaming in, the Earth's daily rotation and its path around the sun, the world distribution of continents and oceans, and land mass elevations. All of these factors interact to produce prevailing winds and ocean currents that influence the global patterns of climate. The physical and chemical development of sediments and soils also depend on climate.

Climate, together with the composition of sediments and soils, influences the growth of primary producers— and, through them, the distribution of ecosystems.

1. Energy from the sun is the initial energy source for nearly all ecosystems on Earth. In addition, solar energy influences the global distribution of ecosystems. It does so by continually providing heat energy that warms the atmosphere and drives the Earth's weather systems.

2. Interactions among global air circulation patterns, ocean currents, and diverse topographic features result in regional variations in patterns of temperature and rainfall. The variations influence the composition of soils and sediments, the growth and distribution of primary producers, and, through them, the distribution of ecosystems.

3. A biome is a large, regional unit of land that can be characterized by the climax vegetation of the ecosystems within its boundaries. Deserts and broadleaf forests are examples. Their distribution corresponds roughly with regional variations in climate, topography, and soil type.

4. The water provinces cover more than 71 percent of the Earth's surface. The oceans hold all but 3 percent of the free-flowing water. Inland seas, estuaries, and diverse bodies of fresh water, including lakes, streams, and rivers, hold the rest.

5. Each freshwater and marine ecosystem has gradients in light availability, temperature, and dissolved gases. The gradients vary daily and with the changing seasons. Primary productivity, and the composition of species, depend on them.

AIR CIRCULATION PATTERNS AND REGIONAL CLIMATES

Each winter Pacific Grove, California, is host to great gatherings of tourists and monarch butterflies (Figure 44.2). Similarly, caribou, Canada geese, sea turtles, and whales undertake vast migrations, moving to and from overwintering grounds. Other animals stay put. Their adaptations in body form, physiology, and behavior help them endure seasonal change. In Canada and the United States, Japan and Italy, flowering plants leaf out, flower, bear fruit, then drop leaves. In the ocean, astronomical numbers of photosynthetic protistans and bacteria show seasonal bursts of primary productivity. Like nearly all organisms, they are exquisitely attuned to regional climates and seasonal change.

Climate starts with incoming rays from the sun. Of the total amount of solar radiation reaching the outer atmosphere, only about half gets through to the Earth's surface. Ozone (O_3) and oxygen (O_2) molecules in the upper atmosphere absorb most of the wavelengths of ultraviolet radiation. Such wavelengths are lethal for most forms of life. Absorption is greatest at altitudes between 17 and 27 kilometers above sea level, where molecules of ozone are the most concentrated (Figure 44.3). Hence the name "ozone layer." Clouds, particles of dust, and water vapor suspended in the atmosphere absorb other wavelengths or reflect them into space.

Radiation penetrating the atmosphere warms the Earth's surface, which gives up heat by radiation and evaporation. The lower atmosphere's molecules absorb some heat and reradiate part of it toward the Earth. The effect is a bit like heat retention in a greenhouse, which lets in the sun's rays while retaining heat that the plants and soil inside are giving up (Section 43.10). Why is this greenhouse effect important? *Heat energy from the sun warms the atmosphere and ultimately drives the Earth's great weather systems.*

The sun's effect varies from one latitude to the next. Its rays are more concentrated at the equator than the poles, so air is heated more at the equator (Figure 44.4). A global pattern of air circulation starts as warm air at the equator rises and spreads north and south. Under the moving air, the Earth rotates faster at the equator than it does at the poles, and this creates worldwide belts of prevailing east and west winds. The differences in solar heating at different latitudes and the modified patterns of air circulation help define the world's major **temperature zones**, which are shown in Figure 44.5*a*.

The differences also help define rainfall patterns at different latitudes. Warm air holds more moisture than cool air. Air warmed at the equator picks up moisture from the seas. It rises to cooler altitudes and gives up moisture as rain, which supports luxuriant growth of equatorial forests. Being drier now, the air moves away

Figure 44.2 Monarch butterflies. These migratory insects gather each winter in trees of California coastal regions and central Mexico. They travel hundreds of kilometers south to those places, which are cool and humid in winter. If they were to stay in their northern breeding grounds, monarchs would risk being killed by more severe climatic conditions.

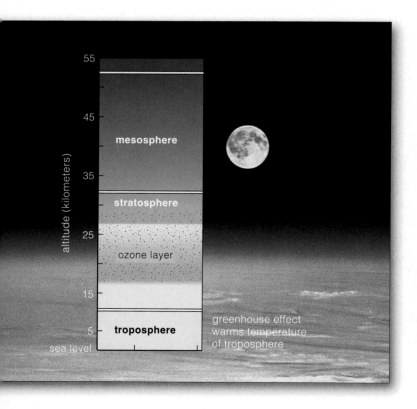

Figure 44.3 Earth's atmosphere. Most of the global air circulation proceeds in the troposphere, where temperature decreases rapidly with altitude. Most of the ultraviolet wavelengths in the sun's rays are absorbed in the upper atmosphere, mainly at the ozone layer. The ozone layer typically ranges between 17 and 27 kilometers above sea level.

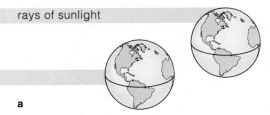

rays of sunlight

a

Figure 44.4 Global air circulation patterns brought about by three interrelated factors. *First*, the sun's rays are less spread out in equatorial regions than in polar regions (**a**). Warm equatorial air rises and spreads northward and southward to produce an initial pattern of air circulation (**b**).

Second, the nonuniform distribution of land creates air pressure variations. Land absorbs and gives up heat faster than the ocean. So depending on what's below them, some parcels of air rise or sink faster than others. Air pressure is lower where air rises, and it is greater where air sinks. The regional pressure differences create winds that disrupt the overall movement of air from the equator to the poles.

Third, the Earth's rotation and overall shape introduce easterly and westerly deflections in wind directions. With each full rotation, the surface turns faster beneath the air masses at the equator (where the Earth's diameter is greatest) and slower beneath air masses at the poles. Thus, a rising air mass cannot move "straight north" or "straight south." Its deflection is the source of prevailing east and west winds (**c**).

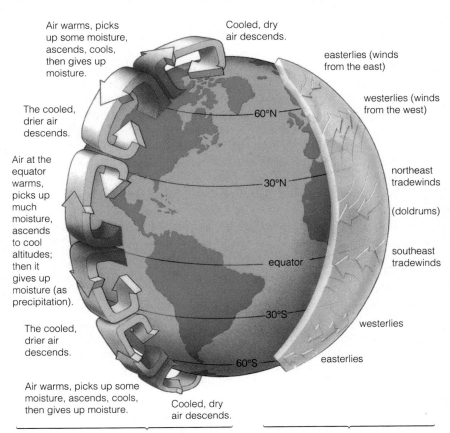

Air warms, picks up some moisture, ascends, cools, then gives up moisture.

Cooled, dry air descends.

easterlies (winds from the east)

westerlies (winds from the west)

The cooled, drier air descends.

Air at the equator warms, picks up much moisture, ascends to cool altitudes; then it gives up moisture (as precipitation).

60°N

30°N

northeast tradewinds

(doldrums)

equator

southeast tradewinds

The cooled, drier air descends.

30°S

westerlies

Air warms, picks up some moisture, ascends, cools, then gives up moisture.

60°S

easterlies

Cooled, dry air descends.

b Initial pattern of air circulation

c Deflections in the paths of air flow near the Earth's surface

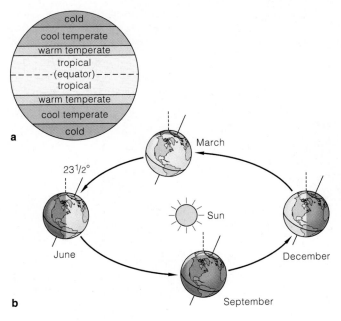

cold

cool temperate

warm temperate

tropical

(equator)

tropical

warm temperate

cool temperate

cold

a

23¹/₂°

March

Sun

June

December

b

September

Figure 44.5 (**a**) World temperature zones. (**b**) Annual variation in incoming solar radiation. The northern end of the Earth's fixed axis tilts toward the sun in June and away from it in December. Thus the equator's position relative to the day–night boundary of illumination varies over the year. Such variations in sunlight intensity and daylength cause seasonal temperature variations in the two hemispheres. Seasonal change is more pronounced with distance from the equator. It is greatest in the interior of continents, away from the ocean's moderating effects.

from the equator. It becomes warmer and drier when it descends at latitudes of about 30°. Deserts tend to form at these latitudes. Farther north and south, air again picks up some moisture and ascends to high altitudes. It creates another moist belt at latitudes of about 60°. Air descends in polar regions, where low temperatures and almost nonexistent precipitation give rise to cold, dry polar deserts.

The amount of solar radiation reaching the surface also varies seasonally because of the Earth's rotation around the sun (Figure 44.5*b*). This leads to seasonal changes in daylength, prevailing wind directions, and temperature. And so primary productivity alternately rises and falls on land and in the seas—and migrations shift many kinds of animals to new locations.

Latitudinal differences in the amount of solar radiation reaching the Earth produce global air circulation patterns. The Earth's rotation and its overall shape affect the patterns to produce latitudinal belts of temperature and rainfall.

Also, the Earth's annual rotation around the sun introduces seasonal changes in winds, temperatures, and rainfall.

Such factors influence the locations of different ecosystems.

THE OCEAN, LANDFORMS, AND REGIONAL CLIMATES

Ocean Currents and Their Effects

The **ocean**—a continuous body of water—covers more than 71 percent of the Earth. The uppermost 10 percent moves in currents that distribute nutrients in marine ecosystems and that influence regional climates. Solar heat and wind friction drive the surface currents.

Latitudinal and seasonal variation in solar heating warm and cool water on a vast scale. The volume of water expands when heated and shrinks when cooled. So sea level is about 8 centimeters (3 inches) higher at the equator than at the poles. The sheer volume of water in this "slope" is enough to get surface waters moving in response to gravity, usually toward the poles. Along the way, surface waters warm air parcels above them. At midlatitudes, about *10 million billion* calories of heat energy per second are transferred to the air!

Rapid mass flows of water—*currents*—arise mostly from the "tugging" of trade winds and westerlies. The Earth's rotation, the distribution of land masses, and the shapes of oceanic basins influence the direction and properties of the currents. Water circulates clockwise in the Northern Hemisphere and counterclockwise in the Southern Hemisphere, as shown in Figure 44.6.

Swift, deep, and narrow currents of nutrient-poor waters move parallel with the east coast of continents. As an example, 55 million cubic meters per second of warm water move northward as the Gulf Stream along the east coast of North America. Slower, shallow, and broad currents paralleling the west coast of continents move cold water toward the equator. As you will see, they move deep, nutrient-rich waters to the surface.

Why are Pacific Northwest coasts cool, foggy, and mild in summer? Winds nearing the coast give up heat to cold coastal water, which the California Current is moving toward the equator. Why are Baltimore and Boston muggy in summer? Air above the Gulf Stream gains heat and moisture, which southerly and easterly winds deliver to those cities. Why are winters milder in London and Edinburgh than in Ontario and central Canada, even though they are all at the same latitude? The North Atlantic Current picks up warm water from the Gulf Stream. As it flows past northwestern Europe, it gives up heat energy to the prevailing winds.

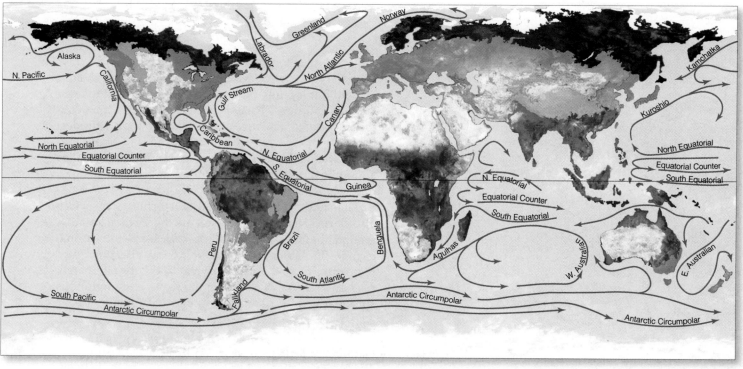

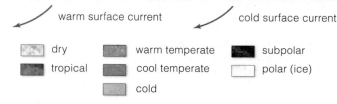

Figure 44.6 Generalized map of major climate zones correlated with surface currents and drifts of the world ocean. Warm surface currents move from the equator toward the poles. Water temperatures differ with latitude and depth. The differences contribute to regional differences in temperature and rainfall. In which direction does a given current flow? It depends on prevailing winds, the Earth's rotation, gravity, the shapes of ocean basins, and the positions of land masses.

warm surface current   cold surface current

- dry
- tropical
- cold
- warm temperate
- cool temperate
- subpolar
- polar (ice)

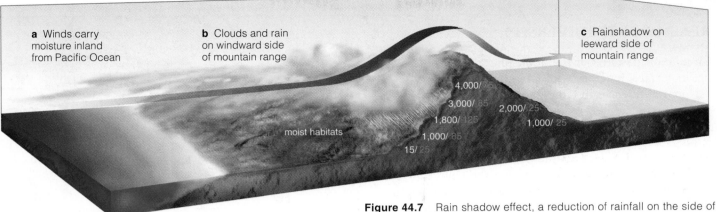

a Winds carry moisture inland from Pacific Ocean

b Clouds and rain on windward side of mountain range

c Rainshadow on leeward side of mountain range

4,000/
3,000/
1,800/
2,000/
1,000/
moist habitats
1,000/
15/

Figure 44.7 Rain shadow effect, a reduction of rainfall on the side of high mountains away from prevailing winds. Plants adapted to arid or semiarid conditions grow in rain shadows. *Blue* numbers show average yearly precipitation in centimeters, as measured at different locations on both sides of the range. *White* numbers signify elevation in meters.

Regarding Rain Shadows and Monsoons

Mountains, valleys, and other aspects of topography influence regional climates. "Topography" refers to a region's physical features, such as elevation. Think of a warm air mass picking up moisture off California's western coast. It moves inland to the Sierra Nevada, a mountain range paralleling the coast. The air cools as it ascends to higher altitudes, then loses moisture as rain (Figure 44.7). Belts of vegetation at different elevations reflect the differences in temperature and moisture. At the western base of the range, we see grasslands with species adapted to semiarid conditions. Higher up we see deciduous and evergreen species adapted to more moisture and cooler air. Higher still, a subalpine belt supports just a few evergreen species that withstand a rigorously cold habitat. Above the subalpine belt, only low, nonwoody plants and dwarfed shrubs can grow.

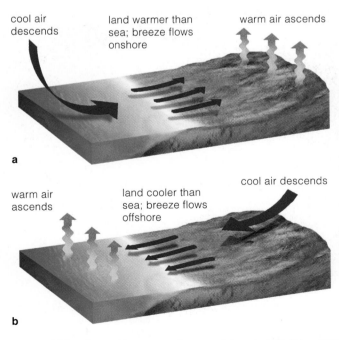

cool air descends

land warmer than sea; breeze flows onshore

warm air ascends

a

warm air ascends

land cooler than sea; breeze flows offshore

cool air descends

b

Figure 44.8 Coastal breezes, afternoon (**a**) and night (**b**).

Air warms and can hold more water after it flows over mountain crests. As it starts its descent, it retains moisture and draws more water out of plants and soil. The result is a **rain shadow**, a semiarid or arid region of sparse rainfall on the leeward side of high mountains. (*Leeward* is the direction not facing a wind; *windward* is the direction from which wind blows.) The Himalayas, Andes, Rockies, and other great mountain ranges cause impressive rain shadows. Similarly, on the windward side of Hawaii's high volcanic peaks are lush tropical forests; arid conditions prevail on the leeward side.

Monsoons are air circulation patterns that influence the continents north or south of warm oceans. The land heats intensely, so low-pressure air parcels form above it. The low pressure draws in moisture-laden air from the ocean, as from the Bay of Bengal into Bangladesh or from the Gulf of Mexico into deserts of the American Southwest. Fantastic summer thunderstorms are one result. Converging trade winds and the equatorial sun cause intense heating and rainfall in a zone called the doldrums. The air circulating north and south creates patterns of wet and dry seasons, as in eastern Africa.

Recurring sea breezes along coastlines are called "mini-monsoons." Here again, water and the adjacent land have different heat capacities. In the morning, the water's temperature does not rise as fast as the land's temperature. When warmed air above the land rises, cooler marine air moves in. After sunset, the land loses heat faster than water. Then, land breezes flow in the reverse direction (Figure 44.8).

Surface ocean currents, in combination with global air circulation patterns, influence regional climates and help distribute nutrients in marine ecosystems.

Air circulation patterns, ocean currents, and landforms interact in ways that influence regional temperatures and moisture levels. Thus they also influence the distribution and dominant features of ecosystems.

REALMS OF BIODIVERSITY

Circulation patterns in the atmosphere and at the ocean surface, together with topography, give rise to regional differences in temperature and moisture. And these differences help explain why deserts, grasslands, forests, and tundra form in some places but not others. It helps us understand the distribution of ecosystems and their species, each adapted to regional conditions.

For example, the information provides clues to why many evolutionarily distant species look alike. Often they emerged through convergent evolution, as Section 18.4 describes. Briefly, their ancestors were not closely related but faced similar selection pressures in similar yet geographically distant places. By natural selection, their body plans underwent similar modifications and ended up looking alike. Ancestors of those cactus and spurge plants in Figure 44.1 both evolved in hot, dry

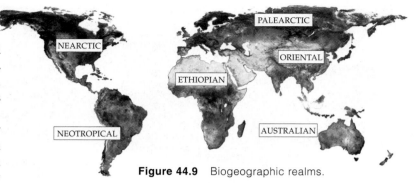

Figure 44.9 Biogeographic realms.

deserts, where water is scarce. Fleshy plant stems that have thick cuticles conserve precious water. And rows of sharp spines deter hungry, thirsty herbivores, which might otherwise chew on juicy plant parts.

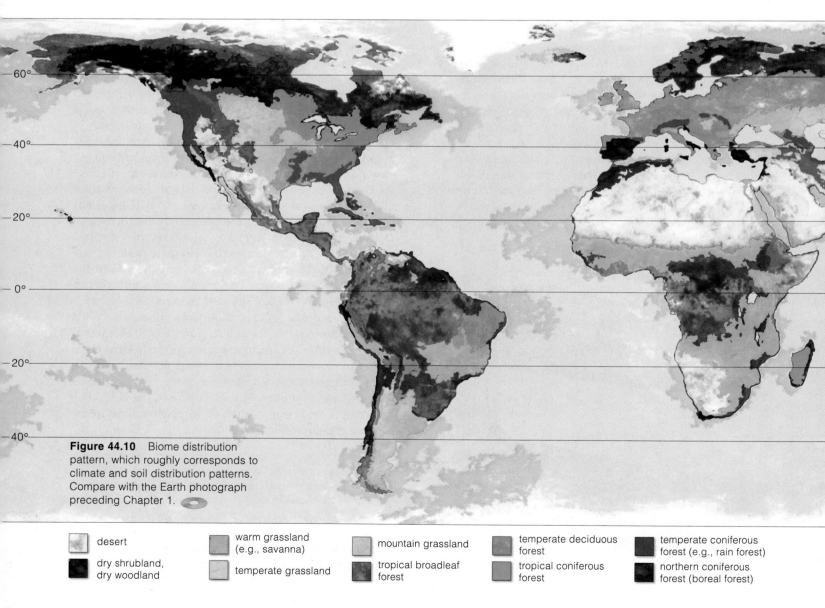

Figure 44.10 Biome distribution pattern, which roughly corresponds to climate and soil distribution patterns. Compare with the Earth photograph preceding Chapter 1.

- desert
- dry shrubland, dry woodland
- warm grassland (e.g., savanna)
- temperate grassland
- mountain grassland
- tropical broadleaf forest
- temperate deciduous forest
- tropical coniferous forest
- temperate coniferous forest (e.g., rain forest)
- northern coniferous forest (boreal forest)

High ← Elevation → Low

ALPINE TUNDRA

MONTANE CONIFEROUS FOREST

DECIDUOUS FOREST

TROPICAL FOREST

TROPICAL FOREST   TEMPERATE DECIDUOUS FOREST   NORTHERN CONIFEROUS FOREST   ARCTIC TUNDRA

High ◄——————— Moisture Availability ———————► Low

Figure 44.11 For North America, changes in plant form along environmental gradients. Gradients in elevation and availability of water help dictate a biome's primary productivity. Other factors, such as mean annual air temperature and soil drainage, also are at work.

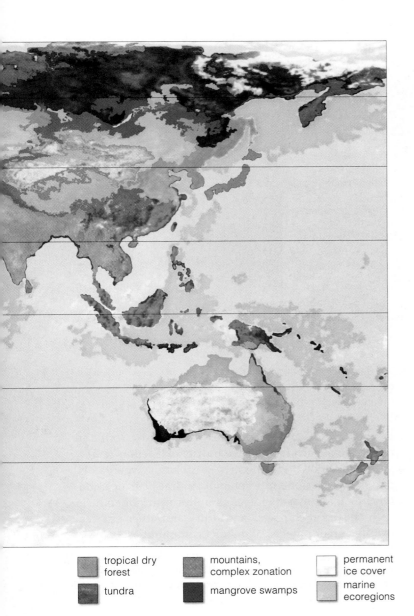

- ☐ tropical dry forest
- ☐ tundra
- ☐ mountains, complex zonation
- ☐ mangrove swamps
- ☐ permanent ice cover
- ☐ marine ecoregions

W. Sclater and, later, Alfred Wallace, bestowed the name **biogeographic realms** on six immense land areas, each with distinctive kinds and numbers of plants and animals (Figure 44.9). Biogeographic realms maintain their identity partly because of climate. They also tend to maintain a distinct identity when mountain ranges, oceans, and other physical barriers restrict gene flow, thus keeping the gene pools of each realm isolated.

A biogeographic realm may be divided into biomes. Each **biome** is a large region of land characterized by habitat conditions and community structure, including endemic species (which evolved nowhere else). Take a look at how Figure 44.10 correlates with environmental factors of the sort shown in Figure 44.11. It helps show that distinctive biomes prevail at certain latitudes and elevations. Biomes dominated by low-growing plants prevail in dry regions, at high elevations, and at high latitudes. Those dominated by tall, leafy plants prevail at tropical latitudes, temperate latitudes, and at lower elevations with warm temperatures and high rainfall. As you will see, soils also affect biome distribution.

Conservationists are working to locate, inventory, and protect "hot spots." These are portions of certain biomes that are the richest in biodiversity and the most vulnerable to species loss (Section 25.5). Twenty-four of the hot spots hold more than half of all land species.

In its Global 200 program, the World Wildlife Fund is targeting ecoregions as well as hot spots. **Ecoregions** are large areas representative of the globally important biomes *and* water provinces vulnerable to extinction. Figure 44.10 identifies the marine ecoregions.

The land and the seas contain realms of biodiversity. The unique identity of each biogeographic realm, biome, and ecoregion is an outcome of habitat conditions, the kinds and numbers of species, and their evolutionary history.

SOILS OF MAJOR BIOMES

Most regions of land have **soils**, which are mixtures of mineral particles and varying amounts of decomposing organic material (humus). As described in Section 27.1, the weathering of hard rocks produces coarse-grained gravel, then sand, silt, and finely grained clay. Water and air infiltrate spaces between particles. How much these particles compact together, and the proportions of each kind, differ within and between regions.

Soils have a layered structure, a *profile*, that reflects their stage of development. Figure 44.12 shows a few examples. Uppermost is topsoil, with the most humus and the most vulnerability to erosion. Topsoil may be less than a centimeter thick on steep slopes and more than a meter thick in grasslands. The lowest layer of soil consists of rocks in varying degrees of weathering.

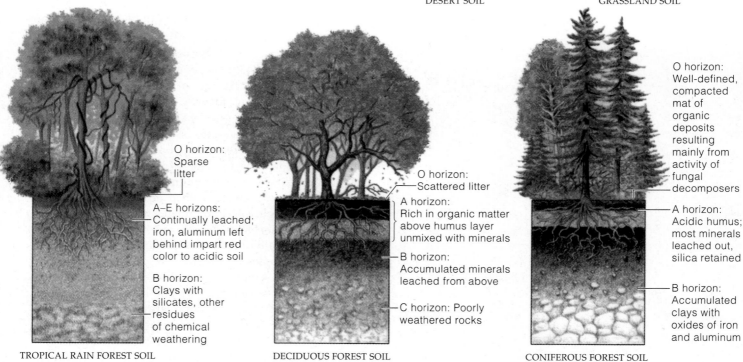

DESERT SOIL

O horizon: Pebbles, little organic matter

A horizon: Shallow, poor soil

B horizon: Leaching results in salinization (accumulated calcium, sodium)

C horizon: Rock fragments from uplands

GRASSLAND SOIL

A horizon: Alkaline, deep, rich in humus

B horizon: Percolating water enriches with calcium carbonates

TROPICAL RAIN FOREST SOIL

O horizon: Sparse litter

A–E horizons: Continually leached; iron, aluminum left behind impart red color to acidic soil

B horizon: Clays with silicates, other residues of chemical weathering

DECIDUOUS FOREST SOIL

O horizon: Scattered litter

A horizon: Rich in organic matter above humus layer unmixed with minerals

B horizon: Accumulated minerals leached from above

C horizon: Poorly weathered rocks

CONIFEROUS FOREST SOIL

O horizon: Well-defined, compacted mat of organic deposits resulting mainly from activity of fungal decomposers

A horizon: Acidic humus; most minerals leached out, silica retained

B horizon: Accumulated clays with oxides of iron and aluminum

The growth of most plants suffers in poorly aerated, poorly draining soils. Remember, loam topsoils have the best mix of sand, silt, and clay for agriculture. They have enough coarse particles to promote drainage and enough fine particles to retain water-soluble mineral ions that serve as nutrients for plant growth. Gravelly or sandy soils promote rapid leaching, which depletes them of water and vital minerals. Clay soils with fine, closely packed particles are poorly aerated and do not drain well. Few plants grow in waterlogged clay soils.

As farmers know, soil affects primary productivity. They grow most crops in cleared, former grasslands. Many burn or clear-cut tropical forests for agriculture. But these biomes have very little topsoil above poorly

Figure 44.12 Soil profiles from a few representative biomes. Compare Section 27.1.

draining sublayers. Clearing exposes the topsoil, and heavy rains leach most of the nutrients (Section 25.6).

We turn now to the deserts, shrublands, woodlands, grasslands, various forests, and tundra. Bear in mind, none of these major biomes is uniform throughout. In each one, local climates, landforms, and other physical features favor patches of distinct communities.

Primary productivity depends on the soil profile and its proportions of sand, silt, clay, gravel, and humus.

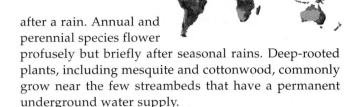

DESERTS

Deserts form on land with less than ten centimeters or so of annual rainfall and high potential for evaporation. Such conditions prevail at latitudes of about 30° north and south. There we find great deserts of the American Southwest, of northern Chile, Australia, northern and southern Africa, and Arabia. Farther north are the high deserts of eastern Oregon, and Asia's vast Gobi and the Kyzyl-Kum east of the Caspian Sea. Rain shadows are the main reason these northern deserts are so arid.

after a rain. Annual and perennial species flower profusely but briefly after seasonal rains. Deep-rooted plants, including mesquite and cottonwood, commonly grow near the few streambeds that have a permanent underground water supply.

More than a third of the Earth's land surfaces is arid or semiarid, without enough rainfall to support crops. Crops in California's Imperial Valley and some other

Figure 44.13 Warm desert near Tucson, Arizona. Primary producers include creosote bushes, multistemmed ocotillos, tall saguaro cacti, and prickly pear cacti with rounded pads.

Deserts do not have lush vegetation. Rain falls in heavy, brief, infrequent pulses that swiftly erode the exposed topsoil. Humidity is so low that the sun's rays easily penetrate the air. They quickly heat the ground's surface, which radiates heat and cools quickly at night.

Although arid or semiarid conditions do not favor large, leafy plants, deserts show plenty of biodiversity. In a patch of Arizona's desert (as in Figure 44.13), you might find creosote and other deep-rooted, evergreen, woody shrubs, fleshy-stemmed, shallow-rooted cacti, tall saguaros, short prickly pears, and ocotillos, which drop leaves more than once a year and grow new ones

deserts need intensive soil management and irrigation. Without drainage programs, crop production declines from waterlogging and salt buildup. Many parts of the world are undergoing **desertification**. The word refers to conversion of grasslands and similarly productive biomes to desertlike wastelands with low biodiversity. We take a closer look at this trend in Section 45.6.

Where the potential for evaporation greatly exceeds sparse rainfall, deserts form. This condition prevails at latitudes 30° north and south and in rain shadows.

DRY SHRUBLANDS, DRY WOODLANDS, AND GRASSLANDS

Dry shrublands and **dry woodlands** prevail in western or southern coastal regions of the continents between latitudes 30° and 40°. These semiarid regions get a bit more rain than deserts, mostly during mild winters. Summers are long, hot, and arid. The dominant plants often have hardened, tough, evergreen leaves.

Dry shrublands get less than 25 to 60 centimeters of rain per year. We see them in California, South Africa, and Mediterranean regions, where they are known by such local names as fynbos and chaparral. California has 2.4 million hectares (6 million acres) of chaparral. In summer, lightning-sparked, wind-driven firestorms sweep through these biomes (Figure 44.14). The shrubs have highly flammable leaves and quickly burn to the ground. However, they are highly adapted to episodes of fire and soon resprout from their root crowns. Trees do not fare as well during firestorms. Shrubs actually "feed" the fires and have the competitive edge.

Dry woodlands dominate where annual rainfall is about 40 to 100 centimeters. The dominant trees can be tall, but they do not form a dense, continuous canopy. Eucalyptus woodlands of southwestern Australia and oak woodlands of California and Oregon are like this.

Grasslands cloak much of the interior of continents in zones between deserts and temperate forests. Warm temperatures prevail in the summer, and winters are exceedingly cold. The 25 to 100 centimeters of annual rainfall keep deserts from forming but aren't enough to support forests. Drought-tolerant primary producers survive strong winds, light and infrequent rainfall, and rapid evaporation. Dominant animals are grazing and burrowing species. Their grazing activities, combined with periodic fires, keep shrublands and forests from encroaching on the fringes of many grasslands.

North America's major grasslands are *shortgrass* and *tallgrass* prairie. Most form on flat or rolling land, as in Figure 44.15a. Roots of perennial plants extend profusely through the topsoil. In the 1930s, shortgrass prairie of the Great Plains was overgrazed and plowed under to grow wheat, which requires more water than this region sometimes gets. Strong winds, prolonged droughts, and unsuitable farming practices turned much of the prairie into the Dust Bowl (Section 45.6). John Steinbeck's *The Grapes of Wrath* and James Michener's *Centennial*, two historical novels, eloquently describe the consequences.

The tallgrass prairie once extended west from temperate deciduous forests (Figures 43.26 and 44.15b). A variety of legumes and composites, including daisies, thrived in the continent's interior, which had richer topsoil and slightly more frequent rainfall. Farmers converted nearly all of the original tallgrass prairie for agriculture, but certain areas are now being restored (Section 42.6).

Between the tropical forests and deserts of Africa, South America, and Australia are broad belts of grasslands with a smattering of shrubs or trees. These are the **savannas** (Figure 44.15c). Rainfall averages 90 to 150 centimeters a year, and prolonged seasonal droughts are common. Where rainfall is low, fast-growing grasses dominate. Acacia and other shrubs grow in regions that get a bit more moisture. Where the rainfall is higher,

Figure 44.14 California chaparral. The dominant plants are multibranched, woody, and typically only a few meters tall. Without periodic fires, they form a nearly impenetrable vegetation cover. The boxed inset shows a firestorm racing through a chaparral-choked canyon above Malibu.

Figure 44.15 (**a**) Rolling shortgrass prairie to the east of the Rocky Mountains. Once, bison were the dominant large herbivore of North American grasslands. At one time, the astoundingly abundant grasses supported 60 million of these hefty ungulates. Ungulates are hooved, plant-eating mammals.

(**b**) A rare patch of natural tallgrass prairie in eastern Kansas.

(**c**) The African savanna—a warm grassland with scattered stands of shrubs and trees. At present, more varieties and greater numbers of large ungulates live here than anywhere else. They include the migratory wildebeests shown in this photograph and giraffes, Cape buffalos, zebras, and impalas.

savannas grade into tropical woodlands with low trees and shrubs, and tall, coarse grasses. *Monsoon* grasslands form in southern Asia where heavy rains alternate with a dry season. Dense stands of tall, coarse grasses form, then die back and often burn during the dry season.

Plants of dry shrublands, dry woodlands, and grasslands are supremely adapted to surviving strong winds, grazing animals, and recurring episodes of drought and fire.

TROPICAL RAIN FORESTS AND OTHER BROADLEAF FORESTS

In forest biomes, tall trees grow close together, forming a fairly continuous canopy over a broad stretch of land. In general, there are three types of trees. Which type prevails in a region depends partly on distance from the equator. Evergreen broadleafs dominate between latitudes 20° north and south. Deciduous broadleafs are common at moist, temperate latitudes where winters are not severe. Evergreen conifers are common at high, colder latitudes and in mountains of temperate zones.

Evergreen broadleaf forests sweep across tropical zones of Africa, the East Indies and Malay Archipelago, Southeast Asia, South America, and Central America. Annual rainfall can exceed 200 centimeters and never

is less than 130 centimeters. One biome, the **tropical rain forest**, depends on regular and heavy rainfall, an annual mean temperature of 25°C, and humidity of at least 80 percent (Figure 44.16). Evergreen trees of this highly productive forest produce new leaves and shed old ones all year long. But litter does not accumulate; decomposition and mineral cycling are notably rapid, given the hot, humid climate. The soils are weathered, humus-deficient, and poor nutrient reservoirs. We will take a closer look at these forests in the next chapter.

Leaving the tropical rain forests, we enter regions where temperatures remain mild but rainfall dwindles during part of the year. This is the start of **deciduous**

Figure 44.16 Tropical rain forests, biomes of great biodiversity. Among millions of known species are jaguars and *Rafflesia*, a leafless, foul-smelling plant with a fly-pollinated flower that grows three meters across. Bromeliads, orchids, and other epiphytes grow on trees, obtaining minerals from organic remains of leaves, insects, and other litter that have dissolved in tiny pockets of water.

SPRING

SUMMER

WINTER

AUTUMN

Figure 44.17 The changing appearance of part of a temperate deciduous forest south of Nashville, Tennessee, in spring, summer, autumn, and winter.

broadleaf forests. Trees of *tropical* deciduous forests drop some or all of their leaves in a pronounced dry season. *Monsoon* forests of India and southeastern Asia have such trees, also. Farther north, in the temperate zone, rainfall is even lower. Winters are cold; water is locked away as snow and ice. *Temperate* deciduous forests, such as those of the southeastern United States, prevail here (Figure 44.17). Decomposition is not as rapid as in the humid tropics, and many nutrients are conserved in accumulated litter on the forest floor.

Complex forests of ash, beech, birch, chestnut, elm, and deciduous oaks once stretched across northeastern North America, Europe, and East Asia. They declined drastically when farmers cleared the land. Pathogenic species introduced to North America killed nearly all chestnuts and many elms (Section 42.8). Now, maple and beech predominate in the northeast. Farther west, oak–hickory forests prevail, then oak woodlands that eventually grade into tallgrass prairie.

In forest biomes, conditions favor dense stands of tall trees that form a continuous canopy over a broad region.

CONIFEROUS FORESTS

Conifers (cone-bearing trees) are primary producers of **coniferous forests**. Most have thickly cuticled, needle-shaped leaves with recessed stomata—adaptations that help them conserve water during droughts and winter. They dominate the boreal forests, montane coniferous forests, temperate rain forests, and pine barrens.

Boreal forests stretch across northern Europe, Asia, and North America. Such forests also are called *taigas*, meaning "swamp forests." Most are in glaciated regions having cold lakes and streams (Figure 44.18a). It rains mostly in summer, and evaporation is low in the cool summer air. The cold, dry winters are more severe in eastern parts of these biomes than in the west, where

surrendered its rights to clear-cut one of the largest intact temperate rain forests outside of tropical regions. The tall trees, some of which are 800 years old, cloak British Columbia's Kitlope Valley.

Southern pine forests dominate the coastal plains of the south Atlantic and the Gulf states. Pine species are adapted to the dry, sandy, nutrient-poor soil and to natural fires or controlled burns. These fires do not get hot enough to damage the trees, and they open up the understory. Forests of pine, scrub oak, and wiregrass grow in New Jersey. Palmettos grow below pines and loblolly in the Deep South. During one recent Florida

ocean winds moderate the climate. Spruce and balsam fir dominate the boreal forests of North America. Pines, birches, and aspens take hold in the burned or logged areas. Where soil is poorly drained, highly acidic bogs dominated by peat mosses, shrubs, and stunted trees prevail. Boreal forests become much less dense to the north, where they grade into arctic tundra.

In the Northern Hemisphere, montane coniferous forests extend southward through the great mountain ranges. Spruces and firs dominate in the north and at higher elevations. They give way to firs and pines in the south and at lower elevations (Figures 3.1 and 44.18b). Some temperate lowlands support coniferous forests. One temperate rain forest that parallels the coast from Alaska on into northern California contains some of the world's tallest trees—Sitka spruce to the far north and redwoods to the south. Logging destroyed much of this biome. One bright note: In 1994, a timber corporation

Figure 44.18 (**a**) Spruce-dominated boreal forest. (**b**) Montane coniferous forest of Yosemite Valley in California's Sierra Nevada. (**c**) Firefighter retreating from flames in a forest of pines, oaks, and palmettos by Daytona Beach, Florida.

heat wave, fires destroyed more than 320,000 acres. Smaller burns had been prohibited in counties where housing and other developments encroached on these forests. Dense, tinder-dry understory had accumulated, and it fed the fires (Figure 44.18c).

Coniferous forests prevail in regions in which a cold, dry season alternates with a cool, rainy season.

ARCTIC AND ALPINE TUNDRA

Tundra is derived from *tuntura*, a Finnish word for the great treeless plain between the polar ice cap and belts of boreal forests in Europe, Asia, and North America. This plain is *arctic* tundra (Figure 44.19*a,b*). Here you find extremely low temperatures, poor drainage, short growing seasons when sunlight is nearly continuous, and poor decomposition of organic matter. The annual precipitation (rain and melting snow) varies regionally but typically is less than 25 centimeters. Lichens and short, hardy, shallow-rooted plants are a base for food webs that include voles, arctic hares, caribou, and other herbivores. Carnivores include arctic foxes, wolves, and polar bears. Many migratory birds nest here in summer.

and it accumulates in soggy masses. About 95 percent of the carbon that photoautotrophs fixed in the arctic tundra is locked up in extensive peat bogs.

A similar type of biome prevails at high elevations in mountains throughout the world. Figure 44.19*c* is one example of this *alpine* tundra. The temperatures at night are typically below freezing. It is just too cold for any trees to grow. However, unlike the arctic tundra, alpine tundra has no permafrost; the soil is thin but well drained. Grasses, heaths, and other small-leafed shrubs form low, compact cushions and mats that can withstand the buffeting of strong winds.

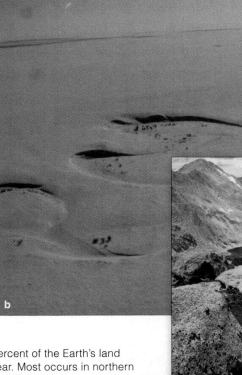

Figure 44.19 Arctic tundra in Russia in summer (**a**) and winter (**b**). About 4 percent of the Earth's land mass is arctic tundra, which snow blankets for as long as nine months of the year. Most occurs in northern Russia and Canada, although we also find some in Alaska and Scandinavia. Small bands of humans have hunted, fished, and herded reindeer in these sparsely populated regions for hundreds of thousands of years. Today, mineral and fossil fuel deposits are being extracted, especially in Russia. These are fragile biomes. Plants and animals must survive extreme cold in winter. They might be vulnerable to industrial pollution. Severe climatic changes that are expected to result from increased global warming may put them at risk. (**c**) Short, compact, hardy plants typical of alpine tundra. Mount Evans, Colorado, is in the distance.

Summers last fifty to sixty days and temperatures average 3°–12°C (37°–54°F), so not much more than the surface soil thaws. Mosquitoes, flies, and other insects proliferate under these conditions. Just beneath the soil surface is a frozen layer, called **permafrost**, that never melts. It is more than 500 meters thick in some places. Permafrost prevents drainage, so the soil above it stays waterlogged. The cool, anaerobic conditions hamper nutrient cycling. Organic matter decomposes slowly,

Even in summer, shaded patches of snow persist in alpine tundra. The thin, fast-draining soil is nutrient-poor, so primary productivity is low.

At high latitudes with short, cool summers and long, cold winters, we find arctic tundra. In high mountains where seasonal changes vary with latitude but where the climate is too cold to support forests, we find alpine tundra.

FRESHWATER PROVINCES

Freshwater and saltwater provinces cover more of the Earth's surface than all biomes combined. They include the ocean, lakes, ponds, wetlands, and those coral reefs described in Section 25.3. "Typical" examples are not that easy to find. Some ponds can be waded across; Lake Baikal in Siberia is more than 1.7 kilometers deep. All aquatic ecosystems have gradients in light penetration, temperature, and dissolved gases, but the values differ greatly. All we can do here is sample the diversity.

Lake Ecosystems

A **lake** is a body of standing fresh water produced by geologic events, as when an advancing glacier carves out the land beneath it. After the glacier retreats, water collects in the exposed basin (Figure 44.20). Over time, erosion and sedimentation alter the dimensions of the lake, which usually ends up filled or drained.

A lake has littoral, limnetic, and profundal zones (Figure 44.21). The littoral extends all around the shore to a depth at which rooted aquatic plants stop growing. Lake diversity is greatest in its shallow, usually well-lit water. Beyond the littoral, the limnetic's open, sunlit water extends to depths where photosynthetic activity is negligible. Cyanobacteria, green algae, and diatoms dominate the lake's *phyto*plankton. They feed rotifers, copepods, and other members of the *zoo*plankton. The profundal extends down through all open water below the depth at which wavelengths of light suitable for photosynthesis can penetrate. Detritus sinks through this zone to the lake's bottom. There, communities of diverse bacterial decomposers live in and upon bottom sediments, and they enrich the water with nutrients.

SEASONAL CHANGES IN LAKES In temperate regions having warm summers and cold winters, lakes show seasonal changes in density and temperature from the surface to the bottom. A layer of ice forms over many of them in midwinter. Water near the freezing point is the least dense and accumulates under the ice. Water at 4°C is the most dense. It collects in deeper layers that are a bit warmer than the surface layer in midwinter.

Figure 44.20 In the Canadian Rockies, the basin of Moraine Lake, formed by glacial action during the most recent ice age.

In spring, daylength increases and the air warms. Lake ice melts, the surface water warms to 4°C, and temperatures turn uniform throughout. Winds over the surface layer cause a **spring overturn**: strong vertical movements carry dissolved oxygen near the surface to the depths, and nutrients released by decomposition move from sediments to the surface.

By midsummer there is a *thermocline*, a midlayer of water where an abrupt change in temperature blocks vertical mixing (Figure 44.22). Being warmer and not as dense, the surface water floats on the thermocline. In cooler water below, decomposers deplete dissolved oxygen. In autumn, the upper layer cools and becomes denser. It sinks, and the thermocline vanishes. During this **fall overturn**, water mixes vertically, so dissolved oxygen moves down and nutrients move up.

Primary productivity corresponds with the seasons. After a spring overturn, the longer daylengths and the cycled nutrients favor higher rates of photosynthesis. Phytoplankton and rooted aquatic plants quickly take up phosphorus, nitrogen, and other nutrients. During a growing season, the thermocline stops vertical mixing. Nutrients locked in the remains of organisms sink to deeper water, so photosynthesis slows. By late summer,

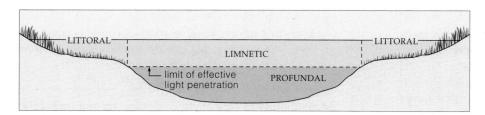

Figure 44.21 Lake zonation. The littoral extends around a lake's edge from the shore to the depth where aquatic plants stop growing. The profundal is all water below the depth of light penetration. Above the profundal are open, sunlit waters of the limnetic zone.

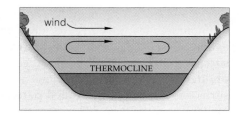

Figure 44.22 Thermal layering, which occurs in many lakes of temperate zones during the summer.

shortages are limiting photosynthesis. A fall overturn cycles nutrients to the surface, which drives a burst of primary productivity. But the shorter daylight hours cannot sustain a long-lasting burst. Not until spring will primary productivity increase once more.

TROPHIC NATURE OF LAKES   The topography, climate, and geologic history of a lake dictate the numbers and kinds of residents, how they are dispersed in the lake, and how they cycle nutrients. Soils of the lake basin and the surrounding regions contribute to the type and amount of nutrients available to support organisms.

Interplays among the climate, soil, basin shape, and metabolic activities of the lake's residents contribute to conditions that range from oligotrophy to eutrophy. Water in *oligotrophic* lakes often is deep, clear, and low in nutrients. Its primary productivity isn't great. Water in *eutrophic* lakes usually is shallow, nutrient rich, and high in primary productivity. Such conditions develop naturally as sediments accumulate in the lake basin. The water gradually becomes less transparent and less deep, compared with an oligotrophic lake, and aquatic communities come to be dominated by phytoplankton. If more sediments build up, the final successional stage will be a filled-in basin—hence no more lake.

Eutrophication refers to any processes that enrich a body of water with nutrients. Section 43.8 described how human activities can bring it about. Here, think of how people caused eutrophication of Lake Washington in Seattle. From 1941 to 1963, phosphate-rich sewage entered the lake. It promoted blooms of cyanobacteria that formed thick, slimy mats in summer, making the lake useless for recreation. With so much phosphorus in the water, nitrogen became the limiting resource. Cyanobacteria are superior competitors for nitrogen; they can fix N_2. They became dominant. The sewage discharges finally were prohibited—and by 1975, the lake neared full recovery. Lower density populations of diatoms and green algae became dominant.

Stream Ecosystems

Flowing-water ecosystems called **streams** start out as freshwater springs or seeps. They grow and merge as they flow downslope, then often combine into a river. Between a river's headwaters and end, we find three

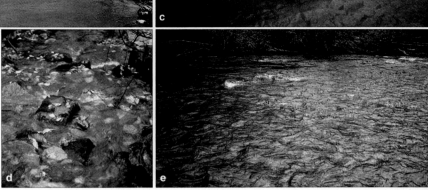

Figure 44.23   Stream habitats in North Carolina and Virginia. (**a**) Pool. (**b**) Pool leading into a riffle. (**c**) A run, Sinking Creek. (**d**) Leaf detritus in a riffle. (**e**) Closer look at a riffle.

kinds of habitats: riffles, pools, and runs (Figure 44.23). *Riffles* are shallow, turbulent stretches where the water flows swiftly over a rough sandy, rocky bottom. *Pools* have deep water flowing slowly over a smooth, sandy, or muddy bottom. *Runs* are smooth-surfaced but fast-flowing stretches over bedrock or rock and sand.

A stream's average flow volume and temperature depend on rainfall, snowmelt, geography, altitude, and even the shade cast by plants. Its solute concentrations are influenced by the streambed's composition as well as by agricultural, industrial, and urban wastes.

Especially in forests, streams import most of the organic matter that supports food webs. Where trees cast shade, they hamper photosynthesis. Most of the organic matter is litter that enters detrital food webs. Aquatic organisms continually take up and release nutrients as water flows downstream. Nutrients move upstream only in the tissues of migratory fishes and other animals. Think of nutrients as spiraling between water and aquatic organisms as the stream flows on its one-way course, usually to a river that flows to the sea.

Ever since cities formed, streams have been sewers for industrial and municipal wastes. The wastes, along with other pollutants from poorly managed farmlands, choked many streams with sediments and chemically poisoned them. Streams, however, are resilient. They can recover impressively with pollution controls.

Freshwater and saltwater provinces are far more extensive than the Earth's biomes. All of the aquatic ecosystems in these water provinces have characteristic gradients in light penetration, temperature, and dissolved gases.

THE OCEAN PROVINCES

Beyond land's end are two vast provinces of the world ocean, which covers nearly three-fourths of the Earth's surface. The *benthic* province includes sediments and rocks of the ocean bottom (Figure 44.24). It begins at continental shelves and extends to deep-sea trenches. The *pelagic* province is the full volume of ocean water. Its neritic zone is all water above continental shelves, and its oceanic zone is the water of the ocean basins.

Walk along the ocean and its vastness may humble you. An immense volume of water extends to a distant horizon—with no mountains or plains or valleys to break it up visually into something less overwhelming. What you do not see are the *submerged* mountains and valleys and plains—which are as varied as the ones that sculpt the dry continents.

Primary Productivity in the Ocean

Photosynthesis proceeds on a stupendous scale in the ocean's upper surface, just as it does on land. Primary productivity varies seasonally, just as it does on land (Section 6.8 and Figure 44.25). Drifting through the water are huge "pastures" of phytoplankton, the basis of food webs that include copepods, shrimplike krill, whales, squids, and fishes. The organic remains and wastes of its communities sink to the bottom of the ocean basin. They support the detrital food webs for most benthic communities.

Near the ocean surface, 70 percent of the primary productivity might be the work of **ultraplankton**—photosynthetic bacteria no more than 2 micrometers (40 millionths of an inch) wide. In subtropical and tropical seas once thought to be practically devoid of producers, 0.035 gram (or 1 ounce) of water holds up to 3 million bacterial cells. In deep water too dark to support photosynthesis, food webs start with **marine snow**. These bits of organic matter drift down to become the food base for staggering biodiversity in midoceanic waters—possibly up to 10 million species. In what may be the greatest of circadian migrations, some species rise thousands of feet to feed in shallower water at night, then move down the next morning. The top carnivores include the siphonophores, a type of soft-bodied cnidarian. At 39 meters, one specimen of *Praya dubia*—a recently discovered siphonophore—is one of the longest existing animals in the world!

Hydrothermal Vents

In the Galápagos Rift, a volcanically active boundary between two crustal plates, we observe communities thriving near **hydrothermal vents**. Near-freezing water seeps down into these fissures in the seafloor and gets heated to extremely high temperatures. As the heated water moves up under pressure, it leaches mineral from rocks before being spewed out. Iron, zinc, copper sulfides, and sulfates of magnesium and calcium are dissolved in the outpourings. They settle out and form rich mineral deposits. Sulfides in these deposits are

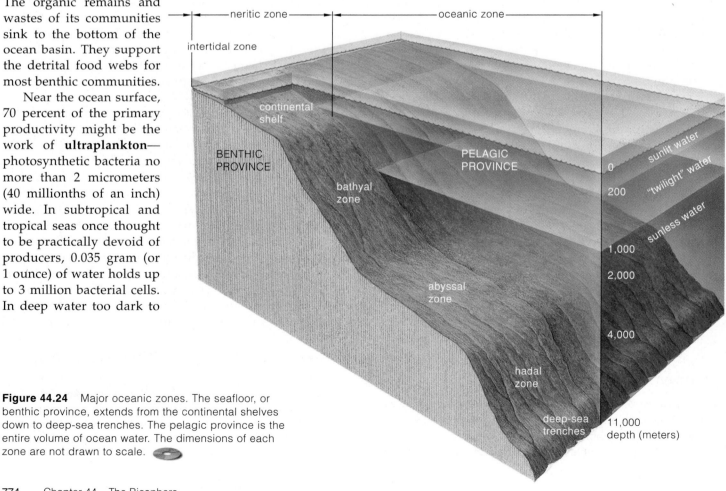

Figure 44.24 Major oceanic zones. The seafloor, or benthic province, extends from the continental shelves down to deep-sea trenches. The pelagic province is the entire volume of ocean water. The dimensions of each zone are not drawn to scale.

Figure 44.25 (**a**) A wave breaking on the ocean shore. (**b**) Near the British Columbia coastline, a whale breaching (leaping out of the water).

(**c**) Seasonal variations in primary production in the ocean, which correspond to latitude. The strongest peak in the *dark-green* graph lines for the north polar and temperate seas corresponds to phytoplankton blooms brought about by a seasonal increase in daylength. The lower peak corresponds to an increase in the availability of nutrients, something like the fall overturn in lakes. In most tropical seas, daylength and nutrient availability do not vary much; neither does primary productivity.

Tube worms (**d**) and crustaceans (**e**) at a hydrothermal vent ecosystem on the ocean floor.

energy sources for chemoautotrophs that are the basis of hydrothermal vent communities. Chemoautotrophic bacteria are the primary producers for food webs that include red tube worms, crustaceans, clams, and fishes (Figure 44.25*d,e*). More hydrothermal vent ecosystems occur in the South Pacific near Easter Island; the Gulf of California, about 150 miles south of the tip of Baja California, Mexico; and the Atlantic. In 1990, a team of United States and Russian scientists found one in Lake Baikal, the world's deepest lake. This lake basin seems to be splitting apart (hence the vents) and may define the beginning of a new ocean.

Did life originate in such nutrient-rich places on the seafloor? Conditions at the surface of the early Earth were about as inhospitable as you might imagine. At the least, cells living on the seafloor would have had

protection from the destructive radiation bombarding the Earth before the oxygen-rich atmosphere formed. Were some descendants of those cells carted closer to the sea surface when the seafloor was uplifted during episodes of crustal crunchings? These are some of the unanswered questions that evolutionary detectives are now asking. Section 6.7 and Chapter 19 include some of their interesting hypotheses.

Although we are most familiar with features of the land, a continuous ocean dominates most of the Earth's surface. Its primary productivity and biodiversity are staggering.

Rich communities that start with chemoautotrophs thrive near hydrothermal vents, which suggests to some that life itself could have originated deep in some ancient ocean.

WETLANDS AND THE INTERTIDAL ZONE

Rich ecosystems develop near the coasts of continents and islands. Like freshwater ecosystems, they differ in their physical and chemical properties, such as light penetration and water temperature, depth, and salinity.

Mangrove Wetlands and Estuaries

In tidal flats at tropical latitudes, we find nutrient-rich, **mangrove wetlands**. "Mangrove" refers to forests in sheltered regions along tropical coasts (Figure 44.26a). Ocean waves do not reach these saltwater ecosystems, so anaerobic sediments and mud accumulate. The term also refers to a particular group of salt-tolerant plants (halophytes), such as red mangrove (*Rhizophora*). Plants have shallow, spreading roots or branching prop roots that extend from the trunk (Figure 44.26b). Many have pneumatophores, root extensions that take up oxygen.

Florida's mangrove wetlands have few mangrove species, whereas those in Southeast Asia have as many as thirty or forty. Seeds of all species germinate while still attached to the parent tree. Seedlings drop to the water and float upright until currents deposit them in the mud of shallow water, where their roots take hold.

Net primary productivity of a mangrove wetland varies, depending partly on the volume and flow rate of water moving in and out with the tides. It depends on salinity and nutrient availability, which is affected by litter accumulation and turnover. But these are all rich ecosystems. Tidal circulation and nutrient inputs from streams and rivers combine to support notable biomass. Particularly along the Malaysian peninsula and the Gulf of Mexico, the nutrient-rich detritus from mangrove wetlands enriches nearby estuaries.

An **estuary** is a partly enclosed coastal region where seawater slowly mixes with nutrient-rich fresh water from rivers, streams, and runoff from nearby land. The

Figure 44.26 (**a**) One mangrove wetland, part of Florida's Everglades. (**b**) Some red mangrove prop roots.

confined conditions, slow mixing of water, and tidal action combine to trap dissolved nutrients. Freshened water continually replenishes the nutrients; that is why estuaries can support productive ecosystems.

Primary producers are phytoplankton, salt-tolerant plants that tolerate submergence at high tide, and algae in mud and on plants. Much of the primary production enters detrital food webs in which bacterial and fungal decomposers are the first to feed. Detritus (and bacteria on and in it) feeds nematodes, snails, crabs, and fishes. Food particles suspended in the slowly moving water feed clams and other filter feeders. Estuaries nurture so many larval and juvenile stages of invertebrates and some fishes that they are often called marine nurseries. Also, many migratory birds use estuaries as rest stops.

Estuaries range from broad, shallow Chesapeake Bay, Mobile Bay, and San Francisco Bay to the narrow, deep fjords of Norway and their equivalents in Alaska and British Columbia. Texas and Florida estuaries lie behind long sandy or muddy spits. New England's salt marshes are familiar estuaries (Figure 44.27).

Many estuarine ecosystems are under assault from raw sewage; agricultural runoff; industrial, urban, and suburban wastes; and upstream diversion of fresh water. Normal conditions, including suitable salinity levels, can't be maintained without inflows of unpolluted fresh water.

Rocky and Sandy Coastlines

Along rocky and sandy coastlines we find ecosystems of the **intertidal zone**, which is not renowned for creature comforts. Waves batter its resident species. Tides alternately

SALT MARSH (estuary)

open ocean   sound   shallow bay   creek

Figure 44.27 Salt marsh of a New England estuary with the open ocean in the distance. *Spartina*, a marsh grass, is the main producer. Its microbe-enriched litter feeds consumers in shallow water and muddy sediments. Compare Section 43.3.

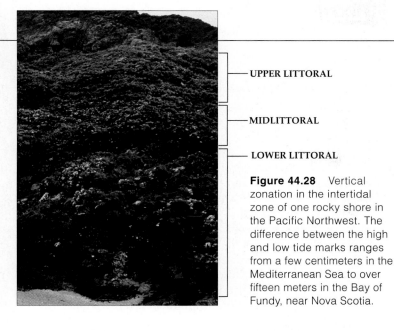

UPPER LITTORAL

MIDLITTORAL

LOWER LITTORAL

Figure 44.28 Vertical zonation in the intertidal zone of one rocky shore in the Pacific Northwest. The difference between the high and low tide marks ranges from a few centimeters in the Mediterranean Sea to over fifteen meters in the Bay of Fundy, near Nova Scotia.

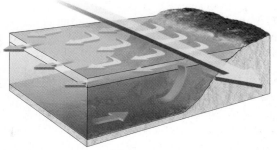

a Wind from the north starts surface ocean water moving.

b Force of Earth's rotation deflects the moving water westward.

c Cold water moves up as replacement.

Figure 44.29 Upwelling along the west coasts in the Northern Hemisphere. Prevailing winds from the north start surface water moving. The force of Earth's rotation deflects the moving water westward. Cold, deeper water moves up to replace it.

submerge and expose them. The higher they are, the more they dry out, freeze in winter, or bake in summer, and the less food comes their way. The lower they are, the more they must compete in limited spaces. At low tides, birds, rats, and raccoons move in and feed on them. High tides bring the predatory fishes.

Generalizing about coastlines isn't easy, for waves and tides continually resculpt them. One feature that rocky and sandy shores do share is vertical zonation.

Rocky shores often have three zones (Figure 44.28). The *upper* littoral is submerged only during the highest tide of the lunar cycle and is sparsely populated. The *mid*littoral is submerged by the highest regular tide and exposed by the lowest. Its tidepools hold red, brown, and green algae, hermit crabs, nudibranchs, sea stars, and fishes. Diversity is greatest in the *lower* littoral, which is exposed only by the lowest tide of the lunar cycle. In all three zones, rapid erosion prevents detritus from accumulating, so grazing food webs prevail.

Waves and currents continually rearrange stretches of loose sediments called *sandy* and *muddy* shores. Few large plants grow in these unstable places, so you won't find many grazing food webs. Detrital food webs start with organic debris from offshore or nearby landforms. Vertical zonation is less clear than along rocky shores. Below the low tide mark in temperate regions are sea cucumbers and blue crabs. Marine worms, crabs, and other invertebrates thrive between high and low tide marks. At night, at the high tide mark, beach hoppers and ghost crabs pop out of their burrows, seeking food.

Upwelling Along the Coasts

In the Northern Hemisphere, prevailing winds from the north parallel the west coasts of continents and tug at the ocean surface. Wind friction gets the surface waters moving. Under the Earth's rotational force, the slow-moving water is deflected westward, away from the coasts. Cold, deep, often nutrient-rich water moves in vertically to replace it (Figure 44.29). When cold, deep water moves up this way, we call it **upwelling**. It happens in equatorial currents and along the coasts of continents, and it cools air masses above it. Example: Fogbanks along the California coast are an outcome of warm air encountering upwellings of cold water.

In the Southern Hemisphere, commercial fishing depends on wind-induced upwelling along Peru and Chile. When prevailing coastal winds blow in from the south and southeast, they tug surface water away from shore. Cold, deeper water that the Humboldt Current delivers to the continental shelf moves to the surface. Tremendous quantities of nitrate and phosphate are pulled up, then carried north by the cold Peru Current. Phytoplankton that depend on these nutrients are the basis of one of the world's richest fisheries.

Every three to seven years, the warm surface waters of the western equatorial Pacific Ocean move eastward. This massive displacement of warm water affects the prevailing wind direction. The eastward flow speeds up so much that it influences the movement of water along the coastlines of Central and South America.

Waters driven toward any coastline will be forced downward and will flow seaward along the continental shelf. Near Peru's coast, a prolonged "downwelling" displaces cooler waters of the Humboldt Current—and so prevents upwelling. The phenomenon, which local fishermen named **El Niño**, has catastrophic effects on productivity, on seabirds that feed on anchovetas and other fishes, and on fishing industries. The next section provides a closer look at the El Niño phenomenon.

Nutrient-rich ecosystems called mangrove swamps and estuaries develop along many coasts. Intertidal zones are not as renowned for creature comforts; waves batter them and tides alternately expose and submerge them.

Rita in the Time of Cholera

We turn now to an application that reinforces a unifying concept of this chapter—that events in the atmosphere, in the ocean, and on land interconnect in ways that can profoundly influence the world of life.

EL NIÑO SOUTHERN OSCILLATION Let's set the stage for our story with a brief look at the *El Niño Southern Oscillation* (ENSO). Massively dislocated rainfall patterns characterize this climatic event, which corresponds to changes in sea surface temperatures and air circulation patterns. "Southern Oscillation" is a recurring seesaw in atmospheric pressure in the western equatorial Pacific, the world's largest reservoir of warm water. More warm air rises here than anywhere else. It is the source of heavy rainfall, which releases heat energy that drives the world's air circulation system.

Between ENSOs, the warm reservoir and heavy rain associated with it move to the west (Figure 44.30). In an ENSO, surface winds prevailing in the western equatorial Pacific pick up speed and "drag" surface waters east (Figure 44.31). As eastward water transport increases, westward transport slows. More warm water in the vast reservoir moves eastward, and so on in a feedback loop between the ocean and the atmosphere. The sea surface temperatures rise, evaporation from the ocean accelerates, and air pressure falls. Humid updrafts trigger violent storms and flooding along coasts and often heavier rain inland. Elsewhere, extended droughts prevail.

As you read earlier, reversal in the westward flow of air and water displaces the cold, deep Humboldt Current and stops upwelling along the western coast of South America. A warm, nutrient-poor current from the east usually reaches the coast around Christmas. Peru's fishermen named the current "El Niño" ("the little one," a reference to the baby Jesus).

Satellites and ocean-bobbing buoys gather data on winds, currents, and sea surface temperatures, which are loaded into supercomputers that simulate global weather patterns. Simulations are getting good: In 1997, scientists predicted the most powerful ENSO of the century a season in advance. Average sea surface temperatures rose nine degrees in the eastern Pacific. Warm water extended 9,660 kilometers west from Peru's coast and 320 kilometers north (Figure 44.32).

Storms that slammed into California, Mexico, Ecuador, and Peru caused heavy flooding and mudslides. An ice storm hit New England and knocked out power for millions in one of Canada's worst natural disasters. Tornadoes devastated parts of the southeastern United States. A numbing heat wave caused hundreds of deaths and triggered wildfires in Mexico and the United States. Australia reeled under a prolonged drought. Heavy rains and floods struck Kenya and Somalia. Monsoons were delayed in Southeast Asia, and fires set to clear tracts of rain forests swept out of control through a million acres. The strongest hurricane ever recorded struck the eastern Pacific; and eight cyclones hit the central Pacific, compared to two the year before.

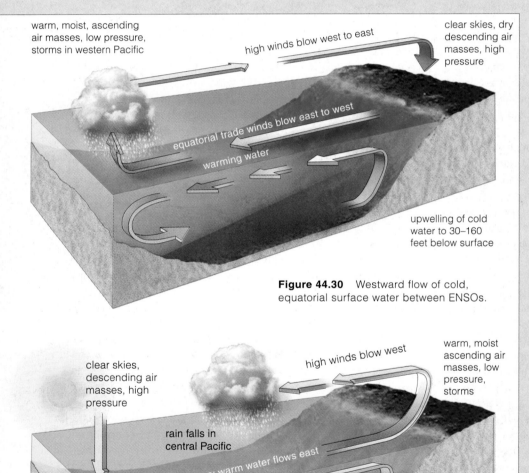

warm, moist, ascending air masses, low pressure, storms in western Pacific

clear skies, dry descending air masses, high pressure

high winds blow west to east

equatorial trade winds blow east to west

warming water

upwelling of cold water to 30–160 feet below surface

Figure 44.30 Westward flow of cold, equatorial surface water between ENSOs.

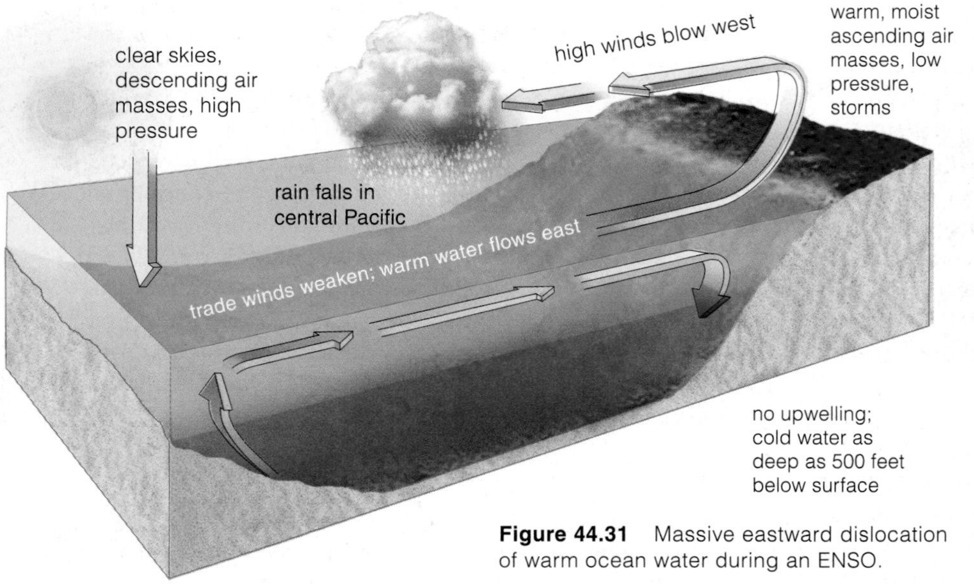

clear skies, descending air masses, high pressure

high winds blow west

warm, moist ascending air masses, low pressure, storms

rain falls in central Pacific

trade winds weaken; warm water flows east

no upwelling; cold water as deep as 500 feet below surface

Figure 44.31 Massive eastward dislocation of warm ocean water during an ENSO.

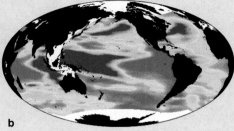

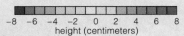

Figure 44.32 (**a**) Sea levels for the 1993–1994 winter between ENSOs. (**b**) Sea levels for the El Niño winter of 1997–1998. Values are based on measured increases in sea surface temperatures.

A CHOLERA CONNECTION During the 1997–1998 El Niño episode, 30,000 cases of cholera were reported in Peru, compared to only 60 from January to August in 1997. For some time, people knew that cholera was linked to water contaminated by *Vibrio cholerae* (Figure 44.33*a*). Epidemics develop after this bacterium infects humans, then enters water in feces expelled during the severe diarrhea that characterizes the disease.

What people did *not* know was where the bacterium lurked between cholera outbreaks. Year after year, no one could find it in humans or in water supplies. Then cholera would emerge simultaneously in places some distance apart—most often in coastal cities, where the urban poor draw water from rivers that enter the sea.

Rita Colwell, a microbiologist who later became the director of the National Science Foundation, suspected that humans did not harbor the pathogen between the outbreaks. Was there an environmental reservoir? Maybe, but nobody had detected any cells of *V. cholerae* in water samples subjected to standard culturing. Then Colwell had a flash of insight: What if no one could find the pathogen because it changes form and enters a dormant, sporelike stage between outbreaks?

During a cholera outbreak in Louisiana, she decided to use labeled antibodies that would bind to a certain protein at the bacterium's surface. Later, antibody tests in Bangladesh pinpointed the bacterium in fifty-one of fifty-two water samples. Standard culture methods had missed it in all but seven of the samples.

V. cholerae thrives in brackish water, rivers, estuaries, and the seas. Colwell knew that plankton (communities of mostly microscopic aquatic species) also thrive in these aquatic environments. She focused her search on the waters near Bangladesh, where outbreaks of cholera are endemic and seasonal. In time Colwell discovered a dormant stage of *V. cholerae* in copepods, marine animals that graze on algae and other members of phytoplankton (Figure 44.33*b*). The abundance of copepods—hence of *V. cholerae*—rise and fall with shifts in the abundance of phytoplankton, the "pastures of the seas."

Colwell already knew about seasonal variations in sea surface temperatures. Remember the saying, Chance favors the prepared mind? She discovered a correlation between seasonal temperature peaks in the Bay of Bengal with cholera cases in regional hospitals. Her correlation held for the 1990–1991 and 1997–1998 El Niño episodes (Figure 44.33*c,d*). Four to six weeks after the sea surface temperatures go up, so do the cases of cholera!

Together with her colleague Anwarul Huq, Colwell is working to identify other factors, such as shifts in salinity and nutrient content, that cause algal blooms, hence rises in copepod population size. Their goal is to integrate the data into a model that can be used to predict precisely where cholera outbreaks will occur and to give advance warning to filter drinking water. Meanwhile, Colwell is advising women in Bangladesh to use four layers of old sari cloth as filters, which remove 99 percent of *V. cholerae* cells from the water (Figure 44.33*e*). They can be rinsed in clean water, sun-dried, and used again and again.

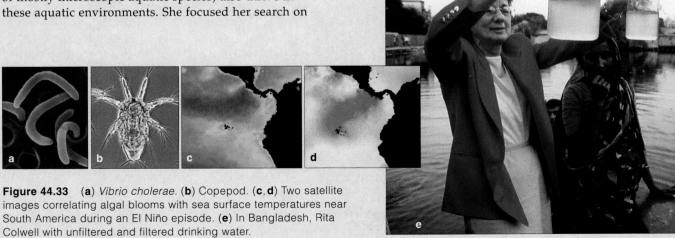

Figure 44.33 (**a**) *Vibrio cholerae.* (**b**) Copepod. (**c,d**) Two satellite images correlating algal blooms with sea surface temperatures near South America during an El Niño episode. (**e**) In Bangladesh, Rita Colwell with unfiltered and filtered drinking water.

SUMMARY

1. The biosphere encompasses the Earth's waters, the lower atmosphere, and the uppermost portions of its crust in which organisms live. Energy flows one way through the biosphere. Materials move through it on a grand scale to influence ecosystems everywhere. *CI*

2. The distribution of species through the biosphere is an outcome of the Earth's history, topography, climate, and interactions among species. *CI, 44.1–44.3*

    a. "Climate" refers to average weather conditions, including temperature, humidity, wind velocity, cloud cover, and rainfall over time. It results from differences in the amount of solar radiation reaching equatorial and polar regions, the Earth's daily rotation and its annual path around the sun, the distribution of continents and seas, and the elevation of land masses. *44.1*

    b. Interacting climatic factors underlie prevailing winds and ocean currents, which shape global weather patterns. Weather affects soil composition and water availability, which affects the growth and distribution of primary producers in ecosystems. *44.2*

3. The world's land masses are classified as six major biogeographic realms. Each is more or less isolated by oceans, mountain ranges, or desert barriers that tend to restrict gene flow between realms. Thus each tends to maintain a characteristic array of species. *44.3*

4. Biomes (a category of major ecosystems on land) are shaped by regional variations in climate, landforms, and soils. Dominant plants reflect prevailing conditions in the biomes called deserts, dry shrublands and dry woodlands, grasslands, broadleaf forests (e.g., tropical forests), coniferous forests, and tundra. *44.3–44.10*

5. Water provinces cover more than 71 percent of the Earth's surface. They include standing fresh water (such as lakes), running fresh water (such as streams), as well as the world oceans and seas. All aquatic ecosystems show gradients in light penetration, water temperature, salinity, and dissolved gases. These factors vary daily and seasonally, and affect primary productivity. *44.10*

6. Wetlands, intertidal zones, rocky and sandy shores, tropical reefs, and regions of the open ocean are major marine ecosystems. Photosynthetic activity is greatest in shallow coastal waters and in regions of upwelling: an upward movement of deep, cool ocean water that often carries nutrients to the surface. *44.11, 44.12*

Review Questions

1. Define biosphere. As part of your answer, also define atmosphere, lithosphere, and hydrosphere. *CI*

2. List the major interacting factors that influence climate. *CI*

3. List some of the ways in which air currents, ocean currents, or both may influence the region where you live. *44.1, 44.2*

4. Indicate where these biomes tend to be located in the world, and describe some of their defining features. *44.3–44.9*
    a. deserts   e. evergreen   h. alpine tundra
    b. dry shrublands   broadleaf forests   i. boreal forest
    c. dry woodlands f. deciduous forests j. montane
    d. grasslands   g. arctic tundra   coniferous forests

5. Define soils, then explain how the composition of regional soils affects ecosystem distribution. *44.4*

6. Describe the littoral, limnetic, and profundal zones of a large temperate lake in terms of seasonal primary productivity. *44.10*

7. Define the two major provinces of the world ocean. Is the open ocean devoid of life? *44.11*

8. Define and characterize the following types of ecosystems: mangrove wetland, estuary, and the intertidal zone. *44.12*

Self-Quiz

1. Solar radiation drives the distribution of weather systems and so influences _____ .
    a. temperature zones   c. seasonal variations
    b. rainfall distribution   d. all of the above

2. The _____ is a shield against ultraviolet wavelengths.
    a. upper atmosphere   c. ozone layer
    b. lower atmosphere   d. greenhouse effect

3. Regional variations in the global patterns of rainfall and temperature depend on _____ .
    a. global air circulation   c. topography
    b. ocean currents   d. all of the above

4. A rain shadow is a reduction in rainfall on the _____ of a mountain range.
    a. windward side   c. highest elevation
    b. leeward side   d. lowest elevation

5. Biogeographic realms are _____ .
    a. land and water provinces   c. divided into biomes
    b. six major land provinces   d. b and c

6. Biome distribution corresponds roughly with regional variations in _____ .
    a. climate   b. soils   c. topography   d. all of the above

7. Dominant plants of _____ are highly adapted to recurring episodes of lightning-sparked fires.
    a. dry shrublands   c. southern pine forests
    b. grasslands   d. all of the above

8. During _____ , deeper, often nutrient-rich water moves to the surface of a body of water.
    a. spring overturns   c. upwellings
    b. fall overturns   d. all of the above

9. Match the terms with the most suitable description.
    ____ boreal forest   a. high productivity; rapid cycling
    ____ permafrost   of nutrients (poor reservoirs)
    ____ chaparral   b. "swamp forest"
    ____ tropical   c. a type of dry shrubland
        rain forest   d. feature of arctic tundra

10. Match the terms with the most suitable description.
    ____ marine snow   a. deep, cool, often nutrient-rich
    ____ upwelling   ocean water moves upward
    ____ eutrophication   b. sediments, rocks of ocean bottom
    ____ estuary   c. partially enclosed mix of seawater
    ____ benthic   and fresh water
        province   d. nutrient enrichment of body
                                of water; reduced transparency,
                                phytoplankton blooms
                            e. basis of midocean food webs

OLIGOTROPHIC LAKE	EUTROPHIC LAKE
Deep, steeply banked	Shallow with broad littoral
Large deep-water volume relative to surface-water volume	Small deep-water volume relative to surface-water volume
Highly transparent	Limited transparency
Water blue or green	Water green to yellow- or brownish-green
Low nutrient content	High nutrient content
Oxygen abundant through all levels throughout year	Oxygen depleted in deep water during summer
Not much phytoplankton; green algae and diatoms dominant	Abundant, thick masses of phytoplankton; and cyanobacteria dominant
Abundant aerobic decomposers favored in profundal zone	Anaerobic decomposers
Low biomass in profundal	High biomass in profundal

b

Figure 44.34 (**a**) Crater Lake, Oregon, a collapsed volcanic cone filled with rainwater and melted snow. Like other volcanoes of the Cascade range, it started forming as a result of tectonic forces that prevailed at the dawn of the Cenozoic. (**b**) Major characteristics of oligotrophic and eutrophic lakes.

Critical Thinking

1. Kangaroos are furry, pouched mammals. Raccoons are furry placental mammals. Speculate on why Australia but not North America is the original home of kangaroos and why the reverse is true of raccoons. Use Section 24.9 and Section 24.13 *Critical Thinking* question 5 when devising an explanation.

2. Reflect on the world distribution of land masses and ocean water, then develop an explanation of why grassland biomes tend to form in the interior of continents, not along their coasts.

3. *Wetlands,* remember, are transitional zones between aquatic and terrestrial ecosystems in which plants are adapted to grow in periodically or permanently saturated soil. They include mangrove swamps and riparian zones (Section 25.6). Wetlands everywhere are being converted for agriculture, home building, and other human activities. Does the great ecological value of wetlands for the nation as a whole outweigh the rights of the private citizens who own many of the remaining wetlands in the United States? Should the private owners be forced to transfer title to the government? If so, who should decide the value of a parcel of land *and* pay for it? Or would such seizure of private property violate laws of the United States?

4. Observe conditions in a lake near your home or a place where you vacation. If lakes aren't part of your life, think about Oregon's Crater Lake instead (Figure 44.34*a*). Using the data in Figure 44.34*b* as a guide, would you conclude Crater Lake is oligotrophic or eutrophic? Will it remain so over time?

5. The United States Forest Service and Geophysical Dynamics Laboratory ran supercomputer programs to predict possible outcomes of a *global warming* trend. (Here you may wish to review Section 43.10.) According to their results, by the year 2030, the United States will experience recurring, devastating forest fires. Severe storms and more frequent rainfall in the western states will accelerate erosion, especially along the coasts and in steep foothills and mountain ranges.

Increases in the deposition of sediments will have adverse effects on rivers, streams, and estuaries. Dry shrublands and woodlands may replace hardwood forests in Minnesota, Iowa, and Wisconsin, and in parts of Missouri, Illinois, Indiana, and Michigan. The breadbasket of the American Midwest will shift north into Canada. Think about where you live now and where you plan to live in the future. How would such changes affect you? Should nations get serious about finding ways to counter global warming? Or do you believe that the scientists who are concerned about this are merely alarmist and that nothing bad will happen? Explain how you have reached your conclusion.

6. Think about the ocean, which covers the majority of the Earth's surface. Its deepest regions are remote from human populations, to say the least. Now think about a modern-day problem—where to dispose of nuclear wastes and other truly hazardous materials. You will be reading about this serious problem in the next chapter. For now, simply be aware that the United States government has stockpiles of very dangerous wastes and has nowhere to put them. Would it be feasible, let alone ethical, to use the deep ocean as a "burying ground" for them? Why or why not?

Selected Key Terms

atmosphere *CI*
biogeographic realm *44.3*
biogeography *CI*
biome *44.3*
biosphere *CI*
boreal forest *44.8*
climate *CI*
coniferous forest *44.8*
deciduous broadleaf forest *44.7*
desert *44.5*
desertification *44.5*
dry shrubland *44.6*
dry woodland *44.6*
ecoregion *44.3*

El Niño *44.12*
estuary *44.12*
eutrophication *44.10*
evergreen broadleaf forest *44.7*
fall overturn *44.10*
grassland *44.6*
hydrothermal vent *44.11*
intertidal zone *44.12*
lake *44.10*
mangrove wetland *44.12*
marine snow *44.11*
monsoon *44.2*
ocean *44.2*

permafrost *44.9*
rain shadow *44.2*
savanna *44.6*
soil *44.4*
southern pine forest *44.8*
spring overturn *44.10*
stream *44.10*
temperature zone *44.1*
tropical rain forest *44.7*
tundra *44.9*
ultraplankton *44.11*
upwelling *44.12*

Readings

Brown, J., and A. Gibson. 1983. *Biogeography*. St. Louis: Mosby.

Dold, C. February 1999. "The Cholera Lesson." *Discover*, 71–76.

Garrison, T. 1996. *Oceanography*. Second edition. Belmont, California: Wadsworth. Accessible; read its lyrical epilogue.

Olson, D. M., and E. Dinerstein, 1998. "The Global 200: A Representation Approach to Conserving the Earth's Most Biologically Valuable Ecoregions." *Conservation Biology* 12(3).

Smith, R. 1996. *Ecology and Field Biology*. Fifth edition. New York: HarperCollins.

On-Line readings at Student Guide for InfoTrac:
www.brookscole.com/biology

45

HUMAN IMPACT ON THE BIOSPHERE

An Indifference of Mythic Proportions

Of all the concepts introduced in the preceding chapter, the one that should be foremost in your mind is this: *The atmosphere, ocean, and land interact in ways that help dictate living conditions throughout the biosphere.* Driven by energy streaming in continually from the sun, these stupendous interactions give rise to globe-spanning temperatures and circulation patterns upon which life ultimately depends. This chapter turns to a related concept of equal importance. Simply put, *we have become major players in these interactions even before we fully comprehend how they work.*

To gain perspective on what is happening, think about something we take for granted—the air around us. The composition of the present-day atmosphere is a result of geologic and metabolic events—most notably, photosynthesis—that began billions of years ago. The first humans, recall, had evolved by 2.4 million years ago. Like us, they breathed oxygen from an atmosphere of ancient origins. Like us, they were protected from the sun's ultraviolet radiation by an ozone shield in the stratosphere. The size of their population was not

much to speak of, and their effect on the biosphere was trivial. About 11,000 years ago, however, agriculture began in earnest, and it laid the foundation for huge increases in population size. Starting a few centuries ago, medical and industrial revolutions expanded that foundation for growth—and human population size skyrocketed in a mere blip of evolutionary time.

Today we are extracting huge amounts of energy and resources from the environment and giving back monumental amounts of wastes. As we do so, we are destabilizing ecosystems everywhere, even though the magnitude of change might not even be recognized when measured on the scale of a lifetime (Figure 45.1).

In a few developed countries, population growth has more or less stabilized, and the resource use per individual has dropped a bit. But resource utilization levels in those places are already high. In developing countries in Central America, Africa, and elsewhere, population sizes and resource consumption are rising fast, even though hundreds of millions of people are already malnourished or starving to death.

1900   1940   1954   1962

Many of the problems sketched out in this chapter are not going to disappear tomorrow. It will take many decades, even centuries, to reverse some trends that are already in motion—and not everyone is ready to make the effort. A few enlightened individuals in Michigan or Alberta or New South Wales can commit themselves to resource conservation and to minimizing pollution. But scattered attempts will not be enough. Individuals of all nations will unite to reverse global trends only when they perceive that the dangers of *not* doing so outweigh the personal benefits of ignoring them.

Does this seem pessimistic? Think of the exhaust fumes released to the air each time you drive a car or truck. Think of oil refineries, food-processing plants, and paper mills that supply you with goods and also release chemical wastes into the nation's waterways. Think of Mexico and other developing countries that produce cheap food by using an unskilled labor force and toxic pesticides. Unregulated pesticide applications poison the people who work the land, the land itself, and sometimes people who buy the exported produce.

Who changes behavior first? We have no answer. We can suggest, however, that a strained biosphere can rapidly impose an answer upon us. As an individual, you might choose to cherish or brood about or ignore any aspect of the world of life. Whatever your choice, the bottom line is that you and all other organisms are in this together. Our lives interconnect to degrees that we are only now starting to comprehend.

Figure 45.1 Tracking human population growth in one small part of the world, based on historical data and satellite imaging. *Red* denotes areas of dense human settlement in and around the San Francisco Bay Area and Sacramento since 1900.

Key Concepts

1. Human population growth has been skyrocketing ever since the mid-eighteenth century. At present, humans have the population size, the technology, and the cultural inclination to use energy and alter the environment at astonishing rates.

2. Pollutants are substances with which ecosystems have had no prior evolutionary experience, in terms of kinds or amounts, so adaptive mechanisms for dealing with them are not in place. In a more restrictive sense, pollutants are substances that accumulate in amounts that have adverse effects on human health, activities, and even survival itself.

3. Many pollutants, including the kinds that contribute to the formation of smog and acid rain, exert regionally harmful effects. The effects of other pollutants, including chlorofluorocarbons that attack the ozone layer in the stratosphere, are global in scale.

4. Conversion of marginally fertile lands for agriculture, highly mechanized deforestation, and other practices required to meet the demands of the growing human population are resulting in a loss of soil fertility and desertification. They also are causing a decline in the quality and quantity of one of the most crucial of all resources—fresh water.

5. Ultimately, the world of life runs on energy inputs from the sun that drive complex interactions among the atmosphere, ocean, and land. As a population, we are intervening in these interactions in ways that may have severe consequences in the near future.

6. We as a species must come to terms with principles of energy flow and principles of resource utilization that govern all systems of life on Earth.

1974

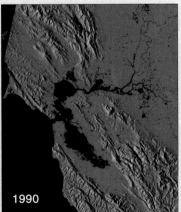
1990

AIR POLLUTION—PRIME EXAMPLES

Let's start this survey of the impact of human activities by defining pollution, a central problem. **Pollutants** are substances with which ecosystems have had no prior evolutionary experience, in terms of kinds or amounts, so adaptive mechanisms that might deal with them are not in place. From the human perspective, pollutants are substances that have accumulated enough to have adverse effects on our health, activities, or survival.

Air pollutants are premier examples. As Table 45.1 shows, they include carbon dioxide, oxides of nitrogen and sulfur, and chlorofluorocarbons. Also among them are photochemical oxidants formed when the sun's rays interact with certain chemicals. The United States alone releases 700,000 metric tons of air pollutants each day. Whether these remain concentrated at the source or are dispersed depends on local climate and topography.

Smog

When weather conditions trap a layer of cool, dense air under a layer of warm air, this is a **thermal inversion**. If the trapped air holds pollutants, winds can't disperse them, and they might accumulate to dangerous levels. Thermal inversions have been key factors in some of the worst air pollution disasters, because they intensify an atmospheric condition called smog (Figure 45.2).

Two types of smog, gray air and brown air, form in major cities. Where winters are cold and wet, **industrial smog** develops as a gray haze over industrialized cities that burn coal and other fossil fuels for manufacturing,

Table 45.1	*Major Classes of Air Pollutants*
Carbon oxides	Carbon monoxide (CO), carbon dioxide (CO_2)
Sulfur oxides	Sulfur dioxide (SO_2), sulfur trioxide (SO_3)
Nitrogen oxides	Nitric oxide (NO), nitrogen dioxide (NO_2), nitrous oxide (N_2O)
Volatile organic compounds	Methane (CH_4), benzene (C_6H_6), chlorofluorocarbons (CFCs)
Photochemical oxidants	Ozone (O_3), peroxyacyl nitrates (PANs), hydrogen peroxide (H_2O_2)
Suspended particles	Solids (dust, soot, asbestos, lead, etc.) and droplets (sulfuric acid, oils, pesticides, etc.)

heating, and generating electric power. Burning releases airborne pollutants, such as dust, smoke, soot, ashes, asbestos, oil, bits of lead and other heavy metals, and sulfur oxides. If not dispersed by winds and rain, the emissions may reach lethal concentrations. Industrial smog caused 4,000 deaths during a 1952 air pollution disaster in London. Before coal burning was restricted, New York, Pittsburgh, and Chicago also were gray-air cities. Today, most industrial smog forms in cities of China, India, and other developing countries as well as cities of coal-dependent countries of eastern Europe.

In warm climates, **photochemical smog** forms as a brown haze over large cities. It gets concentrated when the surrounding land forms a natural basin, as it does around Los Angeles and Mexico City (Figure 45.2). The key culprit is nitric oxide. It escapes from vehicles and reacts with oxygen in air to form nitrogen dioxide. Nitrogen dioxide exposed to sunlight reacts with hydrocarbons and forms photochemical oxidants. Most hydrocarbons are released from spilled or partly burned gasoline. Key oxidants are ozone and peroxyacyl nitrates, or **PANs**. Even traces of PANs sting eyes, irritate lungs, and damage crops.

Acid Deposition

Sulfur and nitrogen oxides are among the worst air pollutants. Coal-burning power plants, metal smelters, and factories emit most sulfur dioxides. Motor vehicles, power plants that burn gas and oil, and nitrogen-rich fertilizers produce nitrogen oxides. In dry weather, fine particles of oxides may briefly remain airborne, then fall as **dry acid deposition**. They form weak solutions of sulfuric and nitric acids when dissolved in atmospheric water. Winds may disperse them far from their source. If they fall to Earth in rain and snow, we call this wet acid deposition, or **acid rain**. The pH of normal rainwater is 5 or so. Acid rain can be 10 to 100 times more acidic, even as potent as lemon

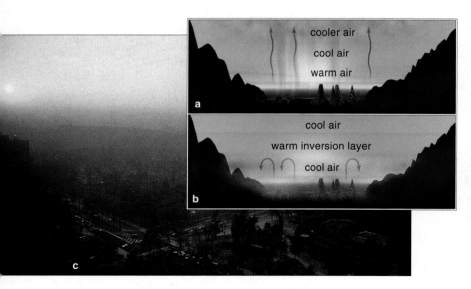

Figure 45.2 (**a**) Normal pattern of air circulation in smog-forming regions. (**b**) Air pollutants trapped under a thermal inversion layer. (**c**) Mexico City on an otherwise bright, sunny morning. Topography, staggering numbers of people and vehicles, and industry combine to make its air among the world's smoggiest. Breathing this city's air is like smoking two packs of cigarettes a day.

cooler air
cool air
warm air
a

cool air
warm inversion layer
cool air
b

c

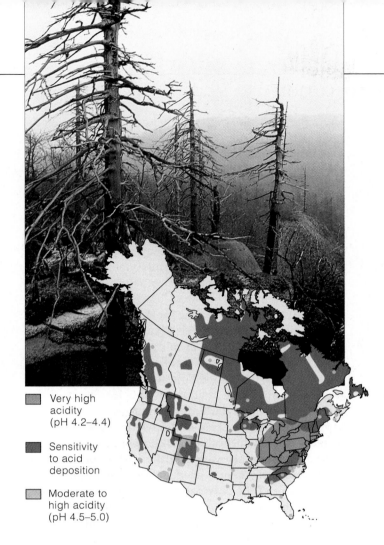

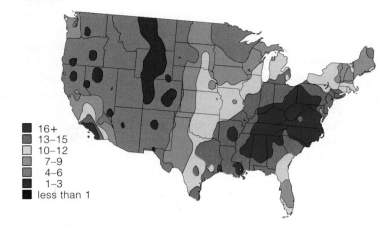

Figure 45.4 Regional differences in the concentrations of fine particles (2.5 μm or less) in micrograms per cubic meter of air, measured in 1992. Data are averages; they understate peak values in individual cities.

16+
13–15
10–12
7–9
4–6
1–3
less than 1

- Very high acidity (pH 4.2–4.4)
- Sensitivity to acid deposition
- Moderate to high acidity (pH 4.5–5.0)

Figure 45.3 Map of average acidities of precipitation and soil sensitivities to acid deposition in 1984 for North American regions. The photograph, from Germany, is an example of how prolonged exposure to air pollutants is contributing to the rapid destruction of forests in much of the world. Weakened trees are more vulnerable to drought, disease, and insect pests.

juice! The deposited acids eat away at metals, marble buildings, rubber, plastics, nylon stockings, and other materials. They can disrupt the physiology of organisms, the chemistry of ecosystems, and biodiversity.

Depending on soil type and vegetation cover, some regions are far more sensitive than others to acid rain (Figure 45.3). Highly alkaline soil can neutralize acids before they enter the streams and lakes of watersheds. Water having a high carbonate content also neutralizes acids. However, many watersheds of northern Europe and southeastern Canada, and of regions throughout the United States have thin soils on top of solid granite. These soils cannot buffer much of the acidic inputs.

Rain in much of eastern North America is thirty to forty times more acidic than it was several decades ago. Crop yields are declining. Fish populations have already disappeared from more than 200 lakes in New York's Adirondack Mountains. By some estimates, fish will vanish from 48,000 lakes in Ontario, Canada, in the next two decades. Pollution from industrial regions

also is changing rainfall acidity enough to contribute to the decline of forest trees and mycorrhizae that support new growth (Chapter 21).

Researchers confirmed long ago that emissions from power plants, factories, and vehicles are the key sources of air pollutants. In 1995, researchers from the Harvard School of Public Health and Brigham Young University reported this finding from a comprehensive air quality study: Whack a year or so off your life span if you live in cities with the dirtiest air—especially air with fine particles of dust, soot, smoke, or acid droplets. Smaller particles are more easily inhaled and can damage lung tissue. Figure 45.4 summarizes air-quality data for the continental United States, gathered in 1992.

At one time the world's tallest smokestack, in the Canadian province of Ontario, accounted for 1 percent by weight of the annual worldwide emissions of sulfur dioxide. But Canada gets more acid deposition from industrialized regions of the midwestern United States than it sends across its southern border. Most of the air pollutants in Scandinavian countries, the Netherlands, Austria, and Switzerland arise in industrialized parts of western and eastern Europe. *Prevailing winds—hence air pollutants—do not stop at national boundaries.*

Pollutants are substances with which ecosystems have had no prior evolutionary experience, in terms of kinds or amounts, so adaptive mechanisms are not in place to deal with them.

Accumulated pollutants can harm organisms, as when they reach levels that adversely affect human health.

Smog formation and acid deposition are two examples of air pollution in specific regions, although prevailing winds often distribute pollutants beyond regional boundaries.

Smog forms mainly as a result of fossil fuel burning in urban and industrialized regions. Airborne acidic pollutants drift down as dry particles or as components of acid rain.

OZONE THINNING—GLOBAL LEGACY OF AIR POLLUTION

Now think about the ozone layer, almost twice as high above sea level as the top of Mount Everest, the highest place on Earth. Each September through mid-October, the layer becomes thinner at high latitudes. The seasonal **ozone thinning** is so pronounced, it was once called an "ozone hole." In 2000, ozone thinning over Antarctica was the most extensive ever recorded; it covered an area greater than North America (Figure 45.5). As in 1997 and 1999, the seasonal ozone loss at high northern latitudes was an alarming 60 percent.

Why is ozone reduction alarming? More ultraviolet radiation reaches the Earth and is triggering far more skin cancers, cataracts, and weakened immune systems. Such radiation also adversely affects photosynthesis. Substantial declines in the oxygen-releasing activity of phytoplankton alone could change the atmosphere's composition (Section 6.9).

CFCs (chlorofluorocarbons) are the major factors in ozone reduction. These compounds of carbon, fluorine, and chlorine are odorless and invisible. You find them in refrigerators and air conditioners (as the coolants), solvents, and plastic foams. CFCs slowly escape into air and resist breakdown. When a free CFC molecule absorbs ultraviolet light, it gives up one chlorine atom. If chlorine reacts with ozone, this yields oxygen and chlorine monoxide—which reacts with free oxygen to release another chlorine atom. Each released chlorine atom converts 10,000+ ozone molecules to oxygen!

Chlorine monoxide levels above Antarctica are 100 to 500 times higher than at midlatitudes. Why? High-altitude ice clouds form there during winter. Winds rotate around the South Pole for most of the winter, like a moving fence that keeps the ice clouds from spreading to other latitudes (Figure 45.5). This also happens on a lesser scale in the Arctic. Ice crystals are surfaces upon which chlorine compounds are swiftly degraded. When air warms in the spring, the chlorine is free to destroy ozone—hence the ozone thinning.

CFCs aren't the only ozone eaters. Methyl bromide, a fungicide, is nastier. It persists in the atmosphere only for a short time. Even so, it will account for about 15 percent of the ozone thinning in the years to come unless production stops.

Some scientists dismiss the threat of chemicals that contain chlorine and bromine. But the vast majority of those who have studied ecosystem modeling programs conclude that these chemicals pose long-term threats—not only to human health and certain crops but also to animal life in general. Substitutes are now available for most uses of CFCs; others are being developed.

Under international agreement, CFC production in developed countries has been phased out. By 2010, developing countries will phase out CFCs. And methyl bromide production will end by 2010. Recent computer models indicate the area of thinning should not get any larger. Assuming the international goals are met, it still will be 50 years before the ozone layer is restored to 1985 levels and another 100 to 200 years to total recovery, to pre-1950 levels. Meanwhile, you and all children and grandchildren of future generations will be living with the destructive effects of air pollutants. And if levels of greenhouse gases continue to rise, the stratosphere could cool enough to increase the size and duration of seasonal ozone depletions over the poles.

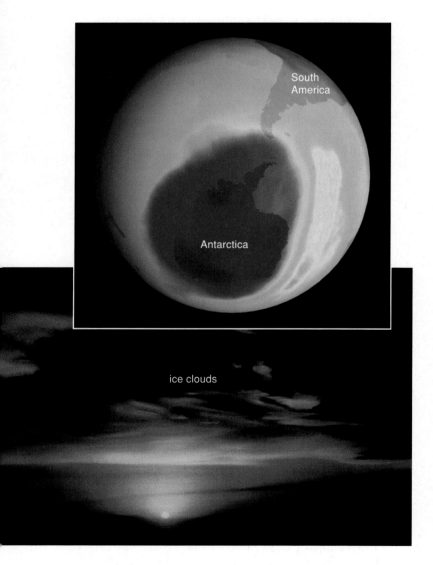

Figure 45.5 Seasonal ozone thinning above Antarctica in the spring of 2000, the largest yet recorded. Ice clouds above Antarctica have a role in the ozone thinning each spring.

Air pollution can have global repercussions, as when CFCs and other compounds contribute to a thinning of the ozone layer that shields life from the sun's ultraviolet radiation.

WHERE TO PUT SOLID WASTES? WHERE TO PRODUCE FOOD?

Oh Bury Me Not, In Our Own Garbage

In natural ecosystems, one organism's wastes serve as resources for others, so by-products of existence are cycled through the system. In developing countries, many resources are scarce. People conserve what they can and discard little. In the United States and some other developed countries, most people use something once, discard it, and buy another. Each year, millions of metric tons of solid wastes are dumped, burned, and buried. Paper products make up half the total volume, which also includes 50 billion nonreturnable cans and bottles. Every week in the United States, paper made from 500,000 trees ends up in Sunday newspapers. If every reader recycled merely one of ten newspapers, 25 million trees a year could be left alone. Recycling paper would reduce the airborne pollutants released during paper manufacturing by 95 percent and require 30 to 50 percent less energy than making new paper.

Our throwaway mentality is unique in the world of life. We are condoning a monumental accumulation of solid wastes in human ecosystems and natural ones. Bury the garbage in landfills? What happens when the space around cities runs out? Besides, landfills "leak" and eventually threaten groundwater supplies. Burn wastes in inefficient incinerators? These spew volumes of pollutants and ashes into the atmosphere.

Recycling is affordable and feasible. At the personal level, individuals can help initiate change by refusing to buy goods that are excessively wrapped, packaged in nondegradable containers, and designed for one-time use. They can participate in curbside recycling, by which they presort recyclable wastes. Such action may be more efficient than building huge resource recovery centers. These require so much trash to turn a profit that owners may end up encouraging the throwaway mentality. They also produce a toxic ash that must be disposed of in landfills, which eventually leak.

Converting Marginal Lands for Agriculture

The human population already uses nearly 21 percent of the Earth's land surfaces for cropland or for grazing. Another 28 percent is said to be potentially suitable for agriculture, but productivity would be so low that the conversion may not be worth the cost (Figure 45.6).

Asia and several other heavily populated regions now experience recurring, severe food shortages. Yet over 80 percent of their productive land is already under cultivation. Scientists have made valiant efforts to improve production on existing cropland. Under the banner of the **green revolution**, their research has been directed toward (1) improving the genetic character of crop plants for higher yields and (2) exporting modern

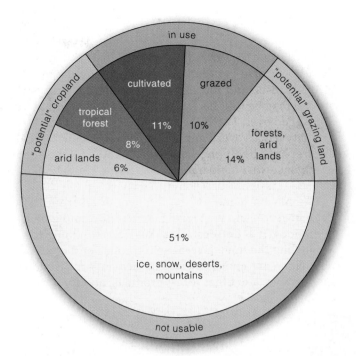

Figure 45.6 Classification of land with respect to its suitability for agriculture. Theoretically, clearing vast tracts of tropical forests and irrigating marginal lands could more than double the world's cropland. Doing so would destroy valuable forest resources, damage the environment, cause severe losses in biodiversity, and possibly cost more than it is worth.

agricultural practices and equipment to the developing countries. Many of these countries rely on *subsistence* agriculture, which runs on energy inputs from sunlight and human labor. They also rely heavily on *animal-assisted* agriculture, with energy inputs from oxen and other draft animals. By contrast, *mechanized* agriculture requires massive inputs of fertilizers, pesticides, and ample irrigation to sustain high-yield crops. It requires fossil fuel energy to drive farm machines. Crop yields are four times as high. But the modern practices use up 100 times more energy and minerals. Also, there are signs that limiting factors are coming into play to slow down further increases in crop yields.

Pressure for increased food production is greatest in parts of Central and South America, Asia, the Middle East, and Africa where human populations are rapidly expanding into marginal lands. Repercussions extend beyond national boundaries, as you will see next.

Our astounding population growth has impact on the Earth's land. We generate huge amounts of solid wastes but reuse or recycle very little. We are expanding into marginal lands for food production, a move that often has high energy costs and environmental consequences.

DEFORESTATION—AN ASSAULT ON FINITE RESOURCES

At one time, tropical forests cloaked regions that were, collectively, twice the size of Europe. For ten thousand years or more, they were home to an estimated 50 to 90 percent of all land-dwelling species. In less than four decades, human populations destroyed more than half of the forests, and most of their spectacular arrays of species may be lost. Each year, another 38 million acres are logged over. That's the equivalent of leveling thirty-four city blocks every minute. Logging extends beyond the tropics. Highly mechanized logging is proceeding in the once-vast temperate forests of the United States, Canada, Europe, Siberia, and elsewhere.

We have a name for the removal of all trees from large tracts of land for logging, agriculture, and grazing operations. It is **deforestation**. Why are we doing this? Paralleling the huge increases in the size of the human population are rapidly increasing demands for wood as fuel, lumber, and other forest products; and for grazing land and cropland. More and more people compete for dwindling resources. As you read in Chapter 25, they do so for economic profit, but also because alternative ways of life simply are not available to most people.

The world's great forests have profound influences on ecosystems. Like enormous sponges, the watersheds of forested regions absorb, hold, and gradually release water. By intervening in the downstream flow of water, they help control soil erosion, flooding, and a build-up of sediments that can clog rivers, lakes, and reservoirs. When the vegetation cover is stripped away, exposed soil becomes vulnerable to leaching of nutrients and to erosion, especially on steep slopes.

Deforestation is now greatest in Brazil, Indonesia, Colombia, and Mexico. If the clearing and destruction continue at present rates, only Brazil and Zaire will still have large tropical forests in 2010. By 2035, most of their forested regions will be cleared, also.

Figures 45.7 and 45.8 are close-up and panoramic views of what is happening in the Amazon basin of South America. Section 22.6 includes one example of heavy deforestation in North America.

In tropical regions, clearing the forests for agriculture sets the stage for long-term losses in productivity. The irony is that tropical forests are one of the worst places to grow crops or raise pasture animals. In intact forests, litter cannot accumulate, for the high temperatures and the heavy, frequent rainfall promote the rapid decomposition of organic wastes and remains. As fast as the decomposers release nutrients, the trees and other plants take them up. Deep, nutrient-rich topsoils simply cannot form.

Well before the advent of highly mechanized logging practices, people were practicing **shifting cultivation** (once referred to as slash-and-burn agriculture). They cut and burn trees,

Figure 45.7 Countries that are allowing the greatest destruction of tropical forests. *Red* denotes regions where 2,000 to 14,800 square kilometers are deforested each year. *Orange* denotes "moderate" deforestation, which encompasses areas of 100 to nearly 2,000 square kilometers.

ANDES

SMOKE FROM FIRES

Figure 45.8 The vast Amazon River basin of South America in September 1988. Smoke from fires that had been deliberately set during the dry season to clear tropical forests, pasturelands, and croplands completely obscured its features. The smoke extended to the Andes Mountains near the western horizon, about 650 miles (1,046 kilometers) away. That smoke cover was the largest that astronauts had ever observed. It extended almost 175 million square kilometers (1,044,000 square miles).

The smoke plume near the center of the photograph covered an area comparable to the immense smoke cover arising from the forest fire in Yellowstone National Park in that year. During the El Niño–induced drought of 1997, fires that had been set deliberately burned out of control. A smoke cover formed again.

Extensive deforestation is not confined to equatorial regions. For instance, in the past century, 2 million acres of redwood forests along the California coast were logged over. Most of the logging in temperate forests has occurred since 1950. Many of the logs are exported to lumber mills overseas, where wages are low.

then till the ashes into the soil. The nutrient-rich ashes can sustain crops for one to several seasons. Afterward, cleared plots are abandoned, for heavy leaching makes the soil infertile. When shifting cultivation is practiced on small, widely separated plots, a forest ecosystem isn't necessarily damaged extensively. But soil fertility plummets with population growth. Larger areas are cleared and plots are cleared again at shorter intervals.

Figure 45.9 Wangari Maathai, a Kenyan who organized the internationally acclaimed Green Belt Movement in 1977. The 50,000 members of this women's group are committed to establishing nurseries, raising seedlings, and planting a tree for each of the 27 million Kenyans. Together with half a million schoolchildren, they had planted more than 10 million trees by 1990. Their success inspired similar programs in more than a dozen countries in Africa. Dr. Maathai's efforts are not appreciated by her own government. Kenyan police have jailed her twice, and in 1992 they severely beat her because of her efforts on behalf of the environment.

Deforestation also affects the rates of evaporation, transpiration, runoff, and maybe regional patterns of rainfall. Trees release 50 to 80 percent of the water vapor above tropical forests. In logged-over regions, annual precipitation drops, and rain swiftly drains away from the exposed, nutrient-poor soil. The regions get hotter and drier—and soil fertility and moisture decline. In time sparse, dry grassland or desertlike conditions may prevail instead of the formerly rich, forested biomes.

Also, tropical forests absorb much of the sunlight reaching equatorial regions. Deforested land is shinier, so to speak, and reflects more incoming energy back into space. And the combined photosynthetic activity of so many trees affects the global cycling of carbon and oxygen. Extensive tree harvesting and burning release carbon stored in the tree biomass to the atmosphere, as carbon dioxide. In this way, deforestation might be a factor in the amplification of the greenhouse effect.

Conservation biologists are attempting to reverse the trend. As three examples, a coalition of 500 groups is dedicated to preserving Brazil's remaining tropical forests. In India, women have already built and installed 300,000 inexpensive, smokeless wood stoves. Over the past decade, the stoves saved more than 182,000 metric tons' worth of trees by reducing demands for fuelwood. In Kenya, women have planted millions of trees to hold soil in place and to provide fuelwood (Figure 45.9).

Once-vast forests helped sustain rapid increases in human population growth. Recent and highly mechanized modes of deforestation are rapidly depleting these finite resources.

You and the Tropical Rain Forest

Developing nations in Latin America, Southeast Asia, and Africa have the fastest-growing populations but not enough food, fuel, and lumber. Of necessity, they turn to their forests for growth-sustaining resources. Most of the forests may disappear within our lifetime. That possibility elicits the most outcries from concerned groups in highly developed nations—which happen to use most of the world's resources, including forest products. You read about economic aspects of this issue in earlier chapters. We invite you to reflect now on a few more aspects.

For purely ethical reasons, many people condemn the destruction of so much biodiversity. Tropical rain forests have the greatest variety and numbers of insects, and the world's largest ones. They are home to the most species of birds and to plants with the largest flowers. Living in the forest canopy and understory are monkeys, tapirs, and jaguars in South America and apes, okapi, and leopards in Africa. Massive vines twist around trees. Orchids, mosses, lichens, and other organisms grow on branches, absorbing minerals that rains deliver to them. Entire communities of microbes, insects, spiders, and amphibians live, breed, and die in small pools of water that collect in furled leaves.

Their disappearance will have repercussions through human life. Only a few strains of crop plants and livestock sustain most human populations, and they are vulnerable to ever evolving pathogens. Tissue-culture specialists and genetic engineers use genes of forest species to develop new or hybrid strains that can make our food base less vulnerable. Geneticists use them to develop more effective antibiotics and vaccines. Aspirin, the most widely used painkiller, is based on a chemical blueprint of an extract from tropical willow leaves. Many ornamental plants and spices and foods, including cocoa, cinnamon, and coffee, originated in the tropics. So did latex, gums, resins, dyes, waxes, and oils for tires, shoes, toothpaste, ice cream, shampoo, compact discs, condoms, and perfumes. Also think about this: Rampant burning of forests is releasing enough air pollutants to change the air we breathe and to help heat the planet during our lifetime.

And so conservation biologists rightly decry the mass extinction, the assaults on species diversity, the depletion of much of the world's genetic reservoir. Yet something else is going on here. Too many of us grow uneasy when we pass through obliterated forests in our own country. Is it because we are losing the comfort of our heritage—a connection with our evolutionary past? Many millions of years ago, our earliest primate ancestors moved into the trees of tropical forests. Through countless generations, their nervous and sensory systems evolved and became highly responsive to information-rich, arboreal worlds.

Does our neural wiring still resonate with rustling leaves, with shafts of light and mosaic shadows? Are we innately attuned to the forests of Eden—or have time and change buried recognition of home?

Figure 45.10 Tropical rain forest in Southeast Asia.

WHO TRADES GRASSLANDS FOR DESERTS?

Long-term shifts in climate often convert grasslands to deserts. Humans do, too. **Desertification** is the name for the conversion of large tracts of natural grasslands to a more desertlike condition. It applies also when the conversion of rain-fed or irrigated croplands causes a 10 percent or greater decline in agricultural productivity. During the past fifty years, 9 million square kilometers worldwide have become desertified.

plants. They also are better at conserving water; they lose little in feces, compared to cattle.

In 1978 a biologist, David Holpcraft, began ranching antelopes, zebras, giraffes, ostriches, and other native herbivores. He also raised cattle as control groups to compare costs and meat yields on the same land. The native herds increased and provided tasty meat. Range conditions did not deteriorate; they improved. Vexing

Figure 45.11 An awe-inspiring dust storm approaching Prowers County, Colorado, in 1934.

The Great Plains of the American Midwest are dry, windy grasslands that are subjected to pronounced, recurring droughts. Extensive conversion of these grasslands to agriculture began in the 1870s. Overgrazing left the ground bare across vast tracts. In May 1934, a cloud of topsoil that blew off the land blanketed the entire eastern portion of the United States, giving the Great Plains a dubious new name—the Dust Bowl. About 3.6 million hectares (9 million acres) of cropland were destroyed. Today, without large-scale irrigation and intensive conservation farming, desertlike conditions could prevail.

Figure 45.12 Desertification in the Sahel, a region of West Africa that forms a belt between the hot, dry Sahara Desert and tropical forests. This savanna country is undergoing rapid desertification as a result of overgrazing, overfarming, and prolonged drought.

Today, at least 200,000 square kilometers are being converted to deserts each year. Long droughts are accelerating the process, just as one did years ago in the Great Plains (Figure 45.11). At present, however, overgrazing of livestock on marginal lands is the source of most large-scale desertification.

In Africa, for instance, there are too many cattle in the wrong places. Cattle require more water than the native wild herbivores do, so they move back and forth between grazing areas and watering holes more often. As they do, they trample grasses and compact the soil surface (Figure 45.12). By contrast, gazelles and other wild herbivores get most (if not all) of their water from

problems remained. Certain tribes in Africa have their own idea of what constitutes "good" meat, and some view cattle as the symbol of wealth in their society.

Without irrigation and conservation practices, grasslands converted for agriculture often end up as deserts.

A GLOBAL WATER CRISIS

The Earth has a tremendous supply of water, but most is just too salty for human consumption or agriculture. Imagine all that water in a bathtub. Withdraw all of the fresh, renewable portion (from lakes, rivers, reservoirs, groundwater, and other sources of surface water) and it would barely fill a teaspoon.

Why not consider **desalinization**—removal of salt from seawater? After all, there are unlimited supplies of seawater. Desalinization processes exist; they distill seawater or force it through membranes, by reverse osmosis. Expensive fuel energy drives these processes, so they may be feasible only in Saudi Arabia and a few other countries that have small population sizes, large reserves of energy, and lots of cash. Actually, in some situations they may be the only alternative to running out of drinkable water. That nearly happened in Santa Barbara and some other cities in California during one prolonged drought. Yet desalinization cannot solve the core problem. It may never be cost-effective for large-scale agriculture, and it makes mountains of salts.

Figure 45.13 Irrigated crops in the Sahara Desert, in Algeria.

Consequences of Heavy Irrigation

Large-scale agriculture accounts for nearly two-thirds of the human population's use of fresh water. In many cases, irrigation water from surface sources is piped into vast fields where water-demanding crops will not grow on their own. At its most extreme, irrigation has turned some hot deserts into lush gardens, although believing these can be maintained over the long term is a bit delusional (Figure 45.13).

Irrigation itself can change the land's suitability for agriculture. Concentrations of mineral salts commonly are high in piped-in water. In regions where soil drains poorly, evaporation may cause **salinization**: a buildup of salt in soil. Salinization can stunt the growth of crop plants, eventually kill them, and decrease yields.

Land that drains poorly also becomes waterlogged. When water accumulates underground, it slowly raises the **water table**, the upper limit at which the ground is fully saturated. When the water table rises close to the ground's surface, soil gets saturated with saline water, which can damage plant roots. Properly managing the water–soil system can reverse the salinization and the waterlogging. The economic cost of doing so is high.

Worse, groundwater overdrafts (the amount nature doesn't replenish) are high (Figure 45.14). For instance, overdrafts have depleted half of the Ogallala aquifer, which supplies irrigation water for 20 percent of the croplands in the United States. So much groundwater has been withdrawn for irrigation in some parts of the San Joaquin Valley in California that the surface of the water table has dropped by as much as six meters.

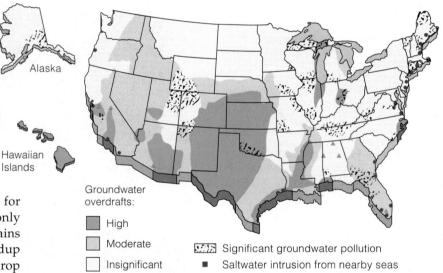

Groundwater
overdrafts:

■ High
□ Moderate   ▦ Significant groundwater pollution
□ Insignificant   ■ Saltwater intrusion from nearby seas

Figure 45.14 Areas of aquifer depletion, saltwater intrusion, and groundwater contamination in the United States.

Water Pollution

Water pollution amplifies the problem of water scarcity. Inputs of sewage, animal wastes, and toxic chemicals from many power-generating plants and factories make water unfit to drink. Runoff from fields puts sediments and pesticides into water, along with phosphates and other plant nutrients that promote algal blooms.

Pollutants collect in lakes, rivers, and bays before reaching the oceans. Many cities throughout the world dump untreated sewage into coastal waters. The cities along rivers and harbors maintain shipping channels by dredging the polluted muck and barging it out to

Figure 45.15 An experimental wastewater treatment facility in Rhode Island. Treatment begins when sewage flows into rows of large water tanks in which water hyacinths, cattails, and other aquatic plants are growing. Decomposers in the tank degrade wastes—which contain nutrients that promote plant growth. Heat from incoming sunlight speeds the decomposition.

From these tanks, water flows through an artificial marsh of sand, gravel, and bulrushes that filter out algae and organic wastes. Then it flows into aquarium tanks, where zooplankton and snails consume microorganisms suspended in the water—and where zooplankton become food for crayfishes, tilapia, and other fishes that can be sold as bait. After ten days, the now-clear water flows into a second artificial marsh for final filtering and cleansing.

sea. They also barge sewage sludge—coarse, settled solids enriched with bacteria, viruses, and toxic metals.

In the United States, about 15,000 facilities partially treat liquid wastes from 70 percent of the population and 87,000 industries. The remaining wastes are mainly from suburban and rural populations. These are treated in lagoons or septic tanks or are directly discharged—untreated—into waterways.

There are three levels of **wastewater treatment**. In *primary* treatment, screens and settling tanks remove sludge, which is dried, burned, dumped in landfills, or treated further. Chlorine often is used to kill pathogens. It does not kill them all, and it produces carcinogens whenever it reacts with certain industrial chemicals.

In *secondary* treatment, microbial populations break down organic matter after primary treatment but before chlorination. Wastewater trickles past microorganisms in gravel beds or is aerated in tanks and seeded with microorganisms. Toxic solutes can poison the microbes. At such times, the facilities shut down until populations of microorganisms become reestablished. Primary and secondary treatments remove most of the suspended solids and oxygen-demanding wastes but not all of the nitrogen, phosphorus, and toxic substances, including heavy metals and pesticides. Then the water is usually chlorinated before being released into the waterways.

Tertiary treatment adequately reduces pollution but is largely experimental and expensive. It is applied to only 5 percent of our nation's wastewater.

In short, most wastewater is not treated adequately. A pattern gets repeated thousands of times along our waterways. Water for drinking is drawn upstream from a city, and wastes from industry and sewage treatment are discharged downstream. It takes no great leap of the imagination to see that water pollution intensifies as rivers flow to the oceans. In Louisiana, waters drained from the central states flow toward the Gulf of Mexico. Its high pollution levels threaten public health as well as ecosystems. Water destined for drinking gets treated to remove pathogens, but treatment does not remove toxic wastes dumped by numerous factories upstream.

This rather bleak picture might be numbing to most of us, but not to biologist John Todd. He constructed experimental wastewater treatment facilities in several greenhouses and artificial lagoons (Figure 45.15). When it works properly, his solar–aquatic treatment system produces water fit to drink. Such natural alternatives cannot work for the large urban areas. But they are an attractive alternative for small towns and rural areas.

The Coming Water Wars

If current rates of human population growth and water depletion hold, the amount of fresh water available for everyone on the planet will soon be 55 to 66 percent less than it was in 1976. Already in this past decade, thirty-three nations have been engaged in conflicts over reductions in water flow, pollution, and silt buildup in major aquifers, rivers, and lakes. The United States and Mexico, Pakistan, India, and Israel and the occupied territories are among the squabblers.

Remember the Persian Gulf War, mainly about oil? Unless we pull off a blue revolution equivalent to the green one, we may be in for upheavals and wars over water rights. Does this sound far-fetched? By building dams and irrigation systems at the headwaters of the Tigris and Euphrates rivers, Turkey can, in the view of one of its dam-site managers, stop the water flow into Syria and Iraq for as long as eight months "to regulate their political behavior." Regional, national, and global planning for the future is long overdue.

Water, not oil, may become the most important fluid of the twenty-first century. National, regional, and global policies for water usage and water rights have yet to be developed.

A QUESTION OF ENERGY INPUTS

Paralleling the J-shaped curve of human population growth is a dramatic rise in total and per capita energy consumption. It is due to increased numbers of energy users and to extravagant consumption and waste. For example, in one city with a near-ideal climate, a major university constructed seven- and eight-story buildings with narrow, sealed windows that cannot be opened to catch prevailing ocean breezes. Neither the buildings nor the windows were designed or aligned to use the abundant sunlight for passive solar heating and breezes for passive cooling. Massive energy-demanding heating and cooling systems were built instead.

Whenever you hear talk of abundant energy supplies, bear in mind that there is a large difference between the *total* and net amounts available. *Net* refers to the amount left over after subtracting the energy necessary to locate, extract, transport, store, and deliver energy to consumers. Some sources, such as direct solar energy, are renewable; but others, including coal, are not (Figure 45.16).

Fossil Fuels

Fossil fuels are the legacy of forests that disappeared many hundreds of millions of years ago, so we can't renew them. Carbon-containing remains of the plants were buried in sediments, compacted, then chemically transformed into coal, petroleum (oil), and natural gas over a great span of time (Section 22.4).

Known reserves of petroleum and natural gas will be depleted in the next century, even with strict conservation. As known reserves run out in accessible areas, we turn to wilderness areas in Alaska and other fragile environments, such as continental shelves. Net energy declines when the cost of extraction and transportation to and from remote areas increases. The environmental effects of extraction and transportation escalate. The 11-million-gallon oil spill from the supertanker *Valdez* off the Alaskan coast is an example. The widespread damage and clean-up costs were enormous.

What about coal? In theory, known reserves may meet the energy needs of the human population for at least a few centuries. But coal burning has been the primary source of air pollution. Most of the known reserves contain low-quality

material with a high sulfur content. Unless sulfur is removed before or after fuel is burned, sulfur dioxides enter the air and contribute to global acid deposition. Fossil fuel burning also releases carbon dioxide and adds to the greenhouse effect.

Extensive strip mining of coal reserves close to the surface invites problems. It reduces the land available for agriculture, grazing, and wildlife. Most strip mining occurs in arid and semiarid regions where the absence of sufficient water supplies and the poor quality of soil make restoration efforts difficult.

Nuclear Energy

NUCLEAR REACTORS In 1945, as Hiroshima burned, people recoiled in horror from the destructive force of nuclear energy. Yet by the 1950s, many considered nuclear energy to be an instrument of progress. Many energy-poor industrialized nations, including France, now rely heavily on nuclear power.

Until recently, construction of more nuclear plants was delayed or canceled in most places. After 1970 in the United States alone, the plans to build 117 nuclear power plants were shelved. Construction of others was stopped before they were finished. At great cost, a few were converted to burn fossil fuels. What happened? People started to question operating costs, efficiency, safety record, and environmental impact of nuclear power.

Yet by 1990, nuclear power was generating electricity at a cost only slightly above that of coal-burning plants. Then, in 2001, California was hit hard by rolling blackouts; and widespread energy shortages put nuclear plants back in the game.

What about safety? Compared to coal-burning plants of the same capacity, a nuclear plant emits less radioactivity and carbon dioxide, and no sulfur dioxide. However, it has the potential for a **meltdown.** Nuclear fuel releases a lot of heat by radioactive decay. In the older plants, water circulating over the nuclear fuel absorbs heat, then it turns into steam that can drive the

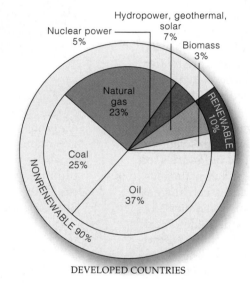

DEVELOPED COUNTRIES

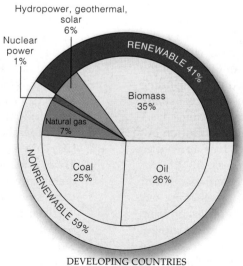

DEVELOPING COUNTRIES

Figure 45.16 Energy consumption in countries that differ greatly in sources of energy and average per capita energy use. The values shown do not take into account energy from the sun, which is the foundation for agriculture.

Figure 45.17 Incident at Chernobyl. (**a**) On April 26, 1986, errors in judgment during a routine test procedure resulted in runaway reactions, explosions, and a full core meltdown at the Chernobyl power plant in Ukraine. Helicopter pilots who were supposed to drop 5,000 tons of lead, sand, clay, and other materials on the blazing core to suppress further release of radioactivity missed the target. Unimpeded and uncovered, nuclear fuel burned for nearly ten days, right on through a six-foot-thick steel and gravel barrier beneath it.

Between 185 and 250 million curies of radioactive matter may have escaped in those ten days alone. Inhaling as little as ten-millionths of a curie of plutonium can cause cancer. Thirty-one people died at once; others died of radiation poisoning in the following weeks. Inhabitants of entire villages were relocated; their former homes were bulldozed under. In time, concrete entombed the 180 tons of partially burned nuclear fuel.

(**b**) Afterward, the number of people opposed to nuclear plants rose sharply, even in France. We get a sense of why opposition increased dramatically by studying maps of the global distribution of radioactive fallout within two weeks of the meltdown. The fallout put 300 to 400 million people at risk for leukemia and other radiation-induced disorders. Throughout Europe, hundreds of millions of dollars were lost when the fallout made crops and livestock unfit for consumption. In 1994, rainwater and air were still moving through 11,000 square meters of holes in the power plant's concrete. By 1998, the rate of thyroid abnormalities in children living immediately downwind from Chernobyl was nearly seven times as high as for those living upwind; their thyroid gland collected iodine radioisotopes from the fallout.

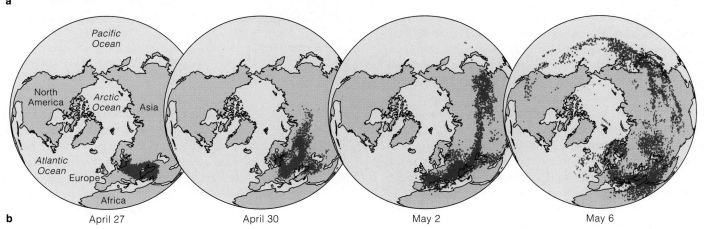

electricity-generating turbines. If the circulating water system leaks, water levels might plummet around the fuel. Fuel may heat past its melting point. Melting fuel could instantly convert the rest of the water to steam. Steam formation and other reactions could blow up the system, releasing radioactive material. An overheated core could melt through the thick concrete slab that contains it, as happened at Chernobyl (Figure 45.17).

New reactors are based on passive cooling and are probably safer. Nuclear fusion reactors, which produce far less radioactive waste, are on the drawing boards.

NUCLEAR WASTE DISPOSAL Unlike coal, nuclear fuel cannot be burned to harmless ashes. After three years or so, the fuel elements are spent, but they still contain uranium fuel as well as hundreds of new radioisotopes that formed in the reactions. The wastes are extremely radioactive and dangerous. They get extremely hot as they undergo radioactive decay, so they are plunged at once into water-filled pools. The water cools them and

keeps radioactive material from escaping. Even after being stored for several months, the isotopes remaining are lethal. Some must be kept isolated for at least 10,000 years. If a certain isotope of plutonium (^{239}Pu) is not removed, wastes must be kept isolated for a quarter of a million years! After nearly fifty years of research, scientists still cannot agree on what is the best way to store high-level radioactive wastes. Even if they could, there is no politically acceptable solution. No one wants radioactive wastes anywhere near where they live.

As if we don't have enough to worry about, after the Soviet Union's breakup, some underpaid workers of a Russian nuclear power plant sold fuel elements on the black market. The buyers? Developing nations that want to make nuclear weapons—and possibly deliver them into the hands of terrorist organizations.

The nuclear genie is out of the bottle, exploitable by the best and worst elements of the human population.

ALTERNATIVE ENERGY SOURCES

Less than thirty years from now, the projected human population size will be such that demands for fossil fuels will increase by 30 percent and for electricity by 265 percent. California is already experiencing massive shortages of electricity. It won't be enough to use and conserve known energy supplies more efficiently. We must develop alternatives, of the sort described here.

Solar–Hydrogen Energy

Each year, incoming sunlight contains about ten times more energy than that in all of the known fossil fuel reserves. That is 15,000 times more energy than our population uses now. Isn't it about time that we start to collect it in earnest, in something besides crops?

For example, when exposed to sunlight, electrodes in "photovoltaic cells" produce an electric current that splits water molecules into oxygen and hydrogen gas (H_2)—which can be used directly as fuel or to generate electricity. The technology to tap such **solar–hydrogen energy** has been around since the 1940s (Figure 45.18a). Such energy is stored efficiently for as long as required. It costs less to distribute H_2 than electricity, and water is the only by-product of using it. Space satellites run on it. Here on Earth, fossil fuels are still "cheaper."

Unlike fossil fuels, however, sunlight and seawater are unlimited resources. The technology's potential to protect the environment is staggering. Environmental scientist G. Tyler Miller Jr. puts it this way: "If we make the transition to an energy-efficient solar–hydrogen age, we can say good-bye to smog, oil spills, acid rain, and nuclear energy, and perhaps to global warming. The reason is simple. As hydrogen burns in air, it reacts with oxygen gas to produce water vapor—not a bad thing to have coming out of tailpipes, chimneys, and smokestacks." Also, if this technology becomes cost-effective for the developing countries, the great forests now being destroyed for timber and fuel might still be around for future generations.

In 1995, photovoltaic cells generated almost 4 million kilowatt-hours of net electricity in the United States (of the total electricity generation of 988.8 million). Thirty states now promote development of solar energy. In 1996, Hawaii, Iowa, and Washingon offered incentives to promote clean, renewable energy technologies.

Wind Energy

As you know, solar energy also is converted into the mechanical energy of winds. Where prevailing winds travel faster than 7.5 meters per second, we find **wind farms**. These arrays of cost-effective turbines exploit wind patterns that arise from latitudinal variations in the intensity of incoming sunlight (Figure 45.18b). One

Figure 45.18 Harnessing solar energy. (**a**) A large array of electricity-producing photovoltaic cells in panels that collect sunlight energy. (**b**) A field of turbines harvesting wind energy.

percent of California's electricity comes from wind farms. Possibly the winds of North and South Dakota alone can meet all but 20 percent of the current energy needs of the United States. Wind energy also has potential for islands and other areas remote from utility grids. One drawback: Winds don't blow on a regular schedule, so they can't be an exclusive or major energy source.

What About Fusion Power?

The sun's gravitational force is enough to compress atomic nuclei to high densities; its temperatures are high enough to force atomic nuclei to fuse. We call this **fusion power**. Similar conditions do not exist on Earth, but maybe we can mimic them. Researchers confine a certain fuel (a heated gas of two isotopes of hydrogen) in magnetic fields, then bombard it with lasers. The fuel implodes, it is compressed to extremely high densities and energy is released. The more energetic the lasers, the greater the compression, and the more the fuel will burn. The bad news is, although the amount of energy released has been steadily increasing, it will be at least fifty years before fusion reactors could be operating, and the costs will probably be high. The good news is, that's about the time fossil fuels may start running out.

Sunlight may end up sustaining the energy needs of the human population in more ways than one.

Biological Principles and the Human Imperative

Molecules, single cells, tissues, organs, organ systems, multicelled organisms, populations, communities, then ecosystems and the biosphere. These are architectural systems of life, assembled in increasingly complex ways during the past 3.8 billion years. We are latecomers to this immense biological building program. Yet within the relatively short span of 10,000 years, our activities have been changing the very character of the land, ocean, and atmosphere, even the genetic character of species.

It would be presumptuous to think we alone have had profound effects on the world of life. As long ago as the Proterozoic era, photosynthetic organisms irrevocably changed the course of biological evolution by gradually enriching the atmosphere with oxygen. During the past as well as the present, competitive adaptations ensured the rise of some groups, whose dominance assured the decline of others. Change is nothing new to this biological building program. What *is* new is the capacity of a species —our own—to comprehend what might be going on.

We now have the population size, the technology, and the cultural inclination to use up energy and modify the environment at frightening rates. Where will rampant, accelerated change lead us? Will feedback controls begin to operate as they do, for example, when population growth exceeds the carrying capacity of the environment? In other words, will negative feedback controls come into play and keep things from getting too far out of hand?

Feedback control will not be enough, for it operates only when deviation already exists. Patterns of resource consumption and growth rates for the human population are founded on an illusion of unlimited resources and a forgiving environment. A prolonged, global shortage of food or the passing of a critical threshold for the global climate can come too fast to be corrected. At some point, deviations may have too great an impact to be reversed.

What about feedforward mechanisms that might serve as early warning systems? For example, when sensory receptors near the surface of skin sense a drop in outside air temperature, each sends messages to the nervous system. That system responds by triggering mechanisms that raise the body's core temperature before the body itself becomes dangerously chilled. If we develop feedforward control mechanisms, maybe we can begin corrective measures before we have altered the environment too significantly.

By themselves, feedforward controls won't work, for they start operating when change is under way. Think of the DEW line—the Distant Early Warning system. It is like a vast sensory receptor, one that can detect the launching of intercontinental ballistic missiles against North America. By the time the system actually detects what it is designed to detect, it may be too late to stop widespread destruction.

It would be naive to assume we can ever reverse who we are at this point in evolutionary time, to de-evolve ourselves culturally and biologically into becoming less complex in the hope of averting disaster. However, there is reason to believe we can avert disaster by using a third kind of control mechanism—a capacity to anticipate events even before they happen. We are not locked into responding only after irreversible change has begun. We have the capacity to anticipate the future—it is the essence of our visions of utopia or of nightmarish hell. *We all have the capacity to adapt to a future that we can partly shape.*

We can, for example, stop trying to "beat nature" and learn to work with it. Individually and collectively, we can work to develop long-term policies at the local, regional, and global levels—policies that take into account biotic and abiotic limits on population growth. Far from being a surrender, this would be one of the most complex and intelligent behaviors of which we are capable.

Having a capacity to adapt and actually using it are not the same thing. We have already put the world of life on dangerous ground because we have not yet mobilized ourselves as a species to work toward self-control.

Our survival depends on predicting possible futures. It depends on preserving, restoring, and even designing and constructing ecosystems that are in harmony with our definition of basic human values and with the biological models available to us. Human values can change; our expectations can and must be adapted to biological reality. *For the principles of energy flow and resource utilization, which govern the survival of all systems of life, do not change.* It is our biological and cultural imperative that we come to terms with these principles, and ask ourselves what our long-term contribution will be to the world of life.

SUMMARY   *Gold* indicates text section

1. Accompanying the extraordinarily rapid growth of the human population are increases in energy demands and in environmental pollution. *CI, 45.10*

2. Pollutants are substances with which ecosystems have had no prior evolutionary experience (in terms of kinds and amounts) and thus have no mechanisms for absorbing or cycling them. Many pollutants result from certain human activities, and they adversely affect the health, activities, or survival of all organisms. *45.1*

3. Smog, a form of air pollution, arises in industrialized and urban regions that rely on fossil fuels. It becomes most concentrated in land basins that promote thermal inversion (a layer of cool, dense air trapped below a warm air layer). Industrial smog forms in industrial coal-burning regions with cold, wet winters. In warm climates, large cities with many fuel-burning vehicles have photochemical smog. Mainly, sunlight makes nitric oxide emitted from vehicles react with hydrocarbons to form photochemical oxidants such as PANs. *45.1*

4. In dry weather, acidic air pollutants (e.g., oxides of nitrogen and sulfur) fall as dry acid deposition. They also dissolve in atmospheric water, then fall to Earth as wet acid deposition, or acid rain. *45.1*

5. Seasonal thinning of the ozone layer at high latitudes has become pronounced as CFCs (chlorofluorocarbons) and other air pollutants rise to the stratosphere, where they deplete ozone and allow more harmful ultraviolet radiation from the sun to reach the Earth. *45.2*

6. Land surfaces are being degraded by:
    a. Passively accepting the dumping, burning, and burial of solid wastes and not making concerted efforts to recycle or reuse materials and to reduce waste. *45.3*
    b. Engaging in rampant deforestation (destruction of vast tracts of forest biomes). *45.4, 45.5*
    c. Contributing to desertification (the conversion of natural grasslands, croplands, and grazing lands, on a large scale, to desertlike conditions). *45.6*

7. Human population growth currently depends upon the expansion of agriculture, made possible by large-scale irrigation and extensive applications of fertilizers and pesticides. Global freshwater supplies are limited, and yet they are being polluted by agricultural runoff (which includes sediments, pesticides, and fertilizers), industrial wastes, and human sewage. *45.7*

8. Fossil fuel energy is nonrenewable, dwindling, and environmentally costly to extract and utilize. Nuclear energy is less polluting, but costs and risks associated with fuel containment and storing radioactive wastes are high. The rapid growth of the human population means that affordable alternatives such as wind energy and solar energy must soon be developed. *45.8, 45.9*

Review Questions

1. Define pollution and list some specific examples of water pollutants. *45.1, 45.7*

2. Distinguish among the following conditions: *45.1*
    a. industrial smog   c. dry acid deposition
    b. photochemical smog   d. wet acid deposition

3. Define CFCs and describe how they apparently contribute to seasonal thinning of the ozone layer in the stratosphere. *45.2*

4. What percent of the Earth's land is under cultivation? What percent is available for new cultivation? *45.3*

5. Define and describe possible consequences of deforestation and of desertification. *45.4, 45.5, 45.6*

6. Which human activity uses the most fresh water? *45.7*

Self-Quiz   ANSWERS IN APPENDIX III

1. Starting a few centuries ago, human population growth has been _____ .
    a. leveling off   c. accelerating
    b. growing slowly   d. not much to speak of

2. Pollutants disrupt ecosystems because _____ .
    a. their components differ from those of natural substances
    b. only humans have uses for them
    c. there are no evolved mechanisms to deal with them
    d. their only effect is on ecosystems, not humans

3. During a thermal inversion, weather conditions trap a layer of _____ air under a layer of _____ air.
    a. warm; cool   c. warm; sooty
    b. cool; warm   d. cool; sooty

4. _____ is a case of regional air pollution.
    a. Smog   c. Ozone layer thinning
    b. Acid rain   d. a and b

5. _____ is a case of air pollution with global effects.
    a. Smog   c. Ozone layer thinning
    b. Acid rain   d. b and c

6. Worldwide, two-thirds of the fresh water used annually goes to _____ .
    a. urban centers   c. treatment facilities
    b. agriculture   d. a and c

7. The upper limit at which the ground is fully saturated with water is called _____ .
    a. groundwater   c. the water table
    b. an aquifer   d. the salinization limit

8. Energy from fossil fuels is _____ ; their extraction and use come at _____ cost to the environment.
    a. renewable; low   c. renewable; high
    b. nonrenewable; low   d. nonrenewable; high

9. Nuclear energy normally pollutes _____ than fossil fuels; it poses _____ dangers than fossil fuels.
    a. less; lesser b. more; greater c. more; fewer d. less; more

10. Match each term with the most suitable description.
    ____ desertification   a. possibly one of our best options
    ____ deforestation   b. soil loss, watershed damage,
    ____ green   altered rainfall patterns follow
         revolution   c. attempt to improve crop
    ____ solar–hydrogen   production on existing land
         energy   d. converting large tracts of natural
    ____ CFCs   grasslands to desertlike state
                             e. invisible, odorless compounds
                                 that contribute to ozone thinning

Figure 45.19 City as ecosystem.

Critical Thinking

1. Kristen, a recent college graduate, is finding her idealism on a collision course with reality. She strongly believes people who live in the United States are obliged to make the world a level playing field for all human beings, with equality in resources, health care, education, economic security, and a pristine environment for all. Yet she also understands that the sheer size of the human population makes this impossible.

Kristen recently said she cannot be party to hard choices and actions that go against her ideals, and she just wants nature to solve the problem for us. Comment on this true story.

2. Investigate where the water for your own city comes from and where it has been. You may find the answer illuminating.

3. Make a list of advantages you personally enjoy as a member of an affluent, industrialized society. List some drawbacks. Do the benefits outweigh the costs? This is not a trick question.

4. It has been said that economic wars, more than military wars, will determine the winners and losers among nations in the near future. The economic growth of certain nations, including some in the former Soviet Union and the Far East, has had devastating impact on the environment. Elsewhere, efforts of conservation biologists to maintain standards of environmental protection add to the cost of goods produced and put practicing nations at serious disadvantage in this global competition.

Should the United States loosen environmental laws to help ensure its economic survival? Why or why not? Should it impose pressure on indifferent nations to encourage reduction in harmful practices? Why or why not?

5. To reduce the backlog of unnamed new species and secure funding, a group of German taxonomists is offering shoppers the opportunity, for $2,500 and up, to name a species (Biopat web site). Example: As a birthday present, investment banker Stan Lai's wife had a Bolivian yellow-striped brown toad named in her husband's honor (*Bufo stanlaii*, or "Stan Lai's toad").

Usually, a species name reflects its appearance or honors its discoverer. But not always; for example, one midge has the name *Dicrotendipes thanatogratus* (the "Grateful Dead" midge). The International Commission on Zoological Nomenclature in London denounces offering taxonomy to the public, which it believes will obscure science and hinder conservation efforts. What does Mrs. Lai think? "Now, every time I hear something about Bolivia, I'll pay attention. After all, I don't want to see our toad die."

Does inviting the public to pay their way into taxonomy undermine the science? Is it a good or bad way to raise public awareness of conservation biology?

6. Populations of every species utilize resources and produce wastes. Use Figure 43.19 as a starting point for a brief essay on the accumulation and uses of energy and materials, including wastes, in a human population of a large city. Contrast your description with the flow of energy and the cycling of materials that proceed in a natural population of some other organism.

7. Many psychologists believe that we spend much of our lives consciously or subconsciously searching for roots that might anchor us in bewildering or frightening times of change. Some philosophers have argued that each of us must find a mountain, river, backyard, or any other place that elicits a sense of sanctuary, of rooted connections with nature. Others believe that forming an emotional connection with nature is based on a romanticized or mystical view of it—and that time would be better spent developing a scientifically sound understanding of nature and the technological means to protect and sustain its resources.

Do you agree with one or the other point of view? Or do you suspect that the two are not mutually exclusive?

Selected Key Terms

acid rain *45.1*	ozone thinning *45.2*
chlorofluorocarbon (CFC) *45.2*	PAN (peroxyacyl nitrate) *45.1*
deforestation *45.4*	photochemical smog *45.1*
desalinization *45.7*	pollutant *45.1*
desertification *45.6*	salinization *45.7*
dry acid deposition *45.1*	shifting cultivation *45.4*
fossil fuel *45.8*	solar–hydrogen energy *45.9*
fusion power *45.9*	thermal inversion *45.1*
green revolution *45.3*	wastewater treatment *45.7*
industrial smog *45.1*	water table *45.7*
meltdown *45.8*	wind farm *45.9*

Readings

Frosh, R. September 1995. "The Industrial Ecology of the Twenty-First Century." *Scientific American* 273(3):178–181.

Kaiser, J. 18 February 2000. "Ecologists on a Mission to Save the World." *Science* 287:1188–1192. Addresses a growing debate over how far environmental scientists should go when interpreting their findings for policy makers.

Miller, G. T., Jr. 2001. *Living in the Environment*. Eighth edition. Belmont, California: Brooks/Cole.

APPENDIX I. CLASSIFICATION SCHEME

The classification scheme that follows is a composite of several that microbiologists, botanists, and zoologists use. Major groupings are agreed upon, more or less. There is not always agreement on what to name a particular grouping or where it may fit within the overall hierarchy. There are several reasons for the lack of total consensus.

First, the fossil record varies in its quality and in its completeness. Therefore, the phylogenetic relationship of one group to others is sometimes open to interpretation. Comparative studies at the molecular level are firming up the picture, but this work is still under way.

Second, ever since the time of Linnaeus, classification schemes have been based on the perceived morphological similarities and differences among organisms. Although some original interpretations are now open to question, we are so used to thinking about organisms in certain ways that reclassification often proceeds slowly.

For example, birds and reptiles traditionally have been placed in separate classes (Reptilia and Aves), yet there are many compelling arguments for grouping the lizards and snakes in one class and the crocodilians, dinosaurs, and birds in a separate class. Some favor six kingdoms but others favor three domains: archaebacteria, eubacteria, and eukaryota (alternatively, archaea, bacteria, and eukarya).

Third, researchers in microbiology, mycology, botany, zoology, and the other fields of biological inquiry have inherited a wealth of literature, based on classification schemes that were developed over time in each of those fields. Many see no good reason to give up the established terminology and thereby disrupt access to the past.

For instance, many microbiologists and botanists use *division*, and zoologists *phylum*, for taxa that are equivalent in the hierarchy of classification. Also, opinions are still polarized with respect to the kingdom Protista, certain members of which could just as easily be grouped with plants, fungi, or animals. Indeed, the term protozoan is a holdover from an earlier scheme in which some single-celled organisms were ranked as simple animals.

Given the problems, why do we even bother imposing artificial frameworks on the history of life? We do this for the same reason that a writer might decide to break up the history of civilization into several volumes, a number of chapters, and many paragraphs. Both efforts are attempts to impart obvious structure to what might otherwise be an overwhelming body of knowledge and to enhance the retrieval of information from it.

Finally, bear in mind that we include this classification scheme primarily for your reference purposes. Besides being open to revision, it also is by no means complete. It does not include the most recently discovered species, as from the mid-ocean. Many existing and extinct organisms of the so-called lesser phyla are not represented here. Our strategy is to focus mainly on organisms mentioned in the text. A few examples of organisms also are listed.

SUPERKINGDOM PROKARYOTA. Prokaryotes. Almost all microscopic species. DNA organized at nucleoid (a region of cytoplasm), not inside a membrane-bound nucleus. All are bacteria, either single cells or simple associations of cells. Autotrophs and heterotrophs. Table A on the following page lists representative types. Reproduce by prokaryotic fission, sometimes by budding and by bacterial conjugation.

The authoritative reference in bacteriology, *Bergey's Manual of Systematic Bacteriology*, calls this "a time of taxonomic transition." It groups bacteria mostly by numerical taxonomy (Section 20.1), not on phylogeny. The scheme presented here reflects strong evidence of evolutionary relationships for at least some bacterial groupings.

KINGDOM EUBACTERIA. Gram-negative, gram-positive forms. Peptidoglycan present in cell wall. Photosynthetic autotrophs, chemosynthetic autotrophs, and heterotrophs.

PHYLUM GRACILICUTES. Typical Gram-negative, thin wall. Autotrophs (photosynthetic and chemosynthetic) and heterotrophs. *Anabaena* and other cyanobacteria. *Escherichia, Pseudomonas, Neisseria, Myxococcus.*

PHYLUM FIRMICUTES. Typical Gram-positive, thick wall. Heterotrophs. *Bacillus, Staphylococcus, Streptococcus, Clostridium, Actinomycetes.*

PHYLUM TENERICUTES. Gram-negative, wall absent. Heterotrophs (saprobes, pathogens). *Mycoplasma.*

KINGDOM ARCHAEBACTERIA. Methanogens, extreme halophiles, extreme thermophiles. Evolutionarily closer to eukaryotic cells than to eubacteria. Strict anaerobes. Distinctive cell wall, membrane lipids, ribosomes, RNA sequences. *Methanobacterium, Halobacterium, Sulfolobus.*

SUPERKINGDOM EUKARYOTA. Eukaryotes. Both single-celled and multicelled species. Cells start out life with a nucleus (encloses the DNA) and usually other membrane-bound organelles. Chromosomes have many histones and other proteins attached.

KINGDOM PROTISTA. Diverse single-celled, colonial, and multicelled eukaryotic species. Existing species are unlike bacteria in characteristics and are most like the earliest, structurally simple eukaryotes. Autotrophs, heterotrophs, or both (Table 20.3). Reproduce sexually and asexually (by meiosis, mitosis, or both). Many related evolutionarily to plants, fungi, and possibly animals.

PHYLUM CHYTRIDIOMYCOTA. Chytrids. Heterotrophs; saprobic decomposers or parasites. *Chytridium.*

PHYLUM OOMYCOTA. Water molds. Heterotrophs. Decomposers, some parasites. *Saprolegnia, Phytophthora, Plasmopara.*

PHYLUM ACRASIOMYCOTA. Cellular slime molds. Heterotrophs with free-living, phagocytic amoeboid cells and spore-bearing stages. *Dictyostelium.*

PHYLUM MYXOMYCOTA. Plasmodial slime molds. Heterotrophs with free-living, phagocytic amoeboid cells and spore-bearing stages. Aggregate into streaming mass of cells that discard plasma membranes. *Physarum.*

Table A Representative Eubacteria and Archaebacteria Grouped on the Basis of Numerical Taxonomy

Some Major Groups	Main Habitats	Characteristics	Representatives
EUBACTERIA			
Photoautotrophs:			
Cyanobacteria, green sulfur bacteria, and purple sulfur bacteria	Mostly lakes, ponds; some marine, terrestrial habitats	Photosynthetic; use sunlight energy, carbon dioxide; cyanobacteria use oxygen-producing noncyclic pathway; some also use cyclic route	*Anabaena, Nostoc, Rhodopseudomonas, Chloroflexus*
Photoheterotrophs:			
Purple nonsulfur and green nonsulfur bacteria	Anaerobic, organically rich muddy soils, and sediments of aquatic habitats	Use sunlight energy; organic compounds as electron donors; some purple nonsulfur may also grow chemotrophically	*Rhodospirillum, Chlorobium*
Chemoautotrophs:			
Nitrifying, sulfur-oxidizing, and iron-oxidizing bacteria	Soil; freshwater, marine habitats	Use carbon dioxide, inorganic compounds as electron donors; influence crop yields, cycling of nutrients in ecosystems	*Nitrosomonas, Nitrobacter, Thiobacillus*
Chemoheterotrophs:			
Spirochetes	Aquatic habitats; parasites of animals	Helically coiled, motile; free-living and parasitic species; some major pathogens	*Spirochaeta, Treponema*
Gram-negative aerobic rods and cocci	Soil, aquatic habitats; parasites of animals, plants	Some major pathogens; some fix nitrogen (e.g., *Rhizobium*)	*Pseudomonas, Neisseria, Rhizobium, Agrobacterium*
Gram-negative facultative anaerobic rods	Soil, plants, animal gut	Many major pathogens; one bioluminescent (*Photobacterium*)	*Salmonella, Escherichia, Proteus, Photobacterium*
Rickettsias and chlamydias	Host cells of animals	Intracellular parasites; many pathogens	*Rickettsia, Chlamydia*
Myxobacteria	Decaying organic material; bark of living trees	Gliding, rod-shaped; aggregation and collective migration of cells	*Myxococcus*
Gram-positive cocci	Soil; skin and mucous membranes of animals	Some major pathogens	*Staphylococcus, Streptococcus*
Endospore-forming rods and cocci	Soil; animal gut	Some major pathogens	*Bacillus, Clostridium*
Gram-positive nonsporulating rods	Fermenting plant, animal material; gut, vaginal tract	Some important in dairy industry, others major contaminators of milk, cheese	*Lactobacillus, Listeria*
Actinomycetes	Soil; some aquatic habitats	Include anaerobes and strict aerobes; major producers of antibiotics	*Actinomyces, Streptomyces*
ARCHAEBACTERIA (ARCHAEA)			
Methanogens	Anaerobic sediments of lakes, swamps; animal gut	Chemosynthetic; methane producers; used in sewage treatment facilities	*Methanobacterium*
Extreme halophiles	Brines (extremely salty water)	Heterotrophic; also, unique photosynthetic pigments (bacteriorhodopsin) form in some	*Halobacterium*
Extreme thermophiles	Acidic soil, hot springs, hydrothermal vents	Heterotrophic or chemosynthetic; use inorganic substances as electron donors	*Sulfolobus, Thermoplasma*

PHYLUM SARCODINA. Amoeboid protozoans. Heterotrophs, free-living or endosymbiotic, some pathogens. Soft-bodied, with or without shell, locomotion by pseudopods. The rhizopods (naked amoebas, foraminiferans), *Amoeba proteus, Entomoeba*. Also actinopods (radiolarians, heliozoans).

PHYLUM CILIOPHORA. Ciliated protozoans. Heterotrophs, predators or symbionts, some parasitic. All have cilia. Free-living, sessile, or motile. *Paramecium, Didinium*, hypotrichs.

PHYLUM MASTIGOPHORA. Animal-like flagellated protozoans. Heterotrophs, free-living, many internal parasites. All with one to several flagella. *Trypanosoma, Trichomonas, Giardia*.

APICOMPLEXA. Heterotrophs, sporozoite-forming parasites. Complex structures at head end. Most familiar members called sporozoans. *Cryptosporidium, Plasmodium, Toxoplasma*.

PHYLUM EUGLENOPHYTA. Euglenoids. Mostly heterotrophs, some autotrophs (photosynthetic), some switch depending on environmental conditions. Most with one short, one long flagellum; red, green, or colorless. *Euglena, Peranema*.

PHYLUM PYRRHOPHYTA. Dinoflagellates. Photosynthetic, mostly, but some heterotrophs. *Fiesteria, Gymnodinium breve*.

PHYLUM CHRYSOPHYTA. Golden algae, yellow-green algae, diatoms. Photosynthetic. Some flagellated, others not. *Mischococcus, Synura, Vaucheria*.

PHYLUM RHODOPHYTA. Red algae. Mostly photosynthetic, some parasitic. Nearly all marine, some in freshwater habitats. *Porphyra, Bonnemaisonia, Euchema*.

PHYLUM PHAEOPHYTA. Brown algae. Photosynthetic, nearly all in temperate or marine waters. *Macrocystis, Fucus, Sargassum, Ectocarpus, Postelsia*.

PHYLUM CHLOROPHYTA. Green algae. Mostly photosynthetic, some parasitic. Most freshwater, some marine or terrestrial. *Chlamydomonas, Spirogyra, Ulva, Volvox, Codium, Halimeda*.

PHYLUM ZYGOMYCOTA. Zygomycetes. Zygosporangia (zygote inside thick wall) formed by sexual reproduction. Bread molds, related forms. *Rhizopus, Philobolus.*

PHYLUM ASCOMYCOTA. Ascomycetes. Sac fungi. Sac-shaped cells form sexual spores (ascospores). Most yeasts and molds, morels, truffles. *Saccharomycetes, Morchella, Neurospora, Sarcoscypha, Claviceps, Ophiostoma, Candida, Aspergillus, Penicillium.*

PHYLUM BASIDIOMYCOTA. Basidiomycetes. Club fungi. Most diverse group. Produce basidiospores inside club-shaped structures. Mushrooms, shelf fungi, stinkhorns. *Agaricus, Amanita, Craterellus, Gymnophilus, Puccinia, Ustilago.*

IMPERFECT FUNGI. Sexual spores absent or undetected. The group has no formal taxonomic status. If better understood, a given species might be grouped with sac fungi or club fungi. *Arthobotrys, Histoplasma, Microsporum, Verticillium.*

LICHENS. Mutualistic interactions between fungal species and a cyanobacterium, green alga, or both. *Lobaria, Usnea, Cladonia.*

PHYLUM RHYNIOPHYTA. Earliest known vascular plants; muddy habitats. Extinct. *Cooksonia, Rhynia.*

PHYLUM PROGYMNOSPERMOPHYTA. Progymnosperms. Ancestral to early seed-bearing plants; extinct. *Archaeopteris.*

PHYLUM PTERIDOSPERMOPHYTA. Seed ferns. Fernlike gymnosperms; extinct. *Medullosa.*

PHYLUM CHAROPHYTA. Stoneworts.

PHYLUM BRYOPHYTA. Bryophytes: mosses, liverworts, hornworts. Seedless, nonvascular, haploid dominance. *Marchantia, Polytrichum, Sphagnum.*

PHYLUM PSILOPHYTA. Whisk ferns. Seedless, vascular. No obvious roots, leaves on sporophyte. *Psilotum.*

PHYLUM LYCOPHYTA. Lycophytes, club mosses. Seedless, vascular. Leaves, branching rhizomes, vascularized roots and stems. *Lepidodendron* (extinct), *Lycopodium, Selaginella.*

PHYLUM SPHENOPHYTA. Horsetails. Seedless, vascular. Some sporophyte stems photosynthetic, others nonphotosynthetic, spore-producing. *Calamites* (extinct), *Equisetum.*

PHYLUM PTEROPHYTA. Ferns. Largest group of seedless vascular plants (12,000 species), mainly tropical, temperate habitats. *Pteris, Trichomanes, Cyathea* (tree ferns), *Polystichum.*

PHYLUM CYCADOPHYTA. Cycads. Gymnosperm group (vascular, bears "naked" seeds). Tropical, subtropical. Palm-shaped leaves, simple cones on male and female plants. *Zamia.*

PHYLUM GINKGOPHYTA. Ginkgo (maidenhair tree). Type of gymnosperm. Seeds with fleshy outer layer. *Ginkgo.*

PHYLUM GNETOPHYTA. Gnetophytes. Only gymnosperms with vessels in xylem and double fertilization (but endosperm does not form). *Ephedra, Welwitchia.*

PHYLUM CONIFEROPHYTA. Conifers. Most common and familiar gymnosperms. Generally cone-bearing species with needle-like or scale-like leaves.

　　Family Pinaceae. Pines, firs, spruces, hemlock, larches, true cedars. *Abies, Cedrus, Pinus, Pseudotsuga.*
　　Family Cupressaceae. Junipers, cypresses. *Cupressus, Juniperus.*
　　Family Taxodiaceae. Bald cypress, redwoods, bigtree. *Metasequoia, Sequoia, Sequoiadendron, Taxodium.*
　　Family Taxaceae. Yews. *Taxus.*

PHYLUM ANTHOPHYTA. Angiosperms (the flowering plants). Largest, most diverse group of vascular seed-bearing plants. Only organisms that produce flowers, fruits.

　Class Dicotyledonae. Dicotyledons (dicots). Some families of some representative orders are listed:
　　Family Nymphaeaceae. Water lilies.
　　Family Papaveraceae. Poppies.
　　Family Brassicaceae. Mustards, cabbages, radishes.
　　Family Malvaceae. Mallows, cotton, okra, hibiscus.
　　Family Solanaceae. Potatoes, eggplant, petunias.
　　Family Salicaceae. Willows, poplars.
　　Family Rosaceae. Roses, apples, almonds, strawberries.
　　Family Fabaceae. Peas, beans, lupines, mesquite.
　　Family Cactaceae. Cacti.
　　Family Euphorbiaceae. Spurges, poinsettia.
　　Family Cucurbitaceae. Gourds, melons, cucumbers, squashes.
　　Family Apiaceae. Parsleys, carrots, poison hemlock.
　　Family Aceraceae. Maples.
　　Family Asteraceae. Composites. Chrysanthemums, sunflowers, lettuces, dandelions.
　Class Monocotyledonae. Monocotyledons (monocots). Some families of several different orders are listed:
　　Family Liliaceae. Lilies, hyacinths, tulips, onions, garlic.
　　Family Iridaceae. Irises, gladioli, crocuses.
　　Family Orchidaceae. Orchids.
　　Family Arecaceae. Date palms, coconut palms.
　　Family Cyperaceae. Sedges.
　　Family Poaceae. Grasses, bamboos, corn, wheat, sugarcane.
　　Family Bromeliaceae. Bromeliads, pineapples, Spanish moss.

PHYLUM PLACOZOA. Marine. Simplest known animal. Two cell layers, no mouth, no organs. *Trichoplax.*

PHYLUM MESOZOA. Ciliated, wormlike parasites, about the same level of complexity as *Trichoplax.*

PHYLUM PORIFERA. Sponges. No symmetry, tissues. *Euplectella.*

PHYLUM CNIDARIA. Radial symmetry, tissues, nematocysts.
　Class Hydrozoa. Hydrozoans. *Hydra, Obelia, Physalia, Prya.*
　Class Scyphozoa. Jellyfishes. *Aurelia.*
　Class Anthozoa. Sea anemones, corals. *Telesto.*

PHYLUM PLATYHELMINTHES. Flatworms. Bilateral, cephalized; simplest animals with organ systems. Saclike gut.
　Class Turbellaria. Triclads (planarians), polyclads. *Dugesia.*
　Class Trematoda. Flukes. *Clonorchis, Schistosoma.*
　Class Cestoda. Tapeworms. *Diphyllobothrium, Taenia.*

PHYLUM NEMERTEA. Ribbon worms. *Tubulanus.*

PHYLUM NEMATODA. Roundworms. *Ascaris, Caenorhabditis elegans, Necator* (hookworms), *Trichinella.*

PHYLUM ROTIFERA. Rotifers. *Asplancha, Philodina.*

PHYLUM MOLLUSCA. Mollusks.
　Class Polyplacophora. Chitons. *Cryptochiton, Tonicella.*

Class Gastropoda. Snails (periwinkles, whelks, limpets, abalones, cowries, conches, nudibranchs, tree snails, garden snails), sea slugs, land slugs. *Aplysia, Ariolimax, Cypraea, Haliotis, Helix, Liguus, Limax, Littorina, Patella.*

Class Bivalvia. Clams, mussels, scallops, cockles, oysters, shipworms. *Ensis, Chlamys, Mytelus, Patinopectin.*

Class Cephalopoda. Squids, octopuses, cuttlefish, nautiluses. *Dosidiscus, Loligo, Nautilus, Octopus, Sepia.*

PHYLUM BRYOZOA.   Bryozoans (moss animals).

PHYLUM BRACHIOPODA.   Lampshells.

PHYLUM ANNELIDA.   Segmented worms.

Class Polychaeta. Mostly marine worms. *Eunice, Neanthes.*

Class Oligochaeta. Mostly freshwater and terrestrial worms, but many marine. *Lumbricus* (earthworms), *Tubifex.*

Class Hirudinea. Leeches. *Hirudo, Placobdella.*

PHYLUM TARDIGRADA.   Water bears.

PHYLUM ONYCHOPHORA.   Onychophorans. *Peripatus.*

PHYLUM ARTHROPODA.

Subphylum Trilobita. Trilobites; extinct.

Subphylum Chelicerata. Chelicerates. Horseshoe crabs, spiders, scorpions, ticks, mites.

Subphylum Crustacea. Shrimps, crayfishes, lobsters, crabs, barnacles, copepods, isopods (sowbugs).

Subphylum Uniramia.
Superclass Myriapoda. Centipedes, millipedes.
Superclass Insecta.
Order Ephemeroptera. Mayflies.
Order Odonata. Dragonflies, damselflies.
Order Orthoptera. Grasshoppers, crickets, katydids.
Order Dermaptera. Earwigs.
Order Blattodea. Cockroaches.
Order Mantodea. Mantids.
Order Isoptera. Termites.
Order Mallophaga. Biting lice.
Order Anoplura. Sucking lice.
Order Homoptera. Cicadas, aphids, leafhoppers, spittlebugs.
Order Hemiptera. Bugs.
Order Coleoptera. Beetles.
Order Diptera. Flies.
Order Mecoptera. Scorpion flies. *Harpobittacus.*
Order Siphonaptera. Fleas.
Order Lepidoptera. Butterflies, moths.
Order Hymenoptera. Wasps, bees, ants.

PHYLUM ECHINODERMATA.   Echinoderms.
Class Asteroidea. Sea stars. *Asterias.*
Class Ophiuroidea. Brittle stars.
Class Echinoidea. Sea urchins, heart urchins, sand dollars.
Class Holothuroidea. Sea cucumbers.
Class Crinoidea. Feather stars, sea lilies.
Class Concentricycloidea. Sea daisies.

PHYLUM HEMICHORDATA.   Acorn worms.

PHYLUM CHORDATA.   Chordates.

Subphylum Urochordata. Tunicates, related forms.

Subphylum Cephalochordata. Lancelets.

Subphylum Vertebrata. Vertebrates.
Class Agnatha. Jawless vertebrates (lampreys, hagfishes).
Class Placodermi. Jawed, heavily armored fishes; extinct.
Class Chondrichthyes. Cartilaginous fishes (sharks, rays, skates, chimaeras).
Class Osteichthyes. Bony fishes.
Subclass Dipnoi. Lungfishes.
Subclass Crossopterygii. Coelacanths, related forms.
Subclass Actinopterygii. Ray-finned fishes.
Order Acipenseriformes. Sturgeons, paddlefishes.
Order Salmoniformes. Salmon, trout.
Order Atheriniformes. Killifishes, guppies.
Order Gasterosteiformes. Seahorses.

Order Perciformes. Perches, wrasses, barracudas, tunas, freshwater bass, mackerels.
Order Lophiiformes. Angler fishes.
Class Amphibia. Mostly tetrapods; embryo in amnion.
Order Caudata. Salamanders.
Order Anura. Frogs, toads.
Order Apoda. Apodans (caecilians).
Class Reptilia. Skin with scales, embryo enclosed in amnion.
Subclass Anapsida. Turtles, tortoises.
Subclass Lepidosaura. *Sphenodon*, lizards, snakes.
Subclass Archosaura. Dinosaurs (extinct), crocodiles, alligators.
Class Aves. Birds. (In some of the more recent schemes, dinosaurs, crocodilians, and birds are grouped in the same category.)
Order Struthioniformes. Ostriches.
Order Sphenisciformes. Penguins.
Order Procellariiformes. Albatrosses, petrels.
Order Ciconiiformes. Herons, bitterns, storks, flamingoes.
Order Anseriformes. Swans, geese, ducks.
Order Falconiformes. Eagles, hawks, vultures, falcons.
Order Galliformes. Ptarmigan, turkeys, domestic fowl.
Order Columbiformes. Pigeons, doves.
Order Strigiformes. Owls.
Order Apodiformes. Swifts, hummingbirds.
Order Passeriformes. Sparrows, jays, finches, crows, robins, starlings, wrens.
Class Mammalia. Skin with hair; young nourished by milk-secreting glands of adult.
Subclass Prototheria. Egg-laying mammals (duckbilled platypus, spiny anteaters).
Subclass Metatheria. Pouched mammals or marsupials (opossums, kangaroos, wombats, Tasmanian devil).
Subclass Eutheria. Placental mammals.
Order Insectivora. Tree shrews, moles, hedgehogs.
Order Scandentia. Insectivorous tree shrews.
Order Chiroptera. Bats.
Order Primates.
Suborder Strepsirhini (prosimians). Lemurs, lorises.
Suborder Haplorhini (tarsioids and anthropoids).
Infraorder Tarsiiformes. Tarsiers.
Infraorder Platyrrhini (New World monkeys).
Family Cebidae. Spider monkeys, howler monkeys, capuchin.
Infraorder Catarrhini (Old World monkeys and hominoids).
Superfamily Cercopithecoidea. Baboons, macaques, langurs.
Superfamily Hominoidea. Apes and humans.
Family Hylobatidae. Gibbon.
Family Pongidae. Chimpanzees, gorillas, orangutans.
Family Hominidae. Existing and extinct human species (*Homo*) and australopiths.
Order Lagomorpha. Rabbits, hares, pikas.
Order Rodentia. Most gnawing animals (squirrels, rats, mice, guinea pigs, porcupines, beavers, etc.).
Order Cetacea. Whales, porpoises.
Order Carnivora. Carnivores.
Suborder Feloidea. Cats, mongooses, hyenas.
Suborder Canoidea. Dogs, weasels, skunks, otters, raccoons, pandas, bears.
Order Pinnipedia. Seals, walruses, sea lions.
Order Proboscidea. Elephants; mammoths (extinct).
Order Sirenia. Sea cows (manatees, dugongs).
Order Perissodactyla. Odd-toed ungulates (horses, tapirs, rhinos).
Order Artiodactyla. Even-toed ungulates (camels, deer, bison, sheep, goats, antelopes, giraffes, etc.).
Order Edentata. Anteaters, tree sloths, armadillos.
Order Tubulidentata. African aardvarks.

APPENDIX II. UNITS OF MEASURE

Metric-English Conversions

Length

English		Metric
inch	=	2.54 centimeters
foot	=	0.30 meter
yard	=	0.91 meter
mile (5,280 feet)	=	1.61 kilometer

To convert	multiply by	to obtain
inches	2.54	centimeters
feet	30.00	centimeters
centimeters	0.39	inches
millimeters	0.039	inches

Weight

English		Metric
grain	=	64.80 milligrams
ounce	=	28.35 grams
pound	=	453.60 grams
ton (short) (2,000 pounds)	=	0.91 metric ton

To convert	multiply by	to obtain
ounces	28.3	grams
pounds	453.6	grams
pounds	0.45	kilograms
grams	0.035	ounces
kilograms	2.2	pounds

Volume

English		Metric
cubic inch	=	16.39 cubic centimeters
cubic foot	=	0.03 cubic meter
cubic yard	=	0.765 cubic meters
ounce	=	0.03 liter
pint	=	0.47 liter
quart	=	0.95 liter
gallon	=	3.79 liters

To convert	multiply by	to obtain
fluid ounces	30.00	milliliters
quart	0.95	liters
milliliters	0.03	fluid ounces
liters	1.06	quarts

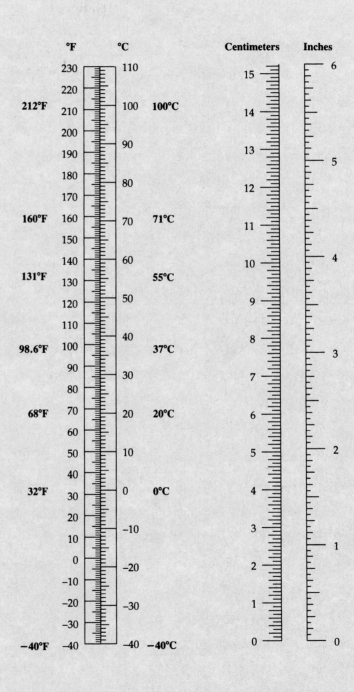

APPENDIX III. ANSWERS TO SELF-QUIZZES *Italicized numbers refer to relevant page numbers*

CHAPTER 1

1. Metabolism — 5
2. Homeostasis — 5
3. cell — 6
4. adaptive — 10
5. mutations — 10
6. d — 4
7. d — 4, 10
8. d — 10, 11
9. d — 13
10. c — 13
11. a — 13
12. c — 10
    e — 11
    d — 13
    b — 12
    a — 12

CHAPTER 2

1. b — 22
2. c — 27
3. e — 28, 29
4. e — 29
5. f — 30, 31
6. acid; base — 30
7. c — 21
   a — 31
   b — 25
   d — 29

CHAPTER 3

1. d — 36
2. e — 37
3. f — 38
4. b — 40
5. d — 42, 46
6. d — 45
7. d — 46
8. c — 42, 43
   e — 46
   b — 41
   d — 46
   a — 38

CHAPTER 4

1. c — 52–53
2. d — 52, 56
3. d — 56, 68
4. False; many cells have an outer wall.
5. c — 52, 70
6. e — 64
   d — 65
   a — 52, 56, 61
   b — 62
   c — 62–63

CHAPTER 5

1. d — 79
2. a — 88–89
3. d — 85
4. b — 86–87
5. c — 81, 84
   g — 81, 85
   a — 81
   d — 81
   e — 81
   b — 82–83
   f — 81

CHAPTER 6

1. carbon dioxide, sunlight energy — 92
2. d — 94
3. b — 94
4. c — 94
5. d — 98
6. e — 100
7. c — 101
8. c — 101
9. d — 98
   e — 98–99
   a — 101, 102
   b — 101
   c — 98

CHAPTER 7

1. d — 110
2. c — 112
3. b — 110, 114
4. c — 111, 116
5. d — 118
6. c — 118
7. b — 118
8. d — 120
9. b — 112
   c — 118
   a — 114–115
   d — 116

CHAPTER 8

1. d — 128
2. b — 128
3. c — 128
4. a — 129
5. c — 130
6. c — 128
7. d — 130–131
8. d — 132
   b — 132
   c — 133
   a — 133

CHAPTER 9

1. d — 139, 140
2. a — 140
3. d — 140
4. b — 141
5. d — 141
6. c — 141
7. c — 141, 142
8. d — 141, 143
9. d — 140
   a — 140
   c — 141
   b — 141

CHAPTER 10

1. a — 155
2. b — 155
3. a — 155
4. b — 155
5. c — 156–157
6. a — 156–157
7. d — 158
8. b — 158
   d — 156
   a — 155
   c — 155

CHAPTER 11

1. c — 172
2. c — 177
3. d — 179
4. a — 179
5. e — 182–183
6. c — 184
7. d — 184
8. c — 172
   e — 183
   d — 184
   b — 182
   a — 170, 173
   f — 176

CHAPTER 12

1. c — 194
2. d — 194
3. d — 194–195
4. c — 195
5. a — 196–197
6. d — 197
7. d — 197
   b — 194–195
   c — 196
   a — 195

CHAPTER 13

1. c — 202
2. b — 202
3. c — 202
4. c — 204
5. a — 204
6. a — 204–205
7. e — 209
   c — 206
   a — 203
   f — 204
   d — 204
   g — 203
   b — 204

CHAPTER 14

1. d — 213
2. d — 213, 214
3. d — 214
4. a — 214
5. c — 214
6. a — 214
7. d — 213, 216
8. b — 216
9. d — 216
10. b — 219
11. b — 219
12. d — 217
    e — 216
    b — 218
    a — 214–215
    c — 217

CHAPTER 15

1. d — 222
2. c — 224
3. Plasmids — 224
4. a — 225
5. b — 225
6. a — 226
7. b — 227
8. d — 228
9. d — 227

c — 230–231
f — 223
e — 229
b — 232
a — 233

CHAPTER 16

1. b — 239
2. d — 244–245
3. populations — 246
4. d — 246
5. c — 255
6. c — 249
7. b — 250
8. c — 253
9. c — 255
   d — 249
   a — 246–247
   b — 256

CHAPTER 17

1. d — 262–263
2. d — 262–263
3. c — 263
4. c — 268
5. b — 268
6. c — 268
   e — 268
   a — 268–269
   d — 269
   b — 269

CHAPTER 18

1. a — 281
2. c — 282–283
3. c — 286
4. c — 286
5. d — 286
6. e — 286
   b — 274
   f — 274–275
   c — 280
   a — 285
   d — 281

CHAPTER 19

1. e — 294–297, 302–307
2. b — 294–295
3. d — 294
4. c — 298
5. d — 298–299
6. d — 298
7. b — 298
   d — 299
   e — 302
   a — 304
   c — 307

CHAPTER 20

1. c — 314
2. c — 316
3. a — 316
4. d — 317
5. a — 318
6. a — 320–321
7. a — 322
8. f — 311, 323
9. a — 326
10. d — 326
11. d — 314

e — 314
b — 317
c — 324
a — 316

CHAPTER 21

1. b — 332
2. a — 334
3. a — 334
4. c — 334
5. c — 335
6. e — 336–337
   c — 336
   d — 334
   a — 334–335
   b — 337
   f — 337

CHAPTER 22

1. d — 342–343
2. c — 343
3. b — 344
4. c — 346
5. b — 343
6. c — 350
   e — 342
   g — 346
   h — 349
   f — 342, 344
   a — 342
   b — 342
   d — 351

CHAPTER 23

1. b — 358
2. c — 359
3. c — 362–363
4. c — 364–365
5. b — 364–367
6. b — 370–371
7. a — 372
8. d — 360
   f — 362
   c — 364
   g — 365
   e — 358, 381
   i — 368
   j — 370
   h — 372
   a — 378
   b — 357, 358

CHAPTER 24

1. d — 384
2. e — 384–385
3. d — 386–387
4. b — 386–387
5. b — 388
6. d — 390
7. f — 392
8. e — 392
9. d — 392, 395, 396
10. d — 394–395
11. e — 396
12. a — 400–401
13. d — 387
    e — 388
    c — 388–389
    b — 390
    g — 392

APPENDIX IV.   ANSWERS TO GENETICS PROBLEMS

CHAPTER 10

1. a. *AB*

    b. *AB, aB*

    c. *Ab, ab*

    d. *AB, Ab, aB, ab*

2. a. All of the offspring will be *AaBB*.

    b. 1/4 *AABB* (25% each genotype)
       1/4 *AABb*
       1/4 *AaBB*
       1/4 *AaBb*

    c. 1/4 *AaBb* (25% each genotype)
       1/4 *Aabb*
       1/4 *aaBb*
       1/4 *aabb*

    d. 1/16 *AABB* (6.25%)
       1/8 *AaBB* (12.5%)
       1/16 *aaBB* (6.25%)
       1/8 *AABb* (12.5%)
       1/4 *AaBb* (25%)
       1/8 *aaBb* (12.5%)
       1/16 *AAbb* (6.25%)
       1/8 *Aabb* (12.5%)
       1/16 *aabb* (6.25%)

3. Yellow is recessive. Because F_1 plants have a green phenotype and must be heterozygous, green must be dominant over the recessive yellow.

4. a. *ABC*

    b. *ABc, aBc*

    c. *ABC, aBc, ABc, aBc*

    d. *ABC, aBC, AbC, abC,*
       ABc, aBc, Abc, abc

5. Because all F_1 plants of this dihybrid cross had to be heterozygous for both genes, then 1/4 (25%) of the F_2 plants will be heterozygous for both genes.

6. a. The mother must be heterozygous for both genes. Both the father and their first child are homozygous recessive for both genes.

    b. The probability that their second child will not be a tongue roller and will have detached earlobes is 1/4 (25%).

7. a. Bill *AaEeSs*, and Marie *AAEESS*

    b. There is a 100% probability that all their children will have Bill's phenotype (flat feet, long eyelashes, and tendency to sneeze).

    c. Zero probability; no child of theirs will have high arches or short eyelashes, and none will be sneezy.

8. a. The mother must be heterozygous $I^A i$. The male with type B blood could have fathered the child if he were heterozygous $I^B i$.

    b. Genotype alone cannot prove the accused male is the father. Even if he happens to be heterozygous, *any* male who carries the *i* allele could be the father, including those heterozygous for type A blood ($I^A i$) or type B blood ($I^B i$) and those with type O blood (*ii*).

9. A mating between a mouse from a true-breeding, white-furred strain and a mouse from a true-breeding, brown-furred strain would provide you with the most direct evidence. Because true-breeding strains typically are homozygous for a trait being studied, all F_1 offspring from this mating should be heterozygous. Record the phenotype of each F_1 mouse, then let them mate with one another. Assuming only one gene locus is involved, these are possible outcomes for the F_2 offspring:

    a. All F_1 mice are brown, and their F_2 offspring segregate 3 brown : 1 white. *Conclusion*: Brown is dominant to white.

    b. All F_1 mice are white, and their F_2 offspring segregate 3 white : 1 brown. *Conclusion*: White is dominant to brown.

    c. All F_1 mice are tan, and the F_2 offspring segregate 1 brown : 2 tan : 1 white. *Conclusion*: The alleles at this locus show incomplete dominance.

10. You cannot guarantee that the puppies won't develop the disorder without more information about Dandelion's genotype. You could do so only if she is a heterozygous carrier, if the male is free of the alleles, and if the alleles are recessive.

11. Fred could use a testcross to find out if his pet's genotype is *WW* or *Ww*. He can let his black guinea pig mate with a white guinea pig having the genotype *ww*.

    If any F_1 offspring are white, then his pet's genotype is *Ww*. If the two parents are allowed to mate repeatedly and all the offspring of the matings are black, there is a high probability that his pet is *WW*. (If, say, ten offspring are all black, then the probability that the male is *WW* is about 99.9 percent. The greater the number of offspring, the more confident Fred can be of his conclusion.)

12. a. 1/2 red, 1/2 pink

    b. All pink

    c. 1/4 red, 1/2 pink, 1/4 white

    d. 1/2 pink, 1/2 white

13. 9/16 walnut comb
3/16 rose comb
3/16 pea comb
1/16 single comb

14. Because both parents are heterozygotes ($Hb^A Hb^S$), the following are the probabilities for each child:

a. 1/4 $Hb^S Hb^S$

b. 1/4 $Hb^A Hb^A$

c. 1/2 $Hb^A Hb^S$

15. A mating of two $M^L M$ cats yields:

1/4 homozygous dominant (MM)

1/2 heterozygous ($M^L M$)

1/4 homozygous recessive ($M^L M^L$)

Because $M^L M^L$ is lethal, the probability that any one kitten among the survivors will be heterozygous is 2/3.

16. a. Both parents must be heterozygotes (Aa). Their children may be albino (aa) or unaffected (AA or Aa).

b. All are aa.

c. The albino father must be aa. They have an albino child, so the mother must be Aa. (If she were AA, they could not have an albino child.) The albino child is aa. The three unaffected children are Aa. There is a 50% chance that any child of theirs will be albino. The observed 3:1 ratio is not surprising, given the small number of offspring.

17. a. All of the offspring will have medium-red color corresponding to the genotype $A^1 A^2 B^1 B^2$.

b. All possible genotypes could appear in the following proportions:

1/16	$A^1 A^2 B^1 B^1$	dark red
1/8	$A^1 A^1 B^1 B^2$	medium-dark red
1/16	$A^1 A^1 B^2 B^2$	medium red
1/8	$A^1 A^2 B^1 B^1$	medium-dark red
1/4	$A^1 A^2 B^1 B^2$	medium red
1/8	$A^1 A^2 B^2 B^2$	light red
1/16	$A^2 A^2 B^1 B^1$	medium red
1/8	$A^2 A^2 B^1 B^2$	light red
1/16	$A^2 A^2 B^2 B^2$	white

CHAPTER 11

1. a. Human males (XY) inherit their X chromosome only from their mother.

b. In males, an X-linked allele will only be found on his one X chromosome. Males can produce two kinds of gametes: one kind with a Y chromosome free of the gene, and the other kind with an X chromosome bearing the X-linked allele.

c. One. Each gamete of a woman who is homozygous for an X-linked allele will have an X chromosome that carries the allele.

d. Two. If a female is heterozygous for an X linked allele, half of the gametes that she produces will contain one of the alleles and the other half will contain the other allele.

2. All of the offspring should be heterozygous for the gene and all should have long wings. However, because some have vestigial wings, the dominant allele might have mutated because of radiation.

3. Because Marfan syndrome is a case of autosomal dominant inheritance and because one parent bears the allele, the probability of any child inheriting the mutant allele is 50%.

4. Because the phenotype appeared in every generation shown in the diagram, this must be a pattern of autosomal dominant inheritance.

5. A daughter could develop this type of muscular dystrophy only if she were to inherit two X-linked recessive alleles—one from her father and one from her mother. However, if a male bears the allele on his X chromosome, it will be expressed, then he will develop the disorder, and most likely he will not father children because of his early death.

6. If no crossover occurs between the two genes, then half the chromosomes will carry alleles AB and half will carry alleles ab.

7. If the alleles are close together, it is highly unlikely that a crossover will separate them from each other during meiosis. The greater the distance between the two loci, the greater the probability that a crossover will separate them from each other.

8. a. Nondisjunction could occur in anaphase I or anaphase II of meiosis.

b. As a result of a translocation, chromosome 21 (which is small) may become attached to the end of chromosome 14. Even though the chromosome number of the new individual would be 46, its somatic cells would contain the translocated chromosome 21, in addition to two normal chromosomes 21.

9. In the mother, a crossover between the two genes at meiosis generates an X chromosome that carries neither mutant allele.

ENERGY-
REQUIRING
STEPS OF
GLYCOLYSIS

(two ATP
invested)

ENERGY-
RELEASING
STEPS OF
GLYCOLYSIS

(four ATP
produced)

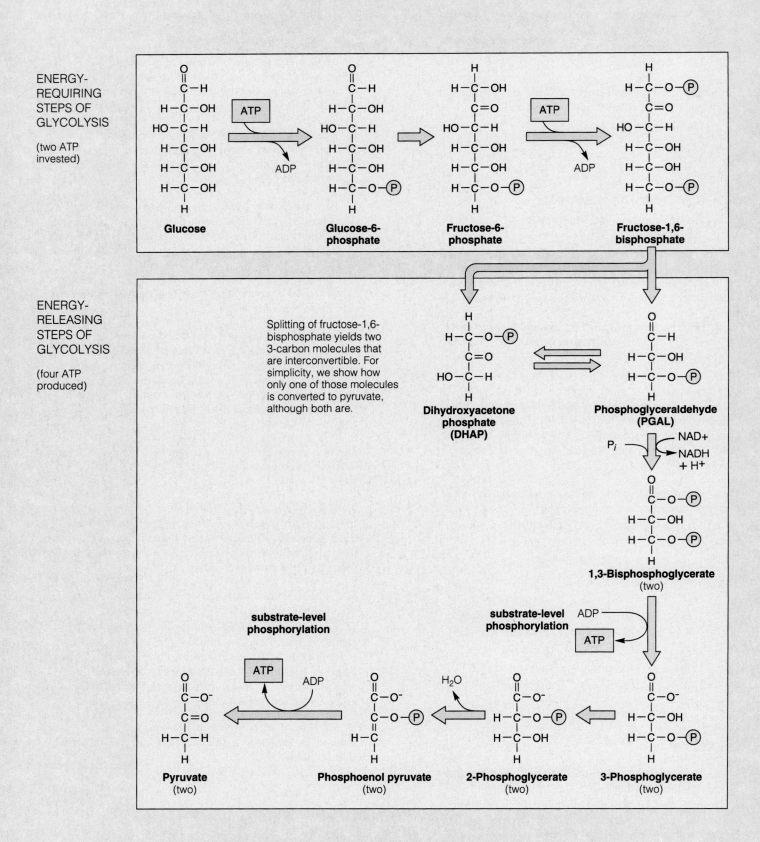

Splitting of fructose-1,6-bisphosphate yields two 3-carbon molecules that are interconvertible. For simplicity, we show how only one of those molecules is converted to pyruvate, although both are.

Figure A   Glycolysis, ending with two 3-carbon pyruvate molecules for each 6-carbon glucose entering the reactions. The *net* energy yield is two ATP molecules (two invested, four produced).

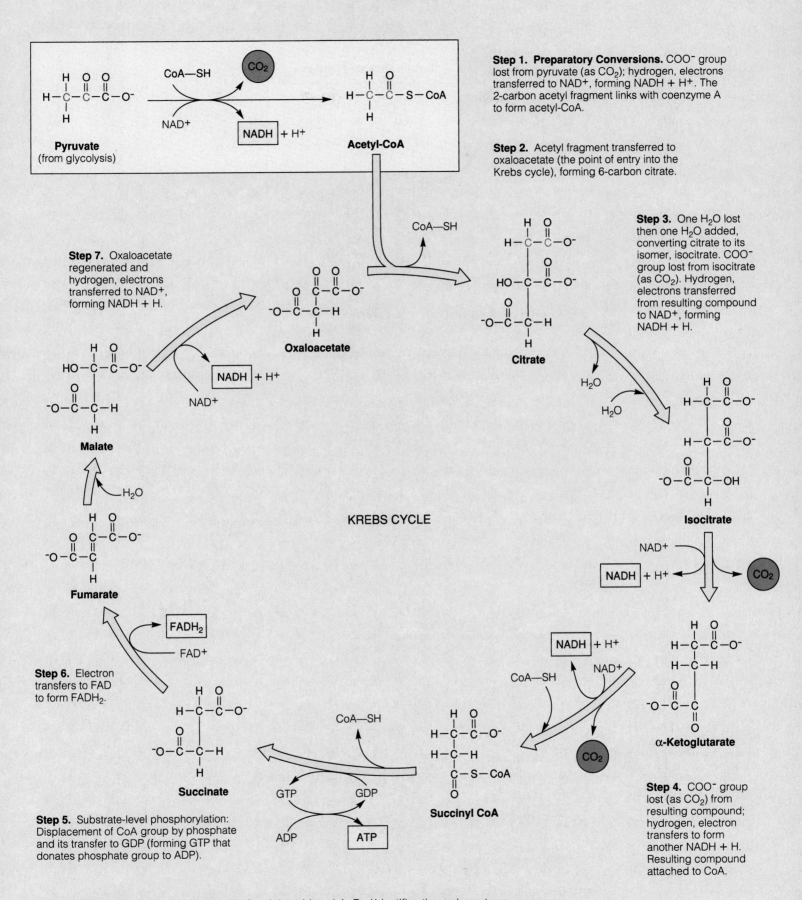

Figure B Krebs cycle (also known as the citric acid cycle). *Red* identifies the carbon atoms entering the cyclic pathway by way of acetyl-CoA.

Step 1. Preparatory Conversions. COO⁻ group lost from pyruvate (as CO_2); hydrogen, electrons transferred to NAD⁺, forming NADH + H⁺. The 2-carbon acetyl fragment links with coenzyme A to form acetyl-CoA.

Step 2. Acetyl fragment transferred to oxaloacetate (the point of entry into the Krebs cycle), forming 6-carbon citrate.

Step 3. One H_2O lost then one H_2O added, converting citrate to its isomer, isocitrate. COO⁻ group lost from isocitrate (as CO_2). Hydrogen, electrons transferred from resulting compound to NAD⁺, forming NADH + H.

Step 4. COO⁻ group lost (as CO_2) from resulting compound; hydrogen, electron transfers to form another NADH + H. Resulting compound attached to CoA.

Step 5. Substrate-level phosphorylation: Displacement of CoA group by phosphate and its transfer to GDP (forming GTP that donates phosphate group to ADP).

Step 6. Electron transfers to FAD to form $FADH_2$.

Step 7. Oxaloacetate regenerated and hydrogen, electrons transferred to NAD⁺, forming NADH + H.

KREBS CYCLE

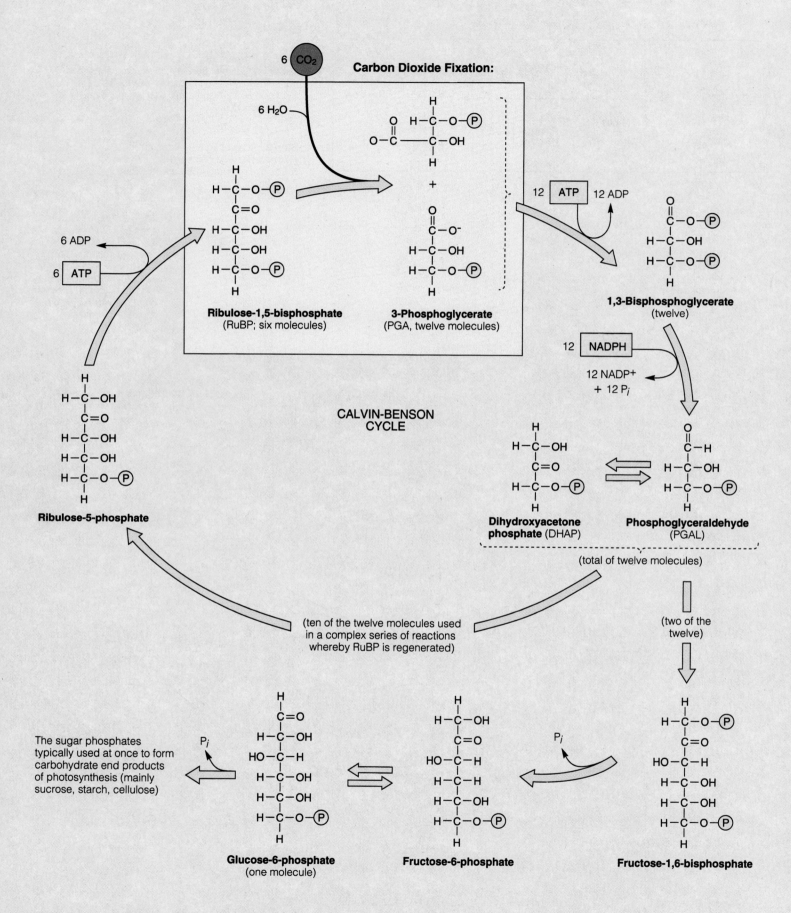

Figure C Calvin–Benson cycle of the light-independent reactions of photosynthesis.

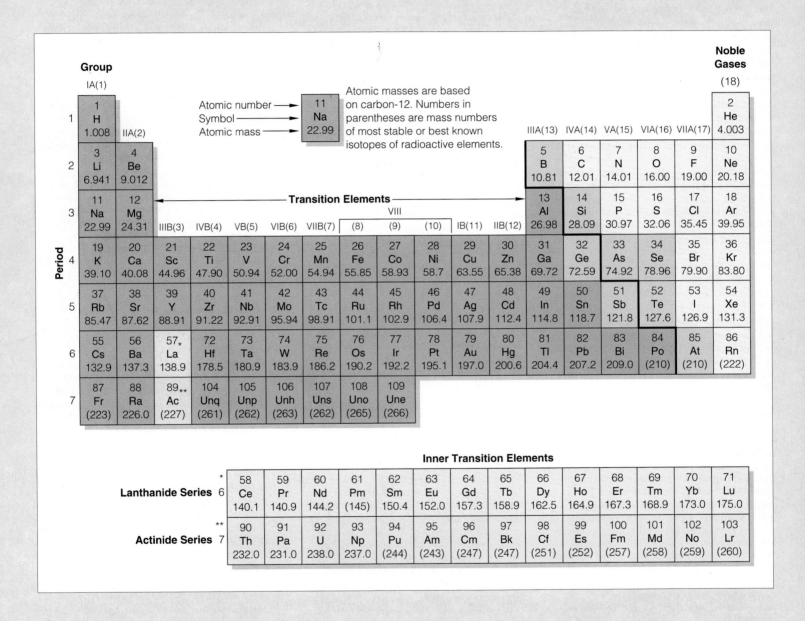

Periodic Table of the Elements

Group

IA(1)

Noble Gases (18)

Atomic number → 11
Symbol → Na
Atomic mass → 22.99

Atomic masses are based on carbon-12. Numbers in parentheses are mass numbers of most stable or best known isotopes of radioactive elements.

Period																		
1	1 H 1.008	IIA(2)											IIIA(13)	IVA(14)	VA(15)	VIA(16)	VIIA(17)	2 He 4.003
2	3 Li 6.941	4 Be 9.012											5 B 10.81	6 C 12.01	7 N 14.01	8 O 16.00	9 F 19.00	10 Ne 20.18
3	11 Na 22.99	12 Mg 24.31	IIIB(3)	IVB(4)	VB(5)	VIB(6)	VIIB(7)	VIII (8)	(9)	(10)	IB(11)	IIB(12)	13 Al 26.98	14 Si 28.09	15 P 30.97	16 S 32.06	17 Cl 35.45	18 Ar 39.95
4	19 K 39.10	20 Ca 40.08	21 Sc 44.96	22 Ti 47.90	23 V 50.94	24 Cr 52.00	25 Mn 54.94	26 Fe 55.85	27 Co 58.93	28 Ni 58.7	29 Cu 63.55	30 Zn 65.38	31 Ga 69.72	32 Ge 72.59	33 As 74.92	34 Se 78.96	35 Br 79.90	36 Kr 83.80
5	37 Rb 85.47	38 Sr 87.62	39 Y 88.91	40 Zr 91.22	41 Nb 92.91	42 Mo 95.94	43 Tc 98.91	44 Ru 101.1	45 Rh 102.9	46 Pd 106.4	47 Ag 107.9	48 Cd 112.4	49 In 114.8	50 Sn 118.7	51 Sb 121.8	52 Te 127.6	53 I 126.9	54 Xe 131.3
6	55 Cs 132.9	56 Ba 137.3	57* La 138.9	72 Hf 178.5	73 Ta 180.9	74 W 183.9	75 Re 186.2	76 Os 190.2	77 Ir 192.2	78 Pt 195.1	79 Au 197.0	80 Hg 200.6	81 Tl 204.4	82 Pb 207.2	83 Bi 209.0	84 Po (210)	85 At (210)	86 Rn (222)
7	87 Fr (223)	88 Ra 226.0	89** Ac (227)	104 Unq (261)	105 Unp (262)	106 Unh (263)	107 Uns (262)	108 Uno (265)	109 Une (266)									

Transition Elements

Inner Transition Elements

Lanthanide Series 6 *	58 Ce 140.1	59 Pr 140.9	60 Nd 144.2	61 Pm (145)	62 Sm 150.4	63 Eu 152.0	64 Gd 157.3	65 Tb 158.9	66 Dy 162.5	67 Ho 164.9	68 Er 167.3	69 Tm 168.9	70 Yb 173.0	71 Lu 175.0
Actinide Series 7 **	90 Th 232.0	91 Pa 231.0	92 U 238.0	93 Np 237.0	94 Pu (244)	95 Am (243)	96 Cm (247)	97 Bk (247)	98 Cf (251)	99 Es (252)	100 Fm (257)	101 Md (258)	102 No (259)	103 Lr (260)

Glossary

ABO blood typing Method of characterizing an individual's blood based on proteins (A, B, or their absence) at surface of red blood cells.

abortion Premature, spontaneous or induced expulsion of the embryo or fetus from uterus.

abscisic acid Plant hormone; induces stomatal closure, bud dormancy, and seed dormancy.

abscission (ab-SIH-zhun) Dropping of leaves, flowers, fruits, or other parts from a plant.

absorption Water and solute uptake by cell or multicelled organism, as when nutrients cross gut lining and enter internal environment.

accessory pigment Photosynthetic pigment; absorbs wavelengths missed by chlorophylls.

acclimatization Long-lasting physiological and behavioral adaptation to a new habitat.

acetylation Attachment of an acetyl group to a compound, such as DNA.

acetylcholine (ACh) Neurotransmitter; acts on brain, spinal cord, glands, and muscles.

acetyl-CoA (uh-SEET-ul) Coenzyme A attached to a two-carbon fragment from pyruvate, which it transfers to oxaloacetate for the Krebs cycle.

acid [L. *acidus*, sour] Any substance that, when dissolved in water, donates hydrogen ions.

acid–base balance State in which extracellular fluid is not too acidic or too basic, an outcome of controls over its concentrations of dissolved ions.

acidity Of a solution, an excess of hydrogen ions relative to hydroxyl ions.

acid rain Wet acid deposition; falling of rain (or snow) rich in sulfur and nitrogen oxides.

acoelomate (ay-SEE-luh-mate) Absence of a fluid-filled cavity between gut and body wall.

actin (AK-tin) Type of cytoskeletal protein.

action potential Abrupt, brief reversal in the resting membrane potential of excitable cells (e.g., neurons, muscle cells).

activation energy For each type of reaction, the minimum amount of collision energy that will drive reactants to an activated state.

active site Crevice in an enzyme molecule where a specific reaction is catalyzed (made to occur far faster than it would spontaneously).

active transport Pumping of a specific solute across a cell membrane against its concentration gradient through a transport protein's interior. Requires an energy boost, as from ATP.

adaptation [L. *adaptare*, to fit] Of evolution, being adapted (or becoming more adapted) to a set of environmental conditions. Of sensory neurons, decline or end of action potentials even if stimulus is maintained at constant strength.

adaptive radiation Macroevolutionary pattern; burst of genetic divergences from a lineage that give rise to many species, each using a novel resource or a new (or newly vacated) habitat.

adaptive trait Any aspect of form, function, or behavior that helps the individual survive and reproduce under prevailing conditions.

adaptive zone Any way of life available for organisms that are physically, ecologically, and evolutionarily equipped to live it.

adenine (AH-de-neen) A purine; a nitrogen-containing base in certain nucleotides.

ADH Antidiuretic hormone. Hypothalamic hormone; promotes water conservation.

adhering junction Junction where a mass of anchored proteins help adjoining cells adhere.

adipose tissue A connective tissue having an abundance of fat-storing cells.

ADP Adenosine diphosphate (ah-DEN-uh-seen die-FOSS-fate). A nucleotide coenzyme.

adrenal gland Endocrine gland. Its cortex secretes glucocorticoids, other hormones; its medulla secretes epinephrine, norepinephrine.

aerobic respiration (air-OH-bik) [Gk. *aer*, air, + *bios*, life] Oxygen-requiring pathway of ATP formation; from glycolysis, to Krebs cycle, to electron transport phosphorylation. Typical net energy yield: 36 ATP per glucose molecule.

African emergence model Holds that *Homo sapiens* originated in Africa, then replaced archaic *Homo* populations in other regions.

age structure Number of individuals in each age category for a population.

agglutination (ah-glue-tin-AY-shun) Defense response; antibodies cause antigen to clump together, which facilitates phagocytosis.

aging Of any multicelled organism showing extensive cell differentiation, the gradual and expected deterioration of the body over time.

AIDS Acquired immune deficiency syndrome. A set of chronic disorders following infection by the human immunodeficiency virus (HIV).

alcohol Organic compound that includes one or more hydroxyl groups (—OH); it dissolves readily in water. Sugars are examples.

alcoholic fermentation Anaerobic ATP-forming pathway. Pyruvate from glycolysis is degraded to acetaldehyde, which accepts electrons from NADH to form ethanol; NAD^+ needed for the reactions is regenerated. Net yield: 2 ATP.

aldosterone (al-DOSS-tuh-rohn) Adrenal cortex hormone; helps control sodium reabsorption.

alga, plural **algae** Photosynthetic protistans, mainly; most are members of phytoplankton.

algal bloom Huge increases in aquatic algal population sizes after nutrient enrichments.

alkylating agent Any substance that transfers methyl or ethyl groups to a compound.

allantois (ah-LAN-twahz) [Gk. *allas*, sausage] Extraembryonic membrane used to exchange gases and store metabolic wastes of embryos of reptiles, birds, some mammals. In humans, it forms urinary bladder, placental blood vessels.

allele (uh-LEEL) One of two or more molecular forms of a gene that arise by mutation and code for different versions of the same trait.

allele frequency For a given gene locus, the relative abundances of each kind of allele among all the individuals of a population.

allergy Hypersensitivity to an allergen (any normally harmless substance that provokes inflammation and often immune responses).

allopatric speciation [Gk. *allos*, different, + L. *patria*, native land]. Speciation model. A physical barrier arises, separates populations or subpopulations of a species, ends gene flow, and favors divergences that end in speciation.

altruism (AL-true-IZ-um) Behavior that lowers an individual's chance of reproductive success but helps others of its species.

alveolus (ahl-VEE-uh-lus), plural **alveoli** [L. *alveus*, small cavity] Cupped, thin-walled outpouching of respiratory bronchiole where lungs and blood exchange O_2 and CO_2.

amino acid (uh-MEE-no) Organic compound with an H atom, amino group, acid group, and R group, all covalently bonded to a carbon atom. Subunit of polypeptide chains.

ammonification (uh-moan-ih-fih-KAY-shun) Of nitrogen cycle, process by which soil fungi and bacteria break down nitrogenous wastes and remains to forms that plants can take up.

amnion (AM-nee-on) An extraembryonic membrane; boundary layer of a fluid-filled sac in which some vertebrate embryos develop.

amniote egg Egg that has extraembryonic membranes and often a shell. Contributed to successful invasion of land by vertebrates.

amphibian Vertebrate with a body plan and reproductive mode somewhere between fishes and reptiles and a mostly bony endoskeleton.

anaerobic electron transport (an-uh-ROW-bik) [Gk. *an*, without, + *aer*, air] Type of prokaryotic ATP-forming pathway. Uses electrons from an organic compound; an inorganic compound (not oxygen) is often the final electron acceptor. Variable but always small net energy yield.

anagenesis Speciation pattern; changes in allele frequencies and morphology accumulate within an unbranched line of descent.

analogous structures (ann-AL-uh-gus) [Gk. *analogos*, similar to one another] Body parts that once differed in evolutionarily distant lineages but converged in structure and function in response to similar environmental pressures.

anaphase (AN-uh-faze) Of mitosis, stage when sister chromatids of each chromosome move to opposite spindle poles. Of anaphase I (meiosis), each duplicated chromosomes and its homologue move to opposite poles. Of anaphase II (meiosis), sister chromatids of each chromosome move to opposite poles.

aneuploidy (AN-yoo-ploy-dee) Having one extra or one less chromosome relative to the parental chromosome number.

angiosperm (AN-gee-oh-spurm) [Gk. *angeion*, vessel, and *spermia*, seed] Flowering plant.

animal Multicelled, motile heterotroph that eats other organisms; has embryonic stages and usually tissues, organs, and organ systems.

Animalia Kingdom of animals.

annelid An invertebrate; a segmented worm (e.g., oligochaete, leech, or polychaete).

annual Plant that lasts one growing season.

anther [Gk. *anthos*, flower] Part of a stamen; pollen forms in it and is dispersed from it.

antibiotic Metabolic product of soil microbes that kills bacterial competitors for nutrients.

antibody [Gk. *anti*, against] Antigen-binding receptor. Only B cells make antibodies, then position them at their surface or secrete them.

anticodon Series of three nucleotide bases in tRNA; can base-pair with an mRNA codon.

antigen (AN-tih-jen) [Gk. *anti*, against, + *genos*, race, kind] Any molecular configuration that certain lymphocytes chemically recognize as nonself and that triggers an immune response.

antigen-presenting cell Cell that processes antigen and displays fragments of it with MHC molecules at the cell surface. Antigen–MHC complexes can stimulate immune responses.

aorta (ay-OR-tah) Of vertebrates, main artery of systemic circulation.

apical dominance Growth-inhibiting effect of a terminal bud on growth of lateral buds.

apical meristem (AY-pih-kul MARE-ih-stem) [L. *apex*, top, + Gk. *meristos*, divisible] Mass of self-perpetuating cells responsible for the primary growth at root tips and shoot tips.

apoptosis (APP-oh-TOE-sis) Programmed cell death of body cells finished with prescribed functions or altered, as by infection or cancer.

Archaebacteria Kingdom of prokaryotes more like eukaryotic cells than eubacteria; includes methanogens, halophiles, and thermophiles.

Archean Eon (3.9–2.5 bya) in which life arose.

archipelago Island chain some distance away from a continent.

area effect Idea that larger islands support more species than smaller ones at equivalent distances from sources of colonizer species.

arteriole (ar-TEER-ee-ole) Type of blood vessel between arteries and capillaries. Collectively, points of selective dilation and constriction.

artery Thick-walled, muscular, rapid-transport vessel that smooths out blood pressure pulses.

arthropod Invertebrate with hard exoskeleton, specialized segments, and jointed appendages.

artificial selection Selection of traits among a population under contrived conditions.

asexual reproduction Any reproductive mode by which offspring arise from a single parent and inherit the genes of that parent only.

atmosphere The volume of gases, airborne particles, and water vapor enveloping Earth.

atmospheric cycle A biogeochemical cycle in which an element occurs in a gaseous phase.

atom Smallest unit of an element that retains the element's properties.

ATP Adenosine triphosphate (ah-DEN-uh-seen try-FOSS-fate). Nucleotide with adenine, ribose, and three phosphate groups that drives most energy-requiring metabolic reactions.

australopith (OHSS-trah-low-pith) [L. *australis*, southern, + Gk. *pithekos*, ape] Early hominid.

autoimmunity Result of inappropriate attacks on normal body cells by lymphocytes.

automated DNA sequencing Recent method that determines the sequence of cloned DNA or PCR-amplified DNA in a few hours.

autonomic nervous system (auto-NOM-ik) All nerves from central nervous system to smooth muscle, cardiac muscle, and glands of viscera.

autosome Any chromosome of a type that is the same in males and females of the species.

autotroph (AH-toe-trofe) [Gk. *autos*, self, + *trophos*, feeder] Any organism that makes its own food with an environmental energy source (e.g., sunlight) and CO_2 as its carbon source.

auxin (AWK-sin) Plant hormone; induces stem lengthening and responses to gravity and light.

axon Neuron's conducting zone.

B cell B lymphocyte; only type of cell that produces and secretes antibodies.

bacterial conjugation Transfer of plasmid DNA from one prokaryotic cell to another.

bacteriophage (bak-TEER-ee-oh-fahj) Category of viruses that infect bacterial cells.

balanced polymorphism Form of selection in which two or more alleles for a trait are being maintained in a population over time.

bark All tissues external to vascular cambium.

Barr body Randomly condensed one of two X chromosomes in cells of female mammals.

basal body A centriole which, after giving rise to microtubules of a flagellum or cilium, remains attached to its base in the cytoplasm.

base Any substance that accepts hydrogen ions (H^+) when dissolved in water; OH^- forms after this. Also, an organic compound with a single or double ring structure containing nitrogen.

base-pair substitution One amino acid has replaced another during protein synthesis.

base sequence Sequential order of bases in a DNA or RNA strand.

behavior Response to external and internal stimuli based on sensory, neural, endocrine, and effector components. Has a genetic basis, can evolve, and can be modified by learning.

behavioral flexibility A capacity to expand on basic activities with novel forms of behavior and learning, especially among primates.

biennial (bi-EN-yul) Flowering plant that completes its life cycle in two growing seasons.

bilateral symmetry Body plan in which left and right halves generally are mirror images.

bile Liver secretion required for fat digestion.

binary fission Asexual reproductive mode; the body of protozoans and some other animals divides in two parts. *See* prokaryotic fission.

binomial system Of taxonomy, assigning a generic and a specific name to each species.

biodiversity [Gk. *bios*, life] All species living in a defined region, ranging from the smallest habitat to the biosphere, in a specified interval.

biogeochemical cycle Slow movement of an element from environmental reservoirs, through food webs, then back to the environment.

biogeography Scientific study of the world distribution of species.

biological clock Internal time-measuring mechanism; adjusts activities seasonally, daily, or both in response to environmental cues.

biological magnification Ever increasing concentration of a nondegradable or slowly degradable substance in body tissues as it is passed along food chains.

biological species concept Defines a species as one or more populations of individuals that are interbreeding under natural conditions, producing fertile offspring, and are isolated reproductively from other such populations. Applies to sexually reproducing species only.

biology The scientific study of life.

bioluminescence Production by an organism of fluorescent light.

biomass Combined weight of all organisms at a given trophic level in an ecosystem.

biome Large subdivision of a biogeographic realm, distinctive in its habitat conditions, community structure, and endemic species.

biosphere [Gk. *bios*, life, + *sphaira*, globe] All regions of the Earth's waters, crust, and atmosphere in which organisms live.

biosynthetic pathway Metabolic pathway by which organic compounds are synthesized.

biotic potential Of population growth for a given species, the maximum rate of increase per individual under ideal conditions.

bipedalism Habitually walking upright on two feet, as by ostriches and hominids.

bird Vertebrate with reptilian ancestry; only organism that produces feathers.

blastocyst (BLASS-tuh-sist) Type of blastula; a surface layer of blastomeres, a cavity filled with their secretions, and an inner cell mass.

blastomere One of the small, nucleated cells that form during cleavage of animal zygote.

blastula Early outcome of cleavage; a number of blastomeres enclose a fluid-filled cavity.

blood Fluid connective tissue of water, solutes, and formed elements (blood cells, platelets). Blood transports substances to and from cells, and helps maintain internal environment.

blood–brain barrier Mechanism that controls which solutes enter cerebrospinal fluid.

blood pressure Fluid pressure, generated by heart contractions, that circulates blood.

bone Vertebrate organ with mineral-hardened connective tissue (bone tissue); helps move body, protects other organs, stores minerals. Some (e.g., breastbone) produce blood cells.

bone remodeling Ongoing mineral deposits and withdrawals from bone that adjust bone strength and maintain levels of calcium and phosphorus in blood.

bony fish Type of fish with an endoskeleton, usually well developed, of bone.

boreal forest A swamp forest (or *taiga*) in glaciated regions with cold lakes and streams.

bottleneck Severe reduction in the size of a population, brought about by intense selection pressure or a natural calamity.

Bowman's capsule Part of a nephron; receives water and solutes being filtered from blood.

brain Of most nervous systems, integrating center that receives and processes sensory input and issues coordinated commands for responses by muscles and glands.

brain stem Most ancient nerve tissue in the vertebrate hindbrain, midbrain, and forebrain.

broadleaf forest Evergreen or deciduous forest of a tropical, subtropical, or temperate region in which tall trees form a continuous canopy (e.g., tropical or deciduous broadleaf forest).

bronchiole Finely branched airway; part of the bronchial tree inside the lung.

bronchus, plural **bronchi** (BRONG-cuss, BRONG-kee) [Gk. *bronchos*, windpipe] Tubular airway; branches from trachea and leads into lungs.

brown alga Mostly marine, photoautotrophic protistan, such as kelp, with chlorophylls a, c_1, and c_2, and carotenoids (e.g., fucoxanthin).

bryophyte Nonvascular land plant requiring free water for fertilization. Haploid dominance in life cycle. Some with cuticle and stomata. A moss, liverwort, or hornwort.

bud Undeveloped shoot, mainly meristematic tissue. Small, protective scales often cover it.

buffer system A weak acid and the base that forms when it dissolves in water. The two work as a pair to counter slight shifts in pH.

bulk Of the vertebrate gut, the volume of undigested material in the small intestine that cannot be decreased by absorption.

bulk flow In response to a pressure gradient, movement of more than one kind of molecule in the same direction in the same medium.

C3 plant Plant that uses three-carbon PGA as the first intermediate for carbon fixation.

C4 plant Plant that uses oxaloacetate (a f carbon compound) as the first in for carbon fixation. CO_2 is fixe cell types; helps counter photore

Calvin–Benson cycle Light-independent cyclic reactions of photosynthesis that form sugars, using ATP energy, NADPH, and CO_2.

CAM plant Type of plant that conserves water by opening stomata only at night, when it fixes carbon dioxide by means of a C4 pathway.

camouflage Coloration, form, patterning, or behavior that helps predators or prey blend with the surroundings and escape detection.

cancer Malignant tumor; mass of altered cells that divide abnormally. Potentially lethal.

capillary, blood [L. *capillus*, hair] Blood vessel with thin endothelial wall and small diameter.

capillary bed Diffusion zone, consisting of great numbers of capillaries, where blood and interstitial fluid exchange substances.

carbamino hemoglobin $HbCO_2$, the form in which 30% of the CO_2 in blood is transported.

carbohydrate Molecule of carbon, hydrogen, and oxygen mostly in a 1:2:1 ratio. Main kinds are monosaccharides, oligosaccharides, and polysaccharides. They are structural materials, energy stores, and transportable energy forms.

carbon cycle An atmospheric cycle. Carbon moves from reservoirs (sediments, rocks, the ocean), through the atmosphere (mostly as CO_2), food webs, and back to the reservoirs.

carbon fixation First of the light-independent reactions. Rubisco, an enzyme, affixes carbon (from CO_2) to RuBP or to another compound for entry into the Calvin–Benson cycle.

carbonic anhydrase Red blood cell enzyme that converts CO_2 to bicarbonate, the form in which 60% of the CO_2 in blood is transported.

carcinogen (kar-SIN-uh-jen) Any substance or agent that can trigger cancer.

cardiac conduction system Specialized cardiac muscle cells that make the heart pump by way of their rhythmic, synchronized excitation.

cardiac cycle (KAR-dee-ak) [Gk. *kardia*, heart, + *kyklos*, circle] Sequence of muscle contraction and relaxation in one heartbeat.

cardiac muscle tissue A contractile tissue that is present only in the heart wall.

cardiac pacemaker Sinoatrial (SA) node; mass of self-excitatory cardiac muscle cells that are the basis of normal rate of heartbeat.

cardiovascular system Organ system that has blood, one or more hearts, and blood vessels.

carnivore Animal that devours other animals.

carnivorous plant Plant that traps, digests, and feeds on insects and other small animals.

carotenoid (kare-OTT-en-oyd) An accessory pigment of photosynthesis.

carpel (KAR-pul) Female reproductive part of a flower. The ovary, stigma, and often a style consist of one or more carpels.

carrying capacity The maximum number of individuals in a population (or species) that a given environment can sustain indefinitely.

cartilage Connective tissue with solid, pliable intercellular material that resists compression.

cartilaginous fish Jawed fish with a cartilage endoskeleton and prominent fins (e.g., sharks).

Casparian strip Narrow, waxy, impermeable band between walls of abutting cells making up root endodermis and exodermis.

catastrophism Idea that abrupt changes in the geologic or fossil record were divinely invoked.

cDNA DNA molecule copied from a mature mRNA transcript by reverse transcription.

cell [L. *cella*, small room] Smallest living unit; organized unit with a capacity to survive and reproduce on its own given DNA instructions, energy sources, and raw materials.

cell count The number of cells of a given type in one microliter of blood.

cell cycle Events by which a cell increases in mass, roughly doubles its cytoplasm, duplicates its DNA, and divides in two. Extends from the time a cell forms until it completes division.

cell differentiation Key development process. Cell lineages become specialized in structure, composition, and function by activating and suppressing part of the genome selectively.

cell junction Site where cells are interacting physically, functionally, or both.

cell plate In a dividing plant cell, a disklike structure that becomes a crosswall with new plasma membrane on both sides.

cell theory All organisms consist of one or more cells, the smallest units with a capacity for independent life that no longer spontaneously arise under existing conditions on Earth.

cell wall A semirigid, permeable structure external to the plasma membrane; helps many cells retain their shape and resist rupturing.

Cenozoic The present era (65 mya to present).

central nervous system Brain and spinal cord.

central vacuole Fluid-filled storage organelle of plant cell; its growth enhances cell surface area.

centriole (SEN-tree-ohl) Structure that gives rise to microtubules of cilia and flagella.

centromere (SEN-troh-meer) A constricted area of a chromosome that has attachment sites for spindle microtubules during nuclear division.

cephalization (SEF-ah-lah-ZAY-shun) [Gk. *kephalikos*, head] Over time, the concentration of sensory structures and nerve cells in a head.

cerebellum (ser-ah-BELL-um) Hindbrain region with reflex centers for maintaining posture and smoothing out limb movements.

cerebral cortex Thin surface layer of cerebral hemispheres; receives, integrates, and stores sensory information; coordinates responses.

cerebrospinal fluid Clear extracellular fluid, inside a system of canals and chambers, that bathes and protects the brain and spinal cord.

cerebrum (suh-REE-bruhm) Forebrain region that generally deals with olfactory input and motor responses. In mammals, it evolved into the most complex integrating center.

channel protein Type of transport protein, open or gated, that spans a cell membrane.

chemical bond A union between the electron structures of two or more atoms or ions.

chemical energy Potential energy of molecules.

chemical synapse (SIN-aps) Thin cleft between a presynaptic neuron and a postsynaptic cell. Neurotransmitter molecules diffuse across it.

chemiosmotic theory (KIM-ee-oz-MOT-ik) Idea that an electrochemical gradient drives ATP formation. H^+ accumulates in a membranous compartment, then ATP forms as it flows out through proteins (ATP synthases) in response to concentration and electric gradients.

chemoautotroph (KEE-moe-AH-toe-trofe) Any prokaryotic cell that synthesizes its own food using carbon dioxide as the carbon source and an inorganic substance as the energy source.

chemoreceptor Sensory receptor that detects ions or molecules dissolved in fluid bathing it.

chlorofluorocarbon (KLORE-oh-FLOOR-oh-car-bun) CFC; a compound of chlorine, fluorine, and carbon; is contributing to ozone thinning.

chlorophyll (KLOR-uh-fill) [Gk. *chloros*, green, + *phyllon*, leaf] Main photosynthetic pigment. Chlorophylls absorb all wavelengths of visible light but not much of green and yellow ones.

chloroplast (KLOR-uh-plast) The organelle of photosynthesis in plants and many protistans.

chordate Animal having a notochord, dorsal hollow nerve cord, pharynx, and gill slits in pharynx wall during at least part of life cycle.

chorion (CORE-ee-on) Type of extraembryonic membrane that becomes part of placenta. Villi form at its surface and facilitate exchanges of substances between the embryo and mother.

chromatid (CROW-mah-tid) Of a duplicated eukaryotic chromosome, one of two DNA molecules (with associated proteins) attached to each other at the centromere.

chromatin A cell's collection of DNA and all of the proteins associated with it.

chromosome (CROW-moe-some) [Gk. *chrōma*, color, + *soma*, body] Of eukaryotic cells, a DNA molecule, duplicated or unduplicated, with many associated proteins. Of prokaryotic cells, a circular DNA molecule.

chromosome number All chromosomes in a given type of cell. *See* haploidy; diploidy.

chrysophyte A category of photosynthetic protistans (e.g., golden algae, yellow-green algae, coccolithophores, diatoms).

cilium (SILL-ee-um), plural **cilia** Short motile or sensory structure of certain eukaryotic cells; its core is a 9 + 2 array of microtubules.

circadian rhythm (ser-KAYD-ee-un) [L. *circa*, about, + *dies*, day] Cycle of physiological events completed every twenty-four hours or so, independently of environmental change.

circulatory system Organ system that moves substances to and from cells, and often helps stabilize body temperature and pH. Typically consists of a heart, blood vessels, and blood.

cladogenesis [Gk. *clad-*, branch] Speciation pattern in which a lineage splits and isolated populations undergo genetic divergence.

cladogram Evolutionary tree diagram with branch points showing relative relationships. Groups closer together share a more recent common ancestor than those farther apart.

classification scheme A way of organizing and retrieving information about species.

cleavage Early stage of animal development. Mitotic cell divisions divide a fertilized egg into many smaller, nucleated cells; original volume of egg cytoplasm does not increase.

cleavage furrow Ringlike depression defining the cutting plane for a dividing animal cell.

cleavage reaction Enzyme action that splits a molecule in two or more parts (e.g., hydrolysis).

climate For a specified region, the prevailing weather conditions (e.g., temperature, cloud cover, wind speed, rainfall, and humidity).

climax community Array of species that has stabilized under prevailing habitat conditions.

climax pattern model Idea that environmental factors often vary in their effects across a large region, so stable communities other than the climax stage may also persist in that region.

cloaca (kloe-AY-kuh) Chamber or duct in last part of gut of some animals; roles in excretion, reproduction, and sometimes respiration.

cloning Making a genetically identical copy of DNA or of an organism.

cloning vector Any plasmid, viral DNA, or some other piece of DNA that researchers use to isolate and amplify DNA of interest.

club fungus Fungus having club-shaped cells that produce and bear spores.

cnidarian (nye-DAR-ee-un) Type of radial invertebrate at the tissue level of organization; the only nematocyst producer.

coal Nonrenewable energy source that formed over 280 million years ago from submerged, undecayed, and compacted plant remains.

codominance In heterozygotes, simultaneous expression of a pair of nonidentical alleles that specify different phenotypes.

codon One of 64 possible base triplets in an mRNA strand. A code word for an amino acid in a polypeptide chain; a few codons also act as START or STOP signals for translation.

coelom (SEE-lum) A peritoneum-lined cavity between the gut and body wall of most animals.

coenzyme Nucleotide that accepts electrons, H atoms that enzymes strip from substrates and transfers them to a different reaction site.

coevolution Joint evolution of two closely interacting species by changes in the selection pressures operating between the two.

cofactor Metal ion or coenzyme; it helps an enzyme catalyze a reaction or transfers atoms, electrons, or functional groups elsewhere.

cohesion Capacity to resist rupturing when placed under tension (stretched).

cohesion–tension theory Of plants, in response to transpiration, columns of water molecules are pulled up through xylem by the collective cohesive strength of their hydrogen bonds.

cohort Group of individuals from a population from the time of birth until the last one dies.

collenchyma (coll-ENG-kih-mah) A simple plant tissue that imparts flexible support during primary growth, as in lengthening stems.

colon (CO-lun) Large intestine.

commensalism Ecological interaction between two (or more) species in which one benefits directly and the other is affected little, if at all.

communication display Social signal, often ritualized with intended changes in functions of common patterns.

communication signal Social cue encoded in stimuli such as specific body coloration, body patterning, odors, sounds, and postures.

community All populations in a habitat. Also, a group of organisms with similar life-styles.

companion cell Specialized parenchyma cell; helps load conducting cells of phloem.

comparative morphology [Gk. *morph*, form] Scientific study of comparable body parts of adults or embryonic stages of major lineages.

competition, interspecific Type of ecological interaction in which individuals of different species compete for a share of resources.

competition, intraspecific Type of ecological interaction in which individuals of the same population compete for a share of resources.

competitive exclusion Theory that two or more species that require identical resources cannot coexist indefinitely.

complement system Set of proteins circulating in inactive form in vertebrate blood. Different kinds promote inflammation, induce lysis of pathogens, and stimulate phagocytes during nonspecific defenses and immune responses.

compound Molecule consisting of two or more elements in unvarying proportions (e.g., H_2O).

concentration gradient A difference in the number of molecules (or ions) of a substance between adjoining regions. Molecules collide constantly and career outward to an adjoining region where they are less concentrated. All substances tend to diffuse down such gradients.

condensation reaction Covalent bonding of two molecules into a larger one; water often forms as a by-product.

conduction In temperature studies, exchange of heat between two touching objects owing to a thermal gradient between them.

cone cell In a vertebrate eye, a photoreceptor that responds to intense light and contributes to sharp daytime vision and color perception.

conifer A gymnosperm; most are evergreen woody trees or shrubs with thickly cuticled needle-like or scale-like leaves; adapted to conserve water in droughts and cold winters.

connective tissue Most abundant, pervasive type of animal tissue. Soft connective tissues —(1) loose, (2) dense, irregular, or (3) dense, regular—differ in their arrangement and number of fibroblasts and fibers, and in the amount of ground substance in which these are embedded. Specialized types include cartilage, bone tissue, adipose tissue, and blood.

conservation biology International systematic survey of the full range of biodiversity, analysis of its evolutionary and ecological origins, and identification of methods to maintain and use it for the benefit of the human population.

consumer [L. *consumere*, to take completely] A heterotroph that feeds on cells or tissues of other organisms (e.g., herbivores, carnivores).

continuous variation Of a population, a more or less continuous range of small differences in a given trait among its individuals.

contractile cell A cell that shortens in response to stimulation and returns to a resting position.

contractile vacuole (kun-TRAK-till VAK-you-ohl) [L. *contractus*, to draw together] Organelle in some protistans; takes up excess water and contracts to expel water through a pore.

control group A group used as a standard for comparison with an experimental group and, ideally, identical with it in all respects except for the one variable being studied.

convection Movement of air or water next to an object that aids conductive heat loss from it.

convergent evolution Dissimilar body parts in evolutionarily distant lineages become similar over time, as independent responses to similar kinds of environmental pressures.

coral reef A formation of accumulated hard parts of corals and other organisms in warm, clear waters between latitudes 25°N and 25°S.

core temperature Internal body temperature, as opposed to near its surface.

cork Periderm component; insulates, protects, and waterproofs woody stems and roots.

cork cambium Lateral meristem that replaces epidermis with cork on woody plant parts.

cornea Transparent cover over front of eye.

corpus callosum (CORE-pus ka-LOW-sum) Band of axons linking two cerebral hemispheres.

corpus luteum (CORE-pus LOO-tee-um) Of female mammals, a glandular structure that forms from cells of a ruptured ovarian follicle; it secretes progesterone and estrogen.

cortex A rindlike layer. In vascular plants, a ground tissue; supports parts and stores food.

cotyledon (KOT-uhl-EE-dun) Seed leaf. One develops in monocot seeds and has digestive roles; two develop in dicot seeds and store food for germination and early growth.

countercurrent flow Movement of two fluids in opposing directions, as in a fish gill.

courtship display Pattern of ritualized social behavior between potential mates. Commonly incorporates frozen postures, exaggerated yet simplified movements, and visual signals.

covalent bond (koe-VAY-lunt) [L. *con*, together, + *valere*, to be strong] Sharing of one or more electrons between atoms or groups of atoms. Nonpolar bonds share them equally. Polar bonds (slightly positive at one end, negative at the other) share them unequally.

creatine phosphate Organic compound that transfers phosphate to ADP in a rapid, short-term, ATP-generating pathway.

cross-bridge formation A reversible, ATP-driven interaction between a sarcomere's actin and myosin filaments; results in a short power stroke that is the basis of muscle contraction.

crossing over At prophase I of meiosis, an interaction in which nonsister chromatids of a pair of homologous chromosomes break at corresponding sites and exchange segments; genetic recombination is the result.

culture Sum of behavior patterns of a social group, passed between generations by way of learning and symbolic behavior.

cuticle (KEW-tih-kull) Body cover. Of plants, a transparent covering of waxes and cutin on outer epidermal cell walls. Of annelids, a thin, flexible coat. Of arthropods, a lightweight exoskeleton hardened with protein and chitin.

cutin Insoluble lipid polymer that functions as the ground substance for plant cuticles.

cycad Slow-growing gymnosperm; its lineage arose in the Permian. Resembles palm trees.

cyclic AMP (SIK-lik) A type of nucleotide (cyclic adenosine monophosphate) with roles in intercellular communication.

cyclic pathway of ATP formation Most ancient photosynthetic pathway. Membrane-bound photosystems give up electrons to transport systems that return them to photosystems. Electron flow across the membrane sets up H^+ gradients that drive ATP formation.

cyst Of many microorganisms, a resting stage with thick outer layers that typically forms under adverse conditions. Of skin, abnormal, fluid-filled sac without an external opening.

cytochrome (SIGH-toe-krome) Iron-containing protein molecule of electron transport systems.

cytokinin (SIGH-toe-KYE-nin) Type of plant hormone; stimulates cell division, promotes leaf expansion, and retards leaf aging.

cytomembrane system Organelles (ER, Golgi bodies) that modify, package, and distribute new protein and lipids; also various vesicles (e.g., lysosomes) that have diverse functions.

cytoplasm (SIGH-toe-plaz-um) All cell parts, particles, and semifluid substances between the plasma membrane and nucleus or nucleoid.

cytoplasmic division Cytokinesis; splitting of a parent cell into daughter cells.

cytoplasmic localization Parceling of a portion of maternal messages in the egg cytoplasm to each blastomere that forms during cleavage.

cytosine (SIGH-toe-seen) Pyrimidine; one of the nitrogen-containing bases in nucleotides.

cytoskeleton Dynamic internal framework that structurally supports eukaryotic cells and organizes and moves a cell and its structures.

decomposer [L. *dis–*, to pieces] Prokaryotic or fungal heterotroph; gets carbon and energy from products, remains, of organisms. Helps cycle nutrients to producers in ecosystems.

decomposition In general, processes by which prokaryotic cells and fungi degrade organic matter (e.g., nitrogenous wastes of organisms).

deductive logic Pattern of thinking; making inferences about specific consequences or predictions that must follow from a hypothesis.

deforestation Removal of all the trees from a large tract of land (e.g., Amazon River Basin).

degradative pathway A stepwise series of metabolic reactions that break down organic compounds to products having lower energy.

deletion At cytological level, loss of a segment from a chromosome. At molecular level, loss of one to a few base pairs from a DNA molecule.

demographic transition model Model that correlates changes in population growth with four stages of economic development.

demographics A population's vital statistics.

denaturation (deh-NAY-chur-AY-shun) Loss of a molecule's three-dimensional shape as weak bonds (e.g., hydrogen bonds) are disrupted.

dendrite (DEN-drite) [Gk. *dendron*, tree] Short, slender extension from cell body of a neuron; commonly a signal input zone.

dendritic cell Type of antigen-presenting cell.

denitrification (DEE-nite-rih-fih-KAY-shun) Conversion of nitrate or nitrite by certain soil bacteria to gaseous nitrogen (N_2) and a small amount of nitrous oxide (N_2O).

density-dependent control Factor that limits population growth by reducing the birth rate or increasing death and dispersal rates (e.g., predation, parasitism, disease, competition).

density-independent factor Factor that causes a population's death rate to rise independently of density (e.g., a severe storm or flood).

dentition (den-TIH-shun) Collectively, the type, size, and number of an animal's teeth.

derived trait A novel feature that evolved but once and is shared only by the descendants of the ancestral species in which it evolved.

dermal tissue system Tissues that cover and protect all exposed surfaces of a plant.

dermis Skin layer beneath the epidermis that consists primarily of dense connective tissue.

desalination Removing salt from seawater.

desert Biome that forms where the potential for evaporation greatly exceeds rainfall, and where soil is thin and vegetation sparse.

desertification (dez-urt-ih-fih-KAY-shun) The conversion of grassland or irrigated or rain-fed cropland to a desertlike condition.

detrital food web Network of food chains in which energy flows mainly from plants through arrays of detritivores and decomposers.

detritivore (dih-TRY-tih-vore) Heterotroph that feeds on decomposing particles of organic matter (e.g., crab, earthworm, roundworm).

deuterostome (DUE-ter-oh-stome) [Gk. *deuteros*, second, + *stoma*, mouth] Bilateral animal in which the anus forms from first indentation in the embryo (e.g., echinoderms, chordates).

development Of multicelled species, emergence of specialized, morphologically distinct body parts according to a genetic program.

diaphragm (DIE-uh-fram) [Gk. *diaphragma*, to partition] Muscular partition between thoracic and abdominal cavities with role in breathing. Also a contraceptive device inserted into the vagina to prevent sperm from entering uterus.

dicot (DIE-kot) [Gk. *di*, two, + *kotylēdōn*, cup-shaped vessel] Dicotyledon. Flowering plant generally characterized by embryos with two cotyledons; net-veined leaves; and floral parts arranged in fours, fives, or multiples of these.

diffusion Net movement of like molecules or ions down their concentration gradient.

digestive system Body sac or tube, often with specialized regions where food is ingested, digested, and absorbed, and where undigested residues are eliminated. Incomplete systems have one opening; complete systems have two.

dihybrid cross An intercross between two F_1 heterozygotes that are identical for two gene loci; the dihybrids are offspring of parents that bred true for different versions of two traits.

dinoflagellate Type of single-celled, flagellated, cellulose-plated protistan. Most are producers of marine phytoplankton; some cause red tides.

dinosaur One of a fabulous group of reptiles that originated in the Triassic and became the dominant land vertebrates for 125 million years.

diploidy (DIP-loyd-ee) Presence of two of each type of chromosome (i.e., pairs of homologous chromosomes) in a cell nucleus at interphase. *Compare* haploidy.

directional selection Mode of natural selection by which allele frequencies underlying a range of phenotypic variation shift in a consistent direction, in response to directional change or to new conditions in the environment.

disaccharide (die-SAK-uh-ride) [Gk. *di*, two, + *sakcharon*, sugar] A common oligosaccharide; two covalently bonded sugar monomers.

disease Outcome of infection when defenses aren't mobilized fast enough and a pathogen's activities interfere with normal body functions.

disruptive selection Mode of natural selection by which the different forms of a trait at both ends of the range of variation are favored and intermediate forms are selected against.

distal tubule Tubular part of nephron where water and sodium are selectively reabsorbed.

distance effect Idea that only species adapted for long-distance dispersal can be potential colonists of islands far from their home range.

diversity of life Sum of all variations in form, function, and behavior in all lineages, from the time of life's origin to the present.

division of labor Of a multicelled organism, the combined contributions of its cells, tissues, organs, organ systems—and even stages of development—to survival and reproduction.

DNA Deoxyribonucleic acid (dee-OX-ee-RYE-bow-new-CLAY-ik) Of cells and many viruses, the molecule of inheritance. H bonds join its two helically twisted nucleotide strands, one of which has instructions (in its base sequence) for synthesizing all of the enzymes and other proteins required to build and maintain cells.

DNA clone Many identical copies of DNA that was inserted into plasmids and later amplified.

DNA fingerprint DNA fragments inherited in a Mendelian pattern that give each individual a unique identity.

DNA ligase (LYE-gaze) Enzyme that seals new base-pairings during DNA replication.

DNA polymerase (poe-LIM-uh-raze) Enzyme of replication and repair that assembles a new strand of DNA on a parent DNA template.

DNA repair Enzyme-mediated process that fixes small-scale alterations in a DNA strand by restoring the original base sequence.

DNA replication Any process by which a cell duplicates its DNA molecules before dividing.

dominance hierarchy Social organization in which some individuals of the group have adopted a subordinate status to others.

dominant allele Of diploid cells, an allele that masks the phenotypic effect of any recessive allele paired with it.

dormancy [L. *dormire*, to sleep] A predictable time of metabolic inactivity for many spores, cysts, seeds, perennials, and some animals.

dosage compensation Any mechanism that balances gene expression between the sexes during critical early stages of development.

double fertilization Of flowering plants only, fusion of a sperm and egg nucleus, plus fusion of another sperm nucleus with nuclei of a cell that gives rise to a nutritive tissue (endosperm).

doubling time Time it takes for a population to double in size.

drug addiction Dependence on a drug, which assumes an "essential" biochemical role in the body following habituation and tolerance.

dry shrubland Biome that forms when annual rainfall is less than 25 to 60 centimeters; short, multibranched woody shrubs dominate.

dry woodland Biome that forms when annual rainfall is about 40 to 100 centimeters; it may have many tall trees but no dense canopy.

duplication Gene sequence repeated several to many hundreds or thousands of times. Even normal chromosomes have such sequences.

echinoderm Type of invertebrate with calcified spines, needles, or plates on body wall. Radial with some bilateral features (e.g., sea stars).

echolocation Use of echoes from self-generated ultrasounds as a navigational mechanism.

ecological succession Processes by which a community develops in sequence, from pioneer species to an end array of species that remain in equilibrium over some region.

ecology [Gk. *oikos*, home, + *logos*, reason] The scientific study of how organisms interact with one another and with their environment.

ecoregion Broad land or ocean region defined by climate, geography, and producer species.

ecosystem Array of organisms, together with their environment, interacting through a flow of energy and a cycling of materials.

ecosystem modeling Analytical method; uses computer programs and models to predict unforeseen effects of ecosystem disturbances.

ectoderm [Gk. *ecto*, outside, + *derma*, skin] The first-formed, outermost primary tissue layer of animal embryos; gives rise to nervous system tissues and integument's outer layer.

ectotherm Poorly insulated animal with low metabolic rates; maintains core temperature by absorbing environmental heat (e.g., from sun).

Ediacaran One of the species with a highly flattened body that arose in the precambrian.

effector Muscle (or gland) which, in response to neural commands, helps cause movement (or chemical change).

effector cell Differentiated cell of a lymphocyte subpopulation that, during immune responses, engages and destroys antigen-bearing agents.

egg Mature female gamete; an ovum.

El Niño Massive eastward movement of warm surface waters of the western equatorial Pacific; displaces cool water off South America. Recurs and disrupts climates throughout the world.

electromagnetic spectrum All wavelengths from radiant energy less than 10^{-5} nm long to radio waves more than 10 km long.

electron Negatively charged unit of matter, with particulate and wavelike properties, that occupies one of the orbitals around the atomic nucleus. Atoms gain, lose, or share electrons.

electron transfer A molecule donates one or more electrons to another, acceptor molecule.

electron transfer phosphorylation (FOSS-for-ih-LAY-shun) Last stage of aerobic respiration; electrons from intermediates flow through a membrane transport system, to O_2. The flow sets up an electrochemical gradient that drives ATP formation at other membrane sites.

electron transport system Organized array of membrane-bound enzymes and cofactors that accept and donate electrons in series. It sets up an electrochemical gradient across the membrane that cells use to do work.

element Fundamental form of matter that has mass, occupies space, and cannot be broken apart into a different form of matter, at least by ordinary physical or chemical means.

embryo (EM-bree-oh) [Gk. *en*, in, + *bryein*, to swell] Of animals, a multicelled body formed by way of cleavage, gastrulation, and other early developmental events. Of plants, a young sporophyte until time of germination.

embryo sac The female gametophyte of flowering plants.

embryonic induction In a growing embryo, release of a gene product from one tissue that affects the development of an adjacent tissue.

emigration Permanent move by a resident out of a population.

emulsification Of chyme, a suspension of fat droplets coated with bile salts.

end product Substance present at the end of a metabolic pathway.

endangered species Endemic (native) species highly vulnerable to extinction.

endergonic reaction (en-dur-GONE-ik) Type of chemical reaction having a net gain in energy.

endocrine gland Ductless gland that secretes hormones, which the bloodstream distributes.

endocrine system Integrative system of cells, tissues, and organs, functionally linked to the nervous system, that exerts control by way of its hormones and other chemical secretions.

endocytosis (EN-doe-sigh-TOE-sis) Cell uptake of substances when part of plasma membrane forms a vesicle around them. Three routes are receptor-mediated endocytosis, bulk transport of extracellular fluid, and phagocytosis.

endoderm [Gk. *endon*, within, + *derma*, skin] Inner primary tissue layer of animal embryos; source of inner gut lining and derived organs.

endodermis Cell layer around root vascular cylinder; influences water and solute uptake.

endometrium (EN-doe-MEET-ree-um) [Gk. *metrios*, of the womb] Inner lining of uterus.

endoplasmic reticulum or **ER** (EN-doe-PLAZ-mik reh-TIK-yoo-lum) Organelle that starts at nucleus and curves through cytoplasm. New polypeptides get side chains in rough ER (has ribosomes on its cytoplasmic side); smooth ER (has no ribosomes) is a site of lipid synthesis.

endoskeleton [Gk. *endon*, within, + *skleros*, hard] Of chordates, an internal framework of cartilage, bone, or both; works with skeletal muscle to position, support, and move body.

endosperm (EN-doe-sperm) Nutritive tissue inside the seed of a flowering plant.

endospore Resting structure formed by some bacteria; encloses a duplicate of the bacterial chromosome and a portion of cytoplasm.

endosymbiosis Continuing physical contact between one species and another species that now lives and reproduces inside its body.

endotherm Animal with high metabolic rates, body form, and behavior to control heat gains and losses over a wide temperature range.

energy Capacity to do work.

energy carrier Molecule that delivers energy from one reaction site to another; mainly ATP.

energy pyramid Pyramidal diagram of an ecosystem's trophic structure; it shows energy losses at each transfer to another trophic level.

enhancer A short DNA base sequence that is a binding site for an activator protein.

entropy (EN-trow-pee) Measure of the degree of disorder in a system (how much energy has become disorganized, usually as dispersing heat, and is no longer available to do work). All systems need energy inputs to counter entropy.

enzyme (EN-zime) A type of protein or one of the few RNAs that catalyze reactions between substances, most often at functional groups.

eosinophil Fast-acting white blood cell; its enzyme secretions digest holes in parasitic worms during an inflammatory response.

epidermis Outermost tissue layer of plants and all animals above sponge level of organization.

epiglottis Flaplike structure between pharynx and larynx; its controlled positional changes direct air into trachea or food into esophagus.

epinephrine (ep-ih-NEF-rin) Adrenal hormone; raises blood levels of sugar and fatty acids, increases heart's rate and force of contraction.

epistasis (eh-PISS-tah-sis) Interaction among the products of two or more gene pairs.

epithelium (EP-ih-THEE-lee-um) Animal tissue that covers external surfaces and lines internal cavities and tubes. One surface is free and the other rests on a basement membrane.

equilibrium, dynamic [Gk. *aequus*, equal, + *libra*, balance] Point at which a reaction runs forward as fast as in reverse, so there is no net change in reactant or product concentrations.

erythrocyte (eh-RITH-row-site) [Gk. *erythros*, red, + *kytos*, vessel] Red blood cell.

erythropoietin Hormone secreted by kidney cells; induces stem cells in bone marrow to divide and give rise to red blood cells.

esophagus (ee-SOF-uh-gus) A muscular tube just after the pharynx in the digestive track of vertebrates and many invertebrates.

essential amino acid Any amino acid that an organism cannot synthesize for itself and must obtain from food.

essential fatty acid Any fatty acid an organism cannot synthesize and must obtain from food.

estrogen (ESS-trow-jen) Female sex hormone; helps oocytes mature, primes uterine lining for pregnancy, helps maintain secondary sexual traits, affects growth and development.

estuary (EST-you-ehr-ee) Partly enclosed coast region where seawater mixes with fresh water and runoff from land, as from rivers.

ethylene (ETH-il-een) Plant hormone; promotes fruit ripening and leaf, flower, fruit abscission.

Eubacteria Kingdom of all prokaryotic cells except archaebacteria.

euglenoid Single-celled, flagellated protistan; photosynthetic but with heterotrophic feeding apparatus (it had phagocytic ancestors).

eukaryotic cell (yoo-CARE-ee-oh-tic) [Gk. *eu*, good, + *karyon*, kernel] Cell having a nucleus and other membrane-bound organelles.

eutrophication Nutrient enrichment of a body of water; typically causes reduced transparency and a phytoplankton-dominated community.

evaporation [L. *e-*, out, + *vapor*, steam] The conversion of a substance from a liquid state to a gaseous state by an input of heat energy.

evaporative heat loss Response to heat stress (e.g., at respiratory surfaces and across skin).

evolution, biological [L. *evolutio*, unrolling] Genetic change in a line of descent. Outcome of microevolutionary events: gene mutation, natural selection, genetic drift, and gene flow.

evolutionary tree Treelike diagram; a branch point means divergence from a shared ancestor and branches signify separate lines of descent.

excretion Removal of excess water, solutes, wastes, and some harmful substances from body by way of a urinary system or glands.

exergonic reaction (EX-ur-GONE-ik) Chemical reaction that shows a net loss in energy.

exocrine gland (EK-suh-krin) [Gk. *es*, out of, + *krinein*, to separate] Glandular structure that secretes products, usually through ducts or tubes, to a free epithelial surface.

exocytosis (EK-so-sigh-TOE-sis) Release of a vesicle's contents at cell surface as it fuses with and becomes part of the plasma membrane.

exodermis Cylindrical sheet of cells inside the root epidermis of most flowering plants; helps control uptake of water and solutes.

exon One of the base sequences of an mRNA transcript that will become translated.

exoskeleton [Gk. *skleros*, hard, stiff] External skeleton (e.g., hardened cuticle of arthropods).

exotic species Species that left its home range and became established in a new community.

experiment, scientific Test that simplifies observation in nature or in the laboratory by manipulating and controlling the conditions under which the observations are made.

exponential growth (EX-po-NEN-shul) Pattern of population growth; the population size expands by ever increasing increments during successive intervals as its reproductive base enlarges; plots out as a J-shaped curve.

external ear Among mammals, one of a pair of sound-collecting flaps on the sides of the head.

extinction Irrevocable loss of a species.

extracellular digestion, absorption Mode of nutrition by which certain organisms grow in or on organic matter, digest it with secreted enzymes, then absorb breakdown products.

extracellular fluid Of most animals, all fluid not in cells; plasma (blood's liquid portion) plus interstitial fluid.

extracellular matrix The ground substance, fibrous proteins, and other materials between cells of animal tissues (e.g., cartilage).

extreme halophile Archaebacterium of unusually saline habitats (e.g., salt lakes).

extreme thermophile Archaebacterium of very hot aquatic habitats (e.g., hot springs).

eye Sensory organ that incorporates a dense array of photoreceptors.

FAD Flavin adenine dinucleotide, a type of nucleotide coenzyme; it transfers electrons and unbound protons (H^+) between reaction sites. At such times it is abbreviated $FADH_2$.

fall overturn Vertical mix of a body of water in fall. Its upper layer cools, gets more dense, sinks. Oxygenated surface water moves down, nutrients from bottom sediments move up.

family pedigree Chart of genetic relationship of family individuals through the generations.

fat Lipid with a glycerol head and one, two, or three fatty acid tails. Unsaturated tails have single covalent bonds in the carbon backbone; saturated tails have one or more double bonds.

fatty acid Molecule with a backbone of up to 36 carbon atoms, a carboxyl group ($-COO^-$ or $-COOH$) at one end, and hydrogen atoms at most or all of the other bonding sites.

feather Of birds, a lightweight structure used in flight, insulation, and behavioral displays.

feedback inhibition Mechanism by which a cellular change resulting from some activity shuts down the activity that brought it about.

fermentation [L. *fermentum*, yeast] Anaerobic pathway of ATP formation, from glycolysis to the transfer of electrons to an intermediate or a breakdown product. NAD^+ is regenerated. Net energy yield: 2 ATP per glucose molecule.

fern A seedless vascular plant of moist or wet habitats; has fronds often divided into leaflets.

fertilization [L. *fertilis*, to carry, to bear] The fusion of a sperm nucleus with the nucleus of an egg, thus forming a zygote.

fetus The stage of animal development after major organ systems formed in the embryo until the time of birth.

fever Any core temperature higher than the set point in the hypothalamic thermostat.

fibrous root system All lateral branchings of adventitious roots arising from a young stem.

Fick's law The larger the surface area and the greater the partial pressure gradient, the faster gases will diffuse across a respiratory surface.

filter feeder Animal that filters food from a current of water directed through a body part (e.g., through a sea squirt's pharynx).

filtration First step in forming urine; pressure of heart contractions filters blood (forces water and all solutes except proteins) into nephrons.

fin Of fishes generally, appendage that helps stabilize, propel, and guide body in water.

first law of thermodynamics The total amount of energy in the universe is constant. Energy cannot be created from nothing and existing energy cannot be destroyed.

fish An aquatic animal of the most ancient, diverse vertebrate lineage; one of the jawless, jawed cartilaginous, or jawed bony species.

fitness Increase in adaptation to environment, as brought about by genetic change.

fixation Loss of all but one kind of allele at a gene locus for all individuals in a population.

fixed action pattern Program of coordinated, stereotyped muscle activity that is completed independently of feedback from environment.

flagellum (fluh-JELL-um), plural **flagella** A motile structure of many free-living eukaryotic cells. Its core has a 9 + 2 array of microtubules. Typically longer than cilia.

flatworm One of the simplest invertebrates having organ systems; a turbellarian, fluke, or tapeworm. Some are notorious parasites.

flower Of angiosperms only, a reproductive structure with nonfertile parts (sepals, petals) and fertile parts (stamens, carpels) attached to a receptacle (modified base of a floral shoot).

fluid mosaic model A cell membrane consists of a lipid bilayer and proteins. Its lipids are mainly structural; they make it impermeable to water-soluble molecules, yet (by packing variations and movements), impart fluidity. Its diverse proteins perform most membrane functions (e.g., transport, signal reception).

food chain Straight-line sequence of steps by which energy stored in autotroph tissues passes on to higher trophic levels.

food pyramid Chart of a purportedly well-balanced diet; continually being refined.

food web Of ecosystems, cross-connecting food chains consisting of producers, consumers, and decomposers, detritivores, or both.

forebrain Most complex portion of vertebrate brain; includes cerebrum (and cerebral cortex), olfactory lobes, and hypothalamus.

forest Biome in which tall trees grow close enough to form a fairly continuous canopy.

fossil Recognizable, physical evidence of an organism that lived in the distant past.

fossil fuel Coal, petroleum, or natural gas; a nonrenewable energy source that formed long ago from remains of swamp forests.

fossilization How fossils form. An organism or evidence of it gets buried in sediments or volcanic ash; water and dissolved inorganic compounds infiltrate it; then chemical changes and pressure from accumulating sediments above transform it to stony hardness.

founder effect A form of bottlenecking. By chance alone, a few individuals that establish a new population have allele frequencies that differ from those of the original population.

free radical Any highly reactive molecular fragment having an unpaired electron.

fruit [L. after *frui*, to enjoy] Flowering plant's mature ovary, often with accessory structures.

FSH Follicle-stimulating hormone produced and secreted by the anterior lobe of pituitary gland; has reproductive roles in both sexes.

functional group An atom or a group of atoms that is covalently bonded to the carbon backbone of an organic compound and that influences its chemical behavior.

functional-group transfer Enzyme-mediated event in which a molecule donates one or more functional groups to another molecule.

Fungi Kingdom of fungi which includes major decomposers plus pathogens and parasites.

fungus Eukaryotic heterotroph; it secretes enzymes that digest food to bits that its cells absorb (extracellular digestion, absorption). The saprobes use nonliving organic matter; parasites feed on living organisms.

fusion power Release of energy from fuel (e.g., heated gas of H isotopes) that implodes when compressed to extremely high densities.

gallbladder Organ that stores bile secreted from liver; its duct connects to small intestine.

gamete (GAM-eet) [Gk. *gametēs*, husband, and *gametē*, wife] Haploid cell, formed by meiotic cell division of a germ cell; required for sexual reproduction. Eggs and sperm are examples.

gamete formation Formation of sex cells (e.g., sperm, eggs) in reproductive tissues or organs.

gametophyte (gam-EET-oh-fite) [Gk. *phyton*, plant] Haploid gamete-producing body that forms during plant life cycles.

ganglion (GANG-lee-on), plural **ganglia** Distinct cluster of cell bodies of neurons.

gap junction Cylindrical arrays of proteins in the plasma membrane that pair up as open channels for signals between adjoining cells.

gastric fluid Highly acidic mix of secretions from the stomach's glandular epithelium (HCl, mucus, pepsinogens, etc.) that act on food.

gastrulation (gas-tru-LAY-shun) Stage of animal development; reorganization of new cells into two or three primary tissue layers.

gel electrophoresis Laboratory technique used to distinguish among molecules. Applied electric field forces them to migrate through a viscous gel and distance themselves from one another by length, size, or electric charge.

gene [German *pangan*, after Gk. *pan*, all; *genes*, to be born] Unit of information for a heritable trait, passed from parents to offspring.

gene control One of the molecular mechanisms that govern when and how fast specific genes will be transcribed and translated, and whether gene products will be activated or inactivated.

gene flow Microevolutionary process; alleles enter and leave a population as an outcome of immigration and emigration, respectively.

gene frequency Abundance of an allele with respect to others at same locus in a population.

gene library Mixed collection of bacteria that house many different cloned DNA fragments.

gene locus A gene's chromosomal location.

gene mutation A small-scale change in the nucleotide sequence of a DNA molecule.

gene pair Two alleles at the same gene locus on a pair of homologous chromosomes.

gene pool All genotypes in a population.

gene therapy Generally, a transfer of one or more normal genes into an organism to correct or lessen adverse effects of a genetic disorder.

genetic abnormality A rare or less common version of a heritable trait.

genetic code [After L. *genesis*, to be born] The correspondence between nucleotide triplets in DNA (then mRNA) and specific sequences of amino acids in a polypeptide chain; the basic language of protein synthesis in cells.

genetic disease Illness in which expression of one or more genes increases susceptibility to infection or weakens immune response to it.

genetic disorder Any inherited condition that causes mild to severe medical problems.

genetic divergence Gradual accumulation of differences in gene pools of populations or subpopulations of a species after a geographic barrier arises and separates them; thereafter, microevolution occurs independently in each.

genetic drift Change in allele frequencies over the generations due to chance alone. Its effect is most pronounced in very small populations.

genetic engineering Deliberately altering the information content of DNA molecules.

genetic equilibrium In theory, a state in which a population is not evolving. These conditions are met: no mutation, very large size, isolation from others of the species, no natural selection (*all* members reproduce, by random mating).

genetic recombination Result of any process that puts new genetic information into a DNA molecule (e.g., by crossing over).

genome All the DNA in a haploid number of chromosomes for a given species.

genotype (JEEN-oh-type) Genetic constitution of an individual; a single gene pair or the sum total of an individual's genes.

genus, plural **genera** (JEEN-US, JEN-er-ah) [L. *genus*, race or origin] A grouping of species that are more closely related to one another in their morphology, ecology, and history than to other species at the same taxonomic level.

geographic dispersal An organism moves out of its home range and becomes established in a new community, as an exotic species.

geologic time scale Time scale for the Earth's history with major subdivisions corresponding to mass extinctions. Now radiometrically dated.

germ cell Animal cell of a lineage set aside for sexual reproduction; gives rise to gametes.

germination (jur-mih-NAY-shun) Of seeds and spores, resumption of growth after dormancy, dispersal from the parent organism, or both.

gibberellin (JIB-er-ELL-un) Plant hormone; promotes elongation of stems, helps seeds and buds break dormancy, and helps flowering.

gill Respiratory organ. Most gills have a thin, moist, vascularized layer for gas exchange.

gill slit Opening in a thin-walled pharynx. Gill slits serve in food-trapping and respiration; jaws evolved from certain gill slit supports.

ginkgo Deciduous gymnosperm; its ancestors were diverse in dinosaur times.

gland Secretory cell or structure derived from epithelium and often still connected to it.

global broiling Theory that an asteroid impact caused the K–T mass extinction by creating a colossal fireball, the debris from which raised global air temperature by thousands of degrees.

global warming A long-term increase in the temperature of the Earth's lower atmosphere.

glomerulus (glow-MARE-you-luss) [L. *glomus*, ball] Bowman's capsule, together with the glomeruli capillaries it cups around.

glottis Opening between two vocal cords.

glucagon (GLUE-kuh-gone) Hormone secreted by pancreas; stimulates cells to convert their stores of glycogen and amino acids to glucose.

glyceride (GLISS-er-eyed) Molecule having one, two, or three fatty acid tails attached to a glycerol backbone; one of the fats or oils.

glycerol (GLISS-er-ohl) [Gk. *glykys*, sweet, + L. *oleum*, oil] Three-carbon compound with three hydroxyl groups; component of fats and oils.

glycogen (GLY-kuh-jen) A highly branched polysaccharide made of glucose monomers; the main storage carbohydrate in animals.

glycolysis (gly-CALL-ih-sis) [Gk. *glykys*, sweet, + *lysis*, breaking apart] Breakdown of glucose or another organic compound to two pyruvate molecules. First stage of aerobic respiration, fermentation, and anaerobic electron transport. Oxygen has no role in glycolysis, which occurs in the cytoplasm of all cells. Two NADH form. Net yield: 2 ATP per glucose molecule.

glycoprotein Protein with linear or branched oligosaccharides covalently bonded to it; most surface proteins of animal cells and many proteins circulating in blood are glycoproteins.

gnetophyte Woody gymnosperm, a vine or shrub; unusual in having vessels in its xylem.

Golgi body (GOHL-gee) Organelle of lipid assembly, polypeptide chain modification, and packaging of both in vesicles for export or for transport to locations in cytoplasm.

gonad (GO-nad) Primary reproductive organ in which animal gametes are produced.

Gondwana Paleozoic supercontinent; with other land masses, it formed Pangea.

graded potential A local signal that alters the resting membrane potential at the input zone of a neuron or other excitable cell; magnitude varies, depending on the stimulus.

gradual model, speciation Idea that species arise by many small morphological changes that accumulate over great spans of time.

grassland Biome that has flat or rolling land, 25–100 centimeters of annual rainfall, warm summers, and a distinct array of grazers; fires recur and regenerate dominant plant species.

gravitropism (GRAV-ih-TROPE-izm) [L. *gravis*, heavy, + Gk. *trepein*, to turn] Directional plant growth in response to gravitational force.

gray matter Unmyelinated axons, dendrites, and cell bodies of neurons, plus neuroglial cells, in the brain and spinal cord.

grazing food web Network of food chains in which energy flows from plants to an array of herbivores, then to carnivores.

green alga Type of protistan biochemically, evolutionarily, and structurally most like plants (e.g., most are photoautotrophs with starch grains and chlorophylls *a*, *b* in chloroplasts).

green revolution In developing countries, use of improved crop strains, modern agricultural equipment and practices (e.g., heavy fertilizer and pesticide application) to raise crop yields.

greenhouse effect Atmospheric gases impede escape of infrared wavelengths (heat) from Earth's sun-warmed surface, absorb them, and radiate much of the heat back toward Earth.

ground meristem (MARE-ih-stem) [Gk. *meristos*, divisible] Primary meristem that gives rise to the ground tissue system of plants.

ground substance Intercellular material in a number of animal tissues; composed of cell secretions and other noncellular components.

ground tissue system Tissues, parenchyma especially, making up most of the plant body.

growth Of multicelled species, increases in the number, size, and volume of cells. Of bacterial populations, increases in the number of cells.

growth factor Protein that plays a role in the body's growth (e.g., by promoting mitosis).

growth ring One of the alternating bands of early and late wood; a "tree ring."

growth, tissue specialization Stage of animal development when organs enlarge and assume specialized functions; continues into adulthood.

guanine Nitrogen-containing base in one of four nucleotide monomers of DNA or RNA.

guard cell One of two adjoining cells defining a stoma, a type of opening where CO_2, O_2, and water vapor cross leaf or stem epidermis.

gut Generally, a sac or tube from which food is absorbed into internal environment. Also, a gastrointestinal tract (from stomach onward).

gymnosperm (JIM-noe-sperm) [Gk. *gymnos*, naked, + *sperma*, seed] Type of vascular plant in which seeds form on exposed surfaces of reproductive structures (e.g., on cone scales).

habitat [L. *habitare*, to live in] Place where an organism or species lives; characterized by its physical and chemical features and its species.

habitat island A region with endemic species (e.g., a forest or lake) located within a "sea" of habitats unsuitable for sustaining them.

habitat loss Reduction in suitable living space and habitat closure through chemical pollution.

hagfish Type of scavenging jawless fish that secretes slimy, smelly mucus.

hair A flexible structure rooted in skin, with a shaft above the skin's surface.

hair cell Mechanoreceptor that responds to being physically bent or tilted.

half-life The time it takes for half of a given quantity of any radioisotope to decay into a different, and less unstable, daughter isotope.

haploidy (HAP-loyd-ee) Presence of only half of the parental number of chromosomes in a spore or gamete, as brought about by meiosis.

hardwood Type of strong, dense wood with many vessels, tracheids, and fibers in xylem.

Hardy–Weinberg rule Allele frequencies stay the same over the generations when there is no mutation, the population is infinitely large and isolated from other populations of the species, mating is random, and all individuals are reproducing equally and randomly.

HDL A high-density lipoprotein in blood; it transports cholesterol to liver for metabolism.

hearing Perception of sound.

heart Muscular pump; its contractions keep blood circulating through the animal body.

heartwood A dry tissue at the core of aging stems and roots that no longer transports water and solutes. It helps the tree defy gravity and is a dumping ground for some metabolic wastes.

heat Thermal energy; a form of kinetic energy.

HeLa cell Cancer cell of a lineage established for research; now used in many laboratories.

helper T cell CD4 lymphocyte; has key roles in antibody- and cell-mediated immune responses. It induces responsive T and B cells to divide and differentiate rapidly into huge armies.

heme group Iron-containing functional group that reversibly binds oxygen; one associated with each of the four chains in hemoglobin.

hemoglobin (HEEM-oh-glow-bin) [Gk. *haima*, blood, + L. *globus*, ball] An iron-containing respiratory protein of red blood cells.

hemostasis (HEE-mow-STAY-sis) [Gk. *haima*, blood, + *stasis*, standing] Process that stops blood loss from damaged blood vessel by way of coagulation, blood vessel spasm, platelet plug formation, and other mechanisms.

herbivore [L. *herba*, grass, + *vovare*, to devour] Plant-eating animal (e.g., snail, deer, manatee).

hermaphrodite (her-MAH-froe-dyte) Individual having both male and female gonads.

heterocyst (HET-er-oh-sist) Cyanobacterial cell, self-modified, that synthesizes a nitrogen-fixing enzyme when nitrogen supplies dwindle.

heterotherm Generally, an endotherm that can lower metabolic rates; activities idle but it cuts energy cost of maintaining its core temperature.

heterotroph (HET-er-oh-trofe) [Gk. *heteros*, other, + *trophos*, feeder] Organism unable to make its own organic compounds; feeds on autotrophs, other heterotrophs, organic wastes.

heterozygous condition (HET-er-oh-ZYE-guss) [Gk. *zygoun*, join together] Having a pair of nonidentical alleles at a gene locus (that is, on a pair of homologous chromosomes).

higher taxon (plural, **taxa**) One of ever more inclusive groupings that reflect relationships among species. Family, order, class, phylum, and kingdom are examples.

hindbrain Medulla oblongata, cerebellum, and pons of vertebrate brain. Includes reflex centers for respiration, blood circulation, and other basic functions; also helps coordinate motor responses and many complex reflexes.

histamine Local signaling molecule that fans inflammation; makes arterioles vasodilate and capillaries more permeable (leaky).

histone Type of protein intimately associated with eukaryotic DNA and largely responsible for organization of eukaryotic chromosomes.

homeostasis (HOE-me-oh-STAY-sis) [Gk. *homo*, same, + *stasis*, standing] State in which physical and chemical aspects of internal environment (blood, interstitial fluid) are being maintained within ranges suitable for cell activities.

homeotic gene In all major animal groups, one of the master genes, the products of which interact with one another and with control elements to map out the overall body plan.

hominid [L. *homo*, man] All species on or near evolutionary road leading to modern humans.

hominoid Apes, humans, and recent ancestors.

homologous chromosome (huh-MOLL-uh-gus) [Gk. *homologia*, correspondence] Of cells with a diploid chromosome number, one of a pair of chromosomes identical in size, shape, and gene sequence, and that interact at meiosis. Nonidentical sex chromosomes (e.g., X and Y) also interact as homologues during meiosis.

homologous structures Of separate lineages, comparable body parts that show underlying similarity even when they may differ in size, shape, or function; outcome of morphological divergence from a shared ancestor.

homozygous condition (HOE-moe-ZYE-guss) For a specified trait, having a pair of identical dominant or recessive alleles at a locus.

hormone [Gk. *hormon*, stir up, set in motion] Signaling molecule secreted by one cell that stimulates or inhibits activities of any cell with receptors for it. Animal hormones are picked up and transported by the bloodstream.

horsetail Seedless vascular plant; has rhizomes, scale-shaped leaves, and hollow photosynthetic stems with silica-reinforced ribs.

host Living organism exploited by a parasite. A definitive host harbors the mature stage of a parasite's life cycle. One or more intermediate hosts harbor immature stages.

hot spot Limited area where human activities are driving many species to extinction.

human A primate of the species *Homo sapiens*.

human genome project Worldwide research project to sequence all 3.2 billion nucleotides in the DNA of human chromosomes.

humus Decomposing organic matter in soil.

hybrid offspring Of a genetic cross, offspring having a pair of nonidentical alleles for a trait.

hybrid zone Where adjoining populations are interbreeding and producing hybrid offspring.

hydrocarbon An organic compound that has only hydrogen bonded to a carbon backbone.

hydrogen bond Weak interaction between a small, highly electronegative atom and an H atom already taking part in a polar covalent bond in the same molecule or a different one.

hydrogen ion Free (or unbound) proton; one hydrogen atom that lost its electron and now bears a positive charge (H^+).

hydrologic cycle Biogeochemical cycle driven by solar energy; water is moved through the atmosphere, on or through land, to the ocean, and back to the atmosphere.

hydrolysis (high-DRAWL-ih-sis) [L. *hydro*, water, + Gk. *lysis*, loosening] Cleavage reaction that breaks covalent bonds and splits a molecule into two or more parts. H^+ and OH^- (from a water molecule) are often attached to the newly exposed bonding sites.

hydrophilic substance [Gk. *philos*, loving] A polar molecule or molecular region that easily dissolves in water (e.g., sugars).

hydrophobic substance [Gk. *phobos*, dreading] A nonpolar molecule or molecular region that strongly resists dissolving in water (e.g., oils).

hydrosphere All of the Earth's water.

hydrostatic pressure Pressure exerted by a volume of fluid against a wall, membrane, or some other structure that encloses the fluid.

hydrostatic skeleton A fluid-filled cavity or cell mass against which contractile cells act.

hydrothermal vent A steaming fissure in the deep ocean floor; has unique ecosystems.

hydroxide ion Ionized compound of one oxygen and one hydrogen atom (OH^-).

hypertonic solution A fluid having a greater solute concentration relative to another fluid.

hypha (HIGH-fuh), plural **hyphae** [Gk. *hyphe*, web] Fungal filament with chitin-reinforced walls; component of a mycelium.

hypodermis Subcutaneous layer with stored fat that helps insulate the body; also anchors skin yet allows it to move somewhat.

hypothalamus [Gk. *hypo*, under, + *thalamos*, inner chamber, or *tholos*, rotunda] Forebrain center of homeostatic control over the internal environment (e.g., salt–water balance, core temperature); influences hunger, thirst, sex, other viscera-related behaviors, and emotions.

hypothesis In science, a possible explanation of a phenomenon, one that has the potential to be proved false by experimental tests.

hypotonic solution A fluid that has a lower solute concentration relative to another fluid.

immigration One or more individuals moves from its population and takes up permanent residence in another population of its species.

immune system Interacting white blood cells that defend the vertebrate body through self/nonself recognition, specificity, and memory.

T and B cell antigen receptors ignore body's own cells yet collectively recognize a billion specific threats. Some B and T cells formed in a primary response are set aside as memory cells for future battles with the same antigen.

immunization Any process that promotes immunity against disease (e.g., vaccination).

immunoglobulin (Ig) One of five classes of antibodies, each with antigen-binding sites as well as other sites with specialized functions.

implantation Event in pregnancy. A blastocyst burrows into endometrium and establishes connections by which a mother will exchange substances with the embryo (and fetus) that develops from the blastocyst's inner cell mass.

imprinting Time-dependent form of learning, usually during a sensitive period for a young animal, triggered by exposure to sign stimuli.

in vitro fertilization Conception outside the body ("in glass" petri dishes or test tubes).

inbreeding Nonrandom mating among close relatives that share many identical alleles.

incomplete dominance Condition in which one allele of a pair is not fully dominant; a heterozygous phenotype somewhere between both homozygous phenotypes emerges.

independent assortment theory Mendelian theory that by the end of meiosis, each pair of homologous chromosomes (and linked genes on each one) are sorted before shipment to gametes independently of how the other pairs were sorted. Later modified to account for the disruptive effect of crossing over on linkages.

indicator species Any species that provides warning of changes in habitat and impending widespread loss of biodiversity.

indirect selection theory Idea that altruistic individuals can pass on their genes indirectly by helping relatives survive and reproduce.

induced-fit model When bound to an active site, a substrate alters an enzyme's shape, thus bringing about a more precise molecular fit between the two that promotes reactivity.

inductive logic Pattern of thinking; deriving a general statement from specific observations.

infection Invasion and multiplication of a pathogen in a host. Disease follows if defenses are not mobilized fast enough; the pathogen's activities interfere with normal body functions.

inflammation, acute Rapid response to tissue injury by phagocytes and certain proteins (e.g., histamine, complement, clotting factors). Signs include localized redness, heat, swelling, pain.

inheritance The transmission, from parents to offspring, of genes that specify structures and functions characteristic of the species.

inhibitor Generally, a substance able to bind with a specific molecule and interfere with its functioning (e.g., a hypothalamic hormone that suppresses anterior pituitary secretion).

insertion Insertion of one to a few bases into a DNA strand. Also, a movable attachment of muscle to bone.

instinctive behavior A behavior performed without having been learned by experience.

insulin Pancreatic hormone that lowers level of glucose in blood by causing cells to take up glucose; promotes protein and fat synthesis and inhibits protein conversion to glucose.

integration, neural [L. *integrare*, coordinate] Moment-by-moment summation of excitatory and inhibitory synapses acting on a neuron.

integrator A control center (e.g., brain) that receives, processes, and stores sensory input, issues commands for coordinated responses.

integument Of animals, protective body cover (e.g., skin). Of seed-bearing plants, one or more layers around an ovule; becomes a seed coat.

integumentary exchange (in-teg-you-MEN-tuh-ree) Respiration across a thin, moist, and often vascularized surface layer of animal body.

intermediate Substance that forms between the start and end of a metabolic pathway.

intermediate filament Cytoskeletal element; mechanically strengthens some animal cells.

internal environment Blood + interstitial fluid.

interneuron Neuron of brain or spinal cord.

internode In vascular plants, the stem region between two successive nodes.

interphase Of a cell cycle, interval between nuclear divisions when a cell increases in mass and roughly doubles the number of its cytoplasmic components. It also duplicates its chromosomes (replicates its DNA) during interphase, but *not* between meiosis I and II.

interstitial fluid (IN-ter-STISH-ul) [L. *interstitus*, to stand in the middle of something] The portion of extracellular fluid that occupies the spaces between animal cells and tissues.

intertidal zone Region between the low and high water marks of a rocky or sandy shore.

intervertebral disk Cartilaginous flex point and shock absorber between vertebrae.

intron A noncoding portion of a pre-mRNA transcript; excised before translation.

inversion Part of a chromosome that became oriented in reverse, with no molecular loss.

invertebrate Any animal without a backbone.

ion, negatively charged (EYE-on) An atom or a molecule that acquired an overall negative charge by gaining one or more electrons.

ion, positively charged Atom or molecule that acquired an overall positive charge by losing one or more electrons.

ion exchange The pH-dependent process by which ions dissociate from soil particles, and other ions dissolved in soil water replace them.

ionic bond Two ions being held together by the attraction of their opposite charge.

ionizing radiation High-energy wavelengths.

isotonic solution A fluid having the same solute concentration as a fluid against which it is being compared.

isotope (EYE-so-tope) One of two or more forms of an element's atoms that differ in the number of neutrons.

jaw Of chordates, either of two cartilaginous or bony parts bordering the mouth, typically bearing teeth used to secure or pummel food. Many invertebrates have similar opposing parts.

joint Area of contact between bones.

J-shaped curve Type of diagrammatic curve that emerges when unrestricted exponential growth of a population is plotted against time.

karyotype (CARE-ee-oh-type) Preparation of metaphase chromosomes sorted by length, centromere location, other defining features.

keratin Tough, water-insoluble protein made by vertebrate epidermal cells (keratinocytes).

key innovation A structural or functional modification to the body that by chance, gives a lineage opportunity to exploit aspects of the environment in more efficient or novel ways.

keystone species A species with a dominant role in shaping community structure.

kidney One of a pair of vertebrate organs that filter ions and other substances from blood; it controls amounts returned to help maintain the internal environment.

kilocalorie 1,000 calories of heat energy; the amount needed to raise the temperature of 1 kilogram of water by 1°C. Used as the unit of measure for the caloric content of foods.

kinase Enzyme (e.g., a protein kinase) that catalyzes a phosphate-group transfer.

kinetic energy Energy of motion.

Krebs cycle A stage of aerobic respiration, in mitochondria only; it (and a few preparatory steps) breaks down pyruvate to CO_2 and H_2O. Many coenzymes accept protons (H^+) and electrons from intermediates and deliver them to the next stage of reactions; 2 ATP form.

lactate fermentation An anaerobic pathway of ATP formation. Pyruvate from glycolysis is converted to three-carbon lactate, and NAD^+ is regenerated. Net energy yield: 2 ATP.

lake A body of standing fresh water in a basin formed by geologic events; characterized by light penetration, temperature, other features.

lamprey Type of predatory, jawless fish that has a cylindrical body, cartilaginous skeleton, and an oral disk with rasping parts.

lancelet A filter-feeding invertebrate chordate with a body tapered at both ends.

large intestine Colon.

larva, plural **larvae** Immature stage between an embryo and adult in many animal life cycles.

larynx (LARE-inks) Tubular airway leading to lungs. Contains vocal cords in some animals.

lateral root Outward branching from the first (primary) root of a taproot system.

LDL Low-density lipoprotein that transports cholesterol; cells take it up, but excess amounts contribute to atherosclerosis.

leaching Removal of some nutrients from soil as water percolates through it.

leaf Chlorophyll-rich plant part adapted for sunlight interception and photosynthesis.

learned behavior Lasting modification of a behavior as a result of experience or practice.

lek Communal display ground for courtship behavior by some animals, including birds.

lens A transparent body that bends light rays so that they converge onto photoreceptors.

leptin An appetite-mediating hormone.

lethal mutation Mutation with drastic effects on phenotype; usually causes death.

LH Luteinizing hormone. Anterior pituitary hormone; reproductive roles in males, females.

lichen (LY-kun) Symbiotic interaction between a fungus and photoautotroph.

life cycle A recurring pattern of genetically programmed events from the time individuals are produced until they themselves reproduce.

life history pattern Patterns of reproduction, survival, and life expectancy for a population.

ligament A strap of dense connective tissue that bridges a joint.

light-dependent reactions The first stage of photosynthesis. Sunlight energy is trapped and converted to chemical energy of ATP, NADPH, or both, depending on the pathway.

light-independent reactions Second stage of photosynthesis; sugar-building reactions that require phosphate-group transfers from ATP, electrons and H atoms from NADPH, and carbon from CO_2. The phosphorylated sugars from these reactions are then converted to end products (e.g., sucrose, cellulose, starch).

lignification Deposition of lignin in secondary wall of many plant cells. Stabilizes and protects wall components, lends strength, rigidity, and water-loss resistance to plant parts. Important factor in the evolution of vascular plants.

lignin Organic compound; three-carbon chain and oxygen atom linked to a six-carbon ring.

limbic system In cerebral hemisphere, centers that govern emotions; has roles in memory.

limiting factor Any essential resource which, in short supply, limits population growth.

lineage (LIN-ee-edge) Line of descent.

linkage group All genes on a chromosome.

lipid A mostly greasy or oily hydrocarbon; strongly resists dissolving in water but readily dissolves in nonpolar substances. All cells use lipids as storage forms of energy, structural materials as in membranes, and cell products.

lipid bilayer Phospholipids, mostly, arranged in two layers; the structural basis of all cell membranes. Hydrophobic tails are sandwiched between the hydrophilic heads; the heads are dissolved in intracellular or extracellular fluid.

liver In vertebrates and many invertebrates, a large gland that stores, converts, and helps maintain blood levels of organic compounds; inactivates most hormone molecules that have completed their tasks; inactivates compounds that can be toxic at high concentrations.

loam Soil with roughly the same proportions of sand, silt, and clay; best for plant growth.

lobe-finned fish Only bony fish with skeletal supporting elements inside a pair of thick, ventral fins; also has paired lungs. Lineage probably included ancestors of amphibians.

local signaling molecule A secretion by one cell type that alters chemical conditions only in localized tissue regions (e.g., prostaglandin).

logistic growth (low-JISS-tik) A population growth pattern. A low-density population slowly increases in size, enters a phase of rapid growth, then levels off in size once the carrying capacity has been reached.

loop of Henle Hairpin-shaped, tubular part of a nephron that reabsorbs water and solutes.

lung Internal, sac-shaped respiratory surface that originally evolved in oxygen-poor aquatic habitats. A few fishes and amphibians, birds, reptiles, and mammals have paired lungs.

lycophyte Seedless vascular plant; needs free water to complete life cycle. Most have leaves, roots, and stems (e.g., club mosses).

lymph (LIMF) [L. *lympha*, water] Tissue fluid that has entered vessels of lymphatic system.

lymph node Lymphoid organ packed with lymphocytes; major site for immune responses.

lymph vascular system Parts of lymphatic system; delivers excess tissue fluid, absorbed fats, and reclaimable solutes to blood.

lymphatic system Supplement to vertebrate circulatory system. Its vessels deliver fluid and solutes from interstitial fluid to blood; its lymphoid organs have roles in body defenses.

lysis [Gk. *lysis*, a loosening] Gross damage to a plasma membrane, cell wall, or both that lets the cytoplasm leak out; causes cell death.

lysosome (LYE-so-sohm) Important organelle of intracellular digestion.

lysozyme Infection-fighting enzyme present in mucous membranes (e.g., of mouth, vagina).

macroevolution Large-scale patterns, trends, and rates of change among higher taxa.

macrophage Phagocytic white blood cell; roles in nonspecific defense and immune responses.

malpighian tubule One of many small tubes that help land-dwelling insects dispose of toxic wastes without losing precious water.

mammal Vertebrate whose females nourish offspring with milk from mammary glands.

mangrove wetland A nutrient-rich ecosystem of tidal flats at tropical latitudes.

mantle Of mollusks, a tissue draped over the visceral mass. Of Earth, a zone of intermediate-density rocks beneath the crust.

marine snow Organic matter, drifting down from ultraplankton to midoceanic water; the basis of food webs and staggering biodiversity.

mass extinction Catastrophic event or phase in geologic time when entire families or other major groups are irrevocably lost.

mast cell Basophil-like cell; secretes histamine.

master gene One of the genes whose products govern the development of an embryo's body according to an inherited plan; all major groups have similar or identical master genes.

mechanoreceptor Sensory cell or nearby cell that detects mechanical energy (changes in pressure, position, or acceleration).

medulla oblongata Hindbrain region with reflex centers for basic tasks (e.g., respiration); coordinates motor responses with complex reflexes (e.g., coughing); also influences brain centers concerned with sleep and arousal.

medusa (meh-DOO-sah) [Gk. *Medousa*, one of three sisters in Greek mythology having snake-entwined hair] Of cnidarian life cycles, a free-swimming, bell-shaped stage, often with oral lobes and tentacles extending below the bell.

megaspore Haploid spore; forms by meiosis in the ovary of seed-bearing plants; one of its cellular descendants develops into an egg.

meiosis (my-OH-sis) [Gk. *meioun*, to diminish] Two-stage nuclear division process that halves the chromosome number of a parental germ cell nucleus, to the haploid number. Basis of gamete formation (and meiospore formation).

memory The capacity to store and retrieve information about past sensory experience.

memory cell B or T cell that formed during an immune response; it remains in a resting phase until a secondary immune response.

menopause (MEN-uh-pozz) [L. *mensis*, month, + *pausa*, stop] The time when the reproductive potential of human females draws to a close.

menstrual cycle A recurring cycle in adult human females. A secondary oocyte is released from an ovary and the uterine lining is primed for pregnancy, all under hormonal control.

meristem A localized region of a plant where undifferentiated cells give rise to cell lineages that form all of the mature tissues.

mesoderm (MEH-zoe-derm) [Gk. *mesos*, middle, + *derm*, skin] Primary tissue layer of all large, complex animals; gives rise to many internal organs and part of the integument.

mesophyll (MEH-zoe-fill) A photosynthetic parenchyma with an abundance of air spaces.

Mesozoic An era (240–65 mya) of spectacular expansion in the range of global diversity.

messenger RNA (mRNA) A single strand of ribonucleotides transcribed from DNA, then translated into a polypeptide chain. The only RNA encoding protein-building instructions.

metabolic pathway (MEH-tuh-BALL-ik) Orderly sequence of enzyme-mediated reactions by which cells maintain, increase, or decrease the concentrations of particular substances.

metabolism (meh-TAB-oh-lizm) [Gk. *meta*, change] All the controlled, enzyme-mediated chemical reactions by which cells acquire and use energy to synthesize, store, degrade, and eliminate substances in ways that contribute to growth, survival, and reproduction.

metamorphosis (me-tuh-MOR-foe-sis) [Gk. *meta*, change, + *morphe*, form] Major changes in body form of certain animals; hormonally controlled growth, tissue reorganization, and remodeling of body parts leads to adult form.

metaphase Of meiosis I, stage when all pairs of homologous chromosomes have become positioned at the spindle equator. Of mitosis or meiosis II, all the duplicated chromosomes are positioned at the spindle equator.

metastasis Abnormal migration of cancer cells, which may establish colonies in other tissues.

methanogen Anaerobic archaebacterium that produces methane gas as by-product.

methylation Attachment of a methyl group to an organic compound; a common gene control.

MHC marker Self-marker protein. Some kinds are on all body cells of the individual; others are unique to macrophages and lymphocytes.

micelle (my-SELL) Of fat digestion, tiny droplet of bile salts, fatty acids, and monoglycerides; role in fat absorption from small intestine.

microevolution Of a population, any change in allele frequencies resulting from mutation, genetic drift, gene flow, natural selection, or some combination of these.

microfilament [Gk. *mikros*, small, + L. *filum*, thread] Two thin, twisted polypeptide chains; cytoskeletal element that helps cell move, and helps produce and maintain cell shapes.

micrograph Photograph of an image that came into view with the aid of a microscope.

microorganism Organism, usually single celled, too small to be observed without a microscope.

microspore Walled haploid spore; becomes a pollen grain in gymnosperms or angiosperms.

microtubular spindle Bipolar array of many microtubules; forms for nuclear division and moves chromosomes apart in controlled ways.

microtubule (my-crow-TUBE-yool) Cylinder of tubulin subunits; cytoskeletal element with roles in cell shape, growth, and motion (e.g., part of cilia, flagella, and spindle apparatus).

microvillus (MY-crow-VILL-us) [L. *villus*, shaggy hair] Slender extension from free surface of certain cells; arrays of many microvilli greatly increase absorptive or secretory surface area.

midbrain Part of vertebrate brain with centers for coordinating reflex responses to visual and auditory input; also relays signals to forebrain.

migration Recurring round trip between two or more regions in response to environmental rhythms (e.g., seasonal change).

mimicry (MIM-ik-ree) Close resemblance in form, behavior, or both between one species (the mimic) and another (its model). Serves in deception, as when an orchid mimics a female insect and so attracts males that pollinate it.

mineral Any element or inorganic compound that formed by natural geologic processes and is required for normal cell functioning.

mitochondrion (MY-toe-KON-dree-on) Double-membrane organelle of ATP formation. Only site of aerobic respiration's second and third stages. May have endosymbiotic origins.

mitosis (my-TOE-sis) [Gk. *mitos*, thread] Type of nuclear division that maintains the parental chromosome number for daughter cells. The basis of growth in size, tissue repair, and often asexual reproduction for eukaryotes.

mixture Two or more elements intermingled in proportions that can and usually do vary.

model Theoretical, detailed description or analogy that helps people visualize something that has not yet been directly observed.

molar Tooth with cusps that help crush, grind, and shear food; one of the cheek teeth.

molecular clock Model used to calculate the time of origin of one lineage relative to others; assumes that the lineage accumulates neutral mutations at predictable rates, measurable as a series of ticks back through time.

molecule Two or more atoms of the same or different elements joined by chemical bonds.

mollusk Only invertebrate with mantle draped over a soft, fleshy body; most have an external or internal shell. Diverse body plans and sizes (e.g., gastropods, bivalves, and cephalopods).

molting Periodic shedding of worn-out or too-small body structures. Permits some animals to grow in size or renew parts (e.g., shells, exoskeletons, hairs, feathers, and horns).

Monera In earlier schemes, one prokaryotic kingdom for archaebacteria and eubacteria.

monocot (MON-oh-kot) Monocotyledon; a flowering plant with one cotyledon in seeds, floral parts usually in threes (or multiples of three), and often parallel-veined leaves.

monohybrid cross Intercross between two F_1 heterozygotes that are identical for one gene locus; they are the offspring of two parents that breed true for different forms of a trait.

monomer Small molecule used as a subunit of polymers, such as sugar monomers of starch.

monosaccharide (MON-oh-SAK-ah-ride) [Gk. *monos*, alone, single, + *sakcharon*, sugar] One of the simple carbohydrates (e.g., glucose).

monsoon Air circulation pattern that moves moisture-laden air arising from warm oceans to continents north or south of them.

morphogen Type of inducer molecule that diffuses through embryonic tissues, activating master genes in sequence and so contributing to mapping out the overall body plan.

morphogenesis (MORE-foe-JEN-ih-sis) [Gk. *morphe*, form, + *genesis*, origin] Programmed, orderly changes in body size, proportion, and shape of an animal embryo through which all specialized tissues and early organs form.

morphological convergence Macroevolutionary pattern. In response to similar environmental pressures over time, evolutionarily distant lineages evolve in similar ways and end up being alike in appearance, functions, or both.

morphological divergence Macroevolutionary pattern. Genetically diverging lineages slowly undergo change from the body form of their common ancestor.

mosaic tissue effect Patches of body tissues in which one or the other of two X chromosomes is being expressed; outcome of X chromosome inactivation in all female mammals.

moss Most common kind of bryophyte.

motor neuron Neuron that relays signals from brain or spinal cord to muscle or gland cells.

motor protein Type of protein (e.g., myosin) attached to microfilaments and microtubules; used in cell movements (e.g., contraction).

motor unit A motor neuron and all muscle cells that form junctions with its endings.

multicelled organism Organism composed of many cells with coordinated metabolic activity; most show extensive cell differentiation into tissues, organs, and organ systems.

multiple allele system Three or more slightly different molecular forms of a gene that occur among individuals of a population.

multiregional model Holds that *Homo sapiens* evolved from *H. erectus* groups that had spread through much of the world by about 1 mya and evolved into regionally distinctive "races."

muscle fatigue Decline in tension of a muscle kept in a state of tetanic contraction as a result of continuous, high-frequency stimulation.

muscle tension Mechanical force exerted by a contracting muscle; resists opposing forces (e.g., gravity or weight of an object being lifted).

muscle tissue Tissue with arrays of cells able to contract under stimulation, then passively lengthen and return to their resting position.

muscle twitch Sequence of muscle contraction and relaxation in response to a brief stimulus.

mushroom Reproductive part of club fungi.

mutagen (MEW-tuh-jen) Environmental agent that can alter DNA's molecular structure.

mutation [L. *mutatus*, a change, + *-ion*, act, result, or process] Heritable change in DNA's molecular structure. Original source of all new alleles and, ultimately, the diversity of life.

mutation rate Of a gene locus, the probability that a spontaneous mutation will occur during or between DNA replication cycles.

mutualism [L. *mutuus*, reciprocal] Symbiotic interaction that benefits both participants.

mycelium (my-SEE-lee-um), plural **mycelia** [Gk. *mykes*, fungus, mushroom, + *helos*, callus] Mesh of tiny, branching filaments (hyphae); the food-absorbing portion of most fungi.

mycorrhiza (MY-coe-RIZE-uh) "Fungus-root." A form of mutualism between fungal hyphae and young plant roots. The plant gives up some carbohydrates and the fungus gives up some of its absorbed mineral ions.

myofibril (MY-oh-FY-brill) One of many striated threadlike structures inside a muscle cell.

myoglobin A pigment with a higher oxygen-binding capacity than hemoglobin; abundant in cardiac and skeletal muscle cells that can contract slowly and resist fatiguing.

myosin (MY-uh-sin) Motor protein with key roles in cell movements.

NAD⁺ Nicotinamide adenine dinucleotide. A nucleotide coenzyme; abbreviated NADH when carrying electrons and H⁺ to a reaction site.

NADP⁺ Nicotinamide adenine dinucleotide phosphate. A phosphorylated nucleotide coenzyme; abbreviated NADPH₂ when it is carrying electrons and H⁺ to a reaction site.

natural killer cell Cytotoxic lymphocyte that touch-kills tumor cells and virus-infected cells.

natural selection Microevolutionary process; the outcome of differences in survival and reproduction among individuals that differ in details of heritable traits.

negative control Use of regulatory proteins to trigger a slowdown in gene expression.

negative feedback mechanism A homeostatic mechanism by which a condition that changed as a result of some activity triggers a response that reverses the change.

nematocyst (NEM-at-uh-sist) [Gk. *nema*, thread + *kystis*, pouch] Cnidarian capsule that has a dischargeable, tube-shaped thread, sometimes barbed; releases a toxin or sticky substance.

nephridium (neh-FRID-ee-um), plural **nephridia** Unit that controls composition and volume of fluid in some invertebrates (e.g., earthworms).

nephron (NEFF-ron) [Gk. *nephros*, kidney] One of the kidney's urine-forming tubules; it filters water and solutes from blood, then selectively reabsorbs adjusted amounts of both.

nerve Sheathed, cordlike bundle of the axons of sensory neurons, motor neurons, or both.

nerve cell One of the cells that receive and integrate signals from sensory receptors, and guide responses to stimuli. *See also* Neuron.

nerve cord A prominent longitudinal nerve; most animals have one, two, or three. Chordate nervous systems arise from a dorsal nerve cord.

nerve net Simple nervous system in epidermis of cnidarians and some other invertebrates; a diffuse mesh of simple, branching nerve cells interacts with contractile and sensory cells.

nervous system Integrative organ system with nerve cells interacting in signal-conducting and information-processing pathways. Detects and processes stimuli, and elicits responses from effectors (e.g., muscles and glands).

nervous tissue Connective tissue composed of neurons and often neuroglia.

net population growth rate per individual (*r*) For population growth equations, a variable combining rates and death rates; assumes that both remain constant in specified interval.

neural tube The embryonic and evolutionary forerunner of brain and spinal cord.

neuroglia (NUR-oh-GLEE-uh) Collectively, cells that structurally and metabolically support neurons. They make up about half the volume of nervous tissue in vertebrates.

neuromuscular junction Synapse between a motor neuron's axon endings and a muscle cell.

neuron (NUR-on) Type of nerve cell; basic communication unit in most nervous systems.

neurotransmitter Any of a class of signaling molecules secreted by neurons. It acts on cells next to it, then is rapidly degraded or recycled.

neutral mutation A mutation with little or no effect on phenotype, so natural selection can't change its frequency in a population.

neutron Unit of matter, one or more of which occupies the atomic nucleus and has mass but no electric charge.

neutrophil Phagocytic white blood cell that acts fast against bacteria.

niche (NITCH) [L. *nidas*, nest] Sum total of all activities and relationships in which individuals of a species engage as they secure and use the resources required to survive and reproduce.

nitrification (nye-trih-fih-KAY-shun) Major nitrogen cycle process. Certain bacteria in soil break down ammonia or ammonium to nitrite, then other bacteria break down the nitrite to nitrate (a form that plants can take up).

nitrogen cycle Atmospheric cycle. Nitrogen moves from its largest reservoir (atmosphere), through the ocean, ocean sediments, soils, and food webs, then back to the atmosphere.

nitrogen fixation Major nitrogen cycle process. Certain bacteria convert gaseous nitrogen to ammonia, which dissolves in their cytoplasm to form ammonium (used in biosynthesis).

node Stem site where one or more leaves form.

noncyclic pathway of ATP formation (non-SIK-lik) [L. *non*, not, + Gk. *kylos*, circle] Light-dependent reactions of photosynthesis, using photolysis, two photosystems, two transport systems. H₂O molecules are split. Released electrons and H atoms are used in ATP and NADPH formation; the oxygen, a by-product, is the basis of Earth's oxygen-rich atmosphere.

nondisjunction Failure of sister chromatids or a pair of homologous chromosomes to separate during meiosis or mitosis. Daughter cells end up with too many or too few chromosomes.

non-ionizing radiation Wavelengths that boost electrons to a higher energy level. DNA easily absorbs one form, ultraviolet light.

notochord (KNOW-toe-kord) Of chordates, a rod of stiffened tissue (not cartilage or bone) that is a supporting structure for the body.

nuclear envelope Outermost portion of the cell nucleus; consists of a double membrane (two lipid bilayers and associated proteins).

nucleic acid (new-CLAY-ik) Single- or double-stranded chain of four kinds of nucleotides joined at their phosphate groups. Nucleic acids (e.g., DNA, RNA) differ in base sequences.

nucleic acid hybridization Any base-pairing between DNA or RNA from different sources.

nucleoid (NEW-KLEE-oid) Portion of bacterial cell interior in which the DNA is physically organized but not enclosed by a membrane.

nucleolus (new-KLEE-oh-lus) [L. *nucleolus*, tiny kernel] In a nondividing cell nucleus, a site for assembling protein and RNA subunits that will later join up as ribosomes in the cytoplasm.

nucleosome (NEW-klee-oh-sohm) A stretch of eukaryotic DNA looped twice around a spool of histone molecules; one of many units that give condensed chromosomes their structure.

nucleotide (NEW-klee-oh-tide) Small organic compound with deoxyribose (a five-carbon sugar), a nitrogenous base, and a phosphate group. Monomer for adenosine phosphates, nucleotide coenzymes, and nucleic acids.

nucleus (NEW-klee-us) [L. *nucleus*, a kernel] Of atoms, a central core of one or more protons and (in all but hydrogen atoms) neutrons. In a eukaryotic cell, the organelle that physically separates DNA from cytoplasmic machinery.

numerical taxonomy Study of the degree of relatedness between an unidentified organism and a known group through comparisons of traits. Used to classify prokaryotic species.

nutrient Element with a direct or indirect role in metabolism that no other element fulfills.

nutrition Processes of selectively ingesting, digesting, absorbing, and converting food into the body's own organic compounds.

nymph Immature, post-embryonic stage of some insect life cycles.

obesity Excess of fat in adipose tissue; caloric intake has exceeded the body's energy output.

ocean A continuous body of water that covers more than 71 percent of the Earth; its currents distribute nutrients in marine ecosystems and influence regional climates.

oligosaccharide (oh-LIG-oh-SAC-uh-RID) Short-chain carbohydrate of two or more covalently bonded sugar monomers (e.g., disaccharides).

omnivore [L. *omnis*, all, + *vovare*, to devour] Animal that eats at more than one trophic level.

oncogene (ON-koe-jeen) Any gene having the potential to induce cancerous transformation.

oocyte Type of immature egg.

oogenesis (oo-oh-JEN-uh-sis) Process by which a germ cell develops into a mature oocyte.

operator Very short base sequence between a promoter and bacterial genes; a binding site for a repressor that can block transcription.

operon Promoter–operator sequence serving more than one bacterial gene; part of a control that adjusts transcription rates up or down.

orbital One of the volumes of space around the atomic nucleus in which one or at most two electrons are likely to be at any instant.

organ Body structure with definite form and function that consists of more than one tissue.

organ formation Developmental stage in which primary tissue layers give rise to cell lineages unique in structure and function. Descendants of those lineages give rise to all the different tissues and organs of the adult.

organ system Organs interacting chemically, physically, or both in a common task.

organelle (or-guh-NELL) Membrane-bound sac or compartment in the cytoplasm having one or more specialized metabolic functions. Most eukaryotic cells have a profusion of them.

organic compound Molecule of one or more elements covalently bonded to some number of carbon atoms.

osmoreceptor Sensory receptor; detects shifts in water volume (solute concentration).

osmosis (oss-MOE-sis) [Gk. *osmos*, pushing] In response to a water concentration gradient, the diffusion of water between two regions that a selectively permeable membrane separates.

osmotic pressure Pressure that operates after hydrostatic pressure develops in an enclosed region (e.g., a cell); it counters water's inward diffusion (stops further rises in fluid volume).

osteoblast Cell that forms and deposits bone, thus storing minerals.

osteoclast Cell that secretes the enzymes that degrade bone tissue to release minerals.

ostracoderm One of the earliest jawless fishes; a bottom-dwelling filter feeder. Extinct.

ovary (OH-vuh-ree) In most animals, a female gonad. In flowering plants, the enlarged base of one or more carpels.

oviduct (OH-vih-dukt) One of a pair of ducts through which eggs travel from an ovary to the uterus. Formerly called a fallopian tube.

ovulation (OHV-you-LAY-shun) Release of a secondary oocyte from an ovary.

ovule (OHV-youl) [L. *ovum*, egg] Tissue mass in a plant ovary that becomes a seed; a female gametophyte with egg cell, nutritious tissue, and a jacket that will become a seed coat.

ovum (OH-vum) Mature secondary oocyte.

oxaloacetate (ox-AL-oh-ASS-ih-tate) A four-carbon compound with roles in metabolism (e.g., the point of entry into the Krebs cycle).

oxidation–reduction reaction A transfer of an electron (and often an unbound proton, or H^+) between atoms or molecules.

oxygen debt Lowered O_2 level in blood when muscle cells have used up more ATP than they have formed by aerobic respiration.

oxyhemoglobin Oxygen bound to hemoglobin (HbO_2) inside red blood cells.

ozone thinning A pronounced and seasonal thinning of the atmosphere's ozone layer.

pain Perception of injury to some body region.

Paleozoic Era (570 to 240 mya) consisting of the Cambrian, Ordovician, Silurian, Devonian, Carboniferous, and Permian.

pancreas (PAN-cree-us) Gland with roles in digestion and organic metabolism. Secretes enzymes and bicarbonate into small intestine; also secretes insulin and glucagon.

pancreatic islet Any of the 2 million or so clusters of endocrine cells of the pancreas.

Pangea Paleozoic supercontinent upon which the first terrestrial plants and animals evolved.

parapatric speciation Idea that neighboring populations can become distinct species while maintaining contact along a common border.

parasite [Gk. *para*, alongside, + *sitos*, food] Organism that lives in or on a host for at least part of its life cycle. It feeds on specific tissues and usually does not kill its host outright.

parasitism Symbiotic interaction in which a species that feeds on its tissues (a parasite) benefits and the other (its host) is harmed.

parasympathetic nerve (PARE-uh-sim-pu-THET-ik) An autonomic nerve. Its signals slow down overall activities and divert energy to basic tasks; they also help make small adustments in internal organ activity by acting continually in opposition to sympathetic nerve signals.

parathyroid gland One of four glands, the secretions of which raise blood calcium levels.

parenchyma (par-ENG-kih-mah) Simple tissue that makes up the bulk of a plant; has roles in photosynthesis, storage, secretion, other tasks.

partial pressure Characteristic of a gas; its contribution to total atmospheric pressure.

passive transport Event in which a transport protein that spans a cell membrane passively permits a solute to diffuse through its interior. Also called facilitated diffusion.

pathogen (PATH-oh-jen) [Gk. *pathos*, suffering, + *genēs*, origin] Any virus, bacterium, fungus, protistan, or parasitic worm that can infect an organism, multiply in it, and cause disease.

pattern formation The ordered, sequential sculpting of embryonic cells into specialized animal tissues and organs; brought about by cytoplasmic localization and, later, inductive interactions among master genes, the products of which map out the basic body plan.

PCR Polymerase chain reaction. A method of enormously amplifying the quantity of DNA fragments cut by restriction enzymes.

peat bog Compressed, soggy, highly acidic mat of accumulated remains of peat mosses.

pedigree Diagram of the genetic connections among related individuals through successive generations; uses standardized symbols.

penis Male copulatory organ by which sperm is deposited in a female reproductive tract.

peptide hormone A hormone that binds to a membrane receptor, thus activating enzyme systems that alter a cell activity. A second messenger in the cell often relays its message.

perennial [L. *per-*, throughout, + *annus*, year] Plant that lives three or more growing seasons.

pericycle (PARE-ih-sigh-kul) [Gk. *peri-*, around, + *kyklos*, circle] One or more cell layers inside the endodermis; gives rise to lateral roots and contributes to secondary growth.

periderm Protective cover that replaces plant epidermis during extensive secondary growth.

peripheral nervous system (per-IF-ur-uhl) [Gk. *peripherein*, to carry around] All nerves leading into and out from the spinal cord and brain. Includes ganglia of those nerves.

peristalsis (pare-ih-STAL-sis) Recurring waves of contraction and relaxation of muscles in the wall of a tubular or saclike organ.

permafrost A water-impermeable, perpetually frozen layer, up to 500 meters thick, beneath the soil surface of the arctic tundra.

peroxisome A vesicle; its enzymes digest fatty acids and amino acids to hydrogen peroxide, which is then converted to harmless products.

PGA Phosphoglycerate (FOSS-foe-GLISS-er-ate) Important intermediate of glycolysis and of the Calvin–Benson cycle.

PGAL Phosphoglyceraldehyde. Intermediate of glycolysis and of the Calvin-Benson cycle.

pH scale Measure of the concentration of free hydrogen ions (H^+) in blood, water, and other solutions. pH 0 is the most acidic, 14 the most basic, and 7, neutral.

phagocytosis (FAG-oh-sigh-TOE-sis) [Gk. *phagein*, to eat, + *kytos*, hollow vessel] Of some cells, engulfment of an extracellular target by way of pseudopod formation and endocytosis.

pharynx (FARE-inks) Among invertebrates, a muscular tube to the gut. In some chordates, a gas exchange organ. In land vertebrates, a dual entrance to the esophagus and trachea.

phenotype (FEE-no-type) [Gk. *phainein*, to show + *typos*, image] Observable trait or traits of an individual that arise from gene interactions and gene–environment interactions.

pheromone (FARE-oh-moan) [Gk. *phero*, to carry, + *-mone*, as in hormone] Hormone-like, nearly odorless exocrine gland secretion. A signaling molecule between individuals of the same species that integrates social behavior.

phloem (FLOW-um) Plant vascular tissue. Live cells (sieve tubes) interconnect as conducting tubes for sugars and other solutes; companion cells help load solutes into the tubes.

phospholipid Organic compound that has a glycerol backbone, two fatty acid tails, and a hydrophilic head of two polar groups (one being phosphate). Phospholipids are the main structural component of cell membranes.

phosphorus cycle A sedimentary cycle; the movement of phosphorus (mainly phosphate ions) from land, through food webs, to ocean sediments, then back to land.

phosphorylation (FOSS-for-ih-LAY-shun) Means of activating molecules. An enzyme attaches inorganic phosphate to a molecule or mediates a phosphate-group transfer, as from ATP.

photoautotroph Photosynthetic autotroph; any organism that synthesizes its own organic compounds using CO_2 for carbon atoms and sunlight for energy. Nearly all plants, some protistans, and a few bacteria do this.

photolysis (foe-TALL-ih-sis) [Gk. *photos*, light, +*-lysis*, breaking apart] Reaction sequence in which photon energy splits water molecules. The released electrons and hydrogen are used in the noncyclic pathway of photosynthesis; the oxygen is released as a by-product.

photon One unit of energy of visible light.

photoperiodism Biological response to change in relative lengths of daylight and darkness.

photoreceptor Light-sensitive sensory cell.

photosynthesis Trapping of sunlight energy, followed by its conversion to chemical energy (ATP, NADPH, or both) and then synthesis of sugar phosphates, which become converted into sucrose, cellulose, starch, and other end products. The main pathway by which energy and carbon enter the web of life.

photosystem Cluster of many light-trapping pigments in a photosynthetic membrane.

phototropism [Gk. *photos*, light, + *trope*, a turning, direction] Change in the direction of cell movement or growth in response to light (e.g., as in a stem bending toward light).

phycobilin (FIE-koe-BY-lin) Type of accessory pigment; notably abundant in red algae and in cyanobacteria.

phylogeny Evolutionary relationships among species, starting with an ancestral form and including branches leading to its descendants.

phytochrome A light-sensitive pigment. Its controlled activation and inactivation affect plant hormone activities that govern leaf expansion, stem branching, stem lengthening, and often seed germination and flowering.

phytoplankton (FIE-toe-PLANK-tun) [Gk. *phyton*, plant, + *planktos*, wandering] Aquatic community of floating or weakly swimming photoautotrophs (e.g., "pastures of the seas").

pigment Any light-absorbing molecule.

pineal gland Light-sensitive endocrine gland; its melatonin secretions affect biological clocks, overall activity, and reproductive cycles.

pioneer species Any opportunistic colonizer of barren or disturbed habitats. Adapted for rapid growth and dispersal.

pith Of most dicot stems, ground tissue inside the ring of vascular bundles.

pituitary gland Endocrine gland that, with the hypothalamus, controls many physiological functions, including activity of many other endocrine glands. Its posterior lobe stores and secretes hypothalamic hormones. Its anterior lobe produces and secretes its own hormones.

placenta (plah-SEN-tuh) Blood-engorged organ of pregnant female placental mammals; made of endometrial tissue and extraembryonic membranes. Permits exchanges between the mother and fetus without intermingling their bloodstreams; sustains the new individual and allows its blood vessels to develop separately.

placoderm One of the earliest fishes with jaws and paired fins; extinct.

placozoan The simplest known animal; tiny, asymmetrically soft-bodied, two cell layers.

plankton [Gk. *planktos*, wandering] Of aquatic habitats, a community of suspended or weakly swimming organisms, mostly microscopic.

plant Generally, a multicelled photoautotroph with well-developed root and shoot systems; photosynthetic cells that include starch grains as well as chlorophylls *a* and *b*; and cellulose, pectin, and other polysaccharides in cell walls.

Plantae Kingdom of plants.

planula Larva, usually with ciliated epidermis. Swims or creeps; may resemble first animals.

plasma (PLAZ-muh) Liquid portion of blood; mainly water in which ions, proteins, sugars, gases, and other substances are dissolved.

plasma membrane Outermost cell membrane; structural and functional boundary between cytoplasm and the fluid outside the cell.

plasmid A small, circular molecule of extra bacterial DNA that carries a few genes and is replicated independently of the chromosome.

plasmodesma (PLAZ-moe-DEZ-muh), plural **plasmodesmata** Plant cell junction; a channel that connects the cytoplasm of adjoining cells.

plate tectonics Theory that great slabs (plates) of the Earth's outer layer float on a hot, plastic mantle. All plates are slowly moving and have rafted continents to new positions over time.

platelet (PLAYT-let) Megakaryocyte fragment that releases substances used in clot formation.

pleiotropy (PLEE-oh-troe-pee) [Gk. *pleon*, more, + *trope*, direction] Positive or negative effects on two or more traits owing to expression of alleles at a single gene locus. Effects may or may not emerge at the same time.

polar body One of four cells that has formed by the meiotic cell division of an oocyte but that does not become the ovum.

pollen grain [L. *pollen*, fine dust] Immature or mature, sperm-bearing male gametophyte of gymnosperms and angiosperms.

pollen sac Chamber inside a flower's anther in which pollen grains develop.

pollen tube Sperm-carrying tube that grows from a germinated pollen grain, through carpel tissues to the egg inside an ovule.

pollination Arrival of a pollen grain on the landing platform (stigma) of a flower's carpel.

pollinator Agent (wind, water, an animal) that puts pollen from male on female reproductive parts of flowers of the same plant species.

pollutant Natural or synthetic substance with which an ecosystem has no prior evolutionary experience, in terms of kinds or amounts; it accumulates to disruptive or harmful levels.

polymer (POH-lih-mur) [Gk. *polus*, many, + *meris*, part] Large molecule of three to millions of monomers of the same or different kinds.

polymerase (puh-LIM-ur-aze) Type of enzyme that catalyzes a polymerization reaction, as in DNA replication and repair.

polymorphism (poly-MORE-fizz-um) [Gk. *polus*, many, + *morphe*, form] The persistence of two or more qualitatively different forms of a trait (morphs) in a population.

polyp (POH-lip) Vase-shaped, sedentary stage of cnidarian life cycles.

polypeptide chain An organic compound of three or more amino acids joined by peptide bonds; atoms of its backbone have this pattern: —N—C—C—N—C—C—. All proteins are composed of one or more polypeptide chains.

polyploidy (POL-ee-PLOYD-ee) Having three or more of each type of chromosome in the nucleus of a eukaryotic cell at interphase.

polysaccharide [Gk. *polus*, many, + *sakcharon*, sugar] Straight or branched chain of many covalently linked sugar units of the same or different kinds. In nature, the most common types are cellulose, starch, and glycogen.

polytene chromosome Insect chromosome consisting of many in-parallel copies of DNA.

pons A hindbrain traffic center for signals between cerebellum and forebrain centers.

population All individuals of the same species that are occupying a specified area.

population density Count of individuals of a population in a habitat (by area or volume).

population distribution Dispersal pattern for individuals of a population through a habitat.

population size The number of individuals that make up the gene pool of a population.

positive control Use of regulatory proteins to promote gene expression.

positive feedback mechanism A homeostatic control mechanism. It sets in motion a chain of events that intensifies change from an original condition; after a limited time, intensification reverses the change.

potential energy A stationary object's capacity to do work owing to its position in space or to the arrangement of its parts.

predation Ecological interaction in which a predator feeds on a prey organism.

predator [L. *prehendere*, to grasp, seize] A heterotroph that eats other living organisms (its prey), does not live in or on them (as parasites do), and may or may not kill them.

prediction Statement about what you should observe in nature if you were to go looking for a particular phenomenon; the if–then process.

pressure flow theory Oganic compounds flow through phloem in response to pressure and concentration and pressure gradients between sources (e.g., leaves) and sinks (e.g., growing parts where they are being used or stored).

pressure gradient A difference in pressure between two adjoining regions.

prey Organism that another organism (e.g., a predator) captures as a food source.

primary growth Plant growth originating at root tips and shoot tips.

primary immune response Defensive actions by white blood cells and their secretions, as elicited by first-time recognition of antigen. Includes antibody- and cell-mediated responses.

primary producer Type of autotroph that secures energy directly from the environment and stores some in its tissues.

primary productivity, gross Of ecosystems, the rate at which primary producers capture and store a given amount of energy in their cells and tissues during a specified interval.

primary productivity, net Of ecosystems, the rate of energy storage in primary producer cells and tissues in excess of rate of aerobic respiration during a specified interval.

primary root First root of a young seedling.

primary wall A thin, flexible plant cell wall of cellulose, polysaccharides, and glycoproteins; allows growing cells to divide or change shape.

primate Mammalian lineage dating from the Eocene; includes prosimians, tarsioids, and anthropoids (monkeys, apes, and humans).

primer Short nucleotide sequence designed to base-pair with any complementary DNA sequence; acts as a START tag for replication.

prion Small infectious protein that causes rare, fatal degenerative diseases of nervous system.

probability The chance that each outcome of a given event will occur is proportional to the number of ways the outcome can be reached.

probe Very short stretch of DNA labeled with a radioisotope; it is designed to base-pair with part of a gene in a DNA sample being studied.

procambium (pro-KAM-bee-um) A meristem that gives rise to primary vascular tissues.

producer Autotroph (self-feeder); it nourishes itself using sources of energy and carbon from the physical environment. Photoautotrophs and chemoautotrophs are examples.

progesterone (pro-JESS-tuh-rown) Of female mammals, a sex hormone secreted by ovaries and the corpus luteum.

proglottid One of many tapeworm body units that bud in sequence, behind the scolex.

prokaryotic cell (pro-CARE-ee-oh-tic) [L. *pro*, before, + Gk. *karyon*, kernel] Archaebacterium or eubacterium; single-celled organism, most often walled; lacks the profusion of membrane-bound organelles observed in eukaryotic cells.

prokaryotic fission Cell division mechanism by which prokaryotic cells alone reproduce.

promoter Short stretch of DNA to which RNA polymerase can bind and start transcription.

prophase Of mitosis, a stage when duplicated chromosomes start to condense, microtubules form a spindle, and the nuclear envelope starts to break up. Duplicated pairs of centrioles (if present) are moved to opposite spindle poles.

prophase I First stage of meiosis I. Each duplicated chromosome starts to condense. It pairs with its homologue; nonsister chromatids usually undergo crossing over and are attached to spindle. One pair of duplicated centrioles (if present) is moved to the opposite spindle pole.

prophase II First stage of meiosis II. Spindle microtubules attach to each chromosome and move them to spindle's equator. A centriole pair (if present) is already at each spindle pole.

protein Organic compound of one or more polypeptide chains folded and twisted into a globular or fibrous shape, overall.

Protista Kingdom of protistans. Chytrids; water molds; slime molds; protozoans; sporozoans; euglenoids; chrysophytes; dinoflagellates; and red, brown, and green algae are major groups.

protistan (pro-TISS-tun) [Gk. *prōtistos*, primal] Photoautotroph or heterotroph (or both) unlike bacteria; some like earliest eukaryotic cells. Has a nucleus, larger ribosomes, mitochondria, ER, Golgi bodies, chromosomes with numerous proteins, and cytoskeletal microtubules. Range in size from microscopic algae to giant kelps.

proto-cell Hypothetic cell-like stage between chemical evolution and the first living cell.

proton Positively charged particle; one or more in the nucleus of each atom. An unbound (free) proton is called a hydrogen ion (H^+).

protostome (PRO-toe-stome) [Gk. *proto*, first, + *stoma*, mouth] Lineage of coelomate, bilateral animals that includes mollusks, annelids, and arthropods. The first indentation to form in protostome embryos becomes the mouth.

protozoan Type of protistan that may be like the single-celled heterotrophs that gave rise to animals. Amoeboid, animal-like, and ciliated protozoans are major categories.

proximal tubule Nephron's tubular portion extending from Bowman's capsule.

pseudopod A dynamically extending lobe of cytoplasm used for motility or engulfment.

puberty Post-embryonic stage; gametes start to mature and secondary sexual traits emerge.

pulmonary circuit Vertebrate cardiovascular route; O_2-poor blood flows from the heart to lungs, gets oxygenated, flows back to heart.

punctuation model, speciation Idea that most morphological changes occur in a brief span when populations start to diverge; speciation is rapid, and the daughter species change little for the next 2–6 million years or so.

Punnett-square method Construction of a simple diagram as a way to predict probable outcomes of a genetic cross.

pupa, plural, **pupae** An immature, post-embryonic stage of many insect life cycles.

purine A nucleotide base with a double ring structure (e.g., adenine or guanine).

pyrimidine (pih-RIM-ih-deen) A nucleotide base with a single ring structure (e.g., cytosine or thymine).

pyruvate (PIE-roo-vate) Organic compound with a backbone of three carbon atoms. Two molecules form as end products of glycolysis.

r Variable in population growth equations that signifies net population growth rate. Birth and death rates are assumed to remain constant and are combined into this one variable.

radial symmetry Animal body plan having four or more roughly equivalent parts around a central axis (e.g., sea anemone).

radiation Radiant energy such as sunlight or a surface warmer than body surface temperature; exposure to it can lead to a heat gain in the animal body.

radioisotope Unstable atom (uneven number of protons and neutrons). It spontaneously emits particles and energy, and so decays into a different atom over a predictable time span.

radiometric dating Method of measuring the proportions of (1) a radioisotope in a mineral trapped long ago in newly formed rock and (2) a daughter isotope that formed from it by radioactive decay in the same rock. Used to assign absolute dates to fossil-containing rocks and to the geologic time scale.

rain shadow A reduction in rainfall on the leeward side of a high mountain range that results in arid or semiarid conditions.

reabsorption In kidneys, diffusion or active transport of water and reclaimable solutes from nephrons into peritubular capillaries; amounts are controlled by ADH and aldosterone. At a capillary bed, osmotic movement of interstitial fluid into a capillary when water concentration between plasma and interstitial fluid differ.

reaction center The only molecule (a special chlorophyll *a*) that can pass electrons out of a photosystem, to a nearby acceptor molecule.

rearrangement, molecular Conversion of one organic compound to another through changes in its internal bonds.

receptor, molecular Protein at which some extracellular substance (e.g., a hormone) docks.

receptor, sensory Sensory cell or specialized cell adjacent to it that can detect a stimulus.

recessive allele [L. *recedere*, to recede] Allele whose expression in heterozygotes is fully or partially masked by expression of its partner. Fully expressed in homozygous recessives.

reciprocal cross A paired cross. In the first cross, one parent displays the trait of interest. In the second, the other parent displays it.

recombinant DNA Any DNA molecule that incorporates one or more nonparental base sequences. Outcome of microbial gene transfer in nature or of recombinant DNA technology.

recombinant DNA technology Procedures by which DNA molecules from different species are isolated, cut up, spliced together, and then enormously amplified to useful quantities.

recombination Any enzyme-mediated reaction that inserts one DNA sequence into another. Generalized recombination uses any pair of homologous sequences between chromosomes as substrates, as occurs at crossing over. Site-specific recombination uses a short stretch of homology between bacterial and viral DNA. A different reaction inserts transposons at new, random sites in bacterial or eukaryotic genomes; no homology is required.

red alga Type of photoautotrophic protistan; most are multicelled and aquatic; an abundance of phycobilins masks their chlorophyll *a*.

red blood cell Erythrocyte. Cell that functions in the rapid transport of oxygen in blood.

red marrow Site of blood cell formation in the spongy tissue of many bones.

reflex [L. *reflectere*, to bend back] Stereotyped, simple movement in response to stimuli. In the simplest reflex arcs, it results from sensory neurons synapsing directly on motor neurons.

refractory period Brief interval following an action potential when a small patch of neural membrane is insensitive to stimulation.

regulatory protein Component of mechanisms that control transcription, translation, and gene products by interacting with DNA, RNA, new polypeptide chains, or proteins (e.g., enzymes).

releaser Hypothalamic hormone; enhances or slows secretions from target cells in anterior lobe of pituitary gland.

renal corpuscle Bowman's capsule plus the glomerular capillaries it cups around.

repressor Protein that binds with an operator on bacterial DNA to block transcription.

reproduction Any process by which a parental cell or organism produces offspring. Among eukaryotes, asexual modes (e.g., binary fission, budding, vegetative propagation) and sexual modes. Prokaryotes use prokaryotic fission only. Viruses cannot reproduce themselves; host organisms execute their replication cycle.

reproductive base The number of actually and potentially reproducing individuals in a population.

reproductive isolating mechanism Heritable feature of body form, functioning, or behavior that prevents interbreeding between two or more genetically divergent populations.

reproductive success Production of viable offspring by the individual.

reptile Carnivorous species belonging to the first vertebrate lineage to escape dependency on free water by way of internal fertilization, efficient kidneys, amniote eggs, and other adaptations. Examples are dinosaurs (extinct), crocodilians, snakes, lizards, and tuataras.

resource partitioning Of two or more species that compete for the same resource, a sharing of the resource in different ways or at different times, which permits them to coexist.

respiration [L. *respirare*, to breathe] For all animals, exchange of environmental oxygen with CO_2 from cells (e.g., by integumentary exchange or by way of a respiratory system).

respiratory cycle One inhalation, exhalation.

respiratory surface Thin, moist epithelium that functions in gas exchange.

respiratory system Type of organ system that exchanges gases between the body and the environment and contributes to homeostasis.

resting membrane potential Of a neuron and other excitable cells, a voltage difference across the plasma membrane that holds steady in the absence of outside stimulation.

restoration ecology Attempts to reestablish biodiversity in ecosystems severely altered by mining, agriculture, and other disturbances.

restriction enzyme One of a class of bacterial enzymes that can cut apart foreign DNA that infects a cell, as by viral attack. Important tool of recombinant DNA technology.

reticular formation A mesh of interneurons extending from the upper spinal cord, through the brain stem, and into the cerebral cortex; it is a low-level pathway of information flow.

retina Dense array of photoreceptors in an eye.

reverse transcription Synthesis of DNA on an RNA template by reverse transcriptase, a viral enzyme. Basis of RNA virus replication cycle and of cDNA synthesis in the laboratory.

Rh blood typing Method of characterizing red blood cells by a certain membrane surface protein. Rh^+ cells have it; Rh^- cells do not.

rhizoid Simple rootlike absorptive structure of some fungi and nonvascular plants.

rhizome Short, branched, mainly horizontal underground stem of some plant species.

ribosomal RNA (rRNA) Type of RNA that combines with proteins to form ribosomes, on which polypeptide chains are assembled.

ribosome Two subunits of rRNA and proteins briefly joined together as a structure on which mRNA is translated into polypeptide chains.

riparian zone Narrow corridor of vegetation along a stream or river.

RNA Ribonucleic acid. Any of a class of single-stranded nucleic acids that function in transcribing and translating the genetic instructions encoded in DNA into proteins.

RNA polymerase Enzyme that catalyzes the assembly of RNA strands on DNA templates.

RNA world One model for prebiotic evolution in which RNA was the template for protein synthesis before the evolution of DNA.

rod cell Vertebrate photoreceptor sensitive to very dim light. Contributes to the coarse perception of movement across visual field.

root Plant part, typically belowground, that absorbs water and dissolved minerals, anchors aboveground parts, and often stores food.

root hair Thin extension of a young, specialized root epidermal. Increases root surface area for absorbing water and minerals.

root nodule Localized swelling on a root of certain legumes and other plants. Develops when nitrogen-fixing bacteria infect the plant, multiply, and become mutualists with it.

roundworm Cylindrical, bilateral, cephalized animal with a false coelom and a complete digestive system. Most cycle nutrients in communities; many are parasites.

rubisco RuBP carboxylase; an enzyme that catalyzes attachment of the carbon atom from CO_2 to RuBP and so starts the Calvin–Benson cycle of the light-independent reactions.

RuBP Ribulose bisphosphate; five-carbon organic compound; used for carbon fixation in the Calvin–Benson cycle, which regenerates it.

ruminant Hoofed, herbivorous mammal with multiple stomach chambers.

sac fungus Fungus that forms sexual spores in sac-shaped cells; e.g., truffles, morels.

salinization Salt buildup in soil through poor drainage, evaporation, and heavy irrigation.

saliva Glandular secretion that mixes with food and starts starch breakdown in mouth.

salt Compound that releases ions other than H^+ and OH^- in solution.

sampling error Use of a sample or subset of a population, an event, or some other aspect of nature for an experimental group that is not large enough to be representative of the whole.

saprobe Heterotroph that obtains energy and carbon from nonliving organic matter and so causes its decay (e.g., many fungal species).

sapwood Of older stems and roots, secondary growth in between the vascular cambium and heartwood; wet, usually pale, not as strong.

sarcomere (SAR-koe-meer) One of many basic units of contraction, defined by Z lines, that subdivide a muscle cell. Shortens in response to ATP-driven interactions between its parallel arrays of actin and myosin components.

sarcoplasmic reticulum (sar-koe-PLAZ-mik reh-TIK-you-lum) System of membrane-bound chambers; threads around muscle cells and takes up, stores, and releases the calcium ions necessary for contraction.

savanna Broad belt of warm grassland with a smattering of shrubs and trees.

scale Of fishes, small, bony plates that protect the body surface without weighing it down.

sclerenchyma (skler-ENG-kih-mah) Simple plant tissue that supports mature plant parts and commonly protects seeds. Most of its cells have thick, lignin-impregnated walls.

second law of thermodynamics A law of nature stating that the spontaneous direction of energy flow is from organized forms to less organized forms; with each conversion, some energy is randomly dispersed in a form (usually heat) not as useful for doing work.

second messenger Molecule within a cell that mediates a hormonal signal.

secondary sexual trait A trait associated with maleness or femaleness but with no direct role in reproduction (e.g., distribution of body hair and body fat). The primary sexual trait is the presence of male or female gonads.

secondary wall Of older plant cells no longer growing but in need of structural support, a wall on the inner surface of the primary wall. Contains lignin in older cells of woody plants.

secretion Release of a substance from cells or glands to the external environment, to the lumen of some organ, or to interstitial fluid.

sedimentary cycle Biogeochemical cycle. An element having no gaseous phase moves from land, through food webs, to the seafloor, then returns to land through long-term uplifting.

seed Mature ovule with an embryo sporophyte inside and integuments that form a seed coat.

seed bank A safe storage facility where genes of diverse plant lineages are being preserved.

segmentation Of animal body plans, a series of units that may or may not be similar to one another in appearance. Of tubular organs, an oscillating movement produced by rings of circular muscle in the tube wall.

segregation, theory of [L. *se-*, apart, + *grex*, herd] Mendelian theory. Sexually reproducing organisms inherit pairs of genes (on pairs of homologous chromosomes), the two genes of each pair are separated from each other at meiosis, and they end up in separate gametes.

selective gene expression Control of which gene products a cell makes or activates during a specified interval. Depends on the cell type, adjustments to changing chemical conditions, external signals, and built-in control systems.

selective permeability Of a cell membrane, a capacity to let some substances but not others cross at certain sites, at certain times owing to its bilayer structure and its transport proteins.

selfish behavior An individual protects or increases its own chance to produce offspring regardless of consequences to its social group.

semen (SEE-mun) The sperm-bearing fluid expelled from a penis during male orgasm.

semiconservative replication [Gk. *hēmi*, half, + L. *conservare*, to keep] Mechanism of DNA duplication. A DNA double helix unzips and a complementary strand is assembled on exposed bases of each strand. Each conserved strand and its new partner wind up together as a double helix; a half-old, half-new molecule.

seminiferous tubule One of the small coiled tubes in testes where sperm start forming.

senescence (sen-ESS-cents) [L. *senescere*, to grow old] Processes leading to the natural death of an organism or to parts of it (e.g., leaves).

sensation Conscious awareness of a stimulus.

sensory adaptation Decrease in response to a stimulus maintained at constant strength.

sensory neuron Type of neuron that detects a stimulus and relays information about it toward an integrating center (e.g., a brain).

sensory system The "front door" of a nervous system; it detects external and internal stimuli and relays information to integrating centers that issue commands for responses.

sex chromosome A chromosome with genes that affect sexual traits. Depending on the species, somatic cells have one or two sex chromosomes of the same or different type (e.g., in mammals, XX females, XY males).

sexual dimorphism Occurrence of female and male phenotypes among the individuals of a sexually reproducing species.

sexual reproduction Production of offspring by meiosis, gamete formation, and fertilization.

sexual selection A microevolutionary process; a type of selection that favors a trait giving an individual a competitive edge in attracting or keeping a mate (favors reproductive success).

shell model Model of electron distribution in which all orbitals available to electrons of atoms occupy a nested series of shells.

shifting cultivation Cutting and burning trees in a plot of land, followed by tilling ashes into soil. Once called slash-and-burn agriculture.

shoot system Aboveground plant parts (e.g., stems, leaves, and flowers).

sieve-tube member One of the cells that join together as phloem's sugar-conducting tubes.

sign stimulus Simple environmental cue that triggers a response to a stimulus, which the nervous system is prewired to recognize.

sink Any part of a plant in which cells are storing or using food (e.g., roots).

sister chromatid Of a duplicated chromosome, one of two DNA molecules (and associated proteins) attached at the centromere until they are separated from each other at mitosis or meiosis; each is then a separate chromosome.

six-kingdom classification scheme A recent phylogenetic scheme that groups all organisms into the kingdoms Eubacteria, Archaebacteria, Protista, Fungi, Plantae, and Animalia.

skeletal muscle Organ with many muscle cells bundled inside a sheath of connective tissue that extends past the muscle, as tendons.

skeletal muscle tissue Striated contractile tissue that is the functional partner of bone.

skin Of vertebrates, the integument plus structures derived from epidermal cells.

sliding filament model Model for how muscles contract, based on the ATP-driven interactions between myosin and actin filaments in each sarcomere of each muscle cell. The sarcomere shortens as actin filaments projecting from its two sides are made to slide toward its center.

small intestine Part of the vertebrate digestive system in which digestion is completed and most dietary nutrients are absorbed.

smog, industrial Polluted, gray-colored air in industrialized cities during cold, wet winters.

smog, photochemical Polluted, brown, smelly air that forms in warm weather in large cities that have many gas-burning vehicles.

smooth muscle tissue Nonstriated contractile tissue found in soft internal organs.

social behavior Diverse interactions among individuals of a species, which display, send, and respond to shared forms of communication that have genetic and learned components.

sodium–potassium pump Type of membrane transport protein that, when activated by ATP, selectively transports potassium ions across a membrane against its concentration gradient, and passively allows sodium ions to cross in the opposite direction.

softwood Wood with tracheids but no vessels or fibers. Weaker, less dense than hardwood.

soil Mixture of mineral particles of variable sizes and decomposing organic material; air and water occupy spaces between particles.

solute (SOL-yoot) [L. *solvere*, to loosen] Any substance dissolved in a solution.

solvent Any fluid (e.g., water) in which one or more substances are dissolved.

somatic cell (so-MAT-ik) [Gk. *somā*, body] Any body cell that is not a germ cell. (Germ cells are the forerunners of gametes.)

somatic nervous system Nerves leading from a central nervous system to skeletal muscles.

somatic sensation Perception of touch, pain, pressure, temperature, motion, or positional changes of body parts.

somite One of many paired segments in a vertebrate embryo that give rise to most bones, skeletal muscles of head and trunk, and dermis.

source Any plant part where photosynthetic cells are making organic compounds.

special sense Vision, hearing, olfaction, or another sensation that arises from a particular location, such as the eyes, ears, or nose.

speciation (spee-see-AY-shun) The formation of a daughter species from a population or subpopulation of a parent species by way of microevolutionary processes. Routes vary in their details and duration.

species (SPEE-sheez) [L. *species*, a kind] One kind of organism. Of sexually reproducing organisms, one or more natural populations in which individuals are interbreeding and are reproductively isolated from other such groups.

specific name A name that, combined with a genus name, designates one species alone.

sperm [Gk. *sperma*, seed] Mature male gamete.

spermatogenesis (sper-MAT-oh-JEN-ih-sis) Formation of mature sperm from a germ cell.

sphere of hydration A clustering of water molecules around the molecules or ions of a substance placed in water owing to positive and negative interactions among them.

sphincter A ring of smooth muscles that can alternatively contract and relax to close off and open a passageway to the body surface.

spinal cord Part of central nervous system, in a canal inside the vertebral column; site of direct reflex connections between sensory and motor neurons; has tracts to and from the brain.

spindle apparatus Dynamic, temporary array of microtubules that moves chromosomes in precise directions during mitosis or meiosis.

spleen A lymphoid organ that is a filtering station for blood, a reservoir of red blood cells, and a reservoir of macrophages.

sponge Animal with no body symmetry, and no tissues, and phagocytic collar cells in its body wall. Lineage dates from precambrian.

spore A reproductive or resting structure of one or a few cells, often walled or coated; used for resisting harsh conditions, dispersal, or both. May be nonsexual or sexual, as formed by way of meiosis. Some bacteria as well as sporozoans, fungi, and plants form spores.

sporophyte [Gk. *phyton*, plant] Vegetative body that produces spore-bearing structures; grows by mitotic cell divisions from a plant zygote.

sporozoan Type of parasitic protistan. Forms a motile infective stage inside specific host cells; some form cysts. One end of the body has a complex of structures used to penetrate hosts.

spring overturn Of many bodies of water (e.g., large lakes in temperate zones) in spring, a downward movement of oxygenated surface water and upward movement of nutrient-rich water from below; fans primary productivity.

S-shaped curve Type of diagrammatic curve that emerges when logistic population growth is plotted against time.

stabilizing selection Mode of natural selection by which intermediate phenotypes in the range of variation are favored and extremes at both ends are eliminated.

stamen (STAY-mun) A male reproductive part of a flower (e.g., anther with stalked filament).

start codon Base triplet in mRNA that serves as the START signal for translation.

statolith A gravity-sensing mechanism based on clusters of particles.

STD One of the sexually transmitted diseases.

stem cell Self-perpetuating, undifferentiated animal cell. Some of its daughter cells are self-perpetuating; others become specialized (e.g., erythrocytes from stem cells in bone marrow).

steroid hormone Lipid-soluble hormone made from cholesterol that acts on a target cell DNA.

sterol (STAIR-all) Lipid with a rigid backbone of four fused carbon rings (e.g., cholesterol). Sterols differ in the number, position, and type of their functional groups.

stigma Sticky or hairy surface tissue on upper part of a carpel (or fused carpels) that captures pollen grains and favors their germination.

stimulus [L. *stimulus*, goad] A specific form of energy (e.g., pressure, light, and heat) that activates a sensory receptor able to detect it.

stoma (STOW-muh), plural **stomata** [Gk. *stoma*, mouth] A gap between two guard cells in leaf or stem epidermis. Opens or closes to control CO_2 movement into a plant and H_2O and O_2 out of it. Stomata help plants conserve water.

stomach Muscular, stretchable sac that mixes and stores ingested food, helps break it apart mechanically and chemically, and controls its expulsion (e.g., into the small intestine).

stop codon Base triplet in mRNA; STOP signal in translation that blocks further additions of amino acids to a new polypeptide chain.

strain One of two organisms with differences that are too minor to classify it as a separate species (e.g., *Escherichia coli* strain 018:K1:H).

stratification Stacked layers of sedimentary rock, built up by gradual deposition of volcanic ash, silt, and other materials over time.

stream A flowing-water ecosystem that starts out as a freshwater spring or seep.

strip logging A way to minimize downside of deforestation; only a very narrow corridor that parallels contours of sloped land is cleared; the upper part is used as a road to haul away logs, then is reseeded from intact forest above it.

strobilus Of many nonflowering plants, a cluster of spore-producing structures.

stroma [Gk. *strōma*, bed] A semifluid matrix between the thylakoid membrane system and two outer membranes of a chloroplast; a zone where sucrose, starch, cellulose, and other end products of photosynthesis are assembled.

stromatolite Fossilized mats of shallow-water microbial communities, mainly cyanobacteria, from Archean to Precambrian. Cell secretions blocked UV radiation but trapped sediments, and new mats grew on old ones; some are half a mile thick and hundreds of miles across.

substrate Reactant or precursor for a specific enzyme-mediated metabolic reaction.

substrate-level phosphorylation The direct, enzyme-mediated transfer of a phosphate group from a substrate to a molecule, as when an intermediate of glycolysis is made to give up a phosphate group to ADP, forming ATP.

succession, primary (suk-SESH-un) [L. *succedere*, to follow after] Ecological pattern by which a community develops in orderly progression, from the time that pioneer species colonize a barren habitat to the climax community.

succession, secondary Ecological pattern by which a disturbed area of a community recovers and moves back toward the climax state.

surface-to-volume ratio Mathematical relation in which volume increases with the cube of the diameter, but surface area increases only with the square. The volume of a cell expanding in diameter only would increase faster than the surface area of the plasma membrane servicing it; a major constraint on cell size and shape.

survivorship curve Plot of age-specific survival of a group of individuals in the environment, from the time of birth until the last one dies.

swim bladder Adjustable flotation device that helps many fishes maintain neutral buoyancy in water; its volume changes as it exchanges gases with blood.

symbiosis (sim-by-OH-sis) [Gk. *sym*, together, + *bios*, life, mode of life] Individuals of one species live near, in, or on those of another species for at least part of life cycle (e.g., in commensalism, mutualism, and parasitism).

sympathetic nerve An autonomic nerve; deals mainly with increasing overall body activities at times of heightened awareness, excitement, or danger; works continually in opposition with parasympathetic nerves to make minor adjustments in internal organ activities.

sympatric speciation [Gk. *sym*, together, + *patria*, native land] A speciation event within the home range of an existing species, in the absence of a physical barrier. Such species may form instantaneously, as by polyploidy.

synaptic integration (sin-AP-tik) Moment-by-moment combining of all excitatory and inhibitory signals arriving at the trigger zone of a neuron or some other excitable cell.

syndrome A set of symptoms that may not individually be a telling clue but collectively characterize a genetic disorder or disease.

systemic circuit (sis-TEM-ik) Of vertebrates, a cardiovascular route in which oxygen-enriched blood flows from the heart through the rest of the body (where it gives up oxygen and takes up carbon dioxide), then back to the heart.

T cell T lymphocyte; a type of white blood cell vital to immune responses (e.g., helper T cells and cytotoxic T cells).

tandem repeat One of many short sequences of DNA, occurring one after the other, in a chromosome. Used in DNA fingerprinting.

taproot system A primary root together with all of its lateral branchings.

target cell Any cell with molecular receptors that can bind a particular signaling molecule.

taxonomy Field of biology that deals with identifying, naming, and classifying species.

TCR Antigen-binding receptor of T cells.

tectum Midbrain's roof. Coordinating center for most sensory inputs and initiating motor responses in fishes and amphibians. In most vertebrates (not mammals), a reflex center that relays sensory input to forebrain.

telophase (TEE-low-faze) Of meiosis I, a stage when one member of each pair of homologous chromosomes has arrived at a spindle pole. Of mitosis and of meiosis II, the stage when chromosomes decondense into threadlike structures and two daughter nuclei form.

temperature A measure of the kinetic energy of ions or molecules in a specified region.

temperature zone Globe-spanning bands of temperature defined by latitude (e.g., cool temperate, warm temperate, equatorial).

tendon A cord or strap of dense connective tissue that attaches a muscle to bone.

territory An area that an animal is defending against competitors for mates, food, water, living space, and other resources.

test A means to determine the accuracy of a prediction, as by conducting experimental or observational tests and by developing models. Scientific tests are conducted under controlled conditions in nature or the laboratory.

testcross Experimental cross to determine whether an individual of unknown genotype that shows dominance for a trait is either homozygous dominant or heterozygous.

testis, plural **testes** A type of primary male reproductive organ (gonad); it produces male gametes and sex hormones.

testosterone (tess-TOSS-tuh-rown) A type of sex hormone necessary for the development and functioning of the male reproductive system of vertebrates.

tetanus (TET-uh-nuss) Of a muscle, a large contraction; repeated stimulation of a motor unit causes muscle twitches to run together. In a disease by the same name, a toxin prevents muscles from being released from contraction.

thalamus (THAL-uh-muss) A forebrain region that is a coordinating center for sensory input and a relay station for signals to the cerebrum.

theory, scientific An explanation of the cause or causes of range of related phenomena. It has been rigorously tested but is still open to tests, revision, and tentative acceptance or rejection.

thermal inversion A layer of dense, cool air trapped beneath a layer of warm air; can hold air pollutants close to the ground.

thermoreceptor Sensory cell or specialized cell next to it that detects radiant energy (heat).

thigmotropism (thig-MOE-truh-pizm) [Gk. *thigm*, touch] Orientation of the direction of growth in response to physical contact with a solid object (e.g., a vine curls around a post).

threshold Of excitable cells (e.g., a neuron or muscle cell), the minimum amount of change in the resting membrane potential that causes an action potential.

thylakoid Internal portion of a chloroplast's membrane system, often folded into flattened sacs, that forms a single compartment. Light-trapping pigments, and enzymes used to form ATP, NADPH, or both, are embedded in it.

thymine A nitrogen-containing base; one of the nucleotides in DNA (not in RNA).

thymus gland Lymphoid organ; its hormones influence T cells, which form in bone marrow but migrate to and differentiate in the thymus.

thyroid gland Endocrine gland; its hormones influence overall growth, development, and metabolic rates of warm-blooded animals.

tight junction Cell junction where strands of fibrous proteins oriented in parallel with a tissue's free surface collectively block leaks between the adjoining cells.

tissue Of multicelled organisms, a group of cells and intercellular substances that function together in one or more specialized tasks.

tissue culture propagation Inducing a new organism to grow from a cell of a parent tissue.

tonicity (toe-NISS-ih-TEE) The relative solute concentrations of two fluids (e.g., cytoplasmic fluid relative to extracellular fluid).

topsoil Uppermost soil layer, the one most essential for plant growth.

total fertility rate (TFR) The average number of children being born to women of a specified population during their reproductive years.

touch-killing Mechanism by which cytotoxic T cells directly release perforins and toxins onto a target cell and cause its destruction.

toxin Normal metabolic product of a species that can hurt or kill a different species.

trace element Any element that represents less than 0.01 percent of body weight.

tracer Substance with an attached radioisotope that researchers can track after delivering it into a cell, body, ecosystem, or some other system. Its emissions are detected as it moves through a pathway or reaches a destination.

trachea (TRAY-kee-uh), plural **tracheae** An air-conducting tube of respiratory systems. Of land vertebrates, the windpipe through which air passes between the larynx and bronchi.

tracheal respiration Of certain invertebrates (e.g., insects), respiration by way of finely branching tubes that start at openings in the integument and dead-end in body tissues.

tracheid (TRAY-kid) One of two types of cells in xylem that conduct water and mineral ions.

tract A cordlike bundle of axons of sensory neurons, motor neurons, or both inside the brain or spinal cord. Comparable to a nerve.

transcription [L. *trans*, across, + *scribere*, to write] First stage of protein synthesis. An RNA strand is assembled on exposed bases of an unwound strand of a DNA double helix. The transcript has a complementary base sequence.

transfer RNA (tRNA) An RNA that binds with and delivers amino acids to a ribosome and that pairs with an mRNA codon during the translation stage of protein synthesis.

translation Stage of protein synthesis when an mRNA's base sequence becomes converted to the amino acid sequence of a new polypeptide chain by rRNA, tRNA, and mRNA interactions.

translocation Of cells, movement of a stretch of DNA to a new chromosomal location with no molecular loss. Of vascular plants, distribution of organic compounds by way of phloem.

transpiration Evaporative water loss from a plant's aboveground parts, leaves especially.

transport protein Type of membrane protein used in active or passive transport of water-soluble substances across the lipid bilayer of a cell membrane. The solutes move across the membrane through the protein's interior.

transposon A stretch of DNA that can move spontaneously and at random to a different location in a genome; may inactivate genes and cause changes in phenotype.

triglyceride (neutral fat) A type of lipid that has three fatty acid tails attached to a glycerol backbone. Triglycerides are the body's most abundant lipids and its richest energy source.

trophic level (TROE-fik) All organisms the same number of transfer steps away from the energy input into an ecosystem.

tropical rain forest A biome characterized by regular, heavy rainfall, an annual mean temperature of 25°C, humidity of 80+ percent, and stunning but threatened biodiversity.

true breeding lineage Of sexually reproducing species, a lineage in which only one version of a trait appears over the generations in all parents and their offspring.

tumor Tissue mass with cells dividing at an abnormally high rate. If benign, cells stay in the home tissue; if malignant, they metastasize.

tundra Biome with poor drainage, extremely low temperatures, short growing season, poor decomposition; arctic tundra is a treeless plain between the polar ice cap and belts of northern boreal forests; alpine tundra occurs at high elevations in mountains throughout the world.

tunicate A baglike, filter-feeding urochordate.

turgor pressure (TUR-gore) Type of internal fluid pressure; it acts against a cell wall when water moves into the cell by osmosis.

ultrafiltration Bulk flow of a small amount of protein-free plasma from a blood capillary when the outward-directed effect of blood pressure exceeds the inward-directed osmotic movement of interstitial fluid.

ultraplankton Photosynthetic bacteria, no more than 2µ wide, that may account for 70 percent of the ocean's primary productivity.

uniformity theory Theory that Earth's surface changes in slow, uniformly repetitive ways except for catastrophes that occur annually, such as large earthquakes and floods. Helped change Darwin's view of evolution. Has since been replaced by plate tectonics theory.

upwelling An upward movement of deep, nutrient-rich water along coasts; it replaces surface waters that move away from shore when the prevailing wind direction shifts.

uracil (YUR-uh-sill) Nitrogen-containing base of a nucleotide in RNA but not DNA. Like thymine, uracil can base-pair with adenine.

urea Waste product formed in the liver from ammonia (derived from protein breakdown) and CO_2; excreted in urine.

ureter One of a pair of tubes that conduct urine from the kidneys to the urinary bladder.

urethra Urine-conducting tube from urinary bladder to an opening at the body's surface.

urinary excretion Mechanism by which excess water and solutes are removed from the body through the urinary system.

urinary system Organ system that adjusts the volume and composition of blood, and thereby helps maintain extracellular fluid.

urine Fluid consisting of any excess water, wastes, and solutes; it forms in kidneys by filtration, reabsorption, and secretion.

uterus (YOU-tur-us) [L. *uterus*, womb] Of a female placental mammal, a muscular, pear-shaped organ in which embryos are contained and nurtured during pregnancy.

vaccination Immunization procedure against a specific pathogen.

vaccine An antigen-containing preparation designed to increase immunity to diseases.

vagina Of female mammalian reproductive system, organ that receives sperm, forms part of birth canal, and channels menstrual flow.

variable Of an experimental test, a specific aspect of an object or event that may differ over time and among individuals. A single variable is directly manipulated in an attempt to support or disprove a prediction.

vascular bundle Array of primary xylem and phloem in multistranded, sheathed cords that thread lengthwise in the ground tissue system.

vascular cambium A lateral meristem that increases stem or root diameter.

vascular cylinder Arrangement of vascular tissues as a central cylinder in roots.

vascular plant Plant with xylem, phloem, and usually well-developed roots, stems, and leaves.

vascular tissue system Xylem and phloem, the conducting tissues that distribute water and solutes through a vascular plant.

vasoconstriction The diameter of one or more blood vessels decreases in response to stimuli.

vasodilation The diameter of one or more blood vessels increases in response to stimuli.

vegetative growth A new plant grows from an extension or fragment of its parent.

vein Of a cardiovascular system, any of the large-diameter vessels that lead back to the heart. Of leaves, one of the vascular bundles that thread through photosynthetic tissues.

venule A small blood vessel that serves as a transitional conducting tube between a small-diameter capillary and a larger diameter vein.

vertebra, plural **vertebrae** One of a series of hard bones that serve as a backbone and that protect the spinal cord.

vertebrate Animal having a backbone.

vesicle (VESS-ih-kul) [L. *vesicula*, little bladder] One of various small, membrane-bound sacs in cytoplasm that function in the transport, storage, or digestion of substances.

vessel member Type of cell in xylem; dead at maturity, but its wall becomes part of a water-conducting pipeline (a vessel).

vestibular apparatus Organ of equilibrium.

vestigial (ves-TIDJ-ul) Applies to a small body part, tissue, or organ abnormally developed or degenerated and unable to function like its normal counterpart (e.g., vestigial wings of mutant fruit flies; "tail bones" of humans).

villus (VIL-us), plural **villi** Any of several fingerlike absorptive structures projecting from the free surface of an epithelium.

viroid Infectious particle of short, tightly folded strands or circles of RNA.

virus A noncellular infectious agent made of DNA or RNA, a protein coat and, in some, an outer lipid envelope; it can be replicated only after its genetic material enters a host cell and subverts the host's metabolic machinery.

viscera All soft organs inside an animal body (e.g., heart, lungs, and stomach).

vision Perception of visual stimuli. Light is focused on a dense layer of photoreceptors that sample details of a light stimulus, which are sent to the brain for image formation.

visual accommodation Adjustments in a lens position or shape that focus light on a retina.

visual field The part of the outside world that an animal sees.

visual signal An observable action or cue that functions as a communication signal.

vitamin Any of more than a dozen organic substances that an organism requires in small amounts for metabolism but that it generally cannot synthesize for itself.

vocal cord Of certain animals, one of the thickened, muscular folds of the larynx that help produce sound waves for vocalization.

warning coloration Of many toxic species and their mimics, avoidance signals (strong colors and patterns) that predators learn to recognize.

water potential Sum of two opposing forces (osmosis and turgor pressure) that can cause a directional movement of water into or out of an enclosed volume (e.g., a walled cell).

water table Upper limit at which ground in a specified region is fully saturated with water.

water–vascular system Of sea stars and sea urchins, a system of many tube feet that are deployed in synchrony for smooth locomotion.

watershed Any specified region in which all precipitation drains into one stream or river.

wavelength A wavelike form of energy in motion. The horizontal distance between the crests of every two successive waves.

wax Organic compound; long-chain fatty acids packed together and attached to long-chain alcohols or carbon rings. Waxes have a firm consistency and repel water.

white blood cell A leukocyte (e.g., eosinophil, neutrophil, macrophage, T or B cell) that, with chemical mediators, counters tissue invasion and tissue damage. Different kinds take part in nonspecific and specific defense responses.

white matter Part of brain and spinal cord; its axons, in glistening white sheaths, specialize in rapid signal transmission.

wild-type allele Allele that occurs normally or with the greatest frequency at a given gene locus among individuals of a population.

wing A body part that functions in flight, as among birds, bats, and many insects.

X chromosome A type of sex chromosome. An XX mammalian embryo becomes female; an XY pairing causes it to develop into a male.

X chromosome inactivation One of the two X chromosomes in somatic cells of mammalian females condenses, which inactivates most of its genes. A dosage compensation mechanism.

X-linked gene A gene on an X chromosome.

X-linked recessive inheritance Recessive condition in which the responsible, mutated gene is on the X chromosome.

xylem (ZYE-lum) [Gk. *xylon*, wood] Of vascular plants, a complex tissue that conducts water and solutes through pipelines of interconnected walls of cells, which are dead at maturity.

Y chromosome Distinctive chromosome in males or females of many species, but not both (e.g., human males XY; human females, XX).

yellow marrow A fatty tissue in the cavities of most mature bones that produces red blood cells when blood loss from the body is severe.

Y-linked gene Gene on a Y chromosome.

yolk Protein- and lipid-rich substance that nourishes embryos in animal eggs.

yolk sac Extraembryonic membrane. In most shelled eggs, it holds nutritive yolk; in humans part becomes a site of blood cell formation and some cells give rise to forerunners of gametes.

zero population growth Of a population, no overall increase or decrease during a specified interval; population size is stabilized.

zooplankton A community of suspended or weakly swimming heterotrophs of aquatic habitats. Most species are microscopic.

zygomycetes Parasitic or saprobic fungus that forms a thick-walled sexual spore (a diploid zygote) in a thin cover (the zygosporangium).

zygote (ZYE-goat) First cell of a new individual, formed by fusion of a sperm nucleus with egg nucleus at fertilization; a fertilized egg.

CREDITS AND ACKNOWLEDGMENTS

This page is an extension of the copyright page.

We thank the following authors, agents, and publishers for permission to use the material listed. We also thank graphics artists for the final rendering of original illustrations developed by Cecie Starr and, more recently, Lisa Starr, in collaboration with general advisers, researchers, and content reviewers.

TABLE OF CONTENTS: **iv** (left to right) Derrell Fowler; Gary Head; model by Dr. David B. Goodin, The Scripps Research Institute / **v** Larry West/FPG / **vi** Dr. Pascale Madaule, France / **vii** (left) From "Multicolor Spectral Karyotyping of Human Chromosomes," by E. Schrock, T. Ried, et al, *Science*, 26 July 1966, 273:495; (right) William E. Ferguson / **viii** (left) Christopher Ralling; (right) Painting, Charles Knight (No. 2425), Courtesy Department of Library Services/American Museum of Natural History / **ix** (left to right) K. G. Murty/Visuals Unlimited; David M. Phillips/Visuals Unlimited; © Sherry K. Pittam; Edward S. Ross / **x** (left to right) Courtesy of Department of Library Services/American Museum of Natural History (Neg. K10273); P. J. Bryant, University of California, Irvine/BPS; Jean Paul Tibbles; Gary Head / **xi** David Cavagnaro/Peter Arnold; David M. Phillips/Visuals Unlimited; John Lotter Gurling/Tom Stack & Associates; David Macdonald / **xii** (left to right) Eric A. Newman; Lennart Nilsson © Boehringer Ingelheim International GmBH; Kenneth Garrett/National Geographic Image Collection; Prof. P. Motta/Dept. of Anatomy/University "La Sapienza," Rome/Science Photo Library/Photo Researchers / **xiii** (left to right) Courtesy The New York Academy of Medicine Library; Lisa Starr with photograph © 2001 PhotoDisc, Inc.; NSIBC/SPL/Photo Researchers; Dr. Douglas Coleman, The Jackson Library / **xiv** (left to right) Gunter Ziesler/Bruce Coleman; Ed Reschke; John H. Gerard; From Lennart Nilsson, *A Child Is Born*, © 1966, 1977 Dell Publishing Company, Inc., / **xv** (left to right) Nigel Cook/Dayton Beach News Journal/Corbis Sygma; Pieter Johnson; Stanley Sessions/Hartwick College; Alan and Sandy Carey / **xvi** Frans Lanting/Minden Pictures

INTRODUCTION NASA Goddard Space Flight Center

CHAPTER 1 **1.1** John McColgan, Bureau of Land Management, Alaska Fire Service / **1.2** Lisa Starr / **1.3** Photograph, (all) Jack de Coningh / **1.4** © Y. Arthrus-Bertrand/Peter Arnold / **1.5** Art, Gary Head; Photographs, (left) Paul DeGreve/FPG; (right) Norman Meyers/Bruce Coleman / **1.6** Art, Lisa Starr; Photographs, (far left) Walt Anderson/Visuals Unlimited; (b) Gregory Dimijian/Photo Researchers; (c) Alan Weaving/Ardea, London / **1.7** (a) R. Robinson/Visuals Unlimited; (b) Tony Brain/SPL/Photo Researchers; (c) M. Abbey/Visuals Unlimited; (d) Edward S. Ross; (e) Gary Head; (f) Edward S. Ross; (g, h) Pat and Tom Leeson/Photo Researchers / **1.8** J. A. Bishop and L. M. Cook / **1.9** Courtesy Derrell Fowler, Tecumseh, OK / **1.10** Art, Gary Head / **1.11** Photographs, (all) Gary Head / **1.12** CNRI/SPL/Photo Researchers / **1.13** Art, Gary Head and Preface, Inc.; Photographs, Dr. Douglas Coleman, The Jackson Laboratory / **1.14** James Carmichael Jr./NHPA / **page 19** © Cabisco/Visuals Unlimited

CHAPTER 2 **2.1** Gary Head for Norman Terry, University of California, Berkeley / **2.2** Jack Carey / **2.3** Lisa Starr / **2.4** (Thyroid scans) Hank Morgan/Rainbow, computer-enhanced by Lisa Starr; Photograph, Gary Head / **2.5** (a) © John Griffin/MediChrome; (b, c) Art, Raychel Ciemma; (d) Dr. Harry T. Chugani, M.D., UCLA School of Medicine / **2.6** Micrograph, Maris and Cramer / **2.7, 2.8** Lisa Starr / **2.9** Lisa Starr and Gary Head / **2.10** (a, b) Art, Lisa Starr; Micrograph, © Bruce Iverson / **page 26** Lisa Starr / **2.11** Vandystadt/Photo Researchers / **2.12** Lisa Starr / **2.13** Photograph, Steve Lissau/Rainbow; Art, Raychel Ciemma / **2.14** © Kennan Ward/Corbis Stock Market; Art, Raychel Ciemma / **2.15** (a) H. Eisenbeiss/Frank Lane Picture Agency; (b) Lisa Starr / **2.16** Lisa Starr / **2.17** Lisa Starr, using photographs © 2000 Photo Disc, Inc. / **2.18** Michael Grecco/Picture Group

CHAPTER 3 **3.1** (left) NASA; (right) Dave Schiefelbein / **page 35** © Ron Sanford/Michael Agliolo/Photo Researchers / **3.2** Art, Gary Head / **page 36** Art, Precision Graphics and Lisa Starr / **3.3** Art, Lisa Starr; Photograph, Tim Davis/Photo Researchers / **3.4, 3.5** Lisa Starr / **3.6** Art and Photograph, Lisa Starr / **3.7** (a, b) Lisa Starr; (c) Photograph, David Scharf/Peter Arnold / **3.8** Art, Precision Graphics / **3.9** Clem Haagner/Ardea, London / **3.10** Art, Precision Graphics; (c) Micrograph, Lewis L. Lainey; (d) Larry Lefever/Grant Heilman Photography; (e) Kenneth Lorenzen / **page 41** Lisa Starr / **3.11** (a) Ron Davis/Shooting Star; (b) Frank Trapper/Corbis Sygma / **3.12–3.15** Art, Precision Graphics / **3.16, 3.17** Lisa Starr / **3.18** (a) CNRI/SPL/Photo Researchers; Art, (b) Robert Demarest; (c) Lisa Starr / **page 46** Art, Gary Head / **3.19** Art, Precision Graphics / **3.20** Lisa Starr / **3.21** Photographs, (a) © Inga Spence/Tom Stack & Associates; (inset) Gary Head; Art, (b–d) Gary Head and Lisa Starr

CHAPTER 4 **4.1** (a) © Bettmann/Corbis; (b) Armed Forces Institute of Pathology; (c) National Library of Medicine; (e) Francis A. Countway Library of Medicine / **4.2** (a) Manfred Kage/Bruce Coleman; (b) George J. Wilder/Visuals Unlimited / **4.3–4.5** Art, Raychel Ciemma / **page 54** Leica Microsystems, Inc., Deerfield, IL. / **4.6** Art, Raychel Ciemma; Micrograph, Driscoll, Youngquist, and Baldeschwieler/CalTech/Science Source/Photo Researchers / **4.7** (a–d) Jeremy Pickett–Heaps, School of Botany, University of Melbourne / **4.8** Art, Raychel Ciemma and Precision Graphics / **4.9** Art, Raychel Ciemma and Preface, Inc.; Micrograph, M. C. Ledbetter, Brookhaven National Laboratory / **4.10** Art, Raychel Ciemma and Preface, Inc.; Micrograph G. L. Decker / **4.11** Micrographs, Stephen Wolfe / **4.12** (a-left) Don W. Fawcett/Visuals Unlimited; (a-right) A. C. Faberge, *Cell and Tissue Research*, 151: 403–415, 1974; (b) Art, Lisa Starr / **4.13** Art, Raychel Ciemma and Precision Graphics / **4.14** Art, Raychel Ciemma; Micrographs (a, b) Don W. Fawcett/Visuals Unlimited / **4.15** Art, (left) Raychel Ciemma; (right) Robert Demarest after a model by J. Kephart; Micrograph, Gary Grimes / **4.16** Micrograph, Keith R. Porter; Art, Raychel Ciemma / **4.17** Art, (above) Lisa Starr; Micrograph, L. K. Shumway / **4.18** Courtesy Dr. Vincenzo Cirulli, Laboratory of Developmental Biology, The Whittier Institute for Diabetes, University of California-San Diego, La Jolla, California / **4.19** Lisa Starr / **4.20** Photographs (a) © Lennart Nilsson; (b) CNRI/SPL/Photo Researchers; Art, Precision Graphics after Stephen Wolfe, *Molecular and Cellular Biology*, Wadsworth, 1993 / **4.21** Ronald Hoham, Dept. of Biology, Colgate University / **4.22** Photographs, (a) Walter Hodge/Peter Arnold; (c) courtesy Burlington Mills; Micrograph, Biophoto Associates/ Photo Researchers; Art, Raychel Ciemma / **4.23** Microscopy, (a) George S. Ellmore; (b) Ed Reschke / **4.24** Art, Raychel Ciemma and Lisa Starr / **4.25** (a) Art, Lisa Starr; (b) Courtesy Dr. G. Cohen–Bazire; (c) K. G. Murti/Visuals Unlimited; (d) R. Calentine / Visuals Unlimited; (e) Gary Gaard, Arthur Kelman / **4.28** © Abraham Menashe

CHAPTER 5 **5.1** (a, b) Models by Dr. David B. Goodin, The Scripps Research Institute / **5.2** Photographs, Gary Head; Art, Raychel Ciemma / **5.3** Evan Cerasoli / **5.4** Photographs, (above) NASA; (below) Manfred Kage/Peter Arnold / **5.5** (a–d) Lisa Starr, using photographs © 1997, 1998 Jet Propulsion Laboratory/California Institute of Technology and NASA, 1972 / **5.6** Art, Raychel Ciemma and Nadine Sokol / **5.7** Lisa Starr / **5.8** Art, Raychel Ciemma / **5.9** Lisa Starr / **5.10** Art, Gary Head / **5.11** Art, Precision Graphics / **5.12** Lisa Starr, using photographs © 1997, 1998 Jet Propulsion Laboratory, California Institute of Technology and NASA, 1972 / **5.13** (a, b) Thomas A. Steitz; (c) Lisa Starr / **5.14, 5.15** Lisa Starr / **5.16** Art, Precision Graphics; (b) © Douglas Faulkner/Sally Faulkner Collection / **5.17, 5.18, 5.19** Lisa Starr / **5.20** (a, b) Art, Raychel Ciemma; (c) M. Abbey/Visuals Unlimited / **5.21** Art, Precision Graphics / **5.22** Art, Raychel Ciemma; Micrographs, M. Sheetz, R. Painter, and S. Singer, *Journal of Cell Biology*, 70:193 (1976) by permission of The Rockefeller University Press / **5.23** Lisa Starr / **5.24, 5.25** Art, Raychel Ciemma / **5.26** Frieder Sauer/Bruce Coleman

CHAPTER 6 **6.1** Art, Raychel Ciemma; Micrograph, Carolina Biological Supply Company / **6.2** Art, Lisa Starr; Photograph, Wernher Krutein/PhotoVault / **6.3** Lisa Starr / **pages 93, 94, 95** Lisa Starr / **6.4** Art, Gary Head and Lisa Starr with photograph from David Neal Parks / **6.5** (a) © Barker-Blakenship/FPG; (b) Art, Precision Graphics / **6.6** Lisa Starr after Stephen L. Wolfe, *Molecular and Cellular Biology*, Wadsworth / **6.7** Larry West/FPG / **6.8–6.11** Lisa Starr / **6.12** E. R. Degginger / **6.13, 6.14** Lisa Starr / **6.15** Art, Lisa Starr; Photographs, (a) © Bill Beatty/Visuals Unlimited; (b) © 2001 PhotoDisc, Inc.; (c, both) © 2001 PhotoDisc, Inc.; Micrographs, (above) © Bruce Iverson, computer-enhanced by Lisa Starr; (below) © Ken Wagner/Visuals Unlimited, computer-enhanced by Lisa Starr / **page 103** Steve Chamberlain, Syracuse University / **6.16** NASA / **6.17–6.19** Lisa Starr / **6.20** (a) Herve Chaumeton/Agence Nature; (b) Douglas Faulkner/Sally Faulkner Collection

CHAPTER 7 **7.1** Stephen Dalton/Photo Researchers, computer-enhanced by Lisa Starr / **7.2** Art, Gary Head and Raychel Ciemma / **7.3** Photographs, (a-left) Gary Head; (a-right) Ed Reschke; (b) Paolo Fioratti; Art, (c) Lisa Starr / **page 112** Lisa Starr / **7.4** (left) Art, Raychel Ciemma and Lisa Starr; (right) Lisa Starr after Ralph Taggart / **7.5** Art, (a) Raychel Ciemma; (b, c) Lisa Starr; Micrograph, Keith R. Porter / **7.6** Photograph, © 2001 PhotoDisc, Inc.; Art on photo, Raychel Ciemma; Art, Lisa Starr / **7.7, 7.8, 7.9** Lisa Starr / **7.10** (a) Lisa Starr; Photographs, (b) Adrian Warren/Ardea, London; (c) David M. Phillips/Visuals Unlimited / **7.11** William Grenfell/Visuals Unlimited / **7.12** Art, Lisa Starr; Photograph, Gary Head / **page 122** (left) Art, Gary Head; (right) Photograph, R. Llewellyn/SuperStock / **page 123** Lisa Starr / **page 125** © Francis Leroy, Biocosmos/Science Photo Library/Photo Researchers

CHAPTER 8 **8.1** (left and right, center) Chris Huss; (right, above and below) Tony Dawson / **8.2** Art, Gary Head / **8.3** Art, Lisa Starr and Raychel Ciemma; Micrographs (a) C. J. Harrison et al, *Cytogenetics and Cell Genetics*, 35: 21–27 © 1983 S. Karger, A. G. Basel; (c) B. Hamkalo; (d) O. L. Miller, Jr., Steve L. McKnight / **8.4** Art, Raychel Ciemma and Gary Head / **8.5** Micrographs, Andrew S. Bajer, University of Oregon / **8.6** Art, Gary Head / **8.7** Art, Raychel Ciemma and Gary Head; Photographs, © Ed Reschke / **8.8** Art, Lisa Starr; Micrograph, © R. Calentine/Visuals Unlimited / **8.9** Art, Raychel Ciemma; Micrograph, © D. M. Phillips/Visuals Unlimited / **8.10** (a–c, e) Lennart Nilsson from *A Child Is Born* © 1966, 1967 Dell Publishing Company, Inc.; (d) Lennart Nilsson from *Behold Man*, © 1974 by Albert Bonniers Förlag and Little, Brown & Company, Boston / **8.11** Photograph, (above) Courtesy of The Family of Henrietta Lacks; Micrograph, Dr. Pascal Madaule, France

CHAPTER 9 **9.1** (a) Jane Burton/Bruce Coleman; (b) Dan Kline/Visuals Unlimited / **9.2** Art, Raychel Ciemma / **9.3** CNRI/SPL/Photo Researchers / **9.4** Art, Raychel Ciemma; Micrographs, with thanks to the John Innes Foundation Trustees, computer-enhanced by Gary Head / **9.5, 9.6** Art, Raychel Ciemma / **9.7** Art, Preface, Inc.; Photograph, (fly) David Maitland/Seaphot Limited / **9.8** Lisa Starr / **9.9** Art, Lisa Starr; Micrograph, David M. Phillips/Visuals Unlimited / **9.10** Art, Raychel Ciemma / **9.11** Art, Precision Graphics / **9.12** © Richard Corman/Corbis Outline

CHAPTER 10 **10.1** (left) ZumaPress; (center) Focus on Sports; (right-above) The Moravian Museum, Brno; (right-below) Fabian/Corbis Sygma / **10.2** Art, Jennifer Wardrip; Photograph, Jean M. Labat/Ardea, London / **10.3** Lisa Starr / **10.4** Art, Precision Graphics / **10.5** Art, Raychel Ciemma and Precision Graphics / **10.6** Art, Precision Graphics / **10.7, 10.8, 10.9, page 158** Art, Raychel Ciemma and Precision Graphics/ **10.10** Photographs, William E. Ferguson / **10.11** Preface, Inc. / **10.12** Art, Preface, Inc.; Micrographs, Stanley Flegler/Visuals Unlimited / **10.13** Art, Preface, Inc.; Photographs, (a, b) Michael Stuckey/Comstock, Inc., (c) Bosco Broyer, photograph by Gary Head / **10.14** David Hosking / **10.15** Ted Somes / **10.16** (above to below) Frank Cezus/FPG; Frank Cezus/FPG; © 2001 PhotoDisc, Inc.; Ted Beaudin/FPG; Stan Sholik/FPG / **10.17** (a) Art, Precision Graphics; Photograph, Daniel Fairbanks, Brigham Young University / **10.18** Art, D. & V. Hennings; Photograph, Jane Burton/Bruce Coleman / **10.19** (above) William E. Ferguson; (below) Eric Crichton/Bruce Coleman / **10.20** Photograph, © Pamela Harper/Harper Horticultural Slide Library; Art, Lisa Starr from Prof. Otto Wilhelm Thomé, Flora von Deutschland Österreich und der Schweiz. 1885, Gera, Germany / **10.21** Evan Cerasoli / **10.22** Leslie Faltheisek/Clacritter Manx / **10.23** Joe McDonald/Visuals Unlimited

CHAPTER 11 **11.1** From "Multicolor Spectral Karyotyping of Human Chromosomes," by E. Schrock, T. Ried, et al, *Science*, 26, July 1966, 273:495. Used by permission of E. Schrock and T. Reid and the American Association for the Advancement of Science, computer-enhanced by Lisa Starr / **11.2** Paul A. Thiessen / **11.3** Art, Raychel Ciemma and Preface, Inc. / **11.4** Art, Raychel Ciemma; Photograph, Charles D. Winters/Photo Researchers; Micrograph,

Omikron/Photo Researchers / **11.5** Art, Precision Graphics and Gary Head; Photographs (left) © 2001 EyeWire; (right) © 2001 PhotoDisc, Inc. / **11.6** (a) Photograph, From Lennart Nilsson, *A Child Is Born*, © 1966, 1977 Dell Publishing Company, Inc.; (b) Robert Demarest after Patten, Carlson & others; (c) Robert Demarest with permission from M. Cummings, *Human Heredity: Principles and Issues*, p. 126 of third Ed. © 1994 by Brooks/Cole. All rights reserved. / **11.7** Art, Raychel Ciemma and Preface Inc.; Photograph, Carolina Biological Supply Company / **11.8** Art, Raychel Ciemma / **11.9** Art, (a, b) Precision Graphics; Photograph, (b) Dr. Victor A. McKusick; (c) Steve Uzzell / **page 180** Art, Precision Graphics / **11.10** Lisa Starr / **11.11** (a) Lisa Starr; (b) Painting, Giraudon/Art Resource, New York / **11.12** (a) Lisa Starr; Photograph, (b) © Bettmann/Corbis; Art, (b) After V. A. McKusick, *Human Genetics*, 2nd Ed., © 1969. Reprinted by permission of Prentice-Hall, Inc., Englewood Cliffs, NJ / **11.13** C. J. Harrison / **11.14** Eddie Adams/AP Photo / **11.15** (a, b) Courtesy G. H. Valentine / **11.16** From "Multicolor Spectral Karyotyping of Human Chromosomes," by E. Schrock, T. Ried, et al, *Science*, 26, July 1966, 273:496. Used by permission of E. Schrock and T. Reid and the American Association for the Advancement of Science / **11.17** Art, Raychel Ciemma / **11.18** (a) Micrograph, Cytogenetics Laboratory, The University of California, San Francisco; Photograph, (a) courtesy of Peninsula Association for Retarded Children and Adults, San Mateo Special Olympics, Burlingame, CA; (b) after Collman and Stoller, *American Journal of Public Health*, 52, 1962; Photograph, (b) by permission of Carole Lafrate / **11.19** Art, Raychel Ciemma / **11.20** Art, Lisa Starr; Photographs, courtesy of Lennart Nilsson from *A Child Is Born*, © 1966, 1977 Dell Publishing Company, Inc. / **11.21** Photograph, (a) Fran Heyl Associates © Jacques Cohen, computer-enhanced by © Pix Elation; Art, (b) Raychel Ciemma / **11.22** Carolina Biological Supply Company / **page 189** © Mitchell Gerber/Corbis

CHAPTER 12 **12.1** A. C. Barrington Brown, © 1968 J. D. Watson / **12.2** Model by Dr. David B. Goodin, The Scripps Research Institute / **12.3** Art, Raychel Ciemma / **12.4** Art, (a, b) Raychel Ciemma; Photograph, (c) Lee D. Simon/Science Source/Photo Researchers / **12.5** Art, Raychel Ciemma / **12.6** Micrograph, Biophoto Associates/SPL/Photo Researchers; Art, Precision Graphics / **12.7** Lisa Starr, incorporating model by Dr. David B. Goodin, The Scripps Research Institute / **12.8** Courtesy of the Cold Springs Harbor Laboratory Archives / **12.9, 12.10** Art, Precision Graphics / **12.11** Photographs, (a) From Daniel Fairbanks; (b) PA News Photo Library

CHAPTER 13 **13.1** (above) Dennis Hallinan/FPG; (below) © Bob Evan/Peter Arnold / **13.2, 13.3** Art, Precision Graphics / **13.4** (a) Art, Palay/Beaubois; (b–e) Lisa Starr / **13.5** Art, Preface, Inc. / **13.6, 13.7** Art, Precision Graphics and Gary Head / **13.8** (a) tRNA model by Dr. David B. Goodin, The Scripps Research Institute; (b, c) Lisa Starr / **13.9** (a) Dr. Thomas A. Steitz; (b) Lisa Starr / **13.10** Lisa Starr / **13.11** (a) Lisa Starr; (b) Precision Graphics / **13.12** Lisa Starr / **13.13** (left) Nik Kleinberg; (right) Peter Starlinger / **13.14** Lisa Starr / **13.15** Courtesy of the National Neurofibromatosis Foundation

CHAPTER 14 **14.1** Photographs, (a) Ken Greer/Visuals Unlimited; (b) Biophoto Associates/Science Source/Photo Researchers; (c) James Stevenson/SPL/Photo Researchers; (right) Gary Head / **page 212** Lisa Starr / **14.2** Art, Raychel Ciemma / **14.3** Brian Matthews, University of Oregon / **14.4** Art, Palay/Beaubois and Hans & Cassady / **14.5** Art, (b) Raychel Ciemma / **14.6** Jack Carey and Lisa Starr / **14.7** Art, Raychel Ciemma; Micrograph, W. Beerman / **14.8** Frank B. Salisbury / **14.9** (a) Lennart Nilsson © Boehringer Ingelheim International GmbH; (b, c) Art, Betsy Palay/Atemis / **14.10** Art, Betsy Palay/Atemis / **14.11** Slim Films / **page 221** Lisa Starr

CHAPTER 15 **15.1** Lewis L. Lainey / **15.2** Science VU/Visuals Unlimited / **15.3** (a) Stanley N. Cohen/Science Source/Photo Researchers; (b) Dr. Huntington Potter and Dr. David Dressler / **page 224** Art, Preface, Inc. / **15.4** Lisa Starr with photograph © 2000 PhotoDisc, Inc. / **15.5, 15.6** Art, Gary Head / **15.7** (above) Damon Biotech, Inc.; (below) Cellmark Diagnostics, Abingdon, UK / **15.8, 15.9** Lisa Starr / **15.10** Stephen Wolfe, *Molecular and Cellular Biology* / **15.11** Art, Lisa Starr; Photograph, Herve Chaumeton/Agence Nature / **15.12** Keith V. Wood / **15.13** Photographs, (a) Dr. Vincent Chiang, School of Forestry and Wood Products, Michigan Technology University; (b) Courtesy Calgene LLC / **15.14** R. Brinster, R. E. Hammer, School of Veterinary Medicine, University of Pennsylvania / **15.15** © Adrian Arbib/Still Pictures / **page 237** S. Stammers/SPL/Photo Researchers

CHAPTER 16 **16.1** Elliot Erwitt/Magnum Photos, Inc. / **16.2** Photographs, (a) Jen & Des Bartlett/Bruce Coleman; (b) Kenneth W. Fink/Photo Researchers; (c) Dave Watts/A.N.T. Photo Library / **16.3** Art, Raychel Ciemma / **16.4** Photographs, (a) Courtesy George P. Darwin, Darwin Museum, Down House; (b) Heather Angel, computer-enhanced by Lisa Starr; (c) Christopher Ralling; (e) Dieter & Mary Plage/Survival Anglia; Art, (d-above) Leonard Morgan; (d-below) Precision Graphics / **16.5** (a) © Philip Boyer/Photo Researchers; (b) Field Museum of Natural History, Chicago. Artist, Charles R. Knight (Neg. CK21T) / **16.6** Photographs, (a) Heather Angel; (b) David Cavagnaro; (c) Dr. P. Evans/Bruce Coleman; (d) Alan Root/Bruce Coleman / **16.7** Courtesy of Down House and The Royal College of Surgeons of England / **page 246** Courtesy Derrell Fowler, Tecumseh, OK / **16.8** Alan Solem / **16.9** Art, Raychel Ciemma and Gary Head / **16.10** Art, Precision Graphics / **16.11** J. A. Bishop and L. M. Cook / **16.12** Art, Precision Graphics / **16.13** Photographs, (a, b) Warren Abrahamson; (c) Forest W. Buchanan/Visuals Unlimited; (d) Kenneth McCrea and Warren Abrahamson; computer-enhanced by Gary Head / **16.14** Art, Precision Graphics / **16.15** Photographs, (a, b) Thomas Bates Smith; (c) art, Preface, Inc. / **16.16** Bruce Beehler / **16.17** Art, Precision Graphics after Ayala and others / **16.18** (above) David Neal Parks; (below) W. Carter Johnson / **16.19** Art, Precision Graphics after computer models developed by Jerry Coyne / **16.20** Art, Raychel Ciemma; Photograph, David Cavagnaro / **16.21** Kjell Sandved/Visuals Unlimited / **page 259** Art, Precision Graphics / **16.22** Art, Precision Graphics using NIH data

CHAPTER 17 **17.1** (a) Photograph, Gary Head with snail courtesy of Larry Reed; (b) Art, adapted from R. K. Selander and D. W. Kaufman, *Evolution*, 29(3), December 31, 1975 / **17.2** Art, Jennifer Wardrip / **17.3** Art, Precision Graphics after F. Ayala and J. Valentine, *Evolving*, Benjamin-Cummings, 1979 / **17.4** Photographs, (a) G. Ziesler/ZEFA; (b) Alvin E. Staffan/Photo Researchers; (c) John Alcock/Arizona State University / **17.5** Photographs, (a, all) Fred McConnaughey/Photo Researchers; (b) Patrice Geisel/Visuals Unlimited; (right) Tom Van Sant/The Geosphere Project, Santa Monica, CA / **17.6** Art, Preface, Inc. / **17.7** Art, Raychel Ciemma / **17.8** Photograph, (above) Policy Programme of WWF Cameroon/www.wwfcameroon.org, computer-enhanced by Lisa Starr; (below) Tom Van Sant/The Geosphere Project, Santa Monica, CA / **17.9** Lisa Starr, after W. Jensen and F. B. Salisbury, *Botany: An Ecological Approach*, Wadsworth, 1972, and others / **17.10** Photographs, (left) © H. Clarke, VIREO/Academy of Natural Sciences; (right) Robert C. Simpson/Nature Stock; Art, Preface, Inc. / **17.11** Art, Precision Graphics / **17.12** Photographs (above) Jack Dermid; (below) Leonard Lee Rue III/FPG; Art, Precision Graphics after P. Dodson, *Evolution: Process and Products*, 3rd Ed., PWS / **17.13** Jen & Des Bartlett/Bruce Coleman

CHAPTER 18 **18.1, 18.2** Lisa Starr, with use of NASA satellite photographs / **18.3** Vatican Museums / **18.4** Photographs, (a) Donald Baird, Princeton Museum of Natural History; (b) A. Feduccia, *The Age of Birds*, Harvard University Press, 1980; (c) H. P. Banks; (d) Jonathan Blair / **18.5** (Globes), Lisa Starr; Chart, Gary Head / **18.6** Art, Preface, Inc. / **18.7** Photograph, (a) © 2001 PhotoDisc, Inc.; Art, (b, c) Lisa Starr / **18.8** Dan Peha/Index Stock Imagery / **18.9** Art, Leonard Morgan / **18.10** Lisa Starr, after A. M. Ziegler, C. R. Scotese, and S. F. Barrett, "Mesozoic and Cenozoic Paleogeographic Maps," and J. Krohn and J. Sundermann (Eds.), *Tidal Friction and the Earth's Rotation II*, Springer–Verlag, 1983; Photograph, Martin Land/Photo Researchers / **18.11** Art, Raychel Ciemma / **18.12** Art, (a) Lisa Starr and Raychel Ciemma, art reference, Natural History Collection, Royal BC Museum; Photographs, (a) Stephen Dalton/Photo Researchers, computer-enhanced by Lisa Starr; (b) Frans Lanting/Minden Pictures, computer-enhanced by Lisa Starr; (c) J. Scott Altenbach, University of New Mexico, computer-enhanced by Lisa Starr / **18.13** Art, Precision Graphics and Gary Head, after E. Guerrant, *Evolution*, 36:699–712; Photographs, Gary Head / **18.14** From T. Storer et al., *General Zoology*, sixth Ed., McGraw-Hill, 1979, reproduced by permission of McGraw-Hill, Inc. / **18.15** Art, Raychel Ciemma / **18.16** Art, Precision Graphics / **18.17** Photographs, (left to right) Kjell B. Sandved/Visuals Unlimited; Jeffrey Sylvester/FPG; Thomas D. Mangelsen/Images of Nature / **18.18** (left to right) Larry Lefever/Grant Heilman Photography; R.I.M. Campbell/Bruce Coleman; Runk & Schoenberger/Grant Heilman Photography; Bruce Coleman / **18.21** Photograph, P. Morris/Ardea London; Art, Raychel Ciemma / **page 291** © 1990 Arthur M. Green

CHAPTER 19 **19.1** Jeff Hester and Paul Scowen/Arizona State University and NASA / **19.2** William K. Hartmann / **19.3** (a) Chesley Bonestell, (b) Art, Raychel Ciemma / **19.4** Art, Precision Graphics / **19.5** Microscopics, (a) Sidney W. Fox; (b) W. Hargreaves and D. Deamer / **19.7** Art, Raychel Ciemma and Precision Graphics / **19.8** Bill Bachmann/Photo Researchers / **19.9** Photographs, (a) Stanley M. Awramik; (b–f) Andrew H. Knoll, Harvard University / **19.10** Micrograph, P. L. Walne and J. H. Arnott, *Planta*, 77: 325–354, 1967 / **19.11** Art, Raychel Ciemma / **19.12** Robert K. Trench / **19.13** Photographs, (a, b) Neville Pledge/South Australian Museum; (c, d) Dr. Chip Clark / **19.14** (a, b) Illustrations by Zdenek Burian, © Jiri Hochman and Martin Hochman; (c) Patricia G. Gensel; (d) © John Barber / **19.15** Painting by Megan Rohn courtesy of David Dilcher; Art, Gary Head; Photographs, (from below clockwise) Runk & Schoenberger/Grant Heilman Photography; Robert and Linda Mitchell; Ed Reschke; Lee Casebere / **19.16** (a) © John Sibbick; (b, c) © Karen Carr/www.karencarr.com / **19.17** NASA Galileo Imaging Team / **19.18** Art, Raychel Ciemma / **19.19** Paintings, (a, b) © Ely Kish; (c) Field Museum of Natural History, Chicago. Artist, Charles R. Knight (Neg. CK8T) / **Globes, page 302–304, 307** Lisa Starr / **19.20** Lisa Starr

CHAPTER 20 **20.1** (a, b) Tony Brain and David Parker/SPL/Photo Researchers; (c) Lee D. Simon/Photo Researchers / **20.2** (a) © Oliver Meckes/Photo Researchers; (b) Courtesy Allan W. H. Bé and David A. Caron; (c) Lewis Trusty/Animals Animals / **20.3** (a) Lisa Starr; (b) CNRI/SPL/Photo Researchers / **20.4** (a) L. J. LeBeau, University of Illinois Hospital/BPS; (b, c) Courtesy of K. Amako and A.Umeda, Kyshu University, Japan / **20.5** © P. Hawtin, University of Southhampton/SPL/Photo Researchers / **20.6** (a) Jack Jones, *Archives of Microbiology*, 136: 254–261, 1983. Reprinted by permission of Springer–Verlag, computer-enhanced by Lisa Starr; (b) © 2000 PhotoDisc, Inc.; (c) NASA; (d) © Alan L. Detrick, Science Source/Photo Researchers / **20.7** (a) Dr. Jeremy Burgess/Science Photo Library/Photo Researchers; (b) Tony Brain/SPL/Photo Researchers; (c) P. W. Johnson and J. MeN. Sieburth, Univ. Rhode Island/BPS / **20.8** Photographs, (above) Edward S. Ross; (center) Courtesy of Wadsworth Center, NY; (deer ticks) Courtesy Pfizer Central Research; Art, Preface, Inc. / **20.9, 20.10** Art, Raychel Ciemma / **20.11** Art, Raychel Ciemma after Stephen L. Wolfe and others / **20.12** (a) K. G. Murti/Visuals Unlimited; (b) Kenneth M. Corbett / **20.13** Art, Palay/Beaubois and Precision Graphics / **20.14** Art, Raychel Ciemma / **20.15** (a) Heather Angel; (b) W. Merrill / **20.16** Art, (a) Leonard Morgan; Photographs, (b) M. Claviez, G. Gerish, and R. Guggenheim; (c) London Scientific Films; (d–f) Carolina Biological Supply Company; (g) Courtesy Robert R. Kay from R. R. Kay, et al., *Development*, 1989 Supplement, pp. 81–90, © The Company of Biologists Ltd., 1989 / **20.17** Edward S. Ross / **20.18** Photographs, (a) M. Abbe/Visuals Unlimited; (b) G. Shih and R. Kessel/Visuals Unlimited; (c) John Clegg/Ardea, London; (d) Neal Ericsson / **20.19** Micrograph, (a) Sidney L. Tamm; Art, (b, c) Raychel Ciemma, redrawn from V. & J. Pearse and M. & R. Buchsbaum, *Living Invertebrates*, The Boxwood Press, 1987. Used by permission / **20.20** Micrographs, (a) Dr. Stan Erlandsen, University of Minnesota; (b) David M. Phillips/Visuals Unlimited / **20.21** (b) Lisa Starr, after Prescott, *Microbiology*, third Ed. / **20.22** Lauren and Homer, Photography by Gary Head / **20.23** Art, Leonard Morgan; Micrograph, Steven L'Hernault / **Table 20.2** Courtesy Saul Tzipori, Div. of Infectious Diseases, Tufts University School of Veterinary Medicine / **20.24** Art, (a-left) Raychel Ciemma; (a-right) Copyright © Richard Triemer, Rutgers; Micrographs, (b) Greta Fryxell, University of Texas, Austin; (c) Emiliania huxleyi photograph by Vita Pariente, scanning electron micrograph taken on a Jeol T330A instrument at the Texas A&M University Electron Microscopy Center; (d) S. Carty, Heidelberg College / **20.25** Photographs, (a-both, b-right) North Carolina State University, Aquatic Botany Lab; (b-left) C. C. Lockwood / **20.26** Art, Raychel Ciemma / **20.27** (a) Art from T. Garrison, *Oceanography: An Invitation to Marine Science*, Brooks/Cole, 1993; Photograph, (b) J. R. Waaland, University of Washington/BPS / **20.28** Photographs, (a) Manfred Kage/Peter Arnold; (b) Ronald W. Hoham, Dept. of Biology, Colgate University; (c) © Linda Sims/Visuals Unlimited; (d) Brian Parker/Tom

Stack & Associates; Art, (d) Raychel Ciemma / **20.29** Art, Raychel Ciemma; Photograph, D. J. Patterson/Seaphot Limited/Planet Earth Pictures / **20.30** Carolina Biological Supply Company

CHAPTER 21 **21.1** © Sherry K. Pittam / **21.2** (a–f) Robert C. Simpson/Nature Stock / **21.3** (a) Robert C. Simpson/ Nature Stock; (b) Jane Burton/Bruce Coleman; (c) Thomas J. Duffy / **21.4** Art, Raychel Ciemma; Micrograph, Garry T. Cole, University of Texas, Austin/BPS / **21.5** (a) Dr. P. Marazzi/SPL/Photo Researchers; (b) Eric Crichton/ Bruce Coleman / **21.6** Photographs, (a) Dr. John D. Cunningham/Visuals Unlimited; (b) © Michael Wood/ mykoweb.com; (c) © North Carolina State University, Department of Plant Pathology; (d) © Fred Stevens/ mykoweb.com; (e) Dennis Kunkel Microscopy, Inc.; (f) Garry T. Cole, University of Texas, Austin/BPS / **21.7** Art, Raychel Ciemma; Micrographs, Ed Reschke / **21.8** (a) After Raven, Evert, and Eichhorn, *Biology of Plants*, 4th Ed., Worth Publishers, New York, 1986; Photograph, (b) © Mark E. Gibson/Visuals Unlimited / **21.9** (a) © 1990 Gary Braasch; (b) F. B. Reeves / **21.10** John Hodgin

CHAPTER 22 **22.1** (a) Pat and Tom Leeson/Photo Researchers; (b) Craig Wood/Visuals Unlimited; (c) Edward S. Ross; (d) © Greg Allikas/ www.orchidworks.com / **22.2** Art, Raychel Ciemma after E. O. Dodson and P. Dodson, *Evolution Process and Product*, third Ed., p. 401, PWS / **22.3** Art, Precision Graphics / **22.4** (a) Field Museum of Natural History, Chicago (Neg. 7500C); (b) Brian Parker/Tom Stack & Associates / **22.5** Photograph, Jane Burton/Bruce Coleman; Art, Raychel Ciemma / **22.6** (a) Fred Bavendam/Peter Arnold; (b) Dr. John D. Cunningham/ Visuals Unlimited / **22.8** (a) Kingsley R. Stern; (b) Ed Reschke/Peter Arnold; (c) William Ferguson; (d) W. H. Hodge; (e) Kratz/ZEFA / **22.9** Art, Raychel Ciemma; Photographs, (inset) A. & E. Bomford/Ardea, London; (below) Lee Casebere / **22.10** Art, Raychel Ciemma; Photographs, (left) Edward S. Ross; (right-above) Robert and Linda Mitchell Photography; (right-below) R. J. Erwin/Photo Researchers / **22.11** © 1994 Robert Glenn Ketchum / **22.12** (a) © Jeff Gnass Photography; (b) Joyce Photographics/Photo Researchers; (c) Kingsley R. Stern / **22.13** (a) Robert and Linda Mitchell; (b) Edward S. Ross; (c) William Ferguson / **22.14** Art, (a) Jennifer Wardrip; Photographs, (b) M. P. L. Fogden/Bruce Coleman; (c) L. Mellichamp/Visuals Unlimited; (d) Peter F. Zika/Visuals Unlimited; (e) Heather Angel / **22.15** Art, Raychel Ciemma / **22.16** (a) George Loun/Visuals Unlimited; (b) Earl Roberge/Photo Researchers; (c) John Mason/ Ardea, London; (d) Dick Davis/Photo Researchers; (e) Rein/ZEFA / **22.17** Bob Cerasoli / **22.18** (left and center) © 1989, 1991 Clinton Webb; (right) Gary Head

CHAPTER 23 **23.1** Courtesy of Department of Library Services, American Museum of Natural History (Neg. K10273) / **23.2** Art, Gary Head; Photograph, Lisa Starr / **23.3** Art, Leonard Morgan / **23.4** Art, Raychel Ciemma / **23.5** (left) After Laszlo Meszoly in L. Margulis, *Early Life*, Jones and Bartlett, 1982 / **23.6** Bruce Hall / **23.7** (a–c) Art, Raychel Ciemma; (d-left) art, Precision Graphics after Bayer and Owre, *The Free-Living Lower Invertebrates*, © 1968 Macmillan; Photographs, (d-right) Don W. Fawcett/Visuals Unlimited; (e) Marty Snyderman/Planet Earth Pictures / **23.8** Art, Raychel Ciemma after Eugene Kozloff / **23.9** (a, b) Art, Raychel Ciemma; Photographs, (c) Courtesy of Dr. William H. Hammer; (d-both) Kim Taylor/Bruce Coleman; (e-both) F. S. Westmorland/Tom Stack & Associates / **23.10** (a) Art, Precision Graphics after T. Storer et al., *General Zoology*, sixth Ed., © 1979 McGraw-Hill; Photographs, (b) Andrew Mounter/Seaphot Limited/Planet Earth Pictures; (c) Christian DellaCorte / **23.11** Art, Raychel Ciemma; Photograph, Kim Taylor/Bruce Coleman / **23.12** (a) Robert and Linda Mitchell Photography; (b) Cath Ellis, University of Hull/SPL/Photo Researchers / **23.13** Art, Lisa Starr; Micrograph, J. Sulston, MRC Laboratory of Molecular Biology / **23.14** Art, Raychel Ciemma / **23.15** Art, Raychel Ciemma; Photograph, Carolina Biological Supply Company / **23.16** (a) © L. Jensen/Visuals Unlimited; (b) Dianora Niccolini / **page 367** Art, Precision Graphics / **23.17** Photographs, (a) Gary Head; (b) Alex Kerstitch; (c) Jeff Foott/Tom Stack & Associates; (d) Frank Park/A.N.T. Photo Library; (e) Bob Cranston; Art, (f) Palay/Beaubois / **23.18** (a, b) Art, Raychel Ciemma / **23.19** J. A. L. Cooke/Oxford Scientific Films / **23.20** (a) © Cabisco/Visuals Unlimited; (b) Jon Kenfield/Bruce Coleman; (c) Art, Precision Graphics after Eugene Kozloff, *Invertebrates* © 1990 by Saunders College Publishing. Reproduced by permission of the publisher; (d) Art, Precision Graphics, adapted from Rasmussen, "Ophelia," Vol. 11, in Eugene Kozloff, *Invertebrates*, 1990 / **23.21** Art, Raychel Ciemma / **23.22** Jane Burton/Bruce Coleman / **23.23** (a) Redrawn from *Living Invertebrates*, V. & J. Pearse/M. &. R. Buchsbaum, The Boxwood Press, 1987. Used by permission; Photographs, (b-above) Jane Burton/ Bruce Coleman; (b-below) Angelo Giampiccolo/FPG; (c) P. J. Bryant, University of California, Irvine/BPS; (d) Ken Lucas/Seaphot Limited/Planet Earth Pictures / **23.24** B. Backes, Arizona State University / **23.25** Photographs, (a) Herve Chaumeton/Agence Nature; (b) Frans Lanting/ Bruce Coleman; (c) Fred Bavendam/Peter Arnold; (d) Agence Nature; Art, Precision Graphics, after D. H. Milne, *Marine Life and the Sea*, Wadsworth, 1995 / **23.26** Art, Precision Graphics / **23.27** (a) Z. Leszczynski/Animals Animals; (b) Steve Martin/Tom Stack & Associates / **23.28** Art, Raychel Ciemma / **23.29** Art, Precision Graphics. From G. Pasteur, "Jean Henri Fabre," *Scientific American*, July 1994. © 1994 by *Scientific American*. All rights reserved / **23.30** Photographs, (a–h) Edward S. Ross; (i) David Maitland/Seaphot Limited/Planet Earth Pictures; (j) C. P. Hickman, Jr. / **23.31** (b) Marlin E. Rice, Iowa State University; (a, c) John Obermeyer, Purdue University, Dept. of Entomology / **23.32** (a) Chris Huss/The Wildlife Collection; (b) John Mason/Ardea, London; (c) Kjell B. Sandved; (d) Ian Took/Biofotos / **23.33** (a, b) Art, L. Calver; (c, d) Photographs, Herve Chaumeton/Agence Nature / **page 379** Inset, Jane Burton/Bruce Coleman / **23.34** Art, Raychel Ciemma / **23.35** Walter Deas/Seaphot Limited/ Planet Earth Pictures / **23.36** John H. Gerard / **23.37** (a) J. Solliday/BPS; (b) Herve Chaumeton/Agence Nature

CHAPTER 24 **24.1** (a) Jean Phillipe Varin/Jacana/Photo Researchers; (b) Tom McHugh/Photo Researchers / **24.2** Art, Gary Head / **24.3** (a) Rick M. Harbo; (b, c) Redrawn from *Living Invertebrates*, V. & J. Pearse and M. & R. Buchsbaum, The Boxwood Press, 1987. Used by permission / **24.4** (a) Art, Raychel Ciemma; (b) Runk & Schoenberger/Grant Heilman Photography / **24.5** Art, D. & V. Hennings and Precision Graphics / **24.6** (a–c) Art, Raychel Ciemma, adapted from A. S. Romer and T. S. Parsons, *The Vertebrate Body*, sixth Ed., Saunders College Publishing, 1986; (d) Photograph, Al Giddings/Images Unlimited / **24.7** (a, b) After D. H. Milne, *Marine Life and the Sea*, Wadsworth, 1995; (c) Photograph, Heather Angel / **24.8** (a) Erwin Christian/ZEFA; (b) Allan Power/Bruce Coleman; (c) Tom McHugh/Photo Researchers / **24.9** (a) Bill Wood/Bruce Coleman; (b) Art, Raychel Ciemma; (c) Robert and Linda Mitchell Photography; (d) Patrice Ceisel/ © 1986 John G. Shedd Aquarium; (e) © Norbert Wu/Peter Arnold; (f) From T. Garrison, *Oceanography: An Invitation to Marine Science*, Wadsworth, 1993 / **24.10** (a) © Marianne Collins; (b, c) art, Laszlo Meszoly and D. & V. Hennings / **24.11** (a) Stephen Dalton/ Photo Researchers; (b) John Serraro/Visuals Unlimited; (c) Jerry W. Nagel; (d) Art, Leonard Morgan/Adapted from A. S. Romer and T. S. Parsons, *The Vertebrate Body*, sixth Ed., Saunders College Publishing, 1986; (e) Juan M. Renjifo/Animals Animals / **24.12** (a) Pieter Johnson; (b) Stanley Sessions, Hartwick College / **24.13** Art, Raychel Ciemma / **24.14** (a) © 1989 D. Braginetz; (b) Z. Leszczynski/Animals Animals / **24.15** (a, c) Art, Raychel Ciemma; Photographs, (a) D. Kaleth/Image Bank; (b) Andrew Dennis/A.N.T. Photo Library; (c) Stephen Dalton/Photo Researchers; (d) Heather Angel / **24.16** Art, Lisa Starr; Photograph, J. L. G. Grande/Bruce Coleman / **24.17** (a) Gerard Lacz/A.N.T. Photo Library; (b) Rajesh Bedi; (c) Thomas D. Mangelsen/Images of Nature / **24.18** (a, b) Art, Raychel Ciemma / **24.19** Photographs, (a) Sandy Roessler/FPG; (b) Leonard Lee Rue III/FPG; (c) © Kevin Schafer/Tom Stack & Associates; (d) Art, Raychel Ciemma after M. Weiss and A. Mann, *Human Biology and Behavior*, 5th Ed., Harper Collins, 1990 / **24.20** Art, Raychel Ciemma / **24.21** Photographs, (a) D. & V. Blagden/A.N.T. Photo Library; (b) courtesy The Zoological Society of San Diego, The San Diego Zoo; (c) Clem Haagner/Ardea, London; (d) Douglas Faulkner/ Photo Researchers; (e) Christopher Crowley / **24.22** Photographs, (a) Larry Burrows/Aspect Picture Library; (b) Bruce Coleman; (c) Tom McHugh/Photo Researchers; (d) Courtesy of Dr. Takeshi Furuichi, Biology, Meiji–Gakuin University-Yokohama; (e) Time Inc. 1965/Larry Burrows Collection / **24.23** Lisa Starr / **24.24** (a) Dr. Donald Johanson, Institute of Human Origins; (b) Louise M. Robbins; (c, d) Kenneth Garrett/National Geographic Image Collection / **24.25** (a) Jean Paul Tibbles; (b, c) John Reader © 1981 / **24.26, 24.27** Lisa Starr / **24.28** Art, Raychel Ciemma / **24.29** Art, Preface, Inc. / **24.30** Lisa Starr, with photograph from NASA / **24.31** Lisa Starr / **24.32** Art, Precision Graphics after National Geographic, February 1997, page 82 / **page 405** Art, D. & V. Hennings

CHAPTER 25 **25.1** Tom Till/© Stone / **25.2** Art, Gary Head / **25.3** (a) Painting by Charles Knight, Department of Library Services/American Museum of Natural History (2425); (b) Mansell Collection/Time, Inc. / **25.4** Lisa Starr / **25.5** Steve Hillebrand, U.S. Fish & Wildlife Service / **25.6** (a) Courtesy of Dan Stanford; (b) Lisa Starr, based on 1940–1984 reports from International Whaling Commission / **25.7** (a) C. B. & D. W. Frith/Bruce Coleman; (b) Douglas Faulkner/Photo Researchers; (c) Douglas Faulkner/ Sally Faulkner Collection / **25.8** (a) From T. Garrison, *Oceanography: An Invitation to Marine Science*, third Ed., Brooks/Cole, 2000. All rights reserved; (b), (above row, left to right) Peter Scoones/Planet Earth Pictures; Jim Doran; (insets) Douglas Faulkner/Sally Faulkner Collection; (center row, left to right) Jeff Rotman; Alex Kerstitch; Douglas Faulkner/Sally Faulkner Collection; (below row, all) Douglas Faulkner/Sally Faulkner Collection / **25.9** Eric Hartmann/Magnum Photos / **25.10** Lisa Starr / **25.11** Lisa Starr, with photographs © 2000 PhotoDisc, Inc. / **25.12** Photographs, Bureau of Land Management / **25.13** Lisa Starr / **page 419** © Gary Head

CHAPTER 26 **26.1** (a) R. Barrick/USGS; (b, c) © 1980 Gary Braasch; (d) Don Johnson/Photo Nats, Inc. / **26.2, 26.4** Art, Raychel Ciemma / **26.5** Micrography, James D. Mauseth, *Plant Anatomy*, Benjamin-Cummings, 1988 / **26.6** (all) Biophoto Associates / **26.7** (a) Jan Robert Factor/Photo Researchers; (b) D. E. Akin and I. L. Risgby, Richard B. Russel Agricultural Research Center, Agricultural Research Service, U.S. Department of Agriculture, Athens, Georgia; (c) Kingsley R. Stern / **26.8** Art, Precision Graphics / **26.9** George S. Ellmore / **26.10** Art, D. & V. Hennings / **26.11** Art, Raychel Ciemma / **26.12** (a) Robert and Linda Mitchell Photography; (b) Roland R. Dute; (c) Gary Head / **26.13** Art, D. & V. Hennings; Micrographs, (a-left) Ray F. Evert; (a-right) James W. Perry; (b-left) Carolina Biological Supply Company; (b-right) James W. Perry / **26.14** Art, D. & V. Hennings / **26.15** (a, above and below) Heather Angel; (b) Gary Head; (c) © 2001 PhotoDisc, Inc. / **26.16** (a) Art, Raychel Ciemma; (b) C. E. Jeffree, et al., *Planta*, 172(1):20–37, 1987. Reprinted by permission of C. E. Jeffree and Springer–Verlag; (c) Dr. Jeremy Burgess/SPL /Photo Researchers / **26.17** (a) Art after Salisbury and Ross, *Plant Physiology*, fourth Ed., Wadsworth; (b) Micrograph, John Limbaugh/Ripon Microslides, Inc.; (c, d) Art, Raychel Ciemma / **26.18** Micrographs, (a-both) Chuck Brown; (b) Carolina Biological Supply Company / **26.19** Photographs, John Limbaugh/Ripon Microslides; Sketch after T. Rost et al., *Botany: A Brief Introduction to Plant Biology*, second Ed., © 1984, John Wiley & Sons / **26.20** Art, (a) Raychel Ciemma, from Rost et al, *Plant Biology*, Wadsworth; (b) Art, Raychel Ciemma / **26.21** Lisa Starr / **26.22** H. A. Core, W. A. Cote, and A. C. Day, *Wood Structure and Identification*, second Ed., Syracuse University Press, 1979 / **26.23** Art, Raychel Ciemma / **26.24** Photographs, Edward S. Ross / **page 435** Art, Lisa Starr; Photograph, © Stephen J. Krasemann/National Audubon Society Collection/Photo Researchers

CHAPTER 27 **27.1** (a, b) Robert and Linda Mitchell Photography; (c) © John N. A. Lott, *Scanning Electron Microscope Study of Green Plants*, St. Louis: C. V. Mosby Company, 1976; (d) Robert C. Simpson/Nature Stock / **27.2** David Cavagnaro/Peter Arnold / **27.3** William Ferguson / **27.4** U.S. Dept. of Agriculture / **27.5** Art, Leonard Morgan; Micrograph, (b) Chuck Brown / **27.6** Jean Paul Reve / **27.7** Photographs, (a) Adrian P. Davies/ Bruce Coleman; (c) Mark E. Dudley and Sharon R. Long; (a, b) Art, Jennifer Wardrip / **27.8** NifTAL Project, Univ. of Hawaii, Maui / **27.9** Micrographs, (a) W. A. Cote of N. C. Brown Center for Ultra-Structure Studies, computer-enhanced by Lisa Starr; (b, c) H. A. Core, W. A. Cote and A. C. Day, *Wood Structure and Identification*, 2nd Ed., Syracuse University Press, 1979 / **27.10** Art, Raychel Ciemma / **27.11** Frank B. Salisbury / **27.12** George S. Ellmore / **27.13** W. Thomson, *American Journal of Botany*, 57(3):316, 1970 / **27.14** Art, Palay/Beaubois / **27.15** T. A. Mansfield / **27.16** Photograph, (left) John Lawlor/ FPG; Scanning micrographs; (a, b) Dr. Jeremy Burgess/ SPL/Photo Researchers / **27.17** Art, Hans & Cassady, Inc. / **27.18** Martin Zimmerman, *Science*, 1961, 133:73–79, © AAAS / **27.19** Art, (left) Palay/Beaubois; (right) Precision Graphics / **27.21** James T. Brock

CHAPTER 28 **28.1** (left) Courtesy Merlin D. Tuttle, Bat Conservation International; (right) John Alcock/Arizona

State University / **28.2** (a) Robert A. Tyrrell; (b, c) Thomas Eisner, Cornell University / **28.3** Art, Raychel Ciemma and Precision Graphics; Photographs © Gary Head / **28.4** John Shaw/Bruce Coleman / **28.5** (a, b) David M. Phillips/Visuals Unlimited; (c) David Scharf/Peter Arnold / **28.6** Art, Raychel Ciemma / **28.7** (a, b) Patricia Schulz; (c, d) Ray F. Evert; (e, f) John Limbaugh/Ripon Microslides, Inc.; Art, (g) Raychel Ciemma / **28.8** Photographs, (a) Gary Head, (b) B. J. Miller, Fairfax, VA/BPS; (c) Richard H. Gross; (d-all) Janet Jones; (e) R. Carr / **28.9** Russell Kaye, © 1993 The Walt Disney Co. Reprinted with permission of Discover Magazine / **28.10** (a) Runk & Schoenberger/Grant Heilman Photography; (b) Kingsley R. Stern / **28.11** From B. Bracegirdle and P. Miles, *An Atlas of Plant Structure*, Heinemann Educational Books, 1977 / **28.12** Art, Raychel Ciemma; (c) Photograph, Herve Chaumeton/Agence Nature / **28.13** Art, Raychel Ciemma; Photograph, (c) Barry L. Runk/Grant Heilman Photography; (d) From Mauseth / **28.14** Art, Raychel Ciemma; Micrograph, Biophot / **28.15** Kingsley R. Stern / **28.16** Sylvan H. Wittwer/Visuals Unlimited / **28.17** (a, b) Micrographs courtesy Randy Moore from "How Roots Respond to Gravity," M. L. Evans, R. Moore, and K. Hasenstein, *Scientific American*, December 1986 / **28.18** John Digby and Richard Firn / **28.19** (a) Frank B. Salisbury; (b, c) Lisa Starr / **28.20** Gary Head / **28.21** Cary Mitchell / **28.23** Frank B. Salisbury / **28.24** Art, Precision Graphics / **28.25** From Rost et al, *Plant Biology*, Wadsworth / **28.26** (a) Jan Zeevart; (b) Plant Industry Station, The Corps Research Division, Agricultural Research Service, U.S. Department of Agriculture / **28.27** N. R. Lersten / **28.28** R. J. Down / **28.29** Gary Head / **page 471** © Kevin Schafer

CHAPTER 29 **29.1** David Macdonald / **29.2** (a) Photograph, Focus on Sports; Micrograph, Manfred Kage/Bruce Coleman; Art, Lisa Starr; (b) Photographs, (left to right) © Ray Simons/Photo Researchers; © Ed Reschke/Peter Arnold; © Don W. Fawcett; Art, Lisa Starr / **29.3** Art, Raychel Ciemma and Lisa Starr / **29.4** Photograph, Gregory Dimijian/Photo Researchers; Art, Raychel Ciemma, adapted from C. P. Hickman, Jr., L. S. Roberts, and A. Larson, *Integrated Principles of Zoology*, ninth Ed., Wm. C. Brown, 1995 / **29.5** Art, Lisa Starr; Photographs (a–c, e) Ed Reschke, (d) Fred Hossler/Visuals Unlimited; (f-above) Prof. P. Motta/Dept. of Anatomy, University "La Sapienza," Rome/SPL/Photo Researchers; (f-below) Ed Reschke / **29.6** Photograph, Roger K. Burnard; Art, (left) Joel Ito; (right) L. Calver / **29.7** Ed Reschke / **29.8** Art, Lisa Starr; Micrographs, (a, c) Ed Reschke; (b) © Biophoto Associates/Photo Researchers, Inc. / **29.9** Art, Robert Demarest / **29.10** (a) Lennart Nilsson from *Behold Man*, © 1974 Albert Bonniers Förlag and Little, Brown and Company, Boston; (b) Kim Taylor/Bruce Coleman / **29.11** Art, L. Calver / **29.12** Art, Palay/Beaubois / **29.14** Art, Preface, Inc.; Photograph, Fred Bruemmer

CHAPTER 30 **30.1** (a) Comstock, Inc.; (b) Kevin Somerville / **30.2** Art, Robert Demarest; Micrograph, Manfred Kage/Peter Arnold / **30.3–30.5** Art, Raychel Ciemma / **30.6** Art, Precision Graphics / **30.7** Art, Kevin Somerville; Micrograph, © Don Fawcett, Bloom and Fawcett, 11th Ed., after J. Desaki and Y. Uehara, Photo Researchers/ **30.8** Art, Lisa Starr; Micrograph, Dr. Constantino Sotelo from *International Cell Biology*, p. 83, 1977. Used by copyright permission of Rockefeller University Press / **30.9** Art, Gary Head / **30.10** Art, Robert Demarest / **30.12** Art, (a) Kevin Somerville and Precision Graphics; (b) Robert Demarest / **30.13** Painting, Sir Charles Bell, 1809, courtesy of Royal College of Surgeons, Edinburgh / **30.14** (a) Art, Raychel Ciemma; (b) Lisa Starr after Eugene Kozloff / **30.15** Art, Raychel Ciemma / **30.16** Art, Kevin Somerville / **30.17** Art, Kevin Somerville and Precision Graphics / **30.18** Art, Robert Demarest and Precision Graphics / **30.19** Art, Robert Demarest; Micrograph, Manfred Kage/Peter Arnold / **30.20** Photographs, (a) Colin Chumbley/Science Source/Photo Researchers; (b) C. Yokochi and J. Rohen, *Photographic Anatomy of the Human Body*, 2nd Ed., Igaku-Shoin, Ltd., 1979 / **30.21** (a) Art, Raychel Ciemma; (b) Photograph, Marcus Raichle, Washington University School of Medicine / **30.22** Art, Kevin Somerville / **30.23** (left) PET scans from E. D. London et al., *Archives of General Psychiatry*, 47:567–574, 1990; Photograph, (c) Ogden Gigli/Photo Researchers

Chapter 31 **31.1** (a) Eric A. Newman; (b) Merlin D. Tuttle, Bat Conservation International / **31.2** Lisa Starr, using photograph © 1997, 1998 Jet Propulsion Laboratory/California Institute of Technology and NASA / **31.3** From Hensel and Bowman, *Journal of Physiology*, 23:564–568, 1960 / **31.4** Art, Palay/Beaubois after Penfield and Rasmussen, *The Cerebral Cortex of Man*, © 1950 Macmillan Publishing Company, Inc. Renewed 1978 by Theodore Rasmussen; Photograph, © Colin Chumbley/Science Source/Photo Researchers / **31.5** Art, Raychel Ciemma / **31.6** Art, Gary Head / **31.7** Art, (a-left, b, d) Robert Demarest; (a-right) Lisa Starr; (c) Precision Graphics / **31.8** Robert E. Preston, courtesy Joseph E. Hawkins, Kresge Hearing Research Institute, University of Michigan Medical School / **31.9** Art, Raychel Ciemma / **31.10** Art after M. Gardiner, *The Biology of Vertebrates*, McGraw-Hill, 1972; Photograph, E. R. Degginger / **31.11** Art, (a) Robert Demarest; (b–d) K. Somerville / **31.12** Chase Swift / **31.13** (a) Micrograph, Lennart Nilsson © Boehringer Ingelheim International GmbH; (b) Art, Raychel Ciemma / **31.14** Micrograph, (a) Omikron/SPL/ Photo Researchers; Art, Robert Demarest / **31.15** Art, Precision Graphics / **31.16** Gerry Ellis/The Wildlife Collection

CHAPTER 32 **32.1** (a, b) Hugo van Lawick/The Jane Goodall Institute / **32.2** Art, Kevin Somerville / **32.3, 32.4** Lisa Starr / **32.5, 32.6** Art, Robert Demarest / **32.7** (a) Mitchell Layton; (b) Syndication International (1986) Ltd.; (c-all) Courtesy of Dr. William H. Daughaday, Washington University School of Medicine, from A. I. Mendelhoff and D. E. Smith, eds., *American Journal of Medicine*, 1956, 20:133 / **32.8** Art, Leonard Morgan / **32.9** Art, (a, b) Raychel Ciemma; Photographs, (c) Gary Head; (d) © Bettmann/Corbis / **32.10** Art, Precision Graphics / **32.11** Biophoto Associates/SPL/Photo Researchers / **32.12** Art, Leonard Morgan / **32.13** (a) John S. Dunning/Ardea, London; (b) Evan Cerasoli / **32.14** The Stover Group/D. J. Fort / **32.15** Art, From R. C. Brusca and G. J. Brusca, *Invertebrates*, © 1990 Sinauer Associates. Used by permission; Photographs, (c) Frans Lanting/Bruce Coleman; (d) Robert and Linda Mitchell Photography / **32.16** Roger K. Burnard

CHAPTER 33 **33.1** John Brandenberg/Minden Pictures / **33.2** Art, L. Calver / **33.3** (a-right) Dr. John D. Cunningham/Visuals Unlimited; (a-left) CNRI/SPL/Photo Researchers; (b) Art, Robert Demarest / **33.4** Michael Keller/FPG / **33.5** Linda Pitkin/Planet Earth Pictures / **33.6** Art, D. & V. Hennings / **33.7** Art, Raychel Ciemma / **33.8** Art, (a-above) Raychel Ciemma; (a-left, b) Joel Ito; Micrograph, Ed Reschke / **33.9** (a, b) Prof. P. Motta/Dept. of Anatomy/University "La Sapienza," Rome/Science Photo Library/Photo Researchers / **33.10** C. Yokochi and J. Rohen, *Photographic Anatomy of the Human Body*, second Ed., Igaku-Shoin, Ltd., 1979 / **33.11** Art, Raychel Ciemma / **33.12** N.H.P.A./A.N.T. Photo Library / **33.13** Art, Robert Demarest / **33.14** Art, Raychel Ciemma / **33.15** Art, Robert Demarest with Gary Head; Photograph, (a) Dance Theatre of Harlem, by Frank Capri; Micrographs, (b, c) © Don Fawcett/Visuals Unlimited, from D. W. Fawcett, *The Cell*, Philadelphia; W. B. Saunders Co., 1966 / **33.16** Art, (above) Robert Demarest; (below) Nadine Sokol and Gary Head / **33.17** Art, Preface, Inc. / **33.18** Art, Robert Demarest / **33.19, 33.20** Art, Gary Head / **33.21** (a) © Sean Sprague/Stock, Boston; (b) Michael Neveux

CHAPTER 34 **34.1** (a) From A. D. Waller, *Physiology: The Servant of Medicine*, Hitchcock Lectures, University of London Press, 1910; (b) Courtesy The New York Academy of Medicine Library / **34.2** Art, Precision Graphics and Gary Head / **34.3** Art, (a, c) Precision Graphics; (b, d) Raychel Ciemma; (e) After M. Labarbera and S. Vogel, *American Scientist*, 1982, 70:54–60 / **34.4** Art, Precision Graphics / **34.5** Art, Palay/Beaubois / **34.6** Photographs, © David Scharf/Peter Arnold / **34.7** Photograph, © 2001 EyeWire, modified by Lisa Starr; Art, Lisa Starr, with art references from Bloodline Image Atlas, University of Nebraska-Omaha/Sherri Wicks, Human Physiology and Anatomy, University of Wisconsin Biology Web Education System, and others / **34.8** Photographs, (a, b) Lester V. Bergman & Associates, Inc.; Art, (c) after F. Ayala and J. Kiger, *Modern Genetics*, © 1980 Benjamin-Cummings / **34.9** Art, Nadine Sokol after G. J. Tortora and N. P. Anagnostakos, *Principles of Anatomy and Physiology*, 6th Ed. © 1990 by Biological Sciences Textbooks, Inc., A & P Textbooks, Inc. and Elia-Sparta, Inc. Reprinted by permission of Harper Collins / **34.10** Art, Precision Graphics / **34.11** Art, Kevin Somerville / **34.12** Art, Raychel Ciemma; Photograph, C. Yokochi and J. Rohen, *Photographic Anatomy of the Human Body*, second Ed., Igaku-Shoin, Ltd., 1979 / **34.13** Art, Precision Graphics / **34.14** Art, (a) Lisa Starr; (b) Raychel Ciemma; Micrographs, Don W. Fawcett / **34.15** Art, Robert Demarest, based on A. Spence, *Basic Human Anatomy*, Benjamin-Cummings, 1982 / **34.16** Art, Precision Graphics / **34.17** Sheila Terry/SPL/Photo Researchers / **34.18** Lisa Starr / **34.19** Art, (a, b-left) Kevin Somerville; (b-right) Lisa Starr, using © 2001 PhotoDisc, Inc. photograph / **34.20** (a) © Ed Reschke; (b) © Biophoto Associates/Photo Researchers; (c) Art, L. Calver / **34.21** Art, L. Calver / **34.22** Art, Precision Graphics / **34.23** Photograph, (a) Prof. P. Motta/Dept. of Anatomy/University "La Sapienza," Rome/Science Photo Library/Photo Researchers; Art, Gary Head / **34.24** Art, (a) Raychel Ciemma; (b) Lisa Starr / **34.25** Art, Raychel Ciemma / **34.26** Lennart Nilsson from *Behold Man*, © 1974 by Albert Bonniers Förlag and Little, Brown and Company, Boston

CHAPTER 35 **35.1** (a) The Granger Collection, New York; (b) Lisa Starr, from Harris, L. J.; Larson, S. B.; Hasel, K. W.; McPherson, A.; *Biochemistry* 36, pp. 1581 (1997). Structure rendered with RIBBONS; (c) Lennart Nilsson © Boehringer Ingelheim International GmbH / **35.2** Robert R. Dourmashkin, courtesy of Clinical Research Centre, Harrow, England / **35.3** Lisa Starr / **35.4** (a) NSIBC/SPL/Photo Researchers; (b) Biology Media/Peter Arnold / **35.5** Art, Raychel Ciemma / **35.6** Art, Lisa Starr; Photograph, © David Scharf/Photo Researchers / **35.7** Art, Lisa Starr; Photograph, © Ken Cavanagh/Photo Researchers, Inc. / **35.8** Art, Precision Graphics / **35.9, 35.10** Lisa Starr / **35.11** Art, Preface, Inc. / **35.12** Art, Raychel Ciemma / **35.13** Art, Lisa Starr / **page 586** Art, Lisa Starr / **35.14** Art, Lisa Starr; Photograph, From Harris, L. J.; Larson, S. B.; Hasel, K. W.; McPherson, A.; *Biochemistry* 36, pp. 1581 (1997). Structure rendered with RIBBONS; / **35.15** Lisa Starr / **35.16** © Dr. A. Liepins/SPL/Photo Researchers / **35.17** (above) Lowell Georgia/Science Source/Photo Researchers; (below) Matt Meadows/Peter Arnold / **35.18** (left) David Scharf/Peter Arnold; (right) Kent Wood/Photo Researchers / **35.19** Ted Thai/Time Magazine / **35.20** © Zeva Olbaum/Peter Arnold / **35.21** (a) Art, Raychel Ciemma after Stephen Wolfe, *Molecular Biology of the Cell*, Wadsworth, 1993; (b) Micrographs, Z. Salahuddin, National Institutes of Health

CHAPTER 36 **36.1** Galen Rowell/Peter Arnold / **36.2** Art, Gary Head and Precision Graphics / **36.4** Photographs, (a) Peter Parks/Oxford Scientific Films; (b) Herve Chaumeton/Agence Nature; Art, Precision Graphics / **36.5** Micrograph, Ed Reschke; Art, Precision Graphics / **36.6** Art, Raychel Ciemma and Preface, Inc. / **36.7** Art, Raychel Ciemma / **36.8** Lisa Starr / **36.9** Micrograph, H. R. Duncker/Justus-Liebig University, Giessen, Germany; Art, Raychel Ciemma / **36.10** Art, Kevin Somerville / **36.11** Photographs, courtesy Kay Elemetrics Corporation; Art, modified from A. Spence and E. Mason, *Human Anatomy and Physiology*, fourth Ed., 1992, West Publishing Company / **36.12** (a) Lisa Starr with photograph © 2000 PhotoDisc, Inc.; Art, (b, c) Lisa Starr; X-Rays, (b, c) SIU/Visuals Unlimited / **36.13** Art, Precision Graphics; Photograph, Giorgio Gualco/Bruce Coleman / **36.14** (a) © R. Kessel/Visuals Unlimited; (b, c) Art, Lisa Starr / **36.15** Art, Leonard Morgan / **36.16** © CNRI/SPL/Photo Researchers / **36.17** (a, b) © O. Auerbach/Visuals Unlimited / **36.18** Photograph, Lennart Nilsson from *Behold Man*, © 1974 by Albert Bonniers Förlag and Little, Brown and Company, Boston / **page 610** Photograph, Courtesy of Ron Romanosky / **36.19** Art, Precision Graphics

CHAPTER 37 **37.1** Hulton Getty Collection/© Stone / **37.2** Art, Gary Head and Precision Graphics / **37.3** Art, Raychel Ciemma / **37.4** Art, adapted from A. Romer and T. Parsons, *The Vertebrate Body*, sixth Ed., Saunders Publishing Company, 1986; Photograph, D. Robert Franz/Plant Earth Pictures / **37.5** Gunter Ziesler/Bruce Coleman, Inc. / **37.6** Art, Kevin Somerville / **37.7** Art, Nadine Sokol / **37.8** Art, Robert Demarest; Photograph, Omikron/SPL/Photo Researchers / **37.9** After A. Vander, et al., *Human Physiology: Mechanisms of Body Function*, fifth Ed., McGraw-Hill, 1990, used by permission; (c) redrawn from *Human Anatomy and Physiology*, fourth Ed., by A. Spence and E. Mason, 1992, Brooks/Cole. All rights reserved / **37.10** Art, Lisa Starr after Sherwood and others.; (b) Microslide, courtesy Mark Nielsen, University of Utah; (e) Micrograph, © D. W. Fawcett/Photo Researchers / **37.11** Art, Raychel Ciemma / **page 621** After A. Vander, et al., *Human Physiology*: Mechanisms of Body Function, fifth Ed., McGraw-Hill, 1990. Used by permission / **37.12** Photograph, Ralph Pleasant/FPG; Art, Precision Graphics / **37.13** Art, Kevin Somerville and Precision Graphics / **37.14** Art, Gary Head; Photographs (above) © 2001 PhotoDisc, Inc.; (center) © Ralph Pleasant/FPG; (below) Elizabeth Hathon/Corbis Stock Market / **37.16** Gary Head / **37.17** Dr. Douglas Coleman, The Jackson Laboratory; Art, Precision Graphics and Gary Head / **37.18** Art, Raychel Ciemma

CHAPTER 38 **38.1** Photographs, (background-left) David Noble/FPG; (below-left) Claude Steelman/Tom Stack & Associates; (right) Gary Head / **38.2** Gary Head and

Precision Graphics / **38.3** Art, Robert Demarest / **38.4** Photograph, (a-left) © Ralph T. Hutchings; Art, Robert Demarest / **38.5** Art, (above) Robert Demarest; (below) Precision Graphics / **38.6** Art, Precision Graphics; Photograph, Gary Head / **38.7** (a, b) From T. Garrison, *Oceanography: An Invitation to Marine Science*, Brooks/Cole, 1993. All rights reserved; (c) Photograph, © Thomas D. Mangelsen/Images of Nature / **38.8** © Bob McKeever/ Tom Stack & Associates / **38.9** Colin Monteath, Hedgehog House, New Zealand/ **38.10** © Bettmann/Corbis

CHAPTER 39 **39.1** Art, Raychel Ciemma; Photographs, (a) Hans Pfletschinger; (c) John H. Gerard; (d, e) © David M. Dennis/Tom Stack & Associates; (f) John Shaw/Tom Stack & Associates / **39.2** (a) Frieder Sauer/ Bruce Coleman (b) © Matjaz Kuntner; (c) Wisniewski/ZEFA; (d) L. L. Rue III; (e) Carolina Biological Supply Company; (f) Fred McKinney/FPG; (g) Gary Head / **39.3** Art, Palay/ Beaubois and Precision Graphics / **39.4** Art, L. Calver; Micrographs, Carolina Biological Supply Company / **39.5** Art, Robert Demarest / **39.6** Art, Precision Graphics; Photograph, Gary Head / **39.7** Art, after V. E. Foe and B. M. Alberts, *Journal of Cell Science*, 1983, 61:32, © The Company of Biologists; Micrograph, J. B. Morrill / **39.8** (Photographic series) Carolina Biological Supply Company; (right photograph) Peter Parks/Oxford Scientific Films/Animals Animals / **39.9** Art, (a) Lisa Starr; (b) Raychel Ciemma after B. Burnside, *Developmental Biology*, 1971, 26:416–441. Used by permission of Academic Press / **39.10** Art, (a, b) Lisa Starr; Photographs, (c) Prof. Jonathon Slack / **39.11** Art, Palay/Beaubois and Preface, Inc. / **39.12** Art, Raychel Ciemma, modified by Lisa Starr / **39.13** Art, (a) Raychel Ciemma; Micrograph, (b) Ed Reschke / **39.14** Art, Raychel Ciemma / **39.15** Art, Precision Graphics / **39.16** Art, Raychel Ciemma / **39.17** Art, (above) Robert Demarest; (below) Raychel Ciemma/ **39.18** Art, Precision Graphics; Photograph, Lennart Nilsson from *A Child Is Born*, © 1966, 1977 Dell Publishing Company, Inc. / **39.19** K. Somerville, Robert Demarest, and Preface, Inc. / **39.20–39.23** Art, Raychel Ciemma / **39.24** Art, Raychel Ciemma; Photographs, Lennart Nilsson, *A Child Is Born*, © 1966, 1977 Dell Publishing Company, Inc. / **39.25** Art, Raychel Ciemma, modified from K. L. Moore, *The Developing Human: Clinically Oriented Embryology*, fourth Ed., Philadelphia: W. B. Saunders Co., 1988 / **39.26** James W. Hanson, M.D. / **39.27** Art, Robert Demarest / **39.28** Art, Raychel Ciemma / **39.29** Art, Raychel Ciemma, adapted from L. B. Arey, *Developmental Anatomy*, Philadelphia, W. B. Saunders Co., 1965; Photograph, Lisa Starr / **39.30** Art, Preface Inc. / **39.31** CNRI/SPL/Photo Researchers / **39.32** (a) Dr. John D. Cunningham/Visuals Unlimited; (b) David M. Phillips/Visuals Unlimited; (c) © Kenneth Greer/Visuals Unlimited / **39.33** Carolina Biological Supply Co. / **page 683** Alan and Sandy Carey

CHAPTER 40 **40.1** (a) Gary Head; (b) Antoinette Jongen/ FPG / **40.2** Art, Precision Graphics / **40.3** (a) E. R. Degginger; (inset) Jeff Foott Productions/Bruce Coleman; (b) © Paul Lally/Stock, Boston / **40.4** Photograph, © Jeff Lepore/Photo Researchers; Art, Precision Graphics / **40.5** (a) Art, Precision Graphics; (b) Micrograph, Stanley Flegler/Visuals Unlimited / **40.6** Art, Precision Graphics / **40.7** Art, Precision Graphics; Photograph, E. Vetter/ZEFA / **40.8** Art, Precision Graphics; Photographs, (a) Jonathan Scott/Planet Earth Pictures; (b) Hal H. Harrison/Photo Researchers; (c) Fred Bavendam/Peter Arnold / **page 693** David Reznick/ University of California-Riverside; computer-enhanced by Lisa Starr / **40.9** Art, Precision Graphics; Photograph, NASA / **40.10** Art, Gary Head after G. T. Miller, Jr., *Environmental Science*, 6th Ed., Brooks/Cole, 1997. All rights reserved. / **40.11** Art, Precision Graphics and Preface, Inc.; Photograph, United Nations / **40.12** Data from Population Reference Bureau after G. T. Miller, Jr., *Living in the Environment*, 8th Ed., Brooks/Cole, 1993. All rights reserved / **40.13** Art, Precision Graphics; Photograph, United Nations / **40.14, 40.15** After G. T. Miller, Jr., *Environmental Science*, sixth Ed., Brooks/Cole, 1997. All rights reserved

CHAPTER 41 **41.1** (left) John Bova/Photo Researchers; (right) Robert Maier/Animals Animals / **41.2** (a) From L. Clark, *Parasitology Today*, 6 (11), Elsevier Trends Journals, 1990, Cambridge, U.K.; (b) Photograph, Jack Clark/ Comstock, Inc. / **41.3** (a) Eugene Kozloff; (b, c) Stevan Arnold, Inc. / **41.4** Photograph, Hans Reinhard/Bruce Coleman; Spectogram, from G. Pohl-Apel and R. Sussinka, *Journal for Ornithologie*, 123:211–214 / **41.5** (a) Evan Cerasoli; (b) From A. N. Meltzoff and M. K. Moore, "Imitation of Facial and Manual Gestures by Human Neonate," *Science*, 1977, 198:75–78. ©1977 by the AAAS / **41.6** Eric Hosking / **41.7** Nina Leen in *Animal Behavior*, Life Nature Library / **page 707** (left) © 2001 PhotoDisc, Inc.; (right) © Ed Reschke / **41.8** Mark Bekoff / **41.9** (above) Edward S. Ross; (below) E. Mickleburgh/Ardea, London / **41.10** Art, D. &. V. Hennings; Photograph, Steven Dalton/ Photo Researchers / **41.11** (a) John Alcock/Arizona State University ; (b) Ray Richardson/ Animals Animals / **41.12** Patricia Caulfield / **41.13** Michael Francis/The Wildlife Collection / **41.14** Frank Lane Agency/Bruce Coleman / **41.15** John Alcock/ Arizona State University / **41.16** Fred Bruemmer / **41.17** (a) John Dominis, *Life Magazine*, © Time, Inc.; (b) A. E. Zuckerman/Tom Stack & Associates / **41.18** (a, b) Kenneth Lorenzen / **41.19** Gregory D. Dimijian/Photo Researchers / **41.20** Yvon Le Maho / **41.21** Lincoln P. Brower / **41.22** F. Schutz / **41.23** John Alcock/Arizona State University

CHAPTER 42 **42.1** (left) Edward S. Ross; (right) Donna Hutchins / **42.2** (a, c) Harlo H. Hadow; (b) Bob and Miriam Francis/Tom Stack & Associates / **42.3** Art, Precision Graphics; Photograph, Clara Calhoun/Bruce Coleman / **42.4** Art, after G. Gause, 1934 / **42.5** Photographs, (a, c) Jane Burton/Bruce Coleman; (b) Heather Angel; Art, (d, e) Precision Graphics from Jane Lubchenco, *American Naturalist*, 112:23–29. © 1978 by The University of Chicago Press / **42.6** Art, after N. Weland and F. Bazazz, *Ecology*, 56: 681–688. © 1975 Ecological Society of America / **42.7** Art, Precision Graphics / **42.8** Art, Precision Graphics; Photograph, Ed Cesar/Photo Researchers / **42.9** © C. James Webb/ Phototake USA / **42.10** (a) James H. Carmichael; (b) Edward S. Ross; (c) W. M. Laetsch / **42.11** Photographs, (all) Edward S. Ross / **42.12** (a, b) Thomas Eisner, Cornell University; (c) Douglas Faulkner/Sally Faulkner Collection; (d) Edward S. Ross/Entymology, 1972, 104:1003–1016 / **42.13** (a–e) Roger K. Burnard; (f) E. R. Degginger; (g) Roger A. Powell / **42.14** Angelina Lax/Photo Researchers / **42.15** (left-inset) John Carnemolla/Australian Picture Library/Westlight; (right) Peter Bird/Australian Picture Library/Westlight / **42.16** Art, Gary Head after W. Dansgaard, et al, *Nature*, 364:218–220, 15 July 1993; D. Raymond et al., *Science*, 259:926–933, February 1993; W. Post, *American Scientist*, 78:310–326, July–August 1990 / **42.17** Art, after S. Fridriksson, *Evolution of Life on a Volcanic Island*, Butterworth: London, 1975; Photograph, Dr. Harold Simon/Tom Stack & Associates / **42.18** (a, b) Art, Gary Head after J. M. Diamond, *Proceedings of the National Academy of Sciences*, 69:3199–3201, 1972; Photographs, (a) © PhotoDisc, Inc.; (c) Frans Lanting/Minden Pictures, computer modified by Lisa Starr/ **42.19** (a, b) Edward S. Ross / **42.20** Heather Angel/Biofotos

CHAPTER 43 **43.1** (a) Wolfgang Kaehler; (b) Frans Lanting/ Minden Pictures / **43.2** Art, Gary Head and Lisa Starr / **43.3** Photograph, Alan and Sandy Carey; Art, Gary Head after R. L. Smith, *Ecology and Field Biology*, fifth Ed. / **43.4, 43.5, 43.6** Lisa Starr with photographs, tallgrass prairie, © Ann B. Swengel/Visuals Unlimited; bacterium, Stanley W. Watson, *International Journal of Systematic Bacteriology*, 21:254–270, 1971; cutworm, © Nigel Cattlin/Holt Studios International/Photo Researchers; gopher and prairie vole, © Tom McHugh/Photo Researchers; sparrow, © Rod Planck/Photo Researchers; yellow spider, Michael Jeffords; frog, John H. Gerard; garter snake, Michael Jeffords; weasel, courtesy Biology Department, Loyola Marymount University; plover, © O. S. Pettingill Jr./Photo Researchers; crow, © Ed Reschke; hawk, © J. Lichter/Photo Researchers; all others © 2000 PhotoDisc, Inc. / **43.7** Art, Precision Graphics and Gary Head; Photographs, (a) D. Robert Franz; (b) Frans Lanting/Bruce Coleman / **43.8** Hans Reinhard/Bruce Coleman / **43.9** Gary Head / **43.10** Gene C. Feldman and Compton J. Tucker/NASA, Goddard Space Flight Center / **43.11, 43.12** Art, Precision Graphics / **43.13** Art, Gary Head; Photograph, Gerry Ellis/The Wildlife Collection / **43.14** Lisa Starr / **43.15** Art, Gary Head / **43.16** Art, Lisa Starr; photograph, © 2000 PhotoDisc, Inc. / **43.17** D. W. Schindler, *Science*, 184:897–899 / **43.18** Art, Lisa Starr after Paul Hertz; Photograph, (fox) © 2000 PhotoDisc, Inc. / **43.19** Lisa Starr with NASA photograph / **43.20** Lisa Starr, with photographs from NASA JSC Digital Image Collection / **43.21** Lisa Starr, compilation of historical records from Vostok and Law Dome ice core samples, Barnola, Raynaud, Lorius, Laboratoire do Glaciologie et de Geophysique de l'Environnement; N.I. Barkov, Arctic and Antarctic Research Institute, and Atmospheric CO_2 concentrations from the South Pole; Keeling and Whorf, Scripps Institute of Oceanography; Jones, Osborn, and Briffa, Climatic Research Unit, University of East Anglia; Parker, Hadley Centre for Climate Prediction and Research; Vostok ice core data set, Chapellaz and Jouzel, NOAA/NGDC Paleoclimatology Program / **43.22** (a) Lisa Starr, compilation of data from Mauna Loa Observatory, Keeling and Whorf, Scripps Institute of Oceanography; photographs, © 2000 PhotoDisc, Inc.; (b) Lisa Starr, compilation of data from Prinn et al, CDIAC, Oak Ridge National Laboratory; World Resources Institute; Khalil and Rasmussen, Oregon Graduate Institute of Science and Technology, CDIAC DB-1010; Leifer and Chan, CDIAC DB-1019; photographs, © 2000 PhotoDisc, Inc. and EyeWire, Inc.; (c, d) Lisa Starr, compilation of data from World Resources Institute; Law Dome ice core samples, Etheridge, Pearman, and Fraser, Commonwealth Scientific and Industrial Research Organisation; Prinn et al, CDIAC, Oak Ridge National Laboratory; Leifer and Chan, CDIAC DB-1019; photograph, © 2000 PhotoDisc, Inc. / **43.23** Art, Lisa Starr, with Gary Head after P. Hertz; Photograph, © 2000 PhotoDisc, Inc. / **43.24** USDA Forest Service / **43.26** After R. Noss, J. M. Scott, and E. LaRoe / **43.27** Courtesy of Matthew Lazzara, Antarctic Meteorological Research Center, University of Wisconsin–Madison

CHAPTER 44 **44.1** (a-above) Edward S. Ross; (a-below) David Noble/FPG; (b-above) Edward S. Ross; (b-below) Richard Coomber/Planet Earth Pictures / **44.2** Edward S. Ross / **44.3** Art, Preface Inc.; Photograph, NASA / **44.4** (b, c) Art, L. Calver/ **44.5** Art, Precision Graphics / **44.6** Lisa Starr, using NASA photograph / **44.7, 44.8** Lisa Starr / **44.9, 44.10** Lisa Starr, after The World Wildlife Fund and other sources, and using NASA photograph / **44.11** Lisa Starr, with photographs © 2000 PhotoDisc, Inc. / **44.12** Art, D. & V. Hennings after Whittaker, Bland and Tilman / **44.13** Harlo H. Hadow / **44.14** (above) Dr. John D. Cunningham/Visuals Unlimited; (inset) AP/Wide World Photos; (below) Jack Wilburn/Animals Animals / **44.15** (a) Kenneth W. Fink/Ardea, London; (b) Ray Wagner/Save the Tall Grass Prairie, Inc.; (c) Jonathan Scott/Planet Earth Pictures / **44.16** (left) © 1991 Gary Braasch; (right) Thase Daniel; (below-left) Adolf Schmidecker/FPG; (below-right) Edward S. Ross / **44.17** (all) Thomas E. Hemmerly / **44.18** (a) Dennis Brokaw; (b) Jack Carey; (c) Nigel Cook/Dayton Beach News Journal/Corbis Sygma / **44.19** (a, b) Dr. Peter Kuhry,Arctic Center, University of Lapland-Finland; (c) Roger N. Clark. All rights reserved / **44.20** D. W. MacManiman / **44.21** Art, Precision Graphics / **44.22** Art, Precision Graphics, modified after E. S. Deevy, Jr., *Scientific American*, October 1951 / **44.23** E. F. Benfield, Virginia Tech / **44.24** Art, L. K. Townsend / **44.25** (a) Ulli Seer/© Stone; (b) McCutcheon/ZEFA; (c) after T. Garrison; (d) J. Frederick Grassle, Woods Hole Institution of Oceanography; (e) Robert Hessler, Woods Hole Institution of Oceanography / **44.26** (a) Courtesy of Everglades National Park; (b) © Beth Davidow/Visuals Unlimited / **44.27** Photograph, E. R. Degginger; Art, D. & V. Hennings / **44.28** Courtesy J. L. Sumich, *Biology of Marine Life*, fifth Ed., W. C. Brown, 1992 / **44.29–44.31** Lisa Starr / **44.32** NOAA/Laboratory for Satellite Altimerty / **44.33** (a) © Dennis Kunkel/ Phototake, NYC; (b) © Roland Birke/OKAPIA/Photo Researchers; (c, d) Courtesy Rita Colwell and A. Huq/UMBI and Byron Wood, Brace Lotitz/NASA; (e) Raghu Rai/Magnum Photos / **44.34** Jack Carey

CHAPTER 45 **45.1** Photograph, Gary Head; Maps, U. S. Geological Survey / **45.2** (a, b) Art, Lisa Starr; (c) Photograph, © United Nations / **45.3** Photograph, Heather Angel; Art, after G. T. Miller, Jr., *Environmental Science: An Introduction*, Brooks/Cole, 1986. All rights reserved; and the Environmental Protection Agency / **45.4** From Center for Air Pollution Impact and Trend Analysis (CAPITA), Washington University, St. Louis / **45.5** (above) NASA Goddard Space Flight Center, Science Visualization Studios; (below) National Science Foundation / **45.6** Data from G. T. Miller, Jr. / **45.7** Photograph, R. Bieregaard/Photo Researchers; Art after G. T. Miller, Jr., *Living in the Environment*, eighth Ed., Brooks/Cole, 1993. All rights reserved / **45.8** NASA / **45.9** William Campbell/Time magazine / **45.10** Gerry Ellis/The Wildlife Collection / **45.11** USDA Soil Conservation Service/Thomas G. Meier / **45.12** Agency for International Development / **45.13** Dr. Charles Henneghien/Bruce Coleman / **45.14** Water Resources Council / **45.15** Ocean Arks International / **45.16** After G. T. Miller, Jr., *Living in the Environment*, tenth Ed., Brooks/Cole. All rights reserved / **45.17** Photograph, (a) AP/Wide World Photos; (b) Art, Precision Graphics after, M. H. Dickerson, "ARAC: Modeling an Ill Wind," in *Energy and Technology Review*, August 1987. Used by permission of University of California, Lawrence Livermore National Laboratory and U.S. Dept. of Energy / **45.18** (a, b) Alex MacLean/Landslides / **page 797** J. McLoughlin/ FPG / **45.19** © 1983 Billy Grimes

Pregnancy, human, 680
  embryonic period, 666–671, 666i–667i
  events leading to, 665
  fetal period, 670–673, 670i–672i
  nutritional considerations, 625, 672–673
  prenatal diagnosis, 186–187
  risks during, 672–673, 672i
Pregnancy test, 667, 677
Pregont, P., 538
Preimplantation diagnosis, 187
Premolar, 396i, 617, 617i
Premotor cortex, 503
Prenatal diagnosis, 186–187, 186i, 187i
Pressure
  air, 598, 598i, 604–605, 759, 761, 778
  hydrostatic, 89, 568i
  negative, in transpiration, 442
  osmotic, 89, 568–569, 568i
  sense of, 511
  in translocation, 446–447, 447i
  turgor, 444, 444i, 445i, 462, 462i
Pressure flow theory, 446–447, 448
Pressure gradient, 80, 598, 604–605, 607i
Preven, 677
Prey, predation and, 724–725, 734
Price, P., 735
Primary cell wall, 68, 68i, 462i
Primary growth, plant, 421, 423, 423i, 434
Primary producer, 738ff., 739i (See also Autotroph; Producer)
Primary productivity, 734, 734i, 738–739, 772–774, 775i, 776
Primary root, 422i, 460i
Primary structure (amino acid), 43, 44, 45
Primary succession, 728, 728i
Primary tissue layer, 481, 648, 649i, 652i
Primate, 183, 283, 283i, 284i, 383, 398i
  evolution, 398–402, 401i, 403, 405, 405i
Primer, 226, 226i
Priming pheromone, 708
Primitive streak, 668, 668i
Prion, 319
PRL, 526, 527i (See also Prolactin)
Probability, 157
Probe, 229, 229i
Procambium, 423, 423i, 426i
Producer, defined, 5, 34, 310, 326, 328, 738
  ecosystem roles, 7, 7i, 9, 34, 96i, 102, 720, 737ff., 738i–753i
Profundal zone, 772, 772i, 781i
Progeria, 179t, 182, 182i
Progesterone, 523i, 526t, 529t, 531, 531i, 661, 661i, 663, 663i, 664i
Progestin, 524t, 677
Proglottid, 365, 366i
Prokaryotic cell (See also Bacterium)
  characteristics, 52, 70–71, 298, 298i–299i, 300i 312–313, 312t
  eukaryotic vs., 8, 72t, 331t
  structure, 70, 70i, 312–313, 312i, 313i
Prokaryotic fission, 128, 128t, 316, 316i
Prolactin (PRL), 217, 523i, 524t, 526, 526t, 527, 639i, 674
Proline, 42i, 204i, 208i
Prometaphase, 132, 136
Promoter, 202, 203i, 214, 215i, 220, 283
Pronghorn antelope, 614–615, 615i
Prop root, 461i, 776, 776i
Prophase I, meiosis, 142i, 144i, 148t, 172i
Prophase II, meiosis, 143i, 149i
Prophase, mitosis, 131i, 132, 132i, 133, 136
Prosimian, 398, 401i
Prostaglandin, 41, 522, 532, 581, 657, 663
Prostate cancer, 657
Prostate gland, 656i, 657, 657i, 658i
Protein (See also specific types)
  chromosomal, 60, 60t, 61, 128–129, 129i, 142, 214, 214i
  function, 35–37, 38, 201, 206
  in human diet, 120, 121i, 625, 625i
  membrane, 52–53, 52i, 53i, 60, 218
  metabolism, 120, 121i, 618t, 619
  regulatory, nature of, 213–216
  structure, 36, 42–45, 43i–45i, 48i, 190, 204i, 208–209
Protein-energy malnutrition, 625
Protein hormone, 524, 524t, 525, 536

Protein kinase, 218
Protein synthesis, 42–44, 43i, 46, 201– 207 (See also Transcription; Translation)
  summary diagram, 210i
Protein-mediated transport, 86–87, 86i, 87i
Proteinoid, 295
Proterozoic, 276i–277i, 299, 300
Protistan (Protista), 299i, 300, 300i, 301
  characteristics, 8i, 72t, 87, 104, 110, 310, 310i, 320i–329i
  classification, 8, 9, 287, 311, 320
  evolution, 298i, 300–301, 320
  groups, 320–324, 326–329, 330, 330t
Proto-cell, 297
Protoderm, 423, 423i, 426i
Proton, 4, 6i, 21, 22, 24, 32t
Protonephridium, 364, 364i
Protostome, 367
Protozoan, 322–323, 322i, 323i, 325t, 330
  biodiversity through time, 418i
Proximal (directional term), 481i
Proximal tubule, 635, 635i, 636, 636i, 637
Prunus, 419i
Pseudocoel, 359
Pseudocoelomate (false coelom), 359i, 365
Pseudomonad, 315
Pseudomonas, 70i, 71i, 325
Pseudopod, 66, 66i, 322
Psilophyta, 346
Psilophyton, 303i
Psilotum, 346i
Psychedelic, 505
Pterosaur, 280i
PTH, 523i, 532, 545
Puberty, 534, 656
Pueraria lobata, 730, 730i
Pulmonary artery, 562i, 563i, 564i, 607i
Pulmonary capillary, 602i, 606
Pulmonary circuit, 557, 562, 562i
Pulmonary vein, 562i, 563i, 564i, 607i
Pulse pressure, 566–567, 567i
Pumpkin, 21i
Punctuation model, speciation, 268
Punnett, R., 163
Punnett-square method, 157, 157i
Pupa, 4i, 5, 376, 376i
Pupil, 511i, 514, 515i
Purine, 194
Pyloric sphincter, 619i
Pyracantha, 119i
Pyrenestes ostrinus, 253, 253i
Pyrimidine, 194
Pyrrhophyta, 326
Pyruvate, 110, 112, 115i, 121i
Python, 241i, 393, 508, 508i, 615i

Q Quadrat, defined, 687
Quadriceps femoris, 547i
Quaking aspen, 458t
Qualitative vs. quantitative variation, 246
Quaternary period, 276i–277i
Queen Victoria, 181, 181i
Quercus, 8 (See also Oak)

R R group, protein, 42, 44
Rabbit, 166, 166i, 731, 731i
Raccoon, 285i, 396i
Radial cleavage, 367, 651, 651i
Radial section, plant, 423i
Radial symmetry, 358–359, 358i, 496
Radiation (See also specific types), 96, 640
Radiation poisoning, 137, 795i
Radiation therapy, 23
Radioactive decay, 22, 277, 277i
Radioactive fallout, 795i
Radioisotope, 22, 23, 32t, 94, 193i, 277, 277i, 795, 795i
Radiolarian, 322, 322i
Radiometric dating, 276i, 277, 277i
Radium, 23, 226
Radius (bone), 543i
Radula, 368, 368i
Rafflesia, 768i
Ragweed, 453, 453i, 591i
Rain shadow, 761, 761i
Rainfall, patterns of, 18, 789
Ramaria, 334i
Rana pipiens, 150i, 644, 644i
Ranching, conservationist, 416–417

Ranunculus, 726
Raphus cucullatus, 409, 409i
"Raptures of the deep," 611
Raspberry, 456t
Rattlesnake, 163, 163i, 393i
Raup, D., 269
Raven, 707
Ray, cartilaginous fish, 388, 388i
Ray, in wood, 432, 433
Reabsorption, 568i, 569, 580, 636–637, 636i
Reactant, defined, 81
Reaction center, photosystem, 98, 98i
Realized niche, 720
Rearrangement, bond, 37
Receptacle (flower), 452, 452i, 457i
Receptor, molecular, 5, 43, 53, 53i, 63, 74, 87, 219, 222, 317, 467i, 571, 653
  bacterial, 315
  lymphocyte, 582, 582i, 583–589, 584i–589, 591
Receptor, sensory, 5, 181, 326, 382, 383, 506–509, 508i, 511i, 513–515, 514i, 518, 567, 602, 607, 617, 619, 637, 640, 646
  neural function and, 488, 490, 492, 492i, 495–496, 495i, 502, 503, 504
  skin, 482, 482i, 511, 511i, 540, 664
Recognition protein, 53, 53i, 219
Recombinant DNA, 222ff., 224i, 225i, 235
Recombination, genetic, 172, 171i, 172, 172i, 197, 223, 233, 246
Recombination, homologous, 233
Rectum, 500i, 616, 616i, 623, 630t
Rectus abdominus, 547i
Recycling, solid wastes, 787
Red alga (Rhodophyta), 97, 107, 107i, 320, 326, 327, 327i, 330t, 412
Red blood cell, 130, 573, 529t, 544
  disorders, 161, 161i
  formation, 544, 544t, 559, 573i, 605, 667
  function, 161, 558–559, 558i, 574
  structure, 4, 55i, 161i, 558, 555i, 568
  tonicity, 88i, 89
Red marrow, 544
Red tide, 326 (See also Algal bloom)
Redwood, 8i, 9i, 433
Reef, coral, 363, 412, 412i, 413i
Reeve, H. K., 714
Reflex, 494, 495i, 506t
Regulatory protein, 214–215, 217
Reindeer, 691i
Relaxin, 673
Release factor, 206
Releasing hormone, 523i, 527
Remission, defined, 560
Renal artery, 563i, 635i, 636i
Renal capsule, 635i
Renal corpuscle, 635, 636i
Renal failure, 638
Renal pelvis, 635i, 638
Renal vein, 563i, 635i
Renewable energy source, 794, 794i, 796i
Renin, 637
Replication (See DNA replication)
Repressor protein, 214–215, 214i, 215i
Reproduction, 4, 126 (See also Asexual reproduction; Sexual reproduction)
Reproductive base, 686, 689
Reproductive isolating mechanism, 262–265, 263i–265i
Reproductive success, 692–693, 707
  of female vs. male, 710–711, 683
Reproductive system, defined, 481i
  flatworm, 364–365, 364i
  human female, 523i, 660–667, 660i–667i
  human male, 523i, 656–659, 656i–659i
  origin of tissues, 481, 648
Reptile (Reptilia), 392, 393, 393i, 403, 513, 515, 542i, 601i, 640
  evolution, 302, 302i, 304–305, 386–387, 392, 498i, 498i, 553, 557, 557i
Resource, environmental
  competition for, 722–723, 722i, 723i
  consumption, human, 696–697, 698–699, 788, 792–793, 794–795, 794i
Resource partitioning, 723, 723i
Respiration, 597, 598 (See also Gas exchange; Respiratory system)
  invertebrate, 599, 599i, 610
  vertebrate, 600–601, 600i, 601i

Respiratory cycle, 604–605, 604i, 605i
Respiratory membrane, 606, 606i
Respiratory surface, 598
Respiratory system, defined, 481i, 602–603
  human, 602–603, 602i, 610
  smoking and, 570, 608–609, 609i
Resting membrane potential, 488
Restoration ecology, 729
Restriction enzyme, 224, 224t, 225
Restriction fragment length polymorphism, 227
Reticular formation, 502
Retina, 511t, 514, 515i, 516, 670i
Retrovirus, 318–319, 592
Reverse transcriptase, 225, 225i, 592
Reynolds, W., 283
Reznick, D., 693
RFLP, 227
Rh blood typing, 561, 561i
Rhea, 240i, 241i
Rheumatic fever, 575
Rheumatoid arthritis, 545, 591
Rhinoceros, 307, 410i
Rhizobium, 314, 441i, 752
Rhizome, 346–347, 458t
Rhizophora, 776
Rhizopus stolonifer, 337, 337i
Rhodophyta, 327
Rhodopsin, 103, 516
RhoGam, 561
Rhyniophyte, 343i
Rhythm method, 676, 677i
Rib bone, 543i
Rib cage, 542i, 543i
Ribeiroia, 391i
Riboflavin, 626t
Ribonucleic acid (See RNA)
Ribose, 38, 295
Ribosomal RNA (See rRNA)
Ribosome, 52, 56t, 60–62, 62i, 63i, 72t
  bacterial (prokaryotic), 70, 70i
  eukaryotic, 57i–59i, 62, 63i, 204, 205, 205i, 206i, 207
Ribulose bisphosphate, 101, 102i
Rickets, 532, 532i, 626t
Riffle, stream, 773, 773i
Riparian zone, 416–417, 416i
Ritonavir, 593
RNA, 4, 5, 46, 46i, 48t, 61, 72t, 201, 202
RNA polymerase, 202, 215, 215i, 217
RNA virus, 317–319, 317i, 318, 325
RNA world, 296, 296i, 297, 297i
Robin, 119i
Rocky Mountain spotted fever, 373
Rod cell, eye, 516, 516i
Rodinia, 299, 308i
Root canal, 617i
Root cap, 422i, 430, 430i, 464
Root, defined, 422
  adventitious, 430, 458, 460i
  function, 422, 423, 431, 433, 437, 440– 443, 440i–443i, 448, 464, 464i
  hormone effects on, 458, 460–464
  lateral, 430, 430i
  primary, 423, 423i, 426i, 430–431, 430i, 431i, 460, 460i, 464
  symbionts, 333–338, 338i, 441, 441i, 452, 423–425, 423i
  tissue components, 421, 423–425, 423i
  woody, 423, 423i, 432–433, 432i, 433i
Root hair, 422i, 430, 430i, 440, 441i
Root nodule, 440–441, 441i
Root system, 342, 421, 430, 430i, 458i
Rootworm, 377i
Ross Ice Shelf, calving from, 755
Rotifer, 358t
Rough endoplasmic reticulum, 57i, 62, 62i, 207
Round window, 512i
Roundworm, 358t, 365, 366, 367i, 569
rRNA, 202, 204–205, 206–207, 206i
RU-486, 677
Rubella, 590i, 672, 682
Rubisco, 101, 102, 105, 217, 217i
RuBP, 101, 102i
Ruffini ending, 511, 511i
Ruminant, 615, 615i
Run, stream, 773, 773i
Runner, plant, 458, 458t
Runoff, 104, 438, 439i, 449, 449i, 776

Index of Applications